An Introduction
To Genetic Analysis

An Introduction to Genetic Analysis

SIXTH EDITION

Anthony J. F. Griffiths

University of British Columbia

Jeffrey H. Miller

University of California, Los Angeles

David T. Suzuki

University of British Columbia

Richard C. Lewontin

Harvard University

William M. Gelbart

Harvard University

W. H. Freeman and Company

NEW YORK

The Cover

Two important genetic principles are depicted on the cover. First, most important cellular functions evolved many hundreds of millions of years ago. Second, the protein products of master control genes regulate the developmental fate of cells and tissues. A human disease called *Aniridia*—no iris in the eye—is represented in the pedigree. The bright spots on the human chromosomes indicate the location of the *Aniridia* gene. The DNA sequence of *Aniridia* reveals that it codes for a protein extremely similar to the protein of the *eyeless* gene in the fruit fly, *Drosophila*. Normally, the *eyeless* gene is expressed only in the tissue that will become the eye. If instead the *eyeless* gene is expressed in other tissues, those now develop into ectopic eyes, establishing the role of *eyeless* as a master control gene. If the human *Aniridia* gene is expressed in different tissues in *Drosophila*, then it also forces those other tissues to develop into eye. Thus, all available evidence points to the strong evolutionary conservation of these genes and the developmental pathways they control.

Library of Congress Cataloging-in-Publication Data

An introduction to genetic analysis / Anthony J. F. Griffiths . . . [et
    al.].—6th ed.
        p.   cm.
    Includes bibliographical references and index.
    ISBN 0-7167-2604-1 (hard cover)
    1. Genetics. 2. Molecular genetics. I. Griffiths, Anthony J. F.
QH430.I62 1996
575.1—dc20   95-44631
                                                              CIP

Printed in the United States of America.

Second printing 1996, RRD

Contents in Brief

Contents

ix

PREFACE

TRADITIONAL STRENGTHS OF
AN INTRODUCTION TO GENETIC ANALYSIS

Genetics has become an indispensable component of almost all research in modern biology and medicine. This position of prominence has been achieved through the powerful merger of classical and molecular approaches. Each analytical approach has its unique strengths: classical genetics is unparalleled in its ability to explore uncharted biological terrain; molecular genetics is equally unparalleled in its ability to unravel cellular mechanisms. It would be unthinkable to teach one without the other, and each is given due prominence in this book. Armed with both approaches, students are able to form an integrated view of genetic principles.

A Balanced Approach

The partnership of classical and molecular genetics has always presented a teaching dilemma: which of the two partners should the student be introduced to first, the classical or the molecular? We believe that students begin much as biologists did at the turn of the century, asking general questions about the laws governing heredity. Therefore the first half of the book introduces the intellectual framework of classical eukaryotic genetics in more or less historical sequence. Although molecular information is provided where appropriate, it is not emphasized in this half. Having acquired the classical framework, the student then proceeds to the second half of the book, which hangs molecular genetics onto this framework. The coverage of genetic mutation is a case in point. In Chapter 7 the student is treated to the classical principles of gene mutation, while Chapter 19 expands this knowledge to include molecular aspects. Because the progression from general to specific is a natural one, this approach makes sense not only in research but also in teaching about research.

Figure 4-19

Focus on Genetic Analysis

True to its title, the theme of this book is genetic analysis. This theme emphasizes our belief that the best way to understand genetics is by learning how genetic inference is made. On almost every page we recreate the landmark experiments in genetics and have the students analyze the data and draw conclusions as if they had done the research themselves. This proactive process teaches students how to think like scientists. The modes of inference and the techniques of analysis are the keys to future exploration.

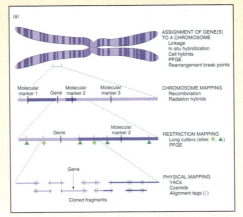

Figure 17-1

Similarly, quantitative analysis is central to the book because many of the new ideas in genetics, from the original conception of the gene to modern techniques such as RFLP mapping, are based on quantitative analysis. The problems at the ends of the chapters provide students with the opportunity to test their understanding in quantitative analyses that effectively simulate the act of doing genetics.

Tools for Analyzing Genetics

Strengths of the previous editions have been retained and reinforced.

KEY CONCEPTS

▶ Recombinant DNA is made by splicing a foreign DNA fragment into a small replicating molecule (such as a plasmid), which will then amplify the fragment along with itself resulting in a molecular "clone" of the inserted DNA fragment.

▶ A collection of DNA clones that together encompass the entire genome is called a *genomic library*.

▶ An individual DNA clone in a library can be detected by using a specific probe for the DNA or its protein product, or by its ability to transform a null mutant.

▶ Restriction enzymes cut DNA at specific target sites, resulting in defined fragments with sticky ends suitable for cloning.

▶ Different-sized DNA fragments produced by restriction enzyme digestion can be fractionated because they migrate to different positions on an electrophoretic gel.

The **Key Concepts** at the chapter openings give an overview of the main principles to be covered in the chapter, stated in simple prose without genetic terminology. These provide a strong pedagogic direction for the reader.

Boxed **Messages** throughout the chapters provide convenient milestones at which the reader can pause and contemplate the material just presented.

Message

At mitosis, nondisjunction, chromosome loss, crossing-over and haploidization all cause a heterozygous pair of alleles to segregate in somatic tissue, resulting in a mosaic expressing the phenotypes of both alleles.

Chapter **Summaries** provide a short distillation of the chapter material and an immediate reinforcement of the concepts. All these items are useful in text review, especially for exam study.

SUMMARY

Genomics is the branch of genetics that deals with the systematic molecular characterization of whole genomes. Some of the methods used are traditional genetic mapping procedures, but in addition specialized techniques have been developed for manipulating the large amounts of DNA in a genome. Genomic analysis is important for two reasons: first, it represents a way of obtaining an overview of the genetic architecture of an organism, and second, it forms a set of basic information that can be

Another end-of-chapter feature is the problems requiring **Concept Maps.** Concept maps grew out of the constructivist movement in education, which asserts that student learning is most effective when new information is brought into direct conflict with previous understanding. The concept map provides a powerful method for visualizing and resolving such conflicts and also aids concept integration.

Concept Map

Draw a concept map interrelating as many of the following terms as possible. Note that the terms are listed in no particular order.

recombinant DNA/probe/in situ hybridization/genetherapy/RFLP/transgenic/mapping/ in vitro mutagenesis/genetic screening

CHAPTER INTEGRATION PROBLEM

Some recessive mutations in maize affect the color, texture, or shape of the kernel. To discuss these mutations in general, we'll use *m* to represent the recessive allele. These mutations can be detected in a straightforward manner by crossing an m^+m^+ male with an *mm* female. Any new mutations are found by visually scanning many seeds, in other words, millions of kernels on thousands of corncobs. Finding a seed with the recessive phenotype shows that the gene mutated in the male. When maize geneticists sow seeds with the mutant phenotype and self the resulting plants, they find one or the other of two kinds of results:

If the mutation was spontaneous, the progeny from the selfing are generally all mutant.

If the mutation followed treatment of the pollen with a mutagen, the progeny from the selfing are generally 3/4 wild type and 1/4 mutant.

Provide an explanation for these two different outcomes.

Solution

As usual, if we restate the results in a slightly different way, it gives us a clue as to what is going on. The difference between outcomes seems to reflect some difference between spontaneous and induced mutation. From a spontaneous mutation, we obtain an individual that seems to be of genotype *mm* because it breeds true for the mutant phenotype. From an induced mutation, we obtain an individual that seems from the 3:1 phenotypic ratio

The **Chapter Integration Problems** are solved problems that emphasize concept integration both within and between chapters. These chapter integration problems help to show how one set of learned skills builds on and interacts with previous ones. They also enable students to develop a holistic perspective as they begin to organize diverse concepts into a coherent body of knowledge.

The problems at the end of each chapter are prefaced by **Solved Problems** that illustrate the ways that geneticists apply principles to experimental data. Research in science education has shown that this application of principles is a process that professionals find second nature, whereas students find it a major stumbling block. The Solved Problems demonstrate this process and prepare the students for solving problems on their own.

The **Problems** themselves continue to be one of the strengths of the book. The problems are generally arranged to start from the simple and proceed to the more difficult. Particularly challenging problems are marked with an asterisk. All problems have been classroom tested. Answers to selected problems are found at the back of the book, and the full set of solutions is in the *Student Companion*, prepared by Diane Lavett (Emory University).

NEW FEATURES OF THE SIXTH EDITION

Revised and Updated

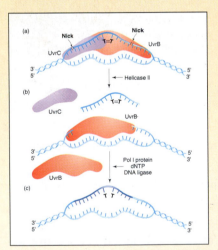

Figure 19-32

All chapters have been revised and updated. Text has been clarified, and new diagrams and photographs have been used throughout. Most of the major revision has been in the chapters that focus on molecular genetic technology (Chapters 14–17) and on the fast-moving field of developmental genetics (Chapters 23–25).

Much of **Chapter 14** (Recombinant DNA Technology) has been rewritten and reorganized. The first part of the chapter focuses on the reasons why geneticists clone DNA and proceeds to a step-by step elucidation of the general methods used. The chapter ends with a description of some other useful techniques in molecular analysis: DNA sequencing (with a new section on automated sequencing), DNA electrophoresis (including Southern and Northern analysis and restriction mapping), and the polymerase chain reaction. Other new topics include pUC plasmids, cloning by functional complementation, positional cloning, cloning by tagging, design of oligonucleotide probes based on amino acid sequence, and ORF analysis. The creation and utilization of DNA libraries has been expanded.

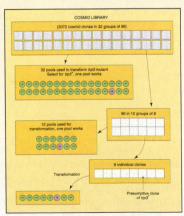

Figure 14-15

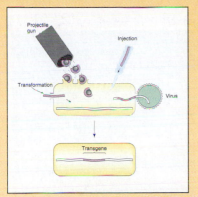

Figure 15-14

Chapter 15 (Applications of Recombinant DNA Technology) has also been extensively revised. Broadly, the new emphasis is on what geneticists can do with cloned genes once they have them. There is new emphasis and explanation of the techniques of reverse genetics, including in vitro mutagenesis, gene disruption, and gene replacement. There is a new section on human gene therapy.

In Chapter 16 (The Structure and Function of Eukaryotic Chromosomes) the description of repetitive DNA has been revised to bring the terminology into line with that used in Chapter 17, for example, minisatellite and microsatellite DNA.

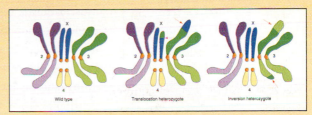

Figure 16-2

Chapter 17 is a new chapter on the topic of Genomics, reflecting current research interest in this area and the interest of the public in the human genome project. The chapter brings together the traditional techniques for genome analysis, such as recombination mapping and somatic cell hybridization, and new approaches designed specifically to manipulate the large amounts of DNA found in eukaryotic chromosomes. New sections deal with fluorescence in situ hybridization, pulsed field gel electrophoresis, mapping by restriction fragment length polymorphisms (RFLPs), mapping by simple sequence length polymorphisms (SSLPs), sperm genotyping, randomly amplified polymorphic DNAs (RAPDs), irradiation and fusion gene transfer (IFGT), contigs, fluorescence-activated sequence-tagged sites, chromosome sorting (FACS), candidate gene approach, and physical mapping of the human Y chromosome.

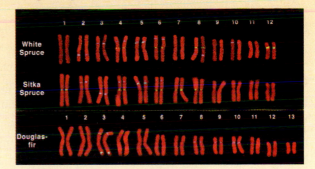

Figure 17-2

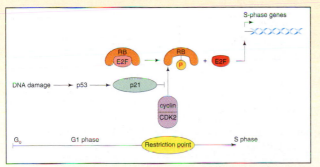

Figure 24-11

Chapters 23–25 are a major revision of Chapters 22–23 in the Fifth Edition to reflect recent advances in the dynamic and changing landscape of developmental genetics. New sections have been added to each of the chapters, and other material has been eliminated to streamline them. Each chapter focuses on a clear theme in developmental genetics. **Chapter 23** discusses how protein activities in a cell are regulated during development and builds on the individual regulatory events to introduce the student to the concept of developmental pathways. New to Chapter 23 is a discussion of epigenetic phenomena, including paramutation in plants and parental imprinting in mammals. In recent years, the dividing line between cell and developmental biology has been blurred. **Chapter 24**

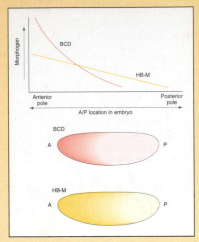

Figure 25-10

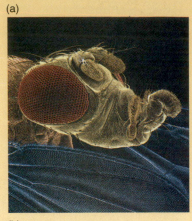

(a)

addresses this by discussing several areas of cell biology that impact on developmental genetics. New topics include the cytoskeleton and cell shape, the cell cycle, and intercellular communication. The chapter culminates with a discussion of the genetics of cancer, bringing together many of the principles discussed in Chapters 23 and 24. Finally, Chapter 25 focuses on the formation of complex biological patterns. It uses the most exquisite case of early *Drosophila* development to exemplify how geneticists are attacking this major problem in developmental biology. Considerable emphasis is placed on how spatial information is incorporated and read out in the developing oocyte and the early embryo and how the principles of cell biology are exploited to pattern the egg. The chapter ends with a consideration of mammalian developmental genetics and the emerging theme that the mechanisms of pattern formation are ancient and highly conserved among animals.

(b)

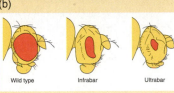

(c)

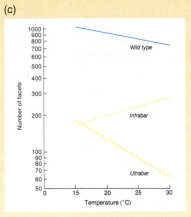

Figure 1-17

Chapter 26, Population Genetics, now precedes the chapter on Quantitative Genetics (Chapter 27), reversing the order in previous editions. Many of the users of the book found that it is a great deal easier to understand quantitative genetics if students first have a grounding in the principles of population genetics, and it is indeed the case that the modern treatment of quantitative genetics grew out of population genetic theory. We have also added a section on the mapping of quantitative trait loci (QTLs) by the new methods of molecular genetics, a subject of considerable interest at present in both human genetics and agriculture.

New Problems

We have added more than 100 new problems, including many problems involving molecular analysis. In addition, most chapters have a new exercise in problem solving called "Unpacking the Problem." This exercise was devised to illustrate the idea that a genetics problem represents only the tip of a vast iceberg of knowledge (we originally considered calling them "iceberg problems"). It is only when the structure of the underlying levels of knowledge is understood that the problem can be solved constructively. The unpacking exercises access this underlying knowledge without actually solving the problem. Some of the component questions in an unpacking exercise might sound trivial, but often they address the kind of fundamental levels of misunderstanding that prevent students from successfully solving problems.

Unpacking the Problem

8. John and Martha are contemplating having children, but John's brother has galactosemia (an autosomal recessive disease), and Martha's great-grandmother also had galactosemia. Martha has a sister who has three children, none of whom has galactosemia. What is the probability that John and Martha's first child will have galactosemia?

In some chapters we expand one specific problem with a list of exercises that help mentally process the principles and other knowledge surrounding

the subject area of the problem. You can make up similar exercises yourself for other problems.

Before attempting a solution, consider some expansion questions such as the following, which are meant only as examples:

a. Can the problem be restated as a pedigree? If so, write one.

b. Can parts of the problem be restated using Punnett squares?

c. Can parts of the problem be restated using branch diagrams?

d. In the pedigree, identify a mating

Course Syllabi

For a two-semester course, the entire text provides an appropriate course structure and syllabus that reflects the range of modern genetics. A syllabus for a one-semester course can be designed around selected chapters. For students familiar with DNA structure and function from introductory biology or cell biology courses, a possible selection of chapters for a one-semester course is Chapters 2, 3, 4, 5, 7, 9, 10, 12, 14, 16, 23, and 25. A one-semester course in molecular genetics could be based on Chapters 10 through 25.

Supplements

A number of useful supplements benefit students and instructors using the textbook. The *Student Companion and Complete Solutions*, by Diane K. Lavett of Emory University, offers worked-out answers to all the problems in the textbook. Dr. Lavett has revised and expanded the section in each chapter devoted to the text's concept maps as a means of helping students build problem-solving skills. Questions and answers about the maps are intended to test student understanding of the connections among the key ideas being taught. Concept maps can also be used by instructors to structure the presentation of chapters.

Figure 1-5

A full-color *overhead transparency set* of about 130 key illustrations from the textbook is available free of charge to qualified adopters. Instructors will find all the approximately 1100 numbered figures and tables in the book, along with their legends, in the *An Introduction to Genetic Analysis CD-ROM*. The images can be viewed on a computer and can be displayed with a projection unit during a lecture. The CD contains special software that allows users to select a sequence of images in advance. The CD also gives the option of showing figures with or without their labels, or with labels that the user has created. The CD is also free to qualified adopters. For more information, instructors should contact their local Freeman representative.

Acknowledgments

Thanks are due to the following people at W. H. Freeman and Company for their considerable support throughout the preparation of this edition: Mary Shuford, Director of Development; Randi Rossignol, Senior Developmental Editor; Philip McCaffrey, Managing Editor; Penny Hull, Project Editor; Vicki Tomaselli, Designer; Denise Wiles and Susan Wein, Illustration Coordinators; Travis Amos, Senior Photo Editor; Larry Marcus, Assistant Photo Editor; Ellen Cash, Production Manager; and Julia DeRosa, Production Coordinator. We also thank the copy editor, William O'Neal; the layout artist, Leon Bolognese; the indexer, Susan Thomas; and the proofreader, Lynn Contrucci.

Tony Griffiths thanks Barbara Moon for showing him the power of concept maps and generally converting him to constructivism. He also thanks Jolie Mayer-Smith for the idea behind the problem-unpacking exercises. Dick Lewontin thanks Rachel Nasca for her work in preparing the manuscript, and Jeffrey Miller and Bill Gelbart thank

Figure 18-33

Kim Anh Miller and Janet and Marnie Gelbart for their constant support. We thank Diane Lavett for critiquing the problem sets and for the solutions to the Chapter Integration Problems for Chapters 11, 15, and 20 and the Solved Problems in Chapters 14 and 15. Finally, we extend our thanks and gratitude to our colleagues who reviewed this edition and whose insights and advice were most helpful:

Professor Kathleen Anderson	San Francisco State University
Dr. M. M. Bentley	University of Calgary
Dr. R. L. Bernstein	San Francisco State University
Dr. Lisa D. Brooks	Brown University
Dr. Andrew Buchman	Pennsylvania State University
Dr. James Cleaver	University of California at San Francisco
Dr. Glen Collier	University of Tulsa
Dr. Victoria Corbin	University of Kansas
Dr. Robert Cohen	University of Kansas
Professor Gray F. Crouse	Emory University
Dr. Thomas Dreesen	Louisiana State University at Baton Rouge
Dr. Kathleen Dunn	Boston College
Professor David Durica	The University of Oklahoma
Dr. Hilary Ellis	Emory University
Dr. Thomas Glover	Hobart and William Smith Colleges
Dr. Robert Grant	Georgia State University
Dr. Charles Hoffman	Boston College
Dr. Robert Kitchen	University of Wyoming
Dr. Richard Kolodner	Dana-Farber Cancer Institute
Dr. Yoke Wah Kow	University of Vermont
Dr. Diane K. Lavett	Emory University
Dr. Marjorie Maguire	University of Texas at Austin
Dr. John Mottinger	University of Rhode Island
Dr. Harold J. Price	Texas A & M University
Dr. R. H. Richardson	University of Texas at Austin
Professor John Scheifelbein	University of Michigan
Dr. David Scicchitano	New York University
Dr. Stephen D'Surney	University of Mississippi
Dr. Christine Tachibana	Pennsylvania State University
Dr. Joanne Tornow	University of Southern Mississippi
Dr. Richard K. Wilson	Washington University Medical School
Dr. Mark Zoeller	ARIAD Pharmaceuticals

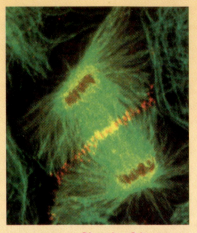

Chapter 24 opener

We believe this edition to be a true celebration of genetics. As authors, we hope that our love of the subject comes through and that the book will stimulate the reader to do some first-hand genetics, whether as professional scientist, student, amateur breeder, or naturalist. Failing this, we hope to impart some lasting impression of the incisiveness, elegance, and power of genetic analysis.

An Introduction
To Genetic Analysis

1

Genetics and the Organism

Genetic variation in the color of corn kernels. Each kernel represents a separate individual with a distinct genetic makeup. The photograph symbolizes the history of humanity's interest in heredity. Humans were breeding corn thousands of years before the rise of the modern discipline of genetics. Extending this heritage, corn today is one of the main research organisms in classical and molecular genetics. (William Sheridan, University of North Dakota; photo by Travis Amos.)

KEY CONCEPTS

▶ Genetics unifies the biological sciences.

▶ Genetics is of direct relevance to human affairs.

▶ Genetics may be defined as the study of genes.

▶ Genetic variation contributes to variation in nature.

▶ As an organism develops, its unique set of genes interacts with its unique environment in determining its characteristics.

hy study genetics? There are two basic reasons. First, genetics has come to occupy a pivotal position in the entire subject of biology; for any serious student of plant, animal, or microbial life, an understanding of genetics is thus essential. Second, genetics, like no other scientific discipline, has become central to numerous aspects of human affairs. It touches our humanity in many different ways. Indeed, genetic issues seem to surface daily in our lives, and no thinking person can afford to be ignorant of its discoveries. In this chapter we take an overview of the science of genetics, showing how it has come to occupy its crucial position. In addition we provide a perspective from which to view the subsequent chapters.

First we need to define what genetics is. Some define it as the study of heredity, but hereditary phenomena have been interesting to humans since before the dawn of civilization. Long before biology or genetics existed as the scientific disciplines we know today, ancient peoples were improving plant crops and domesticated animals by selecting desirable individuals for breeding. They also must have puzzled about the inheritance of traits in humans, and asked such questions as "Why do children resemble their parents?" and "How can various diseases run in families?" But these people could not be called geneticists. Genetics as a set of principles and analytical procedures did not begin until the 1860s when an Augustinian monk named Gregor Mendel (Figure 1-1) performed a set of experiments that pointed to the existence of biological elements called **genes.** The word *genetics* comes from "genes," and genes provide the focus for the subject. Whether geneticists study at the molecular, cellular, organismal, family, population, or evolutionary level, genes are always central in their studies. Simply stated, genetics is the study of genes.

What is a gene? A gene is a section of a threadlike molecule called **deoxyribonucleic acid,** abbreviated **DNA.** DNA, the hereditary material that passes from one generation to the next, dictates the inherent properties of a species. Each cell in an organism has one or two sets of the basic DNA complement, called a **genome.** The genome itself is made up of one or more extremely long molecules of

Figure 1-1 Gregor Mendel. (Moravian Museum, Brno)

DNA that are called **chromosomes.** The genes are the functional regions of the DNA and are simply active segments ranged along the chromosomes (Figure 1-2). In complex organisms chromosomes generally number in the order of tens, but genes number in the order of tens of thousands. Armed with these definitions of genetics and genes, let us go on to see how understanding these subjects has become so important. We will start with the impact of genetics on our own species.

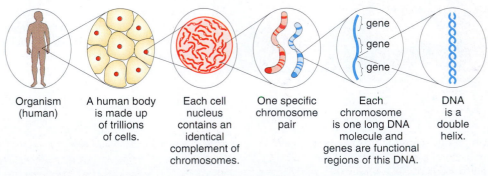

Organism (human) · A human body is made up of trillions of cells. · Each cell nucleus contains an identical complement of chromosomes. · One specific chromosome pair · Each chromosome is one long DNA molecule and genes are functional regions of this DNA. · DNA is a double helix.

Figure 1-2 Successive enlargements of an organism to focus on the genetic material.

Genetics and Human Affairs

Genetics seems to hold a special place in human affairs. Not only is it relevant in the same sense that other scientific disciplines are, but it also has much to tell us about the nature of our humanity, and in this sense it holds a special place among the biological sciences.

Modern society depends on genetics. Take a look at the clothes you are wearing. The cotton of your shirt and jeans came from cotton plants that differ from their wild ancestors because they have been through intensive breeding programs involving the methodical application of standard genetic principles. The same could be said for the sheep that produced the wool for your sweater and coat. Think also about your most recent meal. You can be sure that the rice, the wheat, the chicken, the beef, the pork, and all the other major organisms that feed the planet have been specially engineered with the application of standard genetic procedures (Figure 1-3). The so-called Green Revolution, which dramatically increased crop productivity on a global scale, is a genetic success story about the breeding of highly productive strains of certain major crop species (Figure 1-4).

To maximize crop production, a farmer may plant a vast area with seeds of a single genetic constitution, a practice called *monoculture.* Because pathogenic and predatory organisms attack plant crops, special resistance genes are bred into crop lines to protect them. But the protection is only temporary. Random genetic changes occur constantly in pathogen populations, and such changes sometimes confer new pathogenic ability. This leaves the entire monoculture at risk. For this reason, plant geneticists always have to be one step ahead of the pathogens to prevent widespread epidemics of plant disease or destruction that might have devastating effects on the food supply (Table 1-1 and Figure 1-5).

Fungi and bacteria have also been specially bred for human needs. Yeast is an obvious example; it forms the basis of multibillion dollar industries producing baked goods, alcoholic beverages, and fuel alcohol. Fungi provide the antibiotic penicillin, the immunosuppressant drug cyclosporin that prevents rejection of organ transplants, and a whole range of important industrial compounds such as citric acid and amylase. Bacteria provide such antibiotic chemicals as streptomycin to medical science. Most of the industries using fungi and bacteria have benefited from the application of classic genetic principles, but now we are entering a new age where molecular genetic techniques enable scientists to synthesize entirely novel strains of microbes that are created in a test tube and tailored exclusively to human needs.

Figure 1-3 A geneticist making controlled crosses in corn. Paper bags are used to prevent random cross pollinations and to collect pollen to make the required specific crosses. (Chuck O'Rear/Woodfin Camp.)

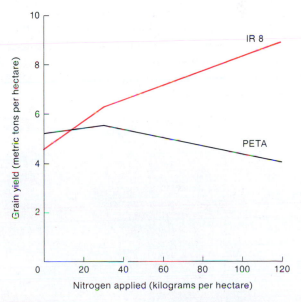

Figure 1-4 The specially bred strain of dwarf rice called *IR 8* owes part of its success to its remarkable response to the application of fertilizer. PETA, an older, nondwarf strain, shows a more typical response. (From Peter R. Jennings, "The Amplification of Agricultural Production." *Scientific American,* September 1976 pp. 180–194.)

Table 1-1 Pest-resistant Strains of Rice

Strain	Year developed	Diseases					Insects		
		Blast fungus	Bacterial blight	Leaf-streak virus	Grassy stunt virus	Tungro virus	Green leafhopper	Brown hopper	Stem borer
IR 8	1966	MR	S	S	S	S	R	S	MS
IR 5	1967	S	S	MS	S	S	R	S	S
IR 20	1969	MR	R	MR	S	R	R	S	MS
IR 22	1969	S	R	MS	S	S	S	S	S
IR 24	1971	S	S	MR	S	MR	R	S	S
IR 26	1973	MR	R	MR	MR	R	R	R	MR

NOTE: The entries describe each strain's susceptibility or resistance to each pest as follows: S = susceptible; MS = moderately susceptible; MR = moderately resistant; R = resistant. (From *Research Highlights,* International Rice Research Institute, 1973, p. 11.)

For example, from such syntheses we now have bacterial strains producing mammalian substances such as insulin for diabetes treatment and growth hormone for treatment of pituitary dwarfism. The engineered insulin product is the genuine human type in contrast to the previous product, which was from cattle and pigs. Some genetically engineered substances are extremely difficult to obtain from other sources; for example, one year's treatment of growth hormone for one child previously had to be extracted from the pituitary glands of approximately 75 human cadavers. In addition, special derivatives of fungal and bacterial strains are being genetically engineered to produce larger amounts of their natural beneficial products.

Molecular genetic engineering was first applied to microbes, but now the same techniques are being applied to plants and animals, resulting in engineered types that could never have been produced with classical genetics. In such an approach, the genome of a plant or animal is changed by exposing cells to fragments of "foreign" DNA carrying desirable genes, often from another species. Such DNA is taken up and can insert itself into one or more of the recipient's chromosomes, in which location it is inherited like any other part of the genome. Cells modified in this way are called **transgenic.** From a transgenic cell a transgenic organism can be produced, all of whose cells contain the additional foreign DNA. Using this technique, a wide range of commercially profitable "designer" plants and animals can be generated. The application of genetic engineering is broadly called **biotechnology.** It is estimated that the biotechnology industry, which has genetics as its basis, will become one of the top money-making industries in the coming decades. Whether this is true or not, the biotechnology industry joins the genetic cornucopia that supports mankind on this planet at its present high population size and living standards.

Another important area of human affairs that is using genetics is forensic science. All members of the human species carry more or less the same set of DNA. But minor variation in the DNA contributes to human variation. If it were possible to compare the entire DNA sequences of individual human beings, then, with the exception of identical twins, every person would be individually distinguishable by a **DNA "fingerprint."** In practice, however, only a tiny piece of the DNA can be examined, so even though a very large number of distinguishable types can be found, many different individuals yield the same fingerprint pattern. Thus, partial "fingerprints," which can be obtained from small samples of blood, semen, or even hair follicles, can narrow down the possible suspects in a criminal case to a small proportion of individuals but not to a specific individual. DNA fingerprints are widely used in police investigations. Even plant DNA fingerprints can be useful. In one reported case, the DNA from two tiny seeds obtained from the back of a suspect's truck clearly established that the truck had been parked under one specific tree, near which a murder victim's body had been found.

Genetics is a crucial component of medicine. A large proportion of human ill health has a genetic basis. For example, it has been estimated that at least 30 percent of pediatric

Figure 1-5 Pure lines of different species of rice specially bred for resistance to various plant parasites. (Patrick Aventurier/Gamma Liaison.)

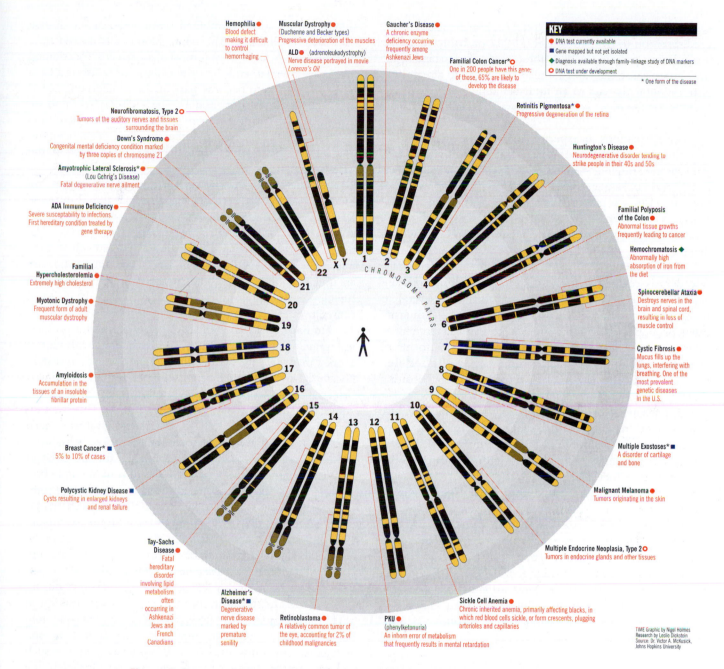

KEY
- DNA test currently available
- Gene mapped but not yet isolated
- Diagnosis available through family-linkage study of DNA markers
- DNA test under development

* One form of the disease

Hemophilia
Blood defect making it difficult to control hemorrhaging

Muscular Dystrophy
(Duchenne and Becker types)
Progressive deterioration of the muscles

ALD (adrenoleukodystrophy)
Nerve disease portrayed in movie *Lorenzo's Oil*

Gaucher's Disease
A chronic enzyme deficiency occurring frequently among Ashkenazi Jews

Familial Colon Cancer*
One in 200 people have this gene; of those, 65% are likely to develop the disease

Retinitis Pigmentosa*
Progressive degeneration of the retina

Neurofibromatosis, Type 2
Tumors of the auditory nerves and tissues surrounding the brain

Down's Syndrome
Congenital mental deficiency condition marked by three copies of chromosome 21

Amyotrophic Lateral Sclerosis*
(Lou Gehrig's Disease)
Fatal degenerative nerve ailment

ADA Immune Deficiency
Severe susceptibility to infections. First hereditary condition treated by gene therapy

Familial Hypercholesterolemia
Extremely high cholesterol

Myotonic Dystrophy
Frequent form of adult muscular dystrophy

Amyloidosis
Accumulation in the tissues of an insoluble fibrillar protein

Breast Cancer*
5% to 10% of cases

Polycystic Kidney Disease
Cysts resulting in enlarged kidneys and renal failure

Tay-Sachs Disease
Fatal hereditary disorder involving lipid metabolism often occurring in Ashkenazi Jews and French Canadians

Alzheimer's Disease*
Degenerative nerve disease marked by premature senility

Retinoblastoma
A relatively common tumor of the eye, accounting for 2% of childhood malignancies

PKU
(phenylketonuria)
An inborn error of metabolism that frequently results in mental retardation

Sickle Cell Anemia
Chronic inherited anemia, primarily affecting blacks, in which red blood cells sickle, or form crescents, plugging arterioles and capillaries

Multiple Endocrine Neoplasia, Type 2
Tumors in endocrine glands and other tissues

Malignant Melanoma
Tumors originating in the skin

Multiple Exostoses*
A disorder of cartilage and bone

Cystic Fibrosis
Mucus fills up the lungs, interfering with breathing. One of the most prevalent genetic diseases in the U.S.

Spinocerebellar Ataxia
Destroys nerves in the brain and spinal cord, resulting in loss of muscle control

Hemochromatosis
Abnormally high absorption of iron from the diet

Familial Polyposis of the Colon
Abnormal tissue growths frequently leading to cancer

Huntington's Disease
Neurodegenerative disorder tending to strike people in their 40s and 50s

CHROMOSOME PAIRS

TIME Graphic by Nigel Holmes
Research by Leslie Dickstein
Source: Dr. Victor A. McKusick, Johns Hopkins University

Figure 1-6 The 23 chromosomes of a human being, showing the positions of genes whose abnormal forms cause some of the better-known hereditary diseases. (*Time*)

hospital admissions have a direct genetic component. However, current research is revealing more and more genetic predispositions to serious conditions as well as milder ailments, so this figure is almost certainly an underestimate.

Genetic ill health can be divided into three major types. The first type is **inherited genetic diseases,** caused by abnormal forms of genes that are passed on from one generation to the next. Many inherited genetic diseases (such as cystic fibrosis, phenylketonuria, and muscular dystrophy) are caused by an abnormal form of a single gene with major effect. The single genes that cause some of the better-known hereditary disorders are shown on the human chro-

mosome maps in Figure 1-6. We live in an exciting era in which many of the single genes responsible for inherited genetic diseases are being isolated and characterized at the molecular level. Some recent examples are the genes responsible for familial Alzheimer's disease and familial breast cancer. The gene *S182* located on chromosome 14 is responsible for 80 percent of early-onset familial Alzheimer's disease, accounting for 10 percent of all cases of Alzheimer's disease. The *S182* gene normally codes for a membrane protein. Familial breast cancer accounts for 5 percent of all breast cancers in our population. Abnormal forms of the genes *BRCA1* and *BRCA2* together account for two-thirds of cases of familial breast cancer. *BRCA1* is also associated with predisposition to ovarian cancer. *BRCA1* probably codes for a regulatory protein. In both these cases, understanding of how genes cause the rarer familial forms of the disease will also undoubtedly lead to an understanding of and effective therapy for the more common "sporadic" forms of the diseases.

Gene isolations in experimental animals can also provide insights that have potential impact on human health. For example, single genes have been isolated that have a major effect on obesity in mice. One gene *ob* codes for a signaling protein that is part of the system for telling the mouse that it has eaten enough food. Another gene *CPE* causes the mouse to slowly become obese and develop susceptibility to diabetes.

The genetic determination of many human hereditary disorders is complex in the sense that the conditions are caused by the interaction of specific forms of several to many genes, all of which also interact with the environment. Examples of such **complex traits** are susceptibilities to heart disease, hypertension, diabetes, various forms of cancer, and infections. However, molecular genetic technology has opened up new avenues for finding and isolating these multiply interacting genes.

The second type of genetic ill health is **somatic genetic disease,** which is caused by the sudden appearance of an abnormal form of a gene in one part of the body. Cancer is the most significant example of this kind of disease. It is little appreciated that cancer, which touches all our lives in some way, is a genetic disease. Although somatic changes are not passed on to the next generation, various predispositions to cancer are inherited as abnormal genes.

The third type of genetic ill health is **chromosomal aberrations,** such as Down syndrome and cri du chat syndrome. These are caused by abnormalities of chromosomal structure or number. Some cases arise de novo in each generation, stemming from cellular "accidents" in the gonads of parents with otherwise normal chromosome sets. Other cases stem from a parent who carries a chromosome aberration.

Genetics has shown that genes are at the root of a large number of diseases, but genetics will also provide relief from suffering in many cases. Already molecular genetic probes are being used to detect defective genes in prospective parents. Furthermore, the defective genes themselves are being isolated and characterized by molecular genetic techniques. For example, the DNA of the gene that causes cystic fibrosis has been isolated and analyzed. From the gene's molecular structure, as revealed by this work, scientists can now investigate the physiological defect that causes the disease. Once the disease is understood, better therapeutic approaches can be designed. Ultimately, there is the hope that direct gene therapy will relieve many genetic diseases. Gene therapy inserts a copy of a normal gene into cells carrying the defective counterpart. The first reported cases of attempted human gene therapy were two children

suffering from severe combined immunodeficiency disease (a hereditary illness of the white blood cells caused by a defective gene). Cells were removed, a correctly functioning gene was inserted, and then the cells were reinserted back into the children to become part of the proliferating blood cell population in their bodies. The children seem to be responding well to the treatment. Since this first case, several more attempts at gene therapy have also been reported, and undoubtedly many more will follow in the years ahead.

Geneticists are also at work studying the human immunodeficiency virus (HIV) that causes acquired immune deficiency syndrome (AIDS). As a natural part of their reproduction, viruses such as HIV insert copies of their genetic material into the chromosomes of the individuals they infect. Therefore, in a sense this too is a genetic disease, and an understanding of how such viral genes integrate and function is an important step in overcoming the diseases caused by these viruses.

Human genetics obviously holds an important position in human affairs, and one measure of this is the amount of money that society is spending on human genetic research. Society rarely allots billions of dollars for single biological projects, but $3 billion has been committed to finance the complete sequencing of the human genome. This is an international collaborative effort by many laboratories, each working on specific chromosomal regions. Already high-resolution physical maps have been obtained for the DNA of two complete chromosomes (the Y chromosome and chromosome 21). Such maps are a necessary prelude to completely sequencing chromosomes. The potential benefits for medicine are immense. But perhaps most exciting is the opportunity finally to see a glimmer of the grand design in the blueprints of what is arguably the most complex structure in the known universe, *Homo sapiens* (Figure 1-7).

Genetics affects one's world view. We each acquire our individual view of the universe and of our own position in that universe gradually, from the beginning of our consciousness. This viewpoint represents our identity as individuals. It drives our attitudes and our actions, and as such determines the kind of people we are and ultimately the kind of society we live in. Any new knowledge has to be accommodated into this world view, or the world view has to be changed to make it fit. Ignorance or rejection of new knowledge leads to closed-mindedness and bigotry. Genetics has provided some powerful new concepts that have radically changed humanity's view of itself and its relation to the rest of the universe.

Probably the best example of how genetics changes a person's world view comes from genetic, cytogenetic, and molecular studies that show we are related not only to apes and other mammals, but, more surprisingly, to all the other living things on the planet, including plants, fungi, and bacteria. Living things share a common system for storing and expressing information and show homology in many structures, even down to the genes themselves. That there is a

Figure 1-7 Two molecular geneticists analyzing DNA sequences. The dark bands are photographic images of radioactive DNA segments. (David Parker/Science Source/Photo Researchers.)

continuous spectrum of relatedness within the living world is a powerful intellectual notion that unifies us with other living organisms. This notion radically affects one's world view. It suggests a view of humanity not as the pinnacle or the center of creation, but as one form equal to other life-forms. Admittedly, this brings us into the domain of philosophy and religion, but that is the point: genetics forces us to consider issues that question how we see ourselves.

Some of the world's biggest and most pressing social issues have an indirect genetic component. For example, some major problems of prejudice and social suffering center on behavioral differences between races and between the sexes. Genetics provides a way of analyzing and thinking about these complex and unresolved issues.

One of the biggest global concerns of biologists is the alarming rate at which we are destroying natural habitats, especially in the tropics, which hold vast reservoirs of plant and animal life. Here again the problem has a crucial genetic component because the issue is one of conserving genetic diversity and genetic resources. An understanding of the full impact of destroying natural habitats requires an understanding of genetics.

Another genetic issue with potential global impact is the genetic health of our populations. Many geneticists are concerned because our human genomes are under assault by an ever-increasing array of environmental agents, especially radiation and chemicals, that are capable of causing random changes in genes. The vast majority of such changes would inevitably be deleterious. In the short run

they might not significantly increase the frequency of inherited disease, but over the long run the changes could accumulate, eventually surfacing as a "genetic time bomb."

Of course, genetic technology, like any other technological advance, brings with it its own share of ethical dilemmas. The Green Revolution, for example, is a success in terms of productivity, but the high levels of fertilizer and pesticide application required for the new plant varieties has raised concerns about water pollution, and ability of farmers in poor countries to buy the expensive fertilizers. The synthesis of new life-forms by genetic engineering has led to a rash of concerns about mankind "playing God," and the morality and legality of patenting designer plants and animals. Access to information about individuals' genetic makeup is also a potential problem in such areas as health insurance and forensics. Already there are credible published claims about specific genes that predispose people to such behavioral patterns as aggressiveness and homosexuality. What would be the impact of knowledge about these genes on the rights of the individuals who carry them? It would be too glib to say that these are not the problems of scientists but of society; scientists must share the responsibility for the societal impact of their discoveries, and all the possible outcomes should be thought out as thoroughly as possible before results are made public.

The above examples show that many of the issues that confront us daily as individuals and as societies intimately involve genetics. Only by understanding the genetic component of these issues can we as global citizens hope to make wise decisions for the future of an ever more complex and unstable world.

Message   Genetic insight helps us understand humanity and human society.

Genetics and Biology

Biology is a huge subject. The planet Earth contains a staggering array of life forms. Already we know about the existence of 286,000 species of flowering plants, 500,000 species of fungi, and 750,000 species of insects. In addition, many more species are still undiscovered. Fifty years ago, the science of biology was divided into separate disciplines, each analyzing life at a different level. There was morphology, physiology, biochemistry, taxonomy, ecology, genetics, and so on, all working largely in separate academic compartments. However, discoveries in genetics have provided unifying themes for the whole of biology, so that now conceptual threads link the disciplines.

The major thematic linking thread is, in reality, a thread, the genetic molecule DNA. We now know that DNA is the basis for all the processes and structures of life. The DNA molecule has a structure that accounts for two of the key properties of life, replication and generation of

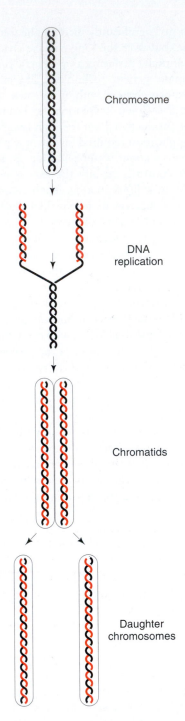

Chromosome

DNA replication

Chromatids

Daughter chromosomes

Figure 1-8 When new cells are made, DNA replication enables a chromosome to become a pair of chromatids, which eventually become daughter chromosomes and pass into the new cells.

form. We will learn that DNA is a double-helical structure that has the inherent property of being able to make copies of itself. DNA copies itself before cell division, and this enables chromosomes to divide into **chromatids** which eventually become **daughter chromosomes** that pass into **daughter cells.** This process of replication and chromatid formation is essentially similar during the division of both asexual and sexual cells, and is diagrammed in Figure 1-8.

(Note, however, that the two types of cell division do have many differences, which we will address in later chapters.) It is this property that enables replicas of cells and organisms to be made and persist through time (Figure 1-9). Hence DNA can be viewed as the thread that connects us with all our evolutionary ancestors. Furthermore, written into the linear sequence of the building blocks of a DNA molecule is a code that contains the instructions for building an organism; we can view this as information, or "that which is necessary to give form." The unique features of a species, whether structures or processes, are seen to be under the influence of DNA. So, underlying the structures studied by morphologists, the reactions studied by physiologists, the homologies studied by evolutionists, and so on, we see the unifying thread of the DNA molecule.

DNA works in virtually the same way in all organisms. This in itself provides another unifying theme, but in addition it means that what is learned in one organism can often be applied in principle to others. For this reason, genetics has made extensive use of model organisms, many of which will appear in the pages of this book. In fact, the advances made in human genetics over the recent decades have been possible in large part because of the advances made with such lowly model organisms as bacteria and fungi.

Genetics has also provided some of the most incisive analytical approaches now being used across the spectrum of the biological disciplines. Foremost is the technique of **genetic dissection.** In this experimental approach, any structure or process can be picked apart, or "dissected," by discovering which genes influence it. For example, in the study of development each abnormal gene that produces a developmental abnormality identifies a component in the normal process of development. Then the larger picture can be assembled by interrelating all the genetically controlled components. Another successful technique is the use of specific genes as markers. Just as you might use brightly colored tags to mark animals or plants in some biological study, geneticists use variant genes to keep track of specific chromosomes, cells, or individuals. This technique has found application across the breadth of the biological disciplines, from cell biology to evolution and ecology. For example, genes for human diseases are now being isolated by virtue of their chromosomal proximity to unrelated marker sequences.

We have seen that molecular genetic engineering has opened up new vistas in applied biotechnology, but the same techniques are just as useful in basic research. With the ability to move genes from organism to organism, scientists have produced plants that glow because they express the phosphorescence genes of fireflies, plants that acquire cold hardiness by expressing antifreeze genes from fish, and giant mice that are expressing the growth hormone genes of rats. Scientists have manipulated genes in yeast to produce totally artificial, functional chromosomes. The ability to isolate a gene in a test tube, to modify its structure in specific ways, and then to reinsert it back into the organism has pro-

Figure 1-9 DNA replication is the basis of the perpetuation of life through time.

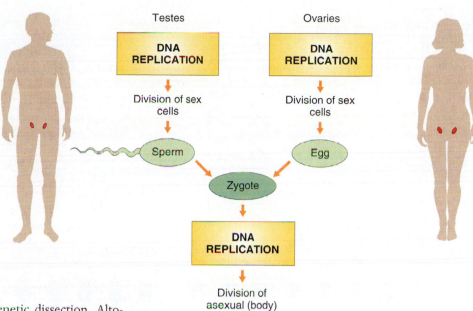

vided the sharpest of scalpels for genetic dissection. Altogether, genetic engineering has revolutionized the biological sciences, and no biologist today can afford to be ignorant of this powerful analytical tool.

Message   Genetic analysis is used not only in the study of heredity, but in all other areas of biology.

Perhaps the biggest biological success story of all is the elucidation of precisely how the genes do their job, in other words, how information becomes form. It is a marvelous story that has developed with amazing rapidity within the span of the careers of scientists now still in their fifties, scientists who never in their wildest dreams imagined that by the 1990s geneticists would be sequencing entire genomes. But today the coding and flow of genetic information in cells are foundations of modern biological thought and baselines from which experimental explorations start off. It is worth summarizing the highlights of how genetic information flows—or, equivalently, how genes act—which has been called the *new paradigm of biology*. Figure 1-10 shows diagrammatically the essentials of gene action in a generalized cell of a **eukaryote.** Eukaryotes are those organisms whose cells have a membrane-bound nucleus. Animals, plants, and fungi are all eukaryotes. Inside the nucleus are the chromosomes, and outside the nucleus are a complex array of membranous structures, including the endoplasmic reticulum and Golgi apparatus, and organelles such as the mitochondria and chloroplasts.

Inside the eukaryotic nucleus, some genes are active more or less constantly, but others have to be turned on and off to suit the needs of the cell or the organism. The signal to activate a gene may come from outside the cell, from, for example, a substance such as a steroid hormone. Or the signal may come from within the cell, for example, from a special regulatory gene whose job it is to turn other genes on and off. The regulatory substances bind to a special region of the gene and initiate the synthesis of copies of the gene's

DNA. These copies are in the form of a molecule called *RNA.* Noncoding regions of the gene called **introns** are cut out at this RNA stage, and the remaining RNA sequence is called **messenger RNA,** or **mRNA.** The mRNA molecules pass out through the nuclear pores into the cytoplasm, and here the information in the sequence of the mRNA is translated into **protein.** Each gene codes for a separate protein, each with specific functions either within the cell (for example, the dark-green protein in Figure 1-10) or for export to other parts of the organism (the pink protein). Proteins are the most important manifestations of form in living organisms: when you look at an organism, what you see is either protein or something that has been made by a protein.

The synthesis of proteins for export (secretory proteins) takes place on the surface of the rough endoplasmic reticulum, a system of large flattened vesicles that is studded on the outside with ribosomes, the molecular machines that synthesize protein. The mRNA passes through the ribosomes, which catalyze the assembly of a string of amino acids that will constitute the protein. Each amino acid is brought to the ribosome by a specific **transfer RNA (tRNA)** molecule that docks onto a specific coding unit of the mRNA.

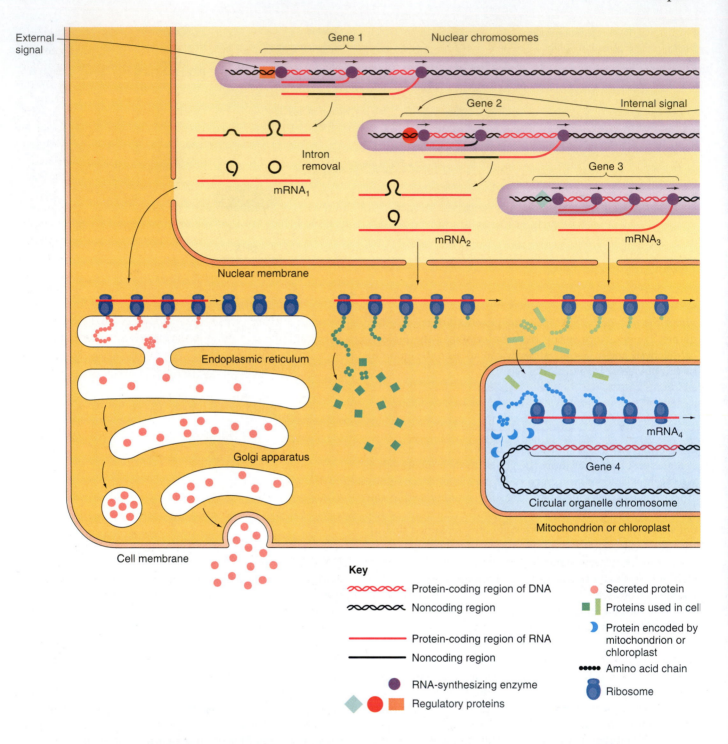

Key

~~~~~ Protein-coding region of DNA

~~~~~ Noncoding region

———— Protein-coding region of RNA

———— Noncoding region

● RNA-synthesizing enzyme

◆ ● ▬ Regulatory proteins

● Secreted protein

■ ▌ Proteins used in cell

◗ Protein encoded by mitochondrion or chloroplast

●●●●● Amino acid chain

▯ Ribosome

The tRNAs are synthesized off special tRNA genes. No tRNAs are ever translated into protein; they recycle constantly, delivering their specific amino acid to the ribosomes. The ribosome itself is made of a complex set of proteins plus several kinds of RNA called *ribosomal RNA (rRNA)*. The genes for rRNA are located in a special chromosomal region called the *nucleolar organizer*. Like tRNA, rRNA is never translated into protein.

The completed amino acid chains are passed into the lumen of the endoplasmic reticulum, where they fold up spontaneously to take on their protein shape. The proteins may be modified at this stage, but eventually are passed into

Figure 1-10 Simplified view of gene action in a eukaryotic cell. The basic flow of genetic information is from DNA to RNA to protein. Four types of genes are shown. Gene 1 responds to external regulatory signals and makes a protein for export; gene 2 responds to internal signals and makes a protein for use in the cytoplasm; gene 3 makes a protein to be transported into an organelle; gene 4 is part of the organelle DNA and makes a protein for use inside its own organelle. Most eukaryotic genes contain introns, regions (generally noncoding) that are cut out in the preparation of functional messenger RNA. Note that many organelle genes have introns and that an RNA-synthesizing enzyme is needed for organelle mRNA synthesis. These details have been omitted from the diagram of the organelle for clarity. (Introns will be explained in detail in subsequent chapters.)

the chambers of the Golgi apparatus, and on into secretory vessels that eventually fuse with the cell membrane and release their contents to the outside.

Proteins destined to function in the cytosol or in mitochondria and chloroplasts are synthesized on ribosomes unbound to membranes. For example, proteins that function as enzymes in glycolysis follow this route. Protein synthesis occurs by the same mechanism using the same kinds of tRNAs. The proteins destined for organelles are specially tagged to target their insertion into the organelle. Mitochondria and chloroplasts have their own small circular DNA chromosomes. Synthesis of proteins encoded by genes on mitochondrial or chloroplast chromosomes takes place on ribosomes inside the organelles themselves. Therefore the proteins in the mitochondria and chloroplasts are of two different origins, either nucleus-coded and imported into the organelle, or organelle-coded and synthesized within the organelle compartment.

Prokaryotes are organisms such as bacteria whose cells have a simpler structure; there is no nucleus nor other membrane-bound structures within these cells. Protein synthesis in prokaryotes is generally similar, using mRNA, tRNA, ribosomes, and the genetic code, but there are some important differences. For example, prokaryotes have no introns, and, furthermore, there are no membrane-bound compartments for the RNA or protein to pass through.

Message The flow of information from DNA to RNA to protein has become a central principle of biology.

Genes and Environment

The net outcome of gene action is that a protein product is made that has one of two basic functions, depending on the gene. First, the protein may be a **structural protein,** contributing to the physical properties of cells or organisms. Examples are microtubule, muscle, and hair proteins. Second, the protein may be an **enzyme** that catalyzes one of the chemical reactions of the cell. Therefore, by coding for proteins, genes determine two important facets of biological structure and function. However, genes cannot act by themselves. The other crucial component in the formula is the environment. The environment influences gene action in many ways, which we will learn about in the subsequent chapters. In the present discussion, however, it is relevant to note that the environment provides the raw materials for the synthetic processes controlled by genes. For example, animals obtain several of the amino acids for their proteins as part of their diet. Again, most of the chemical syntheses in plant cells use carbon atoms taken from the air as carbon dioxide. Finally, bacteria and fungi absorb from their surroundings many substances that are simply treated as carbon and nitrogen skeletons, and their enzymes convert these into the compounds that constitute the living cell. Thus through genes an organism builds the orderly process that we call life out of disorderly environmental materials.

Genetic Determination

From our brief look at gene action, we can see that living organisms mobilize the components of the world around themselves and convert these components into their own living material. An acorn becomes an oak tree, using in the process only water, oxygen, carbon dioxide, some inorganic materials from the soil, and light energy.

An acorn develops into an oak, while the spore of a moss develops into a moss, although both are growing side by side in the same forest (Figure 1-11). The two plants that result from these developmental processes resemble their

Figure 1-11 The genes of a moss direct environmental components to be shaped into a moss, whereas the genes of a tree cause a tree to be constructed from the same components. (Andre Bartschi)

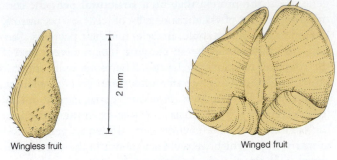

Wingless fruit

2 mm

Winged fruit

Figure 1-12 The fruits of two different forms of *Plectritis congesta*, the sea blush. Any one plant has either all wingless or all winged fruits. In every other way the plants are identical. A simple genetic difference determines the difference in the fruits.

parents and differ from each other, even though they have access to the same narrow range of inorganic materials from the environment. The parents pass to their offspring the specifications for building living cells from environmental materials. These specifications are in the form of genes in the fertilized egg. Because of the information in the genes, the acorn develops into an oak and the moss spore becomes a moss.

Just as genes maintain differences between species such as the oak and moss, they also maintain differences within species. Consider plants of the species *Plectritis congesta,* the sea blush. Two forms of this species are found wherever the plants grow in nature: one form has wingless fruits, and the other has winged fruits (Figure 1-12). These plants will self-pollinate, and we can observe the offspring that result from such "selfs" when these are grown in a greenhouse under uniform conditions: we commonly observe that the selfed progeny of a winged-fruited plant are all winged-fruited and that the selfed progeny from a wingless-fruited plant all have wingless fruits. Because all the progeny were grown in the same environment, we can rule out the possibility that environmental differences cause some plants to bear wingless fruits and others, winged. We can safely conclude that the fruit-shape difference between the original plants, which each passed on to its selfed progeny, results from the different genes they carry.

Plectritis shows two inherited forms that are both perfectly normal. The determinative power of genes is probably more often demonstrated by differences in which one form is normal and the other abnormal. The human inherited disease sickle-cell anemia provides a good example. The underlying cause of the disease is a variant of hemoglobin, the oxygen-transporting protein molecule found in red blood cells. Normal people have a type of hemoglobin called *hemoglobin A,* the information for which is encoded in a gene. A minute chemical change at the molecular level in the DNA of this gene results in the production of a slightly changed hemoglobin, termed *hemoglobin S.* In people possessing only hemoglobin S, the ultimate effect of this small change is severe ill health and usually death. The gene

works its effect on the organism through a complex "cascade effect," as summarized in Figure 1-13.

Observations like these lead to the model of how genes and the environment interact shown in Figure 1-14. In this view, the genes act as a set of instructions for turning more-or-less undifferentiated environmental materials into a specific organism, much as blueprints specify what form of house is to be built from basic materials. The same bricks, mortar, wood, and nails can be made into an A-frame or a flat-roofed house, according to different plans. Such a model implies that the genes are really the dominant elements in

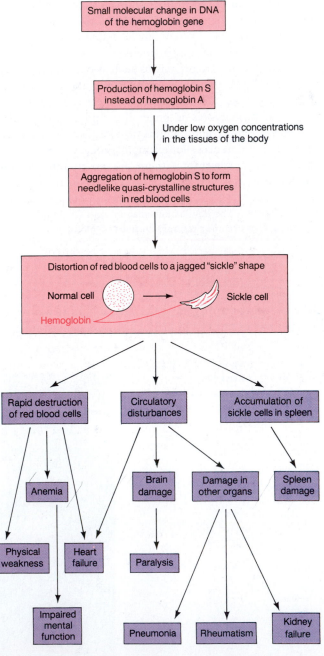

Figure 1-13 Chain of events in human sickle-cell anemia.

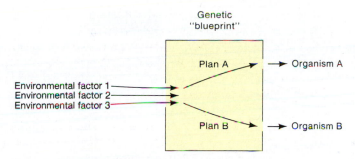

Figure 1-14 A model of determination that emphasizes the role of genes.

the determination of organisms; the environment simply supplies the undifferentiated raw materials.

Environmental Determination

But now consider two monozygotic ("identical") twins, the products of a single fertilized egg that divided and produced two complete individuals with identical genes. Suppose that the twins are born in England but are separated at birth and taken to different countries. If one is reared in China by Chinese-speaking foster parents, she will speak Chinese, while her sister reared in Budapest will speak Hungarian. Each will absorb the cultural values and customs of her environment. Although the twins begin life with identical genetic properties, the different cultural environments in which they live will produce differences between them (and differences from their parents). Obviously, the differences in this case are due to the environment, and genetic effects are of little importance in determining the differences.

This example suggests the model of Figure 1-15, which is the converse of that shown in Figure 1-14. In the model in Figure 1-15, the genes impinge on the system, giving certain general signals for development, but the environment determines the actual course of action. Imagine a set of specifications for a house that simply calls for "a floor that will support 300 pounds per square foot" or "walls with an insulation factor of 15 inches"; the actual appearance and other characteristics of the structure would be determined by the available building materials.

Our different types of example—of purely genetic effect versus that of the environment—lead to two very different models. First, consider the seed example: given a pair of seeds and a uniform growth environment, we would be unable to predict future growth patterns solely from a knowledge of the environment. In any environment we can imagine, if the organisms develop, the acorn becomes an oak and the spore becomes a moss. The model of Figure 1-14 applies here. Second, consider the twins: no information about the set of genes they inherit could possibly enable us to predict their ultimate languages and cultures. Two individuals that are *genetically different* may develop differently in the *same environment*, but two *genetically identical* individuals may develop differently in *different environments*. The model of Figure 1-15 applies here.

In general, of course, we deal with organisms that differ in both genes and environment. If we wish to predict how a living organism will develop, we must first know the genetic constitution that it inherits from its parents. Then we must know the historical sequence of environments to which the developing organism is exposed. Every organism has a developmental history from birth to death. What an organism will become in the next moment depends critically both on the environment it encounters during that moment and on its present state. It makes a difference to an organism not only what environments it encounters but in what sequence it encounters them. A fruit fly (*Drosophila*) develops normally at 20°C. If the temperature is briefly raised to 37°C early in its pupal stage of development, the adult fly will be missing part of the normal vein pattern on its wings. However, if this "temperature shock" is administered just 24 hours later, the vein pattern develops normally.

Our discussion of how genes and the environment interact has brought us to a point where we can appreciate a more general model (Figure 1-16) in which genes and the environment jointly determine (by some rules of development) the actual characteristics of the organism.

Message As an organism transforms developmentally from one stage of its life to another, its genes interact with its environment at each moment of its life history. The interaction of genes and environment determines what organisms are.

Figure 1-15 A model of determination that emphasizes the role of the environment.

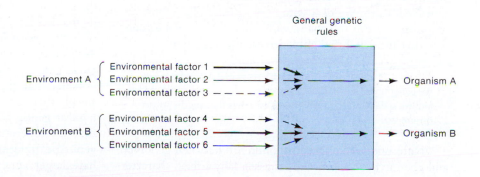

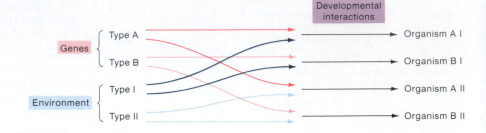

Genotype and Phenotype

In studying how genes and the environment interact to produce an organism, geneticists have developed some useful terms, which are introduced in this section.

A typical organism resembles its parents more than it resembles unrelated individuals. Thus, we often speak as if the individual characteristics themselves are inherited: "He gets his brains from his mother," or "She inherited diabetes from her father." Yet our discussion in the preceding section shows that such statements are inaccurate. "His brains" and "her diabetes" develop through long sequences of events in the life histories of the affected people, and both genes and environment play roles in those sequences. In the biological sense, individuals inherit only the molecular structures of the fertilized eggs from which they develop. Individuals inherit their genes, not the end products of their individual developmental histories.

To prevent such confusion between genes (which are inherited) and developmental outcomes (which are not), geneticists make a fundamental distinction between the genotype and the phenotype of an organism. Organisms share the same genotype if they have the same set of genes. Organisms share the same phenotype if they look or function alike.

Strictly speaking, the genotype describes the complete set of genes inherited by an individual, and the phenotype describes all aspects of the individual's morphology, physiology, behavior, and ecological relationships. In this sense, no two individuals ever belong to the same phenotype, because there is always some difference (however slight) between them in morphology or physiology. Also, except for individuals produced from another organism by asexual reproduction, any two organisms differ at least a little in genotype. In practice, we use the terms *genotype* and *phenotype* in a more restricted sense. We deal with some partial phenotypic description (say, eye color) and with some subset of the genotype (say, the genes that influence eye pigmentation).

Message When we use the terms *phenotype* and *genotype*, we generally mean "partial phenotype" and "partial genotype," and we specify one or a few traits and genes that are the subsets of interest.

Note one very important difference between genotype and phenotype: the genotype is essentially a fixed character of an individual organism; the genotype remains constant throughout life and is essentially unchanged by environmental effects. Most phenotypes change continually throughout the life of an organism as its genes interact with a sequence of environments. Fixity of genotype does not imply fixity of phenotype.

The Norm of Reaction

How can we quantify the relation between the genotype, the environment, and the phenotype? For a particular genotype, we could prepare a table showing the phenotype that would result from the development of that genotype in each possible environment. Such a set of environment-phenotype relationships for a given genotype is called the **norm of reaction** of the genotype. In practice, of course, we can make such a tabulation only for a partial genotype, a partial phenotype, and some particular aspects of the environment. For example, we might specify the eye sizes that fruit flies would have after developing at various constant temperatures; we could do this for several different eye-size genotypes to get the norms of reaction of the species.

Figure 1-17 represents just such norms of reaction for three eye-size genotypes in the fruit fly *Drosophila melanogaster*. The graph is a convenient summary of more extensive tabulated data. The size of the fly eye is measured by counting its individual facets, or cells. The vertical axis of the graph shows the number of facets (on a logarithmic scale); the horizontal axis shows the constant temperature at which the flies develop.

Three norms of reaction are shown on the graph. When flies of the wild-type genotype that is characteristic of flies in natural populations are raised at higher temperatures, they develop eyes that are somewhat smaller than those of wild-type flies raised at cooler temperatures. The graph shows that wild-type phenotypes range from more than 700 to 1000 facets—the wild-type norm of reaction. A fly that has the *ultrabar* genotype has smaller eyes than wild-type flies regardless of temperature during development. Temperatures have a stronger effect on development of *ultrabar* genotypes than on wild-type genotypes, as we see by noticing that the *ultrabar* norm of reaction slopes more steeply than the wild-type norm of reaction. Any fly of the *infrabar* genotype also has smaller eyes than any wild-type fly, but temperatures have the opposite effect on flies of this genotype; *infrabar* flies raised at higher temperatures tend to have larger eyes than those raised at lower temperatures. These norms of reaction indicate that the relationship

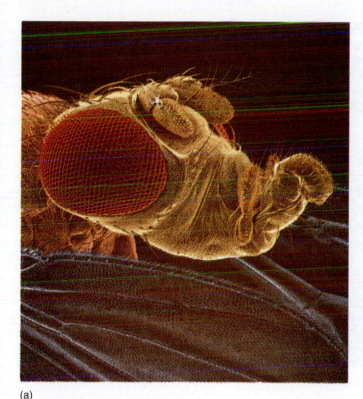

(a)

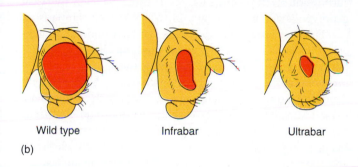

Wild type Infrabar Ultrabar

(b)

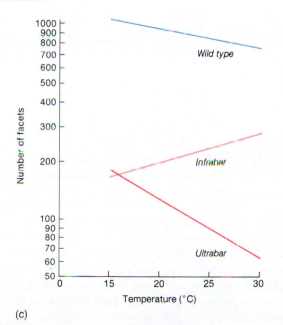

(c)

Figure 1-17 Norms of reaction to temperature for three different eye-size genotypes in *Drosophila*. (a) Closeup showing how the normal eye comprises hundreds of units called *facets*. The number of facets determines eye size. (b) Relative eye sizes of wild-type, *infrabar*, and *ultrabar* flies raised at the higher end of the temperature range. (c) Norm of reaction curves for the three genotypes. (Part a, Don Rio and Sima Misra, University of California, Berkeley.)

between genotype and phenotype is complex rather than simple.

Message A single genotype may produce different phenotypes, depending on the environment in which organisms develop. The same phenotype may be produced by different genotypes, depending on the environment.

If we know that a fruit fly has the wild-type genotype, this information alone does not tell us whether its eye has 800 or 1000 facets. On the other hand, the knowledge that a fruit fly's eye has 170 facets does not tell us whether its genotype is *ultrabar* or *infrabar*. We cannot even make a general statement about the effect of temperature on eye size in *Drosophila*, because the effect is opposite in two different genotypes. We see from Figure 1-17 that some genotypes do differ unambiguously in phenotype, no matter what the environment: any wild-type fly has larger eyes than any *ultrabar* or *infrabar* fly. But other genotypes overlap in phenotypic expression: the eyes of an *ultrabar* fly may be larger or smaller than those of an *infrabar* fly, depending on the temperatures at which the individuals developed.

To obtain a norm of reaction like the norms of reaction in Figure 1-17, we must allow different individuals of identical genotype to develop in many different environments. To carry out such an experiment, we must be able to obtain or produce many fertilized eggs with identical genotypes. For example, to test a human genotype in 10 environments, we would have to obtain genetically identical sibs and raise each individual in a different milieu. Obviously, that is possible neither biologically nor socially. At the present time, we do not know the norm of reaction at any human genotype for any character in any set of environments. Nor is it clear how we can ever acquire such information without the unacceptable manipulation of human individuals.

For a few experimental organisms, special genetic methods make it possible to replicate genotypes and thus to determine norms of reaction. Such studies are particularly easy in plants that can be propagated vegetatively (that is, by cuttings). The pieces cut from a single plant all have the same genotype, so all offspring produced in this way have identical genotypes. Such a study has been done on the yarrow plant, *Achillea millefolium* (Figure 1-18a). The experimental results are shown in Figure 1-18b. Many plants were collected, and three cuttings were taken from each plant. One cutting from each plant was planted at low elevation (30 meters above sea level), one at medium elevation (1400 meters), and one at high elevation (3050 meters). Figure 1-18b shows the mature individuals that developed from the

Figure 1-18 (a) *Achillea millefolium.* (b) Norms of reaction to elevation for seven different *Achillea* plants (seven different genotypes). A cutting from each plant was grown at low, medium, and high elevations. (Part a, Harper Horticultural Slide Library; part b, Carnegie Institution of Washington.)

(a)

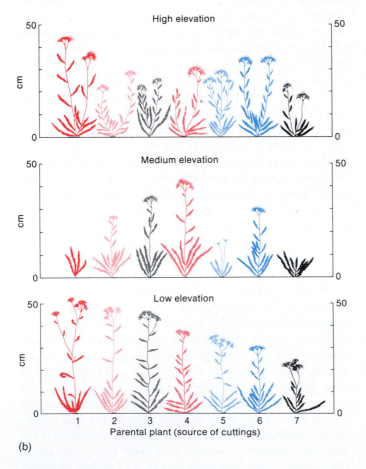

(b)

cuttings of seven plants; each set of three plants of identical genotype is aligned vertically in the figure for comparison.

First, we note an average effect of environment: in general, the plants grew poorly at the medium elevation. This is not true for every genotype, however; the cutting of plant 4 grew best at the medium elevation. Second, we note that no genotype is unconditionally superior in growth to all others. Plant 1 showed the best growth at low and high elevations but showed the poorest growth at the medium elevation. Plant 6 showed the second-worst growth at low elevation and the second-best at high elevation. Once again, we see the complex relationship between genotype and phenotype. Figure 1-19 graphs the norms of reaction derived from the results shown in Figure 1-18b. Each genotype has a different norm of reaction, and the norms cross one another, so that we cannot identify either a "best" genotype or a "best" environment for *Achillea* growth.

We have seen two different patterns of reaction norms. The difference between the wild-type and the other eye-size genotypes in *Drosophila* is such that the corresponding phenotypes show a consistent difference, regardless of the environment. Any fly of wild-type genotype has larger eyes than any fly of the other genotypes, so we could (imprecisely) speak of "large-eye" and "small-eye" genotypes. In this case, the differences in phenotype between genotypes are much greater than the variation within a genotype caused by development in different environments. But the variation for a single *Achillea* genotype in different environments is so great that the norms of reaction cross one another and form no consistent pattern. In this case, it makes no sense to identify a genotype with a particular phenotype except in terms of response to particular environments.

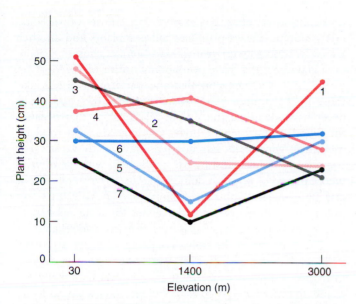

Figure 1-19 Graphic representation of the complete set of results of the type shown in Figure 1-18. Each line represents the norm of reaction of one plant.

Developmental Noise

Thus far, we have assumed that a phenotype is uniquely determined by the interaction of a specific genotype and a specific environment. But a closer look shows some further unexplained variation. According to Figure 1-17, a *Drosophila* of wild-type genotype raised at 16°C has 1000 facets in each eye. In fact, this is only an average value; one fly raised at 16°C may have 980 facets and another may have 1020. Perhaps these variations are due to slight fluctuations in the local environment or slight differences in genotypes. However, a typical count may show that a fly has, say, 1017 facets in the left eye and 982 in the right eye. In another fly, the left eye has slightly fewer facets than the right eye. Yet the left and right eyes of the same fly are genetically identical. Furthermore, under typical experimental conditions, the fly develops as a larva (a few millimeters long) burrowing in homogeneous artificial food in a laboratory bottle and then completes its development as a pupa (also a few millimeters long) glued vertically to the inside of the glass high above the food surface. Surely the environment does not differ significantly from one side of the fly to the other! But if the two eyes experience the same sequence of environments and are identical genetically, then why is there any phenotypic difference between the left and right eyes?

Differences in shape and size are partly dependent on the process of cell division that turns the zygote into a multicellular organism. Cell division, in turn, is sensitive to molecular events within the cell, and these may have a relatively large random component. For example, the vitamin biotin is essential for *Drosophila* growth, but its average concentration is only one molecule per cell. Obviously, the rate of any process that depends on the presence of this molecule will fluctuate as the concentration of biotin varies. But

cells can divide to produce differentiated eye cells only within the relatively short developmental period during which the eye is being formed. Thus, we would expect random variation in such phenotypic characters as the number of eye cells, the number of hairs, the exact shape of small features, and the variations of neurons in a very complex central nervous system—even when the genotype and the environment are precisely fixed. Even such structures as the very simple nervous systems of nematodes vary at random for this reason. Random events in development lead to variation in phenotype; this variation is called **developmental noise.**

Message In some characteristics, such as eye cells in *Drosophila*, developmental noise is a major source of the observed variations in phenotype.

Like noise in a verbal communication, developmental noise adds small random variations to the predictable development governed by norms of reaction. Adding developmental noise to our model of phenotypic development, we obtain something like Figure 1-20. With a given genotype and environment, there is a range of possible outcomes for each developmental step. The developmental process does contain feedback systems that tend to hold the deviations within certain bounds, so that the range of deviation does not increase indefinitely through the many steps of development. However, this feedback is not perfect. For any given genotype, developing in any given sequence of environments, there remains some uncertainty as to the exact phenotype that will result.

Techniques of Genetic Analysis

Our discussion thus far has been based on the wisdom of hindsight. With the wealth of genetic knowledge we now share, we can make generalizations about DNA, genes, phe-

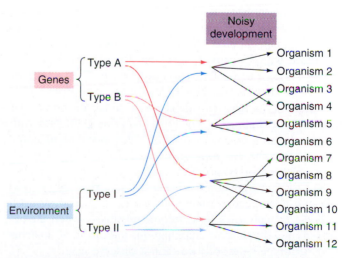

Figure 1-20 A model of phenotypic determination that shows how genes, environment, and developmental noise interact to produce a phenotype.

notypes, and genotypes as though these concepts were self-evident. But obviously this was not always the case; the current wisdom was acquired only after extensive genetic research over the years. Mendel, for example, almost certainly was completely without a conceptual basis for his research at the beginning of his work, but he was able to piece together his genetic principles from the results of his many experiments. This is true of genetic research in general: we start in the unknown, and then ideas and facts emerge out of experimentation. But how does everyday genetics work? We shall explore the answer to this question in much of the rest of this book, but we begin here with an overview of the principles of genetic research.

We have seen that the process of identifying the specific hereditary components of a biological system is called *genetic dissection*. In the same way that an anatomist probes biological structure and function with a scalpel, the geneticist probes biological structure and function armed with genetic variants, usually abnormal ones. If the geneticist is interested in a biological process X, he or she embarks on a search for genetic variants that affect X. Each variant identifies a separate component of X. In much the same way that an auto mechanic can investigate how an internal combustion engine works by pulling out spark-plug leads, or adjusting the gas mixture, the geneticist "tinkers" with a living system by manipulating it genetically.

Message   Genetic dissection is a powerful way of discovering the components of any biological process.

Geneticists use and analyze hereditary variants from the molecular level, probing cellular and organismal processes, all the way up to the population level, where the variants reveal evolutionary processes. The spectrum of variants that has been obtained and studied effectively by geneticists is staggering; as we have seen, there are variants affecting shape, number, biochemical function, and so on. In fact, the general finding has been that genetic variants can be obtained for virtually any biological structure or process of interest to an investigator.

Recall that the genotypes underlying a set of variants of a given biological character fall into one of two categories: (1) those whose norms of reaction do not overlap and (2) those whose norms of reaction do overlap. The techniques of genetic analysis are different for the two categories.

Geneticists have elucidated the cellular and molecular processes of organisms mostly by analyzing genotypes with nonoverlapping norms of reaction. The appeal of such systems is that the environment can virtually be ignored, because each genotype produces a discrete, identifiable phenotype. In fact, geneticists have deliberately sought and selected just such variants. Experiments become much easier—actually, they only become possible—when the observed phenotype is an unambiguous indication of a particular genotype. After all, we cannot observe genotypes; we can visibly distinguish only phenotypes.

The availability of such variants made Mendel's experiments possible. He used genotypes that were established

horticultural varieties. For example, in one set of variants, one genotype always produces tall pea plants and another always produces dwarf plants. Had Mendel chosen variants with genotypes that have overlapping norms of reaction, he would have obtained complex and patternless results like those in the *Achillea* experiment (Figure 1-18b) and could never have identified the simple relationships that became the foundation of genetic understanding. Much of modern molecular genetics is based on research with variants of bacteria that show distinctive characters with genotypes whose norms of reaction do not overlap. For instance, modern DNA manipulation technology relies heavily on selection systems based on bacterial genes for drug resistance; the expression of such genes is very clear-cut and reliable.

The variants in the nonoverlapping category are often strikingly different. Many such variants could never survive in nature, simply because they are so developmentally extreme. In other cases, such as the winged and wingless fruits of *Plectritis,* the strikingly different forms do appear regularly in natural populations.

Reflecting the importance of discrete genetic variants, most of the rest of this book is devoted to the analysis of this kind of variation. The story begins with Mendel's research and theories, proceeds through classical genetics, and ends with the startling discoveries of molecular biology. It must be emphasized that the entire development of this knowledge critically depended on the availability of phenotypes that have simple relationships to genotypes. Because our study of genetic analysis necessarily puts so much emphasis on such simple trait differences, you may get the impression that they represent the most common relationship between gene and trait. They do not. For example, although hundreds of *Drosophila* mutations have been well studied, only about one-quarter are ideally suited for the purposes of careful genetic analysis. Environment and developmental age can be quite important to the expression of even the most useful variants. The widely used mutation *purple* produces an eye color like the normal ruby-red color in very young flies, which darkens to a distinguishable difference only with age. The mutation *Curly*, one of the most important

Figure 1-21 The *curled wing* variant of *Drosophila melanogaster,* caused by a simple genetic change in one gene. In wild-type flies, the wings are flat. (Peter Bryant/BPS.)

tools in genetic analysis in *Drosophila*, results in curled wings at 25°C (Figure 1-21) but in normally straight wings at 19°C.

Simple one-to-one relationships of genotype to phenotype dominate the world of experimental genetics, but in the natural world such relationships are rare. The relationships of genotype to phenotype in nature are almost always one-to-many rather than one-to-one. This explains the rarity of discrete phenotypic classes in natural populations.

Obviously, the analysis of these one-to-many relationships is far more complex. The researcher is confronted with a bewildering range of phenotypes. Special statistical techniques must be used to disentangle the genetic, environmental, and noise components. Chapter 24 deals with such techniques.

Message In general, the relationship between genotype and phenotype cannot be extrapolated from one species to another or even between phenotypic traits that seem superficially similar within a species. A genetic analysis must be carried out for each particular case.

SUMMARY

Genetics is the study of genes at all levels from molecules to populations. As a modern discipline, it began in the 1860s with the work of Gregor Mendel, who first formulated the idea that genes exist. We now know that a gene is a functional region of the long DNA molecule that constitutes the fundamental structure of a chromosome.

Genetics has had a profound impact on human affairs. Much of our food and clothing comes from genetically improved organisms, and molecular genetic engineering is expanding the scope of applied genetics. In addition, a large proportion of human ill health has a genetic component. Genetic research has also provided important insights on the way we see ourselves in relation to the organic world and to the rest of the universe.

Genetics has revolutionized biology by showing that most organisms on the planet work on a common information storage and expression system, centered on DNA—a system in which information flows from DNA to RNA to protein. Genetic dissection, moreover, is an incisive analytical tool used in all the biological sciences.

Genes do not act in a vacuum, but interact with the environment at many levels in producing a phenotype. The relationship of genotype to phenotype across an environmental range is called the *norm of reaction*. Most of the major advances in genetics have come from studying laboratory systems having a simple, one-to-one correspondence of genotype to phenotype. However, in natural populations phenotypic variation generally shows a more complex relationship to genotype and not a one-to-one correspondence.

Concept Map

Each chapter contains an exercise on drawing concept maps. As you draw these maps, you will be organizing knowledge you have just acquired. As we learn, we must place new knowledge into the overall conceptual framework of the discipline. An essential part of this is to discern the interrelationships between new knowledge and previously acquired knowledge. It is not as easy as it sounds, and the mind can sometimes trick us into believing that we have the overall picture. Concept maps are a good way of proving to ourselves that we really do know how the various structures, processes, and ideas of genetics interrelate. The maps can also help in pinpointing gaps in our understanding.

A selection of terms is given, and the challenge is to draw lines or arrows between the ones that you think are related, with a description on the arrow showing just what the relationship is. Draw as many relationships as possible. Force yourself to make connections, no matter how remote they might first seem. The arrow description must be clear enough that another person reading the map can understand what you are thinking. Some-times the maps reveal a simple series of consecutive steps, sometimes a complex network of interactions. There is no one correct answer; there are many correct variations. But of course, you may make some incorrect connections, and finding these misunderstandings is part of the purpose of the exercise.

A sample concept map interrelating the following terms is shown in Figure 1-22.

base-pair substitution / recessive / 3:1 ratio / albino / environment / melanin / metabolic pathway / active site / enzyme / gene

Now draw a concept map for this chapter, interrelating as many of the following terms as possible. Note that the terms are listed in no particular order.

genotype / phenotype / norm of reaction / developmental noise / environment / development / organism

Figure 1-22 Sample concept map.

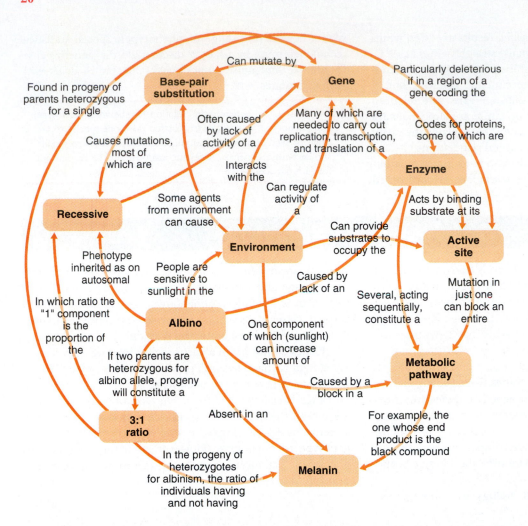

Found in progeny of parents heterozygous for a single

Can mutate by

Particularly deleterious if in a region of a gene coding the

Base-pair substitution

Gene

Often caused by lack of activity of a

Many of which are needed to carry out replication, transcription, and translation of a

Codes for proteins, some of which are

Causes mutations, most of which are

Interacts with the

Enzyme

Recessive

Some agents from environment can cause

Can regulate activity of a

Acts by binding substrate at its

Phenotype inherited as on autosomal

People are sensitive to sunlight in the

Environment

Can provide substrates to occupy the

Active site

In which ratio the "1" component is the proportion of the

Caused by lack of an

Several, acting sequentially, constitute a

Mutation in just one can block an entire

Albino

One component of which (sunlight) can increase amount of

Metabolic pathway

If two parents are heterozygous for albino allele, progeny will constitute a

Caused by a block in a

3:1 ratio

Absent in an

For example, the one whose end product is the black compound

In the progeny of heterozygotes for albinism, the ratio of individuals having and not having

Melanin

GENERAL QUESTIONS

1. Define genetics. Do you think the ancient Egyptian racehorse breeders were geneticists? How might their approaches have differed from those of modern geneticists?

2. List three ways in which genetics has affected modern society.

3. How does DNA dictate the general properties of a species?

4. What are the two features of DNA that suit it for its role as a hereditary molecule? Can you think of alternative types of hereditary molecules that might be found in extraterrestrial life-forms?

5. What is the relevance of norms of reaction to phenotypic variation within a species?

6. What are some causes of phenotypic variation within a species?

7. Is the formula *genotype + environment = phenotype* accurate?

8. List five ethical dilemmas in genetics.

9. What is the goal of genetic dissection?

10. In what sense has genetics unified biology?

2

Mendelian Analysis

The small monastery garden that the monk Gregor Mendel used for his experiments with peas. A statue of Mendel is visible in the background. Today this part of Mendel's monastery is a museum, and the curators have planted red and white begonias in an array that graphically represents some of Mendel's results. (Anthony Griffiths)

KEY CONCEPTS

▶ The existence of genes can be inferred by observing certain progeny ratios in crosses between hereditary variants.

▶ A discrete character difference is often determined by a difference in a single gene.

▶ In higher organisms, each gene is represented twice in each cell.

▶ During sex-cell formation, each member of a gene pair separates into half the sex cells.

▶ During sex-cell formation, different genes are often observed to behave independently of one another.

The gene, the basic functional unit of heredity, is the focal point of the discipline of modern genetics. In all lines of genetic research, the gene provides the common unifying thread to a great diversity of experimentation. Geneticists are concerned with the transmission of genes from generation to generation, with the physical structure of genes, with the variation in genes, and with the ways in which genes dictate the features of a species.

In this chapter we trace how the concept of the gene arose. We shall see that genetics is, in one sense, an abstract science: most of its entities began as hypothetical constructs in the minds of geneticists and were later identified in physical form.

The concept of the gene (but not the word) was first proposed in 1865 by Gregor Mendel. Until then, little progress had been made in understanding heredity. The prevailing notion was that the spermatozoon and egg contained a sampling of essences from the various parts of the parental body; at conception, these essences somehow blended to form the pattern for the new individual. This idea of **blending inheritance** evolved to account for the fact that offspring typically show some characteristics that are similar to those of both parents. However, there are some obvious problems associated with this idea, one of which is that offspring are not always an intermediate blend of their parents' characteristics. Attempts to expand and improve this theory led to no better understanding of heredity.

As a result of his research with pea plants, Mendel proposed instead a theory of **particulate inheritance.** According to Mendel's theory, characters are determined by discrete units that are inherited intact down through the generations. This model explained many observations that could not be explained by the idea of blending inheritance. It also served well as a framework for the later, more detailed understanding of the mechanism of heredity.

The importance of Mendel's ideas was not recognized until about 1900 (after his death). His written work was then rediscovered by three scientists, after each had independently obtained the same kind of results. Mendel's work constitutes the prototype for genetic analysis. He laid down an experimental and logical approach to heredity that is still used today.

Mendel's Experiments

Gregor Mendel was born in the district of Moravia, then part of the Austro-Hungarian Empire. At the end of high school he entered the Augustinian monastery of St. Thomas in the city of Brünn, now Brno of the Czech Republic. His monastery was dedicated to teaching science and to scientific research, so Mendel was sent to university in Vienna in order to obtain his teaching credentials. However, he failed his exams and returned to the monastery at Brünn. There he embarked on the research program of plant hybridiza-

tion that was posthumously to earn him the title of founder of the science of genetics.

Mendel's studies provide an outstanding example of good scientific technique. He chose research material well suited to the study of the problem at hand, designed his experiments carefully, collected large amounts of data, and used mathematical analysis to show that the results were consistent with his explanatory hypothesis. The predictions of the hypothesis were then tested in a new round of experimentation.

Mendel studied the garden pea (*Pisum sativum*) for two main reasons. First, peas were available from seed merchants in a wide array of distinct shapes and colors that could be easily identified and analyzed. Second, peas can either **self** (self-pollinate) or be cross-pollinated. The peas self because the male parts (anthers) and female parts (ovaries) of the flower—which produce the pollen containing the sperm and the ovules containing eggs, respectively—are enclosed by two petals fused to form a compartment called a keel (Figure 2-1). The gardener or experimenter can **cross** (cross-pollinate) any two pea plants at will. The anthers from one plant are removed before they have opened to shed their pollen, an operation called emasculation that is done to prevent selfing. Pollen from the other plant is then transferred to the receptive stigma with a paintbrush or on anthers themselves (Figure 2-2). Thus, the experimenter can readily choose to self or to cross the pea plants.

Other practical reasons for Mendel's choice of peas were that they are cheap and easy to obtain, take up little space, have a relatively short generation time, and produce many offspring. Such considerations enter into the choice of organism for any piece of genetic research. The choice of organism is a crucial decision and is often based on not only scientific criteria but also a good measure of expediency.

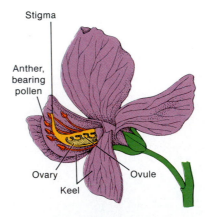

Figure 2-1 A pea flower with the keel cut and opened to expose the reproductive parts. The ovary is also shown in a cutaway view. (After J. B. Hill, H. W. Popp, and A. R. Grove, Jr., *Botany*. Copyright © 1967 by McGraw-Hill.)

Plants Differing in One Character

Mendel chose several *characters* to study. Here, the word **character** means a specific property of an organism; geneticists use this term as a synonym for characteristic or trait.

For each of the characters he chose, Mendel obtained lines of plants, which he grew for two years to make sure they were pure. A **pure line** is a population that breeds true for, or shows no variation in, the particular character being studied; that is, all offspring produced by selfing or crossing within the population show the same form of this character. By making sure his lines bred true, Mendel had made a clever beginning: he had established a fixed baseline for his future studies so that any changes observed following deliberate manipulation in his research would be scientifically meaningful; in effect, he had set up a control experiment.

Two of the pea lines Mendel grew proved to breed true for the character of flower color. One line bred true for purple flowers; the other, for white flowers. Any plant in the purple-flowered line—when selfed or when crossed with others from the same line—produced seeds that all grew into plants with purple flowers. When these plants in turn were selfed or crossed within the line, their progeny also had purple flowers, and so on. The white-flowered line similarly produced only white flowers through all generations. Mendel obtained seven pairs of pure lines for seven characters, with each pair differing in only one character (Figure 2-3).

Each pair of Mendel's plant lines can be said to show a **character difference**—a contrasting difference between

Figure 2-2 One technique of artificial cross-pollination, demonstrated with *Mimulus guttatus,* the yellow monkey flower. To transfer pollen, the experimenter touches anthers from the male parent to the stigma of an emasculated flower, which acts as the female parent. (Anthony Griffiths)

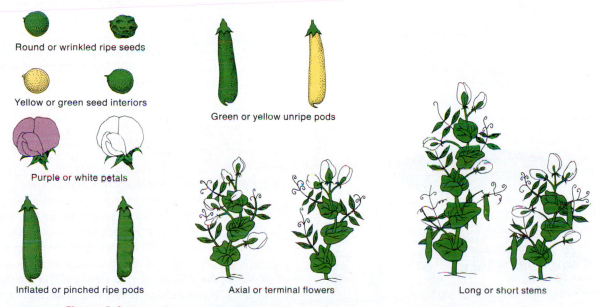

Round or wrinkled ripe seeds

Yellow or green seed interiors

Green or yellow unripe pods

Purple or white petals

Inflated or pinched ripe pods

Axial or terminal flowers

Long or short stems

Figure 2-3 The seven character differences studied by Mendel (After S. Singer and H. Hilgard, *The Biology of People.* Copyright © 1978 by W. H. Freeman and Company.)

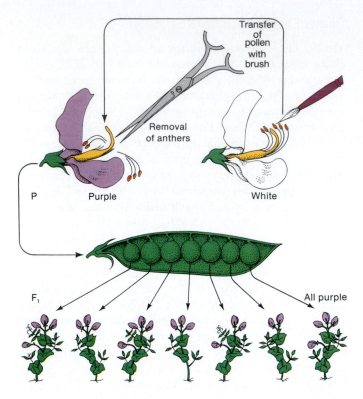

Figure 2-4 Mendel's cross of purple-flowered ♀ × white-flowered ♂.

two lines of organisms (or between two organisms) in one particular character. The differing lines (or individuals) represent different forms that the character may take: they can be called *character forms, character variants,* or **phenotypes.** The term *phenotype* (derived from Greek) literally means "the form that is shown"; it is the term used by geneticists today. Even though such words as gene and phenotype were not coined or used by Mendel, we shall use them in describing Mendel's results and hypotheses.

Figure 2-3 shows seven pea characters, each represented by two contrasting phenotypes. Contrasting phenotypes for a particular character are the starting point for any genetic analysis. This illustrates the point made in Chapter 1 that variation is the raw material for any genetic analysis. Of course, the delineation of characters is somewhat arbitrary; an organism may be "split up" into characters in many different ways. For example, we can state one character difference of the pea plants in at least three ways:

Character	Phenotypes
flower color	purple versus white
flower purpleness	presence versus absence
flower whiteness	absence versus presence

In many cases, the description chosen is a matter of convenience (or chance). Fortunately, the choice does not alter the final conclusions of the analysis, except in the words used.

We turn now to some of Mendel's experiments with the lines breeding true for flower color. In one of his early experiments, Mendel pollinated a purple-flowered plant with pollen from a white-flowered plant. We call the plants from the pure lines the **parental generation** (P). All the plants resulting from this cross had purple flowers (Figure 2-4). This progeny generation is called the **first filial generation** (F_1). (The subsequent generations produced by selfing are symbolized F_2, F_3, and so on.)

Mendel made **reciprocal crosses.** In most plants, any cross can be made in two ways, depending on which phenotype is used as male (♂) or female (♀). For example, the two crosses

phenotype A ♀ × phenotype B ♂

phenotype B ♀ × phenotype A ♂

are reciprocal crosses. Mendel's reciprocal cross in which he pollinated a white flower with pollen from a purple-flowered plant produced the same result (all purple flowers) in the F_1 (Figure 2-5). Mendel concluded that it makes no difference which way the cross is made. If one pure-breeding parent is purple-flowered and the other is white-flowered, all plants in the F_1 have purple flowers. The purple flower color in the F_1 generation is identical with that in the purple-flowered parental plants. In this case, the inheritance obviously is not a simple blending of purple and white colors

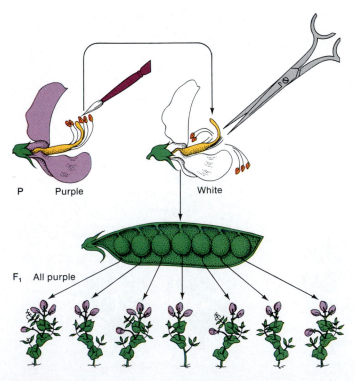

Figure 2-5 Mendel's cross of white-flowered ♀ × purple-flowered ♂.

Table 2-1   Results of All Mendel's Crosses in Which Parents Differed for One Character

	Parental phenotypes	F$_1$	F$_2$	F$_2$ Ratio
1.	Round × wrinkled seeds	All round	5474 round; 1850 wrinkled	2.96 : 1
2.	Yellow × green seeds	All yellow	6022 yellow; 2001 green	3.01 : 1
3.	Purple × white petals	All purple	705 purple; 224 white	3.15 : 1
4.	Inflated × pinched pods	All inflated	882 inflated; 299 pinched	2.95 : 1
5.	Green × yellow pods	All green	428 green; 152 yellow	2.82 : 1
6.	Axial × terminal flowers	All axial	651 axial; 207 terminal	3.14 : 1
7.	Long × short stems	All long	787 long; 277 short	2.84 : 1

to produce some intermediate color. To maintain a theory of blending inheritance, we would have to assume that the purple color is somehow "stronger" than the white color and completely overwhelms any trace of the white phenotype in the blend.

Next, Mendel selfed the F$_1$ plants, allowing the pollen of each flower to fall on its own stigma. He obtained 929 pea seeds from this selfing (the F$_2$ individuals) and planted them. Interestingly, some of the resulting plants were white-flowered; the white phenotype had reappeared. Mendel then did something that, more than anything else, marks the birth of modern genetics: he *counted* the numbers of plants with each phenotype. This procedure had seldom, if ever, been used in studies on inheritance before Mendel's work. Indeed, others had obtained remarkably similar results in breeding studies but had failed to count the numbers in each class. Mendel counted 705 purple-flowered plants and 224 white-flowered plants. He noted that the ratio of 705 : 224 is almost a 3 : 1 ratio (in fact, it is 3.1 : 1).

Mendel repeated the crossing procedures for the six other pairs of pea character differences. He found the same 3 : 1 ratio in the F$_2$ generation for each pair (Table 2-1). By this time, he was undoubtedly beginning to believe in the significance of the 3 : 1 ratio and to seek an explanation for it. In all cases, one parental phenotype disappeared in the F$_1$ and reappeared in one-fourth of the F$_2$. The white phenotype, for example, was completely absent from the F$_1$ generation but reappeared (in its full original form) in one-fourth of the F$_2$ plants.

It is very difficult to devise an explanation of this result in terms of blending inheritance. Even though the F$_1$ flowers were purple, the plants evidently still carried the *potential* to produce progeny with white flowers. Mendel inferred that the F$_1$ plants receive from their parents the ability to produce both the purple phenotype and the white phenotype and that these abilities are retained and passed on to future generations rather than blended. Why is the white phenotype not expressed in the F$_1$ plants? Mendel used the terms **dominant** and **recessive** to describe this phenomenon without explaining the mechanism. In modern terms, the purple phenotype is dominant to the white phenotype

and the white phenotype is recessive to purple. Thus the operational definition of dominance is provided by the phenotype of an F$_1$ established by intercrossing two pure lines. The parental phenotype that is expressed in such F$_1$ individuals is by definition the dominant phenotype.

Mendel went on to show that in the class of F$_2$ individuals showing the dominant phenotype there were in fact two genetically distinct subclasses. In this case he was working with seed color. In peas the color of the seed is determined by the genetic constitution of the seed itself, and not by the maternal parent as in some plant species. This is convenient because the investigator can treat each pea as an individual and can observe its phenotype directly without having to grow up a plant from it as must be done for flower color. This also means much larger numbers can be examined, and it is more convenient to extend studies into subsequent generations. The seed colors Mendel used were yellow and green. He crossed a pure yellow line with a pure green line, and observed that the F$_1$ peas that appeared were all yellow. Symbolically,

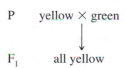

P   yellow × green

F$_1$ all yellow

Therefore, by definition, yellow is the dominant phenotype and green is recessive.

Mendel grew F$_1$ plants from these F$_1$ peas, and then selfed the plants. The peas that developed on the F$_1$ plants represented the F$_2$ generation. He observed that in the pods of the F$_1$ plants $\frac{3}{4}$ of the peas were yellow and $\frac{1}{4}$ were green. Here again in the F$_2$ we see a 3 : 1 phenotypic ratio. Mendel took a sample consisting of 519 yellow F$_2$ peas and grew plants from them. These F$_2$ plants were selfed individually and the peas that developed were noted. Mendel found that 166 of the plants bore only yellow peas, and each of the remaining 353 plants bore a mixture of yellow and green peas in a 3 : 1 ratio. In addition, green F$_2$ peas were grown up and selfed, and were found to bear only green peas. In summary, all the F$_2$ greens were evidently pure breeding like the green parental line, but of the F$_2$ yellows $\frac{2}{3}$ were like the F$_1$ yellows (producing yellow and green seeds in a 3:1 ratio) and $\frac{1}{3}$ were like the pure-breeding yellow parent. Thus, the study of the individual selfings revealed that underlying the 3:1 phenotypic ratio in the F$_2$ generation there was a more fundamental 1:2:1 ratio:

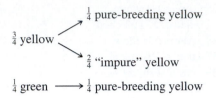

Further studies showed that such 1:2:1 ratios underlay all the phenotypic ratios that Mendel had observed. Thus, the problem really was to explain the 1:2:1 ratio. Mendel's explanation was a classic example of a creative model or hypothesis derived from observation and well suited for testing by further experimentation. He deduced the following explanation of the 1:2:1 ratio.

1. There are hereditary determinants of a particulate nature. (Mendel saw no blending of phenotypes, so he was forced to draw this conclusion.) We now call these determinants *genes*.
2. Each adult pea plant has two genes—a **gene pair**—in each cell for each character studied. Mendel's reasoning here was obvious: the F$_1$ plants, for example, must have had one gene that was responsible for the dominant phenotype and another gene that was responsible for the recessive phenotype, which showed up only in later generations.
3. The members of the gene pairs segregate (separate) equally into the gametes, or eggs and sperm.
4. Consequently, each gamete carries only one member of each gene pair.
5. The union of one gamete from each parent to form the first cell (or zygote) of a new progeny individual is random—that is, gametes combine without regard to which member of a gene pair is carried.

These points can be illustrated diagrammatically for a general case, using *A* to represent the gene that determines the dominant phenotype and *a* to represent the gene for the recessive (as Mendel did). This is similar to the way a mathematician uses symbols to represent abstract entities of various kinds. In Figure 2-6, these symbols are used to illustrate how the above five points explain the 1:2:1 ratio.

The whole model made logical sense of the data. However, many beautiful models have been knocked down under test. Mendel's next job was to test his model. He did this by taking (for example) an F$_1$ plant that grew from a yellow seed and crossing it with a plant grown from a green seed. A 1:1 ratio of yellow to green seeds could be predicted in the next generation. If we let *Y* stand for the gene that deter-

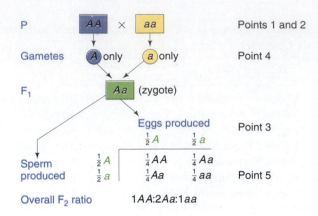

Figure 2-6 Mendel's model of the hereditary determinants of a character difference in the P, F$_1$, and F$_2$. The five points are those listed in the text.

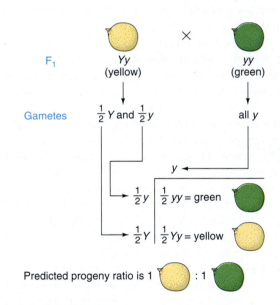

Figure 2-7 Prediction of a 1:1 phenotypic ratio in a cross of an F$_1$ yellow with a green in peas.

mines the dominant phenotype (yellow seeds) and *y* stand for the gene that determines the recessive phenotype (green seeds), we can diagram Mendel's predictions, as shown in Figure 2-7. In this experiment, he obtained 58 yellow (*Yy*) and 52 green (*yy*), a very close approximation to the predicted 1:1 ratio and confirmation of the equal segregation of *Y* and *y* in the F$_1$ individual. This concept of **equal segregation** has been given formal recognition as *Mendel's first law*.

Mendel's First Law   The two members of a gene pair segregate from each other into the gametes, so that half the gametes carry one member of the pair and the other half of the gametes carry the other member of the pair.

Now we need to introduce some more terms. The individuals represented by *A a* are called **heterozygotes,** or sometimes **hybrids,** whereas the individuals in pure lines are called **homozygotes.** In words such as these, *hetero-* means "different" and *homo-* means "identical." Thus, an *AA* plant is said to be **homozygous dominant;** an *aa* plant is homozygous for the recessive gene, or **homozygous recessive.** As we saw in Chapter 1, the designated genetic constitution of the character or characters under study is called the **genotype.** Thus, *YY* and *Yy*, for example, are different genotypes even though the seeds of both types are of the same phenotype (that is, yellow). In such a situation the phenotype can be thought of simply as the outward manifestation of the underlying genotype. Note that underlying the 3 : 1 phenotypic ratio in the F_2 there is a 1 : 2 : 1 genotypic ratio of *YY : Yy : yy*.

Note that strictly speaking the expressions *dominant* and *recessive* are properties of the phenotype. The dominant phenotype is established in analysis by the appearance of the F_1. Obviously, however, a phenotype (which is merely a description) cannot really exert dominance. Mendel showed that the dominance of one phenotype over another is in fact due to the dominance of one member of a gene pair over the other.

Let's pause to let the significance of this work sink in. What Mendel did was to develop an analytic scheme for the identification of genes regulating any biological character or function. Let's take petal color as an example. Starting with two different phenotypes (purple and white) of one character (petal color), Mendel was able to show that the difference was caused by one gene pair. Modern geneticists would say that Mendel's analysis had identified a gene for petal color. What does this mean? It means that in these organisms there is a gene that has a profound effect on the color of the petals. This gene can exist in different forms: a dominant form of the gene (represented by *C*) causes purple petals, and a recessive form of the gene (represented by *c*) causes white petals. The forms *C* and *c* are called **alleles**

(or alternative forms) of that gene for petal color. They are given the same letter symbol to show that they are forms of one gene. We could express this another way by saying that there is a kind of gene, called phonetically a "see" gene, with alleles *C* and *c*. Any individual pea plant will always have two "see" genes, forming a gene pair, and the actual members of the gene pair can be *CC*, *Cc*, or *cc*. Notice that although the members of a gene pair can produce different effects, they obviously both affect the same character.

The related terms gene and allele are potentially confusing, but to clarify the concepts let us jump ahead to think about what they signify at the level of DNA. When alleles like *A* and *a* are examined at the DNA level using modern technology, it is generally found that they are identical for most of their sequence, and differ only at one or a few nucleotides out of the thousands that make up the gene. Therefore, we see that the alleles are truly different versions of the same basic gene. Looked at another way, gene is the generic term and allele is specific.

We have seen that although the terms dominant and recessive are defined at the level of phenotype, the phenotypes clearly reflect the different actions of various alleles. Therefore we can legitimately use the phrases dominant allele and recessive allele as the determinants of dominant and recessive phenotypes. (Note that this usage is more correct than the terms dominant gene and recessive gene, although these terms are sometimes used.)

How does the allele terminology relate to the gene pair concept? A gene pair can consist of identical alleles (in homozygotes) or different alleles (in heterozygotes). When a gene pair segregates as described by Mendel's first law, then it can be identical alleles that segregate, as in homozygotes, or different alleles that segregate, as in heterozygotes. Of course it will only be the heterozygous segregation that can lead to a phenotypic segregation (that is, to two separate phenotypes) in the progeny.

The basic route of Mendelian analysis for a single character is summarized in Table 2-2.

Table 2-2   Summary of the Modus Operandi for Establishing Simple Mendelian Inheritance

Experimental procedure:	1. Choose pure lines showing a character difference (purple versus white flowers).
	2. Cross the lines.
	3. Self the F_1 individuals.
Results:	F_1 is all purple; F_2 is $\frac{3}{4}$ purple and $\frac{1}{4}$ white.
Inferences:	1. The character difference is controlled by a major gene for flower color.
	2. The dominant allele of this gene causes purple petals; the recessive allele causes white petals.
Symbolic interpretation:	

Character	Phenotype	Genotype	Allele	Gene
Flower color	Purple (dominant)	*CC* (homozygous dominant)	*C* (dominant)	Flower-color gene
		Cc (heterozygous)	*c* (recessive)	
	White (recessive)	*cc* (homozygous recessive)		

> **Message** The existence of genes was originally inferred (and is still inferred today) by observing precise mathematical ratios in the descendants of two genetically different parental individuals.

Plants Differing in Two Characters

Mendel's experiments described in the previous section stemmed from two pure-breeding parental lines that differed in one character. As we have seen, such lines produce F_1 progeny that are heterozygous for one gene (genotype Aa). Such heterozygotes are sometimes called **monohybrids.** The selfing or intercross of identical heterozygous F_1 individuals (symbolically $Aa \times Aa$) is called a **monohybrid cross,** and it was this type of cross that provided the interesting 3:1 progeny ratios that suggested the principle of equal segregation. Mendel went on to analyze the descendants of pure lines that differed in *two* characters (using a general symbolism, crosses of the type $AA\,bb \times aa\,BB$ or $AA\,BB \times aa\,bb$), which should produce an F_1 that is a double heterozygote, $Aa\,Bb$, also known as a **dihybrid.** From studying **dihybrid crosses** ($Aa\,Bb \times Aa\,Bb$) he came up with another important principle of heredity.

The two specific characters that he began working with were seed shape and seed color. We have already followed the monohybrid cross for seed color ($Yy \times Yy$), which gave a progeny ratio of 3 yellow:1 green. The seed shape phenotypes were round (determined by allele R) and wrinkled (determined by allele r). The monohybrid cross ($Rr \times Rr$) gave a progeny ratio of 3 round:1 wrinkled (Table 2-1 and Figure 2-8). To perform a dihybrid cross, he started out with two parental pure lines, one with yellow, wrinkled seeds ($YY\,rr$) and the other with green, round seeds ($yy\,RR$). The cross between these produced dihybrid F_1 seeds of genotype $Rr\,Yy$, which he discovered were round and yellow. This showed that the dominance of R over r, and of Y over y, was unaffected by the presence of heterozygosity for the other gene pair in the $Rr\,Yy$ dihybrid. Next Mendel made the dihybrid cross by selfing the dihybrid F_1 to obtain the F_2 generation. The F_2 seeds were of four different types, in the proportions $\frac{9}{16}$ round yellow, $\frac{3}{16}$ round green, $\frac{3}{16}$ wrinkled yellow, and $\frac{1}{16}$ wrinkled green, as shown in Figure 2-9. This rather unexpected 9:3:3:1 ratio seems a lot more complex than the simple 3:1 ratios of the monohybrid crosses. What could be the explanation? Before attempting to explain the ratio, he made dihybrid crosses involving several other combinations of characters, and found that *all* of the dihybrid F_1 individuals produced 9:3:3:1 progeny ratios similar to that obtained for seed shape and color. The 9:3:3:1 ratio was another consistent hereditary pattern that needed to be converted into an idea.

Mendel added up the numbers of individuals in certain F_2 phenotypic classes (the numbers are shown in Figure 2-9) in order to determine if the monohybrid 3:1 F_2 ratios were still present. He noted that in the case of seed

Figure 2-8 Round ($R-$) and wrinkled (rr) peas in a pod of a selfed heterozygous plant (Rr). The phenotypic ratio in this pod happens to be precisely the 3:1 ratio expected on average in the progeny of this selfing. (Recent molecular studies have shown that the wrinkled allele used by Mendel is produced by insertion into the gene of a segment of mobile DNA of the type to be discussed in Chapter 21.) (Madan K. Bhattacharyya)

shape there were 423 round seeds (315 + 108) and 133 wrinkled seeds (101 + 32). This is in fact close to a 3:1 ratio. Next, for seed color there were 416 yellow seeds (315 + 101) and 140 green (108 + 32), also very close to a 3:1 ratio. The presence of these two 3:1 ratios hidden in the 9:3:3:1 ratio undoubtedly gave Mendel the insight he needed to explain the 9:3:3:1 ratio because he realized that it was nothing more than two independent 3:1 ratios combined at random. One way of visualizing the combination of these two ratios is with a branch diagram, as follows:

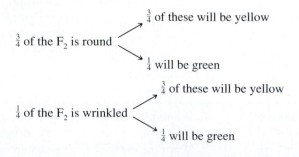

The combined proportions are calculated by multiplying along the branches in the diagram because (for example) $\frac{3}{4}$

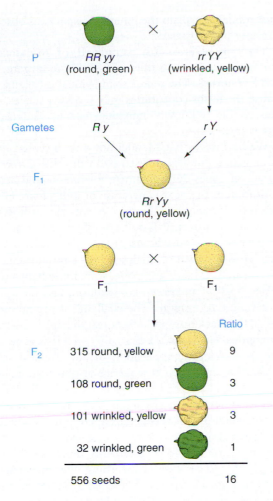

Figure 2-9 The F_2 generation resulting from a dihybrid cross.

Another way of stating the second law is that the process that puts R or r into a gamete is independent of the process that puts Y or y into that same gamete. The second law is an interesting statement that describes the independence of the two gene pairs, but how exactly can we get from this law to the $9:3:3:1$ ratio? Let us consider the gametes produced by the F_1 dihybrid $R\,r\,Y\,y$, because these are the gametes that give rise to the F_2 generation that shows the $9:3:3:1$ ratio. Again we will use the branch diagram to get us started because it illustrates independence visually. Combining Mendel's laws of equal segregation and independent assortment, we can predict that

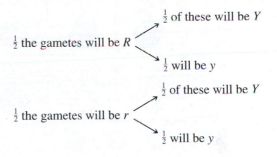

Multiplication along the branches gives us the gamete proportions

$$\tfrac{1}{4}\,R\,Y$$
$$\tfrac{1}{4}\,R\,y$$
$$\tfrac{1}{4}\,r\,Y$$
$$\tfrac{1}{4}\,r\,y$$

of $\tfrac{3}{4}$ is calculated as $\tfrac{3}{4} \times \tfrac{3}{4}$, which equals $\tfrac{9}{16}$. These multiplications give us the four proportions

$$\tfrac{3}{4} \times \tfrac{3}{4} = \tfrac{9}{16} \qquad \text{round yellow}$$
$$\tfrac{3}{4} \times \tfrac{1}{4} = \tfrac{3}{16} \qquad \text{round green}$$
$$\tfrac{1}{4} \times \tfrac{3}{4} = \tfrac{3}{16} \qquad \text{wrinkled yellow}$$
$$\tfrac{1}{4} \times \tfrac{1}{4} = \tfrac{1}{16} \qquad \text{wrinkled green}$$

This is of course the $9:3:3:1$ ratio we are trying to explain. However, is this not merely number juggling? What could the combination of the two $3:1$ ratios mean biologically? The way that Gregor Mendel phrased his explanation does in fact amount to a biological mechanism. In what is now known as *Mendel's second law*, he concluded that *different gene pairs assort independently during gamete formation.*

Mendel's second law During gamete formation the segregation of the alleles of one gene is independent of the segregation of the alleles of another gene. (NOTE: In fact this law applies only to genes that either are on separate chromosomes or are far apart on the same chromosome, as will be shown in Chapter 5.)

and these proportions are a direct result of the application of the two Mendelian laws. However, we still have not arrived at the $9:3:3:1$ ratio. The next step is to recognize that both the male and female gametes will show the same proportions just listed above, because Mendel did not specify different rules for male and female gamete formation. The four female gametic types will be fertilized randomly by the four male gametic types to obtain the F_2, and the best way of showing this graphically is to use a 4×4 grid called a *Punnett square*, which is shown in Figure 2-10. Grids are useful in genetics because their proportions can be drawn according to genetic proportions or ratios being considered, and thereby a visual data representation is obtained. We see in the Punnett square in Figure 2-10, for example, that the areas of the 16 boxes representing the various gametic fusions are each $\tfrac{1}{16}$ of the total area of the grid, simply because the rows and columns were drawn to correspond to the gametic proportions of $\tfrac{1}{4}$ each. As the Punnett square shows, the F_2 contains a variety of genotypes, but there are only four phenotypes and their proportions are in the $9:3:3:1$ ratio. So we see that when we work at the biological level of gamete formation, Mendel's laws explain not only the F_2 phenotypes but also the genotypes underlying them.

Mendel was a thorough scientist; he went on to test his principle of independent assortment in a number of ways.

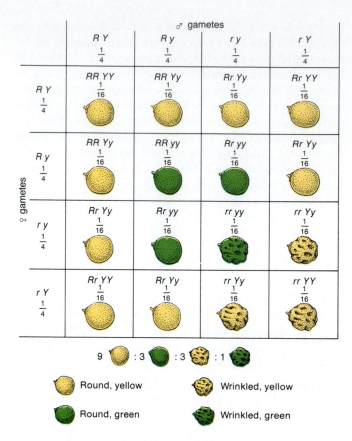

Figure 2-10 Punnett square showing predicted genotypic and phenotypic constitution of the F₂ generation from the dihybrid cross shown in Figure 2-9.

The most direct way zeroed in on the 1:1:1:1 gametic ratio of the F₁ dihybrid $R r Y y$, because this ratio sprang from his principle of independent assortment and was the biological basis of the 9:3:3:1 ratio in the F₂, as we have just demonstrated using the Punnett square. He reasoned that if there was in fact a 1:1:1:1 ratio of $R Y$, $R y$, $r Y$ and $r y$ gametes, if he crossed the F₁ dihybrid to a plant of genotype $r r y y$, which produces only gametes with recessive alleles (genotype $r y$), then the progeny proportions of this cross should be a direct reflection of the gametic proportions of the dihybrid, in other words,

$$\frac{1}{4} R r Y y$$

$$\frac{1}{4} R r y y$$

$$\frac{1}{4} r r Y y$$

$$\frac{1}{4} r r y y$$

This was the result that he obtained, perfectly consistent with his expectations. Similar results were obtained for all the other dihybrid crosses that he had made, and these and other types of tests all showed that he had in fact devised a robust model to explain the inheritance patterns observed in his various pea crosses.

The type of cross just considered, of an individual of unknown genotype to a fully recessive homozygote, is now called a **testcross.** The recessive individual is called a **tester.** Because the tester contributes only recessive alleles, the gametes of the unknown individual can be deduced from progeny phenotypes.

We have seen that Mendel not only invented the concept of genes, but also devised two rules that governed the behavior of genes during gamete formation, and hence the constitution of the next generation of individuals. His work is revered by geneticists because it was the prototypic genetic analysis: in fact, it is still one of the types of analysis that geneticists regularly use today in order to identify genes. Mendel's style of analysis is sometimes called "black box" research. Crosses are made, something happens inside a "black box" of unknown constitution, and then progeny of a certain type emerge. The challenge is to find out what is inside the black box, and of course Mendel was successful in doing this. His work is admired by scientists generally because of its elegant application of the classical scientific method, the experimental sequence of observation, hypothesis, and test. However, like all discoveries that lead to major advances in human understanding, it is the creativity of hypothesis building that sets genius apart. This is where ideas arise of which no one else has conceived, and which make sense of a large range of previously unexplained observations (in Mendel's case, concerning the mechanisms of heredity). Mendel did not know what genes are like, or where they are located. We now know that genes are parts of chromosomes. The pairing, segregation, and independent assortment of chromosomes at meiosis account for Mendel's laws, but this did not become known until the twentieth century, as we shall see in the next chapter.

Methods for Calculating Genetic Ratios

An important part of genetics today is concerned with predicting the types of progeny that emerge from a cross and calculating their expected frequency, in other words, their probability. We have already examined two methods for doing this, Punnett squares and branch diagrams. Punnett squares can be used to show hereditary patterns based on one gene pair, two gene pairs (as in Figure 2-10), or more. Such squares are a good graphic device for representing progeny, but they are time-consuming. Even the 16-compartment Punnett square in Figure 2-10 takes a long time to write out, but in the case of a trihybrid there are 2³, or 8, different gamete types, and the Punnett square has 64 compartments. The branch diagram is easier and is adaptable for phenotypic, genotypic, or gametic proportions. We can illustrate these using a dihybrid $A a B b$.

Progeny genotypes from a self	Progeny phenotypes from a self	Gametes

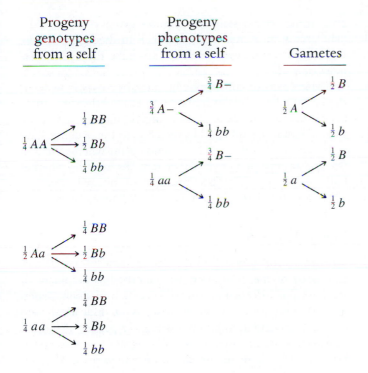

Note that the "tree" of branches for genotypes is quite unwieldy even in this case, which uses two gene pairs, and for three gene pairs there are 3^3, or 27, possible genotypes.

Application of simple statistical rules is the third method for calculating the probabilities (expected frequencies) of specific phenotypes or genotypes coming from a cross. The two statistical rules needed are the **product rule** and the **sum rule,** which we will discuss in that order.

> **Message**   The product rule states that the probability of independent events occurring together is the product of the probabilities of the individual events.

The possible outcomes of rolling dice follow the product rule because the outcome on each separate die is independent of the others. As an example, let us consider two dice and calculate the probability of rolling a pair of 4s. The probability of a 4 on one die is $\frac{1}{6}$ because the die has six sides and only one side carries the 4. This is written

$$p(\text{of a } 4) = \tfrac{1}{6}$$

Therefore, using the product rule, the probability of a 4 appearing on both dice is $\frac{1}{6} \times \frac{1}{6}$, or $\frac{1}{36}$. This is written

$$p(\text{of two 4s}) = \tfrac{1}{6} \times \tfrac{1}{6} = \tfrac{1}{36}$$

> **Message**   The sum rule states that the probability of either of two mutually exclusive events occurring is the sum of their individual probabilities.

In the product rule the focus is on outcomes A *and* B. In the sum rule the focus is on the concept of outcome A *or* B. Dice also can be used to illustrate the sum rule. We have already calculated that the probability of two 4s is $\frac{1}{36}$, and using the same type of calculation, it is clear that the probability of two 5s will be the same, or $\frac{1}{36}$ Now we can calculate the probability of either two 4s or two 5s. Since these outcomes are obviously mutually exclusive, the sum rule can be used to tell us that the answer is $\frac{1}{36} + \frac{1}{36}$, which is $\frac{1}{18}$. This can be written

$$p(\text{two 4s or two 5s}) = \tfrac{1}{36} + \tfrac{1}{36} = \tfrac{1}{18}$$

Now we can consider a genetic example. Assume that we have two plants of genotypes $Aa\,bb\,Cc\,Dd\,Ee$ and $Aa\,Bb\,Cc\,dd\,Ee$, and that from a cross between these plants we want to recover a progeny plant of genotype $aa\,bb\,cc\,dd\,ee$ (perhaps for the purpose of acting as the tester strain in a test cross). To estimate how many progeny plants need to be grown in order to stand a reasonable chance of obtaining the desired genotype, we need to calculate the proportion of the progeny that is expected to be of that genotype. If we assume that all the gene pairs assort independently, then we can do this calculation easily using the product rule. The five different gene pairs are considered individually, like five separate crosses, and then the appropriate probabilities are multiplied together to arrive at the answer.

From $Aa \times Aa$, $\frac{1}{4}$ of the progeny will be aa (see Mendel's crosses); from $bb \times Bb$, $\frac{1}{2}$ the progeny will be bb; from $Cc \times Cc$, $\frac{1}{4}$ of the progeny will be cc; from $Dd \times dd$, $\frac{1}{2}$ the progeny will be dd; from $Ee \times Ee$, $\frac{1}{4}$ of the progeny will be ee. Therefore, the overall probability (or expected frequency) of progeny of genotype $aa\,bb\,cc\,dd\,ee$ will be $\frac{1}{4} \times \frac{1}{2} \times \frac{1}{4} \times \frac{1}{2} \times \frac{1}{4}$, which equals $\frac{1}{256}$. So we learn that hundreds of progeny will need to be isolated in order to stand a chance of obtaining at least one of the desired genotype. This probability calculation can be extended to predict phenotype frequencies or gametic frequencies. Indeed there are thousands of other uses of this method in genetic analysis, and we will encounter many in later chapters.

When Mendel's results were rediscovered in 1900, his principles were tested in a wide spectrum of eukaryotic organisms (organisms with cells that contain nuclei). The results of these tests showed that Mendelian principles were generally applicable. Mendelian ratios (such as 3:1, 1:1, 9:3:3:1, and 1:1:1:1) were extensively reported (Figure 2-11), suggesting that equal segregation and independent assortment are fundamental hereditary processes found throughout nature. Mendel's laws are not merely laws about peas but laws about the genetics of eukaryotic organisms in general. The experimental approach used by Mendel can be extensively applied in plants. However, in some plants and in most animals, the technique of selfing is impossible. This problem can be circumvented by crossing

Figure 2-11 A 9:3:3:1 ratio in the phenotypes of kernels of corn. Each kernel represents a progeny individual. The progeny result from a self on an individual of genotype $Aa\,Bb$, where A = purple, a = yellow, B = smooth, and b = wrinkled. (Anthony Griffiths)

identical genotypes. For example, an F_1 animal resulting from the mating of parents from differing pure lines can be mated to its F_1 siblings (brothers or sisters) to produce an F_2. The F_1 individuals are identical for the genes in question, so the F_1 cross is equivalent to a selfing.

Simple Mendelian Genetics in Humans

Other systems present some special problems in the application of Mendelian methodology. One of the most difficult, yet most interesting, is the human species. Obviously, controlled crosses cannot be made, so human geneticists must resort to scrutinizing records in the hope that informative matings have been made by chance. Such a scrutiny of records of matings is called **pedigree analysis.** A member of a family who first comes to the attention of a geneticist is called the **propositus.** Usually the phenotype of the propositus is exceptional in some way (for example, the propositus might be a dwarf). The investigator then traces the history of the character in the propositus back through the history of the family and draws up a family tree, or pedi-

gree, using certain standard symbols given in Figure 2-12. (The terms autosomal and sex-linked in the figure will be explained later; they are included to make the list of symbols complete.)

Many pairs of contrasting human phenotypes are determined by pairs of alleles inherited in exactly the same manner shown by Mendel's peas. Pedigree analysis can reveal such inheritance patterns, but the clues in the pedigree have to be interpreted differently depending on whether one of the contrasting phenotypes is a rare disorder or whether both phenotypes of a pair are part of normal variation. These are considered separately in the following sections.

Medical Genetics

Medicine is concerned with the disorders of human beings, and many of these disorders are inherited as dominant or recessive phenotypes in a simple Medelian manner. Generally, the propositus, or index case, is an individual who comes to the attention of a physician as a patient. If there is a family history of the disorder, a pedigree is constructed as far as available information allows, and then the pedigree is analyzed. Let us consider the inheritance of recessives first.

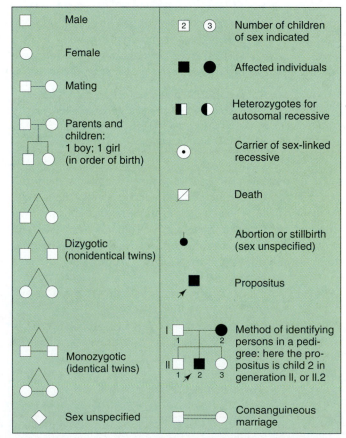

Figure 2-12 Symbols used in human pedigree analysis. (After W. F. Bodmer and L. L. Cavalli-Sforza, *Genetics, Evolution, and Man.* Copyright © 1976 by W. H. Freeman and Company.)

The unusual condition of a recessive disorder is determined by a recessive allele, and the corresponding unaffected phenotype is determined by a dominant allele. For example, the human disease phenylketonuria (PKU) is inherited in a simple Mendelian manner as a recessive phenotype, with PKU determined by an allele p and the normal condition by P. Therefore, sufferers from this disease are of genotype pp, and people who do not have the disease are either PP or Pp. What patterns in a pedigree would reveal such an inheritance? The two key points are that generally the disease appears in the progeny of unaffected parents and that the affected progeny includes both males and females. When we know that both male and female progeny are affected, we can assume that we are dealing with simple Mendelian inheritance, and not a special type of sex-linked inheritance that we will discuss in Chapter 3. The following typical pedigree illustrates the key point that affected children are born to unaffected parents:

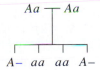

From this pattern we can immediately deduce simple Mendelian inheritance of the recessive allele responsible for the exceptional phenotype (indicated by shading). Furthermore, we can deduce that the parents are both heterozygotes, say Aa: both must have an a allele because each contributed an a allele to each affected child, and both must have an A allele because they are phenotypically normal. We can identify the genotypes of the children (in the order shown) as $A-$, aa, aa, and $A-$. Hence, the pedigree can be rewritten

Notice another interesting feature of pedigree analysis: even though Mendelian rules are at work, Mendelian ratios are rarely observed in families because the sample size is too small. In the above example, we see a 1:1 phenotypic ratio in the progeny of a monohybrid cross. If the couple were to have, say, 20 children, the ratio would be something like 15 unaffected children and 5 with PKU (a 3:1 ratio), but in a sample of four any ratio is possible, and all ratios are commonly found.

The pedigrees of recessive disorders tend to look rather bare, with few shaded symbols. A recessive condition shows up in groups of affected siblings, and the people in earlier and later generations tend not to be affected. To understand why this is so, it is important to have some feel for the genetic structure of populations underlying such rare conditions. By definition, if the condition is rare, most people do not carry the abnormal allele. Furthermore, most of those people who do have the abnormal allele are heterozygous for it rather than homozygous. The basic reason that heterozygotes are much more common than recessive homozygotes is that to be a recessive homozygote, both of your parents must have had the a gene, but to be a heterozygote all you need is one parent with the gene.

Geneticists have a quantitative way to connect the rareness of an allele with the commonness or rarity of heterozygotes and homozygotes in a population. They obtain the relative frequencies of genotypes in a population by assuming that the population is in Hardy-Weinberg equilibrium, to be fully discussed in Chapter 26. Under this simplifying assumption, if the relative proportions of two alleles A and a in a population are p and q, respectively, then the frequencies of the three possible genotypes are given by p^2 for AA, $2pq$ for Aa, and q^2 for aa. A numerical example illustrates this. If we assume that the frequency q of a recessive, disease-causing allele is $\frac{1}{50}$, then p is $\frac{49}{50}$, the frequency of homozygotes with the disease is $q^2 = (\frac{1}{50})^2 = \frac{1}{2500}$, and the frequency of heterozygotes is $2pq = 2 \times \frac{49}{50} \times \frac{1}{50} =$ approximately $\frac{1}{25}$. Hence, for this example, we see that heterozygotes are 100 times more frequent than disease sufferers, and this ratio increases the rarer the allele.

The formation of an affected individual usually depends on the chance union of unrelated heterozygotes. However, inbreeding (mating between relatives) increases the chance that a mating will be between two heterozygotes. An example of a cousin marriage is shown in Figure 2-13. Individuals III-5 and III-6 are first cousins and produce

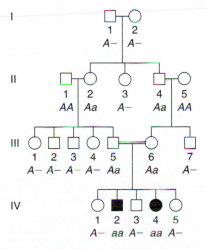

Figure 2-13 Pedigree of a rare recessive phenotype determined by a recessive allele a. Gene symbols normally are not included in pedigree charts, but genotypes are inserted here for reference. Note that individuals II-1 and II-5 marry into the family; they are assumed to be normal because the heritable condition under scrutiny is rare. Note also that it is not possible to be certain of the genotype in some individuals with normal phenotype; such individuals are indicated by $A-$.

two homozygotes for the rare allele. You can see from the figure that an ancestor who is a heterozygote may produce many descendants who are also heterozygotes. Matings between relatives thus run a higher risk of producing abnormal phenotypes caused by homozygosity for recessive alleles than do matings between nonrelatives. It is for this reason that first cousin marriages contribute a large proportion of the sufferers of recessive diseases in the population.

What are some examples of recessive disorders in humans? We have already used PKU as an example of pedigree analysis, but what kind of phenotype is it? PKU is a disease in which the body cannot properly process the amino acid phenylalanine, a component of all proteins in the food we eat. This substance builds up in the body and is converted to phenylpyruvic acid, which interferes with the development of the nervous system, leading to mental retardation. Babies are now routinely tested for this processing deficiency upon birth. If the deficiency is detected, phenylalanine can be withheld by use of a special diet, and the development of the disease can be arrested.

Cystic fibrosis is another disease inherited according to Mendelian rules as a recessive phenotype. This disease has received much recent attention because the allele that causes it was isolated in 1989 and the sequence of its DNA determined. This has led to an understanding of gene function in affected and unaffected individuals, giving hope for more effective treatment. Cystic fibrosis is a disease whose most important symptom is the secretion of large amounts of mucus into the lungs, resulting in death from a combination of effects, but usually precipitated by upper respiratory infection. The mucus can be dislodged by mechanical chest thumpers, and pulmonary infection can be prevented by antibiotics, so with treatment, cystic fibrosis patients can live to adulthood.

Albinism (Figure 2-14) is a rare condition that is inherited in a Mendelian manner as a recessive phenotype in many animals, including humans. The striking "white" phenotype is caused by the inability of the body to make melanin, the pigment that is responsible for most of the black and brown coloration of animals. In humans, such coloration is most evident in hair, skin, and retina, and its absence in albinos (who have the homozygous recessive genotype *a a*) leads to white hair, white skin, and eye pupils that are pink because of the unmasking of the red hemoglobin pigment in blood vessels in the retina.

> **Message** In pedigrees, simple Mendelian inheritance of a recessive disorder is revealed by the appearance of the phenotype in the male and female progeny of unaffected individuals.

What about disorders inherited as dominants? Here the normal allele is recessive, and the abnormal allele is dominant. It might seem paradoxical that a rare disorder can be dominant, but remember that dominance and recessiveness are simply properties of how alleles act, and are not defined in terms of predominance in the population. A good exam-

Figure 2-14 An albino. The phenotype is caused by homozygosity for a recessive allele, say *a a*. The dominant allele *A* determines one step in the chemical synthesis of the dark pigment melanin in the cells of skin, hair, and eye retinas. In *a a* individuals this step is nonfunctional, and the synthesis of melanin is blocked. (© Yves Gellie / Icône)

ple of a rare dominant phenotype with Mendelian inheritance is achondroplasia, a type of dwarfism (see Figure 2-15). Here, people with normal stature are genotypically *d d*, and the dwarf phenotype in principle could be *D d* or *D D*. However, it is believed that the two "doses" of the *D* allele in *D D* individuals produce such a severe effect that this is a lethal genotype. If true, all achondroplastics are heterozygotes.

In pedigree analysis, the major clues for identifying a dominant disorder with Mendelian inheritance are that the phenotype tends to appear in every generation of the pedigree and that affected fathers and mothers transmit the phenotype to both sons and daughters. Again, the representation of both sexes among the affected offspring rules out the sex-linked inheritance mentioned in our discussion of recessive disorders. The phenotype appears in every generation because generally the abnormal allele carried by an individual must have come from a parent in the previous generation. Abnormal alleles can arise de novo by the process of genetic change called mutation. This is relatively rare, but must be kept in mind as a possibility. A typical pedigree for a dominant disorder is shown in Figure 2-16. Once again, notice that Mendelian ratios are not necessarily observed in families. As with recessive disorders, individuals bearing one copy of the rare allele (*A a*) are much more common than those bearing two copies (*A A*), so most affected people are heterozygotes, and virtually all matings involving dominant disorders are *A a* × *a a*. Therefore, when the progeny of

Figure 2-15 The human achondroplasia phenotype, illustrated by a family of five sisters and two brothers. The phenotype is determined by a dominant allele, which we can call *D*, that interferes with bone growth during development. Most members of the human population can be represented as *dd* in regard to this gene. This photograph was taken upon the arrival of the family in Israel after the end of the Second World War. (UPI/Bettmann News Photos)

such matings are totaled, a 1:1 ratio is expected of unaffected (*aa*) to affected individuals (*Aa*).

Huntington's disease is an example of a disease inherited in a Mendelian manner as a dominant phenotype. The phenotype is one of neural degeneration, leading to convulsions and premature death. However, it is a late-onset disease, the symptoms generally not appearing until after the individual has begun to have children (see Figure 2-17). Each child of a carrier of the abnormal allele stands a 50 percent chance of inheriting the allele and the associated disease. This tragic pattern has led to a great effort to find ways of identifying individuals who carry the abnormal allele before they experience the onset of the disease. The application of

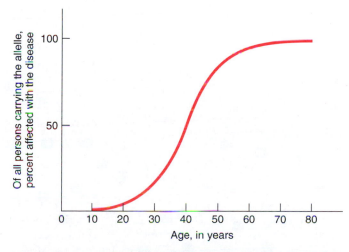

Figure 2-16 Pedigree of a dominant phenotype determined by a dominant allele *A*. In this pedigree, all the genotypes have been deduced.

Figure 2-17 The age of onset of Huntington's disease. The graph shows that people carrying the allele generally do not express the disease until after child-bearing age.

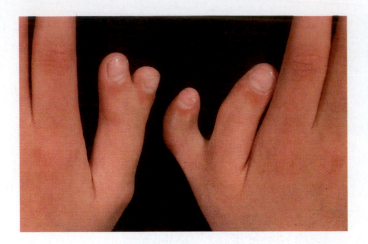

Figure 2-18 Some rare dominant phenotypes of the human hand. (a) (*top*) Polydactyly, a dominant phenotype characterized by extra fingers, toes, or both, determined by an allele *P*. The numbers in the accompanying pedigree (*bottom*) show the number of fingers in the upper lines, and toes in the lower. (Note the variation in expression of the *P* allele, a topic we will cover specifically in Chapter 4.) (b) (*right*) Brachydactyly, a dominant phenotype of short fingers, determined by an allele *B*. Note the very short terminal bones in the fingers compared with those in a normal hand. The pedigree for a family with brachydactyly shows a typical inheritance pattern for a rare dominant condition. All affected individuals are *Bb* and unaffected individuals are *bb*. (a, photo: © Biophoto Associates / Science Source; b, based on C. Stern, *Principles of Human Genetics,* 3d. ed. Copyright © 1973 by W. H. Freeman and Company.)

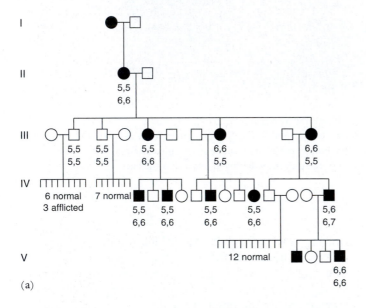

(a)

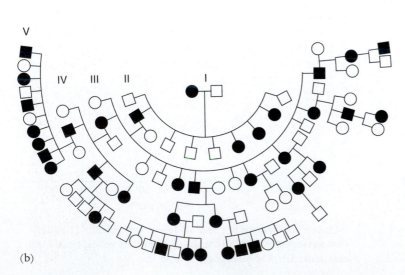

(b)

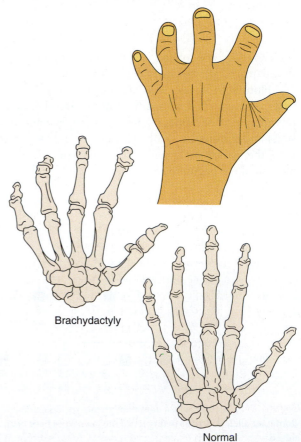

Brachydactyly

Normal

molecular techniques has resulted in a promising screening procedure.

Some other rare dominant conditions are polydactyly (extra digits) and brachydactyly (short digits), shown in Figure 2-18, and piebald spotting, shown in Figure 2-19.

Message Pedigrees of Mendelian dominant disorders show affected males and females in each generation and also show affected men and women transmitting the condition to equal proportions of their sons and daughters.

Normal Phenotypic Variants

All populations of plants and animals show normal variation, and human populations are no exception. Because isolated populations diverge genetically, some human variation corresponds to ethnic differences. We all know, however, that even within ethnic groups there is a vast amount of variation, the type of variation that allows us to distinguish one another. Normal variation is of two types, continuous and discontinuous.

Continuous variation is that shown by measurable characters such as height or weight, in which the phenotypes are arranged along a continuous spectrum. Continuous variation requires special techniques of analysis that we cover in Chapter 26. Discontinuously varying characters, on the other hand, can be classified into distinct phenotypes, such as those analyzed by Mendel. In humans there are many examples of discontinuous characters, such as brown versus blue eyes, dark versus blonde hair, chin dimples versus none, widow's peak versus none, attached versus free earlobes, and so on. Before we look at a sample pedigree, let's consider a useful new term. A set of two or more common, alternative, normal phenotypes is called a **polymorphism,** a word derived from the Greek for "many forms." The alternative phenotypes are called *morphs.* Primarily we will be concerned with sets of two contrasting morphs, the simplest type of polymorphism, called **dimorphism.** So using one of the above examples we would say that earlobes are dimorphic, with attached and free as the two major morphs. The morphs of a polymorphism are often determined by the alleles of one gene, inherited in the simple Mendelian manner described in this chapter.

The interpretation of pedigrees for polymorphisms is somewhat different, because by definition the morphs are common. Let's look at a pedigree for an interesting human dimorphism. Most human populations are dimorphic for the ability to taste the chemical phenylthiocarbamide (PTC). That is, people can either detect it as a foul, bitter taste, or—to the great surprise and disbelief of tasters—cannot taste it at all. From the pedigree in Figure 2-20, we can see that two tasters sometimes produce nontaster children. This makes it clear that the allele that confers the ability to taste is dominant and that the allele for nontasting is recessive. Notice that almost all people who marry into this

family carry the recessive allele either in heterozygous or in homozygous condition. Such a pedigree thus differs from those of rare recessive abnormalities for which it is conventional to assume that all who marry into a family are homozygous normal. As both PTC alleles are common, it is not surprising that all but one of the family members in this pedigree married individuals with at least one copy of the recessive allele.

Polymorphism is an interesting genetic phenomenon. Population geneticists have been surprised at how much polymorphism there is in natural populations of plants and animals generally. Furthermore, even though the genetics of polymorphisms is straightforward, there are very few polymorphisms for which there is satisfactory explanation for the coexistence of the morphs. But polymorphism is rampant at every level of genetic analysis, even down to the DNA level, and indeed polymorphisms observed at the DNA level have been invaluable as landmarks to help geneticists find their way around the chromosomes of complex organisms.

Message Populations of plants and animals (including humans) are highly polymorphic. Contrasting morphs are generally determined by alleles inherited in a simple Mendelian manner.

Simple Mendelian Genetics in Agriculture

Plant breeding methods used by Neolithic farmers were probably the same as those used until the discovery of Mendelian genetics. Basically, the approach was to select superior phenotypes that often appear as rare variants in natural populations and then propagate these over the years. Particularly desirable were pure lines of these favorable phenotypes, because such lines produced constant results over generations of planting. Without the knowledge of Mendelian genetics, how is it possible to develop pure lines? It so happens that self-pollinating plants, such as many crop plants, naturally tend to be homozygous, because selfing reduces heterozygosity in the population as is apparent from considering the genotypes of parents and the genotypes possible among their progeny:

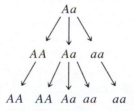

This is illustrated in more detail in Figure 2-21. Thus, pure lines have developed automatically over the years.

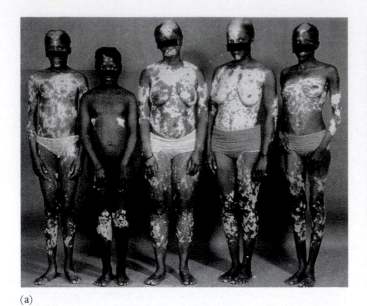
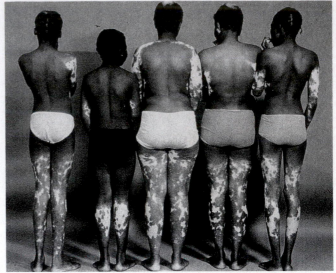

(a)

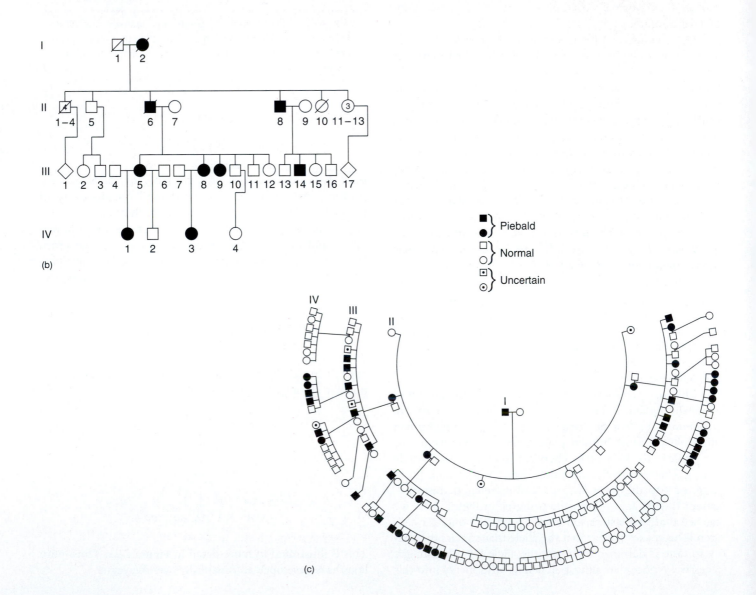

(b)

(c)

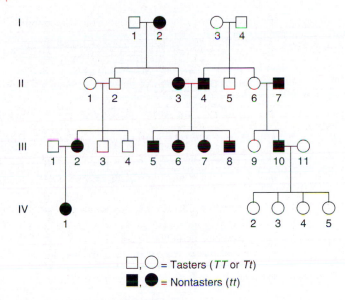

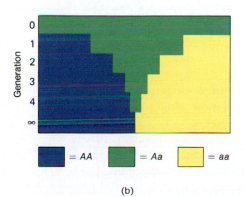

☐, ○ = Tasters (*TT* or *Tt*)

■, ● = Nontasters (*tt*)

Figure 2-20 Pedigree for the ability to taste the chemical PTC.

Figure 2-19 Piebald spotting, a rare dominant human phenotype. Although the phenotype is encountered sporadically in all races, the patterns show up best in those with dark skin. (a) The photographs show front and back views of affected individuals IV-1, IV-3, III-5, III-8, and III-9 from the family pedigree shown in (b). Notice the variation between family members in expression of the piebald gene. A larger pedigree of a Norwegian family is shown in (c). It is believed that the patterns are caused by the dominant allele interfering with the migration of melanocytes (melanin-producing cells) from the dorsal to the ventral surface during development. The white forehead blaze is particularly characteristic and is often accompanied by a white forelock in the hair. The same basic condition is known in mice, and again the melanocytes fail to cover the top of the head and the ventral surface (d). Piebaldism is not a form of albinism; the cells in the light patches have the genetic potential to make melanin, but since they are not melanocytes they are not developmentally programmed to do so. In true albinism, the cells lack the potential to make melanin. (The DNA of the piebald allele has recently been characterized as an allele of *c-kit*, a type of gene called a proto-oncogene, to be discussed in Chapter 7. [(a), (b) from I. Winship, K. Young, R. Martell, R. Ramesar, D. Curtis, and P. Beighton, "Piebaldism: An Autonomous Autosomal Dominant Entity," *Clinical Genetics* 39, 1991, 330; (c) from C. Stern, *Principles of Human Genetics*, 3d ed. Copyright © 1973 by W. H. Freeman and Company; (d) provided by R. A. Fleischman, University of Texas, Southwestern Medical Center, Dallas—also see R. A. Fleischman, D. L. Saltman, V. Stastny, and S. Zneimer, "Deletion of the *c-kit* Protooncogene in the Human Developmental Defect Piebald Trait," *Proceedings of the National Academy of Sciences USA* 88, 1991, 10885.]

Generation	AA	Aa	aa
0	0	100	0
1	25	50	25
2	37.5	25	37.5
3	43.75	12.5	43.75
4	46.875	6.25	46.875
∞	50	0	50

(a)

(b)

Figure 2-21 Repeated generations of selfing increase the proportion of homozygotes. (a) The percentages of the three genotypes are shown through several generations, assuming that all individuals in generation 0 are *A a* and that all individuals reproduce at the same rate. At each generation, *A A* and *a a* breed true, but *A a* individuals produce *A a*, *A A*, and *a a* progeny in a 2 : 1 : 1 ratio. (b) A graphic depiction of how the proportions of genotypes change.

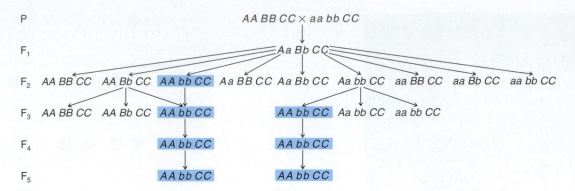

Figure 2-22 The basic technique of plant breeding. A hybrid is generated, and breeding stock is selected in subsequent generations to obtain improved pure lines. For simplicity, we assume here that the parental lines differ in only two genes (*A* and *B*). In this example, the *Aa bb CC* genotype is the one desired; that is, tests on the F_2 generation show this to be the most desirable phenotype.

However, these unsophisticated breeding practices suffered from a major problem: the breeder was forced to rely on favorable natural combinations of genes. With the advent of Mendelian genetics, it became evident that favorable qualities from different lines could be combined through hybridization and subsequent gene reassortment. This procedure forms the basis of modern plant breeding.

The breeding of plants to produce new and improved genotypes works in basically the following way. For naturally self-pollinating plants, such as rice or wheat, two pure lines (each of different favorable genotype) are hybridized by manual cross-pollination, and an F_1 is developed. The F_1 is then allowed to self, and its heterozygous pairs of alleles assort to produce many different genotypes, some of which represent desirable new combinations of the parental genes. A small proportion of these new genotypes will be pure-breeding already; in those that are not, several generations of selfing will produce homozygosity of the relevant genes. Figure 2-22 summarizes this method. An example of the importance of such methodology is in the breeding of highly productive, dwarf lines of rice, one of the world's staple foods. The breeding program is shown in Figure 2-23.

An example of genetic improvement in a species more familiar to most is the tomato. Anyone who has read a recent seed catalog will be familiar with the abbreviations V, F, and N next to a listed tomato variety. These represent, respectively, resistance to the pathogens *Verticillium, Fusarium,* and nematodes—resistance that has been bred into domestic tomatoes, typically from wild relatives. A useful tomato phenotype is determinate growth. The tips of the branches of most tomato species can keep growing (indeterminate growth), but in the determinate lines most branches end in a growth-arresting inflorescence. Determinate plants are bushier and more compact, and they do not need as much staking (Figure 2-24). Determinate growth is caused by a re-

cessive allele *sp* (self-pruning), which has been crossed into modern varieties. The symbol *sp* represents a single gene; many gene symbols are more than one letter. A useful allele of another gene is *u* (uniform ripening); this allele eliminates the green patch or shoulder around the stem on the ripe fruit. Tomato geneticists have produced a large array of different morphological types of tomatoes, many of which have found their way into supermarkets and garden stores (Figure 2-25).

Such examples could be listed for many pages. The point is that simple Mendelian genetics, as described in this chapter, has provided agricultural plant breeding with its rationale and its modern methods. Entire complex genotypes may be constructed from an array of ancestral lines, each showing some desirable feature. When we think of genetic engineering, we think of the molecular techniques of the 1980s and 1990s, but genetic engineering for plant improvement began long ago.

Mendelian genetics also has provided a formal theoretical basis for animal breeding, enabling a greater efficiency than under traditional practices. Most farm animals have been genetically modified in this way. More recent techniques, such as the use of frozen semen, superovulation, artificial insemination, test-tube fertilization, frozen embryos, and surrogate mothers in livestock species, have enabled breeders to amplify the number of offspring of a specific genotype—a number normally limited by the life span of the animal.

Variants and Genetic Dissection

Genetic analysis, as we have seen, must start with parental differences. Without variants, no genetic analysis is possible. Where do these variants—these raw materials for genetic analysis—come from? This is a question that can be an-

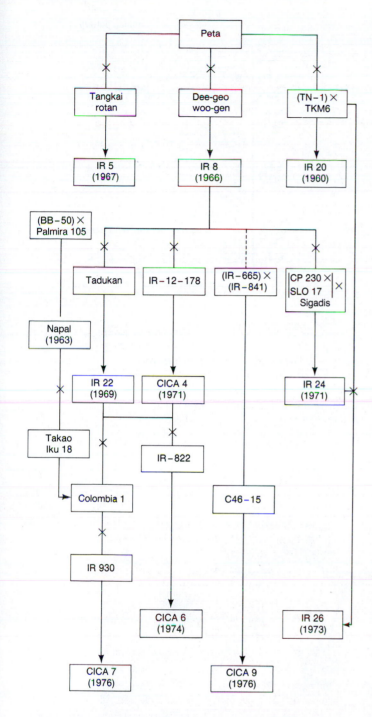

the other hand, for many genes there are two or more common alleles in a population, resulting in genetic polymorphism—the coexistence of genetically determined variant phenotypes in a population. Although the reason for the existence of polymorphisms is usually not easy to discover, for the geneticist they are a useful source of variant alleles for study.

We have seen that the genetic analysis of variants can identify a particular gene that is important for a biological process. This central aspect of modern genetics is called **genetic dissection.** Mendel was the first genetic surgeon. Using genetic analysis, he was able to identify and distinguish among the various components of the hereditary process in a way as convincing as if he had microdissected those components. The fact that the genes he was using were for pea shape, pea color, and so on was largely irrelevant in this connection. Those genes were being used simply as **genetic markers,** which enabled Mendel to trace the hereditary processes of segregation and assortment. A genetic marker is a variant allele that is used to label a biological structure or process throughout the course of an experiment. It is almost as though Mendel were able to "paint" two alleles two different colors and send them through a cross to see how they behaved. Genetic markers are now routinely used in genetics and in all of biology to study all sorts of processes that the marker genes themselves do not directly affect. Markers are often morphological, as in the pea example. But molecular variants (DNA and proteins) are increasingly used as markers too.

Probably without realizing it, Mendel had also invented another aspect of genetic dissection in which the precise genes used *were* important. In this type of analysis, genetic variants are used in such a way that by studying the variant gene function, we can make inferences about the "normal" operation of the gene and the process it controls. Every time a gene is identified by Mendelian analysis, it identifies a component of a biological process.

Once a gene has been identified as affecting, say, petal color in peas, we call it a major gene, but what does this really mean? It is *major* in that it is obviously having a profound effect on the color of the petals. But can we conclude that it is *the* single most important step in the determination of petal color? The answer is no, and the reason may be

swered in full only in later chapters. Briefly, most of the variants like the ones used by Mendel (and by ancient and modern breeders of plants and animals) arise spontaneously in nature or in the breeders' populations without the deliberate action of geneticists.

Let us emphasize again that variants can range from rare to common. Some rare variants are abnormal. Undoubtedly in a natural setting many of them would be weeded out by natural selection, but they can be kept alive by nurture so that the alleles responsible can be studied. On

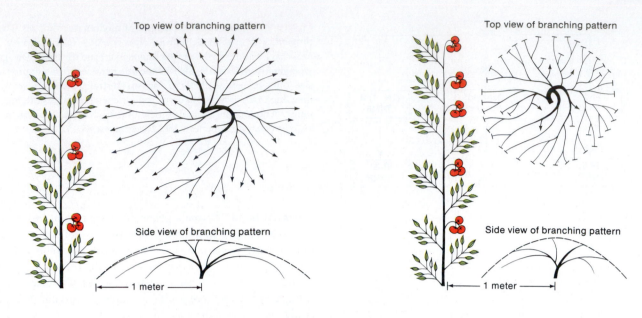

Top view of branching pattern   Top view of branching pattern

Side view of branching pattern   Side view of branching pattern

|← 1 meter →| |← 1 meter →|

Indeterminate (*Sp −*)   Determinate (*sp sp*)

Figure 2-24 Growth characteristics of indeterminate (*Sp −*) and determinate (*sp sp*) tomatoes. Arrows show growth; tomatoes indicate locations of inflorescences. Determinate has progressively fewer leaves between inflorescences, and the determinate stems end with a size-arresting inflorescence. Individual stem branches are shown, plus typical branching patterns for mature plants in top and side views. (In the top view, bars indicate inflorescences that terminate branch growth.) Note the size difference. (From Charles M. Rick, "The Tomato." Copyright © 1978 by Scientific American, Inc. All rights reserved.)

Figure 2-25 Some of the many tomato phenotypes produced by breeders. (Travis Amos)

seen in an analogy. If we were trying to discover how a car engine works, we might pull out various parts and observe the effect on the running of the engine. If a battery cable were disconnected, the engine would stop; this might lead us erroneously to conclude that this cable is *the* most important part of the running of the engine. Other parts are equally necessary, and their removal could also stop or seriously cripple the engine. In a similar way it can be shown (as we will see in Chapter 4) that several genes can be identified, all of which have a major and similar effect on petal coloration.

Mendel's work has withstood the test of time and has provided us with the basic groundwork for all modern genetic study. Yet his work went unrecognized and neglected for 35 years following its publication. Why? There are many possible reasons, but here we shall consider just one. Perhaps it was because biological science at that time could not provide evidence for any real physical units within cells that might correspond to Mendel's genetic particles. Chromosomes had certainly not yet been studied, meiosis had not yet been described, and even the full details of plant life cycles had not been worked out. Without this basic knowledge, it may have seemed that Mendel's ideas were mere numerology.

Message Mendel's work was the prototypical genetic analysis. As such, it is significant for the following reasons:

1. It showed how it is possible to study biological processes by using genetic markers.
2. It showed how the functions of genes themselves can be elucidated from the study of variant alleles.
3. It had far-reaching ramifications in agriculture and medicine.

Above the doorway into the Mendel museum in Brno there is a wistful quip of Mendel's inscribed in Czech, "MÁ DOBA PŘIJDE," meaning "My time will come." Mendel's time did come; the twentieth century saw a massive flowering of research and understanding in heredity, all stemming from his seminal studies in the tiny monastery garden. His hypothetical "factors" (or genes, as we now call them) are a well-understood molecular reality, and even whole genomes are becoming characterized. It is possible to take the latest dramatic research on cloning, gene therapy, transgenics, the human genome project, and so on, and trace them all back through the research literature to that single paper entitled "Experiments on Plant Hybridization," presented in 1865 to the Brünn Natural History Society.

SUMMARY

Modern genetics is based on the concept of the gene, the fundamental unit of heredity. In his experiments with the garden pea, Mendel was the first to recognize the existence of genes. For example, by crossing a pure line of purple-flowered pea plants with a pure line of white-flowered pea plants and then selfing the F_1 generation, which was entirely purple, Mendel produced an F_2 generation of purple plants and white plants in a 3:1 ratio. In crosses such as those of pea plants bearing yellow seeds and pea plants bearing green seeds, he discovered that a 1:2:1 ratio underlies all 3:1 ratios. From these precise mathematical ratios Mendel concluded that there are hereditary determinants of a particulate nature, now known as genes. In higher plant and animal cells, genes exist in pairs. Variant forms of a gene are called alleles. An allele can be either dominant or recessive.

In a cross of heterozygous yellow (Yy) plants with homozygous green (yy) plants, a 1:1 ratio of yellow to green plants was produced. From this ratio Mendel confirmed his so-called first law, which states that two members of a gene pair segregate from each other during gamete formation into equal numbers of gametes. Thus, each gamete carries only one member of each gene pair. The union of gametes to form a zygote is random as regards which allele the gametes carry.

The foregoing conclusions came from Mendel's work with monohybrid crosses. In dihybrid crosses, Mendel found 9:3:3:1 ratios in the F_2, which are really two 3:1 ratios combined at random. From these ratios Mendel inferred that alleles of the two genes in a dihybrid cross behave independently. This concept is Mendel's second law.

Although controlled crosses cannot be made in human beings, human geneticists make use of pedigree analysis to determine inheritance patterns. Mendelian genetics has great significance for humans. Many diseases and other exceptional conditions in humans are determined by recessive alleles inherited in a Mendelian manner; other exceptional conditions are caused by dominant alleles. In addition, Mendelian genetics is widely used in modern agriculture. By combining favorable qualities from different lines through hybridization and subsequent gene reassortment, plant and animal geneticists are able to produce new lines of superior phenotypes.

Finally, Mendel was responsible for the basic techniques of genetic dissection still in use today. One such technique is the use of genes as genetic markers to trace the hereditary processes of segregation and assortment. The other is the study of abnormal variants to discover how genes operate normally.

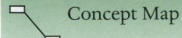

Concept Map

Draw a concept map interrelating as many of the following terms as possible. Note that the terms are listed in no particular order.

independent assortment / genotype / phenotype / dihybrid / testcross / 9:3:3:1 ratio / self / gametes / fertilization / 1:1:1:1 ratio

CHAPTER INTEGRATION PROBLEM

Each chapter has one solved problem in which we stress the integration of concepts from different chapters. Learning in any discipline is a linear process, moving from topic to topic in some kind of appropriate sequence. But of course the discipline itself is not linear, but a set of integrated parts that the professional sees as one whole. We hope that focusing on integration will clarify the overall structure of genetics and that the reader will not see the contents of each chapter in isolation. As we pass from chapter to chapter, the levels of understanding build on the previous ones, and the subject is assembled like the layers of an onion.

Problem

Crosses were made between two pure lines of rabbits that we can call A and B. A male from line A was mated with a female from line B, and the F_1 rabbits were subsequently intercrossed to produce an F_2. It was discovered that $\frac{3}{4}$ of the F_2 animals had white subcutaneous fat, and $\frac{1}{4}$ had yellow subcutaneous fat. Later, the F_1 was examined and was found to have white fat. Several years later, an attempt was made to repeat the experiment using the same male from line A and the same female from line B. This time, the F_1 and all the F_2 (22 animals) had white fat. The only difference between the original experiment and the repeat that seemed relevant was that in the original all the animals were fed on fresh vegetables, and in the repeat they were fed on commercial rabbit chow. Provide an explanation for the difference and a test of your idea.

Solution

The first time the experiment was done, the breeders would have been perfectly justified in proposing that a pair of alleles determine white versus yellow body fat. This is because the data clearly resemble Mendel's results in peas. White must be dominant, so we can represent the white allele as W and the yellow allele as w. The results can then be expressed

P	$WW \times ww$
F$_1$	Ww
F$_2$	$\frac{1}{4} \, WW$
	$\frac{1}{2} \, Ww$
	$\frac{1}{4} \, ww$

No doubt if the parental rabbits had been sacrificed, it would have been predicted that one (we cannot tell which) would have had white fat, and the other yellow. Luckily, this was not done, and the same animals were bred again, leading to a very interesting, different result. Often in science, an unexpected observation can lead to a novel principle, and rather than moving on to something else it is useful to try to explain the inconsistency. So why did the 3:1 ratio disappear? Here are some possible explanations.

First, perhaps the genotypes of the parental animals had changed. This type of spontaneous change affecting the whole animal, or at least its gonads, is very unlikely, because even common experience tells us that organisms tend to be stable to their type. Nevertheless, this is a reasonable idea.

Second, in the repeat, the sample of 22 F_2 animals did not contain any yellow simply by chance ("bad luck"). This again seems unlikely, because the sample was quite large, but it is a definite possibility.

A third explanation draws upon the principle covered in Chapter 1 that genes do not act in a vacuum; they depend on the environment for their effect. Hence, the useful catchphrase arises, "Genotype plus environment equals phenotype." A corollary of this catchphrase is, of course, that genes can act differently in different environments, so

Genotype 1 plus environment 1 equals phenotype 1

and

Genotype 1 plus environment 2 equals phenotype 2

In the present question, the different diets constituted different environments, so a possible explanation of the results is that the recessive allele *w* produces yellow fat only when the diet contains fresh vegetables. This explanation is testable. One way is to repeat the experiment again using vegetables in the food, but the parents might be dead by this time. A more convincing way is to interbreed several of the white-fatted F$_2$ rabbits from the second experiment. According to the original interpretation, about $\frac{3}{4}$ would bear at least one recessive *w* allele for yellow fat, and if their progeny are raised on vegetables, yellow should appear in Mendelian proportions. For example, if we choose two rabbits, *Ww* and *ww*, the progeny would be $\frac{1}{2}$ white and $\frac{1}{2}$ yellow.

If this did not happen, and no yellow progeny appeared in any of the F$_2$ matings, one would be forced back onto explanations one or two. Explanation two can be tested by using larger numbers, and if this explanation doesn't work, we are left with number one, which is difficult to test directly.

As you might have guessed, in reality the diet was the culprit. The specific details illustrate environmental effects beautifully. Fresh vegetables contain yellow substances called xanthophylls, and the dominant allele *W* gives rabbits the ability to break down these substances to a colorless ("white") form. However, *ww* animals lack this ability, and the xanthophylls are deposited in the fat, making it yellow. When no xanthophylls have been ingested, both *W*– and *ww* animals end up with white fat.

SOLVED PROBLEMS

This section in each chapter contains a few solved problems that show how to approach the problem sets that follow. The purpose of the problem sets is to challenge your understanding of the genetic principles learned in the chapter. The best way to demonstrate an understanding of a subject is to be able to use that knowledge in a real or simulated situation. Be forewarned that there is no machinelike way of solving these problems. The three main resources at your disposal are the genetic principles just learned, common sense, and trial and error.

Here is some general advice before beginning. First, it is absolutely essential to read and understand all of the question. Find out exactly what facts are provided, what assumptions have to be made, what clues are given in the question, and what inferences can be made from the available information. Second, be methodical. Staring at the question rarely helps. Restate the information in the question in your own way, preferably using a diagrammatic representation or flowchart to help you think out the problem. Good luck.

1. Consider three yellow, round peas, labeled A, B, and C. Each was grown into a plant and crossed to a plant grown from a green, wrinkled pea. Exactly 100 peas issuing from each cross were sorted into phenotypic classes as follows:

A:	51	yellow, round
	49	green, round
B:	100	yellow, round
C:	24	yellow, round
	26	yellow, wrinkled
	25	green, round
	25	green, wrinkled

What were the genotypes of A, B, and C? (Use gene symbols of your own choosing; be sure to define each one.)

Solution

Notice that each of the crosses is

yellow, round × green, wrinkled

↓

progeny

Because A, B, and C were all crossed to the same plant, all the differences among the three progeny populations must be attributable to differences in the underlying genotypes of A, B, and C.

You might remember a lot about these analyses from the chapter. This is fine, but let's see how much we can deduce from the data. What about dominance? The key cross for deducing dominance is B. Here, the inheritance pattern is

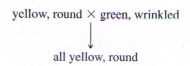

yellow, round × green, wrinkled

↓

all yellow, round

So yellow and round must be dominant phenotypes, because dominance is literally defined in terms of the phenotype of a hybrid. Now we know that the green, wrinkled parent used in each cross must be fully recessive; we have a very convenient situation because it means that each cross is a testcross, which is generally the most informative type of cross.

Turning to the progeny of A, we see a 1:1 ratio for yellow to green. This is a demonstration of Mendel's first law (equal segregation) and shows that for the character of color, the cross must have been heterozygote × homozygous recessive. Letting Y = yellow and y = green, we have

$$Yy \times yy$$

$$\downarrow$$

$$\tfrac{1}{2} Yy \text{ (yellow)}$$
$$\tfrac{1}{2} yy \text{ (green)}$$

For the character of shape, because all the progeny are round, the cross must have been homozygous dominant × homozygous recessive. Letting R = round and r = wrinkled, we have

$$RR \times rr$$

$$\downarrow$$

$$Rr \text{ (round)}$$

Combining the two characters, we have

$$YyRR \times yyrr$$

$$\downarrow$$

$$\tfrac{1}{2} YyRr$$
$$\tfrac{1}{2} yyRr$$

Now, cross B becomes crystal clear and must have been

$$YYRR \times yyrr$$

$$\downarrow$$

$$YyRr$$

because any heterozygosity in pea B would have given rise to several progeny phenotypes, not just one.

What about C? Here, we see a ratio of 50 yellow : 50 green (1:1) and a ratio of 49 round : 51 wrinkled (also 1:1). So both genes in pea C must have been heterozygous, and cross C was

$$YyRr \times yyrr$$

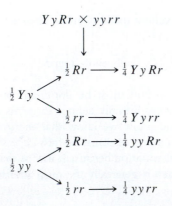

which is a good demonstration of Mendel's second law (independent behavior of different genes).

How would a geneticist have analyzed these crosses? Basically, the same way we just did but with fewer intervening steps. Possibly something like this: "yellow and round dominant; single-gene segregation in A; B homozygous dominant; independent two-gene segregation in C."

2. Phenylketonuria (PKU) is a human hereditary disease resulting from the inability of the body to process the chemical phenylalanine, which is contained in the protein we eat. PKU is manifested in early infancy and, if it remains untreated, generally leads to mental retardation. PKU is caused by a recessive allele with simple Mendelian inheritance.

A couple intends to have children but consults a genetic counselor because the man has a sister with PKU and the woman has a brother with PKU. There are no other known cases in their families. They ask the genetic counselor to determine the probability that their first child will have PKU. What is this probability?

Solution

What can we deduce? If we let the allele causing the PKU phenotype be p and the respective normal allele be P, then the sister and brother of the man and woman, respectively, must have been pp. In order to produce these affected individuals, all four grandparents must have been heterozygous normal. The pedigree can be summarized as follows:

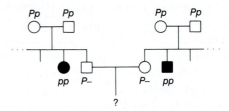

Once these inferences have been made, the problem is reduced to an application of the product rule. The only way the man and woman can have a PKU child is if both of them are heterozygotes (it is obvious that they themselves do not have the disease). Both the grandparental matings are simple Mendelian monohybrid crosses expected to produce progeny in the following proportions:

$$\left. \begin{array}{l} \tfrac{1}{4}\,PP \\ \tfrac{1}{2}\,Pp \end{array} \right\} \quad \text{Normal} \qquad \tfrac{3}{4}$$

$$\tfrac{1}{4}\,pp \qquad \qquad \text{PKU } \tfrac{1}{4}$$

We know that the man and the woman are normal, so the probability of them being a heterozygote is $\tfrac{2}{3}$, because within the $P-$ class, $\tfrac{2}{3}$ are Pp and $\tfrac{1}{3}$ are PP.

The probability of *both* the man and the woman being heterozygotes is $\frac{2}{3} \times \frac{2}{3} = \frac{4}{9}$. If they are both heterozygous, then one-quarter of their children would have PKU, so the probability that their first child will have PKU is $\frac{1}{4}$, and the probability of their being heterozygous *and* of their first child having PKU is $\frac{4}{9} \times \frac{1}{4} = \frac{4}{36} = \frac{1}{9}$, which is the answer.

PROBLEMS

Generally the straightforward problems are at the beginning of a set. Particularly challenging problems are marked with an asterisk.

1. What are Mendel's laws?

2. If you had a fruit fly (*Drosophila melanogaster*) that was of phenotype A, what test would you make to determine if it was *A A* or *A a*?

3. Two black guinea pigs were mated and over several years produced 29 black and 9 white offspring. Explain these results, giving the genotypes of parents and progeny.

4. Look at the Punnett square in Figure 2-10.

a. How many genotypes are there in the 16 squares of the grid?

b. What is the genotypic ratio underlying the 9:3:3:1 phenotypic ratio?

c. Can you devise a simple formula for the calculation of the number of progeny genotypes in dihybrid, trihybrid (etc.) crosses? Repeat for phenotypes.

d. Mendel predicted that within all but one of the phenotypic classes in the Punnett square, there should be several different genotypes. In particular, he performed many crosses to identify the underlying genotypes of the round, yellow phenotype. Show two different ways that could be used to identify the various genotypes underlying the round, yellow phenotype. (Remember, all the round, yellow peas look identical.)

5. You have three dice: one red (R), one green (G), and one blue (B). When all three dice are rolled at the same time, calculate the probability of the following outcomes:

a. 6(R) 6(G) 6(B)

b. 6(R) 5(G) 6(B)

c. 6(R) 5(G) 4(B)

d. no sixes at all

e. 2 sixes and 1 five on any dice

f. 3 sixes or 3 fives

g. the same number on all dice

h. a different number on all dice

6. You have three jars containing marbles, as follows:

jar 1	600 red	and	400 white
jar 2	900 blue	and	100 white
jar 3	10 green	and	990 white

a. If you blindly select one marble from each jar, calculate the probability of obtaining

(1) a red, a blue, and a green
(2) three whites
(3) a red, a green, and a white
(4) a red and two whites
(5) a color and two whites
(6) at least one white

*b. In a certain plant, *R* = red and *r* = white. You self a red *R r* heterozygote with the express purpose of obtaining a white plant for an experiment. What minimum number of seeds do you have to grow to be at least 95 percent certain of obtaining at least one white individual? [HINT: consider your answer to Question 6a(6).]

c. When a woman is injected with an egg fertilized in vitro, the probability of it implanting successfully is 20%. If a woman is injected with five eggs simultaneously, what is the probability that she will become pregnant?

(Part c from Margaret Holm.)

7. a. The ability to taste the chemical phenylthiocarbamide is an autosomal dominant phenotype, and the inability to taste it is recessive. If a taster woman with a nontaster father marries a taster man who in a previous marriage had a nontaster daughter, what is the probability that their first child will be

(1) a nontaster girl
(2) a taster girl
(3) a taster boy

b. What is the probability that their first *two* children will be tasters of any sex?

8. John and Martha are contemplating having children, but John's brother has galactosemia (an autosomal recessive disease), and Martha's great-grandmother also had galactosemia. Martha has a sister who has three children, none of whom has galactosemia. What is the probability that John and Martha's first child will have galactosemia?

 Unpacking the Problem

In some chapters we expand one specific problem with a list of exercises that help mentally process the principles and other knowledge surrounding the subject area of the problem. You can make up similar exercises yourself for other problems.

Before attempting a solution to problem 8, consider some questions such as the following, which are meant only as examples.

a. Can the problem be restated as a pedigree? If so, write one.

b. Can parts of the problem be restated using Punnett squares?

c. Can parts of the problem be restated using branch diagrams?

d. In the pedigree, identify a mating that illustrates Mendel's first law.

e. Define all the scientific terms in the problem, and look up any other terms that you are uncertain about.

f. What assumptions need to be made in answering this problem?

g. Which unmentioned family members must be considered? Why?

h. What statistical rules might be relevant, and in what situations can they be applied? Are there such situations represented in this problem?

i. What are two generalities about autosomal recessive diseases in human populations?

j. What is the relevance of the rareness of the phenotype under study in pedigree analysis generally, and what can be inferred in this problem?

k. In this family, whose genotypes are certain and whose are uncertain?

l. In what way is John's side of the pedigree different from Martha's side? How does this affect your calculations?

m. Is there any information that is irrelevant in the problem as stated?

n. In what way is solving this kind of problem similar to or different from solving problems you have already successfully answered?

o. Can you make up a short story based on the human dilemma in this problem?

Now try to solve the problem. If you are unable, try to identify the obstacle and write a sentence or two

describing your difficulty. Then go back to the expansion questions and see if any of them relate to your difficulty.

9. Holstein cattle normally are black and white. A superb black and white bull, Charlie, was purchased by a farmer for $100,000. The progeny sired by Charlie were all normal in appearance. However, certain pairs of his progeny, when interbred, produced red and white progeny at a frequency of about 25 percent. Charlie was soon removed from the stud lists of the Holstein breeders. Explain precisely why, using symbols.

10. Suppose that a husband and wife are both heterozygous for a recessive gene for albinism. If they have dizygotic (two-egg) twins, what is the probability that both of the twins will have the same phenotype for pigmentation?

11. The plant blue-eyed Mary grows on Vancouver Island and on the lower mainland of British Columbia. The populations are dimorphic for purple blotches on the leaves—some plants have blotches, and others don't. Near Nanaimo, one plant in nature had blotched leaves. This plant, which had not yet flowered, was dug up and taken to a laboratory, where it was allowed to self. Seeds were collected and grown into progeny. One randomly selected (but typical) leaf from each of the progeny is shown in the accompanying figure.

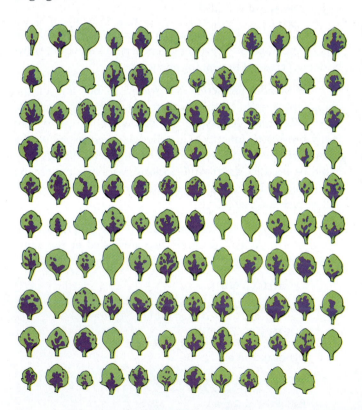

a. Formulate a concise genetic hypothesis to explain these results. Explain all symbols and show all genotypic classes (and the genotype of the original plant).

b. How would you test your hypothesis? Be specific.

12. Can it ever be proved that an animal is *not* a carrier of a recessive allele (that is, not a heterozygote for a given gene)? Explain.

13. In nature, the plant *Plectritis congesta* is dimorphic for fruit shape; that is, individual plants bear either wingless or winged fruits, as shown in the accompanying figure.

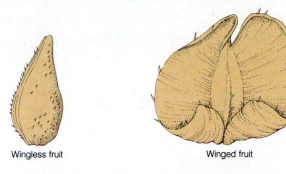

Wingless fruit   Winged fruit

Plants were collected from nature before flowering and were crossed or selfed with the following results:

| | Number of progeny | |
Pollination	Winged	Wingless
winged (selfed)	91	1*
winged (selfed)	90	30
wingless (selfed)	4*	80
winged × wingless	161	0
winged × wingless	29	31
winged × wingless	46	0
winged × winged	44	0
winged × winged	24	0

Interpret these results, and derive the mode of inheritance of these fruit-shape phenotypes. Use symbols. (NOTE: The phenotypes of progeny marked by asterisks probably have a nongenetic explanation. What do you think it is?)

14. The accompanying pedigree is for a rare, but relatively mild, hereditary disorder of the skin.

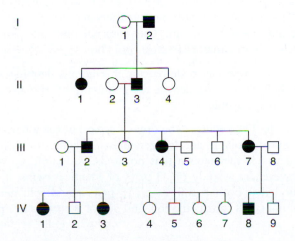

a. Is the disorder inherited as a recessive or a dominant phenotype? State reasons for your answer.

b. Give genotypes for as many individuals in the pedigree as possible. (Invent your own defined allele symbols.)

c. Consider the four unaffected children of parents III-4 and III-5. In all four-child progenies from parents of these genotypes, what proportion is expected to contain all unaffected children?

15. Here are four human pedigrees. The black symbols represent an abnormal phenotype inherited in a simple Mendelian manner.

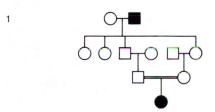

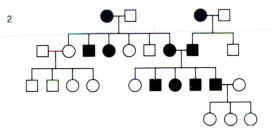

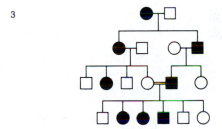

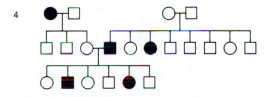

a. For each pedigree, state whether the abnormal condition is dominant or recessive. Try to state the logic behind your answer.

b. In each pedigree, describe the genotypes of as many individuals as possible.

16. Tay-Sachs disease ("infantile amaurotic idiocy") is a rare human disease in which toxic substances accumulate in nerve cells. The recessive allele responsible for the disease is inherited in a simple Mendelian manner. For unknown reasons, the allele is more common in populations of Ashkenazi Jews of Eastern Europe. A woman is planning to marry her first cousin, but the couple discovers that their shared grandfather's sister died in infancy of Tay-Sachs disease.

a. Draw the relevant parts of the pedigree, and show all the genotypes as completely as possible.

b. What is the probability that the cousins' first child will have Tay-Sachs disease, assuming that all people who marry into the family are homozygous normal?

17. The accompanying pedigree was obtained for a rare kidney disease.

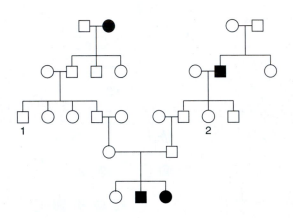

a. Deduce the inheritance of this condition, stating your reasons.

b. If individuals 1 and 2 marry, what is the probability that their first child will have the kidney disease?

18. The accompanying pedigree is for Huntington's disease (HD), a late-onset disorder of the nervous system. The slashes indicate deceased family members.

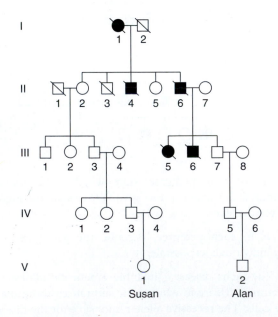

a. Is this pedigree compatible with the mode of inheritance for HD mentioned in the chapter?

b. Consider two newborn children in the two arms of the pedigree, Susan in the left arm and Alan in the right arm. Study the graph in Figure 2-17 and come up with an opinion on the likelihood that they will develop HD. Assume for the sake of the discussion that parents have children at age 25.

19. Consider the accompanying pedigree of a rare autosomal recessive disease, PKU.

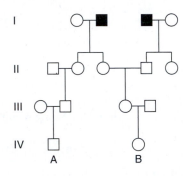

a. List genotypes of as many individuals as possible.

b. If individuals A and B marry, what is the probability their first child will have PKU?

c. If their first child is normal, what is the probability their second child will have PKU?

d. If their first child has the disease, what is the probability their second child will be unaffected?

(Assume all people marrying into the pedigree do not carry the abnormal allele.)

20. A curious polymorphism in human populations has to do with the ability to curl up the sides of the tongue to make a trough ("tongue rolling"). Some people can do this trick, and others simply cannot. Hence it is an example of a dimorphism. Its significance is a complete mystery. In one family, a boy was unable to roll his tongue, but to his great chagrin his sister could. Furthermore, both his parents were rollers, and so were both grandfathers and one paternal uncle and one paternal aunt. One paternal aunt, one paternal uncle, and one maternal uncle could not.

a. Draw the pedigree for this family, defining your symbols clearly, and deduce the genotypes of as many individuals as possible.

b. The pedigree you drew is typical of the inheritance of tongue rolling, and led geneticists to come up with the inheritance mechanism that no doubt you came up with. However, in a study of 33 pairs of identical twins, it was found that both members of 18 pairs could roll, neither member of 8 pairs could roll, and one of the twins in 7 pairs could roll but the other couldn't. Because identical twins are derived from the splitting of one fertilized egg into two embryos, the members of a pair must be genetically identical. How can the existence of the seven discordant pairs be reconciled with your genetic explanation of the pedigree?

21. A rare, recessive allele inherited in a Mendelian manner causes the disease cystic fibrosis. A phenotypically normal man whose father had cystic fibrosis marries a phenotypically normal woman from outside the family, and the couple consider having a child.

    a. Draw the pedigree as far as described.

    b. If the frequency in the population of heterozygotes for cystic fibrosis is 1 in 50, what is the chance that the couple's first child will have cystic fibrosis?

    c. If the first child does have cystic fibrosis, what is the probability that the second child will be normal?

22. In human hair, the black and brown colors are produced by various amounts and combinations of chemicals called *melanins*. However, red hair is produced by a different type of chemical substance about which little is known. Red hair runs in families, and the figure at the bottom of the page shows a large pedigree for red hair.

    a. Does the inheritance pattern in this pedigree suggest that red hair could be caused by a dominant or a recessive allele of a gene that is inherited in a simple Mendelian manner?

    b. Do you think the red hair allele is common or rare in the population as a whole?

(Pedigree from W. R. Singleton and B. Ellis, *Journal of Heredity* 55, 1964, 261.)

23. When many families were tested for the ability to taste the chemical PTC, the matings were grouped into three types and the progeny totaled, with the results shown below:

| | | Children | |
Parents	Number of families	Tasters	Non-tasters
taster × taster	425	929	130
taster × nontaster	289	483	278
nontaster × nontaster	86	5	218

Assuming that PTC tasting is dominant (P) and nontasting is recessive (p), how can the progeny ratios in each of the three types of mating be accounted for?

24. In tomatoes, red fruit is dominant to yellow, two-loculed fruit is dominant to many-loculed fruit, and tall vine is dominant to dwarf. A breeder has two pure lines: red,

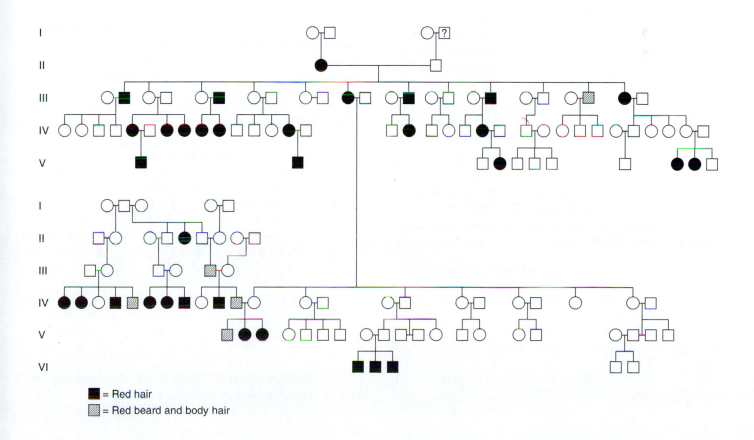

■ = Red hair
▨ = Red beard and body hair

two-loculed, dwarf and yellow, many-loculed, tall. From these he wants to produce a new pure line for trade that is yellow, two-loculed, and tall. How exactly should he go about doing this? Show not only which crosses to make, but also how many progeny should be sampled in each case.

25. In humans, achondroplastic dwarfism and neurofibromatosis are both extremely rare dominant conditions. If a woman with achondroplasia marries a man with neurofibromatosis, what phenotypes could be produced in their children, and in what proportions? (Be sure to define any symbols you use.)

26. In dogs, dark coat color is dominant over albino and short hair is dominant over long hair. Assume that these effects are caused by two independently assorting genes and write the genotypes of the parents in each of the crosses shown below, where D and A stand for the dark and albino phenotypes, respectively, and S and L stand for the short-hair and long-hair phenotypes.

Parental phenotypes	Number of progeny			
	D, S	D, L	A, S	A, L
a. D, S × D, S	89	31	29	11
b. D, S × D, L	18	19	0	0
c. D, S × A, S	20	0	21	0
d. A, S × A, S	0	0	28	9
e. D, L × D, L	0	32	0	10
f. D, S × D, S	46	16	0	0
g. D, S × D, L	30	31	9	11

Use the symbols C and c for the dark and albino coat-color alleles and the symbols S and s for the short-hair and long-hair alleles, respectively. Assume homozygosity unless there is evidence otherwise.

(Problem 26 reprinted by permission of Macmillan Publishing Co., Inc., from *Genetics* by M. Strickberger. Copyright © 1968 by Monroe W. Strickberger.)

27. In tomatoes, two alleles of one gene determine the character difference of purple (P) versus green (G) stems and two alleles of a separate, independent gene determine the character difference of "cut" (C) versus "potato" (Po) leaves. The results for five matings of tomato plant phenotypes are shown below.

Mating	Parental phenotypes	Number of progeny			
		P, C	P, Po	G, C	G, Po
1	P, C × G, C	321	101	310	107
2	P, C × P, Po	219	207	64	71
3	P, C × G, C	722	231	0	0
4	P, C × G, Po	404	0	387	0
5	P, Po × G, C	70	91	86	77

a. Determine which alleles are dominant.

b. What are the most probable genotypes for the parents in each cross?

(Problem 27 from A. M. Srb, R. D. Owen, and R. S. Edgar, *General Genetics,* 2d ed. Copyright © 1965 by W. H. Freeman and Company.)

28. We have dealt mainly with only two genes, but the same principles hold for more than two genes. Consider the cross

$$Aa\,Bb\,Cc\,Dd\,Ee \times aa\,Bb\,cc\,Dd\,ee$$

a. What proportion of progeny will *phenotypically* resemble (1) the first parent, (2) the second parent, (3) either parent, and (4) neither parent?

b. What proportion of progeny will be *genotypically* the same as (1) the first parent, (2) the second parent, (3) either parent, and (4) neither parent?

Assume independent assortment.

29. Most *Drosophila melanogaster* have brown bodies, but members of the species that are homozygous for the recessive allele y have yellow bodies. However, if larvae of pure *YY* lines are reared on food containing silver salts, the resulting adults are yellow. These are called **phenocopies,** environmentally induced copies of genetically determined phenotypes. If you were presented with a single yellow fly, how would you determine if it was a yellow genotype or a yellow phenocopy? (Can you think of examples of phenocopies in humans?)

30. The accompanying pedigree shows the pattern of transmission of two rare human phenotypes, cataract and pituitary dwarfism. Individuals with cataract are shown with a solid *left* half to the symbol; those with pituitary dwarfism are indicated by a solid *right* half.

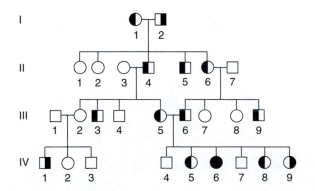

a. What is the most likely mode of inheritance of each of these phenotypes? Explain.

b. List the genotypes of all individuals in generation III as far as possible.

c. If a hypothetical mating occurred between IV-1 and IV-5, what is the probability of the first child's being a dwarf with cataracts? A phenotypically normal child?

(Problem 30 after J. Kuspira and R. Bhambhani,
Compendium of Problems in Genetics. Copyright © 1994
by Wm. C. Brown.)

31. A corn geneticist has three pure lines of genotypes *a a BB CC*, *AA bb CC*, and *AA BB cc*. The phenotypes determined by *a*, *b*, and *c* will all increase the market value of the corn, so naturally he wants to combine them all in one pure line of genotype *a a bb cc*.

a. Outline an effective crossing program that can be used to obtain the *a a bb cc* pure line.

b. At each stage, state exactly which phenotypes will be selected and give their expected frequencies.

c. Is there more than one way to obtain the desired genotype? Which is the best way?

(Assume independent assortment of the three gene pairs. NOTE: Corn will self or cross-pollinate easily.)

32. **a.** A population of annual plants was composed exclusively of individuals of genotype *a a*. One year, a flood introduced many seeds of genotype *AA* and *A a* into the population. Immediately after the introduction there were 55% *AA*, 40% *A a*, and 5% *a a* individuals. Because the region had no insects capable of cross-pollinating this species of plant, all plants routinely self-pollinated. After three generations of selfing, what were the proportions of *AA*, *A a*, and *a a* genotypes?

b. Imagine a plant that has 40% of all its gene pairs heterozygous (*A a*, *B b*, *C c*, etc.). After three generations of selfing, what proportion of gene pairs will still be heterozygous?

c. How does the answer to part b relate to the answer to part a?

3

Chromosome Theory
of Inheritance

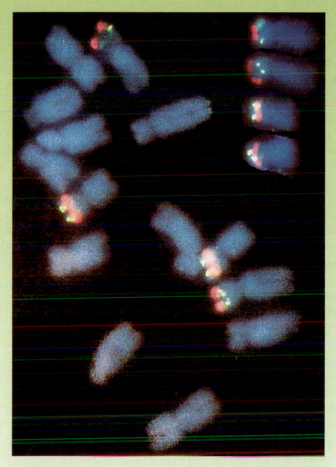

Chromosome set ($2n = 14$) of the wild wheat species *Aegilops umbellulata*. (Inset, top right, shows two aligned pairs.) The chromosomes are shown in mitotic metaphase, so each chromosome is represented by a pair of sister chromatids. The slide has been stained with DAPI, a fluorescent dye that binds generally to DNA, giving a blue-green appearance. Two additional fluorescent probes have been used that bind to the chromosomal regions coding for ribosomal RNA (rRNA). The red probe shows the position of 18S-25S rRNA genes, and the green probe shows 5S rRNA genes. These rRNA gene regions are present twice in the genome, on two different chromosome pairs in inverted order. (From Alexandra Castilho and J. S. Heslop-Harrison.)

KEY CONCEPTS

▶ Genes are parts of chromosomes.

▶ Mitosis is the nuclear division that results in two daughter nuclei, each with genetic material identical with that of the original nucleus.

▶ Meiosis is the nuclear division by which a reproductive cell with two equivalent chromosome sets divides into four meiotic products, each of which has only one set of chromosomes.

▶ Mendel's laws of equal segregation and independent assortment are based on the separation during meiosis of members of each chromosome pair and on the independent meiotic behavior of different chromosome pairs.

▶ Special phenotypic progeny ratios, differing for males and females, mirror the inheritance patterns of genes on the sex chromosomes.

The beauty of Mendel's analysis is that we need not know what genes are or how they control phenotypes to analyze the results of crosses and to predict the outcomes of future crosses using the laws of equal segregation and independent assortment. We can do all this simply by representing abstract, hypothetical factors of inheritance (genes) by symbols—without any concern about their molecular structures or their locations in a cell. Nevertheless, our interest naturally turns to the next obvious question: What structures within cells correspond to these hypothetical genes?

The development of genetics took a major step forward by accepting the notion that the genes, as characterized by Mendel, are parts of specific cellular structures, the chromosomes. This simple concept has become known as the **chromosome theory of heredity.** Although simple, the idea has had profound implications, inextricably uniting the disciplines of genetics and cytology and providing a means of correlating the results of breeding experiments with the behavior of structures that can be actually seen under the microscope. This fusion between genetics and cytology is still an essential part of genetic analysis today and has important applications in medical genetics, agricultural genetics, and evolutionary genetics.

Mitosis and Meiosis

How did the chromosome theory take shape? Evidence gradually accumulated from a variety of sources. One of the first lines of evidence came from observations of how chromosomes behave during the division of a cell's nucleus. The observations leading up to the discovery of the two different types of nuclear division, termed mitosis and meiosis, were as follows. In the interval between Mendel's research and its rediscovery, many biologists were interested in heredity even though they were unaware of Mendel's results, and they approached the problem in a completely different way. These investigators wanted to locate the hereditary material in the cell. An obvious place to look was in the gametes, because they are the only connecting link between generations. Egg and sperm were believed to contribute equally to the genetic endowment of offspring even though they differ greatly in size. Because an egg has a great volume of cytoplasm but a sperm has very little, the cytoplasm of gametes seemed an unlikely seat of the hereditary structures. The nuclei of egg and sperm, however, were known to be approximately equal in size, so the nuclei were considered good candidates for harboring hereditary structures.

What was known about the contents of cell nuclei? It became clear that the most prominent components were the chromosomes, which proved to possess unique properties that set them apart from all other cellular structures. A property that especially intrigued biologists was the constancy of the number of chromosomes from cell to cell within an organism, from organism to organism within any one species, and from generation to generation within that species. The question therefore arose: How is the chromosome number maintained? The question was answered by observing the behavior of chromosomes under the microscope during mitosis and meiosis; from those observations developed the postulation of the chromosome theory—that chromosomes are the structures that contain genes.

Mitosis is the nuclear division associated with the division of somatic cells—cells of the eukaryotic body that are not destined to become sex cells. This kind of division produces a number of genetically identical cells from a single progenitor cell, as, for example, in the division of a fertilized human egg cell to become a multicellular organism composed of trillions of cells. Each single mitosis is associated with a single cell division that produces two genetically identical daughter cells.

Meiosis is the name given to the nuclear divisions in the special cells that are destined to produce gametes. Such a cell is called a **meiocyte.** There are two cell divisions of each meiocyte and two associated meiotic divisions of the nucleus. Hence, each meiocyte generally produces four cells, which we shall call **products of meiosis.** In humans and other animals, meiosis takes place in the gonads, and the products of meiosis are the gametes—sperm (more properly, spermatozoa) and eggs (ova). In flowering plants, meiosis takes place in the anthers and ovaries, and the products of meiosis are **meiospores,** which eventually give rise to gametes. We now turn to the details of these two basic kinds of nuclear division. The following descriptions of the various stages are as general as possible and are applicable to mitosis and meiosis in most organisms in which such divisions take place. Note, however, that the photographic illustrations are all of one organism (a flowering plant, *Lilium regale*) and therefore cannot be completely general for all details of mitosis and meiosis. Hence, a parallel series of idealized drawings is also included.

Mitosis

The **cell cycle** is the series of events from any stage in a cell to the equivalent stage in a daughter cell. For convenience, it is divided into several periods: **M**, **S**, **G1**, and **G2** (Figure 3-1). Mitosis (M) is usually the shortest period of the cycle, lasting for approximately 5 to 10 percent of the cycle. DNA synthesis takes place during the S period. G1 and G2 are gaps between S and M. Together, G1, S, and G2 constitute **interphase,** the time between mitoses. (Interphase used to be called "resting period"; however, cells actively function in many ways during interphase, not the least of which, of course, is in synthesizing DNA.) The chromosomes cannot be seen during interphase (Figure 3-2a), mainly because they are in an extended state and become intertwined with each other like a tangle of yarn.

The net achievement of mitosis is that each chromosome in the nucleus duplicates longitudinally, and then this double structure splits to become two daughter chromosomes, each going to a different daughter nucleus. Mitosis produces two daughter nuclei identical with each other and

Figure 3-1 Stages of the cell cycle. M = mitosis, S = DNA synthesis, G = gap.

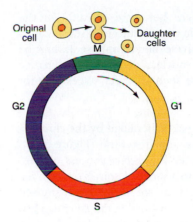

Figure 3-2 Mitosis. The photographs show nuclei of root tip cells of *Lilium regale.* (Modified from J. McLeish and B. Snoad, *Looking at Chromosomes.* Copyright 1958, St. Martin's, Macmillan.)

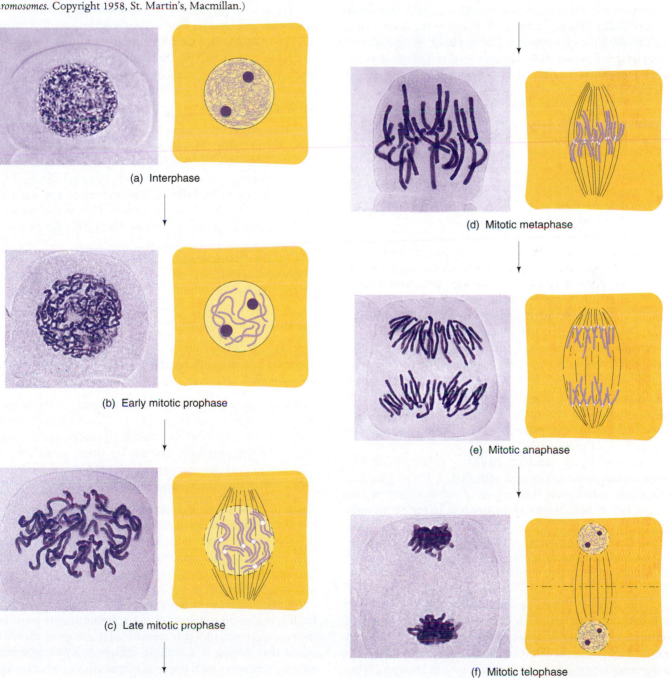

(a) Interphase

(b) Early mitotic prophase

(c) Late mitotic prophase

(d) Mitotic metaphase

(e) Mitotic anaphase

(f) Mitotic telophase

with the nucleus from which they were derived. For the sake of study, scientists divide mitosis into four stages called **prophase, metaphase, anaphase,** and **telophase.** It must be stressed, however, that any nuclear division is a dynamic process on which we impose such arbitrary stages only for our own convenience.

Prophase. The onset of mitosis is heralded by the chromosomes becoming distinct for the first time (Figure 3-2b). They get progressively shorter through a process of contraction, or condensation, into a series of spirals or coils; the coiling produces structures that are more easily moved around (for the same reason, cotton threads are packaged commercially on spools). As the chromosomes become visible, they appear double-stranded, each chromosome being composed of two longitudinal halves called **chromatids** (Figure 3-2c). These "sister" chromatids are joined at a region called the **centromere.** The **nucleoli**—large intranuclear spherical structures—disappear at this stage. The nuclear membrane begins to break down, and the nucleoplasm and cytoplasm become one.

Metaphase. At this stage, the **nuclear spindle** becomes prominent. The spindle is a birdcage-like structure that forms in the nuclear area; it consists of a series of parallel fibers that point to each of two cell poles. The chromosomes move to the equatorial plane of the cell, where the centromeres become attached to a spindle fiber from each pole (Figure 3-2d).

Anaphase. This stage begins when the pairs of sister chromatids separate, one of a pair moving to each pole (Figure 3-2e). The centromeres, which now appear to have divided, separate first. As each chromatid moves, its two arms appear to trail its centromere; a set of V-shaped structures results, with the points of the V's directed at the poles.

Telophase. Now a nuclear membrane re-forms around each daughter nucleus, the chromosomes uncoil, and the nucleoli reappear—all of which effectively re-form interphase nuclei (Figure 3-2f). By the end of telophase, the spindle has dispersed, and the cytoplasm has been divided into two by a new cell membrane.

In each of the resultant daughter cells, the chromosome complement is identical with that of the original cell. Of course, what were referred to as chromatids now take on the role of full-fledged chromosomes in their own right.

Message   Mitosis produces two daughter nuclei that have a chromosomal constitution identical with that of the original nucleus.

The role of chromosomes in directing development was hotly debated in the late nineteenth and early twentieth centuries. Wilhelm Roux's position was that if chromosomes played a role in development, they would have to be differentially partitioned to the various cell lineages. When such partitioning was shown not to be the case, it was concluded, for a while, that chromosomes must therefore not be important for development. The fact that partitioning of chromosomes is *not* required and that the chromosomal material is precisely maintained in development was not proved until later, when nuclear transplantation experiments showed that nuclei from different cell types were functionally interchangeable and, thus, had not lost any developmental potential.

It was evident that mitosis maintains the chromosome number in each nucleus, but this made early investigators puzzle over the joining of two gametes in the fertilization event. They knew that during this process, two nuclei fuse but that the chromosome number nevertheless remains constant. What prevented the doubling of the chromosome number at each generation? This puzzle was resolved by the prediction of a special kind of nuclear division that halved the chromosome number. This special division, which was eventually discovered in the gamete-producing tissues of plants and animals, is called meiosis.

Meiosis

Meiosis is preceded by a premeiotic **S** phase in the meiocyte. Most of the DNA for meiosis is synthesized during this S phase, but some is synthesized during the first prophase of meiosis. Meiosis consists of two cell divisions distinguished as **meiosis I** and **meiosis II.** The events of meiosis I are quite different from those of meiosis II, and the events of both differ from those of mitosis. Each meiotic division is formally divided into prophase, metaphase, anaphase, and telophase. Of these stages, the most complex and lengthy is prophase I, which has its own subdivisions: **leptotene, zygotene, pachytene, diplotene,** and **diakinesis.** Once again, try to imagine those processes as merging dynamically into one another with no clear borders.

Prophase I. Several processes occur at different stages of prophase I, so for convenient study it is divided into five substages.

Leptotene. The chromosomes become visible at this stage as long, thin single threads (Figure 3-3a). The process of chromosome contraction continues during leptotene and throughout the entire prophase. During leptotene small areas of thickening called **chromomeres** develop along each chromosome, which give it the appearance of a necklace of beads.

Zygotene. This is a time of active pairing of the threads (Figure 3-3b), making it apparent that the chromosome complement of the meiocyte is in fact two complete chromosome sets. Thus, each chromosome has a pairing partner, and the two become progressively paired, or **synapsed,** side by side as if by a zipper.

At this point we need to introduce some terminology. Each chromosome pair is called a **homologous pair,** and the two members of a pair are called homologs. It should be noted that pairing is a striking difference from mitosis, in which there is no such process. Furthermore, whereas cells with any number of chromosome sets may undergo mitosis, only cells with two chromosome sets (two **genomes**) undergo meiosis. Cells with two chromosome sets are called **diploid** and are represented symbolically as $2n$, where n is the number of chromosomes in a set. In contrast, cells with only one set (n) are called *haploid*. In most complex organisms such as mammals and flowering plants, the cells of the organism are normally diploid, and the meiocytes are simply a subpopulation of cells that are set aside to undergo meiosis. In haploid organisms, as we shall see later, a diploid meiocyte is constructed as part of the normal reproductive cycle.

How do two homologs find each other during zygotene? The probable answer to this is that the ends of the chromosomes, the **telomeres,** are anchored in the nuclear membrane, and it is likely that homologous telomeres are close, so that the zippering-up process can begin with them. How does the zippering-up work? What is the mechanism whereby two homologs can pair so precisely along their length? Although the mechanism is not precisely understood, we know that it requires an elaborate structure composed of protein and DNA, called a **synaptonemal complex** (Figure 3-4), which is always found sandwiched between homologs during synapsis.

Pachytene. This stage is characterized by thick, fully synapsed threads (Figure 3-3c). Thus, the number of homologous pairs of chromosomes in the nucleus is equal to the number n. Nucleoli are often pronounced during pachytene. The beadlike chromomeres align precisely in the paired homologs, producing a distinctive pattern for each pair.

Diplotene. Although each homolog appeared to be a single thread during leptotene, the DNA had, in fact, already replicated during the premeiotic S phase. This becomes manifest during diplotene as a longitudinal doubleness of each paired homolog (Figure 3-3d). As in mitosis, the divided subunits are called *chromatids.* Hence, since each member of a homologous pair produces two sister chromatids, the synapsed structure now consists of a bundle of four homologous chromatids. At diplotene, the pairing between homologs becomes less tight; in fact, they appear to repel each other, and as they separate slightly, cross-shaped structures called **chiasmata** (singular, chiasma) appear between nonsister chromatids. Each chromosome pair generally has one or more chiasmata. Chiasmata are the visible manifestations of events called crossovers that occurred earlier, probably during zygotene or pachytene. Crossovers represent one major way in which meiosis differs from mitosis—crossovers during mitosis are rare. A crossover is a precise breakage, swapping, and reunion between two nonsister chromatids. Studies performed on abnormal lines of organisms that undergo crossing-over very inefficiently, or not at all, show severe disruption of the orderly events that partition chromosomes into daughter cells at meiosis. Thus, crossing-over obviously helps determine how paired homologs behave, and the presence of at least one crossover per pair is usually essential for proper segregation. Crossovers have another interesting role, which, as we shall see in Chapter 5, is to make new gene combinations, an important source of genetic variation in populations.

Diakinesis. This stage (Figure 3-3e) differs only slightly from diplotene, except for further chromosome contraction. By the end of diakinesis, the long, filamentous chromosome threads of interphase have been replaced by compact units that are far more maneuverable in the movements of the meiotic division.

Metaphase I. The nuclear membrane and nucleoli have disappeared by metaphase I, and each pair of homologs takes up a position in the equatorial plane (Figure 3-3f). At this stage of meiosis, the centromeres do not divide; this lack of division represents a major difference from mitosis. The two centromeres of a homologous chromosome pair attach to spindle fibers from opposite poles.

Anaphase I. As in mitosis, anaphase begins when chromosomes move directionally to the poles. The members of a homologous pair move to opposite poles (Figure 3-3g and h).

Telophase I. This telophase (Figure 3-3i) and the ensuing "interphase," called **interkinesis** (Figure 3-3j), are not universal. In many organisms, these stages do not exist, no nuclear membrane re-forms, and the cells proceed directly to meiosis II. In other organisms, telophase I and the interkinesis are brief in duration; the chromosomes elongate and become diffuse, and the nuclear membrane re-forms. In any case, there is never DNA synthesis at this time, and the genetic state of the chromosomes does not change.

By convention, the two nuclei that result from meiosis I are considered to be haploid. This might seem perverse, because each chromosome is composed of a pair of sister chromatids. Nevertheless, since the sister chromatids remain attached at the centromere, they are counted as one chromosome. Viewed this way, haploidy is best demonstrated by counting centromeres. The key point is that the total number of chromosomes in each cell has been reduced by half. For this reason, meiosis I is called a **reduction division.** In contrast, meiosis II is similar to a mitotic division in that the number of chromosomes (as defined above) is the same in the original and product cells. This type of division is called equational division.

Prophase II. The presence of the haploid number of chromosomes in the contracted state characterizes prophase II (Figure 3-3k).

Metaphase II. The chromosomes arrange themselves on the equatorial plane during metaphase II (Figure 3-3l). Here the chromatids often partly dissociate from each other instead of being closely oppressed as they are in mitosis.

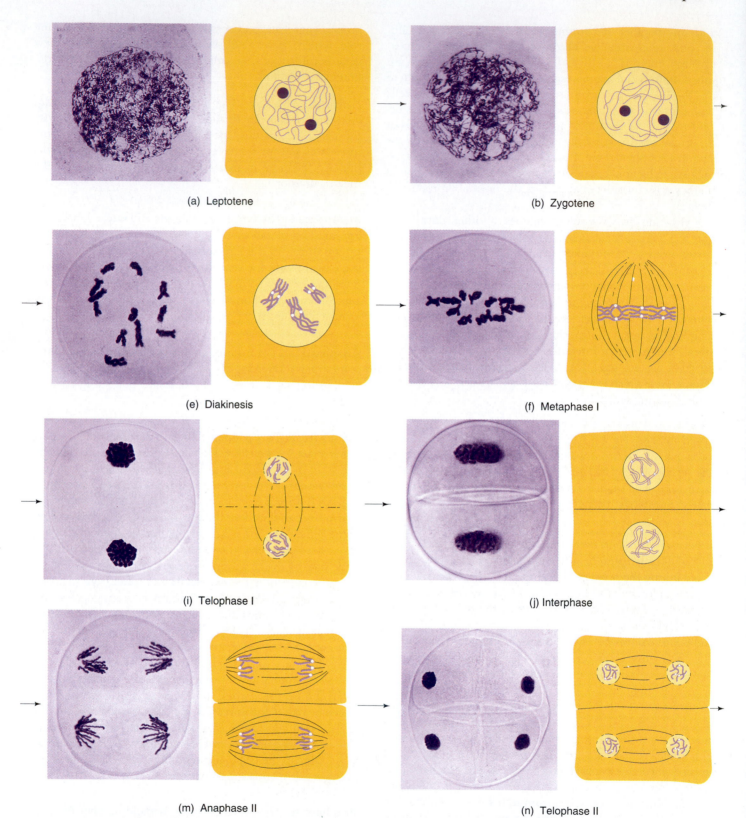

(a) Leptotene

(b) Zygotene

(e) Diakinesis

(f) Metaphase I

(i) Telophase I

(j) Interphase

(m) Anaphase II

(n) Telophase II

Figure 3-3 Meiosis and pollen formation. The photographs are of *Lilium regale*. NOTE: For simplicity, multiple chiasmata are drawn as involving only two chromatids; in reality, all four chromatids can be involved. (Modified from J. McLeish and B. Snoad, *Looking at Chromosomes*. Copyright © 1958, St. Martin's, Macmillan.)

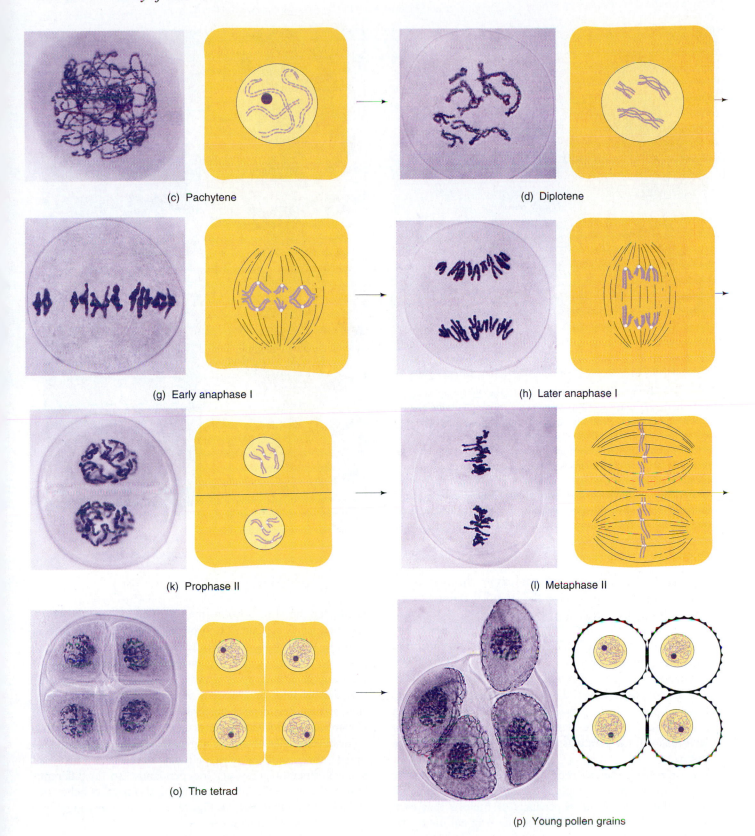

(c) Pachytene

(d) Diplotene

(g) Early anaphase I

(h) Later anaphase I

(k) Prophase II

(l) Metaphase II

(o) The tetrad

(p) Young pollen grains

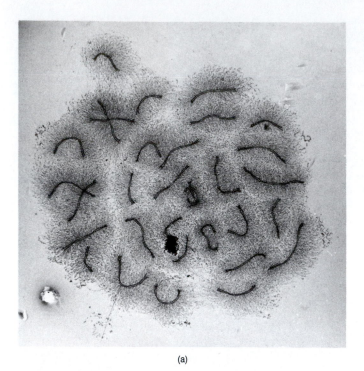

(a)

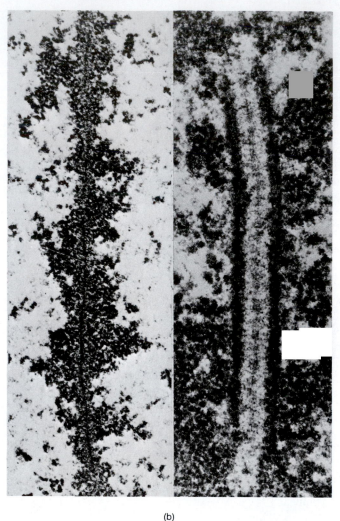

(b)

Figure 3-4 Synaptonemal complexes. (a) In *Hyalophora cecropia*, a silk moth, the normal male chromosome number is 62, giving 31 synaptonemal complexes. In the individual shown here, one chromosome (*center*) is represented three times; such a chromosome is termed *trivalent*. The DNA is arranged in regular loops around the synaptonemal complex. The black, dense structure is the nucleolus. (b) Regular synaptonemal complex in *Lilium tyrinum*. Note (*right*) the two lateral elements of the synaptonemal complex and also (*left*) an unpaired chromosome, showing a central core corresponding to one of the lateral elements. [Parts (a) and (b) courtesy of Peter Moens.]

Anaphase II. Centromeres split and sister chromatids are pulled to opposite poles by the spindle fibers during anaphase II (Figure 3-3m).

Telophase II. The nuclei re-form around the chromosomes at the poles (Figure 3-3n).

The four products of meiosis are shown in Figure 3-3o. In the anthers of a flower, the four products of meiosis develop into pollen grains; these are shown in Figure 3-3p. In other organisms, differentiation produces other kinds of structures from the products of meiosis, such as sperm cells in animals.

In summary, meiosis requires one doubling of genetic material (the premeiotic S phase) and two cell divisions. Inevitably, this must result in products of meiosis that each contain half the genetic material of the original meiocyte.

Message Meiosis generally takes place in a diploid meiocyte and yields four haploid products of meiosis.

A summary of the net events of mitosis and meiosis is shown in Figure 3-5.

The Chromosome Theory of Heredity

Credit for the chromosome theory of heredity—the concept that genes are parts of chromosomes—is usually given to both Walter Sutton (an American who at the time was a graduate student) and Theodor Boveri (a German biologist). In 1902, these investigators recognized independently that the behavior of Mendel's particles during the production of gametes in peas precisely parallels the behavior of chromosomes at meiosis: genes are in pairs (so are chromosomes); the alleles of a gene segregate equally into gametes (so do the members of a pair of homologous chromosomes); different genes act independently (so do different chromosome pairs). After recognizing this parallel behavior (which is summarized in Figure 3-6), both investigators reached the same conclusion.

Message The parallel behavior of genes and chromosomes led to the concept that genes are located on chromosomes.

We saw in Chapter 1 that a major goal of genetics is to explain two apparently conflicting forces of biology: hered-

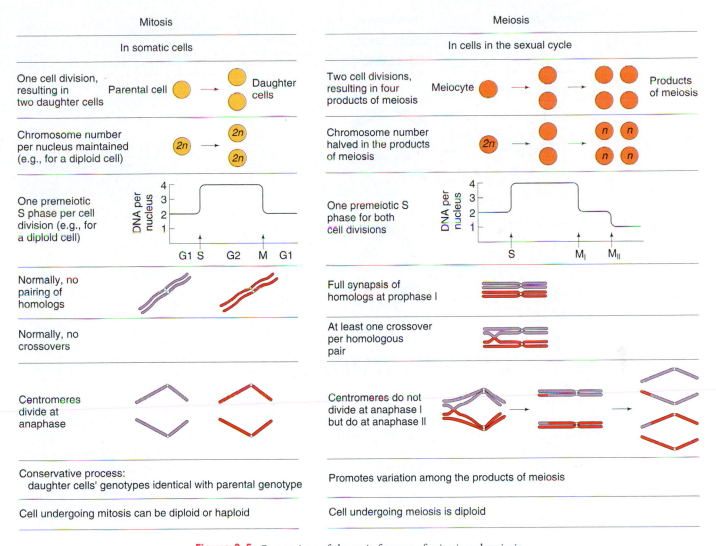

Mitosis	Meiosis
In somatic cells	In cells in the sexual cycle
One cell division, resulting in two daughter cells — Parental cell → Daughter cells	Two cell divisions, resulting in four products of meiosis — Meiocyte → → Products of meiosis
Chromosome number per nucleus maintained (e.g., for a diploid cell) — $2n$ → $2n$ $2n$	Chromosome number halved in the products of meiosis — $2n$ → → n n n n
One premeiotic S phase per cell division (e.g., for a diploid cell) — DNA per nucleus 4 3 2 1 — G1 S G2 M G1	One premeiotic S phase for both cell divisions — DNA per nucleus 4 3 2 1 — S M_I M_{II}
Normally, no pairing of homologs	Full synapsis of homologs at prophase I
Normally, no crossovers	At least one crossover per homologous pair
Centromeres divide at anaphase	Centromeres do not divide at anaphase I but do at anaphase II
Conservative process: daughter cells' genotypes identical with parental genotype	Promotes variation among the products of meiosis
Cell undergoing mitosis can be diploid or haploid	Cell undergoing meiosis is diploid

Figure 3-5 Comparison of the main features of mitosis and meiosis.

ity and variation. The two processes of mitosis and meiosis provide a major clue: mitosis is a conservative process that maintains a genetic status quo, whereas meiosis is a process that generates enormous combinatorial variation, rather like shuffling the gene pack, through independent assortment and (as we shall see later) through crossing-over.

To modern biology students, the chromosome theory may not seem very earthshaking. However, early in the twentieth century, Sutton's and Boveri's hypothesis (which potentially united cytology and the infant field of genetics) was a bombshell. Of course, the first response to publication of the hypothesis was to try to pick holes in it. For years after, there was a raging controversy over the validity of what became known as the Sutton-Boveri chromosome theory of heredity.

It is worth considering some of the objections raised to the Sutton-Boveri theory. For example, at the time, chromosomes could not be detected during interphase (between cell divisions). Boveri had to make some very detailed studies of chromosome position before and after interphase be-

fore he could argue persuasively that chromosomes retain their physical integrity through interphase, even though they are cytologically invisible at that time. It was also pointed out that in some organisms several different pairs of chromosomes look alike, making it impossible to say from visual observation that they are not all pairing randomly, whereas Mendel's laws absolutely require the orderly pairing and segregation of alleles. However, in species in which chromosomes do differ in size and shape, it was verified that chromosomes come in pairs and that these synapse and segregate during meiosis.

In 1913, Elinor Carothers found an unusual chromosomal situation in a certain species of grasshopper—a situation that permitted a direct test of whether different chromosome pairs do indeed segregate independently. Studying grasshopper testes, she found one case in which there was a chromosome pair that had nonidentical members; this is called a **heteromorphic pair,** and the chromosomes presumably show only partial homology. In addition, she found that another chromosome, unrelated to the heteromorphic

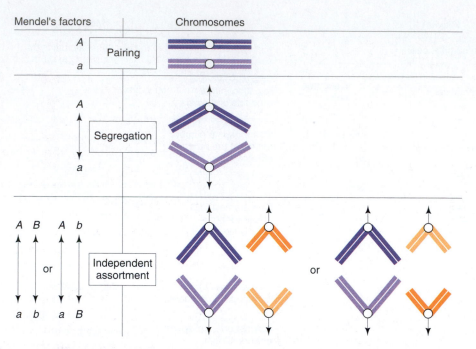

Figure 3-6 Parallels in the behavior of Mendel's genes and chromosomes during meiosis. A dark shade represents one member of a homologous pair; a light shade represents the other member.

pair, had no pairing partner at all. Carothers was able to use these unusual chromosomes as visible cytological markers of the behavior of chromosomes during assortment. By looking at anaphase nuclei, she could count the number of times that each dissimilar chromosome of the heteromorphic pair migrated to the same pole as the chromosome with no pairing partner (Figure 3-7). She observed the two patterns of chromosome behavior with equal frequency. Although these unusual chromosomes obviously are not typical, the results do suggest that nonhomologous chromosomes assort independently.

Other investigators argued that because all chromosomes appear as stringy structures, qualitative differences

between them are of no significance. It was suggested that perhaps all chromosomes were just more or less made of the same stuff. It is worth introducing a study out of historical sequence that effectively counters this objection. In 1922, Alfred Blakeslee performed a study on the chromosomes of jimsonweed (*Datura stramonium*), which has 12 chromosome pairs. He obtained 12 different strains, each of which had the normal 12 chromosome pairs plus an extra representative of one pair. Blakeslee showed that each strain was phenotypically distinct from the others (Figure 3-8). This result would not be expected if there were no genetic significance to the differences between the extra chromosomes.

All these results indicated that the behavior of chromosomes closely parallels that of genes. This of course made the Sutton-Boveri theory attractive, but there was as yet no real proof that genes are located on chromosomes. Further observations, however, did provide such proof, and these began with the discovery of sex linkage.

The Discovery of Sex Linkage

In the crosses discussed thus far, it does not matter which sex of parent is from which strain. Reciprocal crosses (such as strain A ♀ × strain B ♂ and strain A ♂ × strain B ♀) yield similar progeny. The first exception to this pattern was discovered in 1906 by L. Doncaster and G. H. Raynor. They were studying wing color in the magpie moth (*Abraxas*), using two different lines: one with light wings; the other with dark wings. If light-winged females are crossed with dark-winged males, all the progeny have dark wings, showing that the allele for light wings is recessive. However, in the

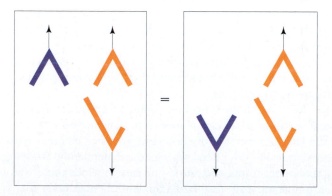

Figure 3-7 Two equally frequent patterns by which a heteromorphic pair and an unpaired chromosome move into gametes, as observed by Carothers.

Figure 3-8 Fruits from *Datura* plants, each having one different extra chromosome. Their characteristic appearances suggest that each chromosome is different. (From E. W. Sinnott, L. C. Dunn, and T. Dobzhansky, *Principles of Genetics*, 5th ed. McGraw-Hill Book Company, New York.)

male × nonbarred male) gave barred males and nonbarred females. Again, the result is that female progeny have their father's phenotype and male progeny have their mother's. Can we find an explanation for these similar results with moths and with chickens?

An explanation came from the laboratory of Thomas Hunt Morgan, who in 1909 began studying inheritance in a fruit fly (*Drosophila melanogaster*). Because this organism has played a key role in the study of inheritance, a brief digression about the creature is worthwhile.

The life cycle of Drosophila is typical of the life cycles of many insects (Figure 3-9). The flies grow vigorously in the laboratory. In the egg, the early embryonic events lead to the production of a larval stage called the first instar. Growing rapidly, the larva molts twice, and the third-instar larva then pupates. In the pupa, the larval carcass is replaced by adult structures, and an image, or adult, emerges from the pupal case, ready to mate within 12 to 14 hours. The adult fly is about 2 millimeters long, so it takes up very little space; thousands of flies can be maintained in vials kept on a single laboratory shelf. The life cycle is very short (12 days at room temperature) in comparison with that of a human, a mouse, or a corn plant; thus, many generations can be raised in a year. Moreover, the flies are extremely prolific: a single female is capable of laying several hundred eggs. Perhaps the beauty of the insect when observed through a microscope added to its early allure as a research organism. In

reciprocal cross (dark female × light male), all the female progeny have light wings and all the male progeny have dark wings. Thus, this pair of reciprocal crosses does not give similar results, and the wing phenotypes in the second cross are associated with the sex of the moths. Note that the female progeny of this second cross are phenotypically similar to their fathers, as the males are to their mothers. How can we explain these results? Before attempting an explanation, let's consider another example.

William Bateson had been studying the inheritance of feather pattern in chickens. One line had feathers with alternating stripes of dark and light coloring, a phenotype called barred. Another line, nonbarred, had feathers of uniform coloring. In the cross barred male × nonbarred female, all the progeny were barred, showing that the allele for nonbarred is recessive. However, the reciprocal cross (barred fe-

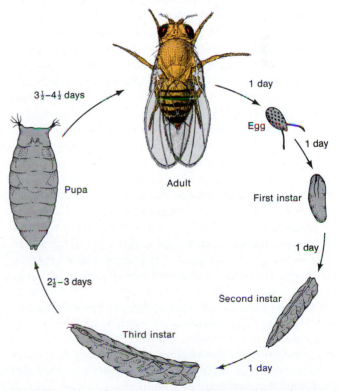

Figure 3-9 Life cycle of *Drosophila melanogaster*, the common fruit fly.

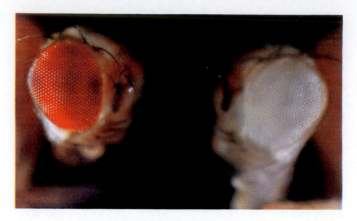

Figure 3-10 Red-eyed and white-eyed *Drosophila*. (Carolina Biological Supply)

Figure 3-11 Segregation of the heteromorphic chromosome pair (X and Y) during meiosis in a *Tenebrio* male. The X and Y chromosomes are being pulled to opposite poles during anaphase I. (From A. M. Srb, R. D. Owen, and R. S. Edgar, *General Genetics*, 2d ed. Copyright © 1965 by W. H. Freeman and Company.)

any case, as we shall see, the choice of *Drosophila* was a very fortunate one for geneticists—and especially for Morgan, whose work earned him a Nobel prize in 1934.

The normal eye color of *Drosophila* is bright red. Early in his studies, Morgan discovered a male with completely white eyes (Figure 3-10). When he crossed this male with red-eyed females, all the F_1 progeny had red eyes, showing that the allele for white is recessive. Crossing the red-eyed F_1 males and females, Morgan obtained a 3 : 1 ratio of red-eyed to white-eyed flies, but all the white-eyed flies were males. Among the red-eyed flies, the ratio of females to males was 2 : 1. What was going on? Clearly the inheritance pattern of the white eye phenotype was related in some way to sex determination, the same conclusion made from the inheritance patterns described above in moths and chickens.

Morgan gathered more data. When he crossed white-eyed females with red-eyed males from a pure line (which is the reciprocal of the cross of the original white male with a normal female), all the females were red and all the males were white. This is very similar to outcomes we still need to explain in the examples of chickens and moths, but note a difference: in chickens and moths the progeny are like the parent of opposite sex when the parental males carry the recessive alleles; in the *Drosophila* cross, this outcome is seen when the female parents carry the recessive alleles.

Before turning to Morgan's explanation of the *Drosophila* results, we should look at some of the cytological information he was able to use in his interpretations. In 1891, working with males of a species of Hemiptera (the true bugs), H. Henking observed that meiotic nuclei contained 11 pairs of chromosomes and an unpaired element that moved to one of the poles during the first meiotic division. Henking called this unpaired element an "X body"; he interpreted it as a nucleolus, but later studies showed it to be a chromosome. Similar unpaired elements were later found in other species. In 1905, Edmond Wilson noted that females of *Protenor* (another Hemipteran) have six pairs of chromosomes, whereas males have five pairs and an unpaired chromosome, which Wilson called (by analogy) the

X chromosome. The females, in fact, have a pair of X chromosomes.

Also in 1905, Nettie Stevens found that males and females of the beetle *Tenebrio* have the same number of chromosomes, but one of the chromosome pairs in males is heteromorphic (that is, the two members differ in size). One member of the heteromorphic pair appears identical with the members of a pair in the female; Stevens called this the *X chromosome*. The other member of the heteromorphic pair is never found in females; Stevens called this the *Y chromosome* (Figure 3-11). She found a similar situation in *Drosophila melanogaster*, which has four pairs of chromosomes, with one of the pairs being heteromorphic in males. Figure 3-12 summarizes the two basic situations: the extra unpaired chromosome and the heteromorphic pair. (You may be wondering about the male grasshoppers studied by Carothers that had both a heteromorphic chromosome pair and an unpaired chromosome. This situation is very unusual, and we needn't worry about it at this point.)

With this background information, Morgan constructed an interpretation of his genetic data. First, it appears that the X and Y chromosomes determine the sex of the fly. *Drosophila* females have four chromosome pairs,

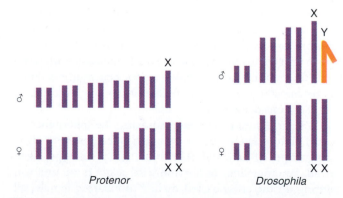

Figure 3-12 The chromosomal constitutions of males and females in two insect species.

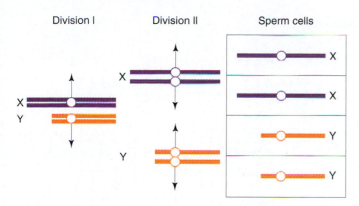

Division I	Division II	Sperm cells

Figure 3-13 Meiotic pairing and the segregation of the X and Y chromosomes into equal numbers of sperm.

Drosophila, Morgan suggested that the heteromorphic chromosomes in chickens and moths be called WZ, with males being ZZ and females being WZ. Thus, if the genes in the chicken and moth crosses are on the Z chromosome, the crosses can be diagrammed as shown in Figure 3-15. The interpretation is consistent with the genetic data. In this case, cytological data provided a confirmation of the genetic hypothesis. In 1914, J. Seiler verified that both chromosomes are identical in all pairs in male moths whereas females have one heteromorphic pair.

> **Message**   The special inheritance pattern of some genes makes it likely that they are borne on the sex chromosomes, which show a parallel pattern of inheritance.

An Aside on Genetic Symbols

In *Drosophila*, a special symbolism for allele designation was introduced to define variant alleles in relation to a "normal" allele. This system is now used by many geneticists and is especially useful in genetic dissection. For a given *Drosophila* character, the allele that is found most frequently in natural populations (or, alternatively, the allele that is found in standard laboratory stocks) is designated as the standard, or **wild type.** All other alleles are then non-wild-type alleles. The symbol for a gene comes from the first non-wild-type allele found. In Morgan's *Drosophila* experiment, this was the allele for white eyes, symbolized by w. The wild-type counterpart allele is conventionally represented by adding a $+$ superscript, so the normal red-eye allele is written w^+.

In a polymorphism, several alleles might be common in nature and all might be regarded as wild-type. In this case, geneticists use appropriate superscripts to distinguish alleles. For example, two alleles of the Adh gene in *Drosophila* are designated Adh^F and Adh^S. This gene encodes the alcohol-metabolizing enzyme alcohol dehydrogenase, and F and S stand for fast and slow movements, respectively, of the enzyme in an electrophoretic gel.

The wild-type allele *can be dominant or recessive* to a non-wild-type allele. For the two alleles w^+ and w, the use of the lowercase letter indicates that the wild-type allele is dominant over the one for white eyes (that is, w is recessive to w^+). As another example, this time concerning the character wing shape, the wild-type phenotype of a fly's wing is straight and flat, and a non-wild-type allele causes the wing to be curled. Because this latter allele is dominant over the wild-type allele, it is written Cy (short for *Curly*), whereas the wild-type allele is written Cy^+. Here note that the capital letter indicates that Cy is dominant over Cy^+. (Also note from these examples that the symbol for a single gene may consist of more than one letter.)

The symbolism specifying wild type is useful because it helps geneticists focus on the procedure of genetic dissection. In many situations, the wild-type allele is defined as being responsible for the normal function, and non-wild-type alleles (whether recessive or dominant) can be regarded as

whereas males have three matching pairs plus a heteromorphic pair. Thus, meiosis in the female produces eggs that each bear one X chromosome. Although the X and Y chromosomes in males are heteromorphic, they seem to synapse and segregate like homologs (Figure 3-13). Thus, meiosis in the male produces two types of sperm, one type bearing an X chromosome and the other bearing a Y chromosome. According to this explanation, union of an egg with an X-bearing sperm produces an XX (female) zygote, and union with a Y-bearing sperm produces an XY (male) zygote. Furthermore, approximately equal numbers of males and females are expected owing to the equal segregation of X and Y.

Morgan next turned to the problem of eye color. Assume that the alleles for red or white eye color are present on the X chromosome, but that there is no counterpart for this gene on the Y chromosome. Thus, females would have two alleles for this gene, whereas males would have only one. This highly unexpected situation proves to fit the data. In the original cross of the white-eyed male with red-eyed females, all F_1 progeny had red eyes, showing that the allele for red eyes is dominant. Therefore, we can represent the two alleles as W (red) and w (white) and diagram the two reciprocal crosses as shown in Figure 3-14.

As we can see from the figure, the genetic results of the two reciprocal crosses are completely consistent with the known meiotic behavior of the X and Y chromosomes. This experiment strongly supports the notion that genes are located on chromosomes. However, it is only a correlation; it does not constitute a definitive proof of the Sutton-Boveri theory.

Can the same XX and XY chromosome theory be applied to the results of the earlier crosses made with chickens and moths? You will find that it cannot. However, Richard Goldschmidt recognized immediately that these results can be explained with a similar hypothesis, making the simple assumption that in this case the *males* have pairs of identical chromosomes whereas the *females* have a heteromorphic pair. To distinguish this situation from the XY situation in

First cross

Second cross

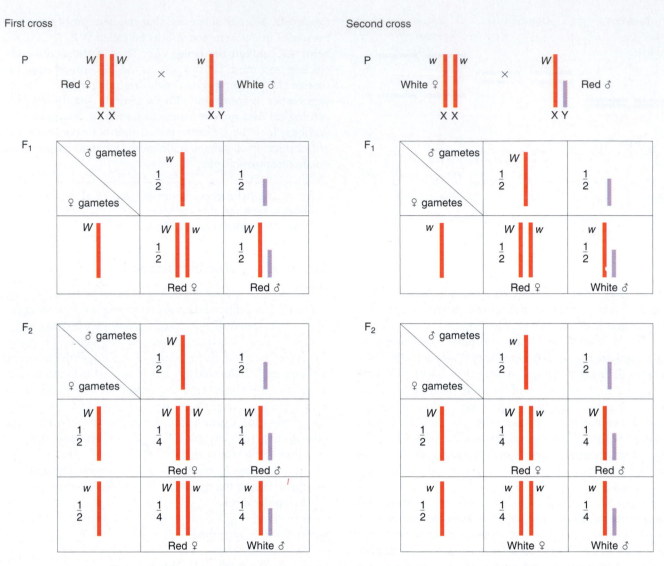

Figure 3-14 Explanation of the different results from reciprocal crosses between red-eyed (red) and white-eyed (white) *Drosophila*.

abnormal. The abnormal alleles then become experimental probes. Geneticists learn how normal functions work by using the abnormal alleles to study the ways in which the normal mechanism can go wrong. Note that Mendel's symbols (*A* and *a*, or *B* and *b*) do not define or emphasize normality. In a pea flower, for example, is purple or white normal? However, the Mendelian symbolism is useful for some purposes; it is used extensively in plant and animal breeding.

A Critical Test of the Chromosome Theory

The correlations between the behavior of genes and the behavior of chromosomes made it very likely that genes are parts of chromosomes. But this was not a critical test of the chromosome theory, and debate continued. The critical analysis came from one of Morgan's students, Calvin Bridges. Bridges's work began with a fruit fly cross we have discussed before. Letting the capital letters X and Y rep-

resent the X and Y chromosomes, we can write the parental genotypes using our new symbolism as X^wX^w (white-eyed) $\times X^{w^+}Y$ (red-eyed). We know that the expected progeny are $X^{w^+}X^w$ (red-eyed females) and X^wY (white-eyed males). When Bridges made the cross on a large scale, he observed a few exceptions among the progeny. About 1 out of every 2000 F_1 progeny was a white-eyed female or a red-eyed male. Collectively, these individuals were called *primary exceptional progeny*. All the primary exceptional males proved to be sterile. However, when Bridges crossed the primary exceptional white-eyed females with normal red-eyed males, in addition to the expected red-eyed female and white-eyed male progeny, a higher proportion of exceptional progeny was observed, 4 percent white-eyed females and red-eyed males that were fertile. These exceptional progeny of primary exceptional mothers were called *secondary exceptional offspring* (Figure 3-16). How did Bridges explain these types of exceptional progeny?

Figure 3-15 Inheritance pattern of genes on the sex chromosomes of two species having the WZ mechanism of sex determination.

Chickens					
First cross	P	Z^BZ^B barred males	$\times$	Z^bW nonbarred females	
	F_1	Z^BZ^b barred males		Z^BW barred females	
Second cross (reciprocal of first)	P	Z^bZ^b nonbarred males	$\times$	Z^BW barred females	
	F_1	Z^BZ^b barred males		Z^bW nonbarred females	

Moths					
First cross	P	Z^LZ^L dark males	$\times$	Z^lW light females	
	F_1	Z^LZ^l dark males		Z^LW dark females	
Second cross (reciprocal of first)	P	Z^lZ^l light males	$\times$	Z^LW dark females	
	F_1	Z^lZ^L dark males		Z^lW light females	

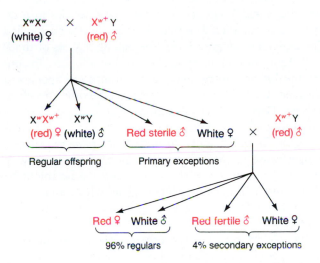

Figure 3-16 *Drosophila* crosses from which primary and secondary exceptional progeny were originally obtained. *Red* = red-eyed, and *white* = white-eyed, *Drosophila*.

male) and $X^{w^+}O$ (red-eyed sterile male). Notice that it is implicit in Bridges's model that the sex of *Drosophila* is determined not by the presence or absence of the Y chromosome but by the number of X chromosomes, with two X chromosomes producing a female and one X chromosome producing a male. What about the sterility of the primary exceptional males? This makes sense if we assume that a male must have a Y chromosome to be fertile.

To summarize, Bridges explained the primary exceptional progeny by postulating rare abnormal meioses that gave rise to viable XXY females and XO males. He went on to test this model in several ways. First he examined microscopically the chromosomes of the primary exceptional progeny, and indeed they were of the type he predicted, XXY and XO. Second he went on to predict the possible chromosome pairings during meiosis in the XXY females, and from this the nature of the secondary exceptional progeny that arose from them. Once again, microscopy confirmed that all his predictions were correct. Therefore, by

It is obvious that the exceptional females—which, like all females, have two X chromosomes—must get both of these chromosomes from their mothers because they are homozygous for *w*. Similarly, exceptional males must get their X chromosomes from their fathers because these chromosomes carry w^+. Bridges hypothesized rare mishaps during meiosis in the female whereby the paired X chromosomes failed to separate during either the first or the second division. This would result in meiotic nuclei containing either two X chromosomes or no X at all. Such a failure to separate is called **nondisjunction;** it produces an XX nucleus and a nullo-X nucleus (containing no X). Fertilization of eggs having these types of nuclei by sperm from a wild-type male produces four zygotic classes (Figure 3-17). It is important to note that the line representing a chromosome in these diagrams is in each case not a single chromosome but a pair of daughter chromatids.

Bridges assumed that XXX and YO zygotes die before development is complete, so the two types of viable exceptional progeny are expected to be X^wX^wY (white-eyed fe-

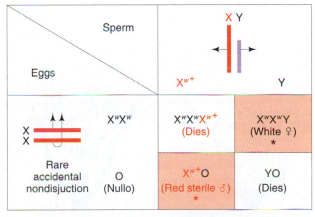

* Primary exceptional progeny

Figure 3-17 Proposed explanation of primary exceptional progeny through nondisjunction of the X chromosomes in the maternal parent. *Red* = red-eyed, and *white* = white-eyed, *Drosophila*.

assuming the chromosomal location of the eye color gene, Bridges was able to predict accurately the nature of several unusual genetic processes.

Message When Bridges used the chromosome theory to predict successfully the nature of certain genetic oddities, the chromosome location of genes was established beyond reasonable doubt.

Sex Chromosomes and Sex Linkage

Mammals, including humans, also have X and Y chromosomes, with males XY and females XX. However, the mechanism of sex determination in mammals differs from that in *Drosophila*. As shown by Bridges, in insects it is the number of X chromosomes that determines sex, with two X's giving a female and one X giving a male, but in mammals it is the presence of the Y that determines maleness and the absence of a Y that determines femaleness. This difference is demonstrated by the sexes of the abnormal chromosome types XXY and XO, as shown in Table 3-1. However, we postpone a full discussion of this topic until Chapter 9.

Vascular plants show a variety of sexual arrangements. Some plants have both male and female sex organs on the same plant. Species with both sex organs combined into the same flower are called **hermaphroditic;** the rose is an exam-

ple. Species in which male and female organs are in separate flowers on the same plant are called **monoecious,** for example, corn. **Dioecious** species, however, have the sexes separate, with female plants bearing flowers containing only ovaries, and male plants bearing flowers containing only anthers (Figure 3-18). Some, but not all, dioecious plants have a heteromorphic pair of chromosomes associated with (and almost certainly determining) the sex of the plant. Of the species with heteromorphic sex chromosomes, a large proportion have an XY system. Critical experiments in a few species suggest a *Drosophila*-like system. Other dioecious plants have no visibly heteromorphic pair of chromosomes; they may still have sex chromosomes, but not visibly distinguishable types.

After studying meiosis in males, cytogeneticists have divided the X and Y chromosomes of some species into regions that pair and regions that do not. The latter are called *differential regions* (Figure 3-19). The pairing regions of the X and Y chromosomes are thought to be homologous. In contrast, the differential region of each chromosome appears to hold genes that have no counterparts on the other sex chromosome. Genes in the differential regions are said to be **hemizygous** ("half-zygous") in males. Genes in the differential region of the X show an inheritance pattern called **X linkage;** those in the differential region of the Y show **Y linkage.** Genes in the pairing region show what might be called **X-and-Y linkage.** In general, genes whose inheritance patterns are different in males and females are said to show **sex linkage.**

We can introduce some other common terminology here. The sex having two identical sex chromosomes (XX or ZZ) is called the **homogametic** sex, because individuals produce only one type of gamete with regard to sex chromosomes. The sex with chromosome combinations XY or ZW is called the **heterogametic** sex, because two different types of gametes are produced. Thus, human and *Drosophila* males, and female birds and moths, are all heterogametic.

Table 3-1 Chromosomal Determination of Sex in *Drosophila* and Humans

Species	Sex chromosomes			
	XX	XY	XXY	XO
Drosophila	♀	♂	♀	♂
Humans	♀	♂	♂	♀

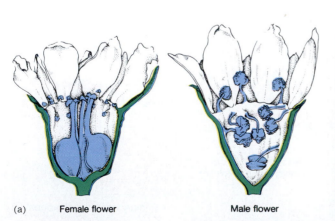

(a) Female flower   Male flower

(b)

Figure 3-18 Two dioecious plant species. (a) *Osmaronia dioica.* (b) *Aruncus dioicus.* Female at left, male at right. [Part (a) Leslie Bohm. Part (b) Anthony Griffiths.]

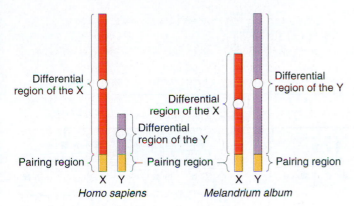

Figure 3-19 Differential and pairing regions of sex chromosomes of humans and of the plant *Melandrium album*. The regions were located by observing where the chromosomes synapsed during meiosis and where they did not.

The nonsex chromosomes (which we might call the "regular" chromosomes) are called **autosomes.** Humans have 46 chromosomes per cell: 44 autosomes plus 2 sex chromosomes. The dioecious plant *Melandrium album* has 22 chromosomes per cell: 20 autosomes plus 2 sex chromosomes.

Genes on the autosome show the kind of inheritance pattern discovered and studied by Mendel. The genes on the differential regions of the sex chromosomes show their own typical patterns of inheritance, as follows.

X-Linked Inheritance

We have already seen an example of X-linked inheritance in *Drosophila*: the inheritance pattern of the white-eye phenotype and its wild-type counterpart, red eye. Of course, eye color is not concerned with sex determination, so we see that genes on the sex chromosomes are not necessarily involved with sexual function. The same is true in humans, where pedigree analysis has revealed many X-linked genes, of which few could be construed as being connected to sexual function. Just as earlier we listed rules for inferring autosomal inheritance patterns, now we list clues for detecting X-linked inheritance in human pedigrees.

Phenotypes with X-linked recessive inheritance typically show the following patterns in pedigrees:

1. Many more males than females show the phenotype under study. This is because a female showing the phenotype can result only from a mating in which both the mother and the father bear the allele (for example, $X^A X^a \times X^a Y$), whereas a male with the phenotype can be produced when only the mother carries the allele. If the recessive allele is very rare, almost all individuals showing the phenotype are males.

The next two clues refer to X-linked recessive phenotypes that are rare in the population. Recall from Chapter 2 our discussion of human phenotypes inherited in a simple

Mendelian manner, which we can now call **autosomal inheritance.** Recall that many such phenotypes are rare disorders, and because of this, certain assumptions could be made about the incidence of certain genotypes in a population. For example, in the case of autosomal recessive phenotypes it is assumed that people marrying into an affected family do not themselves bring the allele in question with them. The same assumption is made in the case of X-linked recessive inheritance. This assumption is particularly relevant to females that enter an affected family, and the point is implicit in the pedigree clues that follow.

2. In the case of a rare X-linked recessive phenotype, none of the offspring of an affected male are affected, but all his daughters are "carriers," bearing the recessive allele masked in the heterozygous condition. Half of the sons borne by these carrier daughters are affected (Figure 3-20). Note that in the case of common X-linked phenotypes, this pattern might be obscured by inheritance of the recessive allele from a heterozygous mother as well as the father.

3. In the case of a rare X-linked recessive phenotype, none of the sons of an affected male show the phenotype under study, nor will they pass the condition to their offspring. The reason behind this lack of male-to-male transmission is that sons obtain their Y chromosome from their fathers, so they cannot normally inherit the father's X chromosome too.

Let us consider some examples of rare X-linked recessive conditions in humans. Perhaps the most familiar example is red-green colorblindness. People with this condition are unable to distinguish red from green, and see them as the same. The genes for color vision have been characterized at the molecular level. Color vision is based on three different kinds of cone cells in the retina, each sensitive to red, green, or blue wavelengths. The genetic determinants for the red and green cone cells are on the X chromosome. As with any X-linked recessive, there are many more males with the phenotype than females.

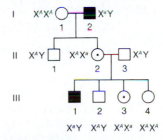

Figure 3-20 Pedigree showing that X-linked recessive alleles expressed in males are then carried unexpressed by his daughters in the next generation, to be expressed again in their sons. Note that III-3 and III-4 cannot be distinguished phenotypically.

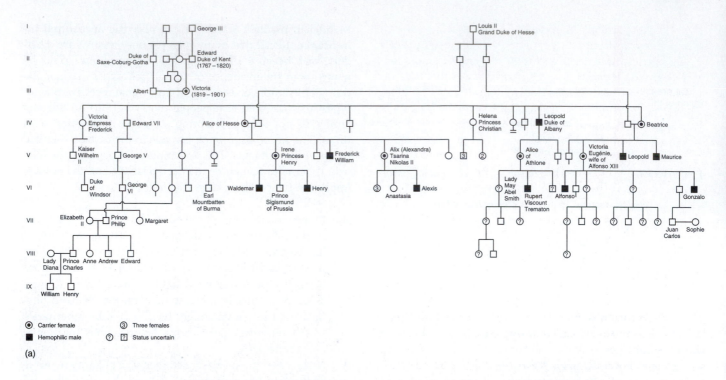

Carrier female　**Three females**

Hemophilic male　**Status uncertain**

(a)

(b)

Figure 3-21 The inheritance of the X-linked recessive condition hemophilia in the royal families of Europe. A recessive allele causing hemophilia (failure of blood clotting) arose in the reproductive cells of Queen Victoria, or one of her parents, through mutation. This hemophilia allele spread into other royal families by intermarriage. (a) This partial pedigree shows affected males and carrier females (heterozygotes). Most spouses marrying into the families have been omitted from the pedigree for simplicity. Can you deduce the likelihood of the present British royal family's harboring the recessive allele? (b) A painting showing Queen Victoria surrounded by her numerous descendants. [(a) Modified from C. Stern, *Principles of Human Genetics,* 3d. ed. Copyright © 1973 by W. H. Freeman and Company. (b), Royal Collection, St. James's Palace. Copyright Her Majesty Queen Elizabeth II.)

Another familiar example is *hemophilia,* the failure of blood to clot. Many proteins must interact in sequence to make blood clot. The most common type of hemophilia is caused by the absence or malfunction of one of these proteins, called *Factor VIII.* The most famous cases of hemophilia are found in the pedigree of interrelated royal families in Europe (Figure 3-21). The original hemophilia allele in the pedigree arose spontaneously (as a mutation) either in the reproductive cells of Queen Victoria's parents or of Queen Victoria herself. The son of the last czar of Russia, Alexis, inherited the allele ultimately from Queen Victoria, who was the grandmother of his mother Alexandra. Nowadays, hemophilia can be treated medically, but it was formerly a potentially fatal condition. It is interesting to note that in the Jewish Talmud there are rules about exemptions to male circumcision that show clearly that the mode of transmission of the disease through unaffected carrier females was well understood in ancient times. For example, one exemption was for the sons of women whose sisters' sons had bled profusely when they were circumcised.

Duchenne muscular dystrophy is a fatal X-linked recessive disease. The phenotype is a wasting and atrophy of muscles. Generally the onset is before the age of 6, with confinement to a wheelchair by 12, and death by 20. The gene for Duchenne muscular dystrophy has now been isolated and characterized, holding out hope for a better understanding of the physiology of this condition and, ultimately, a therapy.

A rare X-linked recessive phenotype that is interesting from the point of view of sexual differentiation is a condition called *testicular feminization syndrome,* which has a frequency of about 1 in 65,000 male births. People afflicted with this syndrome are chromosomally males, having 44 autosomes plus an X and a Y, but they develop as females (Figure 3-22). They have female external genitalia, a blind vagina, and no uterus. Testes may be present either in the labia or in the abdomen. Although many such individuals are happily married, they are, of course, sterile. The condition is not reversed by treatment with male hormone (androgen), so it is sometimes called *androgen insensitivity syndrome.* The reason for the insensitivity is that there is a malfunction in the androgen receptor, so male hormone can have no effect on the target organs that are involved in maleness. In humans, femaleness results when the male-determining system is not functional.

Pedigrees of rare X-linked dominant phenotypes show the following characteristics:

1. Affected males pass the condition on to all their daughters but to none of their sons (Figure 3-23).
2. Females married to unaffected males pass the condition on to half their sons and daughters (Figure 3-24).

There are few examples of X-linked dominant phenotypes in humans. One example is *hypophosphatemia,* a type of vitamin D–resistant rickets.

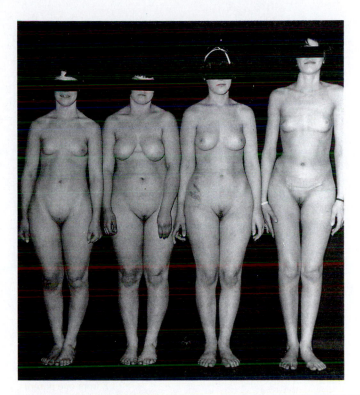

Figure 3-22 Four siblings with testicular feminization syndrome (congenital insensitivity to androgens). All four subjects in this photograph have 44 autosomes plus an X and a Y, but they have inherited the recessive X-linked allele conferring insensitivity to androgens (male hormones). One of their sisters (not shown) who was genetically XX was a carrier and bore a child who also showed testicular feminization syndrome. (Leonard Pinsky, McGill University)

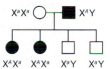

Figure 3-23 Pedigree showing that the daughters of a male expressing an X-linked dominant phenotype will all show the phenotype.

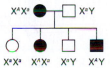

Figure 3-24 Pedigree showing that females affected by an X-linked dominant condition usually are heterozygous and pass the condition to half their sons and daughters.

X Chromosome Inactivation

Early in the development of female mammals, one of the X chromosomes in each cell becomes inactivated. The inactivated X chromosome becomes highly condensed and is visi-

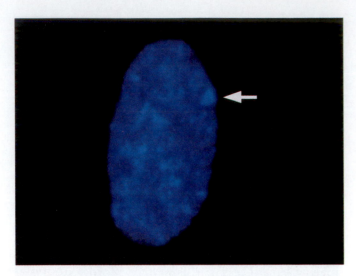

Figure 3-25 A Barr body, a condensed inactivated X chromosome, in the nucleus of a cell of a normal woman. Men have no Barr bodies. The number of Barr bodies in a cell is always equal to the total number of X chromosomes minus one. (Karen Dyer Montgomery)

ble as a darkly staining spot called a **Barr body** (Figure 3-25). Surprisingly, this chromosomal inactivation persists through all the subsequent mitotic divisions that produce the mature body of the animal. The inactivation process is random, affecting either of the X chromosomes. As a result of this inactivation, the adult female body is a mixture, or **mosaic,** of cells with either of the two different X chromosome genotypes (Figure 3-26). During the growth and development of tissues, the mitotic descendants of a progenitor cell often stay next to each other, forming a cluster, so if a female is heterozygous for an X-linked gene that has its effect in that tissue, the two alleles of the heterozygote are expressed in patches, or sectors. A mosaic phenotype familiar to most of us is the coat pigmentation pattern of tortoiseshell and calico cats (Figure 3-27). Such cats are females heterozygous for the alleles *O* (which causes fur to be orange) and *o* (which causes it to be black). Inactivation of the *O*-bearing X chromosome produces a black patch expressing *o*, and inactivation of the *o*-bearing X chromosome produces an orange patch expressing *O*.

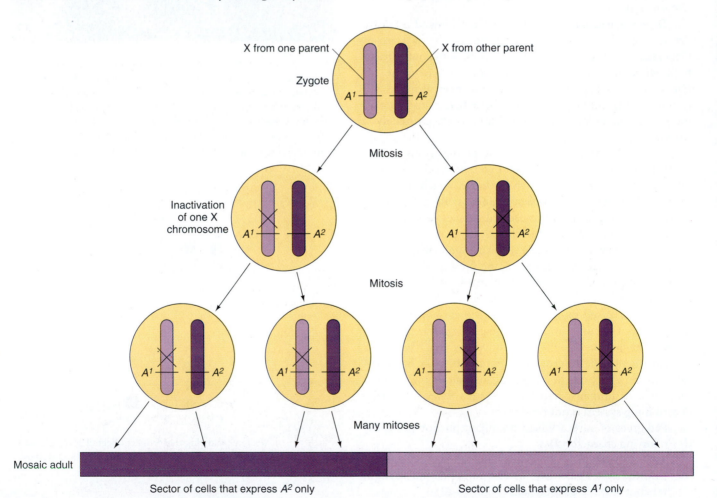

Figure 3-26 X chromosome inactivation in mammals. The zygote of a female mammal heterozygous for an X-linked gene becomes a mosaic adult composed of two cell lines expressing one or the other of the alleles of the heterozygous gene because one or the other X chromosome is inactivated in all cell lines. For simplicity, inactivation is shown at the two-cell stage, but it can take place at other low cell numbers too.

Figure 3-27 A calico cat. Both calico and tortoiseshell cats are females heterozygous for two alleles of an X-linked coat color gene, *O* (orange) and *o* (black). The orange and black sectors are caused by X chromosome inactivation. The white areas are caused by a separate genetic determinant present in calicos, but not in tortoiseshell cats. (Anthony Griffiths)

Although all human females have one of their X chromosomes inactivated in every cell, this is only detectable when a female is heterozygous for an X-linked gene. This is particularly striking when, as in tortoiseshell cats, the phenotype is expressed on the exterior of the body. Such a condition is *anhidrotic ectodermal dysplasia*. Males carrying the responsible allele (let us call it *d*) in its hemizygous condition have no sweat glands. A heterozygous (*Dd*) female has a mosaic of *D* and *d* sectors across her body, as shown in Figure 3-28. Interestingly, the X chromosome location of the gene causing testicular feminization was confirmed when it was shown microscopically that in females heterozygous for the gene, half their fibroblast cells bind androgen but the other half do not.

Y-Linked Inheritance

Genes on the differential region of the human Y chromosome are inherited only by males, with fathers transmitting the region to their sons. However, other than maleness itself, no human phenotype has been conclusively proved to be Y-linked. Hairy ear rims (Figure 3-29) has been proposed as a possibility. The phenotype is extremely rare among the populations of most countries but more common among the populations of India. An Indian geneticist, K. Dronamraju, studied the trait in his own family. Every male in the family descended from a certain male ancestor showed the trait. In other Indian families, however, males seem to transmit the trait to only some of their sons, which is part of the reason that the evidence for Y-linked inheritance is considered to be inconclusive. Of course, maleness itself is Y-linked; a gene that plays a primary role in maleness is the **TDF** gene, which codes for **testis-determining**

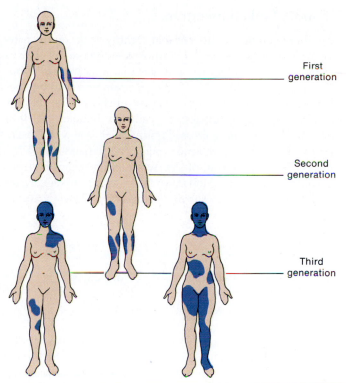

Figure 3-28 Somatic mosaicism in three generations of females heterozygous for sex-linked anhidrotic ectodermal dysplasia (absence of sweat glands). Areas without sweat glands are shown in blue. The extent and location of the different tissues is determined by chance, but each female exhibits the characteristic mosaic pattern.

factor. The TDF gene has been located and mapped on the differential region of the Y chromosome (see Chapters 17 and 23).

The Y chromosome of the fish *Lebistes* carries a gene that determines a phenotype consisting of a pigmented spot at the base of the dorsal fin. This phenotype passes from father to son, and females never carry or express the gene, so it seems to be a clear example of Y linkage.

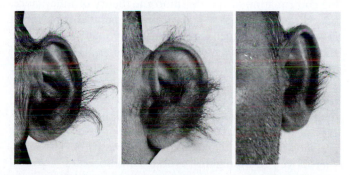

Figure 3-29 Hairy ear rims. This phenotype may be caused by an allele of a Y-linked gene. (From C. Stern, W. R. Centerwall, and S. S. Sarkar, *The American Journal of Human Genetics* 16, 1964, 467. By permission of Grune & Stratton, Inc.)

X-and-Y-Linked Inheritance

Are there hereditary patterns that identify genes on the pairing regions of the X and Y chromosomes? There is a gene in *Drosophila* that is inherited in a way that indicates an X-and-Y location. Curt Stern found that a certain non-wild-type recessive allele in *Drosophila* causes a phenotype of shorter and more slender bristles; he called it bobbed (*b*). If a bobbed X^bX^b female is crossed with a wild-type X^+Y^+ male, all the F_1 progeny are wild-type: X^+X^b females and X^bY^+ males. The same result, of course, would be expected from an autosomal gene. However, the X-and-Y linkage is revealed in the F_2, which shows the following, clear sex-associated pattern:

Sex	Phenotype	Inferred genotype
males	wild type	X^+Y^+ and X^bY^+
females	$\frac{1}{2}$ bobbed	X^bX^b
	$\frac{1}{2}$ wild type	X^+X^b

A combination of cytogenetic and molecular studies has shown that there is a homologous region of the human X and Y chromosomes that pairs and undergoes crossing-over. In fact, it is known that there is an obligatory crossover in this region in every meiosis, so alleles in this vicinity are effectively uncoupled from the unique regions of the X and Y chromosomes and show what is called **pseudoautosomal inheritance.**

> **Message** Inheritance patterns with an unequal representation of phenotypes in males and females can locate the genes concerned to one or both of the sex chromosomes.

The Parallel Behavior of Autosomal Genes and Chromosomes

The patterns of inheritance arising from the normal and the abnormal behavior of the sex chromosomes provide satisfying confirmation of the chromosome theory of inheritance, which was suggested originally by the parallel behavior of genes and autosomal chromosomes at meiosis. Now we should pause and review the segregation and assortment of autosomal genes at meiosis. Autosomal genes represent the great majority of all genes, and are the type most commonly encountered and analyzed. Figure 3-30 is a simplified diagram of the passage of a cell through meiosis. The genotype of the cell is *A a B b*, and the two gene pairs, *A a* and *B b*, are shown on two different chromosome pairs. The hypothetical cell in the figure has four chromosomes: a pair of homologous long chromosomes and a pair of homologous short ones. Such size differences between pairs are common.

Parts 4 and 4′ of Figure 3-30 show that two equally frequent spindle attachments to the centromeres during the first anaphase result in two different allelic segregation patterns. Meiosis then produces four cells of the genotypes shown from each of these segregation patterns. Because segregation patterns 4 and 4′ are equally common, the meiotic product cells of genotypes *A B*, *a b*, *A b*, and *a B* are produced in equal frequencies. In other words, the frequency of each of the four genotypes is $\frac{1}{4}$. This, of course, is the gametic distribution postulated by Mendel for a dihybrid, and is the one we noted along one edge of the Punnett square (see Figure 2-10).

The diagram shows exactly why chromosomal behaviors produce the Mendelian ratios. Notice that Mendel's first law (equal segregation) describes what happens to a pair of alleles when the pair of homologs that carry them separate into opposite cells at the first meiotic division. Notice also that Mendel's second law (independent assortment) results from the independent segregation of different pairs of homologous chromosomes.

The chromosome theory is important in many ways. As we have seen, it accounts for fundamental inheritance patterns observed in crosses of plants and animals. We will encounter many extensions of these patterns in the chapters ahead, and all are well explained by the chromosome theory. The theory is also important because it centers attention on the role of the *cell genotype* in determining the *organismal phenotype*. The phenotype of an organism is determined by the phenotypes of all its individual cells. Cell phenotypes, in turn, are determined by the alleles present on the cells' chromosomes. When we say that an organism is, for example, *A A*, we really mean that each cell of the organism contains a pair of chromosomes both of which carry an *A* allele. The cell genotype determines how the cell functions, thereby controlling the phenotype of the cell, and it is the sum total of all these cell phenotypes that determines the overall phenotype of the organism.

> **Message** The phenotype of an organism is determined by gene action at the cellular level.

Mendelian Genetics and Life Cycles

So far, we have mainly been discussing diploid organisms—organisms with two homologous chromosome sets in each cell. As we have seen, the diploid condition is designated $2n$, where *n* stands for the number of chromosomes in one chromosome set. For example, the pea cell contains two sets of seven chromosomes, so $2n = 14$. The organisms that we encounter most often in our daily lives (animals and flowering plants) are diploid in most of their tissues. Nevertheless, a large portion of the biomass on the earth comprises organisms that spend most of their life cycles in a

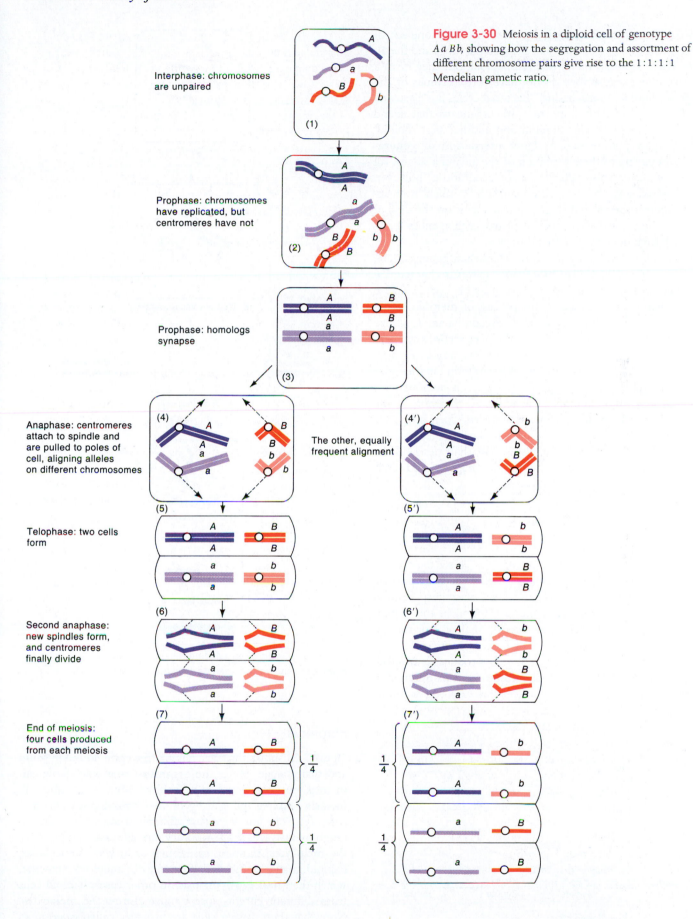

Figure 3-30 Meiosis in a diploid cell of genotype *A a B b*, showing how the segregation and assortment of different chromosome pairs give rise to the 1 : 1 : 1 : 1 Mendelian gametic ratio.

Interphase: chromosomes are unpaired

Prophase: chromosomes have replicated, but centromeres have not

Prophase: homologs synapse

Anaphase: centromeres attach to spindle and are pulled to poles of cell, aligning alleles on different chromosomes

The other, equally frequent alignment

Telophase: two cells form

Second anaphase: new spindles form, and centromeres finally divide

End of meiosis: four cells produced from each meiosis

haploid condition, in which each cell has only one set of chromosomes. Important examples are the fungi and the algae. Bacteria could be considered haploid, but they form a special case because they do not have chromosomes of the type we have been discussing. (Bacterial cycles are discussed in Chapter 10.) Also important are organisms that spend part of their life cycles haploid and another part diploid. Such organisms are said to show **alternation of generations,** referring to the alternation of the 2*n* and *n* stages. All types of plants show alternation of generations; however, the haploid stage of flowering plants and conifers is an inconspicuous specialized structure dependent on the diploid part of the plant. Other types of plants, such as mosses and ferns, have independent haploid stages.

Do all these life cycles show Mendelian genetics? The answer is that Mendelian inheritance patterns characterize any species that has meiosis as part of its life cycle, because Mendelian laws are based on the process of meiosis. All the groups of organisms mentioned, except bacteria, utilize meiosis as part of their cycles. In the next sections we consider the inheritance patterns shown by less familiar eukaryotes and compare them with the patterns found in the more familiar cycles. This is important because it demonstrates the universality of Mendelian genetics. Furthermore, some of the less well known organisms are particularly amenable to genetic analysis and as a result have become used extensively as model genetic organisms. We shall describe the three major types of life cycle, diploid, haploid, and alternating haploid-diploid, in that order.

Diploids

Figure 3-31 summarizes the diploid cycle. This is the cycle of most animals (including humans). The adult body is composed of diploid cells, and meiosis takes place in specialized diploid cells, the meiocytes, that are found in the gonads (testes and ovaries). The products of meiosis are the gametes (eggs or sperm). Fusion of haploid gametes forms a diploid zygote, which, through mitosis, produces a multicellular organism. Mitosis in a diploid proceeds as outlined in Figure 3-32.

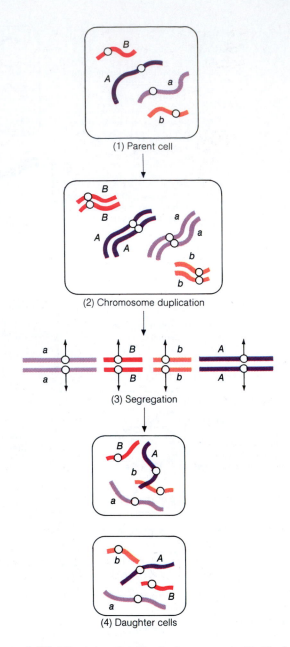

Figure 3-32 Mitosis in a diploid cell of genotype *A a B b.* The heterozygous genes are on separate chromosome pairs.

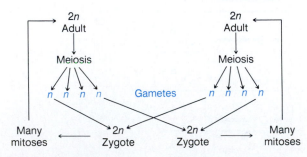

Figure 3-31 The diploid life cycle.

Haploids

Figure 3-33 shows the basic haploid life cycle, found in fungi and many algae. Here, the organism is haploid. How can meiosis possibly occur in a haploid organism? After all, meiosis requires the pairing of *two* homologous chromosome sets. The answer is that all haploid organisms that undergo meiosis create a temporary diploid stage that provides the meiocytes. In some cases such as in yeast, unicellular, haploid, mature individuals fuse to form a diploid meiocyte, which then undergoes meiosis. In other cases, haploid cells from different parents fuse to give rise to the meiocytes. Note that these fusing cells are properly called *gametes,* so

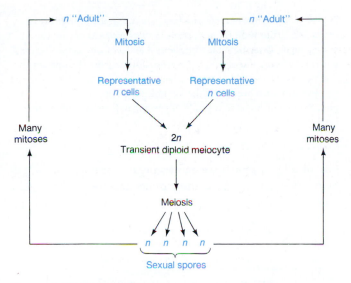

Figure 3-33 The haploid life cycle.

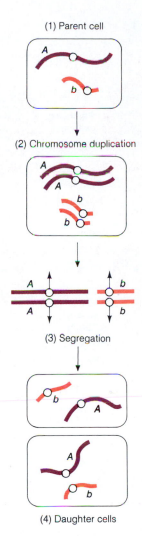

Figure 3-34 Mitosis in a haploid cell of genotype *A b*. The genes are on separate chromosomes.

we see that in these cases gametes arise from mitosis. Meiosis, as usual, produces haploid products of meiosis, which are called **sexual spores.** The sexual spores in some species become new unicellular adults; in other species they each develop through mitosis into a multicellular haploid individual. Notice that a cross between two haploid organisms involves only one meiosis, whereas a cross between two diploid organisms involves a meiosis in each diploid organism. As we shall see, this simplicity makes haploids very attractive for genetic analysis. In haploids, mitosis proceeds as shown in Figure 3-34.

Let's consider a cross in a specific haploid. A convenient organism for demonstration is the orange-colored bread mold *Neurospora.* This fungus is a multicellular haploid in which the cells are joined end to end to form **hyphae,** or threads of cells. The hyphae grow through the substrate and also send up aerial branches that bud off haploid cells known as **conidia** (asexual spores). Conidia can detach and disperse to form new colonies, or alternatively they can act as paternal gametes and fuse with a maternal structure of a different individual (Figure 3-35). However, the different individual must be of the opposite **mating type.** In fungi there are no true sexes, and all haploid cultures develop similarly. However, populations contain distinct genetically determined mating types. In *Neurospora* there are two mating types, called *A* and *a* and the meiotic (sexual) part of the life cycle can take place only if two haploids of different mating type unite. Mating types can be thought of as "physiological sexes," and although this is an inadequate definition, it is a useful phrase that stresses the unseen difference between mating types.

A maternal gamete waits inside a specialized knot of hyphae, and eventually a haploid maternal and a haploid paternal nucleus pair up and divide mitotically to produce numerous pairs. The pairs eventually fuse to form diploid meiocytes. Meiosis occurs, and in each meiocyte four haploid nuclei are produced, which represent the four products of meiosis. For an unknown reason, these four nuclei divide mitotically, resulting in eight nuclei, which develop into eight football-shaped sexual spores called **ascospores.** The ascospores are shot out of a flask-shaped fruiting body that has developed from the knot of hyphae that originally contained the maternal gametic cell. The ascospores can be isolated, each into a culture tube, where each ascospore will grow into a new culture by mitosis (Figure 3-36).

What characters can be studied in such an organism? One character is the color of the conidia. Variants of the normal orange color can be found. Figure 3-37 shows a normal culture and some color variants, including an albino. Another possible character to study is the compactness of the culture, and one pair of contrasting phenotypes are normal *spreading* growth and a densely branching growth called *colonial.*

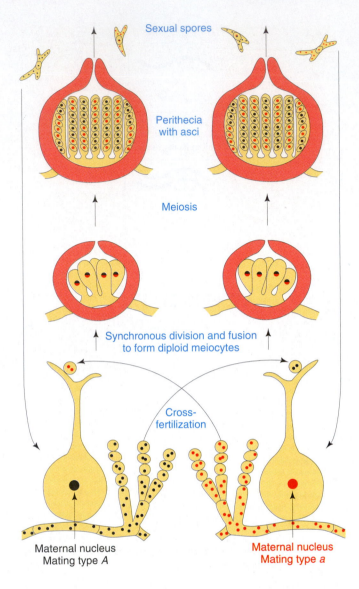

Maternal nucleus
Mating type *A*

Maternal nucleus
Mating type *a*

Figure 3-35 The life cycle of *Neurospora crassa*, the orange bread mold. Self-fertilization is not possible in this species: there are two mating types, determined by the alleles *A* and *a* of one gene. A cross will succeed only if it is *A* × *a*. An asexual spore from the opposite mating type fuses with a receptive hair, and a nucleus travels down the hair to pair with a nucleus in the knot of cells. The *A* and *a* pair then undergo synchronous mitoses, finally fusing to form diploid meiocytes.

pair of alleles, which we can name *al⁺* (orange) and *al* (albino). We can represent the parents and the four progeny types as follows:

Parents

col⁺ al⁺   ×   *col al*

(spreading, orange)   (colonial, albino)

Progeny

col⁺ al⁺ (spreading, orange)

col al (colonial, albino)

col⁺ al (spreading, albino)

col al⁺ (colonial, orange)

The 1:1:1:1 ratio is a result of equal segregation and independent assortment, as illustrated in the following branch diagram:

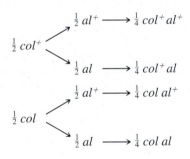

So we see that even in such a lowly organism Mendel's laws are still in operation.

Let's cross a spreading, orange culture with a colonial, albino culture of opposite mating type. We isolate ascospores (progeny) and grow each one into a culture. We would find the following progeny phenotypes and proportions:

25% spreading, orange

25% colonial, albino

25% spreading, albino

25% colonial, orange

In total, half the progeny are spreading and half are colonial. Thus, this phenotypic difference must be determined by the alleles of one gene that have segregated equally at meiosis. We can call these alleles *col⁺* (spreading) and *col* (colonial). The same logic can be applied to the other character: half the progeny are orange and half are albino, so the phenotypic difference in color is also determined by a

Alternating Haploid–Diploid

In an organism with alternation of generations, there are two stages to the life cycle: one diploid and one haploid. One stage is usually more prominent than the other. For example, what we all recognize as a fern plant is the diploid stage, but the organism does have a small, independent, photosynthetic haploid stage that is usually much more difficult to spot on the forest floor. In contrast, the green moss plant is the haploid stage, and the brownish stalk that grows up out of this plant is a dependent diploid stage that is effectively parasitic on it.

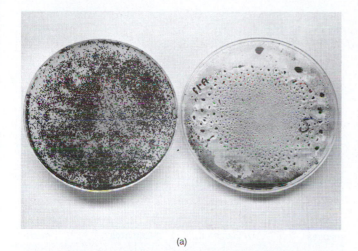

(a)

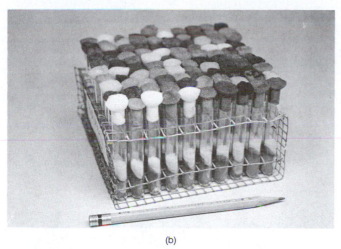

(b)

Figure 3-36 (a) A *Neurospora* cross made in a petri plate (on the left). The many small black spheres are fruiting bodies in which meiosis has occurred; the ascospores (sexual spores) were shot as a fine dust into the condensed moisture on the lid (which has been removed and is to the right of the plate). (b) A rack of progeny cultures, each resulting from one isolated ascospore. (Anthony Griffiths)

In flowering plants, the main green stage is, of course, diploid. The haploid stages of flowering plants are extremely reduced and dependent on the diploid. These haploids are found in the flower. In the anther and the ovary, meiocytes undergo meiosis, and the resulting haploid products of meiosis are called **spores.** The spores undergo a few mitotic divisions to produce a small, multicellular haploid stage. The diploid stage of an organism with alternation of generations is called the **sporophyte,** which means sexual-spore-producing plant, and the haploid stage is called **gametophyte,** which means gamete-producing plant. The male gametophyte of seed plants is known as a pollen grain. Figure 3-38 shows that in flowering plants, cells of the gametophytes act as eggs or sperm in fertilization. The generalized cycle of alternation of generations is shown in Figure 3-39.

In mosses and ferns, the sperm cells are motile and travel from one gametophyte to another in a film of water

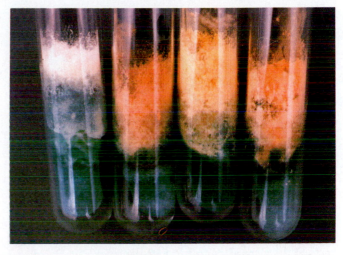

Figure 3-37 Genetically determined color variants of the fungus *Neurospora crassa*. The orange wild-type color is shown on the right, together with albino, yellow, and brown variants. Their genotypes are wild type, $al^+ ylo^+ ad^+$; albino, $al\ ylo^+ ad^+$; yellow, $al^+ ylo\ ad^+$; and brown, $al^+ ylo^+ ad$.

to effect fertilization. Let us consider a cross we might make in a moss. The character to be studied, of course, can pertain to the gametophyte or the sporophyte. Assume that we have a gene whose alleles affect the "leaves" of the gametophyte, with w causing wavy edges and w^+ causing smooth edges. Also assume that a separate gene affects the color of the sporophyte, with r causing reddish coloration and r^+ causing the normal brown coloration. We fertilize a smooth-leaved w^+ gametophyte that also bears the unexpressed allele r by transferring onto it male gametes from a wrinkly leaved w gametophyte, also carrying r^+ (Figure 3-40). Hence, the cross is $w^+ r \times w r^+$.

A diploid sporophyte of genotype $w^+ w\ r^+ r$ develops on the gametophyte, and it is brown because reddish is recessive. Cells of this sporophyte act as meiocytes, and sexual spores (products of meiosis) are produced in the following proportions:

$$25\%\ w^+ r^+$$
$$25\%\ w^+ r$$
$$25\%\ w\ r^+$$
$$25\%\ w\ r$$

Of course, we can directly classify only the leaf character in these gametophytes, and we would have to make the appropriate intercrosses to determine whether each individual is r^+ or r.

Once again, Mendel's laws dictate the inheritance patterns. It is simply a matter of keeping track of the ploidy in each part of the cycle and applying the simple Mendelian ratios.

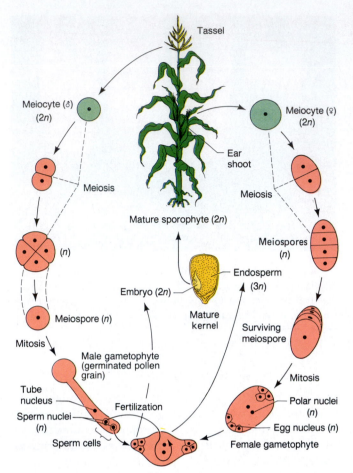

Figure 3-38 Alternation of generations of corn. The male gameto-phyte arises from a meiocyte in the tassel. The female gametophyte arises from a meiocyte in the ear shoot. One sperm cell from the male gametophyte fuses with the egg nucleus of the female gametophyte, and the diploid zygote thus formed develops into the embryo. The other sperm cell fuses with the two polar nuclei in the center of the female gametophyte, forming a triploid ($3n$) cell that generates the en-dosperm tissue surrounding the embryo. The endosperm provides nu-trition to the embryo during seed germination. Which parts of the di-agram represent the haploid stage? Which parts represent the diploid stage?

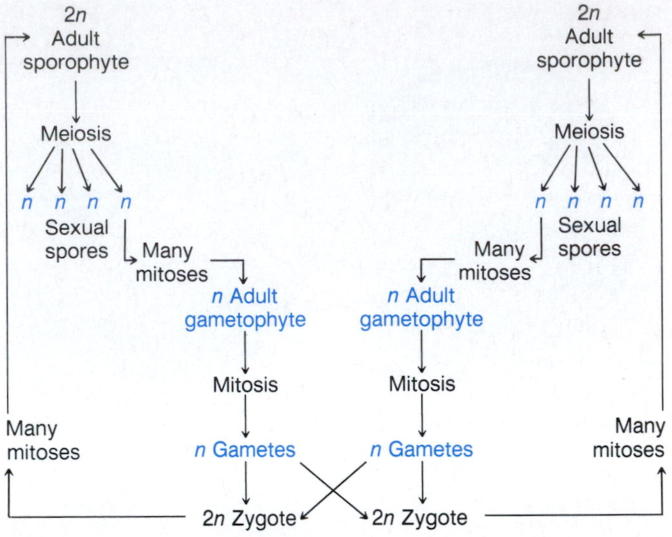

Figure 3-39 The alternation of diploid and haploid stages in the life cycle of plants.

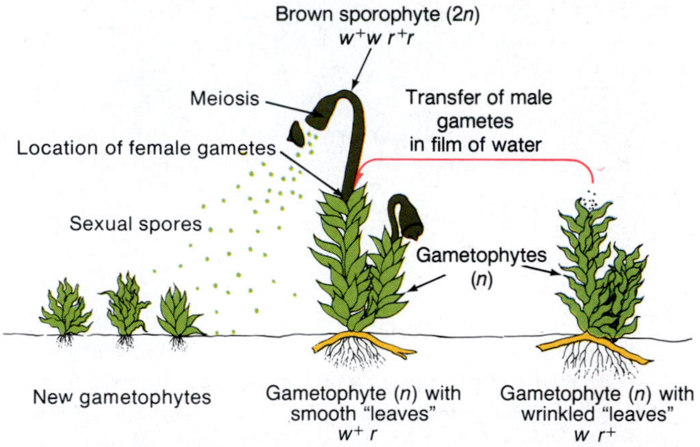

Figure 3-40 Mendelian genetics in a hypothetical cross in a moss. Only the haploid gametophyte express the w^+ and w, and only the diploid sporophyte expresses the r^+ or r alternatives.

Message Mendelian laws apply to meiosis in any organism and may be generally stated as follows:

1. At meiosis, the alleles of a gene segregate equally into the haploid products of meiosis.
2. At meiosis, the alleles of one gene segregate independently of the alleles of genes on other chromosome pairs.

Thus, the theory that genes are located on chromo-somes perfectly explains inheritance patterns in most eu-karyotic organisms. The chromosome theory of inheritance is no longer in doubt; it forms one of the cornerstones of modern biological theory.

SUMMARY

After the rediscovery of Mendelian principles in 1900, scientists set out to discover what structures within cells correspond to Mendel's hypothetical units of heredity, which we now call genes. Recognizing that the behavior of chromosomes during meiosis parallels the behavior of genes, Walter Sutton and Theodor Boveri suggested that genes were located in or on the chromosomes.

In her experiments with a certain species of grasshopper, Elinor Carothers discovered that different chromosome pairs assort independently. This finding provided further evidence that the behavior of chromosomes closely parallels that of genes. Additional evidence for the validity of the chromosome theory of heredity came from the discovery of sex-linked inheritance and the existence of sex chromosomes. In his studies of *Drosophila*, Thomas Hunt Morgan showed that if the relevant alleles are on the X but not the Y, then the inheritance of red or white eye color is consistent with the meiotic behavior of X and Y chromosomes. Finally,

Calvin Bridges's postulation of nondisjunction (the failure of paired chromosomes to separate) during meiosis enabled him to make testable predictions based on the assumption that the gene for eye color is on the X chromosome. The confirmation of these predictions provided unequivocal evidence that genes are located on chromosomes.

We now know which chromosome behaviors produce Mendelian ratios. Mendel's first law (equal segregation) results from the separation of a pair of homologous chromosomes into opposite cells at the first division. Mendel's second law (independent assortment) results from independent behavior of separate pairs of homologous chromosomes.

Because Mendelian laws are based on meiosis, Mendelian inheritance characterizes any organism with a meiotic stage in its life cycle, including diploid organisms, haploid organisms, and organisms with alternating haploid and diploid generations.

Concept Map

Draw a concept map interrelating as many of the following terms as possible. Note that the terms are listed in no particular order.

chromosomes / meiosis / equal segregation / alleles / diploid / independent assortment / haploid / centromere / products of meiosis / Mendelian ratio / spindle fiber attachment

CHAPTER INTEGRATION PROBLEM

Two *Drosophila* flies were mated that had normal (transparent, long) wings. In the progeny, two new phenotypes appeared, dusky wings (having a semiopaque appearance) and clipped wings (with squared ends). The progeny were as follows:

Females	179 transparent, long
	58 transparent, clipped
Males	92 transparent, long
	89 dusky, long
	28 transparent, clipped
	31 dusky, clipped

a. Provide a genetic explanation for these results, showing genotypes of parents and of all progeny classes under your model.

b. Design a test for your model.

Solution

a. The first step is to state any interesting features of the data. The first striking feature is, of course, the appearance of two new phenotypes. We encountered the phenomenon in Chapter 2, and explained it there in terms of recessive alleles masked by their dominant counterparts. So first we might suppose that one or both parental flies have recessive alleles of two different genes. This inference is strengthened by the observation that some progeny express only one of the new phenotypes. If the new phenotypes always appeared together, we might suppose that the same recessive allele determines both.

However, the other striking aspect of the data, which we cannot explain using the Mendelian principles from Chapter 2, is the obvious difference between the sexes; although there are approximately equal numbers of males and females, the males fall into four phenotypic classes but the females comprise only two. This should immediately suggest some kind of sex-linked inheritance. When we study the data, we see that the long and clipped phenotypes are segregating in both males and females but only males have the dusky phenotype. This suggests that the inheritance of wing transparency differs from the inheritance of wing shape. First, long and clipped are found in a 3:1 ratio in both males and females. This can be explained if the parents were both heterozygous for an autosomal gene; we can represent them as Ll, where L stands for long and l stands for clipped.

Having done this partial analysis, we see that it is only the wing transparency inheritance that is associated with sex. The most obvious possibility is that the alleles for transparent (D) and dusky (d) are on the X chromosome, because we have seen in this chapter that gene location on this chromosome gives inheritance patterns correlated with sex. If this suggestion is true, then the parental female must be the one sheltering the d allele, because if the male had the d he would have been dusky whereas we were told that he had transparent wings. Therefore, the female parent would be Dd, and the male D. Let's see if this suggestion works: if it is true, all female progeny would inherit the D allele from their father, so all would be transparent winged. This was observed. Half the sons would be D (transparent) and half d (dusky), which was also observed.

So overall we can represent the female parent as $Dd\,Ll$ and the male parent as $D\,Ll$. Then the progeny would be

Females

$$\frac{1}{2}DD \Big\langle {}^{\frac{3}{4}L- \longrightarrow \frac{3}{8}DD\,L-}_{\frac{1}{4}ll \longrightarrow \frac{1}{8}DD\,ll}$$

$$\frac{1}{2}Dd \Big\langle {}^{\frac{3}{4}L- \longrightarrow \frac{3}{8}Dd\,L-}_{\frac{1}{4}ll \longrightarrow \frac{1}{8}Dd\,ll}$$

$\frac{3}{4}$ transparent, long
$\frac{1}{4}$ transparent, clipped

Males

$$\frac{1}{2}D \Big\langle {}^{\frac{3}{4}L- \longrightarrow \frac{3}{8}D\,L-}_{\frac{1}{4}ll \longrightarrow \frac{1}{8}D\,ll}$$ transparent, long / transparent, clipped

$$\frac{1}{2}d \Big\langle {}^{\frac{3}{4}L- \longrightarrow \frac{3}{8}d\,L-}_{\frac{1}{4}ll \longrightarrow \frac{1}{8}d\,ll}$$ dusky, long / dusky, clipped

b. Generally a good way to test such a model is to make a cross and predict the outcome. But which cross? We have to predict some kind of ratio in the progeny, so it is important to make a cross from which a unique phenotypic ratio can be expected. Notice that using one of the female progeny as a parent would not serve our needs: we cannot say from observing the phenotype of any one of these females what her genotype is. A female with transparent wings could be DD or Dd, and one with long wings could be LL or Ll. It would be good to cross the parental female of the original cross with a dusky, clipped son, because the full genotypes of both are specified under the model we have created. According to our model, this cross is

$$Dd\,Ll \times d\,ll$$

From this we predict:

Females

$$\frac{1}{2}Dd \Big\langle {}^{\frac{1}{2}Ll \longrightarrow \frac{1}{4}Dd\,Ll}_{\frac{1}{2}ll \longrightarrow \frac{1}{4}Dd\,ll}$$

$$\frac{1}{2}dd \Big\langle {}^{\frac{1}{2}Ll \longrightarrow \frac{1}{4}dd\,Ll}_{\frac{1}{2}ll \longrightarrow \frac{1}{4}dd\,ll}$$

Males

$$\frac{1}{2}D \Big\langle {}^{\frac{1}{2}Ll- \longrightarrow \frac{1}{4}D\,Ll}_{\frac{1}{2}ll \longrightarrow \frac{1}{4}D\,ll}$$

$$\frac{1}{2}d \Big\langle {}^{\frac{1}{2}Ll- \longrightarrow \frac{1}{4}d\,Ll}_{\frac{1}{2}ll \longrightarrow \frac{1}{4}d\,ll}$$

SOLVED PROBLEMS

1. A rare human disease afflicted a family as shown in the accompanying pedigree.

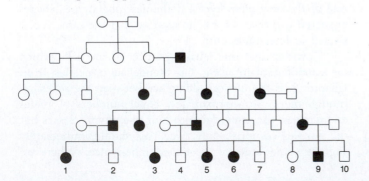

a. Deduce the most likely mode of inheritance.

b. What would the outcomes of the cousin marriages 1 × 9, 1 × 4, 2 × 3 and 2 × 8?

Solution

a. The most likely mode of inheritance is X-linked dominant. We assume that the disease phenotype is dominant because once it is introduced into the pedigree by the male in generation II, it appears in every generation. We assume that the phenotype is X-linked because fathers do not transmit it to their sons. If it were autosomal dominant, father-to-son transmission would be common.

In theory, autosomal recessive could work but it is improbable. In particular, note the marriages between affected members of the family and unaffected outsiders. If the condition were autosomal recessive, the only way these marriages could have affected offspring is if each person marrying into the family were a heterozygote; then the matings would be aa (affected) $\times Aa$ (unaffected). However, we are told that the disease is rare; in such a case, it is highly unlikely that heterozygotes would be so common. X-linked recessive inheritance is impossible because a mating of an affected woman with a normal man could not produce affected daughters. So we can let A represent the disease-causing allele and a represent the normal allele.

b. 1×9: Number 1 must be heterozygous Aa because she must have obtained a from her normal mother. Number 9 must be AY. Hence, the cross is $Aa\,♀ \times AY\,♂$.

Female gametes	Male gametes	Progeny
$\frac{1}{2}A$	$\frac{1}{2}A \longrightarrow$	$\frac{1}{4}AA\,♀$
	$\frac{1}{2}Y \longrightarrow$	$\frac{1}{4}AY\,♂$
$\frac{1}{2}a$	$\frac{1}{2}A \longrightarrow$	$\frac{1}{4}Aa\,♀$
	$\frac{1}{2}Y \longrightarrow$	$\frac{1}{4}aY\,♂$

1×4: Must be $Aa\,♀ \times aY\,♂$.

Female gametes	Male gametes	Progeny
$\frac{1}{2}A$	$\frac{1}{2}a \longrightarrow$	$\frac{1}{4}Aa\,♀$
	$\frac{1}{2}Y \longrightarrow$	$\frac{1}{4}AY\,♂$
$\frac{1}{2}a$	$\frac{1}{2}a \longrightarrow$	$\frac{1}{4}aa\,♀$
	$\frac{1}{2}Y \longrightarrow$	$\frac{1}{4}aY\,♂$

2×3: Must be $aY\,♂ \times Aa\,♀$ (same as 1×4).
2×8: Must be $aY\,♂ \times aa\,♀$ (all progeny normal).

2. Two corn plants are studied; one is Aa, and the other is aa. These two plants are intercrossed in two ways: using Aa as female and aa as male; using aa as female and Aa as male. Recall from Figure 3-38 that the endosperm is $3n$ and is formed by the union of a sperm cell with the two polar nuclei of the female gametophyte.

a. What endosperm genotypes does each cross produce? In what proportions?

b. In an experiment to study the effects of "doses" of alleles, you wish to establish endosperms with genotypes aaa, Aaa, AAa, and AAA (carrying 0, 1, 2, and 3 "doses" of A, respectively). What crosses would you make to obtain these endosperm genotypes?

Solution

a. In such a question, we have to think about meiosis and mitosis at the same time. The meiospores are produced by meiosis; the nuclei of the male and female gametophytes in higher plants are produced by the mitotic division of the meiospore nucleus. We also need to study the corn life cycle to know what nuclei fuse to form the endosperm.

First cross: $Aa\,♀ \times aa\,♂$

Here, the female meiosis will result in spores of which half will be A and half will be a. Therefore, similar portions of haploid female gametophytes will be produced. Their nuclei will be either all A or all a, because mitosis reproduces genetically identical genotypes. Likewise, all nuclei in every male gametophyte will be a. In the corn life cycle, the endosperm is formed from two female nuclei plus one male nucleus, so two endosperm types will be formed as follows.

♀ spore	♀ polar nuclei	♂ sperm	$3n$ endosperm
$\frac{1}{2}A$	A and A	a	$\frac{1}{2}AAa$
$\frac{1}{2}a$	a and a	a	$\frac{1}{2}aaa$

Second cross: $aa\,♀ \times Aa\,♂$

♀ spore	♀ polar nuclei	♂ sperm	$3n$ endosperm
all a	all a and a	$\frac{1}{2}A$	$\frac{1}{2}Aaa$
		$\frac{1}{2}a$	$\frac{1}{2}aaa$

Notice that the phenotypic ratio of endosperm characters would still be Mendelian, even though the underlying endosperm genotypes are slightly different. (Of course, none of these problems arise in embryo characters because embryos are diploid.)

b. This kind of experiment has been very useful in studying plant genetics and molecular biology. In answering the question, all we need to realize is that the two polar nuclei contributing to the endosperm are genetically identical. To obtain endosperms, all of which will be aaa, any $aa \times aa$ cross will work. To obtain endosperms, all of which will be Aaa, the cross must be $aa\,♀ \times AA\,♂$. To obtain embryos, all of which will be AAa, the cross must be $AA\,♀ \times aa\,♂$. For AAA obviously any $AA \times AA$ cross will work. Notice that these endosperm genotypes can be obtained in other crosses, but only in combination with other endosperm genotypes.

PROBLEMS

1. What are the major differences between mitosis and meiosis?

2. When a cell of genotype $Aa\,Bb\,Cc$ having all the genes on separate chromosome pairs divides mitotically, what are the genotypes of the daughter cells?

3. In working with a haploid yeast, you cross a purple (ad^-) strain of mating type a and a white (ad^+) strain of mating type α. If ad^- and ad^+ are alleles of one gene and a and α are alleles of an independently inherited gene on a separate chromosome pair, what progeny do you expect to obtain? In what proportions?

4. The recessive allele s causes *Drosophila* to have small wings and the s^+ allele causes normal wings. This gene is known to be X-linked. If a small-winged male is crossed with a homozygous wild-type female, what ratio of normal to small-winged flies can be expected in each sex in the F_1? If F_1 flies are intercrossed, what F_2 progeny ratios are expected? What progeny ratios are predicted if F_1 females are backcrossed to their father?

5. State where cells divide mitotically and where they divide meiotically in a fern, a moss, a flowering plant, a pine tree, a mushroom, a frog, a butterfly, and a snail.

6. Human cells normally have 46 chromosomes. For each of the following stages, state the number of chromosomes present in a human cell: (**a**) metaphase of mitosis, (**b**) metaphase I of meiosis, (**c**) telophase of mitosis, (**d**) telophase I of meiosis, (**e**) telophase II of meiosis. (In your answers, count chromatids as chromosomes.)

7. Four of the following events are part of both meiosis and mitosis, but one is only meiotic. Which one? (**a**) Chromatid formation, (**b**) spindle formation, (**c**) chromosome condensation, (**d**) chromosome movement to poles, (**e**) chromosome pairing.

8. Suppose that you discover two interesting *rare* cytological abnormalities in the karyotype of a human male. (A karyotype is the total visible chromosome complement.) There is an extra piece (or satellite) on *one* of the chromosomes of pair 4, and there is an abnormal pattern of staining on one of the chromosomes of pair 7. Assuming that all the gametes of this male are equally viable, what proportion of his children will have the same karyotype he has?

9. Suppose that meiosis occurs in the transient diploid stage of the cycle of a haploid organism of chromosome number n. What is the probability that an individual haploid cell resulting from the meiotic division will have a complete parental set of centromeres (that is, a set all from one parent or all from the other parent)?

10. Assuming the sex chromosomes to be identical, name the proportion of all alleles you have in common with (**a**) your mother; (**b**) your brother.

11. A wild-type female schmoo who is graceful (G) is mated to a non-wild-type male who is gruesome (g). Their progeny consist solely of graceful males and gruesome females. Interpret these results and give genotypes.

(Problem 11 from E. H. Simon and H. Grossfield, *The Challenge of Genetics*. Copyright 1971 by Addison-Wesley.)

12. A man with a certain disease marries a normal woman. They have eight children (four boys and four girls); all of the girls have their father's disease, but none of the boys do. What inheritance is suggested? (**a**) Autosomal recessive, (**b**) autosomal dominant, (**c**) Y-linked, (**d**) X-linked dominant, (**e**) X-linked recessive.

13. An X-linked dominant allele causes hypophosphatemia in humans. A man with hypophosphatemia marries a normal woman. What proportion of their sons will have hypophosphatemia? (**a**) $\frac{1}{2}$, (**b**) $\frac{1}{4}$, (**c**) $\frac{1}{3}$, (**d**) 1, (**e**) 0.

14. A condition known as *icthyosis hystrix gravior* appeared in a boy in the early eighteenth century. His skin became very thick and formed loose spines that were sloughed off at intervals. When he grew up, this "porcupine man" married and had six sons, all of whom had this condition, and several daughters, all of whom were normal. For four generations, this condition was passed from father to son. From this evidence, what can you postulate about the location of the gene?

15. Duchenne's muscular dystrophy is sex-linked and usually affects only males. Victims of the disease become progressively weaker, starting early in life.

a. What is the probability that a woman whose brother has Duchenne's disease will have an affected child?

b. If your mother's brother (your uncle) had Duchenne's disease, what is the probability that you have received the allele?

c. If your father's brother had the disease, what is the probability that you have received the allele?

16. The following pedigree is concerned with an inherited dental abnormality, amelogenesis imperfecta.

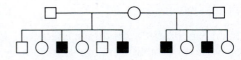

a. What mode of inheritance *best* accounts for the transmission of this trait?

b. Write the genotypes of all family members according to your hypothesis.

17. A sex-linked recessive allele *c* produces a red-green colorblindness in humans. A normal woman whose father was colorblind marries a colorblind man.

a. What genotypes are possible for the mother of the colorblind man?

b. What are the chances that the first child from this marriage will be a colorblind boy?

c. Of the girls produced by these parents, what proportion can be expected to be colorblind?

d. Of all the children (sex unspecified) of these parents, what proportion can be expected to have normal color vision?

18. Male house cats are either black or orange; females are black, orange, or calico.

a. If these coat-color phenotypes are governed by a sex-linked gene, how can these observations be explained?

b. Using appropriate symbols, determine the phenotypes expected in the progeny of a cross between an orange female and a black male.

c. Repeat part b for the reciprocal of the cross described there.

d. Half the females produced by a certain kind of mating are calico, and half are black; half the males are orange, and half are black. What colors are the parental males and females in this kind of mating?

e. Another kind of mating produces progeny in the following proportions: $\frac{1}{4}$ orange males, $\frac{1}{4}$ orange females, $\frac{1}{4}$ black males, and $\frac{1}{4}$ calico females. What colors are the parental males and females in this kind of mating?

19. A man is heterozygous *Bb* for one autosomal gene, and he carries a recessive X-linked allele *d*. What proportion of his sperm will be *bd*? (**a**) 0, (**b**) $\frac{1}{2}$, (**c**) $\frac{1}{8}$, (**d**) $\frac{1}{16}$, (**e**) $\frac{1}{4}$.

20. The accompanying pedigree concerns a certain rare disease that is incapacitating but not fatal.

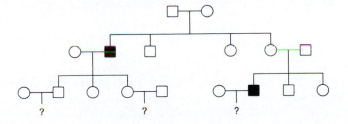

a. Determine the mode of inheritance of this disease.

b. Write the genotype of each individual according to your proposed mode of inheritance.

c. If you were this family's doctor, how would you advise the three couples in the third generation about the likelihood of having an affected child?

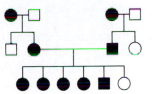

21. Assume that this pedigree is straightforward, with no complications such as illegitimacy.

Phenotype W, found in the individuals represented by the shaded symbols, is rare in the general population. Which of the following patterns of transmission for W are consistent with this pedigree? Which are excluded? (**a**) Autosomal recessive, (**b**) autosomal dominant, (**c**) X-linked recessive, (**d**) X-linked dominant, (**e**) Y-linked.

(Problem 21 from A. M. Srb, R. D. Owen, and R. S. Edgar, *General Genetics*, 2d ed. Copyright © 1965 by W. H. Freeman and Company.)

22. A mutant allele in mice causes a bent tail. Six pairs of mice were crossed. Their phenotypes and those of their progeny are given below. N is normal phenotype; B is bent phenotype. Deduce the mode of inheritance of this phenotype.

	Parents		Progeny	
Cross	♀	♂	♀	♂
1	N	B	All B	All N
2	B	N	$\frac{1}{2}$ B, $\frac{1}{2}$ N	$\frac{1}{2}$ B, $\frac{1}{2}$ N
3	B	N	All B	All B
4	N	N	All N	All N
5	B	B	All B	All B
6	B	B	All B	$\frac{1}{2}$ B, $\frac{1}{2}$ N

a. Is it recessive or dominant?

b. Is it autosomal or sex-linked?

c. What are the genotypes of all parents and progeny?

23. The normal eye color of *Drosophila* is red, but strains in which all flies have brown eyes are available. Similarly, wings are normally long, but there are strains with short wings. A female from a pure line with brown eyes and short wings is crossed with a male from a normal pure line. The F_1 consists of normal females and short-winged males. An

F_2 is then produced by intercrossing the F_1. *Both* sexes of F_2 flies show phenotypes as follows:

$\frac{3}{8}$ red eyes, long wings

$\frac{3}{8}$ red eyes, short wings

$\frac{1}{8}$ brown eyes, long wings

$\frac{1}{8}$ brown eyes, short wings

Deduce the inheritance of these phenotypes, using clearly defined genetic symbols of your own invention. State the genotypes of all three generations and the genotypic proportions of the F_1 and F_2.

 Unpacking the Problem

Before attempting a solution to this problem, try answering the following questions.

a. What does the word "normal" mean in this problem?

b. The words "line" and "strain" are used in this problem: What do they mean and are they interchangeable?

c. Draw a simple sketch of the two parental flies showing their eyes, wings, and sexual differences.

d. How many different characters are involved in this question?

e. How many phenotypes are involved in this problem, and which phenotypes go with which characters?

f. What is the full phenotype of the F_1 females called "normal"?

g. What is the full phenotype of the F_1 males called "short-winged"?

h. List the F_2 phenotypic ratios for each character that you came up with in question d.

i. What do the F_2 phenotypic ratios tell you?

j. What major inheritance pattern distinguishes sex-linked inheritance from autosomal inheritance?

k. Do the F_2 data show such a distinguishing criterion?

l. Do the F_1 data show such a distinguishing criterion?

m. What can you learn about dominance in the F_1? The F_2?

n. What rules about wild-type symbolism can you use in deciding which allelic symbols to invent for these crosses?

o. What does "deduce the inheritance of these phenotypes" mean?

Now try to solve the problem. If you are unable, make a list of questions about the things you do not understand. Inspect the key concepts at the beginning of the chapter and ask yourself which are relevant to your questions. If this doesn't work, inspect the messages of this chapter and ask yourself which might be relevant to your questions.

24. The wild-type (W) *Abraxas* moth has large spots on its wings, but the lacticolor (L) form of this species has very small spots. Crosses were made between strains differing in this character, with the following results:

	Parents		Progeny	
Cross	♀	♂	F_1	F_2
1	L	W	♀ W	♀ $\frac{1}{2}$ L, $\frac{1}{2}$ W
			♂ W	♂ W
2	W	L	♀ L	♀ $\frac{1}{2}$ W, $\frac{1}{2}$ L
			♂ W	♂ $\frac{1}{2}$ W, $\frac{1}{2}$ L

Provide a clear genetic explanation of the results in these two crosses, showing the genotypes of all individuals.

25. A certain gene that governs the activity of the enzyme glucose 6-phosphate dehydrogenase (G6PD) has two common alleles in Mediterranean and African populations. One allele stands for normal G6PD activity, and the other allele, which stands for reduced G6PD activity, confers resistance to malaria.

When the red blood cells of a certain African woman were examined under the microscope, precisely half the cells were found to contain the malarial parasite, whereas the other half appeared normal. Provide a genetic explanation for this finding.

26. Medical literature records the interesting case of a woman whose right breast was larger than the left, who had no pubic hair to the right of the midline, and who suffered from menstrual irregularities. Upon investigation of her family, it was discovered that a brother, a son, and a grandson showed testicular feminization syndrome. One of her daughters had three normal sons. Draw the pedigree from this information, determine if it fits the inheritance mode described in this chapter, and speculate on the cause of the symptoms in the propositus.

27. The pedigree at the top of the facing page is for a rare human disease called *spastic paraplegia*, a nervous disorder in which there is an inability to coordinate voluntary movements.

   a. What mode of inheritance is suggested by this pedigree?

   b. Which individuals must be heterozygous under your model?

(Pedigree from V. A. McKusick, *On the Chromosomes of Man*. Copyright © 1964 by American Institute of Biological Science, Washington, D.C.)

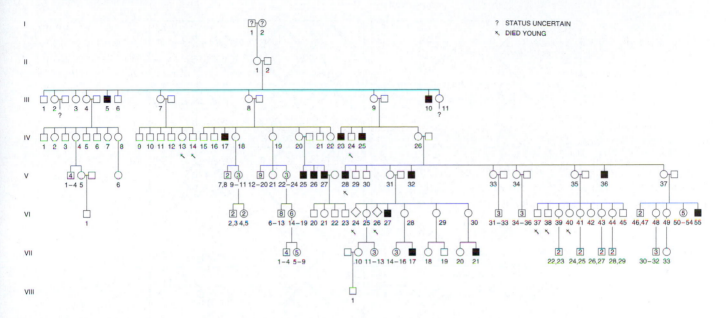

28. The pedigree below shows the inheritance of a rare human disease. Is the pattern best explained as being caused by an X-linked recessive allele or by an autosomal dominant allele with expression limited to males?

(Pedigree modified from J. F. Crow, *Genetics Notes,* 6th ed. Copyright © 1967 by Burgers Publishing Company, Minneapolis.)

29. Pretend that the year is 1868. You are a skilled young lens maker working in Vienna. With your superior new lenses you have just built a microscope that has better resolution than any others available. During your testing of this microscope you have been observing the cells in the testes of grasshoppers, and have been fascinated by the behavior of strange elongated structures you have seen within the dividing cells. One day in the library you read a recent journal paper by G. Mendel on hypothetical "factors" that he claims explain the results of certain crosses in peas. In a flash of revelation you are struck by the parallels between your grasshopper studies and Mendel's, and you resolve to write him a letter. What do you write?

(Based on an idea by Ernest Kroeker.)

∗30. In humans, color vision depends on genes encoding three pigments. The *R* (red pigment) and *G* (green pigment) genes are on the X chromosome, while the *B* (blue pigment) gene is autosomal. A mutation in any one of these genes can cause colorblindness. Suppose that a colorblind man married a woman with normal color vision. All their sons were colorblind, and all their daughters were normal. Specify the genotypes of both parents and all possible children, explaining your reasoning. (A pedigree drawing would probably be helpful.)

(Problem by Rosemary Redfield)

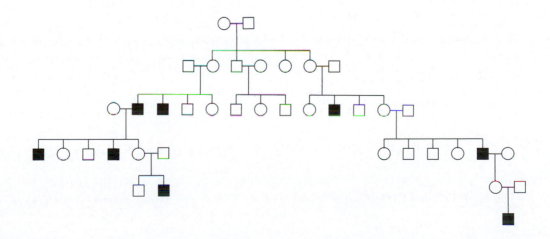

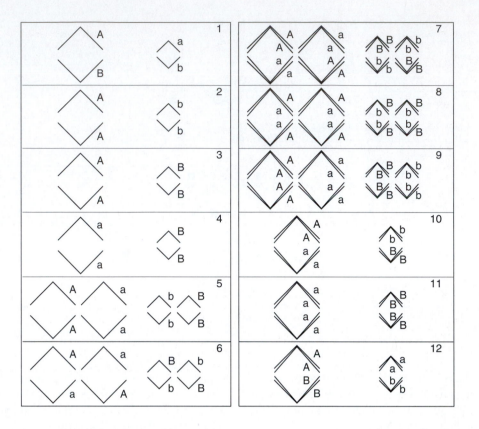

31. The plant *Haplopappus gracilis* is diploid and $2n = 4$. There are one long pair and one short pair of chromosomes. The accompanying diagrams represent anaphases ("pulling apart" stages) of individual cells during meiosis or mitosis in a plant that is genetically a dihybrid (*Aa Bb*) for genes on different chromosomes. The lines represent chromosomes or chromatids, and the points of the V's represent centromeres. In each case, say if the diagram represents a cell in meiosis I, meiosis II, or mitosis. If a diagram shows an impossible situation, say so.

32. A certain type of deafness in humans is inherited as an X-linked recessive. A man who suffers from this type of deafness marries a normal woman and they are expecting a child. They find out that they are distantly related. Part of the family tree follows.

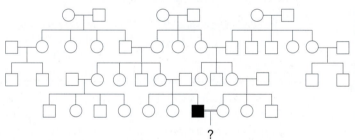

How would you advise the parents about the probability of their child being a deaf boy, a deaf girl, a normal boy, and a normal girl? Be sure to state any assumptions you might make.

4

Extensions of Mendelian Analysis

Variation in shell coloration of the bay scallop (*Argopecten irradians*) cause by three alleles of one gene. Yellow, black, and orange are determined by the alleles p^y, p^b, and p^o, respectively. The groups of small shells represent the proportions obtained upon selfing individuals of genotype $p^y p^b$ (top row), $p^b p^b$ (middle row), and $p^o p^b$ (bottom row). These results demonstrate the allelic relationship and also show that p^y and p^o are both dominant to p^b. (From L. Adamkewicz and M. Castagna, *Journal of Heredity* 79, 1988, 15/BPS.)

KEY CONCEPTS

▶ A gene can have more than two alleles.

▶ Phenotypes of some heterozygotes reveal types of dominance other than full dominance.

▶ Many genes have alleles that can kill the organism.

▶ Most characters are determined by sets of genes that interact with one another and with the environment.

▶ Modified Mendelian ratios reveal gene interactions.

We have seen that Mendel's laws of equal segregation and independent assortment seem to hold across the entire spectrum of eukaryotic organisms. These laws form a base for predicting the outcome of simple crosses. However, it is only a base; the real world of genes and chromosomes is more complex than Mendel's laws suggest, and exceptions and extensions abound. These situations do not invalidate Mendel's laws. Rather, they show that more explanatory elements must be added to the base of equal segregation and independent assortment of alleles to fit these situations into the fabric of genetic analysis. This is the challenge we now must meet. Of course, one extension—sex linkage—has already been explained. This chapter presents an assortment of other extensions, based mainly on the complexities of gene expression. The two following chapters discuss two more major extensions. We shall see that rather than creating a hopeless and bewildering situation, taking these complexities into account reveals a precise and unifying set of principles for the genetic analyst. These principles interlock and support one another in a highly satisfying way that has provided great insight into the mechanisms of inheritance.

Variations on Dominance

Dominance is a good place to start. Mendel reported full dominance (and recessiveness) for all seven genes he studied. However, some examples will illustrate variations on the theme of dominance.

Four-o'clocks are plants native to tropical America. Their name comes from the fact that their flowers open in the late afternoon. When a pure four-o'clock line with red petals is crossed with a pure line with white petals, the F_1 has pink petals. If an F_2 is produced by intercrossing the F_1, the result is

$\frac{1}{4}$ of the plants have red petals

$\frac{1}{2}$ of the plants have pink petals

$\frac{1}{4}$ of the plants have white petals

Because of the $1:2:1$ ratio in the F_2, we can deduce an inheritance pattern based on two alleles of a single gene. However, the heterozygotes (the F_1 and half the F_2) are intermediate in phenotype, suggesting an *incomplete* type of dominance. Inventing allele symbols that have no dominance connotation, we can list the genotypes of the four-o'clocks in this experiment as C_1C_1 (red), C_2C_2 (white), and C_1C_2 (pink). (Subscripts and superscripts are often used to denote alleles.) **Incomplete dominance** describes the general situation in which the phenotype of a heterozygote is intermediate between the two homozygotes on a phenotypic "scale" of measurement. Figure 4-1 gives terms for all the theoretical positions on the scale, but in practice it is difficult to determine exactly where on such a scale the heterozygote is located.

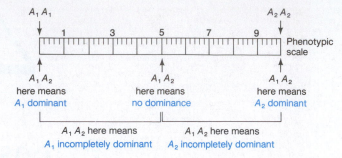

Figure 4-1 Summary of dominance relationships. The ruler represents some sort of phenotypic measurement such as amount of pigment.

The phenotype of the heterozygote is also the key in the phenomenon of **codominance,** in which the heterozygote shows the phenotypes of both the homozygotes. The three MN blood groups found in human populations provide an example. The three blood groups, M, N, and MN correspond to the genotypes L^ML^M, L^NL^N, and L^ML^N, respectively. Blood groups are actually determined by the presence of an immunological antigen on the surface of the red blood cells. As specified by their genotype, people have either antigen M (genotype L^ML^M) or antigen N (genotype L^NL^N), or they have both (genotype L^ML^N). Because the heterozygote has both phenotypes, the two alleles are said to be *codominant.*

The human disease sickle-cell anemia gives interesting insight into dominance. The gene concerned affects the molecule hemoglobin, which transports oxygen and is the major constituent of red blood cells. The three genotypes have different phenotypes, as follows:

Hb^AHb^A: Normal; red blood cells never sickle.

Hb^SHb^S: Severe, often fatal anemia; abnormal hemoglobin causes red blood cells to have sickle shape.

HB^AHb^S: No anemia; red blood cells sickle only under low oxygen concentrations.

Figure 4-2 shows sickle cells. In regard to the presence or absence of anemia, the Hb^A allele is obviously dominant. In regard to blood cell shape, however, there is incomplete dominance. Finally, as we shall now see, in regard to hemoglobin itself there is codominance. The alleles Hb^A and Hb^S actually code for two slightly different forms of hemoglobin, and both these forms are present in the heterozygote, showing that the alleles are codominant. The different hemoglobin forms can be visualized using **electrophoresis,** a technique that separates molecules with different charge and size (Figure 4-3). It so happens that the A and S forms of hemoglobin have different charges, so they can be separated by electrophoresis (Figure 4-4).

We see that homozygous normal people have one type of hemoglobin (A) and anemics have type S, which moves more slowly in the electric field. The heterozygotes have

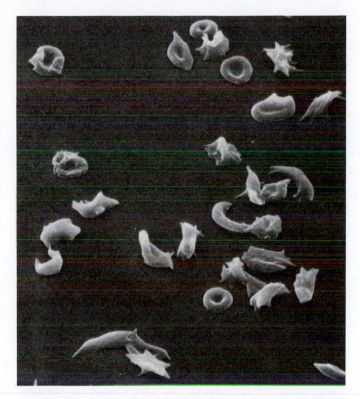

Figure 4-2 Electron micrograph of red blood cells from an individual with sickle-cell anemia. A few rounded cells appear almost normal. (Patricia N. Farnsworth)

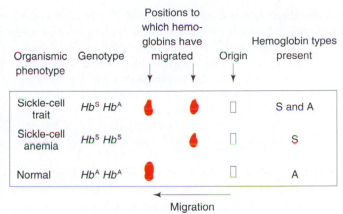

Figure 4-4 Electrophoresis of hemoglobin from an individual with sickle-cell anemia, a heterozygote (called *sickle-cell trait*), and a normal individual. The smudges show the positions to which the hemoglobins migrate on the starch gel.

Message Complete dominance and recessiveness are not essential aspects of Mendel's laws; those laws govern the inheritance patterns of genes, not the functions of the genes.

both types, A and S. In other words, there is codominance at the molecular level.

Sickle-cell anemia illustrates that the terms *incomplete dominance* and *codominance* are somewhat arbitrary. The type of dominance inferred depends on the phenotypic level at which the observations are being made — organismal, cellular, or molecular. Indeed the same caution can be applied to many of the categories that scientists use to classify structures and processes; these categories are devised by humans for their own convenience, and often break down under close scrutiny.

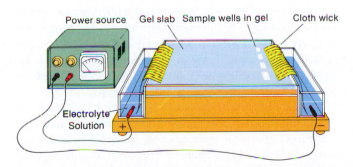

Figure 4-3 Apparatus for electrophoresis. Each sample is placed in a well in a gelatinous slab (a gel). The molecules in the samples migrate different distances on the gel owing to their different electric charges. Several samples are tested at the same time (one in each well). The positions to which the molecules have migrated are later revealed by staining.

Multiple Alleles

Early in the history of genetics, it became clear that it is possible for there to be more than two forms of a gene. Although a diploid organism can possess only two alleles of a gene (and a haploid organism only one), in a population the total number of different alleles for a single gene is often quite large. The existence of these many alleles is called **multiple allelism,** and the set of alleles itself is called an **allelic series.** The concept of allelism is a crucial one in genetics, so we consider several examples. The examples themselves serve also to introduce important areas of genetic investigation.

ABO Blood Group in Humans

The human ABO blood group alleles show multiple allelism. There are four blood types (or phenotypes) in the ABO system, as shown in Table 4-1. The allelic series in-

Table 4-1 ABO Blood Groups in Humans

Blood phenotype	Genotype
O	ii
A	$I^A I^A$ or $I^A i$
B	$I^B I^B$ or $I^B i$
AB	$I^A I^B$

cludes three major alleles—*i*, I^A, and I^B—but, of course, any person has only two of the three alleles (or two copies of one of them). In this allelic series, the alleles I^A and I^B each determine a unique antigen; allele *i* confers inability to produce an antigen. In the genotypes $I^A i$ and $I^B i$, the alleles I^A and I^B are fully dominant to *i*, but they are codominant in the genotype $I^A I^B$.

C Gene in Rabbits

A larger allelic series concerns coat color in rabbits. The alleles in this series are *C* (full color), c^{ch} (chinchilla, a light grayish color), c^h (Himalayan, albino with black extremities), and *c* (albino). Here you can see the value of denoting alleles by superscripts: more than the two symbols *C* and *c* are needed to describe these multiple alleles. In this series, each allele is dominant to the alleles listed after it in the order *C*, c^{ch}, c^h, *c*. Verify this by studying Table 4-2.

Operational Test for Allelism

Now that we have seen two examples of allelic series, it is a good time to pause and ask a question. How do we know that a set of contrasting phenotypes is determined by alleles of only one gene? In other words, what is the operational test for allelism? For now, the answer is simply the observation of Mendelian monohybrid F$_2$ ratios from crosses of all pairwise combinations of pure-breeding lines. For example, consider three pure-line phenotypes in a hypothetical plant species. Line 1 has round spots on the petals; line 2 has oval spots on the petals; and line 3 has no spots on the petals. Suppose that crosses of the three lines yield the following results:

Cross	F$_1$	F$_2$
1 × 2	all round-spotted	$\frac{3}{4}$ round $\frac{1}{4}$ oval
1 × 3	all round-spotted	$\frac{3}{4}$ round $\frac{1}{4}$ unspotted
2 × 3	all oval-spotted	$\frac{3}{4}$ oval $\frac{1}{4}$ unspotted

These results tell us that we are dealing with three alleles of a single gene that affects petal spotting because each possible cross gives a monohybrid F$_2$ ratio. The dominance hierarchy is round > oval > unspotted. We can choose any symbols we wish. Because we don't know which phenotype is the wild type, we could follow the rabbit system and use *S* for the round-spotted allele, s^o for the oval-spotted allele, and *s* for the unspotted allele. Alternatively, we could use S^r for round, S^o for oval, and *s* for unspotted. There are no firm rules about whether to use capital or small letters, particularly for the alleles in the middle of the series that are dominant to some of the alleles but recessive to others.

Table 4-2 *C* Gene in Rabbits

Coat color phenotype	Genotype
Full color	*CC* or Cc^{ch} or Cc^h or *Cc*
Chinchilla	$c^{ch}c^{ch}$ or $c^{ch}c^h$ or $c^{ch}c$
Himalayan	$c^h c^h$ or $c^h c$
Albino	*cc*

What if crosses between pure-breeding lines differing for one character do not produce monohybrid F$_2$ Mendelian ratios? This outcome suggests the interaction of several genes, and this topic is covered later in the chapter.

Clover Chevrons

Clover is the common name for plants of the genus *Trifolium*. There are many species. Some are native to North America, while others grow as introduced weeds. Much genetic research has been done with white clover, which shows considerable variation among individuals in the curious V, or chevron, pattern on the leaves. Figure 4-5 shows that in this species an allelic series determines the different chevron forms (and the absence of chevrons). Study the photographs to determine the type of dominance, if any, of each allele. List the alleles in a way that expresses how they relate to one another in dominance. Are there uncertainties? Does the photographic evidence permit us to say anything about the dominance or recessiveness of allele *v*?

Incompatibility Alleles in Plants

It has been known for millennia that some plants just will not self-fertilize. A single plant may produce both male and female gametes, but no seeds will ever be produced from the fusion of two of these gametes. The same plants, however, will cross with certain other plants, so obviously they are not sterile. This phenomenon is called **self-incompatibility.** Of course, the species of pea used by Mendel was not self-incompatible, for he was able to self his plants with ease. We now know that incompatibility in plants has a genetic basis and that there are several different genetic systems acting in different self-incompatible species. These systems form nice examples of multiple allelism.

One of the most common incompatibility systems is that found in sweet cherries, tobacco, petunias, and evening primroses. In each of these species, one gene, *S*, determines compatibility-incompatibility relations. In any one species, different plants bear different pairs of alleles from a set of many different alleles of the gene. Figure 4-6 shows how this system produces a fully incompatible reaction, a semicompatible reaction, and a fully compatible reaction. If a pollen grain bears an *S* allele that is also present in the ma-

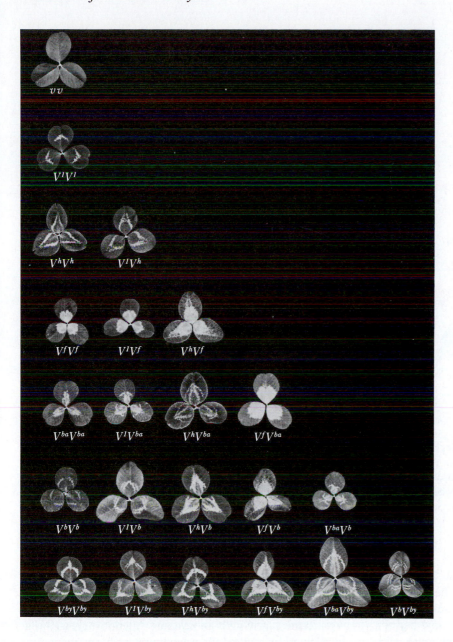

Figure 4-5 Multiple alleles determine the chevron pattern on the leaves of white clover. The genotype of each plant is shown below it. (W. Ellis Davies)

ternal parent, then it will not grow. However, if that allele is not in the maternal tissue, the pollen grain produces a pollen tube containing the male nucleus, and this tube effects fertilization. The number of *S* alleles in a series in one species can be very large (it is more than 50 in evening primroses and clover), and some species have more than 100 alleles. It is generally assumed that this type of system has evolved because it promotes outbreeding.

Since we know that a gene is a chromosomal DNA sequence, we can ask what alleles correspond to at the DNA level. Any alteration in the wild-type gene's DNA sequence will result in a new allele (a new form of the gene). Some of these alterations might cause a change in or an absence of function of the protein coded by that gene, and within this group some would be recognized as alleles producing a detectably different phenotype. So we see that because of the many different possible ways of changing the DNA of a gene, the number of possible alleles is very large, but the number of novel phenotypes produced by these changes is much less.

Message   A gene can have several different states or forms—a situation called *multiple allelism.* The alleles are said to constitute an allelic series, and the members of a series can show any type of dominance to one another.

Lethal Alleles

Normal wild-type mice have coats with a rather dark overall pigmentation. In 1904, Lucien Cuenot studied mice having a lighter coat color called *yellow.* After mating a yellow mouse to a normal mouse from a pure line, Cuenot observed a 1:1 ratio of yellow to normal mice in the progeny.

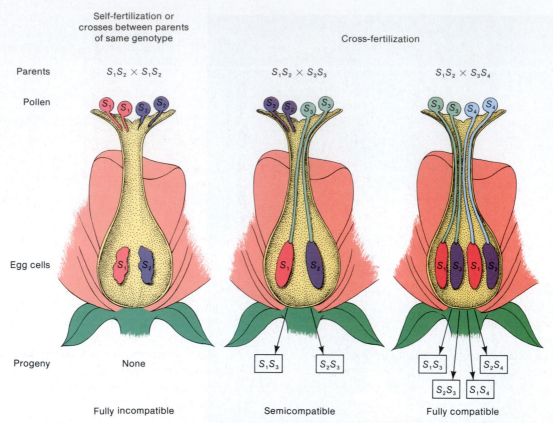

Self-fertilization or
crosses between parents
of same genotype

Cross-fertilization

Parents $S_1S_2 \times S_1S_2$ $S_1S_2 \times S_2S_3$ $S_1S_2 \times S_3S_4$

Pollen

Egg cells

Progeny   None S_1S_3 S_2S_3 S_1S_3 S_2S_4 S_2S_3 S_1S_4

Fully incompatible   Semicompatible   Fully compatible

Figure 4-6 How multiple alleles control self-incompatibility in certain plants. A pollen tube will not grow if the *S* allele that it contains is present in the female parent. This diagram shows only four multiple alleles, but many plant incompatibility systems use far larger numbers of alleles. Such systems promote the exchange of genes between plants by making selfing impossible and crosses between near relations very unlikely. (From A. M. Srb, R. D. Owen, and R. S. Edgar, *General Genetics*, 2d. ed. W. H. Freeman and Company, 1965.)

This observation suggests that a single gene determines these phenotypes, that the yellow mouse was heterozygous for this gene, and that the allele for yellow is dominant to an allele for normal color. However, the situation became more confusing when Cuenot crossed yellow mice with each other. The result was always the same, no matter which individual yellow mice were used:

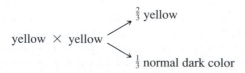

yellow × yellow $\longrightarrow$ $\frac{2}{3}$ yellow

$\frac{1}{3}$ normal dark color

Note two interesting features in these results. First, the 2:1 phenotypic ratio is a departure from Mendelian expectations. Second, because no cross of yellow × yellow ever produced all yellow progeny, as there would be if either parent were a homozygote, it appeared that it is impossible to obtain homozygous yellow mice.

Cuenot explained the results by postulating that all yellow mice are heterozygous for one special gene. A cross between two heterozygotes would be expected to yield a

Mendelian genotypic ratio of 1:2:1. However, if all the mice in one of the homozygous classes died before birth, the live births would then show a 2:1 ratio of heterozygotes to the surviving homozygotes. Cuenot suggested that the allele A^Y for yellow might be dominant to the normal allele A with respect to its effect on color, but A^Y might act as a recessive **lethal** allele with respect to a character we would call *viability*. Thus, a mouse with the homozygous genotype $A^Y A^Y$ dies before birth and is not observed among the progeny. All surviving yellow mice must be heterozygous $A^Y A$, so a cross between yellow mice will always yield the following results:

$$A^Y A \times A^Y A \longrightarrow \begin{array}{l} \frac{1}{4} AA \quad \text{normal} \\ \frac{1}{2} A^Y A \quad \text{yellow} \\ \frac{1}{4} A^Y A^Y \quad \text{die before birth} \end{array}$$

The expected Mendelian ratio of 1:2:1 would be found among the zygotes, but it is altered to a 2:1 ratio in the progeny born because zygotes with a lethal $A^Y A^Y$ genotype do not survive to be counted. This hypothesis was con-

Figure 4-7 A mouse litter from two parents heterozygous for the yellow coat color allele, which is lethal in a double dose. The larger mice are the parents. Not all progeny are visible. (Anthony Griffiths)

firmed by the removal of uteri from pregnant females of the yellow × yellow cross; one-fourth of the embryos were found to be dead. Figure 4-7 shows a typical litter from a cross between yellow mice.

The A^Y allele produces effects on two characters: coat color and survival. Such genes that have more than one distinct phenotypic effect are said to be **pleiotropic** genes. It is entirely possible, however, that both effects of the A^Y pleiotropic allele result from the same basic cause, which promotes yellowness of coat in a single dose and death in a double dose.

The tailless Manx phenotype in cats (Figure 4-8) is also produced by an allele that is lethal in the homozygous state. A single dose of the Manx allele, M^L, severely interferes with normal spinal development, resulting in the absence of a tail in the $M^L M$ heterozygote. In $M^L M^L$ homozygotes, the dou-

ble dose of the gene produces such an extreme developmental abnormality that the embryo does not survive.

There are indeed many different types of lethal alleles. Some lethal alleles produce a recognizable phenotype in the heterozygote, as in the yellow mouse and Manx cat. Some lethal alleles are fully dominant and kill in one dose in the heterozygote. Some confer no detectable effect in the heterozygote at all, and the lethality is fully recessive. Furthermore, lethal alleles differ in the developmental stage at which they express their effects. Human lethals illustrate this very well. It has been estimated that we each carry a small number of recessive lethals in our genomes. The lethal effect is expressed in the homozygous progeny of a mating between two people carrying the same recessive lethal in the heterozygous condition. Some lethals are expressed as deaths in utero, where they either go unnoticed, or are noticed as spontaneous abortions. Other lethals, such as those responsible for Duchenne muscular dystrophy, PKU, and cystic fibrosis, exert their effects in childhood. The time of death can even be in adulthood, as in Huntington's disease. The total of all the deleterious and lethal genes that are present in individuals is called *genetic load,* a kind of genetic burden that the population has to carry.

Exactly how do lethal alleles kill? In some cases it is possible to trace the cascade of events that leads to death. A common situation is that the allele causes a deficiency in some essential chemical reaction. The human disease PKU is a good example of this kind of deficiency. In other cases there is a structural defect. For example, a lethal allele of rats determines abnormal cartilage, and the effect of this abnormality is expressed phenotypically in several different organs, resulting in lethal symptoms, as shown in Figure 4-9. Sickle-cell anemia is another example.

Whether an allele is lethal often depends on the environment in which the organism develops. Whereas certain alleles would be lethal in virtually any environment, others are viable in one environment but lethal in another. For example, many of the phenotypes favored and selected by agricultural breeders would almost certainly be eliminated

Figure 4-8 Manx cat. All such cats are heterozygous for a dominant allele that causes no tail to form. The allele is lethal in homozygous condition. The dissimilar eyes are unrelated to taillessness. (Gerard Lacz/NHPA)

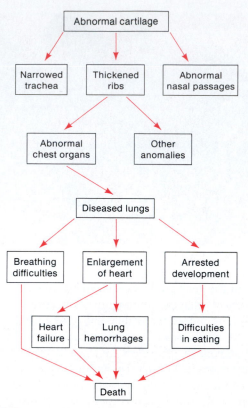

Figure 4-9 Diagram showing how one specific lethal allele causes death in rats. (From I. M. Lerner and W. J. Libby, *Heredity, Evolution, and Society,* 2d. ed. W. H. Freeman and Company, 1976; after H. Gruneberg.)

in nature as a result of competition with the members of the natural population. Modern grain varieties provide good examples; only careful nurturing by the farmer has maintained such phenotypes for our benefit.

In practical genetics, we commonly encounter situations in which expected Mendelian ratios are consistently skewed in one direction by reduced viability owing to one allele. For example, in the cross $Aa \times aa$, we predict a progeny phenotypic ratio of 50 percent $A-$ and 50 percent aa, but we might consistently observe a ratio such as 55% : 45% or 60% : 40%. In such a case, the a allele is said to be *subvital,* or *semilethal,* because the lethality is expressed in only some individuals. Thus, lethality may range from 0 to 100 percent, depending on the gene itself, the rest of the genome, and the environment. We shall return to this topic later.

Several Genes Affecting the Same Character

We saw earlier that the identification of a major gene affecting a character does not mean that this is the only gene af-

fecting that character. An organism is a highly complex machine in which all functions interact to a greater or lesser degree. At the level of genetic determination, the genes likewise can be regarded as cooperating. Therefore, a gene does not act in isolation; its effects depend not only on its own functions but also on the functions of other genes, as well as on the environment. In many cases, genetic analysis can detect the complex interactions of major genes, and we now look at some examples. Typically, the key to an interaction is a **modified Mendelian ratio.**

Coat Color in Mammals

Studies of coat color in mammals reveal beautifully how different genes cooperate in the determination of one character. The mouse is a good mammal for genetic studies because it is small and thus easy to maintain in the laboratory and because its reproductive cycle is short. It is the best-studied mammal with regard to the genetic determination of coat color. The genetic determination of coat color in other mammals closely parallels that of mice, and for this reason the mouse acts as a model system. We shall look at examples from other mammals as our discussion proceeds. At least five major genes interact to determine the coat color of mice: the genes are A, B, C, D, and S.

The A Gene. This gene determines the distribution of pigment in the hair. The wild-type allele A produces a phenotype called *agouti.* Agouti is an overall grayish color with a brindled, or "salt-and-pepper," appearance. It is a common color of mammals in nature. The effect is caused by a band of yellow on the otherwise dark hair shaft. In the nonagouti phenotype (determined by the allele a), the yellow band is absent, so there is solid dark pigment throughout (Figure 4-10).

Figure 4-10 Individual hairs from an agouti and a black mouse. The yellow band on each hair gives the agouti pattern its brindled appearance.

The lethal allele A^Y, discussed in the previous section, is another allele of this gene; it makes the entire shaft yellow. Still another allele is a^t, which results in a "black-and-tan" effect, a yellow belly with dark pigmentation elsewhere. For simplicity, we shall not include these two alleles in the following discussion.

The B Gene. This gene determines the color of pigment. There are two major alleles, *B* coding for black pigment and *b* for brown. The allele *B* gives the normal agouti color in combination with *A*, but gives solid black with *aa*. The genotype *A−bb* gives a streaked brown color called *cinnamon*, and *aa bb* gives solid brown.

The following cross illustrates the inheritance pattern of the *A* and *B* genes:

$$AA\,bb \text{ (cinnamon)} \times aa\,BB \text{ (black)}$$

or $AA\,BB$ (agouti) $\times aa\,bb$ (brown)

$\downarrow$

F_1 all $Aa\,Bb$ (agouti)

$Aa\,Bb$ (agouti) $\times Aa\,Bb$ (agouti)

$\downarrow$

F_2 9 $A−B−$ (agouti)

3 $A−bb$ (cinnamon)

3 $aa\,B−$ (black)

1 $aa\,bb$ (brown)

The breeding of domestic horses seems to have eliminated the *A* allele that determines the agouti phenotype, although certain wild relatives of the horse do have this allele. The color we have called *brown* in mice is called *chestnut* in horses, and this phenotype also is recessive to black.

The C Gene. The wild-type allele *C* permits color expression, and the allele *c* prevents color expression. The *cc* constitution is said to be **epistatic** to the other color genes. The word *epistatic* literally means "standing upon." In other words, in homozygous condition, the *c* allele "stands on," or cancels, the expression of other genes concerned with coat color. The *cc* animals, lacking coat pigment, are called *albinos*. Albinos are common in many mammalian species and have also been reported among birds, snakes, fish (Figure 4-11), and humans (Chapter 2). We have already learned about another allele of the *C* gene: the c^h (Himalayan) allele in rabbits determines that pigment will be deposited only at the body extremities. This same allele has been found in other mammals, including mice, in which the phenotype is also called *Himalayan*, and cats, in which the phenotype is

(a)

(b)

Figure 4-11 Albinism in reptiles and birds. In each case the phenotype is produced by a recessive allele that determines an inability to produce the dark pigment melanin in skin cells. (The normal allele determines ability to synthesize melanin.) (a) Rattlesnake. In this species the normal dark coloration is due entirely to melanin, so the albino allele results in a completely unpigmented appearance. (K. H. Switak/NHPA) (b) Penguin. In this species melanin normally makes dorsal feathers black, but the reddish-orange colors in the head feathers and beak are due to another pigment chemically unrelated to melanin. The recessive albino allele results in no melanin, but the reddish parts are unaffected and retain their normal coloration. (A.N.T./NHPA)

called *Siamese* (Figure 4-12). The allele c^h can be considered a heat-sensitive version of the *c* allele. It is only at the colder body extremities that c^h is functional and can make pigment. In warm parts of the body it behaves just like the albino allele *c*.

(a)

(b)

Figure 4-12 Temperature-sensitive alleles of the *C* gene result in similar phenotypes in several different mammals. These alleles result in very much reduced or no synthesis of the dark pigment melanin in the skin covering warmer parts of the body. At lower temperatures, such as those found at the body extremities, melanin is synthesized, producing darker snout, ears, tail, and feet. (a) Himalayan mouse. (Anthony Griffiths) (b) Siamese cat (Walter Chandoha). Furthermore, Himalayan rabbits (not shown) are often sold as pets. All three are of genotype $c^h c^h$.

Epistatic genes produce interesting modified ratios, as seen in the following cross (in which both parents are aa):

$$BB\,cc\ (\text{albino}) \times bb\,CC\ (\text{brown})$$

or $$BB\,CC\ (\text{black}) \times bb\,cc\ (\text{albino})$$

$$\downarrow$$

F₁   all $Bb\,Cc$ (black)

$$Bb\,Cc\ (\text{black}) \times Bb\,Cc\ (\text{black})$$

$$\downarrow$$

F₂   9 $B-C-$ (black)   9

3 $bb\,C-$ (brown)   3

3 $B-cc$ (albino) ⎫
                               ⎬ 4
1 $bb\,cc$ (albino) ⎭

A phenotypic ratio of 9:3:4 is observed. This ratio indicates a gene interaction of the type called *recessive epistasis*. Recessive epistasis is also well illustrated by the inheritance of the three familiar colors of Labrador retriever dogs. The three colors, black, chocolate, and gold (Figure 4-13), are produced genetically in the following way. The alleles *B* and *b* are equivalent to those in mice and result in black and brown (chocolate), respectively. At another gene, the homozygous constitution *ee* is epistatic to both the *B−* and *bb* alternatives, resulting in the golden color. Therefore, to be black or brown, a dog must have the *E* allele. Whether a golden dog is *B−* or *bb* can be deduced sometimes by observing the color of the nose and lips because the epistasis of *ee* is effective mainly in the coat, an example of tissue-specific gene expression.

It should be pointed out here that the term *epistasis* is often used in a different way (mainly in population genetics) to describe any kind of gene interaction, but we restrict its use to the specific meaning just described.

Message An epistatic allele of one gene eliminates expression of the alternative phenotypes of another gene, and inserts its own phenotype instead.

There is nothing mysterious about epistasis, despite the unusual F₂ ratios. Our example of the epistasis shown by mouse coat color makes perfect sense if we view the *C* allele as necessary for pigment synthesis, and the *B* and *b* alleles as the determinants of what color that pigment will be, with *B* determining black and *b* determining brown. If we then view *c* as an inactive form of *C*, the mechanism of the epistasis becomes clear, because if no pigment is made, then it doesn't really matter whether the determinant for black or for brown is present, because these determinants simply will not be used. The resulting lack of pigment is what we

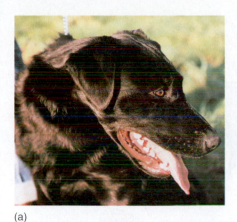
(a)

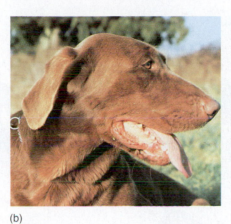
(b)

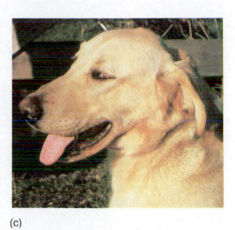
(c)

Figure 4-13 Coat color inheritance in Labrador retrievers. Two alleles *B* and *b*, of a pigment gene determine (a) black and (b) brown, respectively. At a separate gene, *E* allows color deposition in the coat, and *ee* prevents deposition, resulting in (c) the gold phenotype. This is a case of recessive epistasis. Thus the three homozygous genotypes are *BB EE* (a), *bb EE* (b), and *BB ee* or *bb ee* (c). The dog in (c) is most likely *BB ee*—the animal still has the ability to make black pigment, as witnessed by the black nose and lips, but not to deposit this pigment in the hairs. The progeny of a dihybrid cross would produce a 9 : 3 : 4 ratio of black : brown : golden. (Anthony Griffiths)

call *whiteness*. Pigments absorb and reflect specific wavelengths from the visible spectrum, whereas in the absence of pigment there is no specific absorbtion and all light is reflected. We can imagine a sequence of reactions something like the following:

$$\text{no pigment} \xrightarrow{C \text{ allele}} \text{pigment precursor} \nearrow^{B \text{ allele}} \text{black pigment} \searrow_{b \text{ allele}} \text{brown pigment}$$

In general every time one gene is higher, or "upstream," in the genetic chain of command, we would expect there to be an epistatic effect of its inactive allele on the genes lower in the hierarchy of command. Therefore, finding a case of epistasis (as revealed by a modified Mendelian ratio) can suggest hypotheses to the researcher about the sequence in which genes act. This principle can be useful in piecing together developmental pathways.

> **Message**   Epistatic alleles generally act before the genes they cancel, in some biochemical or developmental sequence.

The *D* Gene. The *D* gene controls the intensity of pigment specified by the other coat color genes. The genotypes *DD* and *Dd* permit full expression of color in mice, but *dd* "dilutes" the color, making it look "milky." The effect is due to an uneven distribution of pigment in the hair shaft. Dilute agouti, dilute cinnamon, dilute brown, and dilute black coats all are possible. A gene with such an effect is called a **modifier gene.** In the following cross, we assume that both parents are *aa CC*:

$$BB\,dd \text{ (dilute black)} \times bb\,DD \text{ (brown)}$$

or $$BB\,DD \text{ (black)} \times bb\,dd \text{ (dilute brown)}$$

$\downarrow$

F$_1$ all *Bb Dd* (black)

Bb Dd (black) × *Bb Dd* (black)

$\downarrow$

F$_2$ 9 *B–D–* (black)

3 *B–dd* (dilute black)

3 *bb D–* (brown)

1 *bb dd* (dilute brown)

In horses, the *D* allele shows incomplete dominance. Figure 4-14 shows how dilution affects the appearance of chestnut and bay horses.

The *S* Gene. The *S* gene controls the distribution of coat pigment throughout the body. In effect it controls the presence or absence of spots. The genotype *S–* results in no spots, and *ss* produces a spotting pattern called *piebald* in both mice and horses. This pattern can be superimposed on any of the coat colors discussed earlier—with the exception of albino, of course.

Let us summarize the above discussion of coat color in mice. The normal coat appearance in wild mice is produced by a complex set of interacting genes determining pigment type, pigment distribution in the individual hairs, pigment distribution on the animal's body, and the presence or absence of pigment. Such interactions are deduced from

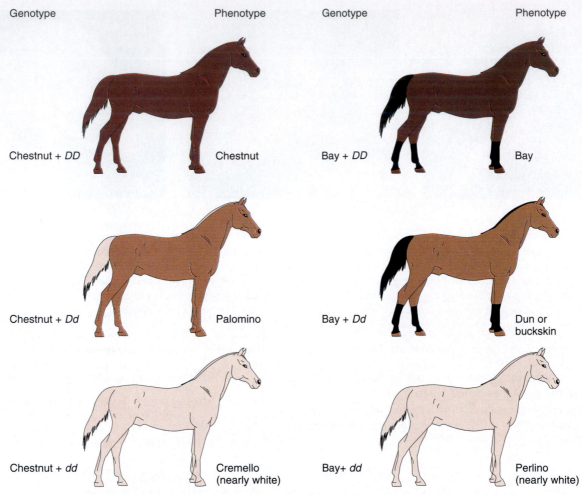

Genotype Phenotype Genotype Phenotype

Chestnut + *DD* Chestnut Bay + *DD* Bay

Chestnut + *Dd* Palomino Bay + *Dd* Dun or buckskin

Chestnut + *dd* Cremello (nearly white) Bay+ *dd* Perlino (nearly white)

Figure 4-14 The modifying effect of the dilution allele on basic chestnut and bay genotypes in horses. Note the incomplete dominance shown by *D*. (From J. W. Evans et al., *The Horse.* W. H. Freeman and Company, 1977.)

crosses in which two or more of the interacting genes are heterozygous for alleles that modify the normal coat color and pattern. Figure 4-15 illustrates some of the pigment patterns in mice. Interacting genes such as these determine most characters in any organism.

More Examples of Gene Interaction

Other organisms show us different types of modified Mendelian ratios and, more important, suggest more interesting mechanisms whereby genes cooperate to program the characteristic features of organisms.

A simple, yet striking, example is the inheritance of skin color in corn snakes (*Elaphe guttata*). The natural color is a black-and-orange, repeating, camouflage pattern, as shown in Figure 4-16a. The phenotype is produced by two

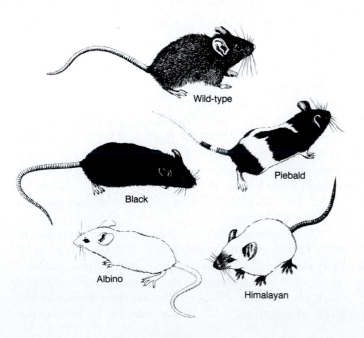

Wild-type

Piebald

Black

Albino

Himalayan

Figure 4-15 Some coat phenotypes in mice.

separate pigments, both of which are under genetic control. One gene determines the orange pigment, and the alleles are O (presence of orange pigment) and o (absence of orange pigment). Another gene determines the black pigment, with alleles B (presence of black pigment) and b (absence of black pigment). The natural pattern is produced by the genotype $O-B-$; a snake that is $oo\,B-$ is black, and a snake that is $O-bb$ is orange (Figure 4-16c). The double homozygous recessive $oo\,bb$ is albino, as shown in the figure. Notice that the faint pink color of the albino is from yet another pigment, the hemoglobin of the blood that is visible through this snake's skin. The albino shows clearly that there is another element to the skin phenotype in addition to pigment, and this is the repeating motif in and around which pigment is deposited.

Since there are two genes in this system, we obtain a typical dihybrid inheritance pattern, and the four unique phenotypes form a $9:3:3:1$ ratio in the F_2. This is not really a modified Mendelian ratio, but it does show clearly the way in which the overall skin appearance is determined by interacting genes. A typical pedigree might be as follows:

$$OO\,bb \text{ (orange)} \times oo\,BB \text{ (black)}$$

$$F_1 \qquad\qquad Oo\,Bb \text{ (natural)}$$

$$Oo\,Bb \text{ (natural)} \times Oo\,Bb \text{ (natural)}$$

$$F_2 \qquad 9\ O-B- \text{ (natural)}$$
$$3\ O-bb \text{ (orange)}$$
$$3\ oo\,B- \text{ (black)}$$
$$1\ oo\,bb \text{ (albino)}$$

A good example of a modified Mendelian ratio that points to an important kind of gene interaction comes from a species of pea that normally has purple petals. Two different white-petaled pure lines of peas are obtained, which we will call lines 1 and 2. When each line is crossed to the purple wild type, the F_1 is purple and the F_2 shows a $3:1$ ratio of purple to white. However, when the two white lines are intercrossed, all the F_1 progeny have purple flowers, and the F_2 shows both purple and white plants in a ratio of $9:7$. How can these results be explained? By now, you should immediately suspect that the $9:7$ ratio is a modification of the Mendelian $9:3:3:1$ ratio. The ratio can be explained by postulating that there are two different genes with similar effects on petal color. Let's represent the alleles by W_1 and w_1 for one gene, and W_2 and w_2 for the other. The cross of the two white lines and subsequent generations can be represented as follows:

$$\text{white line 1} \times \text{white line 2}$$
$$w_1 w_1 W_2 W_2 \qquad W_1 W_1 w_2 w_2$$

$$F_1 \qquad\qquad W_1 w_1 W_2 w_2 \text{ (purple)}$$

$$W_1 w_1 W_2 w_2 \text{ (purple)} \times W_1 w_1 W_2 w_2 \text{ (purple)}$$

$$F_2 \qquad 9\ W_1 - W_2 - \text{ (purple)} \qquad 9$$
$$3\ W_1 - w_2 w_2 \text{ (white)}$$
$$3\ w_1 w_1 W_2 - \text{ (white)} \quad \Big\}\ 7$$
$$1\ w_1 w_1 w_2 w_2 \text{ (white)}$$

The results show that homozygosity for the recessive allele of *either* gene causes a plant to have white petals. To have the purple phenotype, a plant must have at least one dominant allele of both genes. This example illustrates the important genetic process of **complementary gene action,** or **complementation**—in which the two dominant alleles of different genes are cooperating to produce a specific phenotype, in this example, purple petals.

Message Complementation is the production of a wild-type phenotype when two genotypes determining recessive phenotypes come together in the same cell.

The two white lines of peas differ, even though they show exactly the same phenotype. Geneticists often have to deal with independently isolated lines that have the same phenotype. Generally such lines are abnormal recessive phenotypes derived from wild type. The white-petaled peas are just such a case; most species of peas in nature have colored petals, but albino phenotypes arise spontaneously, and are nearly always recessive. When there are two or more such lines, the first obvious question that arises is "Are these different variants determined by recessive alleles of the same gene or not?" The complementation test provides an effective operational test of allelism. If the two recessive phenotypes are intercrossed and a wild-type phenotype is observed in the F_1, the parental genotypes have obviously complemented each other; the recessive alleles must be of different genes. On the other hand, what if we were to cross two independently obtained, recessive, albino lines, and the F_1 and the F_2 were all albino? In this situation the evidence is consistent with the recessive determinants' being alleles of the same gene.

Message When independently derived genotypes producing similar recessive phenotypes fail to complement, the genetic determinants of the phenotypes are alleles of the same gene.

(a)

(b)

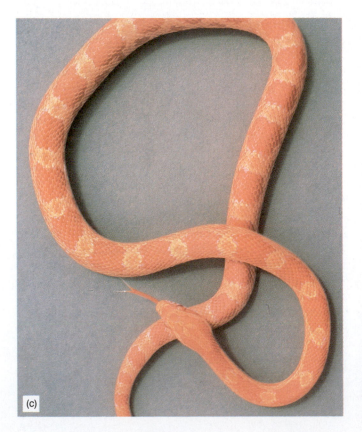

(c)

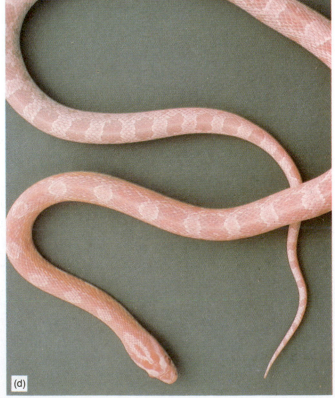

(d)

Figure 4-16 (*facing page*) Analysis of the genes for skin pigment in the corn snake. The wild type (a) has a skin pigmentation pattern made up of a black and an orange pigment. The gene *O* determines an enzyme in the synthetic pathway for orange pigment: when this enzyme is deficient (*oo*), no orange pigment is made and the snake is black (b). Another gene, *B*, determines an enzyme for black pigment: when this enzyme is deficient (*bb*), the snake is orange (c). When both enzymes are deficient, the snake is albino (d). Hence the four homozygous genotypes are *O O B B* (a), *o o B B* (b), *O O b b* (c), and *o o b b* (d). A cross of (a) × (d) or (b) × (c) would give a dihybrid wild-type F₁ and a 9:3:3:1 ratio of the four phenotypes in the F₂. (Anthony Griffiths)

Now we have two tests for allelism: first, the monohybrid 3:1 ratio reveals a dominant and a recessive allele of the same gene, and second, lack of complementation in pairwise crosses of independently isolated lines reveals two recessive alleles of the same gene.

Genetic variants of foxgloves (*Digitalis purpurea*) provide excellent examples of gene interaction in the determination of the overall appearance of an organism. There are three important genes that interact to determine petal coloration. Figure 4-17 shows the phenotypes that we use in this discussion. The first gene determines the ability of the plant to synthesize reddish pigment (a type of anthocyanin).

The *M* allele of this gene stands for ability to synthesize anthocyanin; *m* stands for inability to synthesize anthocyanin, resulting in white petals. The second gene is a modifier gene. One allele, *D*, determines the synthesis of large amounts of anthocyanin (dark reddish), and *d* stands for low amounts (light reddish). Possibly the *D* and *d* alleles regulate the synthesis of pigment by *M*. The third gene affects pigment deposition. The allele *W* prevents pigment deposition in all parts of the petal except in the throat spots, whereas the recessive allele *w* allows deposition of pigment all over the petal. Thus these three genes control the ability to synthesize, the amount synthesized, and the ability for the pigment to be deposited in specific petal cells. We could consider a variety of dihybrid crosses and even a trihybrid cross, but we will consider only one sample cross, a dihybrid cross that illustrates a modified F₂ ratio not yet discussed.

Consider the cross between the two genotypes *M M D D w w* and *M M d d W W*. The phenotype of the first is dark reddish because it has the *D* modifier and the ability to deposit pigment. The second is white with reddish spots because although the plant has the ability to synthesize pigment (conferred by the allele *M*), the *W* allele prevents deposition except in the throat spots. Let us consider the usual type of pedigree but eliminate the *M* allele because it will be homozygous in all individuals.

Figure 4-17 Pigment phenotypes in foxgloves, determined by three separate genes. *M* codes for an enzyme that synthesizes anthocyanin, the reddish pigment seen in these petals; *mm* produces no pigment and produces the phenotype albino with yellowish spots. *D* is an enhancer of anthocyanin, resulting in a darker pigment; *dd* does not enhance. At the third locus, *ww* allows pigment deposition in petals, but *W* prevents pigment deposition except in the spots, and so results in the white, spotted phenotype. Genotypes (and phenotypes) from left to right are *M– W – – –* (white with reddish spots), *mm – – – –* (white with yellowish spots), *M– w w d d* (light reddish), and *M– w w D–* (dark reddish). (Anthony Griffiths)

$(MM)\,DD\,ww \times (MM)\,dd\,WW$

(dark reddish)   (white with reddish spots)

F$_1$ $Dd\,Ww$ (white with reddish spots)

$Dd\,Ww \times Dd\,Ww$

(white with reddish spots)   (white with reddish spots)

F$_2$ 9 $D-W-$ (white with reddish spots)
          3 $dd\,W-$ (white with reddish spots) } 12

          3 $D-ww$ (dark reddish)   3

          1 $dd\,ww$ (light reddish)   1

Overall, a 12:3:1 phenotypic ratio is produced. This kind of interaction is called **dominant epistasis** because, as can be seen from the F$_2$ results, the dominant allele W eliminates the two alternatives expressed by D and d, dark and light reddish, and replaces them with another phenotype, white with reddish spots.

Another important type of gene interaction is **suppression.** A suppressor is an allele that eliminates the phenotypic expression of an allele of a completely separate gene. The term suppression is most often used in connection with reversing the effect of an abnormal allele (a mutation; see Chapter 7), resulting in the normal (wild-type) phenotype. For example, assume that an allele A produces the normal phenotype, whereas a recessive allele a results in abnormality. A recessive suppressor allele s at another gene causes a to be expressed as A, so the $aa\,ss$ genotype will be normal (A-like) phenotype. Suppressor alleles often have no effect on the other allele of the target gene, so in this example the $AA\,ss$ phenotype would be normal.

Suppressors also result in modified Mendelian ratios. Let's look at a real example from *Drosophila*, using a recessive suppressor su of the recessive purple eye color allele pd. A homozygous purple-eyed fly is crossed to a homozygous stock carrying the suppressor.

$pd\,pd\,Su\,Su$ (purple) $\times$ $Pd\,Pd\,su\,su$ (red)

F$_1$ all $Pd\,pd\,Su\,su$ (red)

$Pd\,pd\,Su\,su$ (red) $\times$ $Pd\,pd\,Su\,su$ (red)

F$_2$ 9 $Pd-Su-$ red
         3 $Pd-su\,su$ red } 13 red
         1 $pd\,pd\,su\,su$ red
         3 $pd\,pd\,Su-$ purple   3 purple

The overall ratio in the F$_2$ is 13 red:3 purple.

Suppression is sometimes confused with epistasis. However, note that because a suppressor cancels the expression of an abnormal allele and restores the corresponding normal phenotype, the modified Mendelian F$_2$ ratio can involve only two phenotypes (normal and abnormal), whereas in the case of epistasis the epistatic allele introduces a third phenotype into the ratio.

Suppressor ratios can be confusing because some are identical to those produced by other gene interactions. Suppressors can be recessive or dominant, and their corresponding target alleles also can be recessive or dominant, giving a total of four different kinds of interactions. Writing the suppressor first we can represent these as $R \rightarrow R$, $D \rightarrow D$, $R \rightarrow D$, and $D \rightarrow R$. Of these the first (discussed in detail above) and second both result in the same F$_2$ ratio, 13:3. The third gives a 9:7 ratio in the F$_2$. This ratio seems the same as that discussed earlier under complementation. However, in suppression the ratio is 9 mutant to 7 wild type, whereas in complementation the ratio is 9 wild type to 7 mutant. The last type results in the same ratio as discussed in the next paragraph on duplicate genes, and indeed $D \rightarrow R$ suppression is just another way of looking at the duplicate gene effect.

Message Suppressors cancel the expression of an abnormal (mutant) allele of another gene, resulting in normal phenotype.

In another form of gene interaction, genes may be represented more than once in the genome. The example concerns the genes that control fruit shape in the plant called *shepherd's purse, Capsella bursa-pastoris*. Two different lines have fruits of different shapes: one is "heart-shaped"; the other, "narrow." Are these two phenotypes determined by two alleles of a single gene? A cross between the two lines produces an F$_1$ with heart-shaped fruit; this result is consistent with the hypothesis of determination by a pair of alleles. However, the F$_2$ shows a 15:1 ratio of heart-shaped to narrow, and this ratio suggests a modification of the dihybrid 9:3:3:1 Mendelian ratio. The genetic control of fruit shape can be explained in terms of **duplicate genes** (Figure 4-18). Apparently, heart-shaped fruits result from the presence of at least one dominant allele of *either* gene. The two genes appear to be identical in function. (Contrast this 15:1 ratio with the 9:7 ratio obtained from complementary genes, where *both* dominant genes are necessary to produce a specific phenotype.)

Message Duplicate genes provide alternative genetic determination of a specific phenotype.

Table 4-3 summarizes the various types of gene interactions that modify Mendelian ratios. Note that so far we have considered modified ratios of diploid organisms only. Gene interactions also can be detected in haploid organisms. We have already discussed a gene in the fungus *Neurospora* having an allele *al* that causes the asexual spores to be albino rather than the orange color produced by the al^+ wild-type allele of the same gene (see Figure 3-37). A cross $al \times al^+$ gives $\frac{1}{2}$ al (albino) and $\frac{1}{2}$ al^+ (orange) progeny, conforming to Mendel's first law. At another gene, an allele ylo, gives yellow asexual spores (Figure 3-37), whereas its wild-type allele ylo^+ again stands for orange. The cross $ylo \times ylo^+$ gives $\frac{1}{2}$ ylo (yellow) and $\frac{1}{2}$ ylo^+ (orange) progeny. When an al culture is crossed with a ylo culture, the resulting haploid progeny are $\frac{1}{4}$ yellow, $\frac{1}{4}$ orange, and $\frac{1}{2}$ albino. How can we explain this result? The answer is a kind of epistasis: al and ylo are defective alleles of separate genes whose wild-type alleles are necessary for the normal production of orange pigment. Writing out the genotypes in the cross will demonstrate how the epistasis works:

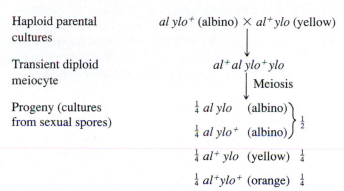

The data clearly suggest that al^+ and ylo^+ are both needed for production of orange pigment and that either al or ylo blocks this production. The epistatic mechanism is revealed by the albino phenotype shown by the $al\ ylo$ genotype, because it suggests another case of an "upstream effect," which we saw in the examples of the mouse allele c and the Labrador retriever allele e. We can postulate from our genetic analysis that al^+ is needed for synthesis of pigment and that the alleles of the other gene determine whether that pigment is orange (ylo^+) or yellow (ylo). We shall see in later chapters that chemical studies generally confirm these ideas that come from genetic analysis.

The various colors of peppers that we see in markets (Figure 4-19) are determined by genes that show a combination of several of the types of interactions we have dis-

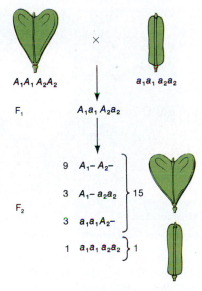

Figure 4-18 Inheritance pattern of duplicate genes controlling fruit shape in shepherd's purse. Either A_1 or A_2 can cause a heart-shaped fruit.

cussed above. The colors shown in the figure are just a few examples from a wider range that is available. The genetic system is four interacting genes. One gene regulates the removal of the green pigment chlorophyll in the fruit. All the types shown in the figure start out green, but a dominant allele Y promotes removal of the chlorophyll, whereas y does not and chlorophyll persists. As the pepper matures, carotenoid pigments are synthesized. One allelic pair determines the type of carotenoid: R determines red, and r yellow. Two recessive alleles of two different genes, c_1 and c_2, have an action similar to each other in that they reduce (down-regulate) the amount of carotenoid synthesized, giving lighter shades. For example, in an $R-$ genotype, c_2 reduces the amount of red pigment, yielding a bright orange appearance, which in this case is not a separate pigment but a reduced level of redness. As another example, in an rr genotype, allele c_2 acts on the yellow pigment to make it a lighter shade called *lemon yellow* (not shown in the figure).

Table 4-3   Some Modified Phenotypic Ratios Produced by Gene Interaction

Type of gene interaction	9 $A-B-$	3 $A-bb$	3 $aa\,B-$	1 $aa\,bb$	Phenotypic ratio
None (four distinct phenotypes)	9	3	3	1	9:3:3:1
Complementary gene action	9	7			9:7
Recessive suppression by aa acting on bb	9	3	4		13:3
Recessive epistasis of aa acting on B and b alleles	9	3	4		9:3:4
Dominant epistasis of A acting on B and b alleles	12		3	1	12:3:1
Duplicate genes	15			1	15:1

Figure 4-19 The colors of peppers are determined by gene interaction, as described in the text. The colors are green, red, yellow, orange, brown, and "white" (actually, pale yellowish green). (Anthony Griffiths)

The two c genes act somewhat like duplicates in that they have comparable effects. Also you can see that the c genes act in a similar manner to the coat color dilution genes in mammals. However, it is also interesting that the effects of c_1 and c_2 are cumulative in that the $c_1 c_1 c_2 c_2$ genotype results in a much more reduced level of carotene than is seen through the action of either allele by itself. The colors shown in Figure 4-19 will serve as examples of some of the 16 possible genotypic combinations. Stated as homozygous genotypes, the green pepper is $rr\,c_1 c_1\, c_2 c_2\, yy$ (chlorophyll, but effectively no carotenoid), the red pepper is $RR\,C_1 C_1\, C_2 C_2\, YY$, the brown is $RR\,C_1 C_1\, C_2 C_2\, yy$ (a combination of green and red), yellow is $rr\,C_1 C_1\, C_2 C_2\, YY$, orange is $RR\,C_1 C_1\, c_2 c_2\, YY$, and the "white" pepper is $rr\,c_1 c_1\, c_2 c_2\, YY$.

The foregoing discussion shows that a gene does *not* determine a phenotype by acting alone; it does so only in conjunction with other genes and with the environment. Although geneticists do routinely ascribe a particular phenotype to a particular allele, we must remember that this is merely a convenient kind of shorthand designed to facilitate genetic analysis. This shorthand arises from the ability of geneticists to isolate individual components of a biological process and to study them separately as part of genetic dissection. Although this logical isolation is an essential aspect of genetics, the message of this chapter is that genes act in concert to produce the overall features of an organism.

Message Different kinds of modified Mendelian dihybrid ratios point to different ways in which genes can interact with each other to determine phenotype.

Penetrance and Expressivity

In the preceding examples, the genetic basis of the dependence of one gene on another has been worked out from clear genetic ratios. In other situations, where the phenotype ascribed to a gene is known to be dependent on other factors but the precise inheritance of those factors has not been established, the terms *penetrance* and *expressivity* may be useful in describing the situation.

We have already encountered penetrance in the discussion of lethal alleles. **Penetrance** is defined as the percentage of individuals with a given genotype who exhibit the phenotype associated with that genotype. For example, an organism may have a particular genotype but may not express the corresponding phenotype because of modifiers, epistatic genes, or suppressors in the rest of the genome or because of a modifying effect of the environment. Penetrance can be used to measure such an effect when it is not known which of these types of modification underlies the effect.

On the other hand, **expressivity** measures the extent to which a given genotype is expressed in an individual at the phenotypic level. Different degrees of expression in different individuals may be due to heterogeneity of the allelic constitution of the rest of the genome, or to environmental factors. Figure 4-20 illustrates the distinction between penetrance and expressivity. Obviously, both penetrance and expressivity are integral to the concept of the norm of reaction, which was discussed in Chapter 1.

Human pedigree analysis and predictions in genetic counseling can often be thwarted by the phenomena of incomplete penetrance and variable expressivity. For example, if a disease-causing allele is not fully penetrant (as often is

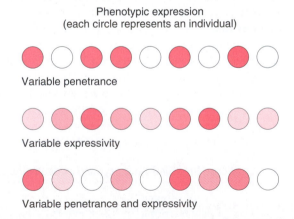

Phenotypic expression
(each circle represents an individual)

Variable penetrance

Variable expressivity

Variable penetrance and expressivity

Figure 4-20 Diagram representing the effects of penetrance and expressivity through a hypothetical character "pigment intensity." In each row, all individuals have the same allele, say *P*, giving them the same "potential to produce pigment." However, effects deriving from the rest of the genome and from the environment may suppress or modify pigment production in an individual.

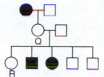

Figure 4-21 Lack of penetrance illustrated by a pedigree for a dominant allele. Individual Q must have the allele (because it was passed on to her progeny), but it was not expressed in her phenotype. An individual such as R cannot be sure that her genotype lacks the allele.

the case), it is difficult to give a clean genetic bill of health to any individual in a disease pedigree (for example, individual R in Figure 4-21). On the other hand, pedigree analysis can sometimes identify individuals who do not express but almost certainly do have a disease genotype (for example, individual Q in Figure 4-21). Similarly, variable expressivity can confound diagnosis. A specific example of variable expressivity is found in Figure 4-22 and also in Figure 2-19 and Problem 11 of Chapter 2.

> **Message** The impact of a gene at the phenotypic level depends not only on its dominance but also on the modifying effects of the rest of the genome and of the environment.

In conclusion, notice that the extensions of Mendelian analysis discussed in this chapter are mainly based on the complexities of gene expression, not on complexities of inheritance. Mendel's principles concerned the inheritance patterns of genes and had little to say about gene expression. When we add the patterns of gene expression to the patterns of gene inheritance, we begin to see the fabric of the modern science of genetics.

The extensions of Mendelian analysis in this chapter and following chapters are not the only ones encountered in routine genetic analysis; they are simply among the most common. The key recognition features for each extension—whether, for example, multiple allelism, incomplete dominance, epistasis, or variable expressivity—occur at the experimental level. The analyst must be constantly on the lookout for these and any other results that might indicate the uniqueness of any given situation. Such signals often lead to the discovery of new phenomena and to the opening up of new research areas.

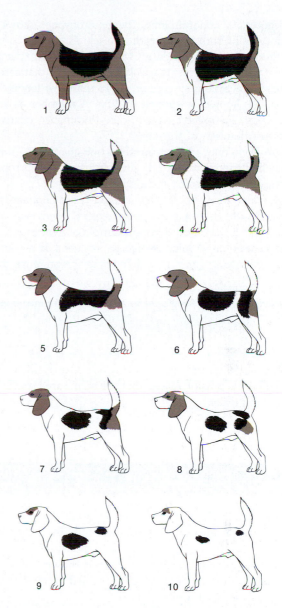

Figure 4-22 Variable expressivity shown by 10 grades of piebald spotting in beagles. Each of these dogs has S^P, the allele responsible for piebald spots in dogs. (Adapted from Clarence C. Little, *The Inheritance of Coat Color in Dogs*, Cornell University Press, 1957, and from Giorgio Schreiber, *Journal of Heredity* 9, 1930, 403.)

SUMMARY

Although Mendel's laws apply to all eukaryotic organisms, these laws are only a starting point for understanding heredity. Genes and chromosomes present many complexities in addition to those unraveled by Mendel.

Mendel observed full dominance in his experiments; we have added examples of incomplete dominance and codominance. In incomplete dominance, the phenotype of a heterozygote is anywhere between the phenotypes of the homozygotes. In codominance, the heterozygote shows the phenotypes of both homozygotes.

From his experiments, Mendel reported particles (genes) with two forms. It was later discovered that across a population a gene may contain more than two forms. This situation is known as *multiple allelism*. A member of an allelic series may exhibit any type of dominance relative to the other members of the series.

An allele may affect more than one character—that is, it may be pleiotropic. The A^Y allele in mice is pleiotropic in affecting both coat color and viability.

A major gene affecting a character is not necessarily the only gene affecting that character; several genes may interact to affect a character. A good example of gene interaction is the coat color of mice, which is produced by a complex set of interacting genes that determine the presence or absence of pigment, pigment type, pigment distribution in the hair, and pigment distribution on the animal.

Gene interaction often produces modified Mendelian ratios in the F_2. Some kinds of interaction have specific names, such as complementary gene action, epistasis, suppression, and duplicate gene action. Genes interact in both diploid and haploid organisms.

Two other important extensions of Mendelian analysis are the concepts of penetrance and expressivity. Penetrance is the percentage of individuals of a specific genotype who express the phenotype associated with that genotype. Expressivity refers to the degree to which a particular genotype is expressed as a phenotype within an individual. Incomplete penetrance and variable expressivity are caused by genetic and environmental variation.

Concept Map

Draw a concept map interrelating as many of the following terms as possible. Note that the terms are listed in no particular order.

dihybrid / environment / modified Mendelian ratio / gene interaction / self-fertilization / 9:3:4 ratio / penetrance / recessive epistasis

CHAPTER INTEGRATION PROBLEM

Most pedigrees show polydactyly (Figure 2-18) to be inherited as a rare autosomal dominant, but the pedigrees of some families do not fully conform to the patterns expected for such inheritance. Such a pedigree is shown below. (The unshaded diamonds stand for the specified number of unaffected individuals of unknown sex.)

a. What irregularity does this pedigree show?

b. What genetic phenomenon does this pedigree illustrate?

c. Suggest a specific gene interaction mechanism that could produce such a pedigree, showing genotypes of pertinent family members.

Solution

a. The normal expectation for an autosomal dominant is for each affected individual to have an affected parent, but this is not seen in this pedigree, and this constitutes the irregularity. What are some possible explanations?

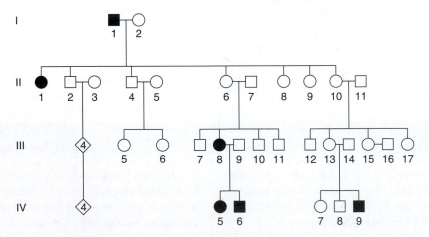

Could some cases of polydactyly be caused by a different gene, one that is an X-linked dominant? This is not a useful suggestion, because we would still have to explain the absence of the condition in individuals II-6 and II-10. Furthermore, postulating recessive inheritance, whether autosomal or sex-linked, requires many people in the pedigree to be heterozygotes, which is inappropriate because polydactyly is a rare condition.

b. This leaves us with the conclusion that polydactyly sometimes must be incompletely penetrant. We have learned in this chapter that some individuals who have the genotype for a particular phenotype do not express it. In this pedigree II-6 and II-10 seem to belong in this category; they must carry the polydactyly gene inherited from I-1 because they transmit it to their progeny.

c. We have seen in the chapter that environmental suppression of gene expression can cause incomplete penetrance, as can suppression by another gene. To give the requested genetic explanation, we must come up with a genetic hypothesis. What do we need to explain? The key is that I-1 passes the gene on to two types of progeny, represented by II-1, who expresses the gene, and by II-6 and II-10, who do not. (From the pedigree we cannot tell if the other children of I-1 have the gene or not.) Is genetic suppression at work? I-1 does not have a suppressor allele because he expresses polydactyly. So the only person a suppressor could come from is I-2. Furthermore, I-2 must be heterozygous for the suppressor gene because at least one of her children does express polydactyly. We have thus formulated the hypothesis that the mating in generation I must have been

$$(I\text{-}1)\,Pp\,ss \times (I\text{-}2)\,pp\,Ss$$

where S is the suppressor, and P is the allele responsible for polydactyly. From this hypothesis, we predict that the progeny will comprise the following four types if the genes assort independently:

Genotype	Phenotype	Example
$Pp\,Ss$	normal (suppressed)	II-6, II-10
$Pp\,ss$	polydactylous	II-1
$pp\,Ss$	normal	
$pp\,ss$	normal	

If S is rare, the matings and progenies of II-6 and II-10 are probably giving

Progeny genotype	Example
$Pp\,Ss$	III-13
$Pp\,ss$	III-8
$pp\,Ss$	
$pp\,ss$	

We cannot rule out the possibilities that II-2 and II-4 have the genotype $Pp\,Ss$ and that by chance none of their descendants are affected.

Note that we have just used concepts from Chapter 1 (environmental effects), Chapter 2 (Mendelian inheritance), Chapter 3 (chromosomal locations of genes), and Chapter 4 (gene interactions).

SOLVED PROBLEMS

1. Beetles of a certain species may have green, blue, or turquoise wing covers. Virgin beetles were selected from a polymorphic laboratory population and mated to determine the inheritance of wing-cover color. The crosses and results were as follows:

Cross	Parents	Progeny
1	blue $\times$ green	all blue
2	blue $\times$ blue	$\frac{3}{4}$ blue : $\frac{1}{4}$ turquoise
3	green $\times$ green	$\frac{3}{4}$ green : $\frac{1}{4}$ turquoise
4	blue $\times$ turquoise	$\frac{1}{2}$ blue : $\frac{1}{2}$ turquoise
5	blue $\times$ blue	$\frac{3}{4}$ blue : $\frac{1}{4}$ green
6	blue $\times$ green	$\frac{1}{2}$ blue : $\frac{1}{2}$ green
7	blue $\times$ green	$\frac{1}{2}$ blue : $\frac{1}{4}$ green : $\frac{1}{4}$ turquoise
8	turquoise $\times$ turquoise	all turquoise

a. Deduce the genetic basis of wing-cover color in this species.

b. Write the genotypes of all parents and progeny as completely as possible.

Solution

a. These data seem complex at first, but the inheritance pattern becomes clear if we consider the crosses one at a time. A general principle of solving such problems, as we have seen, is to begin by looking over all the crosses and by grouping the data to bring out the patterns.

One clue that emerges from an overview of the data is that all the ratios are one-gene ratios: there is no evidence of two separate genes being involved at all. How can such variation be explained with a single gene? The obvious answer is that there is variation for the single gene itself—that is, multiple allelism. Perhaps there are three alleles of one gene; let's call the gene *w* (for wing-cover color) and represent the alleles as w^g, w^b, and w^t. Now we have an additional problem, which is to determine the dominance of these alleles.

Cross (1) tells us something about dominance because the progeny of a blue × green cross are all blue; hence, blue appears to be dominant to green. This conclusion is supported by cross (5), because the green determinant must have been present in the parental stock to appear in the progeny. Cross (3) informs us about the turquoise determinants, which must have been present, although unexpressed, in the parental stock because there are turquoise wing covers in the progeny. So green must be dominant to turquoise. Hence, we have formed a model in which the dominance is $w^b > w^g > w^t$. Indeed, the inferred position of the w^t allele at the bottom of the dominance series is supported by the results of cross (7), where turquoise shows up in the progeny of a blue × green cross.

b. Now it is just a matter of deducing the specific genotypes. Notice that the question states that the parents were taken from a polymorphic population; this means that they could be either homozygous or heterozygous. A parent with blue wing covers, for example, might be homozygous ($w^b w^b$) or heterozygous ($w^b w^g$ or $w^b w^t$). Here, a little trial and error and common sense is called for, but by this stage, the question has essentially been answered and all that remains is to "cross the t's and dot the i's." The following genotypes explain the results. A dash indicates that the genotype may be either homozygous *or* heterozygous in having a second allele further down the allelic series.

Cross	Parents	Progeny
1	$w^b w^b \times w^g -$	$w^b w^g$ or $w^b -$
2	$w^b w^t \times w^b w^t$	$\frac{3}{4} w^b - : \frac{1}{4} w^t w^t$
3	$w^g w^t \times w^g w^t$	$\frac{3}{4} w^g - : \frac{1}{4} w^t w^t$
4	$w^b w^t \times w^t w^t$	$\frac{1}{2} w^b w^t : \frac{1}{2} w^t w^t$
5	$w^b w^g \times w^b w^g$	$\frac{3}{4} w^b - : \frac{1}{4} w^g w^g$
6	$w^b w^g \times w^g w^g$	$\frac{1}{2} w^b w^g : \frac{1}{2} w^g w^g$
7	$w^b w^t \times w^g w^t$	$\frac{1}{2} w^b - : \frac{1}{4} w^g w^t : \frac{1}{4} w^t w^t$
8	$w^t w^t \times w^t w^t$	all $w^t w^t$

2. The leaves of pineapples can be classified into three types: spiny (S), spine tip (ST), and piping (nonspiny) (P). In crosses between pure strains followed by intercrosses of the F_1, the results shown below appeared.

		Phenotypes	
Cross	Parental	F_1	F_2
1	ST × S	ST	99 ST : 34 S
2	P × ST	P	120 P : 39 ST
3	P × S	P	95 P : 25 ST : 8 S

a. Assign gene symbols. Explain these results in terms of the genotypes produced and their ratios.

b. Using the model from part a, give the phenotypic ratios you would expect if you crossed (1) the F_1 progeny from piping × spiny with the spiny parental stock, and (2) the F_1 progeny of piping × spiny with the F_1 progeny of spiny × spiny tip.

Solution

a. First, let's look at the F_2 ratios. We have clear 3:1 ratios in crosses 1 and 2, indicating single-gene segregations. Cross 3, however, shows a ratio that is almost certainly a 12:3:1 ratio. How do we know this? Well, there are simply not that many complex ratios in genetics, and trial and error brings us to the 12:3:1 quite quickly. In the 128 progeny total, the numbers of 96:24:8 are expected, but the actual numbers fit these expectations remarkably well.

One of the principles of this chapter is that modified Mendelian ratios reveal gene interactions. Cross 3 gives F_2 numbers appropriate for a modified dihybrid Mendelian ratio, so it looks as if we are dealing with a two-gene interaction. This seems the most promising place to start; we can go back to crosses 1 and 2 and try to fit them in later.

Any dihybrid ratio is based on the phenotypic proportions 9:3:3:1. Our observed modification groups them as follows:

$$\left. \begin{array}{l} 9\,A-B- \\ 3\,A-bb \end{array} \right\} \quad \text{12 piping}$$

$$3\,aa\,B- \qquad \text{3 spiny tip}$$

$$1\,aa\,bb \qquad \text{1 spiny}$$

So without worrying about the name of the type of gene interaction (we are not asked to supply this anyway), we can already define our three pineapple leaf phenotypes in terms of the proposed allelic pairs *A, a* and *B, b*:

$$\text{piping} = A - (B, b \text{ irrelevant})$$

$$\text{spiny-tip} = aa\,B-$$

$$\text{spiny} = aa\,bb$$

What about the parents of cross 3? The spiny parent must be *aa bb*, and because the *B* gene is needed to produce F_2 spiny-tip individuals, the piping parent must be *AABB*. (Note that we are *told* that all parents are pure, or homozygous.) The F_1 must therefore be *Aa Bb*.

Without further thought, we can write out cross 1 as follows:

$$aa\,BB \times aa\,bb \longrightarrow aa\,Bb \left< \begin{array}{l} \nearrow \frac{3}{4}\,aa\,Bb \\ \searrow \frac{1}{4}\,aa\,bb \end{array} \right.$$

Cross 2 can also be written out partially without further thought, using our arbitrary gene symbols:

$$AA\,-- \times aa\,BB \longrightarrow Aa\,B- \begin{cases} \nearrow \frac{3}{4}\,A--- \\ \searrow \frac{1}{4}\,aa\,B- \end{cases}$$

We know that the F_2 of cross 2 shows single-gene segregation, and it seems certain now that the A, a allelic pair is involved. But the B allele is needed to produce the spiny tip phenotype, so all individuals must be homozygous BB:

$$AABB \times aa\,BB \longrightarrow Aa\,BB \begin{cases} \nearrow \frac{3}{4}\,A-BB \\ \searrow \frac{1}{4}\,aa\,BB \end{cases}$$

Notice that the two single-gene segregations in crosses 1 and 2 do not show that the genes are *not* interacting. What is shown is that the two-gene interaction is not *revealed* by these crosses—only by 3, in which the F_1 is heterozygous for both genes.

b. Now it is simply a matter of using Mendel's laws to predict cross outcomes:

(1) $Aa\,Bb \times aa\,bb \longrightarrow \frac{1}{4}\,Aa\,Bb$

(independent assortment in a standard testcross)
$$\frac{1}{4}\,Aa\,bb$$
$$\frac{1}{4}\,aa\,Bb \quad \text{spiny tip}$$
$$\frac{1}{4}\,aa\,bb \quad \text{spiny}$$

(2) $Aa\,Bb \times aa\,Bb \longrightarrow$

$$\frac{1}{2}\,Aa \begin{cases} \nearrow \frac{3}{4}\,B- \longrightarrow \frac{3}{8} \\ \searrow \frac{1}{4}\,bb \longrightarrow \frac{1}{8} \end{cases} \Big\} \frac{1}{2}\,\text{piping}$$

$$\frac{1}{2}\,aa \begin{cases} \nearrow \frac{3}{4}\,B- \longrightarrow \frac{3}{8} \quad \text{spiny tip} \\ \searrow \frac{1}{4}\,bb \longrightarrow \frac{1}{8} \quad \text{spiny} \end{cases}$$

PROBLEMS

1. If a man of blood group AB marries a woman of blood group A whose father was of blood group O, what different blood groups can this man and woman expect their children to belong to?

2. Erminette fowls have mostly light-colored feathers with an occasional black one, giving a flecked appearance. A cross of two erminettes produced a total of 48 progeny, consisting of 22 erminettes, 14 blacks, and 12 pure whites. What genetic basis of the erminette pattern is suggested? How would you test your hypotheses?

3. Radishes may be long, round, or oval and they may be red, white, or purple. You cross a long, white variety with a round, red one and obtain an oval, purple F_1. The F_2 show nine phenotypic classes as follows: 9 long, red; 15 long, purple; 19 oval, red; 32 oval, purple; 8 long, white; 16 round, purple; 8 round, white; 16 oval, white; and 9 round, red.

a. Provide a genetic explanation of these results. Be sure to define the genotypes, and show the constitution of parents, F_1, and F_2.

b. Predict the genotypic and phenotypic proportions in the progeny of a cross between a long, purple radish and an oval, purple one.

4. In the multiple allele series that determines coat color in rabbits, $C^+ > C^{ch} > C^h$, dominance is from left to right as shown. In a cross of $C^+\,C^{ch} \times C^{ch}\,C^h$, which of the following percentages represents the proportion of progeny that will be Himalayan: **(a)** 100 percent, **(b)** $\frac{3}{4}$, **(c)** $\frac{1}{2}$, **(d)** $\frac{1}{4}$, **(e)** 0 percent?

5. Black, sepia, cream, and albino are all coat colors of guinea pigs. Individual animals (not necessarily from pure lines) showing these colors were intercrossed; the results are tabulated below, where we are using the abbreviations A, albino; B, black; C, cream; and S, sepia, for the phenotypes:

Cross	Parental phenotypes	B	S	C	A
1	B × B	22	0	0	7
2	B × A	10	9	0	0
3	C × C	0	0	34	11
4	S × C	0	24	11	12
5	B × A	13	0	12	0
6	B × C	19	20	0	0
7	B × S	18	20	0	0
8	B × S	14	8	6	0
9	S × S	0	26	9	0
10	C × A	0	0	15	17

a. Deduce the inheritance of these coat colors, using gene symbols of your own choosing. Show all parent and progeny genotypes.

b. If the black animals in (7) and (8) are crossed, what progeny proportions can you predict using your model?

6. In a maternity ward, four babies become accidentally mixed up. The ABO types of the four babies are known to be O, A, B, and AB. The ABO types of the four sets of parents are determined. Indicate which baby belongs to each set of parents: (**a**) AB $\times$ O, (**b**) A $\times$ O, (**c**) A $\times$ AB, (**d**) O $\times$ O.

7. Consider two blood polymorphisms that humans have in addition to the ABO system. Two alleles L^M and L^N determine the M, N, and MN blood groups. The dominant allele R of a different gene causes a person to have the Rh$^+$ (rhesus positive) phenotype, whereas the homozygote for r is Rh$^-$ (rhesus negative). Two men took a paternity dispute to court, each claiming three children to be his own. The blood groups of the men, the children, and their mother were as follows:

Person	Blood group		
husband	O	M	Rh$^+$
wife's lover	AB	MN	Rh$^-$
wife	A	N	Rh$^+$
child 1	O	MN	Rh$^+$
child 2	A	N	Rh$^+$
child 3	A	MN	Rh$^-$

From this evidence, can the paternity of the children be established?

8. On a fox ranch in Wisconsin, a mutation arose that gave a "platinum" coat color. The platinum color proved very popular with buyers of fox coats, but the breeders could not develop a pure-breeding platinum strain. Every time two platinums were crossed, some normal foxes appeared in the progeny. For example, the repeated matings of the same pair of platinums produced 82 platinum and 38 normal progeny. All other such matings gave similar progeny ratios. State a concise genetic hypothesis that accounts for these results.

9. Over a period of several years, Hans Nachtsheim investigated an inherited anomaly of the white blood cells of rabbits. This anomaly, termed the *Pelger anomaly,* involves an arrest of the segmentation of the nuclei of certain white cells. This anomaly does not appear to seriously inconvenience the rabbits.

a. When rabbits showing the typical Pelger anomaly were mated with rabbits from a true-breeding normal stock, Nachtsheim counted 217 offspring showing the Pelger anomaly and 237 normal progeny. What appears to be the genetic basis of the Pelger anomaly?

b. When rabbits with the Pelger anomaly were mated to each other, Nachtsheim found 223 normal progeny, 439 showing the Pelger anomaly, and 39 extremely abnormal progeny. These very abnormal progeny not only had defective white blood cells but also showed severe deformities of the skeletal system; almost all of them died soon after birth. In genetic terms, what do you suppose these extremely de-

fective rabbits represented? Why do you suppose there were only 39 of them?

c. What additional experimental evidence might you collect to support or disprove your answers to part b?

d. In Berlin, about one human in 1000 shows a Pelger anomaly of white blood cells very similar to that described in rabbits. The anomaly is inherited as a simple dominant, but the homozygous type has not been observed in humans. Can you suggest why, if you are permitted an analogy with the condition in rabbits?

e. Again by analogy with rabbits, what phenotypes and genotypes might be expected among the children of a man and woman who both show the Pelger anomaly?

(Problem 9 from A. M. Srb, R. D. Owen, and
R. S. Edgar, *General Genetics,* 2d ed. W. H. Freeman
and Company, 1965.)

10. Two normal-looking fruit flies were crossed and in the progeny there were 202 females and 98 males.

a. What is unusual about this result?

b. Provide a genetic explanation for this anomaly.

c. Provide a test of your hypothesis.

11. You have been given a virgin *Drosophila* female. You notice that the bristles on her thorax are much shorter than normal. You mate her with a normal male (with long bristles) and obtain the following F$_1$ progeny: $\frac{1}{3}$ short-bristled females, $\frac{1}{3}$ long-bristled females, and $\frac{1}{3}$ long-bristled males. A cross of the F$_1$ long-bristled females with their brothers gives only long-bristled F$_2$. A cross of short-bristled females with their brothers gives $\frac{1}{3}$ short-bristled females, $\frac{1}{3}$ long-bristled females, and $\frac{1}{3}$ long-bristled males. Provide a genetic hypothesis to account for all these results, showing genotypes in every cross.

12. A dominant allele H reduces the number of body bristles *Drosophila* flies have, giving rise to a "hairless" phenotype. In the homozygous condition, H is lethal. An independently assorting dominant allele S has no effect on bristle number except in the presence of H, in which case a single dose of S suppresses the hairless phenotype, thus restoring the hairy phenotype. However, S also is lethal in the homozygous (SS) condition.

a. What ratio of hairy to hairless individuals would you find in the live progeny of a cross between two hairy flies both carrying H in the suppressed condition?

b. When the hairless progeny are backcrossed with a parental hairy fly, what phenotypic ratio would you expect to find among their live progeny?

13. A pure-breeding strain of squash that produced disk-shaped fruits (see the accompanying figure) was crossed with a pure-breeding strain having long fruits. The F_1 had disk fruits, but the F_2 showed a new phenotype, sphere, and was composed of the following proportions:

disk	270
sphere	178
long	32

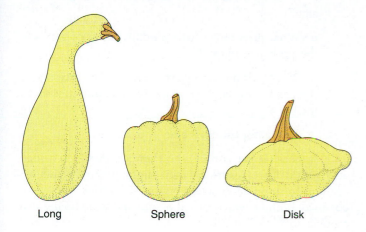

Long	Sphere	Disk

Propose an explanation for these results, and show the genotypes of P, F_1, and F_2 generations.

(Illustration from P. J. Russell, *Genetics*, 3d ed. Harper Collins, 1992.)

14. Because snapdragons (*Antirrhinum*) possess the pigment anthocyanin, they have reddish purple petals. Two pure anthocyanin-less lines of *Antirrhinum* were developed, one in California and one in Holland. They looked identical in having no red pigment at all, manifested as white (albino) flowers. However, when petals from the two lines were ground up together in buffer in the same test tube, the solution, which appeared colorless at first, gradually turned red.

a. What control experiments should an investigator conduct before proceeding with further analysis.

b. What could account for the production of the red color in the test tube?

c. According to your explanation for part b, what would be the genotypes of the two lines?

d. If the two white lines are crossed, what would you predict the phenotypes of the F_1 and F_2 to be?

15. The frizzle fowl is much admired by poultry fanciers. It gets its name from the unusual way that its feathers curl up, giving the impression that it has been (in the memorable words of animal geneticist F. B. Hutt) "pulled backwards through a knothole." Unfortunately, frizzle fowls do not breed true; when two frizzles are intercrossed they always produce 50 percent frizzles, 25 percent normal, and 25 percent with peculiar woolly feathers that soon fall out, leaving the birds naked.

a. Give a genetic explanation for these results, showing genotypes of all phenotypes, and provide a statement of how your explanation works.

b. If you wanted to mass-produce frizzle fowls for sale, which types would be best to use as a breeding pair?

16. Marfan's syndrome is a disorder of the fibrous connective tissue, characterized by many symptoms including long, thin digits, eye defects, heart disease, and long limbs. (Flo Hyman, the American volleyball star, suffered from Marfan's syndrome. She died soon after a game from a ruptured aorta.)

a. Use the pedigree shown below to propose a mode of inheritance for Marfan's syndrome.

b. What genetic phenomenon is shown by this pedigree?

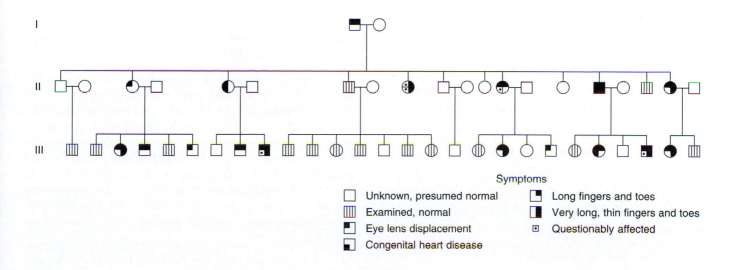

Symptoms

☐ Unknown, presumed normal	☐ Long fingers and toes
▥ Examined, normal	◧ Very long, thin fingers and toes
◪ Eye lens displacement	⊡ Questionably affected
◨ Congenital heart disease	

c. Speculate on a reason for such a phenomenon.

(Illustration from J. V. Neel and W. J. Schull, *Human Heredity*. University of Chicago Press, 1954.)

17. The petals of the plant *Collinsia parviflora* are normally blue, giving the species its common name "blue-eyed Mary." Two pure-breeding lines were obtained from color variants found in nature; the first line had pink petals and the second line had white petals. The following crosses were made between pure lines, with the results shown:

Parents	F_1	F_2
blue × white	blue	101 blue, 33 white
blue × pink	blue	192 blue, 63 pink
pink × white	blue	272 blue, 121 white, 89 pink

a. Explain these results genetically. Define allele symbols you use and show the genetic constitution of parents, F_1, and F_2.

b. A cross between a certain blue F_2 plant and a certain white F_2 plant gave progeny of which $\frac{3}{8}$ were blue, $\frac{1}{8}$ were pink, and $\frac{1}{2}$ were white. What must the genotypes of these two F_2 plants have been?

 Unpacking the Problem

a. What is the character being studied?

b. What is the wild-type phenotype?

c. What is a variant?

d. What are the variants in this problem?

e. What does "in nature" mean?

f. In what way would the variants have been found in nature? (Describe the scene.)

g. At which stages in the experiments would seeds be used?

h. Would the way of writing a cross "blue × white" (for example) mean the same as "white × blue"? Would you expect similar results? Why or why not?

i. In what way do the first two rows in the table differ from the third row?

j. Which phenotypes are dominant?

k. What is complementation?

l. Where does the blueness come from in the progeny of the pink × white cross?

m. What genetic phenomenon does the production of a blue F_1 from pink and white parents represent?

n. List any ratios you can see.

o. Are there any monohybrid ratios?

p. Are there any dihybrid ratios?

q. What does observing monohybrid and dihybrid ratios tell you?

r. List four modified Mendelian ratios that you can think of.

s. Are there any modified Mendelian ratios in the problem?

t. What do modified Mendelian ratios indicate generally?

u. What does the specific modified ratio or ratios in this problem indicate?

v. Draw chromosomes representing the meioses in the parents in the cross blue × white, and meiosis in the F_1.

w. Repeat for the cross blue × pink.

*18. In peas (*Pisum sativum*) the chemical pisatin is associated with defense against parasitic fungi: normal plants are resistant to fungi and contain pisatin. Two pure lines were obtained, both of which lacked pisatin and were highly susceptible to fungal attack. Line 1 was from California and line 2 was from Sweden. The lines were investigated as follows. (NOTE: The normal lines were also pure-breeding.)

Cross	F_1 phenotypes	F_2 phenotypes
line 1 × normal	pisatin	$\frac{3}{4}$ pisatin $\frac{1}{4}$ no pisatin
line 2 × normal	no pisatin	$\frac{3}{4}$ no pisatin $\frac{1}{4}$ pisatin
line 1 × line 2	no pisatin	$\frac{13}{16}$ no pisatin $\frac{3}{16}$ pisatin

a. Propose a model that explains the results of these crosses. Make sure you precisely define any allele symbols you use.

b. Show the genotypes underlying the parents, F_1, and F_2 in each cross.

c. How do lines 1 and 2 differ in the genetic reason for their lacking pisatin?

19. A woman who owned a purebred albino poodle (an autosomal recessive phenotype) wanted white puppies, so she took the dog to a breeder, who said he would mate the female with an albino stud male, also from a pure stock. When six puppies were born they were all black, so the woman sued the breeder, claiming that he replaced the stud male with a black dog, giving her six unwanted puppies. You are called in as an expert witness, and the defense asks you if it is possible to produce black offspring from two pure-breeding recessive albino parents. What testimony do you give?

20. A snapdragon plant that bred true for white petals was crossed to a plant that bred true for purple petals, and all the

F_1 had white petals. The F_1 was selfed. Among the F_2, three phenotypes were observed in the following numbers:

white	240
solid-purple	61
spotted-purple	19
total	320

a. Propose an explanation of these results, showing genotypes of all generations (make up and explain your symbols).

b. A white F_2 plant was crossed to a solid-purple F_2 plant, and the progeny were

white	50%
solid-purple	25%
spotted-purple	25%

What were the genotypes of the F_2 plants crossed?

21. Most flour beetles are black, but several color variants are known. Crosses of pure-breeding parents produced the following results in the F_1 generation, and intercrossing the F_1 from each cross gave the ratios shown for the F_2 generation. The phenotypes are abbreviated Bl, black; Br, brown; Y, yellow; and W, white.

Cross	Parents	F_1	F_2
1	Br × Y	Br	3 Br : 1 Y
2	Bl × Br	Bl	3 Bl : 1 Br
3	Bl × Y	Bl	3 Bl : 1 Y
4	W × Y	Bl	9 Bl : 3 Y : 4 W
5	W × Br	Bl	9 Bl : 3 Br : 4 W
6	Bl × W	Bl	9 Bl : 3 Y : 4 W

a. From these results deduce and explain the inheritance of these colors.

b. Write out the genotypes of each of the parents, the F_1, and the F_2 in all crosses.

22. Two albinos marry and have four normal children. How is this possible?

* 23. Plant breeders obtained three differently derived pure lines of white-flowered *Petunia* plants. They performed crosses and observed progeny phenotypes as follows:

Cross	Parents	Progeny
1	line 1 × line 2	F_1 all white
2	line 1 × line 3	F_1 all red
3	line 2 × line 3	F_1 all white
4	red F_1 × line 1	$\frac{1}{4}$ red : $\frac{3}{4}$ white
5	red F_1 × line 2	$\frac{1}{8}$ red : $\frac{7}{8}$ white
6	red F_1 × line 3	$\frac{1}{2}$ red : $\frac{1}{2}$ white

a. Explain these results, using gene symbols of your own choosing. (Show parental and progeny genotypes in each cross.)

b. If a red F_1 individual from cross 2 is crossed to a white F_1 individual from cross 3, what proportion of progeny will be red?

24. Consider production of flower color in the Japanese morning glory (*Pharbitis nil*). Dominant alleles of either of two separate genes ($A- bb$ or $aa B-$) produce purple petals. $A- B-$ produces blue petals, and $aa bb$ produces scarlet petals. Deduce the genotypes of parents and progeny in the following crosses:

Cross	Parents	Progeny
1	blue × scarlet	$\frac{1}{4}$ blue : $\frac{1}{2}$ purple : $\frac{1}{4}$ scarlet
2	purple × purple	$\frac{1}{4}$ blue : $\frac{1}{2}$ purple : $\frac{1}{4}$ scarlet
3	blue × blue	$\frac{3}{4}$ blue : $\frac{1}{4}$ purple
4	blue × purple	$\frac{3}{8}$ blue : $\frac{4}{8}$ purple : $\frac{1}{8}$ scarlet
5	purple × scarlet	$\frac{1}{2}$ purple : $\frac{1}{2}$ scarlet

25. Corn breeders obtained pure lines whose kernels turn sun-red, pink, scarlet, or orange when exposed to sunlight (normal kernels remain yellow in sunlight). Some crosses between these lines produced the following results. The phenotypes are abbreviated O, orange; P, pink; Sc, scarlet; and SR, sun-red.

		Phenotypes	
Cross	Parents	F_1	F_2
1	SR × P	all SR	66 SR : 20 P
2	O × SR	all SR	998 SR : 314 O
3	O × P	all O	1300 O : 429 P
4	O × Sc	all Y	182 Y : 80 O : 58 Sc

Analyze the results of each cross, and provide a unifying hypothesis to account for *all* the results. (Explain all symbols you use.)

26. Many kinds of wild animals have the agouti coloring pattern, in which each hair has a yellow band around it (see Figure 4-10).

a. Black mice and other black animals do not have the yellow band; each of their hairs is all black. This absence of wild agouti pattern is called *nonagouti*. When mice of a true-breeding agouti line are crossed with nonagoutis, the F_1 is all agouti and the F_2 has a 3:1 ratio of agoutis to non-agoutis. Diagram this cross, letting A represent the allele responsible for the agouti phenotype and a, nonagouti. Show the phenotypes and genotypes of the parents, their gametes, the F_1, their gametes, and the F_2.

b. Another inherited color deviation in mice substitutes brown for the black color in the wild-type hair. Such brown-agouti mice are called *cinnamons*. When wild-type mice are crossed with cinnamons, the F_1 is all wild type and the F_2 has a 3:1 ratio of wild type to cinnamon. Diagram this cross as in part a, letting B stand for the wild-type black allele and b stand for the cinnamon brown allele.

c. When mice of a true-breeding cinnamon line are crossed with mice of a true-breeding nonagouti (black) line, the F_1 is all wild type. Use a genetic diagram to explain this result.

d. In the F_2 of the cross in part c, a fourth color called *chocolate* appears in addition to the parental cinnamon and nonagouti and the wild type of the F_1. Chocolate mice have a solid, rich-brown color. What is the genetic constitution of the chocolates?

e. Assuming that the Aa and Bb allelic pairs assort independently of each other, what would you expect to be the relative frequencies of the four color types in the F_2 described in part d? Diagram the cross of parts c and d, showing phenotypes and genotypes (including gametes).

f. What phenotypes would be observed in what proportions in the progeny of a backcross of F_1 mice from part c to the cinnamon parental stock? to the nonagouti (black) parental stock? Diagram these backcrosses.

g. Diagram a testcross for the F_1 of part c. What colors would result, and in what proportions?

h. Albino (pink-eyed white) mice are homozygous for the recessive member of an allelic pair Cc which assorts independently of the Aa and Bb pairs. Suppose that you have four different highly inbred (and therefore presumably homozygous) albino lines. You cross each of these lines with a true-breeding wild-type line, and you raise a large F_2 progeny from each cross. What genotypes for the albino lines would you deduce from the F_2 phenotypes shown below?

F_2 of line	Wild type	Black	Cinnamon	Chocolate	Albino
			Phenotypes of progeny		
1	87	0	32	0	39
2	62	0	0	0	18
3	96	30	0	0	41
4	287	86	92	29	164

(Problem 26 adapted from A. M. Srb, R. D. Owen, and R. S. Edgar, *General Genetics*, 2d ed. W. H. Freeman and Company, 1965.)

27. An allele A that is not lethal when homozygous causes rats to have yellow coats. The allele R of a separate gene that assorts independently produces a black coat. Together, A and R produce a grayish coat, whereas a and r produce a white coat. A gray male is crossed with a yellow female, and the F_1 is $\frac{3}{8}$ yellow, $\frac{3}{8}$ gray, $\frac{1}{8}$ black, and $\frac{1}{8}$ white. Determine the genotypes of the parents.

28. The genotype $rr\,pp$ gives fowl a single comb, $R-P-$ gives a walnut comb, $rr\,P-$ gives a pea comb, and $R-pp$ gives a rose comb (see the accompanying figure).

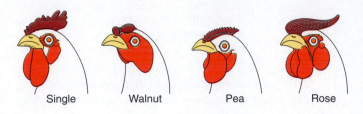

Single   Walnut   Pea   Rose

a. What comb types will appear in the F_1 and in the F_2 in what proportions if single-combed birds are crossed with birds of a true-breeding walnut strain?

b. What are the genotypes of the parents in a walnut × rose mating from which the progeny are $\frac{3}{8}$ rose, $\frac{3}{8}$ walnut, $\frac{1}{8}$ pea, and $\frac{1}{8}$ single?

c. What are the genotypes of the parents in a walnut × rose mating from which all the progeny are walnut?

d. How many genotypes produce a walnut phenotype? Write them out.

29. The production of eye-color pigment in *Drosophila* requires the dominant allele A. The dominant allele P of a second independent gene turns the pigment to purple, but its recessive allele leaves it red. A fly producing no pigment has white eyes. Two pure lines were crossed with the following results:

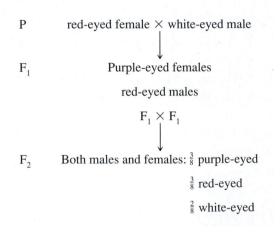

P	red-eyed female × white-eyed male
F_1	Purple-eyed females
	red-eyed males
	$F_1 \times F_1$
F_2	Both males and females: $\frac{3}{8}$ purple-eyed
	$\frac{3}{8}$ red-eyed
	$\frac{2}{8}$ white-eyed

Explain this mode of inheritance and show the genotypes of the parents, the F_1, and the F_2.

30. When true-breeding brown dogs are mated with certain true-breeding white dogs, all the F_1 pups are white. The F_2 progeny from some $F_1 \times F_1$ crosses were 118 white, 32 black, and 10 brown pups. What is the genetic basis for these results?

31. In corn, three dominant alleles, called A, C, and R, must be present to produce colored seeds. Genotypes $A-C-R-$ are colored: all others are colorless. A colored plant is crossed with three tester plants of known genotype. With tester $aa\,cc\,RR$, the colored plant produces 50 percent colored seeds; with *aa CC rr*, it produces 25 percent colored; and with *AA cc rr*, it produces 50 percent colored. What is the genotype of the colored plant?

32. The production of pigment in the outer layer of seeds of corn requires each of the three independently assorting genes *A*, *C*, and *R* to be represented by at least one dominant allele, as specified in problem 31. The dominant allele *Pr* of a fourth independently assorting gene is required to convert the biochemical precursor to a purple pigment, and its recessive allele *pr* makes the pigment red. Plants that do not produce pigment have yellow seeds. Consider a cross of a strain of genotype *AA CC RR pr pr* with a strain of genotype *aa cc rr Pr Pr*.

    a. What are the phenotypes of the parents?

    b. What will be the phenotype of the F$_1$?

    c. What phenotypes, and in what proportions, will appear in the progeny of a selfed F$_1$?

    d. What progeny proportions do you predict from the testcross of an F$_1$?

33. Wild-type strains of the haploid fungus *Neurospora* can make their own tryptophan. An abnormal allele *td* renders the fungus incapable of making its own tryptophan. An individual of genotype *td* grows only when its medium supplies tryptophan. The allele *su* assorts independently of *td*; its only known effect is to suppress the *td* phenotype. Therefore, strains carrying both *td* and *su* do not require tryptophan for growth.

    a. If a *td su* strain is crossed with a genotypically wild-type strain, what genotypes are expected in the progeny, and in what proportions?

    b. What will be the ratio of tryptophan-dependent to tryptophan-independent progeny in the cross of part a?

34. The allele *B* gives mice a black coat, and *b* gives a brown one. The genotype *ee* for another, independently assorting gene prevents expression of *B* and *b*, making the coat color beige, whereas *E*– permits expression of *B* and *b*. Both genes are autosomal. In the following pedigree, black symbols indicate a black coat, pink symbols indicate brown, and white symbols indicate beige.

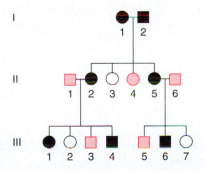

    a. What is the name given to the type of gene interaction in this example?

b. What are the gentoypes of the individuals in the pedigree? (If there are alternative possibilities, state them.)

35. Mice of the genotypes *AA BB CC DD SS* and *aa bb cc dd ss* are crossed. (These gene symbols are explained in the text of this chapter.) The progeny are intercrossed. What phenotypes will be produced in the F$_2$, and in what proportions?

36. Consider the genotypes of two lines of chickens: the pure-line mottled Honduran is *ii DD MM WW*, and the pure-line leghorn is *II dd mm ww*, where:

> *I* = white feathers, *i* = colored feathers
>
> *D* = duplex comb, *d* = simplex comb
>
> *M* = bearded, *m* = beardless
>
> *W* = white skin, *w* = yellow skin

These four genes assort independently. Starting with these two pure lines, what is the fastest and most convenient way of generating a pure line of birds that has colored feathers, simplex comb, yellow skin, and is beardless? Make sure you show

    a. The breeding pedigree

    b. The genotype of each animal represented

    c. How many eggs to hatch in each cross, and why this number

    d. Why your scheme is the fastest and most convenient

37. The following pedigree is for a dominant phenotype governed by an autosomal gene. What does this pedigree suggest about the phenotype, and what can you deduce about the genotype of individual A?

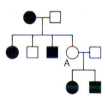

38. The genetic determination of petal coloration in foxgloves is given in Figure 4-17. Consider the following two crosses:

Cross	Parents		Progeny
1	dark reddish	× white with yellowish spots	$\frac{1}{2}$ dark reddish : $\frac{1}{2}$ light reddish
2	white with yellowish spots	× light reddish	$\frac{1}{2}$ white with reddish spots : $\frac{1}{4}$ dark reddish : $\frac{1}{4}$ light reddish

In each case, give the genotypes of parents and progeny with respect to the three genes.

39. A researcher crosses two white-flowered lines of *Antirrhinum plants* as follows and obtains the following results:

$$\text{pure line 1} \times \text{pure line 2}$$
$$\downarrow$$
$$F_1 \quad \text{all white}$$
$$F_1 \times F_1$$
$$\downarrow$$
$$F_2 \quad \text{131 white}$$
$$\text{29 red}$$

a. Deduce the inheritance of these phenotypes, using clearly defined gene symbols. Give the genotypes of the parents, F_1, and F_2.

b. Predict the outcome of crosses of the F_1 to each parental line.

40. In this problem, it is assumed that you know that enzymes are coded by genes—a topic not formally treated until Chapter 12. However, it is a good problem to try here, because it makes a strong connection between gene interactions and cell chemistry.

Assume that two pigments, red and blue, mix to give the normal purple color of petunia petals. Separate biochemical pathways synthesize the two pigments as shown in the top two rows of the diagram below. "White" refers to compounds that are not pigments. (Total lack of pigment results in a white petal.) Red pigment forms from a yellow intermediate that normally is at a concentration too low to color petals.

A third pathway whose compounds do not contribute pigment to petals normally does not affect the blue and red pathways, but if one of its intermediates (white$_3$) should build up in concentration, it can be converted to the yellow intermediate of the red pathway.

In the diagram, A to E represent enzymes; their corresponding genes, all of which are unlinked, may be symbolized by the same letters.

pathway I $\quad \cdots \longrightarrow \text{white}_1 \xrightarrow{\ E\ } \text{blue}$

pathway II $\cdots \rightarrow \text{white}_2 \xrightarrow{\ A\ } \text{yellow} \xrightarrow{\ B\ } \text{red}$
$$\Big\uparrow C$$
pathway III $\quad \cdots \rightarrow \text{white}_3 \xrightarrow{\ D\ } \text{white}_4$

Assume that wild-type alleles are dominant and code for enzyme function, and that recessive alleles represent lack of enzyme function. Deduce which combinations of true-breeding parental genotypes could be crossed to produce F_2 progenies in the following ratios:

a. 9 purple : 3 green : 4 blue

b. 9 purple : 3 red : 3 blue : 1 white

c. 13 purple : 3 blue

d. 9 purple : 3 red : 3 green : 1 yellow

(NOTE: blue mixed with yellow makes green; assume that no mutations are lethal.)

41. The flowers of nasturtiums (*Tropaeolum majus*) may be single (S), double (D), or superdouble (Sd). Superdoubles are female sterile; they originated from a double-flowered variety. Crosses between varieties gave the progenies as listed in the following table, where *pure* means "pure-breeding."

Cross	Parents	Progeny
1	pure S × pure D	All S
2	cross 1 F_1 × cross 1 F_1	78 S : 27 D
3	pure D × Sd	112 Sd : 108 D
4	pure S × Sd	8 Sd : 7 S
5	pure D × cross 4 Sd progeny	18 Sd : 19 S
6	pure D × cross 4 S progeny	14 D : 16 S

Using your own genetic symbols, propose an explanation for the above results, showing (**a**) all the genotypes in each of the six rows above and (**b**) the proposed origin of the superdouble.

*42. In a certain species of fly, the normal eye color is red (R). Four abnormal phenotypes for eye color were found: two were yellow (Y1 and Y2), one was brown (B), and one was orange (O). A pure line was established for each phenotype, and all possible combinations of the pure lines were crossed. Flies of each F_1 were intercrossed to produce an F_2. The F_1's and F_2's are shown within the square below; the pure lines are given in the margins.

		Y1	Y2	B	O
Y1	F_1	all y	all r	all r	all r
	F_2	all y	9 r	9 r	9 r
			7 y	4 y	4 o
				3 b	3 y
Y2	F_1		all y	all r	all r
	F_2		all y	9 r	9 r
				4 y	4 y
				3 b	3 o
B	F_1			all b	all r
	F_2			all b	9 r
					4 o
					3 b
O	F_1				all o
	F_2				all o

a. Define your own symbols and show genotypes of all four pure lines.

b. Show how the F_1 phenotypes and the F_2 ratios are produced.

c. If you understand what a biochemical pathway is, show a biochemical pathway that explains the genetic results, indicating which gene controls which enzyme.

43. In common wheat, *Triticum aestivum*, kernel color is determined by multiply duplicated genes, each with an *R* and an *r* allele. Any number of *R* alleles will give red, and the complete lack of *R* alleles will give the white phenotype. In one cross between a red pure line and a white pure line, the F_2 was $\frac{63}{64}$ red and $\frac{1}{64}$ white.

a. How many *R* genes are segregating in this system?

b. Show genotypes of the parents, the F_1, and the F_2.

c. Different F_2 plants are backcrossed to the white parent. Give examples of genotypes that would give the following progeny ratios in such backcrosses, (1) 1 red:1 white, (2) 3 red:1 white, (3) 7 red:1 white.

***d.** What is the formula that generally relates the number of segregating genes to the proportion of red individuals in the F_2 in such systems?

44. The following pedigree shows the inheritance of deaf-mutism.

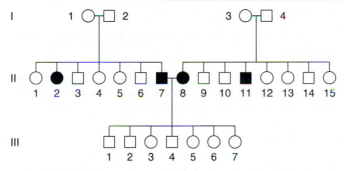

a. Provide an explanation for the inheritance of this rare condition in the two families in generations I and II, showing genotypes of as many individuals as possible using symbols of your own choosing.

b. Provide an explanation for the production of only normal individuals in generation III, making sure your explanation is compatible with the answer to part a.

45. The following pedigree is for blue sclera (bluish thin outer wall to the eye) and brittle bones:

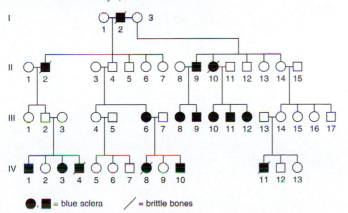

●, ■ = blue sclera   ╱ = brittle bones

a. Are these two abnormalities caused by the same gene or separate genes? State your reasons clearly.

b. Is the gene or genes autosomal or sex-linked?

c. Does the pedigree show any evidence of incomplete penetrance or expressivity? If so, make the best calculations that you can of these measures.

46. Workers of the honeybee line known as *Brown* (nothing to do with color) show what is called "hygienic behavior," that is, they uncap hive compartments containing dead pupae, and then remove the dead pupae. This prevents the spread of infectious bacteria through the colony. Workers of the Van Scoy line, however, do not perform these actions, and therefore this line is said to be "nonhygienic." When a queen from the Brown line was mated with Van Scoy drones, the F_1 were all nonhygienic. When drones from this F_1 inseminated a queen from the Brown line, the progeny behaviors were as follows:

$\frac{1}{4}$ hygienic

$\frac{1}{4}$ uncapping but no removing of pupae

$\frac{1}{2}$ nonhygienic

However, when the nonhygienic individuals were examined further, it was found that if the compartment of dead pupae were uncapped by the beekeeper, then about half the individuals removed the dead pupae, but the other half did not.

a. Propose a genetic hypothesis to explain these behavioral patterns.

b. Discuss the data in terms of epistasis, dominance, and environmental interaction.

(NOTE: Workers are sterile, and all bees from one line carry the same alleles.)

47. In one species of *Drosophila*, the wings are normally round in shape, but you have obtained two pure lines, one of which has oval wings and the other sickle-shaped wings. Crosses between pure lines reveal the following results:

Parents		F_1	
Female	Male	Female	Male
sickle	round	sickle	sickle
round	sickle	sickle	round
sickle	oval	oval	sickle

a. Provide a genetic explanation of these results, defining all allele symbols.

b. If the F_1 oval females from cross 3 are crossed to the F_1 round males from cross 2, what phenotypic proportions are expected in each sex of progeny?

48. Mice normally have one yellow band on their hairs, but variants with two or three bands are known. A female mouse with one band was crossed to a male who had three bands. (Neither animal was from a pure line.) The progeny were

Females	$\frac{1}{2}$	one band
	$\frac{1}{2}$	three bands
Males	$\frac{1}{2}$	one band
	$\frac{1}{2}$	two bands

a. Provide a clear explanation of the inheritance of these phenotypes.

b. Under your model, what would be the outcome of a cross between a three-banded daughter and a one-banded son?

49. In minks, wild types have an almost black coat. Breeders have developed many pure lines of color variants for the mink coat industry. Two such pure lines are platinum (blue-gray) and aleutian (steel gray). These lines were used in crosses, with the following results:

Cross	Parents	F_1	F_2
1	wild × platinum	wild	18 wild, 5 platinum
2	wild × aleutian	wild	27 wild, 10 aleutian
3	platinum × aleutian	wild	133 wild
			41 platinum
			46 aleutian
			17 sapphire (new)

a. Devise a genetic explanation of these three crosses. Show genotypes for parents, F_1, and F_2 in the three crosses, and make sure you show the alleles of each gene you hypothesize in every individual.

b. Predict the F_1 and F_2 phenotypic ratios from crossing sapphire with platinum and aleutian pure lines.

50. In *Drosophila* an autosomal gene determines the shape of the hair, with *B* giving straight and *b* bent hairs. On another autosome, there is a gene of which a dominant allele *I* inhibits hair formation so that the fly is hairless (*i* has no known phenotypic effect).

a. If a straight-haired fly from a pure line is crossed with a fly from a pure-breeding hairless line known to be an inhibited bent genotype, what will the genotypes and phenotypes of the F_1 and the F_2 be?

b. What cross would give the ratio 4 hairless : 3 straight : 1 bent?

51. The following pedigree concerns eye phenotypes in *Tribolium* beetles. The solid symbols represent black eyes, open symbols represent brown eyes, and the cross symbols (×) represent the "eyeless" phenotype, in which eyes are totally absent.

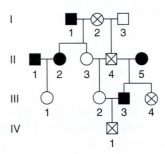

a. From these data deduce the mode of inheritance of these three phenotypes.

b. Using defined gene symbols show the genotype of individual II-3.

52. The normal color of snapdragons is red. Some pure lines showing variations of flower color have been found. When these were crossed, they gave the following results:

Parents	F_1	F_2
1. orange × yellow	orange	3 orange : 1 yellow
2. red × orange	red	3 red : 1 orange
3. red × yellow	red	3 red : 1 yellow
4. red × white	red	3 red : 1 white
5. yellow × white	red	9 red : 3 yellow : 4 white
6. orange × white	red	9 red : 3 orange : 4 white
7. red × white	red	9 red : 3 yellow : 4 white

a. Explain the inheritance of these colors.

b. Write out the genotypes of the parents, the F_1, and the F_2.

5

Linkage I: Basic Eukaryotic Chromosome Mapping

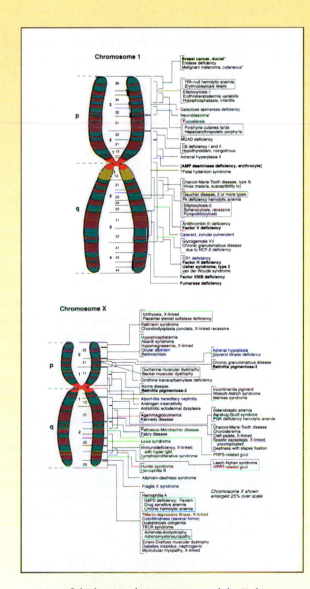

Chromosome 1

Chromosome X

Chromosome X shown enlarged 25% over scale

Map of the human chromosome 1 and the X chromosome. The genes have been positioned using several techniques, including those covered in Chapters 5 and 6. (From *Journal of NIH Research*, 1992.)

KEY CONCEPTS

▶ Two genes close together on the same chromosome pair do not assort independently at meiosis.

▶ Recombination produces genotypes with new combinations of parental alleles.

▶ A pair of homologous chromosomes can exchange parts by crossing-over.

▶ Recombination results from either independent assortment or crossing-over.

▶ Gene loci on a chromosome can be mapped by measuring the frequencies of recombinants produced by crossing-over.

▶ Interlocus map distances based on recombination measurements are roughly additive.

▶ The occurrence of a crossover can influence the occurrence of a second crossover in an adjacent region.

e have already established the basic principles of segregation and assortment, and we have correlated them with chromosome behavior during meiosis. Thus, from the cross $Aa\,Bb \times Aa\,Bb$, we expect a $9:3:3:1$ ratio of phenotypes. As we learned from Bridges's study of nondisjunction (page 68), exceptions to simple Mendelian expectations can direct the experimenter's attention to new discoveries. Just such an exception observed in the progeny of a dihybrid cross provided the clue to the important concepts discussed in this chapter.

Message In genetic analysis, exceptions to predicted behavior often give important new insights.

The Discovery of Linkage

In the early years of this century, William Bateson and R. C. Punnett were studying inheritance in the sweet pea. They studied two genes: one affecting flower color (P, purple, and p, red), and the other affecting the shape of pollen grains (L, long, and l, round). They crossed pure lines $PP\,LL$ (purple, long) $\times pp\,ll$ (red, round), and selfed the F_1 $Pp\,Ll$ heterozygotes to obtain an F_2. Table 5-1 shows the proportions of each phenotype in the F_2 plants.

The F_2 phenotypes deviated strikingly from the expected $9:3:3:1$ ratio. What is going on? This does not appear to be explainable as a modified Mendelian ratio. Note that two phenotypic classes are larger than expected: the purple, long phenotype and the red, round phenotype. As a possible explanation for this, Bateson and Punnett proposed that the F_1 had actually produced more $P\,L$ and $p\,l$ gametes than would be produced by Mendelian independent assortment. Because these were the gametic types in the original pure lines, the researchers thought that physical **coupling** between the dominant alleles P and L and between the recessive alleles p and l might have prevented their indepen-

dent assortment in the F_1. However, the researchers did not know what the nature of this coupling could be.

The confirmation of Bateson and Punnett's hypothesis had to await the development of *Drosophila* as a genetic tool. After the idea of coupling was first proposed, Thomas Hunt Morgan found a similar deviation from Mendel's second law while studying two autosomal genes in *Drosophila*. One of these genes affects eye color (pr, purple, and pr^+, red), and the other affects wing length (vg, vestigial, and vg^+, normal). The wild-type alleles of both genes are dominant. Morgan crossed $pr\,pr\,vg\,vg$ flies with $pr^+pr^+\,vg^+vg^+$ and then testcrossed the doubly heterozygous F_1 females: $pr^+pr\,vg^+vg\,\female \times pr\,pr\,vg\,vg\,\male$.

The use of the testcross is extremely important. Because one parent (the tester) contributes gametes carrying only recessive alleles, the phenotypes of the offspring reveal the gametic contribution of the other, doubly heterozygous parent. Hence, the analyst can concentrate on meiosis in one parent and forget about the other. This contrasts with the analysis of progeny from an F_1 self, where there are two sets of meioses to consider: one in the male parent and one in the female. Morgan's results follow; the alleles contributed by the F_1 female specify the F_2 classes:

$$
\begin{array}{ll}
pr^+\,vg^+ & 1339 \\
pr\,vg & 1195 \\
pr^+\,vg & 151 \\
pr\,vg^+ & \underline{154} \\
 & 2839
\end{array}
$$

Obviously, these numbers deviate drastically from the Mendelian prediction of a $1:1:1:1$ ratio, and they indicate a coupling of genes. The two largest classes are the combinations pr^+vg^+ and $pr\,vg$, originally introduced by the homozygous parental flies. You can see that the testcross clarifies the situation. It directly reveals the allelic combinations in the gametes from one sex in the F_1, thus clearly showing the coupling that could only be inferred from Bateson and Punnett's F_1 self. The testcross also reveals something new: there is approximately a $1:1$ ratio between the two parental types and also between the two nonparental types.

Now let us consider what may be learned by repeating the crossing experiments but changing the combinations of alleles contributed as gametes by the homozygous parents in the first cross. In this cross, each parent was homozygous for one dominant allele and for one recessive allele. Again F_1 females were testcrossed:

$$
\text{P} \qquad pr^+pr^+\,vg\,vg \times pr\,pr\,vg^+vg^+
$$
$$
\downarrow
$$
$$
\text{F}_1 \qquad pr^+pr\,vg^+vg
$$

$$
pr^+pr\,vg^+vg\,\female \times pr\,pr\,vg\,vg\,\male
$$

Table 5-1 Sweet Pea Phenotypes Observed in the F_2 by Bateson and Punnett

Phenotype (and genotype)	Number of Progeny	
	Observed	Expected from $9:3:3:1$ ratio
Purple, long (P–L–)	4831	3911
Purple, round (P–ll)	390	1303
Red, long ($pp\,L$–)	393	1303
Red, round ($pp\,ll$)	1338	435
	6952	6952

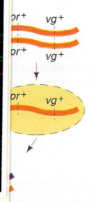

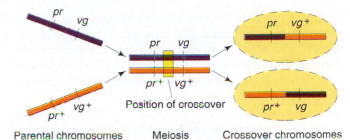

...ined from the

...e to a 1:1:1:1
...classes are those
...r rather than, as
...ssives. But notice
...at were originally
...provide the most
...n the early work
...the term **repul-**
...seemed to them
...alleles "repelled"
...ion in coupling,
..."stick together."
...nomena: coupling

...erning both phe-
...omologous chromo-
...ed from one par-
...me chromosome,
...ous chromosome
...ypothesis also ex-
...chromosome car-
...nd *vg*. Repulsion,
...his case, the domi-
...e recessive allele of
...why allelic combi-
...do we explain the

Figure 5-2 Crossing-over during meiosis. An individual receives one homolog from each parent. The exchange of parts by crossing-over may produce gametic chromosomes whose allelic combinations differ from the parental combinations.

Morgan suggested that when homologous chromosomes pair during meiosis, the chromosomes occasionally exchange parts during a process called **crossing-over.** Figure 5-2 illustrates this physical exchange of chromosome segments. The original arrangements of alleles on the two chromosomes are called the **parental** combinations. The two new combinations are called **crossover products** or **recombinants.**

Morgan's hypothesis that homologs may exchange parts may seem a bit farfetched. Is there any cytologically observable process that could account for crossing-over? We saw in Chapter 3 that during meiosis, when duplicated homologous chromosomes pair with each other, two nonsister chromatids often appear to cross each other. This is diagrammed in Figure 5-3. Recall that the resulting cross-shaped structure is called a *chiasma*. To Morgan, the appearance of the chiasmata visually corroborated the concepts of crossing-over. (Note that the chiasmata seem to indicate that it is chromatids, not unduplicated chromosomes, that cross over. We shall return to this point later.) For the present, let's accept Morgan's interpretation that chiasmata are the cytological counterparts of crossovers and leave the proof until Chapter 20. Note that Morgan did not arrive at this interpretation out of nowhere; he was looking for a *physical* explanation for his *genetic* results. His achievement in correlating the results of breeding experiments with cytological phenomena thus emphasizes the importance of the chromosome theory as a powerful basis for research.

Message Chiasmata are the visible manifestations of crossovers.

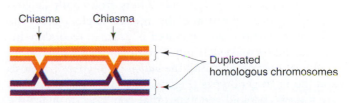

Figure 5-3 Diagrammatic representation of chiasmata at meiosis. Each line represents a chromatid of a pair of synapsed chromosomes.

Figure 5-1 Simple inheritance of two pairs of alleles located on the same chromosome pair.

Data like those just presented, showing coupling and repulsion in testcrosses and in F_1 selfs, are commonly encountered in genetics. Clearly, results of this kind represent a departure from independent assortment. Such exceptions, in fact, constitute a major addition to Mendel's view of the genetic world.

Message   When two genes are close together on the same chromosome pair, they do not assort independently.

The residing of genes on the same chromosome pair is termed **linkage.** Two genes on the same chromosome pair are said to be *linked.* It is also proper to refer to the linkage of specific alleles: for example, in one $A a\, B b$ individual, A might be linked to b; a would then of necessity be linked to B. These terms graphically allude to the existence of a physical entity linking the genes—that is, the chromosome itself. You may wonder why we refer to such genes as "linked" rather than "coupled"; the answer is that the words coupling and repulsion are now used to indicate two different types of linkage conformation in a double heterozygote, as follows:

$$\text{Coupling conformation} \qquad \frac{pr \qquad vg}{pr^+ \qquad vg^+}$$

$$\text{Repulsion conformation} \qquad \frac{pr \qquad vg^+}{pr^+ \qquad vg}$$

In other words, coupling refers to the linkage of two dominant or two recessive alleles, whereas repulsion indicates that dominant alleles are linked with recessive alleles. To ascertain whether a double heterozygote is in coupling or repulsion conformation, an investigator must testcross the double heterozygote or consider the genotypes of its parents.

Recombination

We have already introduced the term *recombination.* This term is widely used in many areas of practical and theoretical genetics, so it is absolutely necessary to be clear about its meaning at this stage. Recombination has been observed in a variety of situations in addition to meiosis, but for the present, let's define it in relation to meiosis. **Meiotic recombination** is any meiotic process that generates a haploid product with a genotype that differs from both haploid genotypes that constituted the meiotic diploid cell. The product of meiosis so generated is called a **recombinant.** This definition makes the important point that we detect recombination by comparing the *output* genotypes of meiosis and the *input* genotypes (Figure 5-4). The input genotypes are the two haploid genotypes that combined to make the genetic constitution of the meiocyte, the diploid cell that undergoes meiosis.

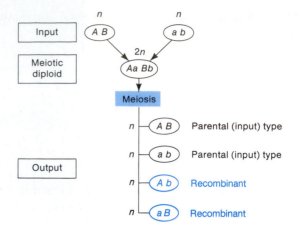

Figure 5-4 Recombinants are those products of meiosis with allelic combinations different from those of the haploid cells that formed the meiotic diploid.

Message   During meiosis recombination generates haploid genotypes differing from the haploid parental genotypes.

Meiotic recombination is a part of both haploid and diploid life cycles; however, detecting recombinants in haploid cycles is straightforward whereas detecting them in diploid cycles is more complex. The input and output types in haploid cycles are the genotypes of individuals and may thus be inferred directly from phenotypes. Figure 5-4 can be viewed as summarizing the simple detection of recombinants in haploid life cycles. The input and output types in diploid life cycles are gametes. Because we must know the input gametes to detect recombinants in a diploid cycle, it is preferable to have pure-breeding parents. Furthermore, we cannot detect recombinant output gametes directly: we must testcross the diploid individual and observe its progeny (Figure 5-5). If a testcross offspring is shown to have been constituted from a recombinant product of meiosis, it too is called a *recombinant.* Notice again that the testcross allows us to concentrate on *one* meiosis and avoid ambiguity. From a self of the F_1 in Figure 5-5, for example, a recombinant $A A\, B b$ offspring cannot be distinguished from $A A\, B B$ without further crosses.

There are two kinds of recombination, and they produce recombinants by completely different methods. But a recombinant is a recombinant, so how can we decide which type of recombination accounts for any particular progeny? The answer lies in the *frequency* of recombinants, as we shall soon see. First, though, let's consider the two types of recombination: interchromosomal and intrachromosomal.

Interchromosomal Recombination

Mendelian independent assortment brings about **interchromosomal recombination** (Figure 5-6). In a testcross the two recombinant classes always make up 50 percent of the

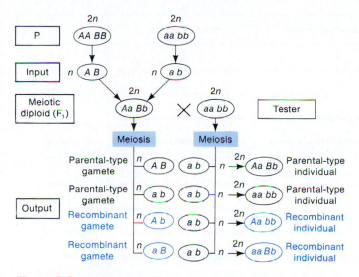

Figure 5-5 The detection of recombination in diploid organisms. Note that Figure 5-4 is actually a part of this figure. Recombinant products of a diploid meiosis are most readily detected in a cross of a heterozygote to a recessive tester.

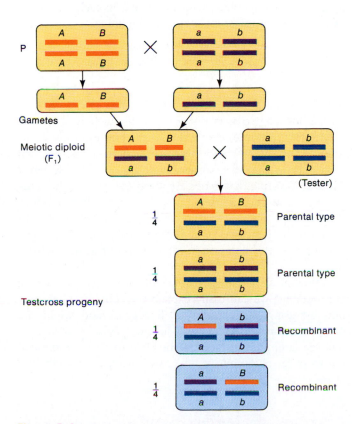

Figure 5-6 Interchromosomal recombination, which always produces a recombinant frequency of 50 percent. This diagram shows two chromosome pairs of a diploid organism with *A* and *a* on one pair and *B* and *b* on the other. Note that we could represent the haploid situation by removing the part marked *P* and the testcross.

progeny; that is, there are 25 percent of each recombinant type among the progeny. If we observe this frequency, we can infer that the two genes under study assort independently.

The simplest interpretation of these frequencies is that the two genes are on separate chromosome pairs. However, as has already been hinted, genes that are far apart on the *same* chromosome pair can act virtually independently and produce the same result. In Chapter 6, we shall see how a large amount of crossing-over effectively unlinks distantly linked genes.

Intrachromosomal Recombination

Crossing-over produces **intrachromosomal recombination.** Any two nonsister chromatids can cross over. (We shall show proof of this in Chapter 6.) Of course, there is not a crossover between two specific genes in all meioses, but when there is, half the products of *that* meiosis are recombinant, as shown in Figure 5-7. Meiosis with no crossover between the genes under study produce only parental genotypes for these genes.

The sign of intrachromosomal recombination is a recombinant frequency of less than 50 percent. The physical linkage of parental gene combinations prevents the independent assortment of two genes that generates the recombinant frequency of 50 percent (Figure 5-8). We saw an example of this situation in Morgan's data (page 124), where the recombinant frequency was $(151 + 154) \div 2839 = 10.7$ percent. This is obviously much less than the 50 percent we would expect with independent assortment. What about recombinant frequencies greater than 50 percent? The answer is that such frequencies are *never* observed, as we shall see in Chapter 6.

Note in Figure 5-7 that crossing-over generates two reciprocal products, which explains why the reciprocal recombinant classes are generally approximately equal in frequency.

> **Message** A recombinant frequency significantly less than 50 percent shows that the genes are linked. A recombinant frequency of 50 percent generally means that the genes are unlinked on separate chromosomes.

The remainder of this chapter deals with intrachromosomal recombination and its causative crossovers.

Linkage Symbolism

Our symbolism for describing crosses becomes cumbersome with the introduction of linkage. We can depict the genetic constitution of each chromosome in the *Drosophila* cross as in the following example:

$$\frac{pr \qquad vg}{pr^+ \qquad vg^+}$$

Figure 5-7 Intrachromosomal recombinants arise from meioses in which nonsister chromatids cross over between the genes under study.

	Meiotic chromosomes	Meiotic products	
Meioses with no crossover between the genes	A ——— B	A ——— B	Parental
	A ——— B	A ——— B	Parental
	a ——— b	a ——— b	Parental
	a ——— b	a ——— b	Parental
Meioses with a crossover between the genes	A ——— B	A ——— B	Parental
	A ⤬ B	A ——— b	Recombinant
	a ⤬ b	a ——— B	Recombinant
	a ——— b	a ——— b	Parental

where each line represents a chromosome; the alleles above are on one chromosome, and those below are on the other chromosome. A crossover is represented by placing an X between the two chromosomes, so that

$$
\frac{pr \qquad vg}{\overline{\text{X}}}\\
pr^{+} \qquad vg^{+}
$$

is the same as

$$
\frac{pr \qquad\qquad vg}{pr^{+}\qquad\qquad vg^{+}}
$$

We can simplify the genotypic designation of linked genes by drawing a single line, with the genes on each side being on the same chromosome; now our symbol is

$$
\frac{pr \qquad vg}{pr^{+} \qquad vg^{+}}
$$

But this is still inconvenient for typing and writing, so let's tip the line to give us $pr\,vg\,/\,pr^{+}\,vg^{+}$, still keeping the genes of one chromosome on one side of the line and those of its homolog on the other. We always designate linked genes on each side in the same order; it is always $a\,b\,/\,a\,b$, never $a\,b\,/\,b\,a$. The rule that genes are always written in the same order permits geneticists to use a shorter notation in which the wild-type allele is written with a plus sign alone. In this notation the genotype $pr\,vg\,/\,pr^{+}\,vg^{+}$ becomes $pr\,vg\,/\,++$. You may see this notation in other books or in research papers.

Now if we reconsider the results obtained by Bateson and Punnett, we can easily explain the coupling phenomenon by using the concept of linkage. Their results are complex because they did not do a testcross. However, we will see later that in fact it is possible to derive estimated numbers for recombinant and parental types in a dihybrid cross.

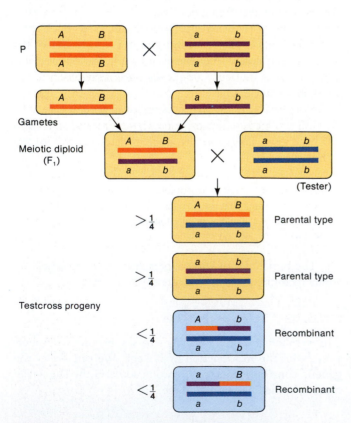

Figure 5-8 Intrachromosomal recombination. Notice that the frequencies of the recombinants add up to less than 50 percent.

Linkage of Genes on the X Chromosome

Until now, we have been considering recombination of autosomal genes. What are the consequences of nonsister chromatids of the X chromosome crossing over between two genes of interest? Recall that a human, or *Drosophila*, female produces male progeny hemizygous for the genes of the X chromosome, so that the genotype of the gamete that a mother contributes to her son is the sole determinant of the son's phenotype. Let's consider an example in which we first observe the F_1 progeny from the mating of two *Drosophila* flies and then the F_2 progeny from intercrossing the F_1. We use here the following symbols: y and y^+ for the alleles governing yellow body and brown body, respectively; w and w^+ for alleles for white eye and red eye; and Y for the Y chromosome.

$$P \quad y w^+ / y w^+ \, ♀ \times y^+ w / Y \, ♂$$

$$F_1 \quad y w^+ / y^+ w \, ♀ \times y w^+ / Y \, ♂$$

The numbers of F_2 males in the phenotypic classes are

$y \, w$	43	recombinant
$y^+ \, w$	2146	parental
$y \, w^+$	2302	parental
$y^+ \, w^+$	22	recombinant
	4513	

Because the F_2 males obtain only a Y chromosome from the F_1 males, these classes reflect perfectly the products of meiosis in the F_1 females. Notice that this eliminates the need for a testcross; we can follow meiosis in a single parent, just as we can in a testcross. The total frequency of the recombinants in this example is $(43 + 22) \div 4513 = 1.4$ percent.

Linkage Maps

The frequency of recombinants for the *Drosophila* autosomal genes we studied (*pr* and *vg*) was 10.7 percent of the progeny—a frequency much greater than that for the linked genes on the X chromosome, studied above. Apparently, the amount of crossing-over between various linked genes differs. Indeed, there is no reason to expect that chromatids would cross over between different linked genes with the same frequency. As Morgan studied more linked genes, he saw that the proportion of recombinant progeny varied considerably, depending on which linked genes were being studied, and he thought that these variations in crossover frequency might somehow reflect the actual distances separating genes on the chromosomes. Morgan assigned the study of this problem to a student, Alfred Sturtevant, who (like Bridges) became a great geneticist. Morgan asked Sturtevant, still an undergraduate at the time, to make some sense of the data on crossing-over between dif-

ferent linked genes. In one night, Sturtevant developed a method for describing relationships between genes that is still used today. In Sturtevant's own words, "In the latter part of 1911, in conversation with Morgan, I suddenly realized that the variations in strength of linkage, already attributed by Morgan to differences in the spatial separation of genes, offered the possibility of determining sequences in the linear dimension of a chromosome. I went home and spent most of the night (to the neglect of my undergraduate homework) in producing the first chromosome map."

As an example of Sturtevant's logic, consider a testcross from which we obtain the following results:

$pr \, vg \, / \, pr \, vg$	165	Parental
$pr^+ \, vg^+ \, / \, pr \, vg$	191	
$pr \, vg^+ \, / \, pr \, vg$	23	Recombinant
$pr^+ \, vg \, / \, pr \, vg$	21	
	400	

The progeny in this example represent 400 female gametes, of which 44 (11 percent) are recombinant. Sturtevant suggested that we can use the percentage of recombinants as a quantitative index of the linear distance between two genes on a genetic map, or **linkage map,** as it is sometimes called.

The basic idea here is quite simple. Imagine two specific genes positioned a certain fixed distance apart. Now imagine random crossing-over along the paired homologs. In some meiotic divisions, nonsister chromatids cross over by chance in the chromosomal region between these genes; from these meioses, recombinants are produced. In other meiotic divisions, there are no crossovers between these genes; no recombinants result from these meioses. Sturtevant postulated a rough proportionality: the greater the distance between the linked genes, the greater the chance that nonsister chromatids would cross over in the region between the genes and, hence, the greater the proportion of recombinants that would be produced. Thus, by determining the frequency of recombinants, we can obtain a measure of the map distance between the genes (Figure 5-9). In fact, we can define one **genetic map unit (m.u.)** as that distance between genes for which one product of meiosis out of 100 is recombinant. Put another way, a **recombinant frequency (RF)** of 0.01 (or 1 percent) is defined as 1 m.u. [A map unit is sometimes referred to as a **centimorgan (cM)** in honor of Thomas Hunt Morgan.]

A direct consequence of the way map distance is measured is that if 5 map units (5 m.u.) separate genes *A* and *B* whereas 3 m.u. separate genes *A* and *C*, then *B* and *C* should be either 8 or 2 m.u. apart (Figure 5-10). Sturtevant found this to be the case. In other words, his analysis strongly suggested that genes are arranged in some linear order.

The place on the map where a gene is located—and on the chromosome—is called the **gene locus** (plural, **loci**). The locus of the eye-color gene and the locus of the wing-

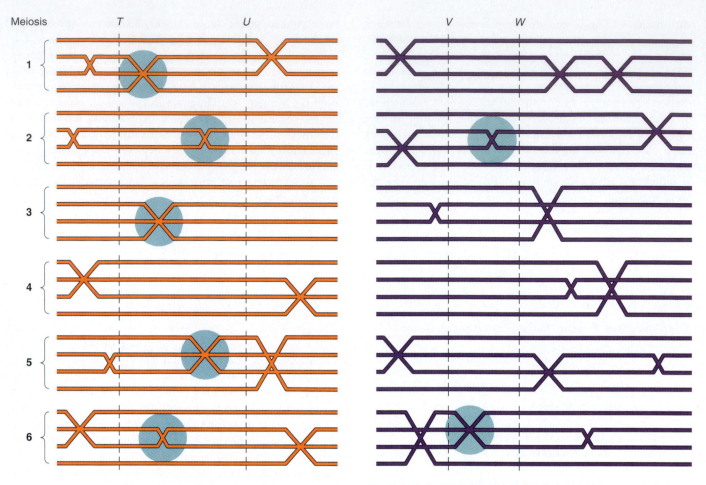

Figure 5-9 Proportionality between chromosome distance and recombinant frequency. During every meiosis, chromatids cross over at random along the chromosome. The two genes *T* and *U* are farther apart on a chromosome than *V* and *W*. Chromatids cross over between *T* and *U* in a larger proportion of meioses than between *V* and *W*, so the recombinant frequency for *T* and *U* is higher than that for *V* and *W*. As we will learn later in the chapter, a crossover can occur between any two nonsister chromatids within a homologous pairing.

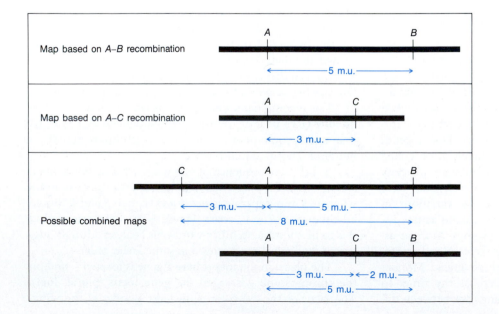

Figure 5-10 Because map distances are additive, calculation of the *A−B* and *A−C* distances leaves us with the two possibilities shown for the *B−C* distance.

length gene, for example, are 11 m.u. apart. The relationship is usually diagrammed this way:

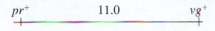

although it could be diagrammed equally well like this:

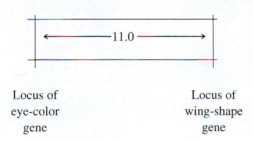

or like this:

Usually we refer to the locus of this eye-color gene in shorthand as the "*pr* locus," after the first discovered non-wild-type allele, but we mean the place on the chromosome where any allele of this gene will be found.

Given a genetic distance in map units, we can predict frequencies of progeny in different classes. For example, in the progeny from a testcross of a female *pr vg / pr⁺ vg⁺* heterozygote, we know there will be 11 percent recombinants, of which $5\frac{1}{2}$ percent will be *pr vg⁺ / pr vg* and $5\frac{1}{2}$ percent will be *pr⁺ vg / pr vg*; of the progeny from a testcross of a female *pr vg⁺ / pr⁺ vg* heterozygote, $5\frac{1}{2}$ percent will be *pr vg / pr vg* and $5\frac{1}{2}$ percent will be *pr⁺ vg⁺ / pr vg*.

There is a strong implication that the "distance" on a linkage map is a physical distance along a chromosome, and Morgan and Sturtevant certainly intended to imply just that. But we should realize that the linkage map is another example of an entity constructed from a purely genetic analysis. The linkage map could have been derived without even knowing that chromosomes existed. Furthermore, at this point in our discussion, we cannot say whether the "genetic distances" calculated by means of recombinant frequencies in any way reflect actual physical distances on chromosomes, although cytogenetic and molecular analysis has shown that genetic distances are, in fact, roughly proportional to chromosome distances. Nevertheless, it must be emphasized that the hypothetical structure (the linkage map) was developed with a very real structure (the chromosome) in mind. In other words, the chromosome theory provided the framework for the development of linkage mapping.

Message Recombination between linked genes can be used to map their distance apart on the chromosome. The unit of mapping (1 m.u.) is defined as a recombinant frequency of 1 percent.

The stage of analysis that we have reached in our discussion is well illustrated by linkage maps of the screwworm (*Cochliomyia hominivorax*). The larval stage of this insect—the worm—is parasitic on mammalian wounds and is a costly pest of livestock in some parts of the world. A genetic system of population control has been proposed, of a type that has been successful in other insects. In order to accomplish this goal, an understanding of the basic genetics of the insect is needed, and one important part of this is to prepare a map of the chromosomes. This animal has six chromosome pairs, and mapping has begun recently.

The job of general mapping starts by finding and analyzing as many variant phenotypes as possible. The adult stage of this insect is a fly, and geneticists have found phenotypic variants among screwworm flies. They found flies of six different eye colors, all different from the brown-eyed, wild-type flies, as Figure 5-11a shows. They also found five variant phenotypes for some other characters. Eleven mutant alleles were shown to determine the 11 variant phenotypes, each at a different autosomal locus. Pure lines of each phenotype were intercrossed to generate dihybrid F_1s, and then these were testcrossed. This revealed the set of four linkage groups shown in Figure 5-11b. Notice that the *ye* and *cw* loci are shown tentatively linked although the recombinant frequency is not significantly different from 50 percent.

A linkage analysis such as the above cannot assign linkage groups to specific chromosomes; this must be done using cytogenetic techniques to be discussed in Chapter 8. In the present example, such cytogenetic techniques have allowed the linkage groups to be correlated with the chromosomes previously numbered as shown in the figure.

Three-Point Testcross

So far, we have looked at linkage in crosses of double heterozygotes to doubly recessive testers. The next level of complexity is a cross of a triple heterozygote to a triply recessive tester. This kind of cross, called a **three-point testcross**, illustrates the standard kind of approach used in linkage analysis. We will consider two examples of such crosses here.

First, we shall focus on three *Drosophila* genes that have the non-wild-type alleles *sc* (short for *scute*, or loss of certain thoracic bristles), *ec* (short for *echinus*, or roughened eye surface), and *vg* (short for *vestigial wing*). We can cross *sc sc ec ec vg vg* triply recessive flies with wild-type flies to generate triple heterozygotes, *sc sc⁺ ec ec⁺ vg vg⁺*. We analyze recombination in these heterozygotes by testcrossing heterozygous females with triply recessive tester males. The results of such a testcross follow. The progeny are listed as gametic genotypes derived from the heterozygous females. Eight gametic types are possible, and these were counted in the following numbers in a sample of 1008 progeny flies:

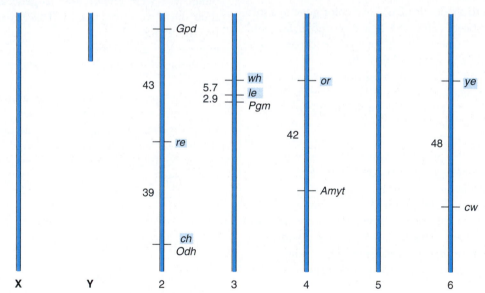

Figure 5-11 (a) Wild-type adult of screwworm and six flies whose eye colors are determined by alleles at six different autosomal loci. (b) Linkage maps of the six eye-color loci (highlighted in blue) and five other loci of screwworms. The numbers between the loci give the recombinant frequencies. (D. B. Taylor, USDA.)

sc	*ec*	*vg*	235
sc⁺	*ec⁺*	*vg⁺*	241
sc	*ec*	*vg⁺*	243
sc⁺	*ec⁺*	*vg*	233
sc	*ec⁺*	*vg*	12
sc⁺	*ec*	*vg⁺*	14
sc	*ec⁺*	*vg⁺*	14
sc⁺	*ec*	*vg*	16
			1008

The systematic way to analyze such crosses is to calculate all possible recombinant frequencies, but it is always worthwhile to inspect the data for obvious patterns before doing this. At first glance, we can note in the above data that there is a considerable deviation from the 1:1:1:1:1:1:1:1 ratio that is expected if the genes are all unlinked. So let's begin to calculate recombinant frequency values, taking the loci a pair at a time. Starting with the *sc* and *ec* loci (ignoring the *vg* locus for the time being), we determine which of the gametic genotypes are recombinant for *sc* and *ec*. Because we know that the heterozygotes were established from *sc ec* and *sc⁺ ec⁺* gametes, we know that the recombinant products of meiosis must be *sc ec⁺* and

$sc^+ ec$. We note from the list that there are $12 + 14 + 14 + 16 = 56$ of these types; therefore, RF = $(56/1008) \times 100 = 5.5$ m.u. This tell us that these loci must be linked on the same chromosome, as follows:

$$sc \xleftarrow{\hspace{1.5cm}} 5.5 \text{ m.u.} \xrightarrow{\hspace{1.5cm}} ec$$

Now let's look at recombination between the *sc* and the *vg* loci. The "input" parental genotypes were *sc vg* and $sc^+ vg^+$, so we must calculate the frequency of $sc\ vg^+$ and $sc^+\ vg$ progeny types (this time, we ignore *ec*). We see that there are $243 + 233 + 14 + 16 = 506$ recombinants; because $506/1008$ is very close to an RF of 50 percent, we conclude that the *sc* and *vg* loci are not linked and are probably not on the same chromosome. We can summarize the linkage relationship as follows:

$$sc \underset{\xleftarrow{\hspace{0.8cm}} 5.5 \text{ m.u.} \xrightarrow{\hspace{0.8cm}}}{\rule{3cm}{0.4pt}} ec \qquad vg$$

Now you should see that the *ec* and *vg* loci must also be unlinked. Adding up the recombinants in the list and calculating the RF will confirm this. (Try it.)

A second example, using some other loci of *Drosophila*, will introduce some more important genetic concepts. Here the non-wild-type alleles are *v* (vermilion eyes), *cv* (crossveinless, or absence of a crossvein on the wing), and *ct* (cut, or snipped wing edges). This time the parental stocks are homozygous doubly recessive flies of genotype $v^+ v^+\ cv\ cv\ ct\ ct$ and homozygous singly recessive flies of genotype $v\ v\ cv^+ cv^+\ ct^+ ct^+$. From this cross, triply heterozygous progeny of genotype $v\ v^+\ cv\ cv^+\ ct\ ct^+$ are obtained, and females of this genotype are testcrossed to triple recessives of genotype $v\ v\ cv\ cv\ ct\ ct$. The female gametic genotypes determining the eight progeny types from this testcross are shown here, with their numbers out of a total sample of 1448 flies:

v	cv^+	ct^+	580
v^+	*cv*	*ct*	592
v	*cv*	ct^+	45
v^+	cv^+	*ct*	40
v	*cv*	*ct*	89
v^+	cv^+	ct^+	94
v	cv^+	*ct*	3
v^+	*cv*	ct^+	5
			1448

Once again, the standard recombination approach is called for, but we must be careful in our classification of parental and recombinant types. Note that the parental input genotypes for the triple heterozygotes are $v^+ cv\ ct$ and

$v\ cv^+ ct^+$; we must take this into consideration when we decide what constitutes a recombinant.

Starting with the *v* and *cv* loci, we see that the recombinants are of genotype *v cv* and $v^+ cv^+$ and that there are $45 + 40 + 89 + 94 = 268$ of these. Out of a total of 1448 flies, this gives an RF of 18.5 percent.

For the *v* and *ct* loci, the recombinants are *v ct* and $v^+ ct^+$. There are $89 + 94 + 3 + 5 = 191$ of these among 1448 flies, so that RF = 13.2 percent.

For *ct* and *cv*, the recombinants are $cv\ ct^+$ and $cv^+ ct$. There are $45 + 40 + 3 + 5 = 93$ of these among the 1448, so that RF = 6.4 percent.

Obviously, all the loci are linked on the same chromosome, because the RF values are all considerably less than 50 percent. Because the *v* and *cv* loci show the largest RF value, they must be farthest apart; therefore, the *ct* locus must be between them. A map can be drawn as follows:

$$v \qquad\qquad ct \qquad\qquad cv$$
$$\xleftarrow{\hspace{1.5cm}} 13.2 \text{ m.u.} \xrightarrow{\hspace{0.5cm}}\!\!\xleftarrow{\hspace{0.5cm}} 6.4 \text{ m.u.} \xrightarrow{\hspace{1cm}}$$

Note several important points here. The first is that we have deduced a different gene order from that of our listing of the progeny genotypes. As the point of the exercise was to determine the linkage relationships of these genes, the original listing was of necessity arbitrary; the order simply was not known before the data were analyzed.

Second, we have definitely established that *ct* is between *v* and *cv* and what the distances are between *ct* and these loci in map units. But we have arbitrarily placed *v* to the left and *cv* to the right; the map could equally well be inverted.

A third point to note is that the two smaller map distances, 13.2 m.u. and 6.4 m.u., add up to 19.6 m.u., which is greater than 18.5 m.u., the distance calculated for *v* and *cv*. Why is this so? The answer to this question lies in the way in which we have analyzed the two rarest classes in our classification of recombination for the *v* and *cv* loci. Now that we have the map, we can see that these two rare classes are in fact double recombinants, arising from two crossovers (Figure 5-12). However, we did not count the $v\ ct\ cv^+$ and $v^+ ct^+ cv$ genotypes when we calculated the RF value for *v* and *cv*; after all, with regard to *v* and *cv*, they are parental combinations ($v\ cv^+$ and $v^+ cv$). In the light of our map, however, we see that this led to an underestimate of the distance between the *v* and *cv* loci. Not only should we have

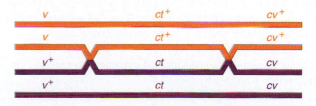

Figure 5-12 An example of a double crossover. Notice that a double crossover produces double recombinant chromatids that have the parental allelic combinations at the outer loci.

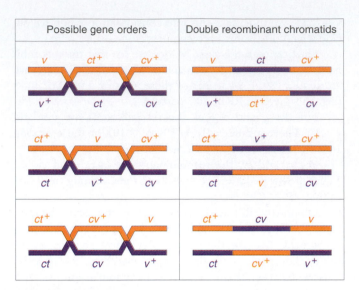

Possible gene orders	Double recombinant chromatids

Figure 5-13 With three genes, only three gene orders are possible. Double crossovers create unique double recombinant genotypes for each gene order. Only the first possibility is compatible with the data in the text.

counted the two rarest classes, we should have counted each of them twice because each represents a double recombinant class! Hence, we can correct the value by adding the numbers $45 + 40 + 89 + 94 + 3 + 3 + 5 + 5 = 284$. Out of the total of 1448, this is exactly 19.6 percent, which is identical with the sum of the two component values.

Now that we have had some experience with the data of this cross, we can look back at the progeny listing and see that it is usually possible to deduce gene order by inspection, without a recombinant frequency analysis. Only three gene orders are possible, each with a different gene in the middle position. It is generally true that the double recombinant classes are the smallest ones. Only one order should be compatible with the smallest classes having been formed by double crossovers, as shown in Figure 5-13. Only one order gives double recombinants of genotype $v\ ct\ cv^+$ and $v^+\ ct^+\ cv$. Notice in passing that the ability to detect double crossover depends on having a heterozygous gene between the two crossovers; if the mothers of these progeny had not been heterozygous $ct\ ct^+$, we could never have identified the double recombinant classes.

Finally, note that linkage maps merely map the loci in relation to each other, using standard map units. We do not know where the loci are on a chromosome—or even which specific chromosome they are on. The linkage map is essentially an abstract construct that can be correlated with a specific chromosome and with specific chromosome regions only by applying special kinds of cytogenetic analyses, as we shall see in Chapter 8.

Message Three (and higher) point testcrosses enable linkage between three (or more) genes to be evaluated in one cross.

Interference

The detection of the double recombinant classes shows that double crossovers must occur. Knowing this, one of the first questions that we may think of is whether the crossovers in adjacent chromosome regions are independent or whether they interact with each other in some way. For example, we might ask if a crossover in one region affects the likelihood of there being a crossover in an adjacent region. It turns out that often the answer is yes, and the interaction is called **interference.**

The analysis can be approached in the following way. If the crossovers in the two regions are independent, then according to the product rule (see page 31), the frequency of double recombinants would equal the product of the recombinant frequencies in the adjacent regions. In the $v-ct-cv$ recombination data, the v-ct RF value is 0.132 and the $ct-cv$ value is 0.064, so double recombinants might be expected at the frequency $0.132 \times 0.064 = 0.0084$ (0.84 percent) if there is independence. In the sample of 1448 flies, $0.0084 \times 1448 = 12$ double recombinants are expected. But the data show that only 8 were actually observed. If this deficiency of double recombinants were consistently observed, it would show us that the two regions are not independent and suggest that the distribution of crossovers favors singles at the expense of doubles. In other words, there is some kind of interference: a crossover reduces the probability of a crossover in an adjacent region.

Interference is quantified by first calculating a term called the **coefficient of coincidence (c.o.c.),** which is the ratio of observed to expected double recombinants and then subtracting this value from 1. Hence

Interference (I) = 1 − c.o.c. =

$$1 - \left[\frac{\text{observed frequency or number of double recombinants}}{\text{expected frequency or number of double recombinants}} \right]$$

In our example

$$I = 1 - \tfrac{8}{12} = \tfrac{4}{12} = \tfrac{1}{3}, \text{ or } 33\%$$

In some regions, there are never any observed double recombinants. In these cases, c.o.c. = 0, so I = 1 and interference is complete. Most of the time, the interference values that are encountered in mapping chromosome loci are between 0 and 1, but in certain special situations, observed doubles exceed expected, giving negative interference values.

Recombination analysis relies so heavily on three-point testcrosses and extended versions of them that it is worth making a step-by-step summary of the analysis, ending with an interference calculation. We shall use numerical values from the data on the v, ct, and cv loci.

1. Calculate recombinant frequencies for each pair of genes:

$$v - cv = 18.5\%$$

$$cv - ct = 6.4\%$$

$$ct - v = 13.2\%$$

2. Represent linkage relationships in a linkage map:

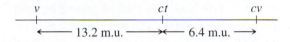

3. Determine the double recombinant classes.
4. Calculate the frequency and number of double recombinants expected if there is no interference:

$$\text{Expected frequency} = 0.132 \times 0.064 = 0.0084$$

$$\text{Expected number} = 0.0084 \times 1448 = 12$$

5. Calculate interference:

$$\text{Observed number of double recombinants} = 8$$

$$\text{Expected number of double recombinants} = 12$$

$$\therefore I = 1 - \tfrac{8}{12} = \tfrac{4}{12} = 0.33, \text{ or } 33\%$$

You may have wondered why we always used heterozygous females for testcrosses in our examples of linkage in *Drosophila*. When $pr\,vg\,/\,pr^{+}\,vg^{+}$ males are crossed with $pr\,vg\,/\,pr\,vg$ females, only $pr\,vg\,/\,pr^{+}\,vg^{+}$ and $pr\,vg\,/\,pr\,vg$ progeny are recovered. This result shows that there is no crossing-over in *Drosophila* males. However, this absence of crossing-over in one sex is limited to certain species; it is not the case for males of all species (or for the heterogametic sex). In other organisms, there is crossing-over in XY males and in WZ females. The reason for the absence of crossing-over in *Drosophila* males is that they have an unusual prophase I, with no synaptonemal complexes.

Incidentally, there is a recombination difference between human sexes as well. Women show higher recombinant frequencies for the same loci than do men. This is shown in a map in a later chapter (Figure 15-37).

Calculating Recombinant Frequencies from Selfed Dihybrids

A testcross is the most convenient way to measure recombinant frequency, but in practice the appropriate tester is not always available. A common situation is that a new phenotype has been identified, and shown by Mendelian analysis to be caused by some genotype such as $a\,a$. In order to map this newly discovered locus, the $a\,a$ individual will be crossed to a variety of genotypes such as bb, where the locus of the B gene is known. In this situation you can see that no $a\,a\,bb$ tester will be available at the outset. In fact $a\,a\,bb$ can be derived only from further breeding experiments.

Nevertheless, it is possible to calculate recombinant frequencies from the self of the dihybrid formed by crossing the two stocks. In this example the parents would be $a\,a\,BB$ and $AA\,BB$, and the dihybrid will be $A\,a\,Bb$. This dihybrid will produce parental gametes, $a\,B$ and $A\,b$, and recombinant gametes $A\,B$ and $a\,b$. Upon selfing, the gametes will fertilize randomly, and initially it seems that the recombinants cannot be identified in the progeny. However, the $a\,a\,bb$ progeny genotype comes to the rescue because this is the only genotype that *must* have been constituted by recombinants, in fact by fertilization of $a\,b$ by another $a\,b$. Therefore, if the frequency of the $a\,b$ products of meiosis is p, then the frequency of $a\,a\,bb$ offspring will be p^2. Therefore, to find p we simply have to take the square root of the frequency of $a\,a\,bb$ progeny. Since we know that the frequency of $a\,b$ will equal the frequency of $A\,B$, we can double p to find the total recombinant frequency.

If the two genes are unlinked, we know that the $a\,a\,bb$ class will be formed at a frequency of $\frac{1}{16}$ (the "1" of the $9:3:3:1$ ratio). In this case the square root of $\frac{1}{16}$ is $\frac{1}{4}$, and on doubling this we would obtain an RF of $\frac{1}{2}$, or 50 percent, as expected. What about linkage? Cases of linkage will produce $a\,a\,bb$ frequencies significantly less than $\frac{1}{16}$. Let's assume we observe that the frequency of $a\,a\,bb$ is 0.01 (1 percent); the frequency of $a\,b$ meiotic products must have been 0.1, or 10 percent, and the RF must be 20 percent, a clear case of linkage. Note also that if we wanted to calculate the parental frequencies, they must be 40 percent each for $a\,B$ and $A\,b$, calculated as 0.5 (100 − 20) percent.

The above method is theoretically correct, but in practice it is inaccurate because it extrapolates from just one of the F_2 phenotypes and furthermore involves taking a square root. A more accurate formulation has been devised that incorporates all the F_2 phenotypes. A statistic called the *product ratio* (z) is calculated, and a recombinant frequency is derived from a table of values of z. In the repulsion dihybrid used above ($A\,b\,/\,a\,B$), the product ratio is calculated as follows, in which the four components of the calculation are the four F_2 phenotypes:

$$z = \frac{(A - B-) \times (a\,a\,bb)}{(A - bb) \times (a\,a\,B-)}$$

The RF values corresponding to selected values of z are shown in Table 5-2. Computer programs are also available that calculate RF values from F_2 data.

Message Recombinant frequency can be calculated indirectly from the progeny of selfed dihybrids.

Examples of Linkage Maps

Linkage maps are an essential aspect of the experimental genetic study of any organism. They are the prelude to any

z	RF
0.001	2.2
0.005	4.9
0.020	9.9
0.040	13.8
0.100	21.1
0.200	28.5
0.300	33.5
0.500	40.3
0.700	45.0

serious piece of genetic manipulation. Many organisms have had their chromosomes intensively mapped in this way. The resultant maps represent a vast amount of genetic analysis generally achieved by collaborative efforts of research groups throughout the world. Figures 5-14 and 5-15 show two examples of linkage maps: one from *Drosophila* and one from the tomato. The *Drosophila* genome is one of the most intensively mapped of all model genetic organisms. The map in Figure 5-14 shows only a fraction of the known loci. Tomatoes, also, have been interesting from the perspectives of both basic and applied genetic research, and the tomato genome is one of the best mapped of plants.

The different panels of Figure 5-15 illustrate some of the stages of understanding through which research arrives at a comprehensive map. First, although chromosomes are visible under the microscope, there is initially no way to lo-

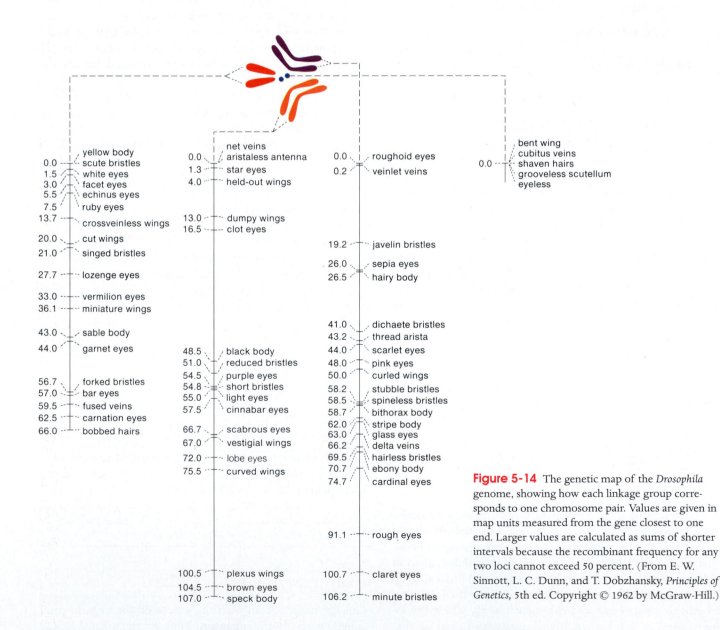

Figure 5-14 The genetic map of the *Drosophila* genome, showing how each linkage group corresponds to one chromosome pair. Values are given in map units measured from the gene closest to one end. Larger values are calculated as sums of shorter intervals because the recombinant frequency for any two loci cannot exceed 50 percent. (From E. W. Sinnott, L. C. Dunn, and T. Dobzhansky, *Principles of Genetics,* 5th ed. Copyright © 1962 by McGraw-Hill.)

cate genes on them. However, the chromosomes can be individually identified and numbered, based on their inherent landmarks such as staining patterns and centromere positions, as has been done in parts a and b. Next, analysis of recombinant frequencies generates a set of linkage groups that must correspond to chromosomes, but specific correlations cannot necessarily be made with the numbered chromosomes. At some stage, as we discussed in the screwworm example, cytogenetic analyses allow the linkage groups to be assigned to specific chromosomes. Part c of Figure 5-15 shows a tomato map made in 1952, with the linkages of the genes known at that time. Each locus is represented by the two alleles used in the original mapping experiments. As more and more loci became known, they were mapped in relation to the loci shown in the figure, so today the map contains hundreds of loci. Some of the chromosome numbers shown in part c are tentative and do not correspond to the modern chromosome numbering system.

The χ^2 Test

In research generally, it is often necessary to compare experimentally observed numbers of items in several different categories with numbers that are predicted on the basis of some hypothesis. If there is a close match then the hypothesis is maintained, while if there is a poor match then the hypothesis is rejected. As part of this process a judgement has to be made about whether the observed numbers are a close enough match to those expected. Very close matches and blatant mismatches generally present no problem in judgement, but inevitably there are grey areas in which the match is not obvious. Genetic analysis often involves the interpretation of numbers in various phenotypic classes. In such cases a statistical procedure called the χ^2 test is commonly used to help in making the decision to hold on to or reject the hypothesis. For example, the test can be used to analyse Mendelian ratios, answering such questions as "Are these experimental results compatible with the 12:3:1 ratio expected on the basis of a hypothesis of dominant epistasis?" Another widespread application is in testing for linkage. The genetic criterion for linkage is the absence of independent assortment. It is not possible to test for linkage directly because a priori we do not have a precise linkage distance to use in the hypothesis, so we are forced to test the hypothesis of the absence of linkage. If the observed results cause rejection of the hypothesis of *no linkage*, then we can infer linkage. This type of **null hypothesis** is generally useful in χ^2 analysis by providing a precise experimental prediction that can be tested.

The χ^2 test is simply a way of quantifying the various deviations expected by chance if a hypothesis is true. Take a simple hypothesis predicting a 1:1 ratio, for example. Even if the hypothesis is true, we would not always expect an exact 1:1 ratio. We can model this with a barrel full of equal numbers of red and blue marbles. If we blindly remove samples

of 100 marbles, on the basis of chance we would expect samples to show small deviations such as 52 red: 48 blue quite commonly, and larger deviations such as 60 red: 40 blue less commonly. The χ^2 test allows us to calculate the probability of such chance deviations from expectations. But, if all levels of deviation are expected with different probabilities even if the hypothesis is true, how can we ever reject a hypothesis? It has become a general scientific convention that a probability of the observations less than or equal to 5% is to be taken as the criterion for rejecting the null hypothesis. The hypothesis might still be true, but we have to make a decision somewhere and the 5% level is the conventional decision line. The logic is that though results this far from expectations are anticipated 5% of the time even when the null hypothesis is true, we will mistakenly reject the null hypothesis in only 5% of cases and we are willing to take this chance of error.

Let's test a specific set of data for linkage using χ^2 analysis. Assume that we have crossed pure-breeding parents of genotypes *AA BB* and *aa bb*, and obtained a dihybrid *Aa Bb*, which we have testcrossed to *aa bb*. A total of 500 progeny are classified as follows (written as gametes from the dihybrid):

140	*A B*
135	*a b*
110	*A b*
115	*a B*

From these data the recombinant frequency is 225/500 = 45%. This seems like a case of linkage because the RF is less than 50% expected from independent assortment. However, it is possible that the two recombinant classes are in the minority merely on the basis of chance, therefore, we need to perform a χ^2 test. The statistic χ^2 is always calculated from actual numbers, not from percentages, proportions, or fractions. Sample size is therefore very important in the χ^2 test, as it is in most considerations of chance phenomena. Sam-

Table 5-3   Contingency table comparing observed and expected results of a testcross to examine linkage between loci A/a and B/b.

locus 2 →		B		b		
	O	*E*	*O*	*E*	Total
locus 1						
A	140	255 × 250/500 = 127.5	110	245 × 250/500 = 122.5	250
a	115	255 × 250/500 = 127.5	135	245 × 250/500 = 122.5	250
Total	255		245		500

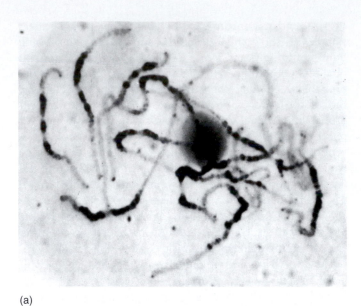

(a)

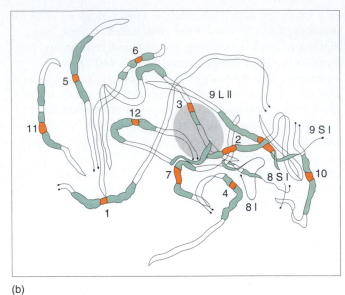

(b)

Figure 5-15 Mapping the chromosomes of tomatoes. (a) Photomicrograph of a meiotic prophase I (pachytene) from anthers, showing the 12 pairs of chromosomes as they appear under the microscope. (b) The currently used chromosome numbering system. The centromeres are colored, and the flanking, densely staining regions (heterochromatin) are shown in black. (c) (*facing page*) A linkage map made in 1952 showing the linkage groupings known at the time. Each locus is flanked by drawings of the variant phenotype that first identified that genetic locus (to the right or above) and the appropriate normal phenotype (to the left or below). Interlocus map distances are shown in map units. (Parts a and b from C. M. Rick, "The Tomato," *Scientific American,* 1978; part c from L. A. Butler.)

ples to be tested generally contain several classes. The letter O is used to represent the observed number in a class, and E the expected number for the same class based on the predictions of the hypothesis. The general formula for calculating χ^2 is as follows:

$$\chi^2 = \text{total of } \frac{(O - E)^2}{E} \text{ for all classes}$$

The problem, then, is to find the expectations, E, for each class. For linkage testing, the phenotypic classes are laid out

Table 5-4   Critical Values of the χ^2 Distribution

df	0.995	0.975	0.9	0.5	0.1	0.05	0.025	0.01	0.005	df
1	.000	.000	0.016	0.455	2.706	3.841	5.024	6.635	7.879	1
2	0.010	0.051	0.211	1.386	4.605	5.991	7.378	9.210	10.597	2
3	0.072	0.216	0.584	2.366	6.251	7.815	9.348	11.345	12.838	3
4	0.207	0.484	1.064	3.357	7.779	9.488	11.143	13.277	14.860	4
5	0.412	0.831	1.610	4.351	9.236	11.070	12.832	15.086	16.750	5
6	0.676	1.237	2.204	5.348	10.645	12.592	14.449	16.812	18.548	6
7	0.989	1.690	2.833	6.346	12.017	14.067	16.013	18.475	20.278	7
8	1.344	2.180	3.490	7.344	13.362	15.507	17.535	20.090	21.955	8
9	1.735	2.700	4.168	8.343	14.684	16.919	19.023	21.666	23.589	9
10	2.156	3.247	4.865	9.342	15.987	18.307	20.483	23.209	25.188	10
11	2.603	3.816	5.578	10.341	17.275	19.675	21.920	24.725	26.757	11
12	3.074	4.404	6.304	11.340	18.549	21.026	23.337	26.217	28.300	12
13	3.565	5.009	7.042	12.340	19.812	22.362	24.736	27.688	29.819	13
14	4.075	5.629	7.790	13.339	21.064	23.685	26.119	29.141	31.319	14
15	4.601	6.262	8.547	14.339	22.307	24.996	27.488	30.578	32.801	15

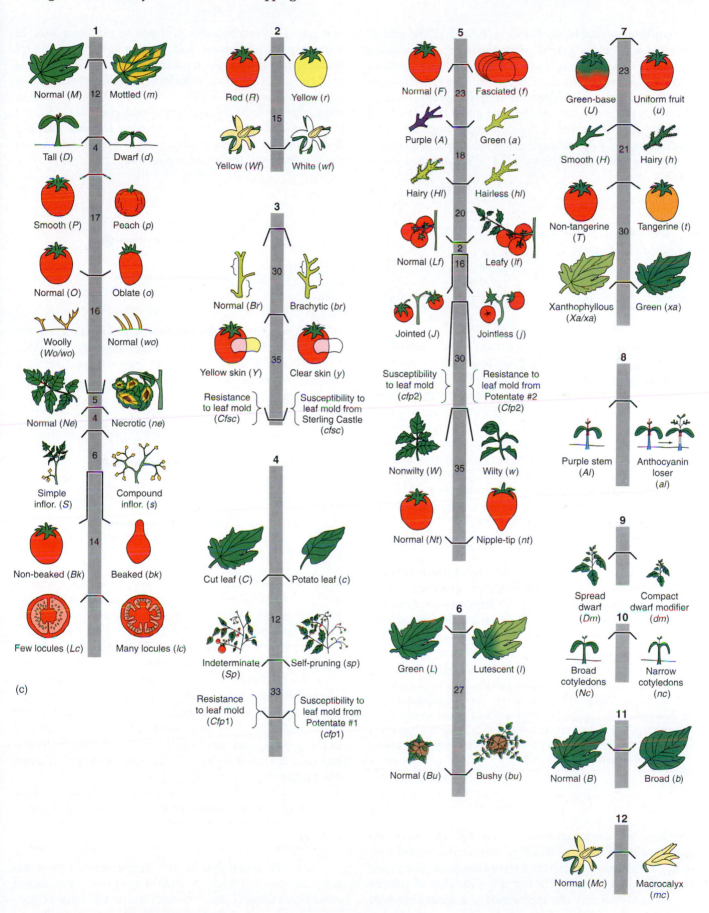

(c)

in a **contingency table,** as shown in Table 5-3. The grid of the table is simply a graphic representation of combining the allele ratios of the two loci at random. The χ^2 analysis tests whether or not the numbers of individuals in the four phenotypic classes of progeny are simply the two allelic ratios combined at random, as expected under the hypothesis of independent assortment. The expected values are calculated simply by subdividing the column totals proportionately to the row totals. The value of χ^2 is calculated as follows:

Class	O	E	$\dfrac{(O - E)^2}{E}$
$A\ B$	140	127.5	1.23
$a\ b$	135	122.5	1.28
$A\ b$	110	122.5	1.28
$a\ B$	115	127.5	1.23
		Total $= \chi^2 =$	5.02

The obtained value of χ^2 is converted into a probability value using a χ^2 table (Table 5-4). To do this we need to compute the number of degrees of freedom (df) used in the χ^2 calculation, which is the number of independent deviations, $O\text{-}E$. But all the deviations in the contingency table turn out to be equal in magnitude (12.5)! This is because the total expected numbers in each row or column have to be equal to the total observed number in each row and column, so there is only one degree of freedom. In general, in contingency tables the degrees of freedom are equal to the number of rows minus one, times the number of columns minus one. In the present example

$$\text{df} = (2 - 1) \times (2 - 1) = 1$$

Therefore, looking along the one degree of freedom line in Table 5-4 we see that the probability of obtaining a deviation from expectations this large (or larger) by chance alone is 0.025, (2.5%). Since this probability is less than 5%, the hypothesis of independent assortment must be rejected. Thus, having rejected the hypothesis of no linkage, we are left with the inference that the loci must be linked.

Message The χ^2 test is used to test observed numbers in a class against the expectations derived from a predictive hypothesis. The test generates the probability of obtaining by chance a specific deviation at least as great as the one observed, assuming that the hypothesis is correct.

The Nature of Crossing-Over

The idea that intrachromosomal recombinants were produced by some kind of exchange of material between homologous chromosomes was a compelling one. But experimentation was necessary to test this idea. One of the first steps was to correlate the appearance of a genetic recombinant with an exchange of parts of chromosomes. Several investigators approached this problem in the same way. In 1931, Harriet Creighton and Barbara McClintock were studying two loci of chromosome 9 of corn: one affecting seed color (C, colored; c, colorless), and the other affecting endosperm composition (Wx, waxy; wx, starchy). Furthermore, the chromosome carrying C and Wx was unusual in that it carried a large, densely staining element (called a *knob*) on the C end and a longer piece of chromosome on the Wx end; thus, the heterozygote was

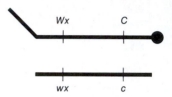

When they compared the chromosomes of genetic recombinants with those of parental—type progeny, Creighton and McClintock found that all the parental types retained the parental chromosomal arrangements, whereas all the recombinants were

or

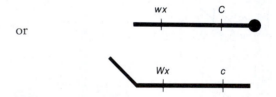

Thus, they correlated the genetic and cytological events of intrachromosomal recombination. The chiasmata appeared to be the sites of the exchange, but the final proof of this did not come until 1978.

But what is the mechanism of chromosome exchange in a crossover event? The short answer is that a crossover results from chromosome breakage and reunion. Two parental chromosomes break at the same position, and then join up again in two nonparental combinations. In Chapter 20 we will discuss positive evidence for this breakage and union process, and also study models of the molecular processes that allow DNA to break and rejoin in such a precise manner.

Message Chromosomes cross over by breaking at the same position and rejoining in two reciprocal nonparental combinations.

You will notice that in our diagrammatic representations of crossing over in this chapter we have shown crossovers taking place at the four-chromatid stage of meiosis. However, just from studying random recombinant products of meiosis, as in a testcross, it is not possible to distinguish this possibility from crossing over at the two-chromosome stage. This matter was settled through the genetic analysis of organisms whose four products of meiosis remain together in groups of four called **tetrads.** These organisms are mainly fungi and unicellular algae. The meiotic products in a single tetrad can be isolated, and this is equivalent to isolating all four chromatids arising from single meioses. Tetrad analyses of crosses involving linked genes clearly show that in many cases tetrads contain four different genotypes with regard to these loci; for example, from the cross

$$A B \times a b$$

some tetrads contain four genotypes

$$A B$$

$$A b$$

$$a B$$

$$a b$$

This result can be explained only by the occurrence of a crossover at the four-chromatid stage because if crossing over occurred at the two-chromosome stage, then there could be only two different genotypes in an individual meiosis, as shown in Figure 5-16.

Tetrad analysis allows the exploration of many other aspects of intrachromosomal recombination, which will be considered in detail in Chapters 6 and 20, but for the present let us use tetrads to answer two more fundamental questions about crossing-over. First, can multiple crossovers involve more than two chromatids? To answer this question, we need to look at double crossovers, and in order to study

double crossovers we need three linked genes. For example, in a cross such as

$$A B C \times a b c$$

there are many different tetrads possible, but some of them can be explained only by double crossovers. Consider the following tetrad as an example:

$$A B c$$

$$A b C$$

$$a B C$$

$$a b c$$

This tetrad must be explained by two crossovers involving three chromatids, as shown in Figure 5-17. Other types of tetrads show that all four of the chromatids can participate in crossing over in the same meiosis. Therefore, two, three, or four chromatids can be involved in crossing-over events in a single meiosis.

If all chromatids can be involved, we can ask if there is any **chromatid interference,** in other words, does the occurrence of a crossover between any two nonsister chromatids affect the likelihood of those two chromatids being involved in another crossover in the same meiosis? Tetrad analysis can answer this question, and shows that generally the distribution of crossovers between chromatids is random; in other words, there is no chromatid interference.

Before we leave the topic of the involvement of chromatids in crossovers, it is worth raising another question that is sometimes asked, which is whether it is possible for crossing-over to occur between sister chromatids. It has been shown in some organisms that indeed there is sister chromatid crossing-over, but since it produces no recombinants, and furthermore since it is not clear if it occurs in all organisms, it is conventional not to represent this type of exchange in crossover diagrams.

Crossing-over is a remarkably precise process. The synapsis and exchange of chromosomes is such that no segments are lost or gained, and four complete chromosomes emerge in a tetrad. A great deal has been learned about the nature of the molecular events in and around the sites of intrachromosomal recombination. Several interesting molecu-

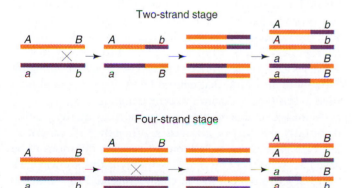

Figure 5-16 Tetrad analysis provides evidence that enabled geneticists to decide whether crossing-over occurs at the two-strand (two-chromosome) or at the four-strand (four-chromatid) stage of meiosis. Because more than two different products of a single meiosis can be seen in some tetrads, crossing-over cannot occur at the two-strand stage.

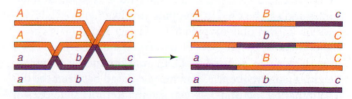

Figure 5-17 One of the several possible types of double-crossover tetrads that are regularly observed. Note that more than two chromatids exchanged parts.

lar processes appear to be involved, including DNA strand breakage, unwinding, hybrid DNA formation, nucleotide excision and polymerization, and DNA strand ligation. We return to these points in Chapter 20.

Linkage Mapping by Recombination in Humans

Humans have thousands of autosomally inherited phenotypes, and it might seem that it should be relatively straightforward to map the loci of the genes causing these phenotypes, using the techniques developed in this chapter. However, progress in mapping these loci was initially slow, for several reasons. First, it is not possible to make controlled crosses in humans, and geneticists had to try to calculate recombinant frequencies from the occasional dihybrids that were produced by chance in human matings. Of course, crosses that were the equivalent of testcrosses were extremely rare. Second, human progenies are generally small, making it difficult to obtain enough data to calculate reliable map distances. Third, the human genome is immense, which means that on average the distances between the known genes are large.

Several technological advances led to rapid improvements in mapping the human genome. First a human-rodent cell-fusion technique (Chapter 17) made it relatively straightforward to assign many human genes to specific chromosomes. This technique also permitted some mapping of genes along a specific chromosome. But perhaps the most dramatic improvements came from the use of DNA markers to act as "milestones" along the chromosomes. It is known that the genomes of all organisms show many sites of neutral variation at the DNA level. In other words although all members of a species have more or less the same genes in the same relative locations, individual organisms within that species differ in which neutral variant sites they contain. These sites are neutral in that they do not have any known effect at the phenotypic level. Sometimes a neutral site is nothing more than a single nucleotide difference within a gene or between genes; sometimes a site is a variable number of tandem repeats of "junk DNA" present between genes. Although these variants are not significant to the phenotype, they are significant to the geneticist because if an individual is heterozygous for one of these sites, the two different DNA types at this site can be considered to be alleles and can be used in mapping just like heterozygous alleles of known phenotypic effect. In fact individuals generally contain large numbers of detectable heterozygous DNA sites, which can be used as **DNA marker loci** in mapping. Recombinant frequencies can be calculated between the different marker loci and between the marker loci and genes of known phenotypic effect. This type of mapping has "fleshed out" the human chromosome map; an example is shown in Figure 5-18. The methodology of mapping DNA markers is covered in Chapters 15 and 17.

Figure 5-18 (*facing page*) Linkage map of human chromosome 1, correlated with chromosome banding pattern. The histogram shows the distribution of all markers available for chromosome 1. Some markers are genes of known phenotype, but the majority are DNA markers based on neutral sequence variation.

A linkage map, based on recombinant frequency analyses of the type discussed in this chapter, is in the center of the figure. It shows only some of the markers available. Map distances are shown in cM (= m.u.). The total length of the chromosome 1 map is 356 cM; it is the longest human chromosome.

The positions of some markers are cross-referenced to a diagram of subregions of chromosome 1 based on standard banding pattern (such a diagram is called an idiogram). These kinds of correlations can only be made using cytogenetic analysis (Chapter 8) and *in situ* hybridization (Chapter 17). Most of the markers shown on the map are molecular, but several genes (highlighted in yellow) are also included:

APOA2	apolipoprotein
ACTN2	actin protein
CRP	C-reactive protein
SPTA1	spectrin protein

(B. R. Jasney et al., *Science*, Sept. 30, 1994.)

The human X chromosome has always been more amenable to mapping by recombination analysis than the autosomes, and the first human chromosome map was for the X chromosome. The reason for this success is that males are hemizygous for X-linked genes, and, just as we did for *Drosophila,* if we look only at male progeny of a dihybrid female, we are effectively sampling her gametic output. In other words we have a close approximation to a testcross. Consider the following situation involving the rare X-linked recessive alleles for defective sugar processing (*g*) and, at another locus, for color blindness (*c*). A doubly affected male (*c g / Y*) marries a normal woman (who is almost certainly *C G / C G*). The daughters of this mating are coupling-conformation heterozygotes. The male children of women of this type will provide an opportunity for geneticists to measure the frequency of recombinants issuing from the maternal meioses (Figure 5-19). A human X chromosome map of some genes causing X-linked phenotypes is shown in Figure 5-20. Note, however, that neutral DNA markers can also be used in this type of X chromosome mapping.

Linkage studies occupy a large portion of the routine day-to-day activities of geneticists. When the first variant allele of a gene is discovered, one of the first questions to be asked is "Where does it map?" Not only is this knowledge a necessary component in the engineering and maintenance of genetic stocks for research use, but it is also of fundamental importance in piecing together an overall view of the architecture of the chromosome—and, in fact, of the entire genome. Linkage is detected in the same way in studies from viruses to humans, and the key is the nonindependent way genes are transmitted from generation to generation. Look for this key when we discuss linkage in following chapters.

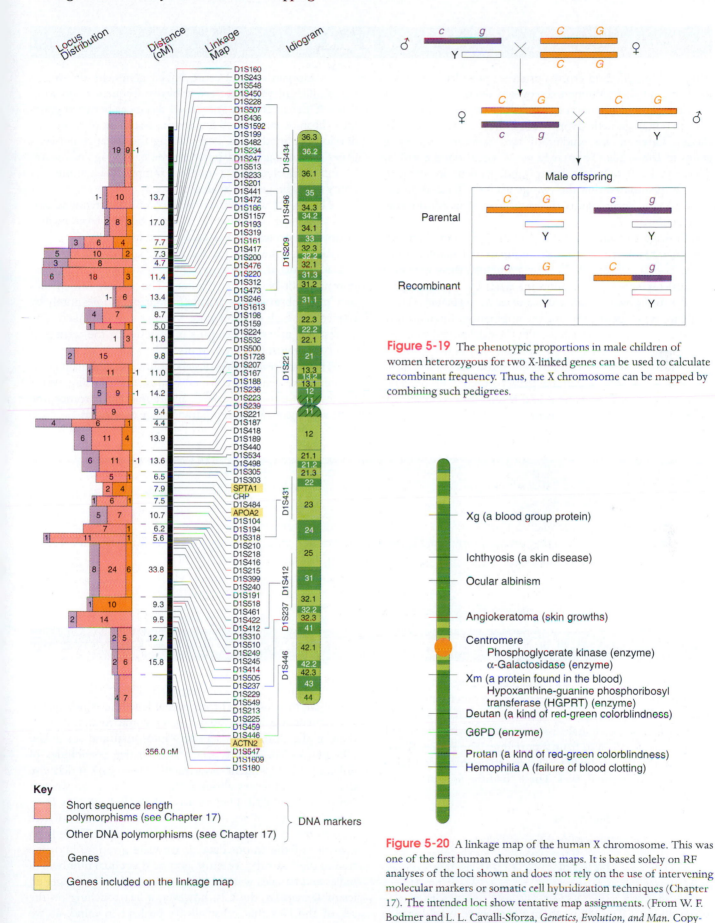

Figure 5-19 The phenotypic proportions in male children of women heterozygous for two X-linked genes can be used to calculate recombinant frequency. Thus, the X chromosome can be mapped by combining such pedigrees.

Figure 5-20 A linkage map of the human X chromosome. This was one of the first human chromosome maps. It is based solely on RF analyses of the loci shown and does not rely on the use of intervening molecular markers or somatic cell hybridization techniques (Chapter 17). The intended loci show tentative map assignments. (From W. F. Bodmer and L. L. Cavalli-Sforza, *Genetics, Evolution, and Man*. Copyright © 1976 by W. H. Freeman and Company.)

SUMMARY

After making dihybrid crosses of sweet pea plants, William Bateson and R. C. Punnett discovered deviations from the $9:3:3:1$ ratio of phenotypes expected in the F$_2$ generation. The parental gametic types outnumbered the other two classes. Later, in his studies of two different autosomal genes in *Drosophila,* Thomas Hunt Morgan found a similar deviation from Mendel's law of independent assortment. Morgan postulated that the two genes were located on the same pair of homologous chromosomes. This relationship is called *linkage.*

Linkage explains why the parental gene combinations stay together but not how the nonparental combinations arise. Morgan postulated that during meiosis there may be a physical exchange of chromosome parts by a process now called *crossing-over,* or *intrachromosomal recombination.* Thus, there are two types of meiotic recombination. Interchromosomal recombination is achieved by Mendelian independent assortment and results in a recombinant frequency of 50 percent. Intrachromosomal recombination occurs when the physical linkage of the parental gene combinations prevents their independent assortment and results in a recombinant frequency of less than 50 percent.

As Morgan studied more linked genes, he discovered many different values for recombinant frequency and wondered if these reflected the actual distances between genes on a chromosome. Alfred Sturtevant, a student of Morgan's, developed a method of determining the distance between genes on a linkage map, based on the percentage of recombinants. A linkage map is another example of a hypothetical entity based on genetic analysis.

The easiest way to measure recombinant frequencies is with a testcross of a dihybrid or trihybrid. However, recombinant frequencies can also be deduced from selfs.

Although the basic test for linkage is deviation from the $1:1:1:1$ ratio of progeny types in a testcross, such a deviation may not be all that obvious. The χ^2 test, which tells how often observations deviate from expectations purely by chance, can help us determine whether loci are linked. The χ^2 test has other applications in genetics in the testing of observed against expected events.

There are several theories about how recombinant chromosomes are generated. We now know that crossing-over is the result of physical breakage and reunion of chromosome parts and occurs at the four-strand stage of meiosis.

Concept Map

Draw a concept map interrelating as many of the following terms as possible. Note that the terms are listed in no particular order.

crossing-over / chromatids / mapping / chromosomes / testcross / linkage / map units / recombination / additivity / independent assortment

CHAPTER INTEGRATION PROBLEM

In *Drosophila,* the alleles *P* and *p* determine red and purple eyes, respectively. The alleles *B* and *b* determine brown and black body, respectively. A geneticist crossed a purple-eyed, brown-bodied female from a pure line with a red-eyed, black-bodied male, also from a pure line. The F$_1$ flies all had red eyes and brown bodies. The geneticist produced an F$_2$ by interbreeding the F$_1$ flies. The F$_2$ had the following composition:

brown, red	684
brown, purple	343
black, red	341
	1368

What do these data tell us about the chromosomal location of the genes?

Solution

When we think about chromosomal location, which of the phenomena we have studied so far come to mind? First, there is the distinction between autosomal and sex-linked inheritance. Then there are the alternative possibilities of linkage or independent assortment. There isn't much else that we should worry about at this stage. How can we sort out these possibilities? Let's start with sex linkage. The characteristic feature of sex linkage is that there is an inheritance pattern that is correlated with sex in some way. There is no evidence of this in our data; as we were given no information to the contrary, we must assume that equal numbers of males and females were obtained in each class. In the genetics of *Drosophila,* there is, however, a fact touching on the sex of the flies that must always be kept in mind; as we learned in this chapter, during meiosis in *Drosophila* males, there is no crossing-over. We will put that notion on the back burner for the moment; it might become useful later.

So we have concluded that both genes must be autosomal. Now we can reconstruct the crosses using gene symbols to restate what we know in a slightly different format.

$$P \quad BBpp \times bbPP$$
$$\downarrow$$
$$F_1 \quad BbPp$$

$$BbPp \times BbPp$$
$$\downarrow$$
$$F_2 \quad B-P-$$
$$B-pp$$
$$bbP-$$

Now we can see better what is so unusual about these data. The mating between the F_1 flies is a dihybrid cross, and from Mendel's second law we expect independently assorting genes, to give four F_2 phenotypic classes in a $9:3:3:1$ ratio. We have only three phenotypic classes, and their ratio is $2:1:1$. Clearly, something else is going on. The missing class is the $bbpp$ genotype, so for one thing, something must be happening to prevent the production of this class.

Remembering the discussion of lethality in Chapter 4, we might speculate that the $bbpp$ genotype is lethal. This is unlikely, however, because that would leave us with the $9:3:3$ part of the ratio, which would reduce to $3:1:1$, not $2:1:1$. But we have just considered another phenomenon that can keep genes from assorting freely to produce specific genotypes, that is, linkage, so let us explore that possibility. For this hypothesis we could write the first cross as

$$P \quad \frac{Bp}{Bp} \times \frac{bP}{bP}$$
$$\downarrow$$
$$F_1 \quad \frac{Bp}{bP}$$

If this is true, what ratios do we expect in the F_2? We might think that to answer this we must know the number of map units that separate the two loci. That is reasonable, but we must start somewhere, so let's assume that 20 map units separate the loci, then see what happens. (Don't forget about trial and error in analysis; scientists use it a lot as part of their calculations.) The expectations for the F_2 can be set up as follows, bearing in mind that there is no crossing-over in the males:

			Sperm		
Eggs		Bp 50%		bP 50%	
BP	10%	$BBPp$	5%	$BbPP$	5%
bp	10%	$Bbpp$	5%	$bbPp$	5%
Bp	40%	$BBpp$	20%	$BbPp$	20%
bP	40%	$BbPp$	20%	$bbPP$	20%

Overall, the phenotypic proportions are

$B-P-$	$5+5+20+20$	$=50\%$
$B-pp$	$5+20$	$=25\%$
$bbP-$	$5+20$	$=25\%$
$bbpp$		$=0\%$

This gives us precisely the ratio that we need, $2:1:1$! How did this happen? Were we just lucky in choosing the 20 percent recombinant frequency? The answer is no; it doesn't matter what frequency we might have chosen, the same result would have prevailed (try it).

In conclusion then, we can say that the ratio was produced by the inheritance of linked genes in repulsion phase in the parental strains, but we cannot determine how far apart they are. Precise mapping would require a testcross.

Although this was a tricky problem, the path we took through it has shown how a variety of different concepts can be used to rule out various possibilities and arrive at a solution that works.

SOLVED PROBLEMS

1. The allele b gives *Drosophila* flies a black body and b^+ gives brown, the wild-type phenotype. The allele wx of a separate gene gives waxy wings and wx^+ gives nonwaxy, the wild-type phenotype. The allele cn of a third gene gives cinnabar eyes and cn^+ gives red, the wild-type phenotype. A female heterozygous for these three genes is testcrossed, and 1000 progeny are classified as follows: 5 wild type; 6 black, waxy, cinnabar; 69 waxy, cinnabar; 67 black; 382 cinnabar; 379 black, waxy; 48 waxy; and 44 black, cinnabar.

a. Explain these numbers.

b. Draw the alleles in their proper positions on the chromosomes of the triple heterozygote.

c. If it is appropriate according to your explanation, calculate interference.

d. If two triple heterozygotes of the above type were crossed, what proportion of progeny would be black, waxy? (Remember, there is no crossing-over in *Drosophila* males.)

Solution

a. One of the general pieces of advice given earlier is to be methodical. Here it is a good idea to write out the genotypes that may be inferred from the phenotypes. The cross is a testcross of type

$$b^+b\ wx^+wx\ cn^+cn \times bb\ wxwx\ cncn$$

Notice that there are distinct pairs of progeny classes in terms of frequency. Already, we can guess that the two largest classes represent parental chromosomes, that the two classes of about 68 represent single crossovers in one region, that the two classes of about 45 represent single crossovers in the other region, and that the two classes of about 5 represent double crossovers. Note that a progeny group may be specified by listing only the non-wild-type phenotypes. We can write out the progeny as classes derived from the female's gametes, grouped as follows:

b^+	wx^+	cn	382
b	wx	cn^+	379
b^+	wx	cn	69
b	wx^+	cn^+	67
b^+	wx	cn^+	48
b	wx^+	cn	44
b	wx	cn	6
b^+	wx^+	cn^+	5
			1000

Writing the classes out this way confirms that the pairs of classes are in fact reciprocal genotypes arising from zero, one, or two crossovers.

At first, because we do not know the parents of the triple heterozygous female, it looks as if we cannot apply the definition of recombination in which gametic genotypes are compared with the two input genotypes that form an individual. But on reflection, the only parental types that make sense in terms of the data presented are $b^+b^+\ wx^+wx^+\ cncn$ and $bb\ wxwx\ cn^+cn^+$ because these are still the most common classes.

Now we can calculate the recombinant frequencies. For

$$b-wx,\ \text{the RF} = \frac{69 + 67 + 48 + 44}{1000} = 22.8\%$$

$$b-cn,\ \text{the RF} = \frac{48 + 44 + 6 + 5}{1000} = 10.3\%$$

$$wx-cn,\ \text{the RF} = \frac{69 + 67 + 6 + 5}{1000} = 14.7\%$$

The map is therefore

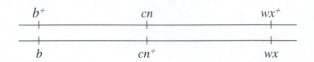

b. The parental chromosomes in the triple heterozygote were

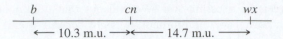

c. The expected number of double recombinants is $0.103 \times 0.147 \times 1000 = 15.141$. The observed number is $6 + 5 = 11$, so interference can be calculated as $I = 1 - 11/15.141 = 1 - 0.726 = 0.274 = 27.4$ percent.

d. Here we are asked to use the newly gained knowledge of the linkage of these genes to predict the outcome of a cross. Note that we are not dealing with a testcross here. We are asked for the expected proportion of black, waxy flies among the progeny of two triple heterozygotes. Because there is no crossing-over in the male, one half of his gametes are $b\ cn^+\ wx$ and the other half are $b^+\ cn\ wx^+$ but the latter are incapable of contributing to the black, waxy phenotype. In the female, two gametic types contribute to black, waxy offspring: $b\ cn^+\ wx$ and $b\ cn\ wx$. The contributions of the parents to the black, waxy phenotype are shown here:

		♀ gametes	
		$b\ cn^+\ wx$	$b\ cn\ wx$
♂ gametes	$b\ cn^+\ wx$	$bb\ cn^+cn^+\ wx\ wx$	$bb\ cn^+cn\ wx\ wx$

The chromosome carrying $b\ cn^+\ wx$ is a nonrecombinant parental type. There is a total of $(382 + 379)/1000 = 76.1$ percent of nonrecombinant parental chromosomes, half of which, or 38.05 percent, carry $b\ cn^+\ wx$. The frequency of $b\ cn\ wx$ is $(11 \div 2)/1000 = 0.55$ percent. In total, then, $38.05 + 0.55 = 38.6$ percent of the female's gametes are capable of forming a black, waxy offspring; half of these fuse with a $b\ cn^+\ wx$ gamete from the male, so the frequency of black, waxy offspring is $38.6/2 = 19.3$ percent.

In summary, the appropriate gametic fusion can be shown this way:

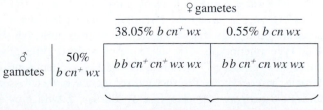

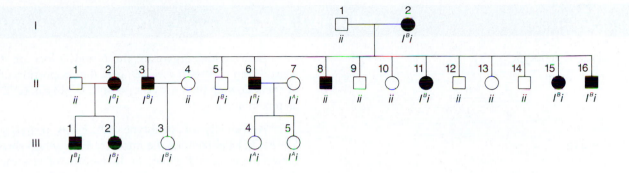

2. A human pedigree shows people affected with the rare nail-patella syndrome (misshapen nails and kneecaps) and also gives the ABO blood group genotype of each individual. Both loci concerned are autosomal. Study the pedigree above.

a. Is the nail-patella syndrome a dominant or recessive phenotype? Give reasons to support your answer.

b. Is there evidence of linkage between the nail-patella gene and the gene for ABO blood type, as judged from this pedigree? Why or why not?

c. If there is evidence of linkage, then draw the alleles on the relevant homologs of the grandparents. If there is no evidence of linkage, draw the alleles on two homologous pairs.

d. According to your model, which descendants represent recombinants?

e. What is the best estimate of RF?

f. If man III-1 marries a normal woman of blood type O, what is the probability that their first child will be blood type B with nail-patella syndrome?

Solution

a. Nail-patella syndrome is most likely dominant. We are told that it is a rare abnormality, so it is unlikely that the unaffected people marrying into the family carry a presumptive recessive allele for nail-patella syndrome. Let N be the causative allele. Then all people with the syndrome are heterozygotes Nn because all (probably including the grandmother, too) result from a mating to an nn normal individual. Notice that the syndrome appears in all three successive generations—another indication of dominant inheritance.

b. There is evidence of linkage. Notice that most of the affected people—those that carry the N allele—also carry the I^B allele; most likely, these alleles are linked on the same chromosome.

c.
$$\frac{n \qquad i}{n \qquad i} \times \frac{N \qquad I^B}{n \qquad i}$$

(The grandmother must carry both recessive alleles in order to produce offspring of genotype ii and nn.)

d. Notice that the grandparental mating is equivalent to a testcross, so the recombinants in generation II are

$$\text{II-5: } n\,I^B/n\,i \quad \text{and} \quad \text{II-8: } N\,i/n\,i$$

whereas all others are nonrecombinants, being either $N\,I^B/n\,i$ or $n\,i/n\,i$.

e. Notice that the grandparental cross and the first two crosses in generation II are identical and, of course, are all testcrosses. Three of the total 16 progeny are recombinant (II-5, II-8, and III-3). This gives a recombinant frequency of RF $= \frac{3}{16} = 18.8$ percent. (We cannot include the cross of II-6 × II-7, because the progeny cannot be designated as recombinant or not.)

f.
$$\text{(III-1 ♂)} \quad \frac{N \qquad I^B}{n \qquad i} \times \frac{n \qquad i}{n \qquad i} \quad \begin{array}{l}\text{(normal} \\ \text{type O ♀)}\end{array}$$

$$\downarrow$$

Gametes

$$81.2\% \begin{cases} N\,I^B & 40.6\% \\ n\,i & 40.6\% \end{cases} \longleftarrow \begin{array}{l}\text{Nail-patella,} \\ \text{blood type B}\end{array}$$

$$18.8\% \begin{cases} N\,i & 9.4\% \\ n\,I^B & 9.4\% \end{cases}$$

The two parental classes are always equal, and so are the two recombinant classes. Hence, the probability that the first child will have nail-patella syndrome and blood type B is 40.6 percent.

PROBLEMS

1. A plant of genotype

$$\frac{A \qquad B}{a \qquad b}$$

is testcrossed to

$$\frac{a \qquad b}{a \qquad b}$$

If the two loci are 10 m.u. apart, what proportion of progeny will be $Aa\,Bb$?

2. The A locus and the D locus are so tightly linked that no recombination is ever observed between them. If $AA\,dd$ is crossed to $aa\,DD$, and the F_1 is intercrossed, what phenotypes will be seen in the F_2 and in what proportions?

3. The R and S loci are 35 m.u. apart. If a plant of genotype

$$\frac{R \qquad S}{r \qquad s}$$

is selfed, what progeny phenotypes will be seen and in what proportions?

4. The cross $EE\,FF \times ee\,ff$ is made, and the F_1 is then backcrossed to the recessive parent. The progeny genotypes are inferred from the phenotypes. The progeny genotypes, written as the gametic contributions of the heterozygous parent, are in the following proportions:

$$E\,F \qquad \tfrac{2}{6}$$

$$E\,f \qquad \tfrac{1}{6}$$

$$e\,F \qquad \tfrac{1}{6}$$

$$e\,f \qquad \tfrac{2}{6}$$

Explain these results.

5. A strain of *Neurospora* with the genotype $H\,I$ is crossed with a strain with the genotype $h\,i$. Half the progeny are $H\,I$, and half are $h\,i$. Explain how this is possible.

6. A female animal with genotype $Aa\,Bb$ is crossed with a double-recessive male ($aa\,bb$). Their progeny include 442 $Aa\,Bb$, 458 $aa\,bb$, 46 $Aa\,bb$, and 54 $aa\,Bb$. Explain these results.

7. If $AA\,BB$ is crossed to $aa\,bb$, and the F_1 is testcrossed, what percent of the testcross progeny will be $aa\,bb$ if the two genes are **(a)** unlinked; **(b)** completely linked (no crossing-over at all); **(c)** 10 map units apart; **(d)** 24 map units apart?

8. In a haploid organism, the C and D loci are 8 m.u. apart. From a cross $C\,d \times c\,D$, give the proportion of each of the following progeny classes: **(a)** $C\,D$, **(b)** $c\,d$, **(c)** $C\,d$, **(d)** all recombinants.

9. A fruit fly of genotype $B\,R/b\,r$ is testcrossed to $b\,r/b\,r$. In 84 percent of the meioses, there are no chiasmata between the linked genes; in 16 percent of the meioses, there is one chiasma between the genes. What proportion of the progeny will be $Bb\,rr$? **(a)** 50 percent, **(b)** 4 percent, **(c)** 84 percent, **(d)** 25 percent, **(e)** 16 percent?

10. A three-point testcross was made in corn. The results and a partial recombination analysis are shown in the following display, which is typical of three-point testcrosses (p = purple leaves, + = green; v = virus-resistant seedlings, + = sensitive; b = brown midriff to seed, + = plain). Study the display and answer the questions below.

P $+ + + + + + \times pp\,vv\,bb$

Gametes $+ + +$ $p\,v\,b$

F_1 $+p\ +v\ +b \times pp\,vv\,bb$ (tester)

Class	Progeny phenotypes	F_1 gametes	Numbers	Recombinant for p-b	p-v	v-b
1	gre sen pla	$+\ +\ +$	3,210			
2	pur res bro	$p\ v\ b$	3,222			
3	gre res pla	$+\ v\ +$	1,024		R	R
4	pur sen bro	$p\ +\ b$	1,044		R	R
5	pur res pla	$p\ v\ +$	690	R		R
6	gre sen bro	$+\ +\ b$	678	R		R
7	gre res bro	$+\ v\ b$	72	R	R	
8	pur sen pla	$p\ +\ +$	60	R	R	
		Total	10,000	1,500	2,200	3,436

a. Determine which genes are linked.

b. Draw a map that shows distances in map units.

c. Calculate interference if appropriate.

Unpacking the Problem

a. Sketch cartoon drawings of the parent (P), F_1, and tester corn plants, and use arrows to show exactly how you would perform this experiment. Show where seeds are collected.

b. Why do all the $+$'s look the same, even for different genes? Why does this not cause confusion?

c. How can a phenotype be purple and brown (for example) at the same time?

d. Is it significant that the genes are written in the order *p-v-b* in the problem?

e. What is a tester and why is it used in this analysis?

f. What does the column marked "progeny phenotypes" represent? In class 1, for example, state exactly what "gre sen pla" means.

g. What does the line marked "gametes" represent, and how is this different from the column marked "F_1 gametes"? In what way is comparison of these two types of gametes relevant to recombination?

h. Which meiosis is the main focus of study? Label it on your drawing.

i. Why are the gametes from the tester not shown?

j. Why are there only eight phenotypic classes? Are there any classes missing?

k. What classes (and in what proportions) would be expected if all the genes are on separate chromosomes?

l. What do the four class sizes (two very big, two intermediate, two intermediate, two very small) correspond to?

m. What can you tell about gene order from inspecting the phenotypic classes and their frequencies?

n. What would be the expected phenotypic class distribution if only two genes are linked?

o. What does the word "point" refer to in a three-point testcross? Does this word useage imply linkage? What would a four-point testcross be like?

p. What is the definition of *recombinant*, and how is it applied here?

q. What do the "recombinant for" columns mean?

r. Why are there only three "recombinant for" columns?

s. What do the R's mean, and how are they determined?

t. What do the column totals signify? How are they used?

u. What is the diagnostic test for linkage?

v. What is a map unit? Is it the same as a centimorgan?

w. In a three-point testcross such as this one, why aren't the F_1 and the tester considered to be parental in calculating recombination? (They *are* parents in one sense.)

x. What is the formula for interference? How are the "expected" frequencies calculated in the coefficient of coincidence formula?

y. Why does part c of the problem say "if appropriate"?

z. How much work is it to obtain such a large progeny size in corn? Approximately how many progeny are represented by one corn cob?

11. An individual heterozygous for four genes, *A a B b C c D d,* is testcrossed to *a a b b c c d d,* and 1000 progeny are classified by the gametic contribution of the heterozygous parent as follows:

a B C D	42
A b c d	43
A B C d	140
a b c D	145
a B c D	6
A b C d	9
A B c d	305
a b C D	310

a. Which genes are linked?

b. If two pure-breeding lines were crossed to produce the heterozygous individual, what were their genotypes?

c. Draw a linkage map of the linked genes, showing the order and the distances in map units.

d. Calculate an interference value, if appropriate.

12. The squirting cucumber, *Echballium elaterium,* has two separate sexes (it is dioecious). The sexes are determined, not by heteromorphic sex chromosomes, but by the alleles of two genes. The alleles at the two loci govern sexual phenotypes as follows: *M* determines male fertility; *m* determines male sterility; *F* determines female sterility; *f* determines female fertility. In populations of this plant, individuals can be male (approximately 50 percent) or female (approximately 50 percent). In addition, a hermaphroditic type is found, but only at a very low frequency. The hermaphrodite has male and female sex organs on the same plant.

a. What must be the full genotype of a male plant? (Indicate linkage relations of the genes.)

b. What must be the full genotype of a female plant? (Indicate linkage relations of the genes.)

c. How does the population maintain an approximately equal proportion of males and females?

d. What is the origin of the rare hermaphrodite?

e. Why are hermaphrodites rare?

* 13. There is an autosomal gene *N* in humans that causes abnormalities in nails and patellae (kneecaps), called the *nail-patella syndrome.* Consider marriages in which one partner has the nail-patella syndrome and blood type A and the other partner has normal nail-patella and blood type O. These marriages produce some children who have both the nail-patella syndrome and blood type A. Assume that unrelated children from this phenotypic group mature, intermarry, and have children. Four phenotypes are observed in the following percentages in this second generation:

nail-patella syndrome, blood type A	66%
normal nail-patella, blood type O	16%
normal nail-patella, blood type A	9%
nail-patella syndrome, blood type O	9%

Fully analyze these data, explaining the relative frequencies of the four phenotypes.

*14. Using the data obtained by Bateson and Punnett (Table 5-1), calculate the map distance (in m.u.) separating the color and shape genes.

15. You have a *Drosophila* line that is homozygous for autosomal recessive alleles *a*, *b*, and *c*, linked in that order. You cross females of this line with males homozygous for the corresponding wild-type alleles. You then cross the F_1 heterozygous males with their heterozygous sisters. You obtain the following F_2 phenotypes (where letters denote recessive phenotypes and pluses denote wild-type phenotypes): 1364 $+++$, 365 abc, 87 $ab+$, 84 $++c$, 47 $a++$, 44 $+bc$, 5 $a+c$, and 4 $+b+$.

*a. What is the recombinant frequency between *a* and *b*? Between *b* and *c*?

b. What is the coefficient of coincidence?

16. R. A. Emerson crossed two different pure-breeding lines of corn and obtained a phenotypically wild-type F_1 that was heterozygous for three alleles that determine recessive phenotypes: *an* determines anther; *br*, brachytic; and *f*, fine. He testcrossed the F_1 to a tester that was homozygous recessive for the three genes and obtained these progeny phenotypes: 355 anther; 339 brachytic, fine; 88 completely wild-type; 55 anther, brachytic, fine; 21 fine; 17 anther, brachytic; 2 brachytic; 2 anther, fine.

a. What were the genotypes of the parental lines?

b. Draw a linkage map for the three genes (include map distances).

c. Calculate the interference value.

17. Chromosome 3 of corn carries three loci having alleles *b* and b^+, *v* and v^+, and *lg* and lg^+. The corresponding recessive phenotypes are abbreviated *b* (for plant-color booster), *v* (for virescent), and *lg* (for liguleless); pluses denote wild-type phenotypes. A testcross of triple recessives with F_1 plants heterozygous for the three genes yields the following progeny phenotypes: 305 $+vlg$, 275 $b++$,

128 $b+lg$, 112 $+v+$, 74 $++lg$, 66 $bv+$, 22 $+++$, and 18 $bvlg$. Give the gene sequence on the chromosome, the map distances between genes, and the coefficient of coincidence.

18. Groodies are useful (but fictional) haploid organisms that are pure genetic tools. A wild-type groody has a fat body, a long tail, and flagella. Non-wild-type lines are known that have thin bodies, or are tailless, or do not have flagella. Groodies can mate with each other (although they are so shy that we do not know how) and produce recombinants. A wild-type groody mates with a thin-bodied groody lacking both tail and flagella. The 1000 baby groodies produced are classified as shown in the following figure. Assign genotypes, and map the three genes.

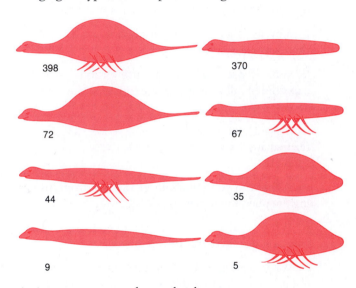

Assign genotypes, and map the three genes.

(Problem 18 from Burton S. Guttman.)

19. Assume that three pairs of alleles are found in *Drosophila*: x^+ and *x*, y^+ and *y*, and z^+ and *z*. As shown by the symbols, each non-wild-type allele is recessive to its wild-type allele. A cross between females heterozygous at these three loci and wild-type males yields the following progeny, classified by the gametic genotypes contributed by the females: 1010 $x^+y^+z^+$ females, 430 xy^+z males, 441 x^+yz^+ males, 39 xyz males, 32 x^+y^+z males, 30 $x^+y^+z^+$ males, 27 xyz^+ males, 1 x^+yz male, and 0 xy^+z^+ males.

a. In what chromosome of *Drosophila* are these genes carried?

b. Draw the relevant chromosomes in the heterozygous female parent, showing the arrangement of the alleles.

c. Calculate the map distances between the genes and the coefficient of coincidence.

20. From the five sets of data given in the following table determine the order of genes by inspection—that is, without calculating recombination values. Recessive phenotypes are symbolized by lowercase letters and dominant phenotypes by pluses.

Phenotypes observed in 3-point testcross	Data set				
	1	2	3	4	5
+ + +	317	1	30	40	305
+ + c	58	4	6	232	0
+ b +	10	31	339	84	28
+ b c	2	77	137	201	107
a + +	0	77	142	194	124
a + c	21	31	291	77	30
a b +	72	4	3	235	1
a b c	203	1	34	46	265

21. From the phenotype data given below for two three-point testcrosses involving (1) *a*, *b*, and *c* and (2) *b*, *c*, and *d*, determine the sequence of the four genes *a*, *b*, *c*, and *d*, and the three map distances between them. Recessive phenotypes are symbolized by lowercase letters and dominant phenotypes by pluses.

(1)		(2)	
+ + +	669	b c d	8
a b +	139	b + +	441
a + +	3	b + d	90
+ + c	121	+ c d	376
+ b c	2	+ + +	14
a + c	2,280	+ + d	153
a b c	653	+ c +	64
+ b +	2,215	b c +	141

22. In *Drosophila*, the allele dp^+ determines long wings and *dp* determines short ("dumpy") wings. At a separate locus e^+ determines gray body and *e* determines ebony body. Both loci are autosomal. The following crosses were made, starting with pure-breeding parents:

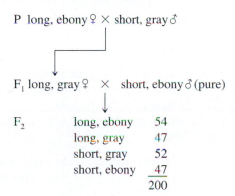

P long, ebony ♀ × short, gray ♂

F₁ long, gray ♀ × short, ebony ♂ (pure)

F₂	long, ebony	54
	long, gray	47
	short, gray	52
	short, ebony	47
		200

Use the χ^2 test to determine if these loci are linked. In doing so, state (**a**) hypothesis, (**b**) calculation of χ^2, (**c**) *p* value, (**d**) what the *p* value means, (**e**) conclusion, (**f**) inferred chromosomal constitutions of parents, F₁, tester, and progeny.

23. Two strains of haploid yeast each carrying two non-wild-type alleles were crossed to determine gene linkage.

Strain A had *asp* and *gal*, which caused it to require aspartate and be unable to utilize galactose; strain B had *rad* and *aro*, which caused it to be sensitive to radiation and to require aromatic amino acids. Haploid progeny were obtained and tested for genotype with the following results (where pluses denote wild-type alleles):

Genotype				Frequency
asp	gal	+	+	0.136
asp	+	+	rad	0.136
asp	gal	+	rad	0.064
asp	+	+	+	0.064
+	gal	aro	+	0.136
+	+	aro	rad	0.136
+	gal	aro	rad	0.064
+	+	aro	+	0.064
asp	gal	aro	+	0.034
asp	+	aro	rad	0.034
asp	gal	aro	rad	0.016
asp	+	aro	+	0.016
+	gal	+	+	0.034
+	+	+	rad	0.034
+	gal	+	rad	0.016
+	+	+	+	
				1.000

a. Calculate the six recombinant frequencies.

b. Draw a linkage map to illustrate positions of the four genetic loci and label in map units.

24. A geneticist puts a female mouse from a line that breeds true for wild-type eye and body color with a male mouse from a pure line having apricot eyes (determined by allele *a*) and gray coat (determined by allele *g*). The mice mate and produce an F₁ that is all wild type. The F₁ mice intermate and produce an F₂ of the following composition:

females	all wild type	
males	wild type	45%
	apricot, gray	45%
	gray	5%
	apricot	5%

a. Explain these frequencies.

b. Give the genotypes of the parents, and of both sexes of the F₁ and F₂.

25. The mother of a family with 10 children has blood type Rh⁺. She also has a very rare condition (elliptocytosis, phenotype E) that causes red blood cells to be oval rather than round in shape but that produces no adverse clinical effects. The father is Rh⁻ (lacks the Rh⁺ antigen) and has normal red cells (phenotype e). The children are 1 Rh⁺ e,

4 Rh$^+$ E, and 5 Rh$^-$ e. Information is available on the mother's parents, who are Rh$^+$ E and Rh$^-$ e. One of the ten children (who is Rh$^+$ E) marries someone who is Rh$^+$ e, and they have an Rh$^+$ E child.

a. Draw the pedigree of this whole family.

b. Is the pedigree in agreement with the hypothesis that the Rh^+ allele is dominant and Rh^- is recessive?

c. What is the mechanism of transmission of elliptocytosis?

*d. Could the genes governing the E and Rh phenotypes be on the same chromosome? If so, estimate the map distance between them, and comment on your result.

26. The father of Mr. Spock, first officer of the starship *Enterprise,* came from planet Vulcan; Spock's mother came from Earth. A Vulcan has pointed ears (determined by allele *P*), adrenals absent (determined by *A*), and a right-sided heart (determined by *R*). All these alleles are dominant to normal Earth alleles. The three loci are autosomal, and they are linked as shown in this linkage map:

If Mr. Spock marries an Earth woman and there is no (genetic) interference, what proportion of their children will have

a. Vulcan phenotypes for all three characters?

b. Earth phenotypes for all three characters?

c. Vulcan ears and heart but Earth adrenals?

d. Vulcan ears but Earth heart and adrenals?

(Problem 26 from D. Harrison, *Problems in Genetics.* Addison-Wesley, 1970.)

27. In a certain diploid plant, the three loci *A*, *B*, and *C* are linked as follows:

One plant is available to you (call it parental plant). It has the constitution $A b c / a B C$.

a. Assuming no interference, if the plant is selfed, what proportion of the progeny will be of the genotype $a b c / a b c$?

b. Again assuming no interference, if the parental plant is crossed with the $a b c / a b c$ plant, what genotypic classes will be found in the progeny? What will be their frequencies if there are 1000 progeny?

*c. Repeat part b, this time assuming 20 percent interference between the regions.

28. From several crosses of the general type $AA BB \times a a bb$ the F_1 individuals of type $Aa Bb$ were testcrossed to $a a bb$. The results are shown below.

Testcross of F_1 from cross	Testcross progeny			
	$Aa Bb$	$a a bb$	$Aa bb$	$a a Bb$
1	310	315	287	288
2	36	38	23	23
3	360	380	230	230
4	74	72	50	44

For each set of progeny, use the χ^2 test to decide if there is evidence of linkage.

29. Certain varieties of flax show different resistances to specific races of the fungus called *flax rust*. For example, the flax variety 77OB is resistant to rust race 24 but susceptible to rust race 22, whereas flax variety Bombay is resistant to rust race 22 and susceptible to rust race 24. When 77OB and Bombay were crossed, the F_1 hybrid was resistant to both rust races. When selfed, the F_1 produced an F_2 containing the phenotypic proportions shown below, where R stands for resistant and S stands for susceptible.

		Rust race 22	
		R	S
Rust race 24	R	184	63
	S	58	15

a. Propose a hypothesis to account for the genetic basis of resistance in flax to these particular rust races. Make a concise statement of the hypothesis, and define any gene symbols you use. Show your proposed genotypes of the 77OB, Bombay, F_1, and F_2 flax plants.

b. Test your hypothesis, using the χ^2 test. Give the expected values, the value of χ^2 (to two decimal places), and the appropriate probability value. Explain exactly what this value means. Do you accept or reject your hypothesis on the basis of the χ^2 test?

(Problem 29 adapted from M. Strickberger, *Genetics.* Macmillan, 1968.)

30. On the next page are two pedigrees in which a vertical bar in a symbol stands for steroid sulfatase deficiency, a horizontal bar for ornithine transcarbamylase deficiency.

a. Is there any evidence in these pedigrees that the genes determining the deficiencies are linked?

b. If the genes are linked, is there any evidence in the pedigree of crossing-over between them?

c. Draw genotypes of these individuals as far as possible.

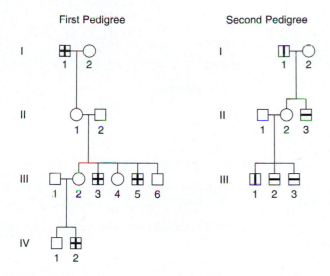

First Pedigree   Second Pedigree

31. The following pedigree shows a family with two rare abnormal phenotypes, blue sclerotics (a brittle bone defect), represented by the black borders to the symbols, and hemophilia, represented by the black centers to the symbols. Individuals represented by completely black symbols have both disorders.

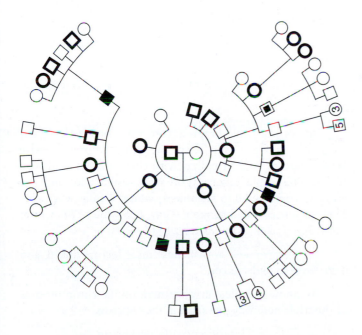

a. What pattern of inheritance is shown by each condition in this pedigree?

b. Provide the genotypes of as many family members as possible.

c. Is there evidence of linkage?

d. Is there any evidence of independent assortment?

e. Can any of the members be judged as recombinants (that is, formed from at least one recombinant gamete)?

32. In the following pedigree, the vertical lines stand for protan colorblindness, and the horizontal lines stand for deutan colorblindness. These are separate conditions causing different misperceptions of colors; each is determined by a separate gene.

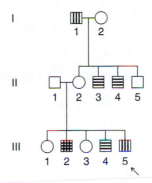

a. Does the pedigree show any evidence that the genes are linked?

b. If there is linkage, does the pedigree show any evidence of crossing-over? Explain both of your answers with the aid of the diagram.

c. Can you calculate a value for the recombination between these genes? Is this an intrachromosomal or an interchromosomal type of recombination?

*** 33.** The human genes for colorblindness and for hemophilia are both on the X chromosome, and they show a recombinant frequency of about 10 percent. Linkage of a pathological gene to a relatively harmless one can be used for genetic prognosis. Here is part of a more extensive pedigree. Blackened symbols indicate that the subjects had hemophilia, and crosses indicate colorblindness. What information could be given to the women III-4 and III-5 as to the likelihood of their having sons with hemophilia?

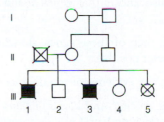

(Problem 33 adapted from J. F. Crow, *Genetics Notes: An Introduction to Genetics.* Burgess, 1983.)

34. A geneticist mapping the genes *A*, *B*, *C*, *D*, and *E* makes two three-point testcrosses. The first cross of pure lines is

$$AA\,BB\,CC\,DD\,EE \times aa\,bb\,CC\,dd\,EE$$

The geneticist crosses the F_1 with a recessive tester and classifies the progeny by the gametic contribution of the F_1:

A B C D E	316
a b C d E	314
A B C d E	31
a b C D E	39
A b C d E	130
a B C D E	140
A b C D E	17
a B C d E	13
	1000

The second cross of pure lines is

$$AA\,BB\,CC\,DD\,EE \times aa\,BB\,cc\,DD\,ee$$

The geneticist crosses the F_1 from this cross with a recessive tester and obtains

A B C D E	243
a B c D e	237
A B c D e	62
a B C D E	58
A B C D e	155
a B c D E	165
a B C D e	46
A B c D E	34
	1000

The geneticist also knows that genes *D* and *E* assort independently.

a. Draw a map of these genes, showing distances in map units wherever possible.

b. Is there any evidence of interference?

35. In the tiny crucifer *Arabidopsis thaliana* a new phenotype "glabrous" (lacking hairs) was discovered, inherited as a recessive (*gg*). It was suspected that the *G* locus was on chromosome 4, so a *gg* plant was crossed to a plant of genotype *yy* (yellow, caused by reduced amounts of chlorophyll), whose locus is known to be on chromosome 4. The F_1 was wild-type in appearance; this was selfed and the following F_2 obtained:

wild-type	514
yellow	237
glabrous	234
yellow, glabrous	15
total	1000

a. Is the *G* locus on chromosome 4?

b. If so, estimate how many map units it is from the *Y* locus.

c. Explain the frequencies of the yellow and the glabrous F_2 classes.

36. In the plant *Arabidopsis thaliana* the loci for pod length (*L* = long, *l* = short) and fruit hairs (*H* = hairy, *h* = smooth) are linked 16 map units apart on the same chromosome. The following crosses were made:

(i) $LL\,HH \times ll\,hh \rightarrow F_1$

(ii) $LL\,hh \times ll\,HH \rightarrow F_1$

If the F_1's from (i) and (ii) are crossed,

a. What proportion of the progeny is expected to be *ll h h*?

b. What proportion of the progeny is expected to be *Ll h h*?

37. In corn, a triple heterozygote was obtained carrying the mutant alleles *s* (shrunken), *w* (white aleurone), and *y* (waxy endosperm), all paired with their normal wild-type alleles. This triple heterozygote was testcrossed, and the progeny contained 116 shrunken, white; 4 fully wild-type; 2538 shrunken; 601 shrunken, waxy; 626 white; 2708 white, waxy; 2 shrunken, white, waxy; and 113 waxy.

a. Determine if any of these three loci are linked, and if so, show map distances.

b. Show the allele arrangement on the chromosomes of the triple heterozygote used in the testcross.

c. Calculate interference, if appropriate.

38. The *A* and *B* loci are linked 10 map units apart. Prove that when a repulsion dihybrid (*A b*/*a B*) is selfed the *z* value will be 0.002 as shown in Table 5-2.

6

Linkage II: Special Eukaryotic Chromosome Mapping Techniques

Some of the color variants of the ascomycete fungus *Aspergillus*, normally dark green. The alleles producing such color variants have been useful markers in the study of mitotic recombination, one of the topics of this chapter. The variant colors are usually pigments that are intermediates in the biochemical reaction sequences terminating in the dark-green pigment. (J. Peberdy, Department of Life Sciences, University of Nottingham, England.)

KEY CONCEPTS

▶ Double and higher multiple crossovers cause underestimates of map distances calculated from recombinant frequencies.

▶ Special mapping formulas provide accurate map distances corrected for the effect of multiple crossovers.

▶ In fungi, analysis of individual meioses can position centromeres on genetic maps.

▶ Analysis of individual meioses provides insight into the mechanisms of gene segregation and recombination during meiosis.

▶ Occasionally chromosomes cross over and assort in diploid cells undergoing mitosis.

I n Chapter 5 we discussed the basic method for mapping eukaryotic chromosomes. According to that method, the investigator measures the *recombinant frequency* in a *random sample* of products of *meiosis* and, from this, determines the degree of linkage between the genes in question. In this chapter we will extend the analysis beyond the three key concepts of the previous sentence:

Recombinant frequency

Random products

Meiosis

First, *recombinant frequency.* You will see that measuring recombinant frequency does not always give an accurate measurement of map distance, and sometimes corrections need to be made. Second, *random products.* Some organisms, because of the way they complete their life cycles, allow the investigator to study the products of individual meioses, and this allows a different view of recombination that is not possible using random meiotic products. Third, *meiosis.* It might come as a surprise to learn that genes can recombine at mitosis. This opens up some important mapping methodologies in those situations.

Mapping chromosomes is one of the central activities of geneticists. The goal of genetics, after all, is to understand the structure, function, and evolution of the genome, and, clearly, knowing the locations of genes is central to this task. The concepts in this chapter are all part of the everyday vocabulary of genetics. In the same way that earlier we extended Mendel's basic rules of segregation, now we have to extend the basic rules of mapping.

Accurate Calculation of Large Map Distances

In Chapter 5 we saw that the basic genetic method of measuring map distance is based on recombinant frequency (RF). A genetic map unit (m.u.) was defined as a recombinant frequency of 1 percent. This is a useful fundamental unit that has stood the test of time and is still used in genetics. However, the larger the recombinant frequency, the less accurate it is as a measure of map distance. In fact, map units calculated from larger recombinant frequencies are smaller than map units calculated from smaller recombinant frequencies, for a given pair of gene loci. We have already encountered this effect in examples from the previous chapter. Typically, when measuring recombination between three linked loci, the sum of the two internal recombinant frequencies is greater than the recombinant frequency between the outside loci. Using such data, what is the most accurate estimate that can be made of map distance between the two outside loci *A* and *C* in the following table?

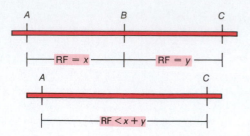

The answer is that $x + y$ is the best estimate, and more accurate than the smaller overall $A - C$ value. This gives us a useful mapping principle, which is as follows:

Message   The best estimates of map distance are obtained from the sum of the distances calculated for shorter subintervals.

However, what if we have no intervening marker loci available to measure recombination in shorter intervals? This would be a situation commonly encountered when beginning to map a new experimental organism, or in cases in which the genome is huge, as it is in human beings. For example, in the above diagram, what if there were no known *B* locus? Would we have to make do with the map distance value obtained directly from the $A - C$ recombinant frequency? Furthermore, what about the shorter intervals themselves? If there were other loci between *A* and *B*, and between *B* and *C*, then we might obtain even better estimates of the *A* to *C* distance. Luckily, there is a way of taking any recombinant frequency and performing a calculation to make it a more accurate measure of map distance, without studying shorter and shorter intervals.

Before we consider the calculation, let's think about the reason why larger RF values are less accurate measures of map distance. We have already encountered the culprit: multiple crossovers. In the previous chapter, we saw that double crossovers often lead to a parental arrangement of alleles and that, therefore, the resulting meiotic products are not counted when measuring recombinant frequency. The same is true for other types of multiple crossovers: triples, quadruples, and so on. So it is easy to see that these automatically lead to an underestimate of map distance, and since the multiples are expected to be relatively more common over longer regions, we can see why the problem is worse for larger recombinant frequencies.

How can we take these multiple crossovers into account when calculating map distances? What we need is a mathematical function that accurately relates recombination to map distance. In other words, what we need is a **mapping function.**

Message   A mapping function is a formula for using recombinant frequencies to calculate map distances corrected for multiple crossover products.

The Poisson Distribution

To derive a mapping function, we need a mathematical tool widely used in genetic analysis because it is useful in describing many different types of genetic processes. This mathematical tool is the **Poisson distribution.** A distribution is merely a description of the frequencies of the different types of classes that arise from sampling. The Poisson distribution describes the frequency of classes containing 0, 1, 2, 3, 4, . . . , i items when the average number of items per sample is small in relation to the total number of items possible. For example, the possible number of tadpoles obtainable in a single dip of a net in a pond is quite large, but most dips yield only one or two or none. The number of dead birds on the side of the highway is potentially very large, but in a sample mile the number is usually small. Such samplings are described well by the Poisson distribution.

Let's consider a numerical example. Suppose we randomly distribute 100 one-dollar bills to 100 students in a lecture room, perhaps by scattering them over the class from some point near the ceiling. The average (or mean) number of bills per student is 1.0, but common sense tells us that it is very unlikely that each of the 100 students will capture one bill. We would expect a few lucky students to grab three or four bills each and quite a few students to come up with two bills each. However, we would expect most students to get either one bill or none. The Poisson distribution provides a quantitative prediction of the results.

In this example, the item being considered is the capture of a bill by a student. We want to divide the students into classes according to the number of bills each captures, then find the frequency of each class. Let m represent the mean number of items (here, $m = 1.0$ bill per student). Let i represent the number for a particular class (say, $i = 3$ for those students who get three bills each). Let $f(i)$ represent the frequency of the i class—that is, the proportion of the 100 students who each capture i bills. The general expression for the Poisson distribution states that

$$f(i) = \frac{e^{-m} m^i}{i!}$$

where e is the base of natural logarithms (e is approximately 2.7) and ! is the factorial symbol (as examples, $3! = 3 \times 2 \times 1 = 6$ and $4! = 4 \times 3 \times 2 \times 1 = 24$; by definition, $0! = 1$). When computing $f(0)$, recall that any number raised to the power of 0 is defined as 1. Table 6-1 gives values of e^{-m} for m values from 0.000 to 1.000. Values for m greater than 1 can be obtained by calculation.

In our example, $m = 1.0$. Using Table 6-1, we compute the frequencies of the classes of students capturing 0, 1, 2, 3, and 4 bills as follows:

$$f(0) = \frac{e^{-1} 1^0}{0!} = \frac{e^{-1}}{1} = 0.368$$

$$f(1) = \frac{e^{-1} 1^1}{1!} = \frac{e^{-1}}{1} = 0.368$$

$$f(2) = \frac{e^{-1} 1^2}{2!} = \frac{e^{-1}}{2 \times 1} = \frac{e^{-1}}{2} = 0.184$$

$$f(3) = \frac{e^{-1} 1^3}{3!} = \frac{e^{-1}}{3 \times 2 \times 1} = \frac{e^{-1}}{6} = 0.061$$

$$f(4) = \frac{e^{-1} 1^4}{4!} = \frac{e^{-1}}{4 \times 3 \times 2 \times 1} = \frac{e^{-1}}{24} = 0.015$$

Figure 6-1 shows a histogram of this distribution. We predict that about 37 students will capture no bills, about 37 will capture one bill, about 18 will capture two bills, about 6 will capture three bills, and about 2 will capture four bills. This accounts for all 100 students; in fact, you can verify that the Poisson distribution yields $f(5) = 0.003$, which makes it likely that no student in this sample of 100 will capture five bills.

Similar distributions may be developed for other m values. Some are shown in Figure 6-2 as curves instead of bar histograms.

Table 6-1	Values of e^{-m} for m Values of 0 to 1*		
m	e^{-m}	m	e^{-m}
0.000	1.00000	0.550	0.57695
0.050	0.95123	0.600	0.54881
0.100	0.90484	0.650	0.52205
0.150	0.86071	0.700	0.49659
0.200	0.81873	0.750	0.47237
0.250	0.77880	0.800	0.44933
0.300	0.74082	0.850	0.42741
0.350	0.70469	0.900	0.40657
0.400	0.67032	0.950	0.38674
0.450	0.63763	1.000	0.36788
0.500	0.60653		

SOURCE: F. James Rohlf and Robert R. Sokal, *Statistical Tables*, 3d ed. W. H. Freeman and Company, 1995.

* Values for m greater than 1 can be obtained from an electronic calculator or by using logarithms.

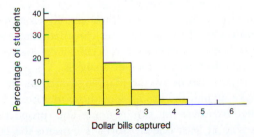

Figure 6-1 Poisson distribution for a mean of 1.0, illustrated by a random distribution of dollar bills to students.

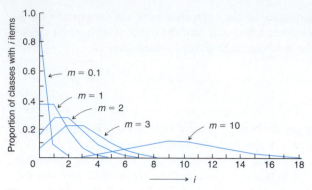

Figure 6-2 Poisson distributions for five different mean values: m is the mean number of items per sample, and i is the actual number of items per sample. (Modified from R. R. Sokal and F. J. Rohlf, *Introduction to Biostatistics.* W. H. Freeman and Company, 1973.)

Derivation of a Mapping Function

The Poisson distribution can also describe the distribution of crossovers along a chromosome during meiosis. In any chromosomal region, the actual number of crossovers is probably small in relation to the total number of possible crossovers in that region. If crossovers are distributed randomly (i.e. there is no interference), then if we knew the mean number of crossovers in the region per meiosis, we could calculate the distribution of meioses with zero, one, two, three, four, and more multiple crossovers. This is unnecessary in the present context because, as we shall see, the only class that is really crucial is the zero class. We want to correlate map distances with observable RF values. Meioses in which there are one, two, three, four, or any finite number of crossovers per meiosis all behave similarly in that they produce an RF of 50 percent among the products of those meioses, whereas the meiosis with no crossovers produce an RF of 0 percent. To see how this can be so, consider a series of meioses in which nonsister chromatids do not cross over, cross over once, and cross over twice, as shown in Figure 6-3. We obtain recombinant products only from meioses with at least one crossover in the region, and always precisely half the products of such meioses are recombinant. We see then that the real determinant of the RF value is the size of the zero crossover class in relation to the rest.

As noted in Figure 6-3, we consider only crossovers between nonsister chromatids; sister-chromatid exchange is thought to be rare at meiosis. If it occurs, it can be shown to have no net effect in most meiotic analyses.

At last, we can derive the mapping function. Recombinants make up half the products of those meioses having at least one crossover in the region. The proportion of meioses with at least one crossover is 1 minus the fraction with zero crossovers. The zero-class frequency will be

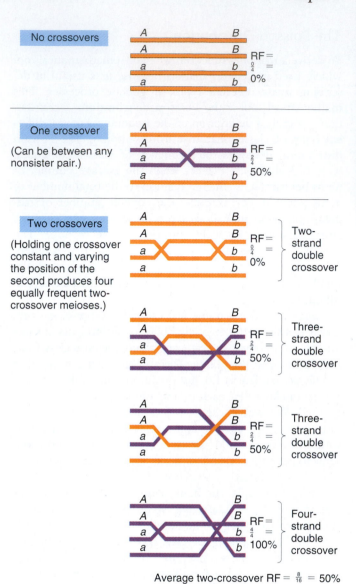

Figure 6-3 Demonstration that the average RF is 50 percent for meioses in which the number of crossovers is not zero. Recombinant chromatids are burgundy. Two-strand double crossovers produce all parental types, so all the chromatids are black. Note that all crossovers are between nonsister chromatids. Try the triple crossover class yourself.

$$\frac{e^{-m} m^0}{0!}$$

which equals

$$e^{-m}$$

so the mapping function can be stated as

$$\mathrm{RF} = \tfrac{1}{2}(1 - e^{-m})$$

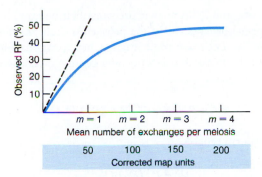

Figure 6-4 The blue line gives the mapping function in graphic form. Where the blue curve and the dashed line coincide, the function is linear and RFs estimate correct map distances well.

This formula relates recombinant frequency to m, the mean number of crossovers. Since the whole concept of genetic mapping is based on the occurrence of crossovers, and proportionality between crossover frequency and the physical size of a chromosomal region, then you can see that m is probably the most fundamental variable in the whole process. In fact, m could be considered to be the ultimate genetic mapping unit.

If we know an RF value, we can calculate m by solving the equation. After obtaining many values of m, we can plot the function as a graph, as in Figure 6-4. Viewing the function plotted as a graph should help us see how it works. First, notice that the function is linear for a certain range corresponding to very small m values. (Remember that m is our best measure of genetic distance.) Therefore, RF is a good measure of distance where the dashed line coincides with the function in Figure 6-4. In this region, the map unit defined as 1 percent RF has real meaning. Therefore, let's use this region of the curve to define corrected map units by considering some small values of m:

For $m = 0.05$, $e^{-m} = 0.95$, and

$$\text{RF} = \tfrac{1}{2}(1 - 0.95) = \tfrac{1}{2}(0.05) = \tfrac{1}{2} \times m$$

For $m = 0.10$, $e^{-m} = 0.90$, and

$$\text{RF} = \tfrac{1}{2}(1 - 0.90) = \tfrac{1}{2}(0.10) = \tfrac{1}{2} \times m$$

We see that $\text{RF} = \tfrac{m}{2}$, and this relation defines the dashed line in Figure 6-4. It allows us to translate m values into corrected map units. Expressing m as a percentage, we see that an m of 100 percent ($= 1$) is the equivalent of 50 corrected map units. Since an m value of 1 is the equivalent of 50 corrected map units, we can express the horizontal axis of Figure 6-4 in our new map units. Now we can see from the graph that two loci separated by 150 corrected map units show an RF of only 50 percent. We can use the graph of the function to convert any RF into map distance simply by

drawing a horizontal line from the RF value to the curve and dropping a perpendicular to the map unit axis—a process equivalent to using the equation $\text{RF} = \tfrac{1}{2}(1 - e^{-m})$ to solve for m.

Let's consider a numerical example of the use of the mapping function. Suppose we get an RF of 27.5 percent. How many corrected map units does this represent? From the function,

$$0.275 = \tfrac{1}{2}(1 - e^{-m})$$

$$0.55 = 1 - e^{-m}$$

Therefore

$$e^{-m} = 1 - 0.55 = 0.45$$

From e^{-m} tables (or by using a calculator), we find that $m = 0.8$, which is the equivalent of 40 corrected map units. If we had been happy to accept 27.5 percent RF as representing 27.5 map units, we would have considerably underestimated the distance between the loci.

Message To estimate map distances most accurately, put RF values through the mapping function. Alternatively, add distances that are each short enough to be in the region where the mapping function is linear.

A corollary of the second statement of this message is that for organisms for which the chromosomes are already well mapped, such as *Drosophila*, a geneticist seldom needs to calculate from the map function to place newly discovered genes on the map. This is because the map is already divided into small, marked regions by known loci. However, when the process of mapping has just begun in a new organism or when the available genetic markers are sparsely distributed, the corrections provided by the function are needed.

Notice that no matter how far apart two loci are on a chromosome, we never observe an RF value of greater than 50 percent. Consequently, an RF value of 50 percent would leave us in doubt about whether two loci are linked or are on separate chromosomes. Stated another way, as m gets larger, e^{-m} gets smaller and RF approaches $\tfrac{1}{2}(1 - 0) = \tfrac{1}{2} \times 1 = 0.5$, or 50 percent. This is an important point: RF values of 100 percent can never be observed, no matter how far apart the loci are.

Analysis of Single Meioses

In Chapter 5 we encountered a type of organism in which the products of single meioses remain together as groups of four cells called a **tetrad**, a term based on the Greek word for "four." Tetrad analysis is possible only in some fungi and single-celled algae. In the mushroomlike fungi (Basidiomycetes), a single meiosis produces a group of four basidiospores—the spores that fall to the earth from the mushroom cap. In ascomycete fungi such as yeasts and molds, the products of a single meiosis are enclosed in a sac called an **ascus.** The spores in the ascus are called *ascospores.* The unicellular alga *Chlamydomonas* also has an ascuslike structure. Basidiospores, ascospores, and the equivalent algal spores are types of **sexual spores.**

Advantages of Haploids for Genetic Analysis

First, let us take note of the fact that fungi and unicellular algae are microbial eukaryotes. This means that just like the prokaryotic microbes, bacteria, they can be propagated easily and cheaply, take up little space, and can be studied in large numbers. These are also useful properties in genetic research. Moreover, recall that fungi and algae are haploid. This has two basic advantages for genetic studies. First, because there is only one chromosome set, dominance and recessiveness normally do not obscure gene expression, and the phenotype is a direct reflection of the genotype. Second, in addition to the usefulness of tetrad analysis, even random meiotic product analysis is easier than in diploids. Recall that in the haploid life cycle the two haploid parental cells unite to form the diploid meiocyte, so for the geneticist this is the only meiosis to worry about. In contrast, in diploids the analyst must consider meioses in both parents. As an example, let's look at how easy it is to study linkage in haploids using random meiotic products. We can make a cross such as $a^+ b^+ \times a\, b$, and test a random sample of meiotic products. This is straightforward because each product develops directly into a haploid progeny individual. We might find the following genotypic frequencies in the progeny:

$$a^+ b^+ \quad 45\%$$
$$a\, b \quad 45\%$$
$$a^+ b \quad 5\%$$
$$a\, b^+ \quad 5\%$$

From such data, it is straightforward to tell which progeny are parental (the first two classes) and which are recombinant (the latter two classes). Then it is simple to compute the RF value as 10 percent, indicating linkage of the two genes 10 map units apart.

It is interesting to note at this point that molecular technology has opened up the ability to analyze haploid products of meiosis in humans. As noted in the previous chapter, neutral variant sites in the DNA are detectable using molecular probes, and "dihybrids" for two such sites can be used to study recombination between them. For example, sperm samples can be diluted so that there is one sperm per test tube. The DNA in one sperm is enough to enable routine screens to be made for neutral variant "alleles." To screen a specific site, the DNA at that site must first be amplified using a special process called the *polymerase chain reaction* (see Chapter 14). Let's assume a man is heterozygous for sites we will call *DNA-P* and *DNA-Q.* His genotype can be designated

$$DNA\text{-}P'\; DNA\text{-}P''\; DNA\text{-}Q'\; DNA\text{-}Q''$$

and his sperm will be

$$DNA\text{-}P'\; DNA\text{-}Q'$$
$$DNA\text{-}P'\; DNA\text{-}Q''$$
$$DNA\text{-}P''\; DNA\text{-}Q'$$
$$DNA\text{-}P''\; DNA\text{-}Q''$$

and their relative frequencies will determine if the *P* and *Q* sites are linked or not. Hence for some loci humans can be analyzed with the same ease as haploids.

Benefits of Analyzing Individual Meioses in Genetics

An important feature of some haploid species is that they produce tetrads, which permit the study of individual meioses. You might be wondering why the study of individual meioses is important; after all, we have just developed some of the basic analytical rules of genetics in the previous chapters without using them. The first answer to this question is that by studying individual meioses the geneticist can make direct observations on the behavior of genes in meiotic processes, with less need for inferences. Consider, for example, the inference that we implicitly make when studying allele segregation in a heterozygote, say *A a.* When we observe equal numbers of *A* and *a* alleles in a random sample of gametes, we attribute this to segregation of the homologous chromosomes carrying *A* and *a* in *individual* meioses. However this is an inference that cannot be confirmed directly in most eukaryotic organisms. Tetrad analysis shows directly that such segregation does take place. This kind of directness is useful in genetic research. In addition, tetrads permit several kinds of studies that are not possible using conventional analysis. In Chapter 5 we encountered some of these studies, specifically those concerning the distribution of crossovers between the four chromatids during the crossover process. Another analysis not possible with random meiotic products is the mapping of centromeres in relation to other loci. Lastly, observations of exceptional allele segregations in tetrad analysis gave rise to one of the central molecular models of crossing-over, the heteroduplex (or hybrid) DNA model; we will discuss this in detail in Chapter 20.

But do fungi and algae have typical eukaryotic chromosomes with typical eukaryotic behavior? The answer is generally yes, so discoveries made in these model systems can be applied with reasonable confidence to plants and animals.

In summary, tetrads provide a direct method for the genetic analysis of individual meioses, and processes can be studied that are not accessible by analyzing random meiotic products. Meiosis is one of the central processes of eukaryotic biology. Although a great deal is known about meiosis, there is still much to be learned. In fact it is curious that despite the wealth of molecular knowledge that is available on how cells carry out their day-to-day affairs, the processes that were some of the first ever to be identified by geneti-

cists (pairing, crossing-over, and segregation) remain among the most mysterious of cellular processes.

Message   Tetrad analysis makes it possible to study individual meioses, a useful ability for the investigation of recombination and assortment.

In some fungi and algae, the cells that represent the four products of meiosis develop directly into a tetrad of four sexual spores (see Figure 6-5). In certain fungi, each of the four products of meiosis undergoes an additional mi-

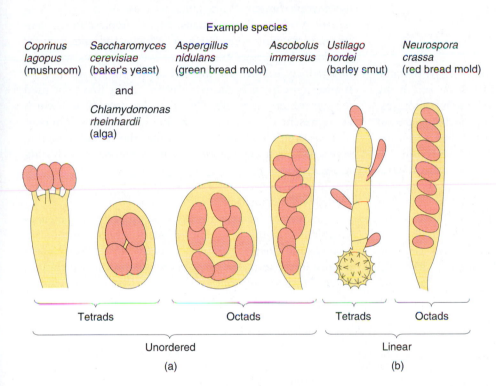

Example species

Coprinus lagopus (mushroom)   *Saccharomyces cerevisiae* (baker's yeast)   *Aspergillus nidulans* (green bread mold)   *Ascobolus immersus*   *Ustilago hordei* (barley smut)   *Neurospora crassa* (red bread mold)

and

Chlamydomonas rheinhardii (alga)

Tetrads   Octads   Tetrads   Octads

Unordered   Linear

(a)   (b)

Figure 6-5   Various forms of tetrads and octads found in different organisms. (a) Unordered. (b) Linear. (c) Normally maturing asci of *Neurospora crassa*. (Namboori B. Raju, *European Journal of Cell Biology*, 1980.)

(c)

totic division, yielding a group of eight cells called an **octad.** However, an octad is simply a double tetrad composed of four spore pairs. The members of a spore pair are identical, being mitotic daughter cells of one of the four products of meiosis.

The sexual spores, whether four or eight in number, can be found in a variety of arrangements. In some species, the spores are found in a jumbled arrangement called an **unordered tetrad,** shown in Figure 6-5a. In other species, the spores are arranged in a striking linear arrangement called a **linear tetrad,** shown in Figure 6.5b. Unordered and linear tetrads each have their specific uses in genetic analysis. We shall deal with the linear types first.

Linear Tetrad Analysis

How are linear tetrads produced? The key fact is that the spindles of the first and second meiotic divisions and of the postmeiotic mitosis are positioned end to end in the long ascus sac and do not overlap. The reason is probably related to the fact that these divisions occur in a tubelike structure, which physically prevents the spindles from overlapping. In any case, the absence of spindle overlap means that the nuclei are laid out in a straight array, and the lineage of each of the eight nuclei of the final octad therefore can be traced back through meiosis, as shown in Figure 6-6.

As you might expect, linear tetrads directly illustrate the segregation and independent assortment of genes at meiosis. Linear tetrads are also ideally suited to a special kind of analysis that is not possible in most organisms: **cen-**

tromere mapping, the locating of centromeres in relation to gene loci on the chromosomes. The centromere is a fascinating region of the chromosome that interacts with the spindle fibers and ensures proper chromosome movement during nuclear division. When this process fails, the daughter cells have abnormal chromosome numbers, which can lead to death or phenotypic abnormality. For example, abnormal chromosome numbers in humans have produced a major class of genetic diseases (which we shall discuss in Chapter 9). Therefore, the centromere is a subject of considerable attention in genetics.

How does centromere mapping work? We shall use the ascomycete fungus *Neurospora crassa* as an example. *Neurospora* produces linear octads. In its simplest form, centromere mapping considers a gene locus and asks how far this locus is from its centromere. We could pick any *Neurospora* locus to illustrate this technique, but we will use the **mating-type locus** because it was one of the first to be used for centromere mapping. This locus has two alleles, which are represented as *A* and *a*, even though neither is dominant or recessive. These alleles determine the *A* and *a* mating types. Although the mating-type phenomenon is interesting in itself, here we are merely using the locus as a genetic marker to illustrate the analysis, and we will not be concerned with its function.

Centromere mapping is based on the fact that a meiosis in which nonsister chromatids cross over between the centromere and a heterozygous locus produces a different allele pattern in the tetrad or octad than a meiosis in which nonsister chromatids do not cross over in that region. Figures

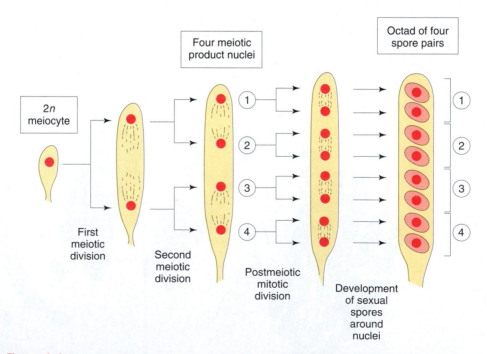

Figure 6-6 Meiosis and postmeiotic mitosis in a linear tetrad. The nuclear spindles do not overlap at any division, so that nuclei never pass each other in the sac. The resulting eight nuclei are laid down in a linear array that can be traced back through the divisions.

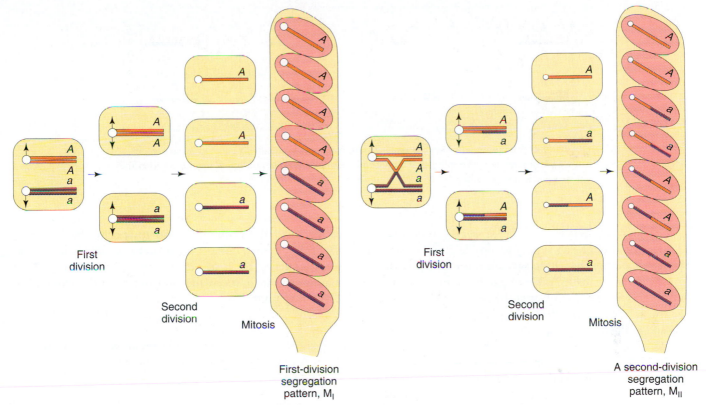

Figure 6-7 *A* and *a* segregate into separate nuclei at the first meiotic division when there is no crossover between the centromere and the locus. The resultant allele pattern in the octad is called a *first-division segregation pattern*.

Figure 6-8 *A* and *a* segregate into separate nuclei at the second meiotic division when there is a crossover between the centromere and the locus.

6-7 and 6-8 show examples of these two possibilities. The simpler pattern, shown in Figure 6-7, arises when there is no crossover. This pattern is typified by all the products at one end of the tetrad or octad carrying one allele and the products at the other end carrying the other allele. You can see from the diagram how this pattern is produced: because there is no spindle overlap, the *A*-bearing nuclei and the *a*-bearing nuclei never pass each other in the ascus. Also notice that although the *A* and *a* alleles are together in the diploid meiocyte nucleus, the first meiotic division cleanly segregates the *A* and *a* alleles, and these alleles remain separate throughout the second division of meiosis. This gives rise to the term **first-division segregation,** and the allele pattern in the spores is called a **first-division segregation pattern,** or M_I **pattern.**

When nonsister chromatids do cross over between the centromere and the locus (see Figure 6-8), the *A* and *a* alleles are still together in the nuclei at the end of the first division of meiosis. There has been no first-division segregation. However, the second meiotic division does move the *A* and *a* alleles into separate nuclei, giving rise to the term **second-division segregation.** The resulting allele pattern in the tetrad or octad is called a **second-division segregation pattern,** or M_{II} **pattern.** This pattern is typified by any

arrangement in which a half tetrad or a half octad contains both alleles.

Now let's look at the following experimental results from a *Neurospora* octad analysis, and interpret them in light of these ideas. Remember, the cross was $A \times a$. The octad patterns obtained were the following:

			Octads		
A	*a*	*A*	*a*	*A*	*a*
A	*a*	*A*	*a*	*A*	*a*
A	*a*	*a*	*A*	*a*	*A*
A	*a*	*a*	*A*	*a*	*A*
a	*A*	*A*	*a*	*a*	*A*
a	*A*	*A*	*a*	*a*	*A*
a	*A*	*a*	*A*	*A*	*a*
a	*A*	*a*	*A*	*A*	*a*
126	132	9	11	10	12

Total = 300

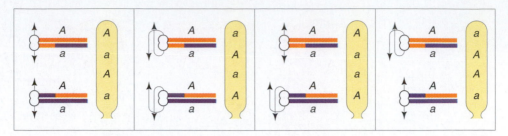

Figure 6-9 Four second-division segregation ascus patterns are equally frequent in linear asci because centromeres attach to spindles at the second meiotic division at random as regards whether they will be pulled "up" or "down."

The first two octads have the M_I segregation pattern, with the first octad being simply an upside-down version of the second. Notice that these two are more or less equal in frequency (126 versus 132). This equality simply reflects that centromeres attach to spindles at the first meiotic division at random as regards whether the arrangement will pull A "up" and a "down" or a "up" and A "down." We can deduce that $126 + 132 = 258$ meioses out of 300, or 86 percent of meioses, had no crossover in the region between the mating-type gene and the centromere.

The remaining four octads all have both A and a present in half octads; hence, by the above definition A and a segregated not at the first but at the second division of meiosis. What accounts for these four different variations on the same basic M_{II} theme? Once again, it is the randomness of spindle attachment, not only at the first but also at the second division of meiosis. This is illustrated in Figure 6-9. The M_{II} patterns in our example total $9 + 11 + 10 + 12 = 42$, or 14 percent, and show that nonsister chromatids crossed over between the mating-type locus and the centromere in 14 percent of the meioses.

We have measured the M_{II} pattern frequency at 14 percent in this example. Does this mean that the mating-type locus is 14 map units from the centromere? The answer is no, but this value can be used to calculate the number of map units. The 14 percent value is a percentage of meioses, and this is not the way map units are defined. Map units are defined in terms of the percentage of recombinant chromosomes issuing from meiosis. Figure 6-10 shows that when a crossover occurs in the region spanned by the centromere and the locus, only half the chromosomes issuing from that

meiosis will be recombinant. So to specify the length of the region in map units, it is necessary to divide the M_{II} pattern frequency by 2. In our example, the distance of the mating-type locus from the centromere is therefore $14 \div 2 = 7$ map units.

> **Message** To calculate the distance of a locus firm its centromere in map units, measure the percentage of tetrads showing second-division segregation patterns for that locus and divide by 2.

The above analysis can be extended to any number of heterozygous pairs segregating in a cross. Let's consider meiosis in a *Neurospora* diploid meiocyte of genotype $a^+ a\ b^+ b$ as an example. We do not know if the two loci are linked, nor do we know how they are located relative to centromeres. There are only limited possibilities, however:

1. The loci are on separate chromosomes.
2. The loci are on opposite sides of the centromere on the same chromosome.
3. The loci are on the same side of the centromere on the same chromosome.

The first two possibilities would show independence of the M_{II} patterns for both loci because a crossover in one arm cannot produce an M_{II} pattern in the other arm. The third possibility is more interesting in that a crossover in the region between the centromere and the locus closest to it would produce coinciding M_{II} patterns for both the two heterozygous loci (barring rare double crossovers). Hence, if a is closer to the centromere than b, for example, we would expect that most asci with M_{II} patterns for a should also show an M_{II} pattern for b (Figure 6-11). However, some M_{II} patterns for gene b will be the result of crossovers in the a-to-b region and will not show a coincident M_{II} pattern for gene a. Such coincidence of M_{II} patterns provides clues about linkage between loci.

> **Message** Coincident second-division segregations can indicate the location and order of genes on the same chromosome arm.

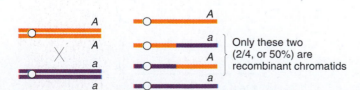

Figure 6-10 Only half the chromatids from a meiosis with a single crossover are recombinant.

Figure 6-12 When a locus is far from the centromere, a large number of crossovers effectively unlink the locus from the centromere.

An intuitive way to calculate the maximum M_{II} frequency is by the following "thought experiment." Consider a heterozygous locus, b^+b, that is so far from the centromere that there are numerous crossovers in the intervening region (Figure 6-12). These many crossovers effectively uncouple the locus from the centromere, and the two b^+ alleles and the two b alleles can end up in the tetrad in any arrangement. We can simulate this by considering how many different ways there are of dropping four marbles (two b^+ and two b) into a test tube (follow this with Figure 6-13). The first marble can be b^+ or b; it makes no difference. Let's assume that it is b^+. We then have two b's and one b^+ left. The next marble determines if the pattern will be M_I or M_{II}. If the marbles are dropped at random into the tube over and over again, one-third of the time the second marble will be b^+; thus, the third and fourth marbles must be b, and an M_I pattern is generated. The other two-thirds of the time, the second marble will be b, generating an M_{II} pattern. Therefore, we see that even with this very large number of crossovers, the $\frac{2}{3}$ frequency of M_{II} patterns can never be exceeded.

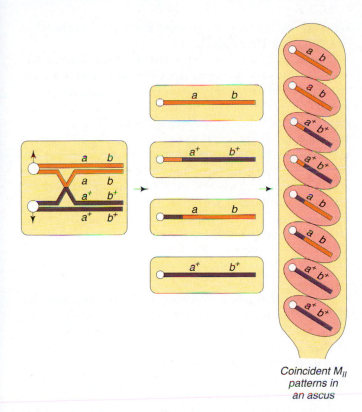

Figure 6-11 When two loci are on the same chromosome arm, a crossover between the centromere and the locus closest to it produces a coincident M_{II} pattern for both loci.

The farther a locus is from the centromere, the greater will be the M_{II} frequency. But the M_{II} frequency never reaches 100 percent; in fact, the theoretical maximum is 67 percent, or $\frac{2}{3}$. The reason for this is that multiple crossovers (especially doubles) become more and more prevalent as the interval becomes larger, and double crossovers can generate M_I patterns as well as M_{II} patterns. For example:

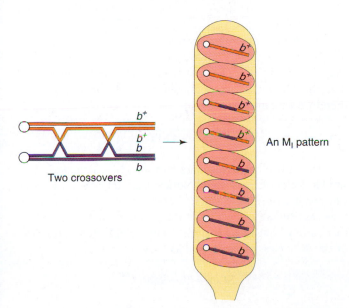

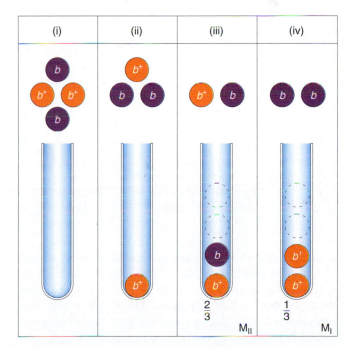

Figure 6-13 Demonstration of the limiting M_I and M_{II} segregation pattern frequencies of $\frac{1}{3}$ and $\frac{2}{3}$ in linear tetrad analysis. (See text for details.)

This raises an enigma. We feel intuitively that this M_{II} maximum should be directly equivalent to the 50 percent RF maximum observed for very large map distances, but $66.7 \div 2 = 33.3$ percent, which would be 33.3 m.u. The villain, once again, is the existence of multiple crossovers. We could derive a map function for M_{II} patterns, but in practice, it is simpler to stick to the analysis of smaller intervals and to recognize that the larger M_{II} frequencies provide increasingly inaccurate estimates of distances between centromeres and loci.

Unordered Tetrad Analysis

In unordered tetrads or octads, the spores are in no particular sequence, so that centromere mapping studies of the type just discussed are not possible. Nevertheless, unordered tetrads are generally useful in analyzing many aspects of meiotic segregation and recombination. Some of these uses will be explored in the problem sets. We will illustrate their use here with one example, which shows how they can be used as an alternative to a mapping function to correct for double crossovers. In this example, we will see again the directness of the tetrad approach, avoiding some of the assumptions of the mathematical method.

Let us assume that we are studying two yeast loci a and b and we want to determine if they are linked and, if so, calculate an accurate map distance. Yeast produces unordered tetrads. How can we use these tetrads to explore linkage? First let's see what different types of unordered tetrad can be expected in such a cross. It turns out that in any dihybrid cross, with or without linkage, there are only three possible unordered tetrad types, and they are defined as follows, using a cross of $a^+ b^+ \times a\, b$:

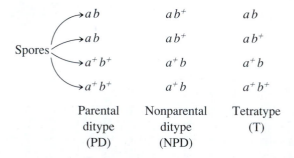

Remember that these asci are unordered, so even though the first type might look like a case of M_I segregation for both loci, that is not the case: the spores could have been written equally well in any order. These asci merely have been classified according to whether they contain two genotypes (*ditypes*) or four genotypes (*tetratypes*, represented by T). Within the ditype class, both genotypes can be either parental (PD) or nonparental (NPD). You might try writing

out a few asci to convince yourself that no types other than PD, NPD, and T are possible.

Now, what do these types tell us about linkage? Remember that only the three types PD, NPD, and T are obtained, no matter whether the loci are linked or not. However, if presented with the frequencies of these three types, one could determine if two loci are linked. You will notice from examining the three types that only the NPD and T types contain recombinants, so these are key classes in determining recombinant frequency. The NPD class contains only recombinants, whereas only half the spores in the T class are recombinant. Therefore, we could write a formula for determining RF using tetrads, where T and NPD represent the percentages of those classes:

$$RF = \tfrac{1}{2}T + NPD$$

If this formula gives a frequency of 50 percent, then we know the loci must be unlinked, and correspondingly if the RF is less than 50 percent, then the genes must be linked and we could use that value to represent the number of map units between them. However, just as with other linkage analyses we have studied earlier in the chapter, this is an underestimate because it does not consider double recombinants and other higher-level crossovers. However, the frequencies of PD, NPD, and T can be used to make a correction for doubles. First we need to understand how the PD, NPD, and T classes are produced in crosses involving linked markers. Let us assume that genes a and b are linked. If we assume that individual meioses can have no crossovers (NCO), a single crossover (SCO), or a double crossover (DCO) in the a-to-b region, then we can represent the classes of unordered asci that emerge from such meioses as shown in Figure 6-14. Of course, triples and higher numbers of crossovers might occur, but we may assume that these are rare and therefore negligible. Notice that Figure 6-14 is merely an extension of Figure 6-3.

The key to the analysis is the NPD class, which arises only from a double crossover involving all four chromatids. Because we are assuming that double crossovers occur randomly between the chromatids, we can also assume that the frequencies of the four DCO classes are equal. This means that the NPD class should contain $\tfrac{1}{4}$ of the DCOs, and therefore we can estimate that

$$DCO = 4NPD$$

The single-crossover class can also be calculated by a similar kind of reasoning. Notice that tetratype (T) asci can result from either single-crossover or double-crossover meioses. But we can estimate the component of the T class that comes from DCO meioses to be 2NPD. Hence, the size of the SCO class can be stated as

$$SCO = T - 2NPD$$

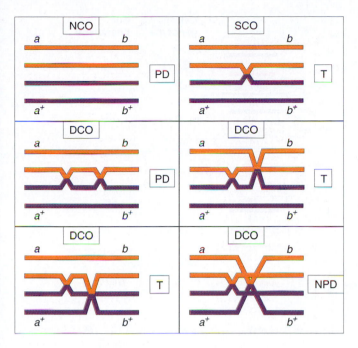

Figure 6-14 The ascus classes produced by crossovers between linked loci. NCO = noncrossover meioses; SCO = single-crossover meioses; DCO = double-crossover meioses.

Now that we have estimated the sizes of the SCO and DCO classes, the noncrossover class can be estimated as

$$NCO = 1 - (SCO + DCO)$$

Thus, we have estimates of the sizes of the NCO, SCO, and DCO classes in this marked region. We can use these values to derive a value for m, the mean number of crossovers per meiosis in this region. We can calculate the value of m simply by taking the sum of the SCO class plus *twice* the DCO class (because this class contains two crossovers). Hence

$$m = (T - 2NPD) + 2(4NPD)$$
$$= T + 6NPD$$

In the mapping-function section, we saw that to convert an m value to map units, it must be multiplied by 50 because each crossover on average produces 50% recombinants. So

$$\text{Map distance} = 50(T + 6NPD) \text{ m.u.}$$

Let's assume that in our hypothetical cross of $a^+ b^+ \times a\, b$, the observed frequencies of the ascus classes are 56 percent PD, 41 percent T, and 3 percent NPD. Using the formula, the map distance between the a and b loci is

$$50[0.41 + (6 \times 0.03)] = 50(0.59)$$
$$= 29.5 \text{ m.u.}$$

Let us compare this value with that obtained directly from the RF. Recall that the formula is

$$RF = \tfrac{1}{2}T + NPD$$

In our example,

$$RF = 0.205 + 0.03$$
$$= 0.235, \text{ or } 23.5 \text{ m.u.}$$

This is 6 m.u. less than the estimate we obtained using the map-distance formula because we could not correct for double crossovers in the RF analysis.

What PD, NPD, and T values are expected when dealing with unlinked genes? The sizes of the PD and NPD classes will be equal as a result of independent assortment. The T class can be produced only from a crossover between either of the two loci and their respective centromeres, and therefore the size of the T class will depend on the total size of these two regions.

The formulas derived above for unordered tetrads can also be applied to linear tetrads.

Message Unordered tetrads can be used to study several aspects of meiotic segregation and recombination, including calculating the frequencies of single and double crossovers, which may be used to calculate accurate map distances.

Mitotic Segregation and Recombination

We normally think of segregation and recombination in connection with meiosis, but these processes do occur (although less frequently) during mitosis. Mitotic segregation and recombination can easily be demonstrated if the genetic system is appropriately chosen.

Mitotic Segregation

In genetics, **segregation** refers to the separation of two alleles constituting a heterozygous genotype into two different cells or individuals. We have seen segregation repeatedly, of course, in our meiotic analyses based on Mendel's first law. However, the alleles of a heterozygote occasionally can be seen to segregate when the heterozygous cell undergoes mitotic division. The following example clarifies the nature of mitotic segregation.

In the 1930s, Calvin Bridges was observing *Drosophila* females that were genotypically M^+M (M is a dominant allele that produces a phenotype of slender bristles). Surprisingly, some females had a patch, or **sector,** of wild-type bristles on a body of predominantly M phenotype. Thus, the alleles of the heterozygote were showing segregation at the

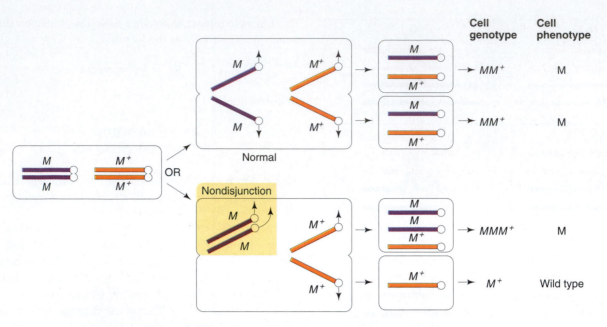

Figure 6-15 Mitotic nondisjunction can lead to phenotypic segregation.

phenotypic level. Bridges concluded that this segregation was the result of **mitotic nondisjunction,** a type of chromosome separation failure that is diagrammed in Figure 6-15. In a similar effect, heterozygotes of autosomal recessive alleles paired with wild-type alleles ($a^+ a$) produce patches of recessive phenotype on backgrounds of wild-type phenotype. These patches also can be explained by mitotic nondisjunction.

Other cases of segregation in the somatic tissue of a heterozygote have been found to be due to **mitotic chromosome loss.** Here one chromosome somehow gets left behind when the daughter nuclei reconstitute after mitotic division (Figure 6-16).

Geneticists find two other terms useful in relation to such phenomena. First, **variegation** is the coexistence of different-looking sectors of somatic tissue, whatever the cause. Second, a **mosaic** is an individual composed of tissues of two or more different genotypes, often recognizable by their different phenotypes. The following sentence gives an example of their usage: the $M^+ M$ *Drosophila* females were variegated, because mitotic nondisjunction had produced mosaics containing sectors of two different genotypes.

Mitotic Crossing-Over

In 1936, Curt Stern found the first case of mitotic segregation caused by **mitotic crossing over.** Working with the *Drosophila* X-linked genes *y* (which codes for yellow body) and *sn* (singed, because it codes for short, curly bristles), Stern made a cross

$$y^+ sn/y^+ sn \times y\ sn^+/Y$$

The female progeny were wild-type in appearance, as expected from their $y^+ sn/y\ sn^+$ genotypes. However, some females had sectors of yellow tissue or of singed tissue; these could be explained by nondisjunction or chromosome loss. Other females showed **twin spots.** A twin spot, in this example, is two adjacent sectors—one of yellow tissue and one of singed tissue—in a background of wild-type tissue (Figure 6-17). Stern noticed that the twin spots were too common to be chance juxtapositions of single spots. Because of this, and because the twin sectors of a twin spot were always adjacent, he reasoned that the twin sectors must be reciprocal products of the same event. That event, he realized, must have begun when, by chance, the homolo-

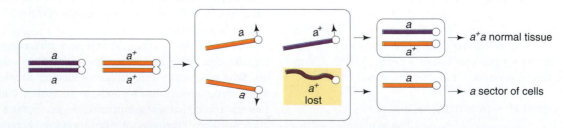

Figure 6-16 Chromosome loss at mitotic division can lead to phenotypic segregation.

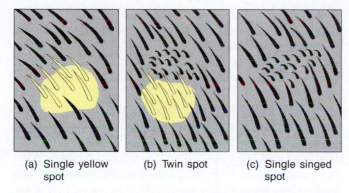

(a) Single yellow spot   (b) Twin spot   (c) Single singed spot

Figure 6-17 Unexpected sectors of body surface phenotypes of *Drosophila* genotype *y sn/y sn*, where *sn* represents singed bristles and *y* represents yellow body.

Mitotic Recombination in Fungi

Some fungi provide convenient systems for the study of mitotic crossing over and chromosome assortment. However, to observe these mitotic phenomena in fungi, the geneticist must generate diploid fungal cells because a haploid cell does not provide the opportunity for two genomes to recombine. Diploids form spontaneously in many fungi. The fungus we shall examine is *Aspergillus nidulans*, a greenish mold. *Aspergillus* is highly suitable for mitotic analysis. The aerial hyphae of this fungus produce long chains of **conidia** (asexual spores). Each conidium has a single nucleus, and the phenotype of any individual spore is dependent only on the genotype of its own nucleus. This makes certain kinds of selective techniques possible. If two haploid strains are mixed, the hyphae fuse and then both types of nuclei are present in a common cytoplasm. Such a strain is called a **heterokaryon**. Genetically, heterokaryons are not diploid, their constitution can best be thought of as *n + n*. Heterokaryons, like other strains, produce uninucleate conidia.

Consider a heterokaryon composed of the following nuclear genotypes:

gous parental chromosomes had come to lie in a pairing conformation, and chromatids of the different homologs must have crossed over between the *sn* locus and the centromere. Figure 6-18 diagrams this crossover. The diagram also shows that a mitotic crossover between the *y* and *sn* loci will give rise to a single yellow spot. Hence both twin and single yellow spots can be explained by a common hypothesis of mitotic crossing over. Note that all heterozygous genes distal to the crossover are rendered homozygous; this is a general characteristic of mitotic crossing over.

Twin spots have been observed in the cells of other diploid organisms, including plants. Mitotic crossing-over seems to take place regularly (although rarely) in all eukaryotes.

The definition of mitotic recombination is similar to that of meiotic recombination: **mitotic recombination** is any mitotic process that generates a diploid daughter cell with a combination of alleles different from that in the diploid parental cell. Compare this definition with that of meiotic recombination on page 126.

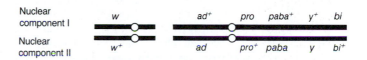

Nuclear component I   *w*   *ad⁺   pro   paba⁺   y⁺   bi*
Nuclear component II   *w⁺*   *ad   pro⁺   paba   y   bi⁺*

The alleles *ad, pro, paba,* and *bi* are all recessive to their wild-type counterparts, and each confers a requirement for a specific chemical supplement to permit growth. Because the heterokaryon has a wild-type allele of each heterozygous gene, it does not require any supplements for growth. The recessive alleles *w* and *y* cause the conidia to be white and yellow, respectively, instead of the wild-type green. (Note that only conidia are ever pigmented; the hyphae are colorless.) Because each conidium produced by a heterokaryon is uninucleate, it has either the yellow or the white phenotype. Therefore, the heterokaryotic colony looks yellowish-white and has a kind of "pepper-and-salt" appearance.

Green sectors appear in some heterokaryons. Green is the normal wild-type color of the fungus, and green coloration in the present example reveals that a diploid nucleus has formed spontaneously and has multiplied to form the green sector. It is the *complementation* (page 170) of the dominant *y⁺* and *w⁺* alleles in the same nucleus that results in the wild-type coloration. Diploid conidia can be removed from a green sector, and a diploid culture can be grown from it. Like the heterokaryon, the diploid cells require no growth supplements because they have the wild-type alleles. When the cultured diploid is fully grown, rare sectors producing either white or yellow conidia can be observed. Some of these sectors are diploid (recognizable because of

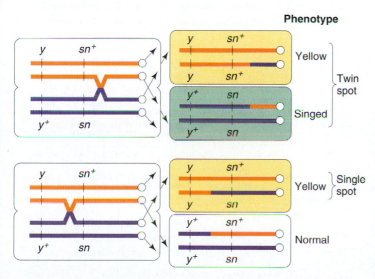

Phenotype

y   sn⁺
y   sn⁺   Yellow

y⁺   sn
y⁺   sn   Singed

Twin spot

y   sn⁺
y   sn   Yellow   } Single spot

y⁺   sn⁺
y⁺   sn   Normal

Figure 6-18 A mitotic crossover can lead to phenotypic segregation of the type shown in Figure 6-17.

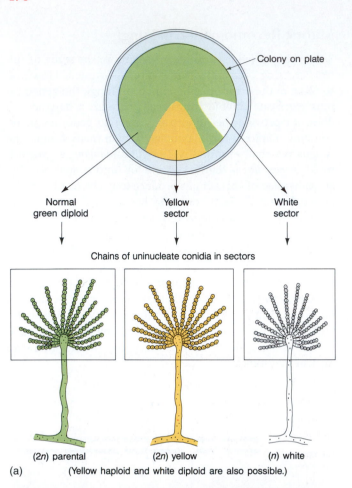

Colony on plate

Normal green diploid   Yellow sector   White sector

Chains of uninucleate conidia in sectors

(2n) parental   (2n) yellow   (n) white

(a)   (Yellow haploid and white diploid are also possible.)

(b)

Figure 6-19 (a) Some sectors showing segregation in an *Aspergillus* diploid of genotype $w^+ w\ y^+ y$, where *w* and *y* cause conidia to be white and yellow, respectively. Haploids are smaller. (b) Photograph of a white and a yellow sector in a diploid colony. (Part b from Etta Käfer)

their large-diameter conidia) and some are haploid (with small-diameter conidia). Two types of sectors are suitable for illustrating mitotic crossing-over and chromosome assortment: these are haploid white sectors, and diploid yellow sectors (Figure 6-19).

Haploid White Sectors. If haploid white sectors are isolated and tested, almost exactly half of them prove to have the genotype $w\ ad^+\ pro\ paba^+\ y^+\ bi$ and half have the genotype $w\ ad\ pro^+\ paba\ y\ bi^+$. Thus, the original diploid nucleus has somehow become haploid, a process known as **haploidization**, presumably through the progressive loss of one member of each chromosome pair. By looking at only white haploid conidia, we automatically select for the *w*-bearing chromosome. In one half of the *w* sectors, the $ad^+\ pro\ paba^+\ y^+\ bi$ chromosome is retained; in the other half, the $ad\ pro^+\ paba\ y\ bi^+$ chromosome is retained (Figure 6-20). In this way, the recessive color alleles can be used to derive linkage information because haploidization results in an outcome similar to independent assortment. The general procedure is first to select the chromosome bearing the color marker and then find out which genes are retained with it and which are independent of it, and in what groupings.

Message Haploidization produces an effect like independent assortment, and can be used to deduce which genes are linked and which assort independently.

Diploid Yellow Sectors. When diploid yellow sectors are tested, they usually prove to contain recombinant chromosomes. For example, one sector type was yellow and also required "paba" (*para*-aminobenzoic acid) for growth. Remember that mitotic crossing-over can make heterozygous loci homozygous, so mitotic crossing-over explains this type (Figure 6-21). Notice that we must follow two spindle fibers to each pole in mitotic analysis. This yellow, paba-requiring diploid arose from a mitotic exchange that obviously took

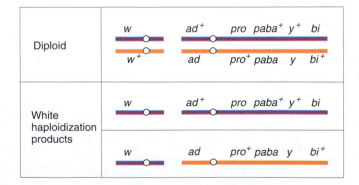

Diploid	w w^+	ad^+ pro paba$^+$ y^+ bi ad pro$^+$ paba y bi$^+$
White haploidization products	w	ad^+ pro paba$^+$ y^+ bi
	w	ad pro$^+$ paba y bi$^+$

Figure 6-20 Genotypes of an *Aspergillus* diploid and two white haploidization products derived from it.

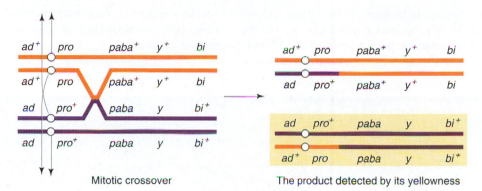

Figure 6-21 A mitotic crossover can produce a diploid yellow sector in the *Aspergillus* diploid shown. Note that the crossover produces homozygosity at all heterozygous loci beyond the crossover.

place in the centromere-to-paba region; other yellow diploids would arise from mitotic exchanges in the paba-to-y region, but these would not require paba for growth. The relative frequencies of these two types provide a kind of mitotic linkage map for that region. Such mapping can be carried out in several fungi. As expected, gene order in mitotic maps corresponds to gene order in meiotic maps, but, unexpectedly, the relative sizes of many of the intervals are found to differ when meiotic and mitotic maps are compared. The reason for this is not known. Note that every heterozygous gene from the point of the crossover to the end of the chromosome arm is made homozygous by a mitotic crossover. This in itself can provide useful mapping information. Because mitotic recombination experiments are done in fast-growing vegetative fungal cultures, the results can be obtained much more quickly than in a meiotic analysis, which requires the slower sexual phase.

> **Message** A mitotic crossover homozygoses all heterozygous genes distal to the crossover, and this principle may be used to deduce gene order. The frequencies of different homozygous classes are a measure of relative map distances.

In this section we have encountered several different processes that can cause a heterozygous allele pair to segregate during mitosis and produce a mosaic of different genotypes. In later chapters we shall encounter other processes that can produce such variegation, but the following message summarizes the sources of variegation we have seen in this chapter.

> **Message** At mitosis, nondisjunction, chromosome loss, crossing-over, and haploidization all cause a heterozygous pair of alleles to segregate in somatic tissue, resulting in a mosaic expressing the phenotypes of both alleles.

SUMMARY

Because of multiple crossovers, map distance is not linearly related to recombinant frequency. A mathematical relationship between map distance and recombinant frequency is called a *mapping function*. The Poisson distribution of crossovers at meiosis can be used to devise a mapping function.

Tetrad analysis is used to study individual meioses. Tetrad analysis is possible in those fungi and single-celled algae in which the four products of each meiosis remain together prior to spore dispersal.

Tetrads may be linear or unordered. Analysis of linear tetrads enables us to map loci in relation to their centromeres and to each other. From crosses in which the parents differ at two loci, the asci in a linear or unordered tetrad may be classified as parental ditype (PD), nonparental ditype (NPD), and tetratype (T). From these classes recombinant frequency can be calculated using the simple formula RF =   A different expression $50(T + 6NPD)$ provides a map distance corrected for the occurrence of double crossovers. In haploid organisms, recombination analysis by random meiotic products is simpler than in diploids.

Although segregation and recombination are normally thought of as meiotic phenomena, alleles do occasionally segregate and recombine during mitosis. Mitotic segregation was first identified in the 1930s, when Calvin Bridges observed patches of M^+ bristles on the body of an M^+M female *Drosophila* that was predominantly *M* in phenotype. He concluded that the patches were the result of abnormal chromosome segregation during mitosis. Around the same time, Curt Stern observed twin spots in *Drosophila* and assumed they must be the reciprocal products of mitotic crossing-over. Mitotic crossing-over homozygoses all heterozygous genes distal to the crossover, and this principle can be used to deduce map position and distance. Some fungi provide convenient systems for the study of mitotic recombination.

Concept Map

Draw a concept map interrelating as many of the following terms as possible. Note that the terms are listed in no particular order.

map distance / map function / linkage map / recombinant frequency / tetrad analysis / centromere mapping / second-division segregation / correction for double crossovers / individual meioses

CHAPTER INTEGRATION PROBLEM

Strains of the haploid fungus *Neurospora* bearing the *am* allele do not grow unless alanine is added to the medium. Normal wild-type strains grow without alanine supplementation. Linked to the *am* locus is a suppressor gene *ssu*. The *ssu* allele suppresses the alanine requirement, and *am ssu* strains do not need alanine supplementation. The *ssu* allele has no effect on the wild-type *am*$^+$ allele, and furthermore the wild-type allele *ssu*$^+$ has no known effect on the *am* locus alleles.

It has been calculated that triple and higher crossovers are negligible in the region between *am* and *ssu*. Meioses thus have no crossover (41 percent), one crossover (39 percent), or two crossovers (20 percent). From this information answer the following questions:

a. In what way is the genetic system described dependent on the environment?

b. What is the recombinant frequency (RF value)?

c. If ascospores from the cross *am ssu* × *am*$^+$ *ssu*$^+$ are sampled at random, what proportion will require alanine to grow?

d. What is the distance between the loci in corrected map units?

e. If asci from the cross *am ssu* × *am*$^+$ *ssu*$^+$ are analyzed as unordered tetrads, what proportion will show the following five phenotypic patterns:

Pattern	Number of ascospores not requiring alanine	Number of ascospores requiring alanine
1	8	0
2	6	2
3	4	4
4	2	6
5	0	8

Solution

a. The only aspect of the environment that is mentioned is alanine as part of the growth medium. If there is alanine in the medium, we cannot identify the two basic phenotypes, the growth phenotype and the no-growth phenotype. Therefore, the ability to study and analyze this interesting case of gene interaction is environmentally dependent, because the starting point for any piece of genetic analysis is variation, and without it there would be no study. Another point that is relevant is that the *am* allele would probably be lethal in a natural environment; it is only because scientists propagate such phenotypes (in this case on special medium—that is, in a special environment) that we have them for study.

b. The chapter has provided us with the formula for calculating the RF from tetrad analysis. We have learned that RF = T/2 + NPD. However, we do not know what the values of T and NPD are, so they have to be calculated. We have been given the frequencies of meioses with 0, 1, and 2 crossovers, and we have learned how PD, T, and NPD are produced from such meioses. We thus have what we need to calculate the required frequencies. All asci with no crossovers are PD, so this contributes 41 percent to the PD class. Single crossovers are also easy; they produce only T asci, so this contributes 39 percent to the T class. Double crossovers are a bit more complicated, but we have seen that they produce $\frac{1}{4}$PD, $\frac{1}{2}$T and $\frac{1}{4}$NPD, which gives us 5 percent PD, 10 percent T, and 5 percent NPD. Adding these, we get 46 percent PD, 49 percent T, and 5 percent NPD. Now we can calculate the RF value, which is $\frac{49}{2}$ + 5 = 29.5 percent.

c. We now expect 29.5 percent recombinants in the ascospores from the cross, comprising the two recombinant classes, *am ssu*$^+$ and *am*$^+$ *ssu*, each being 14.75 percent. The two parental classes are *am ssu* and *am*$^+$ *ssu*$^+$, and we expect them to make up the difference 100 − 29.5 = 70.5 percent, each being 35.25 percent. Which of these four genotypes will require alanine? Obviously, $am^+ ssu^+$ will not; neither will $am\ ssu$. The genotype $am^+ ssu$ will also not require alanine because it has the wild-type allele at the am locus, and we have been told that ssu does not affect that allele. However, $am\ ssu^+$ does not have the suppressor allele, and will require alanine, so the answer is 14.75 percent, the frequency of this class.

d. We have learned in this chapter that if we know the frequency of PD, T, and NPD in an unordered tetrad analysis, we can use these values to correct for the effect of double crossovers, which always tend to cause an underestimation of map distance. The formula is $50(T + 6NPD)$. Applying our calculated values we get $50(0.49 + 0.30)$, which equals 39.5 corrected map units.

e. We have done most of the reasoning to answer this part because we have deduced that the only genotype that

will confer a requirement for alanine is the genotype $am\ ssu^+$. Since we know there are only three ascus types (PD, T, and NPD) in an unordered tetrad analysis of two heterozygous loci, we simply need to deduce how many $am\ ssu^+$ ascospores will be seen in each. In PD asci there will be no $am\ ssu^+$ genotypes, so these asci will give us an $8:0$ ratio. Tetratype (T) asci will contain one spore pair of genotype $am\ ssu^+$, so these will give us a $6:2$ ratio. The NPD class will be composed of four $am\ ssu^+$ ascospores and four $am^+ ssu$ ascospores, giving a $4:4$ ratio. And that covers all the possibilities, so we can answer that the proportions for the table are (1) 46 percent, (2) 49 percent, (3) 5 percent, (4) 0 percent, (5) 0 percent.

In summary, notice the concepts we used from previous chapters: Mendelian segregation, chromosomal inheritance, gene interaction, lethal effects, environmental effects, crossing-over, and mapping.

SOLVED PROBLEMS

1. A cross is made between a haploid strain of *Neurospora*, of genotype $nic^+ ad$ and another haploid strain of genotype $nic\ ad^+$. From this cross, a total of 1000 linear asci are isolated and categorized a follows:

1	2	3	4	5	6	7
$nic^+ ad$	$nic^+ ad^+$	$nic^+ ad^+$	$nic^+ ad$	$nic^+ ad$	$nic^+ ad^+$	$nic^+ ad^+$
$nic^+ ad$	$nic^+ ad^+$	$nic^+ ad^+$	$nic^+ ad$	$nic^+ ad$	$nic^+ ad^+$	$nic^+ ad^+$
$nic^+ ad$	$nic^+ ad^+$	$nic^+ ad$	$nic\ ad$	$nic\ ad^+$	$nic\ ad$	$nic\ ad$
$nic^+ ad$	$nic^+ ad^+$	$nic^+ ad$	$nic\ ad$	$nic\ ad^+$	$nic\ ad$	$nic\ ad$
$nic\ ad^+$	$nic\ ad$	$nic\ ad^+$	$nic^+ ad^+$	$nic^+ ad$	$nic^+ ad^+$	$nic^+ ad$
$nic\ ad^+$	$nic\ ad$	$nic\ ad^+$	$nic^+ ad^+$	$nic^+ ad$	$nic^+ ad^+$	$nic^+ ad$
$nic\ ad^+$	$nic\ ad$	$nic\ ad$	$nic\ ad$	$nic\ ad^+$	$nic\ ad^+$	$nic\ ad^+$
$nic\ ad^+$	$nic\ ad$	$nic\ ad$	$nic\ ad^+$	$nic\ ad^+$	$nic\ ad$	$nic\ ad^+$
808	1	90	5	90	1	5

Map the ad and nic loci in relation to centromeres and to each other.

Solution

What principles can we draw on to solve this problem? It is a good idea to begin by doing something straightforward, which is to calculate the two locus-to-centromere distances. We do not know if the ad and the nic loci are linked, but we do not need to know. The frequencies of the M_{II} patterns for each locus give the distance from locus to centromere. (We can worry about whether it is the same centromere later.)

Remember that an M_{II} pattern is any pattern that is not two blocks of four. Let's start with the distance between the nic locus and the centromere. All we have to do is add the ascus types 4, 5, 6, and 7, because these are all M_{II} patterns

for the nic locus. The total is $5 + 90 + 1 + 5 = 101$, or 10.1 percent. In this chapter, we have seen that to convert this to map units, we must divide by 2, which gives 5.05 m.u.

Now we do the same thing for the ad locus. Here the total of the M_{II} patterns is given by types 3, 5, 6, and 7 and is $90 + 90 + 1 + 5 = 186$, or 18.6 percent, which is 9.3 m.u.

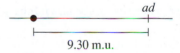

Now we have to put these two together and decide between the following alternatives, all of which are compatible with the above locus-to-centromere distances:

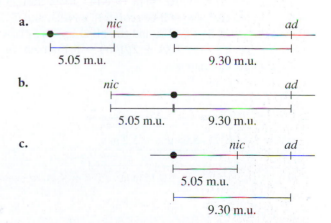

Here a combination of common sense and simple analysis tells us which alternative is correct. First, an inspection of the asci reveals that the most common single type is the one labeled 1, which contains more than 80 percent of all the asci. This type contains only *nic⁺ ad* and *nic ad⁺* genotypes, and they are *parental* genotypes. So we know that recombination is quite low and the loci are certainly linked. This rules out alternative a.

Now consider alternative c; if this were correct, a crossover between the centromere and the *nic* locus would generate not only an M$_{II}$ pattern for that locus but also an M$_{II}$ pattern for the *ad* locus, because it is farther from the centromere than *nic*. The ascus pattern produced by alternative c should be

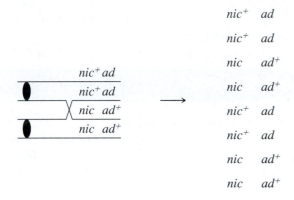

nic⁺	*ad*
nic⁺	*ad*
nic	*ad⁺*
nic	*ad⁺*
nic⁺	*ad*
nic⁺	*ad*
nic	*ad⁺*
nic	*ad⁺*

Remember that the *nic* locus shows M$_{II}$ patterns in asci 4, 5, 6, and 7 (a total of 101 asci); of these, type 5 is the very one we are talking about and contains 90 asci. Therefore, alternative c appears to be correct because ascus type 5 comprises about 90 percent of the M$_{II}$ asci for the *nic* locus. This relationship would not hold if alternative b were correct, because crossovers on either side of the centromere would generate the M$_{II}$ patterns for the *nic* and the *ad* loci independently.

Is the map distance from *nic* to *ad* simply 9.30 − 5.05 = 4.25 m.u.? Close, but not quite. The best way of calculating map distances between loci is always by measuring the recombinant frequency (RF). We could go through the asci and count all the recombinant ascospores, but it is simpler to use the formula RF = $\frac{1}{2}$T + NPD. Even though the asci are linear, they can still be scored PD, NPD, and T. The T asci are classes 3, 4 and 7, and the NPD asci are classes 2 and 6. Hence, RF + [$\frac{1}{2}$(100) + 2]/1000 = 5.2 percent of 5.2 m.u., and a better map is

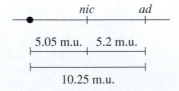

The reason for the underestimate of the *ad*-to-centromere distance calculated from the M$_{II}$ frequency is the occurrence of double crossovers, which can produce an M$_I$ pattern for *ad*, as in ascus type 4:

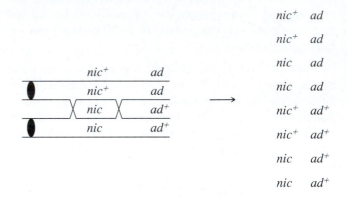

nic⁺	*ad*
nic⁺	*ad*
nic	*ad*
nic	*ad*
nic⁺	*ad⁺*
nic⁺	*ad⁺*
nic	*ad⁺*
nic	*ad⁺*

2. In *Aspergillus*, the recessive chromosome VI alleles *leu1*, *met5*, *thi3*, *pro2*, and *ad2* confer requirements for leucine, methionine, thiamine, proline, and adenine, respectively, whereas their wild-type alleles confer no such requirements. At the tip of chromosome VI there is a locus with a recessive allele *su* that suppresses *ad2*, so that strains expressing *ad2* and *su* require no adenine. A diploid strain is made by combining the following haploid genotypes, where the loci, with the exception of *su*, are written in no particular order:

and
$$su \quad leu1 \quad met5 \quad thi3 \quad pro2 \quad ad2$$
$$su⁺ \quad leu1⁺ \quad met5⁺ \quad thi3⁺ \quad pro2⁺ \quad ad2$$

Diploid asexual spores were spread on a medium containing all supplements except adenine. Most spores did not grow, owing to the recessiveness of *su*, but a few did grow into colonies; 100 of these were removed and tested for the phenotypes of the other markers. The following four classes were found (note that these are diploid phenotypes, not haploid genotypes):

(1)	*su*	leu⁺	met⁻	thi⁺	pro⁺	ad⁻	60
(2)	*su*	leu⁻	met⁻	thi⁺	pro⁺	ad⁻	25
(3)	*su*	leu⁺	met⁺	thi⁺	pro⁺	ad⁻	10
(4)	*su*	leu⁻	met⁻	thi⁻	pro⁺	ad⁻	5

   a. Explain the production of these four classes and their relative amounts.

   b. Why do you think that no colonies expressing *pro2* were recovered?

Solution

   a. The experiment concerns the behavior of diploid cells at mitosis, so the four classes must be explained by a mitotic mechanism. It seems likely that the *su* allele in all classes has been made homozygous *su su*, because only in that condition can it suppress an *ad2 ad2* homozygote to permit growth without adenine. This then was the basis of the original selection of the 100 colonies. But evidently other alleles have become homozygous too—and in different combinations in different classes.

This chapter demonstrates that mitotic crossing-over can produce homozygosity in any recessive allele that is distal to the crossover. Can crossovers in different regions of chromosome VI explain the different classes? Inspection of the classes shows that homozygosity can be produced for either *su* alone (class 3), or *su* and *met* (class 1), or *su, met* and *leu* (class 2), or *su, met, leu,* and *thi* (class 4). This virtually dictates to us that the order must be *su–met–leu–thi*–centromere. The crossovers responsible for homozygosity must then be in the following regions.

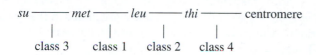

For example, the following crossover is necessary to produce class 2:

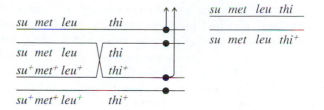

The relative sizes of the classes must reflect the relative distances separating the loci as follows:

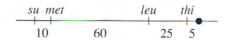

Note that these units are relative proportions and are not the same as meiotic map units.

b. We are told that *pro2* is on chromosome VI, yet it is never made homozygous. Note the importance of the fact that to select for mitotic crossovers, we are making use of an allele *su*, which is at the tip of one arm. Recessive alleles in the other arm would not be simultaneously made homozygous by these crossovers, so *pro2* is probably in the other arm. We have no way of knowing where *ad2* is, because it is homozygous from the outset.

PROBLEMS

1. The *Neurospora* cross *al-2⁺ × al-2* is made. A linear tetrad analysis reveals that the second-division segregation frequency is 8 percent.

    a. Draw two examples of second-division segregation patterns in this cross.

    b. What can the 8 percent value be used to calculate?

2. From the fungal cross *arg-6 al-2 × arg-6⁺ al-2⁺*, what will the spore genotypes be in unordered tetrads that are (**a**) parental ditypes? (**b**) tetratypes? (**c**) nonparental ditypes?

3. For a certain chromosomal region, the mean number of crossovers at meiosis is calculated to be two per meiosis. In that region, what proportion of meioses are predicted to have (**a**) no crossovers? (**b**) one crossover? (**c**) two crossovers?

4. In a *Drosophila* of genotype

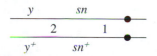

y determines yellow body and *y⁺* determines brown body; *sn* determines singed hairs and *sn⁺* unsinged. What is the

detectable outcome if there is a mitotic crossover during development in (**a**) region 1? (**b**) region 2?

5. Every Friday night, genetics student Jean Allele, exhausted by her studies, goes to the students' union bowling lane to relax. But even there, she is haunted by her genetic studies. The rather modest bowling lane has only four bowling balls: two red and two blue. These are bowled at the pins and are then collected and returned down the chute in random order, coming to rest at the end stop. Over the evening, Jean notices familiar patterns of the four balls as they come to rest at the stop. Compulsively, she counts the different patterns. What patterns did she see, what were their frequencies, and what is the relevance of this matter to genetics? (This is not a trivial question.)

6. **a.** Use the mapping function to calculate the corrected map distance between loci having a recombinant frequency of 20 percent. Remember that an *m* value of 1 is equal to 50 corrected map units.

    b. If you obtain an RF value of 45 percent in one experiment, what can you say about linkage? (The actual numbers you observed were 58 and 52 parental types and 47 and 43 recombinant types out of 200 progeny.)

7. In a haploid yeast, a cross between $arg^-\ ad^-\ nic^+\ leu^+$ and $arg^+\ ad^+\ nic^-\ leu^-$ produces haploid sexual spores, and 20 of these are isolated at random. When the resulting cultures are tested on various media, they give the results shown below, where Arg means arginine; Ad, adenine; Nic, nicotinamide; Leu, leucine; +, growth; and − no growth.

Minimal medium, plus

Culture	Arg, Ad, Nic	Arg, Ad, Leu	Arg, Nic, Leu	Ad, Nic, Leu
1	+	+	−	−
2	−	−	+	+
3	−	+	−	+
4	+	−	+	−
5	−	−	+	+
6	+	+	−	−
7	+	+	−	−
8	−	−	+	+
9	+	−	+	−
10	−	+	−	+
11	−	+	−	+
12	+	−	+	−
13	+	+	−	−
14	+	−	+	−
15	−	+	−	+
16	+	−	−	−
17	+	+	−	−
18	−	−	+	+
19	+	+	−	−
20	−	+	−	+

a. What can you say about linkage among these genes?

b. What is the origin of culture 16?

8. You measure the frequency of recombinants between the linked loci waxy (*wx*) and shrunken (*sh*) on chromosome 9 of corn. The RF is 36 percent. In what proportion of meiocytes would there be

a. no crossovers between the *wx* and *sh* loci?

b. one crossover between the *wx* and *sh* loci?

c. two crossovers between the *wx* and *sh* loci?

d. at least one crossover between the *wx* and *sh* loci?

* 9. In a tetrad analysis, the linkage arrangement of the *p* and *q* loci is as follows:

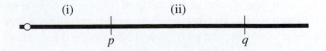

Assume that

- In region (i), there is no crossover in 88 percent of meioses and a single crossover in 12 percent of meioses.

- In region (ii), there is no crossover in 80 percent of meioses and a single crossover in 20 percent of meioses.

- There is no interference (in other words, the situation in one region doesn't affect what is going on in the other region).

What proportions of tetrads will be of the following types? (**a**) M_IM_I, PD; (**b**) M_IM_I, NPD; (**c**) M_IM_{II}, T; (**d**) $M_{II}M_I$, T; (**e**) $M_{II}M_{II}$, PD; (**f**) $M_{II}M_{II}$, NDP; (**g**) $M_{II}M_{II}$, T. (NOTE: Here the M pattern written first is the one that pertains to the *p* locus.) HINT: The easiest way to do this problem is to start by calculating the frequencies of asci with crossovers in both regions, region 1, region 2, and neither region. Then determine what M_I and M_{II} patterns result.

10. The following cross is made in *Neurospora*: $a^+b^+c^+d^+ \times abcd$ (loci *a*, *b*, *c*, and *d* are linked in the order written). Construct crossover diagrams to illustrate how the following unordered ascus patterns could arise:

a. $a^+b^+c\ d^+$
  $a\ b\ c\ d^+$
  $a^+b\ c^+d$
  $a\ b^+c^+d$

b. $a^+b\ c\ d$
  $a^+b^+c^+d$
  $a\ b\ c\ d^+$
  $a\ b^+c^+d^+$

c. $a^+b^+c\ d^+$
  $a^+b\ c^+d$
  $a\ b^+c\ d^+$
  $a\ b\ c^+d$

d. $a^+b\ c^+d$
  $a^+b\ c^+d$
  $a\ b^+c\ d^+$
  $a\ b^+c\ d^+$

e. $a^+b\ c\ d$
  $a\ b\ c\ d^+$
  $a^+b^+c^+d$
  $a\ b^+c^+d^+$

f. $a^+b\ c^+d$
  $a^+b^+c^+d$
  $a\ b\ c\ d^+$
  $a\ b^+c\ d^+$

g. $a^+b\ c^+d^+$
  $a^+b\ c^+d^+$
  $a\ b^+c\ d$
  $a\ b^+c\ d$

h. $a^+b\ c^+d$
  $a\ b^+c\ d^+$
  $a^+b\ c^+d$
  $a\ b^+c\ d^+$

11. A *Neurospora* cross was made between one strain that carried the mating-type allele *A* and the mutant allele *arg-1* and another strain that carried the mating-type allele *a* and the wild-type allele for *arg-1* (+). Four hundred linear tetrads were isolated, and these fell into the following seven classes. (Each class contained several different spore orders within the tetrad.)

1	2	3	4	5	6	7
A arg	A +	A arg	A arg	A arg	A +	A +
A arg	A +	A +	a arg	a +	a arg	a arg
a +	a arg	a arg	A +	A arg	A +	A arg
a +	a arg	a +	a +	a +	a arg	a +
127	125	100	36	2	4	6

a. Deduce the linkage arrangement of the mating-type locus and the *arg-1* locus. Include the centromere or centromeres on any map you draw. Label *all* intervals in map units.

b. Diagram the meiotic divisions that led to tetrad class 6. Label clearly.

 Unpacking the Problem

a. Are fungi generally haploid or diploid?

b. How many ascospores are in the ascus of *Neurospora*? Does your answer match the number represented in this problem? Explain any discrepancy.

c. What is mating type in fungi? How do you think it is determined experimentally?

d. Do the symbols *A* and *a* have anything to do with dominance and recessiveness?

e. What does the symbol *arg-1* mean? How would you test for this genotype?

f. How does this relate to the symbol *+*?

g. What does the expression *wild type* mean?

h. What does the word *mutant* mean?

i. Does the biological function of the alleles shown have anything to do with the solution of this problem?

j. What does the expression *linear tetrad analysis* mean?

k. What more can be learned from linear tetrad analysis that cannot be learned from unordered tetrad analysis?

l. How is a cross made in a fungus such as *Neurospora*? Explain how to isolate asci and individual ascospores. How does the term *tetrad* relate to the terms *ascus* and *octad*?

m. Where does meiosis occur in the *Neurospora* life cycle? (Show this on a diagram of the life cycle.)

n. What does this question have to do with meiosis?

o. Can you write out the genotypes of the two parental strains?

p. Why are only four genotypes shown in each class?

q. Why are there only seven classes? How many ways have you learned for classifying tetrads generally? Which of these classifications can be applied to both linear and unordered tetrads? Can you apply these classifications to the tetrads in this problem? (Classify each class in as many ways as possible.) Can you think of more possibilities in this cross? If so, why are they not shown?

r. What does "Each class contained several different spore orders within the tetrad" mean? Why would these different spore orders not change the class?

s. Why is the class

a +

a +

A arg

A arg

not listed?

t. What does the expression "linkage arrangement" mean?

u. What is a genetic "interval"?

v. Why does the problem state "centromere or centromeres" and not just "centromere"? What is the general method for mapping centromeres in tetrad analysis?

w. What is the total frequency of A + ascospores? (Did you calculate this using a formula or by inspection? Is this a recombinant genotype? If so, is it the only recombinant genotype?)

x. The first two classes are the most common and are approximately equal in frequency. What does this tell you? What is their content of parental and recombinant genotypes?

12. In the fungus *Neurospora*, a strain that was auxotrophic for thiamine (mutant allele *t*) was crossed to a strain that was auxotrophic for methionine (mutant allele *m*). Linear asci were isolated, and these were classified into the following groups:

Spore pair	Ascus types					
1 and 2	t +	t +	t +	t +	t m	t m
3 and 4	t +	t m	+ m	+ +	t m	+ +
5 and 6	+ m	+ +	t +	t m	+ +	t +
7 and 8	+ m	+ m	+ m	+ m	+ +	+ m
Number	260	76	4	54	1	5

a. Determine the linkage relationships of these two genes to their centromere or centromeres and to each other. Specify distances in map units.

b. Draw diagrams to show the origin of each ascus type.

c. If 1000 randomly selected ascospores are plated on a plate of minimal medium, how many are expected to grow into colonies?

13. A geneticist studies 11 different pairs of *Neurospora* loci by making crosses of the type *a b* × *a*⁺ *b*⁺ and then analyzing 100 linear asci from each cross. For the convenience of making a table, the geneticist organizes the data as if all 11 pairs of loci had the same designation—*a* and *b*—as shown below:

Number of asci of type

Cross	a b / a b / a⁺ b⁺ / a⁺ b⁺	a b⁺ / a b⁺ / a⁺ b / a⁺ b	a b / a b⁺ / a⁺ b⁺ / a⁺ b	a b / a⁺ b / a⁺ b⁺ / a b⁺	a b / a⁺ b⁺ / a⁺ b⁺ / a b	a b⁺ / a⁺ b / a⁺ b / a b⁺	a b⁺ / a⁺ b / a⁺ b⁺ / a b
1	34	34	32	0	0	0	0
2	84	1	15	0	0	0	0
3	55	3	40	0	2	0	0
4	71	1	18	1	8	0	1
5	9	6	24	22	8	10	20
6	31	0	1	3	61	0	4
7	95	0	3	2	0	0	0
8	6	7	20	22	12	11	22
9	69	0	10	18	0	1	2
10	16	14	2	60	1	2	5
11	51	49	0	0	0	0	0

For each cross, map the loci in relation to each other and to centromeres.

14. In *Neurospora,* the *a* locus is 5 m.u. from the centromere on chromosome 1. The *b* locus is 10 m.u. from the centromere on chromosome 7. From the cross of $a\,b^+ \times a^+\,b$, determine the frequencies of the following: **(a)** parental ditype asci, **(b)** nonparental ditype asci, **(c)** tetratype asci. **(d)** recombinant ascospores, **(e)** wild-type ascospores. (NOTE: Don't bother with mapping-function complications here.)

15. Three different crosses in *Neurospora* are analyzed on the basis of unordered tetrads. Each cross combines a different pair of linked genes. The results follow:

Cross	Parents	Parental ditypes (%)	Tetra- types (%)	Non- parental ditypes (%)
1	$a\,b^+ \times a^+\,b$	51	45	4
2	$c\,d^+ \times c^+\,d$	64	34	2
3	$e\,f^+ \times e^+\,f$	45	50	5

For each cross, calculate:

a. The frequency of recombinants (RF).

b. The uncorrected map distance, based on RF.

c. The corrected map distance, based on tetrad frequencies.

16. A geneticist crosses two yeast strains differing at the linked loci *ura3* (which governs uracil requirement) and *lys4* (which governs lysine requirement):

$$ura3\ lys4^+ \times ura3^+\ lys4$$

The geneticist isolates and classifies 300 *unordered* tetrads as follows:

ura3 lys4⁺ / ura3⁺ lys4 / ura3 lys4 / ura3⁺ lys4⁺	ura3 lys4 / ura3 lys4 / ura3⁺ lys4⁺ / ura3⁺ lys4⁺	ura3 lys4⁺ / ura3 lys4⁺ / ura3⁺ lys4 / ura3⁺ lys4
138	12	150

a. What is the recombinant frequency?

b. If it is assumed that there may be zero, one, or two (never more) crossovers between these loci at meiosis, what are the percentages of zero, one, and two crossover meioses?

c. What is the distance between the loci in map units *corrected* for double crossovers?

* 17. For an experiment with haploid yeast, you have two different cultures. Each will grow on minimal medium to which arginine has been added, but neither will grow on minimal medium alone. (Minimal medium is inorganic salts plus sugar.) Using appropriate methods, you induce the two cultures to mate. The diploid cells then divide meiotically and form unordered tetrads. Some of the ascospores will grow on minimal medium. You classify a large number of these tetrads for the phenotypes ARG⁻ (arginine-requiring) and ARG⁺ (arginine-independent) and record the following data:

Segregation of ARG⁻ : ARG⁺	Frequency (%)
4 : 0	40
3 : 1	20
2 : 2	40

a. Using symbols of your own choosing, assign genotypes to the two parental cultures. For each of the three kinds of segregation, assign genotypes to the segregants.

b. If there is more than one locus governing arginine requirement, are these loci linked?

* 18. Four histidine loci are known in *Neurospora*. As shown here, each of the four loci is located on a different chromosome.

In your experiment, you begin with an *ad-3* line from which you recover a cell that also requires histidine. Now you wish to determine which of the four histidine loci is involved. You cross the *ad-3 his-?* strain with a wild type (*ad-3⁺ his-1⁺ his-2⁺ his-3⁺ his-4⁺*) and analyze ten unordered tetrads: two are PD, six are T, and two are NPD. From this result, which of the four *his* loci is most probably the one that changed from *his⁺* to *his?*

(Problem 18 from Luke deLange.)

19. The haploid ascomycete fungus *Sordaria* has ascospores that are normally black. Two ascospore color variants (mutants) are isolated. When mutant 1 is crossed to a wild type, the asci produced contain four black spores and four white spores; when mutant 2 is crossed to a wild type, the asci produced contain four black spores and four tan spores. When mutants 1 and 2 are intercrossed, some asci contain four black and four white spores, some asci contain four tan and four white spores, and some asci contain four white and two black and two tan spores. Explain these results giving

    a. Genotypes underlying the three phenotypes

    b. An explanation of the types of asci produced in the crosses to wild type and the intercross

* 20. *Drosophila melanogaster* can have two entire X chromosomes attached to the same centromere:

The joined X chromosomes behave as a single chromosome, called an *attached X*. Work out the inheritance of sex chromosomes in crosses of attached-X-bearing females with normal males. During meiosis, an attached-X chromosome duplicates and segregates as follows:

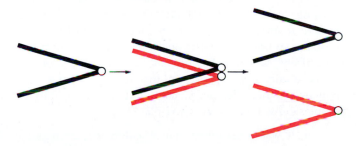

Nonsister chromatids of attached-x chromosomes can cross over. Suppose you have an attached-X chromosome of the following genotype:

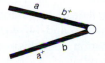

Diagram all the possible genotypic products and their phenotypes (**a**) from single crossovers between *a* and *b* and between *b* and the centromere, and (**b**) from double exchanges between *a* and *b* and between *b* and the centromere. Make sure you consider all possibilities. Describe how such a procedure represents a kind of tetrad analysis.

21. In *Neurospora*, the mating-type locus, which has alleles *a* and *A*, is 5 m.u. from the centromere on chromosome 1, and the cycloheximide-resistance locus, which has alleles *c* and *C*, is 8 m.u. to the other side of the centromere on the same chromosome. What proportion of *unordered* tetrads from the cross *A C × a c* will be of the type

a. *A c*	**b.** *A C*	**c.** *a C*
A C	*A C*	*a C*
a c	*a c*	*A c*
a C	*a c*	*A c*

22. A cross was made in the fungus *Sordaria*,

$$un \; cyh \times un⁺ \; cyh⁺$$

and 100 linear tetrads were isolated. There proved to be six classes in the following proportions:

1	2	3
un cyh	*un cyh*	*un cyh*
un⁺ cyh	*un cyh⁺*	*un cyh*
un cyh⁺	*un⁺ cyh*	*un⁺ cyh⁺*
un⁺ cyh⁺	*un⁺ cyh⁺*	*un⁺ cyh⁺*
15	29	47

4	5	6
un cyh	*un cyh⁺*	*un cyh*
un⁺ cyh⁺	*un⁺ cyh*	*un⁺ cyh⁺*
un cyh	*un cyh⁺*	*un cyh⁺*
un⁺ cyh⁺	*un⁺ cyh*	*un⁺ cyh*
2	2	5

    a. Map the genes in relation to each other and to centromeres. Show distances in uncorrected map units.

    b. Draw six crossover diagrams, one to explain the origin of each class.

    c. Which type of tetrad is missing, and why do you think this is so?

    d. Why is class 6 more common than class 4 or 5?

23. *Neurospora's* chromosome 4 carries the *leu3* locus, located near the centromere; its alleles always segregate at the first division. On the other arm of chromosome 4, the *cys2* locus is 8 m.u. away from the centromere. If we cross a

$leu3^+$ $cys2$ strain and a $leu3$ $cys2^+$ strain, and if we ignore double and other multiple crossovers, what will we expect as the frequencies of the following seven classes of linear tetrad (where $l = leu3$ and $c = cys2$)?

a.	l	c	b.	l	c^+	c.	l	c	d.	l	c
	l	c		l	c^+		l	c^+		l^+	c
	l^+	c^+		l^+	c		l^+	c^+		l^+	c^+
	l^+	c^+		l^+	c		l^+	c		l	c^+

e.	l	c	f.	l	c^+	g.	l	c^+
	l^+	c^+		l^+	c		l^+	c
	l^+	c^+		l^+	c		l^+	c^+
	l	c		l	c^+		l	c

24. Consider the following *Drosophila* alleles:

g codes for gray body (wild type is black)

c codes for curly bristles (wild type is straight)

s codes for stippled body (wild type is smooth)

These autosomal loci are arranged in the order *g-c-s*-centromere. A homozygous gray, stippled strain was crossed to a homozygous curly strain, and the F_1 were generally wild type in their phenotypic appearance, but microscopic examination of the flies revealed that a few rare individuals had one of three basic patterns. The patterns are

a. A gray, stippled sector next to a curly sector

b. A gray sector next to a curly sector

c. Solitary gray sectors

Explain the origin of these three patterns with a single unifying hypothesis.

25. Previous experiments on the fungus *Aspergillus* had shown that the loci *ad, col, phe, pu, sm,* and *w* are all on the same chromosome, but their order was not known. A diploid was constructed with the following genotype: ad col phe pu sm w/ad^+ col^+ phe^+ pu^+ sm^+ w^+ (the gene order is written alphabetically). When this diploid was cultured, white diploid sectors (ww) were observed and isolated. These were found to be of the following *phenotypes* (none of the white diploid sectors expressed *col*):

ad	phe	pu	sm	41%
ad	phe^+	pu	sm^+	30%
ad	phe^+	pu	sm	24%
ad^+	phe^+	pu	sm^+	5%

a. What is the likely origin of these sectors?

b. What is the relative order of the six genes and the centromere?

c. Give relative map distances where possible.

d. Why was *col* not expressed by the white diploid cells?

26. An incompletely dominant allele of soybeans, *Y*, causes yellowish leaves in Y^+ Y heterozygotes. However, heterozygotes regularly show rare patches of green adjacent to a patch of very pale yellow, all in the yellowish background. Propose an explanation for these rare patches.

27. You isolate white haploidized sectors from an *Aspergillus* diploid that is w^+ w a^+ a b^+ b c^+ c and score them for *a, b,* and *c*. You find that 25 percent are w a b c, 25 percent are w a^+ b^+ c^+, 25 percent are w a b^+ c, and 25 percent are w a^+ b c^+. What linkage relations can you deduce from these frequencies? Sketch your conclusions.

28. You know that two loci, *y* and *ribo,* are linked in *Aspergillus,* but you do not know their locations relative to the centromere. You have a diploid culture of genotype y^+ $ribo^+/y$ $ribo$ that is green and that grows without riboflavin supplementation. You notice some yellow sectors in the culture and study them. They are diploid, and you discover that 80 percent of them can grow without riboflavin whereas 20 percent require media supplemented with riboflavin. What is the most likely order of the two loci and the centromere?

29. An *Aspergillus* diploid is pro^+ pro fpa^+ fpa $paba^+$ $paba;$ *pro* is a recessive allele for proline requirement, *fpa* is a recessive allele for fluorophenylalanine resistance, and *paba* is a recessive allele for *para*-aminobenzoic acid (paba) requirement. When conidia are plated on fluorophenylalanine, only resistant colonies develop. Of 154 *diploid* resistant colonies, 35 do not require proline or paba, 110 require paba, and 9 require both.

a. What do these figures indicate?

b. Sketch your conclusions in the form of a map.

c. Some resistant colonies (not the ones described) are haploid. What would you predict their genotype to be?

30. An *Aspergillus* diploid was made by fusing haploid strains of genotype ad^+ leu^+ $ribo^+$ fpa^+ w^+ and ad leu $ribo$ fpa $w.$ The mutant alleles are all recessive and determine requirements for adenine, leucine, and riboflavin; resistance to fluorophenylalanine; and white asexual spores, respectively. When thousands of asexual spores of the diploid strain were spread on medium containing fluorophenylalanine, adenine, leucine, and riboflavin, only 93 colonies grew. These were then isolated.

a. It was determined that 38 colonies were haploid: 22 with the genotype fpa leu $ribo$ ad^+ w^+ and 16 with the genotype fpa leu $ribo$ ad $w.$ What does this result tell us about the linkage of the genes? Summarize with a diagram.

b. The remaining 55 colonies were diploid; of these, 24 required neither leucine nor riboflavin, 17 required riboflavin but not leucine, and 14 required both. What further linkage information does this result provide? Summarize with a diagram.

7

Gene Mutation

The phenotype produced by an unstable mutation in *Zinnia*. The mutation eliminates red pigment, resulting in white tissue. However, during development the mutation frequently reverts back to the normal allele, which permits synthesis of red pigment. Stripes are produced because cell division in the petals takes place mainly on the long axis, so the daughters of the revertant cells tend to be arranged longitudinally. (From M. A. L. Smith, Department of Horticulture, University of Illinois; see *Journal of Heredity* 80, 1989.)

KEY CONCEPTS

▶ Mutation is the process whereby genes change from one allelic form to another.

▶ Forward mutations are changes away from the wild-type allele, and reverse mutations are changes to the wild-type allele.

▶ Mutations can lead to loss of function of a gene, or to new function.

▶ Mutations in germ-line cells can be transmitted to progeny, but somatic mutations cannot.

▶ Selective systems make it easier to obtain mutations.

▶ Mutagens are agents that increase normally low rates of mutation.

▶ Genes mutate randomly, at any time and in any cell of an organism.

▶ A biological process can be dissected genetically if mutations that affect that process can be obtained. Each gene identified by a mutation identifies a separate component of the process.

enetic analysis would not be possible without *variants*—organisms that differ in some particular character. We have considered many examples in which we could analyze differing phenotypes for particular characters. Now we consider the origin of the variants. How, in fact, do genetic variants arise?

The simple answer to this question is that organisms have an inherent tendency to undergo change from one hereditary state to another. Such hereditary change is called **mutation.** Geneticists recognize two different levels at which mutation takes place. In **gene mutation,** an allele of a gene changes, becoming a different allele. Because such a change occurs within a single gene and maps to one chromosomal locus ("point"), a gene mutation is sometimes called a **point mutation.** At the other level of hereditary change—**chromosome mutation**—segments of chromosomes, whole chromosomes, or even entire sets of chromosomes change. Gene mutation is not necessarily involved in such a process; the effects of chromosome mutation are due more to the new arrangements of chromosomes and of the genes they contain. Nevertheless, some chromosome mutations, in particular those proceeding from chromosome breaks, are accompanied by gene mutations caused by the disruption at the breakpoint. In this chapter, we explore gene mutation; in Chapters 8 and 9, we shall consider chromosome mutation.

To consider change, we must have a fixed reference point, or standard. In genetics, the **wild type** provides the standard. Remember that the wild-type allele may be either the form found in nature or the form found in a standard laboratory stock. Any change away from the wild-type allele is called **forward mutation;** any change back to the wild-type allele is called **reverse mutation** (or **reversion** or **back mutation**). For example,

$$\left. \begin{array}{l} a^+ \longrightarrow a \\ D^+ \longrightarrow D \end{array} \right\} \text{forward mutation}$$

$$\left. \begin{array}{l} a \longrightarrow a^+ \\ D \longrightarrow D^+ \end{array} \right\} \text{reverse mutation}$$

The non-wild-type allele of a gene is sometimes called a *mutation.* To use the same word for the process and the product may seem confusing at first, but in practice little confusion arises. Thus, we can speak of a dominant mutation (such as *D* in the example) or a recessive mutation (such as *a*). Bear in mind how arbitrary these gene states are; the wild type of today may have been a mutation in the evolutionary past, and vice versa.

Another useful term is **mutant.** This is, strictly speaking, an adjective and should properly precede a noun. A mutant organism or cell is one whose changed phenotype is attributable to the possession of a mutation. Sometimes the noun is left unstated; in this case, a mutant always means an individual or cell with a phenotype that shows that it bears a mutation.

Two other useful terms are **mutation event,** which is the actual occurrence of a mutation, and **mutation frequency,** the proportion of mutations in a population of cells or individuals.

Somatic versus Germinal Mutation

Genes and chromosomes can mutate in either somatic or germinal tissue, and these changes are called **somatic mutations** and **germinal mutations,** respectively. These two different types are shown diagrammatically in Figure 7-1.

Somatic Mutation

If a somatic mutation occurs in a single cell in developing somatic tissue, that cell is the progenitor of a population of identical mutant cells, all of which have descended from the cell that mutated. A population of identical cells derived asexually from one progenitor cell is called a **clone.** Because the members of a clone tend to stay close to one another during development, an observable outcome of a somatic mutation is often a patch of phenotypically mutant cells called a **mutant sector.** The earlier in development the mutation event, the larger the mutant sector will be (Figure 7-2). Mutant sectors can be identified by eye only if their

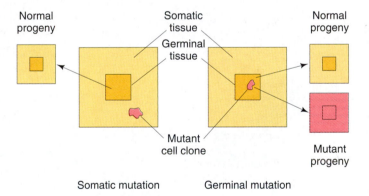

Figure 7-1 Somatic mutations are not transmitted to progeny, but germinal mutations may be transmitted to some or all progeny.

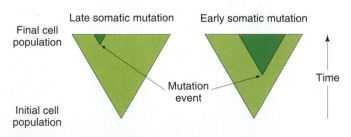

Figure 7-2 Early mutation produces a larger proportion of mutant cells in the growing population than does later mutation.

Figure 7-3 Somatic mutation in the red delicious apple. The mutant allele determining the golden color arose in a flower's ovary wall, which eventually developed into the fleshy part of the apple. The seeds would not be mutant, and would give rise to red-appled trees. (Note that, in fact, the golden delicious apple originally arose as a mutant branch on a red delicious tree.) (Anthony Griffiths)

Figure 7-4 A mutation producing an allele for white petals that arose originally in somatic tissue but eventually became part of germinal tissue and could be transmitted through seeds. The mutation arose in the primordium of a side branch of the rose. The branch grew long and eventually produced flowers. (From Harper Horticultural Slide Library.)

phenotype contrasts visually with the phenotype of the surrounding wild-type cells (Figure 7-3).

In diploids, a dominant mutation is expected to show up in the phenotype of the cell or clone of cells containing it. On the other hand, a recessive mutation will not be expressed, because it is masked by a wild-type allele that is by definition dominant to the recessive mutation. A second mutation could create a homozygous recessive mutation, but this would be rare.

What would be the consequences of a somatic mutation in a cell of a fully developed organism? If the mutation is in tissue in which the cells are still dividing, then there is the possibility of a mutant clone's arising. If the mutation is in a postmitotic cell, that is, one which is no longer dividing, then the impact on phenotype is likely to be negligible. Even when dominant mutations result in a cell that is either dead or defective, this loss of function will be compensated by other normal cells in that tissue. However, mutations that give rise to cancer are a special case. Cancer mutations occur in a special category of genes called **proto-oncogenes,** many of which regulate cell division. When mutated, such cells enter a state of uncontrolled division, resulting in a cluster of cells called a **tumor.** We shall look at some examples later in this chapter. Mechanisms of cancer will be considered in Chapter 24.

Are somatic mutations ever passed on to progeny? This is impossible because somatic cells by definition are those that are never transmitted to progeny. However, note that if we take a plant cutting from a stem or leaf that includes a mutant somatic sector, the plant that grows from the cutting may develop germinal tissue out of the mutant sector. Put another way, a branch bearing flowers can grow out of

the mutant somatic sector. Hence, what arose as a somatic mutation can be transmitted sexually. An example is shown in Figure 7-4.

Any method for the detection of somatic mutation must be able to rule out the possibility that the sector is due to mitotic segregation or recombination (Chapter 6). If the individual is a homozygous diploid, somatic sectoring is almost certainly due to mutation.

Germinal Mutation

A **germinal mutation** occurs in the germ line, special tissue that is set aside during development to form sex cells. If a mutant sex cell participates in fertilization, then the mutation will be passed on to the next generation. Of course, an individual of perfectly normal phenotype and of normal ancestry can harbor undetected mutant sex cells. These mutations can be detected only if they are included in a zygote (Figures 7-5 and 7-6). You will remember from Chapter 3 that the X-linked hemophilia mutation in the European royal families is thought to have arisen in the germ cells of Queen Victoria or one of her parents. The mutation was expressed only in her male descendants.

The experimental detection of germinal mutation depends on the ability to rule out meiotic segregation and recombination as possible causes of phenotypic differences between parents and offspring.

Message Before a new heritable phenotype can be attributed to mutation, both segregation and recombination must be ruled out as possible causes. This is true for both somatic and germinal mutations.

Figure 7-5 Germinal mutation determining white petals in viper's bugloss (*Echium vulgare*). A recessive germinal mutation, *a*, arose in an *A A* blue plant of the previous generation, making its germinal tissue *A a*. Upon selfing, the mutation was transmitted to progeny, some of which were *a a* and expressed the mutant phenotype. (Anthony Griffiths)

Figure 7-6 A mutation to an allele determining curled ears arose in the germ line of a normal straight-eared cat and was expressed in progeny such as the individual shown here. This mutation arose in a population in Lakewood, California, in 1981. It is an autosomal dominant. (From R. Robinson, *Journal of Heredity* 80, 1989, 474.)

Mutant Types

The phenotypic consequences of mutation may be so subtle as to require refined biochemical techniques to detect a difference from the phenotype conferred by the wild-type allele. Alternatively, the mutation may be so severe as to produce gross morphological defects or death. A rough classification follows, based only on the ways in which the mutations are recognized. This classification is not meant to be complete. Furthermore, the various ways of categorizing mutations often overlap.

Morphological Mutations

Morph means "form." Morphological mutations affect the outwardly visible properties of an organism, such as shape, color, or size. Albino ascospores in *Neurospora,* curly wings in *Drosophila,* and dwarf peas are all morphological mutations. Some additional examples of morphological mutants are shown in Figure 7-7.

Lethal Mutations

A new lethal mutant allele is recognized by its effects on the survival of the organism. Sometimes a primary cause of death from a lethal mutation is easy to identify (for example, in certain blood abnormalities). But often the cause of death is hidden, and the mutant allele is recognizable *only* by its effects on viability. An example of a lethal mutation is shown in Figure 7-8.

Conditional Mutations

In the class of conditional mutations, a mutant allele causes a mutant phenotype in only a certain environment, called the **restrictive condition,** but causes a wild-type phenotype in some different environment, called the **permissive condition.** Geneticists have studied many temperature-conditional mutations. For example, certain *Drosophila* mutations are known as "dominant heat-sensitive lethals." Heterozygotes (say, H^+H) are wild-type at 20°C (the permissive condition) but die if the temperature is raised to 30°C (the restrictive condition).

Many mutant organisms are less vigorous than their normal counterparts and thus more troublesome as experimental subjects. For this reason, conditional mutants are useful because they can be grown under permissive conditions and then shifted to restrictive conditions for study. Another advantage of conditional mutations is that they allow the determination of a developmental **sensitive period** at which specific time the gene acts. In these studies organisms carrying some specific conditional mutation are shifted from permissive to restrictive conditions at different times during development. Some shifts will lead to mutants, some to wild types, and from these results the sensitive period of gene action is assessed.

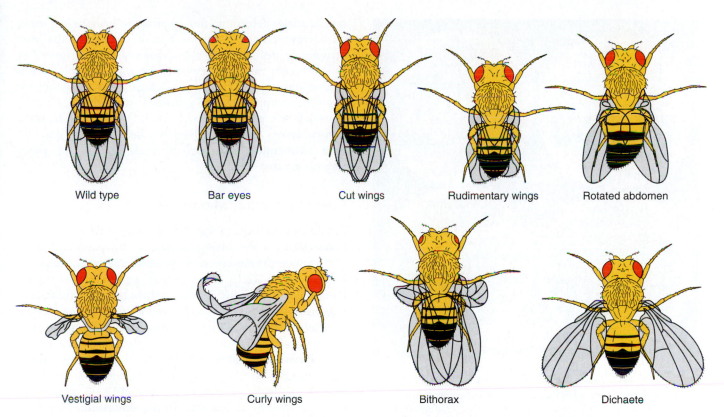

Wild type   Bar eyes   Cut wings   Rudimentary wings   Rotated abdomen

Vestigial wings   Curly wings   Bithorax   Dichaete

Figure 7-7 Eight morphological mutations of *Drosophila,* and the wild type for comparison. Most of the mutant phenotypes are self-explanatory; bithorax is an abnormality of the thorax featuring small wings instead of balancers; the most prominent feature of dichaete is that the wings are held at 45 degrees to the body.

(a)   (b)

Figure 7-8 Phenotypes of (a) the wild type and (b) a mutation affecting plumage of Japanese quail. This mutation arose in a laboratory colony of quail, and could be maintained as an interesting subject for genetic analysis. However, if such a mutation had arisen in nature, it would almost certainly have been lethal. (Janet Fulton)

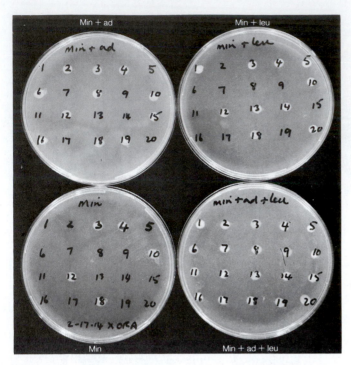

Figure 7-9 Testing strains of *Neurospora crassa* for auxotrophy and prototrophy. In this experiment the test utilizes 20 progeny from a cross of an adenine-requiring auxotroph and a leucine-requiring auxotroph. Genotypically, the cross is *ad leu⁺* × *ad⁺ leu*, and the progeny can carry any of the four possible combinations of these alleles. To test the progeny, the geneticist attempts to grow cells on various kinds of gelled media in petri dishes. The media are minimal medium (Min) with either adenine (*ad, top left*), leucine (*leu, top right*), neither (*bottom left*), or both (*bottom right*). Growth appears as a small circular colony (white in the photograph). Any culture growing on minimal must be *ad⁺ leu⁺*, one growing on adenine and no leucine must be *leu⁺*, and one on leucine and no adenine must be *ad⁺*. All should grow on adenine plus leucine; it is a kind of control to check viability. As examples, culture 8 must be *ad leu⁺*, 9 must be *ad leu*, 10 must be *ad⁺ leu⁺*, and 13 must be *ad⁺ leu*. (Anthony Griffiths)

Biochemical Mutations

Microbial cultures are convenient material for the study of biochemical mutations, which are identified by the loss or change of some biochemical function of the cells. This change typically results in an inability to grow and proliferate. In many cases, however, growth of a mutant cell can be restored by supplementing the growth medium with a specific nutrient. Biochemical mutations have been extensively analyzed in microorganisms. Microorganisms, by and large, are **prototrophic:** they can exist on a substrate of simple inorganic salts and an energy source; such a growth medium is called a **minimal medium.** Biochemical mutants, however, often are **auxotrophic:** they must be supplied certain additional nutrients if they are to grow. For example, one class of biochemically mutant fungi will not grow unless supplied with the nitrogenous base adenine. These are called *ad* mutations, whereas the wild-type allele is *ad⁺*. Mu-

tant *ad* alleles determine the auxotrophic, adenine-requiring phenotype. The practical method of testing for the auxotrophic or protoptrophic phenotype is shown in Figure 7-9.

Although microbial cultures are used for the experimental induction of biochemical mutations, we should note that many human hereditary diseases are biochemical mutations defective in some step of cellular chemistry. The term **inborn errors of metabolism** has been used to describe such biochemical disorders. Phenylketonuria and galactosemia are two examples.

Loss-of-Function Mutations

A wild-type allele has some type of active function or functions specific to the biological role of that particular gene. Mutation events are agents of random change acting at the level of gene (DNA) structure, so generally a mutation event is destructive and deletes or changes crucial functional regions of the gene. Such a change interferes with wild-type function, and the result is a **loss-of-function mutation.** Sometimes loss of function is complete, in which case a **null mutation** is generated. The word *null* means "nothing"; in other words, nothing is left of the wild-type function. Sometimes inactivation is incomplete, in which case the new allele is said to be a **leaky mutation.** This term means that some of the original wild-type function still "leaks through" into the phenotype, and the phenotype is not so clearly mutant as that resulting from a null mutation.

Generally loss-of-function mutations are found to be recessive. In a wild-type diploid cell there are two wild-type alleles of a gene, both making normal gene product. In heterozygotes (the crucial genotypes for testing dominance or recessiveness) the single wild-type allele may be able to provide enough normal gene product to produce a wild-type phenotype. In such cases, loss-of-function mutations are recessive. In some cases the cell is able to "up-regulate" the level of activity of the single wild-type allele so that in the heterozygote the total amount of wild-type gene product is more than half that found in the homozygous wild type. However, some loss-of-function mutations are dominant. In such cases the single wild-type allele in the heterozygote cannot provide the amount of gene product needed for the cells and the organism to be wild-type. The action of loss-of-function mutations is represented diagrammatically in Figure 7-10a and b.

Gain-of-Function Mutations

Because mutation events introduce random genetic changes, most of the time they result in loss of function. The mutation events are like bullets being fired at a complex machine; most of the time they will inactivate it. However, it is conceivable that in rare cases a bullet will strike the machine in such a way that it produces some new function. So it is with mutation events; sometimes the random change by pure chance confers some new function on the gene. In a

The Occurrence of Mutations

Mutation is a biological process that has characterized life from its beginning. As such, it is certainly fascinating and worthy of study. Mutant alleles such as the ones mentioned in the previous sections obviously are invaluable in the study of the process of mutation itself. For instance, they allow us to measure the frequency of mutation. In this connection, they are used as genetic markers, or representative genes: their precise function is not particularly important, except as a way to follow the process.

In modern genetics, however, mutant genes have another role in which their precise function *is* important. We have already referred (in Chapter 2) to genetic dissection as an established approach to biological analysis. Mutant genes can be used as probes to disassemble the constituent parts of a biological function and to examine their workings and interrelationships. Thus, it is of considerable interest to a biologist studying a particular function to have as many mutant forms that affect the function as possible. This has led to "mutant hunts" as an important prelude to any genetic dissection in biology. To identify a genetic variant is to identify a component of the biological process. We explore this idea further later in the chapter.

Message   Mutations serve two research purposes: geneticists use mutations to study the process of mutation itself and to genetically dissect biological functions.

Mutation Detection Systems

The tremendous stability and constancy of form of species from generation to generation suggest that mutation must be a rare process. This supposition has been confirmed, creating a problem for the geneticist who is trying to demonstrate mutation.

The prime need is for a detection system—a set of circumstances in which a mutant allele will make its presence known at the phenotypic level. Such a system ensures that the rare mutations that might occur will not be missed.

One of the main considerations here is that of dominance. The system must be set up so that mutations that create new recessive alleles, which are the most common type, will not be masked by a dominant wild-type allele. (Dominant mutations are less of a problem.) As an example, we can use one of the first detection systems ever set up—the one used by Lewis Stadler in the 1920s to study mutation in corn from *C*, expressed phenotypically as a colored kernel, to *c*, expressed as a white kernel. Here we are dealing with the phenotype of the endosperm of the seed. If you refer to Figure 3-38, you will see that this tissue forms when two identical haploid female nuclei fuse with one haploid nucleus from the male pollen cell. Hence, the tissue has three chromosome sets (it is 3*n*). One dominant *C* allele in combination with two *c* alleles causes the kernel to be colored.

(a)   Null loss-of-function mutation (*m*)

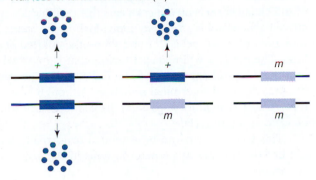

(b)   Leaky loss-of-function mutation (*m'*)

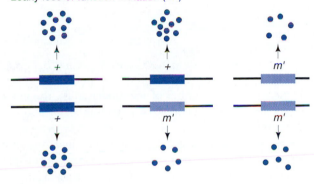

(c)   Gain-of-function mutation (*M*)

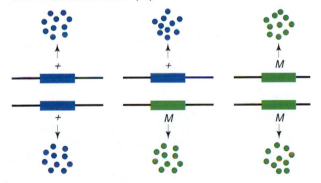

Figure 7-10 (a) Mutation *m* has completely lost its function (it is a null mutation). In the heterozygote, wild-type gene product is still being made, and often this is enough to result in a wild-type phenotype, in which case, *m* will act as a recessive. If the wild-type gene product is insufficient, the mutation will be seen as dominant. (b) Mutation *m'* still retains some function, but in the homozygote there is not enough to produce a wild-type phenotype. (c) Mutation *M* has acquired a new cellular function represented by the gene product colored green. *M* will be expressed in the heterozygote and most likely will act as a dominant. The homozygous mutant may or may not be viable, depending on the role of the + allele.

heterozygote the new function will be expressed, and therefore the gain-of-function mutation most likely will act like a dominant allele and produce some kind of new phenotype. Gain of function is represented in Figure 10c.

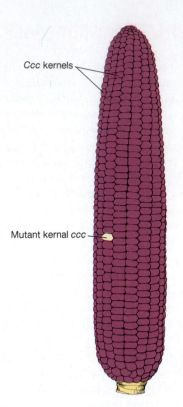

Ccc kernels

Mutant kernal ccc

Figure 7-11 The detection system for mutations at a specific locus of corn. The *C* allele determines the presence of a purple pigment in kernels, whereas *c* results in none. The geneticist makes the cross *cc* ♀ × *CC* ♂, and *C* → *c* mutations in the male germ line show up as unpigmented kernels on the cobs.

Stadler crossed *cc*♀ × *CC*♂ and examined thousands of individual kernels on the corn ears that resulted from this cross. Each kernel is a progeny individual. In the absence of mutation, every kernel is *Ccc* and shows the colored phenotype. Therefore, a white kernel reveals that *C* mutated to *c* in a male reproductive cell of the *CC* parent. The system has thus detected a germinal mutation. Although laborious, this is a very straightforward and reliable method of mutant detection (Figure 7-11).

This basic system can be extended to as many loci as can be conveniently made heterozygous in the same cross. For example

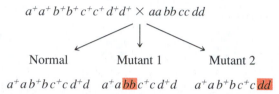

$$a^+ a^+ b^+ b^+ c^+ c^+ d^+ d^+ \times aa\,bb\,cc\,dd$$

Normal   Mutant 1   Mutant 2

$a^+ a\,b^+ b\,c^+ c\,d^+ d$ $a^+ a\,bb\,c^+ c\,d^+ d$ $a^+ a\,b^+ b\,c^+ c\,dd$

By increasing the number of loci under study, the investigator increases the likelihood of detecting a mutation in the experiment.

Stadler's method for detecting germinal mutations is now called the **specific-locus test.** The same principle can be applied to somatic mutations too. Once again, to increase the likelihood of finding and quantifying mutations, the number of loci under study can be increased to any level that is feasible. Let's look at an example in which detectable

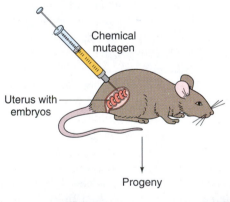

Chemical mutagen

Uterus with embryos

Progeny

Figure 7-12 A detection system for recessive somatic mutations at seven coat color loci in mice. The cross *ln ln pa pa b⁺ b⁺ ch⁺ ch⁺ p⁺ p⁺ d⁺ d⁺ pe pe* × *ln⁺ ln⁺ pa⁺ pa⁺ bb ch ch pp dd pe⁺ pe⁺* results in progeny heterozygous for all seven genes and predominantly wild-type in appearance. Any somatic mutation from wild-type to mutant at any of the loci causes a mutant sector in the coat of the offspring. The mutant colors are leaden (*ln*), pallid (*pa*), brown (*b*), pink-eyed dilution (*p*), dilution (*d*), and pearl (*pe*). The frequency of mutations can be increased by administering to developing embryos a chemical that is known to produce mutations (a chemical mutagen).

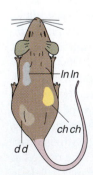

ln ln
ch ch
d d

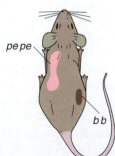

pe pe
b b

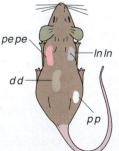

pe pe
d d
ln ln
p p

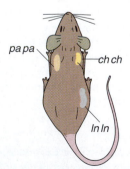

pa pa
ch ch
ln ln

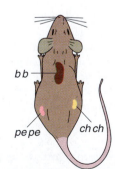

b b
pe pe
ch ch

Genotype of progeny is *ln ln⁺ pa pa⁺ b b⁺ ch ch⁺ p p⁺ d d⁺ pe pe⁺*
Phenotype of progeny is wild type with mutant sectors

mutations affect the coat colors of mice. The phenotypes and genes responsible for them are leaden (*ln*), pallid (*pa*), brown (*b*), chinchilla (*ch*), pink-eyed dilution (*p*), dilution (*d*), and pearl (*pe*). Mice from pure lines of the following genotypes are crossed:

$$ln\,ln\ pa\,pa\ b^+b^+\ ch^+ch^+\ p^+p^+\ d^+d^+\ pe\,pe$$

$$\times ln^+ln^+\ pa^+pa^+\ bb\ ch\,ch\ pp\ dd\ pe^+pe^+$$

and the females bear embryos of the multiply heterozygous genotype

$$ln\,ln^+\ pa\,pa^+\ bb^+\ ch\,ch^+\ pp^+\ dd^+\ pe\,pe^+$$

Because all the mutant alleles are recessive, the coats of the F_1 are expected to be wild-type. However, somatic mutations from wild type to mutant at any of the heterozygous loci cause mutant sectors on the coat (Figure 7-12). The frequency of these mutant sectors increases dramatically if the investigator injects a mutation-inducing chemical into the uterus of the pregnant mother at the eighth day of pregnancy. This exposes the developing embryos to the injected chemical. The progeny show many more mutant sectors than do mice that were not exposed to the mutagen as embryos. The potency of the chemical can be quantified simply by counting the number of mutant sectors and comparing this number with that observed in untreated animals.

In the plant *Tradescantia*, an effective somatic specific-locus test has been developed using a single heterozygous gene. A vegetatively propagated line of *Tradescantia* called *02* is heterozygous *Pp*. These alleles determine the dominant blue and recessive pink pigmentation phenotypes. The pigmentation is expressed in the petals and the stamens. The stamens of this plant have hairs that are chains of single cells. Therefore millions of single somatic cells can easily be scanned under the microscope to look for pink cells, which must have arisen from a $P \rightarrow p$ mutation, converting *Pp* into *pp* (Figure 7-13).

Human geneticists detect germinal mutations by the sudden appearance of a novel phenotype in a pedigree in which there is no previous record of such a phenotype. Dominant mutations are easily detected in this way because they are expected to be expressed in the phenotype. An example is given in Figure 7-14, which shows a pedigree for neurofibromatosis, an autosomal dominant disorder characterized by abnormal skin pigmentation ("café au lait" spots) and by numerous tumors, called *neurofibromas,* that associate with the peripheral or central nervous system. The tumors are visible on the surface of the body as shown in Figure 7-14. The pedigree shows an ancestry free of neurofibromatosis, with the disorder arising suddenly in generation IV in the children of one couple. The mutation must have arisen in the germinal tissue of either the mother or the father. It so happens that the gene mutations causing both neurofibromatosis and achondroplasia have very high mutation frequencies in humans, so a large proportion of cases

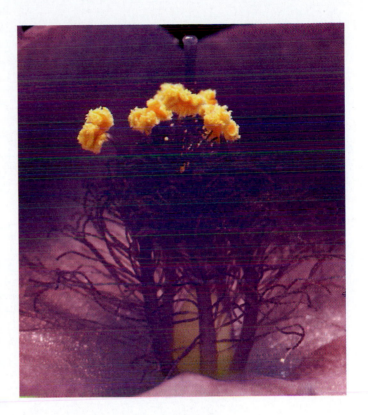

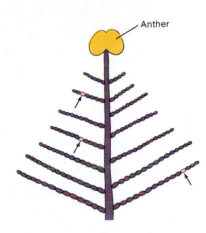

Anther

Figure 7-13 (a) Stamen of a *Tradescantia* plant heterozygous for *P* and *p* alleles. (b) In the chains that constitute the lateral hairs, some cells (marked by arrows) are mutant. The darker color (blue pigmentation) is determined by *P* and the paler color (pink pigmentation) by *pp*. (Part a from Runk/Schoenberger/Grant Heilman.)

are from mutation. At the other end of the spectrum, the gene for Huntington's disease has the lowest mutation frequency, so most cases of this disease are inherited from previous generations. In fact, most cases of Huntington's disease in North America can be traced to two immigrant families.

A human recessive mutation is more difficult to detect. Because of its recessiveness, the mutation is not expressed in the heterozygous state and can go unnoticed for many

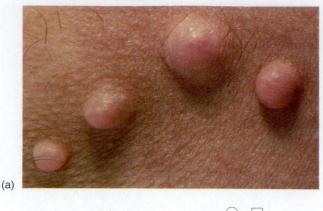

(a)

Figure 7-14 Mutation to neurofibromatosis. (a) Neurofibromas. (b) Pedigree showing that the neurofibromatosis mutation must have been in the germ line of III-8 or III-9. (Part a from Michael English / Custom Medical Stock.)

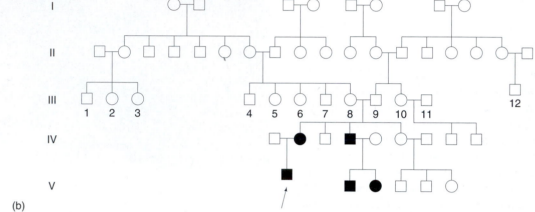

(b)

generations. It will become expressed only through inbreeding or the chance union of two separate heterozygotes. X-linked recessive mutations can be identified more easily than autosomal recessive mutations. We have already discussed the X-linked hemophilia allele present in the royal families of Europe. The disease was not recorded in the pedigree until expressed by one of the sons of Queen Victoria, and therefore the mutation event must have occurred either in the germinal tissue of Queen Victoria herself, or in the germinal tissue of one of her parents. A similar case is shown in the hemophilia pedigree in Figure 7-15. It is possible that the allele was present in the female ancestors, but the absence of hemophilia in a total of 11 male predecessors makes this possibility unlikely.

How can the frequency of autosomal recessive mutation events be measured? The answer begins with the idea that for any particular autosomal mutant phenotype, mutations are constantly occurring, entering the population like water drops from a dripping faucet. On the other hand, the mutations are removed from the population by reverse mutation and by selection, which are like leaks in the water vessel. The losses and gains eventually become equal, in a type of equilibrium. Measuring the frequency of the mutant phenotype at equilibrium provides a way of calculating the frequency of mutation, as we shall see in the population genetics chapter.

Haploids present a great advantage over diploids in mutation studies. The system of detecting mutations in hap-

loids is quite straightforward: any newly arisen recessive allele announces its presence unobscured by a dominant partner allele. In fact, the question of dominance or recessiveness need never arise: there is what amounts to a built-in detection facility. In some cases, a direct identification of mutants is possible. In *Neurospora*, for example, auxotrophic adenine-requiring mutations have been found to map at several loci. Of these, the *ad-3* locus is unique in that cells carrying auxotrophic *ad-3* mutants accumulate a purple pigment when grown on a low concentration of adenine. Thus, auxotrophic mutations of this gene may be detected simply by allowing large populations of single asexual

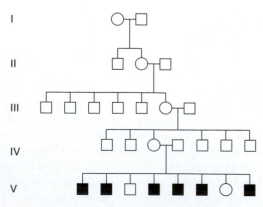

Figure 7-15 Mutation to X-linked recessive hemophilia.

spores to grow into colonies on a medium with limited adenine. The purple mutant colonies can be identified easily among the normal white colonies.

What about other kinds of auxotrophs? Usually, there are no visual pleiotropic effects, as are exhibited with *ad-3* mutations. We will examine the most commonly used detection technique, called *replica plating,* later in this chapter.

How Common Are Mutations?

A detection system must be designed before an investigator can find mutations. If a detection system is available, the investigator can set out on a mutant hunt. One thing will become readily apparent: mutations are, in general, very rare. This is shown in Table 7-1, which presents some data Lewis Stadler collected while working with several corn loci. Mutation studies of this sort are a lot of work! Counting a million of *anything* is no small task. Another feature shown by these data is that different genes seem to generate different frequencies of mutations; a 500-fold range is seen in the corn results. Obviously, one of the prime requisites of mutation analysis is to be able to measure the tendency of different genes to mutate. Two terms are commonly used to quantify mutation: mutation rate and mutation frequency. The **mutation rate** is expressed as the number of mutations occurring in some unit of time. The units commonly used are a cell generation span, an organismal generation span, or a cell division. You can see that these are all biological units of time.

Consider the lineage of cells in Figure 7-16. Obviously, there has been only one mutation event (M), so the numerator of a mutation rate is established. But what can we use as the denominator? The total time for mutation may be represented either by the total number of straight lines in Figure 7-16 (14 total generation spans) or, alternatively, by the total number of cell divisions (7). We might state the mutation rate, for example, as one mutation per seven cell divisions.

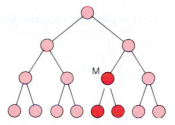

Figure 7-16 A simple cell lineage, showing a mutation at M.

The **mutation frequency** is the frequency at which a specific kind of mutation (or mutant) is found in a population of cells or individuals. The cell population can be gametes, asexual spores, or almost any other cell type. In our example in Figure 7-16, the mutation frequency in the final population of eight cells would be $2/8 = 0.25$.

Some mutation rates and frequencies are shown in Table 7-2.

Selective Systems

The rarity of mutations is a problem if an investigator is trying to amass a collection of a specific type of mutation for genetic study. Geneticists respond to this problem in two ways. One approach is to use a **selective system,** an experimental protocol designed to separate the desired mutant types from wild-type individuals. The other approach is to increase the mutation rate using **mutagens,** agents that have the biological effect of inducing mutations above the background, or spontaneous, rate. "Mutation Induction" will be discussed later in the chapter.

> **Message** Selective systems and mutagens offer two ways to improve the recovery of rare mutations.

Most of the examples of selective systems presented here are in microorganisms. This doesn't mean that selective systems are impossible for studies of more complex organisms, but merely that selection can be used to much better advantage in microbes. A million spores or bacterial cells are easy to produce; a million mice, or even a million fruit flies, require a large-scale commitment of money, time, and laboratory space. Microbes appear frequently in the discussions that follow, so a few words on microbe culture and routine microbial manipulation are appropriate here.

Microbial Techniques

The microbes that we consider in this book are bacteria, fungi, and unicellular algae, all of which are haploid. Fungi and algae are eukaryotic; their DNA is organized into distinct linear chromosomes that are found inside a nucleus enclosed by a nuclear membrane. However, bacteria are prokaryotic. Their DNA is not organized into separate chromosomes like that of eukaryotes. Generally there is one cir-

	Number of gametes tested	Number of mutations	Average number of mutations per million gametes
Gene			
$R \rightarrow r$	554,786	273	492.0
$I \rightarrow i$	265,391	28	106.0
$Pr \rightarrow pr$	647,102	7	11.0
$Su \rightarrow su$	1,678,736	4	2.4
$Y \rightarrow y$	1,745,280	4	2.2
$Sh \rightarrow sh$	2,469,285	3	1.2
$Wx \rightarrow wx$	1,503,744	0	0.0

Table 7-1 Foreward Mutation Frequencies at Some Specific Corn Loci

Table 7-2 **Mutation Rates or Frequencies in Various Organisms**

Organism	Mutation	Value	Units
Bacteriophage T2 (bacterial virus)	Lysis inhibition $r \rightarrow r^+$	1×10^{-8}	*Rate:* mutant genes per gene replication
	Host range $h^+ \rightarrow h$	3×10^{-9}	
Escherichia coli (bacterium)	Lactose fermentation $lac^- \rightarrow lac^+$	2×10^{-7}	*Rate:* mutant cells per cell division
	Histidine requirement $his^- \rightarrow his^+$	4×10^{-8}	
	$his^+ \rightarrow his^-$	2×10^{-6}	
Chlamydomonas reinhardtii (alga)	Streptomycin sensitivity $str^s \rightarrow str^r$	1×10^{-6}	
Neurospora crassa (fungus)	Inositol requirement $inos^- \rightarrow inos^+$	8×10^{-8}	*Frequency* per asexual spore
	adenine reqirement $ad^- \rightarrow ad^+$	4×10^{-8}	
Corn	See Table 7-1		
Drosophila melanogaster (fruit fly)	Eye color $W \rightarrow w$	4×10^{-5}	
Mouse	Dilution $D \rightarrow d$	3×10^{-5}	
Human:			
To autosomal dominants	Huntington's disease	0.1×10^{-5}	
	Nail-patella syndrome	0.2×10^{-5}	
	Epiloia (predisposition to type of brain tumor)	$0.4 – 0.8 \times 10^{-5}$	*Frequency* per gamete
	Multiple polyposis of large intestine	$1 – 3 \times 10^{-5}$	
	Achondroplasia (dwarfism)	$4 – 12 \times 10^{-5}$	
	Neurofibromatosis (predisposition to tumors of nervous system)	$3 – 25 \times 10^{-5}$	
To X-linked recessives	Hemophilia A	$2 – 4 \times 10^{-5}$	
	Duchenne muscular dystrophy	$4 – 10 \times 10^{-5}$	
Bone-marrow tissue-culture cells	Normal $\rightarrow$ azaguanine resistance	7×10^{-4}	*Rate:* mutant cells per cell division

SOURCE: Adapted from R. Sager and F. J. Ryan, *Heredity.* John Wiley, 1961.

cular chromosome, and it is not enclosed in a separate compartment of any kind.

In liquid culture, prokaryotic organisms proliferate as suspensions of individual cells. Each starting cell goes through repeated cell divisions, so that, from each, a series of 2, 4, 8, 16, . . . descendant cells is produced. This exponential growth is limited by the availability of nutrients in the medium, but soon a dense suspension of millions of separate cells is produced. Fungi such as yeasts follow this growth pattern exactly, but in the filamentous fungi the daughter cells remain attached as long chains called *hyphae* or *filaments*. Therefore, a liquid culture started from asexual spores eventually looks like tapioca pudding, with small fuzzy balls of hyphae in suspension, each originating from one spore.

In solid culture, usually on an agar gel surface, descendant cells tend to stay together, so that visible circular colonies are produced. When a suspension of cells is spread on the surface of a plate of culture medium, one colony develops from each original cell in the suspension.

What phenotypes of microorganisms may be examined? Morphological mutations affecting color, shape, and size of colony create useful but limited ranges of phenotypes. Other phenotypes have been far more useful—for example, auxotrophy, resistance to growth inhibitors, and inability to use various organic substances as an energy source.

With this brief introduction to microbial culture in mind, we can now discuss some examples of selective systems.

Reversion of Auxotrophs

For the reversion of auxotrophy to prototrophy, there is a direct selective system. Take an adenine-requiring auxotroph, for example. Grow a culture of this strain on an adenine-containing medium. Then plate a large number of cells on a solid nutritive medium containing no adenine. The only cells that can grow and divide on this medium are adenine prototrophs, which must have arisen by reverse mutation in the original culture (Figure 7-17). For most genes (not just those concerned with nutrition), the rate of reverse mutation is generally lower than the rate of forward mutation. We shall explore the reason for this later.

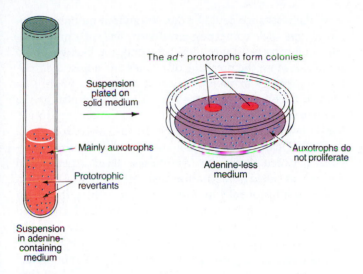

Figure 7-17 Selection for prototrophic revertants of auxotrophic adenine-requiring mutants.

Early studies on the reversion of auxotrophic mutations revealed a complication that must be dealt with in any reversion experiment. Some prototrophic colonies turn out not to be revertants at all but, rather, cases in which a gain-of-function mutation has occurred at a different locus. The second mutation acts as a suppressor of auxotrophy. If we represent the original mutation as *m* and the suppressor as *s*, then the apparent revertant was in fact of genotype *m s*. Suppressor mutations are generally not as common as revertants. They can be distinguished from revertants by crossing to a wild-type strain and observing the reappearance of the original mutant phenotype in the progeny:

$$m\,s \times +\,+$$

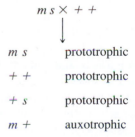

m s	prototrophic
+ +	prototrophic
+ s	prototrophic
m +	auxotrophic

Of course, a true revertant, which would be m^+, would produce only prototrophic progeny when crossed to a wild type.

> **Message**   In any reversion experiment, true revertants must be distinguished from suppressor mutations at a separate locus.

Filtration Enrichment

Filamentous fungi such as *Neurospora* and *Aspergillus* grow as a mass of branching threadlike hyphae. Fungal geneticists can make use of this property when selecting auxotrophic mutants, by using a procedure called **filtration enrichment.**

Let us assume that the investigator is interested in obtaining auxotrophs that specifically require the amino acid leucine. These must result from mutations of the type leu^+ to *leu* at one of several genetic loci. The investigator begins by growing a wild-type culture and, from this, obtaining large numbers of conidia. Most of these conidia will be prototrophic in regard to leucine, as expected of a wild-type culture. However, spontaneous forward mutations occur constantly, and some of the cells will be leucine-requiring auxotrophs bearing a mutant *leu* allele. These are selected as follows. The conidia are placed in a flask of liquid minimal medium and allowed to grow overnight with gentle shaking. Each prototrophic conidium acts as a growth center and produces a spiderlike colony of hyphal threads. However, auxotrophs of all kinds, including the leucine auxotrophs, will not grow but, rather, remain in suspension as live conidia. The contents of the flask are next poured through several layers of fine gauze, which act as a filter. The prototrophic colonies become caught on the fibers of the filter, but the single-celled auxotrophic mutants pass through (Figure 7-18). The cells that pass through the filter are then plated onto minimal medium supplemented with leucine. Each leucine auxotroph will grow and form a visible colony, but all other

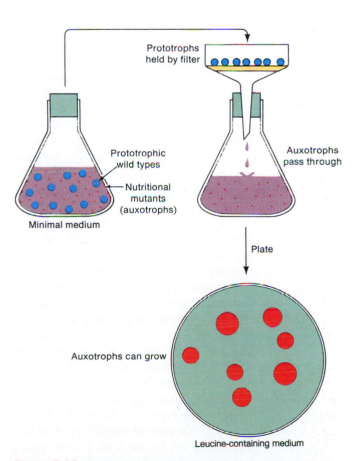

Figure 7-18 The filtration-enrichment method for selecting forward mutations to auxotrophy in filamentous fungi. (This example involves mutation to a leucine requirement.)

kinds of auxotrophs will remain invisible as single cells. Next, the colonies are isolated with a needle and transferred to permanent culture tubes containing supplemented medium. In this way it is relatively easy to amass a large number of auxotrophic mutants.

Penicillin Enrichment

A parallel technique is available for auxotroph selection in bacteria. Many species of bacteria are highly sensitive to the antibiotic penicillin, but only during their proliferating stage. If penicillin is added to a suspension of cells in liquid minimal medium, all the prototrophic cells are killed because they divide, but the auxotrophs survive because they cannot divide without supplementation. After treatment, the penicillin can be removed by washing the cells on a filter. Plating the washed cells on a minimal medium supplemented with some specific chemical such the amino acid leucine will selectively allow leucine auxotrophs to divide and produce visible colonies.

Resistance

Mutations that confer resistance to specific environmental agents not normally tolerated by wild types are easily demonstrated in microorganisms. We shall use an example that was important historically in determining the nature of mutation.

Viruses, like most parasites, are highly specific with regard to the hosts they infect. Some viruses are parasitic on bacteria; this group is called *bacterial viruses,* **bacteriophages,** or just **phages.** An individual phage is so small that it can be seen only with an electron microscope. There is a wide range of different phage types, and each is specific to certain bacterial species. The simple, fast cycles of phages make them ideally suited for genetic analysis. Because of their small size, very large numbers of phages can be used in an experiment, making it easy to detect rare genetic events such as mutations. Also their relatively simple genomes have made them ideal subjects for molecular analysis, and many of the advances in molecular genetics have come from studies on phages. Another important use of phages has been as DNA cloning vectors in recombinant DNA technology. Although they are introduced properly in Chapter 10, we need them at this point to illustrate resistance mutations.

The intestinal bacterium *Escherichia coli* is parasitized by many specific phages. One of these, called T1, was used in early bacterial mutation studies. Phage T1 attacks and kills most *E. coli* cells, liberating a large brood of newly synthesized progeny viruses from each dead bacterial cell. If a plate is spread with large numbers of bacteria (around 10^9) mixed with phages, most of the bacteria will be killed. However, some rare bacterial cells survive and produce colonies that can be isolated.

When T1-resistant bacteria were first observed, their origin was not known, but there were two main hypotheses.

First, the resistance could be due to random mutation from a wild-type allele conferring sensitivity (Ton^s) to a mutant allele conferring resistance (Ton^r). Second, the bacteria could somehow sense the presence of phages and adjust their cellular physiology in such a way that some cells might succeed in becoming resistant. This could be similar to the manner in which bacteria can shift their physiology to utilize a new nutrient in the medium. In 1943, Salvadore Luria and Max Delbrück designed a classic experiment to distinguish between these two hypotheses. Their experimental design was called a **fluctuation test.** Not only was the fluctuation test historically important in determining the origin of phage-resistant bacteria, but it also provided a method for calculating mutation rates that is still used today.

Luria and Delbrück reasoned that the two hypotheses of mutation and physiological change gave different predictions about the numbers of resistant bacteria found in a sample of cultures. Under the mutation hypothesis, in the cultures where a rare mutation event occurs, the mutation event could occur relatively early or relatively late in the growth of the culture because, presumably, mutation is random over time. An early event would produce a larger clone of descendant resistant cells than would a late event. Hence, considerable variation in the number of resistant cells from culture to culture is predicted (Luria and Delbrück used the term *fluctuation* instead of *variation*). Under the physiological change hypothesis, there is no reason to expect such variation; the physiological gear shifting would presumably be quite constant. These different predictions are diagrammed in Figure 7-19.

The practical details of the experiment were as follows. Into each of 20 culture tubes containing 0.2 ml of medium and into one containing 10 ml, they introduced 10^3 *E. coli* cells per milliliter and incubated them until about 10^8 cells per milliliter were obtained. Each of the twenty 0.2-ml cultures was then spread on a plate that had a dense layer of T1 phages. From the 10-ml "bulk" culture, 10 separate 0.2-ml volumes were withdrawn and plated. Many T1-resistant colonies were found, as shown in Table 7-3.

If resistance was due to random mutation during the incubation period, each culture tube would produce a different number of resistant cells; the number would vary depending on how early in the cascade of growing cells the mutation occurred. These resistant cells would each produce a separate colony when plated with the T1 phage. If resistance was due to a physiological adaptation occurring after exposure of the cells to the T1 phage, all cultures would be expected to generate resistant cells at roughly the same rate and little variation from culture to culture would be expected.

Variation from plate to plate was indeed observed in the individual 0.2-ml cultures but not in the samples from the bulk culture (which represent a kind of control experiment). This situation cannot be explained by physiological change, because all the samples spread had the same approximate number of cells. The simplest explanation is random mutation, occurring early in the incubation of the 0.2-ml cultures

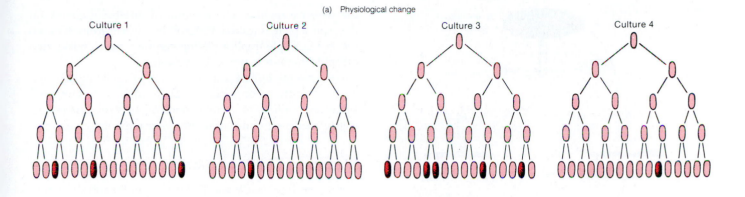

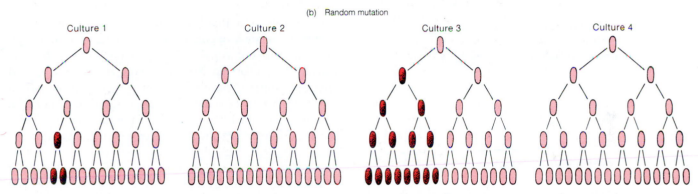

Figure 7-19 Cell pedigrees illustrating the expectations from two contrasting hypotheses about the origin of resistant cells. (a) Physiological change. (b) Random mutation. (From G. S. Stent and R. Calendar, *Molecular Genetics*, 2d ed. W. H. Freeman and Company, 1978.)

(producing a large number of resistant cells and, therefore, a large number of colonies), or late (producing few resistant cells and colonies), or not at all (producing no resistant cells).

This elegant analysis suggests that the resistant cells are selected by the environmental agent (here, phages) rather than produced by it. Can the existence of mutants in a population before selection be demonstrated directly? This demonstration was made possible by the use of a technique called **replica plating**, developed by Joshua and Esther Lederberg in 1952. A population of bacteria was plated on nonselective medium, that is, containing no phages, and from each cell a colony grew. This plate was called the *master plate*. A sterile piece of velvet was pressed down lightly on the surface of the master plate, and this picked up cells wherever there was a colony (Figure 7-20). In this way the velvet picked up a colony "imprint" from the whole plate. On touching the velvet to replica plates containing selective medium (that is, containing T1 phages), cells clinging to the velvet are inoculated onto the replica plates in the same relative positions as the colonies on the original master plate. As expected, rare *Ton*ʳ mutant colonies were found on the replica plates, but the multiple replica plates showed identical patterns of resistant colonies (Figure 7-21). If the muta-

Table 7-3   Results of Luria and Delbrück's Test

Individual cultures		Bulk culture	
Culture number	Number of T1-resistant colonies	Culture number	Number of T1-resistant colonies
1	1	1	14
2	0		
3	3	2	15
4	0		
5	0	3	13
6	5		
7	0	4	21
8	5		
9	0	5	15
10	6		
11	107	6	14
12	0		
13	0	7	26
14	0		
15	1	8	16
16	0		
17	0	9	20
18	64		
19	0	10	13
20	35		
	Mean 11.3		Mean 16.7

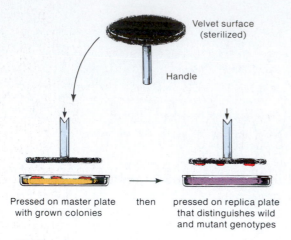

Figure 7-20 Replica plating reveals mutant colonies on a master plate through their behavior on selective replica plates. (From G. S. Stent and R. Calendar, *Molecular Genetics*, 2d ed. W. H. Freeman and Company, 1978.)

tions had occurred *after* exposure to the selective agents, the patterns for each plate would have been as random as the mutations themselves. The mutation events must have occurred before exposure to the selective agent.

Message Mutation is a random process. Any allele in any cell may mutate at any time.

Replica plating has become an important technique of microbial genetics. It is useful for negative selections, in other words, selections for mutants that *fail* to grow under

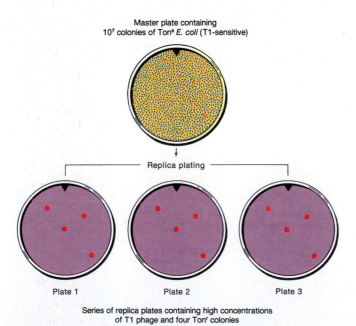

Figure 7-21 Replica plating demonstrates the presence of mutants before selection. The identical patterns on the replicas show that the resistant colonies are from the master. (From G. S. Stent and R. Calendar, *Molecular Genetics*, 2d ed. W. H. Freeman and Company, 1978.)

the selective regime. The position of an absent colony on the replica plate is used to retrieve the mutant from the master. For example, replica plating can be used to select auxotrophic mutants in precisely this way.

In general, replica plating is a way of retaining an original set of strains on a master plate while simultaneously subjecting replicas to various kinds of tests on different selective media or under different environmental conditions.

Measuring Mutation Rate

Let us see how Luria and Delbrück's fluctuation test provides a way of measuring the mutation rate. Consider the 20 cultures that were tested for phage T1 resistance. Here we have a situation that is well described by the Poisson distribution—a great opportunity for mutations to be found, but a large class of samples having no mutations at all in the culture (11 out of the 20). Look back at Figure 7-16, and convince yourself that the number of cell divisions necessary to produce n cells is n minus the original number in the culture (it would be $8 - 1 = 7$ in the example in Figure 7-16). If n is very large, which it is in the fluctuation test cultures, and if the original number of cells was relatively very small, then a sufficiently accurate estimate of the number of cell divisions is given by n itself. If the mutation rate per cell division is μ, then each culture tube may be expected to have contained an average of μn mutation events. The Poisson distribution (see page 157) then tells us that the zero class with no mutants will equal $e^{-\mu n}$. Because we know that the zero class $= 11/20 = 0.55$ and that $n = 0.2 \times 10^8$, we can solve the following equation for μ:

$$0.55 = e^{-\mu(0.2 \times 10^8)}$$

We find that $\mu = 3 \times 10^{-8}$ mutation events per cell division. [Note that in the same data, the mutation frequency is $(1 + 3 + 5 + 5 + 6 + 107 + 1 + 64 + 35)/(20 \times 0.2 \times 10^8)$, which equals $227/(4 \times 10^8)$, or 5.7×10^{-7}.]

Isolating Mutations from Cultures of Plant and Animal Cells

Animal and plant cells can be grown in culture. Like microbial cells, cells taken from certain animal and plant tissues, including cancerous tissue, divide mitotically in culture. Furthermore, many of the techniques used to induce and select mutations in microbial cell cultures can also be used with these plant and animal cells, so **somatic cell genetics** of plants and animals is made possible. The combination of these mutation-selection techniques with sophisticated techniques of cell fusion and hybridization now makes possible a wide variety of in vitro manipulations on plant and animal cells, including the cells of human.

One extensively used cell-culture system uses Chinese hamster ovary (CHO) cells. Many mutant cell types have been isolated from this system and from others. Let's have a look at some examples. In the first place, dominant mutations can be selected. In this context, dominance is defined in terms of the phenotype observed in hybrid cells formed by fusing a mutant cell with a normal cell. Some mutations that confer resistance to a particular drug, such as ouabain, α-amanitin, methotrexate, or colchicine, are dominant. Finding dominant mutations is not surprising: we might expect to find *only* dominant mutations because the cells are diploid and recessives would be hidden.

Nevertheless, recessive mutants are commonly seen, too. Alleles that confer resistance to lectins, 8-azaguanine, or methotrexate or confer the auxotrophic requirement for glycine, proline, or lysine, as well as various temperature-sensitive lethals, are all examples of recessive mutants that have been found. (Note that certain mutant phenotypes are dominant as determined by one mutant allele and recessive as determined by another mutant allele.) How is it possible for recessive mutant phenotypes to show up in diploid mammalian cells in culture? No one has the complete answer to this question. However, there is some evidence that although cultured CHO cells are predominantly diploid, they are hemizygous for very small sections of the genome. Thus, recessive mutations in such regions would be expressed. But whatever accounts for the appearance of such recessive phenotypes, the mutant alleles themselves provide useful markers for somatic cell genetics.

To grow single cells of plants in culture, a geneticist plates cells on a medium containing appropriate combinations of plant hormones. The resulting cell colonies, called **calluses,** can be subdivided for further growth in culture, or they can be transferred to a different medium that promotes differentiation of roots and shoots. Such "plantlets" can then be potted in soil and grown into mature plants. In one application of this technique, geneticists selected cells resistant to a toxin produced by a fungus that is parasitic on certain plants. From calluses growing on the toxin-containing plates, the geneticists produced plants that also showed resistance to the fungal toxin, and hence to fungal disease.

Mutation Induction

The task of finding rare mutations in multicellular organisms is difficult compared with that in microorganisms. So techniques have been designed to open the selective net as wide as possible. In 1928, Hermann J. Muller devised a method of searching for any lethal mutation on the X chromosome in *Drosophila*. He first "constructed" an X chromosome called *ClB*. The *C* stands for a chromosomal rearrangement called an *inversion*, a type we shall learn about in Chapter 8. For the moment, just think of it as standing for crossover suppressor because in a female fly carrying this special *ClB* chromosome and a normal X, the X chromosome chromatids do not cross over. The *ClB* chromosome also bears *l*, a recessive lethal allele, and the allele *B*, which determines the dominant bar-eye phenotype. *C l B* / Y males die because of hemizygosity for the lethal allele, but the chromosome can be maintained in heterozygous *C l B* / $C^+ l^+ B^+$ females. This special *ClB* system allowed Muller to detect lethal mutations anywhere on the X chromosomes in samples of male gametes. The experimental protocol is diagrammed in Figure 7-22. Bar-eyed daughters from the cross of females heterozygous for the *ClB* chromosome and wild-type males are crossed individually with wild-type males. Each bar-eyed daughter lays her eggs in a separate culture vial. When the progeny hatch, the vials are examined for the presence of males. We can see that if there was a new lethal recessive mutation on an X chromosome in one of the original male gametes, then the F_1 female carrying that chromosome will not produce any viable male progeny. Note that only rarely will a new lethal mutation *m* on the X chromosome be an allele of the *l* locus. (Of course, when it is, the *C l B* / 1 female will die.) The absence of males in a vial is easy to determine under low-power magnification and readily allows scoring for the presence of lethal recessive mutations on the X chromosome. Muller found the recessive frequency of such mutations occurring spontaneously to be about 1.5 per thousand chromosomes, still a relatively low value for an entire chromosome.

Figure 7-22 The *ClB* test for new X chromosome mutations in *Drosophila*. The symbol *m* represents a recessive lethal mutation anywhere on the X chromosome. Observing presence versus absence of males in each individual progeny amounts to scanning—by genetic analysis—a sample of gametes from the male parent of the bar-eyed daughter.

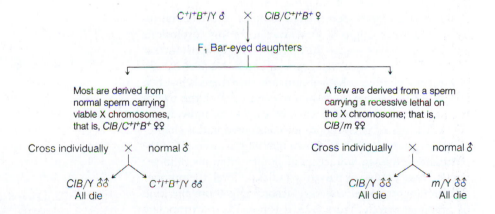

Table 7-4 Relative Efficiencies of Various Types of Radiation in Producing Mutations in *Drosophila*

Type of radiation	Percent of male X chromosomes bearing recessive lethal mutations after a dose of 1000 roentgens*
Visible light (spontaneous)	0.15
X rays (25 Mev)	1.70
β rays, γ rays, hard X rays	2.90
Soft X rays	2.50
Neutrons	1.90
α rays	0.84

*The roentgen (r) is a unit of radiation energy.

Muller then asked whether there were any agents that would increase the rate of mutation. Using the *ClB* test, he measured X-linked lethal frequencies after irradiating males with X rays and observed frequencies that were much higher than in unirradiated controls. His results supplied the first experimental evidence of a mutagen—in this case, the X rays. Recall that a **mutagen** is an agent that causes mutation at higher than spontaneous levels. Mutagens have been invaluable tools not only for studying the mechanism of mutation itself but also for increasing the yield of mutants for other genetic studies.

It is now known that many kinds of radiation increase mutations. Table 7-4 shows the effects of several types of radiation on increasing mutation frequencies in *Drosophila*. To the list in Table 7-4 must be added ultraviolet (UV) radiation, which is also mutagenic. Radiation is often categorized as *ionizing* or *nonionizing,* depending on whether ions are produced in the tissue through which it passes. X rays and γ (gamma) rays, for example, do produce ions, and UV radiation does not.

The harnessing of nuclear energy has become a social issue because of the powerful mutagenic effect of nuclear radiation. Even in the absence of nuclear war, many peacetime uses of nuclear energy—especially as fuel—carry their own hazards. No containment system is infallible, and recent decades have seen many examples of accidental release of nuclear isotopes, for example, through explosions and leaks at power plants. Furthermore, disposal of nuclear wastes is not as easy as had been supposed. Nuclear energy illustrates well the ecological principle that there is no such thing as a free lunch, and the pros and cons of the use of nuclear energy must be weighed by each of us individually.

The kinds of gene mutations considered in this chapter are **point mutations.** It is known that within a certain range of radiation dosage, induction of point mutations is linear; that is, if we double or halve the radiation level, the number of mutants produced will vary accordingly. From the kind of graph shown in Figure 7-23, it is possible to extrapolate to very low radiation levels and to infer very low frequen-

cies of mutation induction. Because exposure to cosmic radiation, radiation from X-ray machines, radioactive fallout from bomb testing, and contamination from nuclear plants is very low, it is possible to conclude that the effects are negligible. Yet even though increases in mutation rates might be low, population numbers are large: every year 200 million new gametes unite to form 100 million babies in the world. In this very large annual "mutation experiment," even low mutation frequencies are potentially translatable into large numbers of mutations.

Radiation doses generally are cumulative. If a population of organisms is repeatedly exposed to radiation, the frequency of mutations induced will be in direct proportion to the total amount of radiation absorbed over time. However, there are exceptions to the cumulative effect. For example, if mice given *x* rads (a biological measure of radiation dose) in one short burst, called an *acute dose,* are compared with those given the same dose gradually over a protracted period of weeks or months, a *chronic dose,* significantly fewer mutations are found in the chronically exposed group. This has been interpreted to mean that there is some form of repair of radiation-induced genetic damage over time.

In 1947 Charlotte Auerbach and J. M. Robson conducted experiments on mustard gases, which had been used as toxins in gas warfare. They found that in addition to being toxic, the mustard gases were also potently mutagenic. This pioneer study provided the impetus for research into

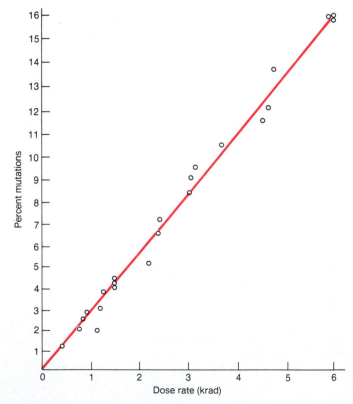

Figure 7-23 Linear relationship between X-ray dose to which *Drosophila melanogaster* were exposed and the percentage of mutations (mainly sex-linked recessive lethals).

the mutagenic effects of other chemicals, and it is now known that many of the chemicals that we are exposed to in our daily environment are mutagens. Cigarette smoke is an obvious example, but there are many others, including certain preservatives, pesticides, herbicides, industrial chemicals, and pollutants. There are even natural chemical mutagens in our foods.

Indeed, some chemicals are so potent that they are now routinely used by geneticists to induce mutations in experimental organisms. The mutagenicity of chemicals is important experimentally because, in many cases, the chemical reactions responsible for mutagenic action are understood, and this knowledge can be used to produce specific types of mutations. Furthermore, many chemicals are much less

toxic to an organism than radiation is and yet result in much higher frequencies of mutation. So, as a research tool, chemical mutagens have been useful in producing the wide array of mutations now available for genetic studies.

Mutation and Cancer

It is now clear that cancer is a genetic disease. Cancer has many causes, but ultimately all these causes exert their effect on a special class of genes called *cancer genes* or **proto-oncogenes**. Many of these proto-oncogenes have now been identified and mapped, as shown in Figure 7-24. Oncogenes normally carry out basic cellular functions, generally related

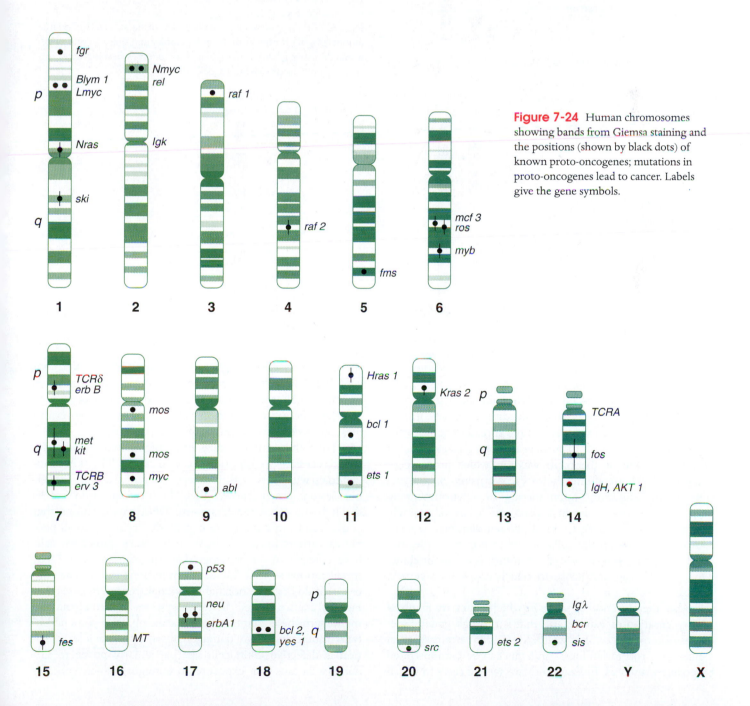

Figure 7-24 Human chromosomes showing bands from Giemsa staining and the positions (shown by black dots) of known proto-oncogenes; mutations in proto-oncogenes lead to cancer. Labels give the gene symbols.

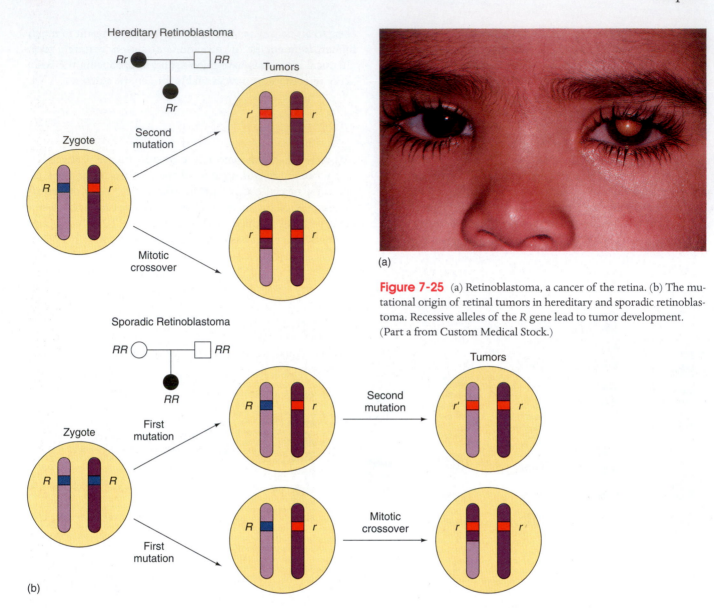

(a)

Figure 7-25 (a) Retinoblastoma, a cancer of the retina. (b) The mutational origin of retinal tumors in hereditary and sporadic retinoblastoma. Recessive alleles of the *R* gene lead to tumor development. (Part a from Custom Medical Stock.)

(b)

to the regulation of cell division. However several types of events can change a proto-oncogene into an oncogene into a state in which it promotes the two main characteristics of cancer, uncontrolled cell division leading to an overgrown group of cells called a **tumor,** and the spread of tumor cells throughout the body to form new tumors, a process called **metastasis.** One of the main ways in which proto-oncogenes can be changed into their cancer-causing (oncogenic) state is by mutation. Spontaneous or environmentally induced mutation occurs in a proto-oncogene of a single cell, which then undergoes multiple cell divisions to form a tumor. Because all the cells of the tumor carry the mutated oncogene, you can see that a tumor is a mutant clone. This, then, is the first sense in which cancer is a genetic disease.

Also, there is another sense in which genes are involved in the causation of cancer, and that is through genetic predisposition to cancer. It is well known that certain types of cancer can "run in families," and this can be demonstrated by pedigree analysis. Indeed there are several cases in which

specific alleles virtually guarantee cancer. How is the concept of predisposition compatible with the mutational basis for cancer? After all, mutation is a random process, so how can there be a predisposition? We will examine two examples, one recessive and one dominant, to clarify.

The first example is xeroderma pigmentosum, an autosomal recessive predisposition to skin cancer. The skin of individuals with this condition becomes covered with cancerous lesions. It is known that the predisposition works through the DNA repair system. DNA modifications that could lead to mutations are constantly occurring in all people as various kinds of chemical mistakes. However, cells have efficient repair enzymes that can remove these DNA modifications so that few or no mutations result. The cancers that do occur in normal individuals represent cases that have by chance eluded the repair system. In xeroderma pigmentosum the enzyme system that normally repairs potentially mutagenic DNA damage caused by ultraviolet radiation is defective. Skin cells are particularly prone to UV damage because of exposure to sunlight, which contains

UV light. Consequently, in the absence of efficient repair, multiple cancers result. It is clear why the predisposition is a recessive phenotype. The mutation results in loss of function, but in a heterozygote the normal allele can do all the DNA repair needed.

The second example of predisposition is retinoblastoma, a retinal cancer. Retinoblastoma arises in two ways. The first is sporadic, with no previous cases in the family history. Generally the tumor is in one eye only (unilateral). The second way is through a hereditary predisposition that pedigree analysis shows is inherited as an autosomal dominant. In this type both eyes are often affected (bilateral). The two types are explained by a common mechanism.

All cases of retinoblastoma (Figure 7-25) are caused by mutations in one gene; let us call this gene R. Cancerous mutant alleles are all recessive, represented r. In hereditary retinoblastoma, the determining element inherited is one r allele, so the people who develop the cancer are all born with genotype Rr. This initially seems at odds with the established dominant inheritance of the condition, but this paradox will become clear. The cancer arises in retinal cells that have become rr. In about 30 percent of cases the second r allele can arise by random mutation, and in about 70 percent of cases the second r allele arises by mitotic cross-

ing-over. Hence, with mutation, a new r allele arises, so the cancerous genotype is really rr'. With mitotic crossing-over, the original r allele is made homozygous, rr. These processes are diagrammed in Figure 7-25b. You can see the way the predisposition works; with one mutant allele already inherited, only one other allele must become mutant to express the disease. This also explains why two eyes are often affected; the probability of the second "r event" is high because each retina is composed of millions of cells.

A person who develops sporadic retinoblastoma inherits the genotype RR. In order for a tumor to develop, two r alleles must arise, either rr from two independent mutations, or first a mutation to Rr then the step to rr by mitotic crossover (Figure 7-25b). Because two events are needed, the "bad luck" of the cancer is usually unilateral. The molecular product of the R gene is most likely a repressor of a cell cycle protein. Mutation inactivates the repressor, and hence cell division becomes unregulated. Other familial predispositions to cancer, including familial predisposition to breast cancer, are expressed in ways similar to the retinoblastoma model. Some genetic aspects of cancer are summarized in Figure 7-26.

These examples show how many genetic principles can be brought to bear on a single practical problem, in this case cancer. Note that in our analysis we have used the concepts of Mendelian inheritance, dominance and recessiveness, pedigree analysis, mitotic crossing-over, and of course our present topic, somatic mutation.

Mutagens in Genetic Dissection

Mutations are the basis of genetic analysis. In the same way that we can learn how the engine of a car works by tinkering with its parts one at a time to see what effect each has, we can see how a cell or an organism works by altering its parts one at a time by inducing mutations. The mutation analysis is a further aspect of the process we have called *genetic dissection* of living systems. As we have seen, the first stage of any mutational dissection is the mutant hunt. The effectiveness of the hunt can be improved dramatically by using mutagens. Let's look at some examples.

First, Table 7-5 shows the relative frequencies of *ad-3* forward mutants in *Neurospora* after various mutagenic treatments. Note that an investigator would have to test about 2.5 million untreated cells to find a single mutant cell but could hope to find a mutant cell among only 450 treated with ICR-170. Table 7-6 provides another example of the improved recovery of mutant cells after treatment with an appropriate mutagen. The term **supermutagen** has been introduced to describe some of these highly potent agents.

For a third example, we turn to *Drosophila*. Until the mid-1960s, most researchers working with *Drosophila* used radiation to induce mutations. Doses of 4000 roentgens produce lethal mutations in perhaps 10 to 11 percent of all X chromosomes of irradiated males. Higher doses, however, make the flies increasingly infertile. Another problem is that

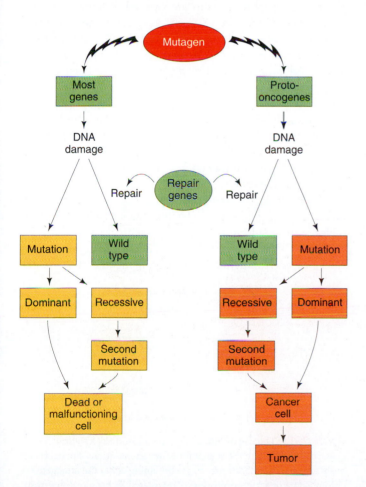

Figure 7-26 Comparison of the various outcomes of somatic mutations in proto-oncogenes and other genes.

Table 7-5 Forward Mutation Frequencies Obtained with Various Mutagens in *Neurospora*

Mutagenic treatment	Exposure time (minutes)	Survival (%)	Number of *ad-3* mutants per 10^6 survivors
No treatment (spontaneous rate)	—	100	~0.4
Amino purine (1–5 mg/ml)	During growth	100	3
Ethyl methane sulfonate (1%)	90	56	25
Nitrous acid (0.05 M)	160	23	128
X rays (2000 r/min)	18	16	259
Methyl methane sulfonate (20 mM)	300	26	350
UV rays (600 erg/mm^2 per min)	6	18	375
Nitrosoguanidine (25 μM)	240	65	1500
ICR-170 acridine mustard (5 μg/ml)	480	28	2287

NOTE: The assay measures the frequency of purple *ad-3* colonies among the white colonies produced by the wild-type *ad-3$^+$*.

many of the mutations induced by X rays involve chromosomal rearrangements, and these can complicate the genetic analysis.

In contrast, the chemical ethyl methane sulfonate (EMS) induces a vastly higher proportion of point mutations. This mutagen is very easily administered simply by placing adult flies on a filter pad saturated with a mixture of sugar and EMS. Simple ingestion of the EMS produces large numbers of mutations. For example, males fed 0.025 M EMS produce sperm carrying lethal alleles on more than 70 percent of all X chromosomes and on virtually every chromosome 2 and chromosome 3. At these levels of mutation induction, it becomes quite feasible to screen for many mutations at specific loci or for defects with unusual phenotypes.

In the *ClB* test, to screen for X-linked recessive lethals it was necessary to test for lethality in the second generation of descendants. However it is possible to screen an X chromosome for viable recessive mutations immediately in the F$_1$ generation, by making use of a special X chromosome rearrangement called an **attached X** (represented as X̂X). An attached X is a compound chromosome formed by the fusing of two separate X chromosomes. It is inherited as a single unit. EMS-treated males are crossed to females bearing an attached X and a Y (X̂X/Y), as diagrammed in Figure 7-27. (If you need to review sex determination in *Drosophila,* turn back to Chapter 3.) The gametes of the female are X̂X and Y. All F$_1$ zygotes carrying sex-linked lethals die, but newly induced, sex-linked mutations governing observable phenotypes are expressed in the males. Because of the effectiveness of the method, it is possible to select a wide range of mutations in a particular region of the genome, an approach known as **saturation mutagenesis.** All F$_1$ males carry a mutagenized paternal X chromosome, each from a different treated sperm.

Geneticists can then take the study of the mutations further: if individual F$_1$ males are crossed with X̂X/Y females, each resulting progeny represents a single "cloned" X. The F$_1$ flies can be raised at 22°C, and then progeny clones of each individual X can be incubated at 29°C and

Table 7-6 The Potency of Various Chemical Mutagens for Inducing *his$^-$* Reverse Mutation in *Salmonella*

Mutagen	Revertants/ nanomole	Ratio of effectiveness*
1,2-Epoxybutane	0.006	1
Benzyl chloride	0.02	3
Methyl methane sulfonate	0.63	105
2-Naphthylamine	8.5	1400
2-Acetylaminofluorene	108	18,000
Aflatoxin B$_1$	7057	1,200,000
Furylfuramide (AF-2)	20,800	3,500,000

*Relative effectiveness of each mutagen, compared against 1,2-epoxybutane.

SOURCE: Adapted from J. McCann and B. N. Ames, in *Advances in Modern Toxicology*, vol. 5. Edited by W. G. Flamm and M. A. Mehlman. Hemisphere Publishing Corp., 1976.

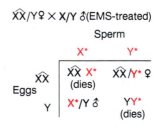

X*, Y* = chromosomes exposed to mutagen

Figure 7-27 The use of attached-X chromosomes (X̂X) in *Drosophila* to facilitate the search for X-linked mutations. Sperm previously exposed to EMS fertilize eggs containing either the attached-X chromosome or a Y chromosome. The treated X chromosomes from the males show up as the hemizygous X's of the sons, revealing phenotypically recessive mutations.

17°C to detect heat-sensitive and cold-sensitive lethals. Such temperature-sensitive (ts) lethals make up about 10 to 12 percent of all EMS-induced lethals. We have already learned how useful such conditional mutants are for genetic analysis. Many laboratories now routinely test for ts mutants in *Drosophila* screening tests. The approach is possible because large numbers of mutants are easily obtained.

The availability of ts lethals has a side benefit in that it simplifies laboratory procedures. A time-consuming task in any large-scale *Drosophila* experiment is separating females from males within 12 hours after emergence from the pupae to prevent undesired mating. (The flies do not mate during the first 12 to 14 hours after emergence.) The task can be eliminated by using ts lethal alleles (represented as l^{ts}) to produce cultures of a single sex at will. For example, the cross $X\hat{X}/Y \; ♀ \times l^{ts}/Y \; ♂$ produces progeny of both sexes at permissive temperatures. If such a culture is shifted to restrictive temperatures before hatching, the l^{ts}/Y males die and only the females hatch. Similarly, homozygosity of an l^{ts} allele in an $X\hat{X}$ chromosome determines that only wild-type males hatch from cultures kept at restrictive temperatures.

> **Message** The mutagen EMS has revolutionized the genetic versatility of *Drosophila* by providing a potent method for the isolation of a wide range of point mutants.

Mutation Breeding

In addition to the mutational dissection of biological systems, geneticists use mutation in other ways. We saw in Chapter 2 that one way of breeding a better crop plant is to make a hybrid and then to select the desired recombinants from the progeny generations. That approach makes use of the variation naturally found between available stocks or isolates from nature. Figure 7-28 shows examples of the outcome of such breeding. Another way to generate variability for selection is to treat with a mutagen. In this way, the variability is increased through human intervention.

A variety of procedures may be used. Pollen may be mutagenized and then used in pollination. Dominant mutations will be expressed in the next generation, and further generations of selfing reveal recessives. Alternatively, seeds may be mutagenized. A cell in the enclosed embryo of a seed may become mutant, and then it may become part of germinal tissue or somatic tissue. If the mutation is in somatic tissue, any dominant mutations will show up in the plant derived from that seed, but this will be the end of the road for such mutations. Germinal mutations will show up in later generations, where they can be selected as appropriate. Figure 7-29 summarizes mutation breeding.

In this chapter, we have seen how gene mutation not only is of biological interest in itself, but can also be put to a variety of experimental and practical uses. In the next chapter, we shall see that much the same kind of statement can be made about chromosome mutation.

Figure 7-28 *Potentilla fruticosa* wild type (*center*) and horticultural varieties arranged in a circle. All the attractive, novel shapes, colors, patterns, and sizes arose through mutations, which were then selected and bred by the horticulturalists. Most flowering plants in our gardens and parks and most food plants in use today have been produced by such a procedure, beginning with either spontaneous or induced mutations. (Anthony Griffiths)

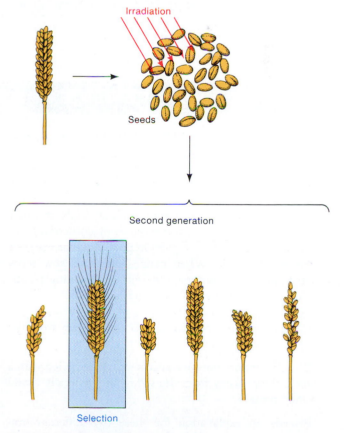

Figure 7-29 Mutation breeding in crops. (From Bjorn Sigurbjornsson, "Induced Mutations in Plants," *Scientific American,* 1970.)

SUMMARY

Genes may mutate in somatic cells or germinal cells. Within these two broad categories, there are several kinds of mutations, including morphological, lethal, conditional, biochemical, and resistance mutations.

Mutations can be used to study the process of mutation itself or to permit genetic dissection of biological functions. In order to carry out such studies, however, it is necessary to have a system for detecting mutant alleles at the phenotypic level. In diploids, dominant mutations are easily detected. On the other hand, the detection of recessive mutations requires the use of special genotypes; otherwise, the mutations may never be manifested in the phenotype. For this reason, detection systems are much more straightforward in haploids, where the question of dominance does not arise.

Mutation events occur randomly at any time and in any cell. Therefore the size of a mutant clone is proportional to the developmental stage at which the mutation event occurred.

Mutations often result in loss of the normal function of the affected allele, either fully (null mutations) or partly (leaky mutations). Loss-of-function mutations are generally recessive, and this can be demonstrated in heterozygotes, in which the wild-type allele provides enough function to result in normal phenotype. Gain-of-function mutations are often dominant.

Somatic mutation in oncogenes is the major cause of cancer. Hereditary predisposition to cancer can be caused by inheriting one cancer-causing allele from a parent, or by possessing defective DNA repair genes. Some environmental chemicals and radiation can influence mutation rates in cancer genes.

Because mutations are rare, geneticists use selective systems or mutagens (or both) to obtain mutations. Selective systems automatically distinguish between mutant and nonmutant states. Mutagens are valuable not only in studying the mechanisms of mutation but also in inducing mutations to be used in other genetic studies. In addition, mutagens are frequently used in crop breeding.

Concept Map

Draw a concept map interrelating as many of the following terms as possible. Note that the terms are listed in no particular order.

mutation / allele / reversion / wild type / recessive / somatic / mutagen / germinal / genetic dissection / progeny

CHAPTER INTEGRATION PROBLEM

Some recessive mutations in maize affect the color, texture, or shape of the kernel. To discuss these mutations in general, we'll use m to represent the recessive allele. These mutations can be detected in a straightforward manner by crossing an $m^+ m^+$ male with an mm female. Any new mutations are found by visually scanning many seeds, in other words millions of kernels on thousands of corncobs. Finding a seed with the recessive phenotype shows that the gene mutated in the male. When maize geneticists sow seeds with the mutant phenotype and self the resulting plants, they find one or the other of two kinds of results:

If the mutation was spontaneous, the progeny from the selfing are generally all mutant.

If the mutation followed treatment of the pollen with a mutagen, the progeny from the selfing are generally $\frac{3}{4}$ wild type and $\frac{1}{4}$ mutant.

Provide an explanation for these two different outcomes.

Solution

As usual, if we restate the results in a slightly different way, it gives us a clue as to what is going on. The difference between outcomes seems to reflect some difference between spontaneous and induced mutation. From a spontaneous mutation, we obtain an individual that seems to be of genotype mm because it breeds true for the mutant phenotype. From an induced mutation, we obtain an individual that seems, from the 3:1 phenotypic ratio upon selfing, to have an $m^+ m$ genotype. Clearly, the latter deduction does not mesh with the observation of a mutant phenotype. This is the paradoxical case. How is it possible for a seed to have a recessive mutant phenotype but to mature into a plant that produces the selfed progeny expected from a heterozygote?

The experimenter should always keep in mind the peculiarities of the organism that is being used. What do we know about corn that might be relevant to this situation? The explanation hinges on mutational events in and around the pollen grain in the anthers of the plant. Recall from

Chapter 3 that the mature pollen grain of maize is not just a cell with a single haploid nucleus, but that there are three important components: a haploid tube nucleus and two haploid nuclei that act as sperm. One of these sperm fuses with one female nucleus to create the zygote, and the other sperm fuses with two other female nuclei to create the triploid endosperm. When mature pollen cells are treated with a mutagen, it is likely that only one of the sperm nuclei mutates in any one cell; the only mutations we see are those in sperm that produce the endosperm, because it is the endosperm that forms the bulk of the mature kernel. This is the solution to the problem, but let's write out the genotypes: The endosperm must be mmm, and the embryo must be m^+m. This is why selfing produces a $3:1$ ratio.

What about the spontaneous mutations? A gene can mutate at any time. There is only a relatively short time between the mitosis that produces the three nuclei of a mature pollen cell and fertilization. A gene is thus most likely to mutate during the relatively long time that precedes the pollen cell's maturation, possibly even a long time before, in tissue that is still somatic. If a gene does mutate during this time to become allele m, the pollen cell's mitoses will produce three nuclei of genotype m. When such a cell unites with a gamete from the mm female, the endosperm is mmm and the embryo is mm.

In summary, we have used concepts from the discussion of germinal and somatic mutation, and of detection systems (all from this chapter); from previous chapters we have used our knowledge of dominance relations, Mendelian genetics, mitosis and meiosis, and organismal life cycles.

SOLVED PROBLEMS

1. A certain plant species normally has the purple pigment anthocyanin dispersed throughout the plant. This makes the green parts of the plant look brown (green plus purple) and the parts of the flower that lack chlorophyll (petals, ovary, anthers) look bright purple. The allele A is essential for anthocyanin production, and, in a homozygote, the recessive alleles a result in a plant that lacks anthocyanin. An interesting new allele called a^u arises, which is unstable. The a^u allele reverts to A at a frequency thousands of times greater than regular (stable) a alleles do.

a. What phenotype would you expect in plants of genotype (1) $a^u a^u$; (2) $a^u a$; (3) $A a^u$?

b. How can you confirm that the reversions are true mutations?

Solution

a. (1) We have seen that mutations tend to be very rare; in the normal course of events, they are seldom observed unless they are seriously sought out. However, an allele that is reverting as frequently as a^u is likely to announce its reversion behavior prominently if a proper detection system is available. In a plant containing billions of somatic cells, many cells will undoubtedly revert at some stage of development. As development proceeds, each initial revertant cell will give rise to a clone of revertant cells that should be visible as a purple or brown sector. Therefore, a plant of $a^u a^u$ genotype should have white flowers with purple sectors of varying size and photosynthetic tissues that are basically green with brown sectors.

(2) Each sector is a clone derived from a single cell with a^u reverting to A. The allele a^u is expected to be dominant over a, because a is essentially inactive and will not prevent a^u from reverting. Hence, $a^u a$ plants will look the same as $a^u a^u$, but with fewer purple or brown sectors because there are half the chances for reversion.

(3) Because A produces pigment in all cells, $A a^u$ will be indistinguishable from AA. Even though a^u is reverting, the reversions will not show up—a detection system is lacking. Such plants will be purple-brown throughout.

b. How can we prove that a pigmented sector is due to a mutation? It is worth mentioning that some spotted patterns (as in, for example, Dalmatian dogs) are not caused by mutant clones but rather by developmental effects. So spots do not necessarily indicate a mutation—a skeptic could invoke some other cause.

The key is that genes are hereditary units and mutant alleles should be transmitted to subsequent generations of cells or individuals. How can we detect the transmission of revertant A alleles to progeny cells or individuals? If we were able to take or "scrape" a few cells out of a sector and grow them into a plant, and that plant turned out to be purple, then the reversion hypothesis would be proved. Well, the technology for doing such an experiment is available. Is there any other way? If revertant sectors appear in the flower, then presumably the germinal tissue (anthers and ovaries) should sometimes be part of a sector. Therefore, we could collect pollen from such a flower and pollinate a plant of genotype aa. If there is some pollen carrying a revertant A, then the progeny from this pollen should be Aa and purple-brown throughout. Finding some progeny of this phenotype would prove the reversion is a true mutation because there is nowhere else the A gene could have come from.

2. A mutation experiment is performed on the *tryp4* locus in yeast. A *tryp4* mutation confers a requirement for the amino acid tryptophan. A *tryp4* mutant allele named *tryp4-1* is known to be revertible; the experiments want to measure reversion frequency in a population of haploid cells. A culture of mating type α and of genotype *tryp4-1* is grown and 10 million cells are plated on a medium lacking tryptophan; 120 colonies are obtained. The genotypes of these prototrophic colonies are checked by crossing each one to a wild-type culture of mating type *a*. Based on the results of these crosses, it is found that the prototrophic colonies were of two types: two-thirds of type 1 and one-third of type 2:

> Type 1α × wild-type *a* ⟶
> > progeny all tryptophan-independent
>
> Type 2α × wild-type *a* ⟶
> > $\frac{3}{4}$ progeny tryptophan-independent
> > $\frac{1}{4}$ progeny tryptophan-requiring

   a. Propose a genetic explanation for the two types.

   b. Calculate the frequency of revertants.

Solution

   a. The technique of plating on a medium lacking tryptophan should be a selection system for *tryp4*$^+$ revertants, because they do not require tryptophan for growth. When backcrossed to a wild type (*tryp4*$^+$), all progeny of revertants should be tryptophan-independent. This behavior is exhibited by type 1 colonies, so these prototrophs are revertants.

Now what about the type 2 colonies? The fact that some progeny are tryptophan-requiring shows that the *tryp4-1* mutation could not have genuinely reverted. Rather, the requirement for tryptophan appears to have been merely masked or suppressed. We have already studied several examples of the suppression of a mutant allele by a new mutation at a separate locus. But even if we had not remembered about these examples, the 3 : 1 ratio in a haploid organism should provide the clue that two independent loci are involved. Let's designate the suppressor mutation as *su* and its inactive wild-type allele as *su*$^+$. Type 2 colonies must be *tryp4-1 su*, and the wild types are *tryp4*$^+$ *su*$^+$. A cross of these strains produces

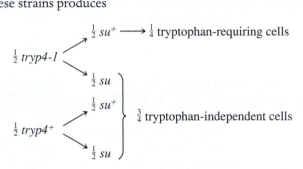

(Notice that *su* has no effect on the *tryp4*$^+$ allele.)

   b. We are told that two-thirds of the colonies are type 1, or true revertants. Thus, there are $120 \times \frac{2}{3} = 80$ revertants out of a total of 10^7, or a revertant frequency of 8×10^{-6} cells.

PROBLEMS

1. A certain species of plant produces flowers with petals that are normally blue. Plants with the mutation *w* produce white petals. In a plant of genotype *ww*, one *w* allele reverts during the development of a petal. What detectable outcome would this reversion produce in the resulting petal?

2. *Penicillium* (a commercially important filamentous fungus) normally can synthesize its own leucine (an amino acid). How would you go about selecting mutants that are leucine-requiring (that cannot synthesize their own leucine)? (NOTE: Like many filamentous fungi, *Penicillium* produces profuse numbers of asexual spores.)

3. How would you select revertants of the yeast allele *pro-1*? This allele confers an inability to synthesize the amino acid proline, which can be synthesized by wild-type yeast and which is necessary for growth.

4. How would you use the replica plating technique to select arginine-requiring mutants of haploid yeast?

5. Using the filtration-enrichment technique, you do all your filtering with a minimal medium and do your final plating on a complete medium that contains every known

nutritional compound. How would you find out what *specific* nutrient is required? After replica plating onto every kind of medium supplement known to science, you still can't identify the nutritional requirement of your new yeast mutant. What could be the reason or reasons?

6. An experiment is initiated to measure the reversion rate of an *ad-3* mutant allele in haploid yeast cells. Each of 100 tubes of a liquid, adenine-containing medium are inoculated with a very small number of mutant cells and incubated until there are 10^6 cells per tube. Then the contents of each tube are spread over a separate plate of solid medium containing no adenine. The plates are observed after one day, and colonies are seen on 63 plates. Calculate the reversion rate of this allele per cell division.

7. Suppose that you want to determine whether caffeine induces mutations in higher organisms. Describe how you might do this (include control tests).

8. In corn, alleles at a single locus determine presence (*Wx*) or absence (*wx*) of amylose in the cell's starch. Cells that have *Wx* stain blue with iodine; those that have only *wx* stain red. Design a system for studying the frequency of rare mutations from $Wx \rightarrow wx$ without using acres of plants. (HINT: You might start by thinking of an easily studied cell type.)

9. Suppose that you cross a single male mouse from a homozygous wild-type stock with several virgin females homozygous for the allele that determines black coat color. The F_1 consists of 38 wild-type mice and five black mice. Both sexes are represented equally in both phenotypic classes. How can you explain this result?

10. A man employed for several years in a nuclear power plant becomes the father of a hemophiliac boy. There is no hemophilia in the extensive pedigrees of the man's ancestors and his wife's ancestors. Another man, also employed for several years in the same plant, has an achondroplastic dwarf child—a condition nowhere recorded in his ancestry or in that of his wife. Both men sue their employer for damages. As a geneticist, you are asked to testify in court. What do you say about each situation? (NOTE: Hemophilia is an X-linked recessive; achondroplasia is an autosomal dominant.)

11. In a large maternity hospital in Copenhagen, there were 94,075 births. Ten of the infants were achondroplastic dwarfs. (Achondroplasia is an autosomal dominant, showing full penetrance.) Only two of the dwarfs had a dwarf parent. What is the mutation frequency to achondroplasia in gametes? Do you have to worry about reversion rates in this problem? Explain.

12. One of the jobs of the Hiroshima-Nagasaki Atomic Bomb Casualty Commission was to assess the genetic consequences of the blast. One of the first things they studied was the sex ratio in the offspring of the survivors. Why do you suppose they did this?

13. Many organisms present examples of unstable recessive alleles that revert to wild types at very high frequencies. Different unstable alleles often revert in different ways. One gene in corn (C), which produces a reddish pigment in the kernels, has several unstable null alleles. The unstable allele c^{m1} reverts late in the development of the kernel and at a very high rate. Another unstable allele, c^{m2}, reverts earlier in the development of the kernel and at a lower rate. Assuming that lack of pigment leaves the cell looking yellowish, what phenotype do you predict in plants of genotype

  a. Cc^{m1}?   **e.** $c^{m1}c$ (c is a stable mutant allele)?

  b. Cc^{m2}?   **f.** $c^{m2}c$?

  c. $c^{m1}c^{m1}$?   **g.** $c^{m1}c^{m2}$?

  d. $c^{m2}c^{m2}$?

14. A haploid strain of *Aspergillus nidulans* carried an auxotrophic *met-8* mutation conferring a requirement for methionine. Several million asexual spores were plated on minimal medium, and two prototrophic colonies grew and were isolated. These prototrophs were crossed sexually to two different strains, with the progeny shown in the body of the following table, where met^+ means methionine is not required for growth, and met^- means methionine is required for growth.

	Crossed to a wild-type strain	Crossed to a strain carrying the original *met-8* allele
Prototroph 1	All met^+	$\frac{1}{2}$ met^+ $\frac{1}{2}$ met^-
Prototroph 2	$\frac{3}{4}$ met^+ $\frac{1}{4}$ met^-	$\frac{1}{2}$ met^+ $\frac{1}{2}$ met^-

  a. Explain the origin of both of the original prototrophic colonies.

  b. Explain the results of all four crosses, using clearly defined gene symbols.

Unpacking the Problem

Before trying to solve this problem, try these questions that are relevant to the experimental system.

a. Draw a labeled diagram that shows how this experiment was done. Show test tubes, plates, etc.

b. Define all the genetic terms in this problem.

c. Many problems show a number next to the auxotrophic mutation's symbol—here the number *8* next to *met*. What does the number mean? Is it necessary to know in order to solve the problem?

d. How many crosses were actually made? What were they?

e. Represent the crosses, using genetic symbols.

f. Is the question about somatic mutation or germinal mutation?

g. Is the question about forward mutation or reversion?

h. Why was there such a small number of prototrophic colonies (two) found on the plate?

i. Why didn't the several million asexual spores grow?

j. Do you think any of the millions of spores that did not grow were mutant? Dead?

k. Do you think the wild type used in the crosses was prototrophic or auxotrophic? Explain.

l. If you had the two prototrophs from the plates and the wild-type strain in three different culture tubes, could you tell them apart just by looking at them?

m. How do you think the *met-8* mutation was obtained in the first place? (Show with a simple diagram. Note, *Aspergillus* is a filamentous fungus.)

n. In Chapter 6 we learned that *Aspergillus* is a convenient organism in which to study mitotic crossing-over and haploidization. Do you think those processes could be involved in this problem? Can they be ruled out as explanations of the results?

o. What progeny do you predict from crossing *met-8* with a wild type? What about *met-8* × *met-8*? What about wild type × wild type?

p. Do you think this is a random meiotic progeny analysis or a tetrad analysis?

q. Draw a simple life cycle diagram of a haploid organism showing where meiosis takes place in the cycle.

r. Consider the $\frac{3}{4} : \frac{1}{4}$ ratio. In haploids, crosses heterozygous for one gene generally give progeny ratios based on halves. How can this idea be extended to give ratios based on quarters?

15. A diploid strain of yeast was heterozygous for several genes as shown below:

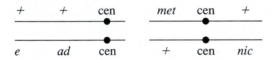

The recessive *e* allele gives an elongated, egg-shaped cell instead of the usual round one, and the recessive *ad* (adenine-requiring) allele causes accumulation of a red pigment in the medium under conditions of limiting adenine (the wild-type allele gives a white color). The *met* and *nic* alleles are recessive auxotrophic "forcing markers," whose job is to force the cells to stay diploid in order to grow on minimal medium. *Cen* represents "centromeres." The original culture was white and grew well on minimal medium, and on limiting adenine produced no red pigment. Cells were irradiated and plated out onto medium with limiting adenine. There were several phenotypes on these plates:

a. Most colonies were white.

b. Several diploid red colonies appeared. The red colonies were examined under the microscope.

(1) Most colonies were composed of round cells.

(2) Some had elongated cells.

c. There were many cells on the plates that did not grow into colonies. Some were transferred to complete medium containing yeast extract (all possible nutrients included).

(1) Some then grew into colonies.

(2) Some did not.

What are the most likely explanations of the origins of these five types?

16. Cells of a haploid wild-type *Neurospora* strain were mutagenized with EMS. Large numbers of these cells were plated and grown up into colonies at 25°C on complete medium (containing all possible nutrients). These plates were used as masters, then replicas were imprinted onto plates of minimal medium and complete medium at both 25°C and at 37°C. There were several mutant phenotypes, as shown in the accompanying sketch. The large circles represent luxuriant growth, the spidery symbols represent weak growth, and a blank means no growth at all. How would you categorize the types of mutants represented by isolates 1 through 5?

	Minimal		Complete	
Strain	25°C	37°C	25°C	37°C
Mutant 1			◯	◯
Mutant 2	◯		◯	
Mutant 3	✳	✳	◯	◯
Mutant 4	◯	✳	◯	✳
Mutant 5	◯		◯	◯
Wild type (control)	◯	◯	◯	◯

17. **a.** You have isolated three tryptophan-requiring mutants of a haploid fungus. You cross them (in pairs) and score 1000 spores from each cross for the ability to grow without added tryptophan.

Cross	Parents	Trp$^+$ / 1000
1	A × B	135
2	A × C	179
3	B × C	50

What can you conclude from these results?

b. You grow up cultures from five of the Trp-requiring spores from cross 2 (you label them "2-1," "2-2," etc.), and you cross each of these cultures with the original mutant B. Again, you score 1000 spores from each cross for the ability to grow without added tryptophan.

Cross	Parents	$Trp^+/1000$
4	2-1 × B	132
5	2-2 × B	13
6	2-3 × B	53
7	2-4 × B	137
8	2-5 × B	47

What can you conclude about each of the strains, 2-1, 2-2, etc.? Your explanation should include each strain's genotype and how it arose.

(Problem 17 by Rosie Redfield.)

18. In order to understand the genetic basis of locomotion in the diploid nematode *Caenorhabditis elegans*, recessive mutations were obtained, all making the worm "twitch" ineffectually instead of moving with its usual smooth gliding motion. These mutations presumably affect the nervous or muscle systems. Twelve homozygous mutants were intercrossed and the hybrids examined to see if they twitched or not. The results were as follows, where a "+" means the hybrid was wild-type (gliding) and "t" means the hybrid twitched:

	1	2	3	4	5	6	7	8	9	10	11	12
1	t	+	+	+	t	+	+	+	+	+	+	+
2		t	+	+	+	t	+	t	+	t	+	+
3			t	t	+	+	+	+	+	+	+	+
4				t	+	+	+	+	+	+	+	+
5					t	+	+	+	+	+	+	+
6						t	+	t	+	t	+	+
7							t	+	+	+	t	t
8								t	+	t	+	+
9									t	+	+	+
10										t	+	+
11											t	t
12												t

a. Explain what the intercrossing experiment was designed to test.

b. How many genes affecting movement have been identified, and which mutations are in which gene? Give the genes symbols of your choice.

c. As illustrative examples, show the relevant genotypes of mutants 1, 2, and 5, and of the $\frac{1}{5}$ and $\frac{1}{2}$ hybrids, using the gene symbols you defined in part b, and explain why different results are obtained in these two hybrids.

19. Note: this problem requires knowledge of the genetic code (see Figure 13-18). In corn, synthesis of purple pigment is controlled by two genes acting sequentially through colorless ("white") intermediates:

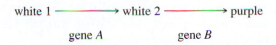

UAG nonsense mutations were obtained in genes A and B, and we can call them a^n and b^n. These both lacked enzyme activity and were recessive to their wild-type alleles A and B. A nonsense mutation is caused by a nucleotide change that results in a stop codon; in this case the same nonsense codon was produced, UAG.

A nonsense suppressor is also available; it is a mutation in a duplicate copy of a tRNA gene. The mutational site is in the anticodon, and this causes the tRNA to insert the amino acid tryptophan at the nonsense codon UAG, thereby allowing the protein synthesis machinery to complete translation. The result is a wild-type phenotype. We will represent the nonsense suppressor allele as T^s and its wild-type allele T^+. Neither T^s nor T^+ has any detectable effect on the wild-type alleles A and B or on any other aspect of the phenotype.

a. Would you expect T^s to be dominant to T^+ or not? Explain.

b. A trihybrid $A\,a^n\,B\,b^n\,T^s\,T^+$ is selfed. If all the genes are unlinked, what phenotypic ratio do you expect in the progeny. Explain carefully. (Assume that absence of purple results in a light color which we can call *white*.)

20. Outline the precise steps that you would take to find the chromosomal locus of a new recessive mutation that you have just found in tomatoes. Show all steps, including what crosses you would make, which lines you would use, how you would obtain the lines, how much field or greenhouse space is needed, how long it would take, etc. (You may refer to the tomato map, which is shown in Figure 5-15.)

21. A *Neurospora* strain unable to synthesize arginine produces a revertant, arginine-independent colony. A cross is made between the revertant and a wild-type strain. What proportion of the progeny from this cross would be arginine-independent if

a. the reversion precisely reversed the original change that produced the *arg*⁻ mutant allele?

b. the revertant phenotype was produced by a mutation in a second gene located on a different chromosome (the new mutation suppresses *arg*⁻)?

c. a suppressor mutation occurred in a second gene located 10 map units from the *arg*⁻ locus on the same chromosome?

22. A certain haploid fungus is normally red owing to a carotenoid pigment. Mutants were obtained that were different colors owing to the presence of different pigments: orange (o^-), pink (p^-), white (w^-), yellow (y^-), and beige (b^-). Each phenotype was inherited as if a single gene mutation governed it. To determine what these mutations signified, double mutants were constructed with all possible combinations, and the results were as follows:

	p^-	w^-	y^-	b^-
o^-	Pink	White	Yellow	Beige
p^-	—	White	Pink	Pink
w^-		—	White	White
y^-			—	Yellow

Interpret this table as follows, using the first entry as an example: the double mutant o^-p^- has a pink phenotype.

a. What do the results of the table mean? Explain in detail.

b. What phenotypes would result if the mutants were paired as heterokaryons in all the combinations possible?

c. What phenotypic proportions would be found in the progeny of a cross between an o^-p^- double mutant and a wild type, if the loci are linked 16 map units apart?

8

Chromosome Mutation I: Changes in Chromosome Structure

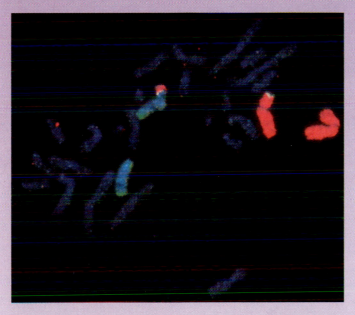

A reciprocal translocation demonstrated by a technique called *chromosome painting*. A suspension of chromosomes from many cells is passed through an electronic device that sorts them by size. DNA is extracted from individual chromosomes, joined to one of several fluorescent dyes, then added to partially denatured chromosomes on a slide. The fluorescent DNA "finds" its own chromosome and binds along its length to paint it. In this preparation a light-blue and a pink dye have been used to paint different chromosomes. The preparation shows one normal pink chromosome, one normal blue, and two that have exchanged their tips. (Lawrence Berkeley Laboratory)

KEY CONCEPTS

▶ Chromosomes have many landmarks that can be used for the detection of rearrangements.

▶ Owing to the strong meiotic pairing affinity of homologous regions, diploids with one standard and one rearranged chromosome set produce pairing structures that have shapes and properties unique to that rearrangement.

▶ A deletion in one chromosome set is generally deleterious as a result of gene imbalance and the unmasking of deleterious alleles in the other chromosome set.

▶ Duplications can lead to gene imbalance but also provide extra material for evolutionary divergence.

▶ Heterozygous inversions show reduced recombination in the region spanned by the inversion, and reduced fertility.

▶ A heterozygous translocation shows 50 percent sterility and linkage of genes on the chromosome involved in the translocation.

Chromosome mutation is the process of change that results in rearranged chromosome parts, abnormal numbers of individual chromosomes, or abnormal numbers of chromosome sets. As with gene mutation, the term *chromosome mutation* is applied both to the process and to the product, so the novel genomic arrangements may be called *chromosome mutations.* Sometimes chromosome mutation can be detected by microscopic examination, sometimes by genetic analysis, and sometimes by both. In contrast, gene mutations are never detectable microscopically on the chromosome; a chromosome bearing a gene mutation looks the same under the microscope as one carrying the wild-type allele.

Many chromosome mutations lead to abnormalities in cell and organismal function. There are two basic reasons for this. First, the chromosome mutations can result in abnormal gene number or position. Second, if chromosome mutation involves chromosome breakage, which is often the case, then the break may occur in the middle of a gene, thereby disrupting its function.

Chromosome mutations are important at several different levels of biology. First, in research, they provide ways of designing special arrangements of genes, uniquely suited to answer certain biological questions. Second, chromosome mutations are important at the applied level, especially in medicine, and in plant and animal breeding. Finally, chromosome mutations have been instrumental in shaping genomes as part of the evolutionary process.

The study of normal and abnormal chromosome sets and their genetic properties is called **cytogenetics,** a discipline that combines cytology with genetics. Cytogenetics uses the normal genome as its standard or wild type. A cytogeneticist must be familiar with the normal features and landmarks of the genome before being able to detect any

Table 8-1   Human Chromosomes

Group	Number	Diagrammatic representation	Relative length*	Centromeric index†
Large chromosomes				
A	1		8.4	48 (M)
	2		8.0	39
	3		6.8	47 (M)
B	4		6.3	29
	5		6.1	29
Medium chromosomes				
C	6		5.9	39
	7		5.4	39
	8		4.9	34
	9		4.8	35
	10		4.6	34
	11		4.6	40
	12		4.7	30
D	13		3.7	17 (A)
	14		3.6	19 (A)
	15		3.5	20 (A)
Small chromosomes				
E	16		3.4	41
	17		3.3	34
	18		2.9	31
F	19		2.7	47 (M)
	20		2.6	45 (M)
G	21		1.9	31
	22		2.0	30
Sex chromosomes				
	X		5.1 (group C)	40
	Y		2.2 (group G)	27 (A)

* Percentage of the total combined length of a haploid set of 22 autosomes.

† Percentage of a chromosome's length spanned by its short arm. The four most metacentric chromosomes are indicated by an (M); the four most acrocentric by an (A).

changes that have occurred. Therefore we begin our discussion with the topography of chromosomes.

The Topography of Chromosomes

In this section we will discuss the features of chromosomes commonly used as landmarks in cytogenetic analysis.

Chromosome Size

The chromosomes of a single genome may differ considerably in size. In the human genome, for example, there is about a three- to fourfold range in size from chromosome 1 (the biggest) to chromosome 21 (the smallest), as shown in Table 8-1. In studying the chromosomes of some species, a cytogeneticist may have difficulty identifying individual chromosomes by size alone, but may be able to group chromosomes of similar size. A change may then be pinpointed in, for example, "one of the chromosomes in size group A."

Centromere Position

The centromere is the structure to which spindle fibers attach. The centromere region usually appears to be constricted, and the position of this constriction defines the ratio between the lengths of the two chromosome arms; this ratio is a useful characteristic (Table 8-1). Centromere positions can be categorized as **telocentric** (at one end), **acrocentric** (off center), or **metacentric** (in the middle). The centromere position determines not only arm ratio but also the shapes of chromosomes as they migrate to opposite poles during anaphase. These anaphase shapes range from a rod to a J to a V (Figure 8-1). In some organisms, such as the lepidoptera, centromeres are "diffuse," so that spindle fibers attach all along the chromosome. When such a chromosome breaks, both parts can still migrate to the poles. In contrast, a break in a chromosome with a single centromere creates a fragment that cannot move to the pole because it has no centromere. Chromosome fragments lacking a centromere are called **acentric** chromosomes (Figure 8-2).

The molecular structure of the centromeres of certain organisms is now known, as is the structure of the tips of the chromosomes, the **telomeres**. Although not morphologically distinct, the telomeres have a unique molecular structure that is crucial to normal chromosome behavior. In yeast, understanding the molecular structure of centromeres and telomeres has made it possible to create entirely new chromosomes. These **yeast artificial chromosomes (YACs)** are composed of a centromere, two telomeres, a DNA replication origin, and in between these any type of DNA that the experimenter wishes to insert there. YACs have become important as DNA carriers (or *vectors*) in the manipulation of the DNA of complex organisms such as humans.

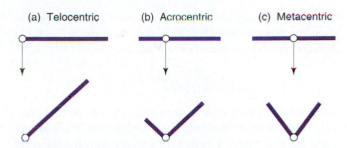

(a) Telocentric   (b) Acrocentric   (c) Metacentric

Figure 8-1 The classification of chromosomes by the position of the centromere. A telocentric chromosome has its centromere at one end; when the chromosome moves toward one pole of the cell during the anaphase of cellular division, it appears as a simple rod. An acrocentric chromosome has its centromere somewhere between the end and the middle of the chromosome; during anaphase movement, the chromosome appears as a J. A metacentric chromosome has its centromere in the middle and appears as a V during anaphase.

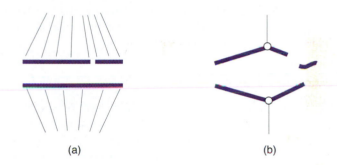

(a)   (b)

Figure 8-2 Spindle fiber attachments to chromosomes. (a) A diffuse centromere with many spindle fibers attaching along the length of a single chromosome. This arrangement is rare. A break in the chromosome does not result in the loss of any chromosome material. (b) Spindle fiber attachment to a single centromere (the common arrangement). In this case, a break in the chromosome results in an acentric fragment, which is lost during cellular division.

Position of Nucleolar Organizers

Nucleoli are intranuclear organelles that contain ribosomal RNA, an important component of ribosomes. Different organisms are differently endowed with nucleoli, which range in number from one to many per chromosome set. The diploid cells of many species have two nucleoli. The nucleoli reside next to secondary constrictions of the chromosomes, called **nucleolar organizers,** which have highly specific positions in the chromosome set. Nucleolar organizers contain the genes that code for ribosomal RNA. Their positions, like those of centromeres, are landmarks for cytogenetic analysis.

Chromomere Patterns

Chromomeres are beadlike, localized thickenings found along the chromosome during prophase of mitosis and meiosis. Homologous chromosomes tend to have homologous sets of chromomeres. Although they can be useful as markers, their molecular nature is not known.

Heterochromatin Patterns

When chromosomes are treated with chemicals that react with DNA, such as Feulgen stain, distinct regions with different staining characteristics are visually revealed. Densely staining regions are called **heterochromatin;** poorly staining regions are said to be **euchromatin.** The distinction reflects the degree of compactness or coiling of the DNA in the chromosome. Heterochromatin can be either *constitutive* or *facultative.* The constitutive type is a permanent feature of a specific chromosome location and is, in this sense, a hereditary feature; Figure 5-15 shows good examples of constitutive heterochromatin in the tomato. The facultative type of heterochromatin, as its name suggests, is sometimes but not always found at a particular chromosomal location. The patterns of heterochromatin and euchromatin along a chromosome are good cytogenetic markers.

Banding Patterns

Special chromosome staining procedures have revealed specific sets of intricate bands (transverse stripes) in many different organisms. The positions and sizes of the bands are highly chromosome-specific. There are Q bands (produced by quinacrine hydrochloride), G bands (produced by Giemsa stain), and R bands (reversed Giemsa). An example of G bands in human chromosomes is shown in Figure 8-3.

A rather specialized kind of banding, which has been used extensively by cytogeneticists for many years, is characteristic of the so-called **polytene** chromosomes in certain organs of the dipteran insects. In 1881, E. G. Balbiani recorded peculiar structures in the nuclei of certain secretory cells of two-winged flies. These structures were long and sausage-shaped and were marked by swellings and cross striations. Unfortunately, Balbiani did not realize they were chromosomes, and his report remained buried in the literature. It was not until 1933 that they were rediscovered, and recognized as a special type of chromosome.

Polytene chromosomes develop in the following way. In secretory tissues, such as the Malpighian tubules, rectum, gut, footpads, and salivary glands of the diptera, the chromosomes replicate their DNA many times without actually separating into distinct chromatids. As a chromosome increases its number of replicas, it elongates and thickens. This bundle of replicas becomes the polytene chromosome. We can look at *Drosophila melanogaster* as an example. This insect has a 2n number of 8, but the special organs contain only four polytene chromosomes. There are four and not eight because during the replication process the homologs unexpectedly become tightly paired. Furthermore, all four polytene chromosomes become joined at a structure called the **chromocenter,** which is a coalescence of the heterochromatic areas around the centromeres of all four chromosome pairs. The chromocenter of *Drosophila* salivary gland chromosomes is shown in Figure 8-4, where *L* and *R* stand for arbitrarily assigned left and right arms.

Along the length of a polytene chromosome, there are transverse stripes called **bands.** Polytene bands are much

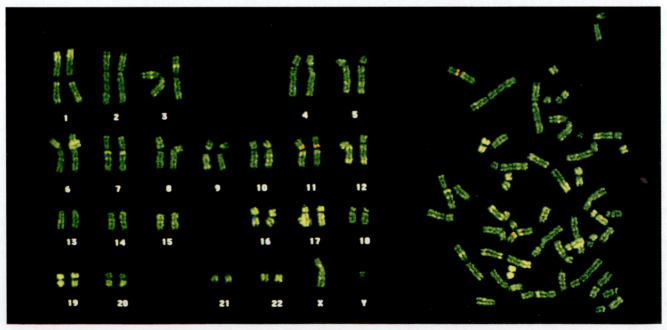

Figure 8-3 Chromosomes of a human male stained with Giemsa to reveal bands.

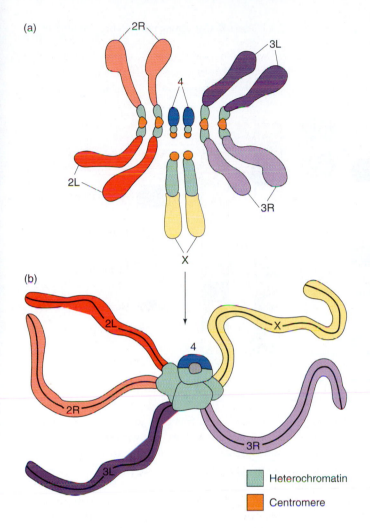

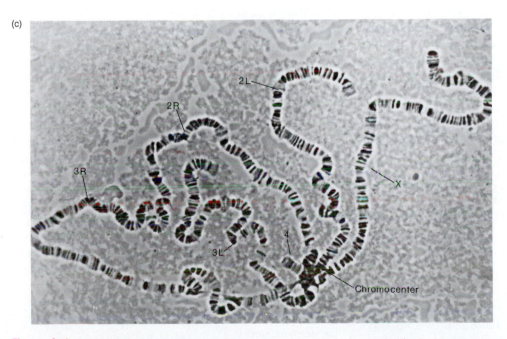

more numerous than Q, G, or R bands, numbering in the hundreds on each chromosome. The bands differ in width and morphology, so that the banding pattern of each chromosome is unique and characteristic of that chromosome. In addition, there are regions that may at times appear swollen (**puffs**) or greatly distended (**Balbiani rings**), and these correspond to regions of RNA synthesis. As we shall see in Chapter 16, recent molecular studies have shown that in any chromosomal region of *Drosophila* there are more genes than there are polytene bands, so there is not a one-to-one correspondence of bands and genes as was once believed. Similarly, the significance of the banding patterns of the chromosomes of vertebrates is not completely clear. However, it is known that the chromosome bands reflect DNA nucleotide composition, and that most of the active genes reside in the light bands.

Using all the available chromosomal landmarks together, cytogeneticists can distinguish each of the chromosomes in many species. As an example, Figure 8-5 shows a map of the chromosomal landmarks of the genome of corn. Notice how the landmarks enable each of the 10 chromosomes to be distinguished under the microscope.

Message   Such features as size, arm ratio, heterochromatin, number and position of thickenings, number and location of nucleolar organizers, and banding pattern identify the individual chromosomes within the set that characterizes a species.

Chromosomal landmarks are useful not only for characterizing normal chromosomes, but also for detecting chromosomal mutations. In this chapter we will discuss mu-

Figure 8-4 Polytene chromosomes form a chromocenter in a *Drosophila* salivary gland. (a) Mitotic metaphase chromosomes, with arms represented by different shades. (b) Heterochromatin coalesces to form the chromocenter. (c) Photograph of polytene chromosomes. (Tom Kaufman)

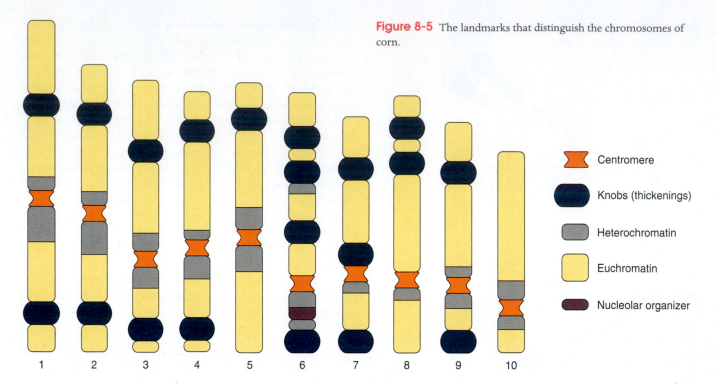

Figure 8-5 The landmarks that distinguish the chromosomes of corn.

Centromere

Knobs (thickenings)

Heterochromatin

Euchromatin

Nucleolar organizer

1 2 3 4 5 6 7 8 9 10

tational changes in chromosome structure, sometimes known as **chromosome rearrangements.** Many rearrangements are recognized by alterations in the chromosomal landmarks. However, there are also genetic ways of detecting rearrangements.

Let's consider two important properties of chromosomes that are useful in understanding structural chromosome mutations. The first is that during prophase I of meiosis, homologous regions of chromosomes have a strong pairing affinity, and if necessary will contort in order to pair. Because of this, many curious structures may be seen in a cell that has one standard set of chromosomes and one aberrant set. Remember that in polytene chromosomes the homologs also pair (even though they are not in meiotic cells), so comparable shapes result. The second important property concerns the ends of broken chromosomes. Changes in structure usually involve chromosome breakage, and the broken chromosome ends are highly "reactive," tending strongly to join with other broken ends. The telomeres (the regular chromosome ends), however, do not tend to join.

Types of Changes in Chromosome Structure

A chromosome rearrangement can occur de novo in a cell of any tissue, and the result will be a heterozygous karyotype consisting of one normal set plus one rearranged set. Rearrangements occurring in somatic tissue may have phenotypic effects in one cell or a somatic sector of cells. Rearrangements occurring in germinal tissue may generate heterozygous meiocytes. In this case the original rearrangement may enter a proportion of the gametes, but in addition various kinds of abnormal meiotic chromosome segregations can occur, resulting in novel types of chromosome rearrangements, and these will also enter the gametes. If a rearrangement takes place de novo within the gamete population, then this too may end up in one of the offspring.

Gametes with rearrangements would most likely unite with normal gametes from the opposite sex, so that the zygote and all the cells of the F_1 individual developing from this zygote would be heterozygous for the rearrangement. The potential phenotypic impact of the arrangement therefore is much more severe. Furthermore, *all* the meiocytes in such an individual will be heterozygous for the rearrangement, and a large proportion of gametes would contain the parental or a derived rearrangement.

Most of the following discussions will concern individuals whose entire body is heterozygous for a rearrangement: in these individuals the rearrangement may have occurred de novo in parental gonads or gametes or may have been derived from a parent whose body was also heterozygous in every cell.

In discussing chromosome rearrangements, it is convenient to use letters to represent different chromosome regions. These letters therefore represent large segments of DNA, each containing many genes.

The simple loss of a chromosomal segment is called a **deletion** or **deficiency.** In the following diagram region B has been deleted:

A B C D E F   →   A C D E F

The presence of two copies of a chromosomal region is called a **duplication:**

A B B C   D E F

A segment of a chromosome can rotate 180 degrees and rejoin the chromosome, resulting in a chromosomal mutation called an **inversion:**

A E D   C B F

Finally, two nonhomologous chromosomes may exchange parts to produce a chromosomal mutation called a **translocation:**

A B C D J K   G H   I E F

We shall consider the properties of each of these chromosomal rearrangements in turn.

Deletions

The process of spontaneously occurring deletion must involve two chromosome breaks to cut out the intervening segment. If the two ends join, and one of them bears the centromere, a shortened chromosome results, which is said to carry a deletion. The deleted fragment is acentric; consequently it is immobile and will be lost. An effective mutagen for inducing chromosomal rearrangements of all kinds is ionizing radiation. This kind of radiation, of which X rays and γ rays are examples, is highly energetic and causes chromosome breaks. The way in which the breaks rejoin determines the kind of rearrangement produced. Two types of deletion are possible. Two breaks can produce an **interstitial deletion,** as shown in Figure 8-6. In principle, a single break can cause a **terminal deletion,** but because of the need for the special chromosome tips (telomeres) it is likely that apparently terminal deletions involve two breaks, one close to the telomere.

The effects of deletions depend on their size. A small deletion within a gene, called an **intragenic deletion,** inactivates the gene and has the same effect as other null mutations of that gene. If the homozygous null phenotype is viable (as for example in human albinism), then the homozygous deletion will also be viable. Intragenic deletions can be distinguished from single nucleotide changes because they are nonrevertible.

For most of this section we will be dealing with **multigenic deletions,** those that remove two to several thousands of genes. These have more severe consequences. If by inbreeding such a deletion is made homozygous (that is, if

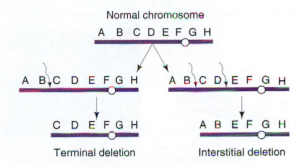

Figure 8-6 Terminal and interstitial deletions. Chromosome can be broken when struck by ionizing radiation (wavy arrows). A terminal deletion is the loss of the end of a chromosome. An interstitial deletion results after two breaks are induced if the terminal portion (AB) rejoins the main body of the chromosome while the acentric fragment (CD) is lost.

both homologs have the same deletion), then the combination is almost always lethal. This suggests that most regions of the chromosomes are essential for normal viability and that complete elimination of any segment from the genome is deleterious. Even individuals heterozygous for a multigenic deletion—those with one normal homolog and one that carries the deletion—may not survive. There are several possible reasons for this. First, a genome has been "finetuned" during evolution to require a specific balance of genes, and the deletion upsets this balance. We shall encounter this balance notion several times in this chapter and the next, because several different types of chromosome mutations upset the ratio or balance of genes in a genome. Second, in many organisms there are recessive lethal and other deleterious mutations throughout the genome. If "covered" by wild-type alleles on the other homolog, these recessives are not expressed. However, a deletion can "uncover" recessives, allowing their expression at the phenotypic level.

Message   The lethality of heterozygous deletions can be explained by genome imbalance and by unmasking of recessive lethal alleles.

Nevertheless, some small deletions are viable in combination with a normal homolog. In these cases the deletion can sometimes be identified by cytogenetic analysis. If meiotic chromosomes are examined in an individual carrying a heterozygous deletion, the region of the deletion can be determined by the failure of the corresponding segment on the normal homolog to pair, resulting in a **deletion loop** (Figure 8-7a). In insects, deletion loops are also detected in the polytene chromosomes, in which the homologs are fused (Figure 8-7b). A deletion can be assigned to a specific chromosome location by determining which chromosome shows the deletion loop, and the position of the loop along the chromosome.

Deletions of some chromosomal regions produce their own unique phenotypes. A good example is a deletion of one specific small chromosome region of *Drosophila.* When

(a) Meiotic chromosome

(b) Polytene chromosomes

Figure 8-7 Looped configurations in a *Drosophila* deletion heterozygote. During the meiotic pairing, the normal homolog forms a loop. The genes in this loop have no alleles to synapse with. Because polytene chromosomes in *Drosophila* have specific banding patterns, we can infer which bands are missing from the homolog with the deletion by observing which bands appear in the loop of the normal homolog. (Part b from William M. Gelbart.)

one homolog carries the deletion, the fly shows a unique notch-wing phenotype, so the deletion acts like a dominant mutation in this regard. But the deletion is lethal when homozygous and therefore acts as a recessive in regard to its lethal effect. The specific dominant phenotypic effect of certain deletions might be caused by one of the chromosome breaks being inside a gene which, when disrupted, will act like a dominant mutation.

What are the genetic properties of deletions? In addition to cytogenetic criteria, there are several purely genetic criteria for inferring the presence of a deletion. These criteria are particularly useful in species whose chromosomes are not easily analyzed cytogenetically.

Two genetic criteria we have encountered already. The first is the failure of the chromosome to survive as a homozygote; however, this effect could also be produced by any lethal mutation. Second, chromosomes with deletions can never revert to a normal condition. This criterion is useful only if there is some specific phenotype associated with the deletion.

A third criterion is that in heterozygous deletions, recombinant frequencies between genes flanking the deletion are lower than in control crosses. This makes intuitive sense because part of the region contains an unpaired chromosomal region, which cannot participate in crossing-over. We will see that inversions have a similar effect on recombinant frequencies, but can be distinguished in other ways.

A fourth criterion for inferring the presence of a deletion is that deletion of a segment on one homolog sometimes unmasks recessive alleles present on the other homolog, leading to their unexpected expression. Consider, for example, the deletion shown in the following diagram:

$$a \quad b \quad c \quad d \quad e \quad f$$

Phenotype is a^+ b c $d^+e^+f^+$

$$a^+ \qquad d^+\ e^+\ f^+$$

In this case none of the six recessive alleles is expected to be expressed, but if *b* and *c* are expressed, then it suggests that a deletion has occurred on the other homolog spanning the b^+ and c^+ loci. Because in such cases it seems that recessive alleles are showing dominance, the effect is called **pseudodominance.**

The pseudodominance effect also can be used in the opposite direction. A known set of overlapping deletions is used to locate the map positions of new mutant alleles. This procedure is called **deletion mapping.** An example from the fruit fly *Drosophila* is shown in Figure 8-8. In this diagram, the recombination map is shown above, marked with distances in map units from the left end. The horizontal bars below the chromosome show the extent of the deletions identified to the left. The mutation prune (*pn*), for example, shows pseudodominance only with deletion 264-38, which determines its location in the 2D-4 to 3A-2 region. However, *fa* shows pseudodominance with all but two deletions, so its position can be pinpointed to band 3C-7.

Deletion analysis makes it possible to compare a linkage map based on recombinant frequency with the chromosome map based on deletion mapping. By and large, where this has been done, the maps correspond well—a satisfying cytological endorsement of a purely genetic creation.

Message Chromosome maps made by analyzing deletion coverage are congruent with linkage maps made by analyzing recombinant frequency.

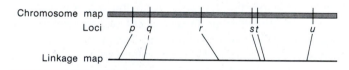

Chromosome map

Loci   *p*   *q*   *r*   *st*   *u*

Linkage map

Moreover, pseudodominance can be used to map a small deletion that cannot be visualized microscopically. Let's consider an X chromosome in *Drosophila* that carries a recessive lethal suspected of being a deletion; we call this chromosome "X*." We can cross X*-bearing females with males carrying recessive alleles of loci on that chromosome. For example, a map of loci in the tip region is

y	*dor*	*br*	*gt*	*swa*	*w*	*rst*	*vt*
0.3	0.3	0.3	0.4	0.2	0.2	0.6	

Suppose we obtain all wild-type flies in crosses between X*/X females and males carrying *y, dor, br, gt, rst,* and *vt* but obtain pseudodominance of *swa* and *w* with X* (that is, X*/*swa* shows the recessive *swa* phenotype and X*/*w* shows the recessive *w* phenotype). Then we have good genetic evidence for a deletion of the chromosome that includes at least the *swa* and *w* loci but not *gt* or *rst*.

Message Deletions are recognized genetically by (1) reduced RF, (2) pseudodominance, (3) recessive lethality, and (4) lack of reverse mutation, and cytologically by (5) deletion loops.

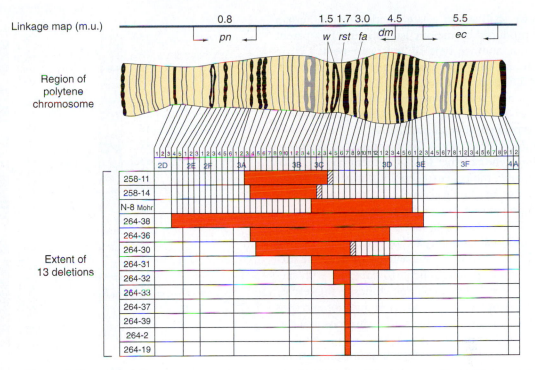

Figure 8-8 Locating genes to chromosomal regions by observing pseudodominance in *Drosophila* heterozygous for deletion and normal chromosomes. The red bars show the extent of the deleted segments in 13 deletions. All recessive alleles spanned by a deletion will be expressed.

Clinicians regularly find deletions in human chromosomes. In most cases the deletions are relatively small, but they nevertheless have an adverse phenotypic effect, even though heterozygous. Deletions of specific human chromosome regions cause unique syndromes of phenotypic abnormalities. An example is the *cri du chat* syndrome, caused by a heterozygous deletion of the tip of the short arm of chromosome 5 (Figure 8-9). It is the convention to call the short arm of a chromosome *p*, and the long arm *q*. The specific bands deleted in cri du chat syndrome are 5p15.2 and 5p15.3, the two most distal bands identifiable on 5p. The most characteristic phenotype in the syndrome is the one that gives it its name, the distinctive catlike mewing cries made by infants with this deletion. Other phenotypic manifestations of the syndrome are microencephaly (abnormally small head) and a moonlike face. Like syndromes caused by other deletions, the cri du chat syndrome also includes mental retardation.

Most human deletions, such as those we have just considered, arise spontaneously in the germ line of a normal parent of an affected person; thus no signs of the deletions are found in the somatic chromosomes of the parents.

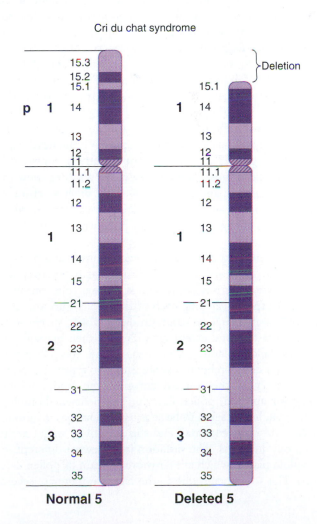

Cri du chat syndrome

Figure 8-9 The cause of the cri du chat syndrome of abnormalities in humans is loss of the tip of the short arm of one of the homologs of chromosome 5.

However, as we shall see in a later section, some human deletions are produced by meiotic irregularities in a parent heterozygous for another type of rearrangement. Cri du chat syndrome, for example, can result from a parent heterozygous for a translocation.

Geneticists have mapped human genes from deletions by using a molecular technique called **in situ hybridization.** This will be explained in detail in Chapter 17, but for now we can cover the basics to show the usefulness of deletions. If an interesting gene or other DNA fragment has been isolated using modern molecular technology, it can be tagged with a radioactive or chemical label, and then added to a chromosomal preparation under the microscope. In such a situation, the DNA recognizes and physically binds to its normal chromosomal counterpart by nucleotide pairing and is recognized as a spot of radioactivity or dye. The precise location of such spots is difficult to correlate with specific bands, but the deletion technique comes to the rescue. If a deletion happens to span the locus in question, no spot will appear when the test is run with the chromosome carrying the deletion, because the region for binding simply is not present (Figure 8-10). By saving cell lines from patients with deletions, geneticists develop test panels of overlapping deletions spanning specific chromosomal regions, and these can be used to pinpoint a gene's position. An example from chromosome 11 is shown in Figure 8-11. The extent of the deletions in the test panel are shown as vertical bars, and the coded DNA fragments under test are shown at the right. By showing that fragment 270, for example, failed to bind to deletions 35, 8, 10, 7, 9, 23, 24, A2, 27A, and 4D, but did bind to the other deletions, it can be inferred that this piece of DNA originally came from the region spanned by 11q13.5 and 11q21.

Chromosome mutations often arise in cancer cells, and we shall see several cases in this chapter and the next. As an example, Figure 8-12 shows some deletions consistently found in solid tumors. Not all the cells in a tumor show the deletion indicated, and often a mixture of different chromosome mutations can be found in one tumor. The contribution of such changes to the cancer phenotype is not understood.

An interesting difference between animals and plants is revealed by deletions. A male animal that is heterozygous for a deletion chromosome and a normal one produces functional sperm carrying each of the two chromosomes in approximately equal numbers. In other words, sperm seem to function to some extent regardless of their genetic content. In diploid plants, on the other hand, the pollen produced by a deletion heterozygote is of two types: functional pollen carrying the normal chromosome, and nonfunctional (or aborted) pollen carrying the deficient homolog. Thus, pollen cells seem to be sensitive to changes in amount of chromosomal material, and this sensitivity might act to weed out deletions. The situation is somewhat different for polyploid plants, which are far more tolerant of pollen deletions. This tolerance is due to the fact that there are several

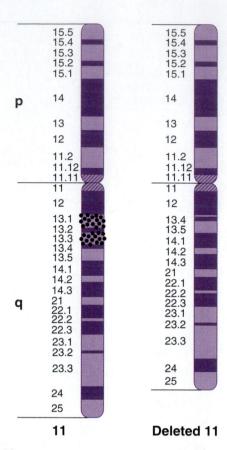

Figure 8-10 Radioactive spots show up only on one chromosome 11, because the other one has a deletion in the region where the radioactive DNA binds.

chromosome sets even in the pollen, and the loss of a segment in one of these sets is less crucial than it would be in a haploid pollen cell. Ovules in either diploid or polyploid plants also are quite tolerant of deletions, presumably because of the nurturing effect of the surrounding maternal tissues.

Duplications

The processes of chromosome mutation sometimes produce an extra copy of some chromosome region. In considering a haploid organism, which has one chromosome set, it is easy to see why such a product is called a *duplication* because the region is now present in duplicate. The duplicate regions can be located adjacent to each other, or one can be in its normal location and the other in a novel location on a different part of the same chromosome or even on another chromosome. In a diploid organism, the chromosome set containing the duplication is generally present together with a standard chromosome set. The cells of such an organism

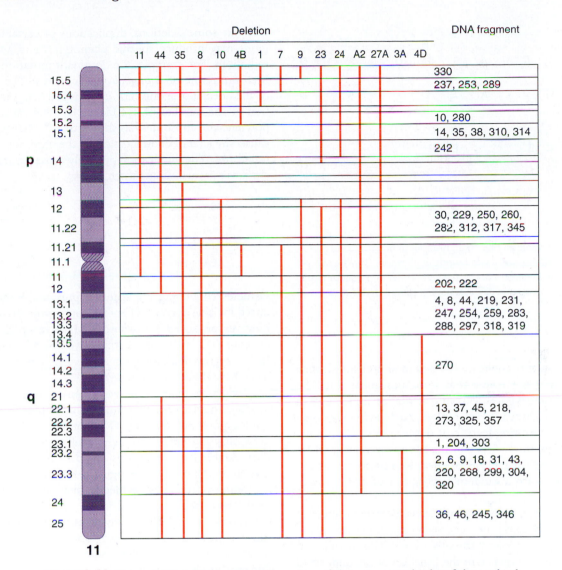

Figure 8-11 Human DNA fragments mapped to regions of chromosome 11 by their failure to bind to particular deletions. The red bars show the extent of the deletions, and the DNA fragments that were mapped are identified at the right. Notice that fragment 270, for example, failed to bind to deletions 35, 8, 10, 7, 9, 23, 24, A2, 27A, and 4D, but did bind to the other deletions. The region spanned by 11q13.5 and 11q21 is missing from all the deletions that did not bind 270; 270 was thus inferred to lie within this region. (From Y. Nakamura.)

will, of course, have three copies of the chromosome region in question, but nevertheless such duplication heterozygotes are generally referred to as duplications because they carry the product of one duplication event.

Duplication heterozygotes also show interesting pairing structures at meiosis or in salivary gland chromosomes. The precise structure that forms depends on the type of duplication. For the present we shall discuss only adjacent duplications. These can be **tandem,** as in the example A B C B C D, or **reverse,** as in the case A B C C B D. The pairing structures in heterozygotes for adjacent duplications are shown in Figure 8-13.

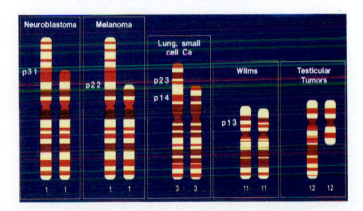

Figure 8-12 Deletions found consistently in several different types of solid tumors in humans. Band numbers represent recurrent breakpoints. (Jorge Yunis)

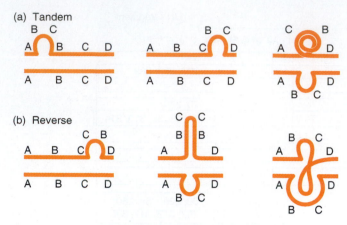

(a) Tandem

(b) Reverse

Figure 8-13 Possible pairing configurations in heterozygotes of a standard chromosome and a side-by-side duplication. Duplicated segments may be (a) in tandem or (b) in reverse order. A particular duplicated segment may assume different configurations in different meioses.

Once a tandem duplication arises in an individual, then by inbreeding it is possible that some descendants will be duplication homozygotes, carrying a total of four copies of the duplicated chromosome region. Such individuals present another interesting feature of duplications, which is the possibility of asymmetrical pairing, as shown in Figure 8-14. The crossing-over at meiosis of such asymmetrically paired regions may create a tandem triplication of the chromosome region.

The extra region of a duplication is free to undergo gene mutation because the necessary basic functions of the region will be provided by the other copy. This provides an opportunity for divergence in the function of the duplicated genes, which could be advantageous in genome evolution. Indeed, from situations in which different gene products with related functions can be compared, such as the globins (to be discussed later), there is good evidence that these products arose as duplicates of one another.

Message Duplications supply additional genetic material capable of evolving new functions.

Like some deletions, duplications of certain genetic regions may produce specific phenotypes and act like gene mutations. For example, the dominant mutation *Bar* on the X chromosome of *Drosophila* produces a slitlike eye instead of the normal oval one. The effect is produced by the reduction of the number of eye facets. Cytological study of the polytene chromosomes has shown that the Bar phenotype is caused by a tandem duplication of the chromosomal region called *16A*. The duplication probably arose from an unequal crossover at meiosis as shown below:

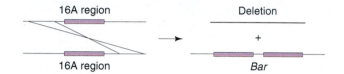

Gametes containing the deletion presumably died or produced inviable zygotes. Gametes containing the duplication, however, produced Bar offspring. Males carrying the *Bar* duplication in a hemizygous state have severely reduced eyes. Heterozygous females with a *Bar* and a normal chromosome have slightly reduced eyes.

Evidence that asymmetric pairing and crossing-over produces higher orders of duplication comes from studying homozygous *Bar* females. Occasionally, such females produce offspring with extremely small eyes called *double Bar*. Each double Bar offspring is found to carry three doses of the *Bar* region in tandem (Figure 8-15).

Some of the best evidence that tandem duplications (and their reciprocal deletions) derive from unequal crossovers comes from studies of the genes that determine the structure of human hemoglobin, the oxygen transport molecule. The hemoglobin molecule is composed of two different kinds of subunits. Furthermore, there are always two of each kind of subunit, giving a total of four subunits. A person has different hemoglobin at different ages—that is, the kinds of subunits differ. For example, a fetus has two α subunits and two γ subunits ($\alpha_2\gamma_2$), whereas an adult has two α subunits and two β subunits ($\alpha_2\beta_2$). The structures of the different subunits are determined by different genes, some of which are linked and some of which are not. The situation is summarized in Figure 8-16.

Figure 8-14 Generation of higher orders of duplications by asymmetric pairing followed by crossing-over in a duplication homozygote.

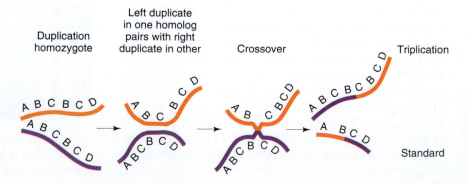

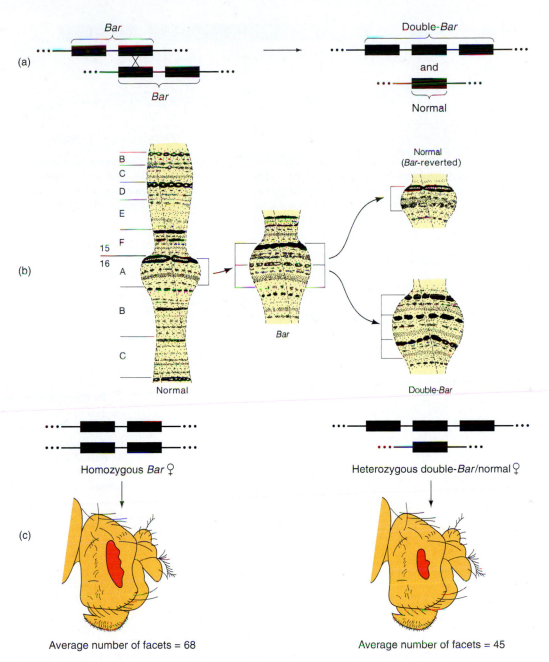

Figure 8-15 Production of double-*Bar* (triplication) and *Bar*-revertant (normal) chromosomes by asymmetric pairing followed by crossing-over in a duplication homozygote. (a) Diagrammatic representation. (b) Cytological representation. (c) Comparison of Bar and double-Bar phenotypes. (Part b from Bridges, *Science* 83, 1936, 210.)

The linked γ-δ-β group provides the data we need on unequal crossover. Some people afflicted with certain thalassemias (a kind of inherited blood disease) have a hemoglobin subunit that is part δ and part β (Lepore hemoglobin) or one that is part γ and part β (Kenya hemoglobin). The origin of these rare hemoglobin subunits can be explained by the unequal crossover models shown in Figure 8-17. Deletion chromosomes determine both Lepore and Kenya hemoglobins, causing their bearers to have the blood diseases. Also diagrammed in Figure 8-17 are the origins of the reciprocal crossover products called *anti-Lepore* and *anti-Kenya*.

Tandem duplications are rare in humans. Most duplications consist of an extra chromosomal arm or part of an arm, generally attached to a nonhomologous chromosome. In discussions that follow about other chromosomal rearrangements—inversions and translocations—we shall learn about the typical origins of such duplications. Like some deletions, duplications of some human chromosomes cause syndromes of phenotypic abnormalities. A person af-

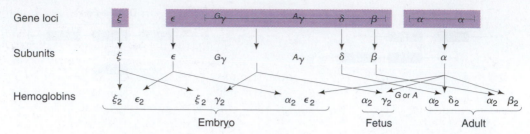

Figure 8-16 Hemoglobin genes in humans. Each Greek letter represents a different gene locus and the hemoglobin subunit determined by that gene. These subunits combine in different ways at different stages in development, as indicated, to yield functioning hemoglobin molecules. Notice that adults have a minor hemoglobin, $\alpha_2\delta_2$ (<2 percent) in addition to $\alpha_2\beta_2$. (From D. J. Weatherall and J. B. Clegg, "Recent Developments in Molecular Genetics of Human Hemoglobin," *Cell* 16, 1979.)

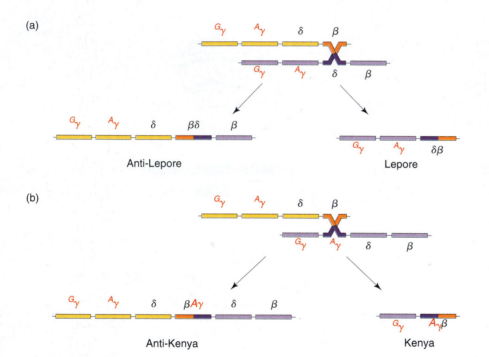

Figure 8-17 Proposed generation of variant human hemoglobin subunits by unequal crossing-over in the γ-δ-β genetic region. Notice that both Lepore and Kenya are deletions. (From D. J. Weatherall and J. B. Clegg, "Recent Developments in Molecular Genetics of Human Hemoglobin," *Cell* 16, 1979.)

flicted with a duplication syndrome has three copies of the duplicated region, whereas other chromosome regions are present in two copies as usual. Humans homozygous for duplications are unknown in medical genetics. Unlike deletions, duplications do not unmask deleterious recessives, so the abnormalities associated with duplications may be attributed solely to the imbalance generated by the extra copy of the chromosome region.

In general, duplications are hard to detect and are rare. However, they are useful tools in research and can be generated from other chromosomal aberrations by manipulations that we shall learn later.

Duplications can also be created using modern DNA technology. Cells can be induced to take up specific fragments of DNA from their own species, and this DNA can insert into the linear structure of a chromosome. It can insert either at its regular locus, replacing the resident sequence, or at a completely different site, called an **ectopic site.** Insertion of the DNA at an ectopic site creates a duplication of the fragment that was taken up. This is a direct and convenient way of making small duplications.

Inversions

If two breaks occur in one chromosome, sometimes the region between the breaks rotates 180 degrees before rejoining with the two end fragments. This creates a chromosomal mutation called an **inversion.** Unlike deletions and duplications, inversions involve no change in the overall amount of the genetic material, so inversions are generally viable and show no particular abnormalities at the phenotypic level. In some cases one of the chromosome breaks is within a gene of essential function, and then that breakpoint acts as a lethal gene mutation linked to the inversion.

(a)

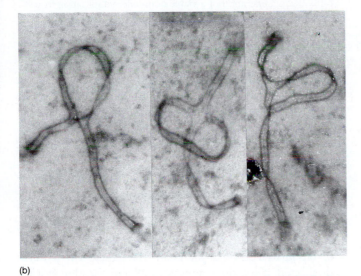

(b)

Figure 8-18 The chromosomes of inversion heterozygotes pair in a loop at meiosis. (a) Diagrammatic representation; each chromosome is actually a pair of sister chromatids. (b) Electron micrograph of synaptonemal complexes at prophase I of meiosis in a mouse heterozygous for a paracentric inversion. Three different meiocytes are shown. (Part b from M. J. Moses, Department of Anatomy, Duke Medical Center.)

In such a case the inversion could not be bred to homozygosity. However, many inversions can be made homozygous, and furthermore inversions can be detected in haploid organisms, so in these cases the breakpoint is clearly not in an essential region.

Most analyses of inversions use heterozygous inversions, diploids in which one chromosome has the standard sequence and one carries the inversion. Microscopic observation of meioses in inversion heterozygotes reveals the location of the inverted segment because one chromosome twists once at the ends of the inversion to pair with the other, untwisted chromosome; in this way the paired homologs form an **inversion loop** (Figure 8-18).

The location of the centromere relative to the inverted segment determines the genetic behavior of the chromosome. If the centromere is outside the inversion, then the inversion is said to be **paracentric,** whereas inversions spanning the centromere are **pericentric:**

How do inversions behave genetically? Crossing-over within the inversion loop of a paracentric inversion connects homologous centromeres in a **dicentric bridge** while also producing an **acentric fragment**—one without a centromere. Then, as the chromosomes separate during anaphase I, the centromeres remain linked by the bridge. This orients the centromeres so that the noncrossover chromatids lie farthest apart. The acentric fragment cannot align itself or move and, consequently, is lost. Tension eventually breaks the bridge, forming two chromosomes with terminal deletions (Figure 8-19). The gametes containing such deleted chromosomes may be inviable, but even if viable, the zygotes they eventually form will be inviable. Hence, a crossover event, which normally generates the recombinant class of meiotic products, instead produces lethal products.

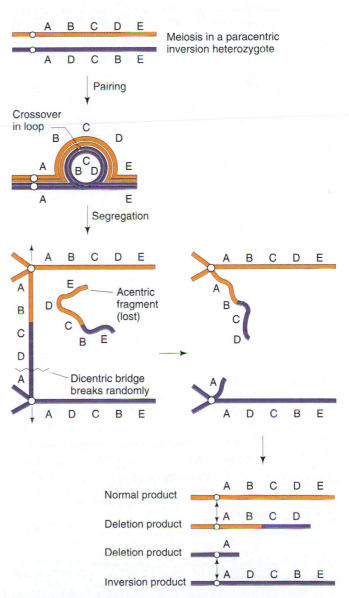

Figure 8-19 Meiotic products resulting from a single crossover within a paracentric inversion loop. Two nonsister chromatids cross over within the loop.

The overall result is a lower recombinant frequency. In fact, for genes within the inversion the RF is zero. For genes flanking the inversion, the RF is reduced in proportion to the relative size of the inversion.

Inversions affect recombination in another way too. Inversion heterozygotes often have mechanical pairing problems in the region of the inversion; this reduces the frequency of crossing-over and hence the recombinant frequency in the region.

The net genetic effect of a pericentric inversion is the same as that of a paracentric one—crossover products are not recovered—but for different reasons. In a pericentric inversion, because the centromeres are contained within the inverted region, the chromosomes that have crossed over disjoin in the normal fashion, without the creation of a bridge. However, the crossover produces chromatids that contain a duplication and a deficiency for different parts of the chromosome (Figure 8-20). In this case, if a nucleus carrying a crossover chromosome is fertilized, the zygote dies because of its genetic imbalance. Again, the result is the selective recovery of noncrossover chromosomes in viable progeny.

Message Two mechanisms reduce the number of recombinant products among the progeny of inversion heterozygotes: elimination of the products of crossovers in the inversion loop and inhibition of pairing in the region of the inversion.

Figure 8-21 Generation of a viable nontandem duplication from a pericentric inversion close to a dispensable chromosome tip.

It is worth adding a note about homozygous inversions. In such cases the homologous inverted chromosomes pair and cross over normally, there are no bridges, and the meiotic products are viable. However, there is an interesting effect, which is that the linkage map will show the inverted gene order.

Geneticists use inversions to create duplications of specific chromosome regions for various experimental purposes. For example, consider a heterozygous pericentric inversion with one breakpoint at the tip (T) of the chromosome, as shown in Figure 8-21. A crossover in the loop produces a chromatid type in which the entire left arm is duplicated; if the tip is nonessential, a duplication stock is generated for investigation. Another way to make a duplication (and a deficiency) is to use two paracentric inversions with overlapping breakpoints (Figure 8-22). A complex loop is formed, and a crossover within the inversion produces the duplication and the deletion. These manipulations are possible only in organisms with thoroughly mapped chromosomes for which large sets of standard rearrangements are available.

We have seen that genetic analysis and meiotic chromosome cytology are both good ways of detecting inversions. As with most rearrangements, there is also the possibility of detection through mitotic chromosome analysis. A key operational feature is to look for new arm ratios. Consider that a chromosome has mutated as follows:

Note that the ratio of the long to the short arm has been changed from about 4 to about 1 by the pericentric inversion. Paracentric inversions do not alter the arm ratio, but they may be detected microscopically if banding or other chromosome landmarks are available.

Message The main diagnostic features of inversions are reduction of recombinant frequency, inversion loops, and reduced fertility from unbalanced or deleted meiotic products, all observed in individuals heterozygous for inversions. Some inversions may be directly observed as an inverted arrangement of chromosomal landmarks.

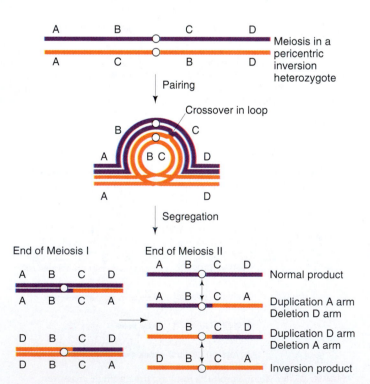

Figure 8-20 Meiotic products resulting from a meiosis with a single crossover within a pericentric inversion loop.

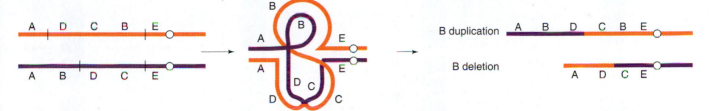

Figure 8-22 Generation of a nontandem duplication by crossing-over between two overlapping inversions.

Inversions are found in about 2 percent of humans. The heterozygous inversion carriers generally show no adverse phenotype, but produce the expected array of abnormal meiotic products from crossing-over in the inversion loop. Let us consider pericentric inversions as an example. Individuals heterozygous for pericentric inversions produce offspring with the duplication-deletion chromosomes predicted; these offspring show varying degrees of abnormalities depending on the lengths of the chromosome regions involved. Some phenotypes caused by duplication-deletion chromosomes are so abnormal as to be incapable of survival to birth, and are lost as spontaneous abortions. However, there is a way to study the abnormal meiotic products that does not depend on survival to term. Human sperm placed in contact with unfertilized eggs of the golden hamster penetrate the eggs but fail to fertilize them. The sperm nucleus does not fuse with the egg nucleus, and if the cell is prepared for cytogenetic examination, the human chromosomes are easily visible as a distinct group (Figure 8-23). This makes it possible to study directly the chromosomal products of a male meiosis and is particularly useful in the study of meiotic products of men who have chromosome mutations.

In one case, a man heterozygous for an inversion of chromosome 3 was subjected to sperm analysis. The inversion was a large one with a high potential for crossing-over in the loop. Four chromosome 3 types were represented in the man's sperm—normal, inversion, and two recombinant types (Figure 8-24). The sperm contained the four types in the following frequencies:

normal	38%
inversion	32%
duplication q, deletion p	17%
duplication p, deletion q	13%

The duplication q–deletion p recombinant chromosome had been observed previously in several abnormal children, but the duplication p–deletion q type had never been seen, and probably zygotes receiving it are too abnormal to survive to term. Presumably, deletion of the larger q fragment has more severe consequences than deletion of the smaller p fragment.

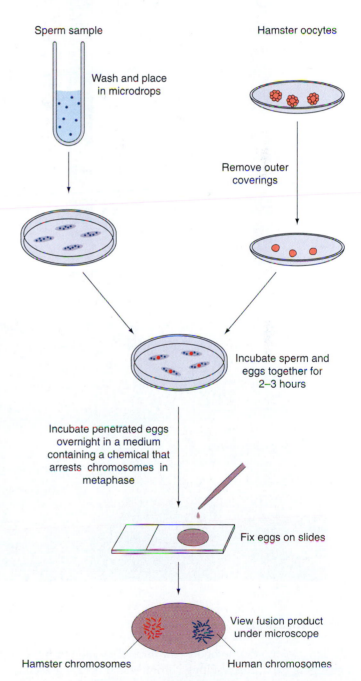

Sperm sample

Hamster oocytes

Wash and place in microdrops

Remove outer coverings

Incubate sperm and eggs together for 2–3 hours

Incubate penetrated eggs overnight in a medium containing a chemical that arrests chromosomes in metaphase

Fix eggs on slides

View fusion product under microscope

Hamster chromosomes

Human chromosomes

Figure 8-23 Human sperm and hamster oocytes are fused to permit study of the chromosomes in the meiotic products of human males. (Adapted from original art by Renée Martin.)

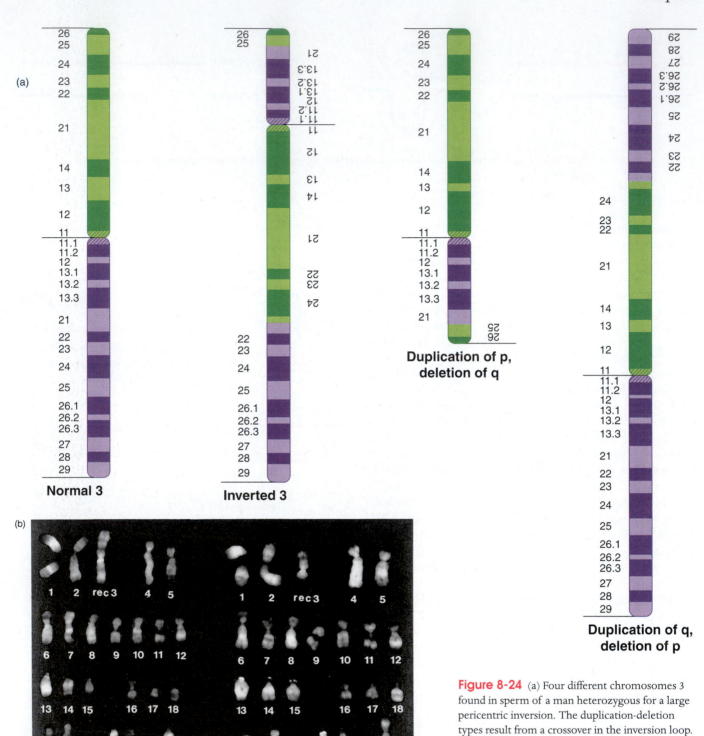

Normal 3

Inverted 3

Duplication of p, deletion of q

Duplication of q, deletion of p

Figure 8-24 (a) Four different chromosomes 3 found in sperm of a man heterozygous for a large pericentric inversion. The duplication-deletion types result from a crossover in the inversion loop. (b) Two complete sperm chromosome sets containing the two duplication-deletion types (labeled rec 3). (Renée Martin)

Translocations

When two nonhomologous chromosomes mutate by exchanging parts, the resulting chromosomal rearrangements are translocations. Here we consider **reciprocal translocations,** the most common type. A segment from one chromosome is exchanged with a segment from another nonho-mologous one, so that two translocation chromosomes are generated simultaneously.

The exchange of chromosome parts between nonhomologs establishes new linkage relationships. These new linkages are revealed if the translocated chromosomes are homozygous and, as we shall see, even when they are heterozygous. Furthermore, translocations may drastically al-

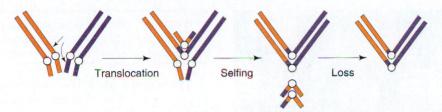

Figure 8-25 Genome restructuring by translocations. Small arrows indicate breakpoints in one homolog of each of two pairs of acrocentric chromosomes. The resulting fusion of the breaks yields one short and one long metacentric chromosome. If, as in plants, self-fertilization (selfing) takes place, an offspring could be formed with only one pair of long and only one pair of short metacentric chromosomes. Under appropriate conditions, the short metacentric chromosome may be lost. Thus, we see a conversion from two acrocentric pairs of chromosomes to one pair of metacentrics.

ter the size of a chromosome as well as the position of its centromere. For example,

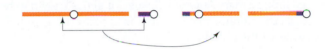

Here a large metacentric chromosome is shortened by half its length to an acrocentric one, and the small chromosome becomes a large one. Examples from natural populations are known in which chromosome numbers have been changed by translocation between acrocentric chromosomes and the subsequent loss of the resulting small chromosome elements (Figure 8-25).

In heterozygotes having two translocated chromosomes and their normal counterparts, there are important genetic and cytological effects. Again, the pairing affinities of homologous regions dictate a characteristic configuration for chromosomes synapsed in meiosis. Figure 8-26, which illustrates meiosis in a reciprocally translocated heterozygote, shows that the configuration is that of a cross.

Remember, the configuration presented in the figure lies on the equatorial plate of the cell at metaphase, with the spindle fibers perpendicular to the page. Thus, the centromeres would actually migrate up out of the page or down under it. Homologous paired centromeres disjoin, whether or not a translocation is present. Because Mendel's second law still applies to different paired centromeres, there are two common patterns of disjunction. The segregation of each of the structurally normal chromosomes with one of the translocated ones (T1 + N2 and T2 + N1) is called **adjacent-1 segregation**. Both meiotic products are duplicated and deficient for different regions. These products are inviable. On the other hand, the two normal chromosomes may segregate together, as do the reciprocal parts of the translocated ones, to produce N1 + N2 and T1 + T2 products. This is called **alternate segregation**. These products are viable. Since the adjacent-1 and alternate segregation patterns are equally frequent, approximately 50 percent of the products are viable and 50 percent inviable. (There is another event, called **adjacent-2**

segregation, in which homologous centromeres migrate to the same pole, but in general this is rare.)

As a result of the equality of adjacent and alternate segregations, half the gametes will be nonfunctional, a condition known as **semisterility**. Semisterility, or "half-sterility," is an important diagnostic for translocation heterozygotes. However, semisterility is defined differently for plants and animals. In plants, the 50 percent unbalanced meiotic products from the adjacent-1 segregation generally abort at the gametic stage (Figure 8-27). In animals, however, the duplication-deletion products are viable as gametes, but lethal to the zygote.

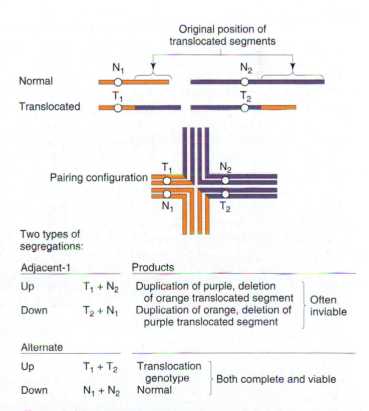

Figure 8-26 The meiotic products resulting from the two most commonly encountered chromosome segregation patterns in a reciprocal translocation heterozygote.

Figure 8-27 Photo of normal and aborted pollen of a semisterile corn plant. The clear pollen grains contain chromosomally unbalanced meiotic products of a reciprocal translocation heterozygote. The opaque pollen grains, which contain either the complete translocation genotype or normal chromosomes, are functional in fertilization and development. (William Sheridan)

Remember also that heterozygotes for the other rearrangements such as deletions and inversions may show some reduction of fertility; but the precise 50 percent reduction in viable gametes or zygotes is usually a reliable diagnostic clue for a translocation.

Translocations are economically important. In agriculture, translocations in certain crop strains can reduce yields considerably owing to the number of unbalanced zygotes that form. On the other hand, translocations are potentially useful: it has been proposed that insect pests could be controlled by introducing translocations into their wild populations. According to the proposal, 50 percent of the offspring of crosses between insects carrying the translocation and wild types would die, and 10/16 of the progeny of crosses between translocation-bearing insects would die.

Message Translocations, inversions, and deletions produce partial sterility by generating unbalanced meiotic products that may themselves die or that may cause zygotes to die.

Genetically, markers on nonhomologous chromosomes appear to be linked if these chromosomes are involved in a translocation and the loci are close to the translocation breakpoint. Figure 8-28 shows a situation in which a translocation heterozygote has been established by crossing an *a a b b* individual with a translocation homozygote bearing the wild-type alleles. We shall assume that *a* and *b* are close to the translocation breakpoint. On testcrossing the heterozygote, the only viable progeny are those bearing the parental genotypes, so linkage is seen between loci that were originally on different chromosomes. In fact, if all four arms of the meiotic pairing structure are genetically marked, recombination studies should result in a cross-shaped linkage map. Apparent linkage of genes known to be on separate nonhomologous chromosomes is a genetic giveaway for the presence of a translocation.

Message Reciprocal translocations are diagnosed genetically by semisterility and by the apparent linkage of genes known to be on separate chromosomes.

Translocations in humans are always carried in the heterozygous state. An example involving chromosomes 5 and 11 is shown in Figure 8-29. The offspring of the person with this particular translocation had a duplication of 11q and a deletion of 5p. These children showed symptoms of both cri du chat syndrome, which is caused by the deletion of 5p, and the syndrome associated with the duplication of 11q. The reciprocal duplication-deficiency chromosome was not observed.

Down syndrome is a constellation of human disorders usually caused by the presence of an extra chromosome 21 that failed to segregate from its homolog at meiosis (see Chapter 9). This common type of Down syndrome (approximately 95% of all cases) is sporadic and shows no recurrence in the family. However, there is a less common type of Down syndrome caused by a special type of translocation called a **Robertsonian translocation,** and this form can recur in a family. A Robertsonian translocation is one that

Figure 8-28 When a translocated fragment carries a marker gene, this marker can show linkage to genes on the other chromosome because the recombinant genotypes (in this case *a⁺ b* and *a b⁺*) tend to be in duplication-deletion gametes and do not survive.

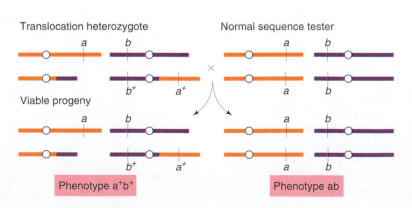

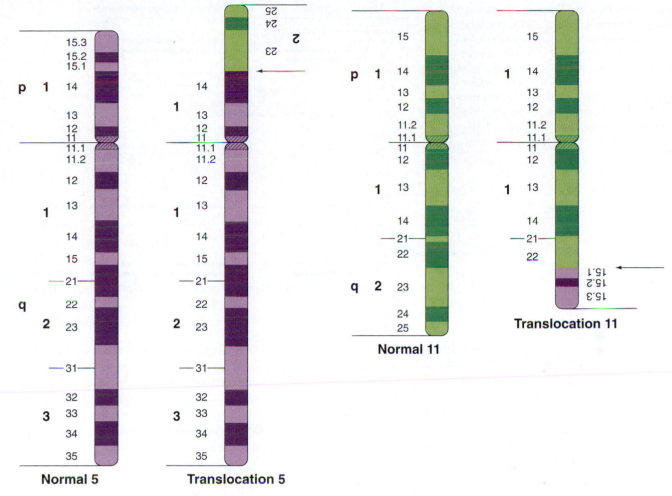

Figure 8-29 A human translocation heterozygote involving reciprocal exchange of 5p and 11q (5p15; 11q23).

combines the long arms of two acrocentric chromosomes, as shown in Figure 8-30. Initially, a small chromosome composed of the two short arms also forms, however, this is generally not present. The material in the short arms must be nonessential because their loss has no effect on phenotype. Translocation Down syndrome is caused by a Robertsonian fusion between chromosomes 21 and 14. The translocated chromosome passes down the generations in unaffected carriers. However, meiotic segregation in the translocation carriers can result in progeny that carry three copies of most of chromosome 21, as shown in Figure 8-30, and these individuals have Down syndrome. The factors responsible for numerous other hereditary disorders have been traced to translocation heterozygosity in the parents.

Translocations also appear in cancer cells, and some examples are shown for solid tumors in Figure 8-31. In solid tumors translocations are not as common as deletions. As with other rearrangements in cancer cells, the involvement with the cancer phenotype generally is not clear. However, in a later section we shall study an example in which the re-

location of a specific proto-oncogene seems to be causally connected to cancer.

As is true for other rearrangements, the translocation breakpoints can sometimes disrupt an essential gene, and the gene is thereby inactivated and behaves as a point mutation. Molecular geneticists can use this effect to pinpoint the location of a human gene and can then proceed to isolate the gene. For example, information from translocations helped in the isolation of the gene for the X-linked recessive disease Duchenne muscular dystrophy. Some rare female cases of Duchenne muscular dystrophy were also heterozygous for translocations between the X chromosome and a variety of different autosomes. The X chromosome breakpoint was always in the band Xp21, so the gene for muscular dystrophy, already known to be X-linked, was evidently in this band and had been disrupted by the break. The hunt for the gene could begin by focusing on that area. In passing, note that the expression of the mutant phenotype in a female must have been because the normal X chromosome was inactivated (Figure 8-32 and page 73).

One specific autosomal breakpoint proved to be useful

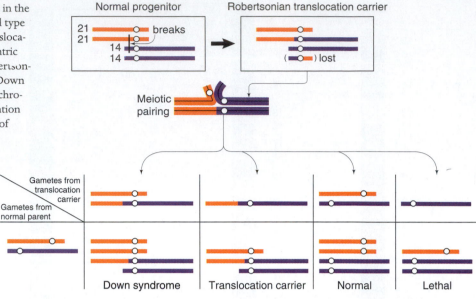

Figure 8-30 How Down syndrome arises in the children of an unaffected carrier of a special type of translocation, called a Robertsonian translocation, in which the long arms of two acrocentric chromosomes have fused. The specific Robertsonian translocation involved in translocation Down syndrome is between the Down syndrome chromosome 21 and chromosome 14. Translocation Down syndrome accounts for less than 5% of cases.

in providing a molecular "tag" for the Duchenne gene. The particular breakpoint that advanced the research was in the ribosomal RNA locus on chromosome 21. (Don't confuse X chromosome band 21 with chromosome 21.) A DNA probe was already available for the ribosomal locus. It was reasoned that, in the X/21 translocation, the Duchenne gene must be disrupted and attached next to the ribosomal RNA gene. The geneticists therefore used the ribosomal probe to isolate a DNA segment that had part of the Duchenne gene on it.

Translocations were also used to isolate the human gene for neurofibromatosis. Once again, the critical chromosomal material came from people who not only had the disease but also carried chromosomal translocations. The translocations all had one of their breakpoints in chromosome 17, in a band close to the centromere. Hence it ap-

peared that this must be the locus of the neurofibromatosis gene, which had been disrupted by the translocation breakage. Subsequent analysis showed that the chromosome 17 breakpoints were not identical, and their positions helped to map the region occupied by the neurofibromatosis gene. Isolation of DNA fragments from this region eventually led to the recovery of the gene itself.

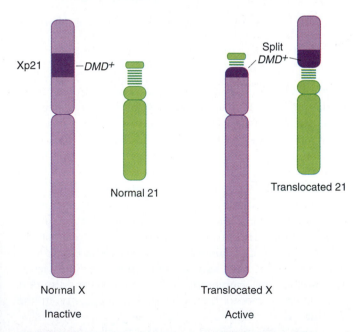

Figure 8-32 Diagram of the chromosomes of a woman with Duchenne muscular dystrophy and heterozygous for a reciprocal translocation between the X chromosome and chromosome 21. The translocation breakpoint disrupted one *DMD*⁺ allele, rendering it nonfunctional, and X chromosome inactivation made its undisrupted partner *DMD*⁺ allele also nonfunctional.

Figure 8-31 Translocations found consistently in several different types of solid tumors in humans. Band numbers indicate breakpoints. (Jorge Yunis)

Figure 8-33 Using translocations with one break-point in heterochromatin to produce a duplication and a deletion. If the upper product of translocation 1 is combined with the upper product of translocation 2 by means of an appropriate mating, a deletion of *b* results. If the lower products of the two translocations are combined, the genotype *a b b c* is produced.

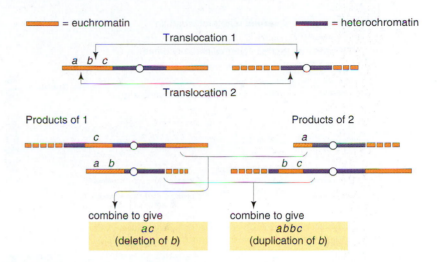

The Use of Translocations in Producing Duplications and Deletions.

Geneticists regularly need to make specific duplications or deletions to answer specific experimental questions. We have seen that they use inversions to do this, and now we shall see how translocations also can be used for the same purpose. Let's take *Drosophila* as an example. For reasons that are still unclear, the densely staining chromosomal regions near the centromere called *heterochromatin* are physically extensive but contain few genes. In fact, for a long time, heterochromatin was considered useless and inert material. In any case, for our purposes, *Drosophila* can tolerate a loss or an excess of heterochromatin with little effect on viability or fertility.

Now let's select two different reciprocal translocations of the same two chromosomes. Each translocation has a breakpoint somewhere in heterochromatin, and each has another breakpoint in euchromatin (nonheterochromatin) on opposite sides of the region we want to duplicate or delete (Figure 8-33). It can be seen that if we have a large collection of translocations having one heterochromatic break and euchromatic breaks at many different sites, then duplications and deletions for many parts of the genome can be produced at will for a variety of experimental purposes. More generally, if one breakpoint of a translocation is near a dispensable tip, then duplication or deletion of this tip can be ignored, and the translocation can be used as a way of generating duplications or deficiencies for the other translocated segment.

Position-Effect Variegation.

In previous chapters, we considered several mechanisms of generating variegation in the somatic cells of a multicellular organism. These mechanisms were somatic segregation, somatic crossover, and somatic mutation. Another cause of variegation is associated with translocations and is called **position-effect variegation.**

The locus for white eye color in *Drosophila* is near the tip of the X chromosome. Consider a translocation in which the tip of an X chromosome carrying w^+ is relocated next to the heterochromatic region of, say, chromosome 4. Position-effect variegation is observed in flies that are heterozygotes for such a translocation and that have the normal X chromosome carrying the recessive allele *w*. The expected eye phenotype is red because the wild-type allele is dominant to *w*. However, in such cases the observed phenotype is a variegated mixture of red and white eye facets. How can we explain the white areas? We could suppose that when the chromosomes broke and rejoined in the translocation, the w^+ allele was somehow changed to a state that made it more mutable in somatic cells; so the white eye tissue reflects cells in which w^+ has mutated to *w*.

In 1972, Burke Judd tested this hypothesis by recombining the w^+ allele out of the translocation and onto a normal X chromosome and by recombining a *w* allele from the normal X chromosome onto the translocation (Figure 8-34). Judd found that when the w^+ allele on the translocation was crossed onto a normal X chromosome and *w* was then inserted into the translocation, the eye color was red; so obviously the w^+ allele was not defective. When he crossed the w^+ allele back onto the translocation, the phenotype was again variegated. Thus, we can conclude that, for some reason, the w^+ is allele in the translocation is not expressed in some cells, thereby allowing the expression of *w*. This kind of variegation is called *position-effect variegation* because the unstable expression of a gene reflects its position in a rearrangement.

Such position effects can affect the genes that cause cancer. For example, most cases of Burkitt's lymphoma, a cancer of certain human antibody-producing cells called *B cells,* are caused by the relocation of a proto-oncogene to a position next to a region that normally enhances the production of antibodies (Figure 8-35). The oncogene is then activated, resulting in cancer.

Message: The expression of a gene can be affected by its position in the genome.

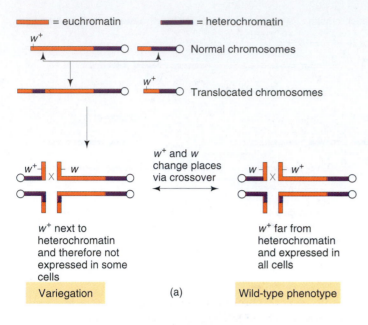

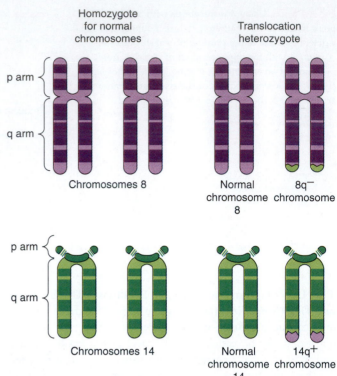

p arm

q arm

Chromosomes 8

Normal chromosome 8

8q⁻ chromosome

p arm

q arm

Chromosomes 14

Normal chromosome 14

14q⁺ chromosome

Homozygote for normal chromosomes

Translocation heterozygote

Figure 8-35 Reciprocal translocations between chromosomes 8 and 14 cause most cases of Burkitt's lymphoma. An oncogene on the tip of chromosome 8 becomes relocated next to an antibody gene enhancer region on chromosome 14.

(b)

Figure 8-34 Position-effect variegation. (a) The translocation of w^+ to a position next to heterochromatin causes the w^+ function to fail in some cells, producing position-effect variegation. (b) A *Drosophila* eye showing the position-effect variegation. (Part b from Randy Mottus.)

Diagnosis of Rearrangements by Tetrad Analysis

In fungi, tetrad analysis can be useful in detecting chromosomal aberrations. Any *Neurospora* genome containing a deletion produces an ascospore that does not ripen to the normal black color and is incapable of growth. The parents of crosses with high proportion of such "aborted" white ascospores usually contain chromosome rearrangements. Duplications are generally recovered as black, viable ascospores.

Specific spore abortion patterns sometimes identify specific rearrangements. For example, a translocation causes three types of asci to be produced in certain proportions. Most asci have either eight black spores (produced from al-

ternate-segregation meioses) or eight white spores (produced from adjacent-1-segregation meioses). A few asci have four black and four white spores, which are produced by crossing-over between either centromere and the translocation breakpoint (Figure 8-36).

Chromosomal Rearrangements in Evolution

At the beginning of the chapter, we noted that chromosomal rearrangements have had evolutionary consequences. This makes it interesting to compare the chromosomes of various species. The most striking aspect of making such comparisons among the chromosomes of closely related primates is the similarities. Figure 8-37 shows the chromosomal banding patterns of humans, chimps, gorillas, and orangutans. Note that a translocation amounting to a fusion of two complete chromosomes produced human chromosome 2.

When more distantly related organisms are compared, more radical rearrangements are seen, as in Figure 8-38, which shows the arrangement of chromosomal loci in three mammals. The intergenus comparisons show many clear examples of translocations. However, note that there is still a considerable level of conservation in the arrangement of genes, which could reflect either a need for certain loci to be adjacent, or the limited amount of time since divergence.

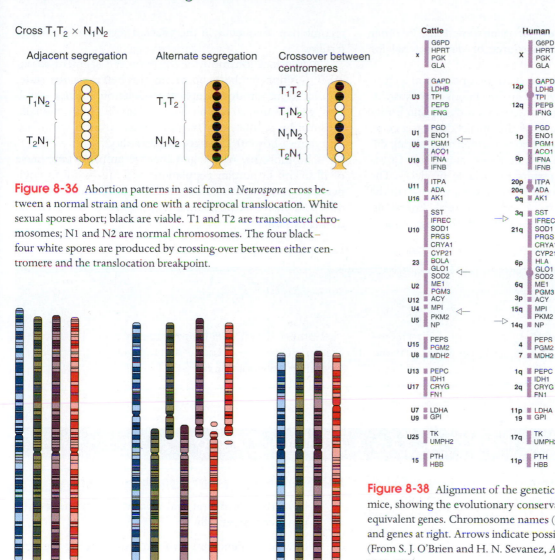

Figure 8-36 Abortion patterns in asci from a *Neurospora* cross between a normal strain and one with a reciprocal translocation. White sexual spores abort; black are viable. T1 and T2 are translocated chromosomes; N1 and N2 are normal chromosomes. The four black–four white spores are produced by crossing-over between either centromere and the translocation breakpoint.

Figure 8-37 Structural similarities and differences of the chromosomes 1 to 3 of humans (H), chimps (C), gorillas (G), and orangutans (O). (Adapted from J. J. Yunis, *Science* 215, 1982, 1525.)

Figure 8-38 Alignment of the genetic maps of cattle, humans, and mice, showing the evolutionary conservation of map positions of equivalent genes. Chromosome names (and arms) are shown at left, and genes at right. Arrows indicate possible translocation points. (From S. J. O'Brien and H. N. Sevanez, *Annual Review of Genetics* 22, 1988, 323.)

SUMMARY

The morphology of chromosomes provides a way of identifying them. Useful features are chromosome size; centromere position; nucleolar organizer position; and chromomere, heterochromatin, and banding patterns.

The chromosome set mutates spontaneously to produce chromosome rearrangements or changes in chromosome number. All these are called *chromosome mutations*.

Four rearrangements of chromosome structure are deletions, duplications, inversions, and translocations. Deletions are missing sections of chromosomes. If the region removed is essential to life, a homozygous deletion is lethal.

Heterozygous deletions can be nonlethal or lethal and can uncover recessive alleles, causing them to be expressed phenotypically. Heterozygous deletions lower recombination between flanking loci. Deletions in one homolog often allow expression of recessive alleles in the other.

Tandem duplications can result from unequal crossover, and nontandem duplications arise from other rearrangements. Duplications can unbalance the genetic material, thereby producing a deleterious phenotypic effect or death in the organism. However, there is good evidence from a number of species, including humans, that duplications can lead to an increased variety of gene functions. In other words, duplications can be a source of new material for evolution.

An inversion is a 180-degree turn of a portion of a chromosome. In the homozygous state, inversions may cause little problem for an organism unless heterochromatin brings about a position effect or one of the breaks disrupts a gene. On the other hand, inversion heterozygotes have pairing difficulties at meiosis, and an inversion loop may result. Crossing-over within the loop results in inviable products. The crossover products of pericentric inversions, which span the centromere, differ from those of paracentric inversions, which do not span the centromere, but both show reduced recombinant frequency in the affected region, and reduced fertility.

A translocation moves a chromosome segment to another position in the genome. In the heterozygous state, translocations produce duplication-deletion meiotic products, which can lead to unbalanced zygotes. New gene linkages can be produced by translocations. Translocation heterozygotes show 50% sterility (semisterility).

Chromosomal rearrangements are an important cause of ill health in human populations and are useful in engineering special strains of organisms in pure and applied biology. Extensive chromosomal rearrangement has occurred during evolution.

Concept Map

Draw a concept map interrelating as many of the following terms as possible. Note that the terms are listed in no particular order.

genome / deletion / loops / duplication / pairing / recombinant frequency / heterozygotes / inversions / lethality / phenotypic ratios

CHAPTER INTEGRATION PROBLEM

Geneticists conducted a *Drosophila* mutant hunt to find recessive lethal mutations on chromosome 2. Ten mutations of this type were obtained. The geneticists then crossed stocks carrying these mutations to produce hybrid zygotes that each bore a lethal chromosome from one stock plus a lethal chromosome from another. All pairwise combinations were crossed, and in every case it was recorded whether the hybrids were viable or not. The results were as follows, where a plus means that the hybrids were viable and a minus means that the combination of the two independently isolated, recessive mutations was lethal.

	1	2	3	4	5	6	7	8	9	10
1	−	+	−	+	+	+	−	+	+	+
2		−	+	+	+	+	+	+	+	−
3			−	+	+	+	−	+	+	+
4				−	+	−	+	+	+	−
5					−	+	+	+	−	−
6						−	+	+	+	−
7							−	+	+	+
8								−	+	+
9									−	−
10										−

Interpret these results as far as possible showing

- how a viable hybrid and an inviable hybrid can be produced genetically.

- how many genes are involved in generating these results.

- what you can say about their location.

- the meaning of any departures from a pattern in the data.

Solution

To begin, let's restate the system under investigation; in each cross the geneticists put together on a pair of homologous chromosomes two recessive lethal mutations that had been independently isolated. We can think of several possible underlying situations.

First, the mutations could be point mutations in the same gene:

Locus of mutation in one isolate

Locus of mutation in another isolate

In this case we would predict that the hybrid genotype would be lethal because we have learned from Chapter 4 that this is how recessive lethals work. So perhaps some of the inviability recorded in the grid can be explained by the pairing of point mutations in the same gene.

Another possibility is that the two mutations are in different genes on chromosome 2:

Because we know the mutations are recessive, which implies that their wild-type alleles are dominant, we would predict that this type would be viable. In fact, in Chapter 4 we called this type of effect *genetic complementation*, based on the idea that the two wild-type alleles are "helping each other out." This then could explain some of the viability in the grid.

Based on these musings, one of the next things that can be done is to find out which mutations are in the same gene, and this should also tell us something about how many genes are involved.

Starting with mutation 1, and looking across the row, we find that it fails to complement with mutations 3 and 7, so we can tentatively infer that mutations 1, 3, and 7 are all in the same gene. Doing the same for all mutations, we arrive at the following groupings, which we write in columns:

1	2	4	5	8
3	10	6	9	
7		10	10	

We certainly see a pattern emerging, one that suggests that five genes had mutated in the stocks. But there is a striking inconsistency to this pattern, and that is the behavior of mutation 10. This mutation fails to complement the mutations in three different groupings. What type of mutation could show this behavior? If we are to believe our hypothesis that five genes are involved, mutation 10 must have damage that spans three separate genes. One of the obvious possibilities is that it is a deletion. Furthermore, we have learned in this chapter that deletions essentially behave like recessive lethals, which is consistent with the geneticists' finding mutation 10 in their search for recessive lethals.

So we could diagram what we have learned about chromosome 2 as follows:

Genes	A	B	C	D	E
	1	2	4	5	8
	3		6	9	
	7				
Deletion					

But note that this is only one of many possible arrangements; we have no data that allow the ordering of *B*, *C*, and *D* within the deleted region, and furthermore we do not know whether *A* and *E* are on the same side or on opposite sides of the deletion.

Note the concepts used in this solution: dominance and recessiveness from Chapter 2, complementation from Chapter 4, gene mutation from Chapter 7, and deletion from the present chapter.

SOLVED PROBLEMS

1. A corn plant is obtained that is heterozygous for a reciprocal translocation and therefore is semisterile. This plant is backcrossed to a chromosomally normal strain that is homozygous for the recessive allele brachytic (*b*), located on chromosome 2. A semisterile F_1 plant is then backcrossed to the homozygous brachytic strain. The progeny obtained show the following phenotypes:

Nonbrachytic		Brachytic	
semisterile	fertile	semisterile	fertile
334	27	42	279

a. What ratio would you expect if the chromosome carrying the brachytic allele is not involved in the translocation?

b. Do you think that chromosome 2 is involved in the translocation? Explain your answer, showing the conformation of the relevant chromosomes of the semisterile F_1 and the reason for the specific numbers obtained.

Solution

a. We should start with the methodical approach and simply restate the data in the form of a diagram, where

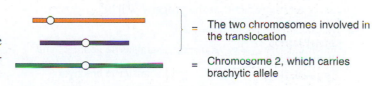

= The two chromosomes involved in the translocation

= Chromosome 2, which carries brachytic allele

To simplify the diagram, we do not show the chromosomes divided into chromatids (although they would be at this stage of meiosis):

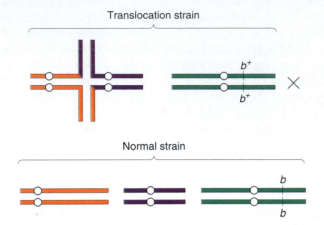

Translocation strain

Normal strain

All the progeny from this cross will be heterozygous for the chromosome carrying the brachytic allele, but what about the chromosomes involved in the translocation? In this chapter, we have seen that only alternate-segregation products survive and that half these survivors will be chromosomally normal and half will carry the two rearranged chromosomes. The rearranged combination will regenerate a translocation heterozygote when it combines with the chromosomally normal complement from the normal parent. These latter types—the semisterile F_1's—are diagrammed below as part of the backcross to the parental brachytic strain:

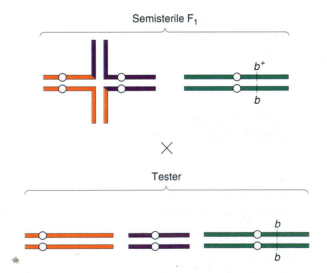

Semisterile F_1

×

Tester

In calculating the expected ratio of phenotypes from this cross, we can treat the behavior of the translocated chromosomes independently of the behavior of chromosome 2. Hence, we can predict that the progeny will be

$\frac{1}{2}$ translocation heterozygotes (semisterile) $\begin{cases} \rightarrow \frac{1}{2} b^+b \rightarrow \frac{1}{4} \text{ semisterile nonbrachytic} \\ \rightarrow \frac{1}{2} bb \rightarrow \frac{1}{4} \text{ semisterile brachytic} \end{cases}$

$\frac{1}{2}$ normal (fertile) $\begin{cases} \rightarrow \frac{1}{2} b^+b \rightarrow \frac{1}{4} \text{ fertile nonbrachytic} \\ \rightarrow \frac{1}{2} bb \rightarrow \frac{1}{4} \text{ fertile brachytic} \end{cases}$

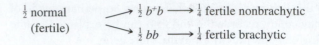

This predicted 1:1:1:1 ratio is quite different from that obtained in the actual cross.

b. Because we observe a departure from the expected ratio based on the independence of the brachytic phenotype and semisterility, it seems likely that chromosome 2 *is* involved in the translocation. Let's assume that the brachytic locus (*b*) is on the solid orange chromosome. But where? For the purpose of the diagram, it doesn't matter where we put it, but it does matter genetically because the position of the *b* locus affects the ratios in the progeny. If we assume that the *b* locus is near the tip of the piece that is translocated, we can redraw the pedigree:

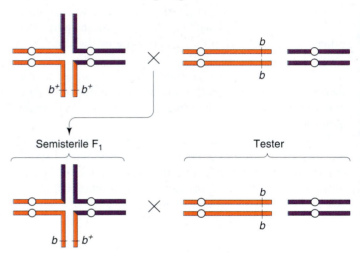

Semisterile F_1 Tester

If the chromosomes of the semisterile F_1 segregate as diagrammed here, we could then predict

$\frac{1}{2}$ fertile brachytic

$\frac{1}{2}$ semisterile nonbrachytic

Most progeny are certainly of this type, so we must be on the right track. How are the two less frequent types produced? Somehow we have to get the b^+ allele onto the normal solid orange chromosome and the *b* allele onto the translocated chromosome. This must be achieved by crossing over between the translocation breakpoint (the center of the cross-shaped structure) and the brachytic locus. To represent this, we must show chromatids, because crossing-over occurs at the chromatid stage:

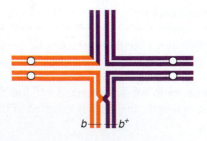

The recombinant chromosomes produce some progeny that are fertile and nonbrachytic and some that are semisterile and brachytic (these two classes together constitute 69 of these out of a total of 682, or a frequency of about 10 percent). We can see that this frequency is really a measure of the map distance (10 m.u.) of the brachytic locus from the breakpoint. (The same basic result would have been obtained if we had drawn the brachytic locus in the part of the chromosome on the other side of the breakpoint.)

2. A maize geneticist is studying recombination between two genes, b and l, which are 18 m.u. apart on the same arm of chromosome 12. She is particularly interested in one pure-breeding line (M) isolated from nature. Crosses within line M give the expected RF of 18 percent between b and l, exactly the same RF obtained when working within the conventional line (P). However, when an appropriately marked stock from M is crossed to an appropriately marked stock from P ($b^+ l^+/b^+ l^+ \times b l/b l$) and the F_1 plants are testcrossed to a $b l/b l$ tester from M, the RF value drops to 2 percent. The same value is obtained when the F_1 is crossed to a $b l/b l$ tester from P. Formulate a model and use it to explain

    a. The RF value within line M

    b. The RF value within line P

    c. The RF value in the $F_1 \times$ M testcross

    d. The RF value in the $F_1 \times$ P testcross

    e. The *origin* of recombinants in the testcrosses

    f. How you would test your model

Solution

In this problem, as in many genetic analyses, the data contain an important clue; once the significance of the clue is realized, the details of the experimental results fall rapidly into place. Here the clue is the drastically reduced recombinant frequency. Not very many genetic mechanisms can cause such a reduction. In this chapter, we have studied two major ones: inversions and deletions. In this problem, however, the deletion hypothesis is highly unlikely for two reasons. First, it is clear that the unusual pure line M is viable, and we have seen that large deletions generally are not viable as homozygotes. Second, the recombinant frequency is normal within line M, and this would not be true for a deletion. So we are left with the basic hypothesis that line M has a large inversion spanning all or most of the $b-l$ region. Initially, we might consider the possibility that one locus is located inside and one locus is located outside the inversion, but this does not explain the values from crosses within the line M (see answer to part a). We may draw diagrams consistent with our model, using R and L to indicate the normal right and left ends of the segment that may be inverted in M and b and l to indicate positions of the loci.

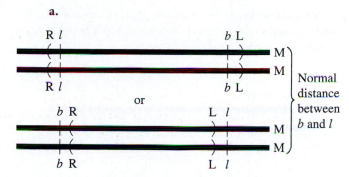

a.

Notice that if only one locus were inside the inversion and the other locus were outside, then a normal map distance would not prevail. For example,

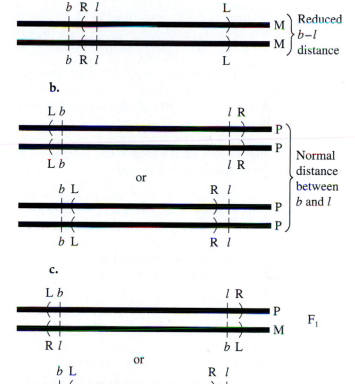

d. Same F_1 as in part c. The tester makes no difference.

e. We have been carrying two basic alternatives. In the first, the recombinants would have to come from double crossovers: one between the loci and one outside the loci but within the inversion. (Note that double crossovers such as two-strand doubles do not create a dicentric bridge.) In the second alternative, the recombinants could come from single crossovers in the small part of the $b-l$ region that is not spanned by the inversion.

f. A simple cytological test is to look for inversion loops at meiosis or, if there are chromosome markers such as constrictions or staining bands, to look for inversion in mitotic chromosomes. For a genetic test, you could map the inverted loci in relation to chromosome markers outside the inversion, as shown here for *r* and *s*:

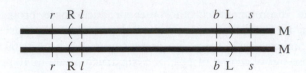

In the normal line P, *b* would map close to *r*; in line M, *b* would map close to *s*.

PROBLEMS

1. List the features (genetic or cytological) that identify and distinguish between the following: **(a)** deletions, **(b)** duplications, **(c)** inversions, **(d)** reciprocal translocations.

2. The two loci *P* and *Bz* are normally 36 m.u. apart on the same arm of a certain plant chromosome. A paracentric inversion spans about one-quarter of this region but does not include either of the loci. What approximate recombinant frequency would you predict for these loci in plants that are **(a)** heterozygous for the paracentric inversion? **(b)** Homozygous for the paracentric inversion?

3. Assume that the following loci in *Drosophila* are linked in the order a–b–c–d–e–f. A fly of genotype *a b c d e f/ a b c d e f* is crossed to a wild-type fly. About half the progeny are fully wild-type in phenotype, but the other half show the recessive phenotype corresponding to *d* and *e*. Propose an explanation for these results.

4. The normal sequence of markers on a certain *Drosophila* chromosome is 123 · 456789, where the dot represents the centromere. Some flies were isolated with chromosome aberrations that have the following structures: **(a)** 123 · 476589, **(b)** 123 · 46789, **(c)** 1654 · 32789, **(d)** 123 · 4566789. Name each type of chromosome mutation and draw diagrams to show how each would synapse with the normal chromosome.

5. A *Neurospora* heterokaryon is established between cells of the genotypes shown in the diagram below. Here *leu, his, ad, nic,* and *met* are all recessive alleles causing specific nutritional requirements for growth. *A* and *a* are the mating-type alleles (one parent must be *A* and the other *a* for a cross to occur). Usually a nucleus carrying *A* will not pair with a nucleus carrying *a* to form a heterokaryon, but the recessive mutant *tol* suppresses this incompatibility and permits heterokaryons to grow on vegetative medium. The

recessive allele *un* prevents the fungus from growing at 37°C (it is a temperature-sensitive allele) and cannot be corrected nutritionally. This heterokaryon grows well on a minimal medium, as do most of the cells derived mitotically from it. However, some rare cells show the following traits:

• They will not grow on a minimal medium unless it is supplemented with leucine.

• When the cells are transferred to a special medium that induces the sexual cycle, a cross does not occur (they will not self).

• They will not grow when moved into a 37°C temperature *even* when supplied with leucine.

• When haploid wild-type cells carrying the mating-type allele *a* are added to these aberrant cells, a cross occurs, but the addition of cells carrying the mating-type allele *A* does not cause a cross.

• From the cross with wild-type *a*, progeny with the genotype of nucleus 1 are recovered, but no alleles from nucleus 2 ever emerge from the cross.

Formulate an explanation for the origin of these strange cells in the original heterokaryon, and account for the observations concerning them.

6. Certain mice called *waltzers* execute bizarre steps in contrast to the normal gait for mice. The difference between normal mice and the waltzers is genetic, with waltzing being a recessive characteristic. W. H. Gates crossed waltzers with homozygous normals and found among several hundred normal progeny a single waltzing mouse (♀). When mated to a waltzing ♂, she produced all waltzing offspring. Mated to a homozygous normal ♂, she produced all

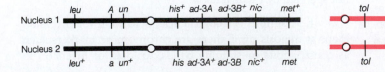

normal progeny. Some ♂♂ and ♀♀ of this normal progeny were intercrossed, and there were no waltzing offspring among their progeny. T. S. Painter examined the chromosomes of waltzing mice that were derived from some of Gates's crosses and that showed a breeding behavior similar to that of the original, unusual waltzing ♀. He found that these individuals had 40 chromosomes, just as in normal mice or the usual waltzing mice. In the unusual waltzers, however, one member of a chromosome pair was abnormally short. Interpret these observations as completely as possible, both genetically and cytologically.

(Problem 6 from A. M. Srb, R. D. Owen, and R. S. Edgar, *General Genetics*, 2d ed. W. H. Freeman and Company, 1965.)

7. *Neurospora crassa* mutations of the *ad-3B* gene are relatively easy to amass because they have a purple coloration in addition to a requirement for adenine. From haploid cultures, 100 spontaneous *ad-3B* mutants were obtained. Cells from each mutant strain were then plated on a minimal medium (containing no adenine) to test for reversion. Even after extensive platings involving a wide array of mutagens, 13 cultures produced no colonies. What kind of mutations are these? Account for both the lack of revertibility and the viability of these strains.

8. Six bands in a salivary gland chromosome of *Drosophila* are shown in the figure below, along with the extent of five deletions (Del1 to Del5):

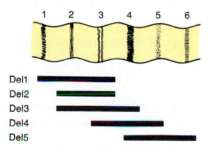

Recessive alleles, *a*, *b*, *c*, *d*, *e*, and *f* are known to be in the region, but their order is unknown. When the deletions are combined with each allele, the following results are obtained:

	a	*b*	*c*	*d*	*e*	*f*
Del1	−	−	−	+	+	+
Del2	−	+	−	+	+	+
Del3	−	+	−	+	−	+
Del4	+	+	−	−	−	+
Del5	+	+	+	−	−	−

In this table, a − means that the deletion is missing the corresponding wild-type allele (the deletion uncovers the recessive) and a + means that the corresponding wild-type allele is still present. Use these data to infer which salivary band corresponds to each gene.

(Problem 8 from D. L. Hartl, D. Friefelder, and L. A. Snyder, *Basic Genetics.* Jones and Bartlett, 1988.)

9. Five recessive lethal mutations in *Drosophila* are all shown to map onto chromosome 2. Flies bearing each of the lethal mutations are then combined pairwise with flies from a stock having six recessive mutations (*h*, *i*, *j*, *k*, *l*, and *m*) that are distributed along chromosome 2 in that order. The appearance of the resulting flies is listed below, where M stands for mutant for any particular phenotype and W stands for wild type.

Lethal mutation	Chromosome 2 marker					
	h	*i*	*j*	*k*	*l*	*m*
1	M	M	W	W	W	W
2	W	W	W	M	M	W
3	W	W	W	W	W	M
4	W	W	W	M	M	M
5	W	W	W	W	W	W

a. What is the probable basis of lethal mutations 1 through 4?

b. What can you say about lethal mutation 5?

10. Two pure lines of corn show different recombinant frequencies in the region from *Pl* to *sm* on chromosome 6. The normal strain (A) shows an RF of 26 percent, and the abnormal strain (B) shows an RF of 8 percent. The two lines are crossed, producing hybrids with reduced fertility.

a. Decide between an inversion and a deletion as the possible cause of the low RF in the abnormal strain. Give your reason.

b. Sketch the approximate relation of the chromosome aberration to the genetic markers.

c. Why do the hybrids show reduced fertility?

11. In a fruit fly of genotype *P Bar Q / p Bar q*, the *P* and *Q* loci represent markers that closely flank the homozygous bar-eye mutation. In a testcross designed to reveal the meiotic products of such a fly, some normal-eye and some double-bar types are recovered at low frequencies. These show the phenotypes associated with the flanking marker combinations *P q* or *p Q*. Explain the following with diagrams:

a. The origin of the rare normal and double-bar types

b. The association of normal and bar-eye types with the flanking marker genotypes

12. A pure line of *Drosophila* is developed that carries a duplication of the X chromosome segment that contains the vermilion-eye gene. The stock is

$$\frac{v^+ \qquad v^-}{v^+ \qquad v^-}$$

and has wild-type color. Females of this stock are mated to vermilion-eyed males that lack the duplication:

$$X\frac{v^+ \qquad v^-}{X \ v^+ \qquad v^-}$$

$$\times$$

$$\frac{X \qquad\qquad v^-}{Y \rule{1em}{0.4pt}}$$

The male offspring all have wild-type eye color, and the female offspring all have vermilion eyes. (a) Explain why these are surprising results in regard to the theory of dominance. Explain the phenotypes of the following: (b) female and male parents, (c) female and male progeny.

13. A fruit fly was found to be heterozygous for a paracentric inversion. However, it was impossible to obtain flies that were homozygous for the inversion even after extensive crosses. What is the most likely explanation of this result?

14. The *Neurospora un3* locus is near the centromere on chromosome 1 and always segregates at the first meiotic division. The *ad3* locus is 10 m.u. to the other side of the same centromere.

a. What linear asci do you predict and in what frequencies in a normal cross of *un3 ad3* × wild type? Assume that only single crossovers or no crossovers occur in the *un3–ad3* region.

b. Most of the time such crosses behave predictably, but in one case a standard *un3 ad3* strain is crossed to a wild type isolated from a field of sugar cane in Hawaii. The results follow:

● *un3 ad3*	● *un3 ad3*
● *un3 ad3*	● *un3 ad3*
● *un3 ad3*	○ ⎫
● *un3 ad3*	○ ⎪
● *un3⁺ ad3⁺*	○ ⎬ abort
● *un3⁺ ad3⁺*	○ ⎭
● *un3⁺ ad3⁺*	● *un3⁺ ad3⁺*
● *un3⁺ ad3⁺*	● *un3⁺ ad3⁺*
(and its upside-down version)	(and other spore pair orders)
80%	20%

Provide an explanation for these results, and state how you could test your idea.

15. A petunia is heterozygous for the following autosomal homologs:

a. Draw the pairing configuration you would see at metaphase I and label all parts of your diagram. Number the chromatids sequentially from top to bottom of the page.

b. A three-strand double crossover occurs with one crossover between the *C* and *D* loci on chromatids 1 and 3, and the second crossover between *G* and *H* loci on chromatids 2 and 3. Diagram the results of these recombination events as you would see them at anaphase I and label your diagram.

c. Draw the chromosome pattern you would see at anaphase II after the crossovers described in part b.

d. Give the genotypes of the gametes from this meiosis that will lead to the formation of viable progeny. Assume that all gametes are fertilized by pollen that has the gene order *A B C D E F G H I*.

16. An inversion heterozygote has one chromosome with the linkage arrangement

$$\rule{1em}{0.4pt}\ \bullet\ a \quad b \quad c \quad d \quad e \quad f \quad g \rule{2em}{0.4pt}$$

and the other possesses an inversion that spans genes *c, d, e,* and *f.*

a. What type of inversion is this?

b. Draw the configuration of these chromosomes at synapsis.

c. If a four-strand double crossover occurs in the areas *c–d* and *e–f*, what are the meiotic products?

d. Repeat for a three-strand double crossover in these regions.

e. Will any of the meiotic products have a normal karyotype? Explain.

17. Orangutans are an endangered species in their natural environment (the islands of Borneo and Sumatra), so a captive-breeding program has been established using orangutans currently held in zoos around the world. One component of this program is research into orangutan cytogenetics. This research has shown that all orangutans from Borneo carry one form of chromosome 2, as shown in the following diagram, and all orangutans from Sumatra carry the other form. Before this cytogenetic difference became known, some matings were carried out between animals from different islands, and 14 hybrid progeny are now being raised in captivity.

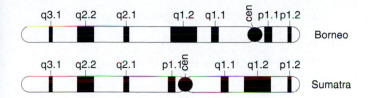

a. What term or terms describe the differences between these chromosomes?

b. Draw the chromosomes 2 of such a hybrid individual, paired during the first meiotic prophase. Be sure to show all the landmarks indicated in the figure, and label all parts of your drawing.

c. In 30 percent of meioses there will be a crossover somewhere in the region between bands p1.1 and q1.2. Draw the gamete chromosomes 2 that would result from a meiosis where a single crossover occurred within band q1.1.

d. What fraction of the gametes produced by a hybrid orangutan will give rise to viable progeny, if only these chromosomes differ between the parents?

(Problem 17 from Rosemary Redfield.)

18. Two groups of geneticists, in California and in Chile, begin work to develop a linkage map of the medfly. They both independently find that the loci for body color (B = black, b = gray) and eye shape (R = round, r = star) are linked 28 m.u. apart. They send strains to each other; a summary of all their findings is shown below:

Cross	F_1	Progeny of F1 $\times$ any $bb\,rr$	
$BB\,RR$ (Calif.) $\times$ $bb\,rr$ (Calif.)	$Bb\,Rr$	$Bb\,Rr$	36%
		$bb\,rr$	36
		$Bb\,rr$	14
		$bb\,Rr$	14
$BB\,RR$ (Chile) $\times$ $bb\,rr$ (Chile)	$Bb\,Rr$	$Bb\,Rr$	36
		$bb\,rr$	36
		$Bb\,rr$	14
		$bb\,Rr$	14
$BB\,RR$ (Calif.) $\times$ $bb\,rr$ (Chile)	$Bb\,Rr$	$Bb\,Rr$	48
or $bb\,rr$ (Calif.) $\times$ $BB\,RR$ (Chile)		$bb\,rr$	48
		$Bb\,rr$	2
		$bb\,Rr$	2

a. Provide a genetic hypothesis that explains the three sets of testcross results.

b. Draw the key chromosomal features of meiosis in the F_1 from a cross of the Californian and Chilean lines.

19. *Drosophila* can be found all over the world. A geneticist studying linkage on chromosome 2 collects pure-breeding strains in the Okanagan Valley (Canada) and in Spain. He crosses flies within each strain and scores the two sets of progeny for traits associated with six loci a–f on chromosome 2. He records the following recombinant frequencies (in percent), where I represents independent assortment, or an RF = 50:

Okanagan

	a	b	c	d	e	f
a	0	12	14	23	3	29
b		0	2	35	15	17
c			0	37	17	15
d				0	20	I
e					0	32
f						0

Spain

	a	b	c	d	e	f
a	0	12	7	30	3	22
b		0	19	18	15	34
c			0	37	4	15
d				0	33	I
e					0	19
f						0

a. Analyze these data with regard to the arrangement of these loci in flies from each geographic location, and provide an explanation for any difference or differences. Draw a map for each set of data, including map distances.

b. Show what recombinant frequencies might be expected in the five chromosomal regions delineated by these loci in Okanagan-Spanish hybrids.

20. An aberrant corn plant gives the following results when testcrossed:

	RF values				
Interval:	d–f	f–b	b–x	x–y	y–p
Control	5	18	23	12	6
Aberrant plant	5	2	2	0	6

(The locus order is centromere–d–f–b–x–y–p.) The aberrant plant is a healthy plant, but it produces far fewer normal ovules and pollen than the control plant.

a. Propose a hypothesis to account for the abnormal recombination values and the reduced fertility in the aberrant plant.

b. Use diagrams to explain the origin of the recombinants according to your hypothesis.

21. The following corn loci are on one arm of chromosome 9, in the order indicated (the distances between them are shown in map units):

$$c \text{——} bz \text{—} wx \text{——} sh \text{—} d \text{——centromere}$$

$$\quad\quad 12 \quad 8 \quad 10 \quad 20 \quad 10$$

C gives colored aleurone; *c*, white aleurone.
Bz gives green leaves; *bz*, bronze leaves.
Wx gives starchy seeds; *wx*, waxy seeds.
Sh gives smooth seeds; *sh*, shrunken seeds.
D gives tall plants; *d*, dwarf.

A plant from a standard stock that is homozygous for all five recessive alleles is crossed to a wild-type plant from Mexico that is homozygous for all five dominant alleles. The F_1 individuals express all the dominant alleles, and when backcrossed to the recessive parent give the following progeny phenotypes:

colored, green, starchy, smooth, tall	360
white, bronze, waxy, shrunk, dwarf	355
colored, bronze, waxy, shrunk, dwarf	40
white, green, starchy, smooth, tall	46
colored, green, starchy, smooth, dwarf	85
white, bronze, waxy, shrunk, tall	84
colored, bronze, waxy, shrunk, tall	8
white, green, starchy, smooth, dwarf	9
colored, green, waxy, smooth, tall	7
white, bronze, starchy, shrunk, dwarf	6

Propose a hypothesis to explain these results. Include the following:

a. A general statement of your hypothesis, with diagrams if necessary

b. Why there are 10 classes

c. An account of the origin of each class, including its frequency

d. At least one test of your hypothesis

22. A *Drosophila* geneticist has a strain of fruit flies that is true-breeding and wild-type. She crosses this strain with a multiply marked X chromosome strain carrying the recessive alleles *y* (yellow), *cv* (crossveinless), *v* (vermilion), *f* (forked), and *car* (carnation), which are evenly spaced along the X chromosome from one end to the other. She collects the heterozygous F_1 female offspring and mates them with *y cv v f B car* males (*B* gives bar, or slitlike eye). The geneticist obtains the following classes among the male offspring:

1. *y   cv   v   f   car*
2. *y⁺ cv⁺ v⁺ f⁺ car⁺*
3. *y   cv⁺ v⁺ f⁺ car*
4. *y⁺ cv   v   f   car⁺*
5. *y   cv   v⁺ f   car*
6. *y⁺ cv⁺ v   f⁺ car⁺*
7. *y   cv   v⁺ f⁺ car*
8. *y⁺ cv⁺ v   f   car⁺*
9. *y   cv⁺ v⁺ f   car*
10. *y⁺ cv   v   f⁺ car⁺*
11. *y   cv   v   f   B car*

a. Account for the results in classes 1 to 10.

b. How can you account for class 11?

c. How would you test your hypothesis?

(Problem 22 from Tom Kaufman.)

*23. Suppose that you are given a *Drosophila* line from which you can get males or virgin females at any time. The line is homozygous for chromosome 2, which has an inversion to prevent crossing-over, a dominant allele (*Cu*) for curled wings, and a recessive allele (*pr*, purple) for dark eyes. The chromosome can be drawn as

You have irradiated sperm in a wild-type male and wish to determine whether recessive lethal mutations have been induced in chromosome 2. How would you go about doing this? (HINT: Remember that each sperm carries a *different* irradiated chromosome 2.) Indicate the kinds and numbers of flies used in each cross.

24. Predict the chromosome shapes that will be produced at anaphase 1 of meiosis in a reciprocal translocation heterozygote undergoing (**a**) alternate segregations and (**b**) adjacent-1 segregations.

25. The *Neurospora* loci *a* and *b* are on separate chromosomes. In a cross of a standard *a b* strain with a wild type obtained from nature, the progeny are as follows: *a b*, 45 percent; *a⁺ b⁺*, 45 percent; *a b⁺*, 5 percent; *a⁺ b*, 5 percent. Interpret these results, and explain the origin of all the progeny types according to your hypothesis.

26. Curly wings (*Cy*) is a dominant mutation on chromosome 2 of *Drosophila*. A *Cy Cy⁺* male is irradiated with X rays and crossed with *Cy⁺ Cy⁺* females. The *Cy Cy⁺* sons are then mated individually with *Cy⁺ Cy⁺* females. From one cross, the progeny are

curly males	146
wild-type males	0
curly females	0
wild-type females	163

What abnormality in chromosome structure is the most likely explanation for these results? Use chromosome diagrams of all strains referred to in the question in your explanation. (HINT: Remember that crossing-over does not occur in male *Drosophila*.)

27. You discover a *Drosophila* male that is heterozygous for a reciprocal translocation between chromosomes 2 and 3. Each chromosomal breakpoint is near the centromere, which is near the center of each of these metacentric chromosomes.

a. Draw a diagram showing how these chromosomes would synapse at meiosis.

b. You find that this fly has the recessive alleles *bw* (brown eye) and *e* (ebony body) on the nontranslocated chromosomes 2 and 3, respectively, and wild-type alleles on the translocated chromosomes. The fly is mated with a female that has normal chromosomes and is homozygous for *bw* and *e*. What type of offspring would you expect and in what ratio? (HINT: Remember that zygotes which have an extra chromosome arm or which are deficient for one chromosome do not survive. There is no crossing-over in *Drosophila* males.)

28. An *insertional* translocation consists of a piece from the center of one chromosome inserted into the middle of another (nonhomologous) chromosome. Thus

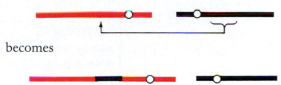

becomes

How will genomes that are heterozygous for such translocations pair at meiosis? What spore abortion patterns will be produced and in what relative proportions in a *Neurospora* cross between a wild type and an insertional translocation? (HINT: Remember that spores with chromosomal duplications survive and become dark, while those with deficiencies are light-spored and inviable.)

29. In corn, the genes for tassel length (alleles *T* and *t*) and rust resistance (alleles *R* and *r*) are known to be on separate chromosomes. In the course of making routine crosses, a breeder noticed that one *Tt Rr* plant gave unusual results in a testcross to the double recessive pollen parent *tt rr*. The results were

progeny:	*Tt Rr*	98
	tt rr	104
	Tt rr	3
	tt Rr	5

corn cobs: only about half as many seeds as usual

a. What are the key features of the data that are different from expected results?

b. State a concise hypothesis that explains the results.

c. Show genotypes of parents and progeny.

d. Draw a diagram showing the arrangement of alleles on the chromosomes.

e. Explain the origin of the two classes of progeny with 3 and 5 members.

Unpacking the Problem

a. What do "a gene for tassel length" and "a gene for rust resistance" mean?

b. Does it matter that the precise meaning of the allelic symbols *T*, *t*, *R*, and *r* is not given? Why or why not?

c. How do the terms *gene* and *allele*, as used here, relate to the concepts of locus and gene pair? (A concept map would be one way of answering this question.)

d. What prior experimental evidence would give the corn geneticists the idea that the two genes are on separate chromosomes?

e. What do you imagine "routine crosses" are to a corn breeder?

f. What term is used to describe genotypes of the type *Tt Rr*?

g. What is a "pollen parent"?

h. What are testcrosses and why do geneticists find them so useful?

i. What progeny types and frequencies might the breeder have been expecting from the testcross?

j. Describe how the observed progeny differ from expectations?

k. What does the approximate equality of the first two progeny classes tell you?

l. What does the approximate equality of the second two progeny classes tell you?

m. What were the gametes from the unusual plant and what were their proportions?

n. Which gametes were in the majority?

o. Which gametes were in the minority?

p. Which of the progeny types seem to be recombinant?

q. Which allele combinations appear to be linked in some way?

r. How can there be linkage of genes supposedly on separate chromosomes?

s. What do these majority and minority classes tell us about the genotypes of the parents of the unusual plant?

t. What is a corn cob?

u. What does a normal corn cob look like (sketch one and label it)?

v. What do the corn cobs from this cross look like (sketch one)?

w. What exactly is a kernel?

x. What effect could lead to the absence of half the kernels?

y. Did half the kernels die? If so, was the female or the male parent the reason for the death?

Now try to solve the problem.

30. The ascomycete fungus *Schizosaccharomyces pombe* produces unordered tetrads. A cross was made between a standard lab strain of genotype *nic-2 leu-3* and a wild-type strain isolated from grapes. Previous genetic studies on lab strains showed that the *nic-2* is on chromosome 1 and *leu-3* is on chromosome 4. Therefore, it was surprising that when approximately 100 tetrads were analyzed, there were only two tetrad types, as shown below, in equal frequency.

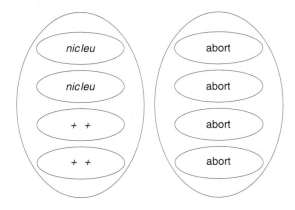

a. What was the expected result of this cross?

b. Provide a genetic explanation for the observed results, showing clearly the genotypes of the parents and ascospores under your model, and explaining why there are only two ascus types of equal frequency.

(NOTE: In this fungus, genomes with deletions or duplication-deletions all abort at meiosis.)

31. In corn, the *g* locus (+ = purple, *g* = green) is on chromosome 2, and the *w* locus (+ = starchy, *w* = waxy) is on chromosome 9. A homozygous wild type was giving unusual results in routine crosses involving these markers. In one analysis it was crossed to green, waxy, and the F_1 looked wild-type, but about half its eggs and pollen aborted. The F_1 was testcrossed, and in 50 progeny there were 28 green, waxy showing normal gamete production, and 22 wild-type showing about 50 percent abortion of gametes.

a. Propose an explanation for these results using diagrams of parents, F_1, and progeny.

b. If the F_1 was selfed, what phenotypes would result and in what proportions?

32. Yellow body in *Drosophila* is caused by a mutant allele *e* of a gene located at the tip of the X chromosome (the wild-type allele causes gray body). In a radiation experiment, a wild-type male was irradiated with X rays and then crossed with a yellow-bodied female. Most of the male progeny were yellow, as expected, but scanning of thousands of flies revealed two gray-bodied (phenotypically wild-type) males. These were crossed to yellow females, with the following results:

	Progeny
gray male 1 × yellow female	females all yellow
	males all gray
gray male 2 × yellow female	$\frac{1}{2}$ females yellow
	$\frac{1}{2}$ females gray
	$\frac{1}{2}$ males yellow
	$\frac{1}{2}$ males gray

a. Explain the origin and crossing behavior of gray male 1.

b. Explain the origin and crossing behavior of gray male 2.

33. Two auxotrophic mutations in *Neurospora*, *ad-3* and *pan-2*, are located on chromosomes 1 and 6, respectively. An unusual *ad-3* line arises in the laboratory, giving the following results:

		Ascospore appearance	RF between *ad-3* and *pan-2*
1.	Normal ad-3 × normal *pan-2*	All black	50%
2.	Abnormal *ad-3* × normal *pan-2*	About $\frac{1}{2}$ black and $\frac{1}{2}$ white (inviable)	1%
3.	Of the black spores from cross 2, about half were completely normal and half repeated the same behavior as the original abnormal *ad-3* strain.		

Explain all three results with the aid of clearly labeled diagrams. (NOTE: In *Neurospora*, ascospores with extra chromosomal material survive and are the normal black color, whereas ascospores lacking any chromosome region are white and inviable.)

34. Corn plants that carry the allele *P* produce purple leaves, and plants homozygous for *p* produce green leaves. The *P* locus is 20 m.u. from the centromere on the long arm of chromosome 9. A homozygous *p p* plant is crossed to a *P P* plant homozygous for a reciprocal translocation that has one breakpoint 30 m.u. from the centromere on the long arm of chromosome 9. The F$_1$ is obtained and backcrossed to the *p p* plant. Predict what proportions of progeny from the backcross will be

a. green, semisterile.

b. green, fully fertile.

c. purple, semisterile.

d. purple, fully fertile.

e. If the F$_1$ is selfed, what proportion of progeny will be green, fully fertile?

35. Chromosomally normal corn plants have a *p* locus on chromosome 1 and an *s* locus on chromosome 5.

P gives dark-green leaves; *p*, pale-green.

S gives large ears; *s*, shrunken ears.

An original plant of genotype *Pp Ss* has the expected phenotype (large ears, dark-green) but gives unexpected results in crosses as follows:

- On selfing, fertility is normal, but the frequency of *p p s s* types is $\frac{1}{4}$ (not $\frac{1}{16}$, as expected).

- When crossed to a normal tester of genotype *p p s s*, the F$_1$ progeny are $\frac{1}{2}$ *Pp Ss* and $\frac{1}{2}$ *p p s s*; fertility is normal.

- When an F$_1$ *Pp Ss* plant is crossed to a normal *p p s s* tester, it proves to be semisterile, but again the progeny are $\frac{1}{2}$ *Pp Ss* and $\frac{1}{2}$ *p p s s*.

Explain these results, showing the full genotypes of the original plant, the tester, and the F$_1$ individuals. How would you test your hypothesis?

36. A corn plant of genotype *pr pr* that has standard chromosomes is crossed with a *Pr Pr* plant that is homozygous for a reciprocal translocation between chromosomes 2 and 5. The F$_1$ is semisterile and phenotypically Pr (a seed color). A backcross to the parent with standard chromosomes gives 764 semisterile Pr; 145 semisterile pr; 186 normal Pr; and 727 normal pr. What is the map distance between the *Pr* locus and the translocation point?

37. A reciprocal translocation of the following types is obtained in *Neurospora*:

The following cross is then made:

Assume that the small, red piece of the chromosome involved in the translocation does not carry any essential genes. How would you select products of meiosis that are duplicated for the translocated part of the solid chromosome?

38. A male rat that is phenotypically normal shows reproductive anomalies when compared with normal male rats, as shown in Table 8-2. Propose a genetic explanation of these unusual results, and say how your idea could be tested.

39. During ascus formation in *Neurospora* any ascospore with a chromosomal deletion aborts and appears white, and any with a duplication survives as a dark spore. What patterns of ascospore abortion would you observe, and in what approximate proportions, in octads from *Neurospora* crosses between a strain with normal chromosomes and one with

Table 8-2

Mating	Embryos (mean no.)			Degeneration (%)
	Implanted in uterine wall	Degenerating after implantation	Normal	
exceptional ♂ × normal ♀	8.7	5.0	3.7	57.5
normal ♂ × normal ♀	9.5	0.6	8.9	6.5

a. a reciprocal translocation?

b. a pericentric inversion?

c. a paracentric inversion?

Be sure to consider the effect of crossing-over in each one.

40. Suppose that you are studying the cytogenetics of five closely related species of *Drosophila*. The figure shown be-

low gives the arrangement of loci (the letters indicate loci that are identical in all five species) and the chromosomal patterns that you find in each species. Show how these species probably evolved from each other, describing the changes occurring at each step. (NOTE: Be sure to compare gene order carefully.)

41. Show how the $\frac{10}{16}$ ratio on page 230 was derived.

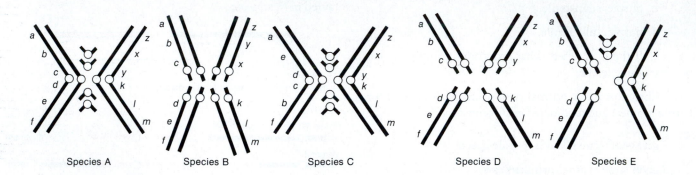

Species A   Species B   Species C   Species D   Species E

9

Chromosome Mutation II: Changes in Number

A geneticist examining an orchid with multiple chromosome sets (a polyploid). Compare the dimensions of the polyploid with those of its diploid parents shown in the foreground. (Runk/Schoenberger/Grant Heilman)

KEY CONCEPTS

▶ Organisms with multiple chromosome sets (polyploids) are generally larger than diploid organisms, but meiotic pairing anomalies make some polyploid organisms sterile.

▶ An even number of polyploid sets is generally more likely to result in fertility. Then the single-locus segregation ratios are different from those of diploids.

▶ Crosses between two different species followed by the doubling of the chromosome number in the hybrid produces a special kind of fertile interspecific polyploid.

▶ Variants in which a single chromosome has been gained or lost generally arise by nondisjunction (abnormal chromosome segregation at meiosis or mitosis).

▶ Such variants tend to be sterile and show the abnormalities attributable to gene imbalance.

▶ When fertile, such variants show abnormal gene segregation ratios for the misrepresented chromosome only.

T he second major type of chromosome mutation is change in chromosome number. There are few aspects of genetics that impinge on human affairs quite so directly as this one. Chromosome numbers change spontaneously as accidents within cells, and this process has been going on as long as there has been life on the planet. In fact, changes in chromosome number have been instrumental in molding genomes during evolution. For examples of such changes, we have to look no farther than the food on our dining tables because many of the plants (and some of the animals) that we eat arose through spontaneous changes in chromosome number during the evolution of those species. Today, breeders emulate this process by manipulating chromosome number to improve productivity or some other useful feature of the organism. But perhaps the main relevance of chromosome numbers is for members of our own species in that a large proportion of genetically determined ill health in humans is caused by abnormal chromosome numbers.

In this chapter we shall investigate the processes that produce new chromosome numbers, the diagnostic tests for detecting such changes, and the properties of cells and individuals carrying the different kinds of variant chromosomal complements. As with any area of cytogenetics, the techniques are a combination of genetics and microscopy. Changes in chromosome number are usually classified into two types, changes in whole chromosome sets and changes in parts of chromosome sets, and these two types are dealt with in the two following sections.

Aberrant Euploidy

The number of chromosomes in a basic set is called the **monoploid number** (x). Organisms with multiples of the monoploid number of chromosomes are called **euploid.** We saw in earlier chapters that eukaryotes normally carry either one chromosome set (haploids) or two sets (diploids). These, then, are both cases of normal euploidy. Euploid types that have more than two sets of chromosomes are called **polyploid.** The polyploid types are named **triploid** ($3x$), **tetraploid** ($4x$), **pentaploid** ($5x$), **hexaploid** ($6x$), and so on. Polyploids arise naturally as spontaneous chromosomal mutations, and as such, they must be considered aberrations because they differ from the previous norm. However, many species of plants and animals have clearly arisen through polyploidy, so evidently evolution can take advantage of polyploidy when it occurs. It is worth noting that organisms with one chromosome set sometimes arise as variants of diploids; such variants are called **monoploid** ($1x$). In some species monoploid stages are part of the regular life cycle, but other monoploids are spontaneous aberrations.

The haploid number (n), which we have already used extensively, refers strictly to the number of chromosomes in gametes. In most animals and many plants that we are familiar with, the haploid number and monoploid number are the same. Hence, n or x (or $2n$ or $2x$) can be used interchangeably. However, in certain plants, such as modern wheat, n and x are different. Wheat has 42 chromosomes, but careful study reveals that it is hexaploid, with six rather similar but not identical sets of seven chromosomes. Hence, $6x = 42$ and $x = 7$. However, the gametes of wheat contain 21 chromosomes, so $n = 21$ and $2n = 42$.

Monoploids

Male bees, wasps, and ants are monoploid. In the normal life cycles of these insects, males develop parthenogenetically — that is, they develop from unfertilized eggs. However, in most species monoploid individuals are abnormal, arising in natural populations as rare aberrations. The germ cells of a monoploid cannot proceed through meiosis normally because the chromosomes have no pairing partners. Thus, monoploids are characteristically sterile. (Male bees, wasps, and ants bypass meiosis in forming gametes; here, mitosis produces the gametes.) If a monoploid cell does undergo meiosis, the single chromosomes segregate randomly, and the probability of all chromosomes going to one pole is $(\frac{1}{2})^{x-1}$, where x is the number of chromosomes. This formula estimates the frequency of viable (whole-set) gametes; obviously a small number if x is large.

Monoploids play an important role in modern approaches to plant breeding. Diploidy is an inherent nuisance when breeders want to induce and select new gene mutations that are favorable and to find new combinations of favorable alleles at different loci. New recessive mutations must be made homozygous before they can be expressed, and favorable allelic combinations in heterozygotes are broken up by meiosis. Monoploids provide a way around some of these problems. In some plant species, monoploids may be artificially derived from the products of meiosis in the plant's anthers. A cell destined to become a pollen grain may instead be induced by cold treatment to grow into an embryoid, a small dividing mass of cells. The embryoid may be grown on agar to form a monoploid plantlet, which can then be potted in soil and allowed to mature (Figure 9-1).

Plant monoploids may be exploited in several ways. In one, they are first examined for favorable traits or allelic combinations, which may arise from heterozygosity already present in the parent or induced in the parent by mutagens. The monoploid can then be subjected to chromosome doubling to achieve a completely homozygous diploid with a normal meiosis, capable of providing seed. How is this achieved? Quite simply, by the application of a compound called **colchicine** to meristematic tissue. Colchicine — an alkaloid drug extracted from the autumn crocus — inhibits the formation of the mitotic spindle, so that cells with two chromosome sets are produced (Figure 9-2). These cells may proliferate to form a sector of diploid tissue that can be identified cytologically.

Another way in which the monoploid may be used is to treat its cells basically like a population of haploid organ-

Figure 9-1 Generating monoploid plants by tissue culture. Pollen grains (haploid) are treated so that they will grow and are placed on agar plates containing certain plant hormones. Under these conditions, haploid embryoids will grow into monoploid plantlets. After being moved to a medium containing different plant hormones, these plantlets will grow into mature monoploid plants with roots, stems, leaves, and flowers.

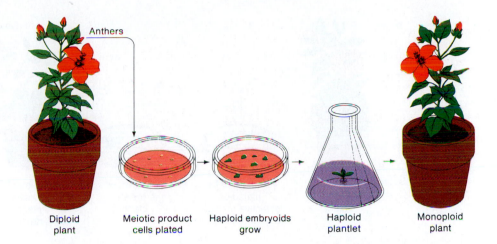

Anthers

Diploid plant   Meiotic product cells plated   Haploid embryoids grow   Haploid plantlet   Monoploid plant

isms in a mutagenesis-and-selection procedure. A population of cells is isolated, their walls are removed by enzymatic treatment, and they are treated with mutagen. They are then plated on a medium that selects for some desirable phenotype. This approach has been used to select for resistance to toxic compounds produced by one of the plant's parasites, and to select for resistance to herbicides being used by farmers to kill weeds. Resistant plantlets eventually grow into haploid plants, which can then be doubled (using colchicine) into a pure-breeding, diploid, resistant type (Figure 9-3).

These are powerful techniques that can circumvent the normally slow process of meiosis-based plant breeding. The techniques have been successfully applied to several important crop plants, such as soybeans and tobacco. This is, of course, another aspect of applied somatic cell genetics in higher organisms.

The anther technique for producing monoploids does not work in all organisms or in all genotypes of an organism. Another useful technique has been developed in barley, an important crop plant. Diploid barley, *Hordeum vulgare,* may be fertilized by pollen from a diploid wild relative called *Hordeum bulbosum.* This results in zygotes with one chromosome set from each parental species. During the ensuing somatic cell divisions, however, the chromosomes of *H. bulbosum* are eliminated from the zygote while all the chromosomes of *H. vulgare* are retained, resulting in a haploid embryo. (The haploidization process appears to be caused by a genetic incompatibility between the chromosomes of the different species.) The chromosomes of the resulting haploids can be doubled with colchicine. This approach has led to the rapid production and widespread planting of several new barley varieties, and it is being used successfully in other species too.

Message To create new plant lines, geneticists produce monoploids with favorable genotypes and then double the chromosomes to form fertile, homozygous diploids.

Polyploids

In the realm of polyploids, we must distinguish between **autopolyploids,** which are composed of multiple sets from within one species, and **allopolyploids,** which are composed of sets from different species. Allopolyploids form

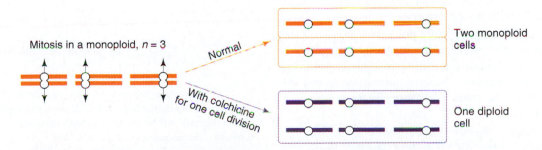

Mitosis in a monoploid, *n* = 3

Normal

Two monoploid cells

With colchicine for one cell division

One diploid cell

Figure 9-2 Using colchicine to generate a diploid from a monoploid. Colchicine added to mitotic cells during metaphase and anaphase disrupts spindle fiber formation, preventing the migration of chromatids after the centromere is split. A single cell is created that contains pairs of identical chromosomes that are homozygous at all loci.

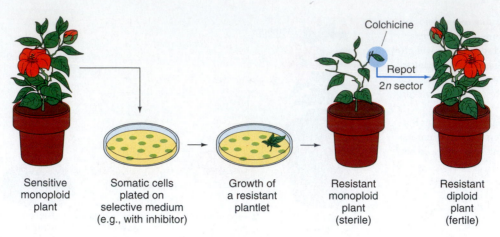

Sensitive monoploid plant → Somatic cells plated on selective medium (e.g., with inhibitor) → Growth of a resistant plantlet → Resistant monoploid plant (sterile) → Resistant diploid plant (fertile)

Figure 9-3 Using microbial techniques in plant engineering. The cell walls of haploid cells are removed enzymatically. The cells are then exposed to a mutagen and plated on an agar medium containing a selective agent, such as a toxic compound produced by a plant parasite. Only those cells containing a resistance mutation that allows them to live within the presence of this toxin will grow. After treatment with the appropriate plant hormones, these cells will grow into mature monoploid plants and, with proper colchicine treatment, can be converted into homozygous diploid plants.

only between closely related species; however, the different chromosome sets are **homeologous** (only partially homologous)—not fully homologous, as they are in autopolyploids.

Triploids

Triploids are usually autopolyploids. They arise spontaneously in nature or are constructed by geneticists from the cross of a 4x (tetraploid) and a 2x (diploid). The 2x and the x gametes unite to form a 3x triploid.

Triploids are characteristically sterile. The problem, like that of monoploids, involves pairing at meiosis. Synapsis or true pairing can take place only between two chromosomes, but one chromosome can pair with one partner along part of its length and with another along the remainder, and this gives rise to an association of three chromosomes. Paired chromosomes of the type found in diploids are called **bivalents.** Associations of three chromosomes are called **trivalents,** and unpaired chromosomes are called **univalents.** Hence in triploids there are two pairing possibilities, resulting in either a trivalent, or a bivalent plus a univalent. Paired centromeres segregate to opposite poles, but unpaired centromeres pass to either pole randomly. We see in Figure 9-4 that the net result of both the pairing possibilities is an uneven segregation, with two chromosomes going in one direction and one in the other. This happens for every chromosome threesome.

If all the single chromosomes pass to the *same* pole, and simultaneously the other two chromosomes pass to the opposite pole, then the gametes formed will be haploid and diploid. The probability of this type of meiosis will be $(\frac{1}{2})^{x-1}$, and this proportion is likely to be low. All other possibilities will give gametes with chromosome numbers in-

termediate between the haploid and diploid number; such genomes are **aneuploid**—"not euploid." It is likely that these aneuploid gametes will not lead to viable progeny—in fact, it is this category that is responsible for the almost complete lack of fertility of triploids. The problem is one of **genome imbalance,** a phenomenon that we will encounter repeatedly in this chapter. For most organisms the euploid chromosome set is a finely tuned set of genes in relative proportions that seem to be functionally significant. Multiples of this set are tolerated because there is no change in the relative proportions of genes. However, the addition of one or more extra chromosomes is nearly always deleteri-

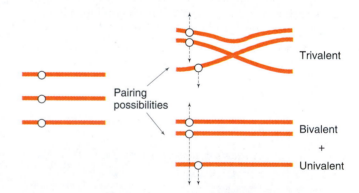

Trivalent

Pairing possibilities

Bivalent
+
Univalent

Figure 9-4 Two possibilities for the pairing of three homologous chromosomes before the first meiotic division in a triploid. Notice that the outcome will be the same in both cases: one resulting cell will receive two chromosomes and the other will receive just one. The probability that the latter cell can become a functional haploid gamete is very small, however, because to do so it would also have to receive only one of the three homologous chromosomes of every other set in the organism. Note that each chromosome is really a pair of chromatids.

ous because the proportions of genes in those extra chromosomes is altered. Although the action of some genes can be regulated to compensate for extra gene "dosage," the overall impact of the extra genetic material seems too great to be overcome by gene regulation. The deleterious effect can be expressed at the level of gametes, making them nonfunctional, or at the level of the zygote, resulting in lethality, sterility, or lowered fitness.

In the case of triploids, it is possible that some haploid or diploid gametes will form, and some may unite to form a euploid zygote, but the likelihood of this is inherently low. Consider bananas. The bananas that are widely available commercially are triploids with 11 chromosomes in each set ($3x = 33$). The probability of a meiosis in which all univalents pass to the same pole is $(\frac{1}{2})^{x-1}$, or $(\frac{1}{2})^{10} = \frac{1}{1024}$, so bananas are effectively sterile. The most obvious expression of the sterility of bananas is that there are no seeds in the fruit that we eat. Another example of the commercial exploitation of triploidy in plants is the production of triploid watermelons. For the same reasons presented in our discussion of bananas, triploid watermelons are seedless, a phenotype favored by some for its convenience.

> **Message** Some types of chromosome mutations are themselves aneuploid; other types produce aneuploid gametes or zygotes. Aneuploidy is nearly always deleterious because of genetic imbalance—the ratio of genes is different from that in euploids, and this interferes with the normal operation of the genome.

Autotetraploids

Autotetraploids arise naturally, by the spontaneous accidental doubling of a $2x$ genome to a $4x$ genome, and also can be induced artificially through the use of colchicine. Autotetraploid plants are advantageous as commercial crops because in plants the larger number of chromosome sets often leads to increased size. Cell size, fruit size, flower size, stomata size, and so on can all be larger in the polyploid (Figure 9-5). Here we see another effect that must be explained by gene numbers. Presumably the amount of gene product (protein or RNA) is proportional to the number of genes in the cell, and this is higher in the cells of polyploids.

> **Message** Polyploid plants are often larger and have larger organs than their diploid relatives.

Because 4 is an even number, autotetraploids can have a regular meiosis, although this is by no means always the case. The crucial factor is how the four homologous chromosomes, one from each of the four sets, pair and segregate. There are several possibilities, as shown in Figure 9-6. Pairings between four chromosomes are called **quadrivalents.** In tetraploids, the two-bivalent and the quadrivalent pairing modes tend to be most regular in segregation, but even here there is no guarantee of a 2:2 segregation. If all

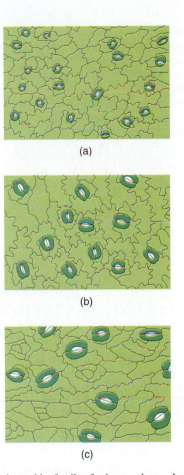

(a)

(b)

(c)

Figure 9-5 Epidermal leaf cells of tobacco plants, showing an increase in cell size, particularly evident in stomata size, with an increase in autopolyploidy. (a) Diploid, (b) tetraploid, (c) octoploid. (From W. Williams, *Genetic Principles and Plant Breeding*. Blackwell Scientific Publications, Ltd.)

chromosome sets segregate 2:2 as they do in some species, then the gametes will be functional and a formal genetic analysis can be developed for such autotetraploids.

Let's consider the genetics of a fertile tetraploid. We can consider an experiment in which colchicine is used to double the chromosomes of an Aa plant to form an $AAaa$ autotetraploid, which we will assume shows 2:2 segregation. We now have a further concern because polyploids such as tetraploids give different phenotypic ratios in their progeny, depending on whether or not the locus in question is tightly linked to the centromere. First, we consider a centromere-linked gene. The three possible pairing and segregation patterns are presented in Figure 9-7; these occur by chance and with equal frequency. As the figure shows, the $2x$ gametes produced are Aa, AA, or aa, in a ratio of 8:2:2, or 4:1:1. If such a plant is selfed, the probability of an $aaaa$ phenotype in the offspring is $\frac{1}{6} \times \frac{1}{6} = \frac{1}{36}$. In other words, a 35:1 phenotypic ratio of $A---:aaaa$ will be observed if A is fully dominant over three a alleles.

If, in the same kind of plant, a genetic locus having the alleles B and b is very far removed from the centromere,

Figure 9-6 Meiotic pairing possibilities in tetraploids. (Each chromosome is really two chromatids.) The four homologous chromosomes may pair as two bivalents or as a quadrivalent. Both possibilities can yield functional gametes. However, the four chromosomes may also pair in a univalent-trivalent combination, yielding nonfunctional gametes. A specific tetraploid can show one or more of these pairings.

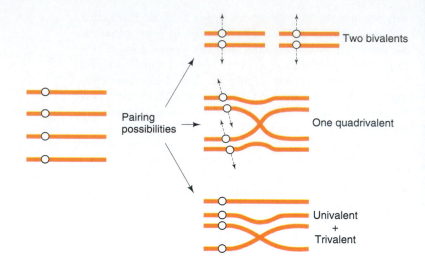

crossing-over must be considered. This forces us to think in terms of chromatids instead of chromosomes; there are four *B* chromatids and four *b* chromatids (Figure 9-8). Because the number of crossovers in such a long region will be large, the genes will become effectively unlinked from their original centromeres. The packaging of genes two at a time into gametes is very much like grabbing two balls at random from a bag of eight balls: four of one kind and four of another. The probability of picking two *b* genes is then

$$\tfrac{4}{8} \text{ (the first one)} \times \tfrac{3}{7} \text{ (the second one)} = \tfrac{12}{56} = \tfrac{3}{14}$$

So, in a selfing, the probability of a *bbbb* phenotype is $\tfrac{3}{14} \times \tfrac{3}{14} = \tfrac{9}{196}$, which is approximately $\tfrac{1}{22}$. Hence, there will be a 21:1 phenotypic ratio of *B---*:*bbbb*. For genetic loci of intermediate position, intermediate ratios will, of course, result.

Allopolyploids

The "classic" **allopolyploid** was synthesized by G. Karpechenko in 1928. He wanted to make a fertile hybrid that would have the leaves of the cabbage (*Brassica*) and the roots of the radish (*Raphanus*). Each of these species has 18 chromosomes, and they are related closely enough to allow intercrossing. A viable hybrid progeny individual was produced from seed. However, this hybrid was functionally sterile because the nine chromosomes from the cabbage parent were different enough from the radish chromosomes that pairs did not synapse and disjoin normally:

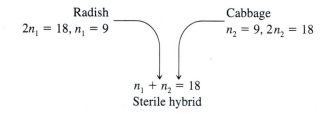

However, one day a few seeds were in fact produced by this (almost) sterile hybrid. On planting, these seeds produced fertile individuals with 36 chromosomes. All these individuals were allopolyploids. They had apparently been derived from spontaneous, accidental chromosome doubling to $2n_1 + 2n_2$ in the sterile hybrid, presumably in tissue that eventually became germinal and underwent meiosis. Thus, in $2n_1 + 2n_2$ tissue, there is a pairing partner for each chromosome and balanced gametes of the type $n_1 + n_2$ are produced. These fuse to give $2n_1 + 2n_2$ allopolyploid progeny, which are also fertile. This kind of allopolyploid is sometimes called an **amphidiploid** which means "doubled diploid" (Figure 9-9). (Unfortunately for Karpechenko, his

	Pairing	Gametes produced by random spindle attachment	
1 2	*A*	1 + 3	*Aa*
		2 + 4	*Aa*
3 4	*A* *a*	1 + 4	*Aa*
	a	2 + 3	*Aa*
1 3	*A* *a*	1 + 2	*AA*
		3 + 4	*aa*
2 4	*A* *a*	1 + 4	*Aa*
		2 + 3	*Aa*
1 4	*A* *a*	1 + 2	*AA*
		3 + 4	*aa*
2 3	*A* *a*	1 + 3	*Aa*
		2 + 4	*Aa*

Figure 9-7 Gene segregation in a tetraploid showing orderly pairing by bivalents. (Each chromosome is really two chromatids.) The locus is assumed to be close to the centromere. Self-fertilization could yield a variety of genotypes, including *aaaa*.

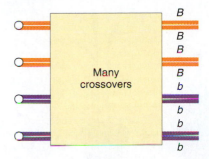

Figure 9-8 Highly diagrammatic representation of a tetraploid meiosis involving a heterozygous locus distant from the centromere. The net effect of multiple crossovers in such a long region will be that the genes become effectively unlinked from their original centromeres. Genes are packaged two at a time into gametes, much as two balls may be grabbed at random from a bag containing eight balls: four of one kind and four of another.

amphidiploid had the roots of a cabbage and the leaves of a radish.)

When the allopolyploid was crossed to either parental species, sterile offspring resulted. The offspring of the cross to radish were $2n_1 + n_2$, constituted from an $n_1 + n_2$ gamete from the allopolyploid and an n_1 gamete from the radish. Obviously, the n_2 chromosomes had no pairing partners, so sterility resulted. Consequently, Karpechenko had effectively created a new species, with no possibility of gene exchange with its parents. He called his new species Raphanobrassica.

Today, allopolyploids are routinely synthesized in plant breeding. Instead of waiting for spontaneous doubling to occur in the sterile hybrid, colchicine is added to induce doubling. The goal of the breeder obviously is to combine some of the useful features of both parental species into one type. This kind of endeavor is very unpredictable, as Karpechenko learned. In fact, only one synthetic amphidiploid has ever been widely used. This is *Triticale*, an amphidiploid between wheat (*Triticum*, $2n = 6x = 42$) and rye (*Secale*, $2n = 2x = 14$). *Triticale* combines the high yields of wheat with the ruggedness of rye. Figure 9-10 shows the procedure for synthesizing *Triticale*.

In nature, allopolyploidy seems to have been a major force in speciation of plants. There are many different examples. One particularly satisfying one is shown by the genus *Brassica*, as illustrated in Figure 9-11. Here three different parent species have hybridized in all possible pair combinations to form new amphidiploid species.

A particularly interesting natural allopolyploid is bread wheat, *Triticum aestivum* ($2n = 6x = 42$). By studying various wild relatives, geneticists have reconstructed a probable evolutionary history of bread wheat (Figure 9-12). In a bread wheat meiosis, there are always 21 pairs of chromosomes. Furthermore, it has been possible to establish that any given chromosome has only one specific pairing partner (homologous pairing)—not five other potential partners (homeologous pairing). The suppression of such homeologous pairing (which would make the species more unstable) is maintained by a gene *Ph* on the long arm of chromosome 5 of the B set. Thus, *Ph* ensures a diploid-like meiotic behavior for this hexaploid species. Without *Ph*, bread wheat could probably never have arisen. It is interesting to speculate as to whether Western civilization could have begun or progressed without this species—in other words, without the *Ph* mutation.

Figure 9-9 The origin of the amphidiploid (*Raphanobrassica*) formed from cabbage (*Brassica*) and radish (*Raphanus*). The fertile amphidiploid arose in this case from spontaneous doubling in the $2n = 18$ sterile hybrid. Colchicine can be used to promote doubling. (From A. M. Srb, R. D. Owen, and R. S. Edgar, *General Genetics*, 2d ed. Copyright 1965 by W. H. Freeman and Company. After G. Karpechenko, Z. *Indukt. Abst. Vererb.* 48, 1928, 27.)

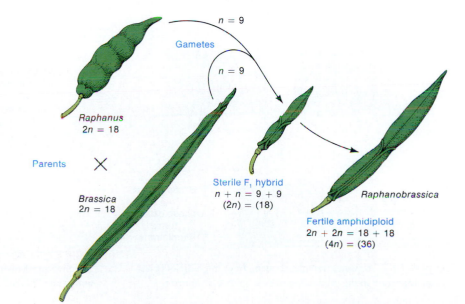

Gametes

$n = 9$

$n = 9$

Raphanus
$2n = 18$

Parents   ✕

Brassica
$2n = 18$

Sterile F₁ hybrid
$n + n = 9 + 9$
$(2n) = (18)$

Raphanobrassica

Fertile amphidiploid
$2n + 2n = 18 + 18$
$(4n) = (36)$

Figure 9-10 Techniques for the production of the amphidiploid *Triticale.* If the hybrid seed does not germinate, then tissue culture (*below*) may be used to obtain a hybrid plant. (From Joseph H. Hulse and David Spurgeon, "Triticale." Copyright © 1974 by Scientific American.)

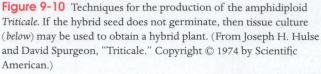

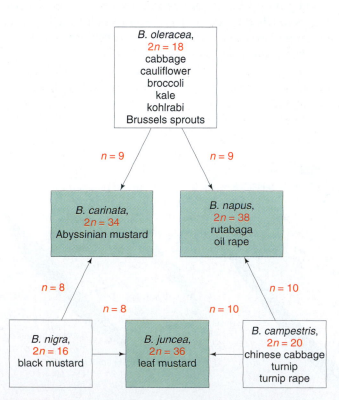

Figure 9-11 A species triangle, showing how amphidiploidy has been important in the production of new species of *Brassica.*

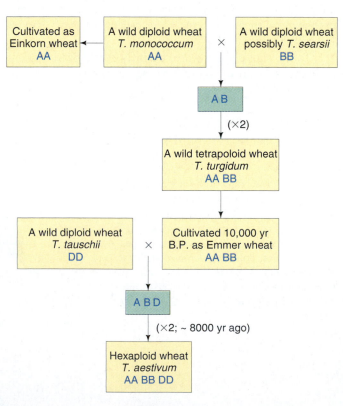

Figure 9-12 Diagram of the proposed evolution of modern hexaploid wheat involving amphidiploid production at two points. A, B, and D are different chromosome sets.

Somatic Allopolyploids from Cell Hybridization

Another innovative approach to plant breeding is to try to make allopolyploid-like hybrids by fusing asexual cells. Theoretically, such a technique would allow us to combine widely differing parental species. The technique does indeed work, but the only allopolyploids that have been produced so far can also be made by the sexual methods we have considered already. In the cell-fusion procedure, cell suspensions of the two parental species are prepared and stripped of their cell walls by special enzyme treatments. The stripped cells are called **protoplasts.** The two protoplast suspensions are then combined with polyethylene glycol, which enhances protoplast fusion. The parental cells and the fused cells proliferate on an agar medium to form colonies (in much the same way as microbes). If these colonies, or *calluses* as they are called, are examined, a fair percentage of them are found to be allopolyploid-like hybrids with chromosome numbers equal to the sum of the parental numbers. Thus, not only do the protoplast cell membranes fuse to form a kind of heterokaryon, but the nuclei fuse, too, to give rise to a $2n_1 + 2n_2$ amphidiploid.

Another good example of an allopolyploid-like hybrid is commercial tobacco, *Nicotiana tabacum,* which has 48 chromosomes. This species of tobacco was originally found in nature as a spontaneous amphidiploid. The two probable parents are *N. sylvestris* and *N. tomentosiformis,* each of which has 24 chromosomes. A sexual cross between *N. tabacum* and either of these two probable parents gives a 36-chromosome hybrid containing 12 chromosome pairs plus 12 unpaired chromosomes. A cross between *N. sylvestris* and *N. tomentosiformis* yields a 24-chromosome hybrid in which there is no pairing at all. Hence, it appears that part of the *N. tabacum* genome is from *N. sylvestris* and part is from *N. tomentosiformis.* This amphidiploid can be re-created either sexually, by using colchicine as described previously, or somatically by cell fusion. When cells of the prospective parental species are fused, a 48-chromosome hybrid cell line is produced, from which plants identical in behavior to *N. tabacum* may be grown. (Note that in the latter method, colchicine is not required, because the fusion product is already amphidiploid.)

The recovery of somatic hybrids may be enhanced if a selective system is available. In one example in *N. tabacum,* two monoploid lines were fused to form a diploid hybrid culture using complementation as the selection system. The first line had whitish, light-sensitive leaves due to a recessive mutation *w.* The other had yellowish, light-sensitive leaves due to a mutation *y* at a separate locus. When the cells were combined in a petri dish, the diploid $w^+w\,y^+y$ calluses could be selected by their resistance to light and their normal green color. The calluses can be grown into plantlets, which then are either grafted onto a mature plant to develop or potted themselves. The protocol for this experiment is illustrated in Figure 9-13.

Two allotetraploids of *Petunia,* one produced by sexual hybridization and the other by somatic hybridization, are compared in Figure 9-14. The figure shows that the two are identical in appearance and produce the same range of progeny types.

Message Allopolyploids can be synthesized by crossing related species and doubling the chromosomes of the hybrid, or by asexually fusing the cells of different species.

Polyploidy in Animals

You may have noticed that most of the discussion of polyploidy so far has concerned plants. Indeed, polyploidy is more common in plants than in animals, but nevertheless there are many cases of polyploid animals. Examples are found in flatworms, leeches, and brine shrimp. In these cases reproduction is by parthenogenesis, the development of a special type of unfertilized egg into an embryo, without the need for fertilization. However, examples are not confined to these so-called lower forms. Polyploid amphibians and reptiles are surprisingly common. These show several modes of reproduction. Polyploid frogs and toads have males and females participating in their sexual cycles, whereas polyploid salamanders and lizards are parthenogenetic.

Some fish are also polyploid, and in two cases it appears that a single polyploid event has given rise to an entire taxonomic family in evolution. This situation contrasts with that in amphibians and reptiles because in those cases the polyploids all have closely related diploid species, and, hence, the polyploid events do not seem to have been important in the evolution of the group as a whole. The Salmonidae family of fishes (containing salmon and trout) is a familiar example of a group that appears to have originated through polyploidy. Salmonids have twice as much DNA as related fish. Different salmonid species have different chromosome numbers, but it has been discovered that the group has an almost invariant number of chromosome arms (some fused in some species) that is twice the number of arms in related groups. Hence, the evidence points to the salmonids' having evolved from a single event that gave rise to a tetraploid.

You might also be interested to know that the sterility of triploids has been commercially exploited in animals as well as plants. Triploid oysters have been developed, and these have a commercial advantage over their diploid relatives. The diploids go through a spawning season, during which they are unpalatable. Triploids, however, because of their sterility, do not spawn and are palatable the whole year round.

Human polyploid zygotes do arise through various kinds of mistakes in cell division. Most die in utero. Occasionally, triploid babies are born, but none survive.

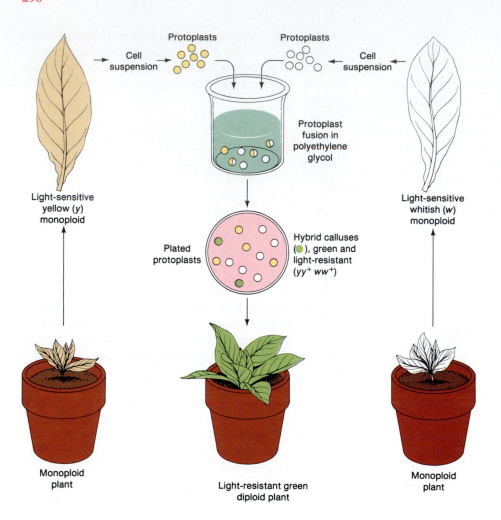

Protoplasts — Cell suspension

Cell suspension — Protoplasts

Protoplast fusion in polyethylene glycol

Plated protoplasts

Hybrid calluses (●), green and light-resistant ($yy^+ \ ww^+$)

Light-sensitive yellow (y) monoploid

Light-sensitive whitish (w) monoploid

Monoploid plant

Light-resistant green diploid plant

Monoploid plant

Figure 9-13 Creating a hybrid of two monoploid lines of *Nicotiana tabacum* by cell fusion. One line has light-sensitive, yellowish leaves, and the other has light-sensitive, whitish leaves. Protoplasts are produced by enzymatically stripping the cell walls from the leaf cells of each strain. The protoplasts can fuse, as indicated; those that fuse as hybrids can be grown into calluses that are light-resistant as a result of recessiveness of the parental genotypes. The calluses, under the appropriate hormone regime, can be grown into green diploid plants.

Aneuploidy

Aneuploidy is the second major category of chromosome mutations involving abnormal chromosome number. An **aneuploid** is a individual whose chromosome number differs from the wild type by part of a chromosome set. Gen-erally the aneuploid chromosome set differs from wild type by only one or a small number of chromosomes. Aneuploids can have a chromosome number either greater or smaller than that of the wild type. Aneuploid nomenclature is based on the number of copies of the specific chrom some involved in the aneuploid state. For example, the ane-uploid condition $2n - 1$ is called **monosomic** (meaning "one chromosome") because there is only one copy of some specific chromosome present instead of the usual two found in its diploid progenitor. The aneuploid $2n + 1$ is called **trisomic,** $2n - 2$ is **nullisomic,** and $n + 1$ is **disomic.**

P. hybrida

P. parodii

F_1 hybrid

Somatic hybrid

1   2   3

4   5   6

Figure 9-14 Sexual hybridization and somatic hybridization pro-duce identical *Petunia* allotetraploids. The two parental lines are illus-trated at the top, the red *P. hybrida* and the white *P. parodii*. The F_1 al-lotetraploid in the center row resulted from crossing the two diploid parents and doubling the chromosome number of the hybrid; the al-lotetraploid next to it resulted from fusion of somatic cells of the two parental lines. The two allotetraploids produce the same range of progeny, illustrated. The numbers identify plants used in further ex-periments. (J. B. Power)

Nullisomics (2n − 2)

Although nullisomy is a lethal condition in diploids, an organism like bread wheat, which behaves meiotically like a diploid although it is a hexaploid, can tolerate nullisomy. The four homoeologous chromosomes apparently compensate for a missing pair of homologs. In fact, all the possible 21 bread wheat nullisomics have been produced; these are illustrated in Figure 9-15. Their appearances differ from the normal hexaploids; furthermore, most of the nullisomics grow less vigorously.

Monosomics (2n − 1)

Monosomic chromosome complements are generally deleterious for two main reasons. First, the missing chromosome perturbs the overall gene balance in the chromosome set. (We encountered this effect earlier.) Second, having a chromosome missing allows any deleterious recessive allele on the single chromosome to be hemizygous and thus to be directly expressed phenotypically. Notice that these are the same effects produced by deletions.

Nondisjunction during mitosis or meiosis is the cause of most aneuploids. Disjunction is the normal separation of homologous chromosomes or chromatids to opposite poles at nuclear division. Nondisjunction is a failure of this disjoining process, and two chromosomes (or chromatids) go to one pole and none to the other. Nondisjunction occurs spontaneously; it is another example of a chance failure of a basic cellular process. Surprisingly the precise molecular processes that fail are not known, but in experimental systems the frequency of nondisjunction can be increased by interfering with microtubule action.

In meiotic nondisjunction, the chromosomes may fail to disjoin at either the first or second division (Figure 9-16). Either way, n + 1 and n − 1 gametes are produced. If an n − 1 gamete is fertilized by an n gamete, a monosomic (2n − 1) zygote is produced. The fusion of an n + 1 and an n gamete yields a trisomic 2n + 1.

In humans, a sex-chromosome monosomic complement of 44 autosomes + 1 X produces a phenotype known as *Turner syndrome*. Affected people have a characteristic, easily recognizable phenotype: they are sterile females, are

Figure 9-15 Ears of the nullisomics of wheat. The symbols under the ears designate the absent chromosome. Although nullisomics are usually lethal in regular diploids, organisms like wheat, which "pretends" to be diploid but is hexaploid, can tolerate nullisomy. Nullisomics, however, are less vigorous growers. (E. R. Sears)

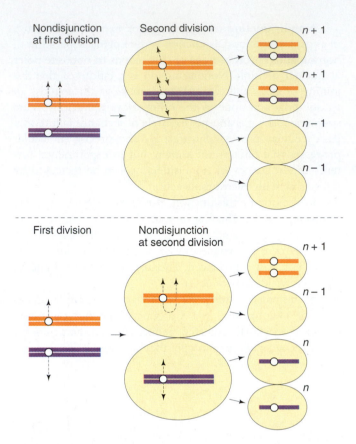

Figure 9-16 The origin of aneuploid gametes by nondisjunction at the first or second meiotic division.

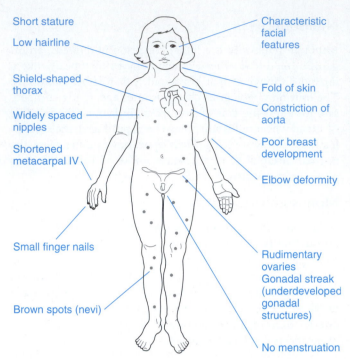

Figure 9-17 Characteristics of Turner syndrome, which results from having a single X chromosome (XO). (Adapted from F. Vogel and A. G. Motulsky, *Human Genetics.* Springer-Verlag, 1982.)

short in stature, and often have a web of skin extending between the neck and shoulders (Figure 9-17). Although their intelligence is near normal, some of their specific cognitive functions are defective. About 1 in 5000 female births have this monosomic chromosomal complement. Monosomics for all human autosomes die in utero.

Geneticists have used viable plant nullisomics and monosomics to identify the chromosomes that carry the loci of newly found recessive mutant alleles. For example, a geneticist may obtain different monosomic lines, each of which lacks a different chromosome. Homozygotes for the new mutant allele are crossed with each monosomic line, and the progeny of each cross are inspected for expression of the recessive phenotype. The phenotype appears in some of the progeny of the parent that is monosomic for the locus-bearing chromosome and thus identifies it. Figure 9-18 shows that this test works because meiosis in the monosomic parent produces some gametes that lack the chromosome bearing the locus. When one of these gametes forms a zygote, the single chromosome contributed by the other homozygous recessive parent determines the phenotype.

Genetic analysis of humans occasionally reveals a similar unmasking of a recessive phenotype by an $n - 1$ gamete. For example, two people whose vision is normal may produce a daughter who has *Turner syndrome* and who is also red-green colorblind. This coincidence is interpreted as follows. First, since the father was not colorblind, the mother must have been heterozygous for the recessive allele, and must have passed this allele on to the Turner daughter. Nondisjunction must have occurred in the father, resulting in an $n - 1$ sperm, which combined with the colorblind allele–bearing egg from the mother.

Message Monosomics show the deleterious effects of genome inbalance, as well as unexpected expression of recessive alleles carried on the monosomic chromosome.

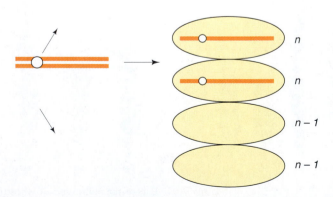

Figure 9-18 Meiosis in which the chromosome of interest is monosomic. Two of the resulting gametes are haploid (n); the other two gametes contain a set lacking a chromosome ($n - 1$).

Trisomics ($2n + 1$)

The trisomic condition is also one of chromosomal imbalance and can result in abnormality or death. However, there are many examples of viable trisomics. You might remember that we studied the trisomics of the Jimson weed *Datura* in Chapter 3 (see Figure 3-8). Furthermore, trisomics can be fertile. When cells from some trisomic organisms are observed under the microscope at the time of meiotic chromosome pairing, the trisomic chromosomes are seen to form a trivalent, an associated group of three, while the other chromosomes form regular bivalents. What genetic ratios might we expect for genes on the trisomic chromosome? Let us consider a gene *A* that is close to the centromere on that chromosome, and let us assume that the genotype is Aaa. Furthermore, if we postulate that two of the centromeres disjoin to opposite poles as in a normal bivalent, and that the other chromosome passes randomly to either pole, then we can predict the three equally frequent segregations shown in Figure 9-19. These segregations result in an overall gametic ratio of $1A:2Aa:2a:1aa$. This ratio and the one corresponding to a trisomic of genotype AAa are observed in practice. If a trisomic tester set is available (much like the nullisomic tester set we discussed earlier), then a new mutation can be located to a chromosome by determining which of the testers gives the special ratio.

There are several examples of viable trisomics in humans. The combination XXY (1 in 1000 male births) results in Klinefelter syndrome, males with lanky builds who are mentally retarded and sterile (Figure 9-20). Another combination, XYY, also occurs in about 1 in 1000 male births. Attempts have been made to link the XYY condition with a predisposition toward violence. This is still hotly debated, although it is now clear that an XYY condition in no way guarantees such behavior. Nevertheless, several enterprising lawyers have attempted to use the XYY genotype as grounds

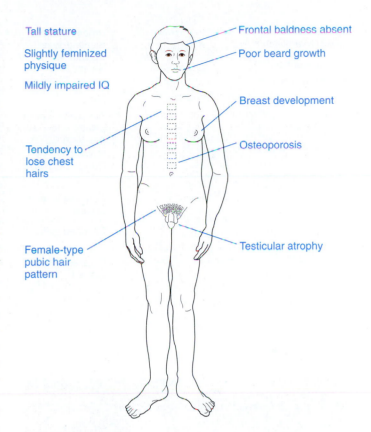

Figure 9-20 Characteristics of Klinefelter syndrome (XXY). (Adapted from F. Vogel and A. G. Motulsky, *Human Genetics*. Springer-Verlag, 1982.)

Labels (clockwise):
- Tall stature
- Slightly feminized physique
- Mildly impaired IQ
- Tendency to lose chest hairs
- Female-type pubic hair pattern
- Frontal baldness absent
- Poor beard growth
- Breast development
- Osteoporosis
- Testicular atrophy

for acquittal or compassion in crimes of violence. The XYY males are usually fertile. Their meioses are of the XY type; the extra Y is not transmitted, and their gametes contain either X or Y, never YY or XY.

The most common type of viable human aneuploid is Down syndrome (Figure 9-21), occurring at a frequency of about 0.15 percent of all live births. We have already encountered the translocation form of Down syndrome in Chapter 8. However, by far the most common type of Down syndrome is trisomy 21 caused by nondisjunction of chromosome 21 in a parent that is chromosomally normal. Like any mechanism, chromosome disjunction is error-prone, and sometimes produces aneuploid gametes. In this type of Down syndrome there is no family history of aneuploidy, unlike the translocation type discussed earlier.

There is a maternal-age effect for Down syndrome; older mothers show greatly elevated risk of having Down syndrome children (Figure 9-22). For this reason, fetal chromosome analysis (by amniocentesis or by chorionic villus sampling) is now recommended for older mothers. Even though the maternal-age effect has been known for many years, the cause of it is still not known. A less pronounced paternal-age effect has also been demonstrated.

The multiple phenotypes that make up Down syndrome include mental retardation, with an IQ in the 20 to

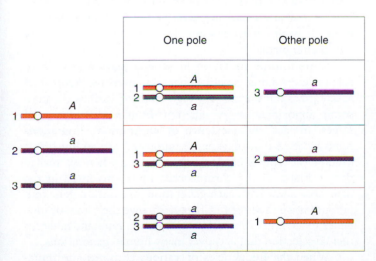

	One pole		Other pole	
	1 _A_		3 _a_	
	2 _a_			
	1 _A_		2 _a_	
	3 _a_			
	2 _a_		1 _A_	
	3 _a_			

Chromosomes:
1 — _A_
2 — _a_
3 — _a_

Figure 9-19 Genotypes of the meiotic products of an Aaa trisomic. Three segregations are equally likely.

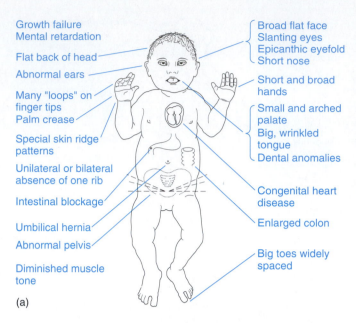

Growth failure
Mental retardation

Flat back of head

Abnormal ears

Many "loops" on
finger tips
Palm crease

Special skin ridge
patterns

Unilateral or bilateral
absence of one rib

Intestinal blockage

Umbilical hernia

Abnormal pelvis

Diminished muscle
tone

Broad flat face
Slanting eyes
Epicanthic eyefold
Short nose

Short and broad
hands

Small and arched
palate
Big, wrinkled
tongue
Dental anomalies

Congenital heart
disease

Enlarged colon

Big toes widely
spaced

(a)

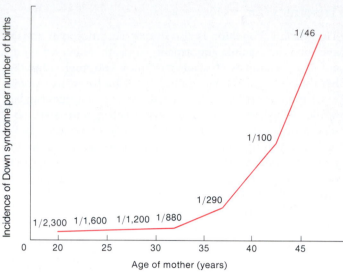

Figure 9-22 Maternal age and the production of Down syndrome offspring. (From L. S. Penrose and G. F. Smith, *Down's Anomaly.* Little, Brown and Company, 1966.)

Figure 9-21 Characteristics of Down syndrome (trisomy 21). (a) Diagrammatic representation of the syndrome in an infant. (b) Athletes with Down syndrome. (Part a adapted from F. Vogel and A. G. Motulsky, *Human Genetics.* Springer-Verlag, 1982; part b from Bob Daemmrich/The Image Works.)

50 range; broad, flat face; eyes with an epicanthic fold; short stature; short hands with a crease across the middle; and a large, wrinkled tongue. Females may be fertile and may produce normal or trisomic progeny, but males have never reproduced. Mean life expectancy is about 17 years, and only 8 percent survive past age 40.

The only other human autosomal trisomics to survive to birth are afflicted with either trisomy 13 (Patau syndrome) or trisomy 18 (Edwards syndrome). Both show severe physical and mental abnormalities. The generalized phenotype of trisomy 13 includes a harelip; a small, malformed head; "rockerbottom" feet; and a mean life expectancy of 130 days. That of trisomy 18 shows "faun-like" ears, a small jaw, a narrow pelvis, and rockerbottom feet; almost all babies with trisomy 18 die within the first few weeks after birth.

Chromosome mutation in general plays a prominent role in determining genetic ill health in humans. Figure 9-23 summarizes the surprisingly high levels of various chromosome abnormalities at different human developmental stages. In fact, the incidence of chromosome mutations ranks close to that of gene mutations in human live births (Table 9-1). This is particularly surprising when we realize that virtually all the chromosome mutations listed in the table arise anew with each generation. In contrast, gene mutations (as we shall see in Chapter 25) owe their level of incidence to a complex interplay of mutation rates and environmental selection that acts over many human generations.

When the frequencies of various chromosome mutations in live births are compared with the corresponding fre-

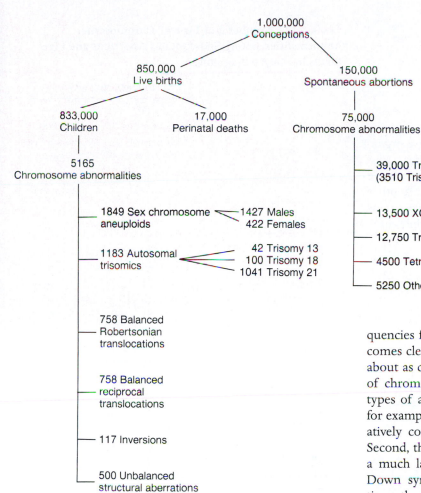

Figure 9-23 The fate of a million implanted human zygotes. (Robertsonian translocations involve fusion or dissociation of centromeres.) (From K. Sankaranarayanan, *Mutation Research* 61, 1979.)

quencies found in spontaneous abortions (Table 9-2), it becomes clear that the chromosome mutations that we know about as clinical abnormalities are just the tip of an iceberg of chromosome mutations. First, we see that many more types of abnormalities are produced than survive to birth; for example, trisomies of chromosome 2, 16, and 22 are relatively common in abortuses, but never survive to birth. Second, the specific aberrations we see surviving are part of a much larger number that do not survive; for example, Down syndrome (trisomy 21) is produced at almost 20 times the frequency we see in live births. This is even more striking in the case of Turner syndrome (XO). It has been estimated that a minimum of 10 percent of conceptions have a major chromosome abnormality; our reproductive success depends on the natural weeding out process that eliminates most of these abnormalities before birth. Incidentally, there is no evidence that would suggest these aberrations are produced by environmental insult to our reproductive systems, or that the frequency of the aberrations is increasing.

Message Trisomics show the deleterious effects of genome inbalance, and produce chromosome-specific modified phenotypic ratios.

Disomics ($n + 1$)

A disomic is an aberration of a haploid organism. In fungi, it is clear that they can result from meiotic nondisjunction. In the fungus *Neurospora* (a haploid), an $n - 1$ meiotic product aborts and does not darken like a normal ascospore; so we may detect MI and MII nondisjunctions by observing asci with 4:4 and 6:2 ratios of normal to aborted spores, respectively, as shown on the next page.

Table 9-1 **Relative Incidence of Human Ill Health due to Gene Mutation and to Chromosome Mutation**

Type of Mutation	Percentage of live births
Gene mutation	
Autosomal dominant	0.90
Autosomal recessive	0.25
X-linked	0.05
Total gene mutation	1.20
Chromosome mutation	
Autosomal trisomies (mainly Down syndrome)	0.14
Other unbalanced autosomal aberrations	0.06
Balanced autosomal aberrations	0.19
Sex chromosomes	
XYY, XX.Y, and other ♂♂	0.17
XO, XXX, and other ♀♀	0.05
Total chromsome mutation	0.61

These diagrams correspond exactly to the outcomes of the chromosomal events shown in Figure 9-16. In these organisms the disomic ($n + 1$) meiotic product becomes a disomic strain directly. The abortion patterns themselves are diagnostic for the presence of disomics in the asci. Another way of detecting disomics in fungi is to cross two strains with homologous chromosomes bearing multiple auxotrophic mutations, for example:

$$ade\ his^+\ nic\ ala^+\ leu\ arg^+$$

$$\times$$

$$ade^+\ his\ nic^+\ ala\ leu^+\ arg$$

From such a cross, large populations of ascospores are plated onto minimal medium. Only ascospores of genotype $+ + + + + +$ can grow and form colonies. Most of these colonies are found to be disomics and not multiple crossover types.

Message Disomics in fungi can be selected from asci showing special spore abortion patterns, or by selecting meiotic progeny that must contain homologous chromosomes from both parents.

Somatic Aneuploids

Aneuploid cells can arise spontaneously in somatic tissue or in cell culture. In such cases, the initial result is a genetic mosaic of cell types.

Human sexual mosaics—individuals whose bodies are a mixture of male and female tissue—are good examples. One type of sexual mosaic, XO/XYY, can be explained by postulating an XY zygote in which the Y chromatids fail to disjoin at an early mitotic division, so that both go to one pole:

Table 9-2 Number and Type of Chromosomal Abnormalities among Spontaneous Abortions and Live Births in 100,000 Pregnancies

	100,000 Pregnancies	
	15,000 spontaneous abortions 7,500 chromosomally abnormal	85,000 live births 550 chromosomally abnormal
Trisomy		
1	0	0
2	159	0
3	53	0
4	95	0
5	0	0
6–12	561	0
13	128	17
14	275	0
15	318	0
16	1229	0
17	10	0
18	223	13
19–20	52	0
21	350	113
22	424	0
Sex chromosomes		
XYY	4	46
XXY	4	44
XO	1350	8
XXX	21	44
Translocations		
Balanced	14	164
Unbalanced	225	52
Polyploid		
Triploid	1275	0
Tetraploid	450	0
Other (mosaics, etc.)	280	49
Total	7500	550

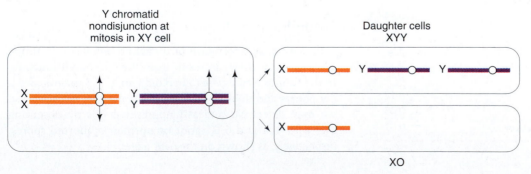

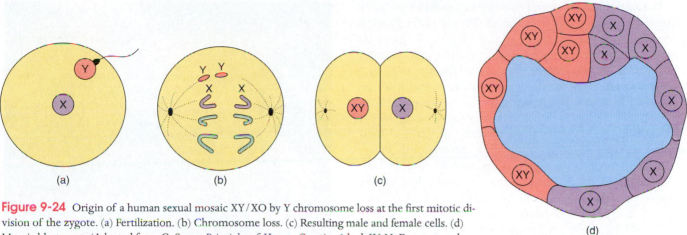

Figure 9-24 Origin of a human sexual mosaic XY/XO by Y chromosome loss at the first mitotic division of the zygote. (a) Fertilization. (b) Chromosome loss. (c) Resulting male and female cells. (d) Mosaic blastocyst. (Adapted from C. Stern, *Principles of Human Genetics,* 3d ed. W. H. Freeman and Company, 1973.)

The phenotypic sex of such individuals depends on where the male and female sectors end up in the body. In the type of nondisjunction being considered, nondisjunction at a later mitotic division would produce a three-way mosaic XY/XO/XYY, which contains a clone of normal male cells. Other sexual mosaics have different explanations; as examples, XO/XY is probably due to early X chromosome loss in a male zygote (Figure 9-24) and XX/XY is probably the result of a double fertilization (fused twins). In general, sexual mosaics are called *gynandromorphs.*

Geneticists working with many species of experimental animals occasionally find gynandromorphs among their stocks. A classic example is the *Drosophila* gynandromorph shown in Figure 9-25. In this case the zygote started out as a female heterozygous for two X-linked genes, white eye and miniature wing ($w^+\,m^+/w\,m$). Loss of the wild-type allele–bearing X chromosome at the first mitotic division resulted in the two cell lines and ultimately in a fly differing from one side to the other in sex, eye color, and size of wing. A similar gynandromorph in the Io moth is shown in Figure 9-26.

> **Message** Mitotic nondisjunction and other types of aberrant mitotic chromosome behavior can give rise to mosaics consisting of two or more chromosomally distinct cell types, including aneuploids.

Figure 9-25 A bilateral gynandromorph of *Drosophila.* The zygote was $w^+\,m^+/w\,m$, but loss of the $w^+\,m^+$ chromosome in the first mitotic division produced a fly that was $\frac{1}{2}$ O/$w\,m$ and male (*left*) and $\frac{1}{2}\,w^+\,m^+/w\,m$ and female (*right*). Mutant allele *w* gives white eye, and *m* gives miniature wing.

Figure 9-26 A bilateral gynandromorph in the Io moth, *Automeris io io.* One half of the body is female and happens to carry the sex chromosome mutation "broken eye"; the other half of the body is male and carries the normal allele of the broken-eye gene. This moth resulted from an event similar to that which produced the *Drosophila* gynandromorph in Figure 9-25. The color and size differences are both sex differences. (T. R. Manley)

Somatic aneuploidy and its resulting mosaics are often observed in association with cancer. People suffering from chronic myeloid leukemia (CML), a cancer of the white blood cells, frequently harbor cells containing the so-called Philadelphia chromosome. This chromosome was once thought to represent an aneuploid condition, but it is now known to be a translocation product in which part of the long arm of chromosome 22 attaches to the long arm of chromosome 9. However, CML patients often show aneuploidy in addition to the Philadelphia chromosome. In one study of 67 people with CML, 33 proved to have an extra Philadelphia chromosome and the remainder had various aneuploidies; the most common aneuploidy was trisomy for the long arm of chromosome 17, which was detected in 28 people. Of 58 people with acute myeloid leukemia, 21 were shown to have aneuploidy for chromosome 8; 16 for chromosome 9; and 10 for chromosome 21. In another study of 15 patients with intestinal tumors, 12 had cells with abnormal chromosomes, at least some with trisomy for chromosome 8, 13, 15, 17, or 21. Of course, such studies merely established correlations, and it is not clear whether the abnormalities are best thought of as a cause or as an effect of cancer.

Message Aneuploids are produced by nondisjunction or some other type of chromosome misdivision at either meiosis or mitosis.

Chromosome Mechanics in Plant Breeding

What we know about chromosome mutation can be used in the type of genetic engineering that produces and maintains

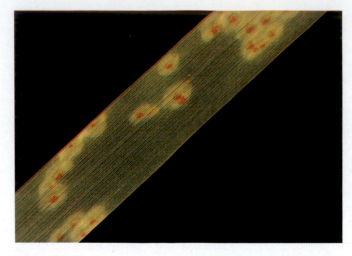

Figure 9-27 A wheat leaf infected by rust fungus, which derives its name from the rust-colored patches of fungal spores produced at the infection centers. Various kinds of rust fungi are pathogens on many important crop species including cereals, pines, and coffee. Rust infection results in billions of dollars of damage yearly, and a large part of plant breeding is directed at producing genetically resistant plant lines. (V. A. Wilmot/Visuals Unlimited)

new crop types in our hungry world. The classic example of this, performed by E. R. Sears in the 1950s, concerns the transfer of a gene for leaf rust resistance from a wild grass, *Aegilops umbellulata*, to bread wheat, which is highly susceptible to this disease, to offset a major problem in the wheat industry (Figure 9-27).

The first problem that Sears encountered was that these two species (Figure 9-28) are not interfertile, so the feat of

(a)

(b)

Figure 9-28 (a) Wheat. (b) *Aegilops umbellulata*. Both whole plants and seed heads are shown. (E. R. Sears)

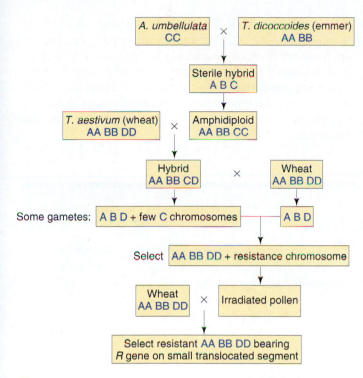

Figure 9-29 Summary of Sears's program for transferring rust resistance from *Aegilops* to wheat. A, B, C, and D represent chromosome sets of diverse origin. *R* represents the genetic determinant (single gene?) of resistance.

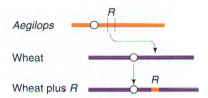

Figure 9-30 Translocation of *Aegilops R* segment to wheat using radiation as a means of breaking the chromosomes.

gene transfer seemed impossible. Sears sidestepped this problem with a bridging cross, in which he crossed *A. umbellulata* to a wild relative of bread wheat called *emmer*, *Triticum dicoccoides*. (Follow the process in Figure 9-29.) *A. umbellulata* is a diploid, $2n = 2x = 14$. We shall call its chromosome sets CC. *T. dicoccoides* is a tetraploid, $2n = 4x = 28$, with sets AA BB. From this cross, the resulting sterile hybrid A B C was doubled into a fertile amphidiploid AA BB CC with 42 chromosomes. This amphidiploid was fertile in crosses with bread wheat (*T. aestivum*), which is represented as $2n = 6x = 42$, AA BB DD. The offspring, AA BB CD, were almost completely sterile

owing to pairing irregularities between the C and D sets. However, backcrosses to wheat did produce a few rare seeds, some of which grew into plants that were resistant to leaf rust, like the wild parent in the first cross. Some of these were almost the desired type, having 43 chromosomes (42 of which were wheat, plus one *Aegilops* chromosome bearing the resistance gene). Thus, in the AA BB CD hybrid, some aberrant form of chromosome assortment had produced a gamete with 22 chromosomes: A B D plus one from the C group.

Unfortunately, the extra chromosome carried just too many undesirable *Aegilops* genes along with the good one, and the plants were weedy and low producers. So the *Aegilops* gene linkage had to be broken. Sears accomplished this by using irradiated pollen from these plants to pollinate wheat. He was looking for translocations of part of the *Aegilops* chromosome tacked onto the wheat chromosomes. These were quite common, but only one turned out to be ideal—a very small, insertional, unidirectional translocation (Figure 9-30). When bred to homozygosity, the resistant plants were indistinguishable from wheat.

The study of chromosome mutation is obviously of great importance—not only in pure biology, where there are ramifications in many areas (especially in evolution), but also in applied biology. The application of such studies is especially important in agriculture and medicine. Rather curiously, very little is known about the mechanisms of chromosome mutation, especially at the molecular level. More details are needed about the normal chemical architecture of chromosomes. We return to this subject in Chapter 16.

SUMMARY

When chromosome mutation changes the number of whole chromosome sets present, aberrant euploids result; when chromosome mutation changes parts of sets, aneuploids result.

Polyploids such as triploids ($3x$) and tetraploids ($4x$) are common in the plant kingdom and are even represented in the animal kingdom. An odd number of chromosome sets makes an organism sterile because there is not a partner for

each chromosome at meiosis, whereas even numbers of sets can produce standard segregation ratios. Allopolyploids (polyploids formed by combining sets from different species) can be made by crossing two related species and then doubling the progeny chromosomes through the use of colchicine, or they can be made through somatic cell fusion. These techniques have important applications in crop breeding, since allopolyploids combine features of both parental species. Polyploidy can result in an organism of larger dimensions; this discovery has permitted important advances to be made in horticulture and in crop breeding.

Aneuploids have also been important in the engineering of specific crop genotypes, although aneuploidy per se usually results in an unbalanced genome with an abnormal phenotype. Examples of aneuploids include monosomics $(2n - 1)$, trisomics $(2n + 1)$, and disomics $(n + 1)$. Aneuploid conditions are well studied in humans. Down syndrome (trisomy 21), Klinefelter syndrome (XXY), and Turner syndrome (XO) are well-documented examples. The spontaneous level of aneuploidy in humans is quite high and produces a major portion of genetically based ill health in human populations. Most instances of aneuploidy result from chromosome nondisjunction at meiosis or chromatid nondisjunction at mitosis. Mitotic nondisjunction can lead to somatic mosaics.

Concept Map

Draw a concept map interrelating as many of the following terms as possible. Note that the terms are listed in no particular order.

euploid / imbalance / nondisjunction / meiosis / gametes / sterility / aneuploid / phenotypic ratios / abortion / mitosis / genes

CHAPTER INTEGRATION PROBLEM

We have lines of mice that breed true for two alternative behavioral phenotypes that we know are determined by two alleles at a single locus: v causes a mouse to move with a "waltzing" gait whereas V determines a normal gait. After crossing the true-breeding waltzers and normals, we observe that most of the F_1 is normal, but unexpectedly, there is one waltzing female. We mate the F_1 waltzer with two different waltzer males, and note that she produces only waltzer progeny. When we mate her with normal males, she produces normal progeny and no waltzers. We mate three of her normal female progeny with two of their brothers and these mice produce 60 progeny, all normal. When, however, we mate one of these same three females with a third brother, we get six normals and two waltzers in a litter of eight. By thinking about the parents of the F_1 waltzer, we can consider some possible explanations of these results:

a. A dominant allele may have mutated to a recessive allele in her normal parent.

b. There may have been a mutation to a dominant suppressor of V in one parent.

c. Meiotic nondisjunction of the chromosome carrying V in her normal parent may have given a viable aneuploid.

d. There may have been a viable deletion spanning V in the meiocyte from her normal parent.

Which of these explanations are possible, and which are eliminated by the genetic analysis? Explain in detail.

Solution

This question will force us to use concepts from Chapter 4 (suppression), Chapter 7 (mutation), Chapter 8 (deletion), and the present chapter (nondisjunction). The best way to answer it is to take the explanations one at a time and see if each fits the results given.

a. Mutation V to v

This hypothesis requires that the exceptional waltzer female be homozygous vv. This is compatible with the results of mating her both to waltzer males, which would—if she is vv—produce all waltzer offspring (vv), and to normal males, which would produce all normal offspring of genotype Vv. However, brother-sister matings within this normal progeny should then produce a 3:1 normal to waltzer ratio. As some of the brother-sister matings actually produced no waltzers, this hypothesis does not explain the data.

b. Suppressor mutation s to S

Here the parents would be $VVss$ and $vvss$, and a germinal mutation in one of them would give the F_1 waltzer the genotype $VvSs$. When we crossed her to a waltzer male, who would be of the genotype $vvss$, we would expect some $Vvss$, that would, of course, be phenotypically normal. However, we saw no normal progeny from this cross, so the hypothesis is already overthrown. Linkage could save the hypothesis temporarily if we assumed that the mutation was in the normal parent, giving a gamete VS. Then the F_1 waltzer would be VS/vs, and if linkage were tight enough, few or no Vs gametes would be produced, the type that are necessary to combine with the *v s* gamete from the male to give *Vv ss* normals. However, if this were true, the cross to the normal males would be *V S/v s* × *V s/V s* and this would give a high percentage of *V S/V s* progeny, which would be waltzers, and of course none were seen.

c. Nondisjunction in the normal parent

This would give a nullisomic gamete that would combine with *v* to give the F₁ waltzer the hemizygous genotype *v*. The subsequent matings would be as follows:

- *v* × *vv* gives *vv* and *v* progeny, all waltzers. This fits.

- *v* × *VV* gives *Vv* and *V* progeny, all normals. This also fits.

- First intercrosses of normal progeny: *V* × *V*. This gives *V* and *VV*, which are all normal. This fits.

- Second intercrosses of normal progeny: *V* × *Vv*. This gives 25 percent each of *VV*, *Vv*, *V* (all normals), and *v* (waltzers). This also fits.

This hypothesis is therefore consistent with the data.

d. Deletion of *V* in normal parent

Let's call the deletion D. The F₁ waltzer would be D *v*, and the subsequent matings would be as follows:

- D *v* × *vv*. This gives *vv* and D *v*, which are all waltzers. This fits.

- D *v* × *VV*. This gives *Vv* and D *V*, which are all normal. This fits.

- First intercrosses of normal progeny: D *V* × D *V*. This gives D *V* and *VV*, which are all normal. This fits.

- Second intercrosses of normal progeny: D *V* × *Vv*. This gives 25 percent of each of *VV*, *Vv*, D *V* (all normals), and D *v* (waltzers). This also fits.

Once again, the hypothesis fits the data provided, so we are left with two hypotheses that are compatible with the results, and further experiments would be necessary to distinguish them. One obvious way would be to examine the chromosomes of the exceptional female under the microscope; aneuploidy should be relatively easy to distinguish from deletion.

SOLVED PROBLEMS

1. There is a controversy about the type of chromosome pairing seen in autotetraploids formed between two geographical races of a certain plant. It is known that chromosomes associate by pairs, but three hypotheses are put forward concerning how pairs form:

a. Pair formation is random.

b. Pairs form only between chromosomes of the same race.

c. Pairs form only between chromosomes of different races.

Consider a locus, *A*, that is closely linked to the centromere. The following cross is made:

Race 1: *AAAA* × Race 2: *aaaa*
↓
Autotetraploid: *AAaa*

The autotetraploid is then selfed. What ratio of phenotypes can be expected under each hypothesis of chromosome pairing? Explain your answer.

Solution

a. Under random pairing, all possible pairing combinations are equally likely. If we label the four homologous chromosomes 1, 2, 3, and 4, then the equally probable combinations are 1-2/3-4, 1-3/2-4, and 1-4/2-3. This kind of situation in an *AAaa* tetraploid was examined in the chapter (see page 253). We saw that *aa* gametes are produced at a frequency of $\frac{1}{6}$ and that the frequency of *aaaa* progeny from selfing must therefore be $\frac{1}{36}$. All other progeny types contain at least one *A* allele. Hence, the expected ratio is 35 *A———* : 1 *aaaa*.

b. In this alternative, the chromosome pairs look like

because both the chromosomes carrying *A* come from race 1 and both the chromosomes carrying *a* come from race 2. This diagram shows how being methodical can help in problem solving, because it makes it very clear that the only possible segregation is that of one *A* and *a* to each pole. Hence, all gametes are *Aa* and, on self-fertilization, all genotypes would be *AAaa*; only the *A* phenotype would be seen.

c. Here the pairing looks like this:

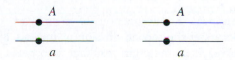

Because segregation of the pairs is independent, the gametic population can be represented as follows:

Pair 1　Pair 2　Gametes

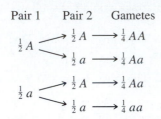

Self-fertilization would produce $aaaa$ progeny at a frequency of $\frac{1}{4} \times \frac{1}{4} = \frac{1}{16}$. Therefore, $\frac{15}{16}$ of the progeny are $A{-}{-}{-}$ and a 15:1 phenotypic ratio is predicted.

In conclusion, the ratio of phenotype A to phenotype a in each case is

a. 35:1

b. 1:0

c. 15:1

2. Owing to the small size of the *Drosophila* chromosome 4, monosomics and trisomics for this chromosome are viable but tetrasomics and nullisomics are not. A fly that is trisomic for chromosome 4 and carries the recessive gene for bent bristles (b) on all copies of chromosome 4 is crossed to a phenotypically normal fly that is monosomic for chromosome 4.

a. What genotypes and phenotypes can be expected in the progeny and in what proportions?

b. If trisomics from these progeny are interbred, what phenotypic ratio can be expected in the next generation? Assume that only one copy of b^+ is needed to produce normal (nonbent) bristles, that unpaired chromosomes pass to either pole at random, and that aneuploid gametes of any kind survive.

Solution

a. The cross is

$$4 \underline{\quad\overset{b}{+}\quad}$$
$$4 \underline{\quad\overset{b}{+}\quad} \times 4 \underline{\quad\overset{b^+}{+}\quad}$$
$$4 \underline{\quad\overset{b}{+}\quad}$$

From the bent-bristled parent, the gametes will be $\frac{1}{2}bb$ and $\frac{1}{2}b$. From the monosomic parent, the gametes will be $\frac{1}{2}b^+$ and $\frac{1}{2}0$ (the latter contain no chromosome 4). Hence, the progeny will be $\frac{1}{4}b^+bb$, $\frac{1}{4}bb$, $\frac{1}{4}b^+b$ and $\frac{1}{4}b$, which provides a phenotypic ratio of 1:1.

b. The trisomics referred to are of genotype b^+bb. If we label these chromosomes 1, 2, and 3, then

$$1 = b$$
$$2 = b$$
$$3 = b^+$$

and segregation produces

$$1, 2\ (bb)/3\ (b^+)$$
$$1, 3\ (bb^+)/2\ (b)$$
$$2, 3\ (bb^+)/1\ (b)$$

Fertilization can be represented as follows:

	$bb(\frac{1}{6})$	$b^+b(\frac{2}{6})$	$b(\frac{2}{6})$	$b^+(\frac{1}{6})$
$bb(\frac{1}{6})$	×	×	bbb	$\frac{1}{36}$
$b^+b(\frac{2}{6})$	×	×	$\frac{2}{36}$	$\frac{2}{36}$
$b(\frac{2}{6})$	bbb $\frac{2}{36}$	$\frac{4}{36}$	bb $\frac{4}{36}$	$\frac{2}{36}$
$b^+(\frac{1}{6})$	$\frac{1}{36}$	$\frac{2}{36}$	$\frac{2}{36}$	$\frac{1}{36}$

The X's represent tetrasomics which, we are told, are not viable. Out of the remaining $\frac{27}{36}$, $\frac{8}{36}$ are of b phenotype; therefore, a ratio of 19:8 is predicted.

PROBLEMS

1. Distinguish between Klinefelter, Down, and Turner syndromes in humans.

2. List two ways in which you could make an allotetraploid between two related diploid plant species $2n = 28$.

3. **a.** Suppose nondisjunction of *Neurospora* chromosome 3 occurs at the second division of meiosis. Show the content of each of the eight resultant ascospores with regard to chromosome 3.

b. *Neurospora* normally has seven chromosomes. How many chromosomes are present in each of the ascospores in part a?

4. **a.** How would you synthesize a pentaploid ($5x$)?

b. How would you synthesize a triploid ($3x$) of genotype Aaa?

c. You have just obtained a rare recessive mutation a^* in a diploid plant, which Mendelian analysis tells you is Aa^*.

From this plant, how would you synthesize a tetraploid ($4x$) of genotype $AAa*a*$?

d. How would you synthesize a tetraploid of genotype $Aaaa$?

e. How would you synthesize a plant that is resistant to a chemical herbicide? (Assume that mutation to this trait is very infrequent, and you are lucky to get one mutant.)

5. Which of the following correctly completes the sentence? Allopolyploids are (**a**) not fertile at all; (**b**) fertile only among themselves; (**c**) fertile with one parent only; (**d**) fertile with both parents only; (**e**) fertile with both parents and themselves.

6. Tetraploid yeast can be created by fusing two diploid cells. These tetraploids undergo meiosis like any other tetraploid and produce four diploid products of meiosis. Assume that homologous chromosomes synapse randomly in pairs and that there is no crossing-over in the region between the *B* locus and the centromere. What nonlinear tetrads are produced by a tetraploid of genotype $BBbb$? What are the frequencies of the ascus types? (NOTE: This question involves tetrad analysis of tetraploid cells instead of the usual diploid cells.)

7. Consider a tetraploid of genotype $AAaa$ in which there is no pairing between chromosomes from the same parent. In another tetraploid of genotype $BBbb$, pairs form only between chromosomes from the same parent. What phenotypic ratios result from selfing in each of these cases? (Assume that one parent carries the dominant allele and that the other carries the recessive allele in each case.)

8. In the plant genus *Triticum* there are many different polyploid species and also of course diploid species. Crosses were made between some different species, and hybrids were obtained. The meiotic pairing was observed in each hybrid; this is recorded in the following table. (A bivalent is two homologous chromosomes paired at meiosis, and a univalent is an unpaired chromosome at meiosis.)

Species crossed to make hybrid	Pairing in hybrid
T. turgidum × *T. monococcum*	7 bivalents + 7 univalents
T. aestivum × *T. monococcum*	7 bivalents + 14 univalents
T. aestivum × *T. turgidum*	14 bivalents + 7 univalents

Explain these results, and in doing so

a. deduce the somatic chromosome number of each species used.

b. state which species are polyploid, and whether they are auto- or allopolyploids.

c. account for the chromosome pairing pattern in the three hybrids.

9. The New World cotton species *Gossypium hirsutum* has a $2n$ chromosome number of 52. The Old World species

G. thurberi and *G. herbaceum* each have a $2n$ number of 26. Hybrids between these species show the following chromosome pairing arrangements at meiosis:

Hybrid	Pairing arrangement
G. hirsutum × *G. thurberi*	13 small bivalents + 13 large univalents
G. hirsutum × *G. herbaceum*	13 large bivalents + 13 small univalents
G. thurberi × *G. herbaceum*	13 large univalents + 13 small univalents

Draw diagrams to interpret these observations phylogenetically, clearly indicating the relationships between the species. How would you go about proving that your interpretation is correct?

(Problem 9 adapted from A. M. Srb, R. D. Owen, and R. S. Edgar, *General Genetics*, 2d ed. W. H. Freeman and Company, 1965.)

10. The genotype of an autotetraploid that is heterozygous at two loci is $FFffGGgg$. Each locus affects a different character, and the two loci are located very close to the centromere on different (nonhomologous) chromosomes.

a. What gametic genotypes are produced by this individual, and in what proportions?

b. If the individual is self-fertilized, what proportion of the progeny will have the genotype $FFFfGGgg$? The genotype $ffffgggg$?

11. Which of the following are *not* caused by meiotic nondisjunction? (**a**) Turner syndrome; (**b**) cri du chat syndrome; (**c**) Down syndrome; (**d**) Klinefelter syndrome; (**e**) XYY syndrome; (**f**) achondroplastic dwarfism; (**g**) Edwards syndrome; (**h**) Marfan syndrome.

12. A woman with Turner syndrome is found to be colorblind. Both her mother and father have normal vision. How can this be explained? Does this outcome tell us whether nondisjunction occurred in the father or in the mother? If the colorblindness gene were close to the centromere (it is not, in fact), would the available facts tell us whether the nondisjunction occurred at the first or at the second meiotic division? Repeat the question for a colorblind man with Klinefelter syndrome.

13. Individuals have been found who are colorblind in one eye but not in the other. What would this suggest if these individuals were (**a**) only or mostly females? (**b**) Only or mostly males? (Assume that this is an X-linked recessive trait.)

14. When geneticists treat human sperm with quinacrine dihydrochloride, about half the sperm show a fluorescent spot thought to be the Y chromosome. About 1.2 percent of normal human sperm show two fluorescent spots. Geneticists examined the sperm of some industrial workmen who were exposed over a period of about one year to the chemical dibromochloropropane. They found the frequency of sperm with two spots to be on average 3.8 percent. Propose an explanation for this result, and explain how you would test it.

15. An individual with Turner syndrome is also found to have the phenotype nystagmus (an eye disorder).

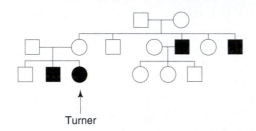

Turner

a. What event or events can explain the coincidence of the two hereditary conditions?

b. In which parent and at which stage did the event or events occur?

16. In British Columbia, the mean age of women giving birth to Down syndrome babies fell from 34 years to 28 years between 1952 and 1972. What are two possible causes of this population trend, and how could the two hypotheses be tested?

17. Several kinds of sexual mosaicism are well documented in humans. Suggest how each of the following examples may have arisen:

a. XX/XO (that is, there are two cell types in the body, XX and XO)

b. XX/XXYY   **c.** XO/XXX

d. XX/XY   **e.** XO/XX/XXX

18. The discovery of chromosome banding in eukaryotes has greatly improved the ability to distinguish various cytogenetic events. Particularly useful are banding polymorphisms, because the "morphs" can be used as chromosome markers. These morphs make chromosomes cytologically distinguishable by virtue of minor variation in band size, position, and so forth. Consider chromosome 21 in humans. Assume that one set of parents is 21^a21^b ♀ × 21^c21^d ♂, where a, b, c, and d represent morphs of a polymorphism for this chromosome. Also assume that fetuses of the following types are produced (where 42A stands for the rest of the autosomes):

1. $42A + 21^b21^b21^c + XY$
2. $42A + 21^a21^b21^d + XX$
3. $42A + 21^b21^d + XY$
4. $42A + 21^a21^c21^c + XX$

5. $42A + 21^a21^b + XY$
6. $42A + 21^a21^c + XYY$
7. $(42A + 21^a + XY)(42A + 21^a21^c21^c + XY)$ (mosaic)
8. $(42A + 21^a21^c + XY)(42A + 21^b21^d + XX)$ (mosaic)

In each case:

a. State the genetic term for the condition.

b. Diagram the events that give rise to the condition.

c. State the individual in which the events took place.

19. Suppose you have a line of mice that has cytologically distinct forms of chromosome 4. The tip of the chromosome can have a knob (called 4^K), or a satellite (4^S), or neither (4). Here are sketches of the three types:

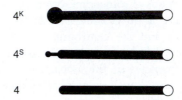

You cross a $4^K 4^S$ female with a 4 4 male, and find most of the progeny are $4^K 4$ or $4^S 4$, as expected. However, you occasionally find some rare types as follows (all other chromosomes are normal):

a. $4^K 4^K 4$ **b.** $4^K 4^S 4$ **c.** 4^K

d. $4/4^K 4^K 4$ mosaic   **e.** $4^K 4/4^S 4$ mosaic

Explain the rare types you have found. Give as precisely as possible the stages at which they originate, and state if they occur in the male parent, the female parent, or the zygote. (Give reasons briefly.)

20. In *Drosophila* a cross (cross 1) is made between two mutant flies, one homozygous for the recessive mutation bent wing (*b*) and the other homozygous for the recessive mutation eyeless (*e*). The mutations *e* and *b* are alleles of two different genes that are known to be very closely linked on the tiny autosomal chromosome 4. All the progeny were wild-type phenotype. One of the female progeny was crossed to a male of genotype *bb ee;* call this *cross 2.* The progeny of cross 2 were mostly of the expected types, but there was also one rare female of wild-type phenotype.

a. Explain what the common progeny are expected to be from cross 2.

b. Could the rare wild-type female have arisen by

(1) crossing-over?

(2) nondisjunction?

Explain.

c. The rare wild-type female was testcrossed to a male of genotype *bb ee* (cross 3). The progeny were

$\frac{1}{6}$ wild-type

$\frac{1}{6}$ bent, eyeless

$\frac{1}{3}$ bent

$\frac{1}{3}$ eyeless

Which of the explanations in part b are compatible with this result? Explain the genotypes and phenotypes of progeny of cross 3 and their proportions.

 Unpacking the Problem

a. Define *homozygous, mutation, allele, closely linked, recessive, wild type, crossing-over, nondisjunction, testcross, phenotype,* and *genotype.*

b. Does this problem involve sex linkage? Explain

c. How many chromosomes does *Drosophila* have?

d. Draw out a clear pedigree summarizing the results of crosses 1, 2, and 3.

e. Draw the gametes produced by both parents in cross 1.

f. Draw the chromosome 4 constitution of the progeny of cross 1.

g. Is it surprising that the progeny of cross 1 are wild-type phenotype? What does this tell you?

h. Draw the chromosome 4 constitution of the male tester used in cross 2, and the gametes he can produce.

i. With respect to chromosome 4, what gametes can the female parent in cross 2 produce in the absence of nondisjunction? Which would be common and which rare?

j. Draw first- and second-division meiotic nondisjunction in the female parent of cross 2, and the resulting gametes.

k. Are any of the gametes from question j aneuploid?

l. Would you expect the aneuploid gametes to give rise to viable progeny? Would these progeny be nullisomic, monosomic, disomic, or trisomic?

m. What progeny phenotypes would be produced by the various gametes considered under i and j?

n. Consider the phenotypic ratio in the progeny of cross 3. Many genetic ratios are based on halves and quarters, but this is based on thirds and sixths. What might this point to?

o. Could there be any significance to the fact that the crosses concern genes on a very small chromosome? When is chromosome size relevant in genetics?

p. Draw out the progeny expected from cross 3 under the two hypotheses, and give some idea of relative proportions.

21. A cross is made in tomatoes between a female plant that is trisomic for chromosome 6 and a normal diploid male plant that is homozygous for the recessive allele for potato leaf (*p p*).

a. A trisomic F_1 plant is backcrossed to the potato-leaved male. What is the ratio of normal-leaved plants to potato-leaved plants when you assume that *p* is on chromosome 6?

b. What is the ratio of normal- to potato-leaved plants when you assume that *p* is not located on chromosome 6?

22. A tomato geneticist attempts to assign five recessive genes to specific chromosomes using trisomics. She crosses each homozygous mutant (*2n*) to each of three trisomics, involving chromosomes 1, 7, and 10. From these crosses, the geneticist selects trisomic progeny (which are less vigorous) and backcrosses them to the appropriate homozygous recessive. The *diploid* progeny from these crosses are examined. Her results follow, where the ratios are wild type : mutant:

Trisomic chromosome	Gene				
	d	*y*	*c*	*h*	*cot*
1	48:55	72:29	56:50	53:54	32:28
7	52:56	52:48	52:51	58:56	81:40
10	45:42	36:33	28:32	96:50	20:17

Which of the genes can the geneticist assign to which chromosomes? (Explain your answer fully.)

23. *Petunia* plants have four loci, *A, B, C,* and *D,* that are known to be very closely linked. A plant of genotype *a B c D / A b C d* is irradiated with gamma rays and then crossed to an *a b c d / a b c d* plant. In the progeny, plants of phenotype A B C D are occasionally found. What are two possible modes of origin, and which is the more likely?

24. A cross in *Neurospora* between strains carrying the multiply marked chromosomes $a\ b^+\ c\ d^+\ e$ and $a^+\ b\ c^+\ d\ e^+$ produces one product of meiosis that grows on minimal medium. (Assume that *a, b, c, d,* and *e* determine nutritional requirements.) When the rare colony grows up, some asexual spores are $a\ b^+\ c\ d^+\ e$ in genotype, some are $a^+\ b\ c^+\ d\ e^+$, and the remainder grow on minimal medium. Explain the origin of the rare product of meiosis and the origin of the three types of asexual spores.

25. Two *Neurospora* auxotrophs are crossed: *pan1* × *leu2*. These loci are linked on the same chromosome arm, with *leu2* between the centromere and *pan1*. Most octads from this cross were of the type expected, but the following unexpected types were also obtained:

a. fully prototrophic
fully prototrophic
fully prototrophic
fully prototrophic
abort
abort
abort
abort

b. pan-requiring
pan-requiring
abort
abort
leu-requiring
leu-requiring
leu-requiring
leu-requiring

c. fully prototrophic
fully prototrophic
abort
abort
leu-requiring
leu-requiring
pan-requiring
pan-requiring

Give explanations for these three types.

∗ 26. The ascomycete *Sordaria brevicollis* has two closely linked complementing markers, b_1 and b_2, that result in buff-colored (light-brown) ascospores. In the cross $b_1 × b_2$, if one or both centromeres divide and separate precociously at the first division of meiosis, what patterns of spore colors are produced in those asci? How can these asci be distinguished from normal asci and from asci in which nondisjunction has occurred? (NOTE: Ascospores in this fungus are normally black, and nullisomic ascospores are white; assume that there is no crossing-over between b_1 and b_2.)

27. Design a test system for detecting agents in the human environment that are potentially capable of causing aneuploidy in eukaryotes.

28. Two alleles determine flower color in the plant *Datura*: *P* determines purple; and *p*, white. The *P* locus is on the smallest chromosome, number 3. In plants trisomic for this chromosome, $n + 1$ pollen is never functional, and only half the $n + 1$ eggs are functional. Assume that *P* is always completely dominant to any number of *p*'s, and that the locus is close to the centromere. What is the expected ratio of purple:white in the progeny of the following crosses?

a. *PPp* ♀ × *PPp* ♂

b. *Ppp* ♀ × *Ppp* ♂

c. *Ppp* ♀ × *pp* ♂

29. There are six main species in the *Brassica* genus: *B. carinata, B. campestris, B. nigra, B. oleracea, B. juncea, B. napus*. You can deduce the interrelationships between these six species from the following table:

Species or F$_1$ hybrid	Chromosome number	Number of bivalents	Number of univalents
B. juncea	36	18	0
B. carinata	34	17	0
B. napus	38	19	0
B. juncea × *B. nigra*	26	8	10
B. napus × *B. campestris*	29	10	9
B. carinata × *B. oleracea*	26	9	8
B. juncea × *B. oleracea*	27	0	27
B. carinata × *B. campestris*	27	0	27
B. napus × *B. nigra*	27	0	27

a. Deduce the chromosome number of *B. campestris*, *B. nigra*, and *B. oleracea*.

b. Show clearly any evolutionary relationships between the six species you can deduce at the chromosomal level.

30. In the fungus *Ascobolus* (similar to *Neurospora*), ascospores are normally black. The mutation *f*, producing fawn ascospores, is in a gene just to the right of the centromere on chromosome 6, whereas mutation *b*, producing beige ascospores, is in a gene just to the left of the same centromere. In a cross of fawn by beige parents ($+ f × b +$) most octads show four fawn and four beige ascospores, but three rare exceptional octads were found as shown below. In the sketch, black is the wild-type phenotype, a vertical line is fawn, a horizontal line is beige, and an empty circle represents an aborted (dead) ascospore.

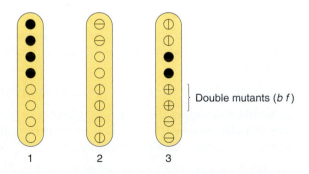

a. Provide reasonable explanations for these three exceptional octads.

b. Diagram the meiosis that gave rise to octad 2.

CHAPTER

10

Recombination in Bacteria and Their Viruses

Bacterial colonies on staining medium. Lac⁻ cells are white. Lac⁺ colonies are stained red. (Jeffrey Miller)

KEY CONCEPTS

▶ The fertility factor (F) permits bacterial cells to transfer DNA to other cells through the process of conjugation.

▶ F can exist in the cytoplasm or can be integrated into the bacterial chromosome.

▶ When F is integrated in the chromosome, chromosomal markers can be transferred during conjugation.

▶ Bacteriophages can also transfer DNA from one bacterial cell to another.

▶ During generalized transduction, random chromosome fragments are incorporated into the heads of certain bacterial phages and transferred to other cells by infection.

▶ During specialized transduction, specific genes near the phage integration sites on the bacterial chromosome are mistakenly incorporated into the phage genome and transferred to other cells by infection.

▶ The different methods of gene transfer in bacteria allow geneticists to make detailed maps of bacterial genes.

Thus far, we have dealt almost exclusively with genes that are packed into chromosomes and enclosed within the nuclei of eukaryotic organisms. However, a very large part of the history of genetics and current genetic analysis (particularly molecular genetics) is concerned with prokaryotic organisms, which have no distinct nuclei, and with viruses. Although **viruses** share some of the definitive properties of organisms, many biologists regard viruses as distinct entities that in some sense are not fully alive. They are not cells; they cannot grow or multiply alone. To reproduce, they must parasitize living cells and use their metabolic machinery. Nevertheless, viruses do have hereditary properties that can be subjected to genetic analysis. Genetic analysis of bacteria and their viruses has yielded key insights into the nature and structure of the genetic material, the genetic code, and mutation.

The **prokaryotes** are the blue-green algae, now classified as "cyanobacteria," and the bacteria. The best studied **viruses**—those that parasitize bacteria—are called **bacteriophages** or, simply, **phages.** Pioneering work with bacteriophages has led to a great deal of recent research on tumor-causing viruses and other kinds of animal and plant viruses.

Compared with eukaryotes, prokaryotic organisms and viruses have simple chromosomes that are not contained within a nuclear membrane. Because they are monoploid, these chromosomes do not undergo meiosis, but they do go through stages analogous to meiosis. The approach to the genetic analysis of recombination in these organisms is surprisingly similar to that for eukaryotes.

The opportunity for genetic recombination in bacteria can arise in several different ways, as this chapter will detail. In the first process we'll examine, **conjugation,** one bacterial cell transfers DNA segments to another cell via direct cell-to-cell contact. A bacterial cell can also pick up a piece of DNA from the environment and incorporate this DNA into its own chromosome; this procedure is called **transformation.** In addition, certain bacterial viruses can pick up a piece of DNA from one bacterial cell and inject it into another, where it can be incorporated into the chromosome, in a process known as **transduction.**

Working with Microorganisms

Bacteria can be grown in a liquid medium or on a solid surface, such as an agar gel, as long as basic nutritive ingredients are supplied. In a liquid medium, the bacteria divide by binary fission: they multiply geometrically until the nutrients are exhausted or until toxic factors (waste products) accumulate to levels that halt the population growth. A small amount of such a liquid culture can be pipetted onto a petri plate containing an agar medium and spread evenly on the surface with a sterile spreader, in a process called **plating** (Figure 10-1). Each cell then reproduces by fission. Because the cells are immobilized in the gel, all the daughter cells re-

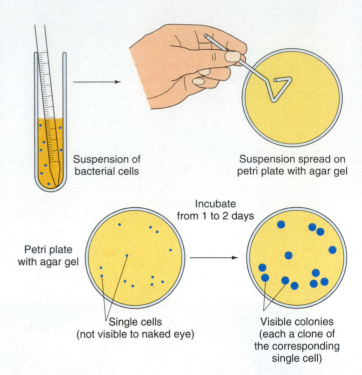

Figure 10-1 Methods of growing bacteria in the laboratory. A few bacterial cells that have been grown in a liquid medium containing nutrients can be spread on an agar medium also containing the appropriate nutrients. Each of these original cells will divide many times by binary fission and eventually give rise to a colony. All cells in a colony, being derived from a single cell, will have the same genotype and phenotype.

main together in a clump. When this mass reaches more than 10^7 cells, it becomes visible to the naked eye as a **colony.** If the initially plated sample contains very few cells, then each distinct colony on the plate will be derived from a single original cell. Members of a colony that share a single genetic ancestor are known as **clones.**

As we discussed in Chapter 7, wild-type bacteria are **prototrophic:** they can grow colonies on **minimal medium**—a substrate containing only inorganic salts, a carbon source for energy, and water. Mutant clones can be identified because they are **auxotrophic:** they will not grow unless the medium contains one or more specific nutrients—say, adenine, or threonine and biotin. Furthermore, wild types are susceptible to certain inhibitors, such as streptomycin, whereas **resistant mutants** can form colonies despite the presence of the inhibitor. These properties allow the geneticist to distinguish different phenotypes among plated colonies.

For many characters, the phenotype of a clone can be determined readily through visual inspection or simple chemical tests. This phenotype can then be assigned to the original cell of the clone, and the frequencies of various phenotypes in the pipetted sample can be determined. Table 10-1 lists some bacterial phenotypes and their genetic symbols. Also refer to Chapter 7 for a further review of microbial genetic techniques.

Table 10-1   Some Genotypic Symbols Used in Bacterial Genetics

Symbol	Character or phenotype associated with symbol
bio⁻	Requires biotin added as a supplement to minimal medium
arg⁻	Requires arginine added as a supplement to minimal medium
met⁻	Requires methionine added as a supplement to minimal medium
lac⁻	Cannot utilize lactose as a carbon source
gal⁻	Cannot utilize galactose as a carbon source
*str*ʳ	Resistant to the antibiotic streptomycin
*str*ˢ	Sensitive to the antibiotic streptomycin

NOTE: Minimal medium is the basic synthetic medium for bacterial growth, without nutrient supplements.

Bacterial Conjugation

This and the following sections describe the discovery of gene transfer in bacteria and explain different types of gene transfer and their use in bacterial genetics. First, we'll consider **conjugation,** a process by which certain bacterial cells can transfer DNA to a second cell with which they make contact. Before we examine the discovery of conjugation, let's summarize what we now know.

The Remarkable Properties of F

The ability to transfer DNA by conjugation is dependent on the presence of a cytoplasmic entity termed the **fertility factor,** or F. Cells carrying F are termed **F⁺**; cells without F are **F⁻**. F is a small, circular DNA element that acts like a minichromosome. It is an example of a class of elements termed plasmids, which are self-replicating extrachromosomal DNA molecules. F contains approximately 100 genes; these give F several important properties:

1. F can replicate its DNA, which allows F to be maintained in a cellular population that is dividing (Figure 10-2a).
2. Cells carrying F produce **pili** (singular, pilus)—minute proteinaceous tubules that allow the F⁺ cells to attach to other cells and maintain contact with them (Figure 10-2b).
3. F⁺ cells can transfer the newly synthesized copy of the circular F genome to a recipient (F⁻) cell that lacks such a genome (Figure 10-2c); note that a copy of F always remains behind in the donating cell. When a donor cell transfers a copy of its cytoplasmic F to an F⁻ cell, the recipient cell also becomes an F⁺ cell, because it now contains a circular F genome.
4. F⁺ cells are usually inhibited from making contact with other F⁺ cells and do not usually transfer the F genome to F⁺ cells.
5. Occasionally, F leaves the cytoplasm and integrates itself into the host bacterial chromosome, as diagrammed in Figure 10-3a. When this occurs, F can also transfer host chromosomal markers to the recipient cell along with its own DNA (Figure 10-3b).

This last process, the linked transfer of chromosomal markers, has some interesting consequences. First, in any population of cells containing the F factor, F will integrate into the chromosomes of a small fraction of cells (Figure 10-3c). These few cells can now transfer chromosomal

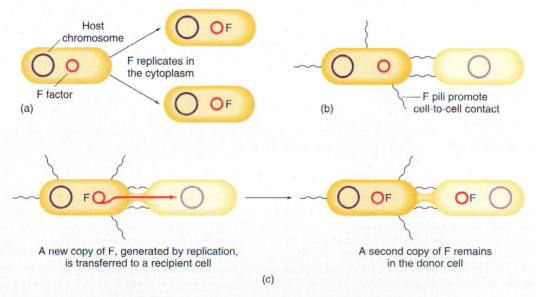

Figure 10-2   Some properties of the fertility (F) factor of *E. coli.*

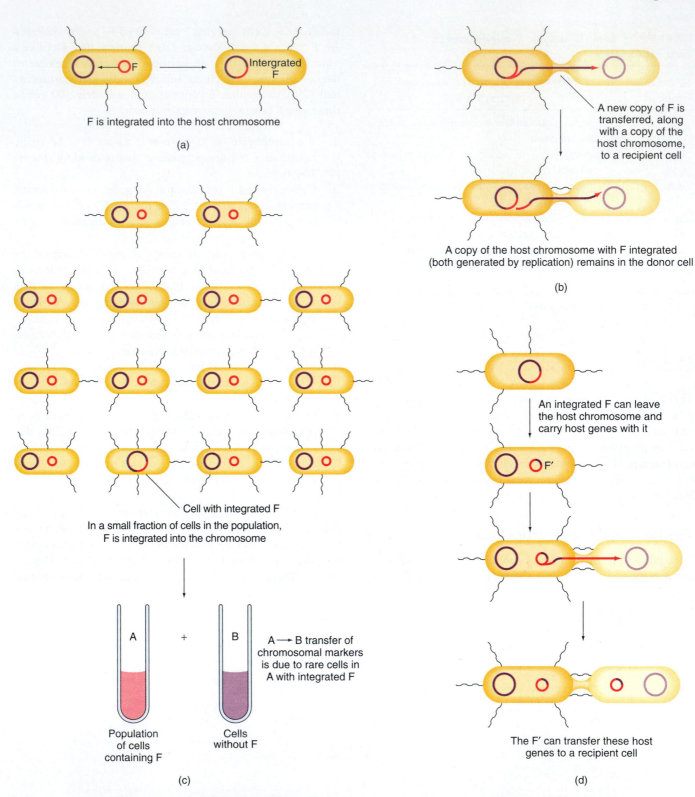

F is integrated into the host chromosome

(a)

In a small fraction of cells in the population, F is integrated into the chromosome

Cell with integrated F

Population of cells containing F

Cells without F

A → B transfer of chromosomal markers is due to rare cells in A with integrated F

(c)

A new copy of F is transferred, along with a copy of the host chromosome, to a recipient cell

A copy of the host chromosome with F integrated (both generated by replication) remains in the donor cell

(b)

An integrated F can leave the host chromosome and carry host genes with it

The F′ can transfer these host genes to a recipient cell

(d)

Figure 10-3 The transfer of *E. coli* chromosomal markers mediated by F. (a) Occasionally, the independent F factor combines with the *E. coli* chromosome. (b) When the integrated F transfers to another *E. coli* cell during conjugation, it carries along any *E. coli* DNA that is attached, thus transferring host chromosomal markers to a new cell. (c) In a population of F$^+$ cells, a few cells will have F integrated into the chromosome; these few cells can transfer chromosomal markers. Therefore, when a population of F$^+$ cells (tube A) is mixed with a population of F$^-$ cells (tube B), a few B cells will show the acquisition of markers from A. (d) Occasionally, the integrated F can leave the chromosome and return to the cytoplasm. In rare cases F can carry host genes with it, incorporating them into the circular F, which is now termed an F′. The F′ can transfer these genes at high efficiency to other cells, since they are part of the F′ genome.

markers to a second strain. Although the level of transfer is small, it is detectable because the transferred chromosomal markers produce genetic recombinants in the second strain of cells. This phenomenon is what led to the initial discovery of gene transfer by conjugation.

It is possible to isolate the specific cells in the bacterial population that have the F factor integrated into the host chromosome and to cultivate pure strains derived from these cells. In such strains, every cell donates chromosomal markers during F transfer, so that the frequency of recombinants for these strains is much higher than it is for cells in the original population, where the F factor is nearly always in the cytoplasm. Therefore, strains with an integrated F factor are termed **high frequency of recombination (Hfr)** strains to distinguish them from normal F⁺ strains, which contain only a few individual Hfr cells that can transfer chromosomal markers and thus display a low frequency of recombination for the population as a whole. Because they transfer chromosomal markers, Hfr strains have proved to be very useful for genetic mapping, as we shall see later on.

F occasionally leaves the chromosome of an Hfr cell and moves back to the cytoplasm in rare cases carrying a few host chromosomal genes with it. This modified F, called **F′**, can now transfer these specific host genes to a recipient cell. Thus, the recipient cell can now contain two copies of the same gene, one on its bacterial chromosome, and one on the newly transferred cytoplasmic F′ factor.

Let's now look at how some of these processes were discovered by pioneers in the field.

The Discovery of Bacterial Gene Transfer

Do bacteria possess any processes similar to sex and recombination? The question was answered in 1946 by the elegantly simple experimental work of Joshua Lederberg and Edward Tatum, who studied two strains of *Escherichia coli* with different nutritional requirements. Strain A would grow on a minimal medium only if the medium were supplemented with methionine and biotin; strain B would grow on a minimal medium only if it were supplemented with threonine, leucine, and thiamine. Thus, we can designate strain A as *met⁻ bio⁻ thr⁺ leu⁺ thi⁺* and strain B as *met⁺ bio⁺ thr⁻ leu⁻ thi⁻*. Note that in this instance no linkage relationship is intended in writing the symbols in a specific order.

In their experiment, Lederberg and Tatum plated bacteria into dishes containing only unsupplemented minimal medium. Some of the dishes were plated only with strain A bacteria, some only with strain B bacteria, and some with a mixture of strain A and strain B bacteria that had been incubated together for several hours in a liquid medium containing all the supplements (Figure 10-4). The plates that received the mixture of strain A and strain B produced growing colonies with a frequency of one in every 10,000,000 cells plated (in scientific notation, 1×10^{-7}), whereas no colonies arose on plates containing either strain

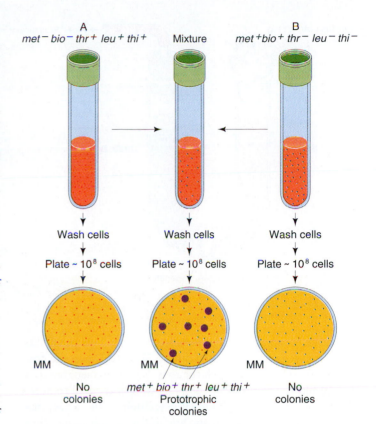

Figure 10-4 Demonstration by Lederberg and Tatum of genetic recombination occurring between bacterial cells. Cells of type A or type B cannot grow on an unsupplemented (minimal) medium (MM), because A and B each carry mutations that cause the inability to synthesize constituents needed for cell growth. When A and B are mixed for a few hours and then plated, however, a few colonies appear on the agar plate. These colonies derive from single cells in which an exchange of genetic material has occurred; they are therefore capable of synthesizing all the required constituents of metabolism.

A or strain B alone. Because only prototrophic (*met⁺ bio⁺ thr⁺ leu⁺ thi⁺*) bacteria can grow on a minimal medium, this observation suggested that some form of recombination of genes had occurred between the genomes of the two strains.

Now, you might object: "What about the possibility that these wild-type colonies were produced by mutation?" The answer is that if the results were due to mutation, then wild-type colonies should have appeared when the strain A and strain B bacteria were plated *by themselves* onto the minimal medium. But they didn't. Note too that more than one mutation would be required to convert each strain to prototrophy (the ability to grow on minimal medium without added nutrients). Therefore, we can conclude that the wild-type colonies were most likely produced by an exchange of genetic material between the two strains.

Requirement for Physical Contact

It could be suggested that the cells of the two strains do not really exchange genes but instead leak substances that the

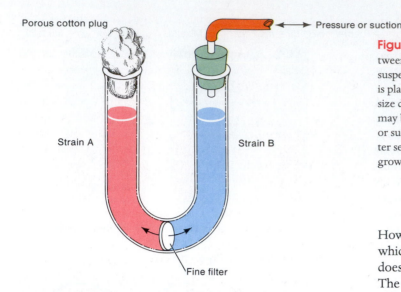

Porous cotton plug

Pressure or suction

Strain A

Strain B

Fine filter

Figure 10-5 Experiment demonstrating that physical contact between bacterial cells is needed for genetic recombination to occur. A suspension of a bacterial strain unable to synthesize certain nutrients is placed in one arm of a U tube. A strain genetically unable to synthesize different required metabolites is placed in the other side. Liquid may be transferred between the arms by the application of pressure or suction, but bacterial cells cannot pass through the center filter. After several hours of incubation, the cells are plated, but no colonies grow on the minimal medium.

other cells can absorb and use for growing. This possibility of "cross-feeding" was ruled out by Bernard Davis. He constructed a U tube in which the two arms were separated by a fine filter. The pores of the filter were too small to allow bacteria to pass through but large enough to allow easy passage of the fluid medium and any dissolved substances (Figure 10-5). Strain A was put in one arm; strain B in the other. After the strains had been incubated for a while, Davis tested the content of each arm to see if cells had become able to grow on a minimal medium, and none were found. In other words, *physical contact* between the two strains was needed for wild-type cells to form. It looked as though some kind of gene transfer was involved and genetic recombinants were indeed produced.

The Discovery of the Fertility Factor (F)

Determining the Direction of Gene Transfer. Having seen that bacteria can exchange genetic material, as can higher organisms, Lederberg and Tatum hypothesized that bacterial genes might be organized into linkage groups, as are the genes of higher organisms. They extended their work with strain A and strain B in attempts to support their hypothesis. Their data clearly showed linkage, because the allele frequencies they obtained from A × B crosses showed great departures from the 1:1 ratio typical of independent gene assortment. However, their results were unclear, in part because they did not know the direction of genetic transfer.

In 1953, William Hayes exploited the properties of the antibiotic streptomycin to determine that genetic transfer occurred in one direction in the types of crosses carried out. Hayes verified the results of Lederberg and Tatum, using a similar cross:

Strain A   Strain B

$met^- thr^+ leu^+ thi^+ \times met^+ thr^- leu^- thi^-$

However, he treated one of the strains with streptomycin, which prevents cell division and subsequently kills cells but does permit mating to continue for a short period of time. The streptomycin can be washed out after it takes effect on the initial strain, so that a second strain will be unexposed to the drug when the strains to be crossed are mixed. When Hayes treated strain A with the streptomycin, washed out the streptomycin, mixed in strain B, and then plated the culture on a minimal medium, he obtained the same frequency of colonies he had obtained in the control experiment with both strains untreated. However, when he treated strain B with streptomycin, washed out the streptomycin, mixed in strain A, and then plated the culture on a minimal medium, he obtained no colonies.

This experiment showed that all the recombinants detected in these genetic transfer experiments took place in strain B. Therefore, the production of colonies of strain B—but not of strain A—on the selective minimal medium was required. The obvious interpretation is that the genetic transfer was not reciprocal but occurred only by transfer from strain A to strain B.

Message The transfer of genetic material in *E. coli* is not reciprocal. One cell acts as the **donor,** and the other cell acts as the **recipient.**

In Hayes's experimental strains, the donor could still transmit genes after exposure to streptomycin, but the recipient could not divide to produce colonies if it had been exposed to streptomycin. This kind of unidirectional transfer of genes was originally analogized to a sexual difference, with the donor being termed "male" and the recipient "female." Although the terms "male" and "female" still persist, it should be stressed that this type of gene transfer is not sexual reproduction. In *bacterial gene transfer,* one organism receives genetic information from a donor; the recipient is changed by that information. In *sexual reproduction,* two organisms donate equally (or nearly so) to the formation of a new organism, but only in exceptional cases is either of the donors changed.

Loss and Regain of Ability to Transfer. By accident, Hayes discovered a variant of his original A strain (male) that would not produce recombinants on crossing with the B strain (female). Apparently, the A males had lost the ability to transfer genetic material and had changed into females. In his analysis of this "sterile" variant, Hayes realized that the fertility of *E. coli* could be lost and regained rather easily.

In further studies, Hayes isolated and cultured a streptomycin-resistant mutant of the sterile A variant. He then mixed these sterile A str^r cells with fertile A male str^s cells. He found that as many as one-third of the A str^r cells had again become fertile, which he judged by their ability to transfer genetic markers to B females. Hayes suggested that maleness, or donor ability, is itself a hereditary state imposed by a **fertility factor (F).** Females lack F and therefore are recipients. Thus, females he designated F⁻, and males F⁺.

Transfer of F during Conjugation. Recombinant genotypes for marker genes occurred relatively rarely in bacterial crosses, Hayes noted, but the F factor apparently was transmitted effectively during physical contact, or **conjugation,** since Hayes recovered male fertile str^r cells very soon after mixing F⁻ str^r cells with F⁺ str^s cells. There seemed to be a kind of **"infectious transfer"** of the F factor that was taking place far more quickly than the exchange of genetic markers by recombination. The physical nature of the F factor was elucidated much later, but these early experiments clearly showed that it involved some kind of independent particle not closely tied to the genetic markers.

Hfr Strains

An important breakthrough came when Luca Cavalli-Sforza obtained a new kind of male from an F⁺ strain (which we now know resulted from the integration of F into the chromosome). On crossing with F⁻ females, this new strain produced 1000 times more recombinants for genetic markers than did a normal F⁺ strain. Cavalli-Sforza designated this derivative an **Hfr** strain to indicate a high frequency of recombination. (The derivative later became known as *Hfr C* to distinguish it from a similar strain, Hfr H, found by Hayes.) In cells resulting from an F⁺ × F⁻ cross, a large proportion of the F⁻ parents were converted to F⁺ by infectious transfer of the F factor. However, in Hfr × F⁻ crosses, virtually none of the F⁻ parents were converted to F⁺ or to Hfr. Thus, infectious transfer of F did not occur in these crosses, since, as we now know, F was no longer in the cytoplasm.

Determining Linkage from Interrupted-Mating Experiments

The exact nature of Hfr strains became clearer in 1957 when Elie Wollman and François Jacob investigated the pattern of transmission of Hfr genes to F⁻ cells during a cross. They crossed Hfr str^s $a^+ b^+ c^+ d^+$ with F⁻ str^r $a^- b^- c^- d^-$. At specific time intervals after mixing, they removed sam-

ples. Each sample was put into a kitchen blender for a few seconds to disrupt the mating cell pairs and then was plated onto a medium containing streptomycin to kill the Hfr donor cells. This is called an **interrupted-mating** procedure. The str^r cells then were tested for the presence of marker alleles from the donor. Those str^r cells bearing donor marker alleles must have been involved in conjugation; such cells are called **exconjugants.** Figure 10-6a shows a plot of the results; azi^r, ton^r, lac^+, and gal^+ correspond to the a^+, b^+, c^+, and d^+ mentioned in our generalized description of the experiment. Figure 10-6b portrays the transfer of markers.

The most striking thing about these results is that each donor allele first appeared in the F⁻ recipients at a specific time after mating began. Furthermore, the donor alleles appeared in a specific sequence. Finally, the maximal yield of cells containing a specific donor allele was smaller for the donor markers that entered later. Putting all these observations together, Wollman and Jacob concluded that gene transfer occurs from a fixed point on the donor chromosome, termed the **origin,** or **O,** and continues in a linear fashion.

Message   The Hfr chromosome, originally circular, unwinds and is transferred to the F⁻ cell in a linear fashion. The unwinding and transfer begin from a specific point at one end of the integrated F, called the **origin,** or **O.** The farther a gene is from O, the later it is transferred to the F⁻; the transfer process most likely will stop before the farthermost genes are transferred.

Wollman and Jacob realized that it would be easy to construct linkage maps from the interrupted-mating results, using as a measure of "distance" the times at which the donor alleles first appear after mating. The units of distance in this case are minutes. Thus, if b^+ begins to enter the F⁻ cell 10 minutes after a^+ begins to enter, then a^+ and b^+ are 10 units apart (Figure 10-7). Like the maps based on crossover frequencies, these linkage maps are purely genetic constructions; at the time, they had no known physical basis.

Chromosome Circularity and Integration of F

When Wollman and Jacob allowed Hfr × F⁻ crosses to continue for as long as two hours before blending, they found that a few of the exconjugants were converted into Hfr. In other words, an important part of F (the terminal portion now known to confer maleness, or donor ability), was eventually being transmitted, but at a very low efficiency, and it apparently was transmitted as the last element of the linear chromosome. We now have the following map, in which the arrow indicates the process of transfer, beginning with O:

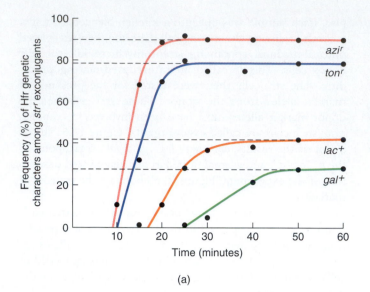

(a)

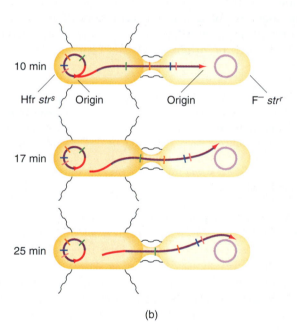

(b)

Figure 10-6 Interrupted-mating conjugation experiments with *E. coli*. F⁻ cells that are *str*ʳ are crossed with Hfr cells that are *str*ˢ. The F⁻ cells have a number of mutations (indicated by the genetic markers *azi, ton, lac,* and *gal*) that prevent them from carrying out specific metabolic steps. However, the Hfr cells are capable of carrying out all these steps. At different times after the cells are mixed, samples are withdrawn, disrupted in a blender to break conjugation between cells, and plated on media containing streptomycin. The antibiotic kills the Hfr cells but allows the F⁻ cells to grow and to be tested for their ability to carry out the four metabolic steps. (a) A plot of the frequency of recombinants for each metabolic marker as a function of time after mating. Transfer of the donor allele for each metabolic step obviously depends on how long conjugation is allowed to continue. (b) A schematic view of the transfer of markers over time. (Part a modified from E. L. Wollman, F. Jacob, and W. Hayes, *Cold Spring Harbor Symposia on Quantitative Biology* 21, 1956, 141.)

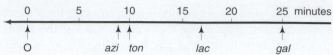

Figure 10-7 Chromosome map from Figure 10-6. A linkage map can be constructed for the *E. coli* chromosome from interrupted-mating studies, using the time at which the donor alleles first appear after mating. The units of distance are given in minutes; arrowhead at left indicates the direction of transfer of the donor alleles.

However, when several different Hfr linkage maps were derived by interrupted-mating and "time-of-entry" studies using different, separately derived Hfr strains, the maps differed from strain to strain:

Hfr H	O *thr pro lac pur gal his gly thi* F
1	O *thr thi gly his gal pur lac pro* F
2	O *pro thr thi gly his gal pur lac* F
3	O *pur lac pro thr thi gly his gal* F
AB 312	O *thi thr pro lac pur gal his gly* F

At first glance, there seems to be a random reshuffling of genes. However, a pattern does exist; the genes are not thrown together at random in each strain. For example, note that in every case the *his* gene has *gal* on one side and *gly* on the other. Similar statements can be made about each gene, except when it appears at one end or the other of the linkage map. The order in which the genes are transferred is not constant. In two Hfr strains, for example, the *his* gene is transferred before the *gly* gene (*his* is closer to O), but in three strains the *gly* gene is transferred before the *his* gene.

How can we account for these unusual results? Allan Campbell proposed a startling hypothesis: suppose that in an F⁺ male, F is a small cytoplasmic element (and therefore easily transferred to an F⁻ cell on conjugation). If the *chromosome* of the F⁺ male were a *ring*, any of the linear Hfr chromosomes could be generated simply by inserting F into the ring at the appropriate place and in the appropriate orientation (Figure 10-8).

Several conclusions—later confirmed—follow from this hypothesis.

1. The orientation in which F is inserted would determine the polarity of the Hfr chromosome, as indicated in Figure 10-8a.
2. At one end of the integrated F factor would be the **origin,** where transfer of the Hfr chromosome begins; the **terminus** at the other end of F would not be transferred unless all the chromosome had been transferred. Since the chromosome often breaks before all of it is transferred, and since the F terminus is what confers maleness, then only a small fraction of the recipient cells would be converted to male cells.

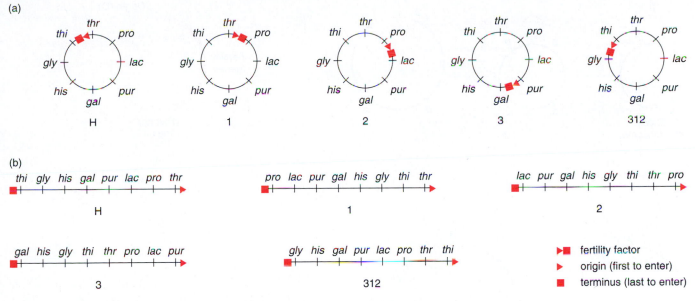

Figure 10-8 Circularity of the *E. coli* chromosome. (a) Through the use of different Hfr strains (H, 1, 2, 3, 312) that have the fertility factor inserted into the chromosome at different points and in different directions, interrupted-mating experiments indicate that the chromosome is circular. The mobilization point (origin) is shown for each strain. (b) The linear order of transfer of markers for each Hfr strain; arrowheads indicate the origin and direction of transfer.

How, then, might F integration be explained? Wollman and Jacob suggested that some kind of crossover event between F and the F^+ chromosome might generate the Hfr chromosome. Campbell then came up with a brilliant extension of that idea. He proposed that if F, like the chromosome, were circular, then a crossover between the two rings would produce a single larger ring with F inserted (Figure 10-9a).

Now suppose that F consists of three different regions, as shown in Figure 10-9b. If the bacterial chromosome had several homologous regions that could match up with the pairing region of F, then different Hfr chromosomes could easily be generated by crossovers at these different sites (Figure 10-9c).

Chromosomal and F circularity were wildly implausible concepts initially, inferred solely from the genetic data; confirmation of their physical reality came only a number of years later. The direct-crossover model of integration was also subsequently confirmed.

Episomes

The fertility factor thus exists in two states: (1) the plasmid state, as a free cytoplasmic element F that is easily transferred to F^- recipients, and (2) the integrated state, as a contiguous part of a circular chromosome that is transmitted only very late in conjugation. The word **episome** (literally, "additional body") was coined for a genetic particle having such a pair of states. A cell containing F in the first state is called an F^+ cell, a cell containing F in the second state is an Hfr cell, and a cell lacking F is an F^- cell.

Infectious elements other than F have been found in *E. coli* and other bacteria. Like F, when these factors exist in the cytoplasmic state, they are termed **plasmids.** Some plasmids (including many used in the recombinant DNA studies described in Chapter 14) cannot integrate into the chromosome, while others, such as temperate phages (see page 297), are episomes that can integrate into the bacterial chromosome as the F factor does. Plasmids are of central importance in genetic engineering; we consider them in more detail later in this book.

Message An **episome** is a genetic factor in bacteria that can exist either as a plasmid in the cytoplasm or as an integrated part of a chromosome. The F factor is an episome.

Mechanics of Transfer

Does an Hfr cell die after donating its chromosome to an F^- cell? The answer is no (unless the culture is treated with streptomycin). The Hfr chromosome replicates while it is transferring a single strand to the F^- cell; this ensures a complete chromosome for the donor cell after mating. The transferred strand is replicated in the recipient cell, and donor genes may become incorporated in the recipient's chromosome through crossovers, creating a recombinant cell. Otherwise, transferred fragments of DNA in the recipient are lost during cell division.

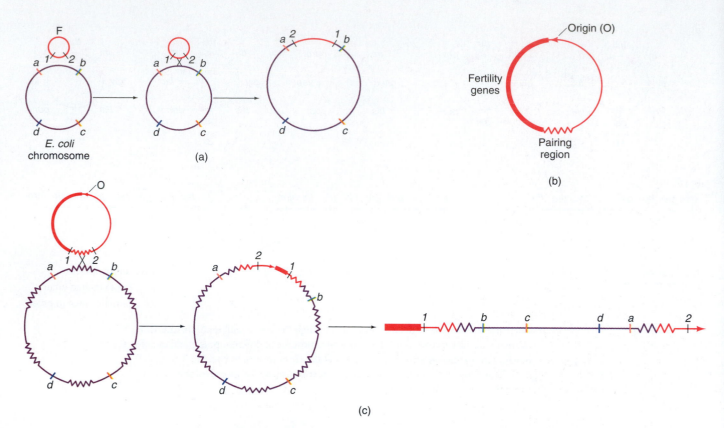

Figure 10-9 Insertion of the F factor into the *E. coli* chromosome. (a) Crossing-over mechanism for the attachment of F. (Hypothetical markers 1 and 2 are shown on F to depict the direction of insertion.) (b) Hypothetical regions in F. The origin (O) is the mobilization point where insertion begins; the pairing region is homologous with a region on the *E. coli* chromosome; fertility genes are responsible for the F⁺ phenotype. (c) Model for the insertion of F. An area of F may share homology with multiple regions of the circular *E. coli* chromosome. Crossover between F and the chromosome could thus insert F into various points on the chromosome. (Jagged lines represent possible regions of homology; the arrowhead indicates the direction of transfer from the origin, O.) In this example, the Hfr cell created by the insertion of F would transfer its genes in the order *a, d, c, b.* What would be the order of transfer if F inserted at the other homologous sites?

We assume that the F⁻ chromosome is also circular, because the recipient F⁻ cell, if it receives the F factor, from an F⁺ cell, is readily converted into an F⁺ cell from which an Hfr cell can be derived.

The picture emerges of a circular Hfr chromosome unwinding a copy of itself, which is then transferred in a linear fashion into the F⁻ cell. How is the transfer achieved? Electron microscope studies show that Hfr and F⁺ cells have fibrous structures, **F pili,** protruding from their cell walls, as shown in Figure 10-10. The F pili facilitate cell-to-cell contact, during which DNA transfer occurs, apparently through pores in the F⁻, although the exact mechanism of transfer is still unclear.

The *E. coli* Conjugation Cycle

We can now summarize the various aspects of the conjugation cycle in *E. coli* (Figure 10-11). We'll review the conjugation cycle in terms of the differences between F⁻, F⁺, and Hfr cells, since these differences epitomize the cycle.

F⁻ strains do not contain the F factor and cannot transfer DNA by conjugation. They are, however, recipients of DNA transferred from F⁺ or Hfr cells by conjugation.

F⁺ cells contain the F factor in the cytoplasm and therefore can transfer F in a highly efficient manner to F⁻ cells during conjugation.

Hfr cells have F integrated into the bacterial chromosome, not in the cytoplasm.

Chromosomal markers are transferred in a strain of F⁺ cells because in any population of F⁺ cells, a small fraction of cells (about 1 in 1000) have been converted to Hfr cells by the integration of F into the bacterial chromosome. Since conjugation experiments are usually carried out by mixing

Figure 10-10 Electron micrograph of a cross between two *E. coli* cells (× 34,300). The pili of the Hfr cell (not present in the F⁻ cell) have been visualized through the addition of viruses that attach specifically to them. (Photo courtesy of Charles Brinton, Jr., and Judith Carnahan. From J. L. Gould and C. G. Gould, *Sexual Selection*. Scientific American Library, New York, 1989.)

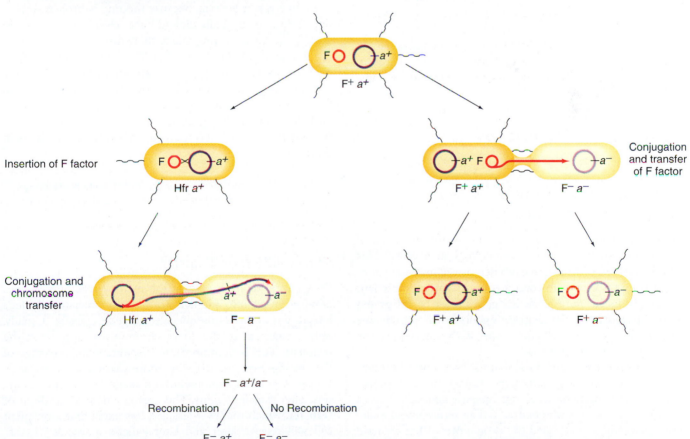

Figure 10-11 Summary of the various events that occur in the conjugational cycle of *E. coli*.

10^7 to 10^8 cells comprising prospective donors and recipients, the population will contain various different Hfr cells derived from independent integrations of F into the chromosome at various different sites. Therefore, when chromosomal markers are transferred by different cells in the population, transfer will start at different points on the chromosome. This results in an approximately equal transfer of markers all around the chromosome, although at a low frequency. This type of F^+-mediated transfer is what Lederberg and Tatum observed when they discovered gene transfer in bacteria.

Each of the Hfr cells in an F^+ population with an integrated F factor can be the source of a new Hfr strain if it is isolated and used to start a clone.

Hfr strains are derived from a clone of Hfr cells in which a specific integration of F into the bacterial chromosome has occurred. Therefore, all the cells in any given Hfr strain have F integrated into the chromosome at exactly the same point.

Hfr populations transfer chromosomal markers to F^- cells at a high frequency compared with F^+ populations, since only a fraction of cells in an F^+ population have F integrated into the chromosome. Further, in any given Hfr strain the markers are transferred from a fixed point in a specific order. This also contrasts with F^+ populations, where the Hfr cells transfer chromosomal markers in no particular fixed order, since the F factor integrates into the chromosome at different points in different F^+ cells.

In an Hfr $\times$ F^- cross, the F^- is not converted to Hfr or to F^+, except in very rare cases, because the Hfr chromosome nearly always breaks before the F terminus is transferred to the F^- cell.

Bacterial Recombination and Mapping the *E. coli* Chromosome

Recombination between Marker Genes after Transfer

Thus far, we have studied only the process of the transfer of genetic information between individuals in a cross. This transfer is inferred from the existence of recombinants produced from the cross. However, before a stable recombinant can be produced, the transferred genes must be integrated or incorporated into the recipient's genome by an exchange mechanism. We now consider some of the special properties of this exchange event.

Genetic exchange in prokaryotes does not take place between two whole genomes (as it does in eukaryotes); rather, it takes place between one *complete* genome, derived from F^-, called the **endogenote,** and an *incomplete* one, derived from the donor, called the **exogenote.** What we have in fact is a partial diploid, or **merozygote.** Bacterial genetics is merozygous genetics. Figure 10-12a is a diagram of a merozygote.

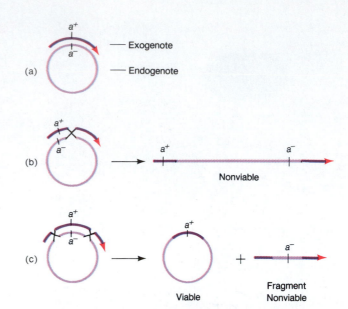

Figure 10-12 Crossover between exogenote and endogenote in a merozygote. (a) The merozygote. (b) A single crossover leads to a partially diploid linear chromosome. (c) An even number of crossovers leads to a ring plus a linear fragment.

A single crossover would not be very useful in generating viable recombinants, because the ring is broken to produce a strange, partially diploid linear chromosome (Figure 10-12b). To keep the ring intact, there must be an even number of crossovers (Figure 10-12c). The fragment produced in such a crossover is only a partial genome, which is generally lost during subsequent cell growth. Hence, both reciprocal products of recombination do not survive—only one does. A further unique property of bacterial exchange, then, is that we must forget about reciprocal exchange products in most cases.

Message In the merozygous genetics of bacteria, we generally are concerned with double crossovers and we do not expect reciprocal recombinants.

The Gradient of Transfer

Only partial diploids exist in the merozygote. Some genes don't even get into the act! To better appreciate this, let's look again at the consequences of gene transfer. Usually, only a fragment of the donor chromosome appears in the recipient. This is because there is spontaneous breakage of the mating pairs, so that the entire chromosome is rarely transferred. The spontaneous breakage can occur at any time after transfer begins. This creates a natural **gradient of transfer,** which makes it less and less likely that a recipient cell will receive later and later genetic markers. ("Later" here refers to markers that are increasingly farther from the origin and hence are donated later in the order of markers transferred.) For example, in a cross of Hfr-donating markers in the order *met, arg, leu,* we would expect a distribution of fragments such as the one represented here:

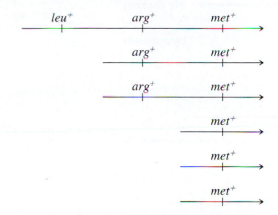

Note that many more fragments contain the *met* locus than the *arg* locus and that the *leu* locus is present on only one fragment. It is easy to see that the closer the marker is to the origin, the greater the chance it will be transferred during conjugation.

The concept of the gradient of transfer is the same as the one described earlier for interrupted matings, except that here we are allowing the natural disruption of mating paris to occur instead of interrupting the pairs mechanically.

Determining Gene Order from the Gradient of Transfer

We can use the natural gradient of transfer to establish the order of genetic markers, provided we select for an early marker that enters before the markers we are ordering. Let's see how this works. Suppose that we use an Hfr strain which donates markers in the order *met, arg, aro, his.* In a cross of an Hfr that is *met⁺ arg⁺ aro⁺ his⁺ strˢ* with an F⁻ that is *met⁻ arg⁻ aro⁻ his⁻ strʳ*, recombinants are selected that can grow on a minimal medium without methionine but with arginine, aromatic amino acids, histidine, and in the presence of streptomycin. Here we are selecting for recombinants in the F⁻ strain that are *met⁺* in a cross in which the *met* locus is transferred as the earliest marker. We can then score for inheritance of the other markers present in the Hfr by testing on supplemental minimal medium lacking, in turn, one of the required nutrients. A typical result would be

$$met^+ = 100\%$$

$$arg^+ = 60\%$$

$$aro^+ = 20\%$$

$$his^+ = 4\%$$

Note how the frequency of inheritance corresponds to the order of transfer. This is because the frequency of inheri-

tance reflects the frequency of transfer. For this method to work, it is crucial that it be applied only to genetic markers that enter after the selected marker—in this case, after *met*.

Higher-Resolution Mapping by Recombinant Frequency in Bacterial Crosses

While interrupted-mating experiments and the natural gradient of transfer can give us a rough set of gene locations over the entire map, other methods are needed to obtain a higher resolution between marker loci that are close together. Here we consider one approach to the problem of using the frequency of recombinants to measure linkage.

Previous attempts to measure linkage in conjugational crosses were hindered by the failure to understand that only fragments of the chromosome are transferred and that the gradient of transfer produces a bias toward the inheritance of early markers. In order to measure linkage and to attach any meaning to a calculated map distance, it is necessary to produce a situation in which every marker has an equal chance at being transferred, so that the recombinant frequencies are dependent only on the distance between the relevant genes.

Suppose that we consider three markers: *met, arg,* and *leu.* If the order is *met, arg, leu,* and if *met* is transferred first and *leu* last, then we really want to set up the situation diagrammed below to calculate map distances separating these markers:

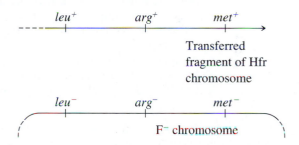

Here, we have to arrange to select the *last* marker to enter, which in this case is *leu.* Why? Because if we select for the last marker, then we know that every cell that received fragments containing the last marker also received the earlier markers, namely *arg* and *met,* on the same fragments. We can then proceed to calculate map distance in the classic manner, where 1 map unit (m.u.) is equal to 1 percent crossovers in the respective interval on the map. In practice, this is done by calculating, among the total recovered recombinants, the percentage of recombinants produced by crossovers between two markers. Let's look at an example.

Sample Cross

In the cross of the Hfr strain just described (*met⁺ arg⁺ leu⁺ strˢ*) with an F⁻ that is *met⁻ arg⁻ leu⁻ strʳ*, we would select *leu⁺* recombinants and then examine them for the *arg* and

(a) Insertion of late marker only

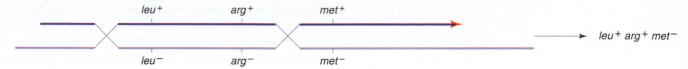

(b) Insertion of late marker and one early marker

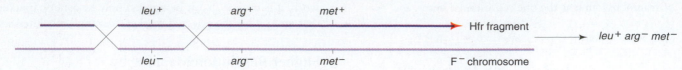

(c) Insertion of late and early markers

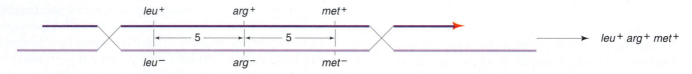

(d) Insertion of late and early markers, but not of marker in between

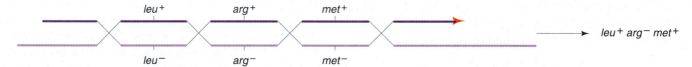

Figure 10-13 Incorporation of a late marker into the F⁻ *E. coli* chromosome. After an Hfr cross, selection is made for the *leu*⁺ marker, which is donated late. The early markers (*arg*⁺ and *met*⁺) may or may not be inserted, depending on the site where recombination between the Hfr fragment and the F⁻ chromosome occurs. If, as indicated in part c, the distance between *leu*⁺ and *arg*⁺ is 5 m.u. and the distance between *arg*⁺ and *met*⁺ is 5 m.u., then crossovers will occur in each of these intervals 5 percent of the time, resulting in the *leu*⁺ *arg*⁻ *met*⁻ (a) and *leu*⁺ *arg*⁺ *met*⁻ (b) recombinants. Crossovers occurring outside of these intervals can result in *leu*⁺ *arg*⁺ *met*⁺ recombinant (c). The *leu*⁺ *arg*⁻ *met*⁺ recombinant requires an additional two crossovers (d).

met markers. In this case, the *arg* and *met* markers are called the **unselected markers.** Figure 10-13 depicts the types of crossover events expected. Note how two crossover events are required to incorporate part of the incoming fragment into the F⁻ chromosome. One crossover must be on each side of the selected (*leu*) marker. Thus, in Figure 10-13, one crossover must be on the left side of the *leu* marker and the second must be on the right side. Suppose that the map distance between each marker is 5 m.u. (5 percent recombination). In 5 percent of the total *leu*⁺ recombinants, the second crossover occurs between *leu* and *arg* (Figure 10-13a); in another 5 percent of the cases, the second crossover occurs between *leu* and *met* (Figure 10-13b). We would then expect 90 percent of the selected *leu*⁺ recombinants to be *arg*⁺ *met*⁺, because the second crossover occurs outside the *leu–arg–met* interval (Figure 10-13c) in 90 percent of the cases. We would also expect 5 percent of the *leu*⁺ recombinants to be *arg*⁻ *met*⁻, resulting from a crossover between *leu* and *arg*, and 5 percent of the *leu*⁺ recombinants to be

arg⁺ *met*⁻, resulting from a crossover between *arg* and *met*. In reality, then, we are simply determining the percentage of the time that the second crossover occurs in each of the three possible intervals.

In a cross such as the one just described, one class of potential recombinants requires an additional two crossover events (Figure 10-13d). In this case, the *leu*⁺ *arg*⁻ *met*⁺ recombinants would require four crossovers instead of two. These recombinants are rarely recovered, because their frequency is sharply reduced compared with the other classes of recombinants.

Deriving Gene Order by Reciprocal Crosses

In many cases, two loci are so close together that it becomes difficult to order them in relation to a third locus. For example, consider three loci linked in this way:

Here *b* and *c* are too closely linked to separate easily. The recombinant frequency (RF) between *a* and *b* under most experimental conditions will be more or less the same as the RF between *a* and *c*. Is the order *a, b, c* or *a, c, b*? One often-used technique for finding out is to make a pair of reciprocal crosses, using the same marker genotypes for the donor in one cross and for the recipient in the other cross. Figure 10-14 shows the crossover events needed to generate prototrophs from the cross $a\ b\ c^+ \times a^+\ b^+\ c$, depending on the order of the loci. If the reciprocal crosses give dramatically different frequencies of wild-type survivors on a minimal medium, then we know that the order is *a, c, b*. If we observe no difference in frequencies between the reciprocal crosses, then the order must be *a, b, c*. Once again, this same principle can be used in other bacterial and phage mapping systems.

Infectious Marker-Gene Transfer by Episomes

Edward Adelberg's work led to the discovery of gene transfer at high frequency by episomes. When he began his recombination experiments in 1959, the particular Hfr strain he used kept producing F$^+$ cells, so the recombination frequencies were not very large. Adelberg called this particular fertility factor F′ to distinguish it from the normal F, for the following reasons:

1. The F′-bearing F$^+$ strain reverted back to an Hfr strain much more frequently than do typical F$^+$ strains.
2. F′ always integrated at the *same place* to give back the original Hfr chromosome. (Remember that randomly selected Hfr derivatives from F$^+$ males have origins at many different positions.)

How could these properties of F′ be explained? The answer came from the recovery of an F′ from an Hfr strain in which the *lac*$^+$ locus was near the end of the Hfr chromosome (since it was transferred very late). Using this Hfr *lac*$^+$ strain, François Jacob and Adelberg found an F$^+$ derivative that transferred *lac*$^+$ to F$^-$ *lac*$^-$ recipients at a very high frequency. Furthermore, the recipients that behaved like F$^+$ *lac*$^+$ occasionally produced F$^-$ *lac*$^-$ daughter cells, at a frequency of 1×10^{-3}. Thus, the genotype of these recipients appeared to be F *lac*$^+$/*lac*$^-$.

Now we have the clue: F′ is a cytoplasmic element that carries a part of the bacterial chromosome. In fact, it is nothing more than F with a piece of the host chromosome incorporated. Its origin and reintegration can be visualized as shown in Figure 10-15. This F′ is known as F′-*lac*, because the piece of host chromosome it picked up has the *lac* gene on it. F′ factors have been found carrying many different chromosomal genes and have been named accordingly. For example, F′ factors carrying *gal* or *trp* are called F′-*gal* and F′-*trp*, respectively. Because F *lac*$^+$/*lac*$^-$ cells are *lac*$^+$ in phe-

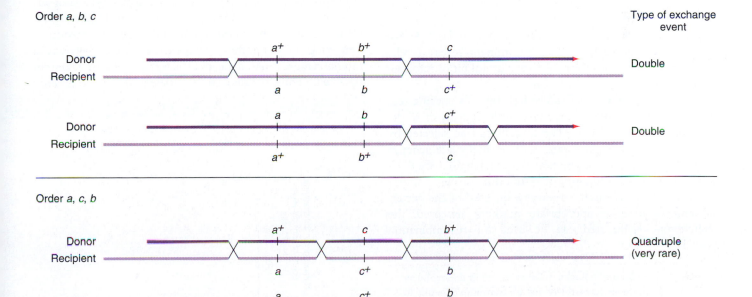

Figure 10-14 Inferring gene order from the relative frequencies of recombinants in reciprocal crosses involving parental genotypes $a\ b\ c^+ \times a^+\ b^+\ c$. If the order is *a, b, c*, then the frequency of $a^+\ b^+\ c^+$ progeny from the reciprocal crosses will be approximately equal. However, if the order is *a, c, b*, then the frequencies will be quite different.

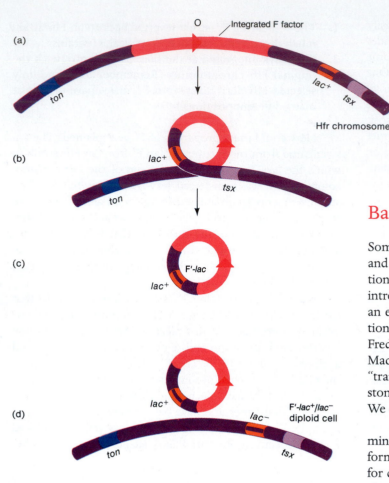

(a)

O   Integrated F factor

ton   lac⁺   tsx

Hfr chromosome

(b)

lac⁺

ton   tsx

(c)

F'-lac

lac⁺

(d)

lac⁺   F'-lac⁺/lac⁻
                          lac⁻   diploid cell

ton   tsx

Figure 10-15 Origin and reintegration of the F′ factor, in this case, F′ *lac*. (a) F is inserted in an Hfr strain between the *ton* and *lac*⁺ alleles. (b,c) Abnormal "outlooping" and separation of F occurs to include the *lac* locus, producing the F′-*lac*⁺ particle. (d) An F *lac*⁺/*lac*⁻ partial diploid is produced by the transfer of the F′-*lac* particle to an F⁻ *lac*⁻ recipient. (From G. S. Stent and R. Calendar, *Molecular Genetics*, 2d ed. Copyright © 1978 by W. H. Freeman and Company, New York.)

Bacterial Transformation

Some bacteria have another method of transferring DNA and producing recombinants that does not require conjugation. The conversion of one genotype into another by the introduction of exogenous DNA (that is, bits of DNA from an external source) is termed **transformation.** Transformation was discovered in *Streptococcus pneumoniae* in 1928 by Frederick Griffith, and in 1944 Oswald T. Avery, Colin M. MacLeod, and Maclyn McCarty demonstrated that the "transforming principle" was DNA. Both results are milestones in the elucidation of the molecular nature of genes. We consider this work in more detail in Chapter 11.

After it was shown that DNA is the agent that determines the polysaccharide character of *S. pneumoniae*, transformation was demonstrated for other genes, such as those for drug resistance (Figure 10-16). The transforming principle, exogenous DNA, is incorporated into the bacterial chromosome by a breakage-and-insertion process analogous to that depicted for Hfr × F⁻ crosses in Figure 10-12. (Note, however, that in *conjugation*, DNA is transferred from one living cell to another through close contact, whereas in *transformation*, isolated pieces of external DNA are taken up by a cell.) Thus, if radioactively labeled DNA from an *arg*⁺

notype, we know that *lac*⁺ is dominant over *lac*⁻. As we shall see later, in Chapter 18, the dominant–recessive relationship between alleles can be a very useful bit of information in interpreting gene function.

Partial diploidy, called **merodiploidy,** for specific segments of the genome can be made with an array of F′ derivatives from Hfr strains. The F′ cells can be selected by looking for the infectious transfer of normally late genes in a specific Hfr strain. The creation of partial diploids by F′ elements is called **sexduction,** or **F′-duction.** Some F′ strains can carry very large parts (up to one-quarter) of the bacterial chromosome; if appropriate markers are used, the merozygotes generated can be used for recombination studies.

Message During conjugation between an Hfr donor and an F⁻ recipient, the genes of the donor are transmitted *linearly* to the F⁻ cell, via the bacterial chromosome, with the inserted fertility factor transferring last.

During conjugation between an F⁺ donor carrying an F′ plasmid and an F⁻ recipient, a specific part of the donor genome may be transmitted *infectiously* to the F⁻ cell, via the plasmid. The transmitted part was originally adjacent to the F locus in an Hfr strain from which the F⁺ was derived.

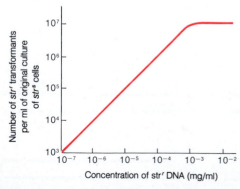

Figure 10-16 The genetic transfer of streptomycin resistance (*str*ʳ) to streptomycin-sensitive (*str*ˢ) cells of *E. coli*. The recovery of *str*ʳ transformants among *str*ˢ cells depends on the concentration of *str*ʳ DNA. (From G. S. Stent and R. Calendar, *Molecular Genetics*, 2d ed. Copyright © 1978 by W. H. Freeman and Company, New York.)

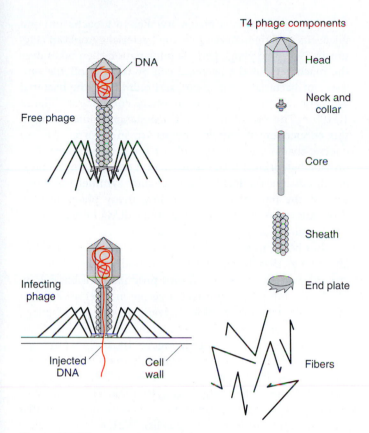

T4 phage components

Figure 10-17 — Phage T4, shown in its free state and in the process of infecting an *E. coli* cell. The infecting phage injects DNA through its core structure into the cell. On the right, a phage has been diagrammatically exploded to show its highly ordered three-dimensional structure. (Modified from R. S. Edgar and R. H. Epstein, "The Genetics of a Bacterial Virus." Copyright © 1965 by Scientific American, Inc. All rights reserved.)

and the frequency of double transformants will equal the product of the single-transformation frequencies. Thus, it should be possible to test for close linkage by testing for a departure from the product rule.

Unfortunately, the situation is made more complex by several factors—the most important of which is that not all cells in a population of bacteria are **competent,** or able to be transformed. Because single transformations are expressed as proportions, the success of the product rule obviously depends on the absolute size of these proportions. There are ways of calculating the proportion of competent cells, but we need not detour into that subject now. You can sharpen your skills in transformation analysis in one of the problems at the end of the chapter, which assumes 100 percent competence.

Bacteriophage Genetics

Infection of Bacteria by Phages

Many bacteriophages play an important role in gene transfer, as we will see below. Most bacteria are susceptible to attack by bacteriophages, which literally means "eaters of bacteria." A phage consists of a nucleic acid "chromosome" (DNA or RNA) surrounded by a coat of protein molecules. One well-studied set of phage strains are identified as T1, T2, and so on. Figures 10-17 and 10-18 show the complicated structure of a phage belonging to the class called **T-even phages** (T2, T4, and so on).

bacterial culture is added to unlabeled *arg⁻* cells, the *arg⁺* transformants (selected by plating on a minimal medium without arginine) can be shown to contain some of the radioactivity. We examine a molecular model of this process in Chapter 20. For now, let's consider transformation simply as a genetic tool.

Linkage Information from Transformation

Transformation has been a very handy tool in several areas of bacterial research. We learn later how it is used in some of the modern techniques of genetic engineering. Here we examine its usefulness in providing linkage information.

When DNA (the bacterial chromosome) is extracted for transformation experiments, some breakage into smaller pieces is inevitable. If two donor genes are located close together on the chromosome, then there is a greater chance that they will be carried on the same piece of transforming DNA and hence will cause a **double transformation.** Conversely, if genes are widely separated on the chromosome, then they will be carried on separate transforming segments

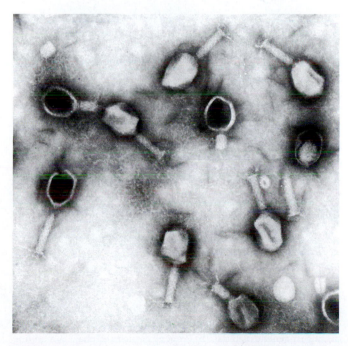

Figure 10-18 — Mature particles of the *E. coli* phage T4 (× 97,500). (Grant Heilman)

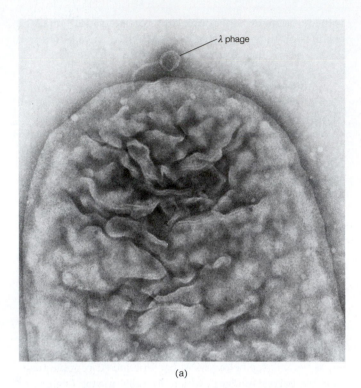

(a)

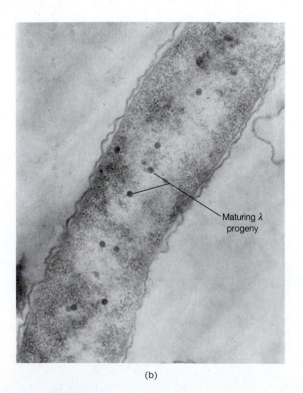

(b)

Figure 10-19 (a) Bacteriophage λ attached to an *E. coli* cell and injecting its genetic material. (b) Progeny particles of phage λ maturing inside an *E. coli* cell. (Jack D. Griffith)

During infection, a phage attaches to a bacterium and injects its genetic material into the bacterial cytoplasm (Figure 10-19a). The phage genetic information then takes over the machinery of the bacterial cell by turning off the synthesis of bacterial components and redirecting the bacterial synthetic material to make more phage components (Figure 10-19b). (The use of the word *information* is interesting in this connection; it literally means "to give form." And of course, that is precisely the role of the genetic material: to provide blueprints for the construction of form. In the present discussion, the form is the elegantly symmetrical structure of the new phages.) Ultimately, many phage descendants are released when the bacterial cell wall breaks open. This breaking-open process is called **lysis.**

But how can we study inheritance in phages when they are so small that they are visible only under the electron microscope? In this case, we cannot produce a visible colony by plating, but we can produce a visible manifestation of an infected bacterium by taking advantage of several phage characters. Let's look at the consequences of a phage's infecting a single bacterial cell. Figure 10-20 shows the sequence of events in the infectious cycle that leads to the release of progeny phages from the lysed cell. After lysis, the progeny phages infect neighboring bacteria. This is an exponentially explosive phenomenon (it causes an exponential increase in the number of lysed cells). Within 15 hours after the start of an experiment of this type, the effects are visible to the naked eye: a clear area, or **plaque,** is present on the opaque lawn of bacteria on the surface of a plate of solid medium (Figure 10-21). Depending on the phage genome, such plaques can be large or small, fuzzy or sharp, and so forth. Thus, **plaque morphology** is a phage character that can be analyzed. Another phage phenotype we can analyze genetically is **host range,** since phages may differ in the spectra of bacterial strains they can infect and lyse. For example, certain strains of bacteria are immune to adsorption (attachment) or injection by phages.

The Phage Cross

Can we cross two phages in the same way we cross two bacterial strains? A phage cross can be illustrated by a cross of T2 phages originally studied by Alfred Hershey. The genotypes of the two parental strains of T2 phage in Hershey's cross were $h^- \, r^+ \times h^+ \, r^-$. The alleles are identified by the following characters: h^- can infect two different *E. coli* strains (which we can call strains 1 and 2); h^+ can infect only strain 1; r^- rapidly lyses cells, thereby producing large plaques; and r^+ slowly lyses cells, thus producing small plaques.

In the cross, *E. coli* strain 1 is infected with both parental T2 phage genotypes at a phage : bacteria concentration (called **multiplicity of infection**) that is high enough to ensure that a large percentage of cells are simultaneously infected by both phage types. This kind of infection (Figure

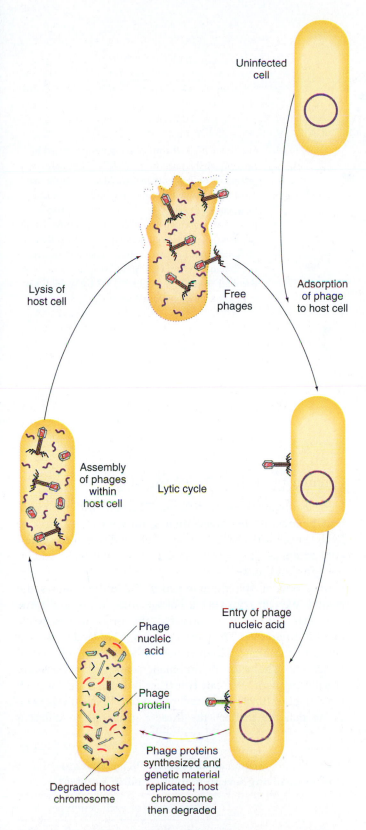

Figure 10-20 A generalized bacteriophage lytic cycle. (Adapted from J. Darnell, H. Lodish, and D. Baltimore, *Molecular Cell Biology.* Copyright © 1986 by W. H. Freeman and Company, New York.)

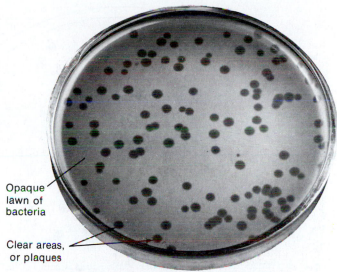

Figure 10-21 The appearance of phage plaques. Individual phages are spread on an agar medium that contains a fully grown "lawn" of *E. coli*. Each phage infects one bacterial cell, producing 100 or more progeny phages that burst the *E. coli* cell and infect neighboring cells. They in turn are exploded with progeny, and the continuing process produces a clear area, or plaque, on the opaque lawn of bacterial cells. (From G. S. Stent, *Molecular Biology of Bacterial Viruses.* Copyright © 1963 by W. H. Freeman and Company, New York.)

10-22) is called a **mixed infection,** or a **double infection.** The phage lysate (the progeny phage) is then analyzed by spreading it onto a bacterial lawn composed of a mixture of *E. coli* strains 1 and 2. Four plaque types are then distinguishable (Figure 10-23 and Table 10-2). These four genotypes can be scored easily as parental ($h^- r^+$ and $h^+ r^-$) and recombinant, and a recombinant frequency can be calculated as follows:

$$RF = \frac{(h^+ r^+) + (h^- r^-)}{\text{total plaques}}$$

If we assume that entire phage genomes recombine, then we are not faced with a merozygous situation, as in a

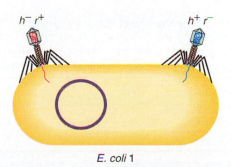

Figure 10-22 A double infection of *E. coli* by two phages.

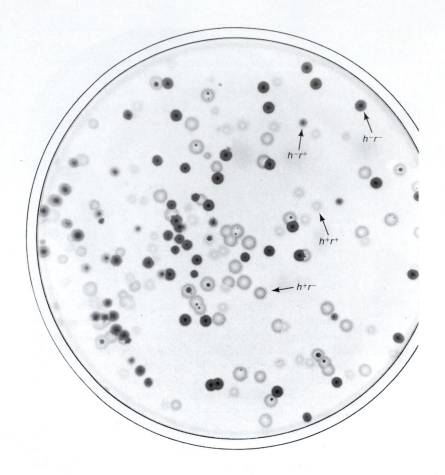

Figure 10-23 Plaque phenotypes produced by progeny of the cross $h^- r^+ \times h^+ r^-$. Enough phages of each genotype are added to ensure that most bacterial cells are infected with at least one phage of each genotype. After lysis, the progeny phages are collected and added to an appropriate *E. coli* lawn. Four plaque phenotypes can be differentiated, representing two parental types and two recombinants. (From G. S. Stent, *Molecular Biology of Bacterial Viruses.* Copyright © 1963 by W. H. Freeman and Company, New York.)

bacterial cross. Presumably, then, single exchanges can occur and produce viable reciprocal products. Nevertheless, phage crosses are subject to complications. First, several rounds of exchange can potentially occur within the host: a recombinant produced shortly after infection may undergo further exchange at later times. Second, recombination can occur between genetically similar phages as well as between different types. Thus, $P_1 \times P_1$ and $P_2 \times P_2$ occur in addition to $P_1 \times P_2$ (P_1 and P_2 refer to phage 1 and phage 2, respectively). For both of these reasons, recombinants from phage crosses are a consequence of a *population* of events rather than defined, single-step exchange events. Nevertheless, *all other things being equal,* the RF calculation does represent a valid index of map distance in phages.

Circularity of the T2 Genetic Map

Hershey obtained several different T2 strains with the rapid-lysis phenotype; he called their genotypes *r1*, *r2*, and so forth (in the order of discovery). Let's indicate three different *r* strains as r_a, r_b, and r_c, each with a mutation in a different gene, and make the cross $r_x^- h^+ \times r_x^+ h^-$, where r_x represents one of the three *r* genes. Table 10-3 shows the results. We can construct a linkage map in just the same way that Sturtevant constructed the original eukaryotic maps in *Drosophila*. The parental types ($r^- h^+$ and $r^+ h^-$) occur with the highest frequencies, although not with equal frequency. The two recombinant classes, however, are equally frequent. We can construct linkage maps for each cross (Figure 10-24a). The different recombination values indicate that the loci for the three *r* genes are in different

Table 10-2   Progeny-Phage Plaque Types from Cross $h^- r^+ \times h^+ r^-$

Phenotype	Inferred genotype
Clear and small	$h^- r^+$
Cloudy and large	$h^+ r^-$
Cloudy and small	$h^+ r^+$
Clear and large	$h^- r^-$

NOTE: Clearness is produced by the h^- allele, which allows infection of *both* bacterial strains in the lawn; cloudiness is produced by the h^+ allele, which limits infection to the cells of strain 1.

Table 10-3   Frequency of Progeny-Phage Types in Crosses Involving Several *r* Mutants and an *h* Mutant

Cross	Percentage of each genotype			
	$r^- h^+$	$r^+ h^-$	$r^+ h^+$	$r^- h^-$
$r_a^- h^+ \times r_a^+ h^-$	34.0	42.0	12.0	12.0
$r_b^- h^+ \times r_b^+ h^-$	32.0	56.0	5.9	6.4
$r_c^- h^+ \times r_c^+ h^-$	39.0	59.0	0.7	0.9

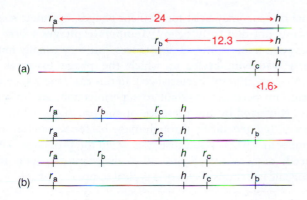

(a)

<1.6>

(b)

Figure 10-24 (a) Distances between gene pairs for each cross given in Table 10-3. (b) The various possible linkage relationships inferred from the distances in (a).

places on the chromosome, so there are four possible linkage maps (Figure 10-24b).

Can we distinguish among these alternatives? First, let's take only r_b, r_c, and h and ask whether the order is r_c, h, r_b or h, r_c, r_b. We can make the cross $r_c^-\, r_b^+ \times r_c^+\, r_b^-$ and compare the RF with the value of 12.3 obtained for the r_b–h interval. From this comparison, we find that the RF is greater than 12.3, so h is located between r_c and r_b (r_c, h, r_b).

Now we ask whether r_a lies on the side of h next to r_b or on the side next to r_c. The data from crosses of r_a with r_b and r_c do not provide a clear-cut answer. After intensive genetic mapping with many different strains of T2, the answer turns out to be that *both* alternative maps are correct. How can both r_a, r_c, h, r_b and r_c, h, r_b, r_a be correct? We have encountered a similar situation before, and the answer is similar. Like the bacterial map, the linkage map of T2 is circular, as shown in Figure 10-25a. After infection, the linear phage genome forms a circle in order to replicate the chromosome. The total genetic length of the T2 linkage map is about 1500 m.u. Figure 10-25b shows a more complex map for another T-even phage.

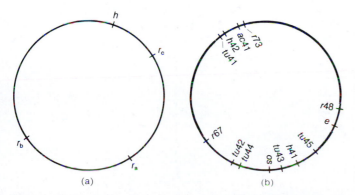

Figure 10-25 Circular maps for T-even phages. (a) Simple linkage map for phage T2. (b) Map of phage T4, inferred from crosses of nonlethal mutants *r* (rapid lysis), *h* (host range), *ac* (acridine resistance), *tu* (turbid plaques), *os* (resistance to osmotic shock), and *e* (lysis defective).

In the bacterial and phage experiments we have considered, the data are initially confusing. However, once the novel conditions for "crossing" are understood, the analysis of recombination and the construction of linkage maps are fairly straightforward procedures.

Message Recombination between phage chromosomes can be studied by bringing the parental chromosomes together in one host cell through mixed infection. Progeny phages can be examined for parental versus recombinant genotypes.

Lysogeny

In the 1920s, long before *E. coli* became the favorite organism of microbial geneticists, some interesting results were obtained in the study of phage infections of *E. coli*. Some bacterial strains were found to be resistant to infection by certain phages, but these resistant bacteria would cause lysis of nonresistant bacteria when the two bacterial strains were mixed together. The resistant bacteria that induced lysis in other cells were said to be **lysogenic bacteria, or lysogens.** When nonlysogenic bacteria were infected with phages derived from a lysogenic strain, a small fraction of the infected cells do not lyse but instead became lysogenic themselves.

Apparently, the lysogenic bacteria could somehow "carry" the phages while remaining immune to their lysing action. Initially, little attention was paid to this phenomenon after some studies seemed to show that the lysogenic bacteria were simply contaminated with external phages that could be removed by careful purification. However, in the mid-1940s, André Lwoff examined lysogenic strains of *Bacillus megaterium* and followed the behavior of a lysogenic strain through many cell divisions. Carefully observing his culture, he separated each pair of daughter cells immediately after division. One cell was put into a culture; the other was observed until it divided. In this way, Lwoff obtained 19 cultures representing 19 generations (19 consecutive cell divisions). All 19 cultures were lysogenic, but tests of the medium showed no free phage at any time during these divisions, thereby confirming that lysogenic behavior is a character that persists through reproduction in the absence of free phage.

On rare occasions, Lwoff observed spontaneous lysis in his cultures. When the medium was spread on a lawn of nonlysogenic cells after one of these spontaneous lyses, plaques appeared, showing that free phages had been released in the lysis. Lwoff was able to propose a hypothesis to explain all his observations: each bacterium of the lysogenic strain contains a noninfective factor that is passed from bacterial generation to generation, but this factor occasionally gives rise to the production of infective phage (without the presence of free phage in the medium). Lwoff called this factor the **prophage** because it somehow seemed to be able to *induce* the formation of a "litter" of infective phage. Later studies showed that a variety of agents, such as ultraviolet light or certain chemicals, could activate the prophage, inducing lysis and infective phage release in a large fraction of a population of lysogenic bacteria.

We now know exactly how Lwoff's observations occur. A lysogenic bacterium contains a prophage, which somehow protects the cell against additional infection, or **superinfection,** from free phages and which is duplicated and passed on to daughter cells during division. In a small fraction of the lysogenic cells, the prophage is induced, or activated, producing infective phage. This process robs the cell of its protection against the phage; it lyses and releases infective phage into the medium, thus infecting any nonlysogenic cells present in the culture.

Phages can be categorized into two types. **Virulent phages** have an infectious cycle that is always **lytic**—for these phages there are no lysogenic bacteria. (Resistant bac-

terial mutants may exist for virulent phages, but their resistance is not due to lysogeny.) **Temperate phages** follow a lytic cycle under some circumstances, but they usually initiate a **lysogenic cycle,** in which the phage exists as a prophage within the bacterial cell. In this case, the lysogenic bacterium becomes resistant to superinfection, an "immunity" conferred by the presence of the prophage, which is transmitted genetically through many bacterial generations. Temperate phages also cause lysis when the prophage is **induced,** or activated. Figure 10-26 diagrams the lytic and lysogenic infectious cycles of a typical temperate phage.

Message   Virulent phages cannot become prophages; they are always lytic. Temperate phages can exist within the bacterial cell as prophages, allowing their hosts to survive as lysogenic bacteria; they are also capable of direct bacterial lysis.

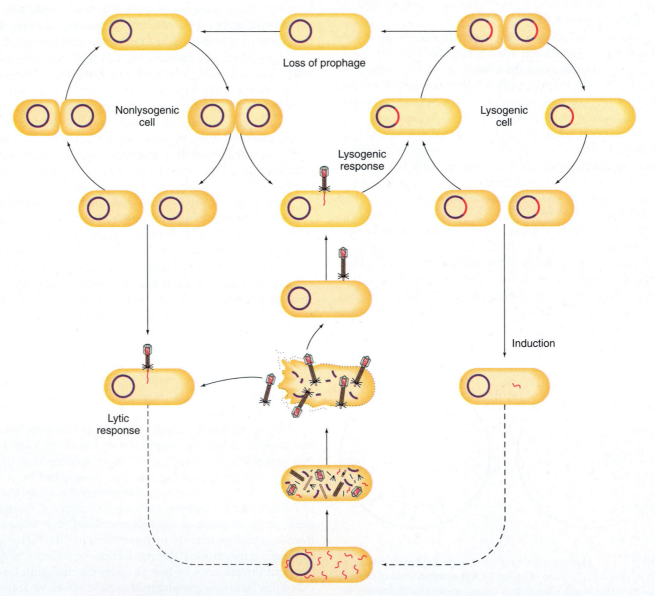

Loss of prophage

Nonlysogenic cell

Lysogenic cell

Lysogenic response

Lytic response

Induction

Figure 10-26   Alternative cell cycles of a temperate phage and its host. (Adapted from A. Lwoff, *Bacteriological Reviews* 17, 1953, 269.)

The Genetic Basis of Lysogeny

What is the nature of the prophage? On induction, the prophage is capable of directing the production of complete mature phage, so all of the phage genome must be present in the prophage. But is the prophage a small particle free in the bacterial cytoplasm—a plasmid—or is it somehow associated with the bacterial genome? Fortuitously, the original strain of *E. coli* used by Lederberg and Tatum (page 279) proved to be lysogenic for a temperate phage called **lambda** (**λ**). Phage λ (Figure 10-27) has become the most intensively studied and best-characterized phage. Crosses between F⁺ and F⁻ cells have yielded interesting results. It turns out that F⁺ × F⁻(λ) crosses yield recombinant lysogenic recipients, whereas the reciprocal cross F⁺(λ) × F⁻ almost never gives lysogenic recombinants.

These results became more understandable when Hfr strains were discovered. In the cross Hfr × F⁻(λ), lysogenic F⁻ exconjugants with Hfr genes are readily recovered. However, in the reciprocal cross Hfr(λ) × F⁻, the early genes from the Hfr chromosome are recovered among the exconjugants, but recombinants for late markers (those expected to transfer after a certain time in mating) are not recovered. Furthermore, lysogenic exconjugants are almost never recovered from this reciprocal cross. What is the explanation? The observations make sense if the λ prophage is behaving like a bacterial gene locus (that is, like part of the bacterial chromosome). In interrupted-mating experiments, the λ prophage always enters the F⁻ cell at a specific time, closely linked to the *gal* locus. Thus, we can assign the λ prophage to a specific locus next to the *gal* region.

In the cross of a lysogenic Hfr with a nonlysogenic (nonimmune) F⁻ recipient, the entry of the λ prophage into the nonimmune cell immediately triggers the prophage into a lytic cycle; this is called **zygotic induction.** But in the cross Hfr(λ) × F⁻(λ), any recombinants are readily recovered (that is, no induction of the prophage, and consequently lysis, occurs) (see Figure 10-28). It would seem that the cytoplasm of the F⁻ cell must exist in two different states (depending on whether or not the cell contains a λ prophage), so that contact between an entering prophage and the cytoplasm of a nonimmune cell immediately induces the lytic cycle. We now know that a cytoplasmic factor specified by the prophage represses the multiplication of the virus, as will be discussed in more detail in Chapter 18. Entry of the prophage into a nonlysogenic environment immediately dilutes this repressing factor, and therefore the virus reproduces. But if the virus specifies the repressing factor, then why doesn't the virus shut itself off again? Clearly it does, because a fraction of infected cells do become lysogenic. There is a race between the λ gene signals for reproduction and those specifying a shutdown. The model of a phage-directed cytoplasmic repressor nicely explains the immunity of the lysogenic bacteria, because any superinfecting phage would immediately encounter a repressor and be inactivated. We discuss this model in more detail in Chapter 18.

Prophage Attachment

How is the prophage attached to the bacterial genome? Allan Campbell proposed in 1962 that λ attaches to the bacterial chromosome by a reciprocal crossover between the circular λ chromosome and the circular *E. coli* chromosome,

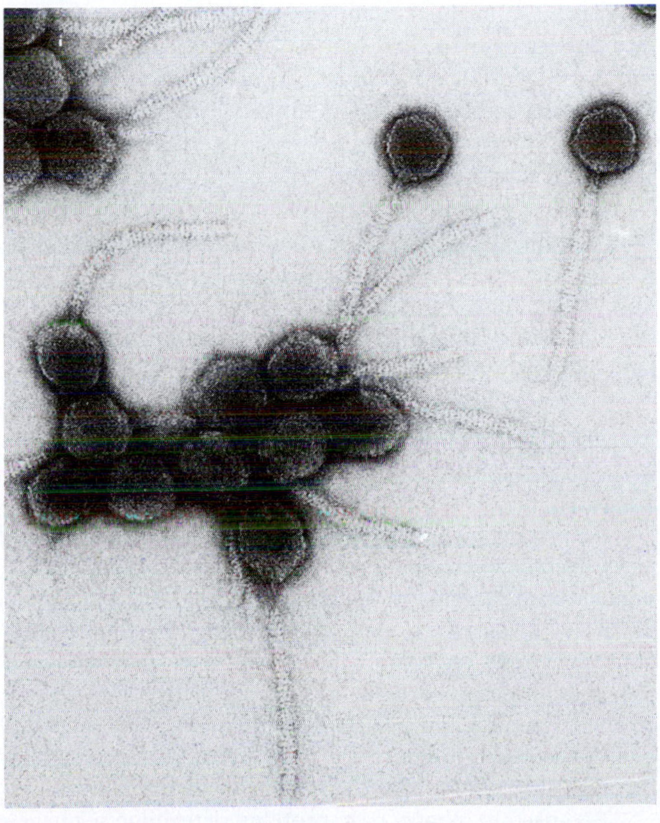

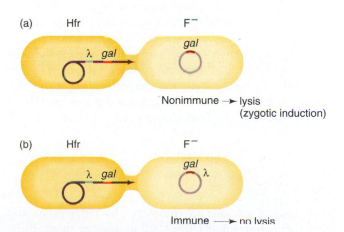

Figure 10-27 Electron micrograph of λ phages. (Jack D. Griffith)

Figure 10-28 Zygotic induction.

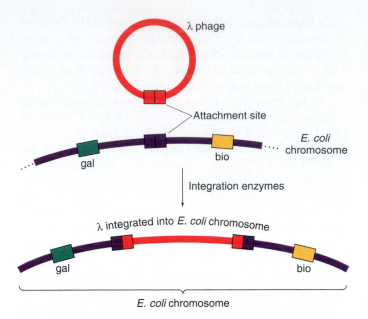

Figure 10-29 Campbell's model for the integration of phage λ into the *E. coli* chromosome. Reciprocal recombination occurs between a specific attachment site on the circular λ DNA and a specific region on the bacterial chromosome between the *gal* and *bio* genes.

as shown in Figure 10-29. The crossover point would occur between a specific site in λ, the **λ attachment site,** and a site in the bacterial chromosome located between the genes *gal* and *bio,* since λ integrates at that position in the *E. coli* chromosome.

One attraction of Campbell's proposal is that it allows predictions that geneticists can test, using phage λ:

1. Integration of the prophage into the *E. coli* chromosome should increase the genetic distance between flanking bacterial markers, as can be seen in Figure 10-29 for *gal* and *bio*. In fact, studies show that time-of-entry or recombination distances between the bacterial genes *are* increased by lysogeny.
2. Deleting bacterial segments adjacent to the prophage site should delete phage genes at least some of the time. Experimental studies also confirm this prediction.

The phenomenon of lysogeny is a very successful way for a temperate phage to avoid eating itself out of house and home. Lysogenic cells can perpetuate and carry the phages around. We consider lysogeny and the integration of the λ phage into the host chromosome in more detail in Chapter 18.

Transduction

Some phages are able to "mobilize" bacterial genes and carry them from one bacterial cell to another through the process of **transduction.** Thus, transduction joins the bat-

tery of modes of genetic transfer in bacteria—along with conjugation, infectious transfer of episomes, and transformation.

There are two kinds of transduction: generalized and specialized. Generalized transducing phage can carry any part of the chromosome, whereas **specialized** transducing phages carry only restricted parts of the bacterial chromosome.

The Discovery of Transduction

In 1951, Joshua Lederberg and Norton Zinder were testing for recombination in the bacterium *Salmonella typhimurium,* using the techniques that had been successful with *E. coli.* The researchers used two different strains: one was *phe⁻ trp⁻ tyr⁻*, and the other was *met⁻ his⁻*. (We won't worry about the nature of these markers except to note that the mutant alleles confer nutritional requirements.) When either strain was plated on a minimal medium, no wild-type cells were observed. However, after mixing the two strains, wild-type cells occurred at a frequency of about 1 in 10^5. Thus far, the situation seems similar to that for recombination in *E. coli.*

However, in this case, the researchers also recovered recombinants from a U-tube experiment, in which cell contact (conjugation) was prevented by a filter separating the two arms. By varying the size of the pores in the filter, they found that the agent responsible for recombination was about the size of the virus P22, a known temperate phage of *Salmonella.* Further studies have supported the suggestion that the vector of recombination is indeed P22. The filterable agent and P22 are identical in properties of size, sensitivity to antiserum, and immunity to hydrolytic enzymes. Thus, Lederberg and Zinder, instead of confirming conjugation in *Salmonella,* had discovered a new type of gene transfer mediated by a virus. They called this process **transduction.** During the lytic cycle, some virus particles somehow pick up bacterial genes that are then transferred to another host, where the virus inserts its contents. Transduction has subsequently been shown to be quite common among both temperate and virulent phages.

Transducing Phages and Generalized Transduction

How are transducing phages produced? In 1965, K. Ikeda and J. Tomizawa threw light on this question in some experiments on the temperate *E. coli* phage P1. They found that when a nonlysogenic donor cell is lysed by P1, the bacterial chromosome is broken up into small pieces. Occasionally, the forming phage particles mistakenly incorporate a piece of the bacterial DNA into a phage head in place of phage DNA. This is the origin of the transducing phage. A similar process can occur when the prophage of P1 is induced.

Because the phage coat proteins determine a phage's ability to attack a cell, transducing phages can bind to a bac-

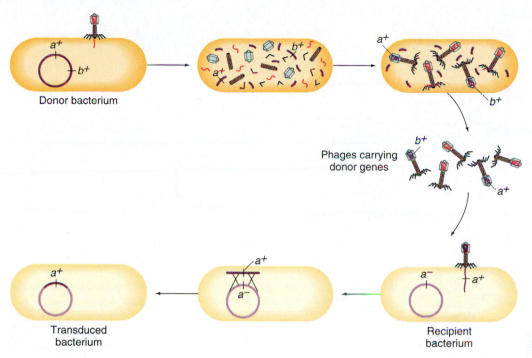

Figure 10-30 The mechanism of generalized transduction. In reality, only a very small minority of phage progeny (1 in 10,000) carries donor genes.

terial cell and inject their contents, which now happen to be donor bacterial genes. When a transducing phage injects its contents into a recipient cell, a merodiploid situation is created in which the transduced bacterial genes can be incorporated by recombination (Figure 10-30). Because any of the host markers can be transduced, this type of transduction is termed **generalized transduction.**

Phages P1 and P22 both belong to a phage group that shows generalized transduction (that is, they transfer virtually any gene of the host chromosome). As prophages, P22 probably inserts into the host chromosome and P1 remains free like a large plasmid. But both transduce by faulty head-stuffing during lysis.

Specialized Transduction

We look now at the process of **specialized transduction,** in which only certain host markers can be transduced.

Lambda (λ) is a good example of a specialized transducer. As a prophage, λ always inserts between the *gal* region and the *bio* region of the *E. coli* host chromosome. In transduction experiments, λ can transduce only the *gal* and *bio* genes. Let's visualize the mechanism of λ transduction.

In Figure 10-31a, we see a schematic representation of the production of a lysogen, or lysogenic bacterium. (The actual recombination between regions of λ and the bacterial chromosome is catalyzed by a specific enzyme system, described more fully in Chapter 20.) The phage and bacterial integration regions are not completely identical, so the integration of λ results in two hybrid integration sites, as shown.

We can induce the lytic cycle with ultraviolet light and produce a lysate (progeny phage population). The normal outlooping of the prophage restores the original phage integration site (Figure 10-31b, i). These phages can integrate normally (as in Figure 10-31a) on subsequent infection of a strain that is not lysogenic for λ.

Very rarely, abnormal outlooping can result in phage particles that now carry the *gal* gene (Figure 10-31b, ii). These particles are **defective** in that some phage genes have been left behind in the host; consequently, they are called **λ*dgal*** (λ-defective *gal*). The λ*dgal* particle has a λ protein coat and tail fibers and can infect bacteria, but it is also defective in its integration site. The hybrid integration site left in λ*dgal* does not provide a correct substrate for the phage-specific enzyme that promotes recombination between the phage and bacterial integration sites. Therefore, efficient integration cannot occur in a single infection of *E. coli* by λ*dgal*. However, coinfection with a wild-type λ phage results in efficient integration of the λ*dgal* phage (Figure 10-13c, i); the second, wild-type phage is often termed a **helper phage.** In general, a helper phage provides something that is required for integration into the host chromosome.

In practice, the lysate of λ produced originally (Figure 10-31b) can be used to infect a *gal*⁻ recipient culture that is nonlysogenic for λ. These infected cells are plated on a minimal medium, and rare *gal*⁺ transductants are the only ones to grow and produce colonies. A small percentage of these

(a) Production of lysogen

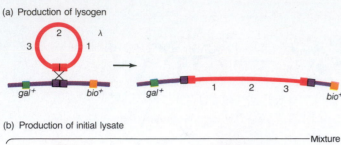

(b) Production of initial lysate

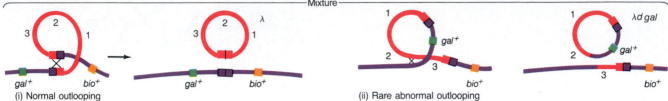

(i) Normal outlooping   (ii) Rare abnormal outlooping

(c) Transduction by initial lysate

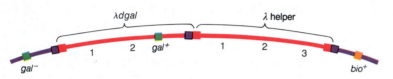

(i) Lysogenic transductants

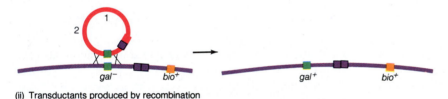

(ii) Transductants produced by recombination

Figure 10-31 Specialized transduction mechanism in phage λ. (a) The production of a lysogenic bacterium takes place by crossing over in a specialized region. (b) The lysogenic bacterial culture can produce normal λ or, rarely, an abnormal particle, λdgal, which is the transducing particle. (c) Transduction by the mixed lysate can produce *gal*⁺ transductants by the coincorporation of λdgal and a λ helper phage or, more rarely, by crossovers flanking the *gal* gene. The purple double squares are bacterial integration sites, the red double squares are λ integration sites, and the pairs of purple and red squares are hybrid integration sites, derived partly from *E. coli* and partly from λ.

transductants result from recombination between the *gal* regions of the λdgal and the recipient chromosome (Figure 10-31c, ii). The vast majority of the transductants (Figure 10-31c, i) are double lysogens: if such a culture is lysed and used as a *gal*⁺ donor in transduction, a very high frequency of transduction is obtained. This lysate is called a **high-frequency transduction (HFT)** lysate. HFT lysates contain a significant fraction of λdgal specialized-transducing phage, since each lysogen already contained a λdgal phage, and induction results in the excision and propagation of both λdgal and helper phage. This is in contrast to the original single lysogen (Figure 10-31a), which on induction yielded λdgal phage (Figure 10-31b, i) at a very low frequency.

Specialized transduction can mobilize only small regions of the bacterial genome that flank the prophage. Transduction by λ can mobilize either the *gal* or the *bio* genes, since these are the genes that can be incorporated into λ by rare, aberrant excision events. Specialized transduction is useful for moving genes from one bacterial strain to another.

Linkage Data from Transduction

Generalized transduction allows us to derive linkage information about bacterial genes when markers are close enough that the phage can pick them up and transduce them in a single piece of DNA. For example, suppose we wanted to find the linkage between *met* and *arg* in *E. coli*. We might set up a cross of a *met*⁺ *arg*⁺ strain with a *met*⁻ *arg*⁻ strain. We could grow phage P1 on the donor *met*⁺ *arg*⁺ strain, allow P1 to infect the *met*⁻ *arg*⁻ strain, and select for *met*⁺ colonies. Then, we could note the percentage of *met*⁺ colonies that became *arg*⁺. Strains transduced to both *met*⁺ and *arg*⁺ are called **cotransductants.**

Linkages usually are expressed as cotransduction frequencies (Figure 10-32). The greater the cotransduction frequency, the closer two genetic markers are.

We can estimate the size of the piece of host chromosome that a phage can pick up by using P1 phage in the following type of experiment:

donor *leu*⁺ *thr*⁺ *azi*ʳ ⟶ recipient *leu*⁻ *thr*⁻ *azi*ˢ

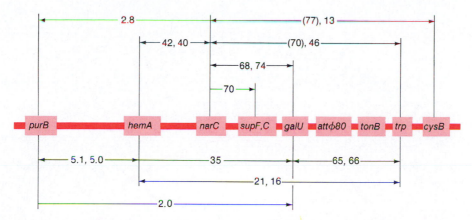

Figure 10-32 Genetic map of the *purB* to *cysB* region of *E. coli* determined by P1 cotransduction. The numbers given are the averages in percent for cotransduction frequencies obtained in several experiments. Where transduction crosses were performed in both directions, the head of each arrow points to the selective marker with the corresponding linkage nearest to each arrow. The values in parentheses are considered unreliable owing to interference from the nonselective marker. (Redrawn from J. R. Guest, *Molecular and General Genetics* 105, 1969, 285.)

We can select for one or more donor markers in the recipient and then (in true merozygous genetics style) look for the presence of the other unselected markers, as outlined in Table 10-4. Experiment 1 in the table tells us that *leu* is relatively close to *azi* and distant from *thr*, leaving us with two possibilities:

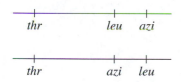

Experiment 2 tells us that *leu* is closer to *thr* than *azi* is, so that the map must be

By selecting for *thr*⁺ and *leu*⁺ in the transducing phage in experiment 3, we see that the transduced piece of genetic material never includes the *azi* locus.

If enough markers were studied to produce a more complete linkage map, we could estimate the size of a transduced segment. Such experiments indicate that P1 cotransduction occurs within approximately 1.5 minutes of

the *E. coli* chromosome map (1 minute equals the length of chromosome transferred by an Hfr in one minute's time at 37°C).

Linkage in the Discovery of Transduction

At this point, we can figure out how Lederberg and Zinder were able to detect transduction in their experiment. Recall that they utilized two multiply marked strains. One strain required phenylalanine, tryptophan, and tyrosine for growth; the second required methionine and histidine. Based on what we now know about P22- and P1-mediated generalized transduction, we would not expect to restore either of these strains to the wild type unless the markers were so closely linked that they could be carried on the same transducing particle. And yet, that is precisely what happened in the experiment. The *met*⁻ *his*⁻ strain was the source of the P22 transducing phage. This strain donated the wild-type markers that replaced the *phe*⁻ *trp*⁻ *tyr*⁻ markers. These markers are all part of the same pathway and are very closely linked. (In fact, we now know that the phenylalanine and tyrosine requirements are caused by the same mutation, although the tryptophan requirement is due to a mutation in a different gene.) If Lederberg and Zinder had used a different set of markers, they might never have discovered transduction!

Message Transduction occurs when newly forming phages acquire host genes and transfer them to other bacterial cells. **Generalized transduction** can transfer any host gene. It occurs when phage packaging accidentally incorporates bacterial DNA instead of phage DNA. **Specialized transduction** is due to faulty separation of the prophage from the bacterial chromosome, so that the new phage includes both phage and bacterial genes. The transducing phage can transfer only specific host genes.

Table 10-4   Accompanying Markers in Specific P1 Transductions

Experiment	Selected marker	Unselected markers
1	*leu*⁺	50% are *azi*ʳ; 2% are *thr*⁺
2	*thr*⁺	3% are *leu*⁺; 0% are *azi*ʳ
3	*leu*⁺ and *thr*⁺	0% are *azi*ʳ

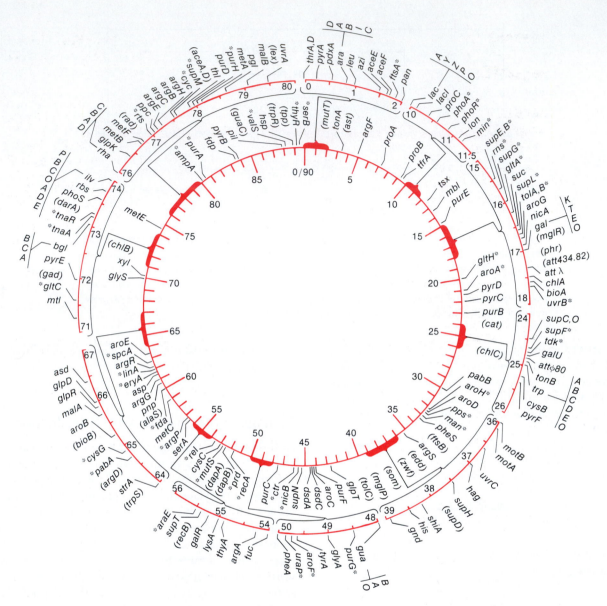

Figure 10-33 The 1963 genetic map of *E. coli*. Units are minutes, based on interrupted-mating experiments, timed from an arbitrarily located origin. (From G. S. Stent, *Molecular Biology of Bacterial Viruses*. Copyright © 1963 by W. H. Freeman and Company, New York.)

Chromosome Mapping

Some very detailed chromosomal maps for bacteria have been obtained by combining the mapping techniques of interrupted mating, recombination mapping, transformation, and transduction. Today, new genetic markers are typically mapped first into a segment of about 10 to 15 map minutes by using a series of Hfr strains that transfer from different points around the chromosome. This allows the selection of markers within the interval to be used for P1 cotransduction.

By 1963, the *E. coli* map (Figure 10-33) already detailed the positions of approximately 100 genes. After 27 years of further refinement, the 1990 map depicts the positions of more than 1400 genes! Figure 10-34 shows a 5-minute portion of the 1990 map (which is adjusted to a scale of 100 minutes). The complexity of these maps illustrates the power and sophistication of genetic analysis at its best.

Bacterial Gene Transfer in Review

1. Gene transfer in bacteria can be achieved through conjugation, transformation, and viral transduction.
2. The inheritance of genetic markers via the conjugative transfer of DNA by Hfr strains, the transformation of portions of the donor chromosome, and generalized transduction all share one important property. Each

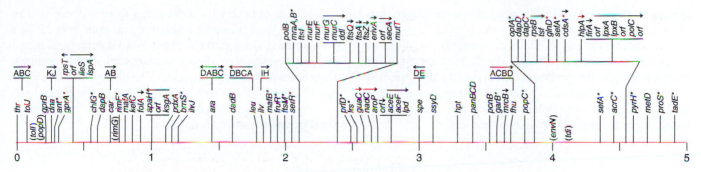

Figure 10-34 Linear scale drawing of a 5-minute portion of the 100-minute 1990 *E. coli* linkage map. Markers in parentheses are not precisely mapped; those marked by asterisks are more precisely mapped than those with parentheses, but are still not exactly known. Arrows above genes and groups of genes indicate the direction of transcription of these loci. (From B. J. Bachmann, "Linkage Map of *Escherichia coli* K-12, Edition 8," *Microbiological Reviews* 54, 1990, 130–197.)

process introduces a DNA fragment into the recipient cell; then a double-crossover event must occur if the fragment is to be incorporated into the recipient genome and subsequently inherited. Unincorporated fragments cannot replicate and are diluted out and lost from the population of daughter cells.

3. The conjugative transfer of F′ factors that carry bacterial genes and the specialized transduction of certain genetic markers are similar processes in that a specific and limited set of bacterial genes in each case is efficiently introduced into the recipient cell. Inheritance does not require normal recombination, as in the case of the in-

heritance of DNA fragments. After the F′ transfer, the F′ factor replicates in the bacterial cytoplasm as a separate entity. The specialized transducing phage DNA is recombined into the bacterial chromosome by a recombination system specific for that phage. In both cases, a partial diploid (merodiploid) results, because each process allows the inheritance of the transferred gene and also of the recipient's counterpart.

4. Gene transfer can be used to map the chromosome. Hfr crosses are first used to localize a mutation to a region of the chromosome. Then, generalized transduction provides a more exact localization.

SUMMARY

Advances in microbial genetics within the past four decades have provided the foundation for recent advances in molecular biology (discussed in the next several chapters). Early in this period, it was discovered that gene transfer and recombination occur between certain different strains of bacteria. In bacteria, however, genetic material is passed in only one direction—from a donor cell (F⁺ or Hfr) to a recipient cell (F⁻). Donor ability is determined by a presence in the cell of a fertility (F) factor acting as an episome.

On occasion, the F factor present in a free state in F⁺ cells can integrate the *E. coli* chromosome and form an Hfr cell. When this occurs, gene transfer and subsequent recombination take place. Furthermore, since the F factor can insert at different places on the host chromosome, investigators were able to show that the *E. coli* chromosome is a single circle, or ring. Interruptions of the transfer at different times has provided geneticists with a new method for

constructing a linkage map of the single chromosome of *E. coli* and other similar bacteria.

Genetic traits can also be transferred from one bacterial cell to another in the form of purified DNA. This process of transformation in bacterial cells was the first demonstration that DNA is the genetic material. For transformation to occur, DNA must be taken into a recipient cell, and recombination between a recipient chromosome and the incorporated DNA then must take place.

Bacteria can also be infected by bacteriophages. In one method of infection, the phage chromosome may enter the bacterial cell and, using the bacterial metabolic machinery, produce progeny phage that burst the host bacteria. The new phages can then infect other cells. If two phages of different genotypes infect the same host, recombination between their chromosomes can take place during this lytic process. Mapping the genetic loci through these recombinational events has led to the discovery that some phage chromosomes also are circular.

In another infection method, lysogeny, the injected phage lies dormant in the bacterial cell. In many cases, this dormant phage (the prophage) incorporates into the host chromosome and replicates with it. Either spontaneously or under appropriate stimulation, the prophage can arise from its latency and can lyse the bacterial host cell.

Phages can carry bacterial genes from a donor to a recipient. In generalized transduction, random host DNA is incorporated alone into the phage head during lysis. In specialized transduction, faulty outlooping of the prophage from a unique chromosomal locus results in the inclusion of specific host genes as well as phage DNA in the phage head.

Figure 10-35 summarizes the processes of conjugation, transformation, and transduction.

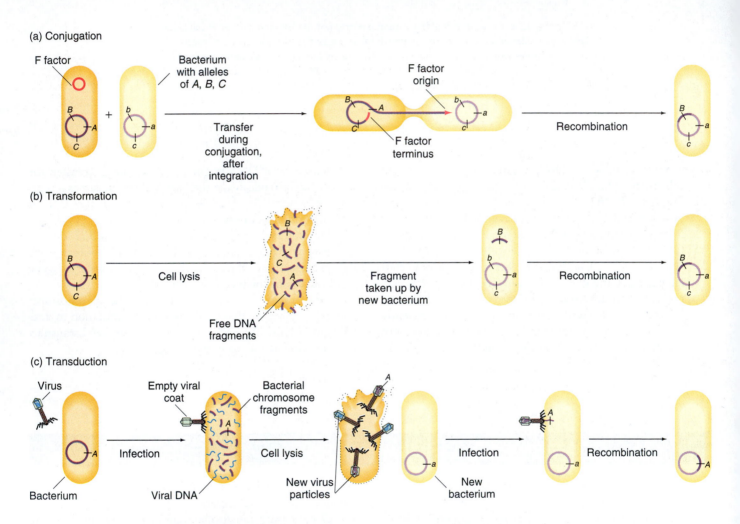

Figure 10-35 Recombination processes in bacteria. Bacterial recombination requires that a bacterial cell receive an allele obtained from another cell. (a) In conjugation, a cytoplasmic element such as the fertility factor (F) integrates into the chromosome of a bacterial cell. During cell-to-cell contact, the integrated factor can transfer part or all of that chromosome to another cell whose chromosome carries alleles of genes on the transferred chromosome. The transferred segment recombines with a homologous segment in the recipient cell's chromosome; in the example shown here, allele *B* thereby replaces allele *b*. (b) In transformation, a DNA segment bearing a particular allele is taken up from the environment by a cell whose chromosome carries a matching allele; the alleles (in our example, *B* and *b*) are then exchanged by homologous recombination. (c) In transduction, after a phage has infected a bacterial cell, one of the newly forming phage particles picks up a bacterial DNA segment instead of viral DNA. When this phage particle infects another cell, it injects its bacterial DNA, which recombines with a homologous segment in the second cell, thereby exchanging any corresponding alleles (in our example, *A* and *a*).

Concept Map

Draw a concept map interrelating as many of the following terms as possible. Note that the terms are listed in no particular order.

bacteria / conjugation / recombination / F plasmid / Hfr / F⁻ / donor / recipient / interrupted mating / chromosome map / pilus / merozygote / gene

CHAPTER INTEGRATION PROBLEM

We saw in Chapter 5 how recombination occurs by breakage and reunion of chromosomes. Suppose a cell were unable to carry out breakage and reunion by generalized recombination (*rec⁻*). How would this cell behave as a recipient in generalized and in specialized transduction? First compare each type of transduction and then determine the effect of the *rec⁻* mutation on the inheritance of genes by each process.

Solution

Generalized transduction involves the incorporation of chromosomal fragments into phage heads, which then infect recipient strains. Fragments of the chromosome are incorporated randomly into phage heads, so that any marker on the bacterial host chromosome can be transduced to another strain by generalized transduction. By contrast, specialized transduction involves the integration of the phage at a specific point on the chromosome and the rare incorporation of chromosomal markers near the integration site

into the phage genome. Therefore, only those markers that are near the specific integration site of the phage on the host chromosome can be transduced.

Inheritance of markers occurs by different routes in generalized and specialized transduction. A generalized transducing phage injects a fragment of the donor chromosome into the recipient. This fragment must be incorporated into the recipient's chromosome by recombination, using the recipient recombination system. Therefore, a *rec⁻* recipient will not be able to incorporate fragments of DNA and cannot inherit markers by generalized transduction. On the other hand, the major route for the inheritance of markers by specialized transduction involves integration of the specialized transducing particle into the host chromosome at the specific phage integration site. This integration, which sometimes requires an additional wild-type (helper) phage, is mediated by a phage-specific enzyme system that is independent of the normal recombination enzymes. Therefore, a *rec⁻* recipient can still inherit genetic markers by specialized transduction.

SOLVED PROBLEMS

1. In *E. coli*, four Hfr strains donate the following genetic markers shown in the order donated:

Strain 1:	Q	W	D	M	T
Strain 2:	A	X	P	T	M
Strain 3:	B	N	C	A	X
Strain 4:	B	Q	W	D	M

All of these Hfr strains are derived from the same F⁺ strain. What is the order of these markers on the circular chromosome of the original F⁺?

Solution

Recall the two-step approach that works well: (1) determine the underlying principle, and (2) draw a diagram. Here the principle is clearly that each Hfr strain donates genetic markers from a fixed point on the circular chromosome and that the earliest markers are donated with the highest frequency. Since not all markers are donated by each Hfr, only the early markers must be donated for each Hfr. Each strain allows us to draw the following circles:

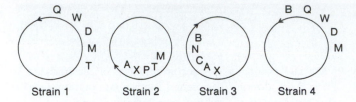

Strain 1   Strain 2   Strain 3   Strain 4

From this information, we can consolidate each circle into one circular linkage map of the order Q, W, D, M, T, P, X, A, C, N, B, Q.

2. In an Hfr × F⁻ cross, *leu*⁺ enters as the first marker, but the order of the other markers is unknown. If the Hfr is wild-type and the F⁻ is auxotrophic for each marker in question, what is the order of the markers in a cross where *leu*⁺ recombinants are selected if 27 percent are *ile*⁺, 13 percent are *mal*⁺, 82 percent are *thr*⁺, and 1 percent are *trp*⁺?

Solution

Recall that spontaneous breakage creates a natural gradient of transfer, which makes it less and less likely for a recipient to receive later and later markers. Because we have selected for the earliest marker in this cross, the frequency of recombinants is a function of the order of entry for each marker. Therefore, we can immediately determine the order of the genetic markers simply by looking at the percentage of recombinants for any marker among the *leu*⁺ recombinants. Because the inheritance of *thr*⁺ is the highest, this must be the first marker to enter after *leu*. The complete order is *leu, thr, ile, mal, trp*.

3. A cross is made between a Hfr that is *met*⁺ *thr*⁺ *pur*⁺ and an F⁻ that is *met*⁻ *thi*⁻ *pur*⁻. Interrupted-mating studies show that *met*⁺ enters the recipient last, so that *met*⁺ recombinants are selected on a medium containing supplements that satisfy only the *pur* and *thi* requirements. These recombinants are tested for the presence of the *thi*⁺ and *pur*⁺ alleles. The following numbers of individuals are found for each genotype:

met⁺ *thi*⁺ *pur*⁺	280
met⁺ *thi*⁺ *pur*⁻	0
met⁺ *thi*⁻ *pur*⁺	6
met⁺ *thi*⁻ *pur*⁻	52

a. Why was methionine (met) left out of the selection medium?

b. What is the gene order?

c. What are the map distances in recombination units?

Solution

a. Methionine was left out of the medium to allow selection for *met*⁺ recombinants, because *met*⁺ is the last marker to enter the recipient. This ensures that all the loci

we are considering in the cross will have already entered each recombinant that we analyze.

b. Here it is helpful to diagram the possible gene orders. Since we know that *met* enters the recipient last, there are only two possible gene orders if the first marker enters on the right: *met, thi, pur* or *met, pur, thi*. How can we distinguish between these two orders? Fortunately, one of the four possible classes of recombinants requires two additional crossovers. Each possible order predicts a different class that arises by four crossovers rather than two. For instance, if the order were *met, thi, pur*, then *met*⁺ *thi*⁻ *pur*⁺ recombinants would be very rare. On the other hand, if the order were *met, pur, thi*, then the four-crossover class would be *met*⁺ *pur*⁻ *thi*⁺. From the information given in the table, it is clear that the *met*⁺ *pur*⁻ *thi*⁺ class is the four-crossover class and therefore that the gene order *met, pur, thi* is correct.

c. Refer to the following diagram:

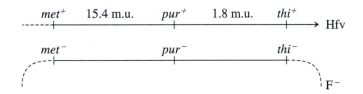

To compute the distance between *met* and *pur*, we compute the percentage of *met*⁺ *pur*⁻ *thi*⁻, which is 52/338 = 15.4 m.u. The distance between *pur* and *thi* is, similarly, 6/338 = 1.8 m.u.

4. Compare the mechanism of transfer and inheritance of the *lac*⁺ genes in crosses with Hfr, F⁺, and F′-*lac*⁺ strains. How would an F⁻ cell that cannot undergo normal homologous recombination (*rec*⁻) behave in crosses with each of these three strains? Would the cell be able to inherit the *lac*⁺ genes?

Solution

Each of these three strains donates genes by conjugation. In the cases of the Hfr and F⁺ strains, the *lac*⁺ genes on the host chromosome are donated. In the Hfr strain, the F factor is integrated into the chromosome in every cell, so that efficient donation of chromosomal markers can occur, particularly if the marker is near the integration site of F and is donated early. The F⁺ cell population contains a small percentage of Hfr cells, in which F is integrated into the chromosome. These cells are responsible for the gene transfer displayed by cultures of F⁺ cells. In the cases of Hfr- and F⁺-mediated gene transfer, inheritance requires the incorporation of a transferred fragment by recombination (recall that two crossovers are needed) into the F⁻ chromosome. Therefore, an F⁻ strain that cannot undergo recombination cannot inherit donor chromosomal markers even though they are transferred by Hfr strains or Hfr cells in F⁺ strains. The fragment cannot be incorporated into the chromosome by recombination. Since these fragments do not possess the ability to replicate within the F$^-$ cell, they are rapidly diluted out during cell division.

Unlike Hfr cells, F$'$ cells transfer genes carried on the F$'$ factor, a process that does not require chromosome transfer.

In this case, the *lac*$^+$ genes are linked to the F$'$ and are transferred with the F$'$ at a high efficiency. In the F$^-$ cell, no recombination is required, because the F$'$-*lac*$^+$ strain can replicate and be maintained in the dividing F$^-$ cell population. Therefore, the *lac*$^+$ genes are inherited even in a *rec*$^-$ strain.

PROBLEMS

1. Describe the state of the F factor in an Hfr, F$^+$, and F$^-$ strain.

2. How does a culture of F$^+$ cells transfer markers from the host chromosome to a recipient?

3. Draw an analogy between gene transfer and integration of the transferred gene into the recipient genome in

    a. Hfr crosses via conjugation and generalized transduction.

    b. F$'$ derivatives such as F$'$*lac* and specialized transduction.

4. Why can generalized transduction transfer any gene, but specialized transduction is restricted to only a small set?

5. A microbial geneticist isolates a new mutation in *E. coli* and wishes to map its chromosomal location. She uses interrupted-mating experiments with Hfr strains and generalized transduction experiments with phage P1. Explain why each technique, by itself, is insufficient for accurate mapping.

6. In *E. coli*, four Hfr strains donate the following markers, shown in the order donated:

strain 1:	M	Z	X	W	C
strain 2:	L	A	N	C	W
strain 3:	A	L	B	R	U
strain 4:	Z	M	U	R	B

All these Hfr strains are derived from the same F$^+$ strain. What is the order of these markers on the circular chromosome of the original F$^+$?

7. Four *E. coli* strains of genotype $a^+ b^-$ are labeled 1, 2, 3, and 4. Four strains of genotype $a^- b^+$ are labeled 5, 6, 7, and 8. The two genotypes are mixed in all possible combinations and (after incubation) are plated to determine the frequency of $a^+ b^+$ recombinants. The following results are obtained, where M = many recombinants, L = low numbers of recombinants, and 0 = no recombinants.

	1	2	3	4
5	0	M	M	0
6	0	M	M	0
7	L	0	0	M
8	0	L	L	0

On the basis of these results, assign a sex type (either Hfr, F$^+$, or F$^-$) to each strain.

8. An Hfr strain of genotype $a^+ b^+ c^+ d^- str^s$ is mated with a female strain of genotype $a^- b^- c^- d^+ str^r$. At various times, the culture is shaken vigorously to separate mating pairs. The cells are then plated on agar of the following three types, where nutrient A allows the growth of a^- cells; nutrient B, of b^- cells; nutrient C, of c^- cells; and nutrient D, of d^- cells (a plus indicates the presence of streptomycin and a nutrient, and a minus indicates its absence):

Agar type	str	A	B	C	D
1	+	+	+	−	+
2	+	−	+	+	+
3	+	+	−	+	+

    a. What donor genes are being selected on each type of agar?

    b. Table 10-5 shows the number of colonies on each type of agar for samples taken at various times after the strains are mixed. Use this information to determine the order of the genes *a*, *b*, and *c*.

    c. From each of the 25-minute plates, 100 colonies are picked and transferred to a dish containing agar with all of the nutrients except D. The numbers of colonies that grow on this medium are 89 for the sample from agar type 1, 51 for the sample from agar type 2, and 8 for the sample

Table 10-5

Time of sampling (minutes)	Number of colonies on agar of type		
	1	2	3
0	0	0	0
5	0	0	0
7.5	100	0	0
10	200	0	0
12.5	300	0	75
15	400	0	150
17.5	400	50	225
20	400	100	250
25	400	100	250

from agar type 3. Using these data, fit gene *d* into the sequence of *a*, *b*, and *c*.

d. At what sampling time would you expect colonies to first appear on agar containing C and streptomycin but no A or B?

(Problem 8 is from D. Freifelder, *Molecular Biology and Biochemistry.* Copyright © 1978 by W. H. Freeman and Company, New York.)

9. You are given two strains of *E. coli.* The Hfr strain is $arg^+ \ ala^+ \ glu^+ \ pro^+ \ leu^+ \ T^s$; the F^- strain is $arg^- \ ala^- \ glu^- \ pro^- \ leu^- \ T^r$. The markers are all nutritional except *T*, which determines sensitivity or resistance to phage T1. The order of entry is as given, with arg^+ entering the recipient first and T^s last. You find that the F^- strain dies when exposed to penicillin (pen^s) but the Hfr strain does not (pen^r). How would you locate the locus for *pen* on the bacterial chromosome with respect to *arg, ala, glu, pro,* and *leu*? Formulate your answer in logical, well-explained steps and draw explicit diagrams where possible.

10. A cross is made between two *E. coli* strains: Hfr $arg^+ \ bio^+ \ leu^+$ × $F^- \ art^- \ bio^- \ leu^-$. Interrupted-mating studies show that arg^+ enters the recipient last, so that arg^+ recombinants are selected on a medium containing *bio* and *leu* only. These recombinants are tested for the presence of bio^+ and leu^+. The following numbers of individuals are found for each genotype:

$arg^+ \ bio^+ \ leu^+$	320
$arg^+ \ bio^+ \ leu^-$	8
$arg^+ \ bio^- \ leu^+$	0
$arg^+ \ bio^- \ leu^-$	48

a. What is the gene order?

b. What are the map distances in recombination units?

11. You make the following *E. coli* cross: Hfr $Z_1^- \ ade^+ \ str^s$ × $F^- \ Z_2^- \ ade^- \ str^r$, in which *str* determines resistance or sensitivity to streptomycin, *ade* determines adenine requirement for growth, and Z_1 and Z_2 are two very close sites having Z^- alleles that cause an inability to use lactose as an energy source. After about an hour, the mixture is plated on a medium containing streptomycin, with glucose as the energy source. Many of the ade^+ colonies that grow are found to be capable of using lactose. However, hardly any of the ade^+ colonies from the reciprocal cross Hfr $Z_2^- \ ade^+ \ str^s$ × $F^- \ Z_1^- \ ade^- \ str^r$ are found to be capable of using lactose. What is the order of the Z_1 and Z_2 sites in relation to the *ade* locus? (Note that the *str* locus is terminal.)

12. Jacob selected eight closely linked *lac*⁻ mutations (called *lac*-1 through *lac*-8) and then attempted to order the mutations with respect to the outside markers *pro* (proline)

Table 10-6

x	y	Cross A	Cross B	x	y	Cross A	Cross B
1	2	173	27	1	8	226	40
1	3	156	34	2	3	24	187
1	4	46	218	2	8	153	17
1	5	30	197	3	6	20	175
1	6	168	32	4	5	205	17
1	7	37	215	5	7	199	34

and *ade* (adenine) by performing a pair of reciprocal crosses for each pair of *lac* mutants:

Cross A   Hfr $pro^- \ lac\text{-}x \ ade^+$ × $F^- \ pro^+ \ lac\text{-}y \ ade^-$

Cross B   Hfr $pro^- \ lac\text{-}y \ ade^+$ × $F^- \ pro^+ \ lac\text{-}x \ ade^-$

In all cases, prototrophs were selected by plating on a minimal medium with lactose as the only carbon source. Table 10-6 shows the number of colonies in the two crosses for each pair of mutants. Determine the relative order of the mutations.

(Problem 12 is from Burton S. Buttman, *Biological Principles.* Copyright © 1971, W. A. Benjamin, Menlo Park, California.)

13. Linkage maps in an Hfr bacterial strain are calculated in units of minutes (the number of minutes between genes indicates the length of time it takes for the second gene to follow the first during conjugation). In making such maps, microbial geneticists assume that the bacterial chromosome is transferred from Hfr to F^- at a constant rate. Thus, two genes separated by 10 minutes near the origin end are assumed to be the same physical distance apart as two genes separated by 10 minutes near the F attachment end. Suggest a critical experiment to test the validity of this assumption.

14. In the cross Hfr $aro^+ \ arg^+ \ ery^r \ str^s$ × $F^- \ aro^- \ arg^- \ ery^s \ str^r$, the markers are transferred in the order given (with aro^+ entering first), but the first three genes are very close together. Exconjugants are plated on a medium containing str (streptomycin, to counterselect Hfr cells), ery (erythromycin), arg (arginine), and aro (aromatic amino acids). The following results are obtained for 300 colonies from these plates isolated and tested for growth on various media: on ery only, 263 strains grow; on ery + arg, 264 strains grow; on ery + aro, 290 strains grow; on ery + arg + aro, 300 strains grow.

a. Draw up a list of genotypes, and indicate the number of individuals in each.

b. Calculate the recombination frequencies.

c. Calculate the ratio of the size of the *arg*-to-*aro* region to the size of the *ery*-to-*arg* region.

15. A particular Hfr strain normally transmits the *pro*⁺ marker as the last one during conjugation. In a cross of this strain with an F⁻ strain, some *pro*⁺ recombinants are recovered early in the mating process. When these *pro*⁺ cells are mixed with F⁻ cells, the majority of the F⁻ cells are converted to *pro*⁺ cells that also carry the F factor. Explain these results.

16. F′ strains in *E. coli* are derived from Hfr strains. In some cases, these F′ strains show a high rate of integration back into the bacterial chromosome of a second strain. Furthermore, the site of integration often is the same site that the sex factor occupied in the original Hfr strain (before production of the F′ strains). Explain these results.

17. You have two *E. coli* strains, F⁻ *str*ʳ *ala*⁻ and Hfr *str*ˢ *ala*⁺, in which the F factor is inserted close to *ala*⁺. Devise a screening test to detect strains carrying F′ *ala*⁺.

18. Five Hfr strains A through E are derived from a single F⁺ strain of *E. coli*. The following chart shows the entry times of the first five markers into an F⁻ strain when each is used in an interrupted-conjugation experiment:

A	B	C	D	E
mal⁺ (1)	*ade*⁺ (13)	*pro*⁺ (3)	*pro*⁺ (10)	*his*⁺ (7)
*str*ˢ (11)	*his*⁺ (28)	*met*⁺ (29)	*gal*⁺ (16)	*gal*⁺ (17)
ser⁺ (16)	*gal*⁺ (38)	*xyl*⁺ (32)	*his*⁺ (26)	*pro*⁺ (23)
ade⁺ (36)	*pro*⁺ (44)	*mal*⁺ (37)	*ade*⁺ (41)	*met*⁺ (49)
his⁺ (51)	*met*⁺ (70)	*str*ˢ (47)	*ser*⁺ (61)	*xyl*⁺ (52)

a. Draw a map of the F⁺ strain, indicating the positions of all genes and their distances apart in minutes.

b. Show the insertion point and orientation of the F plasmid in each Hfr strain.

c. In using each of these Hfr strains, state which gene you would select to obtain the highest proportion of Hfr exconjugants.

19. *Streptococcus pneumoniae* cells of genotype *str*ˢ *mtl*⁻ are transformed by donor DNA of genotype *str*ʳ *mtl*⁺ and (in a separate experiment) by a mixture of two DNAs with genotypes *str*ʳ *mtl*⁻ and *str*ˢ *mtl*⁺. Table 10-7 shows the results.

a. What does the first line of the table tell you? Why?

Table 10-7

Transforming DNA	Percentage of cells transformed to		
	*str*ʳ *mtl*⁻	*str*ˢ *mtl*⁺	*str*ʳ *mtl*⁺
*str*ʳ *mtl*⁺	4.3	0.40	0.17
*str*ʳ *mtl*⁻ + *str*ˢ *mtl*⁺	2.8	0.85	0.0066

Table 10-8

Drug(s) added	Number of colonies	Drug(s) added	Number of colonies
None	10,000	BC	51
A	1156	BC	49
B	1148	CD	786
C	1161	ABC	30
D	1139	ABD	42
AB	46	ACD	630
AC	640	BCD	36
AD	942	ABCD	30

b. What does the second line of the table tell you? Why?

20. A transformation experiment is performed with a donor strain that is resistant to four drugs: A, B, C, and D. The recipient is sensitive to all four drugs. The treated recipient cell population is divided up and plated on media containing various combinations of the drugs. Table 10-8 shows the results.

a. One of the genes obviously is quite distant from the other three, which appear to be tightly (closely) linked. Which is the distant gene?

b. What is the probable order of the three tightly linked genes?

(Problem 20 is from Franklin Stahl, *The Mechanics of Inheritance*, 2d ed. Copyright © 1969, Prentice-Hall, Englewood Cliffs, New Jersey. Reprinted by permission.)

21. Recall that in Chapter 5 we discussed the possibility that a crossover event may affect the likelihood of another crossover. In the bacteriophage T4, gene *a* is 1.0 m.u. from gene *b*, which is 0.2 m.u. from gene *c*. The gene order is *a*, *b*, *c*. In a recombination experiment, you recover five double crossovers between *a* and *c* from 100,000 progeny viruses. Is it correct to conclude that interference is negative? Explain your answer.

22. You have infected *E. coli* cells with two strains of T4 virus. One strain is minute (*m*), rapid-lysis (*r*), and turbid (*tu*); the other is wild-type for all three markers. The lytic products of this infection are plated and classified. Of 10,342 plaques, the following numbers are classified as each genotype:

m r tu	3467	*m + +*	520
+ + +	3729	*+ r tu*	474
m r +	853	*+ r +*	172
m + tu	162	*+ + tu*	965

a. Determine the linkage distances between m and r, between r and tu, and between m and tu.

b. What linkage order would you suggest for the three genes?

c. What is the coefficient of coincidence in this cross, and what does it signify?

(Problem 22 is reprinted with the permission of Macmillan Publishing Co., Inc., from Monroe W. Strickberger, *Genetics.* Copyright © 1968 by Monroe W. Strickberger.)

23. Using P22 as a generalized transducing phage grown on a $pur^+ pro^+ his^+$ bacterial donor, a recipient strain of genotype $pur^- pro^- his^-$ is infected and incubated. Afterwards, transductants for pur^+, pro^+, and his^+ are selected individually in experiments I, II, and III, respectively.

a. What media are used for these selection experiments?

b. The transductants are examined for the presence of unselected donor markers, with the following results:

I		II		III	
$pro^- his^-$	87%	$pur^- his^-$	43%	$pur^- pro^-$	21%
$pro^+ his^-$	0%	$pur^+ his^-$	0%	$pur^+ pro^-$	15%
$pro^- his^+$	10%	$pur^- his^+$	55%	$pur^- pro^+$	60%
$pro^+ his^+$	3%	$pur^+ his^+$	2%	$pur^+ pro^+$	4%

What is the order of the bacterial genes?

c. Which two genes are closest together?

d. On the basis of the order you proposed in part c, explain the relative proportions of genotypes observed in experiment II.

(Problem 23 is from D. Freifelder, *Molecular Biology and Biochemistry.* Copyright © 1978 by W. H. Freeman and Company, New York.)

24. Although most λ-mediated gal^+ transductants are inducible lysogens, a small percentage of these transductants in fact are not lysogens (that is, they contain no integrated λ). Control experiments show that these transductants are not produced by mutation. What is the likely origin of these types?

25. An $ade^+ arg^+ cys^+ his^+ leu^+ pro^+$ bacterial strain is known to be lysogenic for a newly discovered phage, but the site of the prophage is not known. The bacterial map is

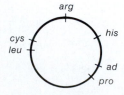

Table 10-9

Medium	Nutrient supplementation in medium						Presence of colonies
	ade	arg	cys	his	leu	pro	
1	−	+	+	+	+	+	N
2	+	−	+	+	+	+	N
3	+	+	−	+	+	+	C
4	+	+	+	−	+	+	N
5	+	+	+	+	−	+	C
6	+	+	+	+	+	−	N

NOTE: + indicates the presence of a nutrient supplement; − indicates supplement not present. N indicates no colonies; C indicates colonies present.

The lysogenic strain is used as a source of the phage, and the phages are added to a bacterial strain of genotype $ade^- arg^- cys^- his^- leu^- pro^-$. After a short incubation, samples of these bacteria are plated on six different media, with the supplementations indicated in Table 10-9. The table also shows whether or not colonies were observed on the various media.

a. What genetic process is at work here?

b. What is the approximate locus of the prophage?

26. You have two strains of λ that can lysogenize *E. coli*; the following figure shows their linkage maps:

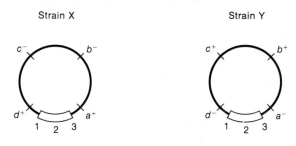

The segment shown at the bottom of the chromosome, designated 1–2–3, is the region responsible for pairing and crossing-over with the *E. coli* chromosome. (Keep the markers on all your drawings.)

a. Diagram the way in which λ strain X is inserted into the *E. coli* chromosome (so that the *E. coli* is lysogenized).

b. It is possible to superinfect the bacteria that are lysogenic for strain X by using strain Y. A certain percentage of these superinfected bacteria become "doubly" lysogenic (that is, lysogenic for both strains). Diagram how this will occur. (Don't worry about how double lysogens are detected.)

c. Diagram how the two λ prophages can pair.

d. It is possible to recover crossover products between the two prophages. Diagram a crossover event and the consequences.

27. You have three strains of *E. coli*. Strain A is F' *cys⁺ trp1/cys⁺ trp1* (that is, both the F' and the chromosome carry *cys⁺* and *trp1*, an allele for tryptophan requirement). Strain B is F⁻ *cys⁻ trp2 Z* (this strain requires cysteine for growth and carries *trp2*, another allele causing a tryptophan requirement; strain B also is lysogenic for the generalized transducing phage Z). Strain C is F⁻ *cys⁺ trp1* (it is an F⁻ derivative of strain A that has lost the F').

a. How would you determine whether *trp1* and *trp2* are alleles of the same locus? (Describe the crosses and the results expected.)

b. Suppose that *trp1* and *trp2* are nonallelic and that the *cys* locus is cotransduced with the *trp* locus. Using phage Z to transduce genes from strain C to strain B, how would you determine the genetic order of *cys, trp1,* and *trp2*?

28. A generalized transducing phage is used to transduce an $a^- b^- c^- d^- e^-$ recipient strain of *E. coli* with an $a^+ b^+ c^+ d^+ e^+$ donor. The recipient culture is plated on various media with the results shown in Table 10-10. (Note that a^- determines a requirement for A as a nutrient, and so forth.) What can you conclude about the linkage and order of the genes?

29. In a generalized transduction system using P1 phage, the donor is *pur⁺ nad⁺ pdx⁻* and the recipient is *pur⁻ nad⁻ pdx⁺*. The donor allele *pur⁺* is initially selected after transduction, and 50 *pur⁺* transductants are then scored for the other alleles present. The results follow:

Genotype	Number of colonies
nad⁺ pdx⁺	3
nad⁺ pdx⁻	10
nad⁻ pdx⁺	24
nad⁻ pdx⁻	13
	50

a. What is the cotransduction frequency for *pur* and *nad*?

b. What is the cotransduction frequency for *pur* and *pdx*?

c. Which of the unselected loci is closest to *pur*?

d. Are *nad* and *pdx* on the same side or on opposite sides of *pur*? Explain. (Draw the exchanges needed to produce the various transformant classes under either order to see which requires the minimum number to produce the results obtained.)

Table 10-10

Compounds added to minimal medium	Presence (+) or absence (−) of colonies
C D E	−
B D E	−
B C E	+
B C D	+
A D E	−
A C E	−
A C D	−
A B E	−
A B D	+
A B C	−

30. In a generalized transduction experiment, phages are collected from an *E. coli* donor strain of genotype *cys⁺ leu⁺ thr⁺* and used to transduce a recipient of genotype *cys⁻ leu⁻ thr⁻*. Initially, the treated recipient population is plated on a minimal medium supplemented with leucine and threonine. Many colonies are obtained.

a. What are the possible genotypes of these colonies?

b. These colonies are then replica-plated onto three different media: (1) minimal plus threonine only, (2) minimal plus leucine only, and (3) minimal. What genotypes could, in theory, grow on these three media?

c. It is observed that 56 percent of the original colonies grow on (1), 5 percent grow on (2), and no colonies grow on (3). What are the actual genotypes of the colonies on (1), (2), and (3)?

d. Draw a map showing the order of the three genes and which of the two outer genes is closer to the middle gene.

* 31. In 1965, Jon Beckwith and Ethan Signer devised a method of obtaining specialized transducing phages carrying the *lac* region. In a two-step approach, the researchers first "transposed" the *lac* genes to a new region of the chromosome and then isolated the specialized transducing particles. They noted that the integration site for the temperate phage φ80 (a relative of phage λ), designated *att80*, was located near one of the genes involved in conferring resistance to the virulent phage T1, termed *tonB*:

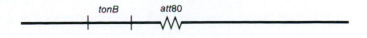

Beckwith and Signer used an F'-*lac* episome that could not replicate at high temperatures in a strain carrying a deletion of the *lac* genes. By forcing the cell to remain *lac*$^+$ at high temperatures, the researchers could select strains in which the episome had integrated into the chromosome, thereby allowing the F *lac* to be maintained at high temperatures. By combining this selection with a simultaneous selection for resistance to T1 phage infection, they found that the only survivors were cells in which the F *lac* had integrated into the *tonB* locus, as shown in the accompanying figure. Can you see why?

This placed the *lac* region near the integration site for phage ϕ80. Describe the subsequent steps that the researchers must have followed to isolate the specialized transducing particles of phage ϕ80 that carried the *lac* region.

11

The Structure of DNA

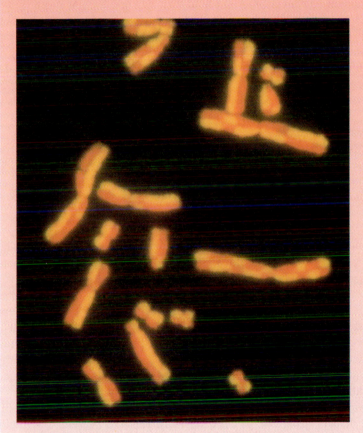

Harlequin chromosomes. (Sheldon Wolff/University of California, San Francisco)

KEY CONCEPTS

▶ Bacterial cells that express one phenotype can be transformed into cells that express a different phenotype; the transforming agent is DNA.

▶ Experiments with labeled T2 phage have established that DNA is the hereditary material.

▶ James Watson and Francis Crick showed that the structure of DNA is a double helix, in which each helix is a chain of nucleotides held together by phosphodiester bonds and in which specific hydrogen bonds are formed by pairs of bases.

▶ The DNA structure suggests that the fidelity of replication can be ensured if the complementary base of each base is specified by hydrogen bonding.

▶ The replication of DNA is semiconservative in that each daughter duplex contains one parental and one newly synthesized strand.

▶ Many of the enzymes involved in DNA synthesis in bacteria have been characterized.

Until now, we have looked at genes as abstract entities that somehow control hereditary traits. Through purely genetic analysis, we have studied the inheritance of different genes. But what about the physical nature of the gene? This question puzzled scientists for many years until it was realized that genes are composed of deoxyribonucleic acid (abbreviated *DNA*) and that DNA has a fascinating structure.

The elucidation of the structure of DNA in 1953 by James Watson and Francis Crick was one of the most exciting discoveries in the history of genetics. It paved the way for the understanding of gene action and heredity in molecular terms. Before we see how the solution of DNA structure was achieved, let's review what was known about genes and DNA at the time that Watson and Crick began their historic collaboration:

1. Genes—the hereditary "factors" described by Mendel—were known to be associated with specific character traits, but their physical nature was not understood.
2. The one-gene–one-enzyme theory (described more fully in Chapter 12) postulated that genes control the structure of proteins.
3. Genes were known to be carried on chromosomes.
4. The chromosomes were found to consist of DNA and protein.
5. Research by Frederick Griffith and, subsequently, by Oswald Avery and his coworkers pointed to DNA as the genetic material. These experiments, described here, showed that bacterial cells that express one phenotype can be transformed into cells that express a different phenotype and that the transforming agent is DNA.

DNA: The Genetic Material

The Discovery of Transformation

A puzzling observation was made by Frederick Griffith in the course of experiments on the bacterium *Streptococcus pneumoniae* in 1928. This bacterium, which causes pneumonia in humans, is normally lethal in mice. However, different strains of this bacterial species have evolved that differ in virulence (in the ability to cause disease or death). In his experiments, Griffith used two strains that are distinguishable by the appearance of their colonies when grown in laboratory cultures. In one strain, a normal virulent type, the cells are enclosed in a polysaccharide capsule, giving colonies a smooth appearance; hence, this strain is labeled *S*. In Griffith's other strain, a mutant nonvirulent type that grows in mice but is not lethal, the polysaccharide coat is absent, giving colonies a rough appearance; this strain is called *R*.

Griffith killed some virulent cells by boiling them and injected the heat-killed cells into mice. The mice survived, showing that the carcasses of the cells do not cause death.

However, mice injected with a mixture of heat-killed virulent cells and live nonvirulent cells did die. Furthermore, live cells could be recovered from the dead mice; these cells gave smooth colonies and were virulent on subsequent injection. Somehow, the cell debris of the boiled S cells had converted the live R cells into live S cells. The process is called **transformation.** Griffith's experiment is summarized in Figure 11-1.

This same basic technique was then used to determine the nature of the *"transforming principle"*—the agent in the cell debris that is specifically responsible for transformation. In 1944, Oswald Avery, C. M. MacLeod, and M. McCarty separated the classes of molecules found in the debris of the dead S cells and tested them for transforming ability, one at a time. These tests showed first that the polysaccharides themselves do not transform the rough cells. Therefore, the polysaccharide coat, although undoubtedly concerned with the pathogenic reaction, is only the phenotypic expression of virulence. In screening the different groups, Avery and his colleagues found that only one class of molecules, DNA, induced transformation of R cells (Figure 11-2). They deduced that DNA is the agent that determines the polysaccharide character and hence the pathogenic character (see pages 290–291 for a description of the mechanism of transformation). Furthermore, it seemed that providing R cells with S DNA was tantamount to providing these cells with S genes.

Message   The demonstration that DNA is the transforming principle was the first demonstration that genes are composed of DNA.

The Hershey-Chase Experiment

The experiments conducted by Avery and his colleagues were definitive, but many scientists were very reluctant to accept DNA (rather than proteins) as the genetic material. The clincher was provided in 1952 by Alfred Hershey and Martha Chase using the phage (virus) T2. They reasoned that phage infection must involve the introduction (injection) into the bacterium of the specific information that dictates viral reproduction. The phage is relatively simple in molecular constitution. Most of its structure is protein, with DNA contained inside the protein sheath of its "head."

Phosphorus is not found in proteins but is an integral part of DNA; conversely, sulfur is present in proteins but never in DNA. Hershey and Chase incorporated the radioisotope of phosphorus (^{32}P) into phage DNA and that of sulfur (^{35}S) into the proteins of a separate phage culture. They then used each phage culture independently to infect *E. coli* with many virus particles per cell. After sufficient time for injection to occur, they sheared the empty phage carcasses (called *ghosts*) off the bacterial cells by agitation in a kitchen blender. They used centrifugation to separate the bacterial cells from the phage ghosts and then measured the radioactivity in the two fractions. When the ^{32}P-labeled

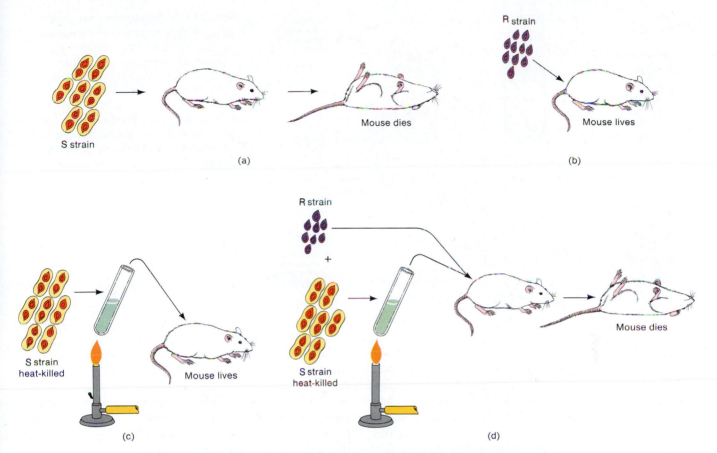

Figure 11-1 The first demonstration of bacterial transformation. (a) Mouse dies after injection with the virulent S strain. (b) Mouse survives after injection with the R strain. (c) Mouse survives after injection with heat-killed S strain. (d) Mouse dies after injection with a mixture of heat-killed S strain and live R strain. The heat-killed S strain somehow transforms the R strain to virulence. Parts a, b, and c act as control experiments for this demonstration. (From G. S. Stent and R. Calendar, *Molecular Genetics*, 2d ed. Copyright 1978 by W. H. Freeman and Company. After R. Sager and F. J. Ryan, *Cell Heredity*. John Wiley, 1961.)

phages were used, most of the radioactivity ended up inside the bacterial cells, indicating that the phage DNA entered the cells. ^{32}P can also be recovered from phage progeny. When the ^{35}S-labeled phages were used, most of the radioactive material ended up in the phage ghosts, indicating that the phage protein never entered the bacterial cell (Fig-

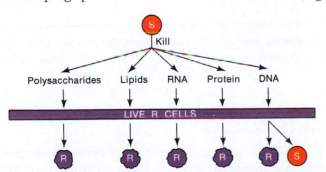

Figure 11-2 Demonstration that DNA is the transforming agent. DNA is the only agent that produces smooth (S) colonies when added to live rough (R) cells.

ure 11-3). The conclusion is inescapable: DNA is the hereditary material; the phage proteins are mere structural packaging that is discarded after delivering the viral DNA to the bacterial cell.

Why such reluctance to accept this conclusion? DNA was thought to be a rather simple chemical. How could all the information about an organism's features be stored in such a simple molecule? How could such information be passed on from one generation to the next? Clearly, the genetic material must have both the ability to encode specific information and the capacity to duplicate that information precisely. What kind of structure could allow such complex functions in so simple a molecule?

The Structure of DNA

Although the DNA structure was not known, the basic building blocks of DNA had been known for many years. The basic elements of DNA had been isolated and deter-

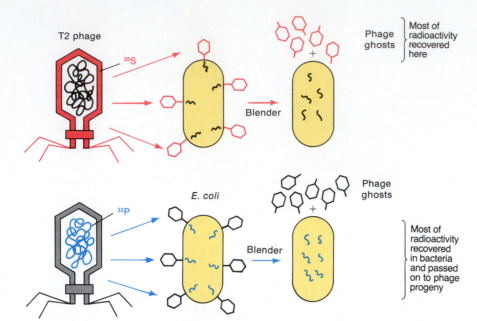

T2 phage

³⁵S

Blender

E. coli

³²P

Blender

Phage ghosts

Most of radioactivity recovered here

Phage ghosts

Most of radioactivity recovered in bacteria and passed on to phage progeny

Figure 11-3 The Hershey-Chase experiment, which demonstrated that the genetic material of phage is DNA, not protein. The experiment uses two sets of T2 bacteriophage. In one set, the protein coat is labeled with radioactive sulfur (^{35}S) not found in DNA. In the other set, the DNA is labeled with radioactive phosphorus (^{32}P) not found in protein. Only the ^{32}P is injected into the *E. coli*, indicating that DNA is the agent necessary for the production of new phages.

mined by partly breaking up purified DNA. These studies showed that DNA is composed of only four basic molecules called **nucleotides,** which are identical except that each contains a different nitrogen base. Each nucleotide contains phosphate, sugar (of the deoxyribose type), and one of the four bases (Figure 11-4). When the phosphate group is not present, the base and the deoxyribose form a **nucleoside** rather than a nucleotide. The four bases are **adenine, guanine, cytosine,** and **thymine.** The full chemical names of the nucleotides are deoxyadenosine 5′-monophosphate (or deoxyadenylate, or dAMP), deoxyguanosine 5′-monophosphate (or deoxyguanylate, or dGMP), deoxycytidine 5′-monophosphate (or deoxycytidylate, or dCMP), and deoxythymidine 5′-monophosphate (or deoxythymidylate, or dTMP). However, it is more convenient just to refer to each nucleotide by the abbreviation of its base (A, G, C, and T, respectively). Two of the bases, adenine and guanine, are similar in structure and are called **purines.** The other two bases, cytosine and thymine, also are similar and are called **pyrimidines.**

After the central role of DNA in heredity became clear, many scientists set out to determine the exact structure of DNA. How can a molecule with such a limited range of different components possibly store the vast range of information about all the protein primary structures of the living organism? The first to succeed in putting the building blocks together and finding a reasonable DNA structure—Watson and Crick in 1953—worked from two kinds of clues. First, Rosalind Franklin and Maurice Wilkins had amassed X-ray diffraction data on DNA structure. In such experiments, X rays are fired at DNA fibers, and the scatter of the rays from the fiber is observed by catching them on photographic film, where the X rays produce spots. The angle of scatter represented by each spot on the film gives information

about the position of an atom or certain groups of atoms in the DNA molecule. This procedure is not simple to carry out (or to explain), and the interpretation of the spot patterns is very difficult. The available data suggested that DNA is long and skinny and that it has two similar parts that are parallel to each other and run along the length of the molecule. The X-ray data showed the molecule to be helical (spiral-like). Other regularities were present in the spot patterns, but no one had yet thought of a three-dimensional structure that could account for just those spot patterns.

The second set of clues available to Watson and Crick came from work done several years earlier by Erwin Chargaff. Studying a large selection of DNAs from different organisms (see Table 11-1), Chargaff established certain empirical rules about the amounts of each component of DNA:

1. The total amount of pyrimidine nucleotides (T + C) always equals the total amount of purine nucleotides (A + G).
2. The amount of T always equals the amount of A, and the amount of C always equals the amount of G. But the amount of A + T is not necessarily equal to the amount of G + C, as can be seen in the last column of Table 11-1. This ratio varies among different organisms.

The Double Helix

The structure that Watson and Crick derived from these clues is a **double helix,** which looks rather like two interlocked bedsprings. Each bedspring (helix) is a chain of nucleotides held together by **phosphodiester bonds,** in which a phosphate group forms a bridge between —OH groups

Purine nucleotides

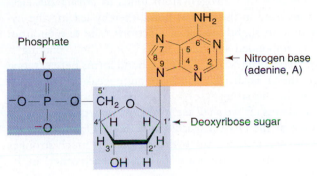

Deoxyadenosine 5'-phosphate (dAMP)

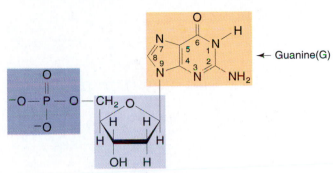

Deoxyguanosine 5'-phosphate (dGMP)

Pyrimidine nucleotides

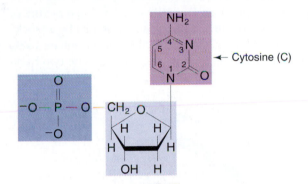

Deoxycytidine 5'-phosphate (dCMP)

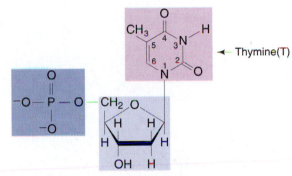

Deoxythymidine 5'-phosphate (dTMP)

Figure 11-4 Chemical structure of the four nucleotides (two with purine bases and two with pyrimidine bases) that are the fundamental building blocks of DNA. The sugar is called *deoxyribose* because it is a variation of a common sugar, ribose, that has one more oxygen atom.

Table 11-1	Molar Properties of Bases* in DNAs from Various Sources					
Organism	Tissue	Adenine	Thymine	Guanine	Cytosine	$\dfrac{A + T}{G + C}$
Escherichia coli (K12)	—	26.0	23.9	24.9	25.2	1.00
Diplococcus pneumoniae	—	29.8	31.6	20.5	18.0	1.59
Mycobacterium tuberculosis	—	15.1	14.6	34.9	35.4	0.42
Yeast	—	31.3	32.9	18.7	17.1	1.79
Paracentrotus lividus (sea urchin)	Sperm	32.8	32.1	17.7	18.4	1.85
Herring	Sperm	27.8	27.5	22.2	22.6	1.23
Rat	Bone marrow	28.6	28.4	21.4	21.5	1.33
Human	Thymus	30.9	29.4	19.9	19.8	1.52
Human	Liver	30.3	30.3	19.5	19.9	1.53
Human	Sperm	30.7	31.2	19.3	18.8	1.62

*Defined as moles of nitrogenous constituents per 100 g-atoms phosphate in hydrolysate.

SOURCE: E. Chargaff and J. Davidson, eds., *The Nucleic Acids.* Academic Press, 1955.

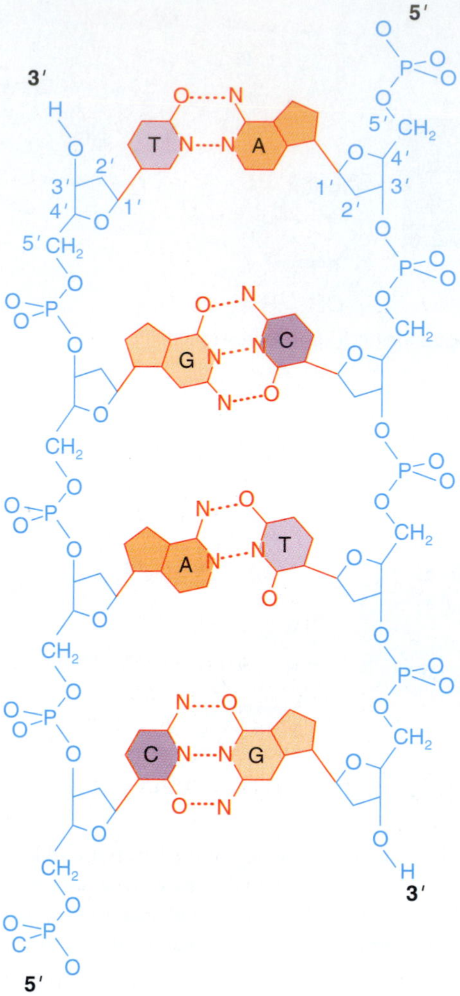

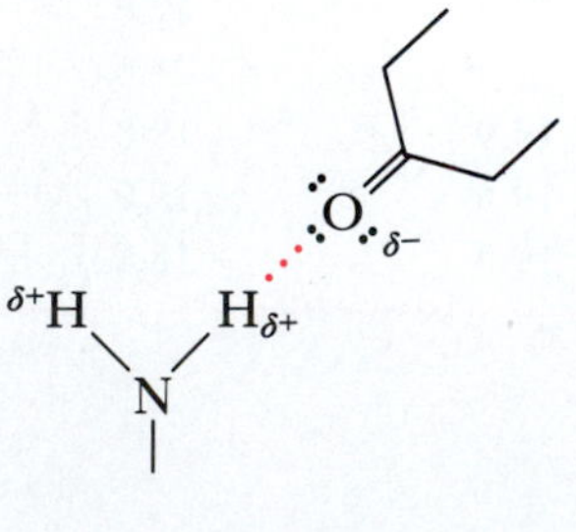

Each hydrogen atom in the NH$_2$ group is slightly positive (δ^+) because the nitrogen atom tends to attract the electrons involved in the N—H bond, thereby leaving the hydrogen atom slightly short of electrons. The oxygen atom has six unbonded electrons in its outer shell, making it slightly negative (δ^-). A hydrogen bond forms between one H and the O. Hydrogen bonds are quite weak (only about 3 percent of the strength of a covalent chemical bond), but this weakness (as we shall see) plays an important role in the function of the DNA molecule in heredity. One further important chemical fact: the hydrogen bond is much stronger if the participating atoms are "pointing at each other" in the ideal orientations.

The hydrogen bonds are formed by pairs of bases and are indicated by dotted lines in Figure 11-5, which shows a part of this paired structure with the helices uncoiled. Each base pair consists of one purine base and one pyrimidine based, paired according to the following rule: G pairs with C, and A pairs with T. In Figure 11-6, a simplified picture of the coiling, each of the base pairs is represented by a "stick" between the "ribbons," or so-called sugar-phosphate backbones of the chains. In Figure 11-5, note that the two back-

Figure 11-5 The DNA double helix, unrolled to show the sugar-phosphate backbones (blue) and base-pair rungs (red). The backbones run in opposite directions; the 5′ and 3′ ends are named for the orientation of the 5′ and 3′ carbon atoms of the sugar rings. Each base pair has one purine base, adenine (A) or guanine (G), and one pyrimidine base, thymine (T) or cytosine (C), connected by hydrogen bonds *(dotted lines)*. (From R. E. Dickerson, "The DNA Helix and How It Is Read." Copyright © 1983 by Scientific American, Inc. All rights reserved.)

on two adjacent sugar residues. The two bedsprings (helices) are held together by **hydrogen bonds,** in which two electronegative atoms "share" a proton, between the bases. Hydrogen bonds occur between hydrogen atoms with a small positive charge and acceptor atoms with a small negative charge. For example,

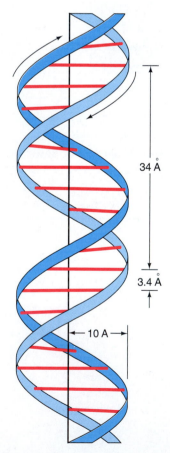

Figure 11-6 A simplified model showing the helical structure of DNA. The sticks represent base pairs, and the ribbons represent the sugar-phosphate backbones of the two antiparallel chains. The various measurements are given in angstroms (1 Å = 0.1 nm).

Pyrimidine + pyrimidine: DNA too thin

Purine + purine: DNA too thick

Purine + pyrimidine: thickness compatible with X-ray data

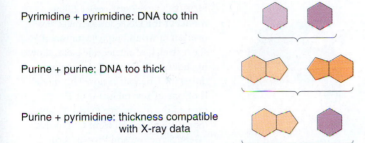

Figure 11-7 The pairing of purines with pyrimidines accounts exactly for the diameter of the DNA double helix determined from X-ray data. (From R. E. Dickerson, "The DNA Helix and How It Is Read." Copyright © 1983 by Scientific American, Inc. All rights reserved.)

bones run in opposite directions; they are thus said to be **antiparallel,** and (for reasons apparent in the figure) one is called the $5' \rightarrow 3'$ strand and the other the $3' \rightarrow 5'$ strand.

The double helix accounted nicely for the X-ray data and also tied in very nicely with Chargaff's data. Studying models they made of the structure, Watson and Crick realized that the observed radius of the double helix (known from the X-ray data) would be explained if a purine base always pairs (by hydrogen bonding) with a pyrimidine base (Figure 11-7). Such pairing would account for the

$(A + G) = (T + C)$ regularity observed by Chargaff, but it would predict four possible pairings: T · · · A, T · · · G, C · · · A, and C · · · G. Chargaff's data, however, indicate that T pairs only with A and C pairs only with G. Watson and Crick showed that only these two pairings have the necessary complementary "lock-and-key" shapes to permit efficient hydrogen bonding (Figure 11-8).

Note that the G–C pair has three hydrogen bonds, whereas the A–T pair has only two. We would predict that DNA containing many G-C pairs would be more stable than DNA containing many A–T pairs. In fact, this prediction is confirmed. We now have a neat explanation for Chargaff's data in terms of DNA structure (Figure 11-9). We also have a structure that is consistent with the X-ray data.

Three-Dimensional View of the Double Helix

In three dimensions, the bases actually form rather flat structures, and these flat bases partially stack on top of one another in the twisted structure of the double helix. This stacking of bases adds tremendously to the stability of the molecule by excluding water molecules from the spaces between the base pairs. (This phenomenon is very much like the stabilizing force that you can feel when you squeeze two plates of glass together underwater and then try to separate them.) Subsequently, it was realized that there were two forms of DNA in the fiber analyzed by diffraction. The A

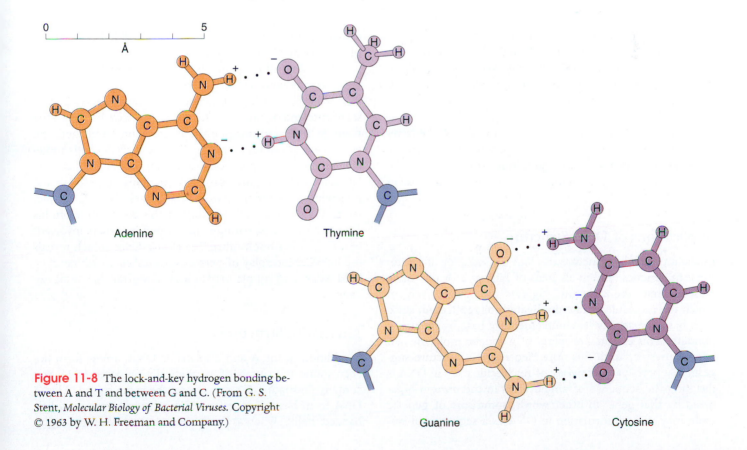

Figure 11-8 The lock-and-key hydrogen bonding between A and T and between G and C. (From G. S. Stent, *Molecular Biology of Bacterial Viruses.* Copyright © 1963 by W. H. Freeman and Company.)

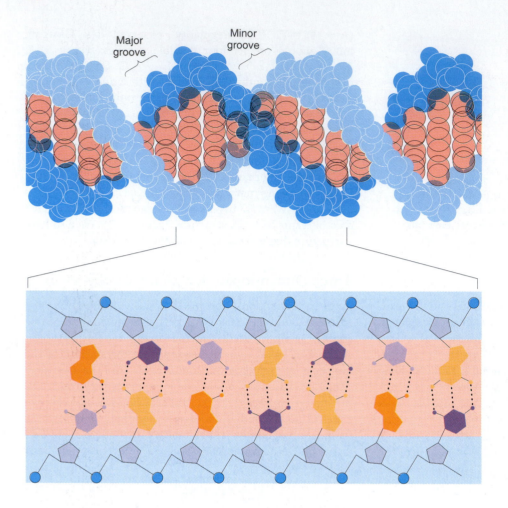

Major groove

Minor groove

form is less hydrated than the **B form** and is more compact. Figure 11-10 shows both the A and the B forms in a schematic three-dimensional drawing. Note the stacking of bases as represented in this diagram. It is believed that the B form of DNA is the form found most frequently in living cells.

The stacking of the base pairs in the double helix results in two grooves in the sugar-phosphate backbones. These are termed the **major** and **minor** grooves and can be readily seen in the space-filling (three-dimensional) model in Figure 11-9.

Implications of DNA Structure

Elucidation of the structure of DNA caused a lot of excitement in genetics (and in all areas of biology) for two basic reasons. First, the structure suggests an obvious way in which the molecule can be **duplicated,** or **replicated,** since each base can specify its complementary base by hydrogen bonding. This essential property of a genetic molecule had been a mystery until this time. Second, the structure suggests that perhaps the *sequence* of nucleotide pairs in DNA is dictating the sequence of amino acids in the protein organized by that gene. In other words, some sort of **genetic code** may write information in DNA as a sequence of nu-

cleotide pairs and then translate it into a different language of amino acid sequences in protein.

This basic information about DNA is now familiar to almost anyone who has read a biology text in elementary or high school, or even magazines and newspapers. It may seem trite and obvious, but try to put yourself back into the scene in 1953 and imagine the excitement! Until then, the evidence that the uninteresting DNA was the genetic molecule seemed disappointing and discouraging. But the Watson-Crick structure of DNA suddenly opened up the possibility of explaining two of the biggest "secrets" of life. James Watson has told the story of this discovery (from his own point of view, strongly questioned by others involved) in a fascinating book called *The Double Helix,* which reveals the intricate interplay of personality clashes, clever insights, hard work, and simple luck in such important scientific advances.

Alternative Structures

In addition to the A and B forms of DNA, a new form has been found in crystals of synthetically prepared DNA that contain alternating G's and C's on the same strand. This **Z DNA** form has a zigzag-like backbone and generates a left-handed helix, whereas both A and B DNA form right-

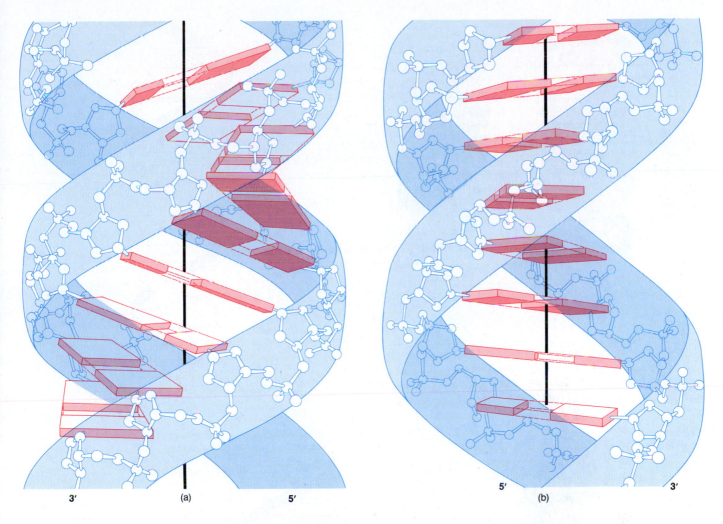

3'   (a)   5'   5'   (b)   3'

Figure 11-10 Schematic drawings of the structures of two forms of DNA. (a) A DNA. (b) B DNA. Analyses of DNA fibers by Struther Arnott and others provided the conceptual basis for these drawings. The sugar-phosphate backbones of the double helix are represented as ribbons and the runglike base pairs connecting them as planks. In A DNA, the base pairs are tilted and are pulled away from the axis of the double helix. In B DNA, on the other hand, the base pairs sit astride the helix axis and are perpendicular to it.

handed helices. Figure 11-11 shows space-filling models of each of the three forms.

Replication of DNA

Semiconservative Replication

Figure 11-12 diagrams the possible mechanism for DNA replication proposed by Watson and Crick. The sugar-phosphate backbones are represented by lines, and the sequence of base pairs is random. Let's imagine that the double helix is like a zipper that unzips, starting at one end (the bottom in this figure). We can see that if this zipper analogy is valid, the unwinding of the two strands will expose single bases on each strand. Because the pairing requirements imposed by the DNA structure are strict, each exposed base will pair

only with its complementary base. Because of this base complementarity, each of the two single strands will act as a **template,** or mold, and will begin to reform a double helix identical with the one from which it was unzipped. The newly added nucleotides are assumed to come from a pool of free nucleotides that must be present in the cell.

If this model is correct, then each daughter molecule should contain one parental nucleotide chain and one newly synthesized nucleotide chain. This prediction has been tested in both prokaryotes and eukaryotes. A little thought shows that there are at least three different ways in which a parental DNA molecule might be related to the daughter molecules. These hypothetical modes are called semiconservative (the Watson-Crick model), conservative, and dispersive (Figure 11-13). In **semiconservative** replication, each daughter duplex contains one parental and one newly

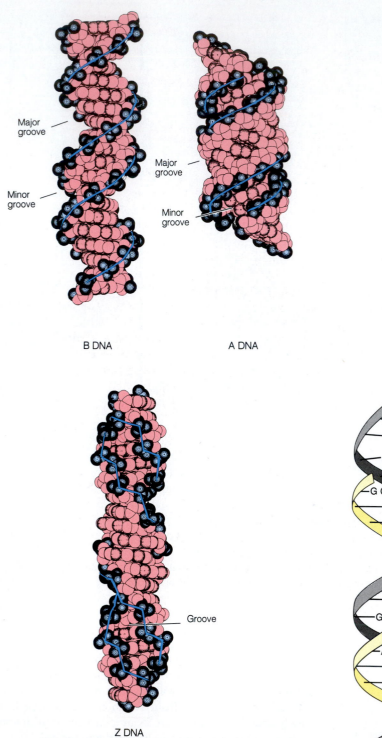

B DNA

A DNA

Major groove

Minor groove

Major groove

Minor groove

Groove

Z DNA

Figure 11-11 Space-filling models of B DNA, A DNA, and Z DNA. (From A. Kornberg and T. A. Baker, *DNA Replication*, 2d ed. Copyright © 1992 by W. H. Freeman and Company.)

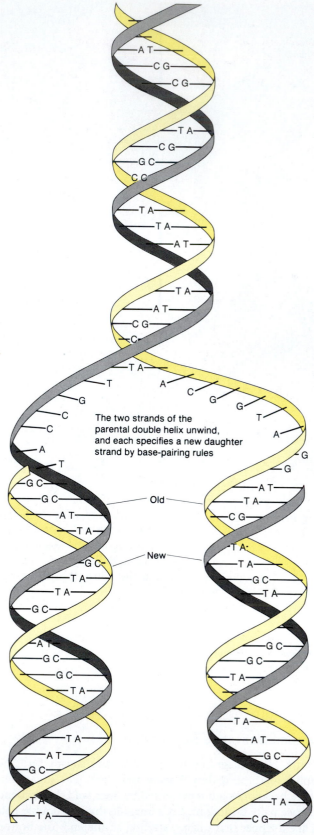

The two strands of the parental double helix unwind, and each specifies a new daughter strand by base-pairing rules

Old

New

Figure 11-12 The model of DNA replication proposed by Watson and Crick is based on the hydrogen-bonding specificity of the base pairs. Complementary strands are shown in different colors.

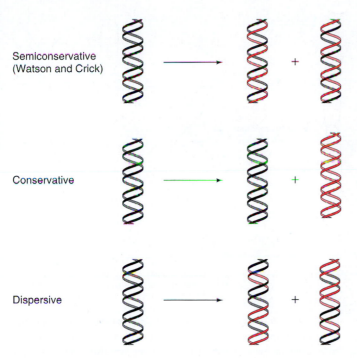

Semiconservative
(Watson and Crick)

Conservative

Dispersive

Figure 11-13 Three alternative patterns for DNA replication. The Watson-Crick model would produce the first (semiconservative) pattern. Red lines represent the newly synthesized strands.

tion also are pushed toward the bottom by centrifugal force. But as they travel down the tube, they encounter the increasing salt concentration, which tends to push them back up owing to the buoyancy of DNA (or its tendency to float). Thus, the DNA finally "settles" at some point in the tube where the centrifugal forces just balance the buoyancy of the molecules in the cesium chloride gradient. The buoyancy of DNA depends on its density (which in turn reflects the ratio of G–C to A–T base pairs). The presence of the heavier isotope of nitrogen changes the buoyant density of DNA. The DNA extracted from cells grown for several generations on ^{15}N medium can readily be distinguished from the DNA of cells grown on ^{14}N medium by the equilibrium position reached in a cesium chloride gradient. Such samples are commonly called *heavy* and *light* DNA, respectively.

Meselson and Stahl found, one generation after the heavy cells were moved to ^{14}N medium, the DNA formed a single band of an intermediate density between the densities of the heavy and light controls. After two generations in ^{14}N medium, the DNA formed two bands: one at the intermediate position, the other at the light position (Figure 11-14). This result would be expected from the semiconservative mode of replication; in fact, the result is compatible *only* with this mode *if* the experiment begins with chromosomes composed of individual double helices (Figure 11-15).

synthesized strand. However, in **conservative** replication, one daughter duplex consists of two newly synthesized strands, and the parent duplex is conserved. **Dispersive** replication results in daughter duplexes that consist of strands containing only *segment* of parental DNA and newly synthesized DNA.

The Meselson-Stahl Experiment

In 1958, Matthew Meselson and Franklin Stahl set out to distinguish among these possibilities. They grew *E. coli* cells in a medium containing the heavy isotope of nitrogen (^{15}N) rather than the normal light (^{14}N) form. This isotope was inserted into the nitrogen bases, which then were incorporated into newly synthesized DNA strands. After many cell divisions in ^{15}N, the DNA of the cells were well labeled with the heavy isotope. The cells were then removed from the ^{15}N medium and put into a ^{14}N medium; after one and two cell divisions, samples were taken. DNA was extracted from the cells in each of these samples and put into a solution of cesium chloride (CsCl) in an ultracentrifuge.

If cesium chloride is spun in a centrifuge at tremendously high speeds (50,000 rpm) for many hours, the cesium and chloride ions tend to be pushed by centrifugal force toward the bottom of the tube. Ultimately, a gradient of Cs^+ and Cl^- ions is established in the tube, with the highest ion concentration at the bottom. Molecules of DNA in the solu-

Autoradiography

The Meselson-Stahl experiment on *E. coli* was essentially duplicated in 1958 by Herbert Taylor on the chromosomes of bean root tip cells, using a cytological technique. Taylor put root cells into a solution containing tritiated thymidine ($[^3H]$-thymidine)—the thymine nucleotide labeled with a radioactive hydrogen isotope called *tritium*. He allowed the cells to undergo mitosis in this solution, so that the $[^3H]$-

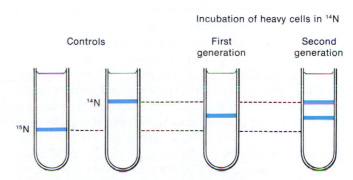

Figure 11-14 Centrifugation of DNA in a cesium chloride (CsCl) gradient. Cultures grown for many generations in ^{15}N and ^{14}N media provide control positions for heavy and light DNA bands, respectively. When the cells grown in ^{15}N are transferred to a ^{14}N medium, the first generation produces an intermediate DNA band and the second generation produces two bands: one intermediate and one light.

Semiconservative

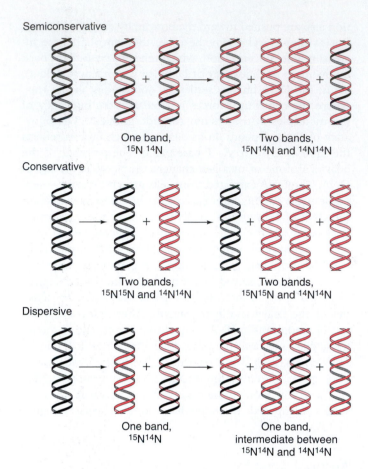

One band,
^{15}N ^{14}N

Two bands,
^{15}N^{14}N and ^{14}N^{14}N

Conservative

Two bands,
^{15}N^{15}N and ^{14}N^{14}N

Two bands,
^{15}N^{15}N and ^{14}N^{14}N

Dispersive

One band,
^{15}N^{14}N

One band,
intermediate between
^{15}N^{14}N and ^{14}N^{14}N

Figure 11-15 Only the semiconservative model of DNA replication predicts results like those shown in Figure 11-14: a single intermediate band in the first generation and one intermediate and one light band in the second generation. (See Figure 11-13 for explanations of symbols.)

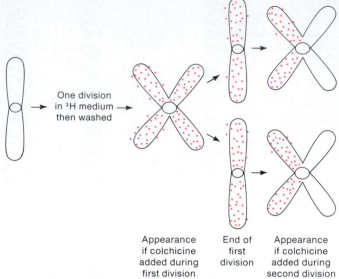

One division
in ^{3}H medium
then washed

Appearance
if colchicine
added during
first division

End of
first
division

Appearance
if colchicine
added during
second division

Figure 11-16 Diagrammatic representation of the autoradiography of chromosomes from cells grown for one cell division in the presence of the radioactive hydrogen isotope ^{3}H (tritium) and then grown in a nonradioactive medium for a second mitotic division. Each dot represents the track of a particle of radioactivity.

thymidine could be incorporated into DNA. He then washed the tips and transferred them to a solution containing nonradioactive thymidine. Addition of colchicine to such a preparation inhibits the spindle apparatus so that chromosomes in metaphase fail to separate and sister chromatids remain "tied together" by the centromere.

The cellular location of ^{3}H can be determined by **autoradiography.** As ^{3}H decays, it emits a beta particle (an energetic electron). If a layer of photographic emulsion is spread over a cell that contains ^{3}H, a chemical reaction takes place wherever a beta particle strikes the emulsion. The emulsion can then be developed like a photographic print, so that the emission track of the beta particle appears as a black spot or grain. The cell can also be stained, so that the structure of the cell is visible, to identify the location of the radioactivity. In effect, autoradiography is a process in which radioactive cell structures "take their own pictures."

Figure 11-16 shows the results observed when colchicine is added during the division in [^{3}H]-thymidine or during the subsequent mitotic division. It is possible to interpret these results by representing each chromatid as a sin-

gle DNA molecule that replicates semiconservatively (Figure 11-17).

Harlequin Chromosomes

Using a more modern staining technique, it is now possible to visualize the semiconservative replication of chromosomes at mitosis without the aid of autoradiography. In this procedure, the chromosomes are allowed to go through two rounds of replication in bromodeoxyuridine (BUdR). The BUdR labeling pattern, shown in Figure 11-18a, is the reciprocal of that of Figure 11-17, since the BUdR is used in both replications, rather than being replaced by normal thymidine for the second replication as in the autoradiographic process. The chromosomes are then stained with fluorescent dye and Giemsa stain; this process distinguishes hybrid chromatids with one BUdR-containing strand and one original strand (dark stain) from those in which both strands contain BUdR (no stain), and generates so-called **harlequin chromosomes** (Figure 11-18b). (Note, in passing, that harlequin chromosomes are particularly favorable for the detection of sister chromatid exchange at mitosis; two examples are seen in Figure 11-18b). Using similar techniques, Taylor showed that chromosome replication at meiosis also is semiconservative.

Chromosome Structure

Figure 11-16 and 11-17 bring up one of the remaining great unsolved questions of genetics: Is a eukaryotic chromosome

Figure 11-17 An explanation of Figure 11-16 at the DNA level. Red lines represent radioactive strands. In the second replication (which occurs in nontritiated solution), both the ^{3}H strand and the nontritiated strand incorporate nonradioactive nucleotides, yielding one hybrid and one nontritiated chromatid.

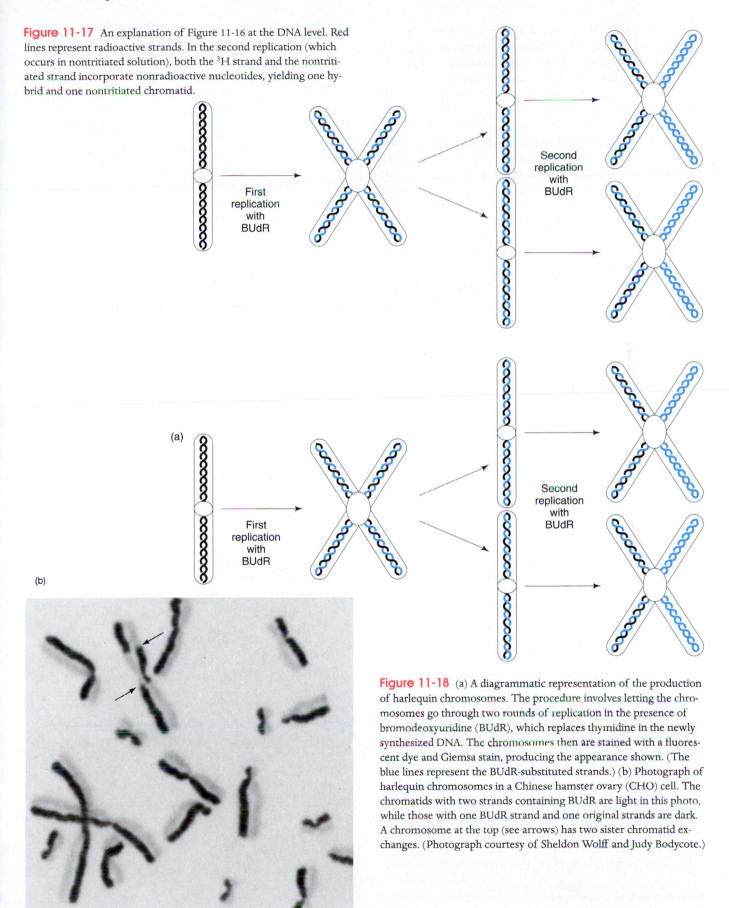

Figure 11-18 (a) A diagrammatic representation of the production of harlequin chromosomes. The procedure involves letting the chromosomes go through two rounds of replication in the presence of bromodeoxyuridine (BUdR), which replaces thymidine in the newly synthesized DNA. The chromosomes then are stained with a fluorescent dye and Giemsa stain, producing the appearance shown. (The blue lines represent the BUdR-substituted strands.) (b) Photograph of harlequin chromosomes in a Chinese hamster ovary (CHO) cell. The chromatids with two strands containing BUdR are light in this photo, while those with one BUdR strand and one original strands are dark. A chromosome at the top (see arrows) has two sister chromatid exchanges. (Photograph courtesy of Sheldon Wolff and Judy Bodycote.)

basically a single DNA molecule surrounded by a protein matrix? Two things strongly suggest that this is, in fact, the case. First, if there were many DNA molecules in the chromosome (whether they were side by side, end to end, or randomly oriented), it would be almost impossible for the chromosome to replicate semiconservatively (with all the label going into one chromatid, as in Taylor's results). Recent studies on isolated chromosomes and long DNA molecules are consistent with the suggestion that *each chromatid is a single molecule of DNA.* That makes a very long molecule. There is enough DNA in a single human chromosome, for example, to stretch out an inch or two and enough DNA in a single nucleus to stretch one meter. (This raises another interesting problem: How is this long molecule packed into the chromosome to permit easy replication?) The second fact supporting a single-molecule hypothesis is that DNA and genes behave as though they are attached end to end in a single string or thread that we call a *linkage group.* All genetic linkage data (Chapter 5) tell us that we need nothing more than a single linear array of genes per chromosome to explain the genetic facts. We examine the structure of the chromosome in Chapter 16.

The Replication Fork

A prediction of the Watson-Crick model of DNA replication is that a replication zipper, or **fork,** will be found in the DNA molecule during replication. In 1963, John Cairns tested this prediction by allowing replicating DNA in bacterial cells to incorporate tritiated thymidine. Theoretically, each newly synthesized daughter molecule should then contain one radioactive ("hot") strand and another nonradioactive ("cold") strand. After varying intervals and varying numbers of replication cycles in a "hot" medium, Cairns extracted the DNA from the cells, put it on a slide, and autoradiographed it for examination under the light microscope. After one replication cycle in [³H]-thymidine, rings of dots appeared in the autoradiograph. Cairns interpreted these rings as shown in Figure 11-19. It is also apparent from this figure that the bacterial chromosome is circular—a fact that also emerged from genetic analysis described earlier (Chapter 10).

During the second replication cycle, the forks predicted by the model were indeed seen. Furthermore, the density of grains in the three segments was such that the interpretation shown in Figure 11-20 could be made. Cairns saw all sizes of these moon-shaped, autoradiographic patterns, corresponding to the progressive movement of the replication zipper, or fork, around the ring. Structures of the sort shown in Figure 11-20 are called **theta (θ) structures.**

Mechanism of DNA Replication

Watson and Crick first reasoned that complementary base pairing provides the basis of fidelity in DNA replication, that is, that each base in the template strand dictates the comple-

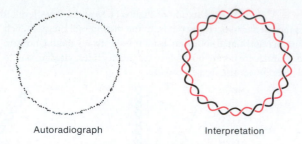

Figure 11-19 *Left:* autoradiograph of a bacterial chromosome after one replication in tritiated thymidine. According to the semiconservative model of replication, one of the two strands should be radioactive. *Right:* interpretation of the autoradiograph. The red line represents the tritiated strand.

mentary base in the new strand. However, we now know that the process of DNA replication is very complex and requires the participation of many different components. Let's examine each of these components and see how they fit together to produce our current picture of DNA synthesis in *Escherichia coli,* the best-studied cellular replication system. In the previous section we introduced the concept of the replication fork. Figure 11-21 gives a detailed schematic view of fork movement during DNA replication; we can refer to this illustration as we consider each component of the process.

DNA Polymerases

In the late 1950s, Arthur Kornberg successfully identified and purified the first DNA polymerase, an enzyme that catalyzes the replication reaction

$$\text{Primer (parental) DNA} + \left\{ \begin{array}{c} \text{dATP} \\ + \\ \text{dGTP} \\ + \\ \text{dCTP} \\ + \\ \text{dTTP} \end{array} \right\} \xrightarrow[\text{polymerase}]{\text{DNA}} \text{progeny DNA}$$

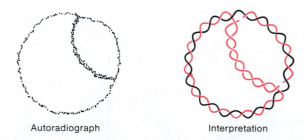

Figure 11-20 *Left:* Autoradiograph of a bacterial chromosome during the second round of replication in tritiated thymidine. In this theta (θ) structure, the newly replicated double helix that crosses the circle could consist of two radioactive strands (if the parental strand was the radioactive one). *Right:* The double thickness of the radioactive tracing on the autoradiogram appears to confirm the interpretation shown here. The red helices represent the "hot" strands.

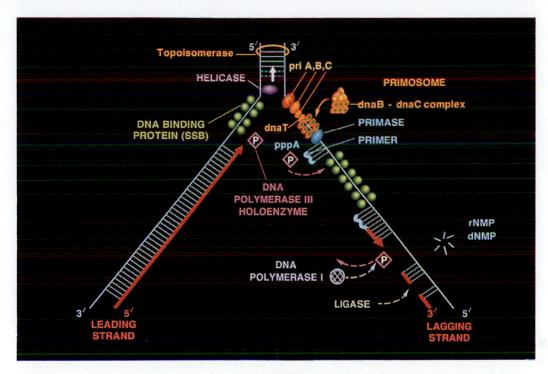

Figure 11-21 DNA replication fork. (From A. Kornberg and T. A. Baker, *DNA Replication*, 2d ed. Copyright © 1992 by W. H. Freeman and Company.)

This reaction works only with the triphosphate forms of the nucleotides (such as deoxyadenosine triphosphate, or dATP). The total amount of DNA at the end of the reaction can be as much as 20 times the amount of original input DNA, so most of the DNA present at the end must be progeny DNA. Figure 11-22 depicts the chain-elongation reaction, or **polymerization** reaction catalyzed by DNA polymerases. We now know that there are three DNA polymerases in *Escherichia coli*. The first enzyme Kornberg purified is called **DNA polymerase I**, or **pol I**. This enzyme has three activities, which appear to be located in different parts of the molecule:

1. A polymerase activity, which catalyzes chain growth in the $5' \rightarrow 3'$ direction
2. A $3' \rightarrow 5'$ exonuclease activity, which removes mismatched bases
3. A $5' \rightarrow 3'$ exonuclease activity, which degrades double-stranded DNA

Subsequently, two additional polymerases, pol II and pol III, were identified in *E. coli*. Pol II may operate to repair damaged DNA, although no particular role has been assigned to this enzyme. Pol III, together with pol I, is involved in the replication of *E. coli* DNA (Figure 11-21). The complete complex, or **holoenzyme,** of pol III contains at least 20 different polypeptide subunits, although the catalytic "core" consists of only three subunits, alpha (α), epsilon (ϵ), and theta (θ). The pol III complex will complete the replication

of single-stranded DNA if there is at least a short segment of duplex already present. The short oligonucleotide that creates the duplex is termed a **primer.**

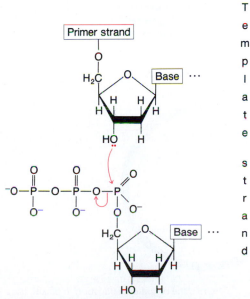

Figure 11-22 Chain-elongation reaction catalyzed by DNA polymerase (From L. Stryer, *Biochemistry*, 4th ed. Copyright © 1995 by Lubert Stryer.)

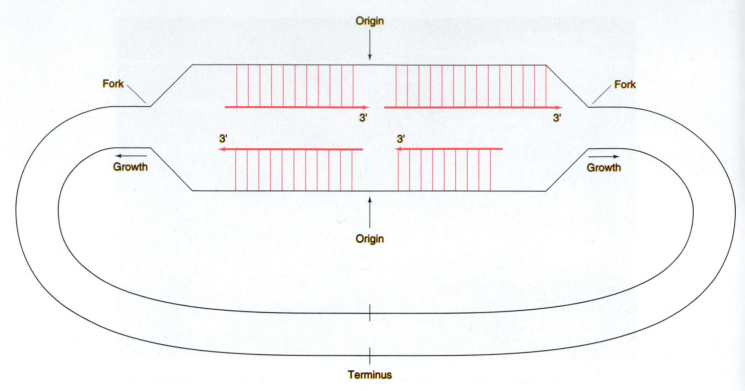

Figure 11-23 Diagrammatic representation of DNA replication proceeding in both directions from the origin. (Modified from A. Kornberg, *DNA Synthesis.* Copyright © 1974 by W. H. Freeman and Company.)

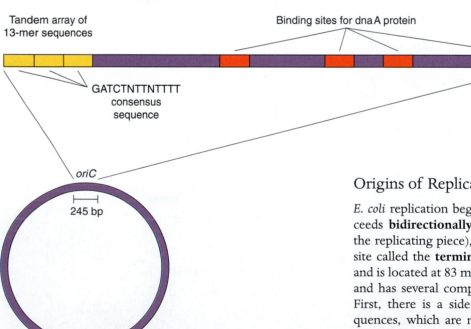

Figure 11-24 *OriC,* the origin of replication in *E. coli,* has a length of 245 bp. It contains a tandem array of three nearly identical 13-nucleotide sequences and four binding sites for DNA protein.

Origins of Replication

E. coli replication begins from a fixed **origin,** but then proceeds **bidirectionally** (with moving forks at both ends of the replicating piece), as shown in Figure 11-23, ending at a site called the **terminus.** The unique origin is termed *oriC* and is located at 83 min on the genetic map. It is 245 bp long and has several components, as illustrated in Figure 11-24. First, there is a side-by-side, or tandem, set of 13-bp sequences, which are nearly identical. There is also a set of binding sites for a protein, the dnaA protein. An initial step in DNA synthesis is the unwinding of the DNA at the origin in response to binding of the dnaA protein. The consequences of bidirectional replication can be seen in Figure 11-25, which gives a larger view of DNA replication.

Priming DNA Synthesis

DNA polymerases can extend a chain but cannot start a chain. Therefore, as mentioned above, DNA synthesis must first be initiated with a primer, or short oligonucleotide, that generates a segment of duplex DNA. The involvement of a primer in DNA replication can be seen in Figure 11-26 (see also Figure 11-21). The replication of the *E. coli* chromosome utilizes RNA primers, which are synthesized either by RNA polymerase or by an enzyme termed **primase.** Primase synthesizes a short (approximately 30-bp-long) stretch of RNA complementary to a specific region of the chromosome. The RNA chain is then extended with DNA by DNA polymerase. *E. coli* primase forms a complex with the template DNA, and additional proteins, such as dnaB, dnaT, pri A, pri B, and pri C. The entire complex is termed a **primosome** (see Figure 11-21).

Leading Strand and Lagging Strand

DNA polymerases synthesize new chains only in the 5′ → 3′ direction and, therefore, because of the antiparallel nature of the DNA molecule, move in a 3′ → 5′ direction *on*

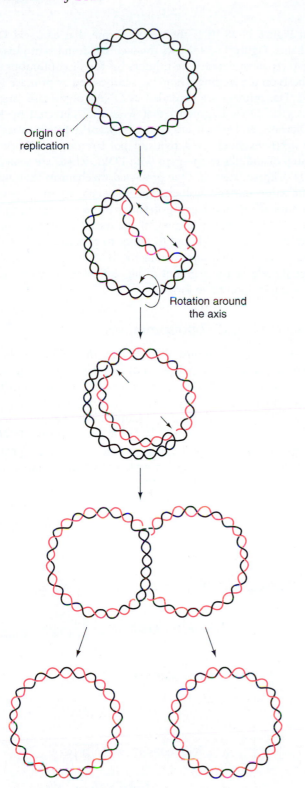

Figure 11-25 Bidirectional replication of a circular DNA molecule.

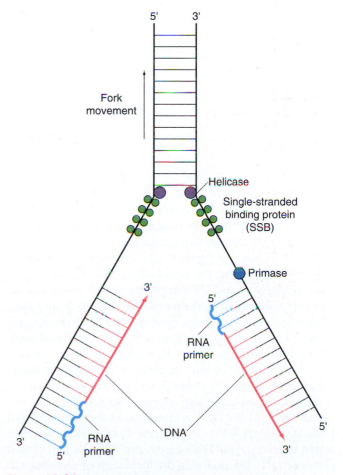

Figure 11-26 Initiation of DNA synthesis by an RNA primer.

in Figure 11-28. It is the only enzyme that can seal DNA chains. Figure 11-29 shows the lagging-strand synthesis and gap repair in detail. The primers for the discontinuous synthesis on the lagging strand are synthesized by primase (step a). The primers are extended by DNA polymerase (step b) to yield DNA fragments that were first detected by Reiji Okazaki, and which are termed **Okazaki fragments.** The $5' \rightarrow 3'$ exonuclease activity of pol I removes the primers (step c) and fills in the gaps with DNA, which are sealed by DNA ligase (step d). One proposed mechanism that allows the same dimeric holoenzyme molecule to participate in both leading- and lagging-strand synthesis is shown in Figure 11-30. Here, the looping of the template for the lagging strand allows a single pol III dimer to generate both daughter strands. After approximately 1000 base pairs, pol III would release the segment of lagging-strand duplex and allow a new loop to be formed.

Helicases and Topoisomerases

Helicases are enzymes that disrupt the hydrogen bonds which hold the two DNA strands together in a double helix. Hydrolysis of ATP drives the reaction. Among *E. coli* helicases are the dnaB protein and the rep protein. The rep protein may help to unwind the double helix ahead of the polymerase (refer to Figure 11-21). The unwound DNA is stabilized by the SSB (single-stranded binding) protein, which binds to the single-stranded DNA and retards reformation of the duplex.

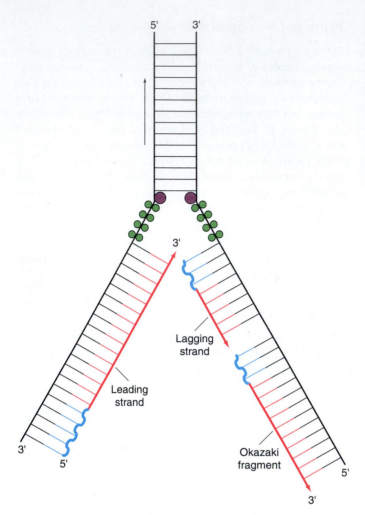

Figure 11-27 DNA synthesis proceeds by continuous synthesis on the leading strand and discontinuous synthesis on the lagging strand.

the template strand. The consequence of this polarity is that while one new strand, the **leading strand,** is synthesized continuously, the other, the **lagging strand,** must be synthesized in short, discontinuous segments, as can be seen in Figure 11-27 (see also Figure 11-21). This is because addition of nucleotides along the template for the lagging strand must proceed toward the template's $5'$ end (since replication *always* moves along the template in a $3' \rightarrow 5'$ direction, so that the new strand can grow $5' \rightarrow 3'$). Thus, the new strand must grow in a direction opposite to the movement of the replication fork. As fork movement exposes a new section of lagging-strand template, a new lagging-strand fragment is begun and proceeds away from the fork until it is stopped by the preceding fragment. In *E. coli,* pol III carries out most of the DNA synthesis on both strands, and pol I fills in the gaps left in the lagging strand, which are then sealed by the enzyme **DNA ligase.** DNA ligases join broken pieces of DNA by catalyzing the formation of a phosphodiester bond between the $5'$ phosphate end of a hydrogen-bonded nucleotide and an adjacent $3'$ OH group, as shown

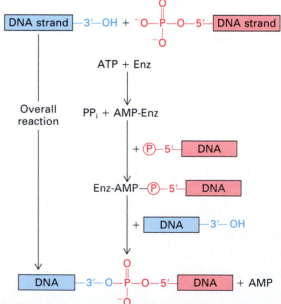

Figure 11-28 The reaction catalyzed by DNA ligase (Enz) joins the $3'$ OH end of one fragment to the $5'$ phosphate of the adjacent fragment (From H. Lodish, D. Baltimore, A. Berk, S. L. Zipursky, P. Matsudaira, and J. Darnell, *Molecular Cell Biology,* 3d ed. Copyright © 1995 by Scientific American Books, Inc.)

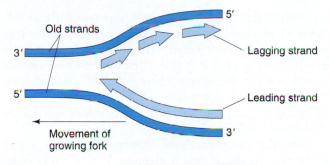

Old strands

3'

5'

Movement of growing fork

Lagging strand

Leading strand

5'

3'

Lagging-strand synthesis

(a) RNA oligonucleotides (primer) copied from DNA.

(b) DNA polymerase elongates RNA primers with new DNA.

(c) DNA polymerase removes 5' RNA at end of neighboring fragment and fills gap.

(d) DNA ligase joins adjacent fragments.

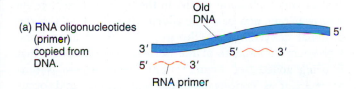

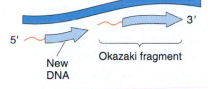

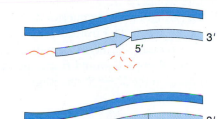

Old DNA

RNA primer

New DNA

Okazaki fragment

Ligation

Figure 11-29 The overall structure of a growing fork (top) and steps in synthesis of the lagging strand. (From H. Lodish, D. Baltimore, A. Berk, S. L. Zipursky, P. Matsudaira, and J. Darnell, *Molecular Cell Biology*, 3d ed. Copyright © 1995 by Scientific American Books, Inc.)

The action of helicases during DNA replication generates twists in the circular DNA that need to be removed to allow replication to continue. We should appreciate that circular DNA can be twisted and coiled, much like the extra coils that can be introduced into a rubber band. This **supercoiling** can be created or relaxed by enzymes termed **topoisomerases,** an example of which is DNA gyrase (Figure 11-31). Topoisomerases can also induce (*catenate*) or remove (*decatenate*) knots, or links in a chain. There are two basic types of isomerases. Type I enzymes induce a single-stranded break into the DNA duplex. Type II enzymes cause a break in both strands. In *Escherichia coli*, topo I and topo III are examples of type I enzymes, whereas gyrase is an example of a type II enzyme.

Untwisting of the DNA strands to open the replication fork causes extra twisting at other regions, and the supercoiling releases the strain of the extra twisting (see Figure 11-32). During replication, gyrase is needed to remove positive supercoils ahead of the replication fork.

Exonuclease Editing

Both DNA polymerase I and DNA polymerase III also possess 3' → 5' exonuclease activity, which serves a "proof-

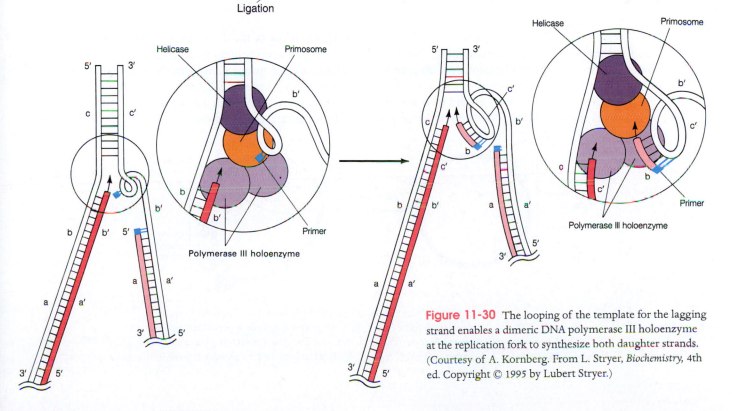

Figure 11-30 The looping of the template for the lagging strand enables a dimeric DNA polymerase III holoenzyme at the replication fork to synthesize both daughter strands. (Courtesy of A. Kornberg. From L. Stryer, *Biochemistry*, 4th ed. Copyright © 1995 by Lubert Stryer.)

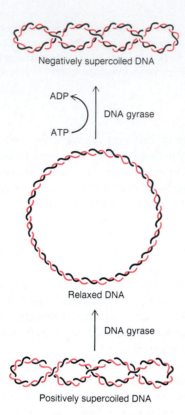

Figure 11-31 DNA gyrase–catalyzed supercoiling. Replicating DNA generates "positive" supercoils, depicted at the bottom of the diagram, as a result of rapid rotation of the DNA at the replication fork. DNA gyrase can nick and close phosphodiester bonds, relieving the supercoiling, as shown here (relaxed DNA). Gyrase can also generate supercoils twisted in the opposite direction, termed *negative* supercoils; this arrangement facilitates the unwinding of the helix. (Modified from L. Stryer, *Biochemistry,* 4th ed. Copyright © 1995 by Lubert Stryer.)

functional ϵ have a higher mutation rate (see Chapter 19). Figure 11-34 shows the excision of a cytosine residue that has erroneously been paired with an adenine. As can be seen, hydrolysis occurs at the 5′ end of the mismatched base; removal of the incorrect base leaves a 3′ OH group on the preceding base, which is then free to continue the growing strand by accepting the correct nucleotide triphosphate (thymidine, in this case).

Note that this exonuclease activity occurs at the 3′ end of the growing strand (and is therefore 3′ → 5′). The coordination of exonuclease activity with strand growth helps to explain why replication occurs in the 5′ → 3′ direction. As we saw earlier, new bases are added when the 3′ OH on the terminal deoxyribose of the growing strand attacks the high-energy phosphate of the nucleotide triphosphate that is being added (see Figure 11-22). Chain growth is thus 5′ → 3′. It is conceivable that replication could occur 3′ → 5′ (in Figure 11-22, the 5′ triphosphate at the bottom would be the last base on the chain, and the 3′ OH that attacks it would be on the free nucleotide triphosphate about to be added to the strand). However, if replication occurred in this direction, exonuclease excisions would occur at the 5′ end of the strand. When a mismatched base was removed, a 5′ OH would be left at the end of the growing strand. The 3′ OH of an incoming nucleotide triphosphate would thus be facing this 5′ OH instead of the high-energy 5′ triphosphate necessary for bond formation. No bond would form and strand growth would stop. Therefore, replication does not occur in the 3′ → 5′ direction.

DNA Replication in Eukaryotes

Although many of the details of DNA replication are similar in both prokaryotes and eukaryotes, there are a number of differences. For instance, in bacteria such as *E. coli* the replication-division cycle usually requires 40 minutes, but in eukaryotes it can vary from 1.4 hours in yeast to 24 hours in cultured animal cells, and may last 100 to 200 hours in some

reading" and "editing" function by searching for mismatched bases that were inserted erroneously during polymerization and excising them. The proofreading activity of pol III is the ϵ subunit, which must be bound to α for full proofreading activity (see Figure 11-33). Strains lacking a

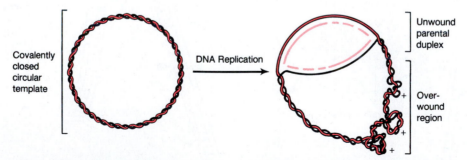

Figure 11-32 Swivel function of topoisomerase during replication. Extra-twisted (positively supercoiled) regions accumulate ahead of the fork as the parental strands separate for replication. A topoisomerase is required to remove these regions, acting as a swivel to allow extensive replication. (From A. Kornberg and T. A. Baker, *DNA Replication,* 2d ed. Copyright © 1992 by W. H. Freeman and Company.)

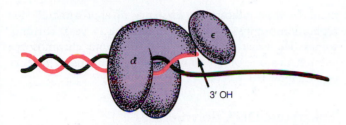

Figure 11-33 Proofreading by the pol III α-ε complex (From A. Kornberg and T. A. Baker, *DNA Replication*, 2d ed. Copyright © 1992 by W. H. Freeman and Company.)

cells. Eukaryotes have to solve the problem of coordinating the replication of more than one chromosome, as well as replicating the complex structure of the chromosome itself (see Chapter 16 for a description of chromosome structure).

The Cell Cycle

The eukaryotic **cell cycle** can be divided into four distinct phases, pictured schematically in Figure 11-35. These are

1. **Gap 1 (G1) phase.** The decision to replicate is made during this phase. Cells synthesize many of the factors and enzymes needed for replication. A commitment to proliferate is made when the cell proceeds past the "restriction point" (R), after which the programmed set of events cannot be reversed. Cells must reach a certain size to be able to enter the division cycle. Cells not entering the division cycle become quiescent (the **G0 phase**).
2. **Synthesis (S) phase.** Initiation of DNA synthesis occurs at multiple origins (see below), and the genome is replicated.

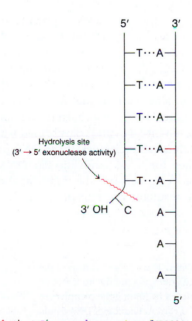

Figure 11-34 3′ → 5′ exonuclease action of DNA polymerase III.

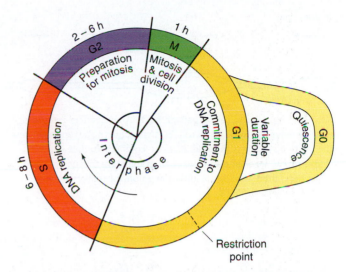

Figure 11-35 Growth cycle of a eukaryotic cell, including a quiescent phase G0. Variation in cell cycle time is due largely to variation in the duration of the G1 phase. (Modified from A. Kornberg and T. A. Baker, *DNA Replication*, 2d ed. Copyright © 1992 by W. H. Freeman and Company.)

3. **Gap 2 (G2) phase.** The cell prepares for entry into mitosis. Chromosomes condense, and cytoplasmic factors important for mitosis are synthesized.
4. **Mitosis (M) phase.** Chromosomes are segregated into daughter cells.

Table 11-2 compares the lengths of cell cycle periods in several eukaryotes.

Origins of Replication

In eukaryotes, replication proceeds from multiple points of origin. This can be demonstrated by a procedure in which a eukaryotic cell is briefly exposed to [³H]-thymidine, in a step called a **pulse** exposure, and then is provided an excess of "cold" (unlabeled) thymidine, in a step called the **chase**; the DNA is then extracted, and autoradiographs are made. Figure 11-36 shows the results of such a procedure, with what appear to be distinct, simultaneously replicating regions along the DNA molecule. Replication appears to be-

Table 11-2	Length of Cell Cycle Periods in Eukaryotes				
	Cell period				Total
Organism	M	G1	S	G2	doubling time
Human (h)	1	8	10	5	~24
Plant (h)	1	8	12	8	~29
Yeast (min)	20	25	40	35	~120

SOURCE: A. Kornberg and T. A. Baker, *DNA Replication*, 2d. ed. W. H. Freeman and Company, 1992.

Autoradiogram

Interpretation

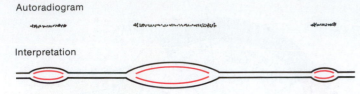

Figure 11-36 A replication pattern in DNA revealed by autoradiography. A cell is briefly exposed to [³H]-thymidine (pulse) and then provided with an excess of nonradioactive ("cold") thymidine (chase). DNA is spread on a slide and autoradiographed. In the interpretation shown here, there are several initiation points for replication within one double helix of DNA.

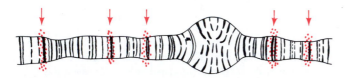

Figure 11-37 Replication pattern in a *Drosophila* polytene chromosome revealed by autoradiography. Several points of replication are seen within a single chromosome, as indicated by the arrows.

gin at several different sites on these eukaryotic chromosomes. Similarly, a pulse-and-chase study of DNA replication in polytene (giant) chromosomes of *Drosophila* by autoradiography reveals many replication regions within single chromosome arms (Figure 11-37). As yet there is no firm proof that these regions are indeed different starting points on a single DNA molecule. However, recent experi-

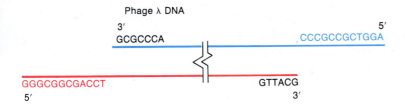

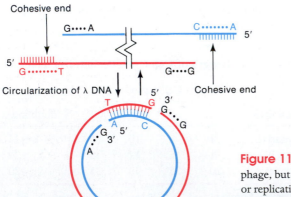

ments in yeast indicate the existence of approximately 400 replication origins distributed among the 17 yeast chromosomes. The exact structure of the eukaryotic chromosome still remains one of the most exciting problems in genetics (see Chapter 16).

Eukaryotic DNA Polymerases

There are at least four DNA polymerases, α, β, γ, and δ, in higher eukaryotes. Polymerase α and δ in the nucleus have roles similar to pol II in *E. coli*. Polymerase β is involved in DNA repair and gap filling. The γ polymerase is found in mitochondria and appears to be involved in replication of mitochondrial DNA.

DNA and the Gene

We have now learned that DNA is the genetic material and consists of a linear sequence of nucleotide pairs. The obvious conclusion is that the allele maps represent a genetic equivalent of the nucleotide-pair sequences in DNA. We can validate this assumption if we can show that *the genetic maps are congruent with DNA maps*. This step was first taken using some elegant genetic and biochemical manipulations.

The DNA of the λ phage turns out to be a linear stretch of DNA that can be circularized because each 5′ end of the two strands has an extra terminal extension of 12 bases that is complementary to the other 5′ end (Figure 11-38). Because they are complementary, these ends can pair to join the DNA into a circle; they are known as *cohesive, or sticky, ends.*

When a linear object such as a DNA molecule is subjected to shear stress (say, by pipetting or stirring), the mechanics of the stress causes breaks to occur, principally in the middle of the molecule. When DNA from the λ phage is sheared in half, the two halves happen to have different G–C : A–T ratios, which means, as we saw earlier, that their buoyant densities differ and they can be separated by centrifugation in CsCl. When the two half-molecules are separated, their genetic content can be assayed through their ability to recombine with normal λ DNA carrying known mutations (Figure 11-39). First, bacteria are simultaneously infected with mutant DNA and transformed with one or the other of the DNA half-molecules. If the DNA fragment contains the genes for which the injected DNA is mutant, crossovers will occur between the fragment and the mutant DNA and the resultant recombination can be detected by

Figure 11-38 Circularization of DNA. The λ DNA is linear in the phage, but once in a host cell, it circularizes as a prelude to insertion or replication. Circularization is achieved by joining the complementary ("sticky") single-stranded ends. (From A. Kornberg, *DNA Synthesis.* Copyright © 1974 by W. H. Freeman and Company.)

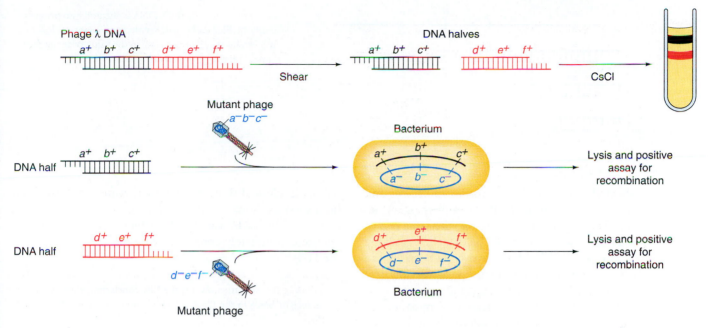

Figure 11-39 When λ DNA is broken, it forms roughly equal halves that happen to have different buoyant densities. Each half can be "rescued" by a multiply mutant phage during simultaneous transformation and infection. From these experiments, it has been shown that only genes from one specific half of the genetic map could be rescued from one specific half of the DNA. (Note that the bacterial chromosome has been omitted here to simplify the diagram.)

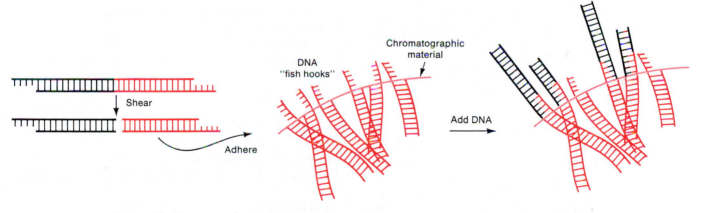

Figure 11-40 One specific sticky end of a DNA molecule can be used as a "fish hook" for the other sticky end plus any genes that may be attached to it. One half of the λ phage DNA is attached to chromatographic material so that the cohesive or sticky ends are exposed. Any DNA passing over this chromatographic material that has single-stranded base sequences complementary to the "fish hook" cohesive ends will stick. Any other DNA will pass through. Later, the "stuck" DNA can be removed from the chromatographic material by heating it to break the hydrogen bonds.

observing plaque-forming phage. If no recombination is detected, the half-molecule does not contain the loci for the mutant genes. A given half-molecule, or DNA fragment of any size, can be tested against a number of mutant markers to see if it recombines with them and thus contains their loci. This process of incorporation of DNA fragments into normal phage DNA and subsequent "rescue" of the affected genes upon lysis is called a **marker-rescue experiment.** *Such experiments demonstrate that one particular half of the DNA*

molecule carries the information of one particular half of the linkage map.

Subsequent experiments have demonstrated that the order of genes on a section of DNA matches the order on linkage maps. Dale Kaiser and his associates separated one of the DNA halves carrying a cohesive end and attached the other half to chromatographic material over which DNA could be passed (Figure 11-40). The single-stranded ends dangled free like fish hooks. When the other half of the

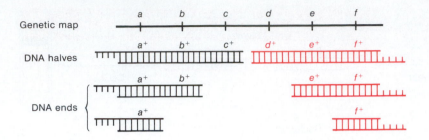

Genetic map

DNA halves

DNA ends

Figure 11-41 As the λ DNA is sheared into progressively smaller and smaller molecules, genes are lost to the fish hooks (Figure 11-40) in the same order as they appear on the genetic map.

DNA was sheared into smaller and smaller molecules and then passed over the sticky ends, the complementary sequences stuck. The pieces that stuck all had the same end of the λ DNA. These stuck pieces could be detached easily by raising the temperature to break the hydrogen bonds in the sticky ends. In this way, a series of fractions (of varying size) of one end of the λ DNA were separated and tested for content by marker rescue. The progressive elimination of markers indicated their respective positions along the DNA

half-molecule and thus revealed their sequence. Kaiser and his associates showed that an unambiguous arrangement of genes on the DNA can be determined and that this sequence is completely congruent with the genetic map (Figure 11-41).

Message   We are now justified in concluding that the sequence of bases in the DNA is indeed congruent with the gene map.

SUMMARY

Experimental work on the molecular nature of hereditary material has demonstrated conclusively that DNA (not protein, RNA, or some other substance) is indeed the genetic material. Using data supplied by others, Watson and Crick created a double helical model with two DNA strands, wound around each other, running in antiparallel fashion. Specificity of binding the two strands together is based on the fit of adenine (A) to thymine (T) and guanine (G) to cytosine (C). The former pair is held by two hydrogen bonds; the latter, by three.

The Watson-Crick model shows how DNA can be replicated in an orderly fashion—a prime requirement for genetic material. Replication is accomplished semiconservatively in both prokaryotes and eukaryotes. One double helix is replicated into two identical helices, each with identical linear orders of nucleotides; each of the two new double

helices is composed of one old and one newly polymerized strand of DNA.

Replication is achieved with the aid of several enzymes, including DNA polymerase, gyrase, and helicase. Replication starts at special regions of the DNA called *origins of replication* and proceeds down the DNA in both directions. Since DNA polymerase acts only in a $5' \rightarrow 3'$ direction, one of the newly synthesized strands at each replication fork must be synthesized in short segments and then joined using the enzyme ligase. DNA polymerization cannot begin without a short primer, which is also synthesized with special enzymes.

The marker-rescue technique has demonstrated that the sequence of genes on a chromosome corresponds exactly to the linear sequence of DNA and has shown that the genetic and chemical entities are one and the same thing.

▱\
◻

Concept Map

Draw a concept map interrelating as many of the following terms as possible. Note that the terms are listed in no particular order.

DNA double helix / nucleotides / hydrogen bonding / semiconservative / $5'$ / replication / chromatid / mitosis / DNA polymerase / meiosis / S phase / gene / $3'$

CHAPTER INTEGRATION PROBLEM

We discussed mitosis and meiosis in Chapter 3. Considering what we have covered in this chapter concerning DNA replication, draw a graph showing DNA content against time in a cell that undergoes mitosis and then meiosis. Assume a diploid cell.

Solution

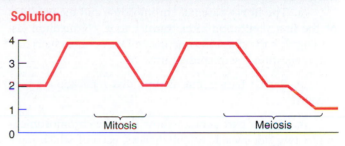

(Solution from Diane K. Lavett.)

SOLVED PROBLEMS

1. If the GC content of a DNA molecule is 56 percent, what are the percentages of the four bases (A, T, G, and C) in this molecule?

Solution

If the GC content is 56 percent, then since G = C, the content of G is 28 percent and the content of C is 28 percent. The content of AT is 100 − 56 = 44 percent. Since A = T, the content of A is 22 percent and the content of T is 22 percent.

2. Describe the expected pattern of bands in a CsCl gradient for *conservative* replication in the Meselson-Stahl experiment. Draw a diagram.

Solution

Refer to Figure 11-15 for an additional explanation. In conservative replication, if bacteria are grown in the presence of ^{15}N and then shifted to ^{14}N, one DNA molecule will be all ^{15}N after the first generation and the other molecule will be all ^{14}N, resulting in one heavy band and one light band in the gradient. After the second generation, the ^{15}N DNA will yield one molecule with all ^{15}N and one molecule with all ^{14}N, whereas the ^{14}N DNA will yield only ^{14}N DNA. Thus, only all ^{14}N or all ^{15}N DNA is generated, again, yielding a light band and a heavy band:

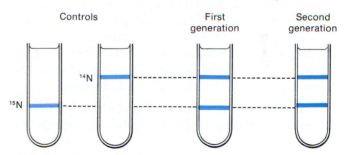

PROBLEMS

1. Describe the types of chemical bonds and forces involved in the DNA double helix.

2. Explain what is meant by the terms *conservative* and *semiconservative replication*.

3. What is meant by a *primer*, and why are primers necessary for DNA replication?

4. What are helicases and topoisomerases?

5. Why is DNA synthesis continuous on one strand and discontinuous on the opposite strand?

6. If thymine makes up 15 percent of the bases in a specific DNA molecule, what percentage of the bases is cytosine?

7. If the GC content of a DNA molecule is 48 percent, what are the percentages of the four bases (A, T, G, and C) in this molecule?

8. *E. coli* chromosomes in which every nitrogen atom is labeled (that is, every nitrogen atom is the heavy isotope ^{15}N instead of the normal isotope ^{14}N) are allowed to replicate in an environment in which all the nitrogen is ^{14}N. Using a solid line to represent a heavy polynucleotide chain and a dashed line for a light chain, sketch the following:

a. The heavy parental chromosome and the products of the first replication after transfer to a ^{14}N medium, assuming that the chromosome is one DNA double helix and that replication is semiconservative.

b. Repeat part a, but assume that replication is conservative.

c. Repeat part a, but assume that the chromosome is in fact two side-by-side double helices, each of which replicates semiconservatively.

d. Repeat part c, but assume that each side-by-side double helix replicates conservatively and that the overall *chromosome* replication is semiconservative.

e. Repeat part d, but assume that the overall chromosome replication is conservative.

f. If the daughter chromosomes from the first division in ^{14}N are spun in a cesium chloride (CsCl) density gradient and a single band is obtained, which of possibilities a to e can be ruled out? Reconsider the Meselson-Stahl experiment: what does it *prove*?

9. R. Okazaki found that the immediate products of DNA replication in *E. coli* include single-stranded DNA fragments approximately 1000 nucleotides in length after the newly synthesized DNA is extracted and denatured (melted). When he allowed DNA replication to proceed for a longer period of time, he found a lower frequency of these short fragments and long single-stranded DNA chains after extraction and denaturation. Explain how this result might be related to the fact that all known DNA polymerases synthesize DNA only in a $5' \rightarrow 3'$ direction.

10. When plant and animal cells are given pulses of [^{3}H]-thymidine at different times during the cell cycle, hetero-chromatic regions on the chromosomes are invariably shown to be "late replicating." Can you suggest what, if any, biological significance this observation might have?

11. On the planet of Rama, the DNA is of six nucleotide types: A, B, C, D, E, and F. A and B are called *marzines,* C and D are *orsines,* and E and F are *pirines.* The following rules are valid in all Raman DNAs:

Total marzines = total orsines = total pirines

$$A = C = E$$

$$B = D = F$$

a. Prepare a model for the structure of Raman DNA.

b. On Rama, mitosis produces three daughter cells. Bearing this fact in mind, propose a replication pattern for your DNA model.

c. Consider the process of meiosis on Rama. What comments or conclusions can you suggest?

12. If you extract the DNA of the coliphage ϕX174, you will find that its composition is 25 percent A, 33 percent T, 24 percent G, and 18 percent C. Does this make sense in terms of Chargaff's rules? How would you interpret this result? How might such a phage replicate its DNA?

13. The temperature at which a DNA sample denatures can be used to estimate the proportion of its nucleotide pairs that are G–C. What would be the basis for this determination, and what would a high denaturation temperature for a DNA sample indicate?

14. In 1960, Paul Doty and Julius Marmur observed that when DNA is heated to 100°C, all the hydrogen bonds between the complementary strands are destroyed and the DNA becomes single-stranded (see the figure below). If the solution is cooled slowly, some double-stranded DNA is

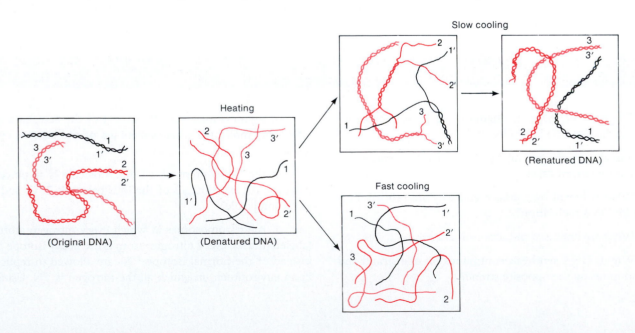

(Original DNA)

Heating

(Denatured DNA)

Slow cooling

Fast cooling

(Renatured DNA)

formed that is biologically normal (for example, it may have transforming ability). Presumably, this **reannealing, or renaturation,** process occurs when two single strands happen to collide in such a way that the complementing base sequences can align and reconstitute the original double helix (see the figure). This reannealing is very specific and precise, making it a powerful tool because stretches of complementary base sequences in *different* DNAs also will anneal after melting and mixing. Thus, the efficiency of annealing provides a measure of similarity between two different DNAs.

Now suppose that you extract DNA from a small virus, denature it, and allow it to reanneal with DNA taken from other strains that carry either a deletion, an inversion, or a duplication. What would you expect to see on inspection with an electron microscope?

15. DNA extracted from a mammal is heat-denatured and then slowly cooled to allow reannealing. The following graph shows the results obtained. These are two "shoulders" in the curve. The first shoulder indicates the presence of a very rapidly annealing part of the DNA—so rapid, in fact, that it occurs before strand interactions take place.

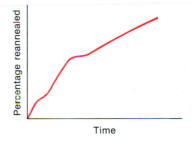

a. What could this part of the DNA be?

b. The second shoulder is a rapidly reannealing part as well. What does this evidence suggest?

16. Design tests to determine the physical relationship between highly repetitive and unique DNA sequences in chromosomes. (HINT: It is possible to vary the size of DNA molecules by the amount of shearing they are subjected to.)

17. Viruses are known to cause cancer in mice. You have a pure preparation of virus DNA, a pure preparation of DNA from the chromosomes of mouse cancer cells, and pure DNA from the chromosomes of normal mouse cells. Viral DNA will specifically anneal with cancer cell DNA, but not with normal-cell DNA. Explore the possible genetic significance of this observation, its significance at the molecular level, and its medical significance.

18. Ruth Kavenaugh and Bruno Zimm have devised a technique to measure the maximal length of the longest DNA molecules in solution. They studied DNA samples from the three *Drosophila* karyotypes shown below. They found the longest molecules in karyotypes (a) and (b) to be of similar length and about twice the length of the longest molecule in (c). Interpret these results.

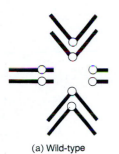

(a) Wild-type

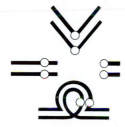

(b) Pericentric inversion

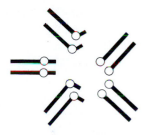

(c) Translocation

19. In the harlequin chromosome technique, you allow *three* rounds of replication in bromodeoxyuridine and then stain the chromosomes. What result do you expect?

12

The Nature of the Gene

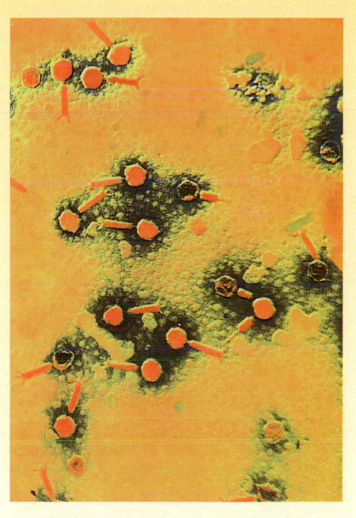

False-color transmission electron micrograph of unidentified phage; magnification, 58,000×. (CNRI / Science Photo Library / Photo Researchers)

KEY CONCEPTS

▶ The one-gene–one-enzyme hypothesis postulates that genes control the structure of proteins.

▶ Studies on hemoglobin demonstrated that a gene mutation can be tied to an altered amino acid sequence in a protein.

▶ The linear sequence of nucleotides in a gene determines the linear sequence of amino acids in a protein.

▶ Fine structural analysis of the *rII* genes in phage T4 showed that the gene consists of a linear array of subelements that can mutate and recombine with one another; these subelements were later correlated with nucleotide pairs.

▶ A gene can be defined as a unit of function by a complementation test.

How Genes Work

What is the nature of the gene, and how do genes control phenotypes? For example, how can one allele of a gene produce a wrinkled pea and another produce a round, smooth pea? Today we realize that all biochemical reactions of a cell are catalyzed by enzymes, which have a specific three-dimensional configuration that is crucial to their function. We now know that genes specify the structures of proteins, some of which are enzymes, and we can even relate the structure of the genetic material to the structure of proteins. Table 12-1 summarizes our current model of the relationship between genotype and phenotype.

How did we arrive at this point? By following two lines of inquiry:

1. What is the physical structure of the genetic material?
2. How does the genetic material exert its effect? (Or, in detail, how does the structure function?)

Chapter 11 recounted the demonstration that DNA is the genetic material and detailed the unraveling of the structure of DNA. In Chapters 12 and 13, we examine how genes function.

The first clues about the nature of primary gene function came from studies of humans. Early in this century, Archibald Garrod, a physician, noted that several hereditary human defects are produced by recessive mutations. Some of these defects can be traced directly to metabolic defects that affect the basic body chemistry—an observation that has led to the suggestion of "inborn errors" in metabolism. For example, phenylketonuria, which is caused by an autosomal recessive allele, results from an inability to convert phenylalanine into tyrosine. Consequently, phenylalanine accumulates and is spontaneously converted into a toxic compound, phenylpyruvic acid. In a different example, the inability to convert tyrosine into the pigment melanin produces an albino. Garrod's observations focused attention on metabolic control by genes.

The One-Gene–One-Enzyme Hypothesis

Clarification of the actual function of genes came from research in the 1940s on *Neurospora* by George Beadle and Edward Tatum, who later received a Nobel prize for their work. Before we describe their actual experiments, let's jump ahead and examine some aspects of **biosynthetic pathways,** based on our current understanding. We now know that molecules are synthesized as a series of steps, each one catalyzed by an enzyme. For instance, a biosynthetic pathway might have four steps, where 1 is the starting material and 5 is the final product:

$$
\begin{array}{ccccc}
A & B & C & D \\
1 \longrightarrow 2 \longrightarrow 3 \longrightarrow 4 \longrightarrow 5
\end{array}
$$

Each step is catalyzed by an enzyme: A, B, C, or D. In turn, each enzyme is specified by a particular gene. We might say that gene *a* specifies enzyme A, gene *b* specifies enzyme B, and so on. Therefore, if we inactivate the gene responsible for an enzyme, we eliminate one required step and the pathway is interrupted.

In the following diagram enzyme B is eliminated owing to a mutation in gene *b*:

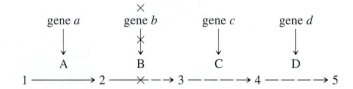

Now the cell cannot carry out the reaction that converts compound 2 to compound 3. It is blocked at compound 2, and cannot go further. But what happens if we add back different intermediate compounds? Suppose that, for example, we feed the cell compound 3 or 4. Can the cell then synthesize the final product, 5? Yes, with either compound 3 or 4 it can, because subsequent steps are not blocked. What if we add more of compound 1? No, adding compound 1 will not allow the synthesis of product 5 because one of the subsequent steps is blocked.

Let's test our understanding of the concept of biosynthetic pathways by looking at a sample problem that illustrates this principle:

Let's assume we've isolated five mutants, 1 to 5, that cannot synthesize compound G for growth. We know the five compounds, A to E, that are required in the biosynthetic pathway, but we do not know the order in which they are synthesized by the wild-type cell. We have tested each

Table 12-1 Model of the Relationship between Genotype and Phenotype

1. The characteristic features of an organism are determined by the phenotype of its parts, which are in turn determined by the phenotype of the component cells.
2. The phenotype of a cell is determined by its internal chemistry, which is controlled by enzymes that catalyze its metabolic reactions.
3. The function of an enzyme depends on its specific three-dimensional structure, which in turn depends on the specific linear sequence of amino acids in the enzyme.
4. The enzymes present in a cell, and structural proteins as well, are determined by the genotype of the cell.
5. Genes specify the linear sequence of amino acids in polypeptides and hence in proteins; thus, genes determine phenotypes.

compound for its ability to support the growth of each mutant, with the following results:

		A	B	C	D	E	G
Mutant	1	−	−	−	+	−	+
	2	−	+	−	+	−	+
	3	−	−	−	−	−	+
	4	−	+	+	+	−	+
	5	+	+	+	+	−	+

Compound tested
(+ = growth; − = no growth)

a. What is the order of compounds A to E in the pathway? How do we approach this type of problem? First, let's work out the underlying principle, and then draw a diagram. The main points are these: a mutation blocks a biosynthetic pathway by canceling out one enzyme needed in the pathway. The mutant therefore lacks one compound needed in the pathway and thus cannot make any of the compounds that come after the blocked step. If we add a compound that is normally synthesized before the block, it will not help matters: the mutant will still be unable to synthesize the rest of the compounds in the pathway. However, the mutant does have the enzymes to make these later compounds; thus, if we add either the blocked compound or any compound that comes after the block, the mutant will grow. So in our testing of mutants with blocks at different steps in the pathway, the compounds that occur latest in the pathway will support the growth of the most mutants, and compounds that occur earliest in the pathway will allow the growth of the fewest mutants. A look at our table shows us that the compound supporting the fewest mutants is compound E, with no mutants supported, followed by A (one mutant), then by C, B, D, and G, the final product. Now we can construct our diagram.

$$E \longrightarrow A \longrightarrow C \longrightarrow B \longrightarrow D \longrightarrow G$$

b. At which point in the pathway is each mutant blocked? Clearly, a mutant that is blocked between E and A cannot be supported by E but can be supported by all the other compounds. Thus, we see that mutant 5 must be blocked in the E–A conversion. We also see that mutant 4 cannot be supported by E or A, so it must be blocked in the A–C conversion. By similar logic, we obtain the order 5–4–2–1–3, which we can insert into the diagram as follows:

$$\overset{5}{E} \longrightarrow \overset{4}{A} \longrightarrow \overset{2}{C} \longrightarrow \overset{1}{B} \longrightarrow \overset{3}{D} \longrightarrow G$$

Now we can comprehend how Beadle and Tatum first worked out this experiment, using one particular biosynthetic pathway of *Neurospora*.

The Experiments of Beadle and Tatum

Beadle and Tatum analyzed mutants of *Neurospora*, a fungus with a haploid genome. They first irradiated *Neurospora* to produce mutations and then tested cultures from ascospores for interesting mutant phenotypes. They detected numerous **auxotrophs**—strains that cannot grow on a minimal medium unless the medium is supplemented with one or more specific nutrients. In each case, the mutation that generated the auxotrophic requirement was inherited as a single-gene mutation: each gave a 1:1 ratio when crossed with a wild type (recall that *Neurospora* is haploid). Figure 12-1 depicts the procedure Beadle and Tatum used.

One set of mutant strains required arginine to grow on a minimal medium. These strains provided the focus for much of Beadle and Tatum's further work. First, they found that the mutations mapped into three different locations on separate chromosomes, even though the same supplement (arginine) satisfied the growth requirement for each mutant. Let's call the three loci the *arg-1*, *arg-2*, and *arg-3* genes. Beadle and Tatum discovered that the auxotrophs for each of the three loci differed in their response to the chemical compounds ornithine and citrulline, which are related to arginine (Figure 12-2). The *arg-1* mutants grew when supplied with ornithine, citrulline, or arginine in addition to the minimal medium. The *arg-2* mutants grew on either arginine or citrulline but not on ornithine. The *arg-3* mutants grew only when arginine was supplied. We can see this more easily by looking at Table 12-2.

It was already known that cellular enzymes often interconvert related compounds such as these. Based on the properties of the *arg* mutants, Beadle and Tatum and their colleagues proposed a biochemical model for such conversions in *Neurospora*:

$$\text{precursor} \xrightarrow{\text{enzyme X}} \text{ornithine} \xrightarrow{\text{enzyme Y}}$$
$$\text{citrulline} \xrightarrow{\text{enzyme Z}} \text{arginine}$$

Note how this relationship easily explains the three classes of mutants shown in Table 12-2. The *arg-1* mutants have a defective enzyme X, so they are unable to convert the precursor into ornithine as the first step in producing arginine. However, they have normal enzymes Y and Z, and so the *arg-1* mutants are able to produce arginine if supplied with either ornithine or citrulline. The *arg-2* mutants lack enzyme Y, and the *arg-3* mutants lack enzyme Z. Thus, a mutation at a particular gene is assumed to interfere with the production of a single enzyme. The defective enzyme, then, creates a block in some biosynthetic pathway. The block can

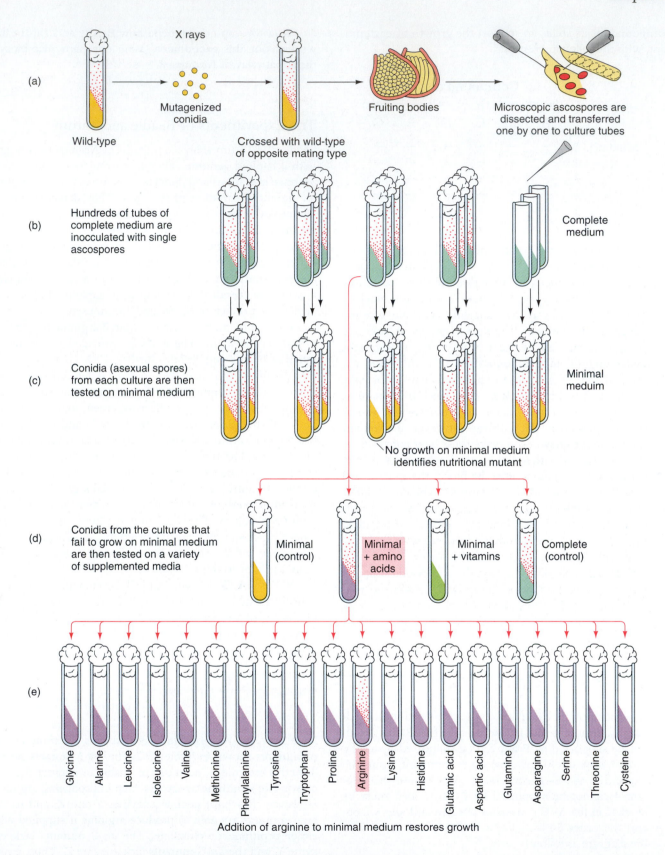

Figure 12-1 The procedure used by Beadle and Tatum. See Chapter 6 for a review of the *Neurospora* life cycle and genetics. (Little, Brown & Co.)

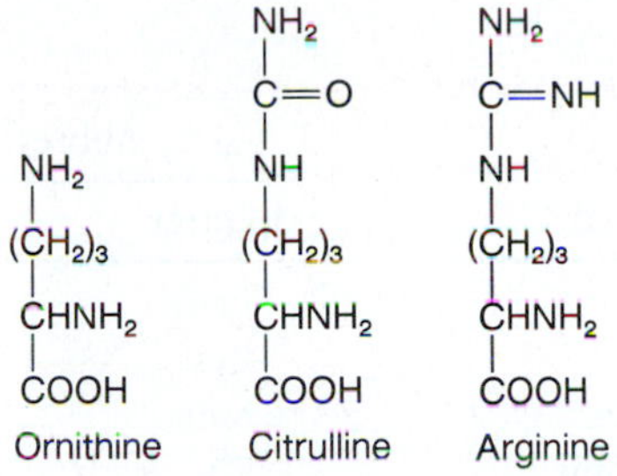

Figure 12-2 Chemical structures of arginine and the related compounds citrulline and ornithine. In the work of Beadle and Tatum, different *arg* auxotrophic mutants of *Neurospora* were found to grow when the medium was supplemented with citrulline, or with ornithine, as an alternative to arginine.

be circumvented by supplying to the cells any compound that normally comes after the block in the pathway.

We can now diagram a more complete biochemical model:

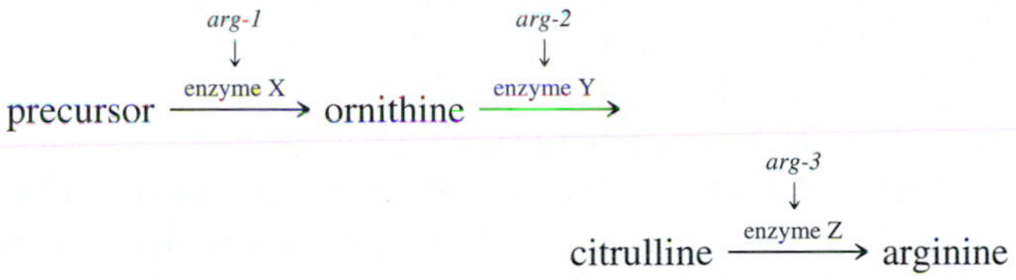

$$\text{precursor} \xrightarrow[\text{enzyme X}]{arg\text{-}1} \text{ornithine} \xrightarrow[\text{enzyme Y}]{arg\text{-}2}$$

$$\text{citrulline} \xrightarrow[\text{enzyme Z}]{arg\text{-}3} \text{arginine}$$

Note that this entire model was inferred from the properties of the mutant classes detected through genetic analysis. Only later were the existence of the biosynthetic pathway and the presence of defective enzymes demonstrated through independent biochemical evidence.

This model, which has become known as the **one-gene–one-enzyme hypothesis,** provided the first exciting insight into the functions of genes: genes somehow were responsible for the function of enzymes, and each gene apparently controlled one specific enzyme. Other researchers obtained similar results for other biosynthetic pathways, and the hypothesis soon achieved general acceptance. It was subsequently refined, as we shall see later in this chapter. Nevertheless, the one-gene–one-enzyme hypothesis be-

came one of the great unifying concepts in biology, because it provided a bridge that brought together the concepts and research techniques of genetics and biochemistry.

Message Genes control biochemical reactions by controlling the production of enzymes.

We should pause to ponder the significance of this discovery. Let's summarize what it established:

1. Biochemical reactions in vivo (in the living cell) occur as a series of discrete, stepwise reactions.
2. Each reaction is specifically catalyzed by a single enzyme.
3. Each enzyme is specified by a single gene.

Gene-Protein Relationships

The one-gene–one-enzyme hypothesis was an impressive step forward in our understanding of gene function, but just *how* do genes control the functioning of enzymes? Enzymes are proteins, and thus we must review the basic facts of protein structure in order to follow the next step in the study of gene function.

Protein Structure

In simple terms, a **protein** is a macromolecule composed of **amino acids** attached end to end in a linear string. The general formula for an amino acid is H_2N—CHR—$COOH$, in which the side chain, or R (reactive) group, can be anything from a hydrogen atom (as in the amino acid glycine) to a complex ring (as in the amino acid tryptophan). There are 20 common amino acids in living organisms (Table 12-3), each having a different R group. Amino acids are linked together in proteins by covalent (chemical) bonds called **peptide bonds.** A peptide bond is formed through a condensation reaction that involves the removal of a water molecule (Figure 12-3).

Several amino acids linked together by peptide bonds form a molecule called a **polypeptide;** the proteins found in living organisms are large polypeptides. For instance, the α chain of human hemoglobin contains 141 amino acids, and some proteins consist of more than 1000 amino acids.

Proteins have a complex structure that is traditionally thought of as having four levels. The linear sequence of the amino acids in a polypeptide chain is called the **primary structure** of the protein. Figure 12-4 shows the linear sequence of tryptophan synthetase (an enzyme) and beef insulin (a hormonal protein).

The **secondary structure** of a protein refers to the interrelationships of amino acids that are close together in the linear sequence. This spatial arrangement often results from the fact that polypeptides can bend into regularly repeating

Table 12-2 Growth of *arg* Mutants in Response to Supplements

Mutant	Ornithine	Citrulline	Arginine
	\(+ = growth; − = no growth\)		
arg-1	+	+	+
arg-2	−	+	+
arg-3	−	−	+

Table 12-3   **The 20 Amino Acids Common in Living Organisms**

Amino Acid	Abbreviation		Amino Acid	Abbreviation	
	3-Letter	1-Letter		3-Letter	1-Letter
Alanine	Ala	A	Leucine	Leu	L
Arginine	Arg	R	Lysine	Lys	K
Asparagine	Asn	N	Methionine	Met	M
Aspartic acid	Asp	D	Phenylalanine	Phe	F
Cysteine	Cys	C	Proline	Pro	P
Glutamic acid	Glu	E	Serine	Ser	S
Glutamine	Gln	Q	Threonine	Thr	T
Glycine	Gly	G	Tryptophan	Trp	W
Histidine	His	H	Tyrosine	Tyr	Y
Isoleucine	Ile	I	Valine	Val	V

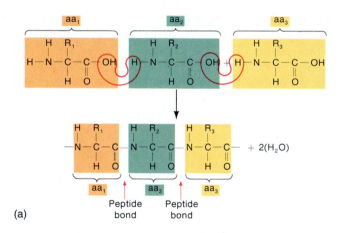

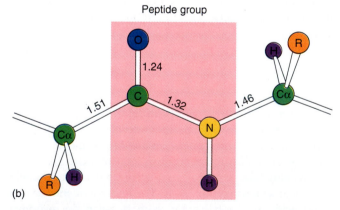

(periodic) structures, generated by hydrogen bonds between the CO and NH groups of different residues. Two of the basic periodic structures are the α helix (Figure 12-5) and the β pleated sheet (Figure 12-6).

A protein also has a three-dimensional architecture, termed the **tertiary structure,** which is generated by electrostatic, hydrogen, and Van der Waals bonds that form between the various amino acid R groups, causing the protein chain to fold back on itself. In many cases amino acids that are far apart in the linear sequence are brought close together in the tertiary structure. Often, two or more folded structures will bind together to form a **quaternary structure;** this structure is **multimeric** because it is composed of several separate polypeptide chains, or monomers.

Figure 12-7 depicts the four levels of protein structure. In Figure 12-7c we can see the tertiary structure of myoglobin. Note how the α helix is folded back on itself to generate the three-dimensional shape of the protein. Figure 12-7d shows the combining of four subunits (two α chains and two β chains) to form the quaternary structure of hemoglobin. Figures 12-8 and 12-9 show the structures of myoglobin and hemoglobin in more detail. The combining of subunits to form a multimeric enzyme can be seen directly in the electron microscope in some cases (Figure 12-10).

Many proteins are compact structures; such proteins are called **globular proteins.** Enzymes and antibodies are among the important globular proteins. Other, unfolded proteins, called **fibrous proteins,** are important components of such structures as hair and muscle.

Figure 12-3 The peptide bond. (a) A polypeptide is formed by the removal of water between amino acids to form peptide bonds. Each aa indicates an amino acid. R_1, R_2, and R_3 represent R groups (side chains) that differentiate the amino acids. R can be anything from a hydrogen atom (as in glycine) to a complex ring (as in tryptophan). (b) The peptide group is a rigid planar unit with the R groups projecting out from the C—N backbone. Standard bond distances (in angstroms) are shown. (Part b from L. Stryer, *Biochemistry,* 4th ed. Copyright ©1995 by Lubert Stryer.)

Message   The linear sequence of a protein folds up to yield a unique three-dimensional configuration. This configuration creates specific sites to which substrates bind and at which catalytic reactions occur. The three-dimensional structure of a protein, which is crucial for its function, is determined solely by the primary structure (linear sequence) of amino acids. Therefore, genes can control enzyme function by controlling the primary structure of proteins.

(a)

                        10   20   30

Met-Glu-Arg-Tyr-Glu-Ser-Leu-Phe-Ala-Gln-Leu-Lys-Glu-Arg-Lys-Glu-Gly-Ala-Phe-Val-Pro-Phe-Val-Thr-Leu-Gly-Asp-Pro-Gly-Ile-

             40   50   60

Glu-Gln-Ser-Leu-Lys-Ile-Ile-Asp-Thr-Leu-Ile-Glu-Ala-Gly-Ala-Asp-Ala-Leu-Glu-Leu-Gly-Ile-Pro-Phe-Ser-Asp-Pro-Leu-Ala-Asp-

             70   80   90

Gly-Pro-Thr-Ile-Gln-Asn-Ala-Thr-Leu-Arg-Ala-Phe-Ala-Ala-Gly-Val-Thr-Pro-Ala-Gln-Cys-Phe-Glu-Met-Leu-Ala-Leu-Ile-Arg-Gln-

           100   110   120

Lys-His-Pro-Thr-Ile-Pro-Ile-Gly-Leu-Leu-Met-Tyr-Ala-Asn-Leu-Val-Phe-Asn-Lys-Gly-Ile-Asp-Glu-Phe-Tyr-Ala-Gln-Cys-Glu-Lys-

           130   140   150

Val-Gly-Val-Asp-Ser-Val-Leu-Val-Ala-Asp-Val-Pro-Val-Gln-Glu-Ser-Ala-Pro-Phe-Arg-Gln-Ala-Ala-Leu-Arg-His-Asn-Val-Ala-Pro-

           160   170   180

Ile-Phe-Ile-Cys-Pro-Pro-Asn-Ala-Asp-Asp-Asp-Leu-Leu-Arg-Gln-Ile-Ala-Ser-Tyr-Gly-Arg-Gly-Tyr-Thr-Tyr-Leu-Leu-Ser-Arg-Ala-

           190   200   210

Gly-Val-Thr-Gly-Ala-Glu-Asn-Arg-Ala-Ala-Leu-Pro-Leu-Asn-His-Leu-Val-Ala-Lys-Leu-Lys-Glu-Tyr-Asn-Ala-Ala-Pro-Pro-Leu-Gln-

           220   230   240

Gly-Phe-Gly-Ile-Ser-Ala-Pro-Asp-Gln-Val-Lys-Ala-Ala-Ile-Asp-Ala-Gly-Ala-Ala-Gly-Ala-Ile-Ser-Gly-Ser-Ala-Ile-Val-Lys-Ile-

           250   260   268

Ile-Glu-Gln-His-Asn-Ile-Glu-Pro-Glu-Lys-Met-Leu-Ala-Ala-Leu-Lys-Val-Phe-Val-Gln-Pro-Met-Lys-Ala-Ala-Thr-Arg-Ser

(b) **A chain**

                     S————S

Gly-Ile-Val-Glu-Gln-Cys-Cys-Ala-Ser-Val-Cys-Ser-Leu-Tyr-Gln-Leu-Glu-Asn-Tyr-Cys-Asn

      5   10   15   21

                S   S

              S

B chain

Phe-Val-Asn-Gln-His-Leu-Cys-Gly-Ser-His-Leu-Val-Glu-Ala-Leu-Tyr-Leu-Val-Cys-Gly-Glu-Arg-Gly-Phe-Phe-Tyr-Thr-Pro-Lys-Ala

     5   10   15   20   25   30

Figure 12-4 Linear sequences of two proteins. (a) The *E. coli* tryptophan synthetase A protein, 268 amino acids long. (b) Bovine insulin protein. Note that the amino acid cysteine can form unique "sulfur bridges," because it contains sulfur.

Determining Protein Sequence

If we purify a particular protein, we find that we can specify a particular ratio of the various amino acids that make up that specific protein. But the protein is not formed by a random hookup of fixed amounts of the various amino acids; each protein has a unique, characteristic sequence. For a small polypeptide, the amino acid sequence can be determined by clipping off one amino acid at a time and identifying it. However, large polypeptides cannot be readily "sequenced" in this way.

Frederick Sanger worked out a brilliant method for deducing the sequence of large polypeptides. There are several different **proteolytic enzymes**—enzymes that can break peptide bonds only between specific amino acids in proteins. Proteolytic enzymes can break a large protein into a number of smaller fragments, which can then be separated

3.6 residues/ turn

Figure 12-5 The α helix, a common basis of secondary protein structure. Each R is a specific side chain on one amino acid. The gray dots are weak hydrogen bonds that bond the CO group of residue *n* to the NH group of residue *n* + 4, thereby stabilizing the helical shape. (From H. Lodish, D. Baltimore, A. Berk, S. L. Zipursky, P. Matsudaira, J. Darnell, *Molecular Cell Biology,* 3d ed. Copyright ©1995 by Scientific American Books, Inc.)

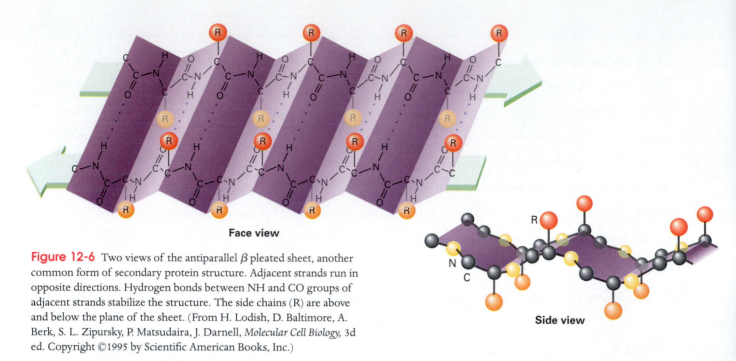

Face view

Side view

Figure 12-6 Two views of the antiparallel β pleated sheet, another common form of secondary protein structure. Adjacent strands run in opposite directions. Hydrogen bonds between NH and CO groups of adjacent strands stabilize the structure. The side chains (R) are above and below the plane of the sheet. (From H. Lodish, D. Baltimore, A. Berk, S. L. Zipursky, P. Matsudaira, J. Darnell, *Molecular Cell Biology,* 3d ed. Copyright ©1995 by Scientific American Books, Inc.)

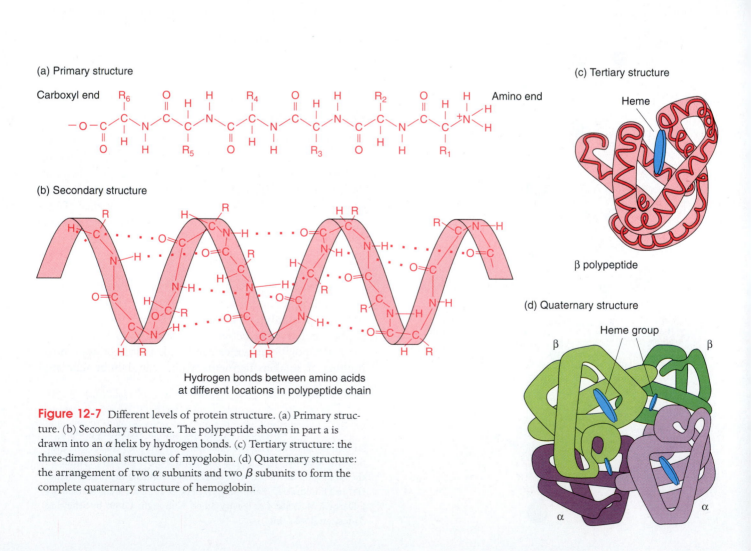

(a) Primary structure

Carboxyl end   Amino end

(b) Secondary structure

Hydrogen bonds between amino acids
at different locations in polypeptide chain

(c) Tertiary structure

Heme

β polypeptide

(d) Quaternary structure

Heme group

Figure 12-7 Different levels of protein structure. (a) Primary structure. (b) Secondary structure. The polypeptide shown in part a is drawn into an α helix by hydrogen bonds. (c) Tertiary structure: the three-dimensional structure of myoglobin. (d) Quaternary structure: the arrangement of two α subunits and two β subunits to form the complete quaternary structure of hemoglobin.

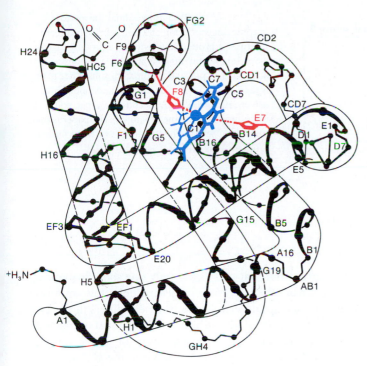

Figure 12-8 Folded tertiary structure of myoglobin, an oxygen-storage protein. Each dot represents an amino acid. The heme group, a cofactor that facilitates the binding of oxygen, is shown in blue. (From L. Stryer, *Biochemistry,* 4th ed. Copyright ©1995 by Lubert Stryer. Based on R. E. Dickerson, *The Proteins,* 2d ed., vol. 2. Edited by H. Neurath. Copyright ©1964 by Academic Press.)

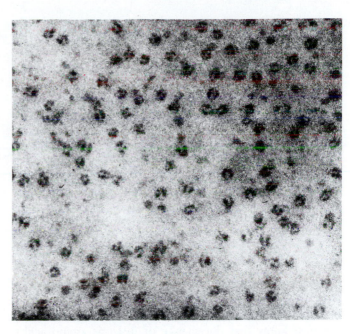

Figure 12-10 Electron micrograph of the enzyme aspartate transcarbamylase. Each small "glob" is an enzyme molecule. Note the quaternary structure: the enzyme is composed of subunits. (Photograph from Jack D. Griffith.)

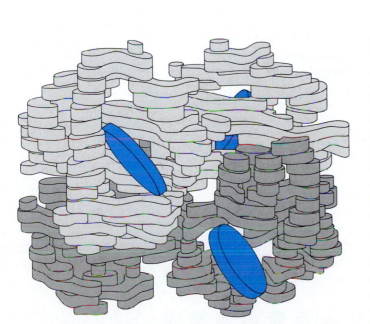

Figure 12-9 A model of the hemoglobin molecule (shown as contoured layers to provide a visualization of the three-dimensional shape). The different shadings indicate the different polypeptide chains, two α and two β subunits, that combine to form the quaternary structure of the protein (see Figure 12-7d). The disks are heme groups—complex structures containing iron. (After M. F. Perutz, "The Hemoglobin Molecule." Copyright ©1964 by Scientific American, Inc. All rights reserved.)

according to their migration speeds in a solvent on chromatographic paper. Because different fragments will move at different speeds in various solvents, **two-dimensional chromatography** can be used to enhance the separation of the fragments (Figure 12-11). In this technique, a mixture of fragments is separated in one solvent; then the paper is turned 90° and another solvent is used. When the paper is stained, the polypeptides appear as spots in a characteristic chromatographic pattern called the **fingerprint** of the protein. Each of the spots can be cut out, and the polypeptide fragments can be washed from the paper. Because each spot contains only small polypeptides, their amino acid sequences can easily be determined.

Using different proteolytic enzymes to cleave the protein at different points, we can repeat the experiment to obtain other sets of fragments. The fragments from the different treatments overlap, because the breaks are made in different places with each treatment. The problem of solving the overall sequence then becomes one of fitting together the small-fragment sequences—almost like solving a tricky jigsaw or crossword puzzle (Figure 12-12).

Using this elegant technique, Sanger confirmed that the sequence of amino acids (as well as the amounts of the various amino acids) is specific to a particular protein. In other words, the amino acid sequence is what makes insulin insulin.

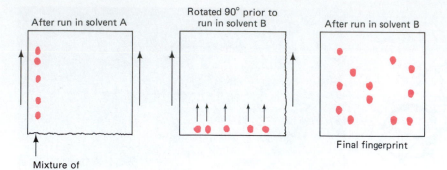

Figure 12-11 Two-dimensional chromatographic fingerprinting of a polypeptide fragment mixture. A protein is digested by a proteolytic enzyme into fragments that are only a few amino acids long. A piece of chromatographic filter paper is then spotted with this mixture and dipped into solvent A. As solvent A ascends the paper, some of the fragments become separated. The paper is then turned 90° and further resolution of the fragments is obtained as solvent B ascends.

Relationship between Gene Mutations and Altered Proteins

We now know that the change of just one amino acid is sometimes enough to alter protein function. This was first shown in 1957 by Vernon Ingram, who studied the globular protein hemoglobin—the molecule that transports oxygen in red blood cells. As we saw in Figures 12-7d and 12-9, hemoglobin is made up of four polypeptide chains: two identical α chains, each containing 141 amino acids, and two identical β chains, each containing 146 amino acids.

Ingram compared hemoglobin A (HbA), the hemoglobin from normal adults, with hemoglobin S (HbS), the protein from people homozygous for the mutant gene that causes sickle-cell anemia, the disease in which red blood cells take on a sickle-cell shape (see Figure 4-2). Using Sanger's technique, Ingram found that the fingerprint of HbS differs from that of HbA in only one spot. Sequencing that spot from the two kinds of hemoglobin, Ingram found that only one amino acid in the fragment differs in the two kinds. Apparently, of all the amino acids known to make up a hemoglobin molecule, a substitution of valine for glutamic acid at just one point, position 6 in the β chain, is all that is needed to produce the defective hemoglobin (Figure 12-13). Unless patients with HbS receive medical attention, this single error in one amino acid in one protein will hasten their death. Figure 12-14 shows how this gene mutation ultimately leads to the pattern of sickle cell disease that we previously saw in Figure 1-13.

Notice what Ingram accomplished. A gene mutation that had been well established through genetic studies was connected with an altered amino acid sequence in a protein! Subsequent studies have identified numerous changes in hemoglobin, and each one is the consequence of a single

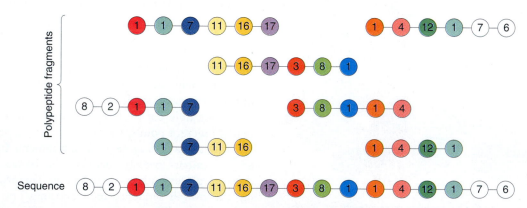

Figure 12-12 Alignment of polypeptide fragments to reconstruct an entire amino acid sequence. Different proteolytic enzymes can be used on the same protein to form different fingerprints, as shown here. The amino acid sequence of each fragment can be determined rather easily, and owing to the overlap of amino acid sequences from different fingerprints, the entire amino acid sequence of the original protein can be determined. Using this procedure, it took Sanger about six years to determine the sequence of the insulin molecule, a relatively small protein.

Normal hemoglobin

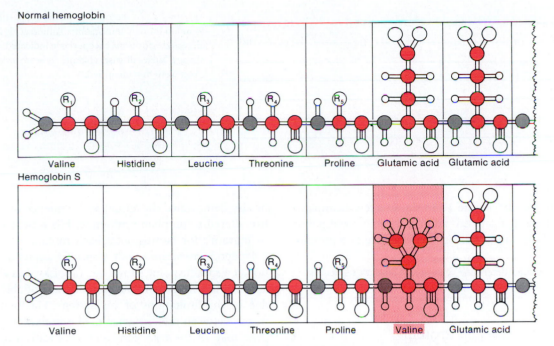

Valine   Histidine   Leucine   Threonine   Proline   Glutamic acid   Glutamic acid

Hemoglobin S

Valine   Histidine   Leucine   Threonine   Proline   Valine   Glutamic acid

Figure 12-13 The difference at the molecular level between normalcy and sickle-cell disease. Shown are only the first seven amino acids; all the rest not shown are identical. (From Anthony Cerami and Charles M. Peterson, "Cyanate and Sickle-Cell Disease." Copyright ©1975 by Scientific American, Inc. All rights reserved.)

Figure 12-14 The compounded consequences of one amino acid substitution in hemoglobin to produce sickle-cell anemia.

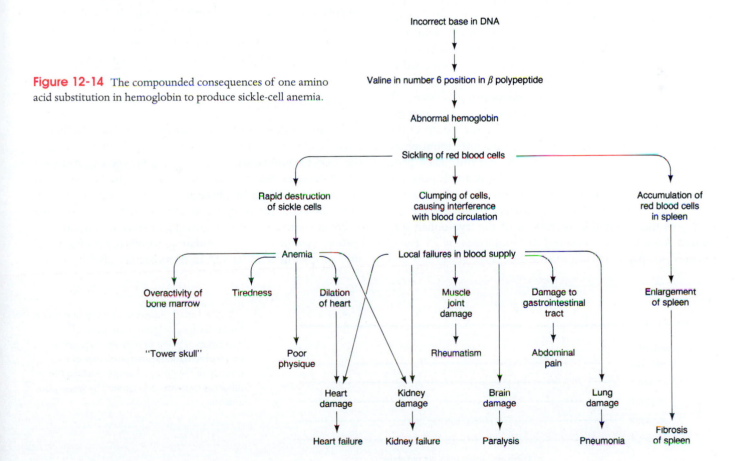

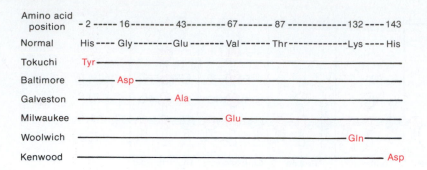

Figure 12-15 Some single amino acid substitutions found in human hemoglobin. Amino acids are normal at all residue positions except those indicated. Each type of change causes disease. (Names indicate areas in which cases were first identified.)

amino acid difference. (Figure 12-15 shows a few examples.) We can conclude that one mutation in a gene corresponds to a change of one amino acid in the sequence of a protein.

Message Genes determine the specific primary sequences of amino acids in specific proteins.

Colinearity of Gene and Protein

Once the structure of DNA was determined by Watson and Crick, it became apparent that the structure of proteins must be encoded in the linear sequence of nucleotides in the DNA. (We'll see in Chapter 13 how this genetic code was deciphered.) Following Ingram's demonstration that one mutation alters one amino acid in a protein, a relationship was sought between the linear sequence of mutant sites in a gene and the linear sequence of amino acids in a protein. (It is possible to map mutational sites within a gene due to studies on genetic fine structure, to be described in detail later in this chapter.)

Charles Yanofsky probed the relationship between altered genes and altered proteins by studying the enzyme tryptophan synthetase in *E. coli*. This enzyme catalyzes the conversion of indole glycerol phosphate into tryptophan. Two genes, *trpA* and *trpB*, control the enzyme. Each gene controls a separate polypeptide; after the A and B polypeptides are produced, they combine to form the active enzyme (a multimeric protein). Yanofsky analyzed mutations in the *trpA* gene that resulted in alterations of the tryptophan synthetase A subunit. He ordered the mutations by P1 transduction (see page 298) to produce a detailed gene map, and

he also determined the amino acid sequence of each respective altered tryptophan synthetase. His results were similar to Ingram's for hemoglobin: each mutant had a defective polypeptide associated with a specific amino acid substitution at a specific point. However, Yanofsky was able to show an exciting correlation that Ingram was not able to observe due to the limitations of his system. He found an exact match between the sequence of the mutational sites in the gene map of the *trpA* gene and the location of the corresponding altered amino acids in the A polypeptide chain. The farther apart two mutational sites were in map units, the more amino acids there were between the corresponding substitutions in the polypeptide (Figure 12-16). Thus, Yanofsky demonstrated **colinearity**—the correspondence between the linear sequence of the gene and that of the polypeptide. Figure 12-17 shows the complete set of data.

Message The linear sequence of nucleotides in a gene determines the linear sequence of amino acids in a protein.

Enzyme Function

How can a single amino acid substitution, such as that in sickle-cell hemoglobin (Figure 12-13), have such a profound effect on protein function and the phenotype of an organism? Take enzymes, for example. Enzymes are known to do their job of catalysis by physically grappling with their substrate molecules, twisting or bending the molecules to make or break chemical bonds. Figure 12-18 shows the gastric digestion enzyme carboxypeptidase in its relaxed position and after grappling with its substrate molecule, glycyltyrosine.

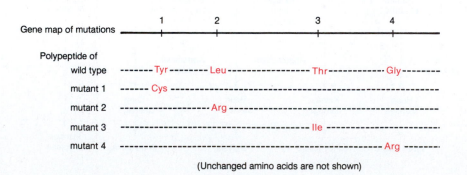

(Unchanged amino acids are not shown)

Figure 12-16 Simplified representation of the colinearity of gene mutations. The genetic map of point mutations (determined by recombinational analysis) corresponds linearly to the changed amino acids in the different mutants (determined by fingerprint analysis).

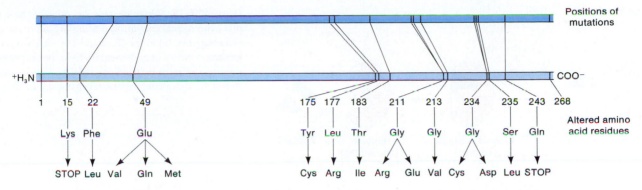

Figure 12-17 Actual colinearity shown in the A protein of tryptophan synthetase from *E. coli*. There is a linear correlation between the mutational sites and the altered amino acid residues. (Based on C. Yanofsky, "Gene Structure and Protein Structure." Copyright ©1967 by Scientific American. All rights reserved.)

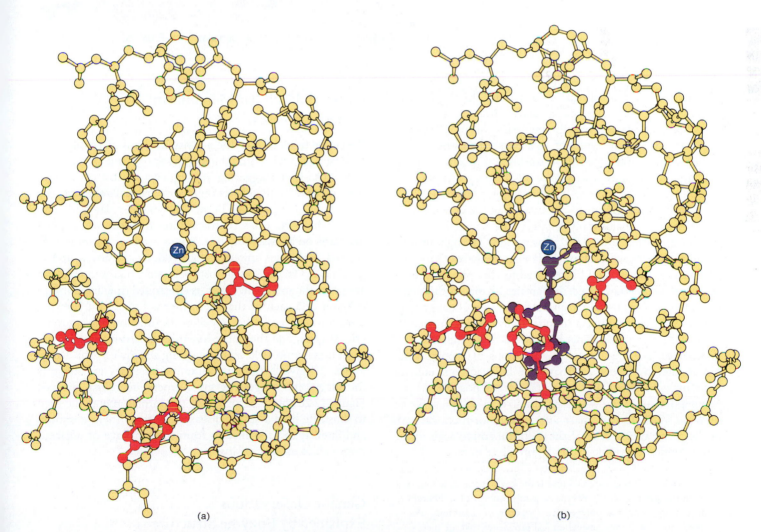

(a)

(b)

Figure 12-18 The active site of the digestive enzyme carboxypeptidase. (a) The enzyme without substrate. (b) The enzyme with its substrate (purple) in position. Three crucial amino acids (red) have changed positions to move closer to the substrate. Carboxypeptidase carves up proteins in the diet. (From W. N. Lipscomb, *Proceedings of the Robert A. Welch Foundation Conferences on Chemical Research* 15, 1971, 140–41.)

(a) LOCK-AND-KEY MODEL

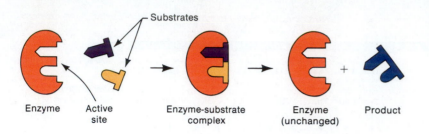

Enzyme | Active site | Enzyme-substrate complex | Enzyme (unchanged) | Product

Figure 12-19 Schematic representation of the action of a hypothetical enzyme in putting two substrate molecules together. (a) In the "lock-and-key" mechanism the substrates have a complementary fit to the enzymes active site. (b) In the induced-fit model, binding of substrates induces a conformational change in the enzyme.

(b) INDUCED-FIT MODEL

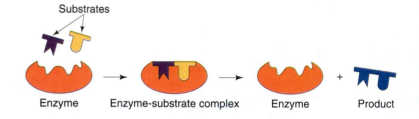

Enzyme | Enzyme-substrate complex | Enzyme | Product

The substrate molecule fits into a notch in the enzyme structure; this notch is called the **active site.**

Figure 12-19 diagrams the general concept. [Note that there are two basic types of reactions performed by enzymes: (1) the breakdown of a substrate into simpler products and (2) as shown in Figure 12-19, the synthesis of a complex product from one or more simpler substrates.]

Much of the globular structure of an enzyme is nonreactive material that simply supports the active site. so we might expect that amino acid substitutions throughout most of the structure would have little effect, whereas very specific amino acids would be required for the part of the enzyme molecule that gives the precise shape to the active site. Hence, the possibility arises that a functional enzyme does not always require a unique amino acid sequence for the *entire* polypeptide. This has proved to be the case: in a number of systems, numerous positions in a polypeptide can be filled by several alternative amino acids, and enzyme function is retained. But at certain other positions in the polypeptide, only the wild-type amino acid will preserve activity; in all likelihood, these amino acids form critical parts of the active sites. Some of these critical amino acids in carboxypeptidase are indicated in Figure 12-18 in red.

Message Protein architecture is the key to gene function. A gene mutation typically results in a substitution of a different amino acid into the polypeptide sequence of a protein. The new amino acid may have chemical properties that are incompatible with the proper protein architecture at that particular position; in such a case, the mutation will lead to a nonfunctional protein.

Genes and Cellular Metabolism: Genetic Diseases

When we think of enzyme activity in terms of cellular metabolism, we realize that the inactivation of one or more enzymes can have staggering consequences. Most of us have been amazed by the charts on laboratory walls showing the myriad of interlocking, branched, and circular pathways along which the cell's chemical intermediates are shunted like parts on an assembly line. Bonds are broken, molecules cleaved, molecules united, groups added or removed, and so on. The key fact is that almost every step, represented by an arrow on the metabolic chart, is controlled (mediated) by an enzyme, and each of these enzymes is produced under the direction of a gene that specifies its function. Change one critical gene, and the entire assembly line can break down.

Humans provide some startling examples. The list in Table 12-4 gives some representative examples and suggests the magnitude of genetic involvement in human disease. Figure 12-20 shows a corner of the human metabolic map to illustrate how a set of diseases, some of them common and familiar to us, can stem from the blockage of adjacent steps in biosynthetic pathways.

Genetic Observations Explained by Enzyme Structure

When the significance of the gene control of cellular chemistry became clear, a lot of other things fell into place. Many genetic observations now could be explained and tied to-

Table 12-4 Representative Examples of Enzymopathies: Inherited Disorders in Which Altered Activity (Usually Deficiency) of a Specific Enzyme Has Been Demonstrated in Humans

Condition	Enzyme with deficient activity*	Condition	Enzyme with deficient activity*
Acatalasia	Catalase	Granulomatous disease	Reduced nicotinamide adenine dinucleotide phosphate (NADPH) oxidase
Acid phosphatase deficiency	Acid phosphatase		
Albinism	Tyrosinase		
Aldosterone deficiency	18-Hydroxydehydrogenase	Hydroxyprolinemia	Hydroxyproline oxidase
		Hyperlysinemia	Lysine-ketoglutarate reductase
Alkaptonuria	Homogentisic acid oxidase		
Angiokeratoma, diffuse (Fabry disease)	Ceramide trihexosidase	Hypophosphatasia	Alkaline phosphatase
		Immunodefiency disease	Adenosine deaminase
Apnea, drug-induced	Pseudocholinesterase		Uridine monophosphate kinase
Argininemia	Arginase		
Argininosuccinic aciduria	Argininosuccinase	Krabbe disease	Galactosylceramide β-galactosidase
Ataxia, intermittent	Pyruvate decarboxylase		
Citrullinemia	Argininosuccinic acid synthetase	Leigh necrotizing encephalomyelopathy	Pyruvate carboxylase
Crigler-Najjar syndrome	Glucuronyl transferase	Maple-sugar urine disease	Keto acid decarboxylase
Cystathioninuria	Cystathionase	Niemann-Pick disease	Sphingomyelinase
Ehlers-Danlos syndrome, type V	Lysyl oxidase	Ornithinemia	Ornithine ketoacid aminotransferase
Farber lipogranulomatosis	Ceramidase	Pentosuria	Xylitol dehydrogenase (L-xylulose reductase)
Galactosemia	Galactose 1-phosphate uridyl transferase	Phenylketonuria	Phenylalanine hydroxylase
Gangliosidosis, GM$_1$; generalized, type I, or infantile form	β-Galactosidase A, B, C	Refsum disease	Phytanic acid oxidase
		Richner-Hanhart syndrome	Tyrosine aminotransferase
Gangliosidosis, GM$_1$; type II, or juvenile form	β-Galactosidase B, C	Sandhoff disease (GM$_2$ gangliosidosis, type II)	Hexosaminidase A, B
Gaucher disease	Glucocerebrosidase	Tay-Sachs disease	Hexosaminidase A
Gout	Hypoxanthine-guanine phosphoribosyl-transferase	Wolman disease	Acid lipase
		Xeroderma pigmentosum	Ultraviolet-specific endonuclease
	Phosphoribosyl pyrophosphate (PRPP) synthetase (increased activity)		

* The form of gout due to increased activity of PRPP is the only disorder listed that is due to *increased* enzymatic activity.

SOURCE: Victor A. McKusick, *Mendelian Inheritance in Man,* 4th ed. Copyright © 1975 by Johns Hopkins University Press.

gether in a single conceptual model. Thus, armed with the understanding of the gene-protein relationship and how enzymes function, we can now re-examine some of the genetic findings discussed in earlier chapters and look at them in terms of the biochemistry involved.

Temperature-Sensitive Alleles. Recall that some mutants appear to be wild-type at normal temperatures but can be detected as mutants at high or low temperatures (see page 184). We now know that such mutations result from the substitution of an amino acid which produces a protein that is functional at normal temperatures, called **permissive** temperatures, but distorted and nonfunctional at high or low temperatures, called **restrictive** temperatures (Figure 12-21).

As we have seen, conditional mutations such as temperature-sensitive mutations can be very useful to geneticists. Stocks of the mutant culture can easily be maintained under

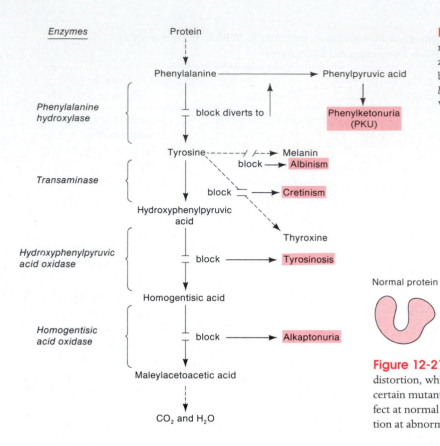

Figure 12-20 One small part of the human metabolic map, showing the consequences of various specific enzyme failures. (Disease phenotypes are shown in colored boxes.) (After I. M. Lerner and W. J. Libby, *Heredity, Evolution, and Society,* 2d ed. Copyright ©1976 by W. H. Freeman and Company.)

Figure 12-21 Schematic representation of protein conformational distortion, which is probably the basis for temperature sensitivity in certain mutants. An amino acid substitution that has no significant effect at normal (permissive) temperatures may cause significant distortion at abnormal (restrictive) temperatures.

permissive conditions, and the mutant phenotype can be studied intensively under restrictive conditions. Such mutants can be very handy in the genetic dissection of biological systems. For example, with a temperature-sensitive allele we can shift to a restrictive temperature at various times during development in order to determine the time at which a gene is active.

Genetic Ratios. Most "classical" (Mendelian) gene-interaction ratios can be explained readily by the one-gene–one-enzyme concept. For example, recall the 9:7 F_2 dihybrid ratio for pea flower pigment (page 103):

P $AA\,bb$ (white) $\times$ $aa\,BB$ (white)

F_1 all $A\,a\,B\,b$ (purple)

F_2 9 $A-B-$ (purple)

3 $A-bb$ (white)

3 $aa\,B-$ (white)

1 $aa\,bb$ (white)

We can easily explain this result if we imagine a biosynthetic pathway that leads ultimately to a purple petal pigment in which there are two colorless (white) precursors:

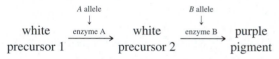

(Can you work out models to explain such ratios as 9:3:4 and 13:3?)

Dominance and Recessiveness. The meaning of dominance and recessiveness also becomes a little clearer in light of the biochemical model. In most cases, dominance represents the presence of enzyme function, whereas recessiveness represents the lack of enzyme function. A heterozygote has one dominant allele that can produce the functional enzyme:

allele $a \longrightarrow$ nonfunctional enzyme (does nothing)

precursor X $\xrightarrow[\text{enzyme A}]{\text{allele } A}$ product Y

If phenotype Y is due to the presence of product Y and phenotype X is due to the absence of product Y, then it is clear that the heterozygote will show phenotype Y and allele A is dominant over allele a.

However, this is not the only possible model. We can build in the concept of a threshold. Suppose that phenotype

Figure 12-22 Hypothetical curves relating enzyme concentration to the amount of product. Two basic situations are possible. In (a) there is a threshold level of the product above which a sharply contrasting phenotype, Y, is observed. Depending on where the product levels in the three possible genotypes occur in relation to this threshold, the heterozygote will show either phenotype X (for example, B_1B_2) or Y (for example, A_1A_2). If allele 2 is the active one, you can see that the active allele (A_2) can be dominant, which is normally the case. Less frequently, the inactive allele (B_1) can be dominant. In (b) there is no threshold, and the heterozygote has an intermediate phenotype. This second situation explains incomplete dominance.

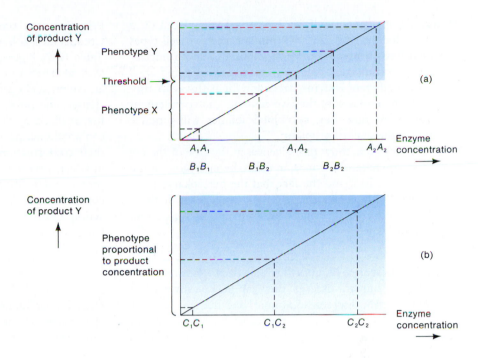

Y is produced only when the concentration of product Y exceeds some threshold level. Further suppose that the homozygote AA produces more enzyme—and hence more product—than the heterozygote Aa. In this case, the phenotype of the heterozygote will depend on the relationship between the threshold and the amount of product Y that is produced by the heterozygote. In Figure 12-22a the heterozygote A_1A_2 does produce enough product Y to exceed the threshold, so the heterozygote has phenotype Y and therefore A_2 is dominant over A_1. However, the situation could be like the one shown for the alleles B_1 and B_2, where the heterozygote B_1B_2 does not exceed the threshold. In this case (which is less common), B_1 is dominant over B_2 and the dominant phenotype is the one involving a lack of product Y.

Figure 12-22b illustrates a situation in which no threshold exists. The heterozygote has an intermediate phenotype—exactly the situation observed in cases of incomplete dominance.

What determines whether a threshold exists and whether a gene will behave like A, B, or C? Probably many interacting factors are involved: other genes, the chemical nature of the product, and (last but not least) the effect of the environment on that particular cell type. Finally, note that Figure 12-22 shows a simple linear relationship between enzyme concentration and the number of active alleles—but this need not be the case, as we'll see in Chapter 18.

Let's go back to the *arg* mutants in *Neurospora,* studied by Beadle and Tatum. (Use Figure 12-23.) Their recessiveness can be demonstrated by making heterokaryons be-

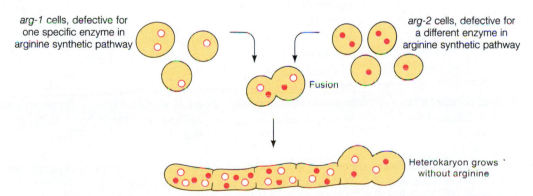

Figure 12-23 Formation of a heterokaryon of *Neurospora,* demonstrating both complementation and recessiveness. Vegetative cells of this normally haploid fungus can fuse, allowing the nuclei from the two strains to intermingle within the same cytoplasm. If each strain is blocked at a different point in a metabolic pathway, as are *arg-1* and *arg-2* mutants, all functions are present in the heterokaryon and the *Neurospora* will grow; in other words, complementation takes place. The growth must be due to the presence of the wild-type gene; hence, the mutant genes *arg-1* and *arg-2* must be recessive.

tween two *arg* mutants. If an *arg-1* mutant is placed on a minimal medium with an *arg-2* mutant, the two cell types fuse and form a **heterokaryon** composed of both nuclear types in a common cytoplasm. Recall that *arg-1* and *arg-2* each lack a different enzyme needed in the pathway to arginine. As a heterokaryon, the two mutants can complement each other, so that their combined abilities will produce arginine in the shared cytoplasm.

How does this show recessiveness in the case of the *arg* mutants? The heterokaryon is, in effect, heterozygous at the *arg-1* and *arg-2* loci, and the fact that the heterokaryon can produce arginine shows that the wild-type genes at these two loci are dominant; hence, the mutant genes *arg-1* and *arg-2* are recessive.

The Prion Diseases. Not only are altered proteins responsible for many human diseases, they can also be involved in spreading certain diseases. In animals and humans, a series of degenerative disorders of the central nervous system result from proteinaceous infectious particles, or **prions.**

Prions cause "scrapie" in sheep and goats, as well as Creutzfeldt-Jakob disease, among others, in humans (see Table 12-5). Figure 12-24 shows a prion protein that can exist in alternative conformations. When a mutation results in an amino acid change that causes a specific conformational change, the prion protein becomes an infectious agent, as first realized by Stanley Prusiner and his co-workers. How can an altered protein influence normal proteins to change their conformation? Figure 12-25 shows one such scheme, in which a scrapie prion protein can bind to a normal prion protein molecule and induce it to refold into the scrapie conformation. This process continues as a sort of chain reaction, until enough altered scrapie prion proteins accumulate to cause toxic effects.

Genetic Fine Structure

Until the beginning of this chapter, our genetic and cytological analysis led us to regard the chromosome as a linear (one-dimensional) array of genes, strung rather like beads

Table 12-5	**Prion Diseases of Humans**			
Disease	Typical Symptoms	Route of Acquisition	Distribution	Span of Overt Illness
Kuru	Loss of coordination, often followed by dementia	Infection (probably through cannibalism, which stopped by 1958)	Known only in highlands of Papua New Guinea; some 2,600 cases have been identified since 1957	Three months to one year
Creutzfeldt-Jakob disease	Dementia, followed by loss of coordination, although sometimes the sequence is reversed	Usually unknown (in "sporadic" disease) Sometimes (in 10 to 15 percent of cases) inheritance of a mutation in the gene coding for the prion protein (PrP) Rarely, infection (as an inadvertent consequence of a medical procedure)	*Sporadic form:* 1 person per million worldwide *Inherited form:* some 100 extended families have been identified *Infectious form:* about 80 cases have been identified	Typically about one year; range is one month to more than ten years
Gerstmann-Sträussler-Scheinker disease	Loss of coordination, often followed by dementia	Inheritance of a mutation in the PrP gene	Some 50 extended families have been identified	Typically two to six years
Fatal familial insomnia	Trouble sleeping and disturbance of autonomic nervous system, followed by insomnia and dementia	Inheritance of a mutation in the PrP gene	Nine extended families have been identified	Typically about one year

SOURCE: S. B. Prusiner, "The Prion Diseases." Copyright © 1995 by Scientific American, Inc. All rights reserved.

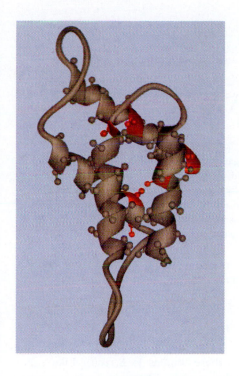
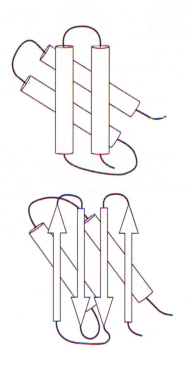

Figure 12-24 Prion protein. The model at the left shows the normal, benign protein PrP with a multiple helix backbone. (See also top right, where cylinders represent helices.) The "scrapie" form (lower right) has β strands, depicted as arrows. Single amino acid substitutions can convert the PrP form to the scrapie form. These are shown in red in the model at the left. (Drawing at left by Fred E. Cohen. Drawing at right by Dimitry Schidlovsky. From S. B. Prusiner, "The Prion Diseases." Copyright ©1995 by Scientific American, Inc. All rights reserved.)

on an unfastened necklace. Indeed, this model is sometimes called the **bead theory.** According to the bead theory, the existence of a gene as a unit of inheritance is recognized through its mutant alleles. All these alleles affect a single phenotypic character, all map to one chromosome locus, all give mutant phenotypes when paired, and all show Mendelian ratios when intercrossed. Several tenets of the bead theory are worth emphasizing:

1. The gene is viewed as a fundamental unit of *structure,* indivisible by crossing-over. Crossing-over occurs between genes (the beads in this model) but never within them.

2. The gene is viewed as the fundamental unit of *change,* or mutation. It changes in toto from one allelic form to another; there are no smaller components within it that can change.

3. The gene is viewed as the fundamental unit of *function* (although the precise function of the gene is not specified in this model). Parts of a gene, if they exist, cannot function.

Yet, how can we reconcile the fact that the gene consists of a series of nucleotides with the view that the gene is the smallest unit of mutation and recombination? Seymour Benzer's work in the 1950s showed that the bead theory was not correct. Benzer was able to use a genetic system in which extremely small levels of recombination could be detected. He demonstrated that whereas a gene can be defined

as a unit of function, a gene can be subdivided into a linear array of sites that are mutable and that can be recombined. The smallest units of mutation and recombination are now known to be correlated with single nucleotide pairs.

Fine-Structure Analysis of the Gene

As we shall see, Benzer's classic analysis of the fine structure of the gene deals with the following material:

1. The life cycle of the bacteriophage
2. Plaque morphology and the *rII* system of phage T4
3. The concept of "selection" in genetic crosses with bacteriophages
4. Deletion mapping
5. Destruction of the bead theory
6. Complementation and the difference between complementation and recombination

The Life Cycle of the Bacteriophage

Benzer chose the bacteriophage genome as the system for his high-resolution genetic studies. Therefore, Benzer's work is much easier to understand if we first review the material covered on pages 291 to 294. Let's summarize some of the essential points. Recall that bacteriophages, or phages, are viruses that attack bacteria. (The bacterial species used

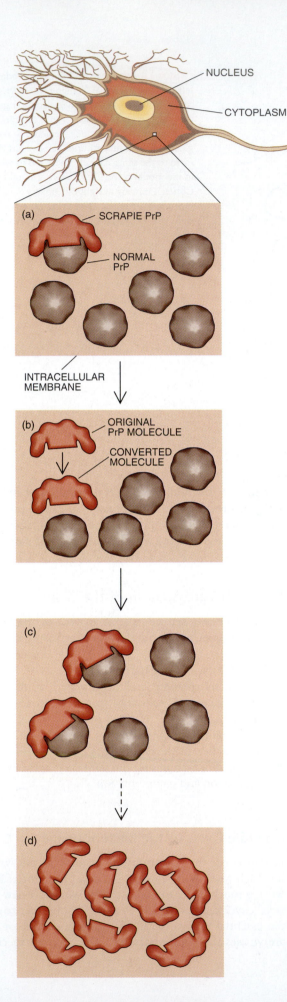

NUCLEUS

CYTOPLASM

(a) SCRAPIE PrP

NORMAL PrP

INTRACELLULAR MEMBRANE

(b) ORIGINAL PrP MOLECULE

CONVERTED MOLECULE

(c)

(d)

Figure 12-25 Propagation of scrapie PrP in neurons of the brain. Propagation probably involves a chain reaction type of conversion of normal PrP molecules by first one molecule of scrapie, then additional ones, as shown in a–d. (By Dimitry Schidlovsky. From S. B. Prusiner, "The Prion Diseases." Copyright ©1995 by Scientific American, Inc. All rights reserved.)

most frequently for these studies is *E. coli*.) Like other viruses, a phage usually consists of a protein coat containing the viral DNA, and tail fibers that allow the virus to attach to the bacterial cell wall. The phage nucleic acid is injected into the cell; this programs the synthesis of proteins that stimulate viral replication and the production of new viral DNA and new coat and tail proteins. After the new phage particles assemble, the bacterial cell is lysed and the particles are released into the surrounding medium. If the infected cell is on a lawn of bacteria immobilized on the agar surface of a petri plate, the new phage progeny will infect neighboring cells, and their progeny in turn will infect other neighboring cells until a small clearing, visible to the naked eye, is produced. This clearing is termed a **plaque** (see Figure 10-21). Different phages make different types of plaques. Phages also differ in the bacterial strains that they infect—a property termed the **host range.** Initially, phages were characterized according to their host range and given numbers in the "T" series. Thus, we have phages T1, T2, T3, T4, T5, and so on. Many experiments relating to the nature of the gene and the genetic code have been carried out with phage T4 (Figure 12-26).

The *rII* System

Benzer sought genetic markers that could be used with bacteriophages. The size and shape of a plaque, he realized, were heritable traits of the virus, and so he began a genetic analysis of plaque morphology. One type of mutant T4 phage produced larger, ragged plaques that were easy to distinguish from wild-type plaques (Figure 12-27). The large plaque size resulted from rapid lysis of the bacteria, so the mutants were termed *r* (for rapid lysis) **mutants.** Benzer analyzed the *r* mutants genetically, and mapped the mutations responsible for the *r* phenotype into two loci: *rI* and *rII*. He then studied the *rII* mutants intensively.

One extraordinary property of *rII* mutants made all of Benzer's work possible: *rII* mutants have a different host range than that of wild-type phages. Two related but different strains of *E. coli*, termed B and K(λ), can be used as different hosts for phage T4. Both bacterial strains can distinguish *rII* mutants from wild-type phages. *E. coli* B allows both to grow, but plaques of different sizes result: wild-type phages produce small plaques, and *rII* mutants produce large plaques. *E. coli* K, an abbreviation for *E. coli* K(λ), does not permit the growth of *rII* mutants, but it does allow wild-type phages to grow. The *rII* mutants are **conditional mutants,** namely, mutants that can grow under one set of conditions but not another. *E. coli* B is said to be **permissive** for *rII* mutants, because it allows phage growth, whereas *E. coli* K is said to be **nonpermissive** for *rII* mutants, because it does not allow phage growth. Figure 12-28 and Table 12-6 show the growth characteristics and plaque morphology of these phages on each host strain.

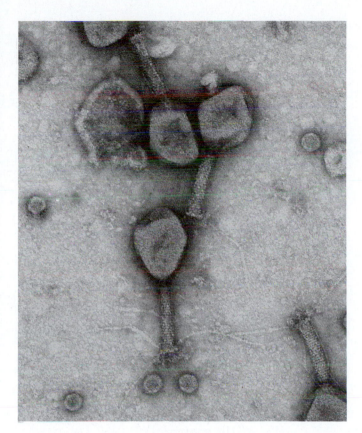

Figure 12-26 Enlargement of the *E. coli* phage T4 showing details of structure: note head, tail, and tail fibers. This was the phage used by Benzer in his experiments on the nature of the *rII* (rapid lysis) gene. (Photograph from Jack D. Griffith.)

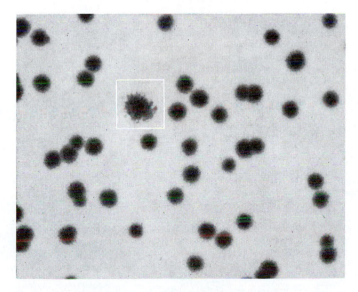

Figure 12-27 A spontaneous mutational event is disclosed by the one mottled plaque (*in the square*) among dozens of normal plaques produced when wild-type phage T4 is plated on a layer of colon bacilli of strain B. Each plaque contains some 10 million progeny descended from a single phage particle. The plaque itself represents a region in which cells have been destroyed. Mutants found in abnormal plaques provide the raw material for genetic mapping. (From S. Benzer, "The Fine Structure of the Gene." Copyright ©1962 by Scientific American, Inc. All rights reserved.)

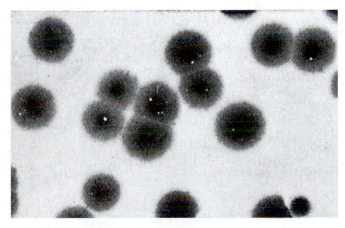

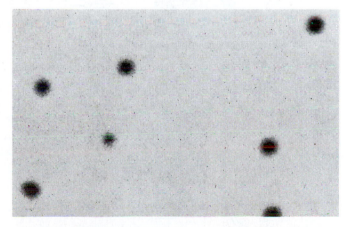

Figure 12-28 Duplicate replatings of a mixed phage population, obtained from a mottled plaque like the one shown in Figure 12-27, give contrasting results, depending on the host. Replated on *E. coli* of strain B (*top*), *rII* mutants produce both large plaques and wild-type phage small plaques. If the same mixed population is plated on strain K (*bottom*), only the wild-type phages produce plaques. (From S. Benzer, "The Fine Structure of the Gene." Copyright ©1962 by Scientific American, Inc. All rights reserved.)

Table 12-6 **Plaque Phenotypes Produced by Different Combinations of *E. coli* and Phage Strains**

T4 phage strain	*E. coli* strain	
	B	K
rII	Large, round	No plaques
rII⁺	Small, ragged	Small, ragged

Selection in Genetic Crosses of Bacteriophages

Benzer crossed various *rII* mutants of the T4 phage and obtained recombination frequencies, which he then used to map mutations within the *rII* gene region. How does one carry out crosses with different phages? In order to do this, phages of two different types are used to infect the same bacterial cell, as described on page 293. If the ratio of phages to bacteria is high enough, then virtually every bacterium will be infected with at least one phage of each type. Once inside the cell, the DNA molecules from the two phages have an opportunity to recombine with one another, generating recombinant phages, which are recovered among the progeny.

Let's see how this works. Suppose we wish to cross two *rII* mutants and recover wild-type recombinants. Because wild-type and *rII* mutants make plaques that can be distinguished from each other, we could cross two different *rII* mutants in *E. coli* B and examine the progeny on *E. coli* B (Figure 12-29, top photograph at lower right), hoping to find small wild-type plaques among the large parental *rII* plaques. If the recombination frequency is high enough to yield 2 to 3 percent or more wild-type plaques, then this method would suffice. However, for recombination that is less frequent than 1 percent, a lot of work would be involved in generating a map of numerous *rII* mutations.

Instead of plating the progeny phages from the cross on *E. coli* B, however, we could plate the progeny on *E. coli* K (Figure 12-29, bottom photograph at right), so that only the wild-type recombinant phages could grow. Even if the recombination frequency is very low (say, 0.01 percent), we could easily detect the recombinant wild-type phages. Why? Because a typical phage lysate (the phage mixture released after lysis of the bacteria) from such an infection (whether it involves a cross or not) contains in excess of 10^9 phages per

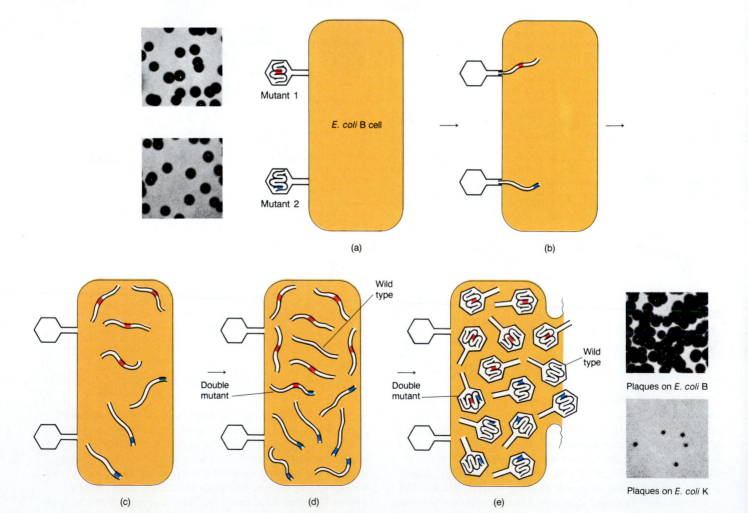

Mutant 1

E. coli B cell

Mutant 2

(a)

(b)

Double mutant

(c)

Wild type

Double mutant

(d)

Wild type

Double mutant

(e)

Plaques on *E. coli* B

Plaques on *E. coli* K

milliliter (ml). If we mix 0.1 ml of such a phage lysate with 0.1 ml of *E. coli* K bacteria, then we will have more than 10^5 (100,000) wild-type recombinant phages infecting the bacteria when the recombination frequency is 0.01 percent. (In practice, increasing dilutions of the phage lysate are used until one yields a countable number of plaques.) Now we can see the power of Benzer's *rII*–*E. coli* B/K system. In a single milliliter, it can find one recombinant or revertant per 10^9 organisms. Contrast this with trying to find one recombinant in 10^9 *Drosophila* or 10^9 mice!

Once we've made our cross, we need to determine the recombinant frequency. First, we count the number of active virus particles, or plaque-forming units (**pfu**), that grew on *E. coli* K (these, remember, are only wild-type recombinant phages), and the number that grew on *E. coli* B (these represent the total progeny phages, since all of the virus particles can grow on strain B). The recombinant frequency can then be calculated as twice the number of pfu on *E. coli* K divided by the number of pfu on *E. coli* B. Why do we use twice the pfu frequency for *E. coli* K? To account for the recombinants that are double mutants and which we cannot detect; such mutants should be present at the same frequency as the wild-type recombinants. Figures 12-29 and 12-30 provide diagrammatic representations of these principles.

Finally, in any cross of this type, we need to plate each parental lysate on *E. coli* K to see how many revertants to wild-type there were in the population. **Back (reverse) mutations** occur at some very low but real frequency. It is important to monitor this frequency and to compare it with our calculated frequency of recombination in order to be sure that recombination—not back reversion of the parental types—has occurred.

In summary, Benzer's use of the *rII* system and two different bacterial hosts provided him with a method for selecting for rare events without having to screen large numbers of plaques. This allowed him to demonstrate that recombination occurred within a gene, as we shall see.

Message   Benzer capitalized on the fantastic resolving power made possible by a system that selects for rare events in rapidly multiplying phages; this allowed him to map a gene in molecular detail.

Intragenic Recombination

Benzer started with an initial sample of eight independently derived *rII* mutant strains, crossing them in all possible combinations of pairs by the double infection of *E. coli* B and subsequent plating onto a lawn of *E. coli* K (Figures 12-29 and 12-30). Using recombinant frequencies, he could map the mutations unambiguously to the right or the left of each other to get what we now call a **gene map** (in this case, the map units are the frequency of *rII*⁺ plaques):

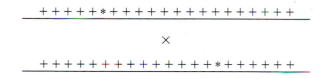

Recombination within a gene, called **intragenic recombination**, seems to be the rule rather than the exception. It can virtually always be found at any locus if a suitable selection system is available to detect recombinants. In other words, a mutant allele can be pictured as a length of genetic material (the gene) that contains a damaged or non-wild-type part—a **mutational site**—somewhere. This partial damage is what causes the non-wild-type phenotype. Different alleles produce different phenotypic effects because they involve damage to different sites within the wild-type allele.

Thus, an allele a^1 can be represented as

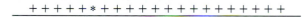

where the asterisk (*) is the mutant site within an otherwise normal gene (denoted by the sites marked +). A cross between a^1 and another mutant allele, a^2, can be represented as

and it is easy to see how the normal, wild-type gene

could be generated by a simple crossover anywhere between the two mutational sites.

Figure 12-29 (*facing page*) The process of recombination permits parts of the DNA of two different phage mutants to be reassembled in a new DNA molecule that may contain both mutations or neither of them. Mutants obtained from two different cultures are introduced into a broth of *E. coli* strain B. Crossing occurs when DNA from each mutant type infects a single bacillus. Most of the DNA replicas are of one type or the other, but occasionally recombination will produce either a double mutant or a wild-type recombinant containing neither mutation. When the progeny of the cross are plated on strain B, all grow successfully, producing many plaques. Plated on strain K, only the wild-type recombinants are able to grow. A single wild-type recombinant can be detected among as many as 100 million progeny. (Modified from S. Benzer, "The Fine Structure of the Gene." Copyright ©1962 by Scientific American, Inc. All rights reserved.)

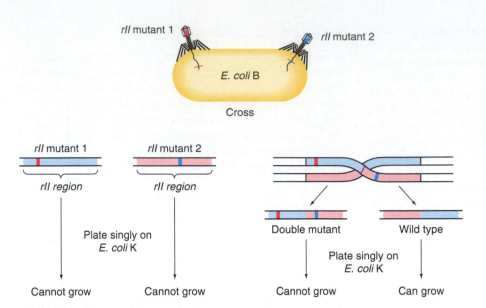

rII mutant 1   rII mutant 2

E. coli B

Cross

rII mutant 1   rII mutant 2

rII region   rII region

Plate singly on
E. coli K

Double mutant   Wild type

Plate singly on
E. coli K

Cannot grow   Cannot grow   Cannot grow   Can grow

Figure 12-30 Selection of intragenic recombinants at the *rII* locus of phage T4. Mutants within the gene *rII* cannot grow on *E. coli* K. When two different phages carrying different alleles of *rII* infect the same bacterial cell, some progeny phages can grow on *E. coli* K; in other words, some progeny have become *rII*⁺. This result indicates that recombination has occurred *within* a single gene and not just between genes.

Benzer showed that, contrary to the classical view, genes were not indivisible but could be subdivided by recombination. Extending his analysis to hundreds of *rII* alleles, Benzer found that the minimal recombinant frequency in a cross between a pair of different mutant alleles was 0.01 percent, even though his analytical system was capable of detecting recombinant frequencies as low as 0.0001 percent if they occurred. This led to the idea that genes were composed of small units, and that recombination could occur between but not within these units. We now know that the single nucleotide pair is the smallest unit of recombination.

> **Message**   A gene is composed of subelements that can recombine. The smallest subunit not divisible by recombination is the single nucleotide pair.

Recombination within genes permits us to construct detailed gene maps, such as the map for the *rII* region shown in Figure 12-31. Making crosses between various mutations provides us with the relative frequencies of intragenic recombinants, and these reveal the order and relative positions of the mutational sites within a gene. It should be noted that recombination *within* genes is the same as recombination *between* genes, except that the scale is different.

Mutational Sites

Using Deletions in Mapping Mutational Sites

Benzer extended his fine-structure analysis to the properties of mutational sites. A useful tool for these studies is **deletion mapping.** Deletions are mutations that result from the elimination of segments of DNA. Deletions can be intercrossed and mapped just like point mutations. The deleted region is represented by a bar. If no wild-type recombinants are produced in a cross between different deletions, then the

bars are shown as overlapping. A typical deletion map might be as follows:

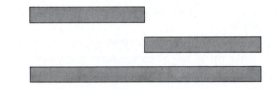

Such deletion maps are useful in delineating regions of the gene to which new point mutations can be assigned. Mutations cannot be converted into wild types in recombination tests against deletion mutations when the DNA corresponding to the wild-type region for that particular mutation is no longer present. Therefore, mutations can be ordered against a set of deletions by rapid tests that do not depend on extensive quantitative measurements.

Deletion Mapping of the *rII* Region

By using deletion mutants, Benzer could rapidly locate new mutational sites in his *rII* gene map. He had found some special *rII* mutants that would not give recombinants when crossed with any of several other types of *rII* mutants, but would give recombinants when crossed to still other *rII* mutants. Benzer realized that his special mutants behaved as if they contained short deletions within the *rII* region, and this model was supported by their lack of reversion to wild types.

Mutants carrying deletions can be used for the rapid location of mutational sites in newly obtained mutants. For example, consider the following gene map, showing 12 identifiable mutational sites:

1 2 3 4 5 6 7 8 9 10 11 12

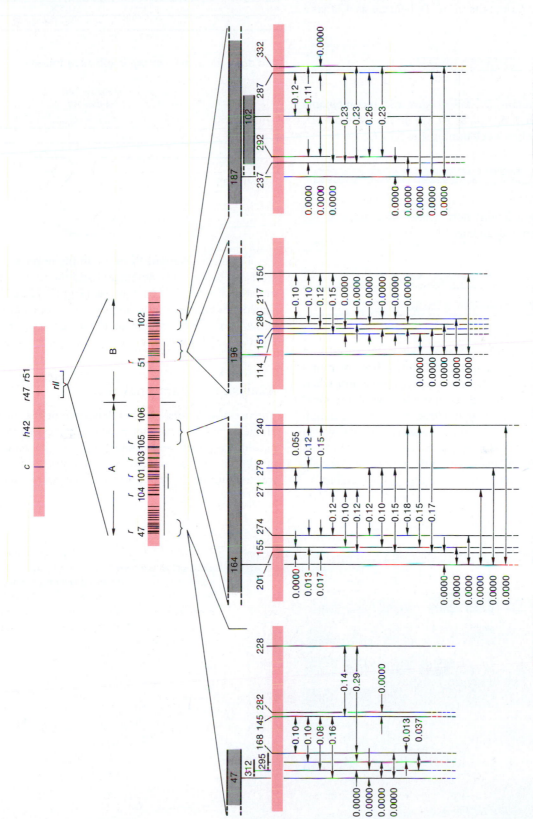

Figure 12-31 Detailed recombination map of the *rII* region of the phage T4 chromosome. The map unit is the percentage of *rII+* recombinants in crosses between the *rII* mutants. Typical regions are progressively enlarged. Numbers on the map represent mutational sites; numbers 47, 64, 196, 187, and 102 represent deletions. Note that the two portions of the *rII* region, A and B, are two different functional units of the region, as described later in the text. (After S. Benzer, *Proceedings of the National Academy of Sciences USA* 41, 1955, 344. From G. S. Stent and R. Calendar, *Molecular Genetics*, 2d ed. Copyright ©1978 by W. H. Freeman and Company.)

Let us suppose that one special mutant, D_1, fails to give *rII*⁺ recombinants when crossed with mutants carrying altered sites *1, 2, 3, 4, 5, 6, 7,* or *8*; therefore, D_1 behaves as if it has a deletion of sites *1* to *8*:

Another special mutant, D_2, fails to give *rII*⁺ recombinants when crossed with *5, 6, 7, 8, 9, 10, 11,* or *12*; therefore, D_2 behaves as if it involves a deletion of sites *5* to *12*:

These overlapping deletions now define three areas of the gene. Let's call them i, ii, and iii:

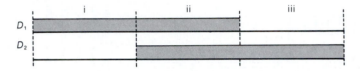

A new mutant that gives *rII*⁺ recombinants when crossed with D_1 but not when crossed with D_2 must have its mutational site in area iii. One that gives *rII*⁺ recombinants when crossed with D_2 but not with D_1 must have its mutational site in area i. A new mutant that does not give *rII*⁺ recombinants with either D_1 or D_2 must have its mutational site in area ii. For example, assume that a mutant in area iii is crossed with D_1. We would envision the cross schematically as follows, where the mutant site is shown with a red bar:

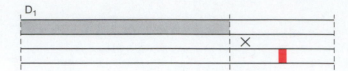

and draw the actual pairing involved as follows:

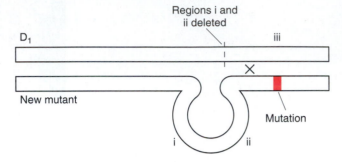

The more deletions there are in the tester set, the more areas can be uniquely designated and the more rapidly new mutational sites can be located (Figure 12-32). Once assigned to a region, a mutation can be mapped against other alleles in the same region to obtain an accurate position. Figure 12-33 shows the complexity of Benzer's actual map.

Analysis of Mutational Sites

The use of deletions enabled Benzer to define the **topology** of the gene—the manner in which the parts are interconnected. His genetic experiments showed the gene to consist of a linear array of mutable subelements. Benzer's next step was to examine the **topography** of the gene—differences in

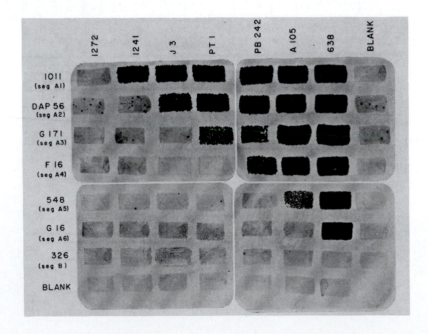

Figure 12-32 Crosses for mapping *rII* mutations. The photograph is a composite of four plates. Each row shows a given mutant tested against the reference deletions of Figure 12-33. The results show each of these mutations to be located in a different segment. Plaques appearing in the blanks are due to revertants present in the mutant stock.

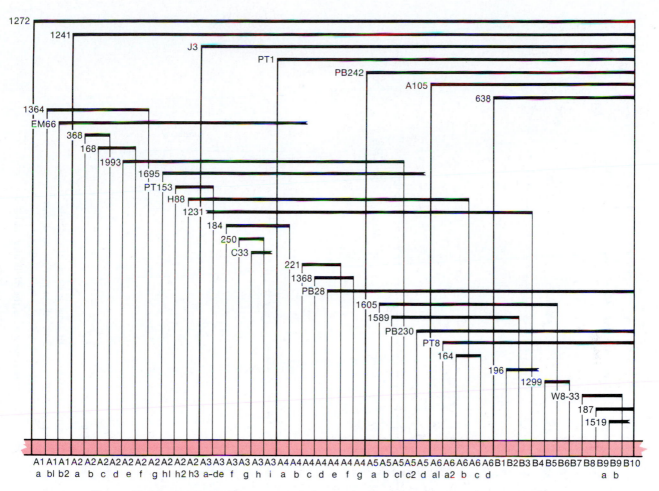

Figure 12-33 Detailed deletion map of the *rII* gene. Each deletion (*horizontal black bar*) has an identification number. Along the bottom are the arbitrary identification numbers of the regions defined by the deletions. Note that some deletions extend out of the *rII* gene. (From G. S. Stent and R. Calendar, *Molecular Genetics,* 2d ed. Copyright ©1978 by W. H. Freeman and Company. After S. Benzer, *Proceedings of the National Academy of Sciences USA 47,* 1961, 403–416.)

the properties of the subelements. Operationally, he determined this by asking whether all the subelements or sites were equally mutable, or whether mutations were prevalent at some sites and rare at others. For this study, it was essential to work with the smallest mutable subelements possible. Instead of multisite mutations (deletions) that exhibited no reversion, Benzer used revertible mutations, since they probably represented small alterations, or point mutations. Also, he discarded mutants with high reversion rates, because high reversion interferes with recombination tests.

Benzer used his deletion mutants to rapidly map the set of point mutations. He first localized each mutation into short deletion segments and then crossed all the point mutations within a segment against one another; any two revertible mutations that failed to recombine with each other represented mutations at the same site.

Figure 12-34 shows the distribution of 1612 spontaneous mutations in the *rII* locus. In Benzer's own words,

"That the distribution is not random leaps to the eye." This extraordinary nonrandom distribution demonstrates that all sites are not equally mutable. Benzer termed sites that are more mutable than other sites **hot spots.** The most prominent hot spot was represented by over 500 repeated occurrences in the collection of 1612 mutations. By examining a Poisson distribution calculated to fit the number of sites having only one or two occurrences (Figure 12-35), Benzer could show that at least 60 sites were truly more mutable than those with only one or two occurrences. Mutations were not observed at all (by chance) in at least 129 sites in this collection, even though these sites were as mutable as those represented by one or two occurrences. When Benzer extended the analysis to include mutagen-induced mutations, results were similar to the spontaneous mutation crosses: there were also hot spots, although often different ones from the spontaneous hot spots, among the mutations generated by mutagenic agents (see Chapter 18).

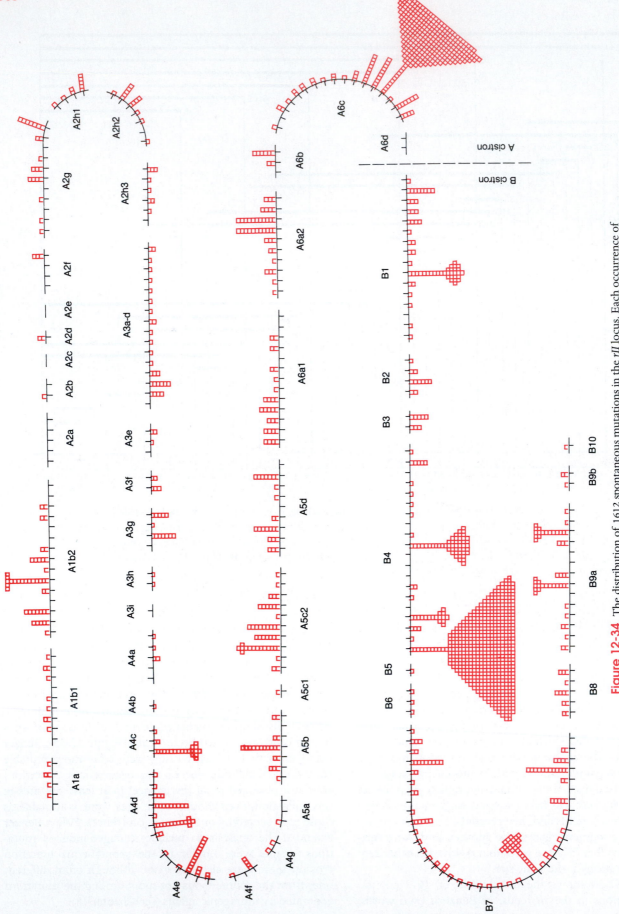

Figure 12-34 The distribution of 1612 spontaneous mutations in the *rII* locus. Each occurrence of an independent mutation is depicted by a square. When mutations are identical, the squares are drawn on top of one another. Although many positions are represented by only a single square, others have a large number of squares; these sites are called *hot spots*. The two subdivisions of the *rII* locus, *A* and *B*, are indicated in this diagram.

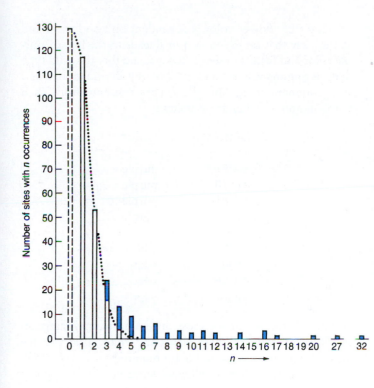

Figure 12-35 Distribution of occurrences of spontaneous mutations at various sites. The dotted curve indicates a Poisson distribution fitted to the number of sites having one and two occurrences. It predicts a minimum estimate for the number of sites of comparable mutability that have zero occurrences due to chance (dashed column at $n = 0$). Blue bars indicate the minimum numbers of sites that have mutation rates significantly higher than the one- and two-occurrence class. (From S. Benzer, *Proceedings of the National Academy of Sciences USA* 47, 1961, 403–416.)

The size of a mutational site was also of interest. The physical size of the *rII* region can be used to calculate the approximate number of base pairs in the region. From this figure, the number of mutational sites was determined to be approximately one-fifth of the number of nucleotide pairs. In other words, the smallest mutable site was five nucleotide pairs or less. The deciphering of the genetic code (Chapter 13), together with the demonstration by Ingram, Yanofsky, and others (described earlier in this chapter) that single amino acid substitutions resulted from single mutations, allowed Benzer to conclude that a mutation could result from the alteration of a single nucleotide pair. (The direct sequencing of DNA, described in Chapter 14, has since confirmed these conclusions in many examples.)

Message The gene can be divided into a linear array of mutable subelements that correspond to individual nucleotide pairs.

Destruction of the Bead Theory

Let's look again at the bead theory in light of Benzer's work. With the aid of deletion mapping, Benzer was able to map an extraordinary number of mutations in the *rII* locus against one another. His experiments have shown that mutations in the same gene can indeed recombine with one another. This result contradicts the bead theory of classical genetics, which held that recombination could occur between genes, but not within genes.

Benzer's analysis of the fine structure of the gene demonstrated that each gene consists of a linear array of subelements, and that these sites within a gene can be altered by mutation and can undergo recombination. This finding also contradicts the bead theory, one tenet of which implies that only the gene as a whole is mutable, not parts of the gene.

Subsequent work by several investigators identified each genetic site as a base pair in double-stranded DNA. Therefore, Benzer's contribution bridged the gulf between classical genetics and the knowledge of the chemical structure of DNA revealed by Watson and Crick. According to the bead theory, the Watson-Crick structure made no sense. However, Benzer's demonstration that genes do indeed have fine structures that can be revealed solely by genetic analysis allowed a fusion of the two disciplines and helped to launch the modern era of molecular genetics. Figure 12-36 illustrates the fine-structure analysis of the *rII* locus and its correspondence with the DNA structure.

Complementation

In another part of his studies, Benzer carried out a series of experiments designed to define the gene in terms of function. Benzer studied the concept of **complementation,** which we discussed briefly on page 358. We should realize that complementation for normally haploid organisms simply represents a situation that we have been dealing with routinely in the genetic analysis of diploids.

Before we describe Benzer's complementation experiments with *rII* mutants, let's first review complementation in diploids.

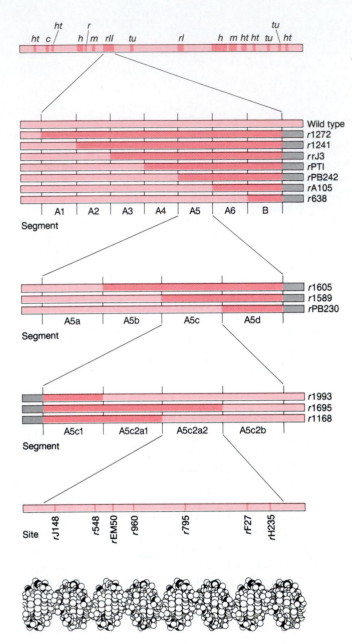

Figure 12-36 Fine-structure analysis of the *rII* locus. This mapping technique localizes the position of a given mutation in progressively smaller segments of the DNA molecule contained in phage T4. The *rII* region represents only a few percent of the entire molecule (*top*). The mapping is done by crossing an unknown mutant with reference mutants having deletions (darker red) of known extent in the *rII* region. Each site represents the smallest mutable unit in the DNA molecule, a single base pair. The molecular segment (*extreme bottom*), estimated to be roughly in proper scale, contains a total of about 40 base pairs. (From S. Benzer, "The Fine Structure of the Gene." Copyright ©1962 by Scientific American, Inc. All rights reserved.)

Complementation in Diploids

Suppose that we are studying a phenotype in a diploid cell that requires the active product of two genes. As an example, let's use the situation for flower color in peas that was detailed in Chapter 4 (see also page 356 in this chapter), in

which purple flower color is dependent on two loci: *A* and *B*. Let's say that mutation in *A* or *B* leads to the loss of purple color and results in white flowers, and that the wild-type allele is dominant in each case. Thus, *A* is dominant to *a* and *B* is dominant to *b*. The phenotypes resulting from each combination of genes are as follows:

Genotype	Color
AA BB	purple
Aa BB	purple
AA Bb	purple
Aa Bb	purple
aa BB	white
aa Bb	white
aa bb	white
AA bb	white
Aa bb	white

What happens when we cross an *AA bb* plant against an *aa BB* plant? Both of these plants are white. (Can you see why? Recall that *both A* and *B* are required for purple flowers.) However, the cross will yield offspring that are *Aa Bb*, and, since both *A* and *B* are dominant to *a* and *b*, respectively, all progeny from this cross will produce purple flowers.

Let's see why this is so in terms of cellular biosynthetic pathways.

Because each parent is homozygous at each locus, we know that all gametes from one parent are *A b* and that all gametes from the other parent are *a B*. The zygote resulting from the union of these two gametes, and all cells descending from this zygote, will complete the two-step cellular pathway that allows the production of purple flower pigment, because each of the two relevant chromosomes supplies one of the functional genes. We can diagram this as follows, where (1) is the starting compound in the cellular pathway, (2) is an intermediate compound, (3) is the final product for purple pigment, and A and B are the enzymes produced by genes *A* and *B*:

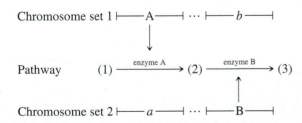

Note that one chromosome set (from one parent) supplies enzyme A but not enzyme B and that the second chromosome set (from the other parent) supplies enzyme B but not enzyme A. Since only one "good" copy of the relevant gene is required to produce each enzyme in this pathway, the two partially defective chromosome sets compensate for their respective deficiencies; that is, they help each other out, or **complement** each other.

Now let's look at complementation in haploids.

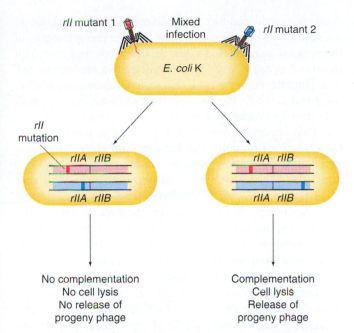

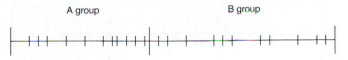

Figure 12-38 Gene map of *rII*.

Figure 12-37 Complementation test: a schematic view of *rII* complementation. Two different mutants of *rII* are used to simultaneously infect *E. coli* K (mixed infection). Normally, an *rII* mutant cannot lyse *E. coli* K or generate progeny phages. However, if the two different mutants can complement each other, then lysis and phage growth will result. If the two *rII* mutants cannot complement each other, then no lysis or phage growth will result.

Complementation in Bacteriophage T4

Benzer wanted to find out whether the entire *rII* region of phage T4 acts as a single functional unit, or whether it is made up of subunits that function independently. Therefore, he tested the mutations that he had mapped in the *rII* region to see whether various pairwise combinations of the mutations would restore the wild-type phenotype. In other words, Benzer looked for complementation in *E. coli* host cells that were temporarily "diploid" for the T4 chromosome. To do this, he carried out a mixed infection with different *rII* mutants (Figure 12-37). His criterion for the wild-type phenotype was the ability to lyse *E. coli* K hosts. (Recall that *rII* mutants cannot do this but that wild-type phages can.)

Complementation tests like the one Benzer conducted are carried out in one cycle of infection; they do not involve the multiple cycles of reinfection required for plaque formation. Samples of the two phages to be tested are spread over a strip of host bacteria on a section of a petri plate at a high ratio of phages to bacteria in order to ensure that essentially every bacterium is infected with both phages. After a period of incubation, the growth or the absence of growth of the bacteria in the strip indicates whether or not the bacteria have lysed as a result of the phage infection.

Pairwise tests of many different mutants allowed Benzer to separate the mutations into two groups, labeled A and B. All mutations in the A group complemented those in

the B group, whereas no mutations in the A group complemented any other mutations in the A group and no mutations in the B group complemented any other mutations in the B group. Benzer found that all mutations in group A mapped in one half of the *rII* locus and that all mutations in group B mapped in the other half of the *rII* locus (Figure 12-38).

The following diagram depicts a model for the *rII* locus based on Benzer's results:

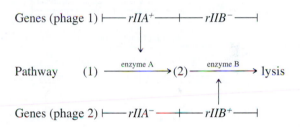

In our model, we assume that two genes, termed *rIIA* and *rIIB*, are involved in lysis. We can envision two steps, (1) and (2), each controlled by an enzyme specified by one of the wild-type *rII* genes. Thus, *rIIA*$^+$ and *rIIB*$^+$ would specify enzymes A and B, respectively, while *rIIA*$^-$ and *rIIB*$^-$ are two different *rII* mutants with defects in either the *rII* A or *rII* B region, respectively.

Because each phage chromosome can program the synthesis of one of the required functions for the cellular pathway, lysis can occur, which is the wild-type phenotype. Therefore, the two phages can complement each other. Notice how this situation is virtually identical with the example given previously for purple flower color in plants. The only difference is that the diploid cells in the case of the plant persist and remain diploid, whereas the double-infection experiment in T4 phage creates a temporary diploidlike cell for the phage chromosome. During the existence of this "diploid," complementation can occur. We should also note how similar complementation with T4 phage in *E. coli* is to complementation in *Neurospora* heterokaryons, described on page 358.

Recombination and Complementation

It is important to understand the difference between recombination and complementation. Recombination represents the creation of new combinations of genes through the physical breakage and rejoining of chromosomes. The progeny from a cross in which recombination has occurred have *new genotypes* that are different from the parental genotypes. Complementation, on the other hand, does not involve any

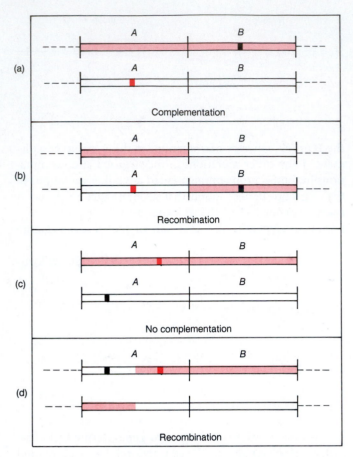

Figure 12-39 (a)

Complementation

(b)

Recombination

(c)

No complementation

(d)

Recombination

Figure 12-39 A comparison of the genetic consequences of complementation and recombination. (a) The mutant pairs can complement each other, since wild-type gene *A* and gene *B* products can mix in the cytoplasm. (b) If recombination occurs, then a rearrangement of the genomes will take place. (c) The mutant pairs cannot complement each other, because neither mutant can contribute a wild-type gene *A* product. (d) A relatively rare recombination event could result in a wild-type phage.

tively, we could conduct an additional genetic analysis: in the case of *rII* mutants, we could analyze the genotypes of the progeny phages. To do this, we first perform a mixed infection of *E. coli* K with the two phages we are testing. We then plate the progeny from this mixed infection on both *E. coli* B and *E. coli* K (Figure 12-40). All phages can grow on B, but only *rII* recombinants can form plaques on K.

Two important details allow our test to work. First, in the case of *rII* mutants, the incidence of recombination is never more than a few percent (see Figure 12-31), and it is rarely that high. Therefore, lysis will occur in only a very small fraction of the infected cells and will not interfere with the interpretation of the test. Second, in our second step, plating the progeny phages, the ratio of phages to bacteria is kept low, so that each bacterium is infected initially by no more than one phage; complementation cannot result in plaque formation in this situation.

Let's look at our test results. If the progeny phages from the mixed infection of *E. coli* K result from complementation, then virtually all the phages will still be *rII* mutants and will not plate on K. However, if recombination is required for phage growth and lysis during the mixed infection, then the analysis of the very low titer of progeny phages will show that about half are *rII*$^+$ recombinants and will plate on K (Figure 12-40).

change in the genotypes of individual chromosomes; rather, it represents the *mixing of gene products*. Complementation occurs during the time that two chromosomes are in the same cell and can each supply a function. Afterward, each respective chromosome remains unaltered. In the case of *rII* mutants, complementation occurs when two different phage chromosomes, with mutations in different *rII* genes, are in the same host cell. However, progeny that result from this complementation carry only the parental genotypes. Figure 12-39 diagrams some of the differences between these two basic processes. The arrangement shown in Figure 12-39a will allow phage growth and lysis (complementation) without a rearrangement of phage chromosomes, whereas the mutant pair shown in Figure 12-39c will not. Recombination can occur in each situation, as shown in parts b and d of the figure.

How can we distinguish operationally between complementation and recombination? To distinguish them defini-

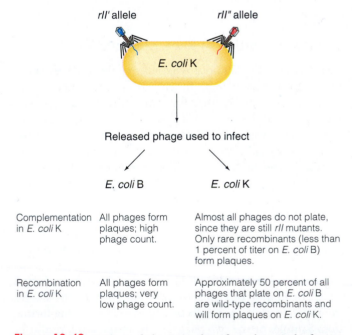

	E. coli B	*E. coli* K
Complementation in *E. coli* K	All phages form plaques; high phage count.	Almost all phages do not plate, since they are still *rII* mutants. Only rare recombinants (less than 1 percent of titer on *E. coli* B) form plaques.
Recombination in *E. coli* K	All phages form plaques; very low phage count.	Approximately 50 percent of all phages that plate on *E. coli* B are wild-type recombinants and will form plaques on *E. coli* K.

Figure 12-40 Analysis of phages resulting from the mixed infection of *E. coli* K. If complementation occurs between two *rII* mutants, then the progeny phages that are released will still be principally *rII* mutants and will fail to plate on *E. coli* K in single-infection experiments. If recombination but no complementation occurs, then there will be a sharply reduced yield of progeny phages due to the rarity of recombination. However, the few resulting phages will consist of approximately 50 percent wild-type recombinants and will form plaques on *E. coli* K in single-infection experiments.

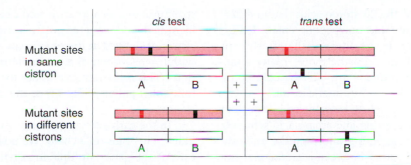

Figure 12-41 The cis-trans test. This test uses complementation to show whether two mutational sites are located within the same functional unit (cistron) in a gene or in different functional units: mutations that lie within the same cistron cannot complement each other. In the trans step, an *E. coli* K cell is infected with two different *rII* mutants, so that the mutations are on different chromosomes. If progeny phages do not grow, complementation has not occurred and the mutants are in the same functional unit, or cistron (*upper right box*). If progeny phages do grow, complementation has occurred and the mutants are in different cistrons (*lower right box*). In the *cis* test, used as a control, an *E. coli* K cell is infected with a T4 double mutant and a wild-type T4, so that both mutations are on a single chromosome (*upper and lower left boxes*). A + represents a successful complementation; a − indicates no complementation.

The Cistron

Mutations that fail to complement one another must affect the same functional unit; mutations that do complement one another must affect different functional units. Benzer called this unit of genetic function a **cistron.**

The cistron gets its name from the **cis-trans test,** which is the term Benzer used to describe a complete complementation test involving mutational sites on the same chromosome (cis) and on opposite chromosomes (trans), as shown in Figure 12-41. The cis portion of the test is really a control; the trans portion is the actual test for complementation. In summary, the cis-trans (or complementation) test is performed to determine whether two mutational sites are located within the same functional unit or in different functional units.

> **Message** A **cistron** is a genetic region within which there is normally no complementation between mutations. The cistron is equivalent to the gene.

Complementation and the Concept of the Gene

Of the three tenets defining a gene according to the classical (bead) theory, the one aspect that has held up and seems most essential is the gene as a unit of function. We now see that the gene is equivalent to the cistron, which we can consider to be a functional unit that can be defined experimentally by a cis-trans complementation test. There are occasional exceptions to this operational definition expressed by the cis-trans test. For example, some genes encode polypeptides that form multimeric proteins; in rare cases these display **intragenic (intracistronic) complementation,** in which individual monomers of mutationally altered proteins are inactive but combine to form the active, functional multimer. Also, some mutations greatly reduce the expression of more than one gene, and complementation would fail to occur between these mutations and mutations in the genes they affect. However, these are rare exceptions and do not upset the basic concept of the gene as a unit of function.

We now know that a cistron is a region of the genetic material that codes for one polypeptide chain. Therefore, the one-gene–one-enzyme hypothesis could be referred to more precisely as the **one-cistron–one-polypeptide hypothesis,** thereby also emphasizing that cistrons, or genes, can code for proteins other than enzymes.

SUMMARY

The work of Beadle and Tatum in the 1940s showed that one gene codes for one protein, often an enzyme. When an enzyme fails to function normally owing to a mutation, a variant phenotype results. These variant phenotypes are often the basis of genetic diseases in any organism, including humans. In order to understand how abnormal enzymes can cause phenotypic change, we need to understand the structure of proteins. Composed of a specific linear sequence of amino acids connected through peptide bonds, the proteins assume specific three-dimensional shapes as a result of the interaction of the 20 amino acids that in different combinations constitute the polypeptide chain. Different areas of this folded chain are sites for the attachment and interaction of substrates. Furthermore, many functional enzymes and other proteins are built by combining several polypeptide chains into a multimeric form.

Specific amino acid changes can be detected in a protein by the technique of fingerprinting and amino acid sequencing. Work of this type by Sanger, Ingram, Yanofsky, and others has demonstrated colinearity between a mutational site on the genetic map and the position of an altered amino acid in a protein.

Further work by Benzer and others illustrated that the gene could be dissected into smaller and smaller pieces. A structural unit of mutation and one of recombination were identified and equated with a single nucleotide pair. A cistron is defined at the phenotypic level as a genetic region within which there is no complementation between mutations. This is the unit that codes for the structure of a single functional polypeptide. Thus, the gene is equivalent to the cistron.

In the early days of genetics, genes were represented as indivisible beads on a chain. We now have a far different picture of the gene. Multiple intragenic mutational sites exist, and recombination may occur anywhere within a gene. In addition, a closer connection between genotype and phenotype was realized when it was established that one cistron is responsible for the synthesis of one polypeptide.

Concept Map

Draw a concept map interrelating as many of the following terms as possible. Note that the terms are listed in no particular order.

allele / mutation / biosynthetic pathway / enzyme / catalysis / auxotroph / dominant / recessive / complementation / cistron / deletion / peptide bond / polypeptide / amino acid sequence / recombination

CHAPTER INTEGRATION PROBLEM

We saw in Chapter 4 that the wild-type hemoglobin allele, Hb^A, displayed dominance to Hb^S (the allele of sickle-cell anemia) with regard to anemia, but codominance with regard to the electrophoretic properties of hemoglobin. Explain the different types of dominance in terms of protein sequence, based on the knowledge of protein structure learned in Chapter 12.

Solution

The protein hemoglobin has two α and two β polypeptide chains, closely fitted together in an elaborate quaternary structure which holds the heme groups that carry oxygen to the tissues. The primary structure of a protein—the linear sequence of its component amino acids—determines the protein's ultimate tertiary and quaternary structures, and hence its activity. Because of the change from glutamic acid to valine at position 6 in the β chain, the hemoglobin of sickle-cell anemia (HbS) is changed structurally, so that it carries less oxygen than normal adult hemoglobin (HbA). Heterogenotes (Hb^AHb^S) still maintain sufficient oxygen-carrying capacity to avoid severe symptoms of anemia; hence, Hb^A shows dominance to Hb^S with regard to anemia. Substituting valine for glutamic acid also changes the electric charge of the molecule, so that HbS moves more slowly than HbA on gel electrophoresis (see Figure 4-4). Thus, the two hemoglobin molecules (HbA and HbS) show codominance with respect to electrophoretic properties, because they both can be resolved and detected by this technique.

SOLVED PROBLEMS

1. Various pairs of *rII* mutants of phage T4 are tested in *E. coli* in both the cis and trans positions, and the "burst size" (the average number of phage particles produced per bacterium) for each test pair is compared. Results for six different *r* mutants—*rM*, *rN*, *rO*, *rP*, *rR*, and *rS*—are as follows (+ indicates the wild-type allele).

Cis genotype	Burst size	Trans genotype	Burst size
rM rN/++	245	*rM*+/+*rN*	250
rO rP/++	256	*rO*+/+*rP*	268
rR rS/++	248	*rR*+/+*rS*	242
rM rO/++	270	*rM*+/+*rO*	0
rM rP/++	255	*rM*+/+*rP*	255
rM rR/++	264	*rM*+/+*rR*	0
rM rS/++	240	*rM*+/+*rS*	240
rN rO/++	257	*rN*+/+*rO*	268
rN rP/++	250	*rN*+/+*rP*	0
rN rR/++	245	*rN*+/+*rR*	255
rN rS/++	259	*rN*+/+*rS*	0
rP rR/++	260	*rP*+/+*rR*	245
rP rS/++	253	*rP*+/+*rS*	0

If we assign the mutation *rO* to the A cistron, what are the locations of the other five mutations with respect to the A and B cistrons? (NOTE: This problem is similar to Problem 21 in the following problem set.)

Solution

In solving problems like this, we first look for the underlying principle and then attempt to draw a diagram to help us work out the solution. The key principle is that in the *trans* position, mutations in the same cistron will not complement one another and thus will yield no progeny phage (no burst), whereas mutations in different cistrons will complement and thus will yield progeny phages of normal burst size. Since *rO* is in the A cistron, we start with the initial diagram

We can now look at the test results in the problem and assign each mutation to a cistron based on whether or not complementation occurs in a mixed infection with *rO* mutants. From the table, it is evident that *N* and *P* complement *O* and must be in a different cistron (the B cistron), whereas *M* does not complement *O* and must be in the same cistron (the A cistron). This leaves us with *R* and *S*, which can be assigned to the A and B cistrons, respectively, based on their

complementation tests with *M*, *N*, and *P*. Therefore, we have the arrangement

We can check this assignment by examining the other crosses to find out if the results are consistent with our answer.

2. The following deletion map shows four deletions (1 to 4) involving the *rIIA* cistron of phage T4:

Five point mutations (*a* to *e*) are tested against these four deletion mutants for their ability (+) or inability (−) to give wild-type (*r*⁺) recombinants; the results are

	a	*b*	*c*	*d*	*e*
1	+	+	+	+	+
2	+	+	+	−	−
3	+	−	+	−	−
4	−	−	+	−	−

What is the order of the point mutations?

Solution

The key principle here is that point mutations can recombine with deletions that *do not* extend past the mutation, but they cannot recombine to yield wild-type phages with deletions that *do* extend past the mutation. For the test results given in the problem, any mutation that recombines with deletion 1 must be to the right of the deletion, any mutation that recombines with deletion 2 must be to the right of deletion 2, and so on. Let's look at point mutation *a*. It recombines with deletions 1, 2, and 3 but not with deletion 4. Therefore, it is to the right of deletions 1, 2, and 3 but not to the right of deletion 4. We can therefore place point mutation *a* in the interval between deletions 3 and 4. Point mutation *b* recombines with deletions 1 and 2 and must be to the right of them. It does not recombine with deletions 3 and 4, so it is in the interval between deletion 2 and 3. Point mutation *c* recombines with all the deletions and is to the right of all deletions, even deletion 4. Finally, both point mutations *d* and *e* recombine only with deletion 1 and must therefore be in the interval between deletions 1 and 2. The solution we have just derived can be summarized as follows:

1 _____ *e,d*

2 _____ *b*

3 _____ *a*

4 _____ *c*

Problem 25 in the following problem set is similar to this sample problem. See if you can apply the reasoning set forth here when you solve Problem 25.

PROBLEMS

1. Describe the primary, secondary, tertiary, and quaternary structures of a protein. Give examples of each type of structure, where possible.

2. What is the one gene–one enzyme hypothesis?

3. What is an auxotroph?

4. What is the exact change in a protein that results in sickle-cell anemia?

5. What experiment did Yanofsky do to demonstrate colinearity between the gene and the protein?

6. What is the difference between complementation and recombination?

7. Complete the following table by filling in the plaque types:

E. coli strain type

T4 phage	B	K
rII	?	?
rII⁺ (wild type)	?	?

8. A common weed, Saint-John's-wort, is toxic to albino animals. It also causes blisters on animals that have white areas of fur. Suggest a possible genetic basis for this reaction.

9. In humans, the disease galactosemia causes mental retardation at an early age because lactose in milk cannot be broken down, and this failure affects brain function. How would you provide a secondary cure for galactosemia?

Would you expect this phenotype to be dominant or recessive?

10. Amniocentesis is a technique in which a hypodermic needle is inserted through the abdominal wall of a pregnant woman and into the amnion, the sac that surrounds the developing embryo, in order to withdraw a small amount of amniotic fluid. This fluid contains cells that come from the embryo (not from the woman). The cells can be cultured; they will divide and grow to form a population of cells on which enzyme analyses and karyotypic analyses can be performed. Of what use would this technique be to a genetic counselor? Name at least three specific conditions under which amniocentesis might be useful. (NOTE: This technique involves a small but real risk to the health of both the woman and the embryo; take this fact into account in your answer.)

11. Table 12-7 shows the ranges of enzymatic activity (in units we need not worry about) observed for enzymes involved in two recessive metabolic diseases in humans. Similar information is available for many metabolic genetic diseases.

a. Of what use is such information to a genetic counselor?

b. Indicate any possible sources of ambiguity in interpreting studies of an individual patient.

c. Reevaluate the concept of dominance in the light of such data.

Table 12-7				
		Range of enzyme activity		
Disease	Enzyme involved	Patients	Parents of patients	Normal individuals
Acatalasia	Catalase	0	1.2–2.7	4.3–6.2
Galactosemia	Gal-1-P uridyl transferase	0–6	9–30	25–40

12. Two albinos marry and have a normal child. How is this possible? Suggest at least two ways. (This question appeared first in Chapter 4. Reconsider it now in light of biosynthetic pathways.)

13. In humans, PKU (phenylketonuria) is a disease caused by an enzyme inefficiency at step A in the following simplified reaction sequence, and AKU (alkaptonuria) is due to an enzyme inefficiency in one of the steps summarized as step B here:

$$\text{phenylalanine} \xrightarrow{A} \text{tyrosine} \xrightarrow{B} CO_2 + H_2O$$

A person with PKU marries a person with AKU. What phenotypes do you expect for their children? (**a**) All normal; (**b**) all having PKU only; (**c**) all having AKU only; (**d**) all having both PKU and AKU; (**e**) some having AKU and some having PKU.

14. Three independently isolated tryptophan-requiring strains of yeast are called *trpB*, *trpD*, and *trpE*. Cell suspensions of each are streaked on a plate supplemented with just enough tryptophan to permit weak growth for a *trp⁻* strain. The streaks are arranged in a triangular pattern so that they do not touch one another. Luxuriant growth is noted at both ends of the trpE streak and at one end of the trpD streak.

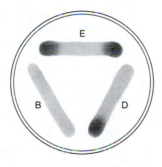

*a. Do you think complementation is involved?

b. Briefly explain the patterns of luxuriant growth.

c. In what order in the tryptophan-synthesizing pathway are the enzymatic steps defective in trpB, trpD, and trpE?

d. Why was it necessary to add a small amount of tryptophan to the medium in order to demonstrate such a growth pattern?

15. In *Drosophila* pupae, certain structures called *imaginal disks* can be detected as thickenings of the skin; after metamorphosis, these imaginal disks develop into specific organs of the adult fly. George Beadle and Boris Ephrussi devised a means of transplanting eye imaginal disks from one larva into another larval host. When the host metamorphoses into an adult, the transplant can be found as a colored eye located in its abdomen. The researchers took two strains of flies that were phenotypically identical in terms of bright scarlet eyes: one due to the sex-linked mutant vermilion (*v*); the other due to cinnabar (*cn*) on chromosome 2. If *v* disks are transplanted into *v* hosts or *cn* disks into *cn* hosts, then the transplants develop as mutant scarlet eyes. Transplanted *cn* or *v* disks in wild-type hosts develop wild-type eye colors. A *cn* disk in a *v* host develops a mutant eye color, but a *v* disk in a *cn* host develops wild-type eye color. Explain these results, and outline the experiments you would propose to test your explanation.

16. In *Drosophila*, the autosomal recessive *bw* causes a dark-brown eye, and the unlinked autosomal recessive *st* causes a bright scarlet eye. A homozygote for both genes has a white eye. Thus, we have the following correspondences between genotypes and phenotypes:

$$st^+ st^+ bw^+ bw^+ = \text{red eye (wild-type)}$$
$$st^+ st^+ bw\, bw = \text{brown eye}$$
$$st\, st\, bw^+ bw^+ = \text{scarlet eye}$$
$$st\, st\, bw\, bw = \text{white eye}$$

Construct a hypothetical biosynthetic pathway showing how the gene products interact and why the different mutant combinations have different phenotypes.

17. Several mutants are isolated, all of which require compound G for growth. The compounds (A to E) in the biosynthetic pathway to G are known, but not their order in the pathway. Each compound is tested for its ability to support the growth of each mutant (1 to 5). In the following table, + indicates growth and − indicates no growth:

| | Compound tested | | | | | |
	A	B	C	D	E	G
Mutant 1	−	−	−	+	−	+
2	−	+	−	+	−	+
3	−	−	−	−	−	+
4	−	+	+	+	−	+
5	+	+	+	+	−	+

a. What is the order of compounds A to E in the pathway?

b. At which point in the pathway is each mutant blocked?

c. Would a heterokaryon composed of double mutants 1,3 and 2,4 grow on a minimal medium? 1,3 and 3,4? 1,2 and 2,4 and 1,4?

18. In *Neurospora* (a haploid), assume that two genes participate in the synthesis of valine. Their mutant alleles are *val-1* and *val-2*, and their wild-type alleles are *val-1⁺* and *val-2⁺*. The two genes are linked on the same chromosome, and a crossover occurs between them, on average, in one of every two meioses.

a. In what proportion of meioses are there no crossovers between the genes?

b. Use the map function to determine the recombinant frequency between these two genes.

c. Progeny from the cross $val\text{-}1\,val\text{-}2^+ \times val\text{-}1^+\,val\text{-}2$ are plated on a medium containing no valine. What proportion of the progeny will grow?

d. The $val\text{-}1\,val\text{-}2^+$ strains accumulate intermediate compound B, and the $val\text{-}1^+\,val\text{-}2$ strains accumulate intermediate compound A. The $val\text{-}1\,val\text{-}2^+$ strains grow on valine or A, but the $val\text{-}1^+\,val\text{-}2$ strains grow only on valine and not on B. Show the pathway order of A and B in relation to valine, and indicate which gene controls each conversion.

19. In a certain plant, the flower petals are normally purple. Two recessive mutations arise in separate plants and are found to be on different chromosomes. Mutation 1 (m_1) gives blue petals when homozygous (m_1m_1). Mutation 2 (m_2) gives red petals when homozygous (m_2m_2). Biochemists working on the synthesis of flower pigments in this species have already described the following pathway:

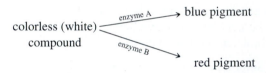

a. Which mutant would you expect to be deficient in enzyme A activity?

b. A plant has the genotype $M_1m_1M_2m_2$. What would you expect its phenotype to be?

c. If the plant in part b is selfed, what colors of progeny would you expect, and in what proportions?

d. Why are these mutants recessive?

20. In sweet peas, the synthesis of purple anthocyanin pigment in the petals is controlled by two genes, B and D. The pathway is

white intermediate $\xrightarrow{\text{gene } B \atop \text{enzyme}}$ blue intermediate $\xrightarrow{\text{gene } D \atop \text{enzyme}}$ anthocyanin (purple)

a. What color petals would you expect in a pure-breeding plant unable to catalyze the first reaction?

b. What color petals would you expect in a pure-breeding plant unable to catalyze the second reaction?

c. If the plants in parts a and b are crossed, what color petals will the F_1 plants have?

d. What ratio of purple:blue:white plants would you expect in the F_2?

21. Various pairs of *rII* mutants of phage T4 are tested in *E. coli* in both the cis and trans positions. Comparisons are made of the "burst size" (the average number of phage particles produced per bacterium). Results for six different *r* mutants—*rU*, *rV*, *rW*, *rX*, *rY*, and *rZ*—are as follows:

Cis genotype	Burst size	Trans genotype	Burst size
rU rV/ + +	250	rU +/ +rV	258
rW rX/ + +	255	rW +/ +rX	252
rY rZ/ + +	245	rY +/ +rZ	0
rU rW/ + +	260	rU +/ +rW	250
rU rX/ + +	270	rU +/ +rX	0
rU rY/ + +	253	rU +/ +rY	0
rU rZ/ + +	250	rU +/ +rZ	0
rV rW/ + +	270	rV +/ +rW	0
rV rX/ + +	263	rV +/ +rX	270
rV rY/ + +	240	rV +/ +rY	250
rV rZ/ + +	274	rV +/ +rZ	260
rW rY/ + +	260	rW +/ +rY	240
rW rZ/ + +	250	rW +/ +rZ	255

If we assign *rV* to the A cistron, what are the locations of the other five *rII* mutations with respect to the A and B cistrons?

(Problem 21 is from M. Strickberger, *Genetics*. Copyright 1968 by Monroe W. Strickberger. Reprinted with permission of Macmillan Publishing Co., Inc.)

22. There is evidence that occasionally during meiosis either one or both homologous centromeres will divide and segregate precociously at the first division rather than at the second division (as is the normal situation). In *Neurospora*, *pan2* alleles produce a pale ascospore, aborted ascospores are completely colorless, and normal ascospores are black. In a cross between two complementing mutations *pan2x* × *pan2y*, what ratios of black:pale:colorless would you expect in asci resulting from the precocious division of (**a**) one centromere? (**b**) Both centromeres? (Assume that *pan2* is near the centromere.)

23. *Protozoon mirabilis* is a hypothetical single-celled haploid green alga that orients to light by means of a red eyespot. By selecting cells that do not move toward the light, 14 white-eyespot mutants (*eye⁻*) are isolated after mutation. It is possible to fuse haploid cells to make diploid individuals. The 14 *eye⁻* mutants are paired in all combinations, and the color of the eyespot is scored in each. Table 12-8 shows the results, where + indicates a red eyespot and − indicates a white eyespot.

a. Mutant 14 obviously is different from the rest. Why might this be?

Table 12-8

	1	2	3	4	5	6	7	8	9	10	11	12	13	14
1	−	+	+	+	−	+	+	−	−	+	+	+	+	−
2	+	−	−	−	+	+	+	+	+	+	+	−	+	−
3	+	−	−	−	+	+	+	+	+	+	+	−	+	−
4	+	−	−	−	+	+	+	+	+	+	+	−	+	−
5	−	+	+	+	−	+	+	−	−	+	+	+	+	−
6	+	+	+	+	+	−	−	+	+	−	−	+	−	−
7	+	+	+	+	+	−	−	+	+	−	−	+	−	−
8	−	+	+	+	−	+	+	−	−	+	+	+	+	−
9	−	+	+	+	−	+	+	−	−	+	+	+	+	−
10	+	+	+	+	+	−	−	+	+	−	−	+	−	−
11	+	+	+	+	+	−	−	+	+	−	−	+	−	−
12	+	−	−	−	+	+	+	+	+	+	+	−	+	−
13	+	+	+	+	+	−	−	+	+	−	−	+	−	−
14	−	−	−	−	−	−	−	−	−	−	−	−	−	−

b. Excluding mutant 14, how many complementation groups are there, and which mutants are in which group?

c. Three crosses are made, with the following results:

Mutants crossed	Number of progeny		
	eye⁺	*eye*⁻	Total
1 × 2	31	89	120
2 × 6	5	113	118
1 × 14	0	97	97

Explain these genetic ratios with symbols.

d. How many genetic loci are involved altogether, and which of the 14 mutants are at each locus?

e. What is the linkage arrangement of the loci? (Draw a map.)

24. You have the following map of the *rII* locus:

A cistron   B cistron

rX   rY   rZ rD   rE   rF

You detect a new mutation, *rW*, and you find that it does not complement any of the mutants in the A or B cistron. You find that wild-type recombinants are obtained in crosses with *rX*, *rY*, *rE*, and *rF* but not with *rZ* or *rD*. Suggest possible explanations for these results. Describe tests you would use to choose between the explanations.

25. The following map shows four deletions (1 to 4) involving the *rIIA* cistron of phage T4:

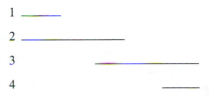

Five point mutations (*a* to *e*) in *rIIA* are tested against these four deletion mutants for their ability (+) or inability (−) to give *r*⁺ recombinants, with the following results:

	a	b	c	d	e
1	+	+	−	+	+
2	+	+	−	−	−
3	−	−	+	−	+
4	+	−	+	+	+

a. What is the order of the point mutants?

b. Another strain of T4 has a point mutation in the *rIIB* cistron. This strain is mixed in turn with each of the *rIIA* deletion mutants, and the mixtures are used to infect *E. coli* K at a multiplicity of infection great enough that each host cell will be infected by at least one *rIIA* and one *rIIB* mutant. A normal plaque is formed with deletions 1, 2, and 3, but no plaque forms with deletion 4. Given that the B cistron is to the right of A, explain the behavior of deletion 4. Does your explanation affect your answer to part a?

26. In a phage, a set of deletions is intercrossed in pairwise combinations. The following results are obtained (a +

indicates that wild-type recombinants are obtained from that cross):

```
        1   2   3   4   5
    1 | −   +   −   +   −
    2 | +   −   +   +   −
    3 | −   +   −   −   −
    4 | +   +   −   −   +
    5 | −   −   −   +   −
```

a. Construct a deletion map from this table.

b. The first geneticists to do a deletion-mapping analysis in the mythical schmoo phage SH4 (which lyses schmoos) came up with this unique set of data:

```
        1   2   3   4
    1 | −   −   +   −
    2 | −   −   −   +
    3 | +   −   −   −
    4 | −   +   −   −
```

Show why this is a unique result by drawing the only deletion map that is compatible with this table. (Don't let your mind be shackled by conventional expectations.)

27. In a haploid eukaryote, four alleles of the *cys-2* gene are obtained. Each allele requires cysteine, and all alleles map to the same locus. The four strains bearing these mutant alleles are crossed to wild types to obtain a set of eight cultures representing the four mutant alleles in association with each mating type. Then the mutant alleles are intercrossed in all pairwise combinations. The haploid meiotic products from each cross are plated on a medium containing no cysteine. In some crosses, cys^+ prototrophs are observed at low frequencies. The results follow, where the numbers represent the frequencies of cys^+ colonies per 10^4 meiotic products plated:

		Mating type A'		
	1	2	3	4
Mating type A″ 1	0	14	2	20
2	14	0	12	6
3	2	12	0	18
4	20	6	18	0

a. Draw a map of the four mutant sites within the *cys-2* gene. Provide a measurement of the relative intersite distances.

b. Do you see any evidence that mutation might be involved in the production of the prototrophs?

* 28. In *Neurospora,* there is a gene controlling the production of adenine, and mutants in this gene are called *ad-3* mutants. The *his-2* locus is 2 m.u. to the left, and the *nic-2* locus is 3 m.u. to the right of the *ad-3* locus (*his-2* controls

Table 12-9

Genotype of *ad-3*⁺ recombinants	Number of *ad-3*⁺ spores picked up		
	Cross 1	Cross 2	Cross 3
his-2 nic-2	0	6	0
his-2⁺ *nic-2*⁺	0	0	0
his-2 nic-2⁺	15	0	5
his-2⁺ *nic-2*	0	0	0
Total ascospores scored	41,236	38,421	43,600

histidine, and *nic-2* controls nicotinamide). Thus, the genetic map is

```
    his-2          ad-3              nic-2
   ├──────────────┼─────────────────┤
        2 m.u.         3 m.u.
```

Three different *ad-3* auxotrophs are detected: *ad-3ᵃ*, *ad-3ᵇ*, and *ad-3ᶜ*. (Use *a*, *b*, and *c* as their labels.) The following crosses are made:

Cross 1   *his-2*⁺ *a nic-2*⁺ × *his-2 b nic-2*

Cross 2   *his-2*⁺ *a nic-2* × *his-2 c nic-2*⁺

Cross 3   *his-2 b nic-2* × *his-2*⁺ *c nic-2*⁺

The ascospores are then plated on a minimal medium containing histidine and nicotinamide, and *ad-3*⁺ prototrophs are picked up. Table 12-9 shows the results obtained. What is the map order of the *ad-3* mutants and the genetic distance between them?

29. In a hypothetical diploid organism, squareness of cells is due to a threshold effect: more than 50 units of "square factor" per cell will produce a square phenotype and less than 50 units will produce a round phenotype. Allele s^f is a functional gene that causes the synthesis of the square factor. Each s^f allele contributes 40 units of square factor; thus, $s^f s^f$ homozygotes have 80 units and are phenotypically square. A mutant allele (s^n) arises; it is nonfunctional, contributing no square factor at all.

a. Which allele will show dominance, s^f or s^n?

b. Are functional alleles necessarily always dominant? Explain your answer.

c. In a system such as this one, how might a specific allele become changed in evolution so that its phenotype shows recessive inheritance at generation 0 and dominant inheritance at a later generation?

* 30. Explain how Benzer was able to calculate the number of sites with zero occurrences, as depicted in Figure 12-35.

(Assume, as Benzer did, that the sites with 0, 1, and 2 occurrences are a set of sites with equal mutability.)

31. In *Collinsia parviflora*, the petal color is normally purple. Four recessive mutations are induced, each of which produces white petals. The four pure-breeding lines are then intercrossed in the following combinations, with the results indicated:

Mutant crosses	F_1	F_2
1 × 2	all purple	$\frac{1}{2}$ purple, $\frac{1}{2}$ white
1 × 3	all purple	$\frac{9}{16}$ purple, $\frac{7}{16}$ white
1 × 4	all white	all white

a. Explain all these results clearly, using diagrams wherever possible.

b. What F_1 and F_2 do you predict from crosses of 2 × 3 and 2 × 4?

*32. A biologist is interested in the genetic control of leucine synthesis in the haploid filamentous fungus *Aspergillus*. He treats spores with mutagen and obtains five point mutations (*a* to *e*), all of which are leucine auxotrophs. He first makes heterokaryons between them to check on their functional relationships. He determines the following results, where + indicates that the heterokaryon grew, and − that the heterokaryon did not grow, on a medium lacking leucine:

	a	*b*	*c*	*d*	*e*
a	−	+	+	+	−
b		−	+	+	+
c			−	+	+
d				−	+
e					−

The biologist then intercrosses the mutations in all possible combinations. From each cross, he tests 500 ascospore progeny by inoculating them onto a medium lacking leucine. The results follow (the numbers represent the number of leucine prototrophs in the 500 progeny):

	a	*b*	*c*	*d*	*e*
a	0	125	128	126	0
b		0	124	2	125
c			0	124	127
d				0	123
e					0

Explain both sets of data genetically. (Note that the two leucine prototrophs from the *b* × *d* cross were found *not* to be due to reversion.)

33. In *Drosophila*, the eye phenotype "star" is caused by recessive mutations (*s*) mapping to one location on the second chromosome. This region is flanked to the left by the *A* locus (allele *A* or *a*) and to the right by the *B* locus (allele *B* or *b*):

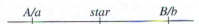

Six independently induced *star* mutations are each made homozygous, with both *A A B B* and *a a b b* constitutions, and the six are intercrossed to study complementation at the *star* locus. The results follow, where + indicates wild eye and s indicates star eye, both being phenotypes of the F_1.

a a s s b b	*A A s s B B*					
	1	2	3	4	5	6
1	s	+	s	s	+	[+]
2		s	+	[+]	s	+
3			s	s	+	+
4				s	+	+
5					s	+
6						s

a. How many cistrons are at the *star* location, and which mutational sites are in each cistron?

b. The heterozygotes in brackets are allowed to produce gametes. In both cases, s^+ recombinant gametes are identified. The gametes are tested for the flanking marker conformation, which is *a B* in both the 1 × 6 heterozygote and the 2 × 4 heterozygote. Order the cistrons in relation to the *A a* and *B b* loci.

34. In Norway in 1934, a mother with two mentally retarded children consulted the physician Asbjørn Følling. In the course of the interview Følling learned that the urine of the children had a curious odor. He later tested their urine with ferric chloride and found out that whereas normal urine gives a brownish color, the children's urine stained green. He deduced that the chemical responsible must be phenylpyruvic acid. Because of chemical similarity to phenylalanine, it seemed likely that this substance had been formed from phenylalanine in the blood, but there was no assay for phenylalanine. However, a certain bacterium could convert phenylalanine to phenylpyruvic acid so that the level of phenylalanine could be measured using the ferric chloride test. It was found that indeed the children had high levels of phenylalanine in their blood, and this was probably the source of the phenylpyruvic acid. Følling noted that in general in families with retarded children whose urine stained green with ferric chloride, there was a proportion of $\frac{1}{4}$ affected children and $\frac{3}{4}$ unaffected. This disease, which came to be known as phenylketonuria (PKU), seemed to have a genetic basis and was inherited like a simple Mendelian recessive.

It became clear that phenylalanine was the culprit and that this chemical built up in the PKU patients and was converted to high levels of phenylpyruvic acid, which then interfered with the normal development of nervous tissue. This led to the formulation of a special diet low in phenylalanine, which could be fed to newborn babies diagnosed with PKU and which allowed normal development to continue without retardation. Indeed it was found that after the child's nervous system had developed, the patient could be taken off the special diet. However, tragically, PKU women who had developed normally with the special diet were found to have babies who were born mentally retarded, and the special diet had no effect on these children.

a. Why do you think the babies of the PKU mothers were born retarded?

b. Explain the reason for the difference in the results of the PKU babies and the babies of PKU mothers.

c. Propose a treatment that might allow PKU mothers to have unaffected children.

d. Write a short essay on PKU integrating concepts at the genetic, diagnostic, enzymic, physiological, pedigree, and population levels.

 Unpacking the Problem

a. Draw the pedigree of the Norwegian family as far as possible. Do you think some siblings might have been unaffected?

b. Could the original children brought to Følling (assume they were 10 years old) have benefitted from the special diet if started at that age?

c. Why were the bacteria needed by Følling?

d. Why was a blood test considered important?

e. Why was there phenylpyruvic acid in the urine?

f. What is the relationship expected between blood concentrations and urine concentrations?

g. Was the green substance in the ferric chloride test important per se?

h. What phenylalanine and phenylpyruvic acid concentrations are expected in the blood and urine of unaffected children?

i. What was the source of the odor in the children's urine?

j. What were the genotypes of the parents?

k. Would families with parents with these genotypes be common?

l. What genotypes of offspring are expected and in what proportions?

m. What would be the genotypes of parents in most families in the population?

n. Why was it inferred that the disease was inherited?

o. Why was it inferred to be inherited as a Mendelian recessive?

p. At which developmental stage(s) is most of the nervous system developing?

q. Why was it believed that PKU adults no longer needed to be on the special diet?

r. What are the relative volumes of the maternal and fetal blood circulation systems?

s. Which types of entities are able to pass the placental barrier (cells, macromolecules, small molecules)?

t. Which PKU-associated substances might pass into the mother from the child? Into the child from the mother?

u. What is an essential amino acid?

v. Is phenylalanine an essential amino acid?

w. How does PKU illustrate the one-gene–one-enzyme concept?

x. Why is PKU recessive?

y. What is the relevance of biochemical pathways to PKU?

z. In the chapter find a chart that shows the position of the PKU-associated chemicals in the chemistry of the cell.

aa. Draw a diagram relating chromosome, gene, mutant site, mRNA, enzyme, substrate, product, in normal individuals and PKU sufferers.

13

DNA Function

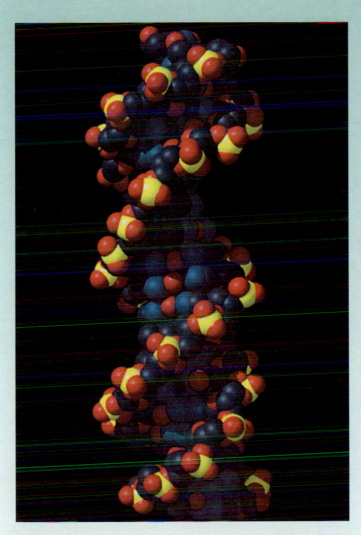

Computer model of DNA. (J. Newdol, Computer Graphics Laboratory, University of California, San Francisco. Copyright © Regents, University of California.)

KEY CONCEPTS

► DNA is transcribed into an mRNA molecule, which is then translated during protein synthesis.

► Translation requires transfer RNAs and ribosomes.

► The genetic code is a nonoverlapping triplet code.

► Special sequences signal the initiation and termination of both transcription and translation.

► In eukaryotes, the initial RNA transcript is processed in several ways to generate the final mRNA.

► Many eukaryotic genes contain segments of DNA, termed *introns,* that interrupt the normal gene coding sequence.

► The primary eukaryotic transcript is spliced in one of a variety of ways to remove the RNA encoded by the intron and to yield the final mRNA.

The genetic information embodied in DNA can either be copied into more DNA during replication or be translated into protein. These are the processes of **information transfer** that constitute DNA function. We considered replication in some detail in Chapter 11. Let's now explore the way in which genetic information is turned into protein. Only a part of this story has been revealed through purely genetic analysis; mutants have been useful as tools for discovering parts that malfunction.

We shall see in this chapter that one of the first clues as to how DNA directs the synthesis of proteins came from bacteriophages, when it was shown that gene expression resulted in the synthesis of RNA molecules from a DNA template, a process termed **transcription.** RNA is synthesized from only one strand of a double-stranded DNA helix. This transcription is catalyzed by an enzyme, **RNA polymerase,** and follows rules similar to those of replication. RNA has similarities to DNA, although ribose is the sugar used in RNA, and uracil replaces thymine. Extraction of RNA from a cell yields several basic varieties: ribosomal, transfer, and messenger RNA, and small amounts of other small RNAs. The three sizes of **ribosomal RNA (rRNA)** combine with an array of specific proteins to form **ribosomes,** which are the machines used for protein synthesis. The ribosome-mediated synthesis of a polypeptide from a messenger RNA molecule is termed **translation. Transfer RNA (tRNA)** comprises a group of rather small RNA molecules, each with specificity for a particular amino acid; they carry the amino acids to the ribosome, where they can be attached to a growing polypeptide.

Messenger RNA (mRNA) molecules, which are produced on the DNA template, contain the information that is translated into proteins. The sequence of bases in mRNA determines the sequence of amino acids. We shall see how this is accomplished and how this **genetic code** was deciphered. Also, we shall see how certain sequences signal the initiation and termination of protein synthesis. In eukaryotes, many mRNA molecules are processed before translation, adding additional features to the mechanism of information transfer.

Transcription

Early investigators had good reason for thinking that information is not transferred directly from DNA to protein. DNA is found in the nucleus (of a eukaryotic cell), whereas protein is known to be synthesized in the cytoplasm. Another nucleic acid, ribonucleic acid (RNA), is necessary as an intermediary.

Early Experiments Suggesting RNA Intermediate

If cells are fed radioactive RNA precursors, then the labeled RNA shows up first of all in the nucleus, indicating that the

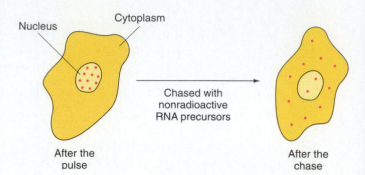

Figure 13-1 RNA synthesized during one short time period is labeled by feeding the cell a brief "pulse" of radioactive RNA precursors, followed by a "chase" of nonradioactive precursors. In an autoradiograph, the labeled RNA appears as dark grains. Apparently, the RNA is synthesized in the nucleus and then moves out into the cytoplasm.

RNA is synthesized there. In a **pulse-chase** experiment, a brief pulse of labeled RNA precursors is given. These are incorporated into RNA molecules. The cells are then transferred to medium with unlabeled RNA precursors. This "chases" the label out of the RNA, since as the RNA breaks down, only the unlabeled precursors are used to synthesize new RNA molecules. The pulse-chase protocol allows one to track a population of RNA molecules, synthesized almost simultaneously, over time. In samples taken after the chase, the labeled RNA is found in the cytoplasm (Figure 13-1). Apparently, the RNA is synthesized in the nucleus and then moves into the cytoplasm, where proteins are synthesized. Thus, it is a good candidate as an information-transfer intermediary between DNA and protein.

In 1957, Elliot Volkin and Lawrence Astrachan made a significant observation. They found that one of the most striking molecular changes when *E. coli* is infected with the phage T2 is a rapid burst of RNA synthesis. Furthermore, this phage-induced RNA "turns over" rapidly, as shown in the following experiment. The infected bacteria are first

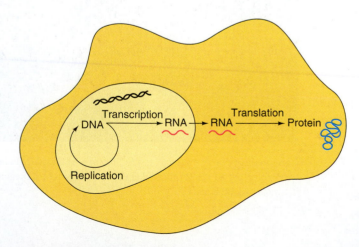

Figure 13-2 The three processes of information transfer: replication, transcription, and translation.

pulsed with radioactive uracil (a specific precursor of RNA); the bacteria are then chased with cold uracil. The RNA recovered shortly after the pulse is labeled, but that recovered somewhat longer after the chase is unlabeled, indicating that the RNA has a very short lifetime. Finally, when the nucleotide contents of *E. coli* and T2 DNA are compared with the nucleotide content of the phage-induced RNA, the RNA is found to be very similar to the phage DNA.

The tentative conclusion from the two experiments described above is that RNA is synthesized from DNA and that it is somehow used to synthesize protein. We can now outline three stages of information transfer (Figure 13-2): *replication* (the synthesis of DNA), *transcription* (the synthesis of an RNA copy of a portion of the DNA), and *translation* (the synthesis of a polypeptide directed by the RNA sequence).

Properties of RNA

Although RNA is a long-chain macromolecule of nucleic acid (as is DNA), it has very different properties. First, RNA is usually single-stranded, not a double helix. Second, RNA has ribose sugar, rather than deoxyribose, in its nucleotides (hence, its name):

Third, RNA has the pyrimidine base **uracil** (abbreviated *U*) instead of thymine. However, uracil does form hydrogen bonds with adenine just as thymine does:

No one is absolutely sure why RNA has uracil instead of thymine or why it has ribose instead of deoxyribose. The most important aspect of RNA is that it is usually single-stranded, but otherwise it is very similar in structure to DNA.

Complementarity and Asymmetry in RNA Synthesis

The similarity of RNA to DNA suggests that transcription may be based on the complementarity of bases, which is

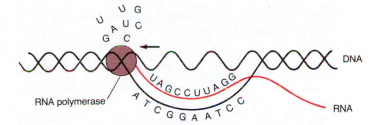

Figure 13-3 Synthesis of RNA on a single-stranded DNA template using free nucleotides. The process is catalyzed by RNA polymerase. Uracil (U) pairs with adenine (A).

also the key to DNA replication. A transcription enzyme, RNA polymerase, could carry out transcription from a DNA template strand in a fashion quite similar to replication (Figure 13-3).

In fact, this model of transcription is confirmed cytologically (Figure 13-4). The fact that RNA can be synthesized with DNA acting as a template is also demonstrated by synthesis in vitro of RNA from nucleotides in the presence of DNA, using an extractable RNA polymerase. Whatever source of DNA is used, the RNA synthesized has an $(A + U)/(G + C)$ ratio similar to the $(A + T)/(G + C)$ ra-

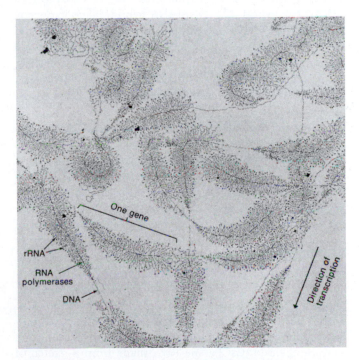

Figure 13-4 Tandemly repeated ribosomal RNA (rRNA) genes being transcribed in the nucleolus of *Triturus viridiscens* (an amphibian). (rRNA is a component of the ribosome, a cellular organelle.) Along each gene, many RNA polymerase molecules are attached and transcribing in one direction. The growing RNA molecules appear as threads extending out from the DNA backbone. The shorter RNA molecules are nearer the beginning of transcription; the longer ones have almost been completed. Hence, the "Christmas tree" appearance. (Photograph from O. L. Miller, Jr., and Barbara A. Hamkalo.)

Table 13-1   Nucleotide Ratios in Various DNAs and in Their Transcripts (in vitro)

DNA source	$\dfrac{(A + T)}{(G + C)}$ of DNA	$\dfrac{(A + U)}{(G + C)}$ of RNA
T2 phage	1.84	1.86
Cow	1.35	1.40
Micrococcus (bacterium)	0.39	0.49

tio of the DNA (Table 13-1). This experiment does not indicate whether the RNA is synthesized from both DNA strands or from just one, but it does indicate that the linear frequency of the A–T pairs (in comparison with the G–C pairs) in the DNA is precisely mirrored in the relative abundance of (A + U) in the RNA. (These points are difficult to grasp without drawing some diagrams; Problem 2 at the end of this chapter provides some opportunities to clarify these notions.)

To test the complementarity of DNA with RNA, investigators can apply the specificity and precision of nucleic acid hybridization. DNA can be denatured and mixed with the RNA formed from it. On slow cooling, some of the RNA strands anneal with complementary DNA to form a DNA-RNA hybrid. The DNA-RNA hybrid differs in density from the DNA-DNA duplex, so its presence can be detected by ultracentrifugation in cesium chloride. Nucleic acids will anneal in this way only if there are stretches of base-sequence complementarity, so the experiment does prove that the RNA transcript is complementary in base sequence to the parent DNA.

Can we determine whether RNA is synthesized from both or only one of the DNA strands? It seems reasonable that only one strand is used, because transcription of RNA from both strands would produce two complementary RNA strands from the same stretch of DNA, and these strands presumably would produce two different kinds of protein (with different amino acid sequences). In fact, a great deal of chemical evidence confirms that transcription usually takes place on only one of the DNA strands (although not necessarily the same strand throughout the entire chromosome).

The hybridization experiment can be extended to explore this problem. If the two strands of DNA have distinctly different purine:pyrimidine ratios, they can be purified separately because they have different densities in cesium chloride (CsCl). The RNA made from a stretch of DNA can be purified and annealed separately to each of the strands to see whether it is complementary to only one. J. Marmur and his colleagues were able to separate the strands of DNA from the *B. subtilis* phage SP8. They denatured the DNA, cooled it rapidly to prevent reannealing of the strands, and then separated the strands in CsCl. They

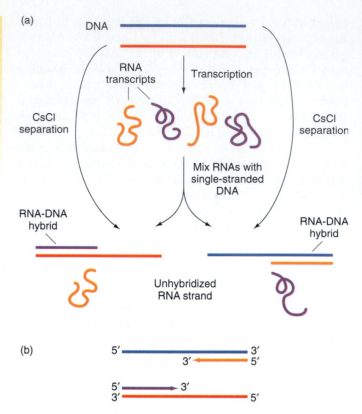

Figure 13-5 (a) DNA-RNA hybridization demonstrates that each RNA transcript is complementary to only one strand of the parent DNA. In this example, each of the two DNA strands is transcribed, but transcription is asymmetrical—only one strand is transcribed at any particular location. (b) A map of this hypothetical genome, showing the direction of transcription for the two transcripts from opposite DNA strands.

showed that the SP8 RNA hybridizes to only one of the two strands, proving that transcription is **symmetrical**—that it occurs only on one DNA strand.

Although, for each gene, RNA is transcribed from only one of the DNA strands, the same DNA strand is not necessarily transcribed throughout the entire chromosome or through all stages of the life cycle. The RNA produced at different stages in the cycle of a phage hybridizes to different segments of the chromosome, showing the different genes that are activated at each stage (Figure 13-5). In λ phage, each of the two DNA strands is partially transcribed at a different stage. In phage T7, however, the same strand is transcribed for both early-acting and late-acting genes. Figure 13-6 shows a sequence of RNA made from the DNA template strand. The DNA strand that is transcribed for a given mRNA is termed the **template strand**. The complementary DNA strand is called the **nontemplate strand**. Note that the mRNA has the same sequence (with U substituted for T) as the nontemplate strand.

Nontemplate strand 5' – `CTGCCATTGTCAGACATGTATACCCCGTACGTCTTCCCGAGCGAAAACGATCTGCGCTGC` – 3' } DNA

Template strand 3' – `GACGGTAACAGTCTGTACATATGGGGCATGCAGAAGGGCTCGCTTTTGCTAGACGCGACG` – 5' }

5' – `CUGCCAUUGUCAGACAUGUAUACCCCGUACGUCUUCCCGAGCGAAAACGAUCUGCGCUGC` – 3'   mRNA

Figure 13-6 The mRNA sequence is complementary to the DNA template strand from which it is synthesized. The sequence shown here is from the gene for the enzyme β-galactosidase, which is involved in lactose metabolism.

RNA Polymerase

In most prokaryotes, a single RNA polymerase species transcribes all types of RNA. Figure 13-7 shows the structure of RNA polymerase from *E. coli*. We can see that the enzyme consists of four different subunit types. The beta (β) subunit has a molecular weight of 150,000 daltons, beta prime (β') 160,000, alpha (α) 40,000, and sigma (σ) 70,000. The σ subunit can dissociate from the rest of the complex, leaving the **core enzyme.** The complete enzyme with σ is termed the **RNA polymerase holoenzyme** and is necessary for correct initiation of transcription, whereas the core enzyme can continue transcription after initiation.

Next, let's look at the three distinct stages of transcription: **initiation, elongation,** and **termination.**

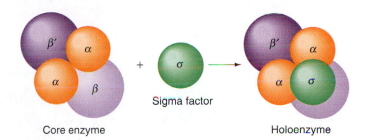

Core enzyme   Sigma factor   Holoenzyme

Figure 13-7 The structure of RNA polymerase. The core enzyme contains two α polypeptides, one β polypeptide, and one β' polypeptide. The addition of the σ subunit allows initiation at promoter sites. (Promoters are discussed in the next section.)

(a) Strong *E. coli* promoters

		−35		−10	+1
tyr tRNA	TCTCAACGTAACAC**TTTACA**GCGGCG••	CGTCATTTGA**TATGATGC**•GCCCC**G**CTTCCCGATAAGGG			
rrn D1	GATCAAAAAAATAC**TTGTGC**AAAAA••	TTGGGATCCC**TATAAT**GCGCCTCC**G**TTGAGACGACAACG			
rrn X1	ATGCATTTTTCCGC**TTGTCTT**CCTGA••	GCCGACTCCC**TATAAT**GCGCCTCC**A**TCGACACGGCGGAT			
rrn (DXE)₂	CCTGAAATTCAGGG**TTGAC**TCTGAAA••	GAGGAAAGCG**TAATATAC**•GCCACC**T**CGCGACAGTGAGC			
rrn E1	CTGCAATTTTTCTA**TTGC**GGCCTGCG••	GAGAACTCCC**TATAAT**GCGCCTCC**A**TCGACACGGCGGAT			
rrn A1	TTTTAAATTTCCTC**TTGTC**AGGCCGG••	AATAACTCCC**TATAAT**GCGCCACC**A**CTGACACGGAACAA			
rrn A2	GCAAAAATAAATGC**TTGAC**TCTGTAG••	CGGGAAGGCG**TATTATGC**•ACACC**C**CGCGCCGCTGAGAA			
λ P_R	TAACACCGTGCGTG**TTGAC**TATTTTA••	CCTCTGGCGGTGA**TAATGG**••**T**TGCATGTACTAAGGAGGT			
λ P_L	TATCTCTGGCGGTG**TTGACAT**AAATA••	CCACTGGCGGTGA**TACTGA**••GCACA**T**CAGCAGGACGCAC			
T7 A3	GTGAAACAAAACGG**TTGACA**ACATGA••	AGTAAACACGG**TACGATGT**•ACCACA**T**GAAACGACAGTGA			
T7 A1	TATCAAAAAGAGTA**TTGACTT**AAAGT••	CTAACCTATAGGA**TACTTA**•CAGCC**A**TCGAGAGGGACACG			
T7 A2	ACGAAAAACAGGTA**TTGACAA**CATGAAGTAACATGCAG**TAAGAT**AC•AAAT**C**GCTAGGTAACACTAG			
fd VIII	GATACAAATCTCCG**TTGTACT**TTGTT••	TCGCGCTTGG**TATAAT**CG•CTGGGG**G**TCAAAGATGAGTG			

Initiation

The regions of the DNA that signal initiation of transcription in prokaryotes are termed **promoters.** (We consider their role in gene control in Chapter 18.) Figure 13-8 shows the promoter sequences from 13 different transcription initiation points on the *E. coli* genome. The bases are aligned according to homologies, or similar base sequences, that appear just before the first base transcribed (designated the "initiation site" in Figure 13-8).

Note in Figure 13-8 that two regions of partial homology appear in virtually each case. These regions have been termed the −35 and −10 regions because of their locations relative to the transcription initiation point. At the bottom of Figure 13-8 an ideal, or consensus, sequence of a promoter is given. Physical experiments have confirmed that

(b) Consensus sequences of promoters

−35 region   15–17 bp   −10 region
TTGACAT   **TATAAT**

Figure 13-8 Promoter sites have regions of similar sequences, as indicated by the yellow region in the 13 different promoter sequences in *E. coli*. Spaces (dots) included to maximize homology at consensus sequences. The gene governed by each promoter sequence is indicated on the left. Numbering is given in terms of the number of bases before (−) or after (+) the RNA synthesis initiation point. Color coding in the consensus sequences at the bottom is as follows: red letters, >75%; boldface black letters, 50–75%; black letters, 40–50%. [From H. Lodish, D. Baltimore, A. Berk, S. L. Zipursky, P. Matsudaira, and J. Darnell, *Molecular Cell Biology*, 3d ed. Copyright ©1995 by W. H. Freeman and Company. See W. R. McClure, *Ann. Rev. Biochem.* 54, 1985, 171 (Consensus Sequences).]

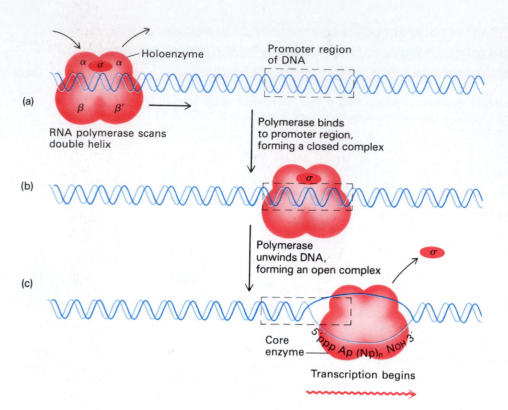

(a)

Holoenzyme

Promoter region
of DNA

RNA polymerase scans
double helix

Polymerase binds
to promoter region,
forming a closed complex

(b)

Polymerase
unwinds DNA,
forming an open complex

(c)

Core
enzyme

5′ ppp Ap (Np)ₙ Nₒₕ 3′

Transcription begins

Figure 13-9 Initiation of transcription. (a) RNA polymerase searches for a promoter site. (b) It recognizes a promoter site and binds tightly, forming a closed complex. (c) The holoenzyme unwinds a short stretch of DNA, forming an open complex. Transcription begins, and the σ factor is released. The RNA transcript is shown, beginning with adenosine triphosphate (pppA), carrying on through an indeterminate number of nucleotides [(Np)ₙ], and ending with Nₒₕ. (Modified from J. Darnell, H. Lodish, and D. Baltimore, *Molecular Cell Biology, 2d ed.* Copyright ©1990 by Scientific American Books, Inc.)

RNA polymerase makes contact with these two regions when binding to the DNA. The enzyme then unwinds DNA and begins the synthesis of an RNA molecule.

The dissociative subunit of RNA polymerase, the σ factor, allows RNA polymerase to recognize and bind specifically to promoter regions. First, the holoenzyme searches for a promoter (Figure 13-9a) and initially binds loosely to it, recognizing the −35 and −10 regions. The resulting structure is termed a *closed promoter complex* (see Figure 13-9b). Then, the enzyme binds more tightly, unwinding bases near the −10 region. When the bound polymerase causes this local denaturation of the DNA duplex, it is said to form an *open promoter complex* (Figure 13-9c). This initiation step, the formation of an open complex, requires the sigma factor.

Elongation

Shortly after initiating transcription, the sigma factor dissociates from the RNA polymerase. The RNA is always synthesized in the 5′ → 3′ direction (Figures 13-10 and 13-11), with nucleoside triphosphates (NTPs) acting as substrates for the enzyme. The equation below represents the addition of each ribonucleotide.

$$\text{NTP} + \text{(NMP)}_n \xrightarrow[\substack{\text{Mg}^{2+} \\ \text{RNA} \\ \text{polymerase}}]{\text{DNA}} \text{(NMP)}_{n+1} + \text{PP}_i$$

The energy for the reaction is derived from splitting the high-energy triphosphate into the monophosphate and releasing the inorganic diphosphates (PP_i), as shown in Figure 13-10. Figure 13-11 gives a physical picture of elongation. Note how a "transcription bubble" must be maintained, since the transcription takes place on a double-stranded template. The bubble must move along the DNA duplex during elongation. Certain sequences may cause stalling or pausing, which becomes critical for termination of transcription (see below).

Termination

RNA polymerase also recognizes signals for chain termination, which involves the release of the nascent RNA and enzyme from the template. There are two major mechanisms for termination in *E. coli.*

In the first the termination is direct. The terminator sequences contain about 40 bp, ending in a GC-rich stretch that is followed by a run of six or more A's on the template strand. The corresponding GC sequences on the RNA are so arranged that the transcript in this region is able to form complementary bonds *with itself,* as can be seen in Figure 13-12. The resulting double-stranded RNA section is called a **hairpin loop.** It is followed by the terminal run of U's that correspond to the A residues on the DNA template. The hairpin loop and section of U residues appear to serve as a signal for the release of RNA polymerase and termination of transcription.

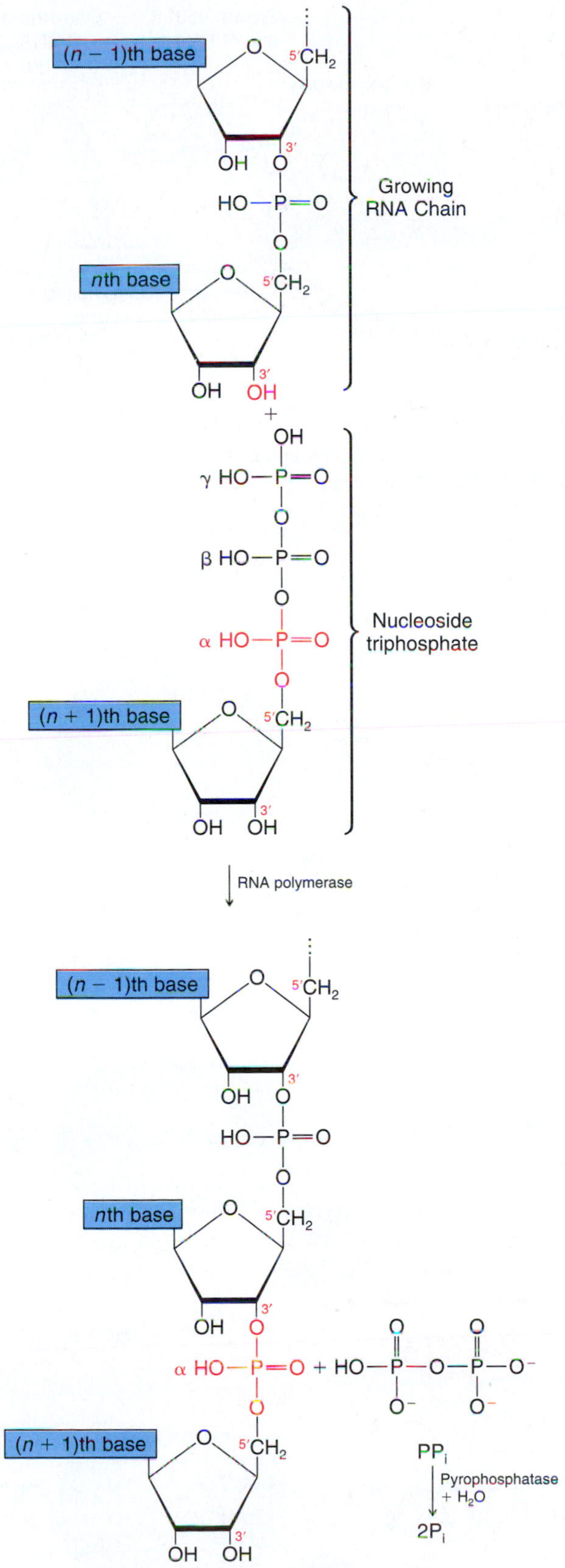

In the second type, the help of an additional protein factor, termed **rho,** is required in order for RNA polymerase to recognize the termination signals. mRNAs with rho-dependent termination signals do not have the string of U residues at the end of the RNA, and usually do not have hairpin loops. A model for rho-dependent termination is shown in Figure 13-13. Rho is a hexamer consisting of six identical subunits, and uses the hydrolysis of ATP to ADP and P_i to drive the termination reaction. The first step in termination is the binding of rho to a specific site on the RNA termed *rut* (Figure 13-13a and b). After binding, rho pulls the RNA off the RNA polymerase, probably by translocating along the mRNA, as depicted in Figure 13-13b and c. The *rut* sites are located just upstream from (that is, 5′ from) sequences at which the RNA polymerase tends to pause.

The efficiency of both mechanisms of termination is influenced by surrounding sequences and other protein factors, as well.

Translation

Although mRNA directs protein synthesis, if you mix mRNA and all 20 amino acids in a test tube and hope to make protein, you will be disappointed. Other components are needed; the discovery of the nature of these components has provided the key to understanding the mechanism of translation. A simple but elegant centrifugation technique helped reveal these additional components.

Application of Sucrose Gradients

The **sucrose density gradient** is created in a test tube by layering successively lower concentrations of sucrose solution, one on top of the other. The material to be studied is carefully placed on top. When the solution is centrifuged in a machine that allows the test tube to swivel freely, the sedimenting material travels through the gradient at different rates that are related to the sizes and shapes of the molecules. Large molecules migrate farther in a given period of time than smaller molecules do. The separated molecules

Figure 13-10 The sequential addition of nucleotides occurs one at a time in the 5′ to 3′ direction. The chain grows by the formation of a bond between the 3′ hydroxyl end of the growing strand and a nucleoside triphosphate, releasing one pyrophosphate ion (PP_i). This results in the net addition of one phosphate, which is incorporated into the backbone of the new strand. DNA grows by reaction with deoxyribonucleoside triphosphates, and RNA grows by reaction with ribonucleoside triphosphates. (From H. Lodish, D. Baltimore, A. Berk, S. L. Zipursky, P. Matsudaira, and J. Darnell, *Molecular Cell Biology,* 3d ed. Copyright ©1995 by Scientific American Books, Inc.)

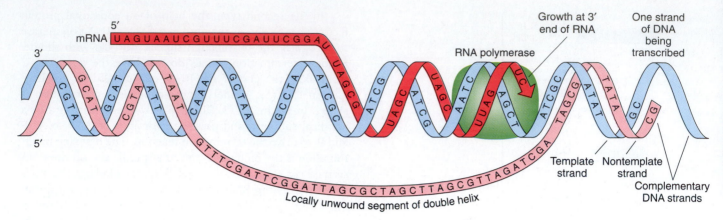

Figure 13-11 Transcription by RNA polymerase. An RNA strand is synthesized in the 5′ → 3′ direction from a locally single-stranded region of DNA. (After E. J. Gardner, M. J. Simmons, and D. P. Snustad, *Principles of Genetics,* 8th ed. Copyright ©1991 by John Wiley and Sons, Inc.)

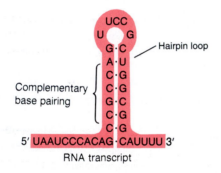

Figure 13-12 The structure of a termination site for RNA polymerase in bacteria. The hairpin structure forms by complementary base pairing *within* the RNA strand.

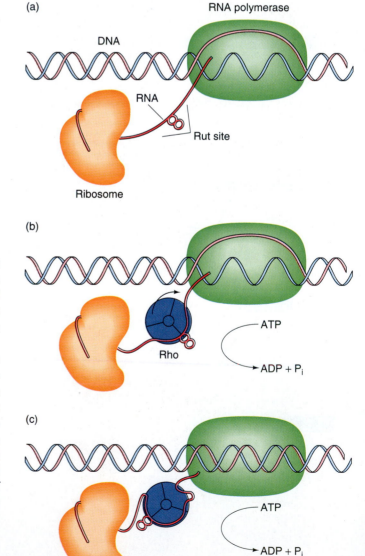

can be collected individually by capturing sequential drops from a small opening in the bottom of the tube (Figure 13-14). The velocity with which a fraction moves the fixed distance to the tube bottom indicates its sedimentation (S) value, which is a measure of the size of the molecules in the fraction.

It is important to note the difference between a sucrose gradient and the CsCl gradient that we considered in Chapter 11. In a CsCl gradient, the molecules being studied have a density somewhere in between the lowest and highest concentrations of CsCl generated in the gradient. Therefore, at equilibrium they will band at a specific point on the gradient. In a sucrose gradient, the molecules being studied are denser than any of the sucrose concentrations used, and at equilibrium they would form a pellet in the bottom of the tube. However, they migrate toward the bottom at vary-

Figure 13-13 A model for rho action on a nascent cotranslated mRNA. (Adapted from J. P. Richardson, *Cell* 64, 1991, 1047–1049.)

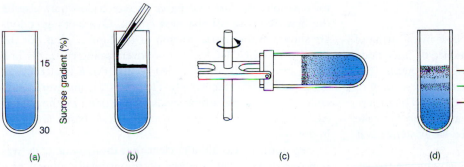

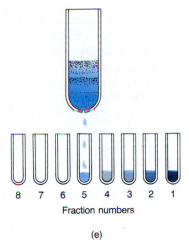

Figure 13-14 The sucrose-gradient technique. (a) A sucrose density gradient is created in a centrifuge tube by layering solutions of differing densities. (b) The sample to be tested is placed on top of the gradient. (c) Centrifugation causes the various components (fractions) of the sample to sediment differentially. (d) The different fractions appear as bands in the centrifuged gradient. (e) The different bands can be collected separately by collecting samples from the bottom of the tube at fixed time intervals. The S value for the fraction is based on its position in the gradient, which is related to the time at which it drips from the bottom of the tube. (From A. Rich, "Polyribosomes." Copyright ©1963 by Scientific American, Inc. All rights reserved.)

ing speeds depending on their size and shape. By comparing the different positions of each molecule in the gradient at a particular time, we can determine the relative sizes of the molecules.

Different Classes of RNA Molecules

Using the separatory powers of the sucrose-gradient technique, it is possible to identify several macromolecules and macromolecular aggregates in a typical protein-synthesizing system. The main components are transfer RNA (tRNA), ribosomes, and messenger RNA (mRNA). Transfer RNA is a class of small (4S) RNA molecules of rather similar type and function. In fact, complete nucleotide sequences have been determined for a large number of tRNA molecules from different organisms.

Ribosomes, which are the sites of protein synthesis, are cellular organelles composed of very complex aggregations of ribosomal proteins and ribosomal RNA (rRNA) components. In *E. coli*, for instance, at least three separate RNA molecules can be distinguished by size in ribosomes: 23S,

16S, and 5S. We do not know precisely how rRNA is bound up with the protein components or how each component functions. Under the electron microscope, a ribosome appears globular. On chemical treatment, it splits into two main globular entities (50S and 30S in *E. coli*). We shall consider the structure of tRNA and rRNA molecules in detail later.

Both rRNA and tRNA molecules will form RNA-DNA hybrids in vitro, indicating that they, like mRNA, are transcribed from the DNA. Table 13-2 summarizes the main types of RNA that can be found in a atypical protein-synthesizing system.

Other components are needed to make protein synthesis work in vitro. These include several enzymes, several protein "factors," and a chemical source of energy. The energy donor is needed because an orderly structure is being created out of a mess of components—a process that requires energy because the system must lose entropy (a measure of disorder or randomness). We will examine the process of protein synthesis after we first consider the genetic code.

Table 13-2 RNA Molecules in *E. coli*

Type	Percentage of cell RNA	Sedimentation coefficient, S	Molecular weight	Number of nucleotides
Ribosomal RNA (rRNA)	80	23	1.2×10^6	3700
		16	0.55×10^6	1700
		5	3.6×10^4	1700
Transfer RNA (tRNA)	15	4	2.5×10^4	75
Messenger RNA (mRNA)	5		Heterogeneous	Varies

SOURCE: Adapted from L. Stryer, *Biochemistry*, 4th ed. Copyright © 1995 by Lubert Stryer.

The Genetic Code

If genes are segments of DNA and if DNA is just a string of nucleotide pairs, then how does the sequence of nucleotide pairs dictate the sequence of amino acids in proteins? The analogy to a code springs to mind at once. The cracking of the genetic code is the story told in this section. The experimentation was sophisticated and swift, and it did not take long for the code to be deciphered once its existence was strongly indicated.

Simple logic tells us that if nucleotide pairs are the *letters* in a code, then a combination of letters could form words representing different amino acids. We must ask how the code is read. Is it overlapping or nonoverlapping? Then we must ask how many letters in the mRNA make up a word, or **codon,** and which specific codon or codons represent each specific amino acid.

Overlapping versus Nonoverlapping Codes

Figure 13-15 shows the difference between an overlapping and a nonoverlapping code. In the example, a three-letter, or **triplet,** code is shown. For the nonoverlapping code, consecutive amino acids are specified by consecutive code words (codons), as shown in the bottom of the figure. For an overlapping code, consecutive amino acids are encoded in the mRNA by codons that share some consecutive bases; for example, the last two bases of one codon may also be the first two bases of the next codon. Overlapping codons are shown in the upper portion of the figure. Thus, for the sequence AUUGCUCAG in a nonoverlapping code, the first three amino acids are encoded by the three triplets AUU, GCU, and CAG, respectively. However, for an overlapping code, the first three amino acids are encoded by the triplets AUU, UUG, and UGC if the overlap is two bases, as shown in the figure.

By 1961, it was already clear that the genetic code was nonoverlapping. The analysis of mutationally altered proteins—in particular, the nitrous acid–generated mutants of tobacco mosaic virus—showed that only a single amino acid changes at one time in one region of the protein. This is predicted by a nonoverlapping code. As you can see from Figure 13-15 an overlapping code predicts that a single base change will alter as many as three amino acids at adjacent positions in the protein.

It should be noted that while the use of an overlapping *code* was ruled out by the analysis of single proteins, nothing precluded the use of alternative reading frames to encode amino acids in two different proteins. In the example here, one protein might be encoded by the series of codons that reads AUU, GCU, CAG, CUU, and so on. A second protein might be encoded by codons that are shifted over by one base and therefore read UUG, CUC, AGC, UUG, and so on. This is an example of storing the information encoding

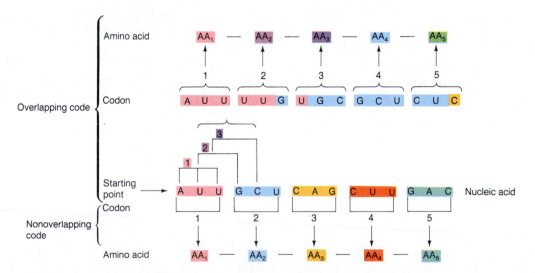

Figure 13-15 The difference between an overlapping and a nonoverlapping code. The case illustrated is for a code with three letters (a triplet code). An overlapping code uses codons that employ some of the same nucleotides as other codons for the translation of a single protein, as shown in the top of the diagram (for the mRNA sequence shown at the bottom of the diagram). In a nonoverlapping code, a protein is translated by reading codons that do not share any of the same nucleotides. Note that the designation of an amino acid as "AA$_2$," "AA$_3$," etc., in the nonoverlapping model does not mean it is necessarily the same amino acid as its numerical counterpart in the overlapping model. This is because the triplets making up the respective codons for the two AA$_3$'s, for example, are different and would more than likely code for different amino acids. Identical amino acid numbers between the two models merely indicate the same amino acid position on the protein chain.

two different proteins in two different reading frames, while still using a genetic code that is read in a nonoverlapping manner during the translation of a *specific* protein. Some examples of such shifts in reading frame have been found.

Number of Letters in the Code

Reading an mRNA molecule from one particular end, only one of four different bases, A, U, G, or C, can be found at each position. Thus, if the words are one letter long, only four words are possible. This cannot be the genetic code, because we must have a word for each of the 20 amino acids commonly found in cellular proteins. If the words are two letters long, then $4^2 = 16$ words are possible; for example, AU, CU, or CC. This vocabulary still is not large enough.

If the words are three letters long, then $4^3 = 64$ words are possible; for example, AUU, GCG, or UGC. This code provides more than enough words to describe the amino acids. We can conclude that the code word must consist of at least three nucleotide pairs. However, if all words are "triplets," then we have a considerable excess of possible words over the 20 needed to name the common amino acids.

Use of Suppressors to Demonstrate a Triplet Code

Convincing proof that a codon is, in fact, three letters long (and no more than three) came from beautiful genetic experiments first reported in 1961 by Francis Crick, Sidney Brenner, and their co-workers, who used mutants in the *rII* locus of T4 phage. Mutations causing the rII phenotype (see Chapter 12) were induced using a chemical called *proflavin,* which was thought to act by the addition or deletion of single nucleotide pairs in DNA. (This assumption is based on experimental evidence not presented here.) The following examples illustrate the action of proflavin on double-stranded DNA.

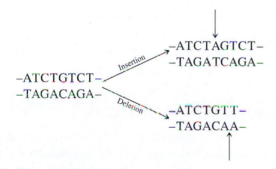

Then, starting with one particular proflavin-induced mutation called *FCO,* Crick and his colleagues again used proflavin to induce "reversions" (reversals of the mutation) that were detected by their wild-type plaques on *E. coli* strain K(λ). Genetic analysis of these plaques revealed that the "revertants" were not identical true wild types, thereby

suggesting that the back mutation was not an exact reversal of the original forward mutation. In fact, the reversion was found to be caused by the presence of a *second mutation* at a different site from—but in the same gene as—that of FCO; this second mutation "suppressed" mutant expression of the original FCO. A **suppressor mutation** counteracts or suppresses the effects of another mutation. Properties of suppressor mutations include the following:

1. A suppressor mutation is at a different site from that of the mutation it counteracts. Therefore, the original mutation can be recovered by genetic crosses between the wild type and the revertant, since the revertant carries both the suppressor and the original mutation.

2. A suppressor mutation may be within the same gene as the mutation it suppresses (internal suppressor), as in the example just given, or it may be in a different gene (external suppressor).

3. Different suppressors may exert their effects in different ways. For instance, some suppressors act at the level of transcription or translation; others alter the physiology of the cell.

The suppressor mutation could be separated from the original forward mutation by recombination, and as we have seen, when this was done, the suppressor was shown to be an *rII* mutation itself (Figure 13-16).

How can we explain these results? If we assume that reading is polarized—that is, if the gene is read from one end only—then the original proflavin-induced addition or deletion could be mutant because it interrupts a normal reading mechanism that establishes the group of bases to be read as words. For example, if each three bases on the resulting mRNA make a word, then the "reading frame" might be established by taking the first three bases from the end as the first word, the next three as the second word, and so on. In that case, a proflavin-induced addition or deletion of a single pair on the DNA would shift the reading frame

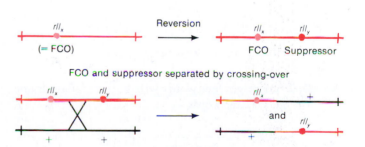

Figure 13-16 The suppressor of an initial *rII* mutation is shown to be an *rII* mutation itself after separation by crossing-over. The original mutant, FCO, was induced by proflavin. Later, when the FCO strain was treated with proflavin again, a revertant was found, which on first appearance seemed to be wild-type. However, it was found that a second mutation within the *rII* region had been induced, and the double mutant *rII*$_x$*rII*$_y$ was shown not to be quite identical with the original wild type.

on the mRNA from that corresponding point on, causing all following words to be misread. Such a **frameshift mutation** could reduce most of the genetic message to gibberish. However, the proper reading frame could be restored by a compensatory insertion or deletion somewhere else, leaving only a short stretch of gibberish between the two. Consider the following example in which three-letter English words are used to represent the codons:

THE FAT CAT ATE THE BIG RAT

Delete C: THE FAT ATA TET HEB IGR AT

Insert A: THE FAT ATA ATE THE BIG RAT

The insertion suppresses the effect of the deletion by restoring most of the sense of the sentence. By itself, however, the insertion also disrupts the sentence:

THE FAT CAT AAT ETH EBI GRA T

If we assume that the FCO mutant is caused by an addition, then the second (suppressor) mutant would have to be a deletion because, as we have seen, this would restore the reading frame of the resulting message (a second insertion would not correct the frame). In the following diagrams, we use a hypothetical nucleotide chain to represent RNA for simplicity. We also assume that the code words are three letters long and are read in one direction (left to right in our diagrams).

1.   Wild-type message

CAU CAU CAU CAU CAU

2.   *rII*$_a$ message: distal words changed (x) by frameshift mutation (words marked √ are unaffected)

Addition ┐

CAU ACA UCA UCA UCA U___

√   x   x   x   x

3.   *rII*$_a$*rII*$_b$ message: few words wrong, but reading frame restored for later words

Deletion ─────────┐

CAU ACA UCU CAU CAU

√   x   x   √   √

The few wrong words in the suppressed genotype could account for the fact that the revertants (suppressed phenotypes) Crick and his associates recovered did not look exactly like the true wild types phenotypically.

We have assumed here that the original frameshift mutation was an addition, but the explanation works just as well if we assume that the original FCO mutation is a deletion and the suppressor is an addition. If the FCO is defined as plus, then suppressor mutations are automatically minus. Experiments have confirmed that a plus cannot suppress a plus and a minus cannot suppress a minus. In other words, two mutations of the same sign never act as suppressors of each other. However, very interestingly, combinations of *three* pluses or *three* minuses have been shown to act together to restore a wild-type phenotype.

This observation provided the first experimental confirmation that a word in the genetic code consists of three successive nucleotide pairs, or a triplet. This is because three additions or three deletions within a gene automatically restore the reading frame in the mRNA if the words are triplets. For example,

Deletions

CAU CAU CAU CAU CAU CAU CAU

CAU ACA UAU CAU CAU CAU

√   x   x   √   √   √

Proof that the genetic deductions about proflavin were correct came from an analysis of proflavin-induced mutations in a gene with a protein product that could be analyzed. George Streisinger worked with the gene that controls the enzyme lysozyme, which has a known amino acid sequence. He induced a mutation in the gene with proflavin and selected for proflavin-induced revertants, which were shown genetically to be double mutants (with mutations of opposite sign). When the protein of the double mutant was analyzed, a stretch of different amino acids lay between two wild-type ends, just as predicted:

Wild type

–Thr–Lys–Ser–Pro–Ser–Leu–Asn–Ala–

Revertant type

–Thr–Lys–Val–His–His–Leu–Met–Ala–

Degeneracy of the Genetic Code

Crick's work also suggested that the genetic code is **degenerate.** That expression is not a moral indictment! It simply means that each of the 64 triplets must have some meaning within the code, so that at least some amino acids must be specified by two or more different triplets. If only 20 triplets are used (with the other 44 being nonsense, in that they do not code for any amino acid), then most frameshift mutations can be expected to produce nonsense words, which presumably would stop the protein-building process. If this were the case, then the suppression of frameshift mutations would rarely, if ever, work. However, if all triplets specify some amino acid, then the changed words would simply result in the insertion of incorrect amino acids into the protein. Thus, Crick reasoned that many or all amino acids must have several different names in the base-pair code; this hypothesis was later confirmed biochemically.

Review

Let's summarize what the work up to this point demonstrates about the genetic code:

1. The code is nonoverlapping.
2. Three bases code for an amino acid. These triplets are termed *codons.*
3. The code is read from a fixed starting point and continues to the end of the coding sequence. We know this because a single frameshift mutation anywhere in the coding sequence alters the codon alignment for the rest of the sequence.
4. The code is degenerate in that some amino acids are specified by more than one codon.

Cracking the Code

The deciphering of the genetic code—determining the amino acid specified by each triplet—is one of the most exciting genetic breakthroughs of the past two decades. Once the necessary experimental techniques became available, the genetic code was broken in a rush.

The first breakthrough was the discovery of how to make synthetic mRNA. If the nucleotides of RNA are mixed with a special enzyme (polynucleotide phosphorylase), a single-stranded RNA is formed in the reaction. No DNA is needed for this synthesis, and so the nucleotides are incorporated at random. The ability to synthesize mRNA offered the exciting prospect of creating specific mRNA sequences and then seeing which amino acids they would specify. The first synthetic messenger obtained, poly(U), was made by reacting only uracil nucleotides with the RNA-synthesizing enzyme, producing –UUUU–. In 1961, Marshall Nirenberg and Heinrich Matthaei mixed poly(U) with the protein-synthesizing machinery of *E. coli* (ribosomes, aminoacyl-tRNAs, a source of chemical energy, several enzymes, and a few other things) in vitro and *observed the formation of a protein!* Of course, the main excitement centered on the question of the amino acid sequence of this protein. It proved to be polyphenylalanine—a string of phenylalanine molecules attached to form a polypeptide. Thus, the triplet UUU must code for phenylalanine:

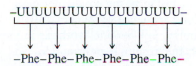

–Phe–Phe–Phe–Phe–Phe–Phe–

Table 13-3 Expected Frequencies of Various Codons in Synthetic mRNA Composed of $\frac{3}{4}$ Uracil and $\frac{1}{4}$ Guanine

Codon	Probability	Ratio*
UUU	$p(UUU) = \frac{3}{4} \times \frac{3}{4} \times \frac{3}{4} = \frac{27}{64}$	1.00
UUG	$p(UUG) = \frac{3}{4} \times \frac{3}{4} \times \frac{1}{4} = \frac{9}{64}$	0.33
UGU	$p(UGU) = \frac{3}{4} \times \frac{1}{4} \times \frac{3}{4} = \frac{9}{64}$	0.33
GUU	$p(GUU) = \frac{1}{4} \times \frac{3}{4} \times \frac{3}{4} = \frac{9}{64}$	0.33
UGG	$p(UGG) = \frac{3}{4} \times \frac{1}{4} \times \frac{1}{4} = \frac{3}{64}$	0.11
GGU	$p(GGU) = \frac{1}{4} \times \frac{1}{4} \times \frac{3}{4} = \frac{3}{64}$	0.11
GUG	$p(GUG) = \frac{1}{4} \times \frac{3}{4} \times \frac{1}{4} = \frac{3}{64}$	0.11
GGG	$p(GGG) = \frac{1}{4} \times \frac{1}{4} \times \frac{1}{4} = \frac{1}{64}$	0.03

* UUU is used as the baseline frequency against which the frequencies of the other codons are measured in establishing the respective ratios. For example, the ratio for UUG is derived from $p(UUG)/p(UUU) = 0.33$.

This type of analysis was extended by mixing nucleotides in a known fixed proportion when making synthetic mRNA. In one experiment, the nucleotides uracil and guanine were mixed in a ratio of 3 : 1. When nucleotides are incorporated at random into synthetic mRNA, the relative frequency at which each triplet will appear in the sequence can be calculated on the basis of the relative proportion of the various nucleotides present (Table 13-3). In the table, note that UUU is used as the baseline frequency against which the other frequencies are measured in determining their respective ratios. For example, UUG, with a probability of $p(UUG) = 9/64$, would be expected only one-third as often as UUU, with its probability of $p(UUU) = 27/64$. Stated alternatively, $p(UUG)/p(UUU) = 9/27 = 1/3 = 0.33$, which is the ratio for UUG given in Table 13-3.

If these codons each code for a different amino acid (that is, are not redundant), we would expect the amino acids generated by this particular mix of guanine and uracil to occur in ratios similar to those of the various codons. Although there is some redundancy among these codons, the ratios of the amino acids actually obtained from this mix of bases (see Table 13-4) are indeed quite similar to the ratios

Table 13-4 Observed Frequencies of Various Amino Acids in Protein Translated from mRNA Composed of $\frac{3}{4}$ Uracil and $\frac{1}{4}$ Guanine

Amino acid	Ratio*
Phenylalanine	1.00
Leucine	0.37
Valine	0.36
Cysteine	0.35
Tryptophan	0.14
Glycine	0.12

* Phenylalanine is used as the baseline concentration against which the concentrations of other amino acids are measured in deriving their respective ratios. Note the correlations with the ratios in Table 13-3.

seen for the codon frequencies in Table 13-3. (In Table 13-4, phenylalanine is used as the baseline in determining ratios.)

From this evidence, we can deduce that codons consisting of one guanine and two uracils (G + 2U) code for valine, leucine, and cysteine, although we cannot distinguish the specific sequence for each of these amino acids. Similarly, one uracil and two guanines (U + 2G) must code for tryptophan, glycine, and perhaps one other. It looks as though the Watson-Crick model is correct in predicting the importance of the precise sequence (not just the ratios of bases). Many provisional assignments (such as those just outline for G and U) were soon obtained, primarily by groups working with Nirenberg or with Severo Ochoa.

Before we consider other code words, we will examine tRNA molecules, which further explain the link between the mRNA codon and amino acid recognition.

tRNA Recognition of the Codon

Is it the tRNA or the amino acid itself that recognizes the portion of the mRNA, or codon, that codes for a specific amino acid? A very convincing experiment has answered this question. In the experiment, cysteinyl-tRNA (tRNA^{Cys}, the tRNA specific for cysteine) charged with cysteine was treated with nickel hydride, which converted the cysteine (while still bound to tRNA^{Cys}) into another amino acid, alanine, without affecting the tRNA:

$$\text{Cysteine} \text{—} \text{tRNA}^{Cys} \xrightarrow{\text{nickel hydride}} \text{alanine} \text{—} \text{tRNA}^{Cys}$$

Protein synthesized with this hybrid species had alanine wherever we would expect cysteine. Thus, the experiment demonstrated that the amino acids are illiterate; they are inserted at the proper position because the tRNA "adapters" recognize the mRNA codons and insert their attached amino acids appropriately. We would expect, then, to find some site on the tRNA that recognizes the mRNA codon by complementary base pairing.

Figure 13-17a shows several functional sites of the tRNA molecule. The site that recognizes an mRNA codon is called the **anticodon;** its bases are complementary and antiparallel to the bases of the codon. Another operationally identifiable site is the amino acid attachment site. The other arms probably assist in binding the tRNA to the ribosome. Figure 13-17b shows a specific tRNA (yeast alanine tRNA). The "flattened" cloverleafs shown in these diagrams are not the normal conformation of tRNA molecules; tRNA normally exists as an L-shaped folded cloverleaf, as shown in Figure 13-17c. These diagrams are supported by very sophisticated chemical analysis of tRNA nucleotide sequences and by X-ray crystallographic data on the overall shape of the molecule. Although tRNA molecules share many structural similarities, each has a unique three-dimensional shape that allows recognition by the correct synthetase, which catalyzes the joining of a tRNA with its specific amino acid to form an aminoacyl-tRNA. (Synthetases will be discussed in

this chapter under "Protein Synthesis.") The specificity of "charging" the tRNAs is crucial to the integrity of protein synthesis.

Where does tRNA come from? If radioactive tRNA is put into a cell nucleus in which the DNA has been partially denatured by heating, the radioactivity appears (by autoradiography) in localized regions of the chromosomes. These regions probably reflect the location of tRNA genes; they are regions of DNA that produce tRNA rather than mRNA, which produces a protein. The labeled tRNA hybridizes to these sites because of the complementarity of base sequences between the tRNA and its parent gene. A similar situation holds for rRNA. Thus, we see that even the one-gene–one-polypeptide idea is not completely valid. Some genes do not code for protein; rather, they specify RNA components of the translational apparatus.

Message   Some genes code for proteins; other genes specify RNA (tRNA or rRNA) as their final product.

How does tRNA get its fancy shape? It probably folds up spontaneously into a conformation that produces maximal stability. Transfer RNA contains many "odd" or modified bases (such as pseudouracil, ψ) in its nucleotides; these play a direct role in folding and also have been implicated in other tRNA functions. You may have noticed some unusual base pairing within the loops of the tRNA in Figure 13-17b; G is hydrogen-bonded to U (instead of C). This apparent mismatching will be discussed in the next section.

New Code Words

Specific code words were finally deciphered through two kinds of experiments. The first involved making "mini mRNAs," each only three nucleotides in length. These, of course, are too short to promote translation into protein, but they do stimulate the binding of aminoacyl-tRNAs (aa-tRNA's) to ribosomes in a kind of abortive attempt at translation. It is possible to make a specific mini mRNA and determine *which* aminoacyl-tRNA it will bind to ribosomes.

For example, the G + 2U problem discussed previously can be resolved by using the following mini mRNAs:

> GUU — stimulates binding of valyl-tRNA
>
> UUG — stimulates binding of leucyl-tRNA
>
> UGU — stimulates binding of cysteinyl-tRNA

Analogous mini RNAs provided a virtually complete cracking of all the $4^3 = 64$ possible codons.

The second kind of experiment that was useful in cracking the genetic code involved the use of *repeating copolymers*. For instance, the copolymer designated $(AGA)_n$, which is a long sequence of AGAAGAAGAAGAAGA, was used to stimulate polypeptide synthesis in vitro. From the sequence of the resulting polypeptides and the possible triplets that could occur in the respective RNA copolymer,

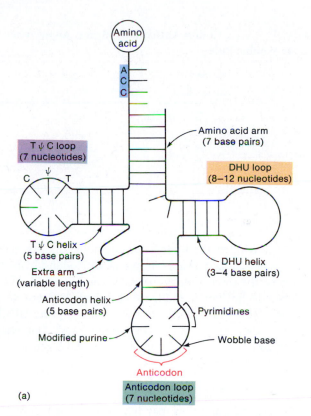

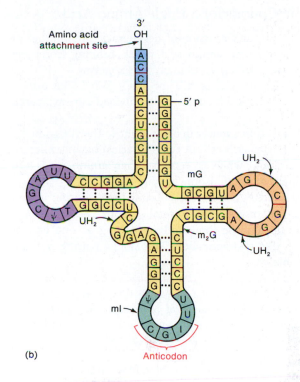

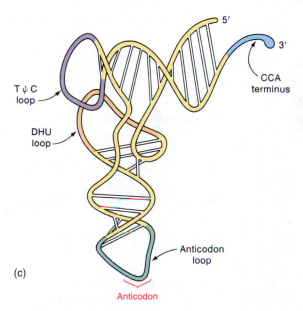

Figure 13-17 The structure of transfer RNA. (a) The functional areas of a generalized tRNA molecule. (b) The specific sequence of yeast alanine tRNA. Arrows indicate several kinds of rare modified bases. (c) The actual three-dimensional structure of yeast phenylalanine tRNA. The symbols ψ, mG, m$_2$G, mI, and DHU (or UH$_2$) are abbreviations for modified bases pseudouridine, methylguanosine, dimethylguanosine, methylinosine, and dihydrouridine, respectively. (Part a from S. Arnott, "The Structure of Transfer RNA," *Progress in Biophysics and Molecular Biology* 22, 1971, 186; parts b and c from L. Stryer, *Biochemistry*, 4th ed. Copyright ©1995 by Lubert Stryer; part c based on a drawing by Dr. Sung-Hou Kim.)

		Second letter				
		U	**C**	**A**	**G**	
First letter	**U**	UUU ⎱ Phe UUC ⎰ UUA ⎱ Leu UUG ⎰	UCU ⎫ UCC ⎬ Ser UCA ⎪ UCG ⎭	UAU ⎱ Tyr UAC ⎰ UAA Stop UAG Stop	UGU ⎱ Cys UGC ⎰ UGA Stop UGG Trp	U C A G
	C	CUU ⎫ CUC ⎬ Leu CUA ⎪ CUG ⎭	CCU ⎫ CCC ⎬ Pro CCA ⎪ CCG ⎭	CAU ⎱ His CAC ⎰ CAA ⎱ Gln CAG ⎰	CGU ⎫ CGC ⎬ Arg CGA ⎪ CGG ⎭	U C A G
	A	AUU ⎱ Ile AUC ⎰ AUA ⎰ AUG Met	ACU ⎫ ACC ⎬ Thr ACA ⎪ ACG ⎭	AAU ⎱ Asn AAC ⎰ AAA ⎱ Lys AAG ⎰	AGU ⎱ Ser AGC ⎰ AGA ⎱ Arg AGG ⎰	U C A G
	G	GUU ⎫ GUC ⎬ Val GUA ⎪ GUG ⎭	GCU ⎫ GCC ⎬ Ala GCA ⎪ GCG ⎭	GAU ⎱ Asp GAC ⎰ GAA ⎱ Glu GAG ⎰	GGU ⎫ GGC ⎬ Gly GGA ⎪ GGG ⎭	U C A G

(Third letter column on right)

Figure 13-18 The genetic code.

many code words could be verified. (This kind of experiment is detailed in Problem 10 at the end of this chapter. In solving it, you can put yourself in the place of H. Gobind Khorana, who received a Nobel prize for directing the experiments.)

Figure 13-18 gives the genetic code dictionary of 64 words. Inspect this figure carefully, and ponder the miracle of molecular genetics. Such an inspection should reveal several points that require further explanation.

Multiple Codons for a Single Amino Acid

As we saw in our discussion of degeneracy, the number of codons for a single amino acid varies, ranging from one (tryptophan = UGG) to as many as six (serine = UCU or UCC or UCA or UCG or AGU or AGC). Why? The answer is complex but not difficult; it can be divided into two parts:

1. Certain amino acids can be brought to the ribosome by several *alternative* tRNA types (species) having different anticodons, whereas certain other amino acids are brought to the ribosome by only one tRNA.
2. Certain tRNA species can bring their specific amino acids in response to several codons, not just one, through a loose kind of base pairing at one end of the codon and anticodon. This sloppy pairing is called **wobble.**

Message The degree of degeneracy for a given amino acid is determined by the number of codons for that amino acid that have only one tRNA each, plus the number of codons for the amino acid that share a tRNA through wobble.

We had better consider wobble first, and it will lead us into a discussion of the various species of tRNA. Wobble is caused by the third nucleotide of an anticodon (at the 5′ end) that is not quite aligned (Figure 13-19). This out-of-line nucleotide sometimes can form hydrogen bonds not only with its normal complementary nucleotide in the third position of the codon but also with a different nucleotide in that position. Crick established certain "wobble rules" that dictate which nucleotides can and cannot form new hydrogen-bonded associations through wobble (Table 13-5). In the table, I (inosine) is one of the rare bases found in tRNA, often in the anticodon.

Figure 13-19 shows the possible codons that one tRNA serine species can recognize. As the wobble rules indicate, G can pair with U or with C. Table 13-6 lists all the codons for serine and shows how different tRNAs can service these codons. This is a good example of the effects of wobble on the genetic code.

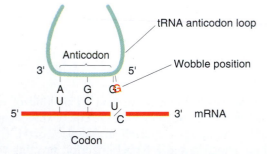

Figure 13-19 In the third site (5′ end) of the anticodon, G can take either of two wobble positions, thus being able to pair with either U or C. This means that a single tRNA species carrying an amino acid (in this case, serine) can recognize two codons—UCU and UCC—in the mRNA.

Table 13-5 Codon-Anticodon Pairings Allowed by the Wobble Rules

5′ end of anticodon	3′ end of codon
G	U or C
C	G only
A	U only
U	A or G
I	U, C, or A

Sometimes there can be an additional tRNA species that we represent as tRNA^Ser4; it has an anticodon identical with any of the three anticodons shown in Table 13-6, but it differs in its nucleotide sequence elsewhere in the tRNA molecule. These four tRNAs are called **isoaccepting tRNAs** because they accept the same amino acid, but they are probably all transcribed from different tRNA genes.

Stop Codons

The second point you may have noticed in Figure 13-18 is that some codons do not specify an amino acid at all. These codons are labeled as **stop** or **termination codons.** They can be regarded as similar to periods or commas punctuating the message encoded in the DNA.

One of the first indications of the existence of stop codons came in 1965 from Brenner's work with the T4 phage. Brenner analyzed certain mutations ($m_1 - m_6$) in a single cistron that controls the head protein of the phage. These mutants had two things in common. First, the head protein of each mutant was a shorter polypeptide chain than that of the wild type. Second, the presence of a suppressor mutation (*su*) in the host chromosome would cause the phage to develop a head protein of normal (wild-type) chain length despite the presence of the *m* mutation (Figure 13-20).

Brenner examined the ends of the shortened proteins and compared them with wild-type protein, recording for

Table 13-6 Different tRNAs That Can Service Codons for Serine

Codon	tRNA	Anticodon
UCU UCC	tRNA^Ser1	AGG + wobble
UCA UCG	tRNA^Ser2	AGU + wobble
AGU AGC	tRNA^Ser3	UCG + wobble

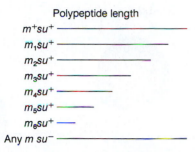

Polypeptide length

m^+su^+ ──────────────

m_1su^+ ──────────────

m_2su^+ ─────────────

m_3su^+ ────────────

m_4su^+ ──────────

m_5su^+ ────────

m_6su^+ ────

Any m su^- ──────────────

Figure 13-20 Polypeptide chain lengths of phage T4 head protein in wild type (*top*) and various amber mutants (*m*). An amber suppressor (*su*) leads to phenotypic development of the wild-type chain.

each mutant the next amino acid that *would* have been inserted to continue the wild-type chain. These amino acids for the six mutations were glutamine, lysine, glutamic acid, tyrosine, tryptophan, and serine. There is no immediately obvious pattern to these results, but Brenner brilliantly deduced that certain codons for each of these amino acids are similar in that each of them can mutate to the codon UAG by a single change in a DNA nucleotide pair. He therefore postulated that UAG is a stop (or termination) codon—a signal to the translation mechanism that the protein is now complete.

UAG was the first stop codon deciphered; it is called the **amber codon.** Mutants that are defective owing to the presence of an abnormal amber codon are called *amber mutants,* and their suppressors are *amber suppressors.* UGA, the **opal codon,** and UAA, the **ochre codon,** are also stop codons and also have suppressors. Stop codons often are called **nonsense codons** because they designate no amino acid. Not surprisingly, stop codons do not act as mini mRNAs in binding aa-tRNA to ribosomes in vitro. We will discuss stop codons and suppressors further after we consider the process of protein synthesis.

Protein Synthesis

We can regard **protein synthesis** as a chemical reaction, and we shall take this approach at first. Then we shall take a three-dimensional look at the physical interactions of the major components.

In protein synthesis as a chemical reaction:

1. Each amino acid (aa) is attached to a tRNA molecule specific to that amino acid by a high-energy bond derived from ATP. The process is catalyzed by a specific enzyme called a **synthetase** (the tRNA is said to be "charged" when the amino acid is attached):

$$aa_1 + tRNA_1 + ATP \xrightarrow{synthetase_1}$$
$$aa_1 - tRNA_1 + AMP + PP_i$$

There is a separate synthetase for each amino acid.

2. The energy of the charged tRNA is converted into a peptide bond linking the amino acid to another one on the ribosome:

$$aa_1 - tRNA_1 + aa_2 - tRNA_2 \xrightarrow[\text{on a ribosome}]{\text{peptidyl transferase}}$$
$$\underbrace{aa_1 - aa_2}_{\substack{\text{Small}\\\text{polypeptide}}} - tRNA_2 + tRNA_1 \text{ (released)}$$

3. New amino acids are linked by means of a peptide bond to the growing chain:

$$aa_3 - tRNA_3 + aa_1 - aa_2 - tRNA_2 \longrightarrow$$
$$\underbrace{aa_1 - aa_2 - aa_3}_{\substack{\text{Larger}\\\text{polypeptide}}} - tRNA_3 + tRNA_2 \text{ (released)}$$

4. This process continues until aa_n (the final amino acid) is added. Of course, the whole thing works only in the presence of mRNA, ribosomes, several additional protein factors, enzymes, and inorganic ions.

The Ribosome

Ribosomes consist of two subunits which in prokaryotes sediment as 50S and 30S particles and which associate to form a 70S particle, as seen in Figure 13-21a. The eukaryotic counterparts are 60S and 40S for the large and small subunits, and 80S for the complete ribosome (Figure 13-21b). Ribosomes contain specific sites that enable them to bind to the mRNA, the tRNAs, and specific protein factors required for protein synthesis. Let's look at a general picture of protein synthesis on the ribosome, and then examine each of the steps in the process in more detail.

Figure 13-22 shows a polypeptide being synthesized on the ribosome. The mRNA binds to the 30S subunit. The tRNAs bind to two sites on the ribosome. These sites overlap the subunits. The **A site** is the entry site for an aminoacyl-tRNA (a tRNA carrying a single amino acid). The peptidyl-tRNA carrying the growing polypeptide chain binds at the **P site.** Each new amino acid is added by the transfer of the growing chain to the new aminoacyl-tRNA, forming a new peptide bond. The deacylated tRNA is then released from the P site, and the ribosome moves one codon farther along the message, transferring the new peptidyl-tRNA to the P site, and leaving the A site vacant for the next incoming aminoacyl-tRNA.

We can separate the process of protein synthesis into three distinct steps. **Initiation, elongation,** and **termination.** Let's examine each of these steps in detail, using prokaryotes as an example.

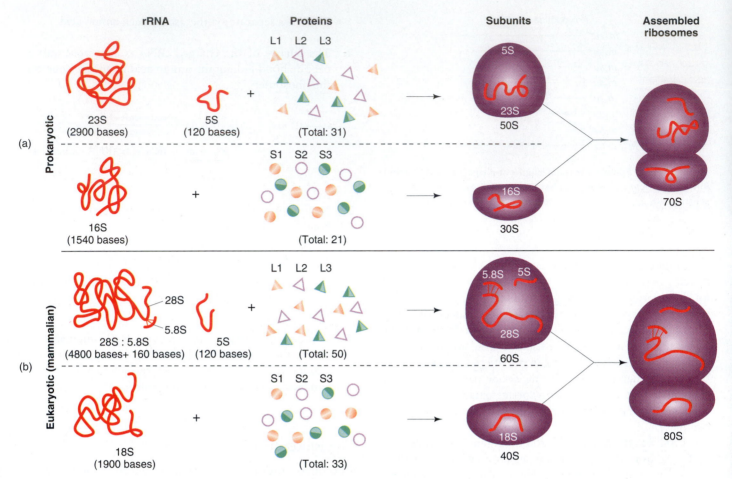

	rRNA	Proteins	Subunits	Assembled ribosomes

(a) Prokaryotic

23S
(2900 bases)   5S
(120 bases)

L1 L2 L3

(Total: 31)

5S
23S
50S

16S
(1540 bases)

S1 S2 S3

(Total: 21)

16S
30S

70S

(b) Eukaryotic (mammalian)

28S
5.8S

28S : 5.8S
(4800 bases+ 160 bases)   5S
(120 bases)

L1 L2 L3

(Total: 50)

5.8S 5S
28S
60S

18S
(1900 bases)

S1 S2 S3

(Total: 33)

18S
40S

80S

Figure 13-21 Ribosomes contain a large and a small subunit. Each subunit contains both rRNA of varying lengths and a set of proteins (designated by different colors). There are two principal rRNA molecules in all ribosomes. Ribosomes from prokaryotes also contain one 120-base-long rRNA which sediments at 5S, whereas eukaryotic ribosomes have two small rRNAs: a 5S RNA molecule similar to the prokaryotic 5S, and a 5.8S molecule 160 bases long. The large subunit proteins are named L1, L2, etc., and the small subunit proteins S1, S2, etc. (From H. Lodish, D. Baltimore, A. Berk, S. L. Zipursky, P. Matsudaira, and J. Darnell, *Molecular Cell Biology,* 3d ed. Copyright ©1995 by Scientific American Books, Inc.)

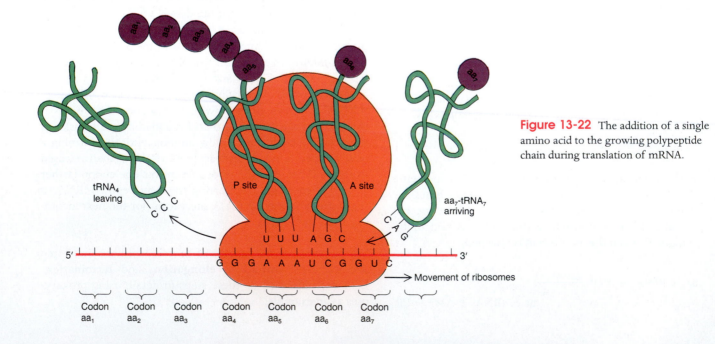

Figure 13-22 The addition of a single amino acid to the growing polypeptide chain during translation of mRNA.

Initiation

Three Steps of Initiation. In addition to mRNA, ribosomes, and specific tRNA molecules, initiation requires the participation of several factors, termed **initiation factors IF1, IF2,** and **IF3.** In *E. coli* and in most other prokaryotic organisms, the first amino acid in any newly synthesized polypeptide is *N*-formylmethionine. It is inserted not by tRNA^Met, however, but by an **initiator tRNA** called tRNA^fMet. This initiator tRNA has the normal methionine anticodon but inserts *N*-formylmethionine rather than methionine (Figure 13-23). In *E. coli*, AUG and GUG, and on rare occasions UUG, serve as initiation codons. When one of these triplets occurs in the initiation position, it is recognized by *N*-formylMet-tRNA and methionine appears as the first amino acid in the chain. Let's examine the steps in initiation in detail.

1. The first step in initiation involves binding of the mRNA to the 30S subunit (Figure 13-24). The binding is stimulated by the initiation factor IF3. When not engaged in protein synthesis, the ribosomal subunits exist in the free form; they assemble into complete ribosomes as a result of the initiation process.

2. The initiation factor IF2 binds to GTP and to the initiator fMet-tRNA, and stimulates the binding of fMet-tRNA to the initiation complex, leading the fMET-tRNA into the P site, as shown in the middle portion of Figure 13-24.

3. A ribosomal protein splits the GTP bound to IF2, helping to drive the assembly of the two ribosomal subunits (Figure 13-24, bottom). At this stage the factors IF2 and IF3 are released. (The exact role of IF1 is not completely clear, although it seems to take part in the recycling of the ribosome.)

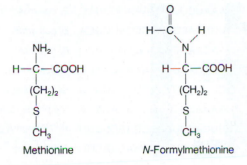

Figure 13-23 The structures of methionine (Met) and *N*-formylmethionine (fMet). A tRNA bearing fMet can initiate a polypeptide chain in prokaryotes but cannot be inserted in a growing chain; a tRNA bearing Met can be inserted in a growing chain but will not initiate a new chain. Both these tRNAs bear the same anticodon complementing the codon AUG.

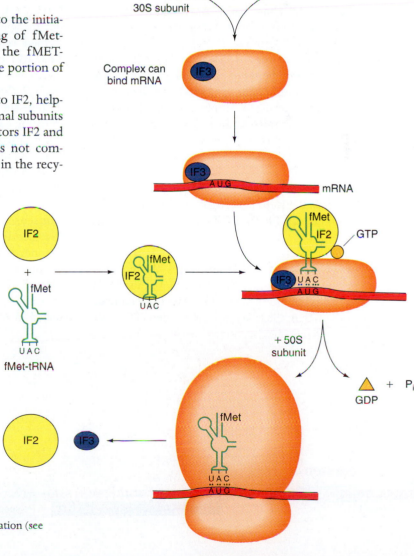

Figure 13-24 The steps involved in initiation of translation (see text).

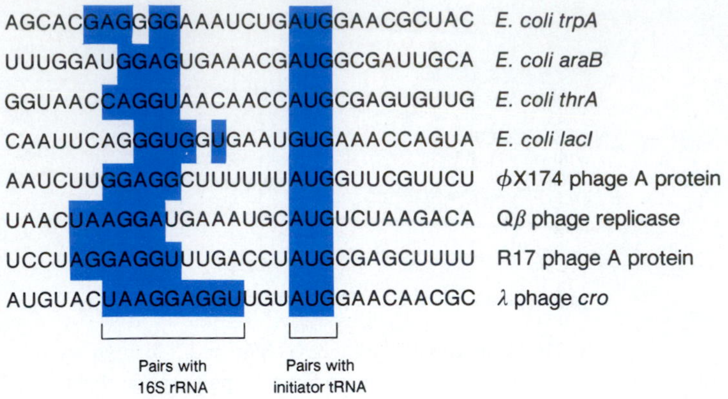

AGCACGAGGGGAAAUCUGAUGGAACGCUAC E. coli trpA
UUUGGAUGGAGUGAAACGAUGGCGAUUGCA E. coli araB
GGUAACCAGGUAACAACCAUGCGAGUGUUG E. coli thrA
CAAUUCAGGGUGGUGAAUGUGAAACCAGUA E. coli lacI
AAUCUUGGAGGCUUUUUUAUGGUUCGUUCU ϕX174 phage A protein
UAACUAAGGAUGAAAUGCAUGUCUAAGACA Qβ phage replicase
UCCUAGGAGGUUUGACCUAUGCGAGCUUUU R17 phage A protein
AUGUACUAAGGAGGUUGUAUGGAACAACGC λ phage cro
Pairs with 16S rRNA
Pairs with initiator tRNA

Figure 13-25 Ribosomal binding site sequences in *E. coli* and its bacteriophages share certain common features, which are shown in the colored regions. The initiation codon (color) is separated by several bases from a short sequence (color) that is complementary to the 3′ end of 16S rRNA. (After L. Stryer, *Biochemistry,* 4th ed. Copyright ©1995 by Lubert Stryer.)

Ribosome Binding Sites. How are the correct initiation codons selected from the many AUG and GUG codons in a mRNA molecule? John Shine and Lynn Dalgarno first noticed that true initiation codons were preceded by sequences that paired well with the 3′ end of 16S rRNA. Figure 13-25 shows some of these sequences. There is a short but variable separation between the Shine-Dalgarno sequence and the initiation codon. Figure 13-26 depicts the base pairing between idealized mRNA and the 16S rRNA that results in ribosome-mRNA complexes leading to protein initiation in the presence of fMet-tRNA.

Elongation

Figure 13-27 details the steps in elongation, which are aided by three protein factors, EF-Tu, EF-Ts, and EF-G. The steps are as follows:

1. The elongation factor EF-Tu mediates the entry of aminoacyl-tRNAs into the A site. To achieve this, EF-Tu first binds to GTP. This activated EF-Tu–GTP complex binds to the tRNA. Next, hydrolysis of the GTP of the complex to GDP helps drive the binding of the aminoacyl-tRNA to the A site, at which point the EF-Tu is released (Figure 13-27a), leaving the new tRNA in the A site (Figure 13-27b).
2. The elongation factor EF-Ts mediates the release of EF-Tu–GDP from the ribosome and also the regeneration of EF-Tu–GTP.
3. In the "translocation" step, the polypeptide chain on the peptidyl-tRNA is transferred to the aminoacyl-tRNA on the A site in a reaction catalyzed by the enzyme peptidyltransferase (Figure 13-27c). The ribosome then translocates by moving one codon farther along the mRNA, going in the 5′ → 3′ direction. This step is mediated by the elongation factor EF-G (Figure 13-27d) and is driven by splitting a GTP to GDP. This action releases the uncharged tRNA from the P site and transfers the newly formed peptidyl-tRNA from the A site to the P site (Figure 13-27e).

Termination

Release Factors. In our previous discussion of the genetic code, we described the three chain-termination codons UAG, UAA, and UGA. Interestingly, these three triplets are not recognized by a tRNA, but instead by protein factors, termed **release factors,** which are abbreviated **RF1** and **RF2.** RF1 recognizes the triplets UAA and UAG, and RF2 the triplets UAA and UGA. A third factor, **RF3,** also helps to catalyze the chain termination. When the peptidyl-tRNA is in the P site, the release factors, in response to the chain-terminating codons, bind to the A site. The polypeptide is then released from the P site, and the ribosomes dissociate into two subunits, in a reaction driven by the hydrolysis of a GTP molecule. Figure 13-28 provides a schematic view of this process.

Nonsense Suppressor Mutations. It is interesting to consider the suppressors of the nonsense mutations that Brenner and coworkers defined. Many of these **nonsense suppressor mutations** are known to alter the anticodon loop of specific tRNAs in such a way as to allow recognition of a nonsense codon in mRNA. Thus, an amino acid is inserted in response to the nonsense codon, and translation

3′ end of 16S rRNA
3′OH
A
U
UCCUCCA
G
A
U
C
5′
3′
GAUUCCUAGGAGGUUUGACCU AUG CGA GCU UUU AGU —mRNA
fMet — Arg — Ala — Phe — Ser —Polypeptide

Figure 13-26 Binding of the Shine-Dalgarno sequence on an mRNA to the 3′ end of 16S rRNA. (After L. Stryer, *Biochemistry,* 4th ed. Copyright ©1995 by Lubert Stryer.)

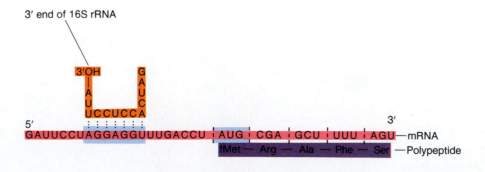

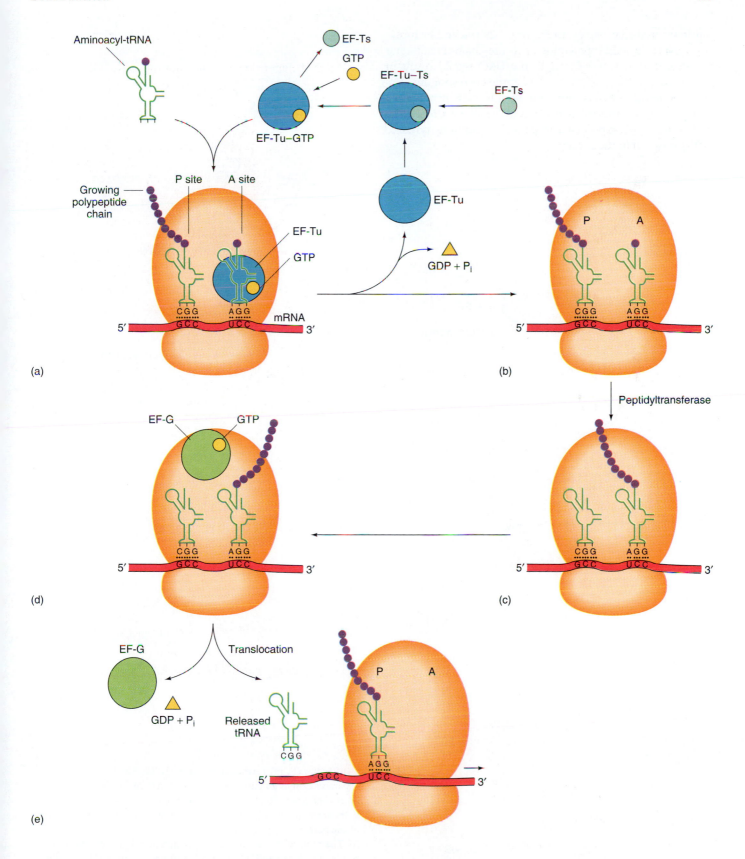

Figure 13-27 The steps involved in elongation (see text).

continues past that triplet. In Figure 13-29 the amber mutation replaces a wild-type codon with the chain-terminating
nonsense codon UAG. By itself, the UAG would result in
prematurely cutting off the protein at the corresponding position. The suppressor mutation in this case produces a
tRNA^Tyr with an anticodon that recognizes the mutant UAG
stop codon. The suppressed mutant thus contains tyrosine
at that position in the protein.

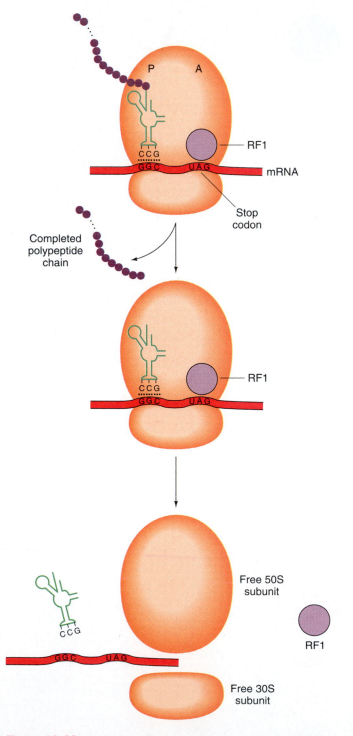

Figure 13-28 The steps leading to termination of protein synthesis
(see text).

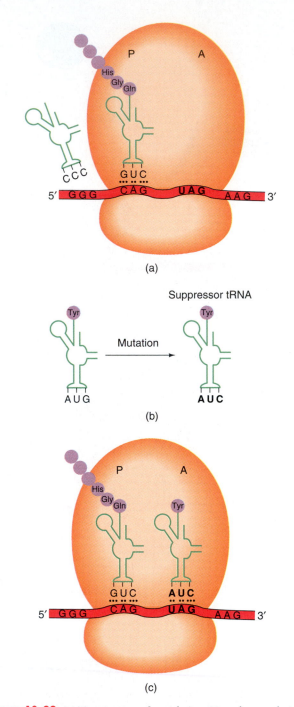

Figure 13-29 (a) Termination of translation. Here the translation
apparatus cannot go past a nonsense codon (UAG in this case), because there is no tRNA that can recognize the UAG triplet. This leads
to the termination of protein synthesis and to the subsequent release
of the polypeptide fragment. The release factors are not shown here.
(b) The molecular consequences of a mutation that alters the anticodon of a tyrosine tRNA. This tRNA can now read the UAG codon.
(c) The suppression of the UAG codon by the altered tRNA, which
now permits chain elongation. (Adapted from James D. Watson, John
Tooze, and David T. Kurtz, *Recombinant DNA: A Short Course.* Copyright ©1983 by W. H. Freeman and Company.)

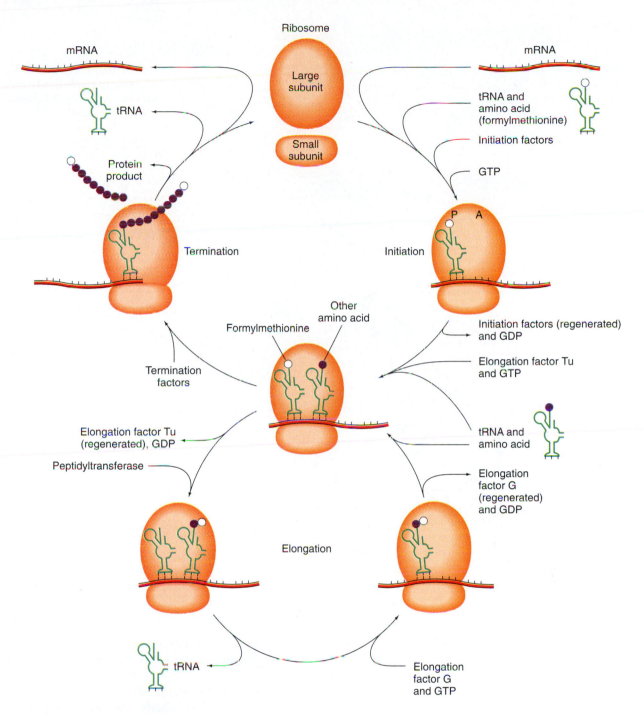

Figure 13-30 The transactions of the ribosome. At initiation, the ribosome recognizes the starting point in a segment of mRNA and binds a molecule of tRNA bearing a single amino acid. In all bacterial proteins, this first amino acid is *N*-formylmethionine. In elongation, a second amino acid is linked to the first one. The ribosome then shifts its position on the mRNA molecule, and the elongation cycle is repeated. When the stop codon is reached, the chain of amino acids folds spontaneously to form a protein. Subsequently, the ribosome splits into its two subunits, which rejoin before a new segment of mRNA is translated. Protein synthesis is facilitated by a number of catalytic proteins (initiation, elongation, and termination factors) and by guanosine triphosphate (GTP), a small molecule that releases energy when it is converted into guanosine diphosphate (GDP). (Adapted from Donald M. Engelman and Peter B. Moore, "Neutron-Scattering Studies of the Ribosome." Copyright ©1976 by Scientific American, Inc. All rights reserved.)

What happens to normal termination signals at the ends of proteins in the presence of a suppressor? Many of the natural termination signals consist of two chain-termination signals in a row. Nonsense suppressors are sufficiently inefficient in translating through chain-terminating triplets, because of competition with release factors, that the probability of suppression at two codons in a row is small. Consequently, very few protein copies are produced that carry many extraneous amino acids resulting from translation beyond the natural stop codon.

Overview of Protein Synthesis

Figure 13-30 provides a summary of the steps in protein synthesis that we have covered in this section. A direct visu-

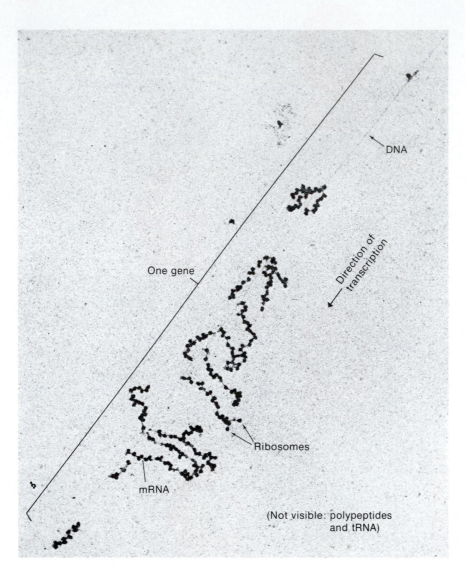

Figure 13-31 A gene of *E. coli* being simultaneously transcribed and translated. (Electron micrograph by O. L. Miller, Jr., and Barbara A. Hamkalo.)

alization of protein synthesis can be seen in the electron micrograph shown in Figure 13-31, which shows the simultaneous transcription and translation of a gene in *E. coli*.

Medicines of the Future Based on Blocking Translation

Short oligonucleotides, stretches of DNA in the 15–25 base range, are now being employed as experimental therapeutic agents to combat AIDS, leukemia, and other diseases. The principle that these so-called antisense oligonucleotides operate on is to combine with the initial portion of a specific messenger RNA, for instance, one of the viral mRNA's for herpes or HIV, and form a double helix, thus blocking the translation of that mRNA. The short antisense DNA is designed to be the exact complement of the mRNA sequence. Figure 13-32 portrays this strategy. Also, sometimes, the double-helical portion of the mRNA formed by the antisense drug can serve as a target for the cellular enzyme ribonuclease H, which then cleaves the message, preventing further translation, as shown in Figure 13-33. Problems of

insolubility and drug breakdown can be partly overcome by replacing one oxygen atom in each phosphate group of the oligonucleotide by a sulfur atom. Although there are still technical obstacles to overcome before these drugs are commonplace, they clearly represent a cutting edge of using molecular tools to combat disease.

Universality of Genetic Information Transfer

Thus far, our discussion has focused on bacteria, but the amazing fact is that the information-transfer and coding processes are virtually identical in all organisms that have been studied. For example, all the different single amino acid substitutions known to occur in human hemoglobin result from single nucleotide-pair substitutions based on the genetic code derived from *E. coli* (Table 13-7). Such observations suggest that the genetic code is shared by all organisms. Furthermore, an information-bearing molecule, such as rabbit red blood cell mRNA, which is predominantly he-

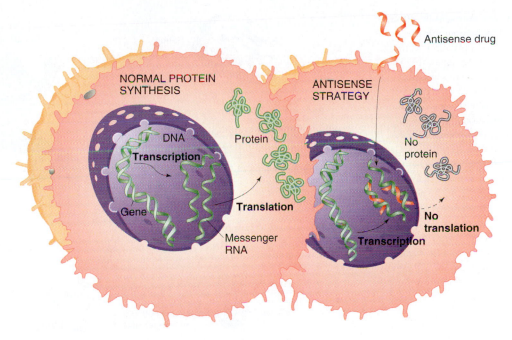

Figure 13-32 The antisense strategy for inhibiting the production of disease-related proteins aims to selectively impede translation. (From J. S. Cohen and M. E. Hogan, "The New Genetic Medicines." Copyright ©1994 by Scientific American, Inc. All rights reserved.)

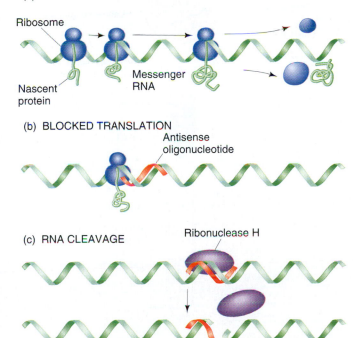

(a) NORMAL TRANSLATION

Ribosome

Messenger RNA

Nascent protein

(b) BLOCKED TRANSLATION

Antisense oligonucleotide

(c) RNA CLEAVAGE

Ribonuclease H

Figure 13-33 The action of an antisense oligonucleotide. (a) Normal translation, which is impeded by an antisense oligonucleotide. (b) The binding of an antisense drug can also induce an enzyme, ribonuclease H, to cut the RNA at the site of drug binding. (c) The cleaved RNA is then rapidly degraded. (From J. S. Cohen and M. E. Hogan, "The New Genetic Medicines." Copyright ©1994 by Scientific American, Inc. All rights reserved.)

Table 13-7 Mutations and Their Inferred Codon Changes

Protein*	Amino acid substitution	Inferred codon change
Hemoglobin	Glu → Val	GAA → GUA
Hemoglobin	Glu → Lys	GAA → AAA
Hemoglobin	Glu → Gly	GAA → GGA
Tryptophan synthetase	Gly → Arg	GGA → AGA
Tryptophan synthetase	Gly → Glu	GGA → GAA
Tryptophan synthetase	Glu → Ala	GAA → GCA
TMV coat protein	Leu → Phe	CUU → UUU
TMV coat protein	Glu → Gly	GAA → GGA
TMV coat protein	Pro → Ser	CCC → UCC

* Mutations are for human hemoglobins, *E. coli* tryptophan synthetase, and tobacco mosaic virus (TMV) coat protein.

SOURCE: L. Stryer, *Biochemistry*, 2d ed. Copyright © 1981 by W. H. Freeman and Company.

moglobin gene transcript, will be translated in an alien environment (such as a frog egg) into rabbit hemoglobin (Figure 13-34). Apparently, the translation apparatus is functionally the same in a wide range of different organisms.

Sequencing techniques at the protein, RNA, and DNA levels (see Chapter 14) have verified that the genetic code is universal in all organisms studied to date, ranging from

Figure 13-34 Rabbit hemoglobin mRNA is translated into rabbit hemoglobin in a frog (*Xenopus*) egg. This translation occurs with the *Xenopus* translation apparatus. The mRNA for rabbit hemoglobin is injected into a *Xenopus* oocyte, which is then incubated in radioactively labeled amino acids. The products of translation are next separated in a chromatography column. Rabbit hemoglobin is identified by its position relative to an unlabeled marker (rabbit) hemoglobin. This experiment is one of many that point toward the uniformity of the genetic molecular mechanism in all forms of life. (Sephadex is a separatory material used in chromatography.) (From C. Lane, "Rabbit Hemoglobin from Frog Eggs." Copyright ©1976 by Scientific American, Inc. All rights reserved.)

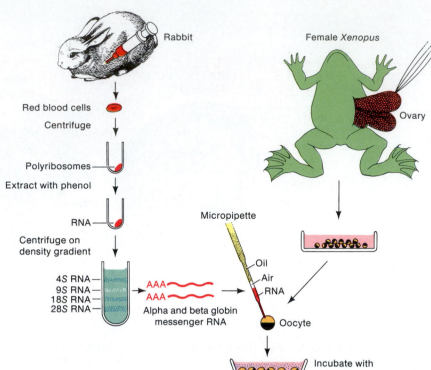

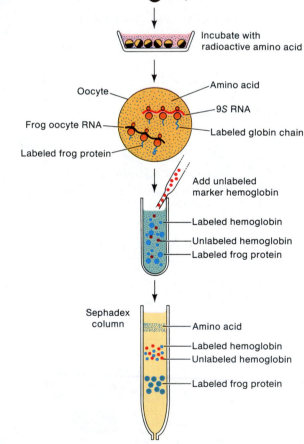

viruses and bacteria to humans. One exception involves mitochondrial DNA (see also Chapter 22). Two codons are translated differently here, owing to the properties of tRNAs that are confined to the mitochondrial system. Thus, whereas AUA is normally translated as isoleucine, it is read as methionine in the mitochondria. Also, the mammalian mitochondria translate UGA as tryptophan, although UGA normally specifies a chain-terminating codon. In yeast, the mitochondria translate UGA as tryptophan, as in mammalian mitochondria, but they translate AUA as isoleucine, as in bacterial (nonmitochondrial) systems. Other exceptions to the universal code are found in the nuclear genome of some protozoans. As we shall see from genetic engineering experiments in Chapter 15, DNA is DNA no matter what its origin. The nature and message of the DNA represent a universal language of life on earth.

Does this interspecific equivalence of parts in the genetic apparatus indicate a common evolutionary ancestry for all life-forms on earth? Or does it simply reflect the fact that this is the only workable biochemical option in the earth environment (biochemical predestination)? Whatever the answer, the wonderful uniformity of the molecular basis of life is firmly established. Minor variations do exist, but they do not detract from the central uniformity of the mechanism we have described.

Message The processes of information storage, replication, transcription, and translation are fundamentally similar in all living systems. In demonstrating this fact, molecular genetics has provided a powerful unifying force in biology. We now know some of the tricks that life uses to achieve persistent order in a randomizing universe.

Eukaryotic RNA

Several aspects of RNA synthesis and processing in eukaryotes are distinctly different from their counterparts in prokaryotes.

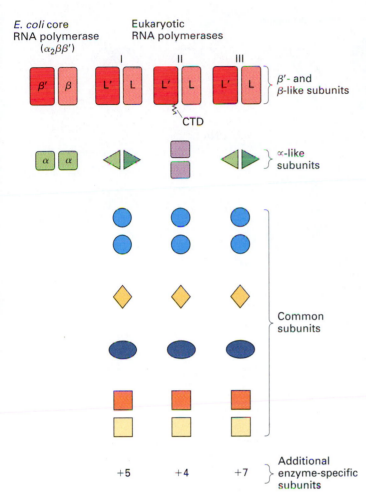

E. coli core
RNA polymerase
($\alpha_2\beta\beta'$)

Eukaryotic
RNA polymerases

β'- and
β-like subunits

CTD

α-like
subunits

Common
subunits

Additional
enzyme-specific
subunits

+5   +4   +7

Figure 13-35 Comparison of the subunit structure of RNA polymerases. The subunits of the three yeast RNA polymerases are compared to that of *E. coli*. Subunits in common are shown with identical symbols and colors. It can be seen that the yeast polymerases have subunits that are related to the *E. coli* subunits as well as having subunits that are unique to yeast and eukaryotes. CTD represents a C-terminal domain that is an essential part of the largest subunit of yeast polymerase II. (From H. Lodish, D. Baltimore, A. Berk, S. L. Zipursky, P. Matsudaira, and J. Darnell, *Molecular Cell Biology*, 3d ed. Copyright ©1995 by Scientific American Books, Inc.)

RNA Synthesis

Whereas a single RNA polymerase species synthesizes all RNAs in prokaryotes, there are three different RNA polymerases in eukaryotic systems:

1. RNA polymerase I synthesizes rRNA.
2. RNA polymerase II synthesizes mRNA. The mRNA molecules are **monocistronic,** whereas many mRNAs are **polycistronic** (code for several cistrons) in prokaryotes.
3. RNA polymerase III synthesizes tRNAs and also small nuclear and cellular RNA molecules.

The eukaryotic polymerases have a more complex subunit structure than prokaryotic polymerases. Figure 13-35 shows the subunits in each of the three yeast RNA polymerases, compared with those for the *E. coli* RNA polymerase. Some of the subunits are similar to the corresponding *E. coli* proteins, but others are not.

RNA Processing

The primary RNA transcript produced in the nucleus usually is processed in several ways before its transport to the cytoplasm, where it is used to program the translation machinery (Figure 13-36). Figure 13-37 depicts these processing

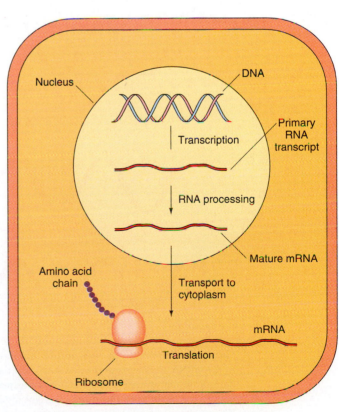

Nucleus

DNA

Transcription

Primary
RNA
transcript

RNA processing

Mature mRNA

Amino acid
chain

Transport to
cytoplasm

mRNA

Translation

Ribosome

Figure 13-36 Gene expression in eukaryotes. The mRNA is processed in the nucleus before transport to the cytoplasm. (From J. E. Darnell, Jr., "The Processing of RNA." Copyright ©1983 by Scientific American, Inc. All rights reserved.)

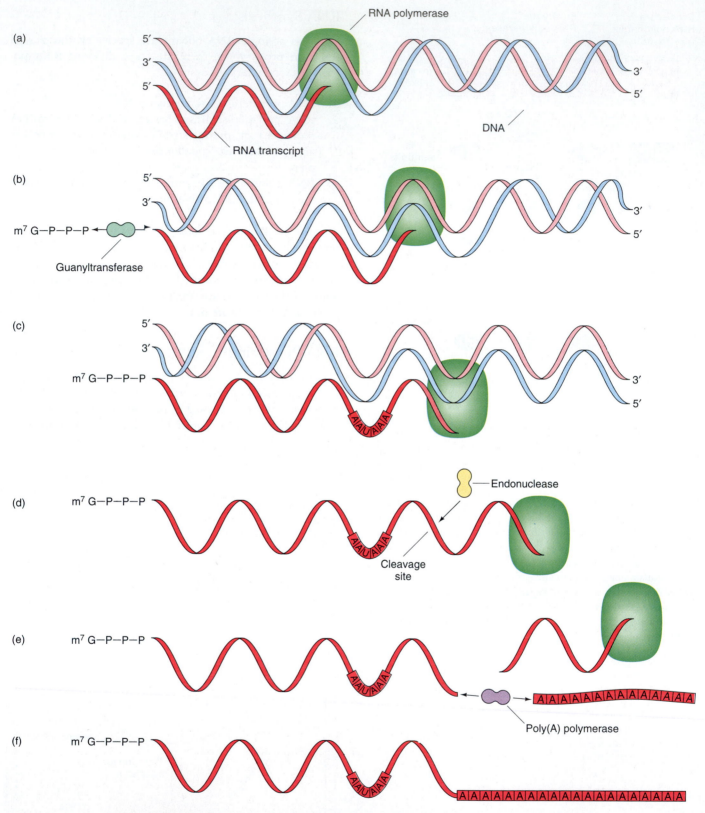

Figure 13-37 Processing of primary mRNA. (a) Transcription is mediated by RNA polymerase. (b) Early in transcription an enzyme, guanyltransferase, adds 7-methylguanosine (m⁷Gppp) to the 5′ end of the mRNA. (c) The sequence AAUAAA, near the 3′ end, helps signal a cleavage event (d) by an endonuclease approximately 20 bp farther downstream. (e) An enzyme, poly(A) polymerase, then adds a poly(A) tail, made up of 150 to 200 adenosine residues, to the site of this cleavage at the 3′ end. (f) This yields the complete primary mRNA. (From J. E. Darnell, Jr., "The Processing of RNA." Copyright ©1983 by Scientific American, Inc. All rights reserved.)

Figure 13-38 Split-gene organization of the gene for the protein ovalbumin. (a) The electron micrograph and (b) its map show the result of an experiment in which a single strand of the DNA incorporating the gene for the egg white protein ovalbumin was allowed to hybridize with ovalbumin mRNA, the molecule from which the protein is translated. The looped-out single-stranded segments of DNA represent the introns (c). The schematic representation of the gene shows the seven introns (black) and eight exons (red) and the number of base pairs in each of the exons; the size of the introns ranges from 251 base pairs for (*B*) to about 1600 (for *G*). (From P. Chambon, "Split Genes." Copyright ©1981 by Scientific American, Inc. All rights reserved.)

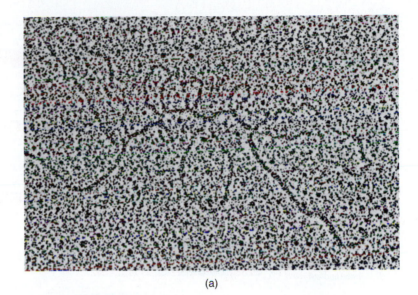

(a)

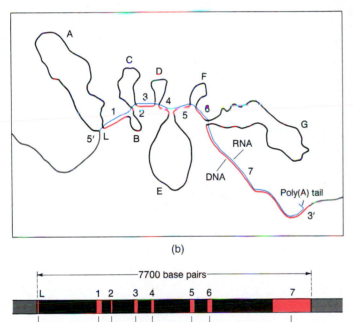

(b)

(c)

cific mRNA molecules. As recombinant DNA techniques (see Chapter 14) facilitated the physical analysis of eukaryotic genes, it became apparent that *primary* RNA transcripts were being shortened by the elimination of internal segments before transport into the cytoplasm. In most higher eukaryotes studied, this was found to be true not only for mRNA but also for rRNA—and even for tRNA in some cases.

Figure 13-38 shows the organization of the gene for chicken ovalbumin, a 386-amino acid polypeptide. The DNA segments that code for the structure of the protein are interrupted by intervening sequences, termed **introns.** In Figure 13-38, these segments are designated with the letters A to G. The primary transcript is processed by a series of "splicing" reactions, much in the same way that a tape-recorded message can be cut and pasted back together. Splicing removes the introns and brings together the coding regions, termed **exons,** to form an mRNA, which now consists of a sequence that is completely colinear with the ovalbumin protein. The exons are indicated by the letter L and numbers 1 to 7 in Figure 13-38. In different genes, introns have been detected that are as large as 2000 base pairs in length. Some genes have as many as 16 introns.

It is clear that splicing occurs after transcription, and in several steps, since RNA transcripts (previously termed "heterogeneous nuclear RNA," or HnRNA) can be isolated that correspond to the entire genetic region (introns + exons), as well as transcripts intermediate in length. In these intermediate-length RNA molecules, certain introns have already been removed, but others are retained. The entire sequence of events for RNA processing and splicing is summarized in Figure 13-39.

events in detail. First a **cap** consisting of a 7-methylguanosine residue linked to the 5′ end of the transcript by a triphosphate bond is added during transcription. Then stretches of adenosine residues are added at the 3′ ends. These **poly(A) tails** are 150 to 200 residues long. Following these modifications, a crucial **splicing** step removes internal portions of the RNA transcript. The uncovering of this process, and the corresponding realization that genes are "split," with coding regions interrupted by "intervening sequences," constitutes one of the most important discoveries in molecular genetics in the last decade.

Split Genes

Studies of mammalian viral DNA transcripts first suggested a lack of correspondence between the viral DNA and spe-

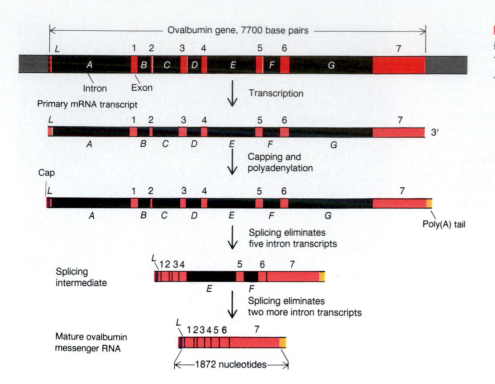

Mechanism of Gene Splicing

There are some sequence homologies at the exon-intron junctures of mRNAs. For instance, there is a GU at the 5′ splice site and an AG at the 3′ splice site of introns in virtually all mRNAs examined. It is thought that splicing enzymes recognize some common configuration of the mRNA and, perhaps with the help of certain **small nuclear ribonucleoprotein** particles, or **snRNPs,** catalyze the cutting and splicing reactions. The snRNPs may help to align the splice sites by hydrogen bonding to the sequences at the exon-intron boundaries (Figure 13-40). In a currently favored model, a lariat-shaped intermediate is spliced to yield the final mRNA. Figure 13-41 portrays the **lariat model.** Here the 5′ splice site is first cleaved (a) in a reaction that creates a loop (b). The newly formed 3′ OH group then cleaves the 3′ splice site (b), releasing the partly looped intron and at the same time joining the two exons (c).

In higher cells, complexes form between the snRNPs, the primary transcript (pre-messenger RNA), and associated factors to form a high-molecular-weight (60S) ribonucleoprotein complex, called a *spliceosome* (Figure 13-42), which catalyzes the splicing transesterification reactions.

Self-Splicing RNA

One exceptional case of RNA splicing occurs in *Tetrahymena,* where Thomas Cech and his co-workers have demon-

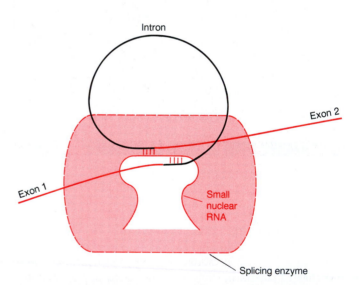

Figure 13-40 Schematic diagram depicting possible participation of small nuclear RNA in the splicing reaction.

strated that the splicing reaction is catalyzed by the RNA molecule itself. Figure 13-43 portrays this reaction. This extraordinary finding is the first demonstration that an RNA molecule can catalyze a specific biological reaction. These RNAs with enzymatic activities have been termed **ribozymes.**

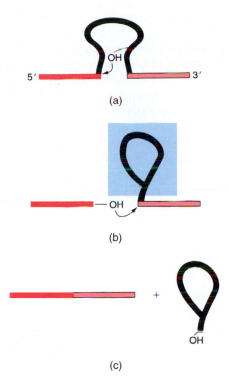

Figure 13-41 Lariat, or loop, formed by a category of introns as they are removed from their RNA molecules. (a) One of the many 2′ hydroxyl groups on the intron attacks the 5′ splice site. (b) Thus the subsequent reaction joins the 5′ end of the intron not to the 3′ end but to the point of the 2-hydroxyl attack a short distance away, yielding a branched structure with a loop: the lariat. (c) The branching is accomplished by the formation of a novel 2′-5′ phosphodiester bond that enables one adenosine nucleotide to form phosphodiester links with three other nucleotides rather than with the usual two. Ligation of the exons frees the lariat from the remainder of the RNA. (From T. Cech, "RNA as an Enzyme." Copyright ©1986 by Scientific American, Inc. All rights reserved.)

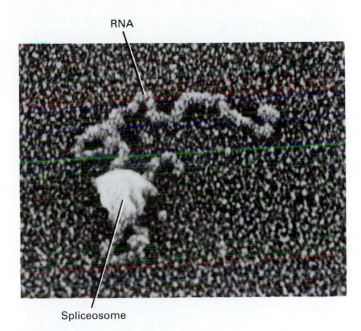

Figure 13-42 Electron micrograph of a spliceosome. (From H. Lodish, D. Baltimore, A. Berk, S. L. Zipursky, P. Matsudaira, and J. Darnell, *Molecular Cell Biology*, 3d ed. Copyright ©1995 by Scientific American Books, Inc.)

Implications of Split Genes

The finding that many eukaryotic genes are interrupted by DNA sequences that are not translated into protein shatters the concept of the gene that we had developed through the end of Chapter 12. Until then we had considered a gene as an uninterrupted sequence of nucleic acid coding for a functional macromolecule (RNA or protein). The gene was co-linear with the protein it encoded. Clearly, this definition of the gene must now be modified, because it no longer holds in all cases. In prokaryotes and, to a large extent, in lower eukaryotes, genes do represent an uninterrupted coding sequence. For many eukaryotic genes, however, the presence of introns interrupts the coding sequence, and the initial RNA transcript must be processed by splicing reactions in order to generate the finished RNA molecule (either mRNA, tRNA, or rRNA).

It is not clear why introns and exons have evolved as such. In some cases, mutations introduced into introns have no noticeable effect on gene expression; in other cases, the occurrence of a mutation in a single intron interferes with expression of its gene. Walter Gilbert has suggested that in many cases, exons encode discrete domains of proteins, and that **exon shuffling** through recombination allows a more rapid evolution of proteins. Figure 13-44 shows all 18 of the exons for the gene encoding the LDL (low-density lipoprotein) receptor protein, which helps transport cholesterol into cells. Note how the functional regions of the protein are grouped in domains which are, in turn, encoded by sets of exons. Many of these domains consist of repeating sequences, each repeat encoded by one exon. The first, fourth, fifth, and sixth domains form a pattern of function found in many other membrane proteins. Furthermore, there is a strong sequence homology between the second domain and the blood protein called *complement factor 9* (*C9*). Likewise, the third domain bears a close resemblance to several other proteins, including epidermal growth factor (EGF) and various blood-clotting factors. These homologies of function and sequence strongly suggest that many such genes are assembled by incorporating, from other loci, exons that have proved successful in performing given functions. Thus, great diversity can be achieved by shuffling around relatively few such sequences.

Protein Processing

Even after mRNA has been successfully translated into its protein product, processing may continue. For example, membrane proteins or proteins that are secreted from the cell are synthesized with a short leader peptide, called a **signal sequence,** at the amino-terminal (N-terminal) end. This is a stretch of 15 to 25 amino acids, most of which are hydrophobic. The signal sequence allows for recognition by factors and protein receptors that mediate transport through the cell membrane; during this process, the signal sequence is cleaved by a peptidase (Figure 13-45). (A similar phenomenon exists for certain bacterial proteins that are secreted.) Moreover, several small peptide hormones, such as corticotropin (ACTH), result from the specific cleavage of a single, large polypeptide precursor.

Protein Splicing

Recently, an extraordinary process involved in splicing out an internal segment of certain proteins has been described in a variety of organisms, including prokaryotes and eukaryotes. This internal segment is termed an **intervening protein sequence,** or **IVPS.** The essential facet of this process is the formation of a new peptide bond between the two sequences flanking the IVPS. This reaction is autocatalytic and can occur in vitro. Figure 13-46 depicts protein splicing in schematic form. Interestingly, all IVPS segments studied so far contain an endonuclease activity, although this activity is unrelated to the protein splicing reaction.

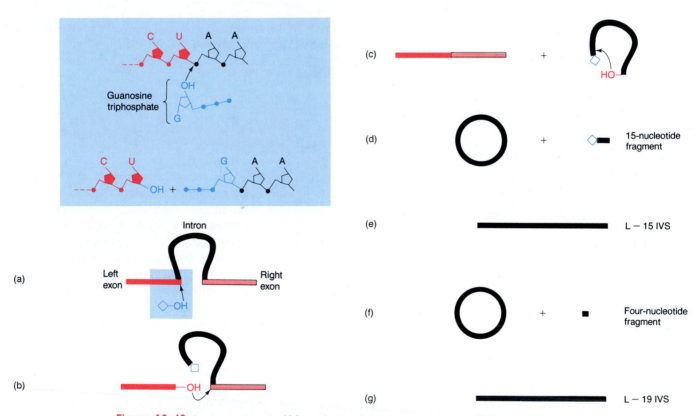

Figure 13-43 Intron removes itself from the *Tetrahymena* rRNA precursor molecule with no assistance from protein enzymes. The cascade of reactions resulting in removal of the intron is unleashed by a free guanosine or guanosine triphosphate molecule (*shown as a diamond*). (a) A hydroxyl group (OH) attached to the guanosine attacks the phosphate at the 5′ end of the intron (*inset*). The phosphodiester bond between the intron and the left exon is broken, and a new bond is formed between the guanosine and the intron. (b) This liberates a new hydroxyl group on the end of the left exon, which begins an attack at the 3′ end of the intron. (c) The bond there is broken and the exons are ligated, or joined, freeing the intron. (d) A similar reaction enables the intron to form a circle, snipping 15 nucleotides off its end in the process. (e) The circle opens into a linear molecule and then closes with the loss of four nucleotides (f). The final, reopened form (g) is known as the L − 19 IVS (linear minus 19 intervening sequence). (From T. Cech, "RNA as an Enzyme." Copyright ©1986 by Scientific American, Inc. All rights reserved.)

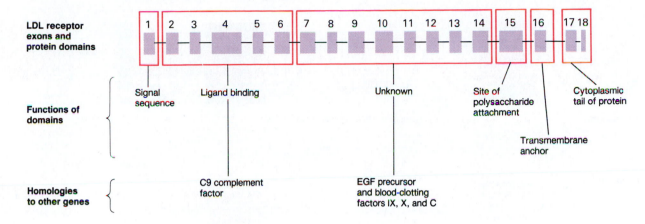

Figure 13-44 The organization of the 18 exons constituting the LDL receptor protein gene into six functional domains, including one signal sequence. (From J. Darnell, H. Lodish, and D. Baltimore, *Molecular Cell Biology*, 2d ed. Copyright ©1990 by Scientific American Books, Inc.)

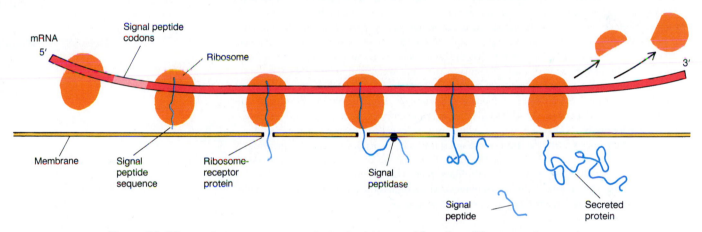

Figure 13-45 Signal sequences. Proteins destined to be secreted from the cell have an amino-terminal sequence that is rich in hydrophobic residues. This signal sequence binds to the membrane and draws the remainder of the protein through the lipid bilayer. The signal sequence is cleaved off the protein during this process by an enzyme called *signal peptidase.* (After J. D. Watson, J. Tooze, and D. T. Kurtz, *Recombinant DNA: A Short Course.* Copyright ©1983 by W. H. Freeman and Company.)

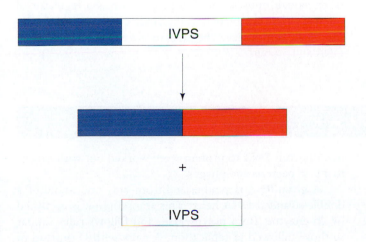

Figure 13-46 Protein splicing results in the removal of an internal segment (IVPS) and the formation of a new peptide bond that links the two regions that originally flanked the IVPS.

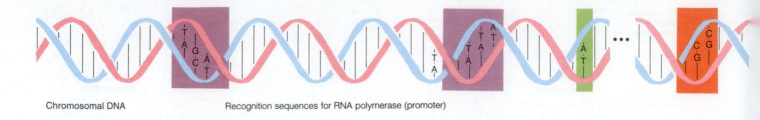

Chromosomal DNA   Recognition sequences for RNA polymerase (promoter)

(a)

Initiation site

Nontemplate strand

TGTTGACA   TATAAT   A   GAGG
ACAACTGT   ATATTA   T   CTCC

Template strand

(b)

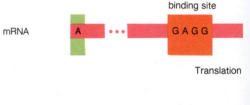

Ribosomal binding site

mRNA   A   GAGG

Translation

Protein chain

(c)

Figure 13-47 Genetic information is stored in the double helix of DNA. (a) Each strand of the helix is a chain of nucleotides, each comprising a deoxyribose sugar and a phosphate group, which form the strand's backbone, as well as one of four bases: adenine (A), guanine (G), thymine (T), or cytosine (C). The information is encoded in the sequence of the bases along a strand. The complementarity of the bases (A always pairs with T and G with C) is the basis of the replication of DNA from generation to generation and of its expression (shown here for bacterial DNA) as protein. Transcription is regulated, in part, by the promoter site upstream from the initiation site. (b) and (c) Expression begins with the transcription of the DNA base sequence from the template strand into a strand of mRNA, which now has the same sequence as the nontemplate strand of the DNA except for the fact that uracil (U) replaces thymine. Transcription into RNA and translation into protein are regulated by special sequences in the DNA and the RNA, respectively. The transcribing enzyme, RNA polymerase, binds to the promoter region before the transcription initiation site; beyond the end of the structural gene, a termination region causes the polymerase to cease transcription. mRNA is translated on cellular organelles called *ribosomes;* each triplet of bases (codon) encodes a particular amino acid and specifies its incorporation into the growing protein chain. A ribosome binding site on the RNA allows translation to begin at a "start" codon, which is always AUG for the amino acid methionine (Met). Translation proceeds until a "stop" codon is reached (UAA is one of three possibilities), which signals the end of translation and the detachment of the completed protein chain from the ribosome.

Review

This chapter and Chapters 11 and 12 have described the development of the central theory of molecular genetics. A linear sequence of nucleotides in DNA is transcribed into a comparable linear sequence of nucleotides in RNA. This RNA sequence, which is processed in several ways in eukaryotes, then serves as a messenger RNA, which is trans-lated into an amino acid sequence in protein by a complex translational apparatus. The protein thus made has importance to the organism either as a structural component (such as hair, muscle, or skin protein) or as a regulator of the body chemistry (such as enzymes or hemoglobin). Figure 13-47 summarizes the structural relationship between DNA and protein.

SUMMARY

We have discovered in earlier chapters that DNA is the genetic material responsible for directing the synthesis of proteins. The first clue as to how DNA accomplishes this feat came from eukaryotes, when it was shown that RNA is synthesized in the nucleus and then transferred to the cytoplasm. However, most of the details of this transfer of infor-mation from DNA to protein were worked out with experiments in bacteria and phages.

A given RNA is synthesized from only one strand of a double-stranded DNA helix. This transcription is catalyzed by an enzyme, RNA polymerase, and follows rules similar to those followed in replication: A pairs with U (instead of

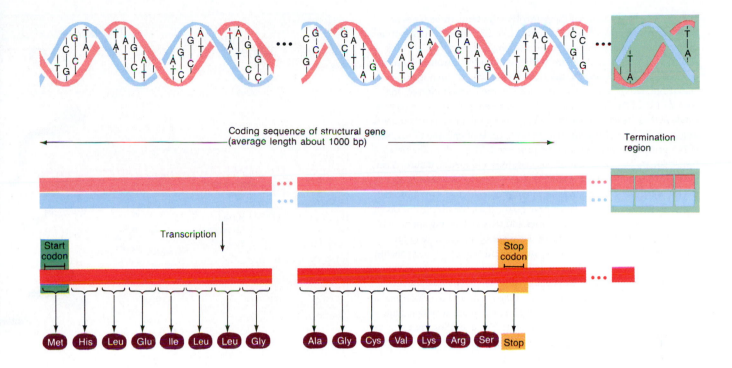

with T, as in DNA), and G pairs with C. Ribose is the sugar used in RNA. Extraction of RNA from a cell yields three basic varieties: ribosomal, transfer, and messenger RNA. The three sizes of ribosomal RNA (rRNA) combine with an array of specific proteins to form ribosomes, which are the machines used for protein synthesis (translation). Transfer RNAs (tRNAs) are a group of rather small RNA molecules, each with specificity for a particular amino acid; they carry the amino acids to the ribosome for attachment to the growing polypeptide.

Messenger RNA (mRNA) molecules are of many sizes and base sequences. These are the molecules that contain information for the structure of proteins. The sequence of codons in mRNA determines the sequence of amino acids that will constitute a polypeptide. Each codon is specific for one amino acid, but several different codons may code for the same amino acid; that is, there is redundancy in the genetic code. In addition, there are three codons for which there are no tRNAs; these stop codons terminate the process of translation. Figure 13-47 summarizes the way in which genetic information is turned into protein.

RNA is processed in eukaryotes before transport to the cytoplasm. Caps and tails are added, and internal portions of the primary transcript are removed. Many genes are therefore "split" in eukaryotes, and the coding segments of a gene are not colinear with the processed mRNA. Table 13-8 and Figure 13-48 summarize some of the differences between RNA synthesis in prokaryotes and eukaryotes.

The processes of information storage, replication, transcription, and translation are fundamentally similar in all living organisms. In demonstrating this similarity, molecular genetics has provided a powerful unifying force in biology.

Table 13-8   **Differences in Gene Expression between Prokaryotes and Eukaryotes**

Prokaryotes	Eukaryotes
1. All RNA species are synthesized by a single RNA polymerase.	1. Three different RNA polymerases are responsible for the different classes of RNA molecules.
2. mRNA is translated during transcription.	2. mRNA is processed before transport to the cytoplasm, where it is translated. Caps and tails are added, and internal portions of the transcript are removed.
3. Genes are contiguous segments of DNA that are colinear with the mRNA that is translated into a protein.	3. Genes are often split. They are not contiguous segments of coding sequences; rather, the coding sequences are interrupted by intervening sequences (introns).
4. mRNAs are often polycistronic.	4. mRNAs are monocistronic.

Figure 13-48 Protein synthesis involves the same three types of RNA in prokaryotic (a) and eukaryotic (b) cells, but with an important difference: in eukaryotes, the protein-coding sequences of DNA (exons) are often separated by intervening sequences (introns) that must be excised from a primary transcript to make mRNA. In both kinds of cell, tRNA, rRNA, and mRNA are made by the transcription of one strand of the DNA double helix. Three different polymerase enzymes catalyze these reactions in eukaryotes, whereas in prokaryotes there is only a single type of polymerase. In both kinds of cell, the tRNA and rRNA primary transcripts must be processed. The ends of the tRNA transcript are cut, and the molecule assumes a looped structure. A single rRNA transcript is cut in several places to form two major types of rRNA, which are then bound to protein molecules to form ribosomal subunits. In prokaryotes, which have no nucleus, the mRNA transcript is generally not processed; ribosomes and tRNAs carrying amino acids begin translating the mRNA into a sequence of amino acids (a protein) as it is being made. In eukaryotes, the nuclear envelope probably facilitates the removal of introns and the splicing of exons from the primary mRNA transcript by protecting it from immediate translation. The mRNA is "read" only after it (like tRNA and the ribosomal subunits) has exited the nucleus through pores in the envelope. Only in the cytoplasm are the mature mRNA, tRNA, and ribosomes united. (From J. E. Darnell, Jr., "RNA." Copyright ©1985 by Scientific American, Inc. All rights reserved.)

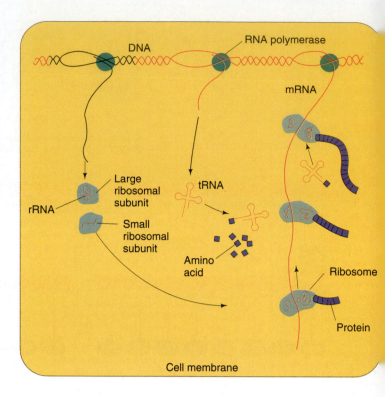

(a) Prokaryote

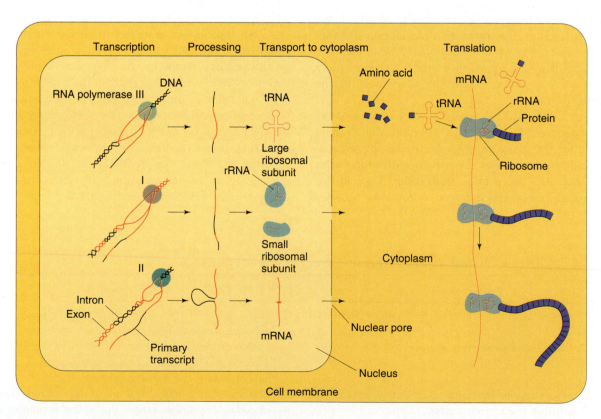

(b) Eukaryote

Concept Map

Draw a concept map interrelating as many of the following terms as possible. Note that the terms are listed in no particular order.

gene / mRNA / tRNA / transcription / splicing / translation / RNA polymerase / ribosome / codon / anticodon / amino acid / intron

CHAPTER INTEGRATION PROBLEM

In Chapter 12 we considered different aspects of protein structure. Refer to Figure 12-8 and explain the effects individually on protein activity of nonsense, missense, and frameshift mutations in a gene encoding a protein.

Solution

Nonsense mutations result in termination of protein synthesis. Only a fragment of the protein is completed. As can be seen from Figure 12-8, fragments of proteins will not be able to adopt the correct configuration to form the active site. Therefore, the protein fragment will be inactive. Missense mutations, on the other hand, result in the exchange of one amino acid for another. If the amino acid is important for the correct folding of the protein, or is part of the active site, or is involved in subunit interaction, then changing that amino acid will often lead to an inactive protein. On the other hand, if the amino acid is on the outside of the protein and not involved in any of these activities or functions, then many such substitutions are acceptable and will not affect activity. Frameshift mutations change the reading frame and result in incorporation of amino acids that are different from those encoded in the wild-type reading frame. Unless the frameshift mutation is at the end of the protein, the resulting altered protein will not be able to fold into the correct configuration and will not have activity.

SOLVED PROBLEMS

1. Using Figure 13-18, show the consequences on subsequent translation of the addition of an adenine base to the beginning of the following coding sequence:

Ⓐ
↓
–CGA–UCG–GAA–CCA–CGU–GAU–AAG–CAU–
– Arg – Ser – Glu – Pro – Arg – Asp – Lys – His –

Solution

With the addition of A at the beginning of the coding sequence, the reading frame shifts, and a different set of amino acids is specified by the sequence, as shown here (note that a set of nonsense codons is encountered, which results in chain termination):

–ACG–AUC–GGA–ACC–ACG–UGA–UAA–GCA–
– Thr – Ile – Gly – Thr – Thr – stop – stop –

2. A single nucleotide addition followed by a single nucleotide deletion approximately 20 bp apart in the DNA causes a change in the protein sequence from

– His – Thr – Glu – Asp – Trp – Leu – His – Gln – Asp –

to

– His – Asp – Arg – Gly – Leu – Ala – Thr – Ser – Asp –

Which nucleotide has been added and which nucleotide has been deleted? What are the original and the new mRNA sequences? (HINT: Consult Figure 13-18.)

Solution

We can draw the mRNA sequence for the original protein sequence (with the inherent ambiguities at this stage):

– His – Thr – Glu – Asp – Trp – Leu – His – Gln – Asp

$$-CA_C^U-ACC-GA_A^A-GA_C^U-UGG-CUC-CA_C^U-CA_G^A-GA_C^U$$
$$\quad\quad\quad A \quad\quad\quad\quad\quad\quad\quad\quad\quad A$$
$$\quad\quad\quad G \quad\quad\quad\quad\quad\quad\quad\quad\quad G$$
$$\quad\quad\quad\quad\quad\quad\quad\quad\quad\quad\quad\quad\quad UUA$$
$$\quad\quad\quad\quad\quad\quad\quad\quad\quad\quad\quad\quad\quad G$$

Because the protein sequence change given to us at the beginning of the problem begins after the first amino acid

(His) owing to a single nucleotide addition, we can deduce that a Thr codon must change to an Asp codon. This change must result from the addition of a G directly before the Thr codon (indicated by a box), which shifts the reading frame, as shown here:

$$-CA{}^{U}_{C}-\boxed{G}AC-UGA-{}^{A}_{G}GA-\overset{}{U}UG-G\overset{}{C}U-UCA-\overset{}{U}CA\overset{\uparrow}{\,}-GA{}^{U}_{C}-$$

– His – Asp – Arg – Gly – Leu – Ala – Thr – Ser – Asp –

Also, because a deletion of a nucleotide must restore the final Asp codon to the correct reading frame, an A or G must have been deleted from the end of the original second-last codon, as shown by the arrow. The original protein sequence permits us to draw the mRNA with a number of ambiguities. However, the protein sequence resulting from the frameshift allows us to determine which nucleotide was in the original mRNA at most of these points of ambiguity. The nucleotide that must have appeared in the original sequence is circled. In only a few cases does the ambiguity remain.

PROBLEMS

1. The two strands of λ phage differ from each other in their GC content. Owing to this property, they can be separated in an alkaline cesium chloride gradient (the alkalinity denatures the double helix). When RNA synthesized by λ phage is isolated from infected cells, it is found to form DNA-RNA hybrids with both strands of λ DNA. What does this tell you? Formulate some testable predictions.

2. The data in Table 13-9 represent the base compositions of two double-stranded DNA sources and their RNA products in experiments conducted in vitro.

   a. From these data, can you determine whether the RNA of these species is copied from a single strand or from both strands of the DNA? How? Drawing a diagram will make it easier to solve this problem.

   b. Explain how you can tell whether the RNA itself is single-stranded or double-stranded.

(Problem 2 is reprinted with permission of Macmillan Publishing Co., Inc., from M. Strickberger, *Genetics*. Copyright ©1968, Monroe W. Strickberger.)

3. Before the true nature of the genetic coding process was fully understood, it was proposed that the message might be read in overlapping triplets. For example, the sequence GCAUC might be read as GCA CAU AUC:

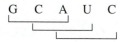

Devise an experimental test of this idea.

4. In protein-synthesizing systems in vitro, the addition of a specific human mRNA to *E. coli* translational apparatus (ribosomes, tRNA, and so forth) stimulates the synthesis of a protein very much like that specified by the mRNA. What does this result show?

5. Which anticodon would you predict for a tRNA species carrying isoleucine? Is there more than one possible answer? If so, state any alternative answers.

6. **a.** In how many cases in the genetic code would you *fail* to know the amino acid specified by a codon if you know only the first two nucleotides of the codon?

   b. In how many cases would you fail to know the first two nucleotides of the codon if you know which amino acid is specified by it?

7. Deduce what the six wild-type codons may have been in the mutants that led Brenner to infer the nature of the amber codon UAG.

8. If a polyribonucleotide contains equal amounts of randomly positioned adenine and uracil bases, what proportion of its triplets would code for (**a**) phenylalanine, (**b**) isoleucine, (**c**) leucine, (**d**) tyrosine?

9. You have synthesized three different messenger RNAs with bases incorporated in random sequence in the following ratios: (**a**) 1U : 5C, (**b**) 1A : 1C : 4U, (**c**) 1A : 1C : 1G : 1U. In a protein-synthesizing system in vitro, indicate the identities

Table 13-9

Species	(A + T)/(G + C)	(A + U)/(G + C)	(A + G)/(U + C)
Bacillus subtilis	1.36	1.30	1.02
E. coli	1.00	0.98	0.80

and proportions of amino acids that will be incorporated into proteins when each of these mRNAs is tested. (Refer to Figure 13-18.)

* **10.** One of the techniques used to decipher the genetic code was to synthesize polypeptides in vitro, using synthetic mRNA with various repeating base sequences—for example, $(AGA)_n$, which could be written out as AGAAGAAGA-AGAAGA. . . . Sometimes the synthesized polypeptide contained just one amino acid (a homopolymer), and sometimes it contained more than one (a heteropolymer), depending on the repeating sequence used. Furthermore, sometimes different polypeptides were made from the same synthetic mRNA, suggesting that the initiation of protein synthesis in the system in vitro does not always start on the end nucleotide of the messenger. For example, from $(AGA)_n$, three polypeptides may have been made: aa_1 homopolymer (abbreviated aa_1-aa_1), aa_2 homopolymer (aa_2-aa_2), and aa_3 homopolymer (aa_3-aa_3). These probably correspond to the following readings derived by starting at different places in the sequence:

AGA AGA AGA AGA . . .

GAA GAA GAA GAA . . .

AAG AAG AAG AAG . . .

Table 13-10 shows the actual results obtained from the experiment done by Khorana.

a. Why do $(GUA)_n$ and $(GAU)_n$ each code for only two homopolypeptides?

b. Why do $(GAUA)_n$ and $(GUAA)_n$ fail to stimulate synthesis?

c. Assign an amino acid to each triplet in the following list. Bear in mind that there often are several codons for a single amino acid and that the first two letters in a codon usually are the important ones (but that the third letter is occasionally significant). Also remember that some very different-looking codons sometimes code for the same amino acid. Try to carry out this task without consulting Figure 13-18.

AUG	GAU	UUG	AAC
GUG	UUC	UUA	CAA
GUU	CUC	AUC	AGA
GUA	CUU	UAU	GAG
UGU	CUA	UAC	GAA
CAC	UCU	ACU	UAG
ACA	AGU	AAG	UGA

To solve this problem requires both logic and trial and error. Don't be disheartened: Khorana received a Nobel prize for doing it. Good luck!

(Problem 10 is from J. Kuspira and G. W. Walker, *Genetics: Questions and Problems.* McGraw-Hill, 1973.)

11. You are studying a gene in *E. coli* that specifies a protein. A part of its sequence is

−Ala−Pro−Trp−Ser−Glu−Lys−Cys−His−

You recover a series of mutants for this gene that show no enzymatic activity. Isolating the mutant enzyme products, you find the following sequences:

Mutant 1
−Ala−Pro−Trp−Arg−Glu−Lys−Cys−His−

Mutant 2
−Ala−Pro−

Mutant 3
−Ala−Pro−Gly−Val−Lys−Asn−Cys−His−

Mutant 4
−Ala−Pro−Trp−Phe−Phe−Thr−Cys−His−

What is the molecular basis for each mutation? What is the DNA sequence that specifies this part of the protein?

12. A single nucleotide addition and a single nucleotide deletion approximately 15 sites apart in the DNA cause a protein change in sequence from

−Lys−Ser−Pro−Ser−Leu−Asn−Ala−Ala−Lys−

to

−Lys−Val−His−His−Leu−Met−Ala−Ala−Lys−

Table 13-10

Synthetic mRNA	Polypeptide(s) synthesized
$(UG)_n$	(Ser-Leu)
$(UG)_n$	(Val-Cys)
$(AC)_n$	(Thr-His)
$(AG)_n$	(Arg-Glu)
$(UUC)_n$	(Ser-Ser) and (Leu-Leu) and (Phe-Phe)
$(UUG)_n$	(Leu-Leu) and (Val-Val) and (Cys-Cys)
$(AAG)_n$	(Arg-Arg) and (Lys-Lys) and (Glu-Glu)
$(CAA)_n$	(Thr-Thr) and (Asn-Asn) and (Gln-Gln)
$(UAC)_n$	(Thr-Thr) and (Leu-Leu) and (Tyr-Tyr)
$(AUC)_n$	(Ile-Ile) and (Ser-Ser) and (His-His)
$(GUA)_n$	(Ser-Ser) and (Val-Val)
$(GAU)_n$	(Asp-Asp) and (Met-Met)
$(UAUC)_n$	(Tyr-Leu-Ser-Ile)
$(UUAC)_n$	(Leu-Leu-Thr-Tyr)
$(GAUA)_n$	None
$(GUAA)_n$	None

NOTE: The order in which the polypeptides or amino acids are listed is not significant except for $(UAUC)_n$ and $(UUAC)_n$.

a. What are the old and the new mRNA nucleotide sequences? (Use Figure 13-18.)

b. Which nucleotide has been added and which has been deleted?

(Problem 12 is from W. D. Stansfield, *Theory and Problems of Genetics*. McGraw-Hill, 1969.)

13. Suppressors of frameshift mutations are now known. Propose a mechanism for their action.

14. Use Figure 13-18 to complete the following table. Assume that reading is from left to right and that the columns represent transcriptional and translational alignments.

C										DNA double helix
			T	G	A					
	C	A		U						mRNA transcribed
					G	C	A			Appropriate tRNA anticodon
		Trp								Amino acids incorporated into protein

15. A mutational event inserts an extra nucleotide pair into DNA. Which of the following do you expect?

a. No protein at all

b. A protein in which one amino acid is changed

c. A protein in which three amino acids are changed

d. A protein in which two amino acids are changed

e. A protein in which most amino acids following the site of the insertion are changed

16. Consider the gene that specifies the structure of hemoglobin. Arrange the following events in the most likely sequence in which they would occur:

a. Anemia is observed.

b. The shape of the oxygen-binding site is altered.

c. An incorrect codon is transcribed into hemoglobin mRNA.

d. The ovum (female gamete) receives a high radiation dose.

e. An incorrect codon is generated in the DNA of the hemoglobin gene.

f. A mother (an X-ray technician) accidentally steps in front of an operating X-ray generator.

g. A child dies.

h. The oxygen-transport capacity of the body is severely impaired.

i. The tRNA anticodon that lines up is one of a type that brings an unsuitable amino acid.

j. Nucleotide-pair substitution occurs in the DNA of the gene for hemoglobin.

17. An induced cell mutant is isolated from a hamster tissue culture because of its resistance to α-amanitin (a poison derived from a fungus). Electrophoresis shows that the mutant has an altered RNA polymerase; *just one* electrophoretic band is in a position different from that of the wild-type polymerase. The cells are presumed to be diploid. What does this experiment tell you about ways in which to detect recessive mutants in such cells?

18. A double-stranded DNA molecule with the sequence shown below produces, in vivo, a polypeptide that is five amino acids long.

TAC ATG ATC ATT TCA CGG AAT TTC TAG CAT GTA
ATG TAC TAG TAA AGT GCC TTA AAG ATC GTA CAT

a. Which strand of DNA is transcribed, and in which direction?

b. Label the 5′ and the 3′ ends of each strand.

c. If an inversion occurs between the second and third triplets from the left and right ends, respectively, and the same strand of DNA is transcribed, how long will the resultant polypeptide be?

d. Assume that the original molecule is intact and that transcription occurs on the bottom strand from left to right. Give the base sequence, and label the 5′ and 3′ ends of the anticodon that inserts the *fourth* amino acid into the nascent polypeptide. What is this amino acid?

14

Recombinant DNA Technology

KEY CONCEPTS

▶ Recombinant DNA is made by splicing a foreign DNA fragment into a small replicating molecule (such as a plasmid), which will then amplify the fragment along with itself resulting in a molecular "clone" of the inserted DNA fragment.

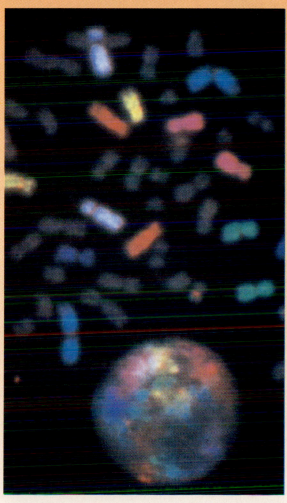

▶ A collection of DNA clones that encompasses the entire genome is called a *genomic library*.

▶ An individual DNA clone in a library can be detected by using a specific probe for the DNA or its protein product, or by its ability to transform a null mutant.

▶ Restriction enzymes cut DNA at specific target sites, resulting in defined fragments with sticky ends suitable for cloning.

▶ Different-sized DNA fragments produced by restriction enzyme digestion can be fractionated because they migrate to different positions on an electrophoretic gel.

▶ DNA or RNA fractionated by size in an electrophoretic gel can be blotted onto a porous membrane, which can be probed to detect the position of specific fragments.

▶ Restriction enzyme target sites can be mapped, providing useful landmarks for DNA manipulation.

▶ A gene can be found by testing overlapping clones radiating outwards from a linked marker.

▶ A cloned gene can be sequenced and used to investigate its cellular function.

▶ With a pair of primers spanning a short DNA sequence, copies of the sequence can be amplified for study.

Chromosome preparation in which several homologous pairs have been painted with chromosome-specific fluorescent probes. An unbroken nucleus also shows specific labeling revealing chromosome positions. The photograph symbolizes the new ways of visualizing the genetic material made possible by using new DNA technologies. (Vysis)

T he goal of genetics is to study the structure and function of genes and genomes. Since Mendel's time, genes have been identified by observing standard phenotypic ratios in controlled crosses. Clues about gene function first came from correlating specific mutations with certain enzyme and other protein deficiencies. Intragenic mapping of mutant sites and comparison of these with amino acid substitutions in the appropriate protein led to a better understanding of how genes work. To these ideas were added discoveries about the nature of DNA and the genetic code, and this led to a fairly comprehensive understanding of basic gene structure and function. However, these were *indirect* inferences about genes; no gene had ever been isolated and its DNA sequence examined directly. Indeed it seemed impossible to isolate an individual gene from the genome. Although it is relatively easy to isolate DNA from living tissue, in the test tube DNA looks like a glob of mucus. How could it be possible to isolate a single gene from this tangled mass of DNA threads? Recombinant DNA technology provides the techniques for doing just that, and today individual genes and other parts of genomes are isolated routinely. Furthermore, using the same technology, geneticists can obtain the nucleotide sequence of an isolated piece of DNA, change it in any specific way the researcher wishes, and reinsert it back into the living cells from which it came—or even into cells of a different organism. Thus, geneticists are able to create novel genotypes that would be impossible using classical genetic techniques. This technology has led not only to a more sophisticated and incisive level of basic genetic analysis, but also to many applications in the areas of medicine, agriculture, and industry.

Making Recombinant DNA

How does recombinant DNA technology work? The organism under study, which will be used to donate DNA for the analysis, is called the **donor organism.** The basic procedure is to extract and cut up DNA from a donor genome into fragments containing one to several genes and allow these fragments to insert themselves individually into opened-up small autonomously replicating DNA molecules such as bacterial plasmids. These small molecules act as carriers, or **vectors,** for the DNA fragments. The vector molecules with their inserts are called **recombinant DNA** because they represent novel combinations of DNA from the donor genome (which can be from any organism) with vector DNA from a completely different source (generally a bacterial plasmid or a virus). The recombinant DNA mixture is then used to transform bacterial cells, and it is common for single recombinant vector molecules to find their way into individual bacterial cells. Bacterial cells are plated and allowed to grow into colonies. An individual transformed cell with a single recombinant vector will divide into a colony with millions of cells, all carrying the same recombinant vector. Therefore an individual colony represents a very large population of identical DNA inserts, and this population is called a **DNA clone.**

Message   Cloning allows the amplification and recovery of a specific DNA segment from large, complex DNA sample such as a genome.

Since the donor DNA was cut into many different fragments, most colonies will carry a different cloned insert. Therefore, the next step is to find a way to select the clone with the insert containing the gene that you are interested in. Once this clone is obtained, the DNA is isolated in bulk and the desired gene is cut out from the vector; in this way a purified and uniform sample of the gene is obtained. Notice that the method works because individual recombinant DNA molecules enter individual bacterial host cells, and then these cells do the job of amplifying the single molecules into large populations of molecules that can be treated like chemical reagents. Figure 14-1 gives a general outline of the approach.

The term *recombinant DNA* must be distinguished from the natural DNA recombinants that result from crossing-over between homologous chromosomes in eukaryotes and prokaryotes. Recombinant DNA in the sense being used in this chapter is an unnatural union of DNAs from nonhomologous sources, usually from different organisms. Some geneticists suggest the alternative name **chimeric DNA,** after the mythological Greek monster Chimera. Down through the ages the Chimera has stood as the symbol of an impossible biological union, a combination of parts of different animals. Likewise, recombinant DNA is a DNA chimera, and would be impossible without the experimental manipulation we call *recombinant DNA technology.*

Isolating DNA

The first step of making recombinant DNA is to isolate donor and vector DNA. General protocols for DNA isolation were available many decades before the advent of recombinant DNA technology. Using such methods, the bulk of DNA extracted from the donor will be genomic DNA, and this generally is the type required for analysis. The procedure used for obtaining vector DNA depends on the nature of the vector. Bacterial plasmids are commonly used vectors, and these must be purified away from the bacterial genomic DNA. A protocol for extracting plasmid DNA by ultracentrifugation is summarized in Figure 14-2. Plasmid DNA forms a distinct band after ultracentrifugation in a cesium chloride density gradient containing ethidium bromide. The plasmid band is collected by punching a hole in the plastic centrifuge tube. Another protocol relies on the observation that at a specific alkaline pH bacterial genomic DNA denatures but plasmids do not. Subsequent neutralization precipitates the genomic DNA, but plasmids stay in solution.

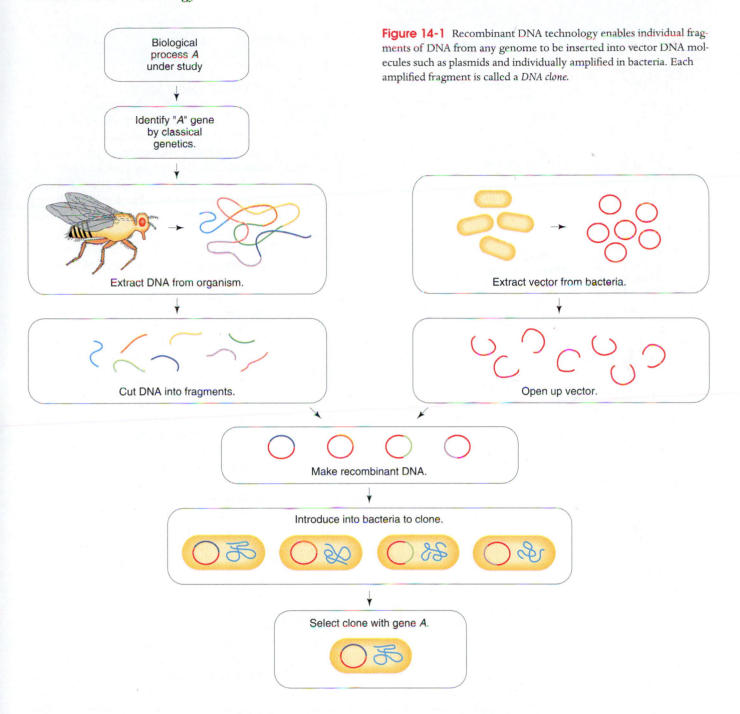

Figure 14-1 Recombinant DNA technology enables individual fragments of DNA from any genome to be inserted into vector DNA molecules such as plasmids and individually amplified in bacteria. Each amplified fragment is called a *DNA clone*.

Cutting DNA

The breakthrough that made recombinant DNA technology possible was the discovery and characterization of **restriction enzymes.** Restriction enzymes are produced by bacteria as a defense mechanism against phages. The enzymes act like scissors, cutting up the DNA of the phage and thereby inactivating it. However, restriction enzymes do not cut randomly, but at specific target sequences, and this is one of the key features that make them suitable for DNA manipulation. Any DNA molecule, from viruses to humans, will contain restriction enzyme target sites purely by chance and

therefore may be cut into defined fragments of size suitable for cloning. Restriction sites are not relevant to the function of the organism, nor would they be cut in vivo because most organisms do not have restriction enzymes. For example, the restriction enzyme *Eco*RI (from *E. coli*) recognizes the following six-nucleotide-pair sequence in the DNA of any organism:

$$5'\text{--GAATTC--}3'$$

$$3'\text{--CTTAAG--}5'$$

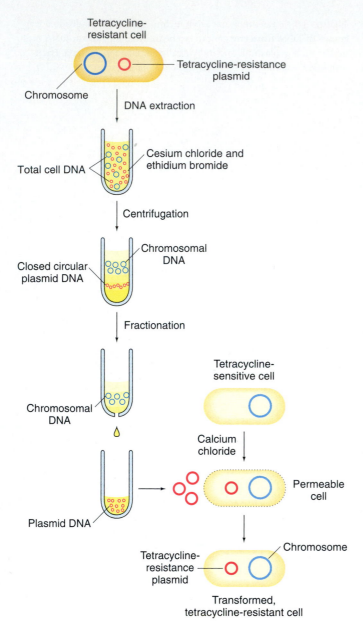

Figure 14-2 Plasmids such as those carrying genes for resistance to the antibiotic tetracycline (*top left*) can be separated from the bacterial chromosomal DNA. Because differential binding of ethidium bromide by the two DNA species makes the circular plasmid DNA denser than the chromosomal DNA, the plasmids form a distinct band on centrifugation in a cesium chloride gradient and can be separated (*bottom left*). They can then be introduced into bacterial cells by transformation (*right*). (Modified from S. N. Cohen, "The Manipulation of Genes." Copyright © 1975 by Scientific American, Inc. All rights reserved.)

$$5'\text{--}G\text{AATTC--}3' \qquad 5'\text{--}G \qquad AATTC\text{--}3'$$
$$3'\text{--}CTTAA\text{G--}5' \longrightarrow 3'\text{--}CTTAA \qquad G\text{--}5'$$

This staggered cut leaves a pair of identical single-stranded "sticky ends." The ends are called *sticky* because they can hydrogen-bond (stick) to a complementary sequence. Figure 14-3 shows *Eco*RI making a single cut in the circular DNA chromosome of the virus SV40: the cut opens up the circle, and the linear molecule formed has two sticky ends. Production of these sticky ends is the second feature of restriction enzymes that makes them suitable for recombinant DNA technology. The principle is simply that if two different DNA molecules are cut with the same restriction enzyme, both will produce fragments with the same complementary sticky ends, making it possible for DNA chimeras to form.

Message   Restriction enzymes have two properties useful in recombinant DNA technology. First they cut DNA into fragments of a size suitable for cloning. Second, many restriction enzymes make staggered cuts generating single-stranded sticky ends conducive to the formation of recombinant DNA.

Dozens of restriction enzymes with different sequence specificities have now been identified, some of which are shown in Table 14-1. You will notice that all the target sequences are palindromes, but some enzymes make staggered cuts like *Eco*RI, whereas others make flush cuts. Even flush cuts, which lack sticky ends, can be altered to render them suitable for making recombinant DNA, as we shall see later.

DNA also can be cut by mechanical shearing. For example, forcing a DNA solution through a hypodermic needle will break up the long chromosome-sized molecules into clonable segments.

Joining DNA

Donor DNA (sometimes called **foreign DNA**) and vector DNA are digested with restriction enzyme and mixed in a test tube in order to allow the ends to join to each other and form recombinant DNA. There are several ways of joining the donor DNA to the vector to create a recombinant DNA molecule. We have seen that many restriction enzymes

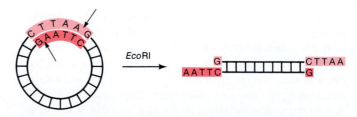

Figure 14-3   The restriction enzyme *Eco*RI is shown cutting a circular DNA molecule bearing one target sequence, resulting in a linear molecule with single-stranded sticky ends.

This type of segment is called a DNA **palindrome,** which means that both strands have the same nucleotide sequence but in antiparallel orientation. The enzyme cuts within this sequence, but in a pair of staggered cuts between the G and the A nucleotides

Table 14-1   Recognition, Cleavage, and Modification Sites of Various Restriction Enzymes

Enzyme	Source organism	Restriction site	Number of cleavage sites in DNA from		
			ϕX174	λ	SV40
EcoRI	Escherichia coli	5′ –G–A–A–T–T–C– / –C–T–T–A–A–G– 5′	0	5	1
EcoRII	E. coli	5′ –G–C–C–T–G–G–C– / –C–G–G–A–C–C–G– 5′	2	>35	16
HindII	Haemophilus influenzae	5′ –G–T–Py–Pu–A–C– / –C–A–Pu–Py–T–G– 5′	13	34	7
HindIII	H. influenzae	5′ –A–A–G–C–T–T– / –T–T–C–G–A–A– 5′	0	6	6
HaeIII	H. aegyptius	5′ –G–G–C–C– / –C–C–G–G– 5′	11	>50	19
HpaII	H. parainfluenzae	5′ –C–C–G–G– / –G–G–C–C– 5′	5	>50	1
PstI	Providencia stuartii	5′ –C–T–G–C–A–G– / –G–A–C–G–T–C– 5′	1	18	2
SmaI	Serratia marcescens	5′ –C–C–C–G–G–G– / –G–G–G–C–C–C– 5′	0	3	0
BamI	Bacillus amyloliquefaciens	5′ –G–G–A–T–C–C– / –C–C–T–A–G–G– 5′	0	5	1
BglII	B. globiggi	5′ –Λ–G–A–T–C–T– / –T–C–T–A–G–A– 5′	0	5	0

NOTE: An asterisk (*) is commonly used to indicate methylation sites, but an "m" is used here to avoid confusion with radioactive labeling.

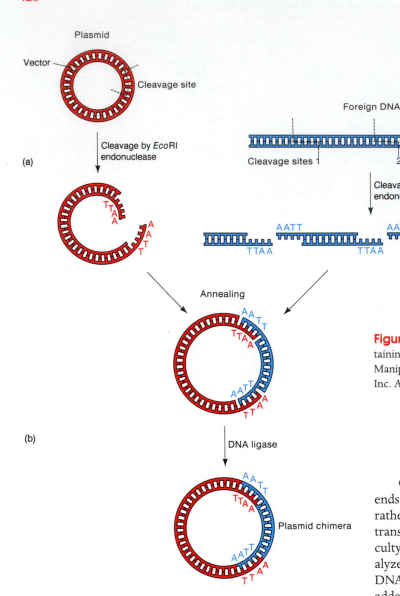

Figure 14-4 Method for generating a chimeric DNA plasmid containing genes derived from foreign DNA. (From S. N. Cohen, "The Manipulation of Genes." Copyright © 1975 by Scientific American, Inc. All rights reserved.)

cleave DNA at a specific sequence and make single-stranded sticky tails. Such strands in the donor DNA then anneal to sticky ends in the vector, which has been cleaved by the same restriction enzyme. An example is shown in Figure 14-4a. In this example, the vector is a plasmid that carries one *Eco*RI restriction site, so digestion with the restriction enzyme *Eco*RI converts the circular DNA to a linear molecule with single-stranded sticky ends. Donor DNA from any other source (say, *Drosophila*) can be treated with *Eco*RI enzyme to produce a population of fragments carrying the same sticky ends. When the two populations are mixed, DNA fragments from the two sources can unite, because duplexes form between their sticky ends. At this stage, although duplexes have formed to generate a population of chimeric molecules, the sugar-phosphate backbones are still not complete at two positions at each junction. However, the fragments can be linked permanently by the addition of the enzyme **DNA ligase,** which creates phosphodiester bonds at the joined ends to make a continuous DNA molecule (Figure 14-4b).

One of the problems of the free availability of sticky ends in solution is that the cut ends of a molecule can rejoin rather than form recombinant DNA. The enzyme terminal transferase from calf thymus offers a way around this difficulty, as illustrated in Figure 14-5. Terminal transferase catalyzes the addition of nucleotide "tails" to the 3′ ends of DNA chains. Thus, if dA (deoxyadenine) molecules are added to, say, vector DNA fragments and dT molecules are added to the donor DNA fragments, only chimeras can form. Any single-stranded gaps (see the figure) created by restriction cleavage are filled by DNA polymerase I and the joints subsequently sealed by DNA ligase. Alternatively, poly(G) and poly(C) tails can be added to the respective fragments at the 3′ ends. Note that in this method it is not the ends produced by restriction cleavage that anneal but rather the artificial sticky ends created by the nucleotide tails. Therefore, in this case, the original restriction ends of the donor DNA can differ from those of the vector.

Amplifying Recombinant DNA

Recombinant plasmid DNA is introduced into host cells by transformation. Once in the host cell, the vector will replicate in the normal way, but now that the donor DNA insert is part of its length, the donor DNA is automatically replicated along with the vector. Each recombinant plasmid that enters a cell will form multiple copies of itself in that cell. Subsequently many cycles of cell division will occur, and the recombinant vectors will undergo more rounds of replica-

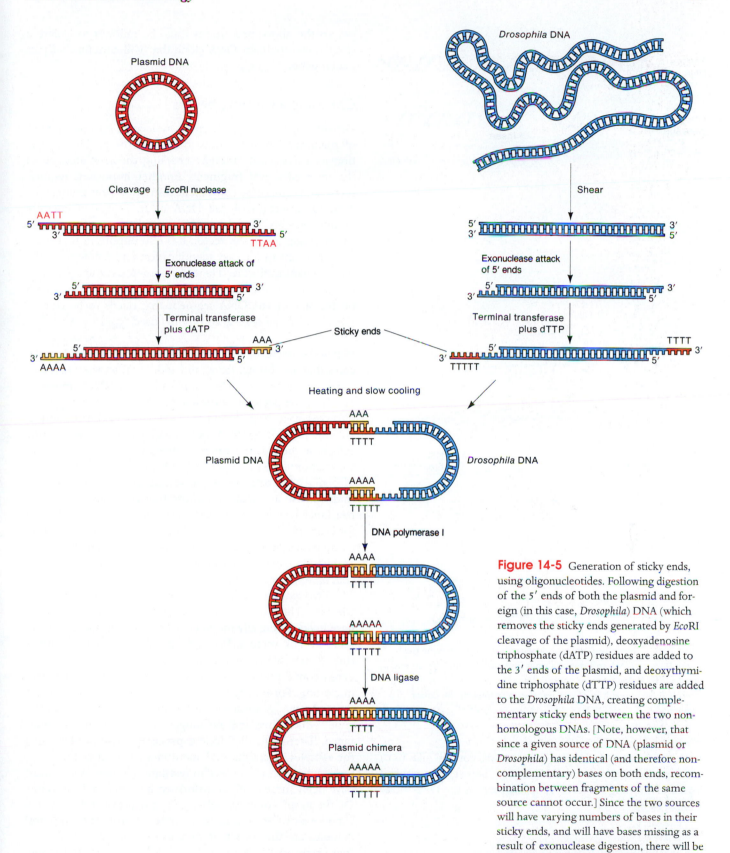

Figure 14-5 Generation of sticky ends, using oligonucleotides. Following digestion of the 5′ ends of both the plasmid and foreign (in this case, *Drosophila*) DNA (which removes the sticky ends generated by *Eco*RI cleavage of the plasmid), deoxyadenosine triphosphate (dATP) residues are added to the 3′ ends of the plasmid, and deoxythymidine triphosphate (dTTP) residues are added to the *Drosophila* DNA, creating complementary sticky ends between the two nonhomologous DNAs. [Note, however, that since a given source of DNA (plasmid or *Drosophila*) has identical (and therefore noncomplementary) bases on both ends, recombination between fragments of the same source cannot occur.] Since the two sources will have varying numbers of bases in their sticky ends, and will have bases missing as a result of exonuclease digestion, there will be gaps in the annealing duplex. These are filled by DNA polymerase I and DNA ligase, to form the completed plasmid chimera. (From S. N. Cohen, "The Manipulation of Genes." Copyright © 1975 by Scientific American, Inc. All rights reserved.)

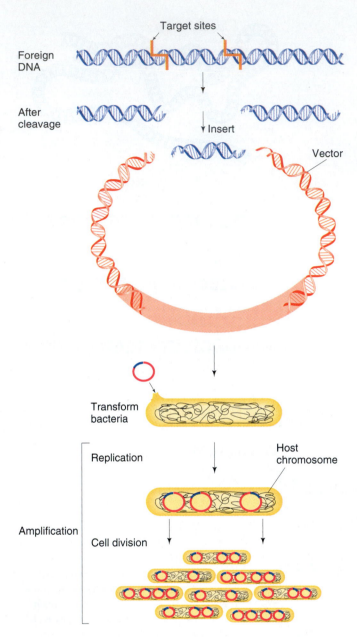

Figure 14-6 How amplification works. Restriction enzyme treatment of donor DNA and vector allows insertion of single fragments into vectors. A single vector enters a bacterial host, where replication and cell division result in a large number of copies of the donor fragment.

tion. The resulting colony of bacteria will contain billions of copies of the single donor DNA insert. This set of amplified copies of the single donor DNA fragment is the DNA clone (Figure 14-6).

Cloning a Specific Gene

The descriptions above are of generic approaches to creating recombinant DNA. However, a geneticist is interested in isolating and characterizing some particular gene of inter-

est, so the above procedures must be tailored to isolate a specific recombinant DNA clone that will contain that particular gene.

Choosing a Cloning Vector

Vectors must be relatively small molecules for convenience of manipulation. They must be capable of prolific replication in a living cell, thereby enabling the amplification of the inserted donor fragment. Another important requirement is that there be convenient restriction sites that can be used for insertion of the DNA to be cloned. Generally, unique sites are useful because then the insert can be targeted to one site in the vector. It is also important that there be a mechanism for easy identification and recovery of the recombinant molecule. There are numerous cloning vectors in current use, and the choice between them often depends on the size of the DNA segment that needs to be cloned. We will consider several commonly used types.

Plasmids. Bacterial plasmids are small circular DNA molecules that are distinct from, and additional to, the main bacterial chromosome. They replicate their DNA independently of the bacterial chromosome. Many different types of plasmids have been found in bacteria. The distribution of any one plasmid within a species is generally sporadic; some cells have the plasmid whereas others do not. In Chapter 10 we encountered the F plasmid, which confers certain types of conjugative behavior. However, the most important plasmids for use as vectors are those that carry genes for drug resistance because drug-resistant phenotypes allow selection for cells transformed by plasmids, and even selection for recombinant DNA. Plasmids are also an efficient means of amplifying cloned DNA because there are many copies per cell, up to several hundreds for some plasmids.

Two extensively used plasmid vectors are shown in Figure 14-7. These are derived from natural plasmids, but both have been genetically modified for convenient use as recombinant DNA vectors. Plasmid pBR322 is simpler in structure; it has two drug-resistance genes, tet^R and amp^R. Both genes contain unique restriction target sites that are useful in cloning. For example, donor DNA could be inserted into the tet^R gene. A successful insertion will inactivate the tet^R gene, which then will no longer confer tetracycline resistance. Therefore, the cloning procedure would be to mix the samples of cut plasmid and donor DNA, transform bacteria, and select for ampicillin-resistant colonies, which must have been successfully transformed by a plasmid molecule. Of the amp^R colonies, only those that prove to be tetracycline-sensitive have inserts; in other words, the amp^R tet^S colonies are the ones that contain recombinant DNA. Further experiments are needed to find the clones with the specific insert required.

The pUC plasmid is a more advanced vector, whose structure allows direct visual selection of colonies containing vectors with donor DNA inserts. The key element is a

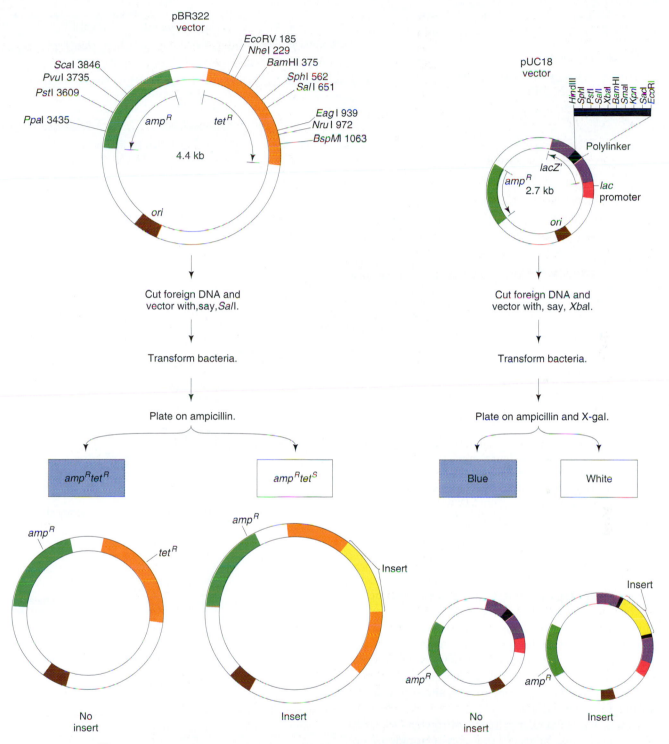

Figure 14-7 Two plasmids designed as vectors for DNA cloning, showing general structure and restriction sites. Insertion into pBR322 is detected by inactivation of one drug-resistance gene (tet^R), indicated by the tet^S (sensitive) phenotype. Insertion into pUC18 is detected by inactivation of the β-galactosidase function of Z′, resulting in an inability to convert the artificial substrate X-gal into a blue dye.

small portion of the *E. coli* β-galactosidase gene. Into this region has been inserted a piece of DNA called a **polylinker,** or **multiple cloning site,** which contains many convenient restriction target sites useful for inserting donor fragments. The polylinker is in frame with the β-galactosidase coding

sequence and does not interfere with its function. The transformation protocol uses recipient cells that contain a β-galactosidase gene lacking the fragment present on the plasmid. An unusual type of complementation occurs in which the partial proteins coded by the two fragments unite to

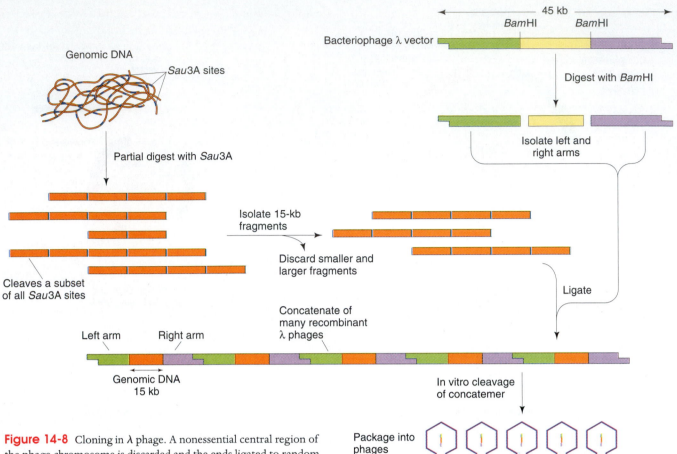

Figure 14-8 Cloning in λ phage. A nonessential central region of the phage chromosome is discarded and the ends ligated to random 15-kb fragments of donor DNA. A linear multimer (concatenate) forms, which is then stuffed into phage heads one monomer at a time using an in vitro packaging system. (From J. D. Watson, M. Gilman, J. Witkowski, and M. Zoller, *Recombinant DNA*, 2d ed. Copyright © 1992 by Scientific American Books.)

form a functional β-galactosidase. A colorless substrate for β-galactosidase called X-gal is added to the medium, and the functional enzyme converts this to a blue dye, which colors the colony blue. If donor DNA is inserted into the polylinker, then the β-galactosidase activity is lost and the colony is white. Hence, selection for white amp^R colonies selects directly for vectors bearing inserts, and such colonies are isolated for further study.

Plasmids that contain large inserts of foreign DNA tend to spontaneously lose the insert; therefore, plasmids are not useful for cloning DNA fragments larger than 20 kb. (DNA length is conventionally measured in kilobases, abbreviated kb. Strictly speaking, this is an incorrect use of the word *base*, because DNA is actually made of nucleotide pairs, and 1 kb really means "1000 nucleotide pairs.")

λ Phage. λ phage is a convenient cloning vector for several reasons. First, λ phage heads will selectively package a chromosome about 50 kb in length, and as will be seen, this property can be used to select for λ molecules with inserts

of donor DNA. The central portion of the phage genome is not required for replication or packaging of lambda DNA molecules in *E. coli,* so the central portion can be cut out using restriction enzymes and discarded. The two "arms" are ligated to restriction-digested donor DNA. The chimeric molecules can be either introduced into *E. coli* using a special type of transformation called *transfection,* or packaged into phage heads in vitro. In this system, DNA and phage head components are mixed together, and infective λ phages form spontaneously. In vitro packaging is another way of allowing selection and amplification of individual re-combinant DNA molecules. In either method, recombinant molecules with 10- to 15-kb inserts are the ones that will be most effectively packaged into phage heads, either in vitro or just prior to bacterial lysis, so the presence of a phage plaque on the bacterial lawn automatically signals the presence of recombinant phage bearing an insert (Figure 14-8). A second useful property of a phage vector is that recombinant molecules are automatically packaged into infective phage particles, which can be conveniently stored and manipulated experimentally.

Cosmids. Cosmids are vectors that are hybrids of λ phages and plasmids, and their DNA can replicate in the cell like a plasmid or be packaged like a phage. However, cosmids can carry DNA inserts about three times larger than those carried by λ itself (up to 45 kb). The key is that most of the λ phage structure has been deleted, but the signal sequences that promote phage headstuffing (*cos* sites) remain. This modified structure enables phage heads to be stuffed with almost all donor DNA. Cosmid DNA can be packaged into phage particles using the in vitro system. Cloning by cosmids is illustrated in Figure 14-9.

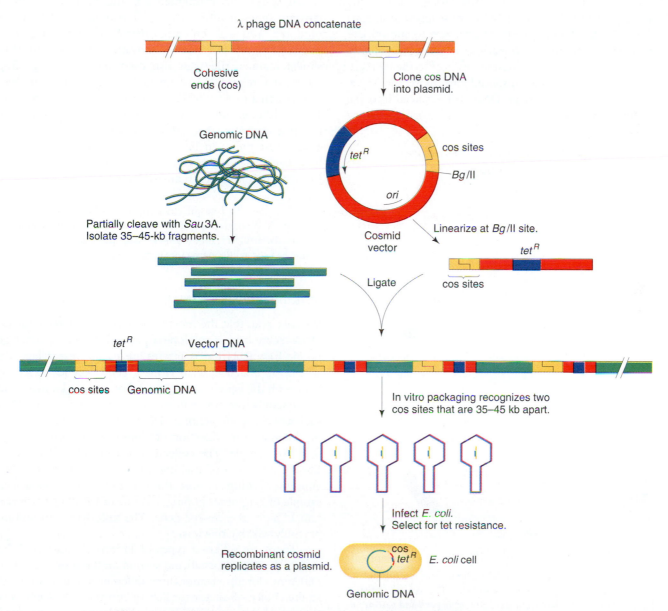

Figure 14-9 Cloning by cosmids. The cosmid is cut at a *Bgl*II site next to the cos site. Donor genomic DNA is cut using *Sau*3A, which gives sticky ends compatible with *Bgl*II. A tandem array of donor and vector DNA results from mixing. Phage is packaged in vitro by cutting at the cos site. The cosmid with insert recircularizes once in the bacterial cell. (From J. D. Watson, M. Gilman, J. Witkowski, and M. Zoller, *Recombinant DNA*, 2d ed. Copyright © 1992 by Scientific American Books.)

Single-Stranded Phages. Some phages contain only single-stranded DNA molecules. On infection of bacteria, the single infecting strand is converted to a double-stranded replicative form, which can be isolated and used for cloning. The advantage of using these phages as cloning vectors is that single-stranded DNA is the very substrate required for the Sanger dideoxy DNA sequencing technique currently in widespread use (page 446). M13 is the phage most widely used for this purpose.

Expression Vectors. Expression vectors are those in which the cloned gene can be transcribed and translated into protein. Thus a foreign protein can be synthesized in a bacterial cell. However, because bacteria cannot process introns, the cloned sequences must be in cDNA form stripped of introns (see below). The cloned gene is inserted next to appropriate bacterial transcription and translation start signals. Some vectors have restriction sites located just next to the *lac* regulatory region, which has been spliced into the vector. These restriction sites permit foreign DNA to be spliced into the vector for expression under the control of the *lac* regulatory system.

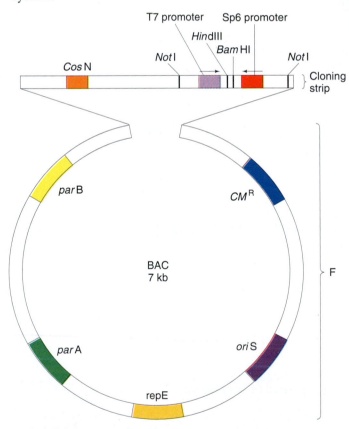

Figure 14-10 Structure of a bacterial artificial chromosome (BAC), used for cloning large fragments of donor DNA. CM^R is a selectable marker for chloramphenicol resistance. *oriS, repE, parA* and *parB* are F genes for replication and regulation of copy number. *CosN* is the *cos* site from λ phage. *Hind*III and *Bam*HI are cloning sites at which foreign DNA is inserted. The two promoters are for transcribing the inserted fragment. The *Not*I sites are used for cutting out the inserted fragment.

YACs. The acronym *YAC* stands for **yeast artificial chromosome.** Because yeast centromeres and telomeres have been cloned and characterized, they can be used to assemble artificial chromosomes that can carry huge amounts of DNA from a donor genome (up to 1000 kb). However, whereas the vectors discussed above use bacteria as hosts, YACs must use yeast cells as hosts. Because of their high carrying capacity, YACs are routinely used to create sets of clones that together encompass entire eukaryotic chromosomes or even entire genomes. YAC cloning is discussed fully in Chapters 15 and 17.

BACs. A relatively new type of vector that is also useful for cloning large DNA fragments is the **bacterial artificial chromosome,** or **BAC.** BACs are based on the 7-kb F factor of *E. coli.* You will remember that F can carry large fragments of *E. coli* DNA as F′ derivatives (Chapter 10). In a similar manner, they can also carry inserts of large fragments of foreign DNA of up to 300 kb, although the average is often around 100 kb. A typical BAC is shown in Figure 14-10. Although the maximum insert sizes are not as large as those of YACs, the BACs have several advantages over YACs. First they can be isolated and manipulated simply with basic bacterial plasmid technology, and second they form less hybrid inserts than do YACs.

Message   Vectors for cloning genes in bacterial hosts are plasmids, λ phages, hybrids of these called *cosmids* and F-based bacterial artificial chromosomes. Yeast artificial chromosomes can be used as cloning vectors in yeast cells.

Making a DNA Library

We have seen that the most important goal of recombinant DNA technology is to clone a particular gene or other genomic fragment that is interesting to the researcher. The specific approach used to clone a gene depends to a large degree on the gene in question and on what is known about it. Generally, the procedures start with a sample of DNA such as eukaryotic genomic DNA. The first step is to make or obtain a large collection of clones made from this original DNA sample. The collection of clones is called a **DNA library.** This step of making a library is sometimes called **shotgun cloning** because the experimenter clones a large sample of fragments hoping that one of the clones will contain a "hit"—the desired gene. The task then is to find out precisely which clone it is.

There are different types of libraries, categorized first on which vector is used, and second on the source of DNA. Different cloning vectors carry different amounts of DNA, so the choice of vector for library construction depends on the size of the genome (or other DNA sample) being made into the library. Plasmid and phage vectors carry smaller amounts, so these are suitable for small genomes. Cosmids carry larger amounts of DNA, and YACs carry the largest amounts of all. Ease of manipulation is another important factor in choosing a vector. A phage library is a suspension of phages. A plasmid or a cosmid library is a suspension of bacteria or a set of defined bacterial cultures stored in culture tubes or microtiter dishes. Similarly, a YAC library is a collection of yeast strains.

The second important decision is whether to make a **genomic library** or a **cDNA library. cDNA,** or **complementary DNA,** is synthetic DNA made from mRNA, using a special enzyme called *reverse transcriptase* originally isolated from retroviruses. Using mRNA as a template, reverse transcriptase synthesizes a single-stranded DNA molecule that can then be used as a template for double-stranded DNA synthesis (Figure 14-11). Because it is made from mRNA, cDNA is devoid of upstream and downstream regulatory sequences and, of course, introns. This means that cDNA from eukaryotes can be translated into functional protein in bacteria, obviously an important feature when cloning and manipulating eukaryotic genes in bacterial hosts.

The choice between genomic DNA and cDNA depends on the situation. If the gene being sought is active in a specific type of tissue in a plant or animal, then it would make sense to use that tissue to prepare mRNA to be converted into cDNA, and then make a cDNA library from that sample. This library should be enriched for the gene in question. A cDNA library is based on the regions of the genome transcribed, so it will inevitably be smaller than a genomic library, which should contain all of the genome. Although genomic libraries are bigger, they do have the benefit of containing genes in their native form, including introns and upstream and downstream regulatory sequences.

In some cases it is possible to narrow down the genomic fraction used in library construction, in order to more easily detect the desired gene. This approach is possible if it is already known which chromosome the gene is on. One technique used in mammalian molecular genetics is to sort the chromosomes with an instrument called a *flow cytometer.* A suspension of chromosomes is passed through the apparatus, which sorts the chromosomes according to size (this procedure is discussed in more detail in Chapter 17). The appropriate chromosomal fraction is then used to make the library. Another technique possible in organisms with small chromosomes is to electrophorese whole chromosomes using pulsed field gel electrophoresis. This procedure is similar to conventional electrophoresis (see below) but uses oscillating fields oriented in several different directions, enabling whole chromosomes (which are huge DNA molecules) to snake through the gel to different positions according to their size. (This procedure is also detailed in Chapter 17.) The appropriate chromosome can be identified on the gel by probing with a chromosome-specific probe (see below). Then the desired chromosome can be cut out, eluted from the gel and used to make a chromosome-specific library.

How can an experimenter determine whether a library is large enough to contain any one unique sequence of interest with a reasonable degree of certainty? There are for-

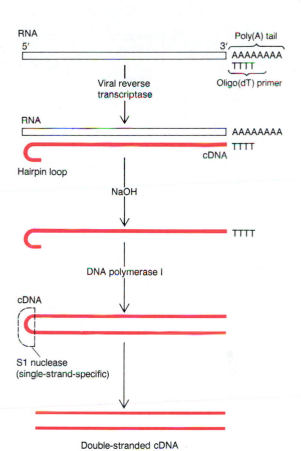

Figure 14-11 The synthesis of double-stranded cDNA from mRNA. A short oligo(dT) chain is hybridized to the poly(A) tail of an mRNA strand. The oligo(dT) segment serves as a primer for the action of reverse transcriptase, which uses the mRNA as a template for the synthesis of a complementary DNA strand. The resulting cDNA ends in a hairpin loop. Once the mRNA strand is degraded by treatment with NaOH, the hairpin loop becomes a primer for DNA polymerase I, which completes the paired DNA strand. The loop is then cleaved by S1 nuclease (which acts only on the single-stranded loop) to produce a double-stranded cDNA molecule. (From J. D. Watson, J. Tooze, and D. T. Kurtz, *Recombinant DNA: A Short Course.* Copyright © 1983 by W. H. Freeman and Company.)

mulas for calculating the minimum number of clones needed, but one can obtain a rough idea of the general order of magnitude of the library simply by taking the total genome size and dividing by the average size of the inserts carried by the vector being used. Generally this number will be at least doubled, but it does provide a ballpark estimate of the magnitude of the job of library construction.

Message The task of isolating a clone of a specific gene begins with making a library of genomic DNA or cDNA, if possible enriched for sequences containing the gene in question.

Finding Specific Clones Using Probes

The library, which might contain as many as hundreds of thousands of cloned fragments, must be screened in order

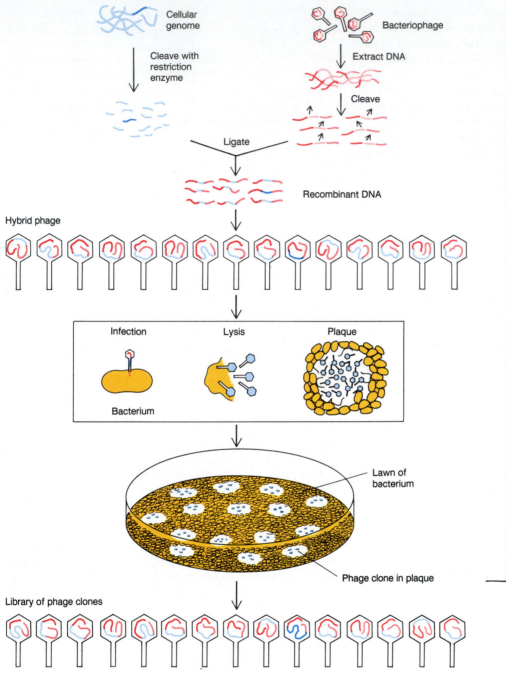

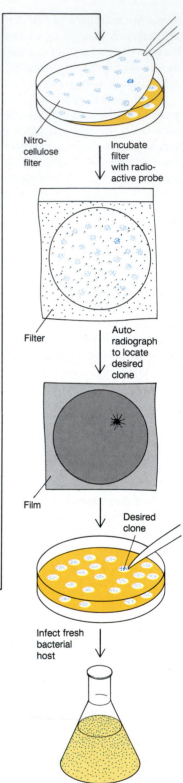

Figure 14-12 (a) A genomic library can be made by cloning genes in λ bacteriophages. When a lawn of bacteria on a petri plate is infected by a large number of different hybrid phages, each plaque in the lawn is inhabited by a single clone of phages descended from the original infecting phage. Each clone carries a different fragment of cellular DNA. The problem now is to identify the clone carrying a particular gene of interest (bright blue) by probing the clones with DNA or RNA known to be related to the desired gene. (b) The plaque pattern is transferred to a nitrocellulose filter, and the phage protein is dissolved, leaving the recombinant DNA, which is then denatured so that it will stick to the filter. The filter is incubated with a radioactively labeled probe: a DNA copy of the messenger RNA representing the desired gene. The probe hybridizes with any recombinant DNA incorporating a matching DNA sequence, and the position of the clone having the DNA is revealed by autoradiography. Now the desired clone can be selected from the culture medium and transferred to a fresh bacterial host, so that a pure gene can be manufactured. (After R. A. Weinberg, "A Molecular Basis of Cancer," and P. Leder, "The Genetics of Antibody Diversity." Copyright © 1983, 1982 by Scientific American, Inc. All rights reserved.)

to find the recombinant DNA molecule containing the gene of interest. This is accomplished by using a specific **probe** that will find and mark the clone for the researcher to identify. Broadly speaking there are two types of probes, those that recognize DNA and those that recognize protein.

Probes to Find DNA. These probes depend on the natural tendency of one single strand of nucleic acid to find and hybridize to a strand of complementary base sequence. A probe that is itself DNA, when denatured (made single-stranded) will therefore find and bind to other similar denatured DNAs in the library. In trying to find a specific clone in a library, there are two steps. First, colonies or plaques of the library on petri dishes are transferred to absorbent membranes (often nitrocellulose filters) by simply laying the membrane on the surface of the medium. The membrane is peeled off, and colonies or plaques clinging to the surface are lysed in situ and the DNA denatured. The next step is to bathe the membrane with a solution of a labeled probe that is specific for the DNA being sought. Generally the probe is itself a cloned piece of DNA that has a sequence homologous to the desired gene. The probe DNA must be denatured; it will then bind only to the DNA of the clone being sought. The position of a positive clone will become clear from the position of the label, often as a spot on an autoradiogram. This technique of probing a library is shown in Figure 14-12.

Where does the DNA to make a probe come from? It seems on the face of it that if you have the DNA to be used as a probe, you don't need to clone it. The DNA can be from several sources. One source is cDNA from tissue that expresses the gene you are interested in. The idea is that because the mRNA of a gene is abundant, if you make cDNAs from this tissue and insert them individually into vectors, then many of them will very likely be for the gene you want. For example, in mammalian reticulocytes 90% of the mRNA is known to be transcribed from the β-globin gene, so this would be a good source of mRNA to make a cDNA probe to find a genomic globin gene. In this case a genomic library would be probed. Of course the need for this kind of analysis depends on which questions are to be asked about the gene. If only the transcribed sequence is of interest, then the cDNA clone itself could provide that information just as well. However, if introns and control region are needed, the genomic clone must be obtained.

Another source of DNA for a probe might be a similar gene from a related organism. For example, if a certain gene has been cloned in the ascomycete fungus *Neurospora*, then it is very likely that this gene could be used as a probe to find the homologous gene from the related fungus *Podospora*. This method depends on the evolutionary conservation of DNA sequences through time. Even though the probe DNA and the DNA of the desired clone might not be identical, they are often similar enough to promote hybridization. The method is jokingly called "clone by phone" because if you can phone a colleague who has a clone of

your gene of interest but from a related organism, then your job of cloning is made relatively easy.

Probe DNA can be synthesized if the protein product of the gene of interest is known and some amino acid sequence has been obtained. Synthetic probes are designed based on knowledge of the genetic code, so an amino acid sequence merely has to be translated backwards to obtain the DNA sequence that encoded it. However, because of the redundancy of the code—in other words, the fact that most amino acids are coded by more than one codon—there are several possible DNA sequences that could have coded for the protein in question. To get around this problem, a short stretch of amino acids with minimal redundancy is selected. The nucleotide sequence is calculated using the codon dictionary. The chemical DNA synthesizing reaction is a stepwise process, so wherever in the sequence there are alternative nucleotides, a mixture of those alternative nucleotides is fed into the reaction and all possible DNA strands are synthesized. Figure 14-13 shows an example in which there are five positions of redundancy, showing 2, 3, 2, 2, and 2 alternatives, respectively. The reaction would make $2 \times 3 \times 2 \times 2 \times 2 = 48$ **oligonucleotide** strands at the same time. This "cocktail" of oligonucleotides would be used as a probe. The correct strand within this cocktail would find the gene of interest. Twenty nucleotides embodies enough specificity to find one unique DNA sequence in the library.

Also, free RNA can be radioactively labeled and used as a probe. This is only possible when a relatively pure population of identical molecules of RNA can be isolated, such as rRNA or fractionated tRNAs.

Probes to Find Proteins. If the protein product of a gene is known and isolated in pure form, then this protein can be

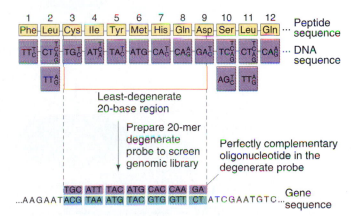

Figure 14-13 A short sequence of a protein is used to design a set of redundant oligonucleotides for use as a probe to recover the gene that coded the protein. One of the set of probes will be a perfect match for the gene. (From H. Lodish, D. Baltimore, A. Berk, S. L. Zipursky, P. Matsudaira, and J. Darnell, *Molecular Cell Biology*, 3d ed. Copyright © 1995 by Scientific American Books, Inc.)

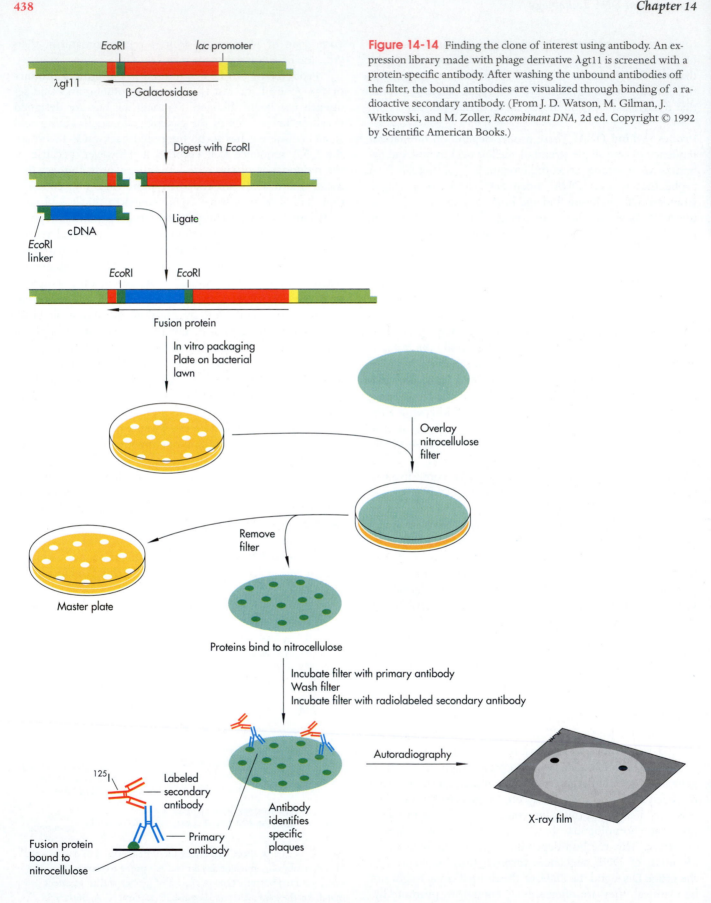

Figure 14-14 Finding the clone of interest using antibody. An expression library made with phage derivative λgt11 is screened with a protein-specific antibody. After washing the unbound antibodies off the filter, the bound antibodies are visualized through binding of a radioactive secondary antibody. (From J. D. Watson, M. Gilman, J. Witkowski, and M. Zoller, *Recombinant DNA*, 2d ed. Copyright © 1992 by Scientific American Books.)

EcoRI   lac promoter

λgt11

β-Galactosidase

Digest with EcoRI

cDNA

EcoRI
linker

Ligate

EcoRI   EcoRI

Fusion protein

In vitro packaging
Plate on bacterial
lawn

Overlay
nitrocellulose
filter

Remove
filter

Master plate

Proteins bind to nitrocellulose

Incubate filter with primary antibody
Wash filter
Incubate filter with radiolabeled secondary antibody

^{125}I

Labeled
secondary
antibody

Primary
antibody

Autoradiography

Antibody
identifies
specific
plaques

X-ray film

Fusion protein
bound to
nitrocellulose

used to detect the clone of the corresponding gene in a library. An antibody to the protein is prepared, and this antibody is used to screen an **expression library.** These libraries are made using expression vectors designed to express high levels of a bacterial protein. To make the library, cDNA is inserted into the vector in frame with the bacterial protein, and the cells will make a fusion protein. A membrane is laid over the surface of the medium, and removed with an imprint of colonies. It is dried and bathed in a solution of the antibody. Positive clones are revealed by making an antibody to the first antibody; the second antibody is labeled by a radioactive isotope or a chemical that will fluoresce or become a colored dye. By detecting the correct protein, the antibody effectively identifies the clone containing the gene that must have synthesized that protein. The process is described in Figure 14-14.

> **Message** A cloned gene can be selected from a library using probes for the gene's DNA sequence or for the gene's protein product.

Finding Specific Clones by Functional Complementation

Cloned genes can also be detected through their ability to confer a missing function on a transformation recipient. This procedure is called **functional complementation.** We have already learned about transformation in prokaryotes (Chapter 10), but eukaryotes can be transformed, too. The procedure differs among eukaryotes, but generally some special treatment of recipient cells is required. For example, to transform fungi, generally the cell walls must be removed enzymatically. Let's assume that you have isolated a mutant that is relevant to some biological process that interests you. For the sake of argument we will assume it is an auxotrophic mutation. If the auxotrophic mutant strain is used as a transformation recipient, DNA from the library is used to transform the recipient and recipient cells plated on minimal medium. Clones that contain the wild-type allele (from the wild-type culture used to make the library) will transform the auxotroph to prototrophy and allow growth on minimal medium.

The reason this transformation method works is that the transforming fragment functionally complements the deficiency caused by the mutant allele in the recipient. It might seem at first that this view of complementation is not the same as the one developed in Chapter 4, the production of a wild-type phenotype from the union of two mutant genomes. However, the transforming vector contributes something the recipient genome lacks (the wild-type allele being sought), and the recipient genome contributes something the vector lacks (the entire remainder of the genome), so a type of complementation is involved.

If the transformation recipient is an organism in which plasmid vectors replicate autonomously (mainly bacteria and yeasts), then the transforming insert can be recovered simply by isolating the plasmid. However, as we shall see, in most eukaryotic organisms the vector cannot replicate and must insert into the genome in order to achieve stable transformation. In these cases, the transforming fragment is relatively inaccessible and must be retrieved from the successful clone in the library. This method uses a library in which the clones are laid out as a collection of numbered bacterial cultures in tubes or microtiter dishes. DNA is isolated in bulk from all the strains in specific subsets of the library, and transformation attempted. By a process of narrowing down the library subsets that successfully transform, the clone with the wild-type allele can be identified (Figure 14-15).

Note that in this protocol DNA cloned in one host (generally a bacterium) is being added to cells of the organism under genetic analysis, often a eukaryote. Therefore, the transforming vector will almost certainly not replicate autonomously in the new host. Transformants are found to contain the vector carrying the wild-type allele inserted into one of the recipient's chromosomes at a location that is different from the mutant locus in the recipient. This is called **ectopic** insertion (Figure 14-16). In other cases the transforming wild-type allele replaces the resident auxotrophic mutation by a double-crossover-like process.

If a eukaryotic gene is cloned on a prokaryotic vector but a specific eukaryotic origin of replication is known and can be added to the vector, the vector will be able to replicate in both bacterial and eukaryotic cells, and insertion into the chromosome is not essential. These types of vectors are called *shuttle vectors*. Without an origin of replication, the exogenous DNA must integrate into the eukaryotic chromosome to effect stable transformation.

> **Message** Cloned genes can be selected by using their DNA to transform and complement null alleles in recipient cells.

Positional Cloning

Positional cloning is a term that can be applied to any method that makes use of information about a gene's position in order to clone it more efficiently. Knowledge about position is used to circumvent the hard work of assaying an entire library to find the clone of interest. Often both probing and complementation form part of positional cloning. A common starting point is the availability of another cloned gene or other marker known to be closely linked to the gene being sought. The linked marker acts as the departure point in a process called **chromosome walking** that will terminate at the target gene. End fragments of the cloned marker are used as probes to select other clones from the library. These probes will detect clones of DNA regions that overlap with the initial clone. Restriction maps (pages 449–451) are made of the DNA of this second set of clones, and again outward fragments are used for a new round of selection of overlapping clones from the library. Hence the walking process

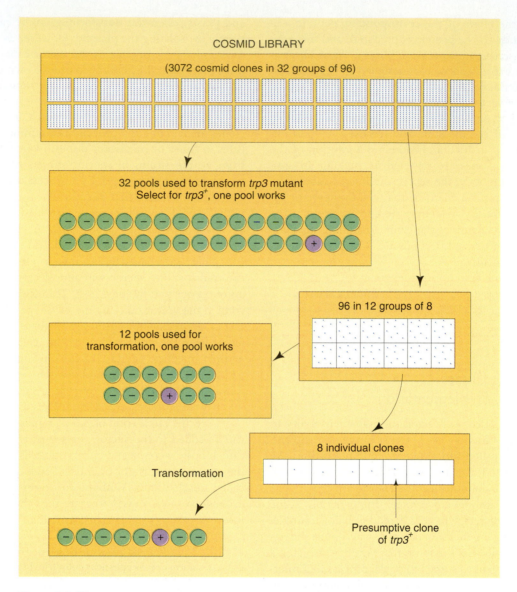

Figure 14-15 Finding a cloned gene by using progressively smaller pooled DNA samples in transformation. In this example, the quest is for the *trp3* gene of *Neurospora* (After J. R. S. Fincham, *Genetic Analysis.* Copyright © 1995 by Blackwell.)

moves outwards in two directions from the start site. Each clone can be sequenced or mapped depending on the intent of the exploration. Figure 14-17 summarizes the procedure of chromosome walking.

Sometimes a cosmid that is known to contain the linked marker will also luckily contain the sought gene, and subcloning and transformation will narrow down the appropriate region of the cosmid. The availability of a large number of neutral DNA markers (restriction fragment length polymorphisms—see page 485) dispersed throughout most genomes has also provided many useful start points. Positional cloning has been particularly useful for cloning human genes, many of which have no known function and cannot be selected by functional complementation. However, ultimately there must be some type of criterion to as-

sess each step of the walk for the gene of interest, and these criteria will depend on the individual gene concerned.

Cloning a Gene by Tagging

Tagging is a cloning method that zeros in on the desired gene directly by inducing a mutation in that gene using a defined DNA sequence as an insertional mutagen. The defined sequence is then used as a tag to recover the gene. When exogenous DNA is added by transformation or by other methods such as injection, it can integrate into the genome and become part of the chromosome. Ectopic integration is random throughout the genome, and apparently no segment of chromosomal DNA is immune to integration. When integration occurs within or near a gene, the in-

Figure 14-16 The possible fates of transforming DNA. A donor wild-type allele A^+ (cloned in a bacterial vector) transforms an A^- recipient by one of three different types of insertion. NOTE: The recipient generally is of the same species as the donor DNA, either prokaryotic or eukaryotic. Two recipient chromosomes are shown, I and II.

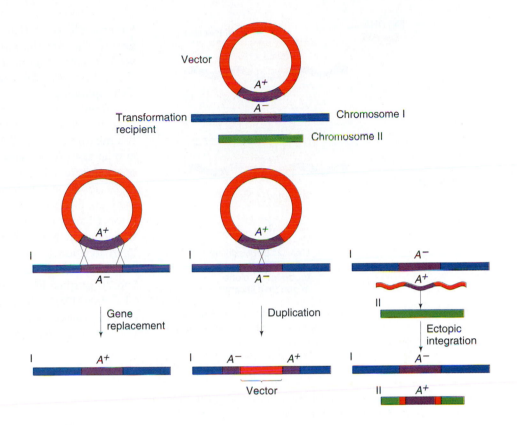

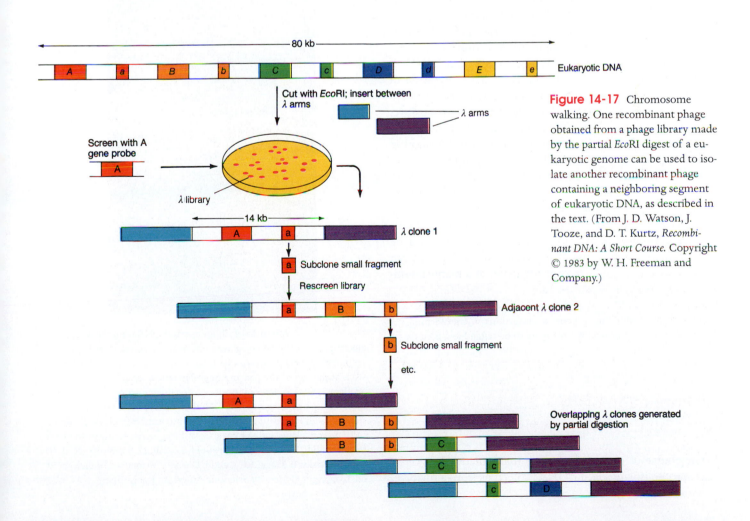

Figure 14-17 Chromosome walking. One recombinant phage obtained from a phage library made by the partial *Eco*RI digest of a eukaryotic genome can be used to isolate another recombinant phage containing a neighboring segment of eukaryotic DNA, as described in the text. (From J. D. Watson, J. Tooze, and D. T. Kurtz, *Recombinant DNA: A Short Course.* Copyright © 1983 by W. H. Freeman and Company.)

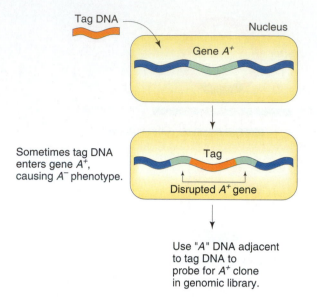

Figure 14-18 Using DNA insertion as a tag for marking and recovering a gene from the genome. The tag DNA can be transforming DNA or an endogenous transposon (movable element).

tegrating fragment acts like a mutagen, disrupting the function of the gene. This property can be used to good advantage. Let's say you use exogenous fragment X for transformation, and recover transformants. Some of these transformants will be mutant for the target gene. The next step is to cross the transformants to determine if the mutant phenotype segregates with fragment X. If it does, it is likely that the mutation was caused by the integration of fragment X. The DNA of this mutant line is used to construct a library, and fragment X can be used as a probe to recover the gene. To recover the intact wild-type gene, a subclone of the disrupted gene sequence is used in another round of probing, this time with a wild-type library. The approach is summarized in Figure 14-18.

A similar approach uses transposons (mobile DNA fragments) as tags instead of transforming fragments (see page 640). Using a line containing an active transposon, mutants for the desired gene are selected. Many of these will be caused by insertion of the transposon. The transposon can then be used as a tag to recover the gene, in a manner similar to that shown in Figure 14-18.

Message Mutating a gene by insertion of transforming DNA or a transposon allows the gene to be tagged as a prelude to its isolation.

Fractionation of DNA by Electrophoresis

In the course of gene and genome manipulation it is often necessary to isolate specific DNA molecules from a mixture.

For example, recall that in cloning using λ phage it is necessary to separate the two chromosome arms from the unwanted central region. There are several ways of fractionating DNA, but the most extensively used method is electrophoresis (see page 93 for a drawing of an electrophoretic apparatus). Electrophoresis separates molecules in an electric field by virtue of their charge, size, and shape. If a mixture of linear DNA molecules is placed in a well cut into an agarose gel, and the well is placed near the cathode of an electric field, the molecules will move through the gel to the anode, at speeds dependent on their size. Therefore if there are distinct size classes in the mixture, these will form distinct bands on the gel. The bands can be visualized by staining the DNA with ethidium bromide, which causes the DNA to fluoresce in UV light (Figure 14-19). If bands are well separated, an individual band can be cut from the gel and the purified DNA sample removed from the gel matrix. Therefore DNA electrophoresis can be either diagnostic (showing what sizes of molecules are present) or preparative (useful in isolating specific DNAs).

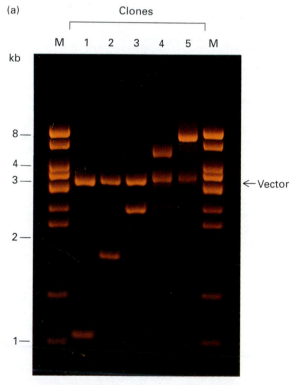

Figure 14-19 Mixtures of different-sized DNA fragments separated electrophoretically on an agarose gel. In this case the samples are five recombinant vectors treated with *Eco*RI. The mixtures are applied to wells at the top of the gel, and fragments move under the influence of an electric field to different positions dependent on size (and, therefore, number of charges). The DNA bands have been visualized by staining with ethidium bromide and photographing under UV light. (*M* represents lanes containing standard fragments acting as markers for estimating DNA length.) (From H. Lodish, D. Baltimore, A. Berk, S. L. Zipursky, P. Matsudaira, and J. Darnell, *Molecular Cell Biology*, 3d ed. Copyright © 1995 by Scientific American Books, Inc.)

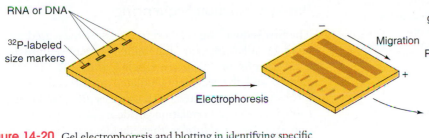

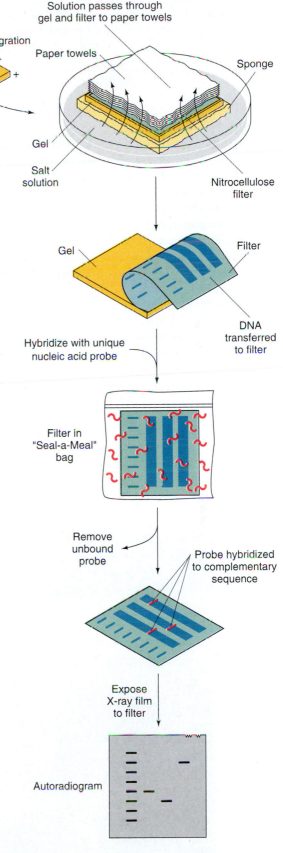

Figure 14-20 Gel electrophoresis and blotting in identifying specific cloned genes. RNA or DNA restriction fragments are applied to an agarose gel and electrophoresed. The various fragments migrate at differing rates according to their respective sizes. The gel is placed in buffer and covered by a nitrocellulose filter and a stack of paper towels. The fragments are denatured to single strands, so that they can stick to the filter. They are carried to the filter by the buffer, which is wicked up by the towels. The filter is then removed and incubated with a radioactively labeled single-stranded probe that is complementary to the targeted sequence. Unbound probe is washed away, and X-ray film is exposed to the filter. Since the radioactive probe has hybridized only with its complementary restriction fragments, the film will be exposed only in bands corresponding to those fragments. Comparison of these bands with labeled markers reveals the number and size of the fragments in which the targeted sequences are found. This information can be used to position the sequences along restriction maps. This procedure is termed *Southern blotting* when DNA is transferred to nitrocellulose and *Northern blotting* when RNA is transferred. (From J. D. Watson, M. Gilman, J. Witkowski, and M. Zoller, *Recombinant DNA*, 2d ed. Copyright © 1992 by Scientific American Books.)

Restriction enzyme digestion of genomic DNA of a eukaryote results in so many fragments that a stained electrophoretic gel shows a smear of DNA. A probe can identify one fragment in this mixture, using a technique developed by E. M. Southern called **Southern blotting,** which has many uses in gene and genomic analysis. After DNA fragments are fractionated on an electrophoretic gel, an absorbent membrane is laid over the gel and the DNA bands are transferred ("blotted") onto the membrane by capillary action. When transferred to the membrane, the DNA bands stay in the same relative positions as on the gel. The membrane is bathed in a labeled probe, and an autoradiogram is used to reveal the presence of any bands on the gel that were homologous to the probe. If appropriate, those bands can be cut out of the gel and further processed.

The Southern blotting technique can be extended to detect a specific *RNA* molecule from a population of different RNAs separated on a gel. This technique has been termed **Northern blotting** to contrast it with the Southern technique for *DNA* analysis. The RNA is blotted onto a membrane and probed in the same way as for Southern blotting. A technique called **Western blotting** has also been developed to transfer *protein* from a gel and then to visualize specific proteins with antibodies.

The procedure of Southern blotting is illustrated in Figure 14-20, and a typical autoradiogram is shown in Figure 14-21.

DNA Sequence Determination

Success in cloning a desired gene is merely the beginning of a second round of analysis in which the aim is to characterize the structure and function of that gene. One important option is to obtain its nucleotide sequence, because this is the ultimate characterization of genetic structure. One of the requirements for DNA sequencing is the ability to obtain defined fragments of DNA. Thus, there is a strong interdependence of DNA cloning and DNA sequencing technology, since DNA cloning provides large samples of defined DNA fragments.

The key to DNA sequencing involves starting with a defined fragment of DNA labeled at one end, then generating a population of molecules that differ in size by one base, and being able to separate these molecules and identify the base in each case. The technique involves acrylamide or agarose gel electrophoresis of single-stranded DNA molecules. The mobility of a strand is inversely proportional to the logarithm of its length. This technique is so sensitive that fragments differing in length by only a single nucleotide can be separated.

Base-Destruction Sequencing

The first sequencing method to be extensively used was devised by Allan Maxam and Walter Gilbert (Figure 14-22). The 5′ ends of DNA are labeled with ^{32}P, the DNA is cleaved and the fragments isolated by size, and then the DNA duplex is separated, with one specific strand being discarded, to yield a population of identical strands labeled on one end. The mixture is then divided into four samples, each of which is subjected to a different chemical reagent that destroys one or two specific bases. The four reagents destroy (1) only G, (2) A and G, (3) T and C, or (4) only C. The loss of a base makes the sugar-phosphate backbone of the DNA chain more likely to break at that point. The reagent concentration is adjusted so that only about 1 in 50 of the target bases is destroyed. This means that within a given sample, on some strands only one nucleotide will be destroyed. For such strands, fragment length will be determined by the distance from the labeled 5′ end to the particular nucleotide that was lost in the sequence. Since this is a random event, in each of the base-destroying reagents, each occurrence of that particular base will yield a labeled fragment of distinctive size. When these pieces are separated in the different lanes of a gel according to which reagent they were in, they can be arranged in order of length, and the base destroyed at each site can be determined by noting in which lane or lanes the band appears. Thus, the sequence of bases in the strand can quite simply be read from the pattern of bands on the gel. Figure 14-22k shows an actual gel

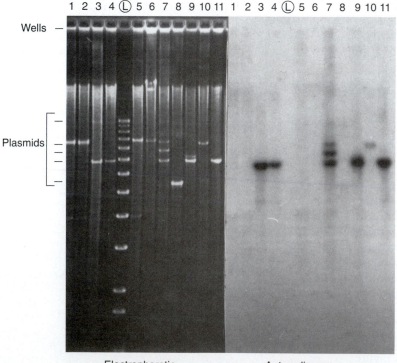

Electrophoretic gel stained with ethidium bromide

Autoradiogram using plasmid from strain 11 as probe

Figure 14-21 An example of a Southern analysis. The DNA from the mitochondria of 11 strains of *Neurospora* from different geographical locations was electrophoresed and the gel stained with ethidium bromide. The bright bands are linear plasmids, and the positions of some of these are marked. The lane marked "L" is a calibration ladder composed of DNA fragments of known size. The sizes of the rungs visible in this figure are (from the bottom up) 1.5 kb, 2 kb, and then one additional kilobase for each rung, up to 12 kb. The plasmid in strain 11 was cloned and used as a probe in a Southern hybridization. The autoradiogram shows that several other plasmids have sequences homologous to the strain 11 plasmid. Indeed strain 7 has a family of three plasmids, all homologous to the strain 11 plasmid. (From Simon Yang.)

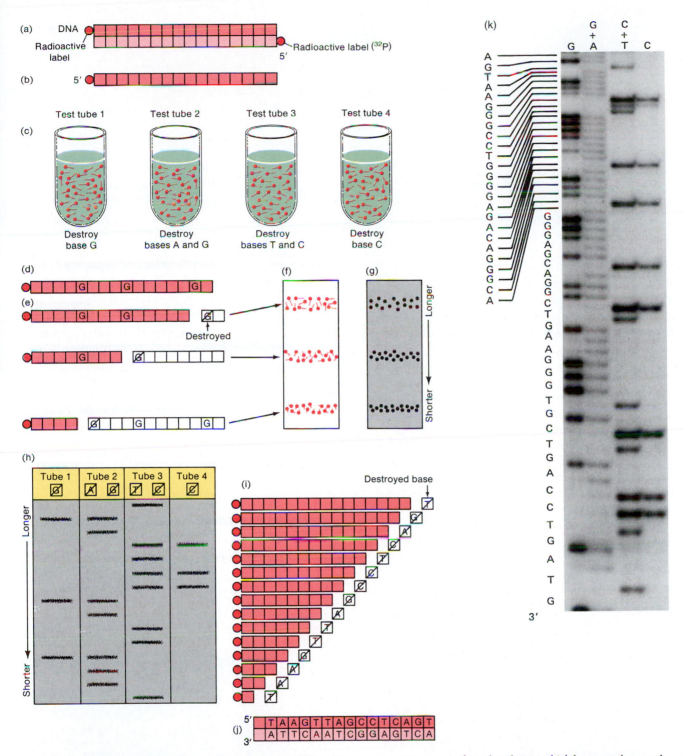

Figure 14-22 DNA sequencing, devised by Maxam and Gilbert. (a) The 5′ end of each strand is labeled with ^{32}P. (b) The strands are separated, and one specific strand is retained. (c) Four equivalent fractions of this strand are created and placed in four test tubes. Each tube is treated with a different reagent that selectively destroys either one or two of the four bases. The concentration is adjusted so that only a small portion of the target bases is attacked, thereby generating a population of fragments of different lengths. (d) A hypothetical example: a strand containing three G bases. (e) After treatment in test tube 1, this strand yields a population of *labeled* fragments of three different lengths. (f) The fragments are separated on a gel. (g) The bands of labeled fragments are identified by autoradiography. (h) Comparison of the bands produced by labeled fragments from the four different treatments provides a display of successively shorter fragments when read from top to bottom on the gels. (i) The appearance of the band on one or two particular gels indicates which base was destroyed to yield the fragment. (j) The sequence of base pairs in the original DNA can then be inferred. (k) Autoradiograph of a Maxam-Gilbert sequencing gel, showing a portion of a sequence of nucleotides located between the two α-globin genes of human DNA. The DNA fragment was labeled at the 5′ end and subjected to degradation by methods that cleave at G, C, G and A, C and T (cleavage reactions need not be specific to a *single* base to provide the necessary information). The sequence that can be read from the gel is indicated at the left. Parts a–j modified from W. Gilbert and L. Villa-Komaroff, "Useful Proteins from Recombinant Bacteria." Copyright © 1980 by Scientific American, Inc. All rights reserved. Part k courtesy of John Hess and C.-K. James Shen, University of California, Davis. From F. J. Ayala and J. A. Kiger, Jr., *Modern Genetics*, 2d ed. Copyright © 1984 by Benjamin/Cummings Publishing Company, 1984.)

Figure 14-23 The structure of 2′, 3′-dideoxy nucleotides, which are employed in the Sanger DNA sequencing method.

used to determine a portion of a sequence of nucleotides located between the two α-globin genes of human DNA.

Dideoxy Sequencing

The sequencing method now most commonly used was developed by Fred Sanger. His method is based on single-stranded DNA and DNA synthesis in the presence of dideoxy nucleotides, which lack a 3′-hydroxyl group (Figure 14-23). The respective dideoxy nucleotide triphosphates (ddNTPs) can be incorporated into a growing chain, but they terminate synthesis because they lack the 3′-hydroxyl necessary to bond with the next nucleotide triphosphate. Four reaction tubes are each prepared with single-stranded DNA template for the sequence of interest, plus DNA polymerase and a section of labeled primer. Each tube receives a small amount of a different ddNTP (ddATP, ddTTP, ddCTP, or ddGTP), together with the four normal deoxynucleotide triphosphates (dNTPs). In any given tube, various chain lengths will be produced, each corresponding to the point at which the respective ddNTP for that tube was incorporated and terminated chain growth. These lengths, in turn, are a clear indication of where the bases complementary to the ddNTPs occur on the template strand. The relative lengths can be visualized by running the four samples in four lanes on an electrophoretic gel, and the positions of the corresponding bases can be determined by reading along the four lanes, as illustrated in Figures 14-24 and 14-25.

Figure 14-24 The dideoxy sequencing method. (a) A labeled primer (designed from the flanking vector sequence) is used to initiate DNA synthesis. Synthesis is arrested randomly by the addition of four different dideoxy nucleotides (ddATP is shown here). (b) The resulting fragments are separated electrophoretically and subjected to autoradiography. (From J. D. Watson, M. Gilman, J. Witkowski, and M. Zoller, *Recombinant DNA*, 2d ed. Copyright © 1992 by Scientific American Books.)

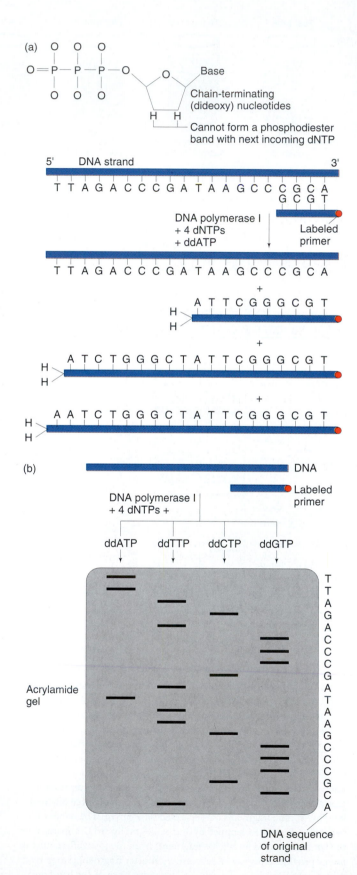

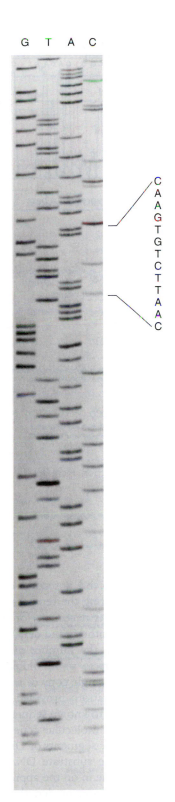

Figure 14-25 Sanger sequencing gel. The inferred sequence is shown at the right. (Photo from Loida Escote-Carlson.)

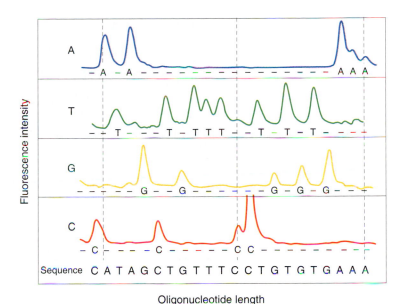

Oligonucleotide length

Figure 14-26 Using dye fluorescence instead of radioactivity as a marker in dideoxy sequencing. Four different dyes are used on primers in four different sets of termination reactions.

Instead of autoradiography, fluorescence detection of tags attached to oligonucleotide primers can be employed. Here, a different color emitter is used for each of the four reactions, but the four mixtures are electrophoresed together. Figure 14-26 shows the sequence of the colors, allowing a direct reading of the nucleotide sequence. Flourescence detection is widely used in automated DNA sequencing machines, which can read out 450 bases in one run, and improved machines promise even longer readouts. Figure 14-27 depicts a readout from an automated sequencing run. Using a set of automated sequencers, in 1995 Craig Venter, Hamilton Smith, and their co-workers announced the first complete DNA sequence of a living organism, the bacterium *Haemophilus influenzae.* They deciphered the entire sequence of the 1.9-megabase genome!

> **Message** A cloned DNA fragment can be sequenced by two different methods that generate sets of labeled single-stranded DNAs differing by one nucleotide, allowing the nucleotide sequence to be read off an electrophoretic gel.

Sometimes the precise location of a gene on a cloned fragment is not known, and sequencing can be used to find that gene. The DNA sequence of the cloned fragment is fed into a computer, which then scans all six reading frames (three in each direction) looking for a gene-sized stretch of DNA beginning with an ATG initiation codon and ending with a stop codon. Figure 14-28 shows such an analysis in which two candidate genes have been identified as long **open reading frames (ORFs).** It is interesting to note that ORF analysis and Mendelian genetics, although poles apart in their approaches, are both ways of identifying genes.

CHAPTER INTEGRATION PROBLEM

In Chapter 13 we studied the structure of tRNA molecules. Suppose that you want to clone the gene from organism X that encodes a certain tRNA. You possess the purified tRNA and also an *E. coli* plasmid that contains a single *Eco*RI cutting site in a *tet*^{R} (tetracycline resistance) gene, and also a gene for resistance to ampicillin (*amp*^{R}). How can you clone the gene of interest?

Solution

You can use the tRNA itself to probe for the DNA containing the gene. One method is to digest the DNA from organism X with *Eco*RI and then mix with the plasmid, which you also have cut with *Eco*RI. After transformation of an *amp*^{S} *tet*^{S} recipient, Amp^R colonies are selected, indicating successful transformation. Of these, select the colonies that are Tet^S. These will contain vectors with inserts, and many of these are needed to make the library. The library is then tested using the tRNA as the probe. You can examine those colonies hybridizing with the probe further to find out how much of the tRNA sequence they contain.

Alternatively, you can run *Eco*RI-digested DNA from organism X on a gel and then identify the correct band by probing with the tRNA. This region of the gel can be cut out and used as a source of highly enriched DNA to clone into the plasmid cut with *Eco*RI.

SOLVED PROBLEMS

1. The restriction enzyme *Hin*dIII cuts DNA at the sequence AAGCTT, and the restriction enzyme *Hpa*II cuts DNA at the sequence CCGG. On average, how frequently will each enzyme cut double-stranded DNA? (In other words, what is the average spacing between restriction sites?)

Solution

We need consider only one strand of DNA, because both sequences will be present on the opposite strand at the same site due to the symmetry of the sequences. The frequency of the six-base-long *Hin*dIII sequence is $(1/4)^6 = 1/4096$, since there are four possibilities at each of the six positions. Therefore, the average spacing between *Hin*dIII sites is approximately 4 kb. For *Hpa*II, the frequency of the four-base-long sequence is $(1/4)^4$, or 1/256. The average spacing between *Hpa*II sites is approximately 0.25 kb.

2. From Table 14-1 determine whether the *Eco*RI enzyme or the *Sma*I enzyme would be more useful for cloning. Explain your answer.

Solution

Both restriction enzymes recognize a six-base-pair sequence, so both would be expected to have approximately the same number of recognition sites per genome. The major difference between the two is that *Eco*RI leaves staggered ends while *Sma*I leaves blunt ends. Staggered ends are much easier to manipulate during cloning because of the base-pairing capacity inherent in them.

(Solution from Diane K. Lavett.)

PROBLEMS

1. A circular bacterial plasmid (pBP1) has a single *Hin*dIII restriction enzyme site in the middle of a tetracycline resistance gene (*tet*^{R}). Fruitfly genomic DNA is digested with *Hin*dIII, and a library is made in pBP1. Probing reveals that clone 15 contains a specific *Drosophila* gene of interest. Clone 15 is studied by restriction analysis with *Hin*dIII and another restriction enzyme *Eco*RV. The ethidium bromide–stained electrophoretic gel shows bands as in the accompanying diagram (the control was an uninserted plasmid pBP1). The sizes of the bands (in kilobases) are shown alongside. (NOTE: Circular molecules do not give intense bands on this type of gel, so you can assume that all bands represent linear molecules.)

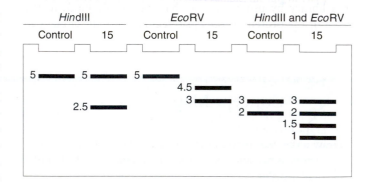

a. Draw restriction maps for plasmid pBP1 with and without the insert, showing the sites of the target sequences and the approximate position of the tet^R gene.

b. If the same tet^R gene cloned in a completely non-homologous vector is made radioactive and used as a probe in a Southern blot of this gel, which bands do you expect to appear radioactive on an autoradiogram?

c. If the same gene of interest from a fly closely related to *Drosophila* has been cloned in a nonhomologous vector, and this is used as a probe for the same gel, what bands do you expect to see on the autoradiogram?

 Unpacking the Problem

a. What are plasmids, and what is their use in recombinant DNA technology?

b. Where would you look to discover a completely new bacterial plasmid?

c. Draw a simple sketch of the pBP1 plasmid showing the tet^R gene.

d. Why do plasmids carry drug-resistance genes?

e. What does a restriction enzyme do, and how is this useful in recombinant DNA technology?

f. Why is it useful that pBP1 has a single *Hind*III site?

g. Why is it useful that pBP1 has a single *Hind*III site inside the tet^R gene?

h. What is the expected effect of cutting pBP1 with *Hind*III?

i. Is there any experimental support shown for the statement that pBP1 has a single *Hind*III site?

j. Explain briefly how the library was made.

k. Explain briefly how the probing was done to find the gene of interest.

l. In what way is clone 15 different from all the other clones?

m. How do you think genomic DNA was obtained from *Drosophila*?

n. What is the functional difference between restriction enzymes?

o. Explain precisely why ethidium bromide is needed in this experiment.

p. See if you can find other references to ethidium bromide in this book, and compare the uses to which it is put.

q. How was the Southern blot done?

r. How was the autoradiogram made?

s. What do the spots on an autoradiogram represent?

t. Compare the use of the word *label* in this question with its use in common parlance.

u. What is a kilobase? What is it a measure of? What are some other ways of measuring the same thing?

v. What causes DNA to move on an electrophoretic gel?

w. What does *nonhomologous* mean in the way it is used in this problem?

x. On the gel, what really is in a band? How might you prove what you say?

y. What is a restriction map a map of?

z. What is represented by the total number of kilobases in each lane? What kind of consistency should you see in the different lanes?

aa. Why are there more bands in the double digest than the single digests?

bb. When donor DNA is inserted at a restriction site, are any new restriction sites created by this process?

cc. When you see a band of the same size in different lanes, does it mean they must be the same DNA fragment? Can you find an example relevant to this point in this problem?

dd. What is the relationship between the number of bands produced and the number of restriction sites in a circular molecule?

2. The restriction enzyme *Eco*RI cuts DNA at the sequence GTTAAC, and the enzyme *Hae*III cuts DNA at the sequence GGCC. On average, how frequently will each enzyme cut double-stranded DNA? (In other words, what is the average spacing between restriction sites?)

3. The bacteriophage ϕX174 has in its head a single strand of DNA as its genetic material. On infection of a bacterial cell, the phage forms a complementary strand on the infective strand to yield a double-stranded replicative form (RF). Design an experiment using ϕX174 to determine whether or not transcription occurs on both strands of the RF double helix.

4. You have a purified DNA molecule, and you wish to map restriction-enzyme sites along its length. After digestion with *Eco*RI, you obtain four fragments: 1, 2, 3, and 4. After digestion of each of these fragments with *Hind*II, you find that fragment 3 yields two subfragments (3_1 and 3_2) and that fragment 2 yields three (2_1, 2_2, and 2_3). After digestion of the entire DNA molecule with *Hind*II, you recover four pieces: A, B, C, and D. When these pieces are treated with *Eco*RI, piece D yields fragments 1 and 3_1, A yields 3_2 and 2_1, and B yields 2_3 and 4. The C piece is identical with 2_2. Draw a restriction map of this DNA.

5. After treating *Drosophila* DNA with a restriction enzyme, the fragments are attached to plasmids and selected as clones in *E. coli*. Using this "shotgun" technique, David Hogness has recovered every DNA sequence of *Drosophila* in a library.

a. How would you identify the clone that contains DNA coding for the protein actin?

b. How would you identify a clone coding for a specific tRNA?

6. You have isolated and cloned a segment of DNA that is known to be a unique sequence in the genome. It maps near the tip of the X chromosome and is about 10 kb in length. You label the 5′ ends with ^{32}P and cleave the molecule with *Eco*RI. You obtain two fragments: one is 8.5 kb long; the other is 1.5 kb. You separate the 8.5 kb fragment into two fractions, partially digesting one with *Hae*III and the other with *Hind*II. You then separate each sample on an agarose gel. By autoradiography you obtain the following results:

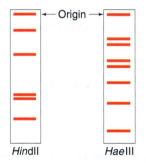

Draw a restriction-enzyme map of the complete 10-kb molecule.

7. As shown at the top of the autoradiographs in the accompanying figure, the Maxam-Gilbert technique has been used for sequencing two different DNA fragments from the *a* mating-type locus in yeast. The bases above each gel are the ones attacked by the reagent. What are the base sequences of these two DNA fragments?

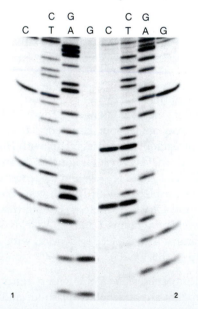

8. Design an experiment to allow the purification of DNA sequences in the Y chromosome.

9. Calculate the average distances (in nucleotide pairs) between the restriction sites in organism X for the following restriction enzymes:

*Alu*I	5′ AGCT 3′	
	3′ TCGA 5′	
*Eco*RI	5′ GAATTC 3′	
	3′ CTTAAG 5′	
*Acy*I	5′ G Pu CG Py C 3′	
	3′ C Py GC Pu G 5′	

(NOTE: Py = any pyrimidine; Pu = any purine.)

10. A linear fragment of DNA is cleaved with the individual restriction enzymes *Hind*III and *Sma*I and then with a combination of the two enzymes. The fragments obtained are

*Hind*III	2.5 kb, 5.0 kb
*Sma*I	2.0 kb, 5.5 kb
*Hind*III and *Sma*I	2.5 kb, 3.0 kb, 2.0 kb

a. Draw the restriction map.

b. The mixture of fragments produced by the combined enzymes is cleaved with the enzyme *Eco*RI, resulting in the loss of the 3-kb fragment (band stained with ethidium bromide on an agarose gel) and the appearance of a band stained with ethidium bromide representing a 1.5-kb fragment. Mark the *Eco*RI cleavage site on the restriction map.

(Problem 10 courtesy of Joan McPherson. From A. J. F. Griffiths and J. McPherson, *100 + Principles of Genetics*. W. H. Freeman and Company, 1989.)

11. A viral DNA fragment carrying a specific gene *V* is introduced into a muscle cell culture by transformation. Following incubation with ^{32}P-labeled ribonucleotides, the virus-encoded RNA product is isolated at two timed intervals. The radiolabeled viral RNA is treated as follows. First, it is hybridized to a specific cDNA previously constructed from viral-gene-*V* mature mRNA. Second, the hybrid is treated with RNase. Finally, the hybrid is denatured and electrophoresed on a gel, which is then subjected to autoradiography. The following results suggest that the pathologic nature of the virus is time-related (the number of nucleotides is indicated on the bands observed):

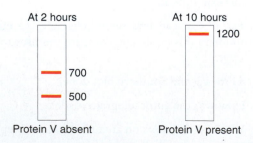

a. What is the size of the mature mRNA for gene *V*?

b. Draw a diagram of each hybrid and indicate what the illustrated bands represent.

c. Why is protein V not produced until after 2 hours?

(Problem 11 courtesy of Joan McPherson. From A. J. F. Griffiths and J. McPherson, *100 + Principles of Genetics.* W. H. Freeman and Company, 1989.)

12. The gene for β-tubulin has been cloned from *Neurospora* and is available. List a step-by-step procedure for cloning the same gene from the related fungus *Podospora*, using as the cloning vector the pBR *E. coli* plasmid shown here, where *kan* = kanamycin and *tet* = tetracycline:

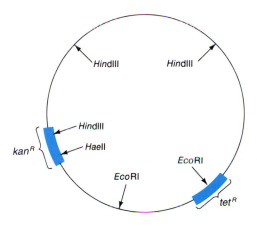

A circular bacterial plasmid containing a gene for tetracycline resistance was cut with restriction enzyme *BglII*. Electrophoresis showed one band of 14 kb.

13. a. What can be deduced from this result?

The plasmid was cut with *EcoRV* and electrophoresis produced two bands, of 2.5 and 11.5 kb.

b. What can be deduced from this result?

Digestion with both enzymes together resulted in three bands of 2.5, 5.5 and 6 kb.

c. What can be deduced from this result?

Plasmid DNA cut with *BglII* was mixed and ligated with donor DNA fragments also cut with *BglII*, to make recombinant DNA molecules. All recombinant clones proved to be tetracycline sensitive.

d. What can be deduced from this result?

One recombinant clone was cut with *BglII* and fragments of 4 and 14 kb were observed.

e. Explain this result.

The same clone was treated with *EcoRV* and fragments of 2.5, 7, and 8.5 were observed.

f. Explain these results showing a restriction map of the recombinant DNA.

14. a. A fragment of mouse DNA with *EcoRI* sticky ends carries the gene *M*. This DNA fragment, which is 8 kb long, is inserted into the bacterial plasmid pBR322 at the *EcoRI* site. The recombinant plasmid is cut with three different restriction enzyme treatments. The patterns of ethidium bromide fragments, following electrophoresis on agarose gels, are shown in this diagram:

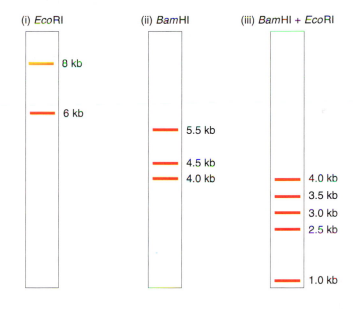

A Southern blot is prepared from gel (iii). Which fragments will hybridize to a probe (^{32}P) of pBR plasmid DNA?

b. Gene *X* is carried on a plasmid consisting of 5300 nucleotide pairs (5300 bp). Cleavage of the plasmid with the restriction enzyme *BamHI* gives fragments 1, 2, and 3, as indicated in the diagram shown below (B = *BamHI* restriction site). Tandem copies of gene *X* are contained within a single *BamHI* fragment. If gene *X* encodes a protein *X* of 400 amino acids, indicate the approximate positions and orientations of the gene *X* copies.

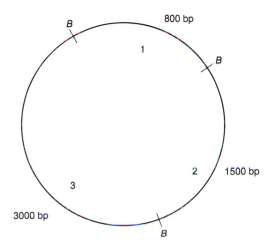

(Problem 14 courtesy of Joan McPherson. From A. J. F. Griffiths and J. McPherson, *100 + Principles of Genetics.* W. H. Freeman and Company, 1989.)

15. In functional complementation in both prokaryotes and eukaryotes prototrophy is often the phenotype selected to detect transformants. Prototrophic cells are used for

c. Mark the *Hpa*II site on the cDNA and the orientation of the two *Hpa* fragments.

(Problem 21 courtesy of Joan McPherson. From
A. J. F. Griffiths and J. McPherson, *100 + Principles
of Genetics.* W. H. Freeman and Company, 1989.)

22. Two children are investigated for the expression of a gene (*D*) that encodes an important enzyme for muscle development. The results of the studies of the gene and its product below:

Individual 1

Probe: ^{32}P gene *D* : Autoradiograms

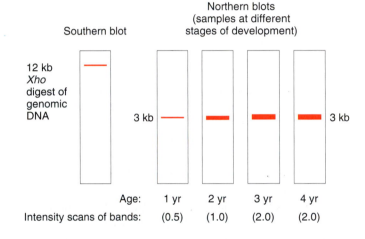

Southern blot / Northern blots (samples at different stages of development)

12 kb *Xho* digest of genomic DNA

3 kb — 3 kb

Age:	1 yr	2 yr	3 yr	4 yr
Intensity scans of bands:	(0.5)	(1.0)	(2.0)	(2.0)

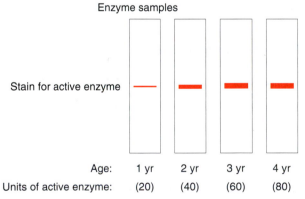

Enzyme samples

Stain for active enzyme

Age:	1 yr	2 yr	3 yr	4 yr
Units of active enzyme:	(20)	(40)	(60)	(80)

For individual 2, the enzyme activity of each stage was very low and could be estimated only at approximately 0.1 unit at age 1, 2, 3, and 4.

Individual 2

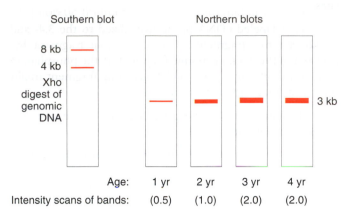

Southern blot / Northern blots

8 kb —
4 kb —
Xho digest of genomic DNA

3 kb

Age:	1 yr	2 yr	3 yr	4 yr
Intensity scans of bands:	(0.5)	(1.0)	(2.0)	(2.0)

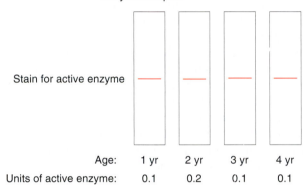

Enzyme samples

Stain for active enzyme

Age:	1 yr	2 yr	3 yr	4 yr
Units of active enzyme:	0.1	0.2	0.1	0.1

a. For both individuals, draw graphs representing the developmental expression of the gene. (Fully label both axes.)

b. How can you explain the very low levels of active enzyme for individual 2? (Protein degradation is only one possibility.)

c. How might you explain the change in the Southern blot for individual 2 compared with individual 1.

d. If only one mutant gene has been detected in family studies of the two children, define the individual children as either homozygous or heterozygous for gene *D*.

(Problem 22 courtesy of Joan McPherson. From
A. J. F. Griffiths and J. McPherson, *100 + Principles
of Genetics.* W. H. Freeman and Company, 1989.)

15

Applications of Recombinant DNA Technology

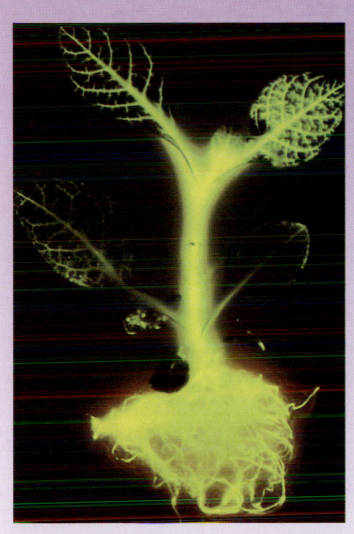

Transgenic tobacco plant expressing the luciferase gene from a firefly. (Keith Wood, Promega, Madison, Wisc.)

KEY CONCEPTS

▶ Cloning a gene allows it to be used for many different molecular applications, including probing of nucleic acids, chromosomes, and cells, and in vitro manipulation to design novel genotypes that would be impossible to make by traditional breeding methods.

▶ In vitro mutagenesis allows single-base substitutions or larger changes to be made at specific positions within a gene.

▶ Reverse genetics works in the opposite direction to traditional genetics; it allows the function of an unknown DNA or protein sequence to be identified at the phenotypic level.

▶ A gene with an easily detectable product can be spliced to the regulatory regions of another gene and thereby act as a reporter for that gene's function.

▶ Recombinant DNA technology is used to make plants and animals that carry genes from other species.

▶ Recombinant DNA technology can be applied to prenatal diagnosis of human genetic disease.

▶ In the chromosomes of an individual, specific restriction sites can be either present or absent, resulting in restriction fragment length polymorphisms, which can be used as loci for genome mapping and in diagnosis of linked disease genes.

I n the previous chapter we examined methods for creating and analyzing recombinant DNA molecules. In the following material we will see how these methods allow geneticists to alter genes at specific sites, and to use these altered genes to modify the characteristics of bacteria, fungi, plants, and animals. Recombinant DNA methodology also opens up new frontiers in the diagnosis and therapy of human genetic diseases.

In Vitro Mutagenesis

For the first time in history, geneticists now can program alterations of the genetic material. The late 1970s and the early 1980s saw the development of techniques that permit alterations of specific base pairs, allowing the introduction of many types of point mutations, deletions, and insertions into segments of cloned DNA. By directing specific mutations into predetermined segments of DNA, we make possible many experimental applications. Mutations can be tailored with exact precision, and the "designer genes" can then be inserted back into the organism and the phenotypic effects observed.

Nucleotide Substitution

Base substitutions can be introduced into cloned DNA after the generation of short, single-stranded regions, as summarized in Figure 15-1. Single-stranded regions are produced following restriction enzyme cleavage by exploiting the exonuclease activity of the *E. coli* DNA polymerase I enzyme. In the absence of deoxynucleotide triphosphates, this enzyme will digest single strands from the DNA duplex in the $3' \rightarrow 5'$ direction. If only one triphosphate is available in the reaction, the digestion will stop when the next base is the same as the one nucleotide available. In the example shown in Figure 15-1, dTTP is the only supplied nucleotide triphosphate, so the digestion following restriction cleavage of a circular DNA molecule leaves a 4-bp single-stranded stretch on one end, and a 5-bp stretch on the other end. In the single-stranded regions, treatment with bisulfite ions will deaminate cytosine to uracil, which leads to $C \rightarrow T$ transitions, as described in Chapter 19. After treatment with bisulfite, the plasmid is recircularized and the gap is filled by polymerase I and ligase.

Alternatively, base analogs can be incorporated by creating short, single-stranded gaps within a duplex (without cleaving both strands). The gap is then repaired using a modified form of DNA polymerase I (see Chapter 10), together with three nucleotide triphosphates and *N*-4-hydroxycytosine in place of T. Both tautomeric forms of *N*-4-hy-

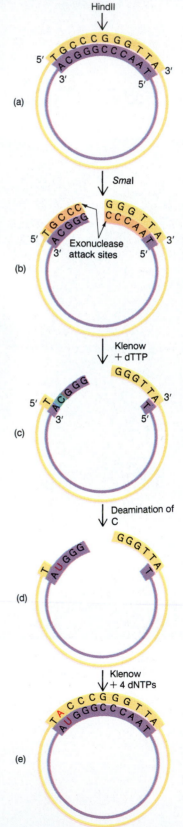

Figure 15-1 Creation of a substitution mutant through the deamination of cytosine. A restriction enzyme site—in this case, *Sma*I (a, b)—is treated with a modified form of DNA polymerase I, termed the *Klenow fragment,* in the presence of dTTP. (c) The Klenow fragment, acting as an exonuclease, performs a $3' \rightarrow 5'$ digestion of single strands until a T residue is encountered. (d) The deamination of C residues on exposed single strands is effected by the addition of bisulfite. The molecule is now repaired by the addition of all four dNTPs and the Klenow fragment. (e) The result in this case is the alteration of a G–C base pair to an A–U pair.

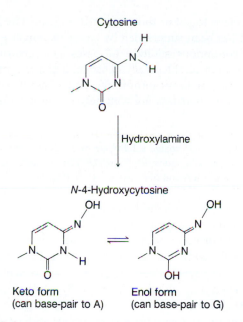

Cytosine

Hydroxylamine

N-4-Hydroxycytosine

Keto form
(can base-pair to A)

Enol form
(can base-pair to G)

Figure 15-2 The formation of *N*-4-hydroxycytosine after hydroxyl-amine treatment.

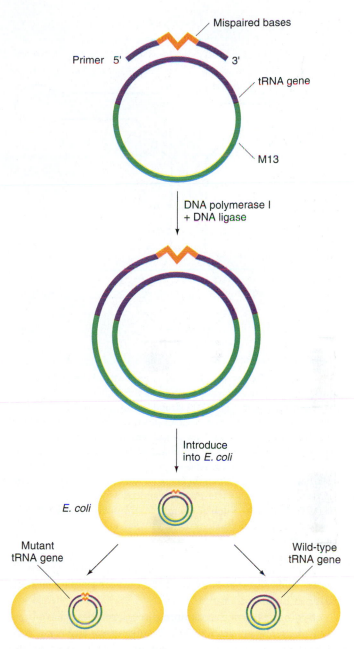

Figure 15-3 Creation of a substitution mutant by use of a synthetic oligonucleotide. The 12- to 15-base oligonucleotide is constructed so that it is complementary to a region of a DNA strand, but with one or two mismatches. When mixed with a clone of the complementary strand, the oligonucleotide will anneal to it even though the match is not exact, so long as the hybridization conditions are not stringent and the mismatches are in the middle of the oligonucleotide segment. The segment then serves as a primer for DNA polymerase I, which synthe-sizes the remainder of the complementary strand. When the resulting double-stranded molecule is introduced into *E. coli*, the molecule replicates to re-create either the mutated sequence or the original wild-type sequence.

droxycytosine (keto and enol) occur (Figure 15-2), allowing pairing with either G or A. The analog is incorporated in place of T but can also pair with G. Therefore, this in vitro method results in T → C transitions.

Less specific changes can be made at single-stranded gaps by carrying out the polymerization reaction in the ab-sence of one of the four deoxynucleotide triphosphates. At a low rate in vitro, polymerases will add nucleotides at ran-dom opposite the base complementary to the missing base.

Site-Directed Mutagenesis

The above methods are limited in their application to spe-cific nucleotide positions. However, Michael Smith has de-vised a powerful method that circumvents this limitation and allows the creation of mutations at any specific site. This technique employs synthetic oligonucleotides con-structed in vitro (see below and Chapter 14).

As a first step, the gene of interest is cloned into a sin-gle-stranded phage vector, such as the phage M13. A syn-thetic oligonucleotide serves as a primer for the in vitro syn-thesis of the complementary strand of the M13 vector (Figure 15-3). Any desired specific base change can be pro-grammed into the sequence of the synthetic primer. Al-though there will be a mispaired base when the synthetic oligonucleotide hybridizes with the complementary se-quence on the M13 vector, one or two mismatched bases can be tolerated when hybridization occurs at a low temper-ature and a high salt concentration. After DNA synthesis is mediated by DNA polymerase in vitro, the M13 DNA is replicated in *E. coli*, in which case many of the resulting phages will be the desired mutant.

The synthetic oligonucleotide can also serve as a la-beled probe to distinguish wild-type from mutant phages. Although the mismatched base will allow the primer to hy-bridize with both types of phage at low temperature, the

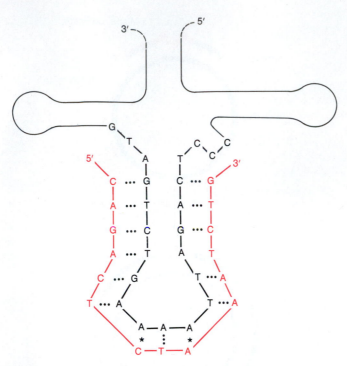

Figure 15-4 The portion of the DNA sense strand (black) for the gene coding for the lysine tRNA, showing the sequences transcribing the anticodon region. (The DNA strand is shown in the shape of the corresponding tRNA it transcribes for illustrative purposes; it does not actually assume this shape.) Base-paired with this is a synthetic oligonucleotide (red) with two mismatched base pairs (*). This can serve as a primer in the formation of a complementary strand as described in Figure 15-3. When the sense strand of this antisense mutant is produced, it codes for a (3′)AUC(5′) anticodon, which recognizes the (5′)UAG(3′) stop codon of the mRNA. Consequently, the tRNA incorporates a lysine at the stop codon, thus serving as an amber (UAG) suppressor.

primer will hybridize only with the complementary mutant phage at high temperature.

Synthetic primers containing mispaired bases have been used to create an amber (UAG) suppressor tRNA by converting the anticodon to a sequence that recognizes the UAG triplet instead of the AAA triplet (lysine codon), as depicted in Figure 15-4. The altered tRNA now inserts lysine in response to the UAG codon.

Gene Synthesis

Oligonucleotides are used not only as primers and as PCR probes (which we discussed in Chapter 14) but also in the construction of synthetic genes. Oligonucleotides were first synthesized in the mid-1960s by Gobind Khorana. He used oligonucleotides to assemble the DNA coding sequence for an alanine tRNA molecule, whose sequence had already been determined by Robert Holley. Khorana synthesized oligonucleotides by adding one base at a time in separate reactions. The oligonucleotides were designed to have overlapping complementary sequences, and he was eventually

able to piece together the entire tRNA gene. The Khorana method has been superseded by easier techniques that rely on the continuous addition of bases to a growing chain fixed to an insoluble resin. The solid-phase synthesis of DNA chains is now automated, and oligonucleotides of lengths up to 60 bases are routinely synthesized in several hours.

> **Message** Automated solid-phase methods permit the rapid synthesis of oligonucleotides, which can be used to make probes or primers and to construct synthetic genes.

The Khorana method was applied by Herbert Boyer's group to make the gene coding for a small human growth-regulating hormone, somatostatin. The hormone is a short polypeptide with the sequence shown in Figure 15-5. Boyer's group synthesized the gene using overlapping fragments. Then they added a triplet specifying methionine and an *Eco*RI cleavage site on the amino end; on the other end, they placed two consecutive stop triplets and a *Bam*HI site (Figure 15-6). The entire gene was inserted into a plasmid carrying the bacterial gene β-galactosidase, within which there is an *Eco*RI site. The other end of the somatostatin gene hybridized with a *Bam*HI site elsewhere in the plasmid. The *E. coli* selected for their possession of the plasmid were found to produce a protein chimera containing part of β-galactosidase fused to somatostatin via a methionine residue. Methionine is cleaved by cyanogen bromide, so the active hormone could be liberated by such treatment (Figure 15-7).

Another example of gene synthesis utilizing synthetic oligonucleotides produced by automated solid-phase synthesis was provided by John Abelson and Jeffrey Miller and their co-workers in 1986. They constructed an artificial phenylalanine tRNA gene, with two changes in the anticodon, so that it would read the UAG (amber) codon and become an amber suppressor tRNA. The gene can be synthesized by combining six oligonucleotides containing the two alterations. These oligonucleotides are synthesized individually, with short overlaps to allow annealing in vitro. Specific restriction-enzyme cutting sites are present at each end of the duplex that is generated. Figure 15-8 portrays the steps in the synthesis. After annealing, ligase is added and the gene is inserted into an expression vector that contains an efficient regulatory region (promoter). Because a nonsense suppressor should result, the presence of an active suppressor tRNA is detected as the ability to suppress a non-

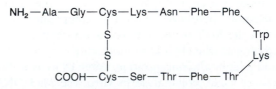

Figure 15-5 The amino acid sequence of the hormone somatostatin.

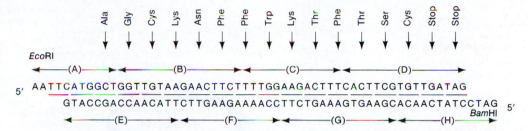

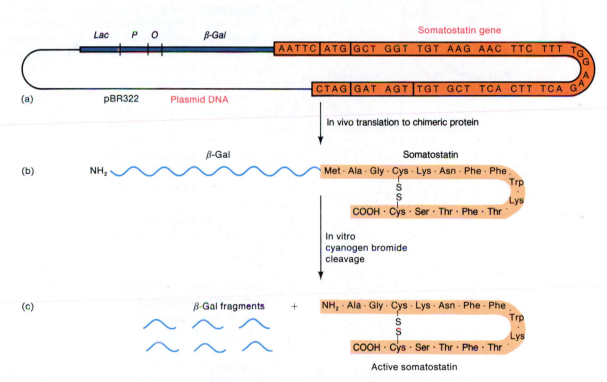

Figure 15-6 The overlapping complementary sequences (indicated by the letters in parentheses) synthesized to produce the somatostatin gene. A triplet specifying methionine is added to the 5′ end of the somatostatin coding region (inferred from the amino acid sequence); adjacent to this, an *Eco*RI restriction sequence is added. A *Bam*HI restriction sequence is added at the other end of the "artificial" gene. (From K. Itakura et al., "Expression in *Escherichia coli* of a Chemically Synthesized Gene for the Hormone Somatostatin," *Science* 198, 1977, 1056–1063. Copyright ©1977 by the American Association for the Advancement of Science.)

Figure 15-7 The production of somatostatin by *E. coli*. (a) The plasmid carrying the synthetic DNA sequence is added to the bacterial cell. (β-Gal = β-galactosidase gene; *Lac, P,* and *O* = regulatory sequences for β-Gal.) (b) A chimeric polypeptide is produced. (c) The desired somatostatin is liberated by treatment with cyanogen bromide. (From K. Itakura et al., "Expression in *Escherichia coli* of a Chemically Synthesized Gene of the Hormone Somatostatin," *Science* 198, 1977, 1056–1063. Copyright ©1977 by the American Association for the Advancement of Science.)

sense mutation in the *lacZ* gene (which encodes β-galactosidase).

Expressing Eukaryotic Genes in Bacteria

Since the original work on the production of somatostatin, many different systems have been developed for overproducing foreign proteins in *E. coli*. One example uses phage T7 RNA polymerase operating on a T7 promoter. Late in infection, phage T7 synthesizes enormous amounts of gene products from several sites termed *late promoters*. Since host chromosomal genes are not synthesized at this point, these are the only proteins synthesized. Figure 15-9 depicts two aspects of this system. The T7 RNA polymerase gene is transcribed from the *lac* promoter, and normally shut off by the *lac* repressor protein (see Chapter 19). Upon the

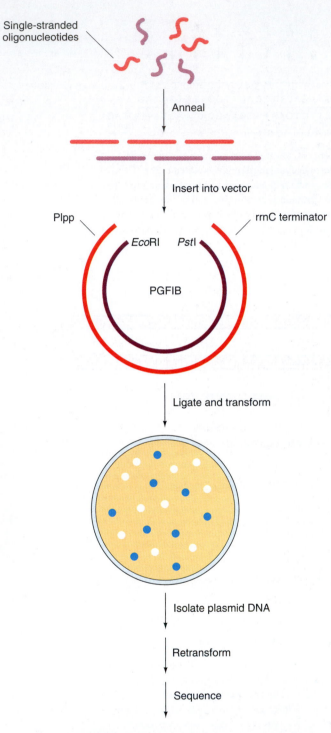

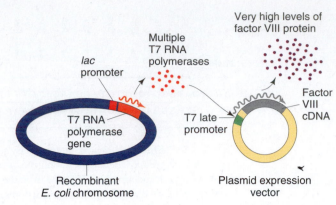

Figure 15-9 Two-step expression vector system based on bacteriophage T7 RNA polymerase and T7 late promoter. The chromosome of a specially engineered *E. coli* cell contains a copy of the T7 RNA polymerase gene under the transcriptional control of the *lac* promoter. When transcription from the *lac* promoter is induced by addition of IPTG, the T7 RNA polymerase gene is transcribed, and the mRNA is translated into the enzyme. The T7 RNA polymerase molecules produced then initiate transcription at a very high rate from the T7 late promoter on the expression vector. Multiple copies of the expression vector are present in such cells, although only one copy is diagrammed here. The large quantity of mRNA transcribed from the cDNA cloned next to the T7 late promoter is translated into abundant protein product. (Modified from H. Lodish, D. Baltimore, A. Berk, S. L. Zipursky, P. Matsudaira, J. Darnell, *Molecular Cell Biology,* 3d ed. Copyright ©1995 by Scientific American Books, Inc.)

addition of the lactose analog IPTG, the repressor is inactivated and large amounts of T7 RNA polymerase are synthesized. The gene of choice, in this example the human blood-clotting factor, factor VIII (defective in hemophiliacs), has been inserted (from a cDNA copy) adjacent to a late T7 promoter. The T7 RNA polymerase then transcribes this gene at high levels, resulting in high production of the factor VIII protein by bacteria.

Now many proteins such as human insulin (Figure 15-10), human growth hormone (Figure 15-11), and a wide range of pharmaceuticals are mass-produced from genetically engineered bacteria and fungi. Figure 15-12 shows an example in which the genetic engineering resulted in a drug that happened to give a strikingly different color to the bacterial colony when streaked onto a petri plate.

Reverse Genetics

"Regular" genetics begins with a mutant phenotype, goes on to show the existence of the relevant gene by classical genetics, and finally clones and sequences the gene to determine its DNA and protein sequence. However, **reverse genetics,** a new approach made possible by recombinant DNA technology, works in the opposite direction. Reverse genetics starts from a protein or DNA for which there is no ge-

Figure 15-8 Synthesis of a tRNA suppressor gene. Six oligonucleotides are annealed and ligated into a vector (pGFIB) constructed for the expression of the synthetic tRNA gene. The vector contains restriction sites (*Eco*RI and *Pst*I) for the insertion of the synthesized segment, as well as a promoter (Plpp) derived from the lipopolysaccharide gene and a transcription terminator (rrnC) derived from ribosomal RNA genes. The plasmid is used to transform cells that carry an amber mutation in the *lacZ* gene. These cells cannot synthesize β-galactosidase (transformations are blue, revealing the presence of active β galactosidase). These transformants are then chosen for further analysis.

netic information and then works backwards to make a mutant gene, finishing with the mutant phenotype.

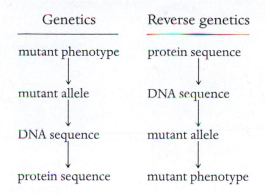

Genetics	Reverse genetics
mutant phenotype	protein sequence
↓	↓
mutant allele	DNA sequence
↓	↓
DNA sequence	mutant allele
↓	↓
protein sequence	mutant phenotype

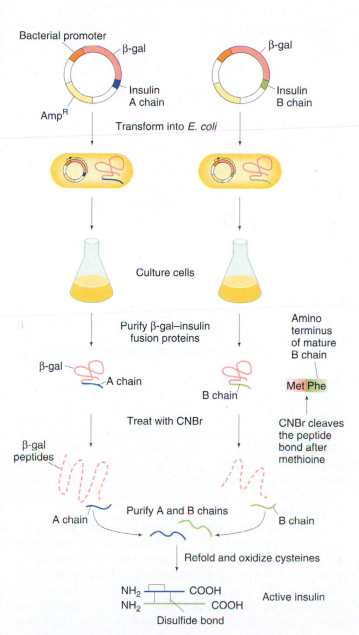

Figure 15-10 Expression of human insulin in *E. coli*. The two chains of insulin are made separately as fusion proteins with β-galactosidase. They are processed chemically, then mixed, and active insulin forms. (Copyright ©1992 by J. D. Watson, M. Gilman, J. Witkowski, and M. Zoller, *Recombinant DNA*, 2d ed. Scientific American Books.)

Often the starting point of reverse genetics is a protein or a gene "in search of a function," or looked at another way, a protein or a gene "in search of a phenotype." For example, a cloned gene sequence can be subjected to some form of in vitro mutagenesis and reinserted back into the organism to determine its phenotypic outcome. In the case of a protein of unknown function, the amino acid sequence can be translated backwards into a DNA sequence, which can then by synthesized in vitro. This sequence is then used as a probe to find the relevant gene, which is cloned and itself subjected to in vitro mutagenesis to determine phenotypic impact. This approach will become even more significant as data accumulate in whole genome–sequencing projects. Sequencing reveals numerous unknown ORFs, which are genes in search of a function.

Important tools for reverse genetics are **gene disruption** and **gene replacement.** Small-scale deletions can be introduced into a cloned gene by cutting out restriction fragments or eroding cut DNA using exonuclease. Base-pair substitutions can be introduced using oligonucleotide-directed mutagenesis. Sometimes a selectable marker is inserted into the middle of a cloned gene, which is then used to transform a wild-type recipient. Selecting for the marker yields some transformants in which the disrupted gene has replaced the in situ wild-type allele (Figure 15-13).

Message   Reverse genetics uses cryptic DNA or protein sequences to discover the normal role of that gene at the phenotypic level.

Recombinant DNA Technology in Eukaryotes

The genomes of eukaryotes are much larger and much more complex than those of bacteria. For example, in contrast to the *E. coli* genome of approximately 4000 kilobases (kb), the simple eukaryote *Neurospora crassa* has a haploid genome size of 27,000 kb; in other words, each one of the seven *Neurospora* chromosomes is approximately the same size as one bacterial genome. In humans, the genome size is 3,000,000 kb. This means that special extensions of recombinant DNA technology must be applied to manipulate these large genomes.

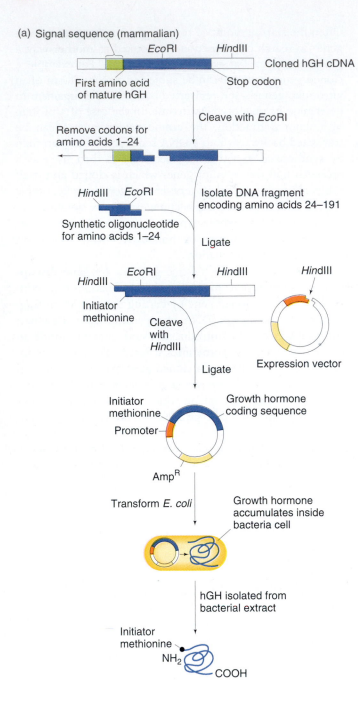

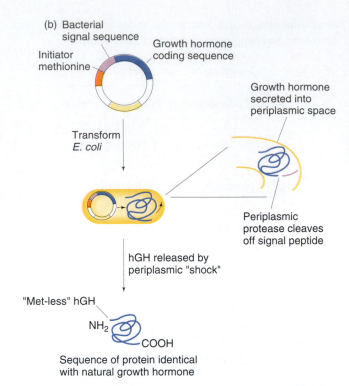

Figure 15-11 Expression of human growth hormone (hGH) in *E. coli*. (a) The human signal sequence is removed, enabling the protein to be produced in bacterial cells. The product contains an extra bacterial methionine. (b) A bacterial signal sequence can be added that targets the protein for secretion to the outside. In this method the product has no extra methionine. (Copyright ©1992 by J. D. Watson, M. Gilman, J. Witkowski, and M. Zoller, *Recombinant DNA*, 2d ed. Scientific American Books.)

Transgenic Organisms in Molecular Genetic Research

A **transgenic organism** is defined as one derived from a cell whose genome has been modified by the addition of exogenous DNA. (We have already considered one type of transgenic organism in our discussion of transformation.) The exogenous DNA can be a manipulated sequence from the same species, or DNA from another species that has some property desirable in the particular experiment. The operational gene in the exogenous DNA sample is called the **transgene.** DNA can be introduced into a cell by a variety of techniques, including transformation, injection, viral infection, or entry on shot tungsten particles (Figure 15-14).

Of course, transgenic organisms are useful in basic research because their genotypes have been genetically engineered to suit them for some specific experimental purpose. One important application is in the use of **reporter genes.** Sometimes it is difficult to detect the activity of a particular gene in the tissue where it normally functions. This problem can be circumvented by splicing the regulatory region of the gene in question (its promoter; see Chapter 19) to the coding region of a gene whose product is easily detectable, known as the *reporter gene.* Wherever and whenever the gene in question is active, the reporter gene will announce that activity in the appropriate tissue. Examples will be given later in the chapter.

Furthermore, transgenic plants, fungi, and animals are now being extensively used for applied research because of the added scope that the approach gives to the design of profitable genotypes. We have already considered the use of transgenic bacteria carrying human and various other genes important in medical and pharmaceutical applications. Broadly speaking the term **biotechnology** is the name given to this area of research, the application of recombinant DNA techniques in commercial enterprises. Patented "de-

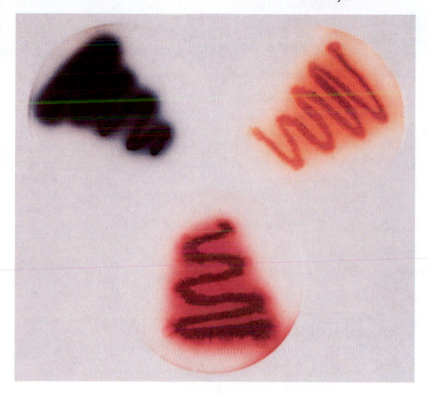

Actinorhodin

Medermycin

Figure 15-12 Genetically engineered bacteria. This figure shows how new antibiotics can be synthesized by genetically engineered bacteria. Different species of *Streptomyces* produce related but different antibiotics. One species produces actinorhodin, which is blue at alkaline pH. A second species produces medermycin, which is brown. When the genes from the second species were cloned and transformed on a plasmid into the first species, some of the transformed bacteria had acquired the capacity to synthesize a new antibiotic, mederrhodin A, which gives a reddish-purple color. The structures of the three compounds are shown here, together with the streaks of the bacteria synthesizing each of the three compounds. The new antibiotic, mederrhodin A, is similar to medermycin but carries a hydroxyl group at the 6 position. The cloned segment probably contains a gene that encodes a hydrolyase that is expressed when transformed into a different species. (Mervyn J. Bibb, John Innes Institute, United Kingdom)

Mederrhodin A

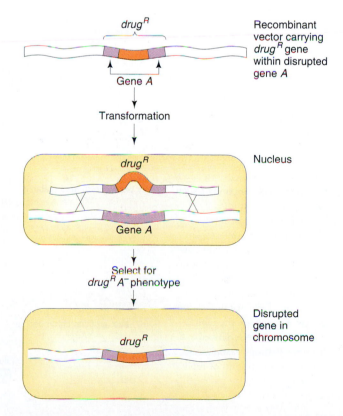

signer organisms" are expected to become a dominant aspect of the economy early in the next century.

Message Transgenesis, the design of a specific genotype by the addition of exogenous DNA to the genome, has increased the scope of breeding in basic genetics and in commercial applications.

Genetic Engineering in Yeast

The yeast *Saccharomyces cerevisiae* has become the *E. coli* of the eukaryotes. One of the main reasons is that the classical

Figure 15-13 Gene disruption by homologous integration of a sequence containing a selectable marker.

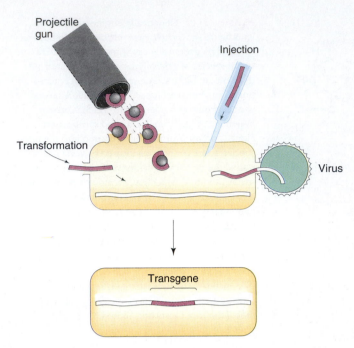

Figure 15-14 Some of the different ways of introducing foreign DNA into a cell.

genetics of yeast is extremely well developed, and the stockpile of thousands of mutants affecting hundreds of different phenotypes is a rich source of genetic markers in the development of yeast as a molecular system. Today the blend of classical genetics and molecular biology is indeed a powerful analytical combination in any organism. In yeast, another important advantage is the occurrence of a circular 6.3-kb natural yeast plasmid, named the 2μ plasmid, after its circumference. This plasmid is transmitted normally to the products of meiosis and mitosis. The 2μ plasmid forms the basis for several specially engineered, sophisticated yeast vectors.

Yeast Vectors. The simplest yeast vectors are derivatives of bacterial plasmids into which a section of yeast DNA has been inserted (Figure 15-15a). When transformed into yeast cells, these plasmids can be inserted into yeast chromosomes by homologous recombination, involving a single or a double crossover (Figure 15-16). Either the entire plasmid is inserted, or the targeted allele is replaced by the allele on the plasmid. Such integrations can be detected by plating

Figure 15-15 Simplified representations of four different kinds of plasmids used in yeast. Each is shown acting as a vector for some genetic region of interest, which has been inserted into the vector. The function of such segments can be studied by transforming a yeast strain of suitable genotype. Selectable markers are needed for the routine detection of the plasmid in bacteria or yeast. Origins of replication are sites needed for the bacterial or yeast replication enzymes to initiate the replication process. (DNA derived from the natural yeast plasmid 2μ has its own origins of replication.)

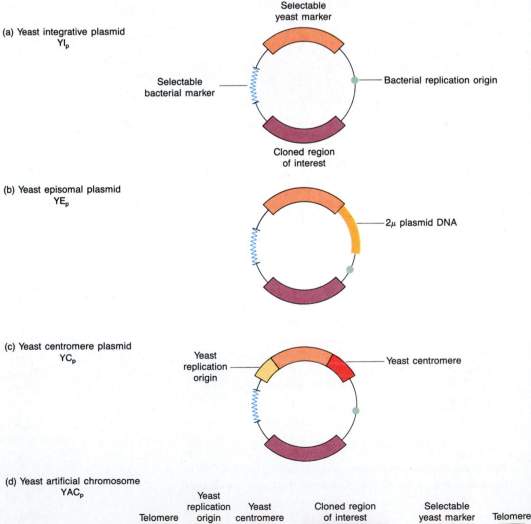

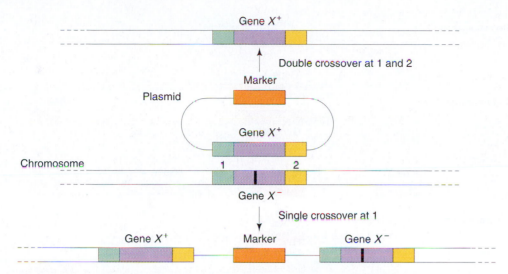

Figure 15-16 Two ways in which a recipient yeast strain bearing a defective X^- can be transformed by a plasmid bearing an active allele (gene X^+). The mutant site of gene X^- is represented as a vertical black bar. Single crossovers at position 2 are also possible but are not shown.

cells on a medium that selects for the allele on the plasmid. Because bacterial plasmids do not replicate in yeast, integration is the only way to generate a stable transgenic phenotype.

If the 2μ plasmid is used as the basic vector and other bacterial and yeast segments are spliced into it (Figure 15-15b), then a construct is obtained that has several useful properties. First, the 2μ segment confers the ability to replicate autonomously in the yeast cell, and insertion is not necessary for a stable transgenic phenotype. Second, genes can be introduced into yeast, and their effects can be studied in that organism; then the plasmid can be recovered and put back into *E. coli,* provided that a bacterial replication origin and a selectable bacterial marker are on the plasmid. Such **shuttle vectors** are very useful in the routine cloning and manipulation of yeast genes.

With any autonomously replicating plasmid, there is the possibility that a daughter cell will not inherit a copy because the partitioning of copies to daughter cells is essentially a random process depending on where the plasmids are in the cell when the new cell wall is formed. However, if the section of yeast DNA containing a centromere is added to the plasmid (Figure 15-15c), then the nuclear spindle that ensures the proper segregation of chromosomes will treat the plasmid in somewhat the same way and partition it to daughter cells more efficiently at cell division. The addition of a centromere is one step toward the formation of an artificial chromosome. A further step has been made by linearizing a plasmid containing a centromere and adding the DNA from yeast telomeres to the ends (Figure 15-15d). If this construct contains yeast replication origins (autonomous replication sequences, ARS), then it constitutes a yeast artificial chromosome (YAC), which behaves in many ways like a small yeast chromosome at mitosis and meiosis. For example, when two haploid cells—one bearing a

$trp^- \ ura^+$ YAC and another bearing a $trp^+ \ ura^-$ YAC are brought together to form a diploid, some tetrads will show the clean segregations expected if these two elements are behaving as regular chromosomes. In other words, two ascospores will show $trp^- \ ura^+$ and the other two will show $trp^+ \ ura^-$ genotypes.

Applications of Yeast Vectors. We have already discussed how an in vitro–mutated gene can be inserted back into an organism to replace the resident wild-type gene and the effects observed. The yeast integrative plasmid (YIp) provides a model of how this can be achieved in a eukaryote. A mutated gene and its flanking regions, carried on an integrating plasmid, provide a region of homology in which a crossover can occur; the entire plasmid then is inserted into the wild-type yeast chromosome at the proper locus (Figure 15-17). Owing to the presence of two copies of the gene in question (one mutated and one wild-type) on one chromosome, pairing can occur by looping and a crossover at a different site can excise the plasmid—this time bearing the wild-type allele and leaving the mutant allele in the normal chromosomal locus.

Selection for Loss of Plasmid. In some cases, we can select for the elimination of the plasmid sequence. For example, if the plasmid selection marker is $lys2^+$ and the recipient yeast chromosome bears $lys2^-$ (Figure 15-17), then insertion produces a strain with $lys2^+$ and $lys2^-$. The chemical α-aminoadipate permits the growth only of $lys2^-$ strains, so strains that have lost the insert can then be selected by plating on this chemical. About half these strains retain the original wild-type gene of interest (gene X^+); the remainder retain the plasmid-borne mutant allele X^-. In other cases, the mutant phenotype of the mutant gene X^- can be selected directly by appropriate platings.

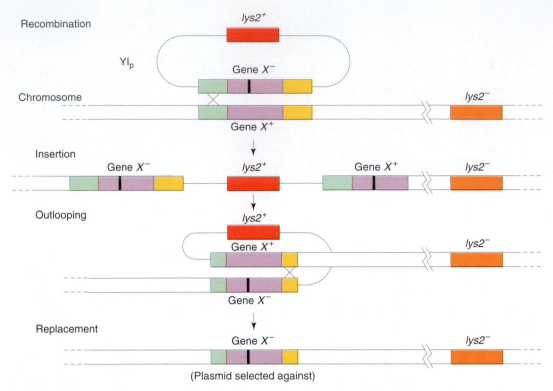

Figure 15-17 A two-step method for replacing an active gene X^+ with a deliberately engineered mutant allele (X^-) for the purpose of observing its effects on phenotype. The $lys2^+$ allele is a yeast marker that can be selected *for* (through the prototrophy that it confers) or *against* (by plating on the chemical α-aminoadipate).

Note that YIp plasmids (Figure 15-15a) carry two yeast elements: the gene under investigation, *X*, and the selectable marker. Both can undergo homologous recombination with their respective chromosomal loci. The specificity of insertion can be targeted more efficiently by making a restriction cut in the plasmid at the gene in question (gene X^+). Since the ends are recombinogenic (recombine preferentially), they direct the plasmid's entry to that specific site (Figure 15-18).

Gene Disruptions. Sometimes, all that is desired is to specifically inactivate a gene of interest. The methods for such disruptions were pioneered in yeast. A way of making such gene disruptions in one step is actually to insert another gene with a selectable function into the middle of a wild-type allele of the gene of interest, carried in a plasmid. A linear derivative of such a construct will then insert specifically at the wild-type locus, automatically disrupting it by virtue of the selectable gene inside it. Thus, if the gene of interest is $lys2^+$ (Figure 15-19), a $his3^+$ gene might be chosen to be inserted into $lys2^+$ carried in a plasmid. A linear form of this would be used to transform a strain of genotype $lys2^+$ $his3^-$, and $his3$ transformants would be selected. Some transformants are found also to be $lys2^-$ by virtue of the replacement of $lys2$ by the disrupted $lys2$ allele from the plasmid.

Studying Regulation. Centromere plasmids can be used to study the regulatory elements upstream (5′) of a gene (Figure 15-20). The relevant coding region and its upstream region can be spliced into a plasmid, which can be selected by a separate yeast marker such as $URA3^+$. The upstream region can be manipulated by inducing a series of deletions, which are achieved by cutting the DNA, using a special exonuclease to chew away the DNA in one direction to different extents, and then rejoining it. The experimental objective is then to determine which of these deletions still permits normal functioning of the gene. Proper function is assayed by transforming the plasmid into a recipient in which the chromosome locus carries a defective mutant allele and then monitoring for return of gene function in the recipient. The results generally define a specific region that is necessary for normal function and regulation of the gene.

In such regulatory studies, it is often more convenient to use a reporter gene instead of the gene of interest. Therefore, if the regulation of gene X is of interest, the upstream regulatory regions of gene X are spliced to the reporter gene. The reporter gene has a phenotype that is easier to monitor than that of gene X, so the normal regulatory signals of gene X are expressed through the reporter. A gene that has been extensively used as a reporter in yeast is the bacterial *lacZ* gene, which codes for the enzyme β-galactosidase. This enzyme normally breaks down lactose, but it can

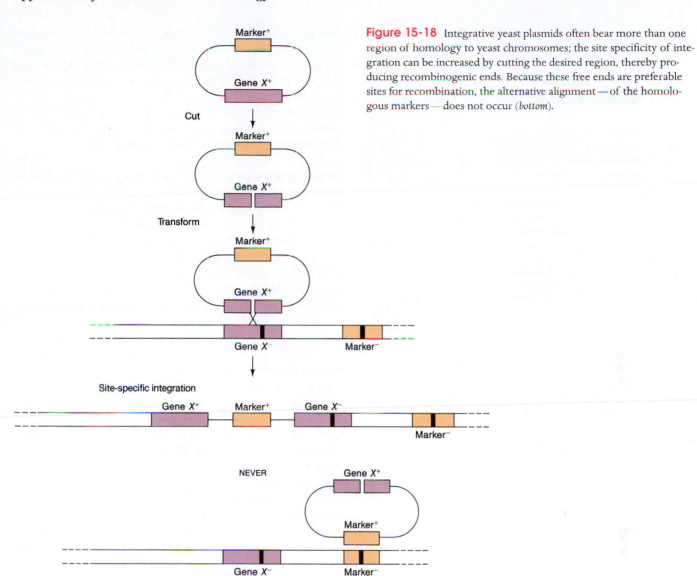

Figure 15-18 Integrative yeast plasmids often bear more than one region of homology to yeast chromosomes; the site specificity of integration can be increased by cutting the desired region, thereby producing recombinogenic ends. Because these free ends are preferable sites for recombination, the alternative alignment—of the homologous markers—does not occur (*bottom*).

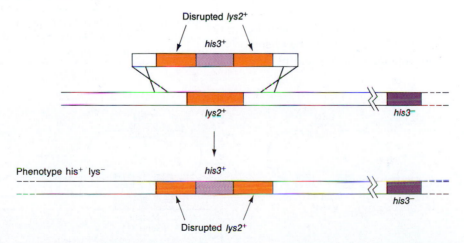

Figure 15-19 A single-step method for replacing a wild-type gene with a disrupted (inactivated) version carried on a linear DNA fragment for the purpose of specifically knocking out gene function. Integrants are selected as *his*[+] in this case.

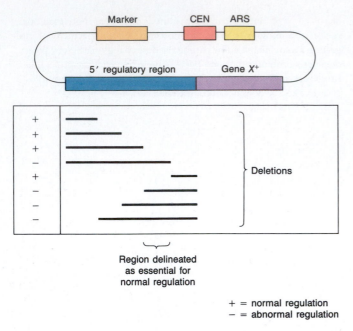

Figure 15-20   The regulation of the yeast gene X^+ can be studied by manipulating its regulatory region through deletion analysis in vitro and then transforming the constructs into a yeast strain bearing a defective allele X^-. (CEN = centromere sequence; ARS = autonomously replicating sequence.)

also break down an analog of lactose, X-gal (5-bromo-4-chloro-indolyl-β,D-galactoside), very efficiently to yield 5-bromo-4-chloro-indigo, which is bright blue. The blue color is expressed as a blue yeast colony whenever it is active. (In Chapter 14 we saw the use of this enzyme to select for integrants into the pUC18 plasmid.) Generally, the fusions are constructed in such a way that the upstream regulatory regions of gene X plus a few codons from the structural gene X are fused in the correct reading frame to the region coding for the enzymatically active portion of β-galactosidase. These constructs can be transformed on a nonintegrative vector.

Retrieval. Autonomously replicating vectors can also be used as retrieval agents. If a particularly interesting phenotype is produced by a specific mutant allele, it is useful to be able to retrieve that allele easily and examine its structure and function. This can be achieved by transformation with a gapped, centromeric plasmid that bears a deleted form of the gene of interest. The gap is repaired using information from the in situ mutant locus. The mutant sequence in the targeted locus is thus introduced into the plasmid. The repaired plasmid is then simply reisolated from the strain and examined at the molecular level (Figure 15-21).

Other Applications.   YACs are finding extensive use as cloning vectors for large sections of eukaryotic (especially human) DNA. Consider, for example, that the size of the region coding blood-clotting factor VIII in humans is known to span about 190 kb and that the gene for Duchenne muscular dystrophy spans more than 1000 kb. Furthermore, the large size of mammalian genomes in general means that libraries made in bacterial vectors would be huge. Yeast artificial chromosomes, as we have seen, can carry much longer inserts, and the library size is correspondingly smaller. We return to this topic in Chapter 17.

Another useful facility in the genetic engineering of yeast is the ability to separate chromosomes using pulsed field gel electrophoresis. Because the chromosomes of fungi are small by eukaryotic standards, whole chromosomes can be visualized by this technique. In more complex eukaryotes, the chromosomes are huge and do not migrate through a gel. However, if they are cut using restriction enzymes that cut at rare target sites (called *long cutter enzymes*), then the chromosomes can be cut into sizes that can conveniently be separated and characterized using pulsed field electrophoresis. (This topic will be covered in Chapter 17.)

The yeast system for genetic engineering is by far the most sophisticated of eukaryotes at present. However, the same techniques are being rapidly applied to other organisms—especially ones with well-defined genetic systems, such as the filamentous fungi *Neurospora* and *Aspergillus* and the nematode *Caenorhabditis elegans*.

Message   Yeast vectors can be integrative, autonomously replicating, or resembling artificial chromosomes, allowing genes to be isolated, manipulated, inserted, and retrieved during molecular genetic analysis.

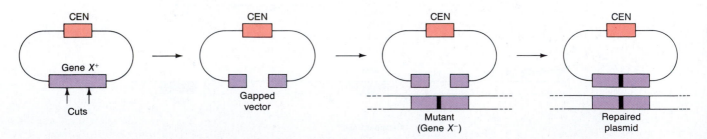

Figure 15-21   A gapped yeast centromeric plasmid is repaired by DNA copied from the homologous chromosome locus. This provides a convenient way of retrieving a mutant sequence of particular interest.

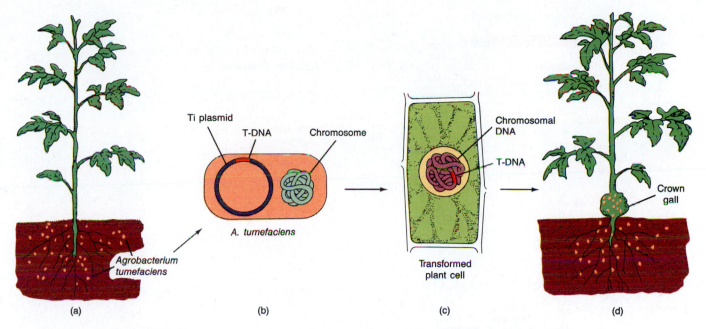

Figure 15-22 In the process of causing crown gall disease, the bacterium *Agrobacterium tumefaciens* inserts a portion of its Ti plasmid—a region called *T-DNA*—into a chromosome of the host plant.

Genetic Engineering in Plants

Because of their economic significance, plants have long been the subject of genetic analysis aimed at developing improved varieties. Recombinant DNA technology has introduced a new dimension to this effort because the genome modifications made possible by this technology are almost limitless. No longer is breeding confined to selecting variants within the species. DNA can now be introduced from other species of plants, animals, or even bacteria!

The Ti Plasmid. The only vectors routinely used to produce transgenic plants are derived from a soil bacterium called *Agrobacterium tumefaciens*. This bacterium causes what is known as *crown gall disease*, in which the infected plant produces uncontrolled growths (tumors, or galls), normally at the base (or crown) of the plant. The key to tumor production is a large (200-kb) circular DNA plasmid—the Ti (tumor-inducing) plasmid. When the bacterium infects a plant cell, a part of the Ti plasmid—a region called the *T-DNA*—is transferred and inserted, apparently more or less at random, into the genome of the host plant (Figure 15-22). The functions required for this transfer are outside the T-DNA on the Ti plasmid. The T-DNA itself carries several interesting functions, including the production of the tumor and the synthesis of compounds called *opines*. Opines are actually synthesized by the host plant under the direction of the T-DNA. The bacterium then uses the opines for its own purposes, calling on opine-utilizing genes on the Ti plasmid. Two important opines are nopaline and octopine; two separate Ti plasmids produce them. The structure of Ti is shown in Figure 15-23.

Using the Ti Plasmid as a Vector. The natural behavior of the Ti plasmid makes it well-suited for the role of a plant vector. If the DNA of interest could be spliced into the T-DNA, then the whole package would be inserted in a stable state into a plant chromosome. This system has indeed been made to work essentially in this way, but with some necessary modifications. Let's follow through one protocol.

Ti plasmids are too large to be easily manipulated and cannot readily be made smaller, since they contain few

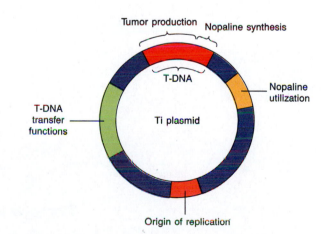

Figure 15-23 Simplified representation of the major regions of the Ti plasmid of *A. tumefaciens*. The T-DNA, when inserted into the chromosomal DNA of the host plant, directs the synthesis of nopaline, which is then utilized by the bacterium for its own purposes. T-DNA also directs the plant cell to divide in an uncontrolled manner, producing a tumor.

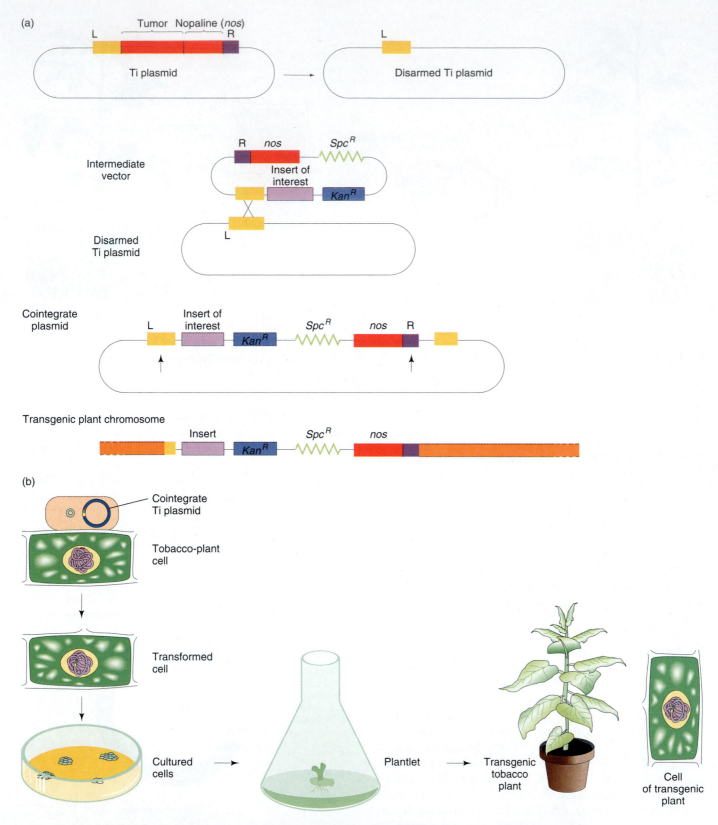

Figure 15-24 (a) To produce transgenic plants, an intermediate vector of manageable size is used to clone the segment of interest. In the method shown here, the intermediate vector is then recombined with an attenuated ("disarmed") Ti plasmid to generate a cointegrate structure bearing the insert of interest and a selectable plant kanamycin-resistance markers between the T-DNA borders, which are all the T-DNA that is necessary to promote insertion. (b) The generation of a transgenic plant through the growth of a cell transformed by T-DNA.

unique restriction sites. Consequently, a smaller, intermediate vector initially receives the insert of interest and the various other genes and segments necessary for recombination, replication, and antibiotic resistance. Once engineered with the desired gene elements, this intermediate vector can then be inserted into the Ti plasmid, forming a cointegrate plasmid that can be introduced into a plant cell by transformation. Figure 15-24a shows one method of creating the cointegrate. The Ti plasmid that will receive the intermediate vector is first attenuated; that is, it has the entire right-hand region of its T-DNA, including tumor genes and nopaline-synthesis genes deleted, rendering it incapable of tumor formation—a "nuisance" aspect of the T-DNA function. It retains the left-hand border (L) of its T-DNA, which will be used as the crossover site for incorporation of the intermediate vector. The intermediate vector has had a convenient cloning segment spliced in, containing a variety of unique restriction sites. The gene of interest has been inserted at this site in Figure 15-25. Also spliced into the vector are a selectable bacterial gene (Spc^R) for spectinomycin resistance; a bacterial kanamycin-resistance gene (Kan^R), engineered for expression in plants; and two segments of T-DNA. One segment carries the nopaline-synthesis gene (nos), plus the right-hand T-DNA border sequence (R). The second T-DNA segment comes from near the left-hand border and provides a section for recombination with a homologous portion of region L, which was retained in the disarmed Ti plasmid. After the intermediate vectors are introduced to *Agrobacterium* cells containing the disarmed Ti plasmids (by conjugation with *E. coli*), plasmid recombinants (cointegrates) can be selected by plating on spectinomycin. The selected bacterial colonies will contain only the Ti plasmid, since the intermediate vector is incapable of replication in *Agrobacterium*.

As Figure 15-25b shows, following spectinomycin selection for the cointegrates, bacteria containing the recombinant double, or "cointegrant," plasmid are then used to infect cut segments of plant tissue, such as punched out leaf disks. If bacterial infection of plant cells occurs, anything between the left and right T-DNA border sequences can be inserted into the plant chromosomes. If the leaf disks are placed on a medium containing kanamycin, the only plant cells that will go through cell division are those that have acquired the Kan^R gene from T-DNA transfer. The growth of such cells results in a clump, or callus, which is an indication that transformation has occurred. These calluses can be induced to form shoots and roots and then be transferred to soil, where they develop into transgenic plants (Figure 15-25b). Often only one T-DNA insert is detectable in such plants, where it segregates at meiosis like a regular Mendelian allele (Figure 15-25). The insert can be detected by a T-DNA probe in a Southern hybridization, or by the detection of the chemical nopaline in the transgenic tissue.

Expression of Cloned DNA. The DNA cloned into the T-DNA can, of course, be any DNA the investigator wants to

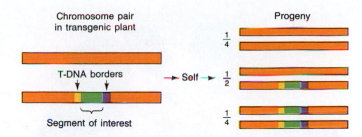

Figure 15-25 T-DNA and any DNA contained within it are inserted into a plant chromosome in the transgenic plant and then transmitted in a Mendelian pattern of inheritance.

test in the plant being used. One particularly striking foreign DNA that has been inserted using T-DNA is the gene for the enzyme luciferase, which is isolated from fireflies. The enzyme catalyzes the reaction of a chemical called *luciferin* with ATP; in this process, light is emitted, which explains why fireflies glow in the dark. A transgenic tobacco plant expressing the luciferase gene will also glow in the dark when watered with a solution of luciferin (see chapter opening photograph). This might seem like an attempt to develop technology to make Christmas trees that do not need lights, but in fact the luciferase gene is useful as a reporter to monitor the function of any gene during development. In other words, the upstream regulatory sequences of any gene of interest can be fused to the luciferase gene and put into a plant via T-DNA. Then the luciferase gene will follow the same developmental pattern as the normally regulated gene does, but the luciferase gene will announce its activity prominently by glowing at various times or in various tissues, depending on the regulatory sequence.

Other genes used as reporters in plants are the bacterial *GUS* (β-glucuronidase) gene, which turns X-gluc blue, and the bacterial *lac* (β-galactosidase) gene, which turns X-gal blue.

An agriculturally important example of inserting foreign DNA via T-DNA is a bacterial gene for resistance to the herbicide glyphosate. This gene confers resistance to the transgenic plant, enabling it to withstand the field application of glyphosate as a weed killer. Another example is "flavrsavr" tomatoes, which carry a gene that delays fruit-ripening, giving the fruit better transportability and longer shelf life. Indeed, most aspects of plant productivity are being subjected to genetic engineering. Not only are the qualities of the plant species itself being manipulated, but plants are also being used as convenient factories to produce proteins coded by foreign genes.

Genetic Engineering in Animals

Some of the animals most extensively used as model systems for DNA manipulation are *Caenorhabditis elegans* (a nematode), *Drosophila*, and mice. Versions of many of the techniques considered so far can also be applied in these animal systems.

Figure 15-26 Transgenic *Drosophila* expressing a bacterial β-galactosidase gene. *Drosophila* was transformed with a construct consisting of the *E. coli lacZ* gene driven by a *Drosophila* heat shock promoter. The resulting flies were heat-shocked, killed immediately, and stained for β-galactosidase activity, detected by production of a blue pigment. Transformed fly is at right, normal fly at left. (John Lis)

Transgenic Animals. There are several ways of producing transgenic animals. One example is the production of transgenic *Drosophila* by the injection of plasmid vectors containing P elements into the fly egg (described in detail in Chapter 21, page 652). Transgenic *Drosophila* provide us with another illustration of the use of the bacterial *lacZ* gene as a reporter in the study of gene regulation during development. The *lacZ* gene is fused to the upstream regulatory region of a *Drosophila* heat shock gene, which is normally activated by high temperatures. This construct is then used to generate transgenic flies. Following heat shock, the flies are killed and bathed in X-gal (pages 470–472). The resulting pattern of blue tissues provides information on the major sites of action of the heat shock gene (Figure 15-26). Another example of transgenesis is the production of transgenic mammals by injecting special plasmid vectors into a fertilized egg (discussed later in this chapter).

In both these cases, because it is the egg that is initially made transgenic, the extra DNA can find its way into germ-line cells, is then passed on to the progeny derived from these cells, and behaves from then on rather like a regular

Figure 15-27 Production of a pharmaceutically important protein in the milk of transgenic sheep. The gene of interest encodes a protein that is of therapeutic importance, such as tissue plasminogen activator used to dissolve blood clots in humans. The gene is placed under control of the β-lactoglobulin promoter, active only in mammary tissue, and is introduced into sheep ova by microinjection of the expression vector into the nucleus. The injected ova are implanted into foster mothers, and progeny expressing the transgene are identified by PCR amplification of chromosomal DNA using primers from the sequence of the gene of interest. Transgenic sheep express this gene only in mammary tissue and secrete high levels of the gene's protein into the milk, from which it can be purified.

nuclear gene. Like plants, animals are being manipulated not only to improve the qualities of the animal itself, but also to act as convenient producers of foreign proteins. For example, mammalian milk is easily obtained, so it is a convenient medium for the collection of proteins that are otherwise more difficult to obtain (Figure 15-27).

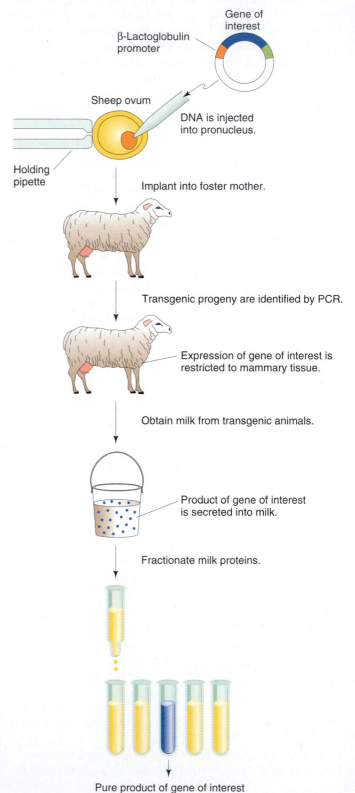

Gene Disruptions and Gene Replacement in Mice. From gene disruptions in yeast, we can gain valuable information about the role of a specific gene by replacing it with a mutated copy. These so-called knockout versions of the organism can then be examined for altered phenotypes. Knockout mice are invaluable models for the study of mutants similar to those found in humans. For instance knockout mice that lack vital DNA repair enzymes (see Chapter 19) have been created to study whether this causes an increase in cancer rates.

Figures 15-28 and 15-29 portray the generation of a knockout mouse. First, a cloned, disrupted gene is used to produce embryonic stem (ES) cells containing a gene knockout (Figure 15-29a). Although recombination of the

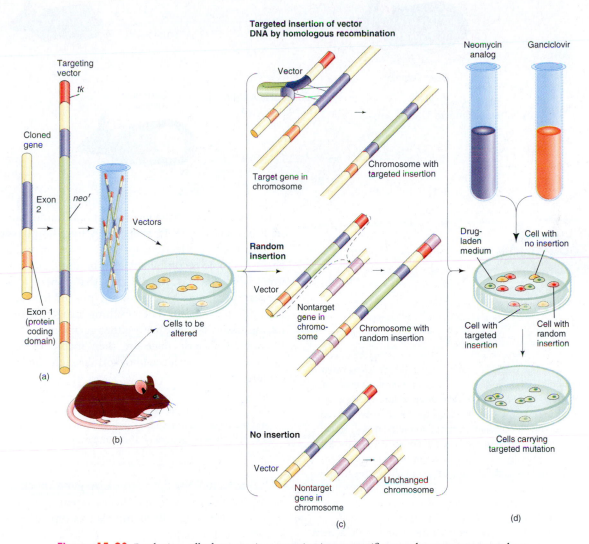

Figure 15-28 Producing cells that contain a mutation in one specific gene, known as a targeted mutation or a gene knockout. (a) Copies of a cloned gene are altered in vitro to produce the targeting vector. The gene shown here has been inactivated by insertion of the *neo*^r gene (*green*) into a protein coding region (*blue*). The *neo*^r gene will serve later as a marker to indicate that the vector DNA took up residence in a chromosome. The vector has also been engineered to carry a second marker at one end: the herpes *tk* gene (*red*). These markers are standard, but others could be used instead. (b) Once a vector, with its dual markers, is complete, it is introduced into cells isolated from a mouse embryo. (c) When homologous recombination occurs (*top*): those regions on the vector (together with any DNA in between) take the place of the original gene, excluding the marker at the tip (*red*). In many cells, though, the full vector (complete with the extra marker) fits itself randomly into a chromosome (*middle*) or does not become integrated at all (*bottom*). (d) To isolate cells carrying a targeted mutation, all the cells are put into a medium containing selected drugs, here a neomycin analog (G418) and ganciclovir. G418 is lethal to cells unless they carry a functional *neo*^r gene, and so it eliminates cells in which no integration of vector DNA has occurred (*yellow*). Meanwhile, ganciclovir kills any cells that harbor the *tk* gene, thereby eliminating cells bearing a randomly integrated vector (*red*). Consequently, virtually the only cells that survive and proliferate are those harboring the targeted insertion (*green*). (Redrawn from M. R. Capecchi, "Targeted Gene Replacement," *Scientific American,* March 1994.)

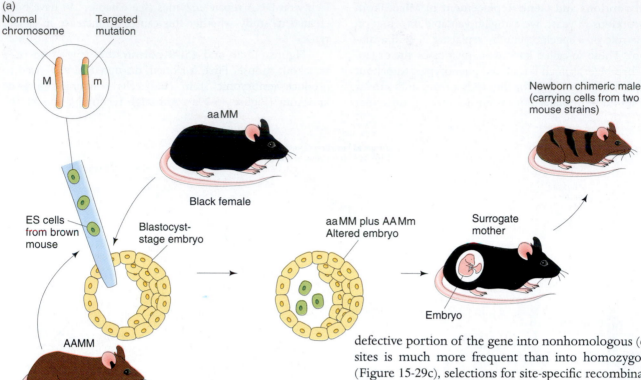

Figure 15-29 Producing a "knockout mouse" carrying the targeted mutation. (a) Embryonic stem (ES) cells (*green at far left*) are isolated from an agouti mouse strain and altered to carry a targeted mutation in one chromosome. The ES cells are then inserted into young embryos, one of which is shown. Coat color of the future newborns are a guide to whether the ES cells have survived in the embryo. Hence, ES cells are typically put into embryos that, in the absence of the ES cells, would acquire a totally black coat. Such embryos are obtained from a black strain that lacks the dominant agouti allele. (See Chapter 4.) The embryos containing the ES cells grow to term in surrogate mothers. Newborns with agouti shading intermixed with black indicate that the ES cells have survived and proliferated in an animal. (Such individuals are called *chimeras* because they contain cells derived from two different strains of mice.) Solid black coloring, in contrast, would indicate that the ES cells had perished, and these mice are excluded. (b) A represents agouti, a black; m is the targeted mutation and M is its wild-type allele. Chimeric males are mated to black (nonagouti) females. Progeny are screened for evidence of the targeted mutation (*green in inset*) in the gene of interest. Direct examination of the genes in the agouti mice reveals which of those animals (*boxed*) inherited the targeted mutation. Males and females carrying the mutation are mated to one another to produce mice whose cells carry the chosen mutation in both copies of the target gene (*inset*) and thus lack a functional gene. Such animals (*boxed*) are identified definitively by direct analyses of their DNA. The knockout results in a curly-tail phenotype. (Redrawn from M. R. Capecchi, "Targeted Gene Replacement," *Scientific American*, March 1994.)

defective portion of the gene into nonhomologous (ectopic) sites is much more frequent than into homozygous sites (Figure 15-29c), selections for site-specific recombinants and against ectopic recombinants can be used, as shown in Figure 15-29d. Second, the ES cells that contain one copy of the disrupted gene of interest are injected into an early embryo (Figure 15-30). The resulting progeny are chimeric, having tissue derived from either recipient or transplanted ES lines. Chimeric mice are then mated to produce homozygous mice with the knockout in each copy of the gene (Figure 15-30).

Message   Fungal, plant, and animal genes can be cloned and manipulated in bacteria, and reintroduced into the eukaryote cell, where generally they integrate into chromosomal DNA.

The technology for mammalian gene knockouts is similar to that necessary to specifically replace a defective allele by a wild-type allele. In mammals this type of procedure is known as *gene therapy*.

Gene Therapy

Gene therapy has become possible because of the development of the technology for making transgenic organisms. The general approach of gene therapy is nothing more than an extension of the clone selection technique of functional complementation (Chapter 14). The functions absent in the recipient as a result of a defective gene are introduced on a vector which inserts into one of the chromosomes of the recipient and thereby generates a transgenic animal that has been genetically cured. The technique is of great relevance to humans because it offers the hope of correcting hereditary diseases. However, gene therapy is also being applied to mammals other than humans.

(b)

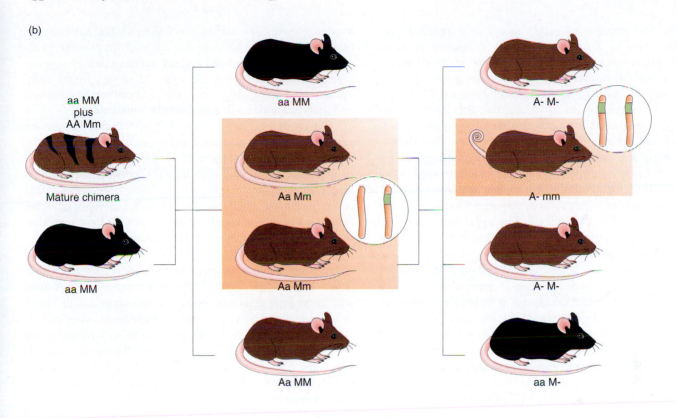

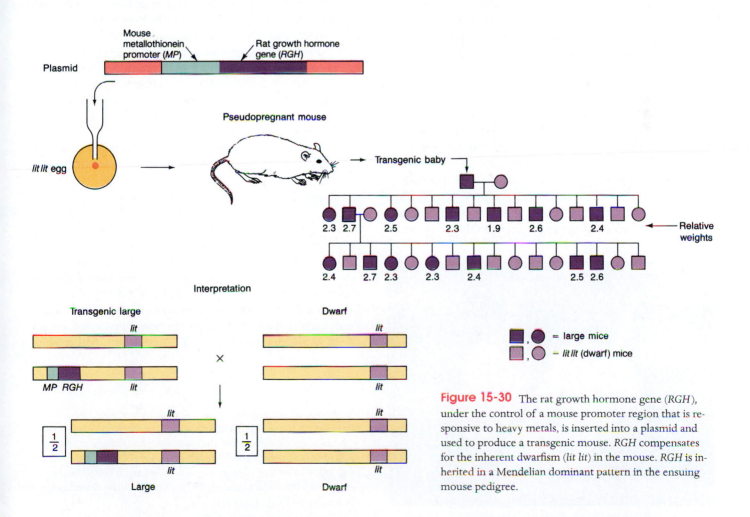

Figure 15-30 The rat growth hormone gene (*RGH*), under the control of a mouse promoter region that is responsive to heavy metals, is inserted into a plasmid and used to produce a transgenic mouse. *RGH* compensates for the inherent dwarfism (*lit lit*) in the mouse. *RGH* is inherited in a Mendelian dominant pattern in the ensuing mouse pedigree.

The first example of gene therapy in a mammal was the correction of a growth hormone deficiency in mice. The recessive mutation little (*lit*) results in dwarf mice. Even though the mouse's growth hormone gene is present and apparently normal, no messenger RNA (mRNA) is produced. The initial step in correcting this deficiency was to inject homozygous *lit lit* eggs with about 5000 copies of a 5-kb linear DNA fragment that contained the rat growth hormone structural gene (*RGH*) fused to a regulator-promoter sequence from a mouse metallothionein gene (*MP*). The normal job of metallothionein is to detoxify heavy metals, so that the regulatory sequence is responsive to the presence of heavy metals in the animal. The eggs were then implanted into pseudopregnant mice, and the baby mice were raised. About 1 percent of these babies turned out to be transgenic, showing increased size when heavy metals were administered during development. A representative transgenic mouse was then crossed to a *lit lit* female. The ensuing pedigree is shown in Figure 15-30. We can see in the figure that mice two to three times the weight of their *lit lit* relatives are produced down through the generations, with the rat growth hormone transgene acting as a dominant marker, always heterozygous in this pedigree. The rat growth hormone transgene also makes *lit*$^+$ mice bigger (Figure 15-31).

The site of insertion of the introduced DNA in mammals is highly variable, and the DNA is generally not found at the homologous locus. Hence, gene therapy most often provides not a genuine correction of the original problem but a masking of it.

Human Gene Therapy

Perhaps the most exciting and controversial application of transgenic technology is in human gene therapy, the treatment and alleviation of human genetic disease by adding exogenous wild-type genes to correct the defective function of mutations. We saw above that the first case of gene therapy in mammals was to "cure" a genotypically dwarf fertilized mouse egg by injecting the appropriate wild-type allele for

normal growth. This technique (Figure 15-32a) has little application in humans because it is currently impossible to diagnose whether a fertilized egg cell carries a defective genotype without destroying the cell. (However, in an early embryo containing only a few cells, one cell can be removed and analyzed with no ill effects on the remainder.)

There are two basic types of gene therapy that can be applied to humans, germinal and somatic. The goal of **germinal gene therapy** (Figure 15-32b) is the more ambitious: to introduce transgenic cells into the germ line as well as into the somatic cell population. Not only should this achieve a cure of the individual treated, but some gametes could also carry the corrected genotype. We have seen that such germinal therapy has been achieved using mice eggs. However, the protocol that is most relevant for application to humans is removal of an early embryo (blastocyst) with a defective genotype from a pregnant mouse and injection with transgenic cells containing the wild-type allele. These cells become part of many tissues of the body, often including the germ line, which will give rise to the gonads. Then the gene can be passed on to some or all progeny, depending on the size of the clone of transgenic cells that lodges in the germinal area. However, no human germinal gene therapy has been performed to date.

Most transforming fragments will insert ectopically throughout the genome. This is a disadvantage in human gene therapy not only because of the possibility of the ectopic insert's causing gene disruption, but also because even if the disease phenotype is reversed, the defective allele is still present and can segregate away from the transgene in future generations. Therefore, for effective germinal gene therapy an efficient targeted gene replacement will be necessary, in which case the wild-type transgene replaces the resident defective copy by a double crossover.

Somatic gene therapy (Figure 15-33c) focuses only on the body, or soma, attempting to effect a reversal of the disease phenotype by treating some somatic tissues in the affected individual. At present it is not possible to render an entire body transgenic, so the method addresses diseases whose phenotype is caused by genes that are expressed predominantly in one tissue. In such cases it is found that not all the cells of that tissue need to become transgenic; a portion of transgenic cells can ameliorate the overall disease symptoms. The method proceeds by removing some cells from an individual with the defective genotype and making these cells transgenic through the introduction of copies of the cloned wild-type gene. These cells are then reintro-

Figure 15-31 Transgenic mouse. The mice are siblings, but the mouse on the left was derived from an egg transformed by injection with a new gene composed of the mouse metallothionein promoter fused to the rat growth hormone structural gene. (This mouse weighs 44 g, and its untreated sibling 29 g.) The new gene is passed on to progeny, in a Mendelian manner, and so is proven to be chromosomally integrated. (R. L. Brinster)

(a) INJECTION OF FERTILIZED EGG

Transgene

Egg

Mosaic soma

(b) GERMINAL THERAPY

Transgenic cell

Blastocyst stage

Mosaic gonad

(c) SOMATIC THERAPY

Transgene

Transgenic clones

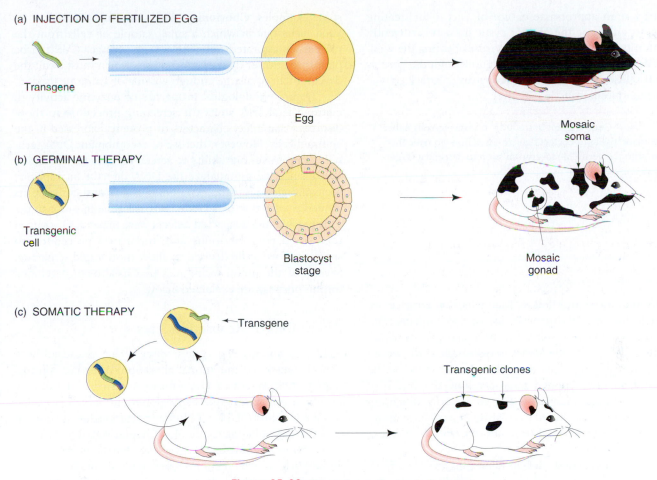

Figure 15-32 Types of gene therapy in the mouse

duced to the patient's body, where they provide normal gene function.

Currently there are two ways of getting the transgene into the defective somatic cells. Both methods use viruses. The older method uses a disarmed retrovirus with the transgene spliced into its genome, replacing most of the viral genes. The natural cycle of retroviruses involves integration of the viral genome at some location in one of the host cell's chromosomes. The recombinant retrovirus will carry the transgene along with it into the chromosome. This is a potential problem because the integrating virus can act as an insertional mutagen and inactivate some unknown resident gene, causing a mutation. Another problem with this type of vector is that a retrovirus attacks only proliferating cells such as blood cells. This procedure has been used for somatic gene therapy of severe combined immunodeficiency disease (SCID), otherwise known as *bubble-boy disease*. This disease is caused by a mutation in the gene coding for the blood enzyme adenosine deaminase (ADA). In the gene therapy, blood stem cells were removed from the bone marrow, the transgene was added, and the transgenic cells were reintroduced into the blood system. Prognosis on these patients is currently good.

Even solid tissues seem to be accessible to somatic gene therapy. One dramatic case involves gene therapy of a patient homozygous for the recessive allele ($LDLR^-$) for low-density-lipoprotein receptor. This mutant allele increases the risk of atherosclerosis and coronary disease. The receptor protein is made in liver cells, so 15 percent of the patient's liver was removed, and the liver cells were dissociated and treated with retrovirus carrying the $LDLR^+$ allele. Transgenic cells were reintroduced back into the body by injection into the portal venous system, which takes blood from the intestine to the liver. The transgenic cells took up residence in the liver. The latest reports are that the procedure seems to be working and the patient's lipid profile has improved.

Another vector used in human gene therapy is adenovirus. This virus normally attacks respiratory epithelia, injecting its genome into the epithelial cells. The viral genome does not integrate into a chromosome but persists extrachromosomally in the cells. This eliminates the problem of insertional mutagenesis by the vector. Another advantage of adenovirus as a vector is that it attacks nondividing cells, making most tissues susceptible in principle. Since cystic fibrosis is a disease of the respiratory epithelium, adenovirus is an appropriate choice of vector for treating this disease, and gene therapy for cystic fibrosis is currently being attempted using adenovirus. Viruses bearing the wild-type cystic fibrosis allele are introduced through the nose as a spray. It is also possible to use adenovirus to attack cells of the nervous system, muscle, and liver.

Message   Gene therapy introduces transgenic cells either into somatic tissue to correct defective function (somatic therapy) or into the germ line for transmission to descendants (germinal therapy).

Recombinant DNA and Screening for Genetic Diseases

Recessive mutations that follow Mendelian inheritance are responsible for over 500 genetic diseases of humans. Homozygous individuals resulting from marriages involving two heterozygotes for the same recessive allele will be affected by the disease. Cells derived from the fetus can be screened at an early enough stage to allow the option of abortion of afflicted individuals. We now know which enzymes or proteins are altered or missing in a number of genetic diseases (refer to the list of inborn errors of metabolism, Table 12-4).

To detect such genetic defects, fetal cells are taken from the amniotic fluid, separated from other components, and cultured to allow the analysis of chromosomes, proteins, enzymic reactions, and other biochemical properties. This process, termed **amniocentesis** (Figure 15-33), can already pinpoint a number of known disorders; Table 15-1 lists

some examples. **Chorionic villus sampling (CVS)** is a related technique in which a small sample of cells from the placenta is aspirated out with a long syringe. CVS can be performed earlier than amniocentesis, which must await the development of a large enough volume of amniotic fluid.

Testing physiological properties or enzymic activity in cultured fetal cells limits the screening procedure to those disorders that affect characters or proteins expressed in the cultured cells. However, the use of recombinant DNA technology improves our ability to screen for genetic diseases in utero, since we can analyze the DNA directly. In principle, we could clone out the gene being tested and compare its sequence with that of a cloned normal gene in order to determine whether suspected defects were present. Of course, this would be a laborious and impractical procedure, so shortcuts have to be devised to allow more rapid screening. Several of the useful techniques that have been developed for this purpose are explained below.

Alterations of Restriction Sites

Sickle-cell anemia is a genetic disease that is caused by a well-characterized mutational alteration. Affecting approximately 0.25 percent of U.S. African Americans, the disease results from an altered hemoglobin, in which a valine residue is substituted for a glutamic acid residue at position 6 in the β-globin chain (see also Chapter 12). The GAG to GTG change eliminates a cleavage site for the restriction enzyme *Mst*II, which cuts the sequence CCTNAGG (where N represents any of the four bases). The change from CCT*GAGG* to CCT*GTGG* can thus be recognized by Southern analysis using labeled β-globin cDNA as a probe, since the DNA derived from individuals with sickle-cell disease will lack one fragment contained in the DNA from normal indi-

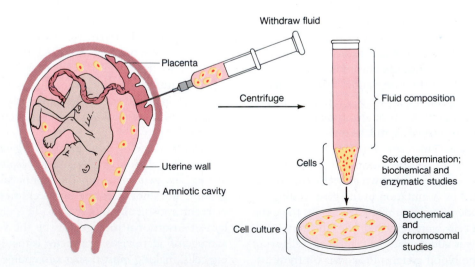

Figure 15-33 Amniocentesis

Table 15-1   Some Common Genetic Diseases

Inborn errors of metabolism	Approximate incidence among live births
1. Cystic fibrosis	1/1600 Caucasians
2. Duchenne muscular dystrophy	1/3000 boys (X-linked)
3. Gaucher's disease (defective glucocerebrosidase)	1/2500 Ashkenazi Jews; 1/75,000 others
4. Tay-Sachs disease (defective hexosaminidase A)	1/3500 Ashkenazi Jews; 1/35,000 others
5. Essential pentosuria (a benign condition)	1/2000 Ashkenazi Jews; 1/50,000 others
6. Classic hemophilia (defective clotting factor VIII)	1/10,000 boys (X-linked)
7. Phenylketonuria (defective phenylalanine hydroxylase)	1/5000 Celtic Irish; 1/15,000 others
8. Cystinuria (mutated gene unknown)	1/15,000
9. Metachromatic leukodystrophy (defective arylsulfatase A)	1/40,000
10. Galactosemia (defective galactose 1-phosphate uridyl transferase)	1/40,000

Hemoglobinopathies	Approximate incidence among live births
1. Sickle-cell anemia (defective β-globin chain)	1/400 U.S. blacks. In some West African populations, the frequency of heterozygotes is 40%.
2. β-thalassemia (defective β-globin chain)	1/400 among some Mediterranean populations

NOTE: Although the vast majority of the over 500 recognized recessive genetic diseases are extremely rare, in combination they represent an enormous burden of human suffering. As is consistent with Mendelian mutations, the incidence of some of these diseases is much higher in certain racial groups than in others.

SOURCE: J. D. Watson, J. Tooze, and D. T. Kurtz, *Recombinant DNA: A Short Course*. Copyright 1983 by W. H. Freeman and Company.

viduals, and in addition, there will be a large (uncleaved) fragment not seen in normal DNA (Figure 15-34).

Probing for Altered Sequences

When a genetic disorder can be attributed to a change in a specific nucleotide, synthetic oligonucleotide probes can identify that change. The best example is α_1-antitrypsin deficiency, which leads to a greatly increased probability of developing pulmonary emphysema. The condition results from a single base change at a known nucleotide position. A synthetic oligonucleotide probe that contains the wild-type sequence in the relevant region of the gene can be used in a Southern analysis to determine whether the DNA contains the wild-type or the mutant sequence. At higher temperatures, a complementary sequence will hybridize, whereas a sequence containing even a single mismatched base will not.

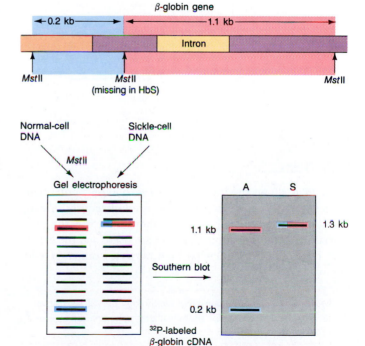

Figure 15-34 Detection of the sickle-cell globin gene by Southern blotting. The base change (A → T) that causes sickle-cell anemia destroys an *Mst*II site that is present in the normal β-globin gene. This difference can be detected by Southern blotting. (Modified from J. D. Watson, J. Tooze, and D. T. Kurtz, *Recombinant DNA: A Short Course*. Copyright © 1983 by W. H. Freeman and Company.)

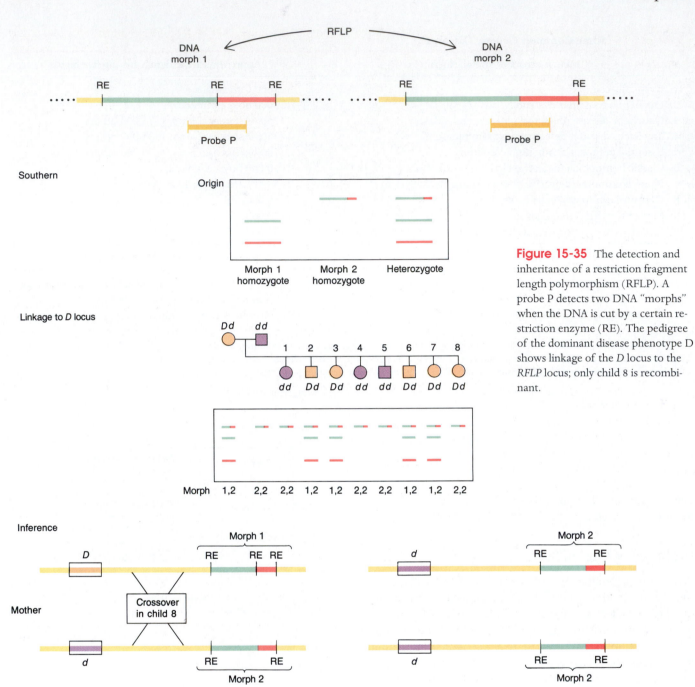

Figure 15-35 The detection and inheritance of a restriction fragment length polymorphism (RFLP). A probe P detects two DNA "morphs" when the DNA is cut by a certain restriction enzyme (RE). The pedigree of the dominant disease phenotype D shows linkage of the *D* locus to the *RFLP* locus; only child 8 is recombinant.

PCR Tests

Since PCR allows the investigator to zero in on a potentially defective DNA sequence, it can be used effectively in diagnosis of diseases in which a specific mutational site is in question. Primers spanning the site are used, and the amplified DNA can be sequenced or otherwise compared with the wild type.

> **Message** Recombinant DNA technology provides sensitive techniques for testing for mutant alleles in people or in embryos in utero.

Linkage to Restriction Sites

The presence of a disease allele can be diagnosed by detecting a restriction site known from crossover analysis to be closely linked. We have seen that any one restriction site is generally found in the same position in all homologous chromosomes throughout a population. However, some restriction sites are not present in all homologous chromosomes in the population, and individuals who show the presence of the site on one homolog and its absence on the other, that is, who are +/− "heterozygotes," provide key diagnostic tools. Their heterozygous loci can be used to study recombination just as in any other heterozygote. The +/− differences in the DNA appear to be neutral; that is, they do not affect DNA function. Usually, they are caused by a single nucleotide difference that happens to create or destroy a restriction site.

Generally, in eukaryotes it is quite easy to find restriction site differences. When used as a probe, most cloned segments of, say, human DNA are capable of detecting restriction-site variation. We might begin with a randomly chosen cloned fragment and use it in Southern hybridizations against restriction-enzyme-digested DNA preparations from a sample of people. If we are unlucky, we will see the same pattern on all autoradiograms: either one band or more, depending on whether the specific restriction enzyme used happened to cut into the region spanned by the probe. But eventually it is likely that we will find a variant—a fragment pattern different from the others, representing a restriction-site difference (Figure 15-35).

The coexistence in a population of two or more alternative phenotypes, generally attributable to the alleles of one gene, is called **polymorphism** (Greek: "presence of many forms"). At the molecular level, the alternative phenotypes can be the restriction fragments of varying number and size. The coexistence of two or more restriction fragment patterns, as revealed by hybridization to a particular probe, is called **restriction fragment length polymorphism (RFLP)**. In family pedigrees containing individuals heterozygous for an RFLP and a disease gene, it is possible to test for linkage between the disease locus and the RFLP locus (Figure 15-36). Once linkage to an RFLP is established, the RFLP genotype can be used to diagnose the genotype at the disease gene locus. Another example of this method is diagrammed in Figure 15-36, in which a *Hpa*I-site polymorphism is very closely linked to the sickle-cell β-globin gene.

Examining linkage to restriction-site changes is simply performing at the DNA level the same type of analysis that is often applied at the phenotypic level, namely, using an identifiable phenotype of an individual to give additional information about another, more cryptic part of the genotype. For instance, colorblindness can be used to yield information about the state of the locus governing the disease hemophilia, also on the X chromosome (see Problem 32 in

Chapter 5). Because the *Cb* (colorblindness) locus is closely linked to the *Hb* (hemophilia) locus, the probability of the presence of specific *Hb* alleles in a colorblind individual can be determined, provided the genotypes of the parents are known. This same principle is illustrated in Figure 15-36; however, instead of using colorblindness, we are using the restriction-enzyme cleavage pattern of the DNA as a means of assessing the genetic condition of a linked region. In diagnosing disease genotypes in a family pedigree, of course, the first step is to find an RFLP linked to the disease gene in question. However, because of the commonness of RFLPs this is quite feasible.

RFLP loci also can be mapped in relation to each other to provide numerous landmarks in chromosome mapping. Using RFLPs as map reference points has been so successful that the entire chromosomal maps of humans and a wide range of other organisms are now liberally sprinkled with RFLP loci (for an example in humans, see Figure 15-37).

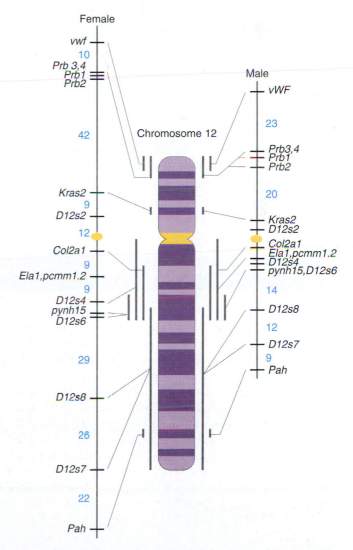

Figure 15-37 The human chromosome 12, showing the location of RFLP marker loci that have been detected in various pedigrees. Recombinant frequencies are shown for meiosis in men and in women; note that crossing-over appears to be more frequent in women.

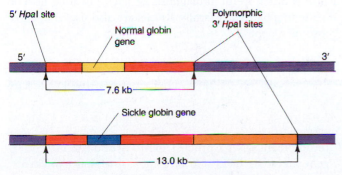

Figure 15-36 *Hpa*I-site polymorphism is diagnostic for the sickle β-globin gene in humans.

These loci are helpful landmarks to help experimenters find their way around the genome. As an example, if a geneticist is interested in cloning and studying a gene that causes a human disease, finding a linked RFLP can be a useful starting point. If the RFLP is less than about 1 map unit away (1 m.u. in humans has been calibrated to represent approximately 1000 kb), then the DNA identified by the RFLP probe identifies a chromosomal locus that could be used as a starting point to clone the disease-causing gene. These techniques are discussed further in Chapter 17.

We have considered simple two-allele restriction fragment length dimorphism as examples of RFLPs. However, some probes pick up multiple alleles. The family in Figure 15-38 illustrates such a situation.

Message   Restriction fragment length polymorphisms (RFLPs) provide neutral heterozygous marker loci for chromosome mapping in experimental genetics, and for diagnosis of linked disease alleles in human families.

It is worth comparing the process of making a restriction map (*restriction mapping,* page 449–451), with the process of *RFLP mapping*. Restriction maps are based on physical analysis of DNA, whereas RFLP maps are based on recombination analysis of matings. Note also that restriction mapping is based on restriction sites with no variation, whereas RFLP mapping is based on restriction-site variation between chromosomes. Most restriction maps are short-range (fine-scale) maps, although long-range maps can be constructed with rare-cutting restriction enzymes. In contrast, RFLP mapping generally produces long-range (coarse-scale) maps. RFLP mapping of whole genomes will be covered in detail in Chapter 17.

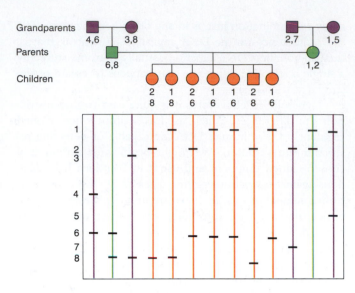

Figure 15-38 A pedigree illustrating the inheritance of alleles of an RFLP locus with a variable number of tandem repeats. In this polymorphism a specific base sequence is repeated a variable number of times between two restriction sites. Thus, it is not the restriction sites per se that vary, but rather the number of repeats between the restriction sites. When these sections are cut by the restriction enzyme, the resulting fragment lengths depend on the number of repeats, and the fragments will migrate in distinctive patterns on an electrophoretic gel, as seen in the autoradiograph presented here below the family tree. This family shows eight alleles, that is, eight fragments of different lengths at this locus.

SUMMARY

Methods are now available to manipulate eukaryotic DNA in microbial vectors, to change specific base pairs in a gene, and then to reintroduce the DNA into the eukaryotic organism from which it came. Biotechnology is the application of recombinant DNA techniques to animals, plants, and microbes important in commerce. Transgenic organisms, modified by insertion of specific exogenous DNA, have found a central place in molecular genetics because they allow highly specific genome modifications to be tailored to the needs of the experiment. Human gene therapy is a special application of transgenic technology: germinal therapy aims at incorporating some transgenic cells into the germ line so that they can be passed on to descendants, and somatic therapy introduces a proportion of genetically modified cells into the body. Amniocentesis and CVS both use recombinant DNA techniques for diagnosis. Linkage to an RFLP is often useful in diagnosis of alleles causing genetic disease in a particular family. RFLPs are also useful in chromosome mapping.

Concept Map

Draw a concept map interrelating as many of the following terms as possible. Note that the terms are listed in no particular order.

recombinant DNA / probe / in situ hybridization / gene therapy / RFLP / transgenic / mapping / in vitro mutagenesis / genetic screening

CHAPTER INTEGRATION PROBLEM

In Chapter 2 we learned how pedigrees can be used to trace a family's genetic history, and in Chapter 14 we examined Southern blots. We can incorporate these ideas into the concepts discussed in this chapter to help find information about human genetic diseases. Huntington's disease (HD) is a lethal neurodegenerative disorder that exhibits autosomal dominant inheritance. Because the onset of symptoms is usually not until the third, fourth, or fifth decade of life, patients with HD usually have already had their children, and some of them inherit the disease. There has been little hope of a reliable pre-onset diagnosis until recently, when a team of scientists searched for and found a cloned probe (called G8) that revealed a DNA polymorphism (actually a tetramorphism) relevant to HD. The probe and its four hybridizing DNA types are shown here; the vertical lines represent *Hind*III cutting sites:

a. Draw the Southern blots expected from the cells of people who are homozygous (*AA*, *BB*, *CC*, and *DD*) and all who are heterozygous (*AB*, *AC*, and so on). Are they all different?

b. What do the DNA differences result from in terms of restriction sites? Do you think they are probably trivial or potentially adaptive? Explain.

c. When human-mouse cell lines were studied, the G8 probe bound only to DNA containing human chromosome 4. What does this tell you?

d. Two families showing HD—one from Venezuela, and one from the United States—are checked to determine their G8-hybridizing DNA type. The results are shown in the pedigree at the top of the next page, where solid black symbols indicate HD and slashes indicate family members who were dead in 1983. What linkage associations do you see, and what do they tell you?

e. If a 20-year-old member of the Venezuelan family needs genetic counseling, what test would you devise and what advice would you give for each outcome? Repeat this for the U.S. family.

f. How might these data be helpful in finding the primary defect of HD?

g. Could these results be useful in counseling other HD families? Explain.

h. Are there any exceptional individuals in the pedigrees? If so, account for them.

Solution

a.

	AA	BB	CC	DD	AB	AC	AD	BC	BD	CD
17.5	—	—			—	—	—	—	—	—
15.0			—	—	—	—	—	—	—	—
8.4	—	—	—	—	—	—	—	—	—	—
4.9		—	—					—	—	—
3.7	—		—		—	—	—	—	—	—
2.3	—	—	—	—	—	—	—	—	—	—
1.2	—		—		—	—	—	—	—	—

A D and *B C* are identical. The rest are different.

b. The differences in restriction sites come from differences in DNA sequence. There is no evidence on which to base a judgment of either trivial or potentially adaptive differences.

c. The sequence that gave rise to the G8 probe is located on chromosome 4.

d. In the Venezuelan family there is a strong association of the disease with RFLP morph C, suggesting tight linkage. In the U.S. family there is a weak association of this disease with another morph, A. However, this again suggests linkage. The difference between the families might be explained by statistical variation.

e. In each case, test for the relevant polymorphism by digesting with *Hind*III and probing with G8.

In the family from Venezuela, there is one crossover individual (VI-5) among the 20 that carry the *C* polymorphism. Therefore, the G8 probe is 100% $(\frac{1}{20})$ = 5 m.u. from the Huntington's disease gene. If the person tests positive for the *C* polymorphism, there is a 95 percent chance of having the gene for Huntington's disease and a 5 percent chance of not having it. If the person tests negative for the *C* polymorphism, there is a 5 percent chance that he has the Huntington's disease gene and a 95 percent chance that he does not have the gene.

The situation in the family from the United States is unclear in comparison with the situation in the family from Venezuela. In the American family, the *A* polymorphism is present in four people with no family history of Huntington's disease who married into the family in question. Their presence makes it impossible to identify crossover individuals unambiguously. It can be assumed that the G8 probe is identifying a polymorphism that is approximately 5 m.u. from the Huntington locus, just as it did in the family from

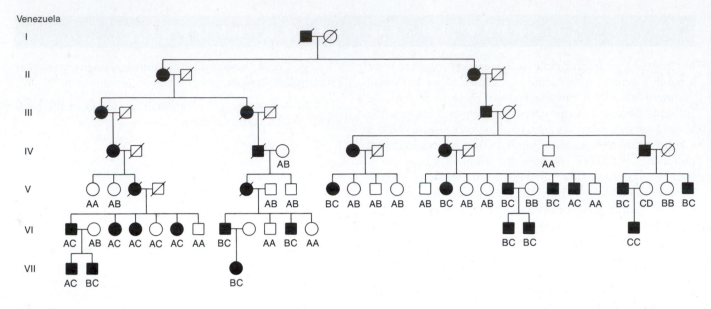

Venezuela

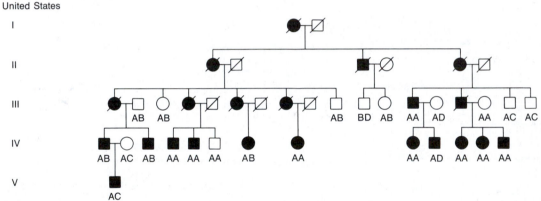

United States

Venezuela. However, the conclusions from testing of the individual in question would vary with the polymorphism genotype of his affected parent.

For instance, if the affected parent were AA and the individual in question were $A-$, the chance of the person's having inherited the Huntington's disease gene would be 50 percent. If, however, the affected parent were AD and the individual in question were $A-$, the chance of having inherited the Huntington's disease gene would be 95 percent, unless the unaffected parent also carried A. In that case, the risk would be 50 percent.

f. The G8 probe can be used to identify the region in which the Huntington's disease gene is located. The locus can be isolated by means of chromosome walking. The gene can be transcribed and translated, and the protein product can be identified.

g. The family must show the RFLP. Then linkage of HD to one of the RFLP morphs would have to be established.

h. One exceptional person was identified already: Venezuela VI-5, who is a crossover between the polymorphism and the Huntington's disease gene. In the U.S. pedigree there is no individual who is an obligate crossover.

(Solution from Diane K. Lavett.)

SOLVED PROBLEMS

1. DNA studies are performed on a large family that shows a certain autosomal dominant disease of late onset (approximately 40 years of age). A DNA sample from each family member is digested with the restriction enzyme *Taq*I and run on an electrophoretic gel. A Southern blot is then performed, using a radioactive probe consisting of a portion of human DNA cloned in a bacterial plasmid. The autoradiogram is shown at the top of the next page, aligned with the family pedigree. Affected members are shown in black.

Pedigree

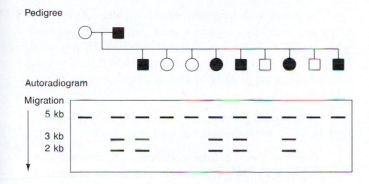

Autoradiogram

a. Analyze fully the relationship between the DNA variation, the probe DNA, and the gene for the disease. Draw the relevant chromosome regions.

b. How do you explain the last son?

c. Of what use would these results be in counseling people from this family who subsequently married?

Solution

a. All individuals have the 5-kb band, indicating that the band sequence is not involved with the gene in question. All affected individuals, except the last son, have the 3-kb and 2-kb bands. These two bands are not seen in unaffected individuals. The suggestion is that the two bands are close to or part of the gene in question.

b. The last affected son indicates that the two bands are not part of the gene in question. He represents a crossover between the gene in question and the two bands.

c. The 2-kb and 3-kb bands are closely linked to the dominant allele. Their presence in an individual would indicate a high risk of developing the disorder, while their absence would indicate a low risk of developing the disorder. Exact risk cannot be stated until the map units between the two bands and the gene in question are determined. Although a rough estimate of map units can be made from the pedigree, the sample size is too small to make the estimate reliable.

(Solution from Diane K. Lavett.)

2. A yeast plasmid carrying the yeast *leu2*$^+$ gene is used to transform nonrevertible haploid *leu2*$^-$ yeast cells. Several *leu*$^+$-transformed colonies appear on a leucineless medium. Thus, *leu2*$^+$ DNA presumably has entered the recipient cells, but now you have to decide what has happened to it inside these cells. Crosses of transformants to *leu2*$^-$ testers reveal that there are three types of transformants, A, B, and C, reflecting three different fates of the *leu2*$^+$ in the transformation. The results are

$$\text{Type A} \times leu2^- \longrightarrow \tfrac{1}{2}\, leu^-$$
$$\tfrac{1}{2}\, leu^+, \times \text{standard } leu2^+$$
$$\longrightarrow \tfrac{3}{4}\, leu^+$$
$$\tfrac{1}{4}\, leu^-$$

$$\text{Type B} \times leu2^- \longrightarrow \tfrac{1}{2}\, leu^-$$
$$\tfrac{1}{2}\, leu^+, \times \text{standard } leu2^+$$
$$\longrightarrow 100\%\, leu^+$$
$$0\%\, leu^-$$

$$\text{Type C} \times leu2^- \longrightarrow 100\%\, leu^+$$

What three different fates of the *leu2*$^+$ DNA do these results suggest? Be sure to explain *all* the results according to your hypotheses. Use diagrams if possible.

Solution

If the yeast plasmid remains unintegrated, then it replicates independently of the chromosomes. During meiosis, the daughter plasmids would be distributed to the daughter cells, resulting in 100 percent transformation. This was observed in type C.

If one copy of the plasmid is inserted, when crossed with a *leu2*$^-$ line, the resulting offspring would have a ratio of 1 *leu*$^+$: 1 *leu*$^-$. This is seen in type A and type B.

When the resulting *leu*$^+$ cells are crossed with standard *leu2*$^+$ lines, the data from type A cells suggest that the inserted gene is segregating independently of the standard *leu2*$^+$ gene, and data from type B cells suggest that the inserted gene is located in the same locus as the standard *leu2*$^+$ allele. Therefore, the gene in type A cells did not insert at the *leu2* site, and the gene in type B cells did.

(Solution from Diane K. Lavett.)

PROBLEMS

1. Transgenic tobacco plants were obtained in which the vector Ti plasmid was designed to insert the gene of interest plus an adjacent kanamycin resistance gene. The inheritance of chromosomal insertion was followed by testing progeny for kanamycin resistance. Two plants typified the results obtained generally. When plant 1 was backcrossed to wild-type tobacco, 50 percent of the progeny were kanamycin-resistant and 50 percent were sensitive. When plant 2 was backcrossed to the wild type, 75 percent of the progeny were kanamycin-resistant, and 25 percent were sensitive. What must have been the difference between the two transgenic plants? What would you predict about the situation regarding the gene of interest?

2. In *Neurospora,* which has seven chromosomes, the following chromosomal rearrangements were obtained in different strains:

a. A paracentric inversion of chromosome 1 (the largest chromosome)

b. A pericentric inversion of chromosome 1

c. A reciprocal translocation in which about half of chromosome 1 was exchanged with about half of chromosome 7 (the smallest chromosome)

d. A unidirectional insertional translocation, in which a part of one chromosome was inserted into another

e. A disomic ($n + 1$)

f. A monosomic ($2n - 1$)

g. A tandem duplication of a large part of chromosome 1

From all these strains and a normal wild type, DNA was isolated carefully to avoid mechanical breakage, and the samples were subjected to pulsed field gel electrophoresis. Predict the bands you would expect to see in each case.

3. In a bacterial vector, you have cloned a plant gene that codes for a photosynthesis protein. Now you wish to find out if this gene is active in roots and other nonphotosynthetic tissue. How would you go about this? Describe the experimental details as well as you can.

4. In *Neurospora* you have two chromosome 5 probes that detect RFLP loci approximately 20 m.u. apart. In Southern blots of a *Pst*I digest of strain 1, probe A picks up two fragments of 1 and 2 kb, and probe B picks up two fragments of 4 and 1.5 kb. In Southern blots of strain 2, probe A detects one band of 3 kb, and probe B detects one band of 5.5 kb. Draw the Southern banding patterns you would see in *Pst*I digests of individual ascospore cultures from asci, using both probes simultaneously. Be sure to state how many different ascus types you expect, and be sure to account for the occurrence of crossovers.

5. A cystic fibrosis mutation in a certain pedigree involves a single nucleotide-pair change. This change destroys an *Eco*RI restriction site normally found in this position. How would you use this information in counseling individuals in this family about their likelihood of being carriers? State the precise experiments needed. Assume that you detect that a woman in this family is a carrier, and it transpires she is married to an unrelated man who is also a heterozygote for cystic fibrosis, but in his case it is a different mutation in the same gene. How would you counsel this couple about the risks of a child's having cystic fibrosis?

6. In yeast, you have sequenced a piece of wild-type DNA and it clearly contains a gene, but you do not know what gene it is. Therefore, to investigate further, you would like to find out its mutant phenotype. How would you use the cloned wild-type gene to do this? Show your experimental steps clearly.

7. How would you use pulsed field gel electrophoresis to find out what chromosome a cloned gene is on?

8. Bacterial glucuronidase converts a colorless substance called *X-gluc* into a bright-blue indigo pigment. The gene for glucuronidase also works in plants if given a plant promoter region. How would you use this gene as a reporter gene to find out in which tissues a plant gene you have just cloned is normally active? (Assume X-gluc is easily taken up by the plant tissues.)

9. In mouse *Hind*III restriction digests, a certain probe picks a simple RFLP consisting of two alternative alleles of 1.7 kb and 3.8 kb. A mouse heterozygous for a dominant allele for bent tail and also heterozygous for the above RFLP is mated to a wild-type mouse that shows only the 3.8-kb fragment. Forty percent of the bent-tail progeny are homozygous for the 3.8-kb allele, and 60 percent are heterozygous for the 3.8- and the 1.7-kb forms.

a. Is the bent-tail locus linked to the RFLP locus? Draw the parental and progeny chromosomes to illustrate your answer.

b. What RFLP types do you predict among the wild-type offspring, and in what proportions?

10. Probes *A* and *B* are used to detect RFLP's in a study of two different haploid populations of yeast. In population 1, morph *A* gives morph *A1* and gene *B* gives morph *B1*; in population 2, gene *A* gives *A2* and gene *B* gives *B2*. These morphs are distinguished by the *Hind*III fragments bound by the probe (size are in kb).

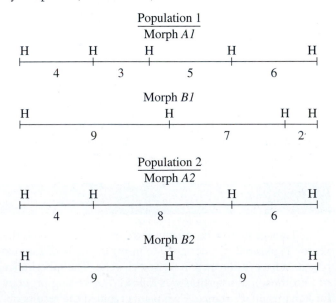

Meiotic products are examined, and the DNA fragments hybridizing to probes *A* and *B*, are given for each type:

Spore type	Frequency	DNA (*Hind*III fragments) Probe A	Probe B
1	15%	4, 3, 5, 6	9, 9
2	15%	4, 8, 6	9, 7, 2
3	35%	4, 3, 5, 6	9, 7, 2
4	35%	4, 8, 6	9, 9

a. How were the different spore DNAs produced?

b. Draw the appropriate chromosomal region(s).

(Problem 10 courtesy of Joan McPherson.)

11. The plant *Arabidopsis thaliana* was transformed using the Ti plasmid into which a kanamycin-resistance gene had been inserted in the T-DNA region. Two kanamycin-resistant colonies (A and B) were selected, and plants were regenerated from these. The plants were allowed to self, and the results were as follows:

Plant A selfed $\longrightarrow$ $\frac{3}{4}$ progeny resistant to kanamycin

$\frac{1}{4}$ progeny sensitive to kanamycin

Plant B selfed $\longrightarrow$ $\frac{15}{16}$ progeny resistant to kanamycin

$\frac{1}{16}$ progeny sensitive to kanamycin

a. Draw the relevant plant chromosomes in both plants.

b. Explain the two different ratios.

12. Two different circular yeast plasmid vectors (YP1 and YP2) were used to transform leu$^-$ cells to leu$^+$. The resulting leu$^+$ cultures from both experiments were crossed to the same leu$^-$ cell of opposite mating type. Typical results were as follows:

YP1 leu$^+$ × leu$^-$ $\longrightarrow$ all progeny leu$^+$ and the DNA of all these showed positive hybridization to a probe specific to the vector YP1

YP2 leu$^+$ × leu$^-$ $\longrightarrow$ $\frac{1}{2}$ progeny leu$^+$ and hybridize to vector probe specific to YP2

$\frac{1}{2}$ progeny leu$^-$ and do not hybridize to YP2 probe

a. Explain the different action of these two plasmids during transformation.

b. If total DNA is extracted from transformants from YP1 and YP2 and digested with an enzyme that cuts once within the vector (and not the insert), predict the results of electrophoresis and Southern analyses of the DNA, using the specific plasmid as a probe in each case.

13. A linear 9-kb *Neurospora* plasmid mar1 seems to be a neutral passenger, causing no apparent harm to its host. It has a restriction map as follows:

*Bgl*II   *Xba*I   *Bgl*II

However, it was suspected that this plasmid sometimes integrates into genomic DNA. To test this idea, the central large *Bgl*II fragment was cloned into a pUC vector, and this was used as a probe in a Southern analysis of *Xba*I-digested genomic DNA from a mar1-containing strain. Predict what the autoradiogram would look like if

a. the plasmid never integrates.

b. the plasmid occasionally integrates into genomic DNA.

14. The accompanying pedigree shows a man affected with dominantly inherited Huntington disease. Both of his at-risk children have elected prenatal diagnosis, although they do not wish their status to be evaluated. The Southern blot has been typed for a two-allele system (with allelic fragments at 4.9 kb or 3.7 and 1.2 kb) that shows 4 percent recombination with Huntington disease. For each of the two fetal samples tested calculate the chance that the fetus will inherit the allele for Huntington.

(Problem 14 from Steve Wood.)

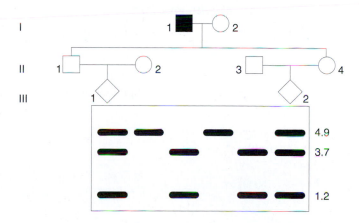

15. A *Neurospora* geneticist is interested in the genes that control hyphal extension. He decides to clone a sample of these genes. It was known from previous mutational analysis of *Neurospora* that a common type of mutant has a small-colony ("colonial") phenotype on plates, caused by abnormal hyphal extension. Therefore, he decides to do a tagging experiment using transforming DNA to produce colonial mutants by insertional mutagenesis. He transforms *Neurospora* cells using a bacterial plasmid carrying a gene for benomyl resistance (*ben-R*) and recovers resistant colonies on benomyl-containing medium. Some colonies show the colonial phenotype being sought, so a sample of these is isolated and tested. The colonial isolates prove to be of two types as follows:

Type 1   *col ben-R* × wild type (+ *ben-S*)

    Progeny $\frac{1}{2}$ *col ben-R*

             $\frac{1}{2}$ + *ben-S*

Type 2   *col ben-R* × wild type (+ *ben-S*)

    Progeny $\frac{1}{4}$ *col ben-R*

             $\frac{1}{4}$ *col ben-S*

             $\frac{1}{4}$ + *ben-R*

             $\frac{1}{4}$ + *ben-S*

a. Explain the difference between these two types of results.

b. Which type should he use to try to clone the genes affecting hyphal extension?

c. How should he proceed with the tagging protocol?

d. If a probe specific for the bacterial plasmid is available, which progeny should be hybridized by this probe?

 Unpacking the Problem

a. What is hyphal extension, and why do you think anyone would find it interesting?

b. How does the general approach of this experiment fit in with the general genetic approach of mutational dissection?

c. Is *Neurospora* haploid or diploid, and is this relevant to the problem?

d. Is it appropriate to transform a fungus (a eukaryote) with a bacterial plasmid? Does it matter?

e. What is transformation, and in what way is it useful in molecular genetics?

f. How are cells prepared for transformation?

g. What is the fate of transforming DNA if successful transformation occurs?

h. Draw a successful plasmid entry into the host cell, and also draw a representation of a successful stable transformation.

i. Does it make any difference to the protocol to know what benomyl is? What role is the benomyl-resistance gene playing in this experiment? Would the experiment work with another resistance marker?

j. What does the word *colonial* mean in the present context? Why did the experimenter think that finding and characterizing colonial mutations would help understand hyphal extension?

k. What kind of "previous mutational studies" do you think are being referred to?

l. Draw the appearance of a typical petri plate after transformation and selection. Pay attention to the appearance of the colonies.

m. What is tagging? How does it relate to insertional mutagenesis? How does insertion cause mutation?

n. How are crosses made and progeny isolated in *Neurospora?*

o. Is recombination relevant to this question? Can an RF value be calculated? What does it mean?

p. Why are there only two colonial types? Is it possible to predict which will be more common?

q. What is a probe? How are probes useful in molecular genetics? How would the probing experiment be done?

r. How would you obtain a probe specific to the bacterial plasmid?

16

The Structure and Function of Eukaryotic Chromosomes

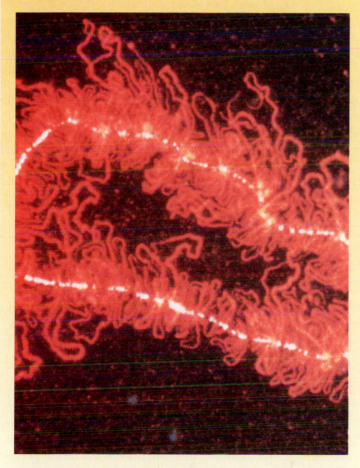

Lampbrush chromosomes. The chromosomes of some animals take on this lampbrush appearance during meiotic diplotene in females. The lampbrush structure is thought to reflect the underlying organization of all chromosomes, a central scaffold (here stained brightly) and projecting lateral loops (stained red) formed by a folded continuous strand of DNA with associated histone proteins. (M. Roth and J. Gall)

KEY CONCEPTS

▶ A chromosome contains only one long DNA molecule.

▶ DNA winds around protein spools, and the spooled unit then coils, loops, and supercoils, forming a chromosome.

▶ The degree to which DNA is packaged affects gene action.

▶ A large proportion of active and inactive eukaryotic DNA is present in multiple copies.

▶ Most of the multiple-copy DNA has no known function.

▶ Different types of multiple-copy DNA can be dispersed throughout the genome, or arrayed in tandem.

▶ Centromeres have specialized molecular sequences that attach to spindles, and telomeres have specialized sequences that permit complete replication of chromosome ends.

▶ Replication and transcription both occur with the protein spools in place.

In our discussions so far, we have treated the eukaryotic chromosome as a fat line on a piece of paper. For much of genetic analysis this is an adequate representation, and for much of this century it in fact reflected the general level of understanding of chromosome structure. However, as technology has advanced, the fat line has gradually become a molecular reality. In this chapter, we will discuss the architecture of the eukaryotic genome, examining its topology at the molecular level. Many of the chromosomal landmarks we have analyzed at the level of light microscopy, such as bands, centromeres, and nucleolar organizers, need to be reexamined using higher-resolution analyses. Once this picture is developed, we will examine the problems of replication and transcription of DNA at the chromosomal level. The analytical procedures that are used in this subject area are a powerful combination of genetics, molecular biology, and light and electron microscopy.

Genome Size in Eukaryotes

In comparison with prokaryotes, the genomes of eukaryotes contain a lot of DNA. Table 16-1 shows some haploid genome sizes of eukaryotes that have become model genetic organisms, and also shows the size of the *E. coli* genome for comparison. Some idea of the proportional sizes of prokaryote and eukaryote genomes can be obtained from the observation that the smallest of the seven chromosomes of the fungus *Neurospora* (a eukaryote) has approximately the same amount of DNA as the entire genome of *E. coli*.

Table 16-1 also shows that there is considerable variation of genome size within eukaryotes. Although there is a rough proportionality between the complexity of the organism and the genome size, there are some wild exceptions; for example, some amphibians contain up to 10^{11} nucleotide pairs per haploid genome, almost 100 times more than most mammals! We shall see that this disproportional-

ity (called the *C value paradox*, because the amount of DNA is called the *C value*) results mainly from the large amount of variation in repetitive DNA of no apparent function.

How do genome sizes in nucleotide pairs relate to the genetic units that we have studied so far, the map unit or centimorgan unit? In *Neurospora crassa* the genome size is 2700 kilobases (kb), whereas the total map size of all the chromosomes (all the genetic intervals added up) is 1000 map units. From this we arrive at the relationship that 1 map unit is approximately 2.7 kb. A similar calculation in humans shows that in our species 1 map unit is approximately 1 megabase, or 1000 kb.

Can we obtain some feel for how big these genomes are in more familiar units such as millimeters and meters? The chromosome of the prokaryote *E. coli* is about 1.3 mm of DNA. In stark contrast, a human cell contains about 2 m of DNA (1 m per chromosome set). The human body consists of approximately 10^{13} cells, and therefore contains a total of about 2×10^{13} m of DNA. Some idea of the extreme length of this DNA can be obtained by comparing it with the distance from earth to the sun, which is 1.5×10^{11} m. You can see that the DNA in your body could stretch to the sun and back about 50 times. This peculiar fact makes the point that the DNA of eukaryotes is obviously efficiently packed. In fact, the packing occurs at the level of the nucleus, where the 2 m of DNA in a human cell is packed into 46 chromosomes, all in a nucleus 0.006 mm in diameter. In this chapter we have to translate what we have learned about the structure and function of eukaryotic genes into the "real world" of the nucleus. Instead of envisioning replication and transcription machinery moving along the airy-looking straight lines that we have used to represent genes in previous chapters, we must now come to grips with the fact that these processes take place in what must be very much like the inside of a densely wound ball of wool.

One DNA Molecule per Chromosome

If eukaryotic cells are broken, and the contents of their nuclei are examined under the electron microscope, the chromosomes appear as masses of spaghettilike fibers with diameters of about 30 nm. Some examples are shown in the electron micrograph in Figure 16-1. In the 1960s, Ernest DuPraw studied such chromosomes carefully and showed that there are no ends protruding from the fibrillar mass. This suggests that each chromosome is one, long, fine fiber folded up in some way. If the fiber somehow corresponds to a DNA molecule, then we arrive at the idea that each chromosome is one densely folded DNA molecule.

In 1973, Ruth Kavenoff and Bruno Zimm performed experiments that showed this was most likely the case. They studied *Drosophila* DNA by using a viscoelastic recoil technique, which measures the size of DNA molecules in solution by measuring their elastic recoil properties. Put simply, the procedure is analogous to stretching out a coiled spring

Table 16-1 Haploid Genome Sizes (in Nucleotide Pairs) of Model Genetic Organisms

Organism	Group	Genome Size
T4	bacteriophage	2.0×10^5
E. coli	bacterium	4.2×10^6
Saccharomyces cerevisiae	yeast	1.8×10^7
Neurospora crassa	mold	2.7×10^7
Dictyostelium discoideum	slime mold	5.4×10^7
Caenorhabditis elegans	nematode	1.0×10^8
Arabidopsis thaliana	plant	1.0×10^8
Drosophila melanogaster	insect	1.4×10^8
Mus musculus	mammal	3.0×10^9
Homo sapiens	mammal	3.3×10^9
Zea mays	plant	5.4×10^9

Figure 16-1 Electron micrograph of metaphase chromosomes from a honeybee. The chromosomes each appear to be composed of one continuous fiber 30 nm wide. (From E. J. DuPraw, *Cell and Molecular Biology.* Copyright © 1968 by Academic Press.)

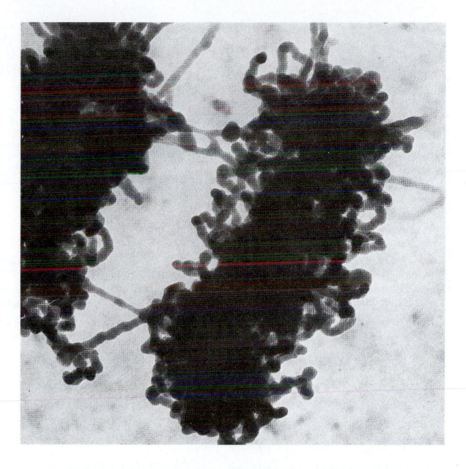

and measuring how long it takes to return to its fully coiled state. DNA is stretched by spinning a paddle in the solution and is then allowed to recoil into its relaxed state. The recoil time is known to be proportional to the size of the largest molecules present. In their study of *Drosophila melanogaster,* which has four pairs of chromosomes, Kavenoff and Zimm obtained a value of 41×10^9 daltons for the largest DNA molecule in the wild-type genome. Then they studied two chromosomal rearrangements (Figure 16-2). One, a translocation involving chromosome 3, resulted in a chromosome that was one-third larger than the largest wild-type chromosome. Their viscoelastic measurement in this case was 58×10^9 daltons, also about one-third larger. The other rearrangement, a pericentric inversion, altered the arm ratio for chromosome 3 but not its size. The viscoelastic measurement was 42 times 10^9 daltons, not significantly differ-

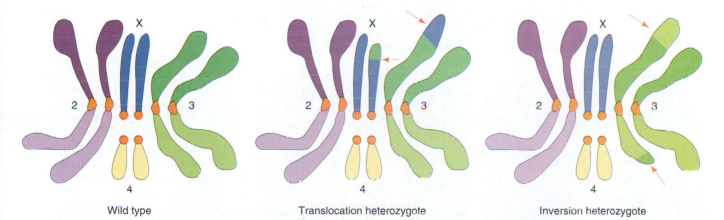

Figure 16-2 Genomes of *Drosophila* used to study DNA lengths in chromosomes. The chromosomes of three females are shown, one a normal wild type, one heterozygous for an X;3 translocation, and one heterozygous for a pericentric inversion of chromosome 3. Arrows mark breakpoints. Only the translocation produced a DNA molecule that was significantly longer.

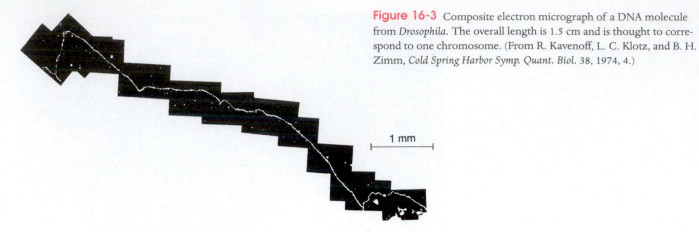

Figure 16-3 Composite electron micrograph of a DNA molecule from *Drosophila*. The overall length is 1.5 cm and is thought to correspond to one chromosome. (From R. Kavenoff, L. C. Klotz, and B. H. Zimm, *Cold Spring Harbor Symp. Quant. Biol.* 38, 1974, 4.)

1 mm

ent from that of the wild type. In other words, making a chromosome bigger made the DNA of that chromosome bigger by exactly the same proportion, and rearranging the chromosome but keeping the same length had no effect on DNA size. It looked like the chromosome was indeed one strand of DNA, continuous from one end through the centromere, to the other end. Kavenoff and Zimm also were able to piece together electron micrographs of DNA molecules about 1.5 cm long, each presumably corresponding to a *Drosophila* chromosome (Figure 16-3).

Today, geneticists can demonstrate directly that certain chromosomes contain single DNA molecules by using pulsed field gel electrophoresis, a technique for separating very long DNA molecules by size (see Chapters 15 and 17). If the DNA of an organism with relatively small chromosomes, such as *Neurospora*, is run for long periods of time in this apparatus, the number of bands that separate on the gel is equal to the number of chromosomes (seven, in the case of *Neurospora*). If each chromosome contained more than one DNA molecule, the number of bands would be ex-

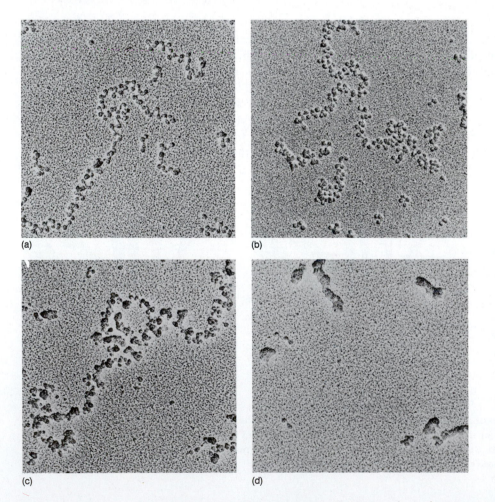

(a)

(b)

(c)

(d)

Figure 16-4 Condensation of chromatin with increasing salt concentration is demonstrated in electron micrographs made by Fritz Thoma and Theo Koller. At a very low salt concentration, as in (a), chromatin forms a loose fiber about 10 nm thick; nucleosomes (discussed in text) are connected by short stretches of DNA. At a concentration with an ionic strength closer to that of normal physiological conditions, as in (d), chromatin forms a thick fiber some 30 nm thick. The origin of this solenoid can be deduced by an examination of chromatin at increasing intermediate ionic strengths, as in (b) and (c). It arises from a shallow coiling of the nucleosome filament. The chromatin is enlarged here about 80,000 diameters.

pected to be greater than the number of chromosomes. Such separations cannot be made for organisms with large chromosomes (such as humans and *Drosophila*), because the DNA molecules are too large to move through the gel; nevertheless, all the evidence points to the general principle that a chromosome contains one DNA molecule.

Message Each eukaryotic chromosome contains a single, long, folded DNA molecule.

The Role of Histone Proteins in Packaging DNA

We have seen that because the length of a chromosomal DNA molecule is much greater than the length of a chromosome, there must be an efficient packaging system. What are the mechanisms that pack DNA into chromosomes? How is the very long DNA thread converted into the

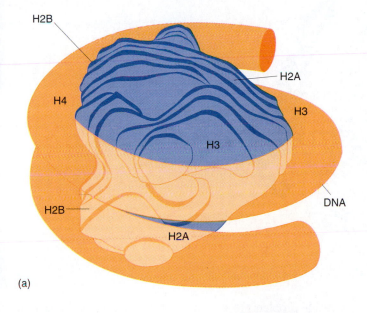

(a)

relatively thick, dense rod that is a chromosome? The overall mixture of material that chromosomes are composed of is given the general name **chromatin.** It is DNA and protein. If chromatin is extracted and treated with differing concentrations of salt, different degrees of compaction, or condensation, are observed under the electron microscope (Figure 16-4). With low salt concentrations, a structure about 10 nm in diameter is seen that resembles a bead necklace. The string between the beads of the necklace can be digested away with the enzyme DNase, so the string can be inferred to be DNA. The beads on the necklace are called **nucleosomes,** and these can be shown to consist of special chromosomal proteins, called **histones,** and DNA. Histone structure is remarkably conserved across the gamut of eukaryotic organisms, and nucleosomes are always found to contain an octamer of two units each of histones H2A, H2B, H3, and H4. The DNA is wrapped twice around the octamer as shown in Figure 16-5a. When salt concentrations are higher, the nucleosome bead necklace gradually assumes a coiled form called a **solenoid** (Figure 16-5b). This solenoid produced in vitro is 30 nm in diameter, and probably corresponds to the in vivo spaghettilike structures we first encountered in Figure 16-1. The solenoid is thought to be stabilized by another histone, H1, that runs down the center of the structure, as the figure shows.

We see then that to achieve its first level of packaging, DNA winds onto histones, which act somewhat like spools. Further coiling results in the solenoid conformation. However, it takes one more level of packaging to convert the solenoids into the three-dimensional structure we call the *chromosome.*

Higher-Order Coiling

Many cytogenetic studies show that chromosomes appear to be coiled, and Figure 16-6 shows a good example from the nucleus of a protozoan. Whereas the diameter of the

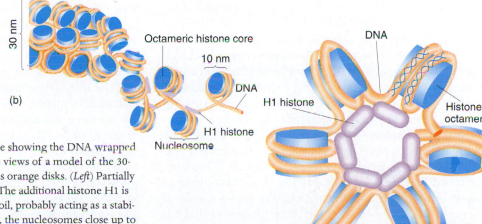

Figure 16-5 (a) Model of a nucleosome showing the DNA wrapped twice around a histone octamer. (b) Two views of a model of the 30-nm solenoid showing histone octamers as orange disks. (*Left*) Partially unwound lateral view. (*Right*) End view. The additional histone H1 is shown running down the center of the coil, probably acting as a stabilizer. With increasing salt concentrations, the nucleosomes close up to form a solenoid with six nucleosomes per turn. (From H. Lodish, D. Baltimore, A. Berk, S. L. Zipursky, P. Matsudaira, J. Darnell, *Molecular Cell Biology,* 3d ed. Copyright © 1995 by Scientific American Books.)

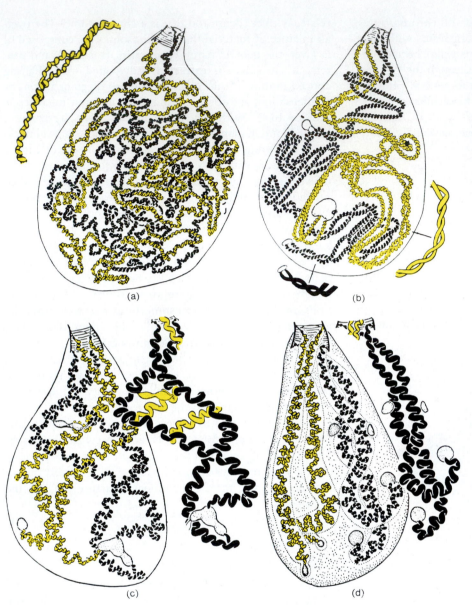

(a)

(b)

(c)

(d)

Figure 16-6 Drawings of chromosomes in meiotic prophase in a protozoan, demonstrating different degrees of coiling and supercoiling visible with the light microscope. Two large chromosomes are shown: one colored and the other black; (a) to (d) is a progression. (a) Coiling is seen, though duplication becomes apparent. (b) Duplication is well advanced. (c) Supercoiling is beginning. (d) Supercoiling is well advanced. (From L. R. Cleveland, "The Whole Life Cycle of Chromosomes and Their Coiling Systems," *Trans. Am. Philosophical Soc.* 39, 1949, 1.)

solenoids is 30 nm, the diameter of these coils is the same as the diameter of the chromosome during cell division, often about 700 nm. What produces these supercoils? One clue comes from observing mitotic metaphase chromosomes from which the histone proteins have been removed chemically. After such treatment, the chromosomes have a densely staining central core of nonhistone protein called the **scaffold,** as shown in Figure 16-7 and in the electron micrograph on the opening page of this chapter. Projecting laterally from this protein scaffold are loops of DNA. At high magnifications, it is clear from electron micrographs that each DNA loop begins and ends at the scaffold. It has been discovered that the central scaffold in metaphase chromosomes is largely composed of the enzyme topoisomerase II (see page 331). You will remember from Chapter 11 that this enzyme has the ability to pass a strand of DNA through another cut strand. Evidently, this central scaffold manipulates the vast skein of DNA during replication, preventing many

possible problems of unwinding DNA strands at this crucial stage. In any case, it is well established that there is a scaffold in eukaryotic chromosomes, and it seems to be a major organizing device for these chromosomes.

Now to return to the question of how the supercoiling of the chromosome is produced. The best evidence suggests that the solenoids arrange in loops emanating from the central scaffold matrix, which itself is in the form of a spiral. We see the general idea in Figure 16-8, which shows a representation of loosely coiled interphase chromosomes and the more tightly coiled metaphase chromosomes. How do the loops attach to the scaffold? There appear to be special regions along the DNA called **scaffold attachment regions,** or SARs. The evidence for these is as follows. When histoneless chromatin is treated with restriction enzymes, the DNA loops are cut off the scaffold, but special regions of DNA remain attached to the scaffold. These are resistant to exonuclease digestion and have been shown to have protein bound

Figure 16-7 Electron micrograph of a metaphase chromosome from a cultured human cell. Note the central core, or scaffold, from which the DNA strands extend outward. No free ends are visible at the outer edge. At even higher magnification, it is clear that each loop begins and ends near the same region of the scaffold. (From W. R. Baumbach and K. W. Adolph, *Cold Spring Harbor Symp. Quant. Biol.*, Cold Spring Harbor Laboratory, Cold Spring Harbor, N.Y., 1977.)

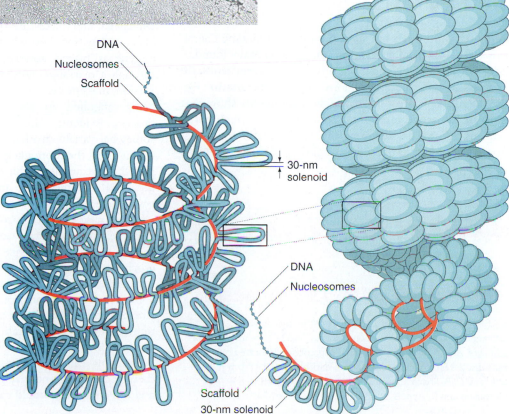

DNA

Nucleosomes

Scaffold

30-nm solenoid

DNA

Nucleosomes

Scaffold

30-nm solenoid

Figure 16-8 Model for chromosome structure. On the left is shown a more relaxed supercoil, as at interphase. On the right, much tighter coiling is shown, representing metaphase: here the loops are so densely packed, only their tips are visible. At the free ends the solenoids are shown uncoiled to give an approximation of relative scale.

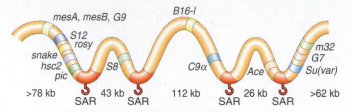

Figure 16-9 Some loop domains that have been mapped in *Drosophila*. Gene loci and scaffold attachment regions (SARs) are shown.

to them. When the protein is digested away, the DNA regions that bind the protein have been shown in *Drosophila* to contain sequences that are known to be specific for topoisomerase binding. This makes it likely that these regions are the SARs that glue the loops onto the scaffold. Some specific loops that have been mapped in *Drosophila* are shown in Figure 16-9. The size of the loops ranges between 4.5 and 112 kb. The SARs are only in nontranscribed regions of the DNA.

Message   In the progressive levels of chromosome packing

1. DNA winds onto nucleosome spools.
2. The nucleosome chain coils into a solenoid.
3. The solenoid loops, and the loops attach to a central scaffold.
4. The scaffold plus loops arrange into a giant supercoil.

Heterochromatin and Euchromatin

When chromosomes are stained with DNA-binding chemicals such as the Feulgen reagent, they are generally found to have both densely staining regions called **heterochromatin** and less densely staining regions called **euchromatin** (Figure 16-10). Chromosome mapping experiments show that most of the active genes that are detectable through mutation are located in euchromatin. Euchromatin stains less densely because it is packed less tightly, and the general idea

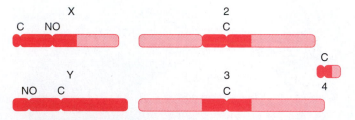

Figure 16-10 Position of heterochromatin in *Drosophila melanogaster*. C represents the centromeres, and NO the nucleolar organizers. A secondary chromosomal constriction is located on chromosome 2. Heterochromatin is more darkly colored. (From A. Hilliker and C. B. Sharp, *Chromosome Structure and Function*, J. P. Gustafson and R. Appels, eds. Plenum, 1988, pp. 91–95.)

is that this is the state most compatible with transcription and gene activity. Heterochromatin in most organisms is found flanking the centromeres, but some whole chromosomes, such as the *Drosophila* Y, are heterochromatic.

The most extensive studies of the genetics of heterochromatin have been done in *Drosophila*. For example, Arthur Hilliker has isolated lethal mutations in genes that map in heterochromatin. Although the function of most of these genes is unknown, the fact that mutation in heterochromatin can be lethal demonstrates that the heterochromatin must show some gene activity. The proportion of *Drosophila* heterochromatin that maps as genes is only 1/100 of the proportion of *Drosophila* euchromatin that maps as genes; there are obviously great stretches of DNA between the functional genes. For now, note that there is clearly a major distinction in the gene activity of the two kinds of chromatin. We will have more to say about the stretches between genes later.

In thinking about the architecture of chromosomes, a key question is how heterochromatin differs from euchromatin, and more important, how this difference is maintained. First, the difference, as we have seen, is the degree of compaction. It is not known if this difference is similar to the difference between interphase and metaphase chromosomes, in other words, a difference in the tightness of coiling as shown in Figure 16-8, or if it resembles other structural differences. The question of how this difference between euchromatin and heterochromatin is maintained is also difficult to answer. In fact the answer is not known at present, but the way in which the problem is being addressed is interesting, as it illustrates well the genetic approach to such a biological question. The investigation is in *Drosophila,* and makes use of the phenomenon of position-effect variegation. You will recall from page 233 that when a gene is placed abnormally close to heterochromatin through a rearrangement such as a translocation or an inversion, in some cells the gene is inactivated, presumably by being engulfed in the heterochromatin condensation process. Evidently, the edge of the heterochromatin is somewhat variable from cell to cell, and the closer a gene is to this edge, the more likely it is to become occasionally engulfed. An example of this effect in mouse is shown in Figure 16-11. The mosaic of active and inactive tissue is called *position-effect variegation* (see Chapter 8 for a full discussion).

Thomas Grigliatti reasoned that since heterochromatin is of relatively constant location across individuals of a species, then the difference between euchromatic and heterochromatic regions must be genetically determined. The way to find out what the heterochromatin-controlling genes are, and what types of proteins they code, would be to isolate mutations that either suppress position-effect variegation [called *Su(var)* mutations] or enhance position-effect variegation [*E(var)* mutations]. The suppressors would presumably cause the heterochromatin to shrink in size, thereby engulfing and inactivating a marker gene less often, and the enhancers would cause the heterochromatin to en-

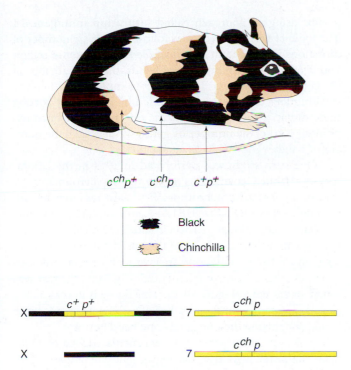

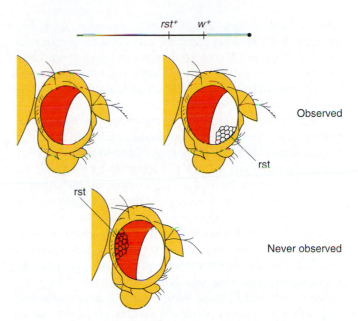

Figure 16-11 Position effect variegation in a female mouse with an insertional translocation of a segment from chromosome 7 containing alleles of two different coat color loci. The heterochromatin formed during X chromosome inactivation sometimes engulfs the c^+ allele, giving a $c^{ch}p^+$ phenotype that is beige, and sometimes engulfs both wild-type alleles, giving a $c^{ch}p$ that is white. (Modified from J. R. S. Fincham, *Genetic Analysis: Principles, Scope, and Objectives*. Blackwell, London, 1994.)

Figure 16-12 When the wild-type allele of the $Su(var)$ gene is present in three copies, the heterochromatin sometimes engulfs the rst (roughest) locus, giving the mutant rough-looking phenotype. Note that rst is never inactivated when w is active, showing that the inactivation effect must spread along the chromosome. (From T. A. Grigliatti, *Methods in Cell Biology* 35, 1991, 587.)

large and engulf a marker gene more often. Many examples of both types of mutations were produced after mutagenesis. Furthermore these mutations not only affected the proportion of cells that showed marker gene inactivation, but as predicted also affected the amount of heterochromatin visible near the marker gene's chromosome band when viewed under the microscope.

Some interesting genetic effects were found. For example, it was shown that deletion of one wild-type $Su(var)^+$ allele, leaving only one functional $Su(var)^+$, suppressed variegation just as the $Su(var)$ mutant allele does. But conversely, if an extra wild-type $Su(var)^+$ allele was added through duplication (making a total of three), then inactivation of the target gene was enhanced, as it is in a fly carrying an $E(var)$ mutation. Thus there seems to be a gene dose effect, where more $Su(var)^+$ alleles mean the production of more heterochromatin. The results also suggest that normally the amount of heterochromatin-forming gene product is in limited supply. Another interesting discovery was that the gene inactivation sometimes spread past the marker gene to engulf another locus. For example, in normal strains gene w (for *white eye*) is occasionally engulfed by the heterochromatin next to it, but the gene rst (for *roughest*) on the other side of w never is. In lines with three $Su(var)^+$ alleles, how-

ever, w is inactivated in 75 percent of cells, and rst is also inactivated in 30 percent (Figure 16-12). This points clearly to an enhanced spreading effect.

The $Su(var)$ and $E(var)$ genes are now being isolated and sequenced, and their products will undoubtedly be a range of proteins that are used in chromatin assembly and condensation.

Message Euchromatin contains most of the active genes. Heterochromatin is more condensed, and the edge of the heterochromatin sometimes engulfs genes, thereby inactivating them.

Chromosome Bands

One of the basic chromosomal banding patterns is that produced by Giemsa reagent, a DNA stain applied after mild proteolytic digestion of the chromosomes. This reagent produces patterns of light-staining (G-light) regions and dark-staining (G-dark) regions (see page 525). The patterns are consistent within species. In the complete set of 23 human chromosomes there are approximately 850 G-dark bands visible at metaphase of mitosis. These have provided a useful way of subdividing the various regions of chromosomes, and each band has been assigned a specific number.

The difference between dark- and light-staining regions was believed to be caused by differences in the relative proportions of bases: the G-light bands being relatively GC-rich, and the G-dark bands AT-rich. However, it is now thought that the differences are too small to account for banding patterns. The crucial factor appears to be chromatin packing density: the G-dark regions are packed denser, with tighter coils, and this provides a higher density of DNA to take up the stain.

In addition, various other correlations have been made. For example, deoxynucleotide labeling studies showed that G-light bands are early-replicating. Furthermore if polysomal (poly*ribo*somal) mRNA (representing genes being actively transcribed) is used to label chromosomes in situ, then most label binds to the G-light regions, suggesting that these contain most of the active genes. From such an analysis it was presumed that the density of active genes is higher in the G-light bands, and this notion has been tested directly as follows. Some restriction enzymes recognize target sequences that are long and hence uncommon along the DNA. Consequently, these "long-cutters" generate relatively large DNA fragments. Most mammalian genes are known to be close to CpG-rich islands of unknown function. If long-cutters with several GC pairs in their target sites are used to digest genomic DNA, then each resulting fragment consists roughly of a gene and all the untranscribed DNA that follows it until the next CpG island (and the next gene). The genome can thus be cut into a series of fragment lengths roughly reflecting the relative spacing between genes. After separating these size fractions electrophoretically, it was shown that the shorter fragments (representing closely spaced genes) hybridize preferentially to the G-light regions.

Our view of chromosome banding is based largely on how chromosomes stain when they are in mitotic metaphase. Does this banding pattern also exist during interphase? There is evidence from in situ labeling with probes that there is banding in interphase, but the degree to which this corresponds to metaphase banding is not known. Nevertheless, the domains revealed by metaphase banding obviously must still be in the same relative position in interphase, whether banded or not.

It is interesting to ask if the relative positions of bands are of any functional significance. We saw in Chapter 8 that as mammals evolved, many blocks of bands were conserved throughout the various lineages. Unfortunately, however, this does not tell us whether this reflects some kind of essential functional grouping of the bands, or whether there simply has not been enough time for the scissors of evolution to cut and rearrange the groups of bands.

Possibly the most spectacular banding, which has had a perennial fascination for geneticists, is that of the giant salivary gland chromosomes of *Drosophila* larvae. Each of these huge chromosomes shows a clear and unique array of alternating dark and light transverse bands. One of the obvious questions that has been asked is what the bands represent. An early suggestion was that each band corresponds to a separate gene. This idea was tested by several research groups using an approach called **saturation mutagenesis.** The logic of this approach is as follows. If a large number of mutations is induced in a well-studied chromosome region with a certain known number of bands, then presumably all the genes in that area will be subject to mutation. When the total number of genes is determined by genetically sorting these mutations into groups, if the one gene – one band hypothesis is true, the number of genes should equal the number of bands.

One such study was performed in 1979 in the laboratory of Arthur Chovnick, using the region surrounding the eye-color gene *rosy* on chromosome 3. The region is technically defined as 87D2-4 to 87E12-F1, an interval of 23 visible polytene bands. The geneticists isolated 153 recessive lethals falling within this region. Then they used complementation tests and a set of deletions to group the mutations into genes. A total of 21 complementation groups (genes) was found, as shown in Figure 16-13. This figure is in reasonably close agreement with the number of polytene chromosome bands, supporting the one gene – one band notion.

However, in more recent experiments, 315 kb of DNA from the same chromosomal region and encompassing 12 of its bands, has been cloned in fragments. These cloned fragments were used to detect RNA transcripts in Northern hybridizations. The experiment revealed over 40 distinct mRNAs transcribed at various developmental stages, as shown in Figure 16-12c. Hence these molecular studies show that most likely there are more genes than bands. It is not known why all these genes were not detected in the saturation mutagenesis.

Before we leave studies of polytene chromosomes, we can address the nature of these chromosomes at the molecular level. It has been calculated that the euchromatic region of the polytene X chromosome of *Drosophila hydei* contains about 1024 copies of the DNA of that region. This must have arisen from 2^{10} DNA duplications, to give a parallel array of molecules that provide the great bulk of the chromosome. These parallel DNAs would of course also be organized on histones. A model for such a structure, using a much smaller number of parallel DNAs for illustration purposes, is shown in Figure 16-14.

Centromeric DNA

One of the prominent landmarks of a chromosome is its centromere, a constriction that becomes visible during nuclear division and appears to act as the attachment point for fibers of the nuclear spindle. Recombinant DNA techniques allow us to study centromere structure at the molecular level. The prototype study was done in the yeast *Saccharomyces cerevisiae* by Louis Clarke and John Carbon, and the first centromere to be isolated and sequenced was that of chromosome 3. The study began by cloning the genes

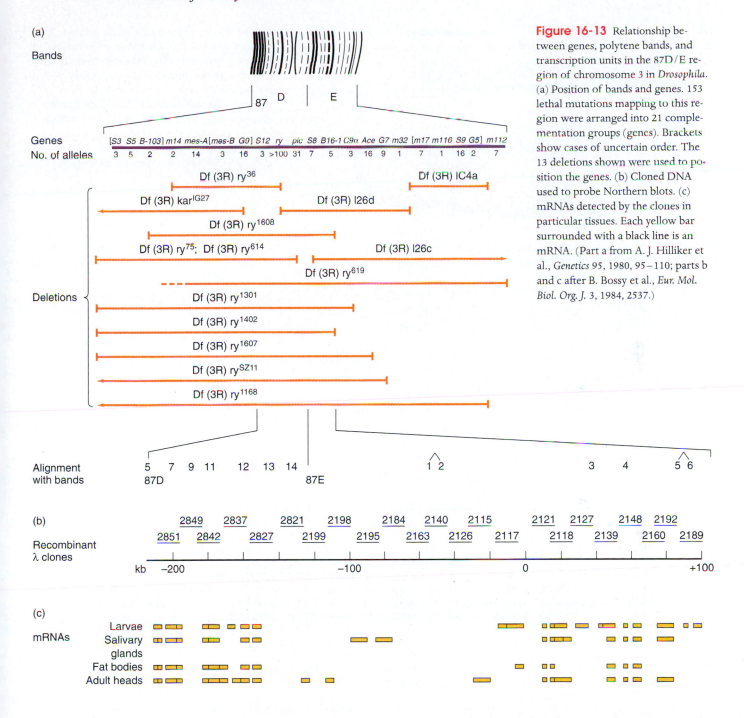

Figure 16-13 Relationship between genes, polytene bands, and transcription units in the 87D/E region of chromosome 3 in *Drosophila*. (a) Position of bands and genes. 153 lethal mutations mapping to this region were arranged into 21 complementation groups (genes). Brackets show cases of uncertain order. The 13 deletions shown were used to position the genes. (b) Cloned DNA used to probe Northern blots. (c) mRNAs detected by the clones in particular tissues. Each yellow bar surrounded with a black line is an mRNA. (Part a from A. J. Hilliker et al., *Genetics* 95, 1980, 95–110; parts b and c after B. Bossy et al., *Eur. Mol. Biol. Org. J.* 3, 1984, 2537.)

LEU2 and CDC10, which are closely linked on opposite sides of the chromosome 3 centromere (CEN3). A chromosome walk was undertaken from one to the other. The intervening fragments were tested by splicing them individually into an artificially created yeast plasmid to see if they conferred ability to segregate the plasmid properly at meiosis, allowing it to behave in effect like the extra chromosome of a disomic. Without any centromeric sequence, the plasmid was found in all four ascospores, but in the presence of certain fragments from the LEU2–CDC10 region, the plasmid was present in two ascospore but not in the other two.

Clearly this 2:2 segregation indicates passage of the plasmid to one pole at the first meiotic division, and then a chromatidlike separation at the second division, precisely as expected of the extra chromosome of an aneuploid. Subsequent sequencing of the centromeres of chromosome 3 and other chromosomes revealed a common pattern as shown in Figure 16-15. Deletion and in vitro mutagenesis showed that centromeric regions I, II, and III were required for proper segregation. These regions probably act as an attachment for proteins that bind to the spindle fiber, as shown in Figure 16-16.

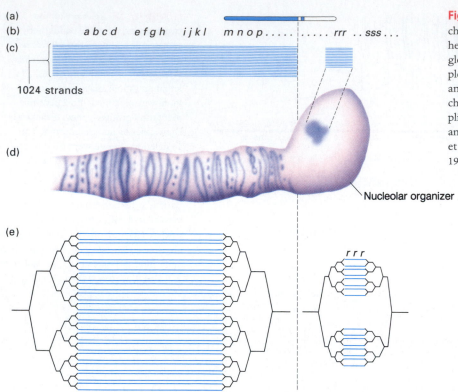

(a)

(b) *a b c d   e f g h   i j k l   m n o p . . . . rrr . . sss . . .*

(c)

1024 strands

(d)

Nucleolar organizer

(e)

r r r

Figure 16-14 Organization of the polytene X chromosome in *Drosophila hydei*. (a) Distribution of heterochromatin, mainly in the right arm. (b) Single-copy genes *a* through *p*, ribosomal genes *r*, simple-sequence repetitive DNA *s*. (c) Relative amounts of amplification of DNA. (d) Polytene chromosome with bands. (e) Model for DNA amplification. Note lower level of amplification for *r*, and no amplification of *s*. (Part e from C. D. Laird et al., *Cold Spring Harbor Symp. Quant. Biol.*, 38, 1973, 311.)

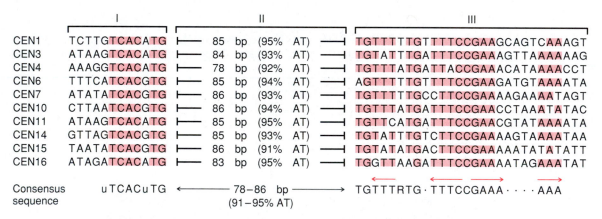

	I		II		III
CEN1	TCTTGTCACATG	85 bp	(95% AT)	TGTTTTTGTTTTCCGAAGCAGTCAAAGT
CEN3	ATAAGTCACATG	84 bp	(93% AT)	TGTATTTGATTTCCGAAAGTTAAAAAAG
CEN4	AAAGGTCACATG	78 bp	(92% AT)	TGTTTATGATTACCGAAACATAAAACCT
CEN6	TTTCATCACGTG	85 bp	(94% AT)	AGTTTTTGTTTTCCGAAGATGTAAAATA
CEN7	ATATATCACGTG	86 bp	(93% AT)	TGTTTTTGCCTTCCGAAAAGAAAATAGT
CEN10	CTTAATCACGTG	86 bp	(94% AT)	TGTTTATGATTTCCGAACCTAAATATAC
CEN11	ATAAGTCACATG	85 bp	(95% AT)	TGTTCATGATTTCCGAACGTATAAAATA
CEN14	GTTAGTCACGTG	85 bp	(93% AT)	TGTATTTGTCTTCCGAAAAGTAAAATAA
CEN15	TAATATCACGTG	86 bp	(91% AT)	TGTATATGACTTCCGAAAAATATATATT
CEN16	ATAGATCACATG	83 bp	(95% AT)	TGGTTAAGATTTCCGAAAATAGAAATAT

Consensus sequence	uTCACuTG	← 78–86 bp → (91–95% AT)	TGTTTRTG · TTTCCGAAA · · · · AAA

Figure 16-15 Centromere sequences for 10 yeast chromosomes. Three basic elements I, II, and III are found in all. Common nucleotides are shaded and written below as a consensus sequence, where U = A or G, and R = A or T. Arrows above the consensus sequence show an inverted repeat that might be important in protein binding. (Modified from L. Clarke and J. Carbon, *Ann. Rev. Genetics* 19, 1985, 29.)

What happens to the centromere DNA during meiosis I, when the centromere does not divide? It has been postulated that the centromere imposes some kind of block on DNA replication, giving the structure shown in Figure 16-17. This is supported by the fact that centromeres cannot act as origins of replication when engineered into test plasmids. Replication must enter from outside the centromere and completes through the centromere only at anaphase II.

Sequence Organization

From the mRNA analysis of the *Drosophila* chromosomal region shown in Figure 16-12, it is clear that there are nontranscribed regions between the transcriptional units. This observation raises the entire question of how the genes are arranged on chromosomes. What proportion of chromosomal DNA comprises active or potentially active genes? What

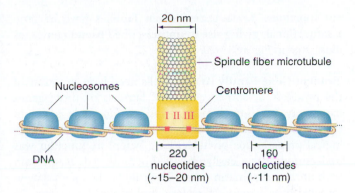

Figure 16-16 Attachment of the spindle fiber microtubule to the centromere in a yeast chromosome. (Modified from L. Clarke and J. Carbon, *Ann Rev. Genetics* 19, 1985, 29.)

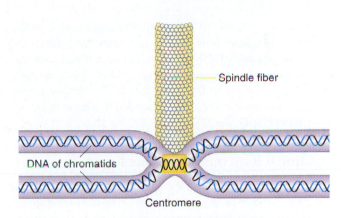

Figure 16-17 Model for arrangement of DNA strands in the centromere and the sister chromatids at the first meiosis anaphase.

is the nature of the DNA between the genes? In this section we will answer some of these questions. As in most aspects of research, the situation turns out to be more complex than expected. But at the same time such unexpected results are always an integral part of the excitement of science.

Some of the first informative results came from studies in which DNA samples from eukaryotic nuclei were heated to separate the two strands of the double helix, and then allowed to cool. Whenever this is done, complementary sequences, such as those that are initially hydrogen-bonded together in a double helix, eventually find each other through random movements of the molecules in solution. This process is called **reannealing.** But surprisingly, the DNA started coming together much faster than was expected on the basis of separated strands of single genes finding each other. The kinetics of reannealing required postulation of a class of DNA that was present in many copies in the genome. This became known as **repetitive DNA.** Of course, the single-copy genes reanneal eventually, but they take much longer. Now it is known that there are several classes of repetitive DNA. In fact, we can classify eukaryotic DNA as shown in Figure 16-18.

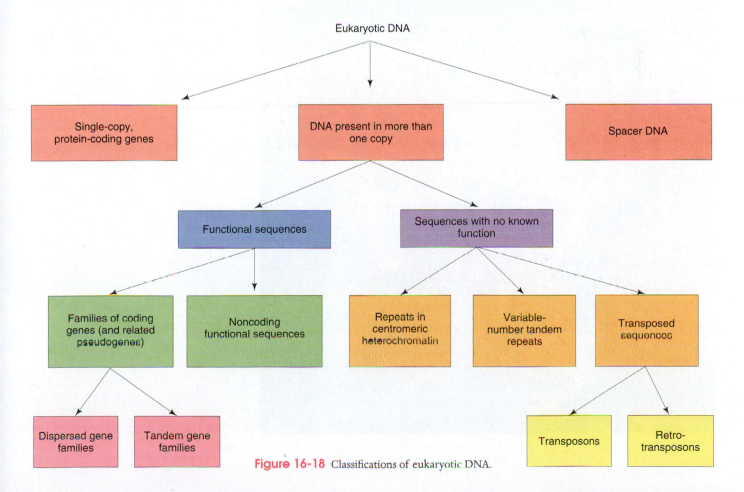

Figure 16-18 Classifications of eukaryotic DNA.

If we are to come to grips with the architecture of the eukaryotic genome, it is clearly necessary to understand how these classes of DNA are arranged to make up the chromosomes. It turns out that the kind of DNA we have spent most time on so far in the book, the unique single-copy genes (Figure 16-19), are embedded in a diverse assembly of repetitive DNAs.

Functional Repetitive Sequences

Dispersed Gene Families. Several types of proteins are coded by families of homologous genes spread throughout the genome. Such families may comprise only a few genes or very many, as some examples illustrate: actins, 5 to 30; keratins more than 20; myosin heavy chain, 5 to 10; tubulins, 3 to 15; insect eggshell proteins, 50; globins, up to 5; immunoglobulin variable region, 500; ovalbumin, 3; and histones, 100 to 1000. The exact DNA sequences of the genes within a family may diverge as have the genes for human hemoglobins that we studied on page 222, and the different homologous genes may come to have slightly differ-

ent functions. Some genes within families have become nonfunctional, giving rise to untranscribed **pseudogenes,** as illustrated in Figure 16-20.

Tandem Gene Family Arrays. Cells need large amounts of the products of some genes, and families of these genes have evolved into tandem arrays. A good example is seen in the nucleolar organizer (NO), which was observed in cytological preparations of nuclei long before its function was understood. It was easily observed because it does not stain with normal chromatin stains. The role of the NO has been revealed by a variety of genetic and molecular studies. In the 1960s, it was suggested that NOs might be tandem arrays of genes that code for rRNA. Ferruccio Ritossa and Sol Spiegelman tested this hypothesis by constructing *Drosophila melanogaster* strains having different numbers of NOs per cell. In this species, the NOs are located in the heterochromatin of the X chromosome and in the short arm of the Y chromosome. (In *Drosophila,* the X chromosome is always depicted with the centromere on the right. We assume this orientation of the chromosome in discussing positions on

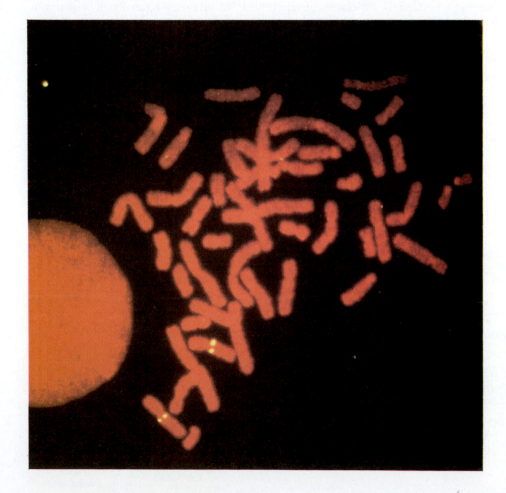

Figure 16-19 Chromosomes probed in situ with a fluorescent probe specific for a gene present in a single copy in each chromosome set, in this case a muscle protein. Only one locus shows a fluorescent spot corresponding to the probe bound to the muscle protein gene. (From Peter Lichter et al., *Science* 247, 1990, 64.)

Figure 16-20 The human α-like and β-like globin gene families are each organized into a single cluster that includes functional genes and pseudogenes; the latter are denoted here by ψ. (After B. Lewin, *Genes.* John Wiley, 1983.)

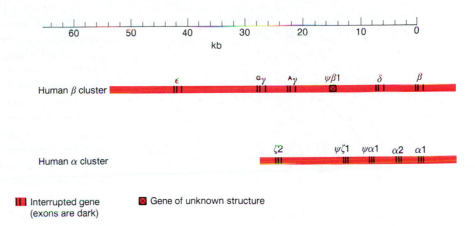

Human β cluster

Human α cluster

‖ Interrupted gene (exons are dark)   ⊠ Gene of unknown structure

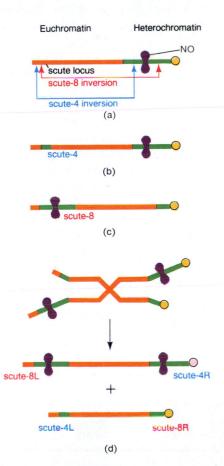

Euchromatin   Heterochromatin

scute locus
scute-8 inversion
scute-4 inversion

(a)

scute-4

(b)

scute-8

(c)

scute-8L   scute-4R

+

scute-4L   scute-8R

(d)

Figure 16-21 The scute-4 and scute-8 inversions. (a) Breakpoints for the two inversions on the *Drosophila* X chromosome. (b) The result of the scute-4 inversion. (c) The result of the scute-8 inversion. (d) A crossover between scute-4 and scute-8 in a heterozygous female yields one crossover product with two nucleolar organizers and another product with no nucleolar organizer. These products are identified as scute-8L scute-4R and scute-4L scute-8R, respectively. The notation scute-8L scute-4R indicates that the left part of the chromosome is derived from the scute-8 chromosome and the right part is derived from the scute-4 chromosome.

it.) Several mutant strains of *Drosophila* exist with inversions of the X chromosome in which the left breakpoint is near the *scute* locus and the right breakpoint is in the heterochromatic region; these inversion mutants are named for the recessive scute phenotype they confer (a reduced number of bristles). The inversion scute-8 has its right breakpoint between the NO and the centromere; the inversion scute-4 has its right breakpoint on the distal side of NO (Figure 16-21). From females heterozygous for scute-4 and scute-8, crossover products can be recovered that carry either two or zero NOs. Different numbers of NOs per cell then can be engineered by appropriate genetic combinations (Table 16-2).

By annealing radioactive ribosomal RNA to known amounts of DNA, F. M. Ritossa and S. Spiegelman measured the amount of DNA homologous to rRNA. They found a linear relationship between the number of NOs per cell and the amount of 18S and 28S rRNA that hybridized (Figure 16-22). This result demonstrates that the rRNA loci are located in the NOs. Subsequently, hybridization in situ has confirmed the NO location of the DNA corresponding

Table 16-2 Chromosomal Complements with Various Numbers of NOs per Cell

Chromosomes	Number of NOs per cell
scute-4L scute-8R/Y ♂	1
scute-4L scute-8R/wild-type/X ♀	1
wild-type X/Y ♂	2
wild-type/wild-type/X/X ♀	2
scute-8L scute-4R/wild-type/X ♀	3
scute-8L scute-4R/scute-8L scute-4R	4

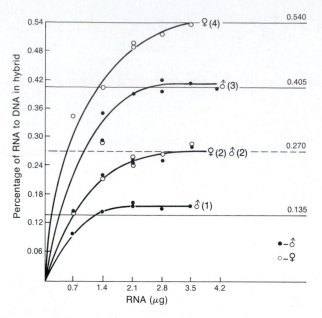

Figure 16-22 The amount of rRNA that hybridizes to a constant amount of DNA. The plateau is reached when all DNA complementary to the rRNA is hybridized. Each curve is obtained using DNA samples isolated from individuals with a particular number of the nucleolar organizer (NO) regions per cell (in parentheses). (From F. M. Ritossa and S. Spiegelman, *Proc. of the Natl. Acad. of Sci. USA 53*, 1965, 737.)

to rRNA. Furthermore, the sizes of the 18S and 28S rRNAs are now known, as are the percentages of the total DNA that hybridize to them. Thus it was estimated that each *Drosophila* X chromosome has about 200 genes for rRNA. Obviously, such redundancy is one way of ensuring a large amount of rRNA per cell.

Message   The nucleolar organizer, which is cytologically distinct, is a tandem array of genes that code for ribosomal RNA.

We now know that the NOs on the *Drosophila* X and Y chromosomes contain 250 and 150 tandem copies of rRNA genes, respectively. One human NO has about 250 copies. In Figure 13-4 we encountered an electron micrograph of the transcription of rRNA tandem arrays in an amphibian.

Another example of a tandem array is the genes for tRNA. In humans there are about 50 chromosomal sites corresponding to the different tRNA types, and at each site there are between 10 to 100 copies.

Finally, the genes for histones themselves are arranged in tandem arrays in some species (Figure 16-23). For histone arrays, as for the other tandemly repeated arrays, sequencing analysis has shown that the multiple copies are identical. Because one might expect that mutation would lead to some differences at noncrucial sites, it seems there is some mechanism for maintaining constancy across the members of the array. The most likely mechanism is some kind of gene conversion, a process that will be discussed in Chapter 20.

Noncoding Functional Sequences. Telomeres, the tips of chromosomes, have tandem arrays of simple DNA sequences that do not code for an RNA or a protein product, but nevertheless have a definite function. For example, in the ciliate *Tetrahymena*, there is repetition of the sequence of TTGGGG, and in humans it is TTAGGG. The telomeric repeats are there to solve a functional problem that is inherent in the replication of linear DNA molecules. Figure 16-24 shows the problem: for the leading strand the polynucleotide addition can always extend to the end because it is automatically primed from behind. However, at the tip the lagging strand reaches a point where its system of RNA priming cannot work, and an unpolymerized section remains. This would result in chromosome shortening. However, an enzyme called *telomerase* adds the simple repeat units to the ends. It seems likely that these additional units bend back on themselves in a hairpin structure made possi-

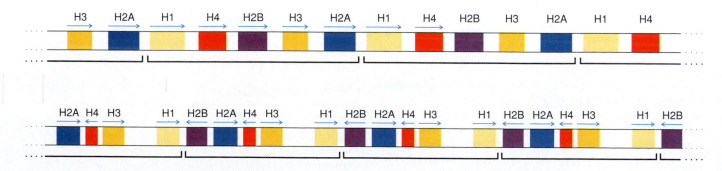

Figure 16-23 Tandem repeats of histone genes in sea urchin and fruit fly. Only a small fraction of the repeats is shown. Arrows indicate direction of transcription.

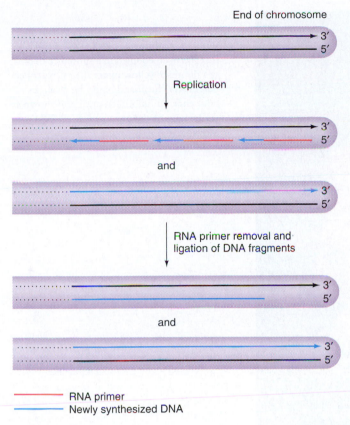

End of chromosome

Replication

and

RNA primer removal and ligation of DNA fragments

and

——— RNA primer
——— Newly synthesized DNA

Figure 16-24 The replication problem at chromosome ends. There is no way of priming the last section of the lagging strand, and a shortened chromosome would result.

ble by an unusual type of G–G hydrogen bonding (Figure 16-25). This provides a 3′ end for filling in the remaining section. The addition of extra telomeric repeats is under careful genetic control, as demonstrated by the fact that yeast mutants have been isolated that are defective in this control. It is interesting to note that through recombinant DNA technology it has been possible to transfer telomeres from other organisms onto yeast artificial chromosomes. Generally they work! However, any additional units that are added to the ends in vivo are yeast units; apparently the

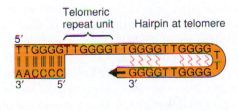

Telomeric repeat unit   Hairpin at telomere

5′
TTGGGGTTGGGGTTGGGGTTGGGG
AACCCC   GGGGTTGGGG
3′   5′   3′

‖ ‖‖ Normal hydrogen bonds
≀   Unusual G-to-G hydrogen bonds

Figure 16-25 Model for the way that telomeric repeats solve the replication problem at chromosome ends. By folding itself into a hairpin secured by an unusual type of hydrogen bonding between G's, the DNA gains a priming site.

telomerase does not use the ends of the DNA as a template, and its specificity resides within itself. Figure 16-26 demonstrates the positions of the telomeric DNA through in situ hybridization. An age-dependent decline in telomere length has been found in several somatic tissues in humans. In addition, human fibroblasts in culture show progressive telomere shortening up to their eventual death. Such observations have led to the telomere theory of aging, and the validity of this theory is now under test.

Sequences with No Known Function

For some repetitive DNA, no function is known. This category contains DNA sequences that generally have many more copies in the genome than the functional sequences discussed above. The combined size of this class is surprising; for example, it has been calculated that about 20 percent of the human genome consists of nonfunctional repetitive sequences of one kind or another. We will consider three types.

Highly Repetitive Centromeric DNA. After genomic DNA has been centrifuged in a cesium chloride density gradient, satellite bands often are visible, distinct from the main DNA band. Upon isolation, such **satellite DNA** is found to consist of multiple tandem repeats of short nucleotide sequences, stretching up to hundreds of kilobases in length. When probes are prepared from such simple-sequence DNA and used in chromosomal in situ labeling experiments, the great bulk of the satellite DNA is found to reside in the heterochromatic regions flanking the centromeres. There can be either one or several basic repeating units, but usually they are less than 10 bases long. For example, in *Drosophila melanogaster*, the sequence AATAACATAG is found in tandem arrays around all centromeres. Similarly, in the guinea pig, the shorter sequence CCCTAA is arrayed flanking the centromeres. In situ labeling of a mouse satellite DNA is shown in Figure 16-27.

Because the centromeric repeats are a nonrepresentative sample of the genomic DNA, the G+C content can be significantly different from the rest of the DNA. This is why the DNA forms a separate satellite band in a cesium chloride gradient. There is no demonstrable function for centromeric repetitive DNA, nor is there any understanding of its relation to heterochromatin or to genes in the heterochromatin. Some organisms have staggering amounts of this DNA; for example, as much as 50 percent of kangaroo DNA can be centromeric satellite DNA.

VNTRs. A special class of tandem repeats shows variable number at different loci and in different individuals. These are called **VNTRs** (**variable number tandem repeats,** sometimes called **minisatellite DNA**). The VNTR loci in humans are 1- to 5-kb sequences consisting of variable numbers of a repeating unit 15 to 100 nucleotides long. If a VNTR probe is available, and the total genomic DNA is cut

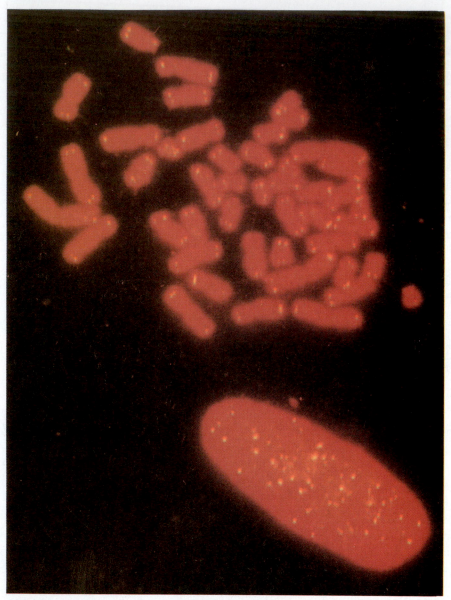

Figure 16-26 Chromosomes probed in situ with a telomere-specific DNA probe that has been coupled to a substance that can fluoresce yellow under the microscope. Each sister chromatid binds the probe at both ends. An unbroken nucleus is shown at the bottom of the photograph. (Robert Moyzis)

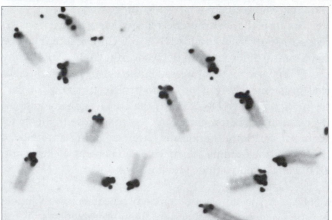

Figure 16-27 Autoradiographic localization of simple-sequence mouse DNA to the centromeres. A radioactive probe for simple-sequence DNA was added to chromosomes whose DNA had been denatured. (Note that all mouse chromosomes have their centromeres at one end.) (From M. L. Pardue and J. G. Gall, *Science* 168, 1970, 1356.)

with a restriction enzyme that has no target sites within the VNTR arrays, then a Southern blot reveals a large number of different-sized fragments that are bound by the probe. Because of the variability in the number of tandem repeats from individual to individual, the set of fragments that shows up on the Southern autoradiogram is highly individualistic. In fact these patterns are called **DNA fingerprints** (Figure 16-28). They are used extensively in forensic medicine. In a criminal investigation, blood or semen can be used to prepare DNA fingerprints, and the patterns compared with those of a suspect. Minute samples of tissue such as follicle cells clinging to single hairs can be used by amplifying VNTR sequences through the polymerase chain reaction.

One of the first probes that detected human VNTRs was obtained in 1985 by Alec Jeffries from a quadruple repeat of a 33-bp sequence found within the first intron of the

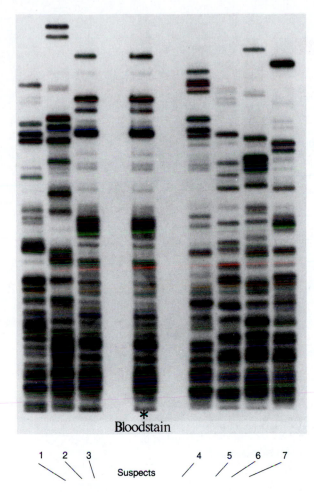

Bloodstain

1 2 3   4 5 6 7
                Suspects

Figure 16-28 DNA fingerprints from a bloodstain at the scene of a crime and from the blood of seven suspects. (Cellmark Diagnostics, Germantown, Md.)

myoglobin gene (Figure 16-29a). Once some of the VNTRs in the DNA fingerprint were cloned and sequenced, it was found that the reason for hybridization of the probe to the VNTRs was a 13-bp core sequence that was common to all the repeats and to the probe. The longer sequences into which the core was embedded were not necessarily similar.

DNA fingerprints can be made in many different species of organisms, so VNTRs must be common, and the technique has great value in testing genetic individuality.

Another class of dispersed repetitive DNA is composed of dinucleotide repeats. This class of repetitive DNA is called **microsatellite DNA.** Although not normally included in the VNTR class, microsatellites are in fact dispersed regions composed of variable numbers of tandemly repeated dinucleotides. Because the number of repeats varies between individuals, this type of DNA has been most useful in providing a dense array of molecular markers for mapping the huge expanses of the human genome, as we shall see in Chapter 17.

Transposed Sequences. A large proportion of eukaryote genomes is composed of repetitive elements that have prop-

agated within the genome by making copies of themselves, which can move into new locations. These are collectively called transposed sequences, and they are described in detail in Chapter 21. Those elements that move as DNA are called **transposons.** In Chapter 21 we consider examples of transposons in bacteria, corn, and Drosophila. Many genomes show multiple copies of such elements or truncated versions of them dispersed throughout the genome.

Another general class of transposed sequences is the **retrotransposons,** DNA sequences that propagate themselves through the action of reverse transcriptase, the enzyme that can make a DNA strand from RNA. One type within this category is repetitive sequences whose structures are related to retroviruses. They move by reverse transcription of their RNA transcripts into DNA which then inserts throughtout the genome. Examples of retroviral-like retrotransposons are the copia elements of *Drosophila* (5-kb sequences present at about 50 copies per genome), the Ty elements of yeast (6-kb elements with about 30 full-length copies per genome and about 100 solo copies of their terminal duplicated regions, called delta sequences), and the **LINES** (**long interspersed elements**) of mammals (1 to 5-kb elements present in 20,000 to 40,000 copies per human genome. Figure 16-30 shows the general structures of several retrotransposons.

The Alu repetitive sequence in the human genome, so named because it contains an *Alu* restriction enzyme target site, is an example of a class of retrotransposons that are not related to retroviruses. The human genome contains hundreds of thousands of whole and partial Alu sequences, scattered between genes and within introns and making up about 5 percent of human DNA. The full Alu sequence is about 200 nucleotides long and bears remarkable resemblance to 7SL RNA, an RNA that is part of a complex involved in secretion of newly synthesized polypeptides through the endoplasmic reticulum. Presumably the Alu sequences have originated as reverse transcripts of these RNA molecules. Short interspersed repeats such as Alu sequences are collectively called **SINEs** (for **short intespersed elements**).

Other examples of this class of moderately repetitive elements are the many scattered **pseudogenes** that have clearly been created by the reverse transcription process because they do not contain the introns that are found in the original functional gene.

Spacer DNA

The final category of DNA is **spacer DNA.** This is basically what is left after all the recognizable units have been identified. Needless to say, little is known about spacer DNA. Possibly its only function is to space, but no studies have yet been performed to delete such DNA and observe the consequences.

Message Single-copy genes are embedded in a complex array of tandem and dispersed types of repetitive DNA, most of which have no known function.

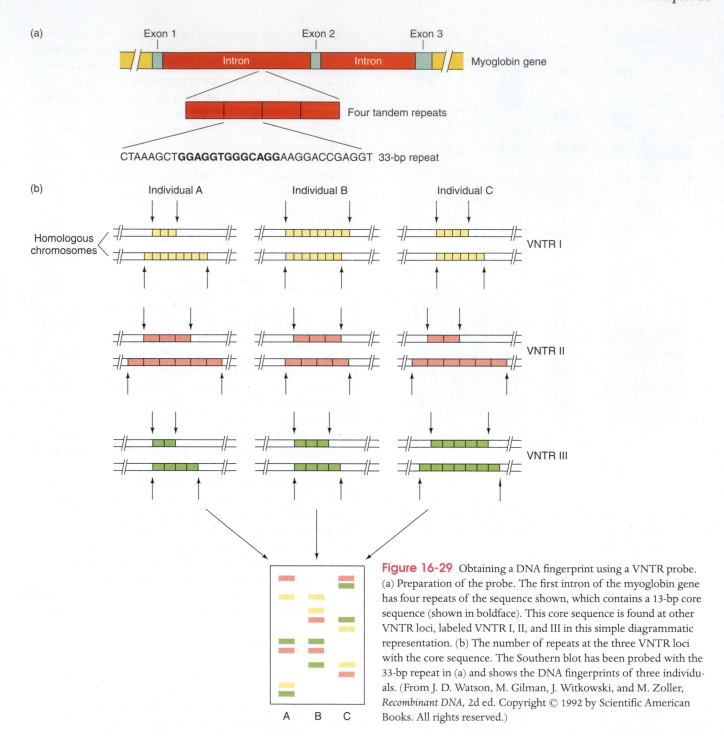

(a)

Exon 1   Exon 2   Exon 3

Intron   Intron   Myoglobin gene

Four tandem repeats

CTAAAGCT**GGAGGTGGGCAGG**AAGGACCGAGGT 33-bp repeat

(b)

Individual A   Individual B   Individual C

Homologous chromosomes

VNTR I

VNTR II

VNTR III

A   B   C

Figure 16-29 Obtaining a DNA fingerprint using a VNTR probe. (a) Preparation of the probe. The first intron of the myoglobin gene has four repeats of the sequence shown, which contains a 13-bp core sequence (shown in boldface). This core sequence is found at other VNTR loci, labeled VNTR I, II, and III in this simple diagrammatic representation. (b) The number of repeats at the three VNTR loci with the core sequence. The Southern blot has been probed with the 33-bp repeat in (a) and shows the DNA fingerprints of three individuals. (From J. D. Watson, M. Gilman, J. Witkowski, and M. Zoller, *Recombinant DNA,* 2d ed. Copyright © 1992 by Scientific American Books. All rights reserved.)

The existence of DNA with no known function has presented a dilemma for geneticists. Previous ideas on the power of natural selection would have predicted that nonfunctional DNA should be purged by selection. It might be imagined that nonfunctional DNA is a genetic burden if only because of the additional energy that the organism needs to put into its synthesis. However, this notion seems inadequate. Perhaps the seemingly functionless DNA does

have a function, possibly as a kind of genetic ballast that gives the chromosome bulk to permit efficient partitioning at cell division, or perhaps to separate the functional elements (genes) for their efficient regulation. Alternatively, the repetitive elements that make up most of the nonfunctional DNA might have hit upon a way of avoiding detection by the forces of natural selection. Such DNA has been termed **selfish DNA,** a type of DNA that exists only for the

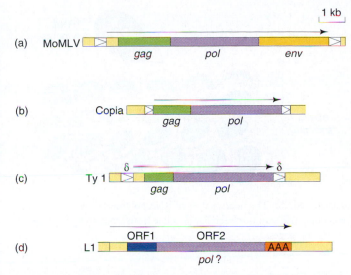

Figure 16-30 The general structure of some retroelements found in eukaryotic genomes. *Gag, pol,* and *env* are retroviral genes. ORF = open reading frame. AAA is a poly(A) tail. Unshaded triangles show direct terminal repeats. Arrows show direction of transcription. (a) A retrovirus, Moloney murine leukemia virus of mice. (b) Copia (*Drosophila* retrotransposon). (c) Ty1 (yeast retrotransposon). (d) L1, a human LINE. (Modified from J. R. S. Fincham, *Genetic Analysis: Principles, Scope and Objectives*. Blackwell, London, 1994.)

purpose of existing and is never exposed to the rigors of phenotype.

Figure 16-31 shows a stylized diagram of the overall organization of a hypothetical eukaryotic chromosome, summarizing much of the above discussion. However, we can now study the architecture of a real eukaryotic chromosome because several complete chromosomes of yeast have been fully sequenced. The arrangement of the genes and other elements on one of these chromosomes (number 3) is shown in Figure 17-14. The entire chromosome is 315 kb long. Analysis of the complete sequence showed that there are 182 open reading frames (shown in dark blue). Of these supposed 182 protein-coding genes, less than a quarter had been discovered by classical genetic techniques, and these are shown in yellow in the figure. The possible functions of the unknown open reading frames are being explored by comparing them with other DNA sequences in the computer data bank, and by looking for mutant phenotypes that

might be caused by gene disruptions. The delta sequences are the solitary terminal repeats mentioned earlier, derived from Ty transposons. Note that yeast has virtually no interspersed repetitive DNA, so the genes are seen to be relatively close together, separated by short spacer sequences.

Replication and Transcription of Chromatin

We have seen that the central scaffold of the chromosome contains topoisomerase II, which must surely play a role in replication, but the details of how polymerases and associated factors gain access to the center of the chromatin is not clear. There are many other crucial questions that cannot be answered fully. A major one is how the replication machinery can pass through the nucleosomes. Nucleosomes appear not to leave the DNA in this process, but may disassemble in situ. There is still controversy about the arrangement of "old" and "new" nucleosomes on the two daughter DNA molecules, but in balance most of the evidence favors a dispersive model, in which each daughter molecule gets some old and some new nucleosomes (Figure 16-32).

Similar problems exist for understanding transcription. If RNA polymerase is drawn to the same scale as the solenoid structure, it is seen to be approximately the same size (Figure 16-33). Therefore it seems unlikely that the enzyme could do its job without a great deal of loosening of the DNA-histone complex. Indeed there is evidence that when a gene is being transcribed in interphase, it becomes sensitive to DNase attack, as would be predicted if there was relaxing of the nucleosome spools to expose unwound DNA. This effect can be demonstrated in chick cells that are synthesizing globin proteins. If the nuclei of such cells are exposed to DNase, subsequent restriction enzyme digestion following the removal of histones reveals no intact globin genes. However, in cells where the gene was not being expressed, a clear globin gene fragment is obtained at the end of the procedure. The histones evidently protect DNA from DNase attack. One possible conformation of the loosened arrays of histones during transcription is shown in Figure 16-34.

Message In chromatin the transcription and replication processes act on DNA with nucleosomes still bound but reorganized in some way.

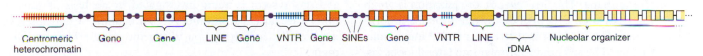

Figure 16-31 General depiction of a eukaryotic chromosomal landscape. A small region of a chromosome is shown that happens to have five protein-coding genes, one end of a nucleolar organizer, and one end of centromeric heterochromatin. Various kinds of repetitive DNAs are shown. (Each chromosome would normally have several thousand genes.)

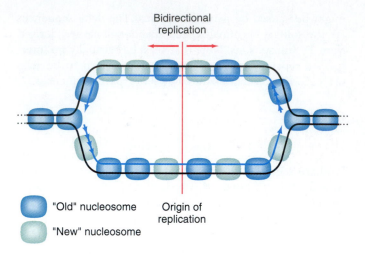

Bidirectional
replication

"Old" nucleosome   Origin of
                        replication

"New" nucleosome

Figure 16-32 The dispersive model for the association of newly formed nucleosomes with the daughter DNA molecules at the replication fork.

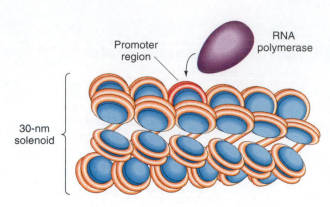

Figure 16-33 The relative sizes of RNA polymerase and the 30-nm solenoid. Transcription seems impossible from DNA so tightly wound on nucleosomes.

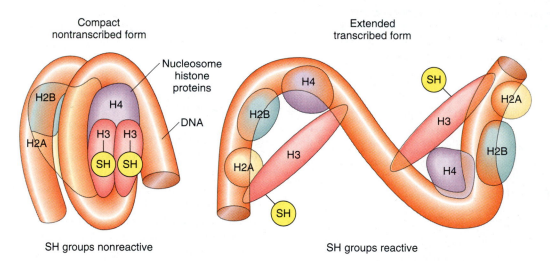

Figure 16-34 Compact form of nucleosomes and a proposed unfolded transcriptional form. (After C. P. Pryor et al., *Cell* 34, 1983, 1033.)

SUMMARY

Each eukaryotic chromosome represents one DNA molecule continuous from one end through the centromere to the other end. This DNA is efficiently packed to achieve the chromosome shape. The first level of organization is winding of the DNA onto octamers of histone proteins to form nucleosomes. The nucleosomal strand is coiled into a solenoid structure. The solenoid is thrown into lateral loops fastened to a central scaffold at specific scaffold attachment regions. This whole structure is arranged in large supercoils. The compactness of the supercoils at metaphase makes it easy for chromosomes to segregate and reveals a banding pattern when stained with Giemsa reagent. The light-stain-

ing bands contain most of the active genes. In interphase the supercoils are less compact. The difference between euchromatin and heterochromatin is maintained genetically, and this system is being investigated by analysis of mutations that alter position-effect variegation. Centromeres have specialized structures to permit spindle fiber attachment.

Eukaryotic chromosomes contain a large proportion of repetitive DNA. Some is in tandem arrays and some is interspersed throughout the genome. Some repetitive DNA, both tandemly repeated and dispersed, consists of genes of known function, whereas a large proportion has no known function. Centromeric heterochromatin contains highly repetitive short tandem repeats. An important type of dispersed repeats is transposable elements of several types. Telomeres have specialized repetitive structures to permit DNA replication to proceed to the end of the chromosome.

Replication and transcription both take place through nucleosomes, but the nucleosomal strands partially unwind and histones might partially disassemble. Nucleosome replication is dispersive, with new and old nucleosomes dispersed on both leading and lagging strands.

Concept Map

Draw a concept map interrelating as many of the following terms as possible. Note that the terms are listed in no particular order.

histones / chromosome / transcription / position-effect variegation / DNA / nucleosomes / bands / loops / scaffold / heterochromatin

CHAPTER INTEGRATION PROBLEM

A yellow-bodied *Drosophila* stock (yy) is transformed with a vector containing a fully functional wild-type allele y^+, which codes for brown body color. Two transgenic lines are obtained: one has a fully brown body, and the other has some brown sectors and some yellow sectors. Give a possible explanation for the difference between the two transgenic flies, and propose a test of your idea.

Solution

Evidently the *y* allele is recessive; otherwise, no brown tissue would be expected in the transformed individuals (Chapter 2). Therefore the transgenic flies, which probably contain two *y* alleles and one y^+ allele, are expected to be brown if the y^+ allele is fully dominant over two recessive *y*

alleles. We saw in the chapter on aneuploidy that this kind of dominance is often encountered (Chapter 9). Indeed it is found in one line in the present experiment. The other line shows variegation, and this should give us a clue to the location of the transgene. If the transgene is inserted into a chromosome next to heterochromatin, it could show position-effect variegation. No other mechanism for variegation (for example, somatic crossing-over, X chromosome inactivation) makes sense in this case. The position of the transgene in both lines can be mapped using standard linkage analysis, as covered in Chapter 5. We predict the transgene to be at a locus next to heterochromatin in the yellow and brown line and in the other line we expect it to be in euchromatin, possibly at the true *yellow* locus, but more likely at a new site.

SOLVED PROBLEM

Investigators have found that *Drosophila* E(var) mutations are rarer than Su(var) mutations. Why do you think this is so?

Solution

We have learned in this chapter that E(var) mutations enhance position-effect variegation by increasing the spread of

heterochromatin. Therefore we might postulate that the euchromatic regions adjacent to heterochromatin contain vital genes that would be condensed in such a large proportion of cells in flies with the E(var) mutation that the vital function would be lost. Such *flies* would be nonviable.

PROBLEMS

1. Two men in a paternity dispute claim to be the father of a certain child. Forensic geneticists use DNA fingerprints to settle the case, and the relevant ones are in the figure at

the top of the next page (M is the mother, C the child, and F1 and F2 are the two men who claim parentage):

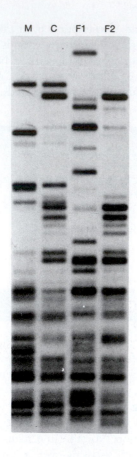

M   C   F1   F2

a. Who is the father? Explain.

b. Attribute as many bands as possible to one or the other parent.

c. Are there any bands that cannot be attributed to a parent?

d. If the child did show a band that neither parent possessed, how might this be explained? (Such situations do sometimes occur.)

e. Explain what a band actually represents, using diagrams.

(DNA fingerprint from Cellmark Diagnostics, Germantown, Md.)

2. Suppose that an essential wild-type allele in *Drosophila* is translocated next to heterochromatin on one homolog, and that the other homolog carries the gene in its normal chromosomal position, but the allele is recessive and lethal. Predict the organismal phenotype in (a) a *Su(var)* mutant, (b) an *E(var)* mutant, and (c) a normal background. Assume first that the gene product is diffusible throughout the body. Then make the alternative assumption that it is used only in cells where it is synthesized.

3. Predict the in situ chromosomal hybridization pattern in humans using probes consisting of

a. a single copy gene.

b. a fragment of telomere sequence.

c. rDNA.

d. a SINE.

e. satellite DNA.

4. Assume that a SINE occurs on average every 50 kb, and you cut DNA with a restriction enzyme that cuts on average every 100 kb. You denature the DNA with heat, then cool it slowly to promote reannealing. Predict what you might see with the electron microscope.

5. Forensic geneticists prepared DNA fingerprints from the blood of a rape victim, from semen taken from her body, and from samples obtained from three suspects. The resultant Southern blots are shown below:

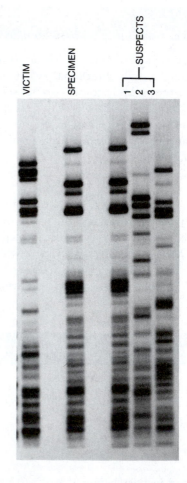

a. Is there a clear case of guilt of one of the suspects?

b. Can any of the suspects be absolved of guilt? Explain your answers.

(DNA fingerprints from Cellmark Diagnostics, Germantown, Md.)

6. A VNTR probe revealed simple DNA fingerprints in the family shown at the top of the facing page:

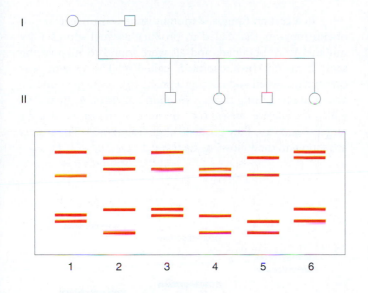

| 1 | 2 | 3 | 4 | 5 | 6 |

From these data, what can you say about the number of VNTR loci, their linkage, and the number and size of the alleles?

7. The genes for eye color (*w* = white, *+* = red) and eye texture (*r* = rough, *+* = smooth) are located at the tip of the X chromosome in *Drosophila*. A set of inversions of X chromosomes carrying all wild-type alleles was paired individually with normal X chromosomes carrying the recessive alleles *w* and *r*. The resulting genotypes (female) were Normal *w r* / Inversion *+ +*. The eyes of these females were examined, and the results were as shown in the following table:

| | Percent of eye facets that were | | |
	White	Rough	White and rough
1 $\underline{\frac{w^+\ r^+}{(\quad)}}$	50	0	10
2 $\underline{\frac{w^+\ r^+}{(\quad)}}$	10	0	2
3 $\underline{\frac{w^+\ r^+}{(\quad)}}$	0	30	0
4 $\underline{\frac{w^+\ r^+}{(\quad)}}$	12	0	4

NOTE: ( ) = inversion breakpoint.

a. Explain the different frequencies obtained for each inversion.

b. In what way are the explanations of the zero values in inversions 3 and 4 different?

c. What frequencies would you predict in the three columns for a female of the type Inversion 1 *+ +* / Inversion 2 *+ +*?

8. Transposons can be used as mutagens. They inactivate genes by inserting into them or near them. In one experiment in yeast, single copies of a transposon were added transgenically to haploid cells containing no transposon. The single copies were allowed to integrate at random throughout the genome. Then the strains that resulted from this procedure were examined to determine their phenotype. It was found that 70 percent of these had an abnormal phenotype that could be attributed to the transposon insertion, but 30 percent appeared fully wild type.

a. How would you make sure that the strains with the wild-type phenotype actually do have a transposon insertion?

b. How would you make sure that the abnormal phenotypes were in fact caused by the transposon insertion?

c. How do you explain the 30 percent that are unaffected phenotypically?

9. A *Drosophila* female was heterozygous for a reciprocal translocation between the X chromosome and chromosome 3. Her eyes were of wild-type color (red). She was crossed to a male of mixed ancestry who had white eyes. The female progeny were as follows:

$\frac{1}{2}$ red-eyed

$\frac{1}{4}$ variegated eyes with 60% white facets

$\frac{1}{4}$ variegated eyes with 20% white facets

When the females with 20 percent white facets were mated to their father, some of their female progeny had variegated eyes with only 2 percent white facets. Provide a hypothesis to explain the females with

a. 60% white facets.

b. 20% white facets.

c. 2% white facets.

d. Account for the proportions of the first two types in the progeny of the cross.

 Unpacking the Problem

a. Define the terms *variegation, facet, reciprocal translocation, wild type, heterozygous.*

b. What does the phrase "of mixed ancestry" convey?

c. Make up a sentence including the words *variegation, translocation, facet,* and *wild type.*

d. Review what you know about the location of the gene for white eyes in *Drosophila*.

e. What is the X chromosome constitution of male and female *Drosophila?*

f. What must the eye-color genotype of the male parent be?

g. What must the eye-color genotype of the female parent be?

h. In a "regular" cross of the genotypes in questions f and g, what are the expected progeny proportions in each sex?

i. What are three causes of variegation, and which of these seems most appropriate in this question?

j Is the fact that the translocation involved chromosome 3 specifically relevant to this question or not?

k. What might explain the equal segregation of red to variegated in the female progeny?

l. What might explain the equal segregation of 60% to 20% white facets?

m. In a fly with 60% white facets, what is the phenotype of the other 40% of facets?

n. There are three classes of variegation: high, intermediate, and low frequencies of white facets. What fundamental Mendelian allele combination could explain the existence of three distinct classes of this type?

o. Is it significant that it was females with the lower of the two white-facet frequencies that gave rise to progeny with even lower white-facet frequencies when crossed to their father?

p. What could the father be contributing to the level of white facets found in female descendants?

10. In a certain fungus, a mutant hunt was undertaken to obtain mutants that failed to produce asexual spores. Fifty mutants were obtained, and all were found to map to the same locus on chromosome 5, called *sp*. The *sp* gene was eventually cloned and used as a probe to survey genomic digests obtained using three restriction enzymes, A, B, and C, which are all rare cutters (i.e., because of the rarity of their target sequences, they produce large fragments). The results are shown in the following diagram:

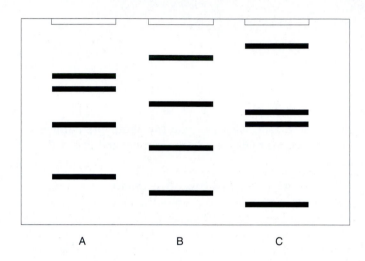

a. Provide an explanation for why the probe does not bind to only one fragment.

b. Why does the probe bind to four fragments in each case.

c. Do you think the enzymes cut the DNA in the middle of the *sp* gene?

17

Genomics

Pigs involved in the international pig genome mapping project "PiGMaP." Two pure parental strains, Large White and Meisham, differ by many molecular and morphological narkers. These lines are crossed and the F_1 hybrids are interbred to produce an F_2. In the F_2 the various patterns of marker assortment are used to determine linkage and to map the genome. In this picture the reclining sow is an F_1 hybrid and the variously colored nursing piglets are F_2 individuals. (Reproduced by kind permission of the Roslin Institute, Edinburgh.)

KEY CONCEPTS

▶ Genomics is the molecular characterization of whole genomes.

▶ Genomic analysis begins by using several different techniques to assign genes to specific chromosomes.

▶ The second level of analysis is low-resolution chromosome mapping, based mainly on meiotic recombination analysis. In analysis of humans, special cell hybrids can also be used.

▶ Neutral DNA variation has provided numerous heterozygous molecular marker loci for use in fine-scale recombination mapping.

▶ The highest level of resolution in genomic mapping analysis is physical mapping of cloned DNA fragments.

▶ The arrangement of cloned DNA fragments into overlapping sets is facilitated by special molecular procedures for tagging the clones.

▶ Ultimately, whole chromosomal and genomic sequences can be obtained by sequencing members of the overlapping clone set.

▶ Genomics provides an information base for isolating specific genes of interest, including human disease genes.

Genomics is the subdiscipline of genetics concerned with the cloning and molecular characterization of whole genomes. The goal of genetics generally is to understand the structure, function, and evolution of genomes, and indeed all the analysis in this book is focused on this pursuit. The goal of genomics, the subject of this chapter, is clearly within this general statement of goals, so why does it merit special attention? The answer is that although there is considerable overlap and complementarity between the analytical procedures within genetics, special experimental techniques have been devised to carry out the difficult task of manipulating whole genomes, which contain huge amounts of DNA. Such special techniques are the focus of this chapter.

Genome projects are now in progress in a range of different organisms, including humans (the "human genome project"), mice, *Drosophila,* a yeast, *Caenorhabditis elegans* (a nematode), *Arabidopsis thaliana* (a plant), and several bacteria. The genomes of several viruses and one bacterium have already been fully sequenced, as have several entire yeast chromosomes. Several mitochondrial and chloroplast genomes have also been fully sequenced. Because of the scope of these tasks, many of the projects are international ventures, involving hundreds of geneticists collaborating and sharing data about different regions of the genome. Often groups or even whole nations specialize in analyzing certain specific chromosomes. Clearly, biologists agree on the importance of this kind of research, and it is worth asking why this is so before we consider some of the details of the techniques.

First, obtaining a set of ordered clones of an entire chromosome or an entire genome is an invaluable baseline for future molecular studies. These clones can be used for finding and manipulating individual genes and also as a starting point for genomic sequencing. The DNA of a genome is the blueprint of a species, the information needed to build a living cell and a living organism. Hence knowledge of its specific nucleotide sequence will eventually provide answers to one of the basic questions of biology: Why is an organism the way it is and different from other organisms? Of course, just knowing the sequence of an entire genome will not in itself answer all the questions, but it is the beginning of a second round of analysis, which will attempt to understand the functions of individual genes and noncoding sequences and how they interact.

Although a great deal is now known about the structure and function of individual genes, little is known about the principles by which functional and nonfunctional DNA regions are organized into whole genomes. This is another level of inquiry that can be addressed through analysis of whole sequences. Part of this question will obviously be evolutionary: How were genomes assembled through the processes of evolution? Are the precise positions of genes on chromosomes unimportant, or must genes be in specific locations in order to ensure proper function? It is known that there is conservation of gene position between related taxa (Figure 8-38), but the significance of this has not been established. Another question concerns the numbers and types of genes needed to specify an individual species and how the numbers and types differ between taxa. There are undoubtedly many other regularities to genome structure and function that cannot be addressed without full sequences from several organisms.

Finally, the human genome project has particular significance to us as humans, not only because of the insights it can provide about human nature, but also because of the unique applications in medicine. Although progress is rapid in cloning and characterizing human disease genes, there is still a long way to go. For example, it has been estimated that less than 5 percent of human genes with known phenotypic expression have been mapped, and of these only a small proportion has been characterized at the molecular level.

Message Characterizing whole genomes is important to the fundamental understanding of the design principles of living organisms and for the discovery of new genes such as those that are involved in human genetic disease.

Genomic analysis generally proceeds through several steps of increasing resolution:

1. Assigning loci of genes and molecular markers to specific chromosomes
2. Calculating map distances between loci on a chromosome
3. Physically mapping individual DNA fragments (cloned or uncloned)

These genomic techniques are useful not only for making general chromosome maps, but also for finding the location of a specific gene on an established map for the purpose of isolating that gene. These general approaches are diagrammed in Figure 17-1.

Assigning Loci to Specific Chromosomes

Several different methods are useful in assigning genes or markers to individual chromosomes. Some of these have been covered elsewhere in this book and are included here for completeness. Other techniques are described for the first time.

Linkage to Known Loci

In well-studied organisms, it is a simple matter to cross a strain carrying the "new" unmapped allele to a set of strains carrying markers spread throughout the genome, each one of known chromosomal location. Meiotic recombinant frequencies of less than 50 percent reveal linkage and therefore assign the new allele to a particular chromosome. Such data

(a)

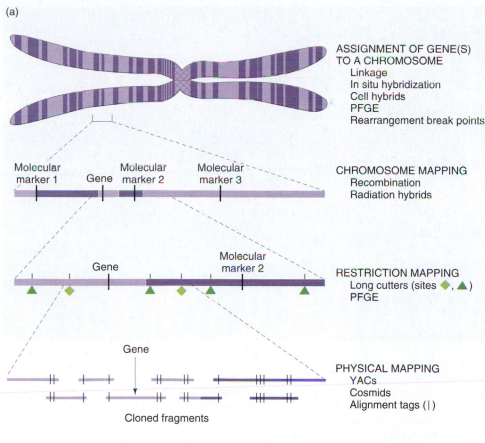

ASSIGNMENT OF GENE(S)
TO A CHROMOSOME
   Linkage
   In situ hybridization
   Cell hybrids
   PFGE
   Rearrangement break points

Molecular
marker 1   Gene   Molecular
marker 2   Molecular
marker 3

CHROMOSOME MAPPING
   Recombination
   Radiation hybrids

Gene   Molecular
marker 2

RESTRICTION MAPPING
   Long cutters (sites ◆, ▲)
   PFGE

Gene

PHYSICAL MAPPING
   YACs
   Cosmids
   Alignment tags (|)

Cloned fragments

Figure 17-1 Overview of the general approaches of genomics. (a) General scheme for making a genome map using analyses at increasing levels of resolution. (b) Finding one specific gene, the breast cancer gene BRCA1, by using the genomic map at increasing levels of resolution. (Part b copyright © 1994 by the New York Times Company. Reprinted by permission.)

(b)

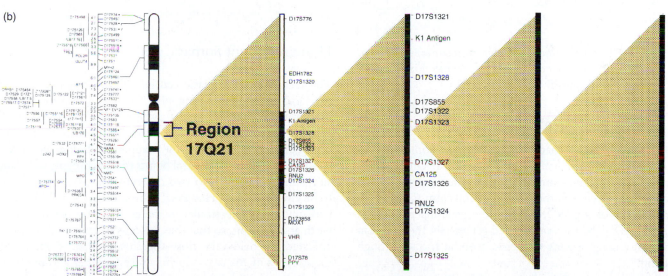

Region 17Q21

Chromosome 17:	Segment of 17Q21:	Smaller segment	BAC283:	BRCA1:
Millions of base pairs	2 million pairs	of 17Q21: 650,000 pairs	130,000 pairs	80,000 pairs

often permit a rough assignment of chromosomal position, perhaps to a chromosomal arm or even to a band.

In Situ Hybridization

If a cloned gene is available, then this can be used to make a labeled probe for in situ hybridization. If the individual chromosomes of the genomic set are recognizable through their banding patterns, size, arm ratio, or other cytological feature, then the new gene can be assigned to the chromosome it hybridizes to. The locus of hybridization also reveals a rough chromosomal position.

Commonly used probe labels are radioactivity and fluorescence. In the process of **fluorescent in situ hybridization (FISH),** the clone is labeled with a fluorescent dye, and a partially denatured chromosome preparation is bathed in

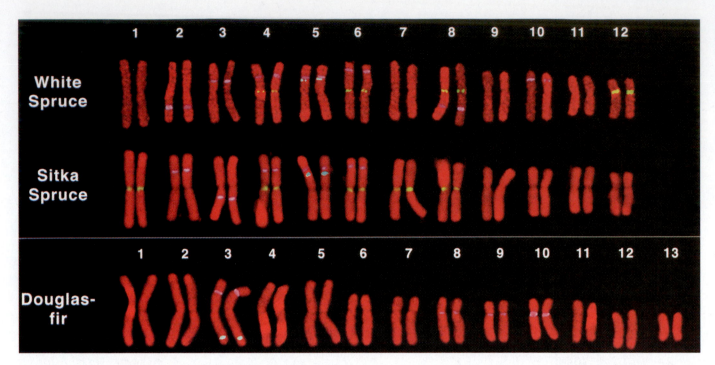

Figure 17-2 Fluorescence in situ hybridization (FISH) analysis of the chromosomes of three conifer-ous tree species. The probes used were clones of 5S rDNA (blue), the region coding for 18S, 5.8S, and 26S rDNA (pink), and in the cases of white and Sitka spruce a type of satellite DNA code-named SGR-31 (green). In this case the FISH analysis was undertaken to locate the relevant loci, use the loci as markers to identify specific chromosomes, and make evolutionary comparisons between the species. (Photo from Garth Brown, Vindhya Amarasinghe, and John Carlson.)

the probe. The probe binds to the chromosome in situ, and the location of the cloned fragment is revealed by a bright fluorescent spot (Figure 17-2).

Rearrangement Breakpoints

We saw in Chapter 8 that mutant alleles are sometimes as-sociated with chromosomal rearrangements. Often this is because the chromosome break that gives rise to the re-arrangement occurs in the middle of (or close to) the cod-ing sequence of a gene, rendering the gene inactive. If the breakpoints can be seen or mapped genetically, then this can be used to assign a gene to a chromosome and to a rough position.

Pulsed Field Gel Electrophoresis (PFGE)

If chromosomes are small enough to be separated by PFGE (Figure 17-3), the DNA bands on the gel can be used to lo-cate new genes by hybridization. First, correlations must be made to establish which DNA band corresponds to which chromosome. Size, translocations between known chromo-somes, and hybridization to probes of known location are useful for this purpose. Then a new cloned gene can be used as a probe to a Southern blot of the PFGE gel, and hence its chromosomal locus determined.

Human-Rodent Somatic Cell Hybrids

The technique of somatic cell hybridization is extensively used in human genome mapping, but it can in principle be used in many different animal systems. The procedure uses cells growing in culture. A virus called the *Sendai virus* has a useful property that makes the mapping technique possible. Each Sendai virus has several points of attachment, so it can simultaneously attach to two different cells if they happen to be close together. However, a virus is very small in com-parison with a cell (similar to the comparison between the earth and the sun), so that the two cells to which the virus is attached are held very close together indeed. In fact, the membranes of the two cells may fuse together and the two cells become one—a binucleate heterokaryon.

If suspensions of human and mouse cells are mixed to-gether in the presence of Sendai virus that has been inacti-vated by ultraviolet light, the virus can mediate fusion of the cells from the different species (Figure 17-4). Once the cells fuse, the nuclei subsequently fuse to form a uninucle-ate cell line composed of both a human and a mouse chro-mosome set. Because the mouse and human chromosomes are recognizably different in number and shape, in the hy-brid cells the two sets can be readily distinguished. How-ever, over the course of subsequent cell divisions, for un-known reasons the human chromosomes are gradually

(a)

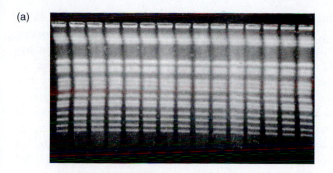

(b)

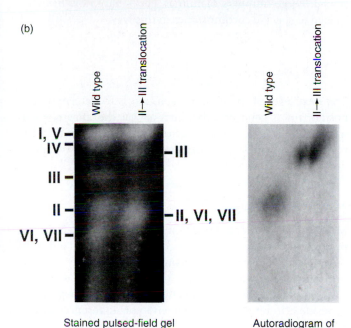

Stained pulsed-field gel

Autoradiogram of
Southern blot using
a cloned gene probe

eliminated from the hybrid at random. Perhaps this process is analogous to haploidization in the fungus *Aspergillus* (Chapter 6).

The loss of human chromosomes can be arrested to encourage the formation of a stable partial hybrid in the following way. The mouse cells used are mutant for some biochemical function, so if the cells are to grow, the missing function must be supplied by the human genome. This selective technique results in the maintenance of hybrid cells that have a complete set of mouse chromosomes and a small number of human chromosomes, which vary in number and type from hybrid to hybrid but which always include the human chromosome carrying the wild-type allele defective in the mouse genome.

Let's look at the specific genes that make the selective system work. In cells, DNA can be made either de novo ("from scratch") or through a salvage pathway that uses molecular skeletons already available. The selective technique involves the application of a chemical, aminopterin, that blocks the de novo synthetic pathway, confining DNA synthesis to the salvage pathway. Two essential salvage enzymes are relevant to the system, as shown in the following two reactions:

Figure 17-3 (a) Pulsed field gel electrophoresis of uncut chromosomal DNA from different strains of yeast. Sixteen bands are resolved. Since there are 22 chromosomes in yeast, some have obviously comigrated in the gel. (All lanes were loaded identically.) (b) Using pulsed-field gel electrophoresis to separate *Neurospora* chromosomes and to locate a cloned gene onto one chromosomal segment. (*Left*) Chromosomes of wild type and a translocation in which a 100-kb section of chromosome II has been iserted into chromosome III. (*Right*) Autoradiogram using a probe consisting of a cloned gene of previously unknown location. The results show that the gene is located on the region of chromosome II that is moved in the translocation. (Part a from Bio-Rad Laboratories; part b from Myron Smith.)

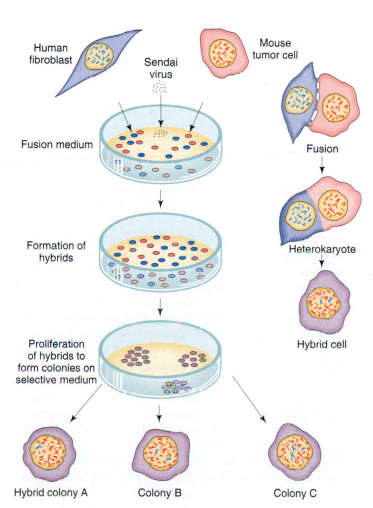

Figure 17-4 Cell-fusion techniques applied to human and mouse cells produce colonies, each of which contains a full mouse genome plus a few human chromosomes (blue). A fibroblast is a cell of fibrous connective tissue. (From F. H. Ruddle and R. S. Kucherlapati, "Hybrid Cells and Human Genes." *Scientific American*, 1974.)

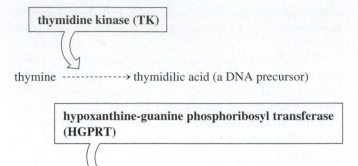

The mouse cell line to be fused is genetically unable to make TK because it is homozygous for the allele tk^-, whereas the human cell line is genetically unable to make HGPRT because it is homozygous at another locus for the allele $hgprt^-$. So the genotypes of the two fusing cell lines are

mouse: $tk^- tk^- hgprt^+ hgprt^+$

human: $tk^+ tk^+ hgprt^- hgprt^-$

Because each is deficient for one enzyme, neither the mouse nor the human cells are able to make DNA individually. In the hybrid cells, however, the tk^+ allele complements the $hgprt^+$ allele, so that the cells can make both enzymes. Therefore, DNA is synthesized and the cells can proliferate. Most human chromosomes are eliminated from the hybrid cell cultures because their loss has no effect on the cultures' ability to grow. But to continue to grow in medium containing hypoxanthine, aminopterin, and thymidine (*HAT medium*), a hybrid culture must retain at least one of the human chromosomes that carries the tk^+ allele.

Luckily, the progressive elimination of the human chromosomes from the fused cell lines can be followed under the microscope because mouse chromosomes can easily be distinguished from human chromosomes. Moreover, chromosome stains such as quinacrine and Giemsa reveal a pattern of banding within the chromosomes. The size and the position of these bands vary from chromosome to chromosome, but the banding patterns are highly specific and invariant for each chromosome. Thus, it is relatively easy to identify the human chromosomes that are present in any hybrid cell (Figure 17-5). Different hybrid cells are grown separately into lines; eventually a bank of lines is produced that contains, in total, all the human chromosomes.

With a complete bank of chromosomes, we can begin to assign genes or markers to chromosomes. If the human chromosome set is homozygous for a human molecular marker, such as an allele that controls a cell-surface antigen, drug resistance, a nutritional requirement, a specific protein, or a DNA marker, then the presence or absence of this genetic marker in each line of hybrid cells can be correlated with the presence or absence of certain human chromo-

somes in each line. Data of this sort are presented in Table 17-1, in which "$+$" means presence and "$-$" means absence of the genetic marker. We can see that in the different hybrid cell lines, genetic markers 1 and 3 are always present or absent together. We can conclude, then, that they are linked. Furthermore, the presence or absence of genes 1 and 3 is correlated with the presence or absence of chromosome 2, so we can assume that these genes are located on chromosome 2. By the same reasoning, gene 2 must be on chromosome 1, but the location of gene 4 cannot be assigned. Large numbers of human genes have now been localized to specific chromosomes in this way.

Chromosome Mapping

The next level of increasing resolution is to determine the position of a gene or other marker on the chromosome map. There are several approaches, and once again they apply both previously discussed techniques and newly introduced ones.

Meiotic Mapping

Meiotic linkage mapping used in genomics is based on the principles of mapping we covered in Chapters 5 and 6, in other words, on analyzing recombinant frequency in dihybrid and multihybrid crosses. In experimental organisms such as yeast, *Neurospora, Drosophila,* and *Arabidopsis,* the genes determining qualitative phenotypic differences can be mapped in a straightforward way because of the ease with which controlled experimental crosses (such as testcrosses) can be made. Therefore in these organisms the chromosome maps built over the years appear full of genes with known phenotypic effect, all mapped to their individual loci.

This is not the case for humans, because there is a lack of informative crosses, progeny sample sizes are too small for accurate statistical determination of linkage, and the human genome is enormous. In fact it was a difficult task even to assign a human disease gene to an individual autosome by linkage analysis. The invention of human-rodent cell hybrid mapping techniques was an important advance because it led to the assignment of a large number of human genes to specific chromosomes, and in some cases to an approximate position within a chromosome. (We will deal with these techniques in the next section.)

Even in those organisms in which the maps appeared to be "full" of loci of known phenotypic effect, measurements showed that the chromosomal intervals between genes had to contain large amounts of DNA. These gaps were inaccessible to linkage analysis because there were no markers in those regions. What was needed were large numbers of additional genetic markers that could be used to fill in the gaps to provide a higher-resolution map. This need was met by the discovery of various kinds of **molecular markers.** A molecular marker is a site of heterozygosity for some type

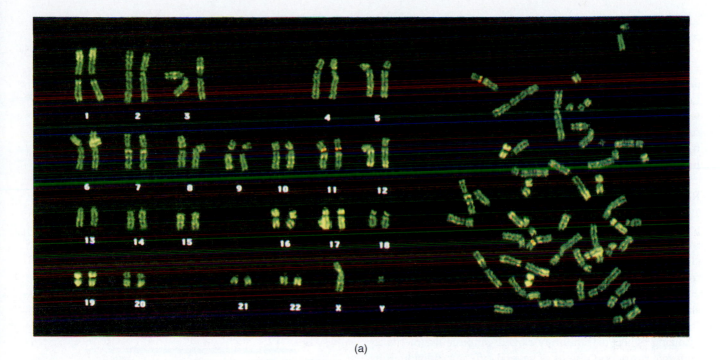

(a)

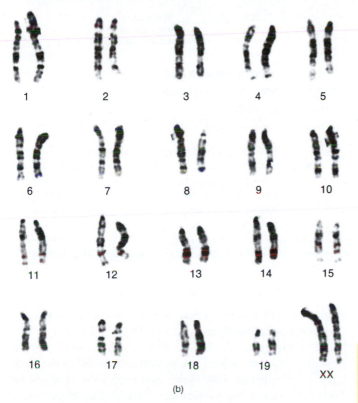

(b)

Figure 17-5 (a) Stained human chromosomes. Under the microscope the chromosomes appear as a jumbled cluster, as shown to the right. This array is photographed, and the individual chromosomes are cut out of the photograph and then grouped by size and banding pattern, as shown at left. The chromosome set of a male is shown. (b) The chromosomes of a female mouse, shown for comparison. To the experienced eye, the mouse chromosomes can be easily distinguished from human chromosomes, as required in the human-rodent cell hybrid technique of gene localization. (Part a from David Ward, Yale University School of Medicine; part b from Jackson Laboratory, Bar Harbor, Me.)

of **neutral DNA variation.** Neutral variation is that which is not associated with any measurable phenotypic variation. Such a "DNA locus," when heterozygous, can be used in mapping analysis just as with a conventional heterozygous allele pair. Note that in mapping, the biological significance of the DNA marker is not important in itself; the heterozygous site is merely a convenient reference point that will be

Table 17-1   Comparison of Five Hybrid Lines

		Hybrid cell lines				
		A	B	C	D	E
Human genes	1	+	−	−	+	−
	2	−	+	−	+	−
	3	+	−	−	+	−
	4	+	+	+	−	−
Human chromosomes	1	−	+	−	+	−
	2	+	−	−	+	−
	3	−	−	−	+	+

useful in finding one's way around the genome. In this way markers are being used just as milestones were used by travelers in previous centuries. Milestones marked off the individual miles between two cities (say, Bristol and London), and each had a written inscription, such as "Bristol 25 miles, London 75 miles." Such information told travelers exactly where they were on the map. They were interested in getting to Bristol or London and were not interested in the milestones (markers) per se. However, the travelers would have been lost without them.

Restriction Fragment Length Polymorphisms. In the 1980s, the first examples were discovered of neutral DNA variation that could be used as markers in mapping. These markers were **restriction fragment length polymorphisms (RFLPs).** It is known that RFLPs can have several causes at the DNA level; but often they are caused simply by the presence or absence of a restriction enzyme target site—even a single base difference can be responsible (see Chapter 15). RFLPs are identified by a rather hit-or-miss method of hybridizing panels of randomly cloned genomic fragments to genomic restriction digests of several different individuals in a family or a population. Because RFLPs are a relatively common type of variation in nature, this method succeeds in finding RFLPs in most organisms.

RFLP mapping is often performed on a defined set of strains or individuals that become "standards" for mapping that species. For example, in the fungus *Neurospora,* the two wild-type strains Oak Ridge and Mauriceville are known to show many RFLP differences, so these have become the standards used in RFLP mapping. The RFLPs can be mapped relative to one another or to genes of known phenotypic expression. For example, a cross can be made of the type *ad* $1^{OR} 2^{OR} \times ad^+ 1^M 2^M$, where *ad* stands for "adenine requirement," and 1 and 2 are RFLP loci with either the Oak Ridge (OR) or Mauriceville (M) "alleles." Progeny are tested for all three loci and recombinant frequencies calculated in the usual way. Detailed maps of the RFLP loci are now available. Most mutants in *Neurospora* have been obtained in Oak Ridge wild-type strains, so it is a simple matter to map the mutant alleles to RFLPs simply by crossing the mutant Oak Ridge strain to the wild-type Mauriceville strain. An example of mapping a phenotypic mutant using RFLP markers is shown in Figure 17-6.

Similar standard strains have been established in other organisms. Furthermore, an analogous approach has been used in the human genome mapping project by collecting DNA from a defined set of individuals in 61 families with an average of eight children per family, and making this DNA available across the globe to provide a standard for DNA marker mapping.

DNA Markers Based on Variable Numbers of Short-Sequence Repeats. In mapping the human genome, RFLPs were useful but have now been largely replaced in routine analysis by markers based on variation in the number of

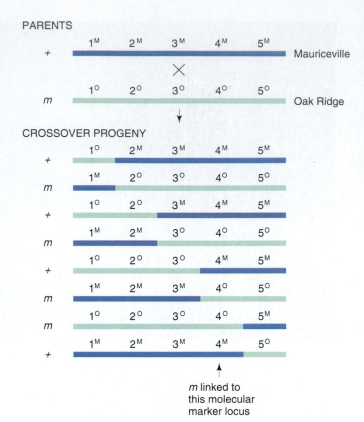

Key

1^M–5^M Mauriceville molecular markers

1^O–5° Oak Ridge molecular markers

All detected as RFLPS

m mutant allele of unknown location

Figure 17-6 Mapping a gene (*m*) by RFLP analysis in *Neurospora.* The two parental strains show many RFLPs ranged along the chromosomes; their loci are labeled 1 through 5. The two parental strains are from Oak Ridge (Tennessee) and Mauriceville (Texas), and their RFLP alleles are labeled O and M. Many different progeny types are recovered, and some of the more common types are shown. The results show that the + allele always segregates with 4^M, and the *m* allele always segregates with 4°, suggesting linkage of *m* to RFLP locus 4.

short tandem repeats. These markers are termed **simple-sequence length polymorphisms (SSLPs).** These have two basic advantages over RFLPs. First, in the case of RFLPs, usually only two "alleles" or morphs are found in the pedigree or population. This limits their usefulness; it would be better to have a larger number of alleles that could act as specific tags for a larger variety of homologous chromosomal regions. The SSLPs fill this need because multiple allelism is much more common, and as many as 15 alleles have been found for one locus. Second, the heterozygosity for RFLPs can be low; in other words, if one allele is relatively uncommon in relation to the other, the proportion of heterozygotes (the crucial individuals useful in mapping) will be low. However SSLPs, in addition to having more alleles, also show much higher levels of heterozygosity, and this

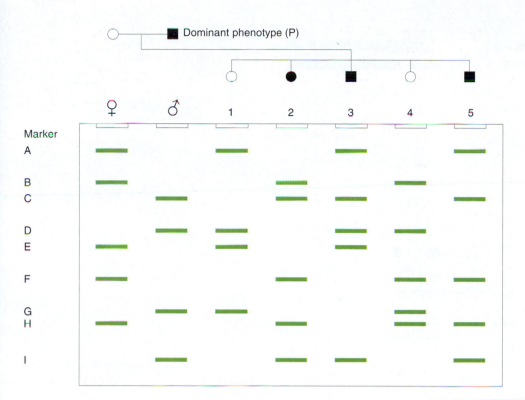

Figure 17-7 Using DNA finger-print bands as molecular markers in mapping. Simplified fingerprints are shown for parents and five progeny. Examples illustrate methods of linkage analysis. Molecular markers can be mapped to one another or to a locus with known phenotypic expression.

ANALYSIS EXAMPLES

F and H   Always inherited together — linked?

A and B   In progeny always either A *or* B — "allelic"?

A and D   Four combinations; A and D, A, D or neither — unlinked?

F, H, and E   Always *either* F and H *or* E — closely linked in trans?

Allele *P*   Possibly linked to I and C.

makes them more useful in mapping. Two types of SSLPs are now routinely used in human mapping.

Minisatellite Markers   In Chapter 16 we saw that human VNTR loci contain variable numbers of a 33-bp, tandemly repetitive DNA element at different locations throughout the genome. This repetitive element forms the basis of DNA fingerprinting. A DNA fingerprint is an array of bands in a Southern hybridization of a restriction digest, as shown in Figure 16-26. The individual bands of the DNA finger-print represent different-sized DNA sequences at many different chromosomal positions. If parents differ for a particular band, then this difference becomes a heterozygous ("plus/minus") locus that can be used in mapping. A simple example is shown in Figure 17-7.

The probe that detects minisatellite DNA variation in humans also turns out to generate DNA fingerprints in a surprising range of other organisms, so the minisatellite technique can be used in mapping in these species, too.

Microsatellite Markers   Microsatellite DNA is a class of repetitive DNA based on dinucleotide repeats. The most common type is repeats of CA and its complement GT, as in the following example:

5′ . . . . . . C-A-C-A-C-A-C-A-C-A-C-A-C-A-C-A . . . 3′

3′ . . . . . . G-T-G-T-G-T-G-T-G-T-G-T-G-T-G-T . . . 5′

Probes to detect these markers are made with the help of the polymerase chain reaction (PCR; see Chapter 14). First, digestion of human DNA with the restriction enzyme *Alu*I results in fragments with an average length of 400 bp, and these are cloned into an M13 phage vector. Phages with $(CA)_n/(GT)_n$ inserts are identified by hybridizing with a $(CA)_n/(GT)_n$ probe. Positive clones are sequenced, and PCR primer pairs are designed based on sequences flanking the repetitive tract in the different clones:

	Primer 1	
Alu ──────	──────▶ $(CA)_n$	────── *Alu*
Alu ──────	$(GT)_n$ ◀──────	────── *Alu*
	Primer 2	

The primers are used to amplify DNA using genomic DNA as a substrate. An individual primer pair will amplify its own repetitive tract and any size variants of it in DNAs from different individuals. A high proportion of PCR primer pairs

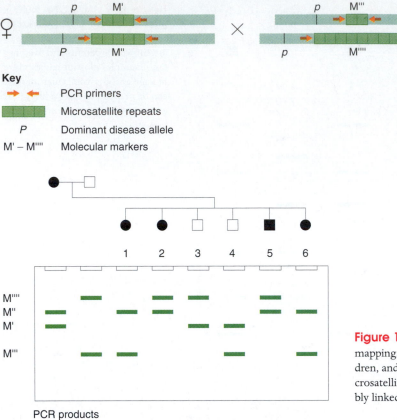

Key

→ ←   PCR primers

▮   Microsatellite repeats

P   Dominant disease allele

M' – M''''   Molecular markers

PCR products

Figure 17-8 Using microsatellite repeats as molecular markers for mapping. A hybridization pattern is shown for a family with six children, and this is interpreted above using four different-sized microsatellite "alleles," M' through M'''', one of which (M'') is probably linked in cis configuration to the disease allele *P*.

reveal at least three marker "alleles" of different-sized amplification products. An example of the microsatellite mapping technique is shown in Figure 17-8. Thousands of primer pairs can be made that likewise detect thousands of marker loci. The latest microsatellite marker map of human chromosome 1 was shown in Figure 5-18.

Note some differences in the convenience of RFLP and SSLP analyses. RFLP analysis requires a specific cloned probe to be on hand in the laboratory for the detection of each individual marker locus. Microsatellite analysis requires a primer pair for each marker locus, but these primer sequences can easily be shared around the world, distributed by electronic mail and rapidly constructed using a DNA synthesizer. Minisatellite analysis requires just one probe that detects the core sequence of the repetitive element at loci anywhere in the genome.

Together, the discovery of RFLP and SSLP markers has enabled the construction of a human genetic map with centimorgan (cM or map unit, m.u.) density. Although such resolution is a remarkable achievement, a centimorgan is still a huge segment of DNA, estimated in humans at approximately 1 megabase (1 Mb = 1 million base pairs, or 1000 kb).

Message   Meiotic recombination analysis of loci of genes with known phenotypic effect, RFLP markers, and SSLP markers has resulted in a map of the human genome that is saturated down to the 1-centimorgan (map unit) level.

RFLP analysis can be used in most organisms. SSLP analysis is being applied to other mammals such as the mouse, and in principle the potential application is even wider.

Other Meiotic Mapping Techniques. Several other approaches to meiotic mapping of molecular markers have been devised.

Molecular Genotyping of Sperm In Chapter 6 we saw that individual human or other animals' sperm can be genotyped at the molecular level, essentially reducing animal meiotic analysis to a simple fungal ascosopore analysis. This genotyping must be done by PCR analysis because the DNA content of an individual sperm is too small for direct manipulation. Individual sperm are isolated into separate tubes by dilution, the sperm are disrupted, and then their DNA is tested for specific marker alleles using appropriate PCR primers.

Homozygosity Analysis This approach, which is useful in human genome mapping, is based on pedigrees in which a recessive Mendelian allele has been brought to homozygosity through inbreeding, as in Figure 17-9. The recessive allele can be traced to one particular ancestor. This ancestor and his or her mate can then be analyzed for a large number of molecular markers such as RFLPs and SSLPs. If the individual carrying the homozygous recessive genotype is also homozygous by descent for one of the molecular markers,

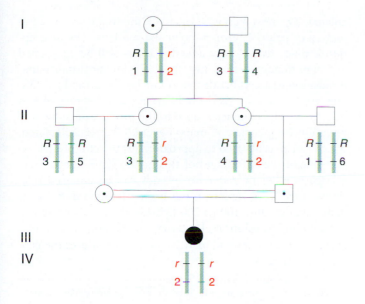

1–6 are alleles of a molecular marker.

Figure 17-9 Using homozygosity by descent to detect linkage. The individual in generation IV became homozygous for the autosomal recessive allele *r* because of inbreeding (a cousin marriage). This person was found to be also homozygous for a molecular marker allele 2, which must have come from the woman in generation I. Most likely the alleles *r* and 2 were linked on the same chromosome in the original woman and have been inherited through three generations as a "package."

then this is strong evidence (but not proof) that the molecular marker is linked to the Mendelian allele because, otherwise, the molecular marker would segregate independently and consequently would be less likely to have become homozygous by descent.

Randomly Amplified Polymorphic DNAs (RAPDs) A single PCR primer designed at random will often by chance amplify several different regions of the genome. The single sequence "finds" DNA bracketed by two inverted copies of the primer sequence. The result is a set of different-sized amplified bands of DNA (Figure 17-10). In a cross, some of the amplified bands may be unique to one parent, in which case they can be treated as heterozygous (plus / minus) loci and used as molecular markers in mapping analysis. Notice too that the set of amplified DNA fragments (called a *RAPD*, pronounced "rapid") is yet another type of DNA fingerprint that can be used to characterize an individual. Such identity tags can be very useful in routine genetic analysis or in population studies.

Chromosome Mapping Using Human-Rodent Hybrid Cells

Earlier in the chapter we discussed the use of human-rodent cell hybrids to assign genes to chromosomes. This tech-

nique can be extended to obtain mapping data. One extension of the hybrid cell technique is called **chromosome-mediated gene transfer.** First, samples of individual human

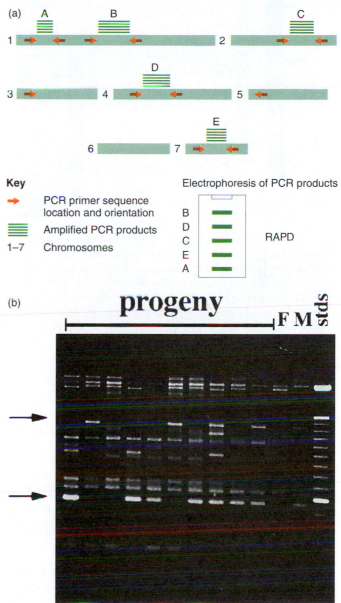

Figure 17-10 (a) Randomly amplified polymorphic DNA analysis (RAPD analysis) provides molecular chromosome markers. If another strain lacks one of the bands, this can be considered as a heterozygous marker locus and used in mapping, as shown in the example in part b. (b) RAPD analysis of a cross in a species of tree. A 10-nucleotide primer was used to amplify regions of genomic DNA in the parental trees (M = male, F = female) and 10 progeny. DNA standards for size calibration are shown in the right lane, marked "stds." The arrows point to two bands that represent two loci in a "dihybrid" test-cross. The two alleles of both these loci are expressed as the presence and the absence of a RAPD band. Of the two parents only the male showed these bands, but the male must have been heterozygous for the presence (+ allele) and absence (− allele) for bands at both loci. The male could be designated $1^+1^-\ 2^+2^-$, and the female $1^-1^-\ 2^-2^-$. The progeny show various parental and nonparental combinations of these alleles. (John E. Carlson)

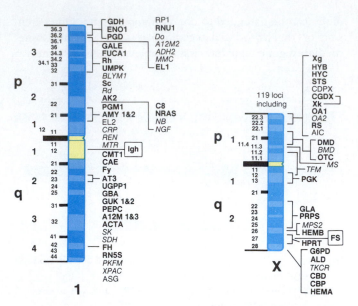

Figure 17-11 Maps of the human chromosome 1 and the X chromosome. The loci are labeled with code names. Many code names are based on the protein specified by the gene; others represent marker loci identified only as DNA variants. Such loci with some molecular tag can be mapped well by somatic cell hybridization methods. Some of these loci are interesting in themselves, and some are useful as chromosomal landmarks to which other loci may be mapped. For autosomes such as chromosome 1, approximately $\frac{2}{3}$ of the loci have been assigned by somatic hybridization, and the remainder from pedigree analysis. Compare these maps with those in the Chapter 5 opener, which list genes that determine various disorders and other phenotypes that map to these chromosomes. Protein variants are the basis of most of these conditions. p and q designate the short and long arms, respectively. (From V. A. McKusick, *Genetic Maps*. Vol. 2. Edited by S. J. O'Brien. Cold Spring Harbor, 1982.)

chromosomes are isolated by flow sorting (discussed later in the chapter). Then a sample of one specific chromosome under study is added to rodent cells. The human chromosomes are engulfed by the rodent cells and whole chromosomes or fragments become incorporated into the rodent nucleus. Correlations are then made between the human chromosomes or fragments present and human markers. The closer two human markers are on a chromosome, the more often they are transferred together. Some results of these kinds of mapping are shown in Figure 17-11.

A second method, called **irradiation and fusion gene transfer** (**IFGT**) is an extension of the above technique, designed to generate a higher-resolution map of molecular markers along a chromosome. The procedure is to irradiate human cells with 3000 rads of X rays to fragment the chromosomes, then fuse the irradiated cells with the rodent cells to form a panel of different hybrids. In this case the hybrids have an assortment of fragments of human chromosomes, as diagrammed in Figure 17-12. Most of the fragments are seen to be embedded into the rodent chromosomes, but truncated human chromosomes can also be found. First, the retention of various molecular markers in the hybrids is cal-

culated. The next step is to calculate the frequency of coretention of pairs of human molecular markers. The assumption is made that closely linked markers will be coretained at a high frequencies because there is a low probability that a radiation-induced break will occur between the loci. Distant markers and markers on different chromosomes should be retained at frequencies close to the product of individual retention frequencies. A mapping unit cR_{3000} is calculated, which can be calibrated to approximately 0.1 cM (m.u.).

One of the advantages of this method is in sample size; a large number of radiation hybrids can be amassed relatively easily. Nevertheless, it has been estimated that a standard panel of only 100 to 200 hybrids would be enough to obtain a high-resolution cR_{3000} map of the human genome, in other words, a tenfold increase in resolution over the centimorgan map.

Message Correlation of retention of human markers and chromosomes in hybrid rodent-human cell lines allows chromosomal assignment of the markers. Coretention of different human markers in X-irradiated hybrids allows high-resolution mapping of the chromosomal loci of the markers.

Physical Mapping of Genomes

How can the resolution of the mapping process be increased even further? This increased resolution is accomplished by manipulating DNA fragments directly; such frag-

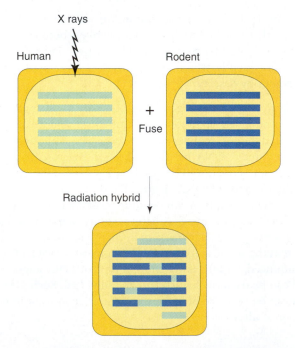

Figure 17-12 Making radiation hybrids using irradiation by X-rays. Fragments of human chromosomes integrate into rodent chromosomes. A panel of different radiation hybrids is analyzed for cotransfer of human markers, which can indicate linkage.

ments can be either cloned or not. Since DNA is the physical material of the genome, the procedures are generally called **physical mapping.** One goal of this procedure is to produce a set of clones that together encompass an entire chromosome or an entire genome. This achieves two purposes. First the genetic markers carried on the clones can be aligned and hence contribute to the overall genome mapping process. Second, once the contiguous clones are obtained, they represent an ordered library of DNA sequences that can be used for future genetic analysis, for example, to intensively characterize one specific genetic region of interest, or to find a previously uncharacterized gene.

Intact genomic DNA is difficult to manipulate. Some smaller genomes such as those of fungi can be analyzed by pulsed field gel electrophoresis (PFGE). After a lengthy run, the chromosomes separate into specific bands that can be identified using chromosome-specific probes (refer to Figure 17-2). However, in most genomes the individual chromosomes are too big and too numerous for this simple procedure. The long-cutter restriction enzymes such as *Not*I can cleave the chromosomal DNA into more manageable fragments, many of which can be separated by PFGE. It would seem that a broad-scale restriction map (see chapter 15) could be constructed using several such enzymes. This approach has been successful, but the rarity of suitable long-cutter enzymes, and a nonrandom distribution of the target sequences of these enzymes has limited the approach.

The technology is developing for direct determination of restriction cut sites on whole chromosomes under the microscope. One procedure, developed in yeast, is called **optical mapping.** A sample of deproteinized yeast chromosomes of one type (collected after pulsed field gel electrophoresis) is placed in a molten gel, and as the gel cools, the chromosomes are stretched by promoting flow of the gel fluid. Once set in the gel in the stretched state, the chromosomes are treated with restriction enzymes and the target sites become visible under the microscope as gaps that increase in size over time. The length of the fragments can be measured optically and, hence, an ordered restriction map obtained. This method has an advantage over conventional restriction mapping that sorts fragment sizes by gel migration, because only small numbers of chromosomes are needed, whereas gel electrophoresis requires bulk DNA preparations.

Another useful physical technique that can be used in *Drosophila* and in humans is to actually cut out specific portions of chromosomes under the microscope using micro tools and a micromanipulator. These chromosomal pieces contain very little DNA, but this can be amplified using PCR.

However, cloned genomic DNA fragments are of more general application and usefulness. Cosmids, YACs (yeast artificial chromosomes), and BACs have been the most useful vectors for cloning whole genomes. However, the task is a daunting one. Even so-called small genomes are still very large. Consider, for example, the 100-Mb genome of the tiny nematode *Caenorhabditis elegans;* since a cosmid insert is around 40 kb, it would require at least 2500 cosmids to embrace this genome, and many more to narrow the number down to such a set. YACs can contain on the order of 1 Mb, so here the task is somewhat simpler.

Cloning a whole genome begins by amassing a large number of randomly cloned inserts, such as cosmid clones. The contents of these clones must be characterized in some way and overlaps determined. A set of overlapping clones is called a **contig.** In the early phases of a genome project, contigs are numerous, and represent cloned "islands" of the genome. But as more and more clones are characterized, contigs enlarge and merge into one another, and eventually the project should end up with a set of contigs that equals the number of chromosomes.

Message Genomic cloning proceeds by assembling clones into overlapping groups called *contigs*. As more data accumulate, the contigs become equivalent to whole chromosomes.

Several different techniques are used to characterize the clones, and we shall consider three.

Clone Fingerprints

The genomic insert carried by a vector has its own unique sequence that can be used to generate a DNA fingerprint. For example, a multiple restriction enzyme digestion can generate a set of bands that are a unique "fingerprint" of that clone. The different sets generated by separate clones can be aligned either visually or using a computer program, to determine if there is any overlap between the inserted DNAs. In this way the contig can be built up.

Sequence-Tagged Sites (STSs)

If unique short sequences of large cloned inserts could be identified, then these could be used to align the various clones into contigs. For example, one could find that clone A has sequences 1 and 2, and clone B has 2 and 3, so clones A and B must overlap in the region of sequence 2. This approach is now possible. The practical procedure once again is to amass a large set of random clones with small genomic inserts (say, in λ phage) and sequence short regions of each. From these sequences, pairs of PCR primers are designed that will amplify the short specific sequence of DNA flanked by the primers. These short DNA sequences are known as **sequence-tagged sites (STSs).** Even though initially it is not known where in the genome these STSs are located, a panel of STSs can be used to characterize clones with large genomic inserts (such as YAC clones), and from this the clones that are shown to have specific STSs in common must have overlapping inserts and therefore can be aligned into contigs. An example of this process is shown in Figure 17-13.

DATA (STS content of YACs)

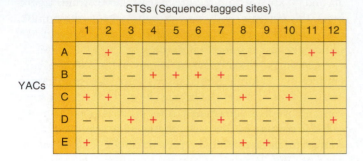

CONTIG

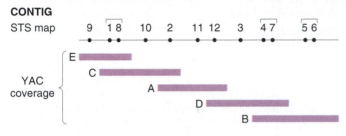

� Order uncertain

Figure 17-13 Using sequence-tagged sites (STSs) to order overlapping clones (YACs, in this example) into a contig. Six different YACs are tested to determine which STSs they contain (*above*), and these data are used to assemble a physical map (*below*).

Random Sequencing

The genome sequencing project for the bacterium *Hemophilus influenzae* used a different but analogous strategy for clone alignment. Genomic DNA was mechanically sheared and shotgun cloning was used to obtain a large number of random clones that were presumed to overlap each other in numerous ways. Primers based on adjacent vector DNA were used to sequence short regions at the ends of the cloned *Hemophilus* inserts. Then these short sequences were used (much like sequence-tagged sites) to align the genomic clones. Because so many random short sequences were obtained, together they encompassed the great majority of the *Hemophilus* genome. Gaps were filled in by "primer walking," that is, using the end of a cloned sequence as a primer to sequence into adjacent uncloned fragments. (The organization of the full genomic sequence is shown in Figure 17-18.)

The combination of these physical methods has resulted in the cloning of whole chromosomes and whole genomes of several experimental organisms. For example, the *C. elegans* genome is now available as sets of closmid or YAC contigs. Also, the DNA of the contigs has been arranged on nitrocellulose filters in ordered arrays, so to find out where a specific piece of DNA of interest lies in the genome, that DNA is used as a probe onto the contig filters, and a positive hybridization signal will announce the precise location of the DNA (see Figure 17-14). Such arrays have been called *polytene filters* because they can be used in somewhat the same way that a polytene chromosome is used for in situ hybridization.

Cloning and Mapping the Human Y Chromosome

The two smallest human chromosomes, chromosome 21 and the Y chromosome, have been fully cloned in overlapping sets of YAC clones (contigs). As part of this procedure, a high-resolution STS map has been obtained, and these sites can act as landmarks for future genetic research on these chromosomes. We will examine the cloning of the Y chromosome because it illustrates several of the techniques we have discussed. The STS map of the Y was in fact obtained by two different methods, YAC alignment, and deletion analysis.

YAC Alignment. First, human chromosomes were flow-sorted by **fluorescence-activated chromosome sorting (FACS)** (Figure 17-15). In this procedure metaphase chromosomes are stained with two dyes, one of which binds to AT-rich regions, and the other to GC-rich regions. Cells are disrupted to liberate whole chromosomes into liquid suspension. This suspension is converted into a spray in which the concentration of chromosomes is such that each spray droplet contains one chromosome. The spray passes through laser beams tuned to excite the fluorescence. Each chromosome produces its own characteristic fluorescence signal, which is recognized electronically, and two deflector plates direct the droplets containing the specific chromosome needed into a collection tube.

This process yielded a sample of Y chromosomes, from which λ clones were made. From clones that did not contain repetitive DNA, STS primers were designed. In all, 160 primer pairs were made. A Y chromosome YAC library of 10,368 clones was obtained in which the average insert size was 650 kb. From these numbers it was estimated that each point on the Y chromosome had been sampled an average of 4 times. The YAC clones were divided into 18 pools of 576 YACs, and the pools were screened with the STS primers. Subdivision of positive pools led rapidly to the assignment of a particular STS to specific YACs. The total STS content of each YAC was assessed and overlaps between the YACs determined in the same way that was shown in the generalized example in Figure 17-15.

Deletion Analysis. Various types of Y chromosome deletions occur naturally. For example, some XX males contain truncated fragments of the Y, while some XY females have deletions of the region containing the maleness (testis-determining) gene (see Chapter 23). These Y deletions have been maintained in cell culture and formed the basis for aligning the Y chromosome STSs. Each deletion was tested for STS content. Because by nature the deletions were nested sets, the STS content could be used not only to de-

Chromosomes and aligned YACs

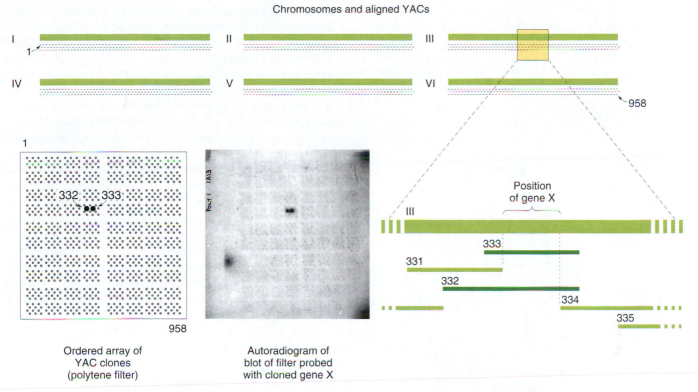

Ordered array of
YAC clones
(polytene filter)

Autoradiogram of
blot of filter probed
with cloned gene X

Figure 17-14 Using an ordered array of YACs to locate the map position of a newly cloned gene in *Caenorhabditis elegans*. The YAC clones are placed in order on a polytene filter. DNA from this filter is blotted and probed with the cloned gene giving the autoradiogram with two positive spots corresponding to two adjacent YACs numbered 332 and 333. Because of the hybridization pattern obtained, the location of the cloned gene can be narrowed down to a small region on chromosome III. (Autoradiogram from Alan Coulson.)

velop an STS map but also to map the coverage of the deletions. The principle is illustrated in Figure 17-16. The STS map produced by YAC alignment and by deletion analysis were identical.

Message Clones can be arranged into contigs by matching DNA fingerprints, by matching short sequences within cloned segments, and by analyzing deletions.

Genetic Analysis Using Physical Maps

A high-resolution physical map is an important resource for future genetic studies on any organism. This is particularly true for studies on positional cloning (see Chapter 14), the success of which is entirely dependent on the availability of a high-resolution map and other positional information about a candidate genetic region. Many of the successes in cloning human disease genes have used positional cloning. This is because although in principle a knowledge of a gene's function can be used in the process of cloning (functional cloning; for an example of functional complementation, see Chapter 14), for most human disease genes there is no knowledge of function, so positional cloning is the alternative.

Isolating Human Disease Genes by Positional Cloning

Figure 17-1 summarizes the general approach of positional cloning in humans. Pedigree and other meiotic analyses narrow down the gene locus to within a centimorgan of some molecular marker locus, and then physical mapping and cloning narrow down the region further. Often a cytogenetic rearrangement such as a deletion (Chapter 8) or restriction fragment length change due to mutation by trinucleotide repeats (Chapter 19) will provide important clues about precise location. Indeed, most cases of positional cloning performed so far have used information from rearrangements. Once the search has narrowed sufficiently, the gene must be found in the DNA sequence by a variety of methods including transcript analysis of the cloned region. Finally, the candidate gene must be sequenced and a mutant site found to confirm that this is indeed the gene sought.

As an example, we will follow the methods used to identify the genomic sequence of the cystic fibrosis (CF) gene. Cystic fibrosis is one of the more common genetic diseases, affecting about 1 in 2500 people. It is inherited as an autosomal recessive. Approximately 1 person in 25 is heterozygous for the mutant allele. The search for this impor-

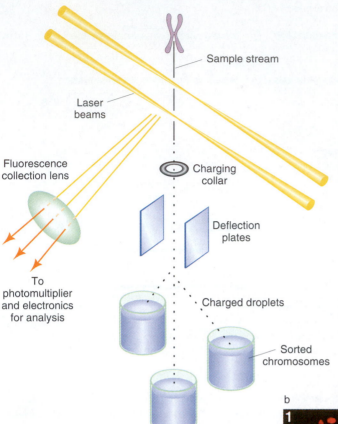

tant gene was an early tour de force of positional cloning, involving the international collaboration of several research teams. The search was particularly difficult because no chromosome rearrangements were available to help narrow down the location. The analysis provided a model for other searches for genes of unknown function.

The major symptoms of cystic fibrosis are well known. Most notably they are serious respiratory and digestive problems and extremely salty sweat. However, no primary biochemical defect was known at the time the gene was isolated. Linkage to molecular markers had located the gene to the long arm of chromosome 7, between bands 7q22 and 7q31.1. The *CF* gene was thought to be inside this region, flanked by the gene *met* (a proto-oncogene, see Chapter 24) on one end and a molecular marker D788 at the other end. But between these markers lay 1.5 centimorgans (map units) of DNA, a vast uncharted terrain of 1.5 Mb. Additional markers within the region were obtained by using new probes derived from a chromosome 7 library made by flow sorting.

Figure 17-15 (a) Chromosome purification using flow sorting. Chromosomes stained with a fluorescent dye are passed through a laser beam. Each time, the amount of fluorescence is measured and the chromosome deflected accordingly. The chromosomes are then collected as droplets. (b) Using flow sorting to obtain DNA to be used as probes for "painting" mouse chromosomes. The painted chromosomes appear yellow. Because several mouse chromosomes are the same size, some paints label more than one chromosome. These studies are useful in identifying chromosomes and in studying translocations (see the Chapter 8 opening figure). (Part a courtesy of the Los Alamos National Laboratory, Los Alamos, N. M. Part b courtesy of Pamela H. Rabbitts, Medical Research Council, Clinical Oncology and Radiotherapeutics Unit, Cambridge, England.)

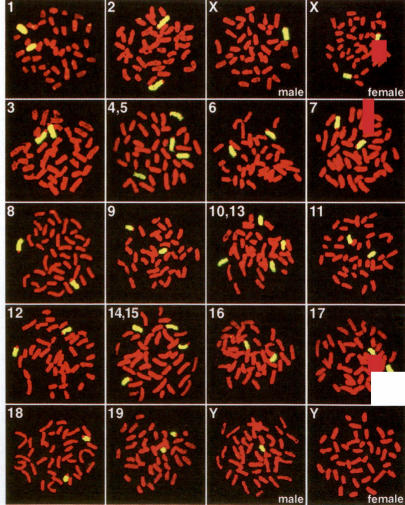

a

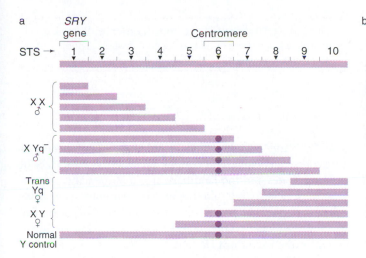

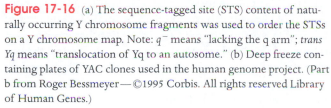

b

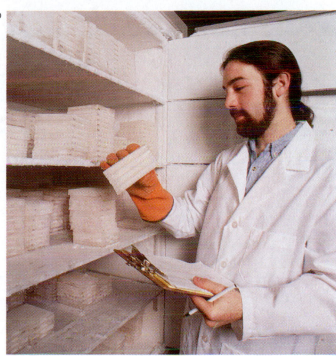

Figure 17-16 (a) The sequence-tagged site (STS) content of naturally occurring Y chromosome fragments was used to order the STSs on a Y chromosome map. Note: *q⁻* means "lacking the q arm"; *trans Yq* means "translocation of Yq to an autosome." (b) Deep freeze containing plates of YAC clones used in the human genome project. (Part b from Roger Bessmeyer—©1995 Corbis. All rights reserved Library of Human Genes.)

However, the two key techniques that were used to traverse the huge genetic distances were chromosome walking (page 439) and a related technique called **chromosome jumping.** This latter technique provides a way of jumping across potentially unclonable areas of DNA, and also generates widely spaced landmarks along the sequence that can be used as initiation points for multiple bidirectional chromosomal walks.

Chromosome jumping is illustrated in Figure 17-17. In this procedure, large fragments are created by partial restriction cleavage of the DNA of the region believed to contain the gene of interest. Each DNA fragment is then circularized, thus bringing the beginning and end of the fragment together. This junction is cut out and cloned into a phage vector, which, together with the other junction segments, makes up a *jumping library*. A probe from the beginning of the stretch of DNA under investigation can be used to screen the jumping library to find the clone that contains the beginning sequence. Once found, the other end of the junction sequence is excised and used to screen the library again to make a second jump. From each jump position, chromosome walks can be made in both directions in order to search for genelike sequences.

A restriction map of the overall region was obtained with rare-cutting restriction enzymes, and the restriction sites used to position and orient the sequences obtained from jumping and walking. Once enough sequencing had been done to cover representative parts of the overall region, then the hunt for any genes along this stretch began. Genes were sought by several techniques. First, it was known that genes in humans generally are preceded at the 5′ end by clusters of cytosines and guanines, called *CpG is-*

lands, and several of these were found. Second, it was reasoned that a gene would show homology to the DNA of other animals, because of evolutionary conservation, so candidate sequences were used to probe what were called *zoo blots* of genomic DNA from a range of animals. Third, genes should have appropriate start and stop signals. Fourth, genes should be transcribed, and transcripts should be found.

Ultimately, a strong candidate gene was found spanning 250 kb of the region. Cystic fibrosis symptoms are expressed in sweat glands, so from cultured sweat gland cells, cDNA was prepared, and a 6500-nucleotide cDNA homologous to the candidate gene was detected. Upon sequencing this cDNA in normal and CF patients, it was found that the cDNA of the patients showed a deletion of three base pairs, eliminating a phenylalanine from the protein. Therefore it was very likely that this was the CF coding sequence. Thus the CF gene had been found. From its sequence, an amino acid sequence was inferred, and from this the three-dimensional structure of the protein was predicted. This protein showed structural similarities to ion-transport proteins in other systems, suggesting that a transport defect is the primary cause of CF. When used to transform mutant cell lines from CF patients, the wild-type gene restored normal function—the final confirmation that the isolated sequence was in fact the CF gene.

As more and more of the genome is mapped at high resolution (as shown above for the Y chromosome), then positional cloning will become much easier than in the case of the CF gene. Often intensive cloning and sequencing of a region will reveal the presence of genes of unknown function. If a disease gene is mapped to that general area, then

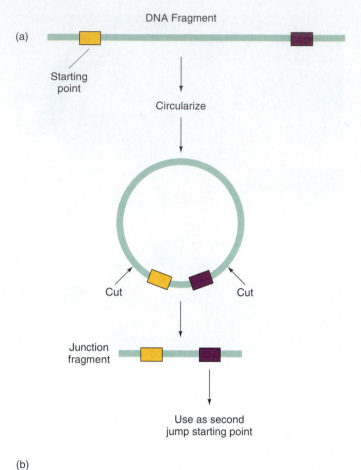

DNA Fragment

(a)

Starting point

Circularize

Cut   Cut

Junction fragment

Use as second jump starting point

(b)

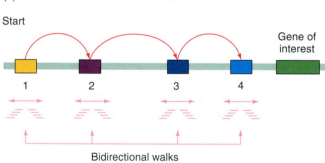

Start

Gene of interest

1   2   3   4

Bidirectional walks

Figure 17-17 Manipulating cloned genomic fragments for chromosome jumping, a modified type of chromosome walking that can bypass regions difficult to clone, such as those containing repetitive DNA (see text).

these genelike sequences become **candidate genes** for the disease gene without the need for extensive chromosome walking and jumping. This procedure is termed the **candidate gene approach** to disease gene identification. Knowledge about the disease such as biochemical defect and pattern of tissue expression can be matched to the sequence domains and tissue expression of the candidate gene. The method works in the opposite direction too; the domains and tissue expression of randomly sequenced genes often suggest a possible disease gene.

Message Positional cloning is made easier by the availability of overlapping clones that represent the whole genome or a whole chromosome.

Genome Sequencing

The final and most detailed level of genomic analysis is genome sequencing. The contigs provide the basic sets of ordered clones and the landmarks for a sequencing project. Several smaller viral and organellar genomes have been fully sequenced. However, the first (and to date only) full genomic sequence to be obtained for a free-living organism is that of the bacterium *Hemophilus influenzae,* whose host is humans. The full sequence was 1.83 Mb in size and contained 1743 known genes and ORFs. A total of 58 percent of the total has been assigned to a function based either on studies of previously obtained mutants or on matches to other known genes on the genomic database. The complete sequence is represented in Figure 17-18, which is color coded by genes of related function, as explained in the figure legend.

Several complete chromosomes of the yeast *Saccharomyces cerevisiae* have been fully sequenced, and it seems certain that this will be the first complete genomic sequence to be obtained for a eukaryotic organism. The genes of one of the fully sequenced chromosomes, chromosome 3, are shown in Figure 17-19.

The potential benefits of sequencing genomes are immense. In basic research the sequence provides a database that acts as the jumping-off point for all future genetic analysis in that organism. All unassigned genes are genes in search of a function, and reverse genetics can be used to disrupt such a gene in order to determine its normal functional role, in the classical style of genetic dissection. In the analysis of humans, unassigned genes can become candidate genes that might prove to be instrumental in causing human ill health or disease. If the map position of a disease gene becomes known, then unassigned genes known to be in that location are automatically candidates for causing that disease phenotype. Furthermore, questions about how many and what types of genes an organism has lead to fundamental issues concerning the types of functions that are needed to code for a living organism in a variety of ecological niches. Also, of course, there are questions that arise about evolution and how one organism has obtained its set of genes and how it can make do without genes that other organisms absolutely need. In other words, an entirely new area of research will be possible, the design features of genomes.

In commercial research, knowledge of DNA sequence can be productive. For example, pharmaceutical companies have recognized the potential applications of finding new genes through sequencing because these genes may suggest molecular targets that might lead to the development of

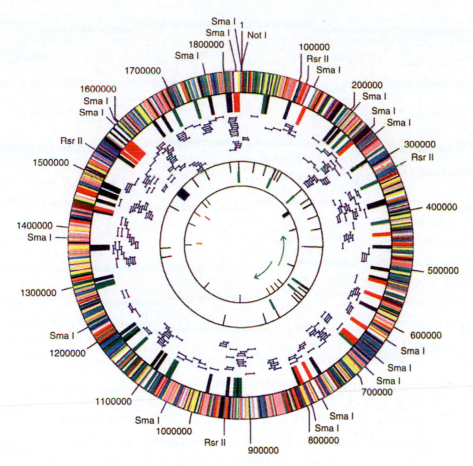

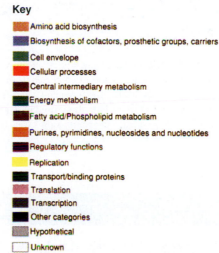

Figure 17-18 The architecture of the genome of the bacterium *Hemophilus influenzae,* based on the complete genomic sequence reported in 1995. The circles have the following meanings, starting with the outer circle:

- Restriction sites and genomic positions, starting from the *Not*I site
- Genes color coded by their functional groups (a key is shown in the figure)
- Regions of high GC content: red >42%, blue >40%. Regions of high AT content: black >66%, green >64%.
- Coverage of some clones used in sequencing
- Ribosomal operons (green), tRNAs (black), mu-like prophage (blue)
- Positions of simple tandem repeats
- Replication origins (green arrows) and termination sequences (red)

(Art by Dr. Anthony R. Kerlavage; Institute for Genomic Research. From *Science,* 1995, 269, 449–604.)

Key

- Amino acid biosynthesis
- Biosynthesis of cofactors, prosthetic groups, carriers
- Cell envelope
- Cellular processes
- Central intermediary metabolism
- Energy metabolism
- Fatty acid/Phospholipid metabolism
- Purines, pyrimidines, nucleosides and nucleotides
- Regulatory functions
- Replication
- Transport/binding proteins
- Translation
- Transcription
- Other categories
- Hypothetical
- Unknown

new drugs. Likewise, agriculturalists recognize that the same is true in their discipline, and genes might be found that have significance in plant or animal improvement. It is not only single-gene phenotypes that are under scrutiny. Many inherited conditions are determined by the interaction of several to many genes. Furthermore, many desirable plant or animal phenotypes are likewise determined by such *polygenes.* The availability of large numbers of molecular markers throughout the genome makes it feasible through linkage analysis to start looking for and piecing together the interacting genes. Hence an analysis becomes possible that was impossible by classical genetic analysis.

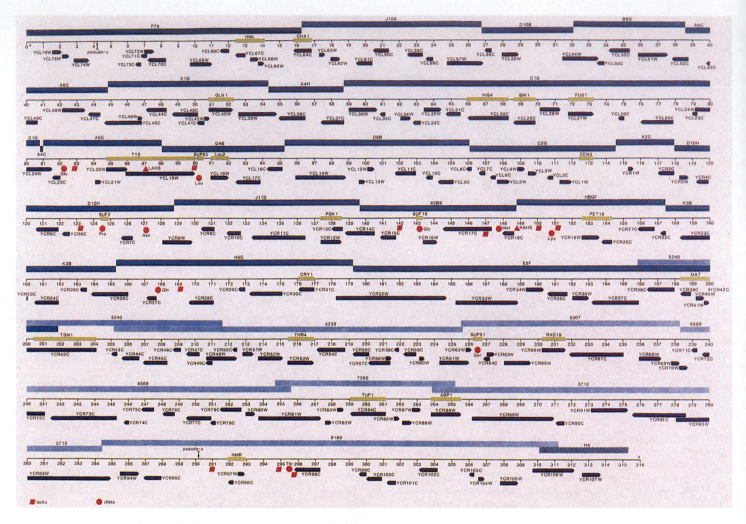

Figure 17-19 The genetic landscape of chromosome 3 in yeast, determined by sequencing the entire chromosome. Genes previously detected from mutant phenotypes are shown in yellow. Open reading frames, which likely are protein-coding genes, were detected by sequence analysis and are shown below the line in dark blue. The clones used in sequencing are shown above the line in light blue. Delta sequences are derived from the transposon Ty. (S. G. Oliver et al., *Nature* 357, 1992, 38–46.)

SUMMARY

Genomics is the branch of genetics that deals with the systematic molecular characterization of whole genomes. Some of the methods used are traditional genetic mapping procedures, but in addition specialized techniques have been developed for manipulating the large amounts of DNA in a genome. Genomic analysis is important for two reasons: first, it represents a way of obtaining an overview of the genetic architecture of an organism, and second, it forms a set of basic information that can be used to find new genes such as those that are involved in human disease. Genomic analysis generally proceeds from low-resolution analysis to techniques with higher resolution. Initially genes must be assigned to chromosomes, and this can be achieved with a variety of techniques including linkage to standard markers, in situ hybridization, pulsed field gel electrophoresis, and human-rodent cell hybrids. The arrangement of loci along a chromosome can be determined using various types of meiotic recombination mapping and (in humans) X-irradiated hybrids. Particularly useful are molecular markers that can fill in the gaps between genes of known phenotypic association. RFLPs, SSLPs, and RAPDs all provide heterozygous loci that can be used as molecular marker loci in mapping. The highest level of genomic resolution is physical mapping of DNA fragments. Most useful are fragments that have been cloned in vectors that carry large DNA inserts, such as cosmids and YACs. The goal of physical mapping is to produce a set of overlapping clones that encompass an entire chromosome or an entire genome. Sequence-tagged sites are particularly useful in aligning overlapping cosmids into contigs. As more clones are characterized, contigs grow to the size of entire chromosomes. Genomic maps have been used in the positional cloning of human disease genes of unknown function. The maps have provided suitable starting points for chromosome walking and jumping. Genomic sequencing often reveals genes that have never been associated with a phenotype; these must be investigated by doing gene disruptions to check for a possible mutant phenotype.

Concept Map

Draw a concept map interrelating as many of the following terms as possible. Note that the terms are listed in no particular order.

genomics / contig / physical map / YAC / RFLP / SSLP / STS / FISH / recombinant frequency / molecular marker

CHAPTER INTEGRATION PROBLEM

A *Neurospora* geneticist has just isolated a new mutation that causes aluminum insensitivity (*al*) in a strain of Oak Ridge background (see page 526). She wishes to clone the gene by positional cloning and therefore needs to map it. For reasons that we don't need to go into, she suspects it is located near the tip of the right arm of chromosome 4. Luckily there are three RFLP markers (1, 2, and 3) available in that vicinity, so the following cross is made:

al (Oak Ridge background) × *al*⁺ (Mauriceville background)

One hundred progeny are isolated and tested for *al* and the six RFLP alleles 1^O, 2^O, 3° 1^M, 2^M, and 3^M. The results were as follows, where O and M represent the RFLP alleles, and *al* and + represent *al* and *al*⁺:

RFLP 1	O	M	O	M	O	M
RFLP 2	O	M	M	O	O	M
RFLP 3	O	M	M	O	M	O
al locus	*al*	+	*al*	+	*al*	+
Total of genotype	34	36	6	4	12	8

    a. Is the *al* locus in fact in this vicinity?

    b. If so, which RFLP is it closest to?

    c. How many map units separate the three RFLP loci?

Solution

This is a mapping problem, but with the twist that some of the markers are traditional types (that we have encountered in previous chapters) but others are molecular markers (in this case, RFLPs). Nevertheless, the principle of mapping is the same as we used before; in other words, it is based on recombinant frequency. In any recombination analysis we must be clear of the genotype of the parents before we can classify progeny into recombinant classes. In this case we know that the Oak Ridge parent must contain all O alleles, and Mauriceville all M alleles; therefore, the parents were

$$al\ 1^O\ 2^O\ 3^O \times\ +\ 1^M\ 2^M\ 3^M$$

and knowing this makes determining recombinant classes easy. We see from the data that the parental classes are the two most common (34 and 36). We first of all notice that the *al* alleles are tightly linked to RFLP 1 (all progeny are *al* 1^O or + 1^M). Therefore, the *al* locus is definitely on this part of the chromosome 4. There are $6 + 4 = 10$ recombinants between RFLP 1 and 2, so these loci must be 10 map units apart. There are $12 + 8 = 20$ recombinants between RFLP 2 and 3; that is, they are 20 map units apart. There are $6 + 4 + 12 + 8 = 30$ recombinants between RFLP 1 and 3, showing that these must flank RFLP 2. Therefore, the map is

al RFLP 1   10 m.u.   RFLP 2   20 m.u.   RFLP 3

There are evidently no double recombinants, which would have been of the type M O M and O M O.

    Notice that there is really no new principle at work in the solution of this problem; the real challenge is to understand the nature of RFLPs and to translate this into genotypes from which to study recombination. If you still don't understand RFLPs, you might ask yourself how exactly the RFLP alleles are tested experimentally.

SOLVED PROBLEM

Duchenne muscular dystrophy (DMD) is an X-linked recessive human disease affecting muscles. Six small boys had DMD, together with various other disorders, and they were found to have small deletions of the X chromosome, as shown below:

```
       1   2   3   4   5   6   7   8   9   10  11  12  13
    ●●●━━━━━━━━━━━━━━━━━━━━━━━      ━━━━━━━●●●● Δ i
    ●●●━━━━━━━━━━        ━━━━━━━━━━━━━━━━●●●● Δ ii
    ●●●━━━━━━━━        ━━━━━━━━━━━━━━━━━━━●●●● Δ iii
    ●●●━━━━━━━        ━━━━━━━━━━━━━━━●●●● Δ iv
    ●●●━━━━━         ━━━━━━━━━━━━━━●●●● Δ v
    ●●●━━━         ━━━━━━━━━━━●●●● Δ vi
```

a. On the basis of this information, which chromosomal region most likely contains the gene for DMD?

b. Why did the boys show other symptoms in addition to DMD?

c. How would you use DNA samples from these three boys and DNA from unaffected boys to obtain an enriched sample of DNA containing the gene for DMD, as a prelude to cloning the gene?

Solution

a. The only region that all the deletions are lacking is the chromosomal region labeled 5, so this presumably contains the gene for DMD.

b. The other symptoms probably result from the deletion of the other regions surrounding the DMD region.

c. If the DNA from all the DMD deletions is denatured (i.e., its strands separated) and bound to some kind of filter, the normal DNA can be cut by shearing or by restriction enzyme treatment, denatured, and passed through the filter containing the deleted DNA. Most DNA will bind to the filter, but the region 5 DNA will pass through. This process can be repeated several times. The filtrate DNA can be cloned and then used in a FISH analysis to see if it binds to the DMD X chromosomes. If not, it becomes a candidate for the DMD-containing sequence.

PROBLEMS

1. From in situ hybridizations, it was known that five different YACs containing genomic fragments hybridized to one specific chromosome band of the human genome. Genomic DNA was digested with a long-cutter restriction enzyme, and radioactively labeled YACs were each hybridized to blots of the digest. The autoradiogram was as follows:

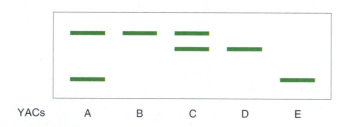

a. Use these results to order the three hybridized restriction fragments.

b. Show the locations of the YACs in relation to the three genomic restriction fragments in part a.

 Unpacking the Problem

a. State two types of hybridization used in genetics. What type of hybridizations are used in this question, and what is the molecular basis for such hybridizations? (Draw a rough sketch of what happens at the molecular level during hybridization.)

b. How are in situ hybridizations done in general? How would the specific in situ hybridizations in this question be done (as in the first sentence)?

c. What is a YAC?

d. What are chromosome bands, and what procedure is used to produce them? Sketch a chromosome with some bands and show how the in situ hybridizations would look.

e. How would it have been shown that five different YACs hybridize to one band?

f. What is a genomic fragment? Would you expect the five YACs to contain the same genomic fragment or different ones? How do you think these genomic fragments were produced (what are some general ways of fragmenting DNA)? Does it matter how the DNA was fragmented?

g. What is a restriction enzyme?

h. What is a long cutter? If you do not know what a long cutter is, what do you think it might be, and does your guess make sense of this part of the problem? If not, refer to discussions of long cutters in the chapter.

i. Why were the YACs radioactively labeled? (What does it mean to radioactively label something?)

j. What is an autoradiogram?

k. Write a sentence that uses the words *DNA, digestion, restriction enzyme, blot, autoradiogram*.

l. Explain exactly how the pattern of dark bands shown in the question was obtained.

m. Approximately how many kilobases of DNA are in a human genome?

n. If human genomic DNA is digested with a restriction enzyme, roughly how many fragments would be produced? Tens? Hundreds? Thousands? Tens of thousands? Hundreds of thousands?

o. Would all these DNA fragments be different? Mostly different?

p. If these fragments are separated on an electrophoretic gel, what would you see if you added a DNA stain to the gel?

q. How does your answer to the previous question compare with the number of autoradiogram bands in this question?

r. Part a of the problem mentions "three hybridized restriction fragments"—point to these in the diagram.

s. Would there actually be any restriction fragments on an autoradiogram?

t. Which YACs hybridize to one restriction fragment and which YACs hybridize to two DNA fragment?

u. How is it possible for a YAC to hybridize to two DNA fragments? Suggest two explanations, and decide which makes more sense in this question. Does the fact that all the YACs in this question bind to one chromosome band (and apparently nothing else) help you in deciding? Could a YAC hybridize to more than two fragments?

v. Distinguish the use of the word *band* by cytogeneticists (chromosome microscopists) from the use of the word *band* by molecular geneticists. In what way do these uses come together in this question?

2. Three genes *LEU2*, *ADE3*, and *MATa* were cloned in yeast. A *Neurospora* geneticist wanted to find out if *Neurospora* had these three genes and, if so, wanted to clone the *Neurospora* equivalents. As a first step to this analysis, he hybridized the clones as radioactive probes to PFGE preparations of *Neurospora* chromosomes, and the results are shown at the top of the next column. Which genes are present in *Neurospora*, and what chromosomes are they on?

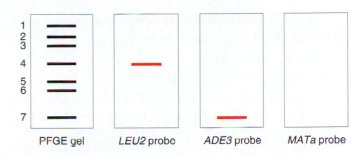

3. A *Neurospora* geneticist wanted to clone the gene *cys-1*, which was believed to be near the centromere on chromosome 5. Two RFLP markers (RFLP1 and RFLP2) were available in that vicinity, so he made the cross

Oak Ridge *cys-1* × Mauriceville *cys-1⁺*

Then 100 ascospores were tested for RFLP and *cys-1* genotypes, and the results were

RFLP 1	O	M	O	M	O	M
RFLP 2	O	M	M	O	M	O
cys locus	cys	+	+	cys	cys	+
Total of genotype	40	43	2	3	7	5

a. Is *cys-1* in this region of the chromosome?

b. If so, draw a map of the loci in this region, labeled with map units.

c. What would be a suitable next step in cloning the *cys-1* gene?

4. In a certain haploid fungus, there had been extensive genetic analysis, including genetic mapping, and four linkage groups had been developed, suggesting four chromosomes. However, the chromosomes were very small and difficult to see under the microscope, so it was not known if there really were four chromosomes. The advent of PFGE technology showed that there are four chromosomes. However, the linkage groups still needed to be assigned to these chromosomes. To begin this process, a cloned gene *P* was used in a Southern analysis with a PFGE preparation from wild type and from two translocations that were known from genetic studies to involve linkage groups 1 and 4 in one case and 2 and 4 in the other. The results follow:

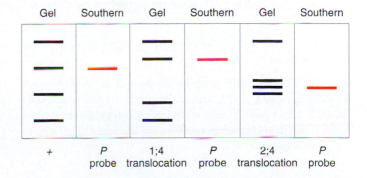

a. From these data, determine which of the four real chromosomes (bands) the gene is on.

b. Determine which chromosome (band) corresponds to each of the four linkage groups.

5. A cloned gene from *Arabidopsis* is used as a radioactive probe against DNA samples from cabbage (in the same plant family) digested by three different restriction enzymes. In the case of enzyme 1, there were three radioactive bands on the autoradiogram, for enzyme 2 there was one band, and for enzyme 3 there were two bands. How can these results be explained?

6. Five YAC clones of human DNA (YAC-A through YAC-E) were tested for sequence-tagged sites STS1 through STS7. The results are shown in the following table, in which a "+" shows that the YAC contains that STS.

STS

YAC	1	2	3	4	5	6	7
A	+	−	+	+	−	−	−
B	+	−	−	−	+	−	−
C	−	−	+	+	−	−	+
D	−	+	−	−	+	+	−
E	−	−	+	−	−	−	+

a. Draw a physical map showing the STS order.

b. Align the YACs into a contig.

7. Seven human-rodent radiation hybrids were obtained and tested for six different human genome molecular markers A through F. The results are shown below, where a "+" shows the presence of a marker.

Radiation hybrids

Markers	1	2	3	4	5	6	7
A	−	+	−	−	+	+	−
B	+	−	+	−	−	−	−
C	+	−	+	+	−	+	−
D	−	+	−	+	+	+	−
E	+	−	−	+	+	−	+
F	+	−	−	+	+	−	+

a. What marker linkages are suggested by these results?

b. Is there any evidence of markers being on separate chromosomes?

8. A RAPD primer amplified two bands in *Aspergillus nidulans* haploid strain 1, and no bands in *Aspergillus nidulans* strain 2 (which was from a different country). These strains were crossed, and seven progeny were analyzed. The results were as follows:

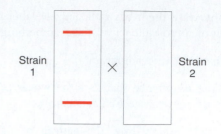

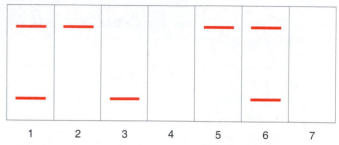

a. Draw diagrams that explain the difference between the parents.

b. Explain the origin of the progeny and their relative frequencies.

c. What would a nonparental ditype tetrad look like with regard to this particular amplification?

9. A *Caenorhabditis* contig for one region of chromosome 2 is as follows, where A through H are cosmids:

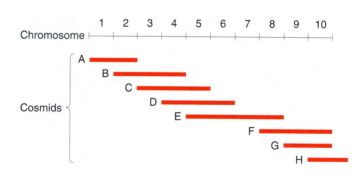

a. A cloned gene *pBR322-x* hybridized to cosmids C, D, and E. What is the approximate location of this gene *x* on the chromosome?

b. A cloned gene *pUC18-y* hybridized only to cosmids E and F. What is its location?

c. Explain exactly how it is possible for both probes to hybridize to cosmid E?

10. A geneticist succeeds in maintaining three colonies of human-mouse hybrid cells. The only human chromosomes retained by the hybrid cells are those indicated by pluses below:

Hybrid colony	Human chromosome							
	1	2	3	4	5	6	7	8
A	+	+	+	+	−	−	−	−
B	+	+	−	−	+	+	−	−
C	+	−	+	−	+	−	+	−

The geneticist tests each of the colonies for the presence of five enzymes (α, β, γ, δ, and ϵ) with the following results: α is active only in colony C, β is active in all three colonies, γ is active only in colonies B and C, δ is active only in colony B, and ϵ shows no activity in any colony. What can the geneticist conclude about the locations of the genes responsible for these enzyme activities?

11. Consider the following set of eight hybridized human-mouse cell lines:

Cell line	Human chromosome								
	1	2	6	9	12	13	17	21	X
A	+	+	−	q	−	p	+	+	+
B	+	−	p	+	−	+	+	−	−
C	−	+	+	+	p	−	+	−	+
D	+	+	−	+	+	−	q	−	+
E	p	−	+	−	q	−	+	+	q
F	−	p	−	−	q	−	+	+	p
G	q	+	−	+	+	+	+	−	−
H	+	q	+	−	−	q	+	−	+

Each cell line may carry an intact (numbered) chromosome (+), only its long arm (q), only its short arm (p), or it may lack the chromosome (−).

The following human enzymes were tested for their presence (+) or absence (−) in cell lines A–H:

Enzyme	Cell line							
	A	B	C	D	E	F	G	H
steroid sulfatase	+	−	+	+	−	+	−	+
phosphoglucomutase-3	−	−	+	−	+	−	−	+
esterase D	−	+	−	−	−	−	+	+
phosphofructokinase	+	−	−	−	+	+	−	−
amylase	+	+	−	+	+	−	−	+
galactokinase	+	+	+	+	+	+	+	+

Identify the chromosome carrying each enzyme locus. Where possible, identify the chromosome arm.

(Problem 11 from L. A. Snyder, D. Freifelder, and D. L. Hartl, *General Genetics.* Jones and Bartlett, 1985.)

12. A certain disease is inherited as an autosomal dominant *N*. It was noted that some patients carry reciprocal translocations in which one of the chromosomes involved is always chromosome 3 and the break is always in band 3q3.1. Four molecular probes (a through d) are known to hybridize in situ to this band but their order is not known. In the translocations, only probe c hybridizes to chromosome 3 carrying a portion of another chromosome, and probes a, b, and d always hybridize to the translocated fragment.

a. Draw diagrams that illustrate the meaning of these findings.

b. How would you use this information for positional cloning of the normal allele *n*?

c. Once *n* is cloned, how would you use this clone to investigate the nature of the mutations in disease patients who do not have translocations?

13. The gene for the autosomal dominant disease shown in this pedigree is thought to be on chromosome 4, so five RFLPs (1–5) mapped on chromosome 4 were tested in all family members. The results are also shown in the diagram; the superscripts represent different alleles of the RFLP loci.

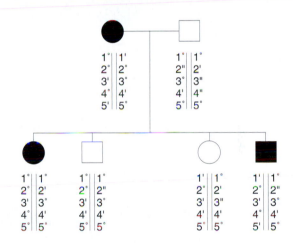

a. Explain how this experiment was carried out.

b. Decide which RFLP locus is closest to the disease gene (explain your logic).

c. How would you use this information to clone the disease gene?

14. Let's say you undertake a study of hybridized human and mouse cells because you want to map part of human chromosome 17. Three loci on this chromosome, *a, b,* and *c,* are concerned with making the compounds a, b, and c—all of which are essential for growth and present in mice as well as humans. You fuse mouse cells that are phenotypically $a^- b^- c^-$ with $a^+ b^+ c^+$ human cells. Assume that you find a hybrid in which the only human component is the right arm of chromosome 17 (17R), translocated by some unknown mechanism to a mouse chromosome. The hybrid can make the compounds a, b, and c. Then you treat cells with adenovirus, which causes chromosome breaks. Assume that you can isolate 200 lines in which bits of the translocated 17R have been clipped off. You test these lines for the ability to make a, b, and c, and obtain the following results:

Number	Can make
0	a only
0	b only
12	c only
0	a and b only
80	b and c only
0	a and c only
60	a, b, and c
48	nothing

a. How would these different types arise?

b. Are *a, b,* and *c* all located on the right arm of chromosome 17? If so, draw a map indicating their relative positions.

c. How would banding patterns help you in your analysis?

18

Control of Gene Expression

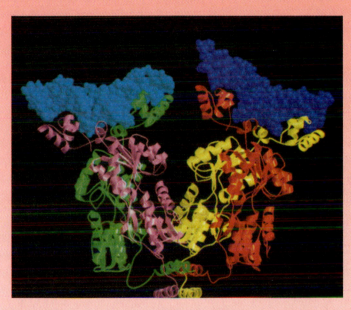

View of the *lac* repressor-DNA complex, as determined by X-ray crystallography. Here the repressor tetramer is shown binding to two operators. Each dimer binds to one operator. The operators are 21 base pairs long here and are shown in dark and light blue. The monomers are colored green, violet, red, and yellow. Note how the amino-terminal headpiece on one monomer (violet) crosses the other monomer (green) to bind the first part of the operator, while the second monomer headpiece (green) crosses over to bind the second part of the operator. Thus, the recognition helix from each of two subunits binds to the consecutive parts of the same operator. (Figure supplied by M. Lewis, Geoffrey Chang, N. C. Horton, M. A. Kercher, H. C. Pace, M. A. Schumacher, R. G. Brennan, and P. Lu, University of Pennsylvania).

KEY CONCEPTS

▶ Gene regulation is most often mediated by proteins that react to environmental signals by raising or lowering the transcription rates of specific genes.

▶ Negative control in prokaryotes is exemplified by the *lac* system, in which a repressor protein blocks transcription by binding to a site on the DNA termed the *operator.*

▶ Positive control in prokaryotes occurs when protein factors are required to activate transcription.

▶ Many regulatory proteins have common structural features.

▶ Transcriptional control of eukaryotes is also mediated by trans-acting protein factors, which bind to specific regulatory sequences.

▶ Additional regulatory sites on the DNA, termed *enhancers,* modulate gene expression in eukaryotes by interacting with specific regulatory proteins.

ntil now, we have discussed what genes are and how genetic change occurs and is inherited. Also, we saw in Chapter 13 how genes are expressed by being transcribed into RNA molecules, many of which are translated into proteins. But how does the cell *regulate* the expression of all its genes? We can see that control of gene expression is crucial to an organism. In higher cells, specific cell types have differentiated to the point that they are highly specialized. An eye cell in humans synthesizes the proteins important for eye color but does not produce the detoxification enzymes that are synthesized in liver cells. Each cell type has arranged to express only some of its genes.

Bacteria also have a need to regulate the expression of their genes. Enzymes involved in sugar metabolism provide an example. Metabolic enzymes are required to break down different carbon sources to yield energy. However, there are many different types of compounds that bacteria could use as carbon sources, including sugars such as lactose, glucose, maltose, rhamnose, raffinose, melibiose, galactose, and xylose. Several enzymes allow each of these compounds to enter the cell and to catalyze different steps in sugar breakdown. If a cell were to synthesize simultaneously all the enzymes it might possibly need, it would cost the cell much more energy to produce the enzymes than it could ever derive from breaking down any of the prospective carbon sources. Therefore, the cell has devised mechanisms to repress all the genes encoding enzymes that are not needed and to activate those genes at a time when the enzymes are needed. Clearly, to do this, two requirements must be met:

1. A method must be found for turning on or off each specific gene or group of genes.

2. The cell must be able to recognize situations in which it should activate or repress a specific gene or group of genes.

Let's review the current model for gene control and then discuss its development.

Basic Control Circuits

The first system we will focus on is concerned with lactose metabolism in *Escherichia coli*. We have now learned a lot about how this system works. Figure 18-1 shows a physical model for the control of the lactose enzymes.

The metabolism of lactose requires two enzymes: a permease to transport lactose into the cell and β-galactosidase to cleave the lactose molecule to yield glucose and galactose. Permease and β-galactosidase are encoded by two contiguous genes, *Z* and *Y*, respectively. A third gene, the *A* gene, encodes an additional enzyme, termed *transacetylase,* but this enzyme is not required for lactose metabolism, and we will not concentrate on it for now. All three genes are transcribed into a single or **polycistronic** messenger RNA (mRNA) molecule. Thus, it can be seen that by regulating the production of this mRNA, the regulation of the synthesis of all three enzymes can be coordinated. A fourth gene, the *I* gene, which maps near but not directly adjacent to the *Z*, *Y*, and *A* genes, encodes a **repressor** protein, so named because it can block the expression of the *Z*, *Y*, and *A* genes. The repressor binds to a region of DNA near the beginning of the *Z* gene and near the point at which transcription of the polycistronic mRNA begins. The site on the DNA to which the repressor binds is termed the **operator.** One necessary property of the repressor is that it

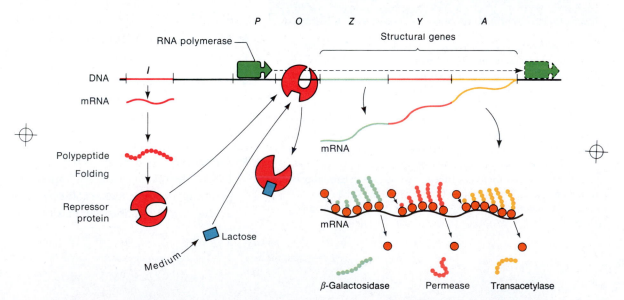

Figure 18-1 Regulation of the *lactose* operon. The *I* gene continually makes repressor. The repressor binds to the *O* (operator) region, blocking the RNA polymerase bound to *P* from transcribing the adjacent structural genes. When lactose is present, it binds to the repressor and changes its shape so that the repressor no longer binds to *O*. The RNA polymerase is then able to transcribe the *Z*, *Y*, and *A* structural genes, so the three enzymes are produced.

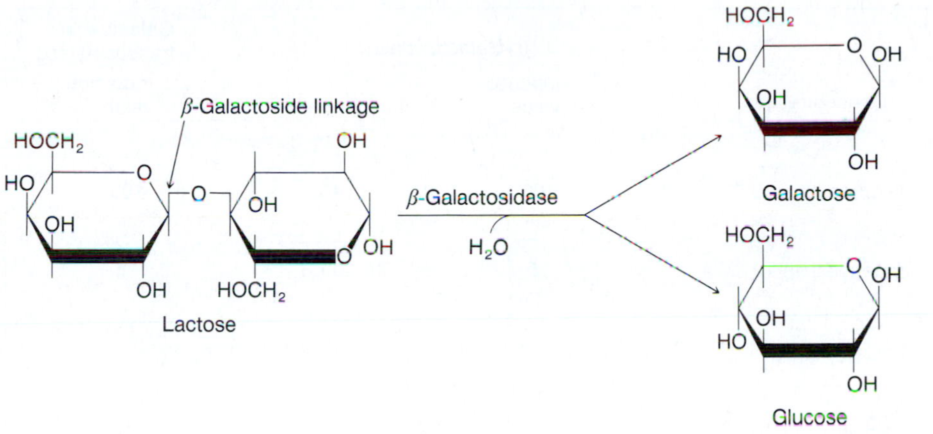

Figure 18-2 The metabolism of lactose. The enzyme β-galactosidase catalyzes a reaction in which water is added to the β-galactosidase linkage to break lactose into separate molecules of galactose and glucose. The enzyme lactose permease is required to transport lactose into the cell.

be able to recognize a specific short sequence of DNA—namely, a specific operator. This ensures that the repressor will bind only to the site on the DNA near the genes that it is controlling and not to other random sites all over the chromosome. By binding to the operator, the repressor prevents the initiation of transcription by RNA polymerase. Normally, RNA polymerase binds to specific regions of the DNA at the beginning of genes or groups of genes, termed **promoters** (see Chapter 13), so that it can initiate transcription at the proper starting points. The *POZYA* segments shown in Figure 18-1 constitute an **operon,** which is a genetic unit of coordinate expression.

The *lac* repressor is a molecule with two recognition sites—one that can recognize the specific operator sequence for the *lac* operon and another that can recognize lactose and certain analogs of lactose. When the repressor binds to lactose derivatives, it undergoes a conformational change; this slight alteration in shape changes the operator binding site so that the repressor loses affinity for the operator. Thus, in response to binding lactose derivatives, the repressor falls off the DNA. This satisfies the second requirement for such a control system—the ability to recognize conditions under which it is worthwhile to activate expression of the *lac* genes. The relief of repression for systems such as *lac* is termed **induction;** derivatives of lactose that inactivate the repressor and lead to expression of the *lac* genes are termed **inducers.**

Other bacterial systems operate by using protein "activator" molecules, which must bind to DNA as a prerequisite of transcription. Still additional mechanisms of control require proteins that allow the continuation of transcription in response to intracellular signals. Before we examine some of these control circuits in detail, let's review the classic work that initially described bacterial control systems, for these studies are landmarks in the use of genetic analysis.

Discovery of the *lac* System: Negative Control

The first major breakthrough in understanding gene control came in the 1950s with the detailed genetic analysis, by François Jacob and Jacques Monod, of the enzymes concerned with lactose metabolism in *E. coli* and of λ phage immunity. Jacob and Monod used the lactose metabolism system of *E. coli* (see Figure 18-2) to attack the problem of enzyme induction (originally termed *adaptation*)—that is, the appearance of a specific enzyme only in the presence of its substrates. This phenomenon had been observed in bacteria for many years. How could a cell possibly "know" precisely which enzymes to synthesize? How could a particular substrate induce the appearance of a specific enzyme?

For the *lac* system, such an induction phenomenon could be illustrated when, in the presence of certain galactosides termed *inducers,* cells produced over 1000 times more of the enzyme β-galactosidase, which cleaves β-galactosides, than they produced when grown in the absence of such sugars. What role did the inducer play in the induction phenomenon? One idea was that the inducer was simply activating a pre-β-galactosidase intermediate that had accumulated in the cell. However, when Jacob and Monod followed the fate of radioactively labeled amino acids added to growing cells either before or after the addition of an inducer, they could show that induction represented the synthesis of new enzyme molecules. Kinetic studies established that these molecules could be detected as early as three minutes after addition of an inducer. Also, withdrawal of the inducer brought about an abrupt halt in the synthesis of the new enzyme. Therefore, it became clear that the cell has a mechanism for turning on and off gene expression in response to environmental signals. What could this mechanism be?

Jacob and Monod wanted to determine the correlation between the catalytic center on the enzyme being induced and the molecular structure of the inducer. At that time, the notion was prevalent that the inducer (the small β-galactoside molecule) could instruct the formation of the catalytic center, in some way serving as a mold for the active site of the enzyme. It was found that although only galactosides serve as inducers, there is no correlation between the inducing capacity of a compound and its affinity for β-galactosidase. Figure 18-3 shows some of the researchers' actual data. Note for instance how phenylethylthiogalactoside has a very high affinity for β-galactosidase and yet is a poor in-

Compound	Concentrations	β-Galactosidase			Galactoside transacetylase
		Induction value	V	$1/K_m$	Induction value
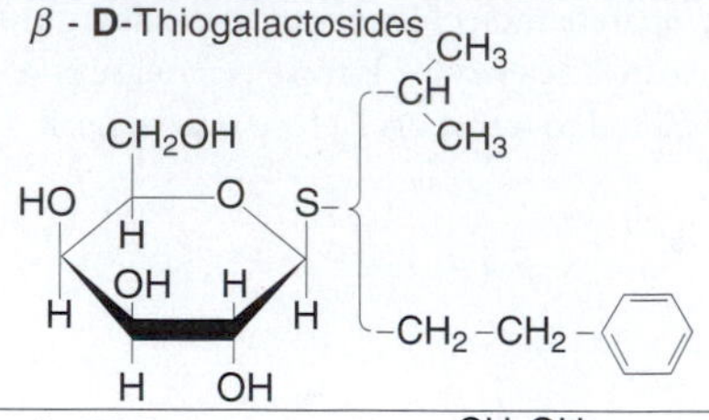 β - **D**-Thiogalactosides	(isopropyl) 10^{-4}M	100	0	140	100
	(phenylethyl) 10^{-3}M	5	0	10,000	3
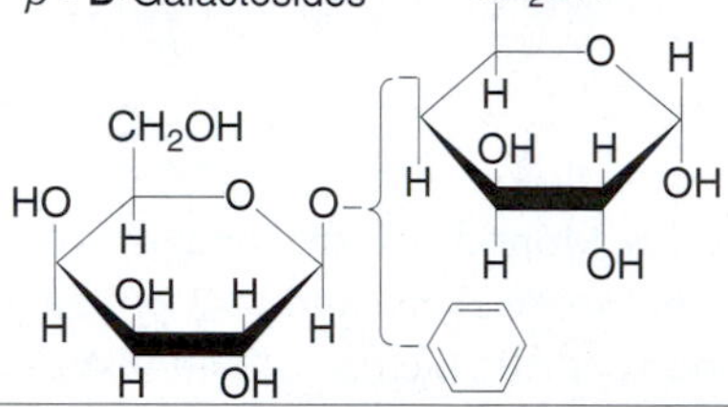 β - **D**-Galactosides	(lactose) 10^{-3}M	17	30	14	12
	(phenyl) 10^{-3}M	15	100	100	11

Figure 18-3 Induction of β-galactosidase and galactoside transacetylase by various galactosides. The data are taken directly from the original paper in 1961 by Jacob and Monod. The induction values represent the specific activities of each enzyme, given as the percentage of the maximum induced value. The V column refers to the maximal substrate activity of each compound with respect to β-galactosidase. The $1/K_m$ column expresses the affinity of each compound with respect to β-galactosidase. Both values are given as percentages of the values observed with the reference compound phenylgalactoside. (From F. Jacob and J. Monod, *Journal of Molecular Biology 3*, 1961, 318.)

ducer. Some strong inducers, such as the synthetically prepared isopropyl-β-D-thiogalactoside (IPTG; see Figure 18-4), are not even substrates for the enzyme. The different stereospecificities for inducing β-galactosidase versus for binding it are clues that the element involved in controlling β-galactosidase synthesis is distinct from β-galactosidase itself.

Genes Controlled Together

When Jacob and Monod induced β-galactosidase, they also induced the enzyme permease, which is required to transport lactose into the cell. The analysis of mutants indicated that each enzyme was encoded by a different gene. The enzyme transacetylase (with a dispensable and as-yet-un-

Figure 18-4 Structure of the inducer of the *lac* operon, IPTG. The β-D-thiogalactoside linkage is not cleaved by β-galactosidase, allowing manipulation of the intracellular concentration of this inducer.

known function) also was characterized and later shown to be encoded by a separate gene, although it was induced together with β-galactosidase and permease (see the right-hand column in Figure 18-3). Therefore, Jacob and Monod could identify three **coordinately controlled genes:** the Z gene encoding β-galactosidase, the Y gene encoding permease, and the A gene encoding transacetylase. Mapping defined the Z, Y, and A genes as being closely linked on the chromosome. Later studies of these and other coordinately controlled genes showed that in many cases a polycistronic mRNA molecule is produced by a contiguous set of genes. Transcription of this mRNA and its translation into protein proceeded in the same direction (see also Chapter 13). This enlightens us about an additional class of mutation referred to as **polar mutations.** Polar mutations not only affect the gene within which they map but also reduce or eliminate the expression of all genes that are farther down the line, or "distal" to the gene containing the mutation. Polar mutations exert their effects by interfering with either continued transcription or translation of the polycistronic mRNA. The first type of polar mutation characterized was the chain-terminating nonsense mutation located early in the Z gene, which lowers the expression of permease and transacetylase. Other polar mutations turned out to result from the insertion of DNA into the middle of genes (see Chapter 21). It should be noted, however, that only some of the mutations in a gene are polar.

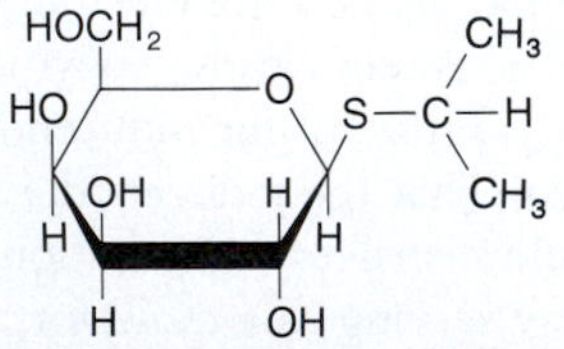

The *I* Gene

Further genetic analysis shed more light on the control circuit. Jacob and Monod characterized a new class of mutant, which synthesized all three enzymes at full levels, even in the absence of an inducer. For the first time, a mutant had been found with a defect not in the *activity* of an enzyme but in the control of enzyme *production*. These **constitutive** (always expressed in an unregulated fashion) **mutants** were found to have mutations mapping close to but distinct from the *Z*, *Y*, and *A* genes, permitting the definition of the *I* locus as the region controlling the inducibility of the *lac* enzymes. I^+ cells synthesize full levels of the *lac* enzymes only in the presence of an inducer, whereas I^- cells synthesize full levels in the presence or absence of an inducer. Figure 18-5 depicts the *lac* region defined by these experiments.

The Repressor

Arthur Pardee, together with Jacob and Monod, carried out an experiment which argued that the *I* gene determines the synthesis of a repressor molecule that blocks activation of the *lac* genes. This experiment, known as the *Pajamo experiment* (a composite of the authors' names), is depicted in Figure 18-6. Conjugation between an Hfr donor and an F⁻ recipient results in a transfer of a fragment of the Hfr chromosome into the F⁻. When the fragment carries the *lac* genes, the recipient cells become diploid for the *lac* genes. This partial diploid state persists for several hours, during which time measurements of gene activity can be carried out. In their experiment, the F⁻ recipients were $I^- Z^-$. They could not synthesize active β-galactosidase. Shortly after introduction of an $I^+ Z^+$ chromosome segment from the donor Hfr, the recipient cells gained the capacity to synthesize β-galactosidase. As can be seen from Figure 18-6, however, after about an hour, the synthesis of β-galactosidase stopped. If an inducer such as IPTG was added at this point, then the synthesis of β-galactosidase continued. In other words, after about an hour, β-galactosidase synthesis became inducible. The simplest interpretation of these data was that the *I* gene product was absent in the cytoplasm of the $I^- Z^-$ recipient, so that when the $I^+ Z^+$ chromosome was introduced, β-galactosidase could begin to be synthesized at the maximal rate. However, the *I* gene product could also be synthesized, and after an hour the concentration of I product in the cytoplasm was sufficient to be able to block, or "repress," β-galactosidase syn-

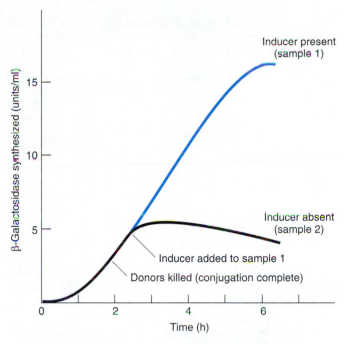

Figure 18-6 The Pajamo experiment. After the conjugation of $I^+ Z^+$ (donor) cells with $I^- Z^-$ (recipient) cells in the absence of inducer, the donors were killed. The merozygotes ($I^+ Z^+ / I^- Z^-$) produced β-galactosidase even without inducer up to about 2 h (sample 2), and then enzyme synthesis ceased. Only if inducer was added (sample 1), did enzyme synthesis continue. [See A. B. Pardee, F. Jacob, and J. Monod, *Journal of Molecular Biology* 1, 1959, 165.)

thesis. The addition of IPTG inactivated the *I* gene product, allowing induction.

The discovery of F′ factors (see Chapter 10) carrying the *lac* region, allowed the construction of stable partial diploids, and facilitated more direct complementation tests. Recall the complementation tests performed by Seymour Benzer (Chapter 12), in which complementation occurring in the trans position implies the action of a diffusable product. Tests with I^+ and I^- genes showed that I^+ is dominant over I^- in the trans position, strengthening the conclusion that the *I* gene product acted through the cytoplasm as a repressor. Table 18-1 shows the effect of various combinations of mutations, in the induced and noninduced state, on the production of β-galactosidase and permease.

A piece of evidence in support of the repressor model was the characterization of I^s mutations. Although mapping within the *I* gene, these mutations prevented induction of the *lac* enzymes by lactose or by the synthetic inducer IPTG (Figure 18-4). Moreover, they were dominant in trans to both an I^+ and an I^- allele (Table 18-2). The I^s mutation eliminates response to an inducer, presumably by altering the stereospecific binding site and destroying inducer binding. Therefore, even in the presence of IPTG, these molecules can still block *lac* enzyme synthesis. This would also explain their dominance, since the I^s repressor would be active, even in the presence of the wild-type repressor that

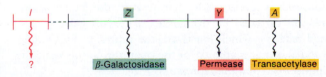

Figure 18-5 The *I* locus: the region controlling the inducibility of the *lac* enzymes.

Table 18-1　　Synthesis of β-Galactosidase and Permease in Haploid and Heterozygous Diploid Strains

Strain	Genotype	β-Galactosidase Noninduced	β-Galactosidase Induced	Permease Noninduced	Permease Induced
1	$I^+ Z^+ Y^+$	−	+	−	+
2	$I^- Z^+ Y^+$	+	+	+	+
3	$I^+ Z^- Y^+/FI^- Z^+ Y^+$	−	+	−	+
4	$I^- Z^- Y^+/FI^+ Z^+ Y^-$	−	+	−	+
5	$I^- Z^- Y^+/FI^- Z^+ Y^+$	+	+	+	+
6	$\nabla(I, Z, Y)/FI^- Z^+ Y^+$	+	+	+	+

NOTE: Bacteria were grown in glycerol as a carbon source and induced by IPTG. The presence of the maximal level of the enzyme is indicated by "+"; the absence or very low level of an enzyme is indicated by "−"; "∇" indicates deletion.

was inactivated by the inducer. The I^s mutations clearly pointed to a direct interaction between the I gene product and the inducer.

Message　The I^- mutation affects the *DNA-binding region* of the repressor, thus preventing binding and allowing transcription, even in the *absence* of inducer; the I^s mutation affects the *inducer-binding region* of the repressor, thus repressing transcription, even in the *presence* of inducer.

The Operator and the Operon

The specificity of the repressor, which results in turning off *lac* enzyme synthesis, suggests a stereospecific complex with an element that Jacob and Monod termed the *operator*. The operator was postulated to be a region of DNA near the beginning of the set of genes it controlled. The researchers sought mutations in the operator that would allow synthesis of the *lac* enzymes even in the presence of an active repressor. These mutations should be dominant in the cis position. Whereas trans dominance reflects a diffusible product, cis

Table 18-2　Synthesis of β-Galactosidase and Permease by the Wild Type and by Stains Carrying Different Alleles of the I Gene

Genotype	Inducer	β-Galactosidase	Permease
$I^+ Z^+ Y^+$	None	−	−
	IPTG	+	+
$I^s Z^+ Y^+$	None	−	−
	IPTG	−	−
$I^s Z^+ Y^+/FI^+$	None	−	−
	IPTG	−	−
$I^s Z^+ Y^+/FI^-$	None	−	−
	IPTG	−	−

NOTE: Bacteria were grown in glycerol with and without the inducer IPTG. Presence of the indicated enzyme is represented by "+"; absence or low levels, by "−."

dominance reflects the action of an element that affects only the genes directly adjacent to it. No diffusible product is altered by the mutation. By selecting for constitutivity (unrepressed synthesis) in cells with two copies of the *lac* region, to eliminate the effects of single I^- mutations, such mutations were detected and labeled O^c, for **operator constitutive.** As Table 18-3 indicates, strains carrying these mutations are capable of synthesizing maximal amounts of enzyme in the presence of IPTG, but can also synthesize 10 to 20 percent of these levels in the absence of an inducer. The O^c mutations are indeed dominant in the cis position, as shown in Table 18-3. Mapping experiments have pinpointed the operator locus between I and Z.

As we have seen, the *OZYA* segment constitutes a genetic unit of coordinate expression that Jacob and Monod termed the *operon*. Figure 18-7 depicts a simplified operon model for the *lac* system. The *lac* operon is said to be under the **negative control** of the *lac* repressor, since the repressor normally blocks expression of the *lac* enzymes in the absence of an inducer.

Let's review the model in Figure 18-7, as postulated by Jacob and Monod. The Z and Y genes code for the structure of two enzymes required for the metabolism of the sugar lactose, β-galactosidase and permease, respectively. The A gene codes for transacetylase. All three genes are linked together on the chromosome. Their transcription into a polycistronic (single) mRNA provides the basis for coordinate control at the level of mRNA synthesis. The synthesis of the polycistronic *lac* mRNA can be blocked by the action of a repressor protein molecule, which binds to an operator region near the start point for transcription. The repressor is the product of the I gene. Therefore, mutations in the I gene that prevent the synthesis of a functional repressor result in unrepressed, or constitutive, synthesis of the *lac* enzymes. Repression can also be overcome by certain galactosides, termed *inducers,* which inactivate the repressor by binding to it and altering its affinity for the operator. In this manner, the inducer can pull the repressor off the DNA.

We can now understand the properties of some of the diploids used for complementation tests, in light of the

Table 18-3 Synthesis of β-Galactosidase and Permease in Haploid and Heterozygous Diploid Operator Mutants

Genotype	β-Galactosidase		Permease	
	Noninduced	Induced	Noninduced	Induced
$O^+ Z^+ Y^+$	−	+	−	+
$O^+ Z^+ Y^+/FO^+ Z^- Y^+$	−	+	−	+
$O^c Z^+ Y^+$	+	+	+	+
$O^+ Z^+ Y^-/FO^c Z^+ Y^+$	+	+	+	+
$O^+ Z^+ Y^+/FO^c Z^- Y^+$	−	+	+	+
$O^+ Z^- Y^+/FO^c Z^+ Y^-$	+	+	−	+
$I^s O^+ Z^+ Y^+/FI^+ O^c Z^+ Y^+$	+	+	+	+

NOTE: Bacteria were grown in glycerol with and without the inducer IPTG. The presence and absence of enzyme are indicated by + and −, respectively.

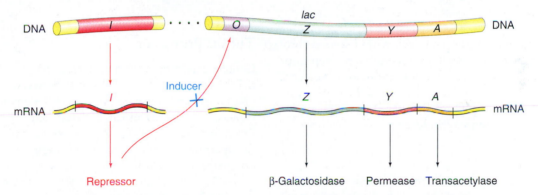

Figure 18-7 A simplified *lac* operon model. The three genes Z, Y, and A are coordinately expressed. The product of the *I* gene, the repressor, blocks the expression of the Z, Y, and A genes by interacting with the operator (O). The inducer can inactivate the repressor, thereby preventing interaction with the operator. When this happens, the operon is fully expressed. Mutations in *I* or *O* can also result in expression of the three *lac* enzymes, even in the absence of an inducer.

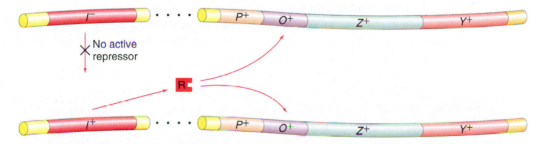

Figure 18-8 I^- mutations are recessive to wild-type. Although no active repressor is synthesized from the I^- gene, the wild-type (I^+) gene provides a functional repressor that binds to both operators in a diploid cell and blocks *lac* operon expression (in the absence of an inducer).

operon model. Figure 18-8 shows how I^- mutations are recessive to wild-type, because one functional *I* gene is all that is needed to produce a repressor that can bind to both operators in a diploid. On the other hand, Figure 18-9 shows a diploid cell carrying one copy of a wild-type *I* gene and one copy of an I^s gene. The I^s mutation alters the inducer binding site so that repressor no longer binds to inducer, although it still recognizes the operator. These diploid cells will be Lac⁻, because the altered repressor will always bind to the operator, even in the presence of an inducer, and block synthesis of the *lac* enzymes.

Operator mutations (O^c) are cis-dominant, because they are dominant only for genes directly linked to them on the same chromosome, as diagrammed in Figure 18-10. Thus, if

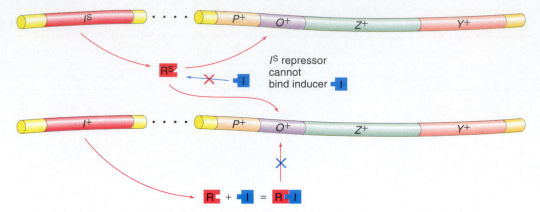

Figure 18-9 I^s mutations are dominant to wild-type. Even though the wild-type repressor is inactivated by an inducer, the I^s repressor is not and can bind to both operators in a diploid cell. Therefore, no enzyme is produced in an I^s/I^+ diploid, even in the presence of an inducer.

an altered operator is in the same cell with a second chromosome that contains a wild-type operator, the repressor will recognize the wild-type operator and repress the genes linked to it, but will not recognize the altered operator. Therefore, the genes linked to the altered operator are expressed even in the absence of inducer.

Allostery

The *lac* repressor is a protein with two different binding sites. One site recognizes the inducer molecule; the second site recognizes the *lac* operator sequence on the DNA. Interaction of the repressor with the inducer lowers the affinity of the repressor for the operator. This change in affinity for operator is mediated by a conformational change in the repressor protein. In one conformation, the repressor binds the operator well; in a second conformation, it does not. Proteins that function this way are termed **allosteric** proteins. Allosteric transitions, the change from one conformation to another, occur in many different proteins.

The *lac* Promoter

Genetic experiments demonstrated that an element essential for *lac* transcription is located between *I* and *O* in the operon model for the *lac* system. This element, termed the *promoter* (**P**), serves as an initiation site for transcription. Promoter mutations affect the transcription of all genes in the operon in a similar manner. Promoter mutations are cis-dominant, as would be expected for a site on the DNA that serves as a recognition element for transcription initiation, since each promoter governs transcription only for those genes in the operon adjacent to it on the *same* DNA molecule. As outlined in Chapter 13, in vitro experiments have demonstrated that RNA polymerase binds to the promoter region and that repressor binding to the operator can block RNA polymerase from binding to the promoter. Mutant analysis, physical experiments, and comparisons with other promoters have identified two binding regions for RNA polymerase in a typical prokaryotic promoter. Figure 18-11 summarizes this body of information (see also Figure 13-8).

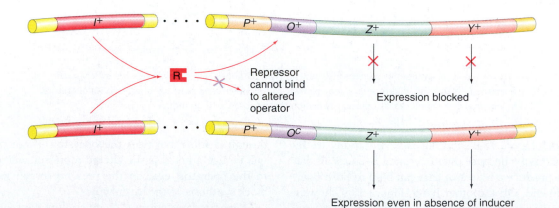

Figure 18-10 O^c mutations are dominant in the cis position. Because a repressor cannot bind to O^c operators, genes linked to an O^c operator are expressed even in the absence of an inducer. However, genes linked to an O^+ operator are still subject to repression.

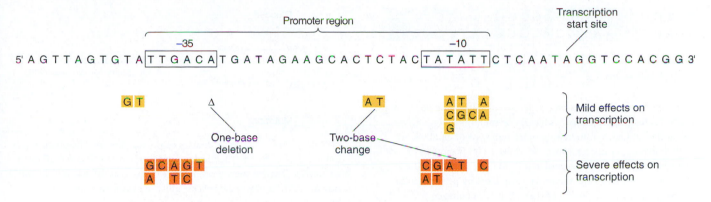

Figure 18-11 Specific DNA sequences are important for efficient transcription of *E. coli* genes by RNA polymerase. The boxed sequences at approximately 35 and 10 nucleotides before the transcription start site are highly conserved in all *E. coli* promoters. Mutations in these regions have mild (gold) and severe (orange) effects on transcription. The mutations may be changes of single nucleotides or pairs of nucleotides, or a deletion (Δ) may occur. (From J. D. Watson, M. Gilman, J. Witkowski, and M. Zoller, *Recombinant DNA*, 2d ed. Copyright © 1992 by Scientific American Books.)

Message The *lac* operon is a cluster of structural genes that specify enzymes involved in lactose metabolism. These genes are controlled by the coordinated actions of cis-dominant promoter and operator regions. The activity of these regions is, in turn, determined by a repressor molecule specified by a separate regulator gene. Figure 18-1 integrates all this information into a single picture.

Characterization of the *lac* Repressor and the *lac* Operator

Several genetic experiments have argued strongly that the repressor is a protein—instead of, for instance, an RNA molecule—the most compelling of which was the discov-

ery of suppressible nonsense mutations in the *I* gene, since the resulting nonsense codons exert their effect by signaling termination of the polypeptide chain during translation. The decisive experiment, however, was provided by Walter Gilbert and Benno Müller-Hill, who in 1966 isolated and purified the repressor by monitoring the binding of the radioactively labeled inducer IPTG. They demonstrated that the repressor is a protein consisting of four identical subunits, each with a molecular weight of approximately 38,000 daltons. Each molecule contains four IPTG-binding sites. (A more detailed description of the repressor is given later in the chapter.) In vitro, repressor binds to DNA containing the operator (see Figure 18-12) and comes off the DNA in the presence of IPTG. Gilbert and his co-workers have shown that the repressor can protect specific bases in

Figure 18-12 The Lac repressor (large whitish sphere) bound to DNA at the operator region of the operon. (Photograph courtesy of Jack D. Griffith.)

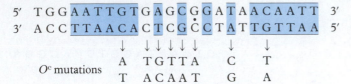

5′ TGG AATTGTGAGCGGATAACAATT 3′
3′ ACC TTAACACTCGCCTATTGTTAA 5′

O^c mutations

	A	TGTTA	C	T
	T	ACAAT	G	A

Figure 18-13 The DNA base sequence of the lactose operator and the base changes associated with eight *O^c* mutations. Regions of twofold rotational symmetry are indicated by color and by a dot at their axis of symmetry. (From W. Gilbert, A. Maxam, and A. Mirzabekov, in N. O. Kjeldgaard and O. Malløe, eds., *Control of Ribosome Synthesis*. Academic Press, 1976. Used by permission of Munksgaard International Publishers, Ltd., Copenhagen.)

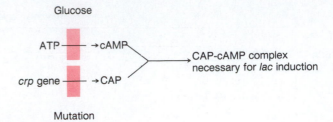

Figure 18-14 Catabolite control of the *lac* operon. The operon is inducible by lactose to the maximal levels when cAMP and CAP form a complex. The *lac* operon cannot be expressed at full levels if formation of cAMP is blocked by excess glucose or if formation of CAP is blocked by mutation of the *crp* gene. (CAP = catabolite activator protein; cAMP = cyclic adenosine monophosphate; *crp* = structural gene responsible for synthesizing CAP.)

the operator from chemical reagents. These experiments provide crucial proofs of the mechanism of repressor action formulated by Jacob and Monod.

Gilbert used the enzyme DNase to break apart the DNA bound to the repressor. He was able to recover short DNA strands that had been shielded from the enzyme activity by the repressor molecule and that presumably represented the operator sequence. This sequence was determined, and each operator mutation was shown to involve a change in the sequence (Figure 18-13). These results confirm the identity of the operator locus as a specific sequence of 17 to 25 nucleotides situated just before the structural Z gene. They also show the incredible specificity of repressor-operator recognition, which is disrupted by a single base substitution. When the sequence of bases in the *lac* mRNA (transcribed from the *lac* operon) was determined, the first 21 bases on the 5′ initiation end proved to be complementary to the operator sequence Gilbert had determined.

Catabolite Repression of the *lac* Operon: Positive Control

There is an additional control system superimposed on the repressor-operator system. This system exists because cells have specific enzymes that favor glucose uptake and metab-

olism. If both lactose *and* glucose are present, synthesis of β-galactosidase is not induced until all the glucose has been utilized. Thus, the cell conserves its metabolic machinery (that, for example, induces the *lac* enzymes) by utilizing any existing glucose before going through the steps of creating new machinery to exploit the lactose. The operon model outlined previously will not account for the suppression of induction by glucose, so we must modify it.

Studies indicate that in fact some catabolic breakdown product of glucose (no exact identity is yet known) prevents activation of the *lac* operon by lactose, so this effect was originally called **catabolite repression.** The effect of the glucose catabolite is exerted on an important cellular constituent called *cyclic adenosine monophosphate (cAMP)* in a way we shall see shortly.

When glucose is present in high concentrations, the cAMP concentration is low; as the glucose concentration decreases, the concentration of cAMP increases correspondingly. The high concentration of cAMP is necessary for activation of the *lac* operon. Mutants that cannot convert ATP to cAMP cannot be induced to produce β-galactosidase, because the concentration of cAMP is not great enough to activate the *lac* operon. In addition, there are other mutants

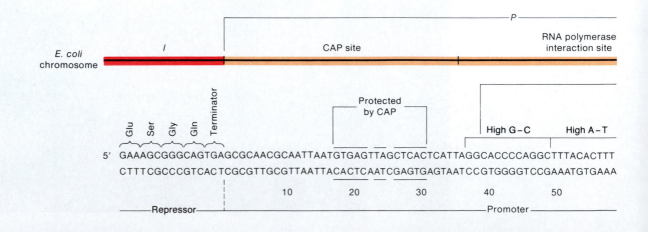

```
GTGAGTTAGCTCAC
            •
CACTCAATCGAGTG
```

Figure 18-15 The DNA base sequence to which the CAP-cAMP complex binds. Regions of twofold rotational symmetry are indicated by the colored boxes and by a dot at their axis of symmetry.

that do make cAMP but cannot activate the *lac* enzymes, because they lack yet another protein, called *CAP (catabolite activator protein),* made by the *crp* gene. The CAP protein forms a complex with cAMP, and it is this complex that activates the *lac* operon (Figure 18-14).

How does catabolite repression fit into our model for the structure and regulation of the *lac* operon? Recall the technique that Gilbert used to identify the operator base sequence. In a similar experiment, the CAP-cAMP complex was added to DNA, and the DNA then was subjected to digestion by the enzyme DNase. The surviving strands are presumably those shielded from digestion by an attached CAP-cAMP complex. The sequence of these strands (Figure 18-15) clearly is different from the operator sequence (Figure 18-13), but it also has a rotational twofold symmetry.

The entire *lac* operon can be inserted into λ phage in such a way that the initiation of transcription is prompted by the phage gene adjacent to *lac*. In this case, the transcribed product carries a complementary copy of the base sequence from the *lac* control regions (sequences not transcribed in the *lac* mRNA). We already know the amino acid sequences for the repressor and β-galactosidase, so these sequences can be identified, and the remaining sequences can be assigned to the control regions (Figure 18-16). We can also fit the known repressor, CAP-cAMP, and RNA polymerase binding sites into the detailed model, as shown in Figures 18-16 and 18-17. In Figure 18-17d, the DNA is shown as being bent when CAP binds. This may aid RNA polymerase binding to the promoter. There is also evidence to suggest that CAP makes direct protein-protein contacts with RNA polymerase, through the RNA polymerase α sub-

unit, that are important for the CAP activation effect (see Figure 18-17e).

Glucose control is accomplished because a glucose-breakdown product inhibits formation of the CAP-cAMP complex required for the RNA polymerase to attach at the *lac* promoter site. Even when there is a shortage of glucose catabolites and CAP-cAMP forms, the mechanism for lactose metabolism will be created only if lactose is present. This level of control is accomplished because lactose must bind to the repressor protein to remove it from the operator site and permit transcription of the *lac* operon. Thus, the cell conserves its energy and resources by producing the lactose-metabolizing enzymes only when they are both needed and useful.

Whereas inducer-repressor control of the *lac* operon is an example of negative control, in which expression is normally blocked, the CAP-cAMP system is an example of **positive control,** because its expression requires the presence of an activating signal—in this case, the interaction of the CAP-cAMP complex with the CAP region. Figure 18-18 distinguishes between these two basic types of control systems.

> **Message** The *lac* operon has an added level of control so that the operon remains inactive in the presence of glucose even if lactose is also present. High concentrations of glucose catabolites produce low concentrations of cyclic adenosine monophosphate (cAMP), which must form a complex with CAP to permit the induction of the *lac* operon.

By using different combinations of controlling elements, bacteria have evolved numerous strategies for regulating gene expression. Some examples follow.

Dual Positive and Negative Control: The Arabinose Operon

The metabolism of the sugar arabinose is catalyzed by three enzymes encoded by the *araB, araA,* and *araD* genes. Figure 18-19 depicts this operon. Expression is activated at *araI*, the

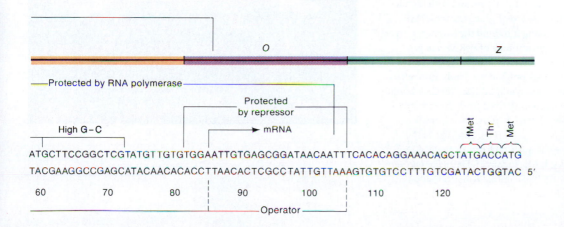

Figure 18-16 The base sequence and the genetic boundaries of the control region of the *lac* operon, with partial sequences for the structural genes. (After R. C. Dickson, J. Abelson, W. M. Barnes, and W. S. Reznikoff, "Genetic Regulation: The *Lac* Control Region," *Science* 187, 1975, 27. Copyright ©1975 by the American Association for the Advancement of Science.)

(a) Glucose present (cAMP); no lactose; no *lac* mRNA

Repressor

(b) Glucose present (cAMP low); lactose present

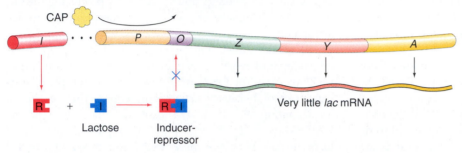

Very little *lac* mRNA

Lactose   Inducer-
repressor

(c) No glucose (cAMP high); lactose present

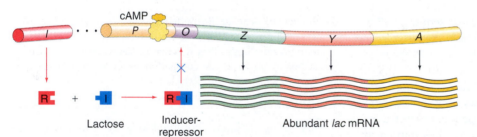

Lactose   Inducer-   Abundant *lac* mRNA
repressor

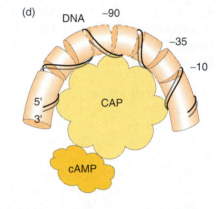

(d)   DNA   −90
                      −35
                      −10

5'   CAP
3'

cAMP

Figure 18-17 Negative and positive control of the *lac* operon by the *lac* repressor and catabolite activator protein (CAP), respectively. (a) In the absence of lactose to serve as an inducer, the *lac* repressor is able to bind the operator; regardless of the levels of cAMP and presence of CAP, mRNA production is repressed. (b) With lactose (or its metabolite *allolactose*) present to bind the repressor, the repressor is unable to bind the operator; however, only small amounts of mRNA are produced because the presence of glucose keeps the levels of cAMP low, and thus the cAMP-CAP complex does not form and bind the promoter. (c) With the repressor inactivated by lactose and with high levels of cAMP present (owing to the absence of glucose), the cAMP binds the CAP, activating it and enabling it to bind the promoter; the *lac* operon is thus activated, and large amounts of mRNA are produced. (d) When CAP binds the promoter, it creates a bend greater than 90° in the DNA. Apparently, RNA polymerase binds more effectively when the promoter is in this bent configuration. (e) CAP bound to its DNA recognition site. The figure is derived from the structural analysis of the CAP-DNA complex. [Parts a–d redrawn from B. Gartenberg and D. M. Crothers, *Nature* 333, 1988, 824. (See H. N. Lie-Johnson et al., *Cell* 47, 1986, 995.) Adapted from H. Lodish, D. Baltimore, A. Berk, S. L. Zipursky, P. Matsudaira, J. Darnell, *Molecular Cell Biology*, 3d ed. Copyright ©1995 by Scientific American Books. Part e from L. Schultz and T. A. Steitz.]

(e)

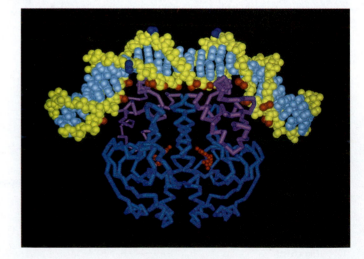

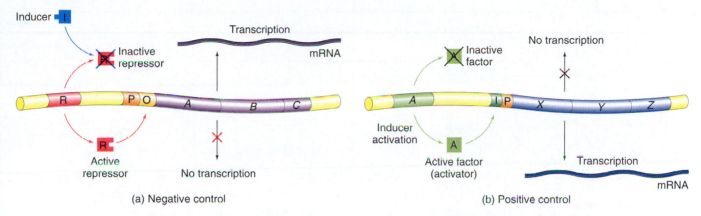

(a) Negative control   (b) Positive control

Figure 18-18 Comparison of positive and negative control. The basic aspects of negative and positive control are depicted. (a) In negative control, an active repressor (encoded by the *R* gene in the example shown here) blocks gene expression of the *A,B,C* operon by binding to an operator site (*O*). An inactive repressor allows gene expression. The repressor can be inactivated either by an inducer or by mutation. (b) In positive control, an active factor is required for gene expression, as shown for the *X,Y,Z* operon here. Small molecules can convert an inactive factor into an active one, as in the case of cyclic AMP and the CAP protein. An inactive positive control factor results in no gene expression. The activator binds to the control region of the operon, termed *I* in this case. (The positions of both *O* and *I* with respect to the promoter, *P*, in the two examples are arbitrarily drawn.)

Control gene   Control sites   Structural genes

Figure 18-19 Map of the *ara* region. The *BAD* genes together with the *I* and *O* sites constitute the *ara* operon.

initiator region. Within this region, the product of the *araC* gene, when bound to arabinose, can activate transcription, perhaps by directly affecting RNA polymerase binding in the *araI* region, which contains the promoter for the *araB, araA,*

and *araD* genes. This represents positive control, since the product of the regulatory gene (*araC*) must be active in order for the operon to be expressed. An additional positive control is mediated by the same CAP-cAMP system that regulates *lac* expression.

In the presence of arabinose, both CAP protein and binding of the C product to the initiator region are required to allow RNA polymerase to bind to the promoter for the *araB, araA,* and *araD* genes (Figure 18-20a). In the absence of arabinose, the *araC* product assumes a different conformation and actually represses the *ara* operon by binding both to *araI* and to an operator region, *araO,* thereby forming a loop (Figure 18-20b) that prevents transcription. Thus, the

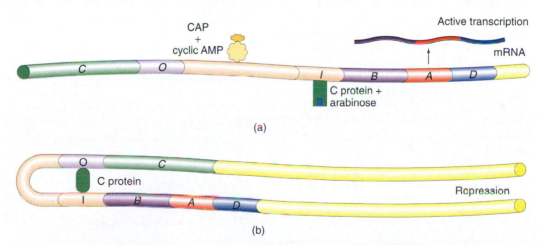

Figure 18-20 Dual control of the *ara* operon. (a) In the presence of arabinose, the *araC* protein binds to the *araI* region, and when bound to cyclic AMP, the CAP protein binds to a site adjacent to *araI*. This stimulates the transcription of the *araB, araA,* and *araD* genes. (b) In the absence of arabinose, the *araC* protein binds to both the *araI* and *araO* regions, forming a DNA loop. This prevents transcription of the *ara* operon.

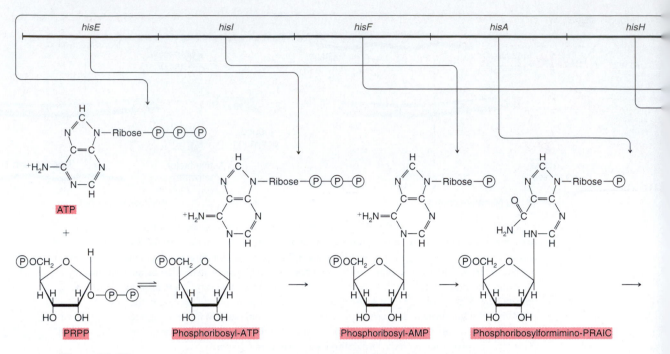

Figure 18-21 The histidine (*his*) gene cluster and the metabolic pathway that it controls. Note that the sequence of genes in the cluster generally corresponds to the sequence of steps that each gene catalyzes in the pathway for the synthesis of histidine. The fact that the final gene in the sequence (*hisG*) catalyzes the first step in the reaction sequence is probably not a coincidence; this pattern is commonly found. (℗ = phosphate; PRPP = phosphoribosylpyrophosphate; PRAIC = phosphoribosylaminoimidazole carboxamide.)

araC protein has two conformations that promote two opposing functions at two alternative binding sites. The conformation is dependent on whether the inducer, arabinose, is bound to the protein.

Metabolic Pathways

Coordinate control of genes in bacteria is widespread. In the early 1960s, when Milislav Demerec studied the distribution of loci affecting a common biosynthetic pathway, he found that the genes controlling steps in the synthesis of the amino acid tryptophan in *Salmonella typhimurium* are clustered together in a restricted part of the genome. Demerec then looked at the distribution of genes involved in a number of different metabolic pathways. Analyzing auxotrophic mutations representing 87 different cistrons, he found that 63 could be located in 17 functionally similar clusters. A **cluster** is defined as two or more loci that control related functions, where the loci are carried on a single transducing fragment and are not separated by an unrelated gene.

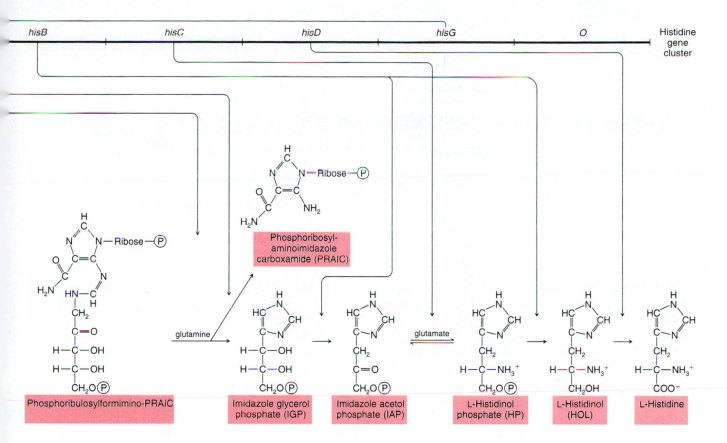

Furthermore, in cases where the sequence of catalytic activity is known, there is a remarkable congruence between the sequence of genes on the chromosome and the sequence in which their products act in the metabolic pathway. This congruence is strikingly illustrated by the histidine cluster in *Salmonella,* extensively studied in the early 1960s by Philip Hartman and Bruce Ames (Figure 18-21), and by the tryptophan cluster in *E. coli* (Figure 18-22), characterized during the same period by Charles Yanofsky.

Message   Genes involved in the same metabolic pathway are frequently tightly clustered on prokaryotic chromosomes, often in the same sequence as the reactions that they control. Furthermore, the genes within a cluster often are expressed at the same time.

The Tryptophan Genes: Negative Control with Superimposed Attenuation

The *lac* operon is an example of an inducible system in the sense that the synthesis of an enzyme is induced by the presence of its substrate. Repressible systems also exist, in which an excess of product leads to a shutdown of the pro-

Figure 18-22   The genetic sequence of cistrons in the *trp* operon of *E. coli* and the sequence of reactions catalyzed by the enzyme products of the *trp* structural genes. The products of genes *trpD* and *trpE* form a complex that catalyzes specific steps, as do the products of genes *trpB* and *trpA*. Tryptophan synthetase is a tetrameric enzyme formed by the products of *trpB* and *trpA*. It catalyzes a two-step process leading to the formation of tryptophan. (PRPP = phosphoribosylpyrophosphate; CDRP = 1-(*o*-carboxyphenylamino)-1-deoxyribulose 5-phosphate.) (After S. Tanemura and R. H. Bauerle, *Genetics* 95, 1980, 545–559.)

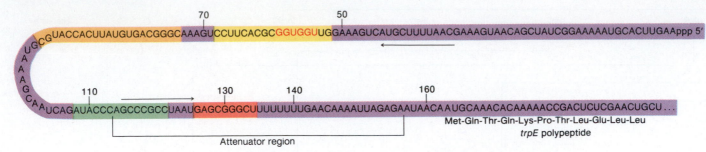

Figure 18-23 The leader sequence, showing the attenuator segment of the *trp* operon, along with the beginning of the *trpE* structural sequence (showing the amino acid sequence of the *trpE* polypeptide). For reference purposes, the color scheme for the various segments is repeated in Figures 18-25 and 18-26. (Modified from G. S. Stent and R. Calendar, *Molecular Genetics*, 2d ed. Copyright ©1978 by W. H. Freeman and Company. Based on unpublished data provided by Charles Yanofsky.)

duction of the enzymes involved in synthesizing that product. Such a control system has been identified for a cluster of genes controlling enzymes in the pathway for tryptophan production. Synthesis of tryptophan is shut off when there is an excess of tryptophan in the medium. Jacob and Monod suggested that the cluster of five *trp* cistrons in *E. coli* forms another operon, differing from the *lac* operon in that the tryptophan repressor will bind to the *trp* operator only when it *is* bound to tryptophan (Figure 18-22). (Recall that the *lac* repressor binds to the operator *except* when it is bound to lactose.) A second control pathway also modulates tryptophan biosynthesis at the level of enzyme activity. This is termed **feedback inhibition.** Here, the first enzyme in the pathway, encoded by the *trpE* and *trpD* genes, is inhibited by tryptophan itself.

As with the *lac* operon, further analysis of the *trp* operon has revealed yet another level of control superimposed on the basic repressor-operator mechanism. In studying mutant strains (carrying a mutation in *trpR*, the repressor locus) that continue to produce *trp* mRNA in the presence of tryptophan, Yanofsky found that removal of tryptophan from the medium leads to almost a tenfold increase in *trp* mRNA production in these strains, even though the Trp repressor was inactive, and thus could not account for the increase through normal derepression of the operator because of low tryptophan levels. Furthermore, Yanofsky identified the region responsible for this increase by isolating a totally constitutive mutant strain that produces *trp* mRNA at this tenfold maximal level, even in

the presence of tryptophan, and he showed that this mutation has a deletion located between the operator and the *trpE* cistron (see the map in Figure 18-22).

Yanofsky was able to isolate the polycistronic *trp* operon mRNA. On sequencing it, he found a long sequence, termed the **leader sequence,** of approximately 160 bases at the 5′ end before the first triplet in the *trpE* gene. The deletion mutant that always produces *trp* mRNA at maximal levels has a deletion extending from base 130 to base 160 (Figure 18-23). Yanofsky called the element inactivated by the deletion the **attenuator,** because its presence apparently leads to a reduction in the rate of mRNA transcription when tryptophan is present. Figure 18-24 shows the position of these elements in the *trp* operon. But what is the role of the leader sequence in bases 1 to 130? A surprising observation provides the key to solving this problem.

While studying mRNAs transcribed by the *trp* operon (using *trpR⁻* mutants), Yanofsky discovered that even in the presence of high levels of tryptophan (which should cause the attenuator region to reduce the rate of transcription tenfold), the first 141 bases of the leader sequence were always transcribed at the maximal rate, though the full-length mRNA occurred, as expected, at levels only one-tenth as great. Another way of stating this is that, even in the presence of high concentrations of tryptophan, the first 141 bases are transcribed in maximal numbers, but because of some attenuating mechanism in this region only 1 in 10 of the mRNAs can be transcribed farther (to completion). This suggests that the end of the attenuator acts as a mRNA

Figure 18-24 Diagram of the *trp* operon showing the promoter (*P*), operator (*O*), and attenuator (*A*) control sites and the genes for the leader sequence (*L*) and enzymes of the tryptophan pathway (*E, D, C, B,* and *A*). (Adapted from L. Stryer, *Biochemistry*, 4th ed. Copyright ©1995 by Lubert Stryer.)

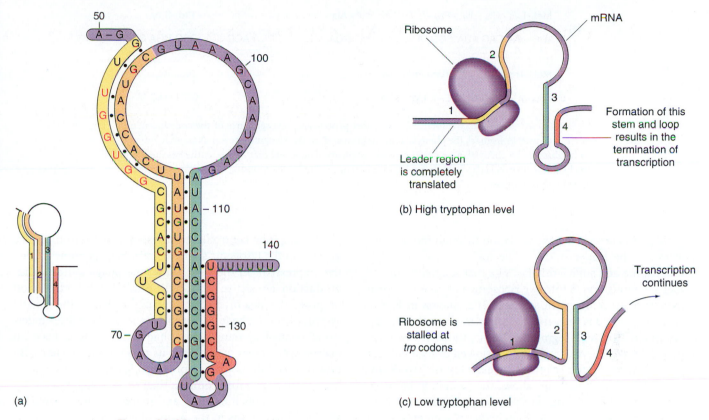

Figure 18-25 Model for attenuation in the *trp* operon. (a) Proposed secondary structures in *E. coli* terminated *trp* leader RNA. Four regions can base-pair to form three stem-and-loop structures. (b) When tryptophan is abundant, segment 1 of the *trp* mRNA is fully translated. Segment 2 enters the ribosome (although it is not translated), which enables segments 3 and 4 to base-pair. This base-paired region somehow signals RNA polymerase to terminate transcription. In contrast, when tryptophan is scarce (c), the ribosome is stalled at the codons of segment 1. Segment 2 interacts with 3 instead of being drawn into the ribosome, and so segments 3 and 4 cannot pair. Consequently, transcription continues. (After D. L. Oxender, G. Zurawski, and C. Yanofsky, *Proc. Natl. Acad. Sci.* 76, 1979, 5524.)

chain terminator that, *in the presence of tryptophan,* halts transcription of 9 out of 10 mRNAs. In the absence of tryptophan, this attenuator is somehow deactivated, and every mRNA goes to completion—hence, the tenfold increase. In those *trpR⁻* mutants in which the attenuator is also deleted, there is no block to extension of the mRNA, so transcription is carried through in every case (that is, the maximal rate of production occurs), regardless of whether tryptophan is present or not.

What causes the interference with termination at the attenuator in the absence of tryptophan? Figure 18-25 pre-

sents a model based on alternative secondary structures formed by the mRNA in the leader region. The model proposes that one of the two conformations favors transcription termination and that the other favors elongation. Translation of part of the leader sequence would promote the conformation that favors termination.

It is known that a portion of the leader sequence near the beginning is in fact translated and yields a short peptide of 14 amino acids. There are two tryptophan codons in the translated stretch of the leader mRNA (Figure 18-26). When excess tryptophan is present, there is a sufficient supply of

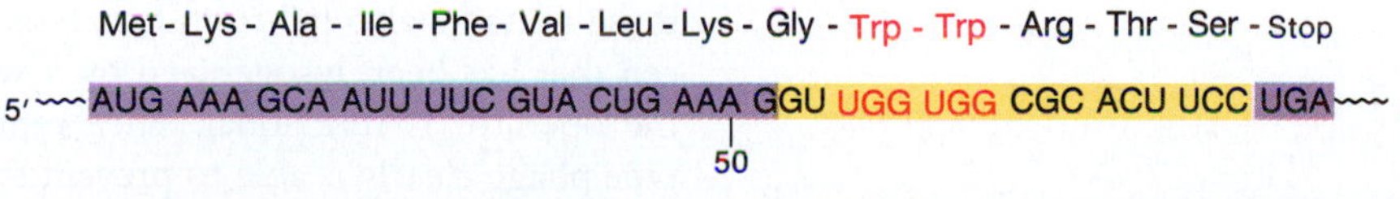

Met - Lys - Ala - Ile - Phe - Val - Leu - Lys - Gly - Trp - Trp - Arg - Thr - Ser - Stop

5′ ∿∿ AUG AAA GCA AUU UUC GUA CUG AAA GGU UGG UGG CGC ACU UCC UGA ∿∿ 3′

50

Figure 18-26 The translated portion of the *trp* leader region, shown with the corresponding sequence of the leader mRNA. Translation of the leader sequence ends at the stop codon.

(a)

Met - Lys - His - Ile - Pro - Phe - Phe - Phe - Ala - Phe - Phe - Phe - Thr - Phe - Pro - Stop

5' AUG AAA CAC AUA CCG UUU UUC UUC GCA UUC UUU UUU ACC UUC CCC UGA 3'

(b)

Met - Thr - Arg - Val - Gln - Phe - Lys - His - His - His - His - His - His - His - Pro - Asp -

5' AUG ACA CGC GUU CAA UUU AAA CAC CAC CAU CAU CAC CAU CAU CCU GAC 3'

Figure 18-27 Amino acid sequence of the leader peptide and base sequence of the corresponding portion of mRNA from (a) the phenylalanine operon and (b) the histidine operon. Note that 7 of the 15 residues in the phenylalanine operon leader are phenylalanine and that 7 consecutive residues from the histidine operon leader peptide are histidine. (From L. Stryer, *Biochemistry*, 4th ed. Copyright © 1995 by Lubert Stryer.)

Trp-tRNA to allow efficient translation through the relevant portion of the leader mRNA. The mRNA passes through the ribosome at a sufficiently fast rate that segment 2 of the leader section is drawn into the ribosome before it can form a loop-stem structure with segment 3, as shown in Figure 18-25b. Segment 3 is thus able to form a transcription-termination stem-loop with segment 4. In conditions of low tryptophan levels, however, translation is slowed in segment 1 at the Trp codons by the relative unavailability of Trp-tRNA. As Figure 18-25c shows, this allows the stem-loop between segments 2 and 3 to form, thus preventing the formation of the transcription-terminating loop between segments 3 and 4. Consequently, under conditions of low tryptophan, transcription is not stopped by the attenuator. In this manner, an additional tenfold range of tryptophan biosynthetic enzymes is superimposed on the normal range that is achieved by repressor-operator interaction. The analysis of numerous point mutations in the *trp* leader sequence that favor or disfavor the respective secondary structures lends strong support to Yanofsky's attenuation model.

Several operons for enzymes in biosynthetic pathways have attenuation controls similar to the one described for tryptophan (Figure 18-27). For instance, the leader region of the *his* operon, which encodes the enzymes of the histidine biosynthetic pathway, contains a translated region with seven consecutive histidine codons. Mutations at outside loci that result in lowering levels of normal charged *his* tRNA produce partially constitutive levels of the enzymes encoded by the *his* operon.

Message   The *trp* operon is regulated by a negative repressor-operator control system that represses the synthesis of tryptophan enzymes when tryptophan is present in the medium. A second level of control involves an attenuator region where termination of transcription is induced by the presence of tryptophan.

The λ Phage: A Complex of Operons

At the time they proposed the operon model, Jacob and Monod suggested that the genetic activity of temperate phages might be controlled by a system analogous to the *lac* operon. In the lysogenic state, the prophage genome is inactive (repressed). In the lytic phase, the phage genes for reproduction are active (induced). Since Jacob and Monod proposed the idea of operon control for phages, the genetic system of the λ phage has become one of the better-understood control systems. This phage does indeed have an operon-type system controlling its two functional states. By now, you should not be surprised to learn that this system has proved to be more complex than initially suggested.

Allan Campbell induced and mapped many conditionally lethal mutations in the λ phage (Figure 18-28), providing clear evidence for the clustering of genes with related functions. Furthermore, mutations in the *N*, *O*, and *P* genes prevent most of the genome from being expressed after phage infection, with only those loci lying between *N* and *O* being active. We shall soon see the significance of this observation.

When normal bacteria are infected by wild-type λ phage particles, two possible sequences may follow: (1) a phage may be integrated into the bacterial chromosome as an inert prophage (thus lysogenizing the bacterial cell), or (2) the phage may produce the necessary enzymes to guide the production of the products needed for phage maturation and cell lysis. When wild-type phage particles are placed on a lawn of sensitive bacteria, clearings (plaques) appear where bacterial cells are infected and lysed, but these plaques are turbid because lysogenized bacteria (which are resistant to phage superinfection) grow within the plaques.

Mutant phages that form clear plaques can be selected as a source of phages that are unable to lysogenize cells. Such *clear* (*c*) mutants prove to be analogous to *I* and *O* mutants in *E. coli*. For example, conditional mutants for a site called *cI* are unable to establish a lysogenic state under restrictive conditions, suggesting either a defective repressor or a defective repressor binding site (operator). However, these mutants also fail to induce lysis after superinfecting a cell that has been lysogenized by a wild-type prophage, so the operator is functional, since repressor from the wild-type phage clearly is able to prevent the mutant from entering the lytic phase. Apparently, the *cI* mutation produces a defective repressor in the phage control system.

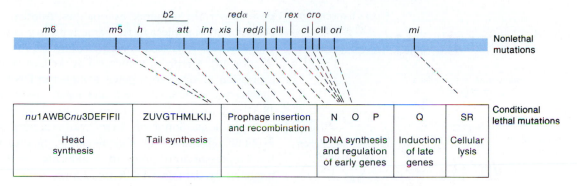

Figure 18-28 The genetic map of the λ phage. The positions of nonlethal and conditionally lethal mutations are indicated, and the characteristic clusters of genes with related functions are shown. (From A. Campbell, *The Episomes.* Harper & Row, 1969.)

Virulent mutant λ phage have also been isolated that do not lysogenize cells but that do grow and enter the lytic cycle after superinfecting a lysogenized cell. Thus, these mutant phage are insensitive to the λ repressor. Genetic mapping has revealed *two* operators, designated O_L and O_R, located on the left and right of *cI*, respectively. When the λ phage infects a nonimmune host, RNA polymerase begins transcription from the two promoters, P_L and P_R. This begins a race to determine whether the phage will express its virulent functions and enter the lytic cycle or whether it will repress these functions and establish lysogeny. Figure 18-29 depicts part of the map of λ and shows the extent of some of the different transcripts. P_L governs the leftward transcript that encodes the anti-terminator protein N, plus a number of lytic functions. P_R governs the rightward transcript, including the *cro, cII,* and *Q* genes, among others.

These transcripts normally terminate at specific critical points, before key genes necessary for lytic development can be transcribed. However, if the N protein is synthesized, it prevents termination of the transcript (see Figure 18-29), and the lytic functions are expressed. Transcription from P_L and P_R can be blocked by repressor binding to the operators O_L and O_R. In the former case this would prevent synthesis of the N protein and thus also block the extension of transcripts from P_L and P_R.

The critical point in establishing lysogeny is the synthesis of the λ repressor, encoded by the *cI* gene. The *cI* gene has two promoters: one, P_E, for establishing lysogeny, and the second, P_M, for maintaining lysogeny. Transcription initiated at each promoter requires an activator protein. For P_M the activator is the *cI*-encoded repressor itself (see below). Therefore, in order to synthesize repressor in a cell without

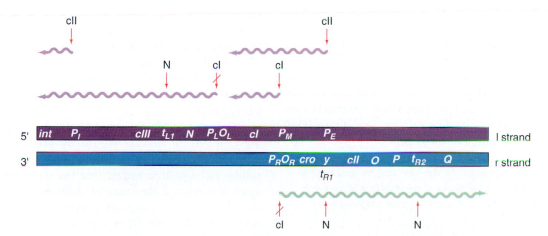

Figure 18-29 Partial genetic map of λ phage, showing the primary genes and proteins of regulation. The wavy lines represent mRNA and show the origin, direction, and extent of transcription. A vertical arrow at the *beginning* of an mRNA marks the site where the indicated gene product activates a promoter. If the arrow has a slash through it, the gene product represses transcription. Where an arrow marks a site *within* an mRNA, the indicated gene product prevents termination of transcription at the site. t_{L1} is a potential terminator for P_L, and t_{R1} is a potential terminator for P_R, when N does not act on them, as the diagram indicates. (Adapted from D. Wulff and M. Rosenberg, in R. W. Hendrix et al., eds., *Lambda II.* Cold Spring Harbor Laboratory, Cold Spring Harbor, N.Y., 1983.)

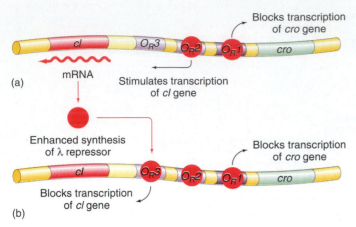

Figure 18-30 Self-regulation of cI repressor levels. (a) Binding of cI repressor to O_R1 both represses transcription of *cro* and potentiates the binding of a cI molecule at O_R2, which in turn stimulates transcription of cI repressor. (b) When cI repressor levels are high, the third site, O_R3, is bound by cI repressor. Since this overlaps the *cI* promoter, further production of cI repressor is blocked until its levels drop and the O_R3 site is again free. (Adapted from L. Stryer, *Biochemistry*, 4th ed. Copyright ©1995 by Lubert Stryer.)

preexisting repressor molecules, P_E must be used. P_E is activated by the *cII* product. Thus, *cII* must be transcribed to establish lysogeny. One additional protein is required to establish lysogeny, the *integrase*, encoded by *int*. The *int* gene is transcribed from its own promoter, P_I (Figure 18-29), which also requires activation by the *cII* gene product.

In the lysogenic state, the λ repressor controls its own synthesis from the promoter P_M by interacting with the three operators O_R1, O_R2, and O_R3, as diagrammed in Figure 18-30. These operators have different affinities for the λ repressor, with $O_R1 > O_R2 > O_R3$. When there is too little λ repressor, only the O_R1 site, which blocks *cro* expression, is filled. (The *cro* protein prevents synthesis of λ repressor by binding to O_R3.) The binding of repressor to O_R1 facilitates the binding of a second molecule of repressor to O_R2. When λ repressor binds to O_R2, it actually serves to activate transcription of the *cI* gene. However, when too much λ repressor is present, then binding to the low-affinity O_R3 blocks further transcription of the *cI* gene.

Message   The regulation of the lytic and lysogenic states of the λ phage provides a model for interacting control systems that may be useful for interpreting gene regulation in eukaryotes.

Multioperon Repression

Repressors can simultaneously control several or even a large number of operons. For instance, the Trp repressor simultaneously regulates the *trp* operon, the *aroH* gene, and the *trp* repressor gene itself. The three respective operators

show important sequence homologies. Another multicomponent system subject to the same repressor is the SOS repair pathway (discussed in conjunction with mutagenesis in Chapter 19). A repressor, encoded by the *lexA* gene, reduces expression of numerous genes involved in DNA repair, including the *recA, uvrA, uvrB,* and *umuC* genes (see Figure 18-31 and page 599). The LexA repressor also regulates its own synthesis. Some of these LexA-repressed genes still maintain a residual level of expression in the presence of the LexA repressor. In response to an inducing signal, somehow triggered by damaged DNA that blocks replication, a protease activity of the protein encoded by *recA* is activated. This results in the cleavage of the LexA repressor, allowing high levels of synthesis of the SOS functions. UV light and other agents that damage DNA induce the SOS functions. Interestingly, the RecA protease also cleaves the λ repressor (the product of the λ *cI* gene), as well as the repressors of several other phages related to λ. This is the physical basis for the classical UV-light induction of λ lysogens.

Structure of Regulatory Proteins

Protein sequence analyses and structural comparisons indicate that a number of DNA-binding regulatory proteins share important features. Many consist of a DNA-binding domain, located at one end of the protein, which protrudes from the main "core" of the protein. In certain cases, the core protein contains the inducer binding site. This arrangement (Figure 18-32) holds for the Lac, λ cI, and λ Cro repressors, as well as for the CAP protein.

It has been postulated that protruding α helices fit into the major groove of the DNA. These models have now been verified by the elucidation of the three-dimensional structure of the *lac* repressor binding to a synthetic operator (Figure 18-33). Here two α helices from the repressor protein interact with the two consecutive major grooves of the DNA of the operator site. The helices are connected by a turn in the protein secondary structure. This **helix-turn-helix motif** (Figure 18-34) is common to many bacterial regulatory proteins.

Striking partial-sequence homologies of the DNA-binding domains of several regulatory proteins suggest that other such proteins bind to their own operator sites in a similar fashion. It is hoped that the determination of the three-dimensional structure of different repressor-operator complexes by X-ray crystallography will allow the elucidation of the rules for DNA sequence recognition by regulatory proteins.

Transcription: Gene Regulation in Eukaryotes—An Overview

Activating specific genes in higher eukaryotic cells plays a key role during embryological development and tissue dif-

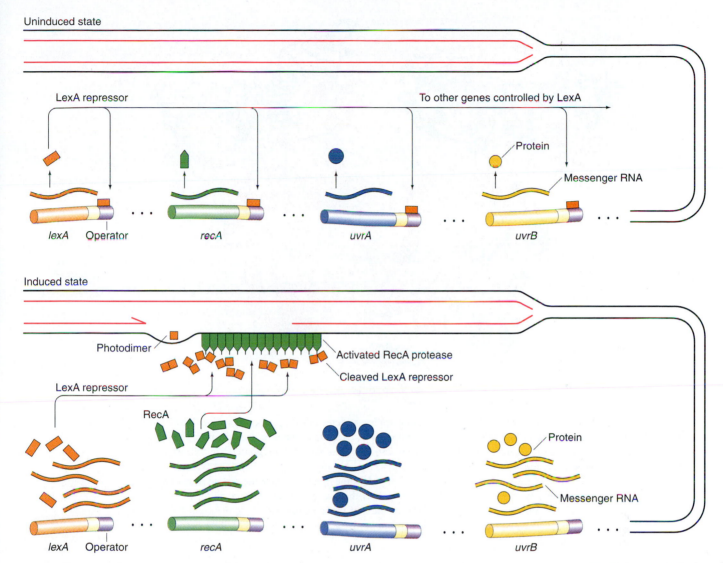

Figure 18-31 Regulatory system based on LexA and RecA. This system is quiescent during normal growth in the absence of damage to DNA (*top*). The LexA repressor binds to the operators of *lexA*, *recA*, *uvrA*, *uvrB*, and some other genes, keeping the synthesis of messenger RNA and protein at the low level characteristic of uninduced cells. Damage to DNA sufficient to produce a postreplication gap activates the SOS response (*bottom*). The RecA protein binds to the single-stranded DNA opposite the gap; its protein-cleaving activity is thereby activated, and the LexA repressor is cleaved. In the absence of a functional repressor, the LexA-controlled genes are switched on, and protein is synthesized at an increased rate. (From P. Howard-Flanders, "Inducible Repair of DNA." Copyright ©1981 by Scientific American, Inc. All rights reserved.)

ferentiation. Often, the turning on or off of sets of genes is not reversible, unlike the situation in bacteria, where gene activity is continually modulated in response to physiological conditions. There are exceptions, for instance, single-cell eukaryotes such as yeast or certain cells in multicellular organisms that do respond to changing environmental conditions. Liver cells, for example, induce detoxification enzymes in response to certain chemicals (see page 602).

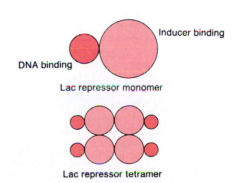

Figure 18-32 Schematic diagram showing the arrangement of domains in the Lac repressor. All mutations affecting DNA and operator binding result in alterations in the amino-terminal end of the protein, whereas mutants defective in inducer binding or aggregation result in alterations in the remaining portion of the protein.

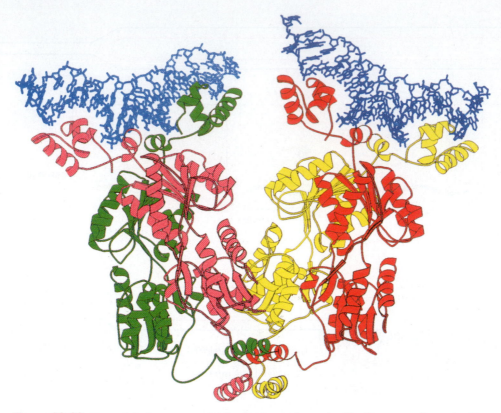

Figure 18-33 View of the *lac* repressor-DNA complex, as determined by X-ray crystallography. Here the repressor tetramer is shown binding to two operators. Each dimer binds to one operator. The operators are 21 base pairs long here and are shown in dark and light blue. The monomers are colored green, violet, red, and yellow. Note how the amino-terminal headpiece on one monomer (violet) crosses the other monomer (green) to bind the first part of the operator, while the second monomer headpiece (green) crosses over to bind the second part of the operator. Thus, the recognition helix from each of two subunits binds to the consecutive parts of the same operator. (Figure supplied by M. Lewis, Geoffrey Chang, N. C. Horton, M. A. Kercher, H. C. Pace, M. A. Schumacher, R. G. Brennan, and P. Lu, University of Pennsylvania).

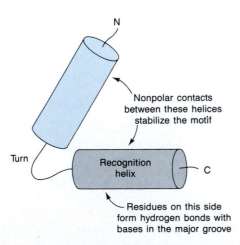

Figure 18-34 Helix-turn-helix motif of DNA-binding proteins. Each monomer of these dimeric proteins contains a helix-turn-helix; the two units are separated by 34Å — the pitch of the DNA helix. (From L. Stryer, *Biochemistry*, 4th ed. Copyright ©1995 by Lubert Stryer.)

Most genes are controlled at the level of transcription, and the mechanisms are similar in concept to those found for bacteria, in that cytoplasmic, or **trans-acting, factors** interact with control points on the DNA, or **cis-acting regulatory sequences.** However, the details are often more complicated. Regulatory sites on the DNA can be located far away from the transcription start site, and a battery of factors are sometimes needed to bring about complete activation of certain genes.

Cis Control of Transcription

As mentioned in Chapter 13, there are three different RNA (ribonucleic acid) polymerases in eukaryotic systems. All mRNA molecules are synthesized by RNA polymerase II. In order to achieve maximal rates of transcription, RNA polymerase II requires several sets of cis-acting control sequences: promoters and additional sequence elements termed **promoter-proximal elements** and **enhancers** in mammalian cells and **upstream activating sequences (UASs)** in yeast. Each of these elements is recognized by

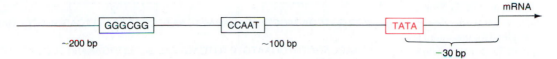

Figure 18-35 Promoter in higher eukaryotes. The TATA box is located approximately 30 base pairs (bp) from the mRNA start site. Usually, two or more promoter-proximal elements are found 100 and 200 bp upstream of the mRNA start site. The CCAAT box and the GC box are shown here. Other upstream elements include the sequences GCCACACCC and ATGCAAAT.

trans-acting factors, which serve as positive control elements. Properly functioning promoters are required for transcription initiation; promoter-proximal elements and enhancers maximize the rate of transcription for promoters. These sites were originally distinguished by the distance from the point of initiation at which they operate. Promoters and promoter-proximal elements function near the initiation site, whereas enhancers can usually function at great distances from this point. More recent work has shown that the internal structure of these elements is more similar than originally imagined.

Promoters

In the same manner that a comparison of promoter sequences in bacteria (Figure 13-8) led to an understanding of important elements in prokaryotic promoters, a compilation of DNA sequences upstream of mRNA start sites in eukaryotes has revealed conserved sequence elements. Figure 18-35 gives a schematic view of sequence elements that have been identified as part of eukaryotic promoters. The

TATA box is involved in directing RNA polymerase II to begin transcribing approximately 30 base pairs (bp) downstream in mammalian systems and 60 to 120 bp downstream in yeast. The TATA box works most efficiently together with other upstream sequences, promoter-proximal elements, located within 100 bp and 200 bp from the start of transcription. The **CCAAT** box serves as one of these sequences, and a GC-rich segment often serves as a second sequence (Figure 18-35). Site-directed mutational analysis has demonstrated that altering bases in the TATA box or in the promoter-proximal elements lowers in vivo transcription rates. In one version of this experiment by Richard Myers, Kit Tilly, and Tom Maniatis (Figure 18-36), a "saturation mutagenesis" in vivo alters, in series, each of the bases in the promoter for the β-globin gene. The effect on transcription rates is measured for changes at each base in the promoter sequence. Base changes outside the TATA box and the promoter-proximal elements have no effect on levels of transcription, whereas an alteration in either of these elements severely lowers transcription rates. Some genes that are expressed at very low rates do not contain TATA boxes at all, but do contain CG-rich sequences upstream

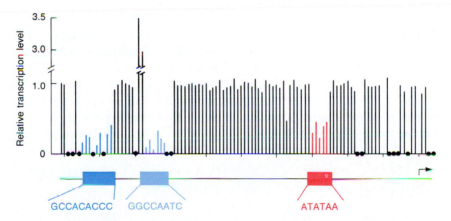

Figure 18-36 Consequences of point mutations in the promoter for the β-globin gene. Point mutations through the promoter region were analyzed for their effects on transcription rates. The height of each line represents the transcription level, relative to a wild-type promoter, that results from promoters with base changes at that point. A level of 1.0 means that the rates are equal to the wild-type rate; reductions in transcription rates yield levels less than 1.0. Almost every nucleotide throughout the promoter was tested, except for the points shown with black dots. The diagram below the bar graph shows the position of the TATA box and two promoter-proximal elements of the promoter. Only the base substitutions that lie within the three promoter elements change the level of transcription. (From T. Maniatis, S. Goodbourn, and J. A. Fischer, *Science* 236, 1987, 1237–1245.)

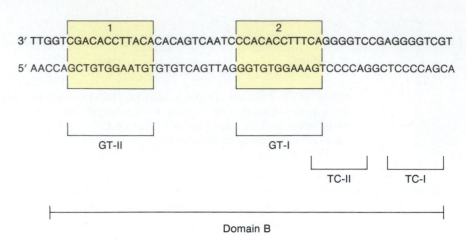

Figure 18-37 Organization of the SV40 enhancer. Boxed sequences 1 to 5 indicate the sequences that are required for maximum levels of enhancer activity. The brackets below each sequence indicate the location of sequence motifs that are repeated within the enhancer. Functional domains A and B are indicated at the bottom of the figure. (From T. Maniatis, S. Goodbourn, and J. A. Fischer, *Science* 236, 1987, 1237–1245.)

from the transcription start sites. Here, transcription begins at different, nearby, points.

Unlike prokaryotic promoters, eukaryotic promoters do not provide sufficient recognition signals for RNA polymerase to initiate transcription in vivo. The TATA box and the upstream sequences must each be recognized by regulatory proteins, which bind to these sites and activate transcription, by enabling RNA polymerase to bind and initiate properly. We will describe some of these trans-acting factors shortly.

Enhancers

Enhancers are cis-acting sequences that can greatly increase transcription rates from promoters on the same DNA molecule. Enhancers are similar to promoters in that they are organized as a series of cis-acting sequences that are recognized by trans-acting factors. However, enhancers can act at longer distances than promoter elements can, sometimes as long as 50 kb. Also, they can operate either upstream or downstream from the promoter they are enhancing. Enhancer elements are intricately structured. Figure 18-37 shows the DNA sequence of the SV40 enhancer. There are two basic domains in the enhancer, which, together, contain five sequence elements required for maximal enhancement of transcription. Different sequence motifs serve as recognition sites for specific trans-acting factors. Figure 18-38 depicts the organization of the SV40 controlling elements. The en-

hancer increases the rates of transcription of both early and late mRNA. Sequences at the end of the enhancer serve as promoter-proximal elements for the late mRNA promoter.

Another interesting property of enhancers is that they can be tissue-specific, operating in some cell types but not in others. For example, antibody genes are flanked by powerful enhancers that operate only in the B cells of the immune system. Other enhancers respond to external signals, as in heat shock genes, which are expressed after a shift to high temperature, or the metallothionein genes, which encode heavy metal detoxification proteins that are expressed in the presence of cadmium or zinc. Some enhancers are activated by steroid hormones, as discussed later. Enhancers are important elements in differential gene activity in higher eukaryotes and will be discussed further in Chapter 25.

Trans Control of Transcription

Trans-Acting Transcription Factors

A large number of transcription factors have now been identified in eukaryotic cells. Some of these bind to enhancer regions. Others bind to cis-acting elements in the promoter. Trans-acting factors (TAFs) help RNA polymerase II to initiate transcription, and together with RNA polymerase they form a preinitiation complex, as pictured in Figure 18-39a, which shows the assembly of the initiation complex. The transcription factors for RNA polymerase II

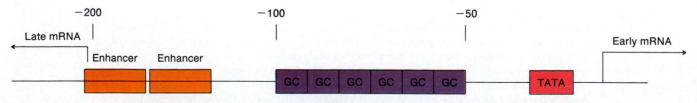

Figure 18-38 SV40 transcriptional control elements. The TATA box, the promoter-proximal elements (the GC boxes), and the two enhancer elements are required for maximal rates of early RNA transcription. The enhancer elements also stimulate transcription of late mRNAs. (Modified from W. S. Dynan and R. Tjian, *Nature* 316, 1985, 774–777.)

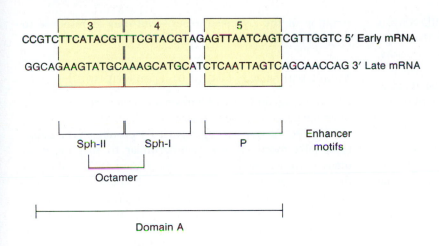

CCGTCTTCATACGTTTCGTACGTAGAGTTAATCAGTCGTTGGTC 5' Early mRNA

GGCAGAAGTATGCAAAGCATGCATCTCAATTAGTCAGCAACCAG 3' Late mRNA

Sph-II   Sph-I   P   Enhancer motifs

Octamer

Domain A

(a)

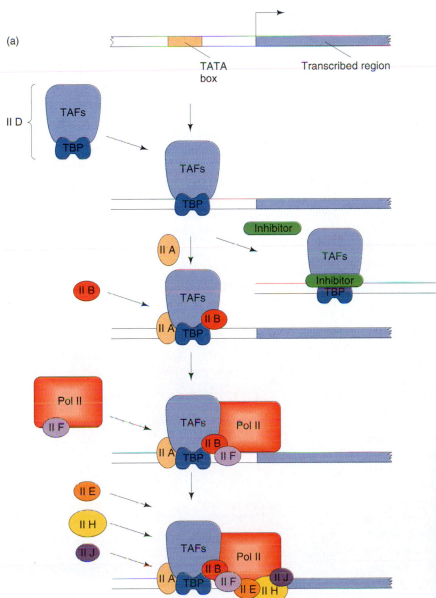

TATA box   Transcribed region

II D   TAFs   TBP

TAFs   TBP

Inhibitor

II A   TAFs   TAFs   Inhibitor   TBP

II B   II A   TBP   TAFs   II B

Pol II   II F   TAFs   Pol II   II A   TBP   II B   II F

II E   II H   II J   TAFs   Pol II   II A   TBP   II B   II F   II E   II H   II J

Figure 18-39 (a) Assembly of the RNA polymerase II initiation complex begins with the binding of TFIID to the TATA box. TFIID is composed of one TATA box–binding subunit called *TBP* (dark blue) and more than eight other subunits (TAFs), represented by one large symbol (light blue). Inhibitors (green) can bind to the TFIID-promoter complex, blocking the binding of other general transcription factors. Binding of TFIIA to the TFIID-promoter complex (to form the D-A complex) prevents inhibitor binding. TFIIB then binds to the D-A complex, followed by binding of a preformed complex between TFIIF and RNA polymerase II (red). Finally, TFIIE, TFIIH, and TFIIJ must add to the complex, in that order, for transcription to be initiated. (b) TATA-binding protein (blue) is a remarkably symmetrical, saddle-shaped molecule. Its underside rides on DNA (yellow) and seems to bend it. This bending may somehow facilitate assembly of the complex that initiates transcription. Coactivators, which are not depicted, are thought to bind tightly to the upper surface. (Part a from H. Lodish, D. Baltimore, A. Berk, S. L. Zipursky, P. Matsudaira, and J. Darnell, *Molecular Cell Biology*, 3d ed. Copyright ©1995 by Scientific American Books; part b courtesy of J. L. Kim and S. K. Berley, from *Nature, 365*, 1993, 520–527.

(b)

(TFII) are distinguished by letters. The factor TFIID consists of a TATA box–binding protein (TBP) and more than eight additional subunits (TAFs). The TFII transcription factors, along with TBP, are often referred to as "basal" or "general" transcription factors, since they are the minimal requirement for RNA polymerase II to initiate transcription at a promoter. Figure 18-39b shows the structure of the TATA box–binding protein binding to DNA. Several proteins that recognize the CCAAT box have been identified from mammalian cells. These factors can distinguish between subunits of CCAAT elements. GC boxes are recognized by additional factors. For instance, the Spl protein recognizes the upstream GC boxes in the SV40 promoter (Figure 18-38). Factors that bind to TATA box sequences have been identified in *Drosophila* and yeast.

Two examples of trans-acting factors in yeast that operate on the enhancerlike upstream activating sequences (UASs) are the GCN4 and the GAL4 proteins. GCN4 activates the transcription of many genes involved in amino acid biosynthetic pathways. In response to amino acid starvation, GCN4 protein levels rise and, in turn, increase the levels of the amino acid biosynthetic genes. The UASs recognized by GCN4 contain the principal recognition sequence element ATGACTCAT. GAL4 is involved in activating UASs for the promoters of genes involved in galactose metabolism. GAL4 binds to four sites with similar 17-bp sequences within these UAS elements.

Domain Structure of Transcription Factors

Several interesting aspects of trans-acting proteins have emerged from recent studies. It is clear that many of these proteins have two separate domains: one of which recognizes and binds to cis-acting DNA sequences, and the second of which activates transcription. This can be seen in the GAL4 protein, which consists of 881 amino acids. A fragment of the protein containing only the first 147 amino acids can bind to DNA at specific UAS sites but cannot activate transcription. However, fragments containing later sequences in the protein cannot bind to DNA but can activate transcription when attached to fragments that bind DNA. Also, when the 144 carboxyl-terminal residues of GAL4 are fused to the DNA-binding segment of the human estrogen receptor, this hybrid protein can activate transcription at the binding site normally recognized by the estrogen receptor. This experiment shows not only that several different proteins have similar domain structures but also that the mechanisms of activation are similar in diverse systems. In a related experiment, a hybrid was constructed between two different steroid hormone–receptor molecules: the glucocorticoid receptor and the progesterone receptor. Here the DNA-binding region of the glucocorticoid receptor was substituted for the progesterone receptor binding region, but the activating portion of the progesterone receptor was left intact. The fused protein responded by binding to glucocorticoid regulatory sequences and stimulating the transcription of the cis-linked genes—but only in response to progesterone.

Structural Motifs Found in Transcription Factors

DNA-binding domains of eukaryotic transcription factors often fall into one of several structural families.

One set of trans-acting regulatory proteins has a **helix-turn-helix** motif (Figure 18-40) similar to that found in many bacterial regulatory proteins (Figure 18-34). Many of these proteins are involved in *Drosophila* development, particularly in determining the spatial arrangement of body segments (see pages 766–769). The genes that encode these proteins, called *homeotic proteins,* contain a similar sequence in the region that encodes the DNA-binding domain. This sequence, named a **homeotic box,** encodes the recognition helix of the helix-turn-helix segment for these regulatory proteins. As Figure 18-40 shows, there are three helices. Helix 3 is the recognition helix and makes important contacts with the DNA. Helices 1 and 2 make contacts with other proteins.

A second common structural motif arises from a cysteine- and histidine-rich region in a DNA-binding domain that complexes zinc. The protrusions in the protein structure resemble fingers (Figure 18-41); for this reason, the structures have been termed **zinc fingers.** Zinc fingers have been found in steroid receptor proteins and among transcription factors for many mRNA promoters.

Leucine zipper proteins can form dimers through a hydrophobic interface formed by leucine residues spaced 7 amino acids apart, as shown in Figure 18-42. Flanking the leucine zipper is a DNA-binding domain containing many lysine and arginine residues. Dimer formation is required for DNA binding. Figure 18-43 shows a leucine zipper pro-

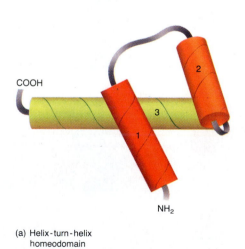

(a) Helix-turn-helix homeodomain

Figure 18-40 Structural model for the helix-turn-helix DNA-binding domain. (From J. D. Watson, M. Gilman, J. Witkowski, and M. Zoller, *Recombinant DNA,* 2d ed. Copyright © 1992 by Scientific American Books.)

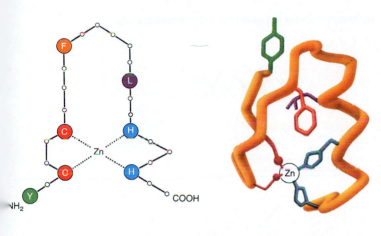

(b) C₂H₂ zinc finger

Figure 18-41 Structural model for the zinc finger DNA-binding domain. (From J. D. Watson, M. Gilman, J. Witkowski, and M. Zoller, *Recombinant DNA*, 2d ed. Copyright © 1992 by Scientific American Books.)

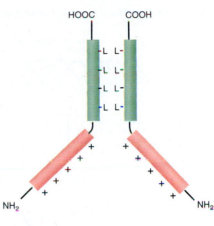

(c) Leucine zipper

Figure 18-42 Structural model for the leucine zipper DNA-binding domain. (From J. D. Watson, M. Gilman, J. Witkowski, and M. Zoller, *Recombinant DNA*, 2d ed. Copyright © 1992 by Scientific American Books.)

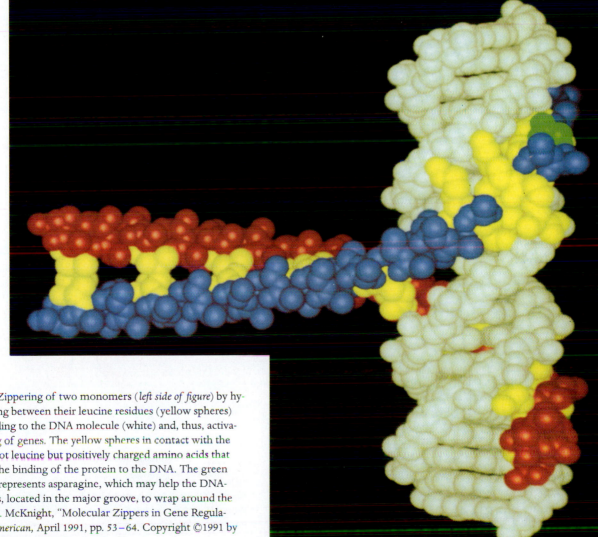

Figure 18-43 Zippering of two monomers (*left side of figure*) by hydrophobic bonding between their leucine residues (yellow spheres) can facilitate binding to the DNA molecule (white) and, thus, activation and silencing of genes. The yellow spheres in contact with the DNA represent not leucine but positively charged amino acids that may strengthen the binding of the protein to the DNA. The green cluster of atoms represents asparagine, which may help the DNA-binding segments, located in the major groove, to wrap around the DNA. (From S. L. McKnight, "Molecular Zippers in Gene Regulation," *Scientific American*, April 1991, pp. 53–64. Copyright ©1991 by Scientific American, Inc. All rights reserved.)

HETERODIMERS INCREASE NUMBER OF USABLE MOTIFS

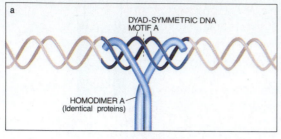

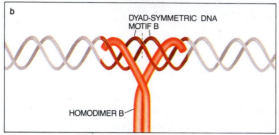

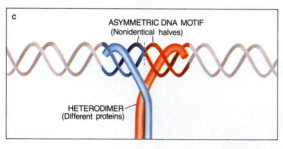

HETERODIMERS INCREASE VARIETY OF PROTEIN COMBINATIONS

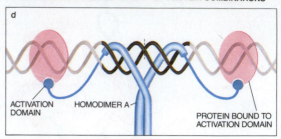

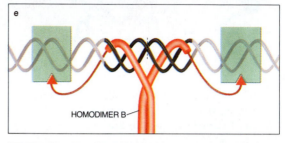

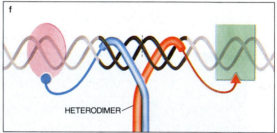

Figure 18-44 Zippered pairs of heterodimers have two possible advantages for an organism. (1) The subunits of homodimers tend to recognize only dyad-symmetric (twinlike) DNA sequences (panels a and b). (c) Heterodimers, the subunits of which may bind to different sequences, free the DNA from such constraints of symmetry and thus increase the number of sequence motifs available for gene regulation. (2) On the other hand, the increased diversity afforded by heterodimers could arise at the protein level. The subunits of homodimers (panels d and e) each bind identical activating proteins. (f) Heterodimers, however, can bind two different proteins, thus increasing the adaptive strategies for turning genes on and off. (From S. L. McKnight, "Molecular Zippers in Gene Regulation," *Scientific American,* April 1991, pp. 16–42. Copyright ©1991 by Scientific American, Inc. All rights reserved.)

tein binding to DNA. A number of proto-oncogenes (see Chapter 24), such as c-*jun* and c-*fos,* encode leucine zipper proteins. **Heterodimers,** zippered pairs of nonidentical proteins, also can form, and offer benefits to the organism by increasing the number of usable DNA sequences and the variety of protein combinations that can be used to turn genes on and off, as illustrated by Figure 18-44. For instance, the heterodimers of c-Jun and c-Fos proteins bind DNA much more strongly than **homodimers** of c-Fos or c-Jun alone.

Helix-loop-helix proteins (Figure 18-45) also form dimers, but not via a leucine zipper. The dimer interface

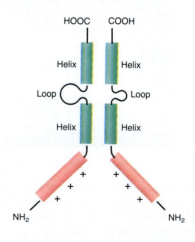

(d) Helix-loop-helix

Figure 18-45 Stuctural model for the helix-loop-helix DNA-binding domain. (From J. D. Watson, M. Gilman, J. Witkowski, and M. Zoller, *Recombinant DNA,* 2d ed. Copyright © 1992 by Scientific American Books.)

consists of two helices linked by a loop. Some proto-onco-genes (for example, *c-myc*) encode helix-loop-helix proteins, as do genes involved in differentiation.

Mechanism of Action of Transcription Factors

Transcription factors that bind to DNA can be shown to increase transcription in vivo and in vitro. Figure 18-46 shows one type of experiment used to demonstrate increased tran-

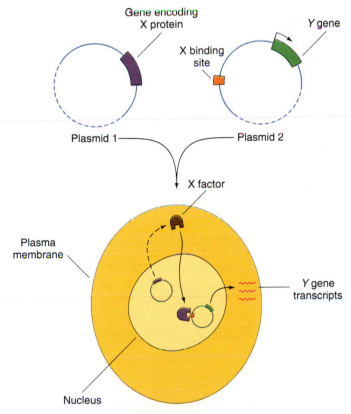

Figure 18-46 An in vivo test for transcription factor activity. A host cell receives two engineered plasmids. One contains a gene encoding protein X, the putative transcription factor being investigated, in a form that will allow expression of protein X at high levels. The second plasmid contains a test gene Y with a binding site for X. The host cell lacks both gene Y and the gene for protein X. Following introduction of the plasmids, the host cell is tested for the presence of transcripts for gene Y. If gene Y mRNA levels increase in the presence of X, the protein is considered a positively acting regulator; if levels go down, it is viewed as negatively acting. The use of plasmids that produce a mutated or rearranged factor can elucidate important domains in the protein. (From H. Lodish, D. Baltimore, A. Berk, S. L. Zipursky, P. Matsudaira, and J. Darnell, *Molecular Cell Biology*, 3d ed. Copyright © 1995 by Scientific American Books.)

scription in vivo. Here, two different plasmids are used. One plasmid encodes the trans-acting factor X, and the second contains the gene Y to be transcribed, together with the cis-controlling elements recognized by X. After introducing both plasmids into the same cell, transcripts from gene Y are measured. In vitro systems utilize purified protein factors and measure RNA polymerase transcription from DNA containing a gene with binding sites for the transcription factor.

Most models for the action of trans-acting factors at a distance from the genes they activate involve some type of DNA looping reminiscent of the model proposed for the bacterial *ara* operon (Figure 18-20). Figure 18-47 details a model for activation of the preinitiation complex (recall Figure 18-39). In this model multiply bound activators interact with transcription factors via "coactivators," protein adaptor molecules which relay signals from activators to the transcription factors. Some of the TAFs in TFIID (Figure 18-39) are examples of coactivators. This model, arising from experiments by B. Franklin Pugh, Robert Tjian, and their co-workers, is further illustrated in Figure 18-48, which highlights the interaction between the activator, in this case the protein Spl, and a coactivator.

Regulation of Transcription Factors

Some transcription factors are synthesized only in specific tissues. The activities of the factors themselves are also regulated in different cell types.

Steroid Hormones

In a number of cases, transcription factors are activated by hormones. These molecules are usually synthesized in one type of cell and then transported to the cells on which they act. Protein hormones, such as insulin, principally act by binding to cell-surface receptors and causing a signal to be transmitted across the plasma membrane, stimulating changes inside the cell. In some cases, the proteins are transported into the cells themselves. Small hormones and, in particular, steroid hormones enter cells by virtue of their lipid solubility and bind to and regulate specific transcription factors in the nucleus. In this respect, they are formally analogous to inducers in certain bacterial systems.

One example is the female sex hormone estrogen. In chicken oviducts, the egg white protein ovalbumin is specifically synthesized in response to estrogen, which causes increased transcription of the ovalbumin gene. The estrogen molecule activates transcription by binding to a protein receptor molecule that first recognizes the estrogen molecule in the cytoplasm and transports it to the nucleus. The receptor molecule is a transcription factor which then binds to the DNA at an enhancer site. The steroid-receptor activa-

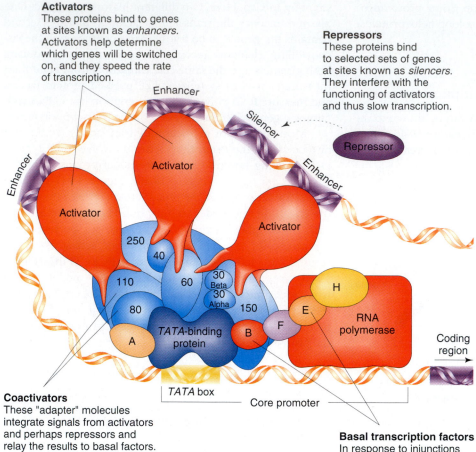

Activators
These proteins bind to genes at sites known as *enhancers*. Activators help determine which genes will be switched on, and they speed the rate of transcription.

Repressors
These proteins bind to selected sets of genes at sites known as *silencers*. They interfere with the functioning of activators and thus slow transcription.

Coactivators
These "adapter" molecules integrate signals from activators and perhaps repressors and relay the results to basal factors.

Basal transcription factors
In response to injunctions from activators, these factors position RNA polymerase at the start of the protein-coding region of a gene and send the enzyme on its way.

Figure 18-47 The molecular apparatus controlling transcription in human cells consists of four kinds of components. Basal transcription factors (red and orange shapes at bottom), generally named by single letters, are essential for transcription but cannot by themselves increase or decrease its rate. That task falls to regulatory molecules known as activators (red) and repressors (gray); these can vary from gene to gene. Activators, and possibly repressors, communicate with the basal factors through coactivators (blue)—proteins that are linked in a tight complex to the TATA-binding protein (TBP), the first of the basal transcription factors to land on a regulatory region of genes known as the *core promoter*. Coactivators, which include some of the TAF's in TFIID (see Fig. 18-39), are named according to their molecular weights (in kilodaltons). (From Robert Tjian, "Molecular Machines That Control Genes," *Scientific American,* February 1995, pp. 55–61. Copyright ©1995 by Scientific American, Inc. All rights reserved.)

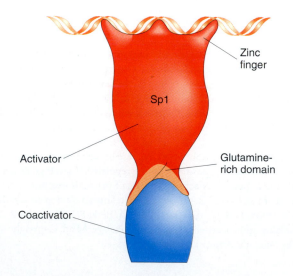

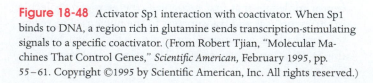

Figure 18-48 Activator Sp1 interaction with coactivator. When Sp1 binds to DNA, a region rich in glutamine sends transcription-stimulating signals to a specific coactivator. (From Robert Tjian, "Molecular Machines That Control Genes," *Scientific American,* February 1995, pp. 55–61. Copyright ©1995 by Scientific American, Inc. All rights reserved.)

Figure 18-49 The action of steroid hormones at enhancer sequences. A glucocorticoid hormone (blue) binds to a soluble receptor protein. This complex, in turn, binds to enhancer sequences, such as the one found in the DNA of the mouse mammary tumor virus (MMTV), and enables them to stimulate the transcription of hormone-responsive genes. (From L. Stryer, *Biochemistry*, 4th ed. Copyright ©1995 by Lubert Stryer.)

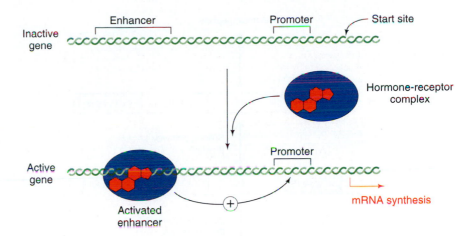

tion of genes has been well studied in the mouse mammary tumor virus (MMTV) system. This virus replicates very well in cells with high levels of glucocorticoid steroids. Figure 18-49 diagrams the action of these steroids on MMTV transcription.

Message Eukaryotic genes are activated by two cis-acting control sequences—promoters and enhancers—which are recognized by trans-acting proteins. The trans-acting proteins allow the RNA polymerase to initiate transcription and to achieve maximal rates of transcription.

5-Methylcytosine Regulation

Genes which are transcriptionally inactive contain 5-methylcytosine (5-methyl-C) residues (Figure 18-50), usually as part of p5-methyl-CpG sequences, whereas genes which are active are free of 5-methyl-C. The methyl groups are added after replication of the DNA susceptible to methylation. Several percent of the cytosines in DNA from mammals are methylated. Although no direct experiments have proved that methylation controls transcription, the correlation between methylation and inactivation of transcription, together with the inherited tissue-specific pattern of methylation, points to a key role. Experiments with a cytosine

analog, 5-azacytodine, which cannot be methylated at the C-5 position, reveal that the addition of 5-azacytodine can activate expression of certain previously unexpressed genes.

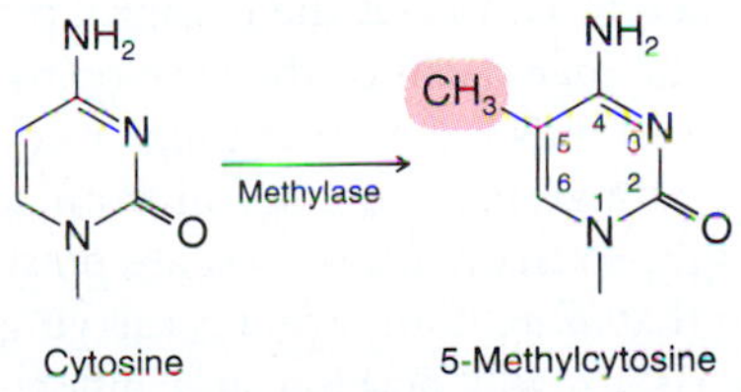

Post-Transcriptional Control

Differential Processing

We saw in Chapter 13 how many eukaryotic transcripts are processed. Poly(A) tails are added, and the primary transcript is spliced to yield the final mRNA. Figure 18-51 details how the same transcript can yield two different mRNAs, depending on which of several different splicing pathways is utilized. Thus, from the same transcript one splicing pathway results in the production of the calcitonin peptide in the thyroid, whereas a second splicing pathway ultimately generates the CGRP peptide in the hypothalamus. Differential poly(A) addition can also result in different mRNAs from the same primary transcript. Table 18-4 provides some interesting examples of differential processing of pre-mRNA.

Control of mRNA Degradation

Clearly the concentration of a given mRNA in the cytoplasm depends not only on its rate of synthesis, but also on its stability. By controlling the rate of mRNA turnover, cells have an additional opportunity to fine-tune mRNA levels. Poly(A) tails help to stabilize mRNAs. Without poly(A) tails, messengers are rapidly degraded. Other factors affecting mRNA stability in specific cases include hormones, low-molecular-weight compounds, and specific sequences at the untranslated 3′ end of the mRNA. For example, one class of mRNAs with short half-lives contains repeats of sequences of the form AUUUA at their 3′ ends, as shown in Figure 18-52a. Transplanting one of these sequences into the more stable β-globin gene reduces the half-life of the resulting mRNA from more than 10 hours to less than 2 hours, as seen in Figure 18-52b.

Figure 18-50 Cytosine can be converted to 5-methylcytosine by methylases.

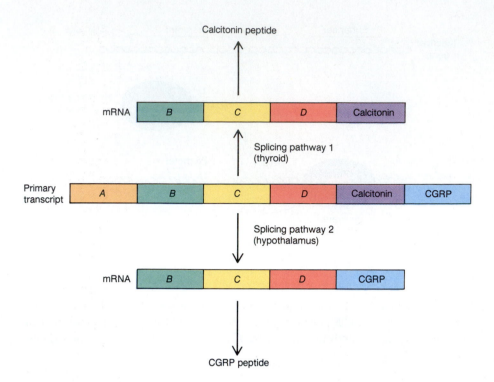

Control of Ubiquitous Molecules in Eukaryotic Cells

Several classes of molecules are found in abundance in every eukaryotic cell: histones, the translational apparatus, membrane components, and so on. Several strategies have evolved to maintain a sufficient amount of these products in the eukaryotic cell. These strategies include continuous and repeated transcription throughout the cell cycle, repetition of genes, and extrachromosomal amplification of specific gene sequences.

Genetic Redundancy

Genes coding for rRNA and tRNA have been extensively analyzed because the availability of purified gene products provides a probe with which to recover DNA that is complementary. Simple DNA-RNA hybridization shows that each *Xenopus* chromosome carrying a nucleolus organizer (NO) has 450 copies of the DNA coding for 18S and 28S rRNA. In contrast, there are 20,000 copies in each nucleus of the genes coding for 5S rRNA, and these genes are not located in the NO region. Donald Brown and his collaborators during the 1970s analyzed the NO DNA extensively and showed that it carries tandem duplications of a large (40S) transcription unit separated from its neighbors by nontranscribed spacers. The initial transcript is then processed to release the smaller rRNAs finally found in the ribosome.

Max Birnstiel and his associates have shown that the five histone genes in sea urchins also are found as a tightly linked unit (see Figure 16-23). There are several hundred copies of a repeating unit consisting of the genes in the sequence *H4–H2B–H3–H2A–H1*. Transcription begins at *H4* and ends at *H1*; each gene is transcribed separately from its other histone neighbors. As we have seen in earlier chapters, tandem duplications can pair asymmetrically between homologous chromosomes or within the same chromosome; exchanges then increase or reduce the number of copies. It remains to be seen how (or whether) the number of tandem repeats is kept constant in a particular species.

Gene Amplification

Specific gene amplification has been studied extensively in the case of the rRNA genes in amphibian oocytes. As noted, each chromosome of *Xenopus* carries about 450 copies of the DNA coding for 18S and 28S rRNA. However, the oocyte contains up to 1000 times this number of copies of these genes. The oocyte is a very specialized cell that is loaded with the nutritive material needed to maintain the embryo until the tadpole stage without any ingestion of food. The cytoplasm also is prepared with the translational apparatus needed to carry out the complex program of differentiation in the many cells of the developing embryo.

During oogenesis, the maturing oocyte increases greatly in size as material is poured in from adjacent nurse cells. (Amazingly, so many ribosomes are present in the egg at fertilization that *an an* homozygotes carrying no genes for 18S and 28S rRNA will develop and differentiate to the twitching tadpole stage before death.) The oocyte nucleus also contributes material as the cell proceeds to the first

Table 18-4   Examples of Differential Processing of Pre-mRNA

Gene products	Product differences	Basis of differential control	
		Poly(A) choice	Splicing variation
Mammals:			
Calcitonin and CGRP (rat)	Two proteins, calcitonin (Ca^{2+}-regulating hormone) and calcitonin gene–related peptide (CGRP), which probably functions in taste, are formed preferentially in thyroid and brain, respectively.	+	+
Kininogens (cow)	Two prehormones which both yield bradykinin, which controls blood pressure, differ in their protease susceptibility.	+	
Immunoglobulin heavy chains (mouse)	These differ at their carboxyl ends in different antibody molecules.	+	
Troponin (rat)	Two different muscle proteins appear in different rat skeletal muscles.		+
Myosin light chains (rat)*	Two proteins, 140 and 180 amino acids long, respectively, appear in different "fast twitch" muscles.		+
Preprotachykinin (cow)	Two prehormones that release substance P, a neuropeptide that affects smooth muscle, are formed; the larger prehormone also releases substance K, a neuropeptide of unknown function.		+
Fibronenctin (rat)	The structural protein has at least six forms.		+
Drosophila:			
Tropomyosin	Two muscle proteins differ between embryos and adults.		+
Myosin heavy chains	Two muscle proteins differ between embryos and adults; a third form is found in pupae and adults.	+	
Ultrabithorax	Different gene products that control thoracic segment development are found in early embryos and larvae.	+	
Glycinamide ribotide transformylase	This enzyme is encoded by a long mRNA; a shorter mRNA that shares several exons encodes a second polypeptide.	+	

* The rat myosin light-chain proteins are processed from two different overlapping primary transcripts with different 5' ends. The resulting mRNAs also contain different exons within the region of overlap, plus four exons that are similar. An almost-identical situation exists in chicken myosin genes.

SOURCE: J. Darnell, H. Lodish, and D. Baltimore, *Molecular Cell Biology*, 2d. ed. Copyright 1990 by Scientific American Books, Inc.

meiotic prophase, where further development is arrested. The amphibian meiotic chromosomes are enormous, with numerous lateral loops representing regions of intense genetic activity (Figure 18-53). In the oocyte nucleus, there are hundreds of extrachromosomal nucleoli of varying size. Each nucleolus contains a different-sized ring of rDNA that has been replicated and released from the chromosome. The steps involved in this process of DNA amplification are completely unknown. The DNA rings actively produce

rRNA that is assembled into ribosomes, which are stored in the nucleolus until their release during meiosis.

Other examples of genetic amplification are known. Specific puffs of the polytene chromosomes in dipterans are found to produce excess DNA rather than mRNA. Also, George Rudkin showed in the early 1960s that the polytene chromosomes from *Drosophila* salivary glands themselves may represent specific amplifications of euchromatic DNA about a thousandfold, whereas the proximal heterochro-

(a) 3′ Sequences of unstable mRNAs

GMCSF UAAUAUUUAUAUAUUUAUAUUUUUAAAAUAUUUAUUUAUUUAUUUAUUUAA

IFNβ UUUUGAAAUUUUUAUUAAAUUAUGAGUUAUUUUUUAUUUAUUUAAAUUUUAUUUU

IL-1 UUAUUUUUUAAUUAUUAUUUAUAUAUGUAUUUAUAAAUAUAUUUAAGAUAAU

TNF AUUAUUUAUUAUUUUAUUUUAUUAUUUUAUUUAUUUAUUUA

c-Fos GUUUUUAAUUUAUUUAUUAAGAUGGAUUCUCAGAUAUUUAUAUUUUUAUUUU

c-Myc UAAUUUUUUUUUAUUUAAGUACAUUUUGCUUUUUAAAGUUGAUUUUUUUCU

(b) Activity of A(U)$_n$A repeats

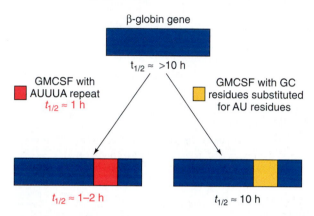

β-globin gene

$t_{1/2} \approx$ >10 h

GMCSF with
AUUUA repeat
$t_{1/2} \approx$ 1 h

GMCSF with GC
residues substituted
for AU residues

$t_{1/2} \approx$ 1–2 h

$t_{1/2} \approx$ 10 h

Figure 18-52 Effect of a 3′ untranslated sequence on mRNA stability. (a) Sequences of 3′ untranslated regions of mRNAs encoding human granulocyte-monocyte–stimulating factor (GMCSF), fibroblast interferon β (IFNβ), interleukin 1 (IL-1), tumor necrosis factor (TNF), and two proto-oncogene proteins (c-Fos and c-Myc). Characteristic A(U)$_n$A sequences are shown in red; some of these are overlapping. Numerous other proto-oncogene and cell growth factor mRNAs contain these sequences. (b) Effect of A(U)$_n$A sequences on mRNA half-life. Cultured cells were transfected with recombinant β-globin DNA containing a 62-bp segment of GMCSF DNA encoding either the A(U)$_n$A-rich segment shown in part a or an altered segment in which many AU residues were replaced with GC residues. The destabilizing effect of the A(U)$_n$A sequences was clearly demonstrated by the decrease in half-life ($t_{1/2}$) of the corresponding mRNAs. (See G. Shaw and R. Kamen, *Cell* 46, 1986, 659. From J. Darnell, H. Lodish, and D. Baltimore, *Molecular Cell Biology*, 2d ed. Copyright ©1990 by Scientific American Books.)

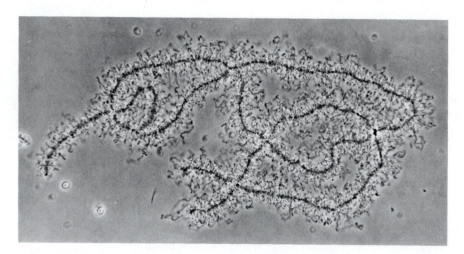

Figure 18-53 The lampbrush chromosomes of an amphibian oocyte. (Photograph courtesy of J. Gall.)

matic regions may be replicated only a few times. Thus, in addition to the mechanism for replicating chromosomal DNA normally for cell division, there exists a means for amplifying specific loci and not others.

Message Cellular mechanisms exist to ensure an adequate supply of the gene products that are vital to all cells. These mechanisms include increasing the number of gene copies per chromosome (redundancy) and increasing the number of gene copies in the cell (amplification).

SUMMARY

The operon model explains how prokaryotic genes are controlled through a mechanism that coordinates the activity of a number of related genes. In negative control, the initiation of transcription is controlled at the operator by a repressor with binding affinities to the operator that may be altered by inducer molecules. Inactivation of the repressor—the negative control element—is required for active transcription. In positive control, transcription initiation requires the activation of a factor. Sometimes one control system is superimposed on another. For instance, superimposed on the repressor-operator system for the *lac* operon is the cAMP-CAP positive control system. Modulation of transcription by an attenuator sequence supplements the repressor-operator control in the *trp* operon. In λ phage, multiple binding sites for repressors, as well as repressors of repressors, add additional levels of control.

A major problem in transcriptional control in eukaryotes is understanding how thousands of promoters can be regulated to yield desired levels of mRNA. It is now clear that promoters are governed both by the number and type of cis-acting control elements and by the action of regulatory proteins that recognize these elements. Also modulating the levels of transcription are enhancers—cis-acting sequences that, when recognized by functional regulatory proteins, can greatly increase the activity of promoters. In some aspects, transcriptional control in eukaryotes is similar to that found in prokaryotes: namely, trans-acting factors recognize cis-acting (or cis-dominant) sites in both cases. The mechanism of action may even be similar in many cases, as evidenced by the looping out of DNA in the *ara* system in bacteria and by enhancer action in eukaryotes. Also, structural motifs of some regulatory proteins share a common thread as in the case of the helix-turn-helix motif in bacterial regulatory proteins and in the homeotic proteins in *Drosophila*. However, other structural motifs appear to be unique to classes in eukaryotic regulatory proteins, as exemplified by the zinc finger and leucine zipper motifs for many trans-acting factors. Eukaryotic mRNA levels are controlled in several different ways, including synthesis and degradation.

Concept Map

Draw a concept map interrelating as many of the following terms as possible. Note that the terms are listed in no particular order.

environment / promoter / operator / gene / operon / RNA polymerase / mRNA / enhancer / trans-acting factors / regulation

CHAPTER INTEGRATION PROBLEM

In Chapter 10 we learned how Hfr strains transfer a segment of the chromosome into a recipient strain, and in this chapter we saw how Jacob, Pardee, and Monod initially exploited this phenomenon to create temporary partial diploids. Explain what happens to *lacZ* expression when an Hfr transfers $I^+ Z^+$ genes into a recipient which is $I^- Z^-$, under conditions which prevent protein synthesis in the donor strain. How would this change if we use a recipient that is $I^+ Z^-$ and transfer the same $I^+ Z^+$ genes from the Hfr? Draw a diagram showing the results of these two experiments.

Solution

As shown in Figure 18-6, the introduction of $I^+ Z^+$ genes into the cytoplasm of the cells containing no repressor results in an initial burst of β-galactosidase synthesis, which lasts until enough repressor can build up in the cytoplasm to block further β-galactosidase synthesis (see the diagram that follows). In the case of a recipient that is $I^+ Z^-$, there is already repressor in the cytoplasm, so there is no initial burst of β-galactosidase synthesis.

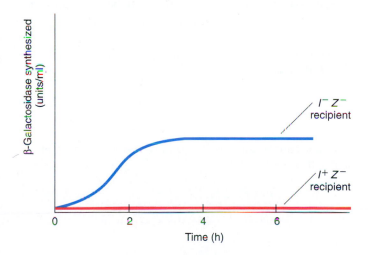

SOLVED PROBLEMS

This set of four solved problems, which are similar to Problem 4 at the end of this chapter, are designed to test understanding of the operon model. Here we are given several diploids and are asked to determine whether Z and Y gene products are made in the presence or absence of an inducer. Use a table similar to the one in Problem 4 as a basis for your answers.

1. $$\frac{I^-P^-O^cZ^+Y^+}{I^+P^+O^+Z^-Y^-}$$

Solution

One way to approach these problems is first to consider each chromosome separately and then to construct a diagram. The following figure diagrams this diploid:

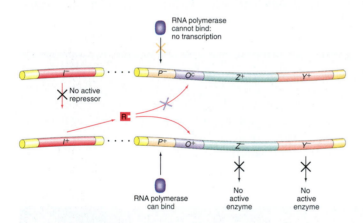

The first chromosome is P^-, so no *lac* enzyme can be synthesized off it. The second chromosome (O^+) can be transcribed, and this transcription is repressible. However, the structural genes linked to the good promoter are defective; thus, no active Z product or Y product can be produced. The symbols to add to your table are "$----$."

2. $$\frac{I^+P^-O^+Z^+Y^+}{I^-P^+O^+Z^+Y^-}$$

Solution

The first chromosome is P^-, so no enzyme can be synthesized off it:

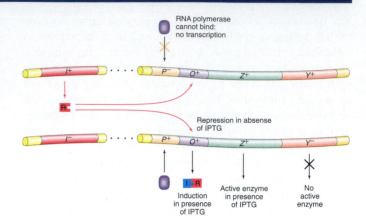

The second chromosome is O^+, so transcription will be repressed by the repressor supplied from the first chromosome, which can act *in trans* through the cytoplasm. However, only the Z gene from this chromosome is intact. Therefore, in the absence of an inducer, no enzyme will be made; in the presence of an inducer, only the Z gene product, β-galactosidase, will be produced. The symbols to add to the table are "$-+--$."

3. $$\frac{I^+P^+O^cZ^-Y^+}{I^+P^-O^+Z^+Y^-}$$

Solution

Because the second chromosome is P^-, we need consider only the first chromosome:

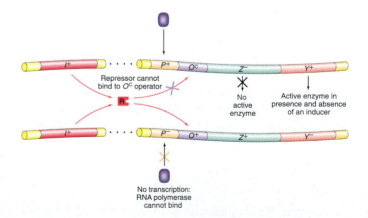

This chromosome is O^c, so that enzyme is made in the absence of an inducer, although only the Y gene is active. The entries in the table should be "$--++$."

4. $\dfrac{I^s P^+ O^+ Z^+ Y^-}{I^- P^+ O^c Z^- Y^+}$

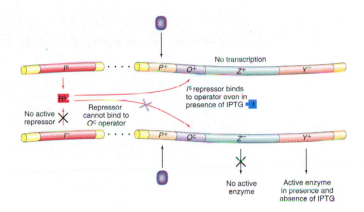

Solution

In the presence of an I^s repressor, all wild-type operators will be shut off, both with and without an inducer (see the accompanying drawing). Therefore, the first chromosome will be unable to produce any enzyme. However, the second chromosome has an altered (O^c) operator and can produce enzyme in both the absence and presence of an inducer. Only the Y gene is active on this chromosome, so the entries in the table should be "$--++$."

PROBLEMS

1. Explain why I^- mutations in the *lac* system are normally recessive to I^+ mutations and why I^+ mutations are recessive to I^s mutations.

2. What do we mean when we say that O^c mutations in the *lac* system are cis-dominant?

3. The genes shown in Table 18-5 are from the *lac* operon system of *E. coli*. The symbols a, b, and c represent the repressor (I) gene, the operator (O) region, and the structural gene (Z) for β-galactosidase, although not necessarily in that order. Furthermore, the order in which the symbols are written in the genotypes is not necessarily the actual sequence in the *lac* operon.

    a. State which symbol (a, b, or c) represents each of the *lac* genes I, O, and Z.

    b. In Table 18-5, a superscript minus sign on a gene symbol merely indicates a mutant, but you know there are some mutant behaviors in this system that are given special mutant designations. Use the conventional gene symbols for the *lac* operon to designate each genotype in the table.

        (Problem 3 is from J. Kuspira and G. W. Walker,
Genetics: Questions and Problems. Copyright ©1973
by McGraw-Hill.)

4. The map of the *lac* operon is

$$\underline{I \quad\quad POZY}$$

The promoter (P) region is the start site of transcription through the binding of the RNA polymerase molecule before actual mRNA production. Mutationally altered promoters (P^-) apparently cannot bind the RNA polymerase mole-

Table 18-5

Genotype	Activity ($+$) or Inactivity ($-$) of Z Gene	
	Inducer Absent	Inducer Present
$a^- b^+ c^+$	$+$	$+$
$a^+ b^+ c^-$	$+$	$+$
$a^+ b^- c^-$	$-$	$-$
$a^+ b^- c^+/a^- b^+ c^-$	$+$	$+$
$a^+ b^+ c^+/a^- b^- c^-$	$-$	$+$
$a^+ b^+ c^-/a^- b^- c^+$	$-$	$+$
$a^- b^+ c^+/a^+ b^- c^-$	$+$	$+$

cule. Certain predictions can be made about the effect of P^- mutations. Use your predictions and your knowledge of the lactose system to complete Table 18-6. Insert a "$+$" where an enzyme is produced and a "$-$" where no enzyme is produced.

5. In a haploid eukaryotic organism, you are studying two enzymes that perform sequential conversions of a nutrient A supplied in the medium:

$$A \xrightarrow{E_1} B \xrightarrow{E_2} C$$

Treatment of cells with mutagen produces three different mutant types with respect to these functions. Mutants of type 1 show no E_1 function; all type 1 mutations map to a single locus on linkage group II. Mutations of type 2 show

Table 18-6

Part	Genotype	β-Galactosidase		Permease	
		No Lactose	Lactose	No Lactose	Lactose
Example	$I^+ P^+ O^+ Z^+ Y^+/I^+ P^+ O^+ Z^+ Y^+$	−	+	−	+
(a)	$I^- P^+ O^c Z^+ Y^-/I^+ P^+ O^+ Z^- Y^+$				
(b)	$I^+ P^- O^c Z^- Y^+/I^- P^+ O^c Z^+ Y^-$				
(c)	$I^s P^+ O^+ Z^+ Y^-/I^+ P^+ O^+ Z^- Y^+$				
(d)	$I^s P^+ O^+ Z^+ Y^+/I^- P^+ O^+ Z^+ Y^+$				
(e)	$I^- P^+ O^c Z^+ Y^-/I^- P^+ O^+ Z^- Y^+$				
(f)	$I^- P^- O^+ Z^+ Y^+/I^- P^+ O^c Z^+ Y^-$				
(g)	$I^+ P^+ O^+ Z^- Y^+/I^- P^+ O^+ Z^+ Y^-$				

no E_2 function; all type 2 mutations map to a single locus on linkage group VIII. Mutants of type 3 show no E_1 or E_2 function; all type 3 mutants map to a single locus on linkage group I.

 a. Compare this system with the *lac* operon of *E. coli,* pointing out the similarities and the differences. (Be sure to account for each mutant type at the molecular level.)

 b. If you were to intensify the mutant hunt, would you expect to find any other mutant types on the basis of your model? Explain.

6. In *Neurospora,* all mutants affecting the enzymes carbamyl phosphate synthetase and aspartate transcarbamylase map at the *pyr-3* locus. If you induce *pyr-3* mutations by ICR-170 (a chemical mutagen), you find that either both enzyme functions are lacking or only the transcarbamylase function is lacking; in no case is the synthetase activity lacking when the transcarbamylase activity is present. (ICR-170 is assumed to induce frameshifts.) Interpret these results in terms of a possible operon.

7. In 1972, Suzanne Bourgeois and Alan Jobe showed that a derivative of lactose, allolactose, is the true natural inducer of the *lac* operon, rather than lactose itself. Lactose is converted to allolactose by the enzyme β-galactosidase. How does this result explain the early finding that many Z^- mutations that are not polar still do not allow the induction of *lac* permease and transacetylase by lactose?

8. Certain *lacI* mutations eliminate operator binding by the *lac* repressor but do not affect the aggregation of subunits to make a tetramer, the active form of the repressor. These mutations are partially dominant to wild-type. Can you explain the partially I^- phenotype of the I^-I^+ heterodiploids?

9. Explain the fundamental differences between negative control and positive control.

10. Mutants that are *lacY⁻* retain the capacity to synthesize β-galactosidase. However, even though the *lacI* gene is still intact, β-galactosidase can no longer be induced by adding lactose to the medium. How can you explain this?

11. What analogies can you draw between transcriptional trans-acting factors that activate gene expression in eukaryotes and prokaryotes? Give an example.

12. Compare the arrangement of cis-acting sites in the control regions of eukaryotes and prokaryotes.

13. Explain how models for bacterial operons such as *ara* relate to eukaryotic trans-acting proteins and their mechanism of action.

14. It is now known that the Lac repressor of *E. coli* has a leucine zipper at the carboxyl-terminal region of the protein which is required to allow dimers to associate into tetramers. It is also known that a weak operator sequence, O_2, exists early in the Z gene, to which the repressor can also bind, in addition to the normal O_1. The elements are as shown here:

Using Figures 18-20 and 18-32 for reference, devise a model that explains why mutations in *lacI* that eliminate the leucine zipper reduce the ability of the repressor to block *lac* operon transcription completely. Draw a diagram.

15. One interesting mutation in *lacI* results in repressors with 100-fold increased binding to both operator and nonoperator DNA. These repressors display a "reverse" induction curve, allowing β-galactosidase synthesis in the *absence* of inducer (IPTG), but partially repressing β-galactosidase expression in the *presence* of IPTG. How can you explain this? Note that when IPTG binds repressor, it does not completely destroy operator affinity, but rather it reduces affinity 1000-fold. Also, as cells divide and new operators are generated by the synthesis of daughter strands, repressor must find the new operators by searching along the DNA, rapidly binding to and dissociating from nonoperator sequences.

19

Mechanisms of Genetic Change I: Gene Mutation

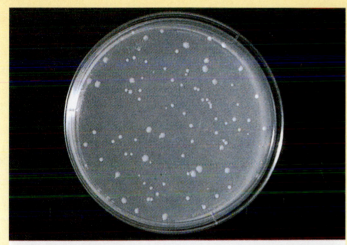

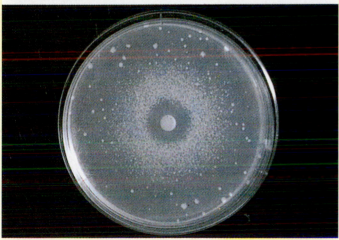

KEY CONCEPTS

► Mutations can occur spontaneously owing to several different mechanisms, including errors of DNA replication and spontaneous damage to the DNA.

► Mutagens are agents that increase the frequency of mutagenesis, usually by altering the DNA.

► Potentially mutagenic and carcinogenic compounds can be detected easily by mutagenesis tests with bacterial systems.

► Biological repair systems eliminate many potentially mutagenic alterations in the DNA.

► Cells lacking certain repair systems have higher than normal mutation rates.

Salmonella test (Ames test) for mutagens. (*Top*) Petri plate containing about 10^9 bacteria that cannot synthesize histidine (the small number of visible colonies are spontaneous revertants). (*Bottom*) Plate containing a disc with a mutagen, which produces a large number of revertants that can synthesize histidine; revertants appear as a ring of colonies around the disc. (Kristien Mortelmans, SRI International Menlo Park, Calif.)

G enetic change can result from a number of processes. Consider an individual organism that represents a variant from an established control population. What mechanism could have produced the change that created this variant individual? The following mechanisms provide four possibilities:

1. Gene mutation
2. Recombination
3. Transposable genetic elements
4. Chromosomal rearrangements

In previous chapters, we have considered genetic changes merely as useful and interesting phenomena. Let's now examine, at the molecular levels, the processes that lead to each type of genetic change. This chapter explores the mechanisms of gene mutation; Chapter 20 describes the events leading to recombination; and Chapter 21 introduces the concept of transposable genetic elements, which can bring about genetic change by moving from one chromosome location to another. Many chromosomal rearrangements are large deletions, duplications, or inversions that can result from recombination and from transposable elements.

Table 19-1 Gene Mutations at the Molecular Level	
Type of Mutation	Result and Example(s)
Forward mutations	
Single-nucleotide-pair (base-pair) substitutions	
At DNA level	
Transition	Purine replaced by a different purine, or pyrimidine replaced by a different pyrimidine: AT ⟶ GC GC ⟶ AT CG ⟶ TA TA ⟶ CG
Transversion	Purine replaced by a pyrimidine, or pyrimidine replaced by a purine: AT ⟶ CG AT ⟶ TA GC ⟶ TA GC ⟶ CG TA ⟶ GC TA ⟶ AT CG ⟶ AT CG ⟶ GC
At protein level	
Silent mutation	Triplet codes for same amino acid: AGG ⟶ CGG both code for Arg
Neutral mutation	Codon specifies different but functionally equivalent amino acid: AAA ⟶ AGA changing basic Lys to basic Arg (will not alter protein function in many cases)
Missense mutation Nonsense mutation	Codon specifies a different and nonfunctional amino acid. Codon signals chain termination: CAG ⟶ UAG changing from a codon for Gln to an amber termination codon
Single-nucleotide-pair addition or deletion: frameshift mutation Addition or deletion of several to many nucleotide pairs	Any addition or deletion of base pairs that is not a multiple of 3 changes the reading frame in DNA segments that code for proteins, resulting in different amino acids from that point on and frequently chain termination.
Reverse mutations	
Exact reversion	AAA (Lys) $\xrightarrow{\text{foreward}}$ GAA (Glu) $\xrightarrow{\text{reverse}}$ AAA (Lys) Wild-type Mutant Wild-type

The Molecular Basis of Gene Mutations

Gene mutations can arise spontaneously or they can be induced. **Spontaneous mutations** are naturally occurring mutations and arise in all cells. **Induced mutations** are produced when an organism is exposed to a mutagenic agent, or mutagen; such mutations typically occur at much higher frequencies than spontaneous mutations do.

To understand the mechanisms of gene mutation requires analysis at the level of DNA and protein molecules.

Preceding chapters have described models for DNA and protein structure and have discussed the nature of mutations that alter these structures. Table 19-1 draws together this information to provide an explanation of the nature of gene mutation at the molecular level.

Molecular genetics techniques can be used to determine the sequence of large segments of DNA and also the sequence changes resulting from mutations. This has greatly increased our understanding of the pathways that lead to mutagenesis and has even helped to unravel the mysteries of mutational hot spots—genetic sites with a penchant for mutating.

Table 19-1 *(continued)*

Type of Mutation	Result and Example(s)
Equivalent reversion	UCC (Ser) $\xrightarrow{\text{foreward}}$ UGC (Cys) $\xrightarrow{\text{reverse}}$ AGC (Ser) Wild-type — Mutant — Wild-type
	CGC (Arg, basic) $\xrightarrow{\text{foreward}}$ CCC (Pro, not basic) $\xrightarrow{\text{reverse}}$ CAC (His, basic) Wild-type — Mutant — Pseudo-wild-type
Intragenic suppressor mutations	
Frameshift of opposite sign at second site within gene	CAT CAT CAT CAT CAT CAT Addition of base alters reading frame $\longrightarrow (+)$ $(-) \longleftarrow$ Deletion of base restores correct reading frame CAT XCA TAT CAT CAT CAT
Second-site missense mutation	A second distortion that restores a more or less wild-type protein conformation after a primary distortion.
Extragenic suppressor mutations	
Nonsense suppressors	A gene (for example, for tyrosine tRNA) undergoes a mutational event in its anticodon region that enables it to recognize and align with a mutant nonsense codon (say, amber UAG) to insert an amino acid (here, tyrosine) and permit completion of the translation.
Missense suppressors	Usually caused by change in tRNA anticodon. One missense suppressor in *E. coli* is an abnormal tRNA that carries glycine but inserts it in response to arginine codons. Although all wild-type arginine codons are mistranslated, the observed mutations are not lethal, probably due to the low efficiency of abnormal substitution.
Frameshift suppressors	Very few examples have been found; in one, a four-nucleotide anticodon in a single tRNA can read a four-letter codon caused by a single-nucleotide-pair insertion.
Physiological suppressors	A defect in one chemical pathway is circumvented by another mutation (for example, one that permits more efficient transport of a compound produced in smaller quantities owing to the original mutation).

Much work until now on the molecular basis of mutation has been carried out in single-celled bacteria and their viruses. However, many mutations leading to inherited diseases in humans have been analyzed recently. We shall review some of the recent findings of these studies. We shall also consider biological repair mechanisms, since repair systems play a key role in mutagenesis, operating to lower the final observed mutation rates. For example, in *Escherichia coli,* with all repair systems functioning, base substitutions occur at rates of 10^{-10} to 10^{-9} per base pair per cell per generation.

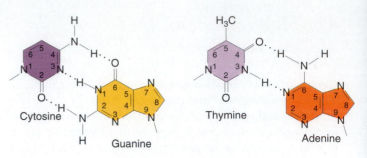

Figure 19-1 Base pairs in DNA. The normal Watson-Crick base pairs are shown.

Spontaneous Mutations

Spontaneous mutations arise from a variety of sources, including errors in DNA replication, spontaneous lesions, and transposable genetic elements. The first two are discussed below; the third is examined in Chapter 21.

Errors in DNA Replication

An error in DNA replication can occur when an illegitimate nucleotide pair (say, A–C) forms during DNA synthesis, leading to a base substitution.

Each of the bases in DNA can appear in one of several forms, called **tautomers,** which are isomers that differ in the positions of their atoms and in the bonds between the atoms. The forms are in equilibrium. The **keto** form of each base is normally present in DNA (Figure 19-1), whereas the **imino** or **enol** forms of the bases are rare. The ability of the wrong tautomer of one of the standard bases to mispair and cause a mutation during DNA replication was first noted by Watson and Crick when they formulated their model for the structure of DNA (Chapter 11). Figure 19-2 demonstrates some possible mispairs resulting from

changes of one tautomer to another, termed **tautomeric shifts.**

Mispairs can also occur when one of the bases becomes **ionized.** This type of mispair may occur more frequently than mispairs involving imino and enol forms of bases.

Transitions. All the mispairs described above lead to **transition mutations,** in which a purine is substituted for a purine or a pyrimidine for a pyrimidine (Figure 19-3). The bacterial DNA polymerase III (Chapter 11) has an editing capacity that recognizes such mismatches and excises them, thus greatly reducing the observed mutations. Another repair system (described later in this chapter) corrects many of the mismatched bases that escape correction by the polymerase editing function.

Transversions. In **transversion mutations,** a pyrimidine is substituted for a purine, or vice versa. Transversions cannot be generated by the mismatches depicted in Figure 19-2. With bases in the DNA in the normal orientation, creation of a transversion by a replication error would require, at some point during replication, mispairing of a purine with a purine or a pyrimidine with a pyrimidine. Although the dimensions of the DNA double helix render such mispairs energetically unfavorable, we now know from X-ray diffraction studies that G–A pairs, as well as other purine-purine pairs, can form.

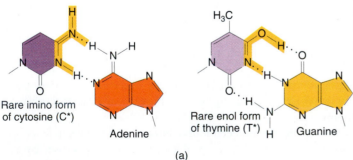

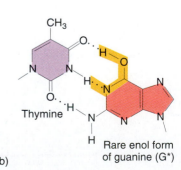

Figure 19-2 Mismatched bases. (a) Mispairs resulting from rare tautomeric forms of the pyrimidines; (b) mispairs resulting from rare tautomeric forms of the purines.

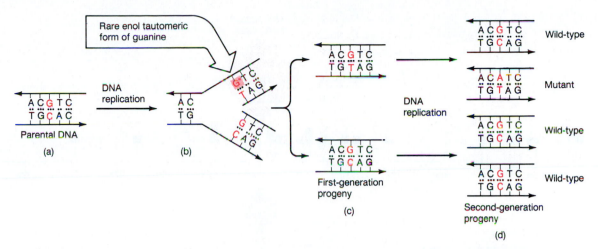

Figure 19-3 Mutation via tautomeric shifts in the bases of DNA. (a) In the example diagrammed, a guanine undergoes a tautomeric shift to its rare enol form (G*) at the time of replication. (b) In its enol form, it pairs with thymine. (c) and (d) During the next replication, the guanine shifts back to its more stable keto form. The thymine incorporated opposite the enol form of guanine, seen in (b), directs the incorporation of adenine during the subsequent replication, shown in (c) and (d). The net result is a GC → AT mutation. If a guanine undergoes a tautomeric shift from the common keto form to the rare enol form at the time of incorporation (as a nucleoside triphosphate, rather than in the template strand diagrammed here), it will be incorporated opposite thymine in the template strand and cause an AT → GC mutation. (From E. J. Gardner and D. P. Snustad, *Principles of Genetics*, 5th ed. Copyright © 1984 by John Wiley & Sons, New York.)

Frameshift Mutations. Replication errors can also lead to **frameshift mutations.** Recall from Chapter 13 that such mutations result in greatly altered proteins.

In the mid-1960s, George Streisinger and his co-workers deduced the nucleotide sequence surrounding different sites of frameshift mutations in the lysozyme gene of phage T4. They found that these mutations often occurred at repeated sequences and formulated a model to account for frame-shifts during DNA synthesis. In the Streisinger model (Figure 19-4), frameshifts arise when loops in single-stranded regions are stabilized by the "slipped mispairing" of repeated sequences.

In the 1970s, Jeffrey Miller and his co-workers examined mutational *hot spots* in the *lacI* gene of *E. coli*. **Hot spots** are sites in a gene that are much more mutable than other sites. The *lacI* work showed that certain hot spots result from repeated sequences, just as predicted by the Streisinger model. Figure 19-5 depicts the distribution of spontaneous muta-

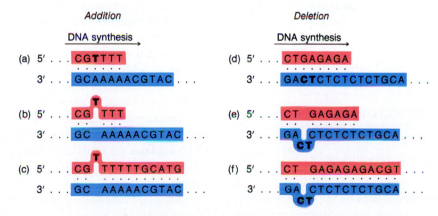

Figure 19-4 A simplified version of the Streisinger model for frameshift formation. (a) to (c) During DNA synthesis, the newly synthesized strand slips, looping out one or several bases. This loop is stabilized by the pairing afforded by the repetitive-sequence unit (the A bases, in this case). An addition of one base pair, A–T, will result at the next round of replication in this example. (d) to (f) If instead of the newly synthesized strand, the template strand slips, then a deletion results. Here the repeating unit is a CT dinucleotide. After slippage, a deletion of two base pairs (C–G and T–A) would result at the next round of replication.

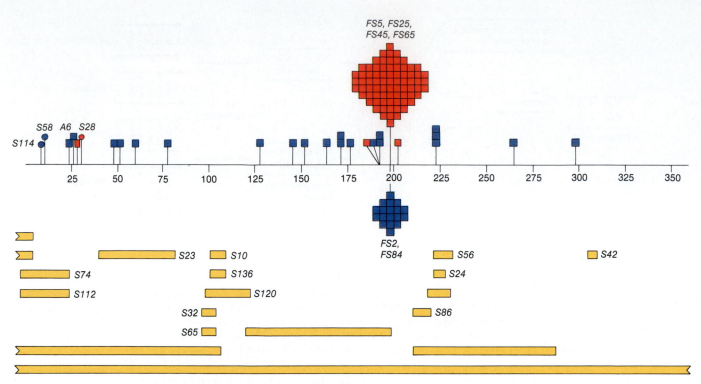

Figure 19-5 The distribution of 140 spontaneous mutations in *lacI*. Each occurrence of a point mutation is indicated by a box. Red boxes designate fast-reverting mutations. Deletions (gold) are represented below. The *I* map is given in terms of the amino acid number in the corresponding *I*-encoded *lac* repressor. Allele numbers refer to mutations that have been analyzed at the DNA sequence level. The mutations *S114* and *S58* (circles) result from the insertion of transposable elements (see Chapter 21). *S28* (red circle) is a duplication of 88 base pairs. (From P. J. Farabaugh, U. Schmeissner, M. Hofer, and J. H. Miller, *Journal of Molecular Biology* 126, 1978, 847.)

tions in the *lacI* gene. Compare this with the distribution in the *rII* genes of T4 seen by Benzer (see Figure 12-34). Note how one or two mutational sites dominate the distribution in both cases. In *lacI*, a four-base-pair sequence repeated three times in tandem in the wild type is the cause of the hot spots (for simplicity, only one strand of the double strand of DNA is indicated):

$$—GTCTGG \; CTGG \; CTGG \; CTGG \; C$$

FS5, FS25, FS45, FS65

wild type   5′—GTCTGG CTGG CTGG C—3′

FS2, FS84

$$GTCTGG \; CTGG \; C$$

The major hot spot, represented here by the mutations *FS5*, *FS25*, *FS45*, and *FS65*, results from the addition of one extra set of the four bases CTGG to one strand of the DNA. This hot spot reverts at a high rate, losing the extra set of four bases. The minor hot spot, represented here by the mutations *FS2* and *FS84*, results from the loss of one set of the four bases CTGG. This mutant does not readily regain the lost set of four base pairs.

How can the Streisinger model explain these observations? The model predicts that the frequency of a particular frameshift depends on the number of base pairs that can form during the slipped mispairing of repeated sequences. The wild-type sequence shown for the *lacI* gene can slip out one CTGG sequence and stabilize this by forming nine base pairs. (Can you work this out by applying the model in Figure 19-4 to the sequence shown for *lacI*?) Whether a deletion or an addition is generated depends on whether the slippage occurs on the template or on the newly synthesized strand, respectively. In a similar fashion, the addition mutant can slip out one CTGG sequence and stabilize this with 13 base pairs (verify this for the *FS5* sequence shown for *lacI*), which explains the rapid reversion of mutations such as *FS5*. However, there are only five base pairs available to stabilize a slipped-out CTGG in the deletion mutant, accounting for the infrequent reversion of mutations such as *FS2* in the sequence shown for *lacI*.

Deletions and Duplications. Large **deletions** (more than a few base pairs) represent a sizable fraction of spontaneous mutations, as shown in Figure 19-5. The majority, although not all, of the deletions occur at repeated sequences. Figure 19-6 shows the results for the first 12 deletions analyzed at

| | S74, S112 | 75 bases | |

CAATTCAGGGTGGTGAATGTGAAACC------CGCGTGGTGAACCAGG

Site (no. of bp)	Sequence repeat	No. of bases deleted	Occurences	
20 to 95	GTGGTGAA	75	2	S74, S112
146 to 269	GCGGCGAT	123	1	S23
331 to 351	AAGCGGCG	20	2	S10, S136
316 to 338	GTCGA	22	2	S32, S65
694 to 707	CA	13	1	S24
694 to 719	CA	25	1	S56
943 to 956	G	13	1	S42
322 to 393	None	71	1	S120
658 to 685	None	27	1	S86

Figure 19-6 Deletions in *lacI*. Deletions occurring in *S74* and *S112* are shown at the top of the figure. As indicated by the gold bars, one of the sequence repeats (aqua) and all the intervening DNA is deleted, leaving one copy of the repeated sequence. All mutations were analyzed by direct DNA sequence determination. (From P. J. Farabaugh, U. Schmeissner, M. Hofer, and J. H. Miller, *Journal of Molecular Biology* 126, 1978, 847.)

the DNA sequence level, presented by Miller and his co-workers in 1978. Further studies have shown that hot spots for deletions involve the longest sequences that are repeated. **Duplications** of segments of DNA have been observed in many organisms. Like deletions, they often occur at sequence repeats.

How do deletions and duplications form? Several mechanisms could account for this. Deletions may be generated as replication errors. For example, an extension of the Streisinger model of slipped mispairing (Figure 19-4) could explain why deletions predominate at short repeated sequences. Alternatively, deletions and duplications could be generated by recombinational mechanisms (to be described in Chapter 20).

Spontaneous Lesions

In addition to replication errors, **spontaneous lesions,** naturally occurring damage to the DNA, can also generate mutations. Two of the most frequent spontaneous lesions result from depurination and deamination.

Depurination, the more common of the two, involves the interruption of the glycosidic bond between the base and deoxyribose and the subsequent loss of a guanine or an adenine residue from the DNA (Figure 19-7). A mammalian cell spontaneously loses about 10,000 purines from its DNA during a 20-hour cell generation period at 37°C. If these lesions were to persist, they would result in significant genetic damage because, during replication, the resulting **apurinic sites** cannot specify a base complementary to the original purine. However, as we shall see later in the chapter, efficient repair systems remove apurinic sites. Under certain

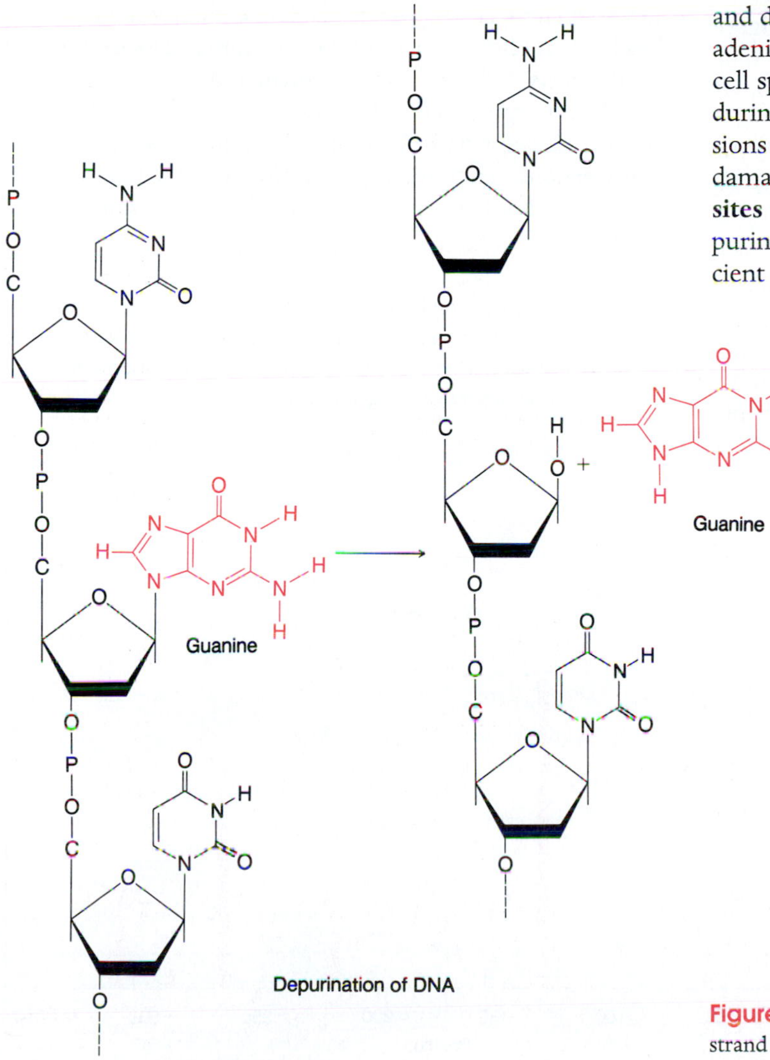

Depurination of DNA

Figure 19-7 The loss of a purine residue (guanine) from a single strand of DNA. The sugar-phosphate backbone is left intact.

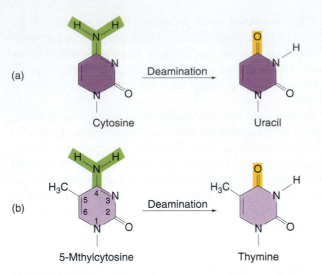

(a)

Cytosine → Deamination → Uracil

(b)

5-Mthylcytosine → Deamination → Thymine

Figure 19-8 Deamination of (a) cytosine and (b) 5-methylcytosine.

conditions (to be described later) a base can be inserted across from an apurinic site; this will frequently result in a mutation.

The **deamination** of cytosine yields uracil (Figure 19-8a). Unrepaired uracil residues will pair with adenine during replication, resulting in the conversion of a G–C pair to an A–T pair (a **GC → AT transition**). In 1978 it was found that deaminations at certain cytosine residues are the cause of one type of mutational hot spot. DNA sequence analysis of GC → AT transition hot spots in the *lacI* gene has shown that 5-methylcytosine residues occur at the position of each hot spot. (Certain bases in prokaryotes and eukaryotes are methylated; see page 421.) Some of the data from this *lacI* study are shown in Figure 19-9. The height of each bar on the graph represents the frequency of mutations at each of a number of sites. It can be seen that the po-

sition of 5-methylcytosine residues correlates nicely with the most mutable sites.

Why are 5-methylcytosines hot spots for mutations? One of the repair enzymes in the cell, uracil-DNA glycosylase, recognizes the uracil residues in the DNA that arise from deaminations and excises them, leaving a gap that is subsequently filled in (a process to be described later in the chapter). However, the deamination of 5-methylcytosine (Figure 19-8b) generates thymine (5-methyluracil), which is not recognized by the enzyme uracil-DNA glycosylase and thus is not repaired. Therefore, C → T transitions generated by deamination are seen more frequently at 5-methylcytosine sites, since they escape this repair system.

A consequence of the frequent mutation of 5-methylcytosine to thymine is the underrepresentation of CpG dinucleotides in higher cells, since this sequence is methylated to give 5-methyl-CpG, which is gradually converted to TpG.

Oxidatively damaged bases represent a third type of spontaneous lesion implicated in mutagenesis. Active oxygen species, such as superoxide radicals ($O_2 \cdot$), hydrogen peroxide (H_2O_2), and hydroxyl radicals ($OH \cdot$), are produced as by-products of normal aerobic metabolism. These can cause oxidative damage to DNA, as well as to precursors of DNA (such as GTP), which results in mutation and which has been implicated in a number of human diseases. Figure 19-10 shows two products of oxidative damage. The 8-oxo-7-hydrodeoxyguanosine (8-oxodG, or "GO") product frequently mispairs with A, resulting in a high level of G → T transversions. Thymidine glycol blocks DNA replication if unrepaired but has not yet been implicated in mutagenesis.

Message Spontaneous mutations can be generated by different processes. Replication errors and spontaneous lesions generate most of the base-substitution and frameshift mutations. Replication errors may also cause some deletions that occur in the absence of mutagenic treatment.

Figure 19-9 5-Methylcytosine hot spots in *E. coli*. Nonsense mutations occurring at 15 different sites in *lacI* were scored. All result from the GC → AT transition. The asterisks (*) mark the position of 5-methylcytosines. Open bars depict sites at which the GC → AT change could be detected but at which no mutations occurred in this particular collection. It can be seen that 5-methylcytosine residues are hot spots for the GC → AT transition. Of 50 independently occurring mutations, 44 were at the four 5-methylcytosine sites and only 6 were at the 11 unmethylated cytosines. (From C. Coulondre et al., *Nature* 274, 1978, 775.).

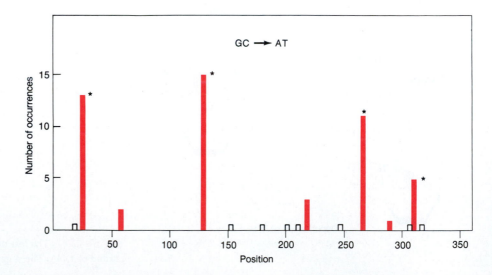

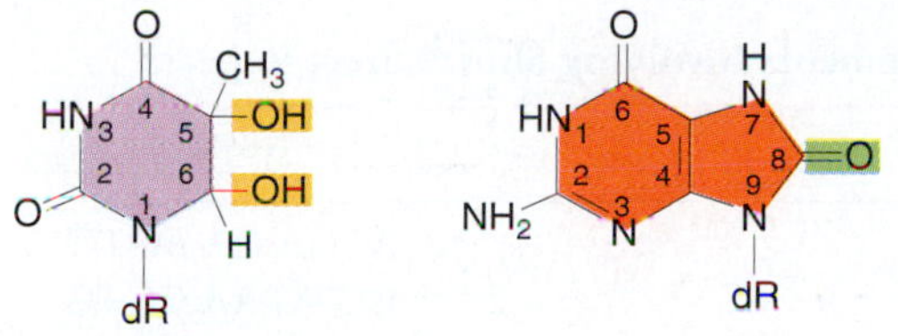

Thymidine glycol

8-Oxo-7-hydrodeoxyguanosine
(8-oxodG)

Figure 19-10 DNA damage products formed after attack by oxygen radicals. dR = deoxyribose.

Spontaneous Mutations and Human Diseases

DNA sequence analysis has revealed the mutations responsible for a number of human hereditary diseases. The previously discussed studies of bacterial mutations allow us to suggest mechanisms that cause these human disorders.

A number of these disorders are due to **deletions or duplications** involving repeated sequences. For example, mitochondrial encephalomyopathies are a group of disorders affecting the central nervous system or the muscles (Kearns-Sayre syndrome). They are characterized by dysfunction of oxidation phosphorylation (a function of the mitochondria) and by changes in mitochondrial structure. These disorders have been shown to result from deletions that occur between repeated sequences. Figure 19-11 depicts one of these deletions. Note how similar it is in form to the spontaneous *E. coli* deletions shown in Figure 19-6. A second example is Fabry disease. This inborn error of glycosphingolipid catabolism results from mutations in the X-linked gene encoding the enzyme α-galactosidase A. Many of these mutations are gene rearrangements, resulting from either deletions or duplications between short direct repeats. Table 19-2 shows the short repeats at the breakpoints of rearrangements leading to Fabry disease, and also some rearrangements in the globin genes, resulting in anemias and thalassemias, that have been analyzed. All these deletions occurred either by a slipped mispairing mechanism, such as pictured in Figure 19-4, or by recombination between the repeated sequences.

A common mechanism that is responsible for a number of genetic diseases is the **expansion of a three-base-pair repeat,** as in the case of fragile X syndrome (Figure 19-12). This syndrome is the most common form of inherited mental retardation, occurring in close to 1 of 1500 males and 1 of 2500 females. It is evidenced cytologically by a fragile site in the X chromosome that results in a break in vitro.

The inheritance of fragile X syndrome is unusual in that 20 percent of the males with a fragile X chromosome are phenotypically normal but transmit the affected chromosome to their daughters, who also appear normal. These males are said to be "normally transmitting males" (NTMs). However, the sons of the daughters of the NTMs frequently display symptoms. The fragile X syndrome results from mutations in a $(CGG)_n$ repeat in the coding sequence of the *FMR-1* gene. Patients with the disease show specific methylation, induced by the mutation, at a nearby CpG cluster, resulting in reduced *FMR-1* expression.

Why do symptoms develop in some persons with a fragile X chromosome and not in others? The answer seems to lie in the number of CGG repeats in the *FMR-1* gene. Humans normally show a considerable variation in the number of CGG repeats in the *FMR-1* gene, ranging from 6 to 54, with 29 repeats representing the most frequent allele. [The variation of CGG repeats produces a corresponding variation in the number of arginine residues (CGG is an arginine codon) in the *FMR-1*-encoded protein.] Both NTMs and their daughters have a much larger number of repeats, ranging from 50 to 200. These increased repeats have been termed **premutations.** All premutation alleles are unstable. The males and females with symptoms of the disease, as well as many carrier females, have additional insertions of DNA, suggesting repeat numbers of 200 to 1300. It has been shown that the frequency of expansion increases with the size of the DNA insertion (and thus, presumably, with the number of repeats). Apparently, the number of repeats in the premutation alleles found in NTMs and their daughters is above a certain threshold, and thus is much more likely to expand to a full mutation than is the case for normal individuals.

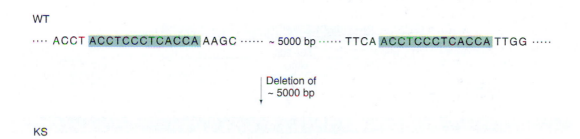

Figure 19-11 Sequences of wild-type (WT) mitochondrial DNA, and deleted DNA (KS) from a patient with Kearns-Sayre/chronic external opthalmoplegia plus syndrome. The 13-base boxed sequence is identical in both WT and KS, and serves as a breakpoint for the DNA deletion. A single base (**bold** type) is altered in KS, aside from the deleted segment.

Table 19-2   Breakpoint Sequences in Mammalian Germinal Rearrangements Involving Short Direct Repeats

Rearrangement	5' Breakpoint	3' Breakpoint
Hb Leiden	TC**CT GA** GGAG	AGGA **GA** AGTC
Hb Lyon	GGGC **AA** GGTG	GGTG **AA** CGTG
Hb Freiburg	TGAA **GT** TGGT	GTTG **GT** GGTG
Hb Niteroi	GGTT **CTTTG** AGTC	AGTC **CTTTG** GGGA
Hb Gun Hill	AGTG **AGCTGCA CTG**T	GACA **AGCTGCA** CGTG
Hb Tochigi	TATG **GG** CAAC	CTAA **GG** TGAA
Hb St. Antoine	TGAT **GGC CTGG**	GC**CT GGC** TCAC
Hb Coventry	TAAT **GCCC TGGC**	**CCTG GCCC** ACAA
γδβ-Thal 1	TC**CC AG** CACT	GAAA **AG** TCTG
Hb BK	GGTA **TCT GG**AG	AATT **TCT** ATTA
Indian HPFH	CGCG **CCACT G**CAC	ATCC **CCACT** ATAT
Dutch β⁰-Thal	AACC **AAATTT** GCAC	GAGA **AAATTT** TTGC
Turkish β-Thal	GTCT **ACCC** TTGG	TTGG **ACCC AG**AG
−(α)²⁰·⁵	CCTA **GGC** AACA	TAAG **GGC** CACG
RB 1	AGCT **TTTATAC** TTGA	TGAA **TTTAAAC** ATAA
Pro-α2(I)	TTTC **TTTC** TAAG	GTGG **TTTC** CCTG
Fabry Family A	GAAC **CCA G**AAC	AGCT **CCA** CCTC
Fabry Family B	CATC **AAG** GGTA	TTAA **AAG** ACTT
Fabry Family D	AGCT **TAGACA G**GTA	AATG **TAGATA** AAGA
Fabry Family E	CAG**C AGAACT G**GGG	GGAA **AGAACT** TTGA
Fabry Family J	TTTA **AC CAG**G	TAAT **AC** ATTT
	GAAA **AT** TTTA	ATGT **AT** AGGC

NOTE: For each gene rearrangement, the 5' and 3' breakpoint regions are shown. Spaces have been inserted between the direct repeats (in **bold**) and the four adjacent 5' and 3' nucleotides which may be involved in the recombinational event. CTGG/CCAG tetranucleotides and CTG/CAG trinucleotides are indicated by highlighting. Hb = hemoglobin; Thal = thalassemia.

SOURCE: R. Kornreich, D. F. Bishop, and R. J. Desnick, *Journal of Biological Chemistry* 2, 1990, 9319.

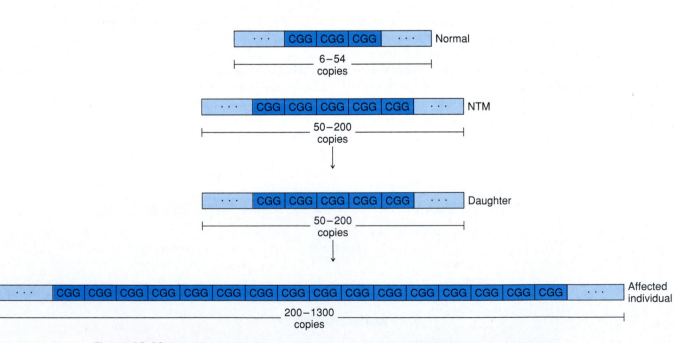

Figure 19-12 Expansion of the CGG triplet in the *FMR-1* gene seen in the fragile X syndrome. Normal individuals have from 6 to 54 copies of the CGG repeat, while individuals from susceptible families display an increase (premutation) in the number of repeats: normally transmitting males (NTMs) and their daughters are phenotypically normal but display 50 to 200 copies of the CGG triplet; the number of repeats expands to some 200 to 1300 in individuals showing full symptoms of the disease.

The proposed mechanism for these repeats is a slipped mispairing during DNA synthesis, just as shown previously (see diagram on page 587) for the *lacI* hot spot involving a one-step expansion of the four-base-pair sequence CTGG. However, the extraordinarily high frequency of mutation at the three-base-pair repeats in the fragile X syndrome suggests that in human cells, after a threshold level of about 50 repeats, the replication machinery cannot faithfully replicate the correct sequence, and large variations in repeat numbers result.

A second inherited disease, X-linked spinal and bulbar muscular atrophy (known as *Kennedy's disease*), also results from the amplification of a three-base-pair repeat, in this case a repeat of the CAG triplet. Kennedy's disease, which is characterized by progressive muscle weakness and atrophy, results from mutations in the gene that codes for the androgen receptor. Normal individuals have an average of 21 CAG repeats in this gene, while affected patients have repeats ranging from 40 to 52.

Myotonic dystrophy, the most common form of adult muscular dystrophy, is yet another example of sequence expansion causing a human disease. Susceptible families display an increase in severity of the disease in successive generations; this is caused by the progressive amplification of a CTG triplet at the 3' end of a transcript. Normal individuals possess, on average, five copies of the CTG repeat; mildly affected individuals have approximately 50 copies, and severely affected people have more than 1000 repeats of the CTG triplet. Additional examples of triplet expansion are still appearing, for instance, Huntington's disease, which has recently been added to the list.

Induced Mutations

Mutational Specificity

When we observe the distribution of mutations induced by different mutagens, we see a distinct specificity that is characteristic of each mutagen. Such **mutational specificity** was first noted in the *rII* system by Benzer in 1961. Specificity arises from a given mutagen's "preference" both for a certain *type* of mutation (for example, GC → AT transitions) and for certain mutational *sites* (hot spots). Figure 19-13 shows the mutational specificity in *lacI* of three mutagens described later: ethyl methanesulfonate (EMS), ultraviolet (UV) light, and aflatoxin B$_1$ (AFB$_1$). The graphs show the distribution of base-substitution mutations that create chain-terminating UAG codons. Figure 19-13 is similar to Figure 12-34, which shows the distribution of mutations in *rII*, except that the specific sequence changes are known for each *lacI* site, allowing the graphs to be broken down into each category of substitution.

Figure 19-13 reveals the two components of mutational specificity. First, each mutagen shown favors a specific category of substitution. For example, EMS and UV favor GC → AT transitions, whereas AFB$_1$ favors GC → TA

transversions. This preference is related to the different mechanisms of mutagenesis. Second, even within the same category, there are large differences in mutation rate. This can be seen best with UV light for the GC → AT changes. Some aspect of the surrounding DNA sequence must cause these differences. In some cases, the cause of mutational hot spots can be determined by DNA sequence studies, as previously described for 5-methylcytosine residues and for certain frameshift sites (Figures 19-5 and 19-9). In many examples of mutagen-induced hot spots the precise reason for the high mutability of specific sites is still unknown. However, high lesion frequency at some sites and reduced repair at certain sites are sometimes causes of hot spots.

Mechanisms of Mutagenesis

Mutagens induce mutations by at least three different mechanisms. They can replace a base in the DNA, alter a base so that it specifically mispairs with another base, or damage a base so that it can no longer pair with any base under normal conditions.

Incorporation of Base Analogs. Some chemical compounds are sufficiently similar to the normal nitrogen bases of DNA that they occasionally are incorporated into DNA in place of normal bases; such compounds are called **base analogs.** Once in place, these analogs have pairing properties unlike those of the normal bases; thus, they can produce mutations by causing incorrect nucleotides to be inserted opposite them during replication. The original base analog exists in only a single strand, but it can cause a nucleotide-pair substitution that is replicated in all DNA copies descended from the original strand.

For example, **5-bromouracil (5-BU)** is an analog of thymine that has bromine at the C-5 position in place of the CH$_3$ group found in thymine. This change does not involve the atoms that take part in hydrogen bonding during base pairing, but the presence of the bromine significantly alters the distribution of electrons in the base. The normal structure (the keto form) of 5-BU pairs with adenine, as shown in Figure 19-14a. 5-BU can frequently change to either the enol form or an ionized form; the latter pairs in vivo with guanine (Figure 19-14b). Thus, the nature of the pair formed during replication will depend on the form of 5-BU at the moment of pairing (Figure 19-15). 5-BU causes transitions almost exclusively, as predicted in Figures 19-14 and 19-15.

Another analog widely used in research is **2-aminopurine (2-AP)**, which is an analog of adenine that can pair with thymine but can also mispair with cytosine when protonated, as shown in Figure 19-16. Therefore, when 2-AP is incorporated into DNA by pairing with thymine, it can generate AT → GC transitions by mispairing with cytosine during subsequent replications. Or, if 2-AP is incorporated by mispairing with cytosine, then GC → AT transitions will result when it pairs with thymine. Genetic studies have shown that 2-AP, like 5-BU, is very specific for transitions.

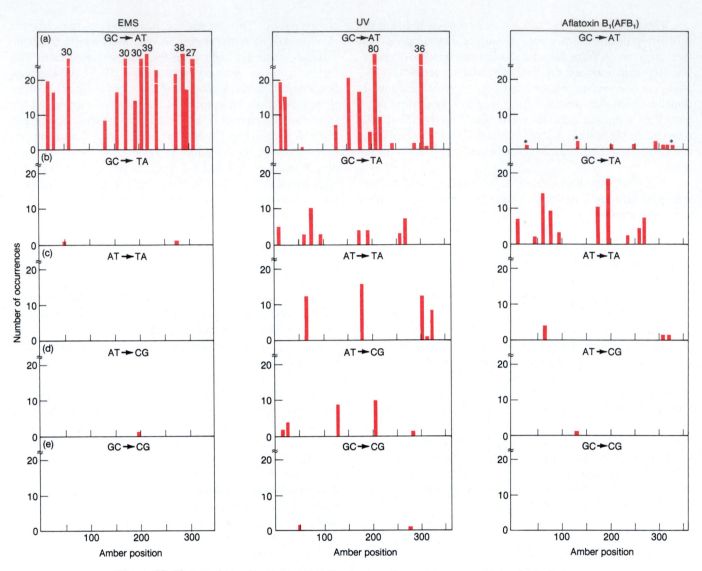

Figure 19-13 Specificity of mutagens. The distribution of mutations among 36 sites in the *lacI* gene is shown for three mutagens: EMS, UV light, and aflatoxin B₁. The height of each bar represents the number of occurrences of mutations at the respective site. Some hot spots are shown off-scale, with the number of occurrences indicated directly above the respective peak. For instance, in the UV-generated collection, one site resulting from a GC → AT transition is represented by 80 occurrences. Each mutational site represented in the figure generates an amber (UAG) codon in the corresponding mRNA. The mutations are arranged according to the type of base substitution involved. Asterisks mark the position of 5-methylcytosines. (Redrawn from C. Coulondre and J. H. Miller, *Journal of Molecular Biology* 117, 1977, 577; and P. L. Foster et al., *Proceedings of the National Academy of Sciences USA* 80, 1983, 2695.)

Figure 19-14 Alternative pairing possibilities for 5-bromouracil (5-BU). 5-BU is an analog of thymine that can be mistakenly incorporated into DNA as a base. It has a bromine atom in place of the methyl group. (a) In its normal keto state, 5-BU mimics the pairing behavior of the thymine it replaces, pairing with adenine. (b) The presence of the bromine atom, however, causes a relatively frequent redistribution of electrons, so that 5-BU can spend part of its existence in the rare ionized form. In this state, it pairs with guanine, mimicking the behavior of cytosine and thus inducing mutations during replication.

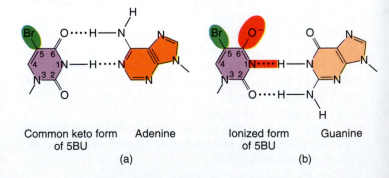

Common keto form   Adenine
of 5BU

(a)

Ionized form   Guanine
of 5BU

(b)

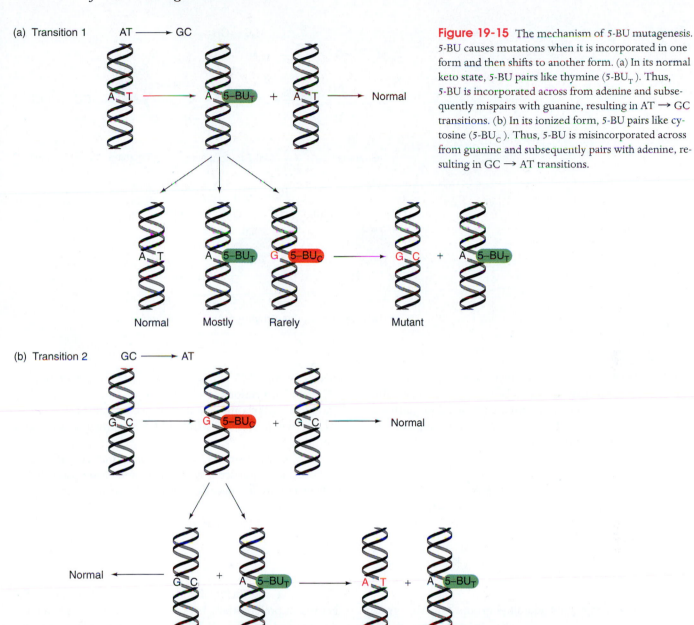

(a) Transition 1   AT ⟶ GC

Normal   Mostly   Rarely   Mutant

(b) Transition 2   GC ⟶ AT

Normal   Mostly   Mutant

Figure 19-15 The mechanism of 5-BU mutagenesis. 5-BU causes mutations when it is incorporated in one form and then shifts to another form. (a) In its normal keto state, 5-BU pairs like thymine (5-BU$_T$). Thus, 5-BU is incorporated across from adenine and subsequently mispairs with guanine, resulting in AT → GC transitions. (b) In its ionized form, 5-BU pairs like cytosine (5-BU$_C$). Thus, 5-BU is misincorporated across from guanine and subsequently pairs with adenine, resulting in GC → AT transitions.

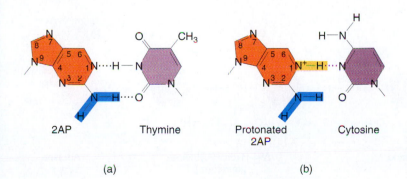

2AP   Thymine   Protonated 2AP   Cytosine

(a)   (b)

Figure 19-16 Alternative pairing possibilities for 2-aminopurine (2-AP), an analog of adenine. Normally, 2-AP pairs with thymine (a), but in its protonated state it can pair with cytosine (b).

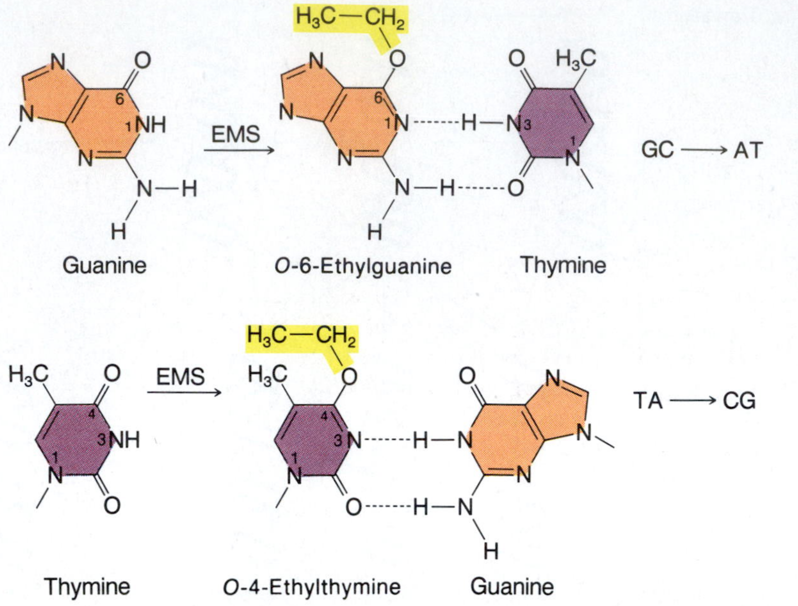

Figure 19-17 Alkylation-induced specific mispairing. The alkylation (in this case, EMS-generated ethylation) of the *O*-6 position of guanine, and the *O*-4 position of thymine, can lead to direct mispairing with thymine and guanine, respectively, as shown here. In bacteria, where mutations have been analyzed in great detail, the principal mutations detected are GC → AT transitions, indicating that the *O*-6 alkylation of guanine is most relevant to mutagenesis.

Specific Mispairing. Some mutagens are not incorporated into the DNA but instead alter a base, causing specific mispairing. Certain **alkylating agents,** such as **ethyl methanesulfonate (EMS)** and the widely used **nitrosoguanidine (NG),** operate via this pathway:

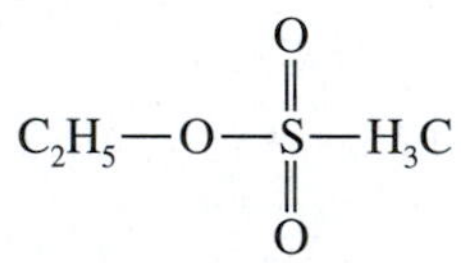

EMS 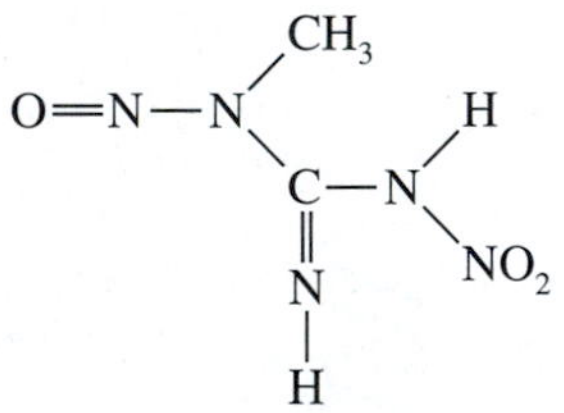　　　　　　　NG

Although such agents add alkyl groups (an ethyl group in the case of EMS and a methyl group in the case of NG) to many positions on all four bases, mutagenicity is best correlated with an addition to the oxygen at the 6 position of guanine to create an *O*-6-alkylguanine. This leads to direct mispairing with thymine, as shown in Figure 19-17 and would result in GC → AT transitions at the next round of replication. As expected, determinations of mutagenic specificity for EMS and NG show a strong preference for GC → AT transitions (see the data for EMS shown in Figure 19-13). Alkylating agents can also modify the bases in dNTPs (where *N* is any base), which are precursors in DNA synthesis.

Hydroxylamine (HA) is a specific inducer of GC → AT transitions, particularly in phage and *Neurospora;* the effects of HA are less specific in *E. coli*. Its structure is

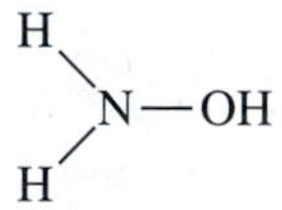

The relative specificity of HA is very probably due to the fact that it preferentially hydroxylates the amino nitrogen at cytosine C-4, creating *N*-4-hydroxycytosine, which can mispair with adenine (Figure 19-18). *N*-4-Hydroxycytosine prepared in vitro has the same ability to pair like thymine and cause mutations (see page 461), which strongly supports the proposed mechanism.

Other examples of mutagens that alter bases specifically are those that deaminate cytosine to uracil (see Figure

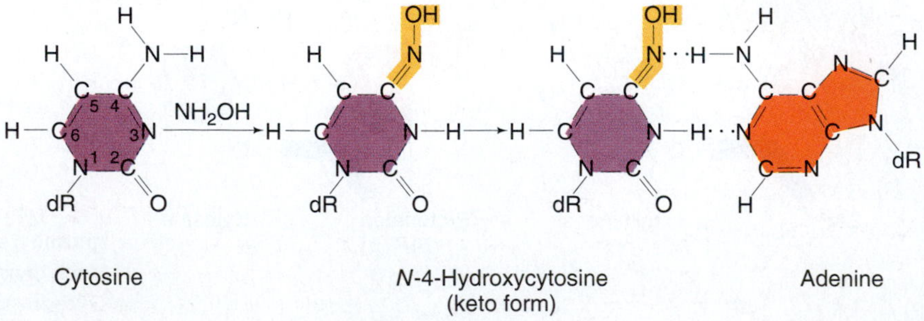

Figure 19-18 A possible explanation for the GC → AT specificity of hydroxylamine (HA) in some organisms. Cytosine is modified to *N*-4-hydroxycytosine, which can mispair with adenine, resulting in a GC → AT transition. dR = deoxyribose.

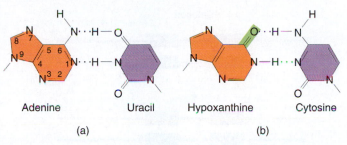

Figure 19-19 Nitrous acid (NA) mutagenesis. (a) NA deaminates cytosine to form uracil, which bonds like thymine. (b) NA deaminates adenine to form hypoxanthine, which bonds like guanine. These altered bonding patterns can lead to mutations. For example, AT may become GC, or GC may become AT. (Modified from E. Freese, in *Structure and Function of Genetic Elements,* Brookhaven Symposia in Biology No. 12, Brookhaven National Laboratory, Upton, N.Y., 1959.)

19-8). For instance, **bisulfite ions** convert cytosine to uracil, as does **nitrous acid (NA).** The uracil residue, if unrepaired, will pair with adenine instead of guanine, generating a C → T transition. Nitrous acid also deaminates adenine to generate hypoxanthine, which can form A–C mispairs (Figure 19-19).

The **intercalating agents** form another important class of DNA modifiers. This group of compounds includes **proflavin, acridine orange,** and a class of chemicals termed **ICR compounds** (Figure 19-20a). These agents are planar molecules, which mimic base pairs and are able to slip themselves in (**intercalate**) between the stacked nitrogen bases at the core of the DNA double helix (Figure 19-20b). In this intercalated position, the agent can cause single-nucleotide-pair insertions or deletions. Intercalating agents may also stack between bases in single-stranded DNA; in so doing, they may stabilize bases that are looped out during frameshift formation, as depicted in the Streisinger model (Figure 19-4). A model for the actions of intercalating agents is shown in Figure 19-21.

Loss of Specific Pairing. A large number of mutagens damage one or more bases so that specific pairing is no longer possible. The result is a replication block, because DNA synthesis will not proceed past a base that cannot specify its complementary base by hydrogen bonding. This is for a good reason: the insertion of bases across from noncoding lesions would lead to frequent mutations.

The **bypass** of such replication blocks by inserting bases requires the activation of a special system in bacterial cells, the **SOS system** (Figure 19-22). The name *SOS* comes from the idea that this system is induced as an emergency response to prevent cell death in the presence of significant DNA damage. SOS induction is a last resort, allowing the cell to trade a certain level of mutagenesis for ultimate survival. (This "induction" is really the activation of gene expression, described more fully in Chapter 18).

Exactly how the SOS bypass system functions is not clear, although in *E. coli* it is known to be dependent on at least three genes, *recA* (which is also involved in general recombination, as described in Chapter 20), *umuC*, and *umuD*. Current models for SOS bypass suggest that the *umuC* and *umuD* proteins combine with the polymerase III DNA replication complex to loosen its otherwise strict specificity and permit replication past noncoding lesions.

Mutagens are dependent on the SOS system for their action when they cause the loss of specific pairing, thereby generating noncoding lesions and blocking replication. The category of SOS-dependent mutagens is important, since it includes most carcinogens, such as ultraviolet (UV) light and aflatoxin B$_1$ (discussed below) and benzo(a)pyrene (see Figure 19-26, discussed later).

How is the SOS system involved in the recovery of cells with mutations induced by certain mutagens? Does the SOS system lower the fidelity of DNA replications so much (to permit the bypass of noncoding lesions) that many replication errors occur, even for undamaged DNA? If this hypothesis were correct, most mutagens generated by different SOS-dependent mutagens would be similar, rather than spe-

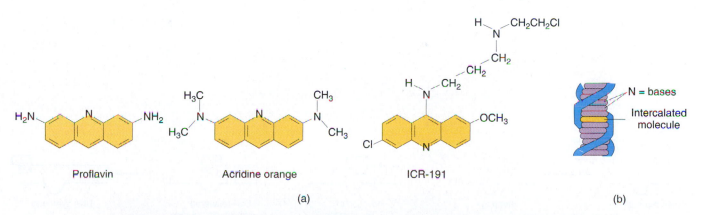

Figure 19-20 Intercalating agents. (a) Structures of the common agents proflavin, acridine orange, and ICR-191. (b) An intercalating agent slips between the nitrogenous bases stacked at the center of the DNA molecule. This occurrence can lead to single-nucleotide-pair insertions and deletions. (From L. S. Lerman, *Proceedings of the National Academy of Sciences USA* 49, 1963, 94.)

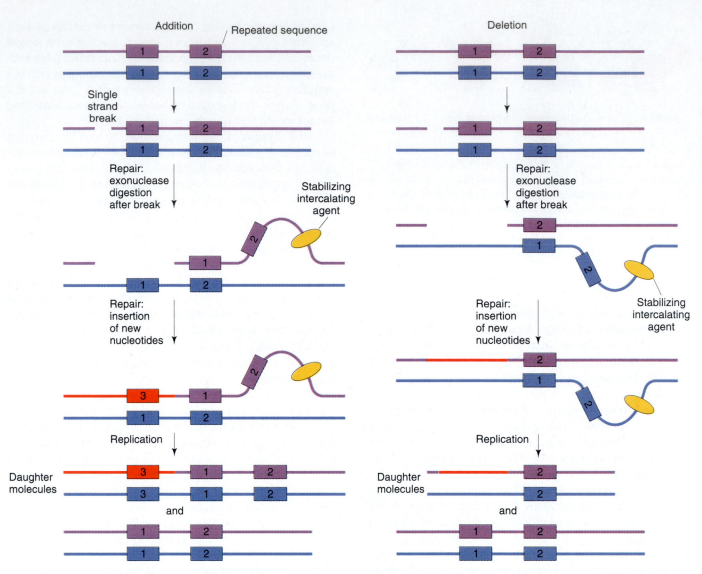

Figure 19-21 Model showing how intercalating agents can cause short deletions or insertions. Here we assume that the agents are active only during DNA processing (repair or recombination), when they insert into a single-stranded loop and stabilize it. (See also Figures 19-4 and 19-20.) For an addition to occur, the loop can form only if there is a short repeat length of complementary sequence. New DNA is shown in red.

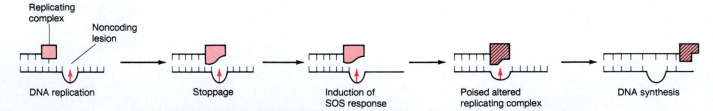

Figure 19-22 The SOS bypass system. This highly schematic diagram represents the stoppage of DNA replication in response to a noncoding lesion. The SOS system relieves this blockage, perhaps by altering the replicating complex.

(a)

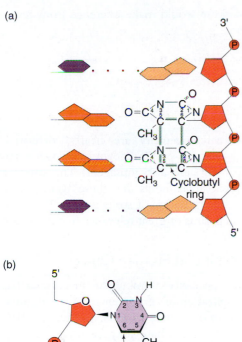

(b)

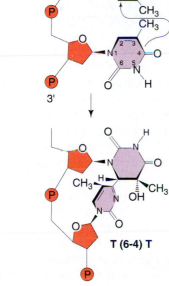

Figure 19-23 (a) Structure of a cyclobutane pyrimidine dimer. Ultraviolet light stimulates the formation of a four-membered cyclobutane ring (orange) between two adjacent pyrimidines on the same DNA strand by acting on the 5,6 double bonds. (b) Structure of the 6-4 photoproduct. The structure forms most prevalently with 5′-C-C-3′ and 5′-T-C-3′, between the C-6 and the C-4 positions of two adjacent pyrimidines, causing a significant perturbation in local structure of the double helix. (Part a adapted from E. C. Friedberg, *DNA Repair.* Copyright © 1985 by W. H. Freeman. Part b adapted from J. S. Taylor et al.,

cific to each mutagen. Most mutations would result from the action of the SOS system itself on undamaged DNA. The mutagen, then, would play the indirect role of inducing the SOS system. Studies of mutational specificity, however, have shown that this is not the case. Instead, a series of different SOS-dependent mutagens have markedly different specificities. You can see this by comparing the specificities

of UV light and aflatoxin B_1 in Figure 19-13. Each mutagen induces a unique distribution of mutations. Therefore, the mutations must be generated in response to specific damaged base pairs. The type of lesion differs in many cases. Some of the most widely studied lesions include UV photoproducts, apurinic sites, and bulky chemical additions on specific bases.

Ultraviolet (UV) light generates a number of photoproducts in DNA. Two different lesions that occur at adjacent pyrimidine residues—the cyclobutane pyrimidine photodimer and the 6-4 photoproduct (Figure 19-23)—have been most strongly correlated with mutagenesis. These lesions interfere with normal base pairing; hence, induction of the SOS system is required for mutagenesis. The insertion of incorrect bases across from UV photoproducts occurs at the 3′ position of the dimer, and more frequently for 5′-CC-3′ and 5′-TC-3′ dimers. The C → T transition is the most frequent mutation, but other base substitutions (transversions) and frameshifts are also stimulated by UV light, as are duplications and deletions. The mutagenic specificity of UV light is illustrated in Figure 19-13.

Aflatoxin B_1 (AFB_1) is a powerful carcinogen. It generates **apurinic sites** following the formation of an addition product at the N-7 position of guanine (Figure 19-24). Studies with apurinic sites generated in vitro have demonstrated a requirement for the SOS system and have also shown that the SOS bypass of these sites leads to the preferential insertion of an adenine across from an apurinic site. This predicts that agents that cause depurination at guanine residues should preferentially induce GC → TA transversions. Can you see why the insertion of an adenine across from an apurinic site derived from a guanine would generate this substitution at the next round of replication? Figure 19-13 shows the genetic analysis of many base substitutions induced by AFB_1. You can verify that most of the substitutions are indeed GC → TA transversions.

AFB_1 is a member of a class of chemical carcinogens that represent **bulky addition products** when they bind covalently to DNA. Other examples include the diol epoxides of **benzo(a)pyrene,** a compound produced by internal combustion engines. For many different compounds it is not yet clear which DNA addition products play the principal role in

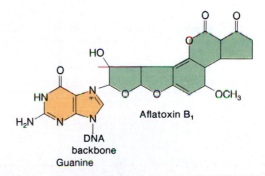

Figure 19-24 The binding of metabolically activated aflatoxin B_1 to DNA.

mutagenesis. In some cases, the mutagenic specificity suggests that depurination may represent an intermediate step in mutagenesis; in others, the question of which mechanism is operating is completely open.

Message Mutagens induce mutations by a variety of mechanisms. Some mutagens mimic normal bases and are incorporated into DNA, where they can mispair. Others damage bases and either cause specific mispairing or destroy pairing by causing nonrecognition of bases. In the latter case, a bypass system, the SOS system, must be induced in order to allow replication past the lesion.

Reversion Analysis

Testing for the reversion of a mutation can tell us something about the nature of the mutation or the action of a mutagen. For example, if a mutation cannot be reverted by action of the mutagen that induced it, then the mutagen must have some relatively specific unidirectional action. In the case of a mutation induced by hydroxylamine (HA), for instance, it would be reasonable to expect that the original mutation is GC → AT, which of course cannot be reverted by another specific GC → AT event. Similarly, mutations that can be reverted by proflavin are in all likelihood frameshift mutations; thus mutations induced by nitrous acid (NA), which are transitions, should not be revertible by proflavin.

Transversions cannot be induced by the agents mentioned above, but they are known definitely to be common among spontaneous mutations, as shown by studies of DNA and protein sequencing. Thus, in the reversion test, if a mutation reverts spontaneously but does not revert in response to a transition mutagen or a frameshift mutagen, then, by elimination, it is probably a transversion.

Table 19-3 summarizes some reversion expectations based on simple assumptions from reversion analysis. Recall that mutagen specificities depend on the organism, the genotype, the gene studied, and the action of biological repair systems. Note also that the kinds of logic employed in the reversion test rely heavily on the assumption that the reversion events are not due to suppressors or transposable el-

ements; either of these would make inference from reversion more difficult.

The Relationship between Mutagens and Carcinogens

Mutagenicity and carcinogenicity are clearly correlated. One study showed that 157 of 175 known carcinogens (approximately 90 percent) are also mutagens. The **somatic mutation theory** of cancer holds that these agents cause cancer by inducing the mutation of somatic cells. Thus, understanding mutagenesis is of great relevance to our society.

Induced Mutations and Human Cancer

Understanding the specificity of mutagens in bacteria has led to the direct implication of certain environmental mutagens in the causation of human cancers. Ultraviolet (UV) light and aflatoxin B_1 (AFB_1) have long been suspected of causing skin cancer and liver cancer, respectively. Now, DNA sequence analysis of mutations in a human cancer gene has provided direct evidence of their involvement. The gene in question is termed *p53*, and is one of a number of **tumor-suppressor genes**—genes which encode proteins that suppress tumor formation. (We will learn more about these genes in Chapter 24). A sizable proportion of human cancer patients have mutated tumor-suppressor genes.

Liver cancer is prevalent in southern Africa and East Asia, and a high exposure to AFB_1 in these regions has been correlated with the high incidence of liver cancer. When *p53* mutations in cancer patients were analyzed, G → T transversions, the signature of AFB_1-induced mutations, were found in liver cancer patients from South Africa and East Asia, but not in patients from these regions with lung, colon, or breast cancer. On the other hand, *p53* mutations in liver cancer patients from areas of low AFB_1 exposure did not result from G → T transversions. These findings, together with the results from the mutagenic specificity studies of AFB_1 (see Figure 19-13), allow us to conclude that AFB_1-induced mutations are a prime cause of liver cancer in South Africa and East Asia.

Table 19-3   Reversion Tests

	Reversion Mutagen			Spontaneous
Mutation	NA	HA or EMS	Proflavin	Reversion
Transition (GC → AT)	+	−	−	+
Transition (AT → GC)	+	+	−	+
Transversion	−	−	−	+
Frameshift	−	−	+	+

NOTE: + indicates a measurable rate of reversion due to a given mutagen. NA = nitrous acid; HA = hydroxylamine; EMS = ethyl methanesulfonate.

Sequencing *p53* mutations has also strengthened the link between UV and human skin cancers. The majority of invasive human squamous cell carcinomas analyzed so far have *p53* mutations, all of them involving mutations at dipyrimidine sites, most of which are C $\longrightarrow$ T substitutions when the C is the 3' pyrimidine of a TC dimer. This is the profile of UV-induced mutations. In addition, several tumors have *p53* mutations resulting from a CC $\longrightarrow$ TT double base change, which is found most frequently among UV-induced mutations.

The modern environment exposes each individual to a wide variety of chemicals in drugs, cosmetics, food preservatives, pesticides, compounds used in industry, pollutants, and so on. Many of these compounds have been shown to be carcinogenic and mutagenic. Examples include the food preservative AF-2, the food fumigant ethylene dibromide, the antischistosome drug hycanthone, several hair-dye additives, and the industrial compound vinyl chloride; all are potent, and some have subsequently been subjected to government control. However, hundreds of new chemicals and products appear on the market each week. How can such vast numbers of new agents be tested for carcinogenicity before much of the population has been exposed to them?

The Ames Test

Many test systems have been devised to screen for carcinogenicity. These tests are time-consuming, typically involving laborious research with small mammals. More rapid tests do exist that make use of microbes (such as fungi or bacteria) and test for mutagenicity rather than carcinogenicity. The most widely used test was developed in the 1970s by Bruce Ames, who worked with *Salmonella typhimurium*. This **Ames test** uses two auxotrophic histidine mutations, which revert by different molecular mechanisms (Figure 19-25). Further properties were genetically engineered into these strains to make them suitable for mutagen detection. First, they carry a mutation that inactivates the excision-repair system (described later). Second, they carry a mutation that eliminates the protective lipopolysaccharide coating of wild-type *Salmonella*, to facilitate the entry of many different chemicals into the cell.

Bacteria are evolutionarily a long way removed from humans. Can the results of a test on bacteria have any real significance in detecting chemicals that are dangerous for humans? First, we have seen that the genetic and chemical nature of DNA is identical in all organisms, so that a compound acting as a mutagen in one organism is likely to have some mutagenic effects in other organisms. Second, Ames devised a way to simulate the human metabolism in the bacterial system. In mammals, much of the important processing of ingested chemicals occurs in the liver, where externally derived compounds normally are detoxified or broken down. In some cases, the action of liver enzymes can create a toxic or mutagenic compound from a substance that was not originally dangerous (Figure 19-26). Ames in-

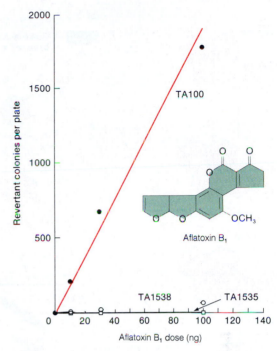

Figure 19-25 Ames test results showing the mutagenicity of aflatoxin B$_1$, which is also a potent carcinogen. TA100, TA1538, and TA1535 are strains of *Salmonella* bearing different *his* auxotrophic mutations. The TA100 strain is highly sensitive to reversion through base-pair substitution. The TA1535 and TA1538 strains are sensitive to reversion through frameshift mutation. The test results show that aflatoxin B$_1$ is a potent mutagen that causes base-pair substitutions but not frameshifts. (From J. McCann and B. N. Ames, in W. G. Flamm and M. A. Mehlman, eds., *Advances in Modern Technology*, Vol. 5. Copyright © by Hemisphere Publishing Corp., Washington, D.C.)

corporated mammalian liver enzymes in his bacterial test system, using rat livers for this purpose. Figure 19-27 outlines the procedure used in the Ames test.

Of course, chemicals detected by this test can be regarded not only as potential carcinogens (sources of somatic mutations) but also as possible causes of mutations in germinal cells. Because the test system is so simple and inexpensive, many laboratories throughout the world now routinely test large numbers of potentially hazardous compounds for mutagenicity and potential carcinogenicity.

Biological Repair Mechanisms

As we have seen in the previous discussions, there are many potential threats to the fidelity of DNA replication. Not only is there an inherent error rate for the replication of DNA, but there are also spontaneous lesions that can provoke additional errors. Moreover, mutagens in the environment can damage DNA and greatly increase the mutation rate.

Living cells have evolved a series of enzymatic systems that repair DNA damage in a variety of ways. Failure of these systems can lead to a higher mutation rate. A number

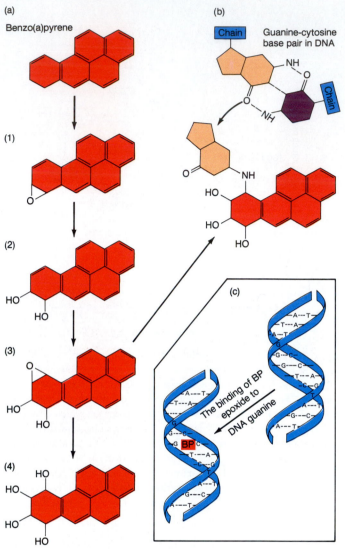

Figure 19-26 The metabolic conversion of benzo(a)pyrene into a mutagen (and a carcinogen). Benzo(a)pyrene (BP) goes through several steps (a) as it is made more water-soluble prior to excretion. One of the intermediates in this process, a diol epoxide (3), is capable of reacting with guanine in DNA (b). This leads to a distortion of the DNA molecule (c) and mutations. Benzo(a)pyrene is therefore a mutagen for any cell that has the enzymes that produce this intermediate. (After I. B. Weinstein et al., *Science*, 193, 1976, 592; from J. Cairns, *Cancer: Science and Society*. Copyright © 1978 by W. H. Freeman and Company, New York.)

of human diseases can be attributed to defects in DNA repair, as we'll see later. Let's first examine some of the characterized repair pathways, and then consider how the cell integrates these systems into an overall strategy for repair.

We can divide repair pathways into several categories.

Avoidance of Errors before They Happen

Some enzymatic systems neutralize potentially damaging compounds before they even react with DNA. One example of such a system involves the detoxification of superoxide radicals produced during oxidative damage to DNA: the enzyme **superoxide dismutase** catalyzes the conversion of the superoxide radicals to hydrogen peroxide, and the enzyme **catalase,** in turn, converts the hydrogen peroxide to water.

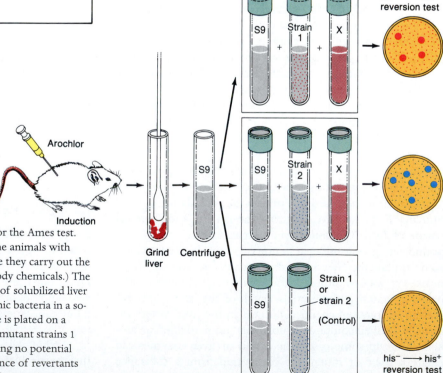

Figure 19-27 Summary of the procedure used for the Ames test. First rat liver enzymes are mobilized by injecting the animals with Arochlor. (Enzymes from the liver are used because they carry out the metabolic processes of detoxifying and toxifying body chemicals.) The rat liver is then homogenized, and the supernatant of solubilized liver enzymes (S9) is added to a suspension of auxotrophic bacteria in a solution of the potential carcinogen (X). This mixture is plated on a medium containing no histidine, and revertants of mutant strains 1 and 2 are looked for. A control experiment containing no potential carcinogen is always run simultaneously. The presence of revertants indicates that the chemical is a mutagen and possibly a carcinogen as well.

Another error-avoidance pathway depends on the protein product of the *mutT* gene: this enzyme prevents the incorporation of 8-oxodG (see Figure 19-10), which arises by oxidation of dGTP into DNA by hydrolyzing the triphosphate of 8-oxodG back to the monophosphate.

Direct Reversal of Damage

The most straightforward way to repair a lesion, once it occurs, is to reverse it directly, thereby regenerating the normal base. Reversal is not always possible, since some types of damage are essentially irreversible. In a few cases, however, lesions can be repaired in this way. One case involves a mutagenic photodimer caused by UV light (see Figure 19-23). The cyclobutane pyrimidine photodimer can be repaired by a **photolyase** that has been found in bacteria and lower eukaryotes, but not in humans. The enzyme binds to the photodimer and splits it, in the presence of certain wavelengths of visible light, to generate the original bases (Figure 19-28). This enzyme cannot operate in the dark, so other repair pathways are required to remove UV damage. A photolyase has also been detected in plants and Drosophila that reverses the 6-4 photoproducts.

Alkyltransferases are also enzymes involved in the direct reversal of lesions. They remove certain alkyl groups that have been added to the O-6 positions of guanine (Figure 19-17) by such agents as NG and EMS. The methyltransferase from *E. coli* has been well studied. This enzyme transfers the methyl group form O-6-methylguanine to a cysteine residue on the protein (Figure 19-29). When this

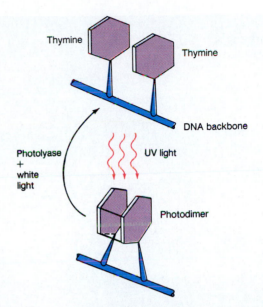

Figure 19-28 Repair of a UV-induced pyrimidine photodimer by a photoreactivating enzyme, or photolyase. The enzyme recognizes the photodimer (here, a thymine dimer) and binds to it. When light is present, the photolyase uses its energy to split the dimer into the original monomers. (After J. D. Watson, *Molecular Biology of the Gene*, 3d ed. Copyright © 1976 by W. A. Benjamin).

happens, the enzyme is inactivated, so this repair system can be saturated if the level of alkylation is high enough. The control of alkyltransferase gene expression is discussed on page 610.

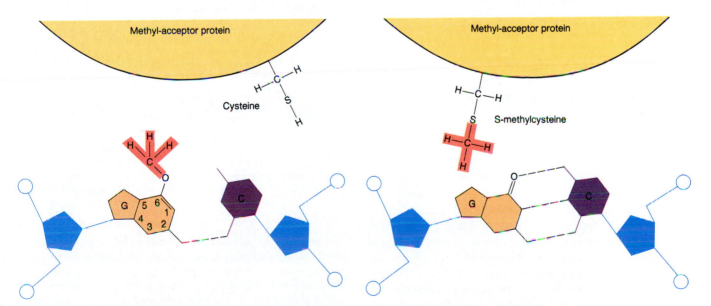

Figure 19-29 Direct reversal of DNA damage by an alkyltransferase. Methylation of a guanine residue by nitrosoguanidine (NG) is repaired by this novel process. The NG adds a methyl group (CH_3) at various sites in the DNA, including an oxygen atom at position 6 of guanine (*left*). This disrupts the hydrogen bonding of guanine to a cytosine. The repair is accomplished by a methyl-acceptor protein, one of the enzymes known as *alkyltransferases*. A cysteine residue on the protein acts as the methyl acceptor: it binds the CH_3 group, thereby restoring the guanine to its original state (*right*). (From P. Howard-Flanders, "Inducible Repair of DNA." Copyright © 1981 by Scientific American, Inc. All rights reserved.)

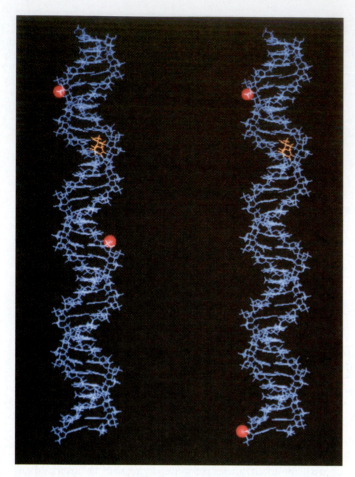

Figure 19-30 Excinuclease incision patterns by *E. coli* (*left*) and human enzymes. The orange points indicate the incision patterns of a lesion, in this case a thymine dimer, which is shown in yellow. (Courtesy of J. E. Hearst in A. Sancar, *Science*, 266, 1974, 1954.)

Excision-Repair Pathways

General Excision Repair. Also termed *nucleotide excision repair*, this system involves the breaking of a phosphodiester bond on either side of the lesion, on the same strand, resulting in the excision of an oligonucleotide. This leaves a gap that is filled in by repair synthesis, and a ligase seals the breaks. In prokaryotes, 12–13 nucleotides are removed, while in eukaryotes 27–29 nucleotides are eliminated. Figure 19-30 depicts the incision pattern in each case.

In *E. coli*, the products of the *uvrA, B,* and *C* genes constitute the excinuclease. Their action is shown in Figure 19-31, right panel). The UvrA protein, which recognizes the damaged DNA, forms a complex with UvrB (Figure 19-31a), and leads the uvrB subunit to the damage site before dissociating (Figure 19-31b). The UvrC protein then binds to UvrB. Each of these subunits makes an incision (Figure 19-31c). The short DNA 12-mer is unwound and released by another protein, helicase II (Figure 19-31d). DNA pol I repair synthesis and ligase replace the missing patch (Figure 19-31e and f). Figure 19-32 shows a more detailed view of these excision events.

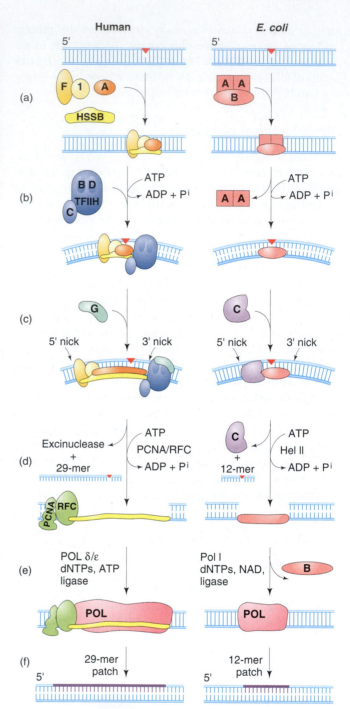

Figure 19-31 Comparison of nucleotide excision repair in humans and *E. coli*. The basic steps in the molecular mechanisms are similar in both organisms. (a) Recognition of damage. Note that in humans, several factors, including subunits of the general transcription factor TFIIH, are required. (b) The helix is distorted in an ATP-dependent manner, and a recognition protein-DNA complex is formed. (c) A nuclease binds and makes two nicks. (d) The excised segment and the excinuclease are released. This is aided by an additional helicase, Hel II in E. coli, and by PCNA (proliferating cell nuclear antigen) and the RFC replication protein in humans. (e) Subsequent repair synthesis proteins bind. (f) The patch is repaired by gap filling and ligation. (Adapted from A. Sancar, *Science*, 266, 1974, 1955.)

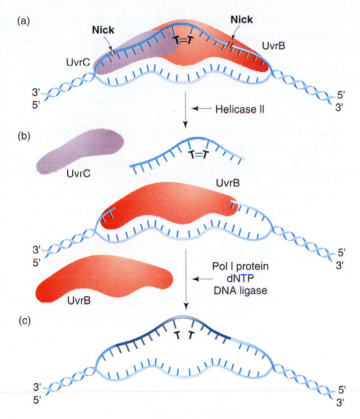

(a)

(b)

(c)

Figure 19-32 Schematic view of events following incision by UvrABC endonuclease in *E. coli*. Here DNA helicase II mediates the release of a segment of the DNA bounded by two nicks in the same strand of DNA. The UvrC protein is also displaced at this point. The subsequent repair synthesis displaces UvrB. (From E. C. Friedberg, G. C. Walked, and W. Seide, *DNA Repair and Mutagenesis*. Copyright © 1995 by the American Society for Microbiology.)

The human excinuclease is considerably more complex than the bacterial counterpart, and involves at least 17 proteins (Figure 19-31, left panel). However, the basic steps are the same as in *E. coli*. Here, some of the proteins involved are XPA, XPB, XPC, XPD, XPF, XPG, and ERCC1, as well as the replication protein HSSB. XPB and XPD (B and D in the figure) are two of the eight subunits of the general transcription factor TFIIH. XPA recognizes the damage and leads TFIIH to the damage site. TFIIH then unwinds the helix, allowing the excinucleases XPG and XPF to operate (see Figure 19-31).

Coupling of Transcription and Repair. The involvement of TFIIH, a transcription factor, in excision repair also underscores the fact that transcription and repair are coupled. In both eukaryotes and prokaryotes there is a preferential repair of the transcribed strand of DNA for actively expressed genes. Figure 19-33 portrays a mechanism for this coupling.

Specific Excision Pathways. Certain lesions are too subtle to cause a distortion large enough to be recognized by the *uvrABC*-encoded general excision-repair system and its counterparts in higher cells. Thus, additional excision pathways are necessary.

DNA Glycosylase Repair Pathway (Base Excision Repair) **DNA glycosylases** do not cleave phosphodiester bonds, but instead cleave *N*-glycosidic (base-sugar) bonds, liberating the altered base and generating an apurinic or an apyrimidinic site, both called **AP sites,** since they are biochemically equivalent (see Figure 19-7). This initial step is shown in Figure 19-34. The resulting AP site is then repaired by an AP endonuclease repair pathway (described in the next section).

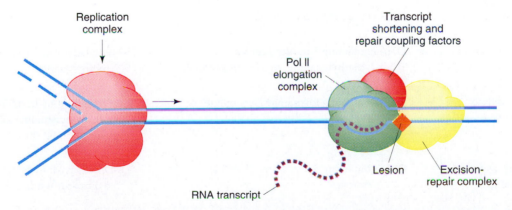

Figure 19-33 Nucleotide excision repair is coupled to transcription. This model for coupled repair is mammalian cells shows RNA polymerase (green) pausing when encountering a lesion. It undergoes a conformational change, allowing the DNA strands at the lesion site to reanneal. Protein factors aid in coupling, in bringing TFIIH and other factors to the site to carry out the incision, excision, and repair reactions. Then transcription can continue normally. (From P. C. Hanawalt, *Science*, 266, 2994, 2957.)

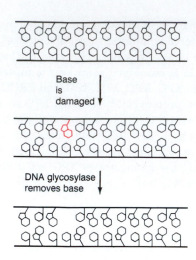

Figure 19-34 Action of DNA glycosylases. Glycosylases remove altered bases and leave an AP site. The AP site is subsequently excised by the AP endonucleases diagrammed in Figure 19-35. (After B. Lewin, *Genes.* Copyright © 1983 by John Wiley.)

Numerous DNA glycosylases exist. One, uracil-DNA glycosylase, removes uracil from DNA. Uracil residues, which result from the spontaneous deamination of cytosine (Figure 19-8), can lead to a C → T transition if unrepaired. It is possible that the natural pairing partner of adenine in DNA is thymine (5-methyluracil), rather than uracil, in order to allow the recognition and excision of these uracil residues. If uracil were a normal constituent of DNA, such repair would not be possible.

There is also a glycosylase that recognizes and excises hypoxanthine, the deamination product of adenine. Other glycosylases remove alkylated bases (such as 3-methyladenine, 3-methylguanine, and 7-methylguanine), ring-opened purines, oxidatively damaged bases, and, in some organisms, UV photodimers. New glycosylases are still being discovered. Table 19-4 shows some of the characterized glycosylases.

AP Endonuclease Repair Pathway All cells have endonucleases that attack the sites left after the spontaneous loss of single

Table 19-4   DNA Glycosylases

Enzyme	Substrate	Products
Ura-DNA glycosylase	DNA containing uracil	Uracil + AP sites
Hmu-DNA glycosylase	DNA containing hydroxymethyluracil	Hydroxymethyluracil + AP sites
5-mC-DNA glycosylase	DNA containing 5-methylcytosine	5-methylcytosine + AP sites
Hx-DNA glycosylase	DNA containing hypoxanthine	Hypoxanthine + AP sites
Thymine mismatch-DNA glycosylase	DNA containing G–T mispairs	Thymine + AP sites
MutY-DNA glycosylase	DNA containing G–A mispairs	Adenine + AP sites
3-mA-DNA glycosylase I	DNA containing 3-methyladenine	3-Methyladenine + AP sites
3-mA-DNA glycosylase II	DNA containing 3-methyladenine, 7-methylguanine, or 3-methylguanine	3-Methyladenine, 7-methylguanine, or 3-methylguanine + AP sites
FaPy-DNA glycosylase	DNA containing formamidopyrimidine moieties, or 8-hydroxyguanine	2,6-Diamino-4-hydroxy-5-*N*-methylformamido-pyrimidine and 8-hydroxyguanine + AP sites
5,6-HT-DNA glycosylase (endonuclease III)	DNA containing 5,6 hydrated thymine-moieties	5,6-Dihydroxydi-hydrothymine or 5,6 dihydrothymine + AP sites
PD-DNA glycosylase	DNA containing pyrimidine dimers	Pyrimidine dimers in DNA with hydrolyzed 5′ glycosyl bonds + AP sites

Source: E. C. Friedberg, G. C. Walker, and W. Siede, *DNA Repair and Mutagenesis.* ASM Press, Washington, D.C.

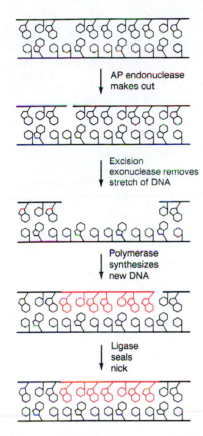

Figure 19-35 Repair of AP (apurinic or apyrimidinic) sites. AP endonucleases recognize AP sites and cut the phosphodiester bond. A stretch of DNA is removed by an exonuclease, and the resulting gap is filled in by DNA polymerase I and DNA ligase. (After B. Lewin, *Genes*. Copyright © 1983 by John Wiley.)

purine or pyrimidine residues. The **AP endonucleases** are vital to the cell, since, as noted earlier, spontaneous depurination is a relatively frequent event. These enzymes introduce chain breaks by cleaving the phosphodiester bonds at AP sites. This initiates an excision-repair process mediated by three further enzymes—an exonuclease, DNA polymerase I, and DNA ligase (Figure 19-35).

Due to the efficiency of the AP endonuclease repair pathway, it can be the final step of other repair pathways. Thus, if damaged base pairs can be excised, leaving an AP site, the AP endonucleases can complete the restoration to the wild type. This is what happens in the DNA glycosylase repair pathway.

The GO System Two glycosylases, the products of the *mutM* and *mutY* genes, work together to prevent mutations arising from the 8-oxodG, or GO, lesion in DNA (Figure 19-10). Together with the product of the *mutT* gene mentioned above, these glycosylases form the GO system. When GO lesions are generated in DNA by spontaneous oxidative damage, a glycosylase encoded by *mutM* removes the lesion (Figure 19-36). Still, some GO lesions persist and mispair with adenine. A second glycosylase, the product of the *mutY* gene, removes the adenine from this specific mispair, leading to

restoration of the correct cytosine by repair synthesis (mediated by DNA polymerase I) and allowing subsequent removal of the GO lesion by the *mutM* product.

The *mutT* product prevents incorporation of GO across from A; otherwise, this would be converted by the *mutY*-encoded glycosylase to a GO lesion across from an AP site, which would result in an AT → CG mutation (Figure 19-36c). (Incorporation of GO across from C could be repaired by the *mutM* product—Figure 19-36c). The human counterparts of the *mutI*, *mutY* and *mutM* gene products have been detected.

Postreplication Repair

Mismatch Repair. Some repair pathways are capable of recognizing errors even after DNA replication has already occurred. One such system, termed the **mismatch repair system,** can detect mismatches that occur during DNA replication. Suppose you were to design an enzyme system that could repair replication errors. What would this system have to be able to do? At least three things:

1. Recognize mismatched base pairs.
2. Determine which base in the mismatch is the incorrect one.
3. Excise the incorrect base and carry out repair synthesis.

The second point is the crucial property of such a system. Unless it is capable of discriminating between the correct and the incorrect bases, the mismatch repair system could not determine which base to excise. If, for example, a G–T mismatch occurs as a replication error, how can the system determine whether G or T is incorrect? Both are normal bases in DNA. But replication errors produce mismatches on the newly synthesized strand, so it is the base on this strand that must be recognized and excised.

To distinguish the old, template strand from the newly synthesized strand, the mismatch repair system in bacteria takes advantage of the normal delay in the postreplication methylation of the sequence

$$5'—G—A—T—C—3'$$
$$3'—C—T—A—G—5'$$

The methylating enzyme is **adenine methylase,** which creates 6-methyladenine on each strand. However, it takes the adenine methylase several minutes to recognize and modify the newly synthesized GATC stretches. During that interval, the mismatch repair system can operate because it can now distinguish the old strand from the new one by the methylation pattern. Methylating the 6-position of adenine does not affect base pairing, and it provides a convenient tag that can be detected by other enzyme systems. Figure 19-37 shows the replication fork during mismatch correction. Note that only the old strand is methylated at GATC sequences right after replication.

Figure 19-36 The GO system. (a) 8-oxo-7-hydrodeoxyguanosine, GO. This is the structure of the predominant tautomeric form of the GO lesion. (b) The GO lesions, generated by oxidative damage, can be removed by the *mutM* gene product (MutM protein) and subsequent repair can restore the original G–C base pair. Translesion synthesis by replicative DNA polymerases is frequently inaccurate, leading to the misincorporation of A opposite the GO lesion. The *mutY* glycosylase (MutY) removes the misincorporated adenine from the A–GO mispairs. Repair polymerases are much less error-prone during translesion synthesis and can lead to a C–GO pair, a substrate for MutM. (c) Oxidative damage can also generate 8-oxo-dGTP from dGTP in the deoxynucleotide pool. MutT is active on 8-oxo-dGTP and hydrolyzes it to 8-oxo-dGMP, effectively removing the triphosphate from the deoxynucleotide pool. If 8-oxo-dGTP were not removed, inaccurate replication could result in the misincorporation of 8-oxo-dGTP opposite template A residues, leading to A–GO mispairs. MutY could be involved in the mutation process because it is active on the A–GO substrate and would remove the template A, leading to the AT → CG transversions that are characteristic of a *mutT* strain. The 8-oxo-dGTP could also be incorporated opposite template cytosines, resulting in a damaged C–GO pair that could be corrected by MutM. (From M. Michaels and J. H. Miller, *Journal of Bacteriology* 174, 1992, 6321.)

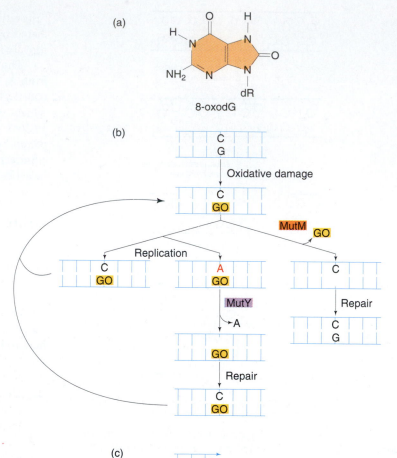

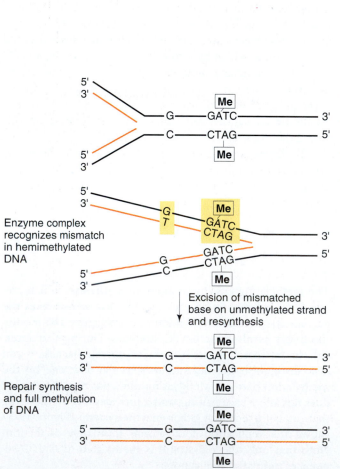

Enzyme complex recognizes mismatch in hemimethylated DNA

Excision of mismatched base on unmethylated strand and resynthesis

Repair synthesis and full methylation of DNA

Figure 19-37 Model for mismatch repair in *E. coli*. Because DNA is methylated by enzymatic reactions that recognize the A in a GATC sequence, directly after DNA replication the newly synthesized strand will not be methylated. The "hemimethylated" DNA duplex serves as a recognition point for the mismatch repair system in discerning the old from the new strand. Here a G:T mismatch is shown. The mismatch repair system can recognize and bind to this mismatch, determine the correct (old) strand since it is the methylated strand of a hemimethylated duplex, and then excise the mismatched base from the new strand. Repair synthesis restores the normal base pair. (Adapted from E. C. Friedberg, *DNA Repair.* Copyright © 1985 by W. H. Freeman and Co., New York.)

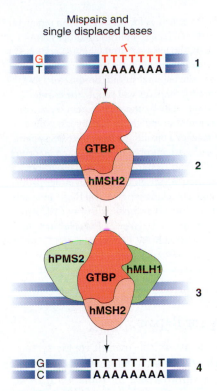

Figure 19-39 Mismatch repair in humans. (1) Mispairs and misaligned bases occur during replication. (2) The G:T binding protein (GTBP) and the human MutS homolog (hMSH2) recognize the incorrect matches. (3) Two additional proteins, hPMS2 and hMLH1, are recruited and form a larger repair complex. (4) The mismatch is repaired after removal, DNA synthesis, and ligation. (From P. Karran, *Science, 268,* 1995, 1857.)

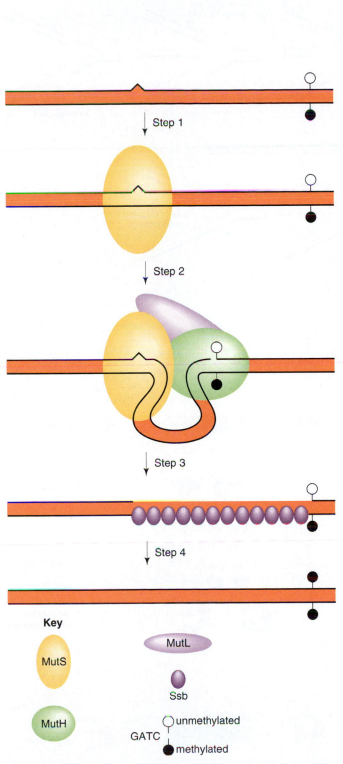

Key

MutS

MutL

Ssb

MutH

GATC — ○ unmethylated
GATC — ● methylated

Figure 19-38 Steps in *E. coli* mismatch repair. (1) MutS binds to mispair. (2) MutH and MutL are recruited to form a complex. MutH cuts the newly synthesized (unmethylated) strand, and exonuclease degradation goes past the point of the mismatch, leaving a patch. (3) Single-strand binding protein (Ssb) protects the single-stranded region across from the missing patch. (4) Repair synthesis and ligation fill in the gap. (From J. Jiricny, *Trends in Genetics,* Vol. 10. Elsener Trends Journals, Cambridge, U.K., 1995.)

Once the mismatched site has been identified, the mismatch repair system corrects the error. Figure 19-38 depicts a model of how the mismatch repair system carries out the correction in *E. coli.*

The mismatch repair system has also been characterized in humans. Two of the proteins, hMSH2 and hMLH1, are very similar to their bacterial counterparts, MutS and MutL respectively. Figure 19-39 depicts how the hMSH2 protein, together with the G:T-binding protein (GTBP), binds to the mismatches and then recruits the other components of the system, HPMS2 and hMLH1, to effect repair of the mismatch.

Recombinational Repair. The *recA* gene, which is involved in SOS bypass (Figure 19-22), is also involved in postreplication repair. Here the DNA replication system stalls at a UV photodimer or other blocking lesions and then restarts past the block, leaving a single-stranded gap. In recombinational repair, this gap is patched by DNA cut from the sister molecule (Figure 19-40a). This process seems to lead to few errors. SOS bypass, by contrast, is highly mutagenic, as described earlier. Here the replication system continues past the lesion (Figures 19-22 and 19-40b), accepting noncomplementary nucleotides for new strand synthesis.

Figure 19-40 Schemes for postreplication repair. (a) In recombinational repair, replication jumps across a blocking lesion, leaving a gap in the new strand. A *recA*-directed protein then fills in the gap, using a piece from the opposite parental strand (because of DNA complementarity, this filler will supply the correct bases for the gap). Finally, the *recA* protein repairs the gap in the parental strand. (b) In SOS bypass, when replication reaches a blocking lesion, the SOS system inserts the necessary number of bases (often incorrect ones) directly across from the lesion and replication continues without a gap. Note that with either pathway the original blocking lesion is still there and must be repaired by some other repair pathway. (Adapted from A. Kornberg and T. Baker, *DNA Replication*, 2d ed. Copyright © 1992 by W. H. Freeman and Co., New York.)

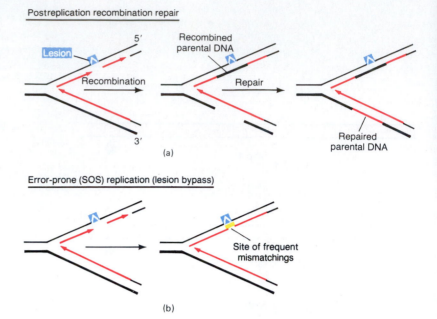

Strategy for Repair

We can now assess the overall repair system strategy used by the cell. The many different repair systems available to the cell are summarized in Table 19-5. It would be convenient if enzymes could be used to directly reverse each specific lesion. However, sometimes that is not chemically possible, and not every possible type of DNA damage can be anticipated. Therefore, a general excision repair system is used to remove any type of damaged base that causes a recognizable distortion in the double helix. When lesions are too subtle to cause such a distortion, specific excision systems, glycosylases, or removal systems are designed. To eliminate replication errors, a postreplication mismatch repair system operates; finally, postreplication recombinational systems eliminate gaps across from blocking lesions that have escaped the other repair systems.

A number of repair pathways are induced in response to damage, such as the SOS system (see Figure 19-22), and many of the proteins involved in repairing alkylation damage were discussed previously. This latter control system, often termed the *adaptive response*, is pictured in Figure 19-41.

Figure 19-41 Model for the regulation of genes encoding repair enzymes for alkylation damage. The four genes shown constitute the Ada regulon. Common sequences in the promoters of two transcription units, the "Ada Box," are shown. When *E. coli* is treated with methylating agents, several positions on the DNA are alkylated, including the O-6 position of guanine, the O-4 position of thymine, and the phosphates in the sugar phosphate backbone. The Ada protein transfers these methyl groups to cysteine residues in either the N-terminal (in the case of transfer from phosphates) or the C-terminal end of the protein. When the N-terminal cysteine receptor has a bound methyl group, the protein undergoes a conformational change and activates transcription from promoters of genes in the Ada regulon, resulting in increased production of the alkylation damage repair enzymes. (Adapted from I. Lindahl et al., *Ann. Rev. Biochem.*, 57, 1988 133–157.)

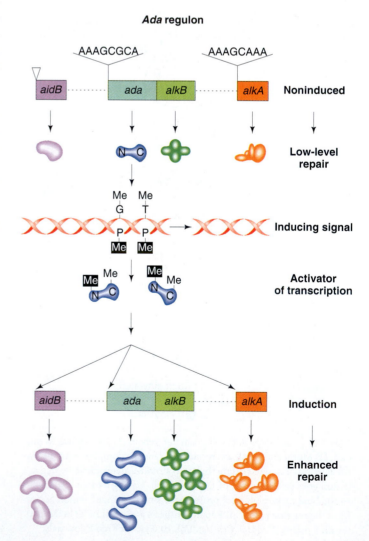

Table 19-5   Repair Systems in *E. coli*

General Mode of Operation	Example	Type of Lesion Repaired	Mechanism
Detoxification	Superoxide dismutase	Prevents formation of oxidative lesion	Converts peroxides to hydrogen peroxide, which is neutralized by catalase
Direct removal of lesions	Alkyltransferases	*O*-6-alkylguanine	Transfers alkyl group from *O*-6-alkylguanine to cysteine residue on transferase
	Photolyase	6-4 photoproduct	Breaks 6-4 bond and restores bases to normal
	Photolyase	UV photodimers	Splits dimers in the presence of white light
General excision	*uvrABC*-encoded exonuclease system	Lesions causing distortions in double helix, such as UV photoproducts and bulky chemical additions	Makes endonucleolytic cut on either side of lesion; resulting gap is repaired by DNA polymerase I and DNA ligase
Specific excision	AP endonucleases	AP sites	Makes endonucleolytic cut; exonuclease creates gap, which is repaired by DNA polymerase I and DNA ligase
	DNA glycosylases	Deaminated bases (uracil, hypoxanthine), certain methylated bases, ring-opened purines, oxydatively damaged bases; and certain other modified bases	Removes base, creating AP site, which is repaired by AP endonucleases
	GO system	8-oxodG	A glycosylase removes 8-oxodG from DNA; another glycosylase converts any remaining 8-oxodG–A mispairs to 8-oxodG–C pairs, and the first glycosylase then removes the 8-oxodG
Postreplication	Mismatch repair system	Replication errors resulting in base-pair mismatches	Recognizes newly synthesized strand by detecting nonmethylated adenine residues in 5′-GATC-3′ sequences; then excises bases from the new strand when a mismatch is detected
	Recombinational repair	Lesions that block replication and result in single-stranded gaps	Recombinational exchange
	SOS system	Lesions that block replication	Allows replication bypass of blocking lesion, resulting in frequent mutations across from lesion

Mutators

As the preceding description of repair processes indicates,
normal cells are programmed for error avoidance. The repair processes are so efficient that the observed base-substitution rate is as low as 10^{-10} to 10^{-9} per base pair per cell
per generation in *E. coli*. However, mutant strains with increased spontaneous mutation rates have been detected.
Such strains are termed **mutators.** In many cases, the mutator phenotype is due to a defective repair system. In humans, these repair defects often lead to serious diseases.

In *E. coli,* the mutator loci *mutH, mutL, mutU,* and *mutS*
affect components of the postreplication mismatch repair
system (see Figure 19-38), as does the *dam* locus, which
specifies the enzyme *deoxyadenosine* methylase. Strains that
are *dam* cannot methylate adenines at GATC sequences (see
Figure 19-37), and so the mismatch repair system can no
longer discriminate between the template and the newly
synthesized strands. This leads to a higher spontaneous mutation rate.

Mutations in the *mutY* locus result in $G-C \rightarrow T-A$
transversions, since many $G-A$ mispairs and all 8-oxodG$-A$
mispairs are unrepaired (see the GO system, above). The
mutM gene encodes a glycosylase that removes 8-oxodG.
Strains lacking *mutM* are mutators for the $G-C \rightarrow T-A$
transversion. Strains which are *mutT$^-$* have elevated rates of
the $A-T \rightarrow C-G$ transversion, since they lack an activity

that prevents the incorporation of 8-oxodG across from
adenine.

Strains that are *ung$^-$* are missing the enzyme uracil
DNA glycosylase. These mutants cannot excise the uracil resulting from cytosine deaminations and as a result have elevated levels of $C \rightarrow T$ transitions. The *mutD* locus is responsible for a very high rate of mutagenesis (at least three
orders of magnitude higher than normal). Mutations at this
locus affect the proofreading functions of DNA polymerase
III (Chapter 11).

Repair Defects and Human Diseases

Several human genetic diseases are known or suspected to
involve repair defects. These often lead to an increased incidence of cancer. Table 19-6 summarizes information about
these diseases, which are usually autosomal recessive disorders.

Xeroderma pigmentosum (XP) results from a defect in
any of the eight genes (complementation groups) involved
in nucleotide excision repair (see Figure 19-31, left panel).
People suffering from this disorder are extremely prone to
UV-induced skin cancers (Figure 19-42) as a result of exposure to sunlight, and have frequent neurological abnormalities. The difference in UV photosensitivity between normal
and diseased cells is evident from the survival curves in Figure 19-43.

Bloom syndrome patients have increased DNA rearrangements. Patients show retarded growth and dilated
blood vessels (telangiectases) in the skin, particularly on the
face. They have a greatly increased cancer rate, with most

Table 19-6   **Human Diseases with DNA-Repair Defects**

Disease	Sensitivity	Cancer Susceptibility	Complementation Groups	Symptoms
Ataxia telangiectasia	γ irradiation	Lymphomas	5	Ataxia, dilation of blood vessels (telangiectases) in skin and eyes, chromosome aberrations, immune dysfunction
Bloom syndrome	Mild alkylating agents	Carcinomas, leukemias, lymphomas	1	Photosensitivity, facial telangiectases, chromosome alterations
Cockayne syndrome	Ultraviolet light		2	Dwafism, retinal atrophy, photosen-sitivity, progeria, deafness, trisomy 10
Fanconi anemia	Cross-linking agents	Leukemias	3	Hypoplastic pancytopenia, congenital anomalies
Xeroderma pigmentosum	Ultraviolet, chemical mutagens	Skin carcinomas and melanomas	8	Skin and eye photosensitivity, keratoses
HNPCC		Colon, ovary	4	Early development of tumors

NOTE: Other human hereditary disorders that may be related to DNA-repair defects include dyskeratosis congenita (Zinsser-Cole-Engman syndrome),
progeria (Hutchinson-Gilford syndrome), and trichothiodystrophy. HNPCC-hereditary nonpolyposis colorectal cancer.

SOURCE: Modified from A. Kornberg and T. Baker, *DNA Replication,* 2d ed. Copyright © 1992 bt W. H. Freeman and Co., New York.

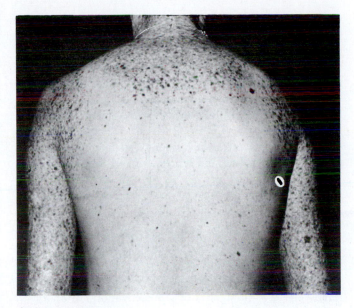

Figure 19-42 Skin cancer in xeroderma pigmentosum. This recessive hereditary disease is caused by a deficiency in one of the excision-repair enzymes, which leads to the formation of skin cancers on exposure of the skin to the UV rays in sunlight. (Photograph courtesy of Dirk Bootsma, Erasmus University, Rotterdam.)

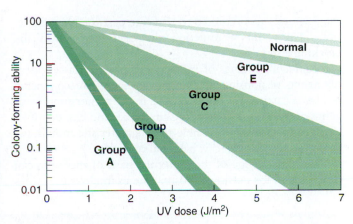

Figure 19-43 Hypersensitivity to UV radiation of XP cells in culture. Here the cells from a number of complementation groups are shown. There is a variation between complementation groups, but all are more sensitive to UV radiation than normal cells. (Adapted from J. E. Cleaver and K. H. Kraemer, *Xeroderma pigmentosum,* in C. R. Scriver et al., eds., *The Metabolic Basis of Inherited Disease.* Copyright © 1989 by McGraw-Hill Book Co., New York.)

patients displaying cancer before the age of 30. **Werner syndrome** also results in a mutator phenotype and shows an increased number of chromosomal deletions. The syndrome is characterized by premature features of normal aging. However, the exact defect has not been identified.

Defects in repair of UV-induced damage are a characteristic of **Cockayne syndrome (CS)** which leads to dwarfism and premature aging. CS cells are deficient in the preferential repair of actively transcribed DNA (see Figure 19-33), owing to a defect in one of the proteins that couples transcription and repair.

Fanconi's anemia (FA) results in a variety of abnormalities, such as short stature, hyperpigmentation, and radial aplasia, and also a reduced level of blood constituents (leukocytes, erythrocytes, and platelets). FA cells appear to be defective in the repair of interstrand cross-links, although it is not yet certain whether altered repair is the prime cause of abnormalities for this disease.

Ataxia telangiectasia (AT) patients have impaired nervous and immune systems, and a higher frequency of lymphomas, most occurring before the age of 20. Severe cases involve progressive mental retardation and muscular uncoordination. Cells from AT patients have increased sensitivity to ionizing radiation and increased chromosomal rearrangements. It is not clear yet what the exact nature of the repair defect in AT cells is.

Hereditary nonpolyposis colorectal cancer (HNPCC), is one of the most common inherited predispositions to cancer, affecting as many as 1 in 200 people in the western world. Recent studies have shown that this syndrome results from a loss of the mismatch repair system. Most HNPCC results from a defect in genes that are the human counterparts (and homologs) of the bacterial MutS and MutL proteins (recall Figure 19-38). The inheritance of HNPCC is autosomal dominant. Cells with one functional copy of the mismatch repair genes have normal mismatch repair activity, but tumor cell lines arise from cells that have lost the one functional copy, and thus are mismatch repair–deficient. These cells display high mutation rates that eventually result in tumor growth and proliferation.

SUMMARY

Gene mutations can arise through many different processes. Spontaneous mutations can result from replication errors or from spontaneous lesions, such as those generated by deamination or depurination. (Recombination and transposable elements can also result in altered genes, as described in Chapters 20 and 21.) Mutagens can increase the frequency of mutations. Some of these agents act by mimicking a base and then mispairing during DNA replication. Others alter bases in the DNA and convert them to derivatives that mispair. A third class of mutagens, which includes most carcinogens, damages DNA in such a way that replication is blocked. The activation of an enzymatic pathway, termed the *SOS system,* is required to replicate past the blocking lesions. This results in mutations that occur across from the blocking lesion.

Molecular studies of mutagenesis in bacteria can identify agents responsible for human cancers. DNA sequencing has been carried out on mutations in the tumor-suppressor genes of patients with liver or skin cancer and has identified changes that closely resemble the characteristic mutational changes caused by specific bacterial mutagens.

Our knowledge of the molecular basis of mutation can be exploited for useful purposes. One example is the Ames test, which utilizes mutant bacterial strains to test compounds in the environment for mutagenic activity. Due to the correlation between mutagenicity and carcinogenicity, the identification of potential carcinogens in the environment can be achieved by this rapid assay.

Repair enzymes present in living cells greatly minimize genetic damage, thus averting many mutations. Mutant organisms lacking certain repair enzymes have higher than normal mutation rates. In humans, repair deficiency leads to a variety of diseases and cancer susceptibilities.

Concept Map

Draw a concept map interrelating as many of the following terms as possible. Note that the terms are listed in no particular order.

mutation / auxotroph / transition / frameshift / transversion / mutagen / photoreactivation / mutator / SOS / genetic disease / amino acid

CHAPTER INTEGRATION PROBLEM

In Chapter 13 we learned that amber and ochre codons were two of the chain-terminating nonsense triplets. Based on the specificity given in the test for aflatoxin B_1 and ethyl methanesulfonate (EMS), describe whether each mutagen would be able to revert amber (UAG) and ochre (UAA) codons back to wild-type.

Solution

EMS induces primarily $G-C \rightarrow A-T$ transitions. UAG codons could not be reverted back to wild-type, since only the UAG $\rightarrow$ UAA change would be stimulated by EMS and that generates a nonsense (ochre) codon. UAA codons would not be acted on by EMS. Aflatoxin B_1 induces primarily $G-C \rightarrow T-A$ transversions. Only the third position of UAG codons would be acted on, resulting in a UAG $\rightarrow$ UAU change (on the mRNA level), which produces tyrosine. Therefore, if tyrosine were an acceptable amino acid at the corresponding site in the protein, aflatoxin B_1 could revert UAG codons. Aflatoxin B_1 would not revert UAA codons because no $G-C$ base pairs appear at the corresponding position in the DNA.

SOLVED PROBLEMS

1. Explain why mutations induced by acridines in phage T4, or by ICR-191 in bacteria, cannot be reverted by 5-bromouracil.

Solution

Acridines and ICR-191 induce mutations by deleting or adding one or more base pairs. This results in a frameshift. However, 5-bromouracil induces mutations by causing the substitution of one base for another. This cannot compensate for the frameshift resulting from the ICR-191 and acridines.

2. A mutant of *E. coli* is highly resistant to mutagenesis by a variety of agents, including ultraviolet light, aflatoxin B_1, and benzo(a)pyrene. Explain one possible cause of this mutant phenotype.

Solution

The mutant might lack the SOS system and perhaps carry a defect in the *umuC* gene. Such strains would not be able to bypass replication-blocking lesions of the type caused by the three mutagens listed. Without the processing of premutational lesions, mutations would not be recovered in viable cells.

PROBLEMS

1. Differentiate between the following pairs:

    a. Transitions and transversions

    b. Silent and neutral mutations

    c. Missense and nonsense mutations

    d. Frameshift and nonsense mutations

2. Why are frameshift mutations more likely than missense mutations to result in proteins that lack normal function?

3. Describe the Streisinger model for frameshift formation. Show how this model can explain mutational hot spots in the *lacI* gene of *E. coli*.

4. Diagram two different mechanisms for deletion formation. How do DNA sequencing experiments suggest these possibilities?

5. Describe two spontaneous lesions that can lead to mutations.

6. Compare the mechanism of action of 5-bromouracil (5-BU) with ethyl methanesulfonate (EMS) in causing mutations. Explain the specificity of mutagenesis for each agent in light of the proposed mechanism.

7. Compare the two different systems involved in the repair of AP sites and in the removal of bulky chemical adducts.

8. Describe the repair systems that operate after replication.

9. Normal ("tight") auxotrophic mutants will not grow at all in the absence of the appropriate supplement to the medium. However, in mutant hunts for auxotrophic mutants, it is common to find some mutants (called "leaky") that grow very slowly in the absence of the appropriate supplement but normally in the presence of the supplement. Propose an explanation for the molecular nature of the leaky mutants.

10. Strain A of *Neurospora* contains an *ad-3* mutation that reverts spontaneously at a rate of 10^{-6}. Strain A is crossed with a newly acquired wild-type isolate, and *ad-3* strains are recovered from the progeny. When 28 different *ad-3* progeny strains are examined, 13 lines are found to revert at the rate of 10^{-6}, but the remaining 15 lines revert at the rate of 10^{-3}. Formulate a hypothesis to account for these findings, and outline an experimental program to test your hypothesis.

11. **a.** Why is it impossible to induce nonsense mutations (represented at the mRNA level by the triplets UAG, UAA, and UGA) by treating wild-type strains with mutagens that cause only AT $\rightarrow$ GC transitions in DNA?

    b. Hydroxylamine (HA) causes only GC $\rightarrow$ AT transitions in DNA. Will HA produce nonsense mutations in wild-type strains?

    c. Will HA treatment revert nonsense mutations?

12. Several auxotrophic point mutants in *Neurospora* are treated with various agents to see if reversion will occur. The following results were obtained ("+" indicates reversion):

Mutant	5-BU	HA	Proflavin	Spontaneous reversion
1	–	–	–	–
2	–	–	+	+
3	+	–	–	+
4	–	–	–	+
5	+	+	–	+

   a. For each of the five mutants, describe the nature of the original mutation event (not the reversion) at the molecular level. Be as specific as possible.

   b. For each of the five mutants, name a possible mutagen that could have caused the original mutation event. (Spontaneous mutation is not an acceptable answer.)

   c. In the reversion experiment for mutant 5, a particularly interesting prototrophic derivative is obtained. When this type is crossed to a standard wild-type strain, the progeny consists of 90 percent prototrophs and 10 percent auxotrophs. Provide a full explanation for these results, including a precise reason for the frequencies observed.

13. You are using nitrous acid to "revert" mutant *nic-2* alleles in *Neurospora*. You treat cells, plate them on a medium without nicotinamide, and look for prototrophic colonies. You obtain the following results for two mutant alleles. Explain these results at the molecular level, and indicate how you would test your hypotheses.

   a. With *nic-2* allele 1, you obtain no prototrophs at all.

   b. With *nic-2* allele 2, you obtain three prototrophic colonies, and you cross each separately with a wild-type strain. From the cross prototroph A $\times$ wild-type, you obtain 100 progeny, all of which are prototrophic. From the cross prototroph B $\times$ wild-type, you obtain 100 progeny, of which 78 are prototrophic and 22 are nicotinamide-requiring. From the cross prototroph C $\times$ wild-type, you obtain 1000 progeny, of which 996 are prototrophic and 4 are nicotinamide-requiring.

14. Devise imaginative screening procedures for detecting the following:

a. Nerve mutants in *Drosophila*.

b. Mutants lacking flagella in a haploid unicellular alga.

c. Supercolossal-sized mutants in bacteria.

d. Mutants that overproduce the black compound melanine in normally white haploid fungus cultures.

e. Individual humans (in large populations) whose eyes polarize incoming light.

f. Negatively phototrophic *Drosophila* or unicellular algae.

g. UV-sensitive mutants in haploid yeast.

20

Mechanisms of Genetic Change II: Recombination

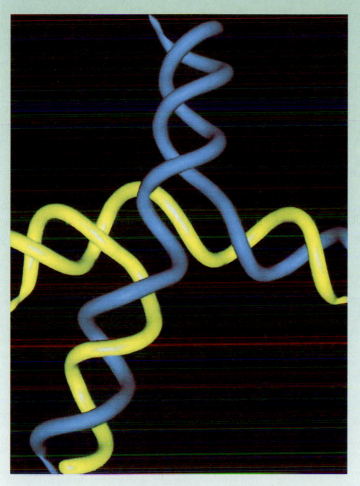

Computer model of exchange site. (Julie Newdol, Computer Graphics Laboratory, University of California, San Francisco. Copyright by Regents, University of California.)

KEY CONCEPTS

▶ Recombination occurs at regions of homology between chromosomes through the breakage and reunion of DNA molecules.

▶ Models for recombination, such as the Holliday model, involve the creation of a heteroduplex branch, or cross bridge, that can migrate and the subsequent splicing of the intermediate structure to yield different types of recombinant DNA molecules.

▶ The Holliday model can be applied to explain genetic crosses.

▶ Many of the enzymes involved in recombination in bacteria have been identified.

▶ Specific recombination systems catalyze recombination events only between certain sequences.

A normal crossover (a reciprocal interchromosomal recombination event) is an extraordinary process in which the genetic material from one parental chromosome and the genetic material from the other parental chromosome are "cut up and pasted together" during each meiosis. Usually, a remarkable precision is attained. Many clues about how this occurs have stimulated the formulation of several models. We can now dissect many of the steps in recombination in terms of specific enzyme reactions.

General Homologous Recombination

Some recombination events occur only at specific sequences. We consider these special events toward the end of the chapter. Here, we focus on the process that results in recombination at any large region of homology between chromosomes.

The Breakage and Reunion of DNA Molecules

Throughout our analysis of linkage, we have implicitly assumed that crossing-over occurs by some process of breakage and reunion of chromatids. The experiments discussed in Chapter 5 provide good *indirect* evidence in favor of breakage and reunion. Furthermore, there is good genetic and cytological evidence that crossing-over occurs during the prophase stage of meiosis, rather than during interphase when chromosomal DNA is replicating. A small amount of DNA synthesis does occur during prophase, but certainly chromosome replication is not associated with crossing-over. One of the first direct proofs that chromosomes (albeit viral chromosomes) can break and rejoin came from experiments on λ phage done in 1961 by Matthew Meselson and Jean Weigle.

Meselson and Weigle simultaneously infected *E. coli* with two strains of λ. One strain, which had the genetic markers *c* and *mi* at one end of the chromosome, was "heavy" because the phages were produced from cells grown in heavy isotopes of carbon (^{13}C) and nitrogen (^{15}N). The other strain was $c^+ mi^+$ for the markers and had "light" DNA because it was harvested from cells grown on the normal light isotopes ^{12}C and ^{14}N. The two DNAs (chromosomes) can be represented as shown in Figure 20-1a. The multiply infected cells were then incubated in a light medium until they lysed.

The progeny phages released from the cells were spun in a cesium chloride density gradient. A wide band was obtained, indicating that the viral DNAs ranged in density from the heavy parental value to the light parental value, with a great many intermediate densities (Figure 20-1b). Interestingly, some recombinant phages were recovered with density values very close to the heavy parental value. They were of genotype $c\ mi^+$, and they must have arisen through an exchange event between the two markers (Figure 20-1c). The heavy density of the chromosome would be expected because only the small tip of the chromosome carrying the mi^+ allele would come from the light parental chromosome. When heavy $c^+ mi^+$ phages were crossed with light $c\ mi$, the heavy recombinants were found to be $c^+ mi$, and the light recombinants were found to be $c\ mi^+$, as expected. These results can be explained in only one way: the recombination event must have occurred through the physical breakage and reunion of DNA. Of course, we have to be careful about extrapolating from viral to eukaryotic chromosomes. However, this evidence shows that the breakage and reunion of DNA strands is a chemical possibility.

Chiasmata: The Crossover Points

In Chapter 5, we made the simple assumption that chiasmata are the actual sites of crossovers. Mapping analysis in-

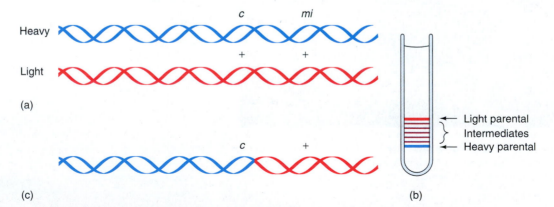

Figure 20-1 Evidence for chromosome breakage and reunion in λ phages. (a) The chromosomes of the two λ strains used to multiply infect *E. coli*. (b) Bands produced when progeny phages are spun in a cesium chloride density gradient. The fact that intermediate densities are obtained indicates a range of chromosome compositions with partly light and partly heavy components. (c) The chromosome of the heavy $c\ mi^+$ progeny resulting from crossover between the two markers. The density of this crossover product confirms that the crossover involved a physical breakage and reunion of the DNA.

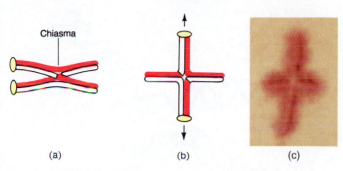

Figure 20-2 Crossing-over between dark- and light-stained nonsister chromatids in a meiosis in the locust. (a) Representation of the chiasma. (b) The best stage for observing is when the centromeres have pulled apart slightly, forming a cross-shaped structure with the chiasma at the center. (c) Photograph of the stage shown in (b). (Photo courtesy of C. Tease and G. H. Jones, *Chromosoma* 69, 1978, 163–178.)

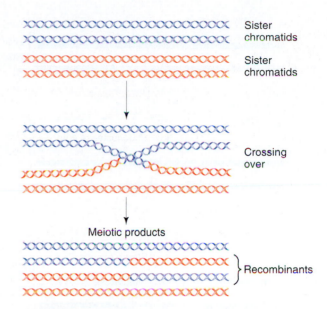

Figure 20-3 The molecular event of recombination may be schematically represented by two double-stranded molecules breaking and rejoining.

directly supports this idea: since an average of one crossover per meiosis produces 50 genetic map units, there should be correlation between the size of the genetic map of a chromosome and the observed mean number of chiasmata per meiosis. The correlation has been made in well-mapped organisms.

However, the harlequin chromosome–staining technique (see Chapter 11) has made it possible to test the idea directly. In 1978, C. Tease and G. H. Jones prepared harlequin chromosomes in meioses of the locust. Remember that the harlequin technique produces sister chromatids: one dark and the other light. When a crossover occurs, it can involve two dark, two light, or dark and light nonsister chromatids as shown in Figure 11-18b. This last situation is crucial because mixed (part dark and part light) crossover chromatids are produced. Tease and Jones found that the dark-light transition occurs right at the chiasma—proving beyond reasonable doubt that these are the crossover sites and settling a question that had been unresolved since the early part of the century (Figure 20-2).

The Holliday Model

Any model for recombination must explain the basic sequence of events depicted in Figure 20-3, in which two homologous chromosomes are **broken** and **rejoined,** with an **exchange** of both strands, and are **resolved** to yield two intact, separate chromosomes. The most plausible recombination model was originally formulated by Robin Holliday to explain the phenomenon of gene conversion (see page 623). The **Holliday model** is typical of the various recombination models that have been proposed for both eukaryotes and prokaryotes. The concept of hybrid, or **heteroduplex,** DNA put forth in this model provides a useful way of expressing the present state of knowledge about crossing-over. Figure 20-4 depicts the Holliday model in its entirety. In some of the figures pertinent to this discussion of recombination,

the DNA strands have been numbered so that they can be followed from one figure to another. The model is based on the creation of a branch, or **cross bridge,** its migration along the two heteroduplex strands, termed **branch migration,** and the subsequent splicing of the intermediate structure in one of two ways to yield different types of recombinant molecules. We will work through the Holliday model in Figure 20-4.

Enzymatic Cleavage and the Creation of Heteroduplex DNA

Looking at 20-4a, we can see that two homologous double helices are aligned, although note that they have been rotated so that the bottom strand of the first helix has the same polarity as the top strand of the second helix ($5' \rightarrow 3'$ in this case). Then a nuclease cleaves the two strands that have the same polarity (Figure 20-4b). The free ends leave their original complementary strands and undergo hydrogen bonding with the complementary strands in the homologous double helix (Figure 20-4c). Ligation produces the structure shown in Figure 20-4d. This partially heteroduplex double helix is a crucial intermediate in recombination. It has been termed the **Holliday structure.**

Branch Migration

The Holliday structure creates a cross bridge, or branch, that can move, or migrate, along the heteroduplex (Figure 20-4de). This phenomenon of branch migration is a distinc-

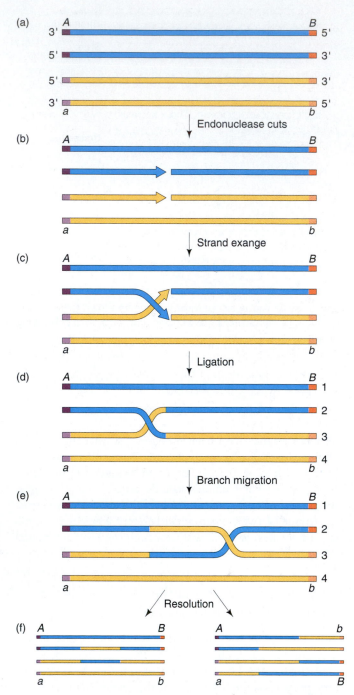

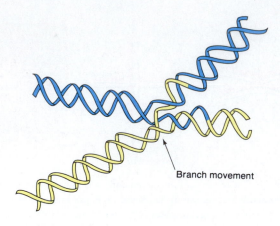

Figure 20-5 Branch migration, the movement of the crossover point between DNA complexes. (After T. Broker, *Journal of Molecular Biology* 81, 1973, 1; from J. D. Watson et al., *Molecular Biology of the Gene*, 4th ed. Copyright ©1987 by Benjamin Cummings.)

tive property of the Holliday structure. Figure 20-5 portrays a more realistic view of this structure as it might appear during branch migration.

Resolution of Holliday Structure

The Holliday structure can be resolved by cutting and ligating either the two originally exchanged strands (Figure 20-4f, *left*) or the originally unexchanged strands (Figure 20-4f, *right*). The former generates a pair of duplexes that are parental, except for a stretch in the middle containing one strand from each parent. If the two parents had different alleles in this stretch, then the DNA will be heteroduplex. The latter resolution step generates two duplexes that are recombinant, with a stretch of heteroduplex DNA.

Figure 20-6 demonstrates one way that we can easily visualize how the Holliday structure can be converted to the recombinant structures with which we are familiar. In 20-6a, we can see the structure that we arrived at in Figure 20-4e drawn out in an extended form. Compare 20-4e and 20-6a until you are convinced that these two structures are indeed equivalent. If we rotate the bottom portion of this structure, as shown in Figure 20-6b, we can generate the form depicted in Figure 20-6c. This last form can be converted back to two unconnected double helices by enzymatically cleaving only two strands. As indicated in 20-6c, cleavage can occur in either of two ways, each of which generates a different product (Figure 20-6d). These cleaved structures can be viewed more simply (Figure 20-6e). Repair synthesis produces the final recombinant molecules (Figure 20-6f). Note the two different types of recombinants.

Application of the Holliday Model to Genetic Crosses

Let's examine how the model shown in Figures 20-4, 20-5, and 20-6 relate to a genetic cross. We can set up a hypotheti-

Figure 20-4 A prototype mechanism for genetic recombination. (a) Two homologous double helices are shown. Each pair represents a chromatid and the two pairs represent two nonsister chromatids. The helices are aligned so that the bottom strand of the first helix has the same polarity as the top strand of the second helix. (b) Two parallel or two antiparallel strands are cut. (c) The free ends become associated with the complementary strands in the homologous double helix. (d) Ligation creates partially heteroduplex double helices. This is the Holliday structure. (e) Migration of the branch point occurs by continued strand transfer by the two polynucleotide chains involved in the crossover. (f) Resolution can occur in one of two ways. These will be described in detail later. (Modified from H. Potter and D. Dressler, *Cold Spring Harbor Symposium on Quantitative Biology* 43, 1979, 970. Cold Spring Harbor Laboratory, Cold Spring Harbor, N.Y.)

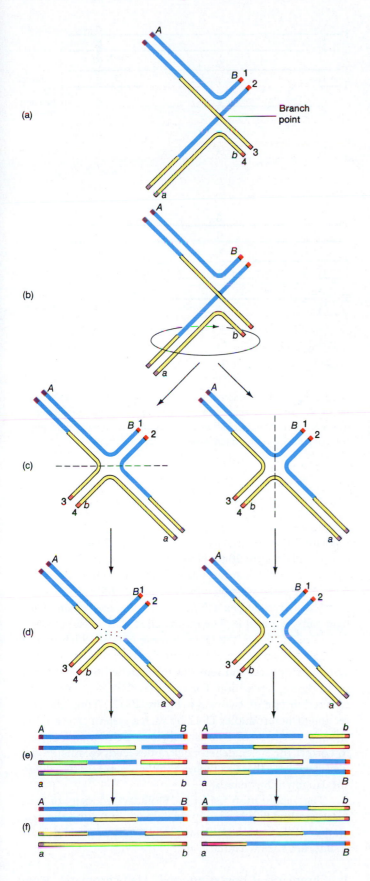

(a)

(b)

(c)

(d)

(e)

(f)

Branch point

Figure 20-6 (a) The Holliday structure shown in an extended form. (b) The rotation of the structure shown in (a) can yield the form depicted in (c). Resolution of the structure shown in (c) can proceed in two ways, depending on the points of enzymatic cleavage, yielding the structures shown in (d). The dotted lines show which segments will rejoin to form recombinant strands for each particular cleavage scheme. The strands are shown linearly in (e) and can be repaired to the forms shown in (f). (From H. Potter and D. Dressler, *Cold Spring Harbor Symposium on Quantitative Biology* 43, 1970, 970. Cold Spring Harbor Laboratory, Cold Spring Harbor, N.Y.)

cal cross of $+ \times m$, in which the $+$ site corresponds to a G–C nucleotide pair and the m site corresponds, after a transition mutation, to an A–T nucleotide pair (Figure 20-7). In Figures 20-7 through 20-10, each set of double lines represents a double helix and constitutes a chromatid. Sister chromatids are the same color. A hypothetical flanking marker is also indicated at the right end of the molecule. The four chromatids are represented as four DNA double helices. The arrows indicate the directions of the DNA strands ($5' \rightarrow 3'$). Also shown is a hypothetical fixed breakage point (perhaps a recognition site for an endonuclease enzyme; its significance will become clear). The fixed breakage points of two nonsister molecules now become the site of phosphodiester-bond breakage (Figure 20-8), and the broken strands unravel away from the fixed breakage point for some distance. The unraveled strands have the same directionality (arrows), so they can rejoin at each other's breakage points through the action of the enzyme ligase (Figure 20-9). We now have two sets of heteroduplex DNA (the double strands that include both black and red parts) containing the two illegitimate purine-pyrimidine pairs G–T and C–A. (We shall return to these later.) Note that the structure generated in Figure 20-9 can be rotated to generate a structure like that depicted in Figure 20-6c.

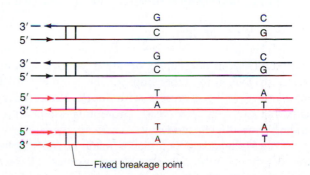

Fixed breakage point

Figure 20-7 The four sets of two lines each represent double helices. Each double helix represents a meiotic chromatid in a cross $+ \times m$. Black lines represent one parent, and red lines represent the other. This figure illustrates the normal synapsis arrangement of the four chromatids during prophase of the first meiotic division. Arrows represent the direction of the antiparallel DNA strands. A fixed breakage point, perhaps a recognition site for an endonuclease enzyme, is also represented.

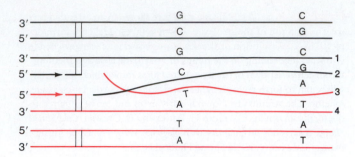

Figure 20-8 The breakage and unwinding of equivalent strands. The broken strands unravel away from the fixed breakage point for some distance and can rejoin at each other's breakage point by means of a ligase enzyme.

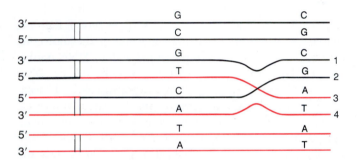

Figure 20-9 Formation of a half-chromatid chiasma and two stretches of heteroduplex DNA extending between the breakpoint and the half-chromatid chiasma. The heteroduplex DNA contains illegitimate base pairs (in this example, GT and CA).

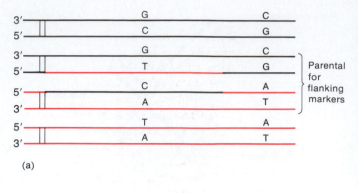

(a)

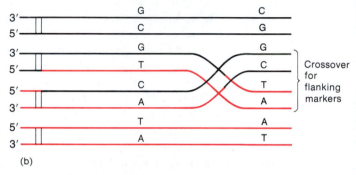

(b)

Figure 20-10 Resolution of the half-chromatid chiasma. (a) In half the cases, resolution will yield a parental conformation for flanking markers. (b) In the other half of the cases, resolution will yield a recombinant conformation (crossover) for flanking markers.

The tangle at the exchange (branch) point is now resolved by "nicking." We can assume that strands 2 and 3 break and rejoin in half the cases (Figure 20-10a) and that strands 1 and 4 break and rejoin in the other half (Figure 20-10b). Thus, we see that the result can be either a crossover or a noncrossover situation (Figure 20-6f) with respect to flanking markers, but that two regions of heteroduplex DNA do exist in either case.

The Holliday model involved symmetry, in that heteroduplex DNA in one duplex occurs in concert with heteroduplex DNA in the other duplex. The symmetric structure worked well for crosses with fungi, but could not explain all the data from yeast, prompting Meselson and Radding to modify the model.

In the Holliday model a nick is made in one strand in each of the two homologous chromosomes. However, Meselson and Radding's proposed model, shown in Figure 20-11, generates the Holliday structure with one single-stranded cut in only one chromosome [step (a)], followed by DNA synthesis (b). After the nick, the displaced single strand invades the second duplex (c), generating a loop, which is excised (d). After ligation to produce a Holliday structure, followed by branch migration (e), a heteroduplex is generated in each chromosome. Resolution of this intermediate (f) occurs exactly as seen in Figure 20-4 (or after rotation, as in Figure 20-6). Note the lack of symmetry in the heteroduplex DNA at resolution following step (f) of Figure 20-11, compared with Figures 20-4f and 20-6f.

The heteroduplex DNA, which contains mismatched base pairs, is unstable. The mismatched nucleotides can produce a distortion in the DNA at these points, which then is recognized by a repair system, much as the mismatch repair system recognizes and removes mismatched base pairs that occur during replication. For example, the G–T pair can be repaired in one of two ways (Figure 20-12). Thus, the correction of heteroduplex DNA by such a system can result in either a wild-type or a mutant allele in a chromatid. Failure to correct the mismatched nucleotides leads to a heteroduplex chromatid that (on replication) will generate two different daughter chromatids.

The application of the Holliday model, or the Meselson-Radding model, to this type of cross and the generation of mismatched alleles nicely explain several genetic results. Let's see what these are.

1: Chromatid Conversion and Half-chromatid Conversion. Much information about recombination in eukaryotes has come from studies of fungi. There is one good rea-

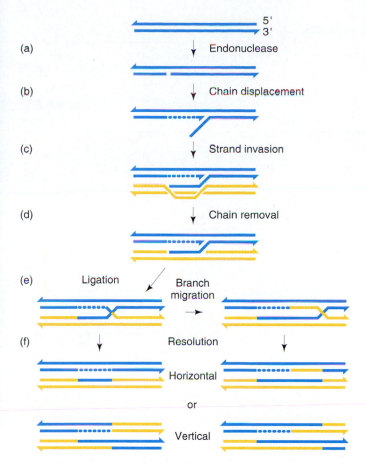

Figure 20-11 reference content:

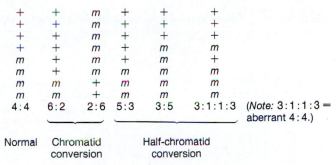

+	+	m	+	+	+
+	+	m	+	+	+
+	+	m	+	+	+
+	+	m	+	m	m
m	+	m	+	m	+
m	+	m	m	m	m
m	m	+	m	m	m
m	m	+	m	m	m
4:4	6:2	2:6	5:3	3:5	3:1:1:3

(Note: 3:1:1:3 = aberrant 4:4.)

| Normal | Chromatid conversion | | Half-chromatid conversion | | |

Figure 20-13 Rare aberrant allele ratios observed in a cross of type $+ \times m$ in fungi. (Ascus genotypes are represented here.) When the Mendelian ratio of 4:4 is not obtained, some of the alleles in the cross have been converted to the opposite allele. In some asci, it appears that the entire chromatid has been converted (6:2 or 2:6 ratios). In others, it appears that only half-chromatids have been converted (5:3, 3:5, or 3:1:1:3 ratios).

Figure 20-11 The Meselson-Radding model. (a) A duplex is cut on one chain. (b) DNA polymerase displaces one chain. (c) The resulting single chain displaces its counterpart in the homolog. (d) This displaced chain is enzymatically digested. (e) Ligation completes the formation of a Holliday junction, which is genetically asymmetric in that only one of the two duplexes has a region of potentially heteroduplex DNA. If the junction migrates, heteroduplex DNA can arise on both duplexes. (f) Resolution of the junction occurs as in the Holliday model. (From F. W. Sthal, "The Holliday Junction on Its Thirtieth Anniversary," *Genetics* 138, 1994, 241–246.)

son for this: the ascus. All four products of a single meiosis can be recovered and examined, so that records of each meiosis can be kept quite precisely in terms of the total genetic information being lost or gained (see Chapter 3).

Non-Mendelian allele ratios are detectable in some asci. Mendel would have predicted 4:4 segregations for all monohybrid crosses, but other ratios are obtained very rarely (in 0.1 to 1 percent of all asci, depending on the fungus species, but 0.1 to 4 percent in yeast). Figure 20-13 gives the most common aberrant ratios obtained. It appears as though some genes in the cross have been "converted" to the opposite allele (Figure 20-14). The process therefore has become known as **gene conversion;** it can occur only where there is heterozygosity for two different alleles of a gene. In some asci (with a 6:2 or 2:6 ratio), the entire chromatid in meiosis seems to have converted; this process is called **chromatid conversion.** In other asci (with a 5:3 or

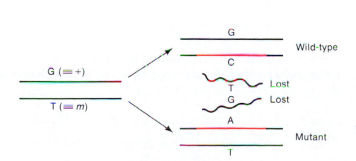

Figure 20-12 Correction of mispaired nucleotides in heteroduplex DNA to either wild-type or mutant pairs. Heteroduplex DNA containing mismatched base pairs is unstable. The mispaired nucleotides can produce a distortion in the DNA at these points that will be recognized by the mismatch repair system described in Chapter 19.

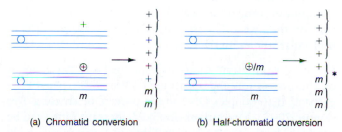

(a) Chromatid conversion   (b) Half-chromatid conversion

Figure 20-14 Gene conversions are inferred from the patterns of alleles observed in asci. (a) In a chromatid conversion, the allele on one chromatid seems somehow to have been converted to an allele like those on the other chromatid pair. The converted allele is shown by the symbol ⊕. One spore pair is of the opposite genotype from that expected in Mendelian segregation. (b) In a half-chromatid conversion, one spore pair (*) has nonidentical alleles. Somehow, one chromatid seems to be "half-converted," giving rise to one spore of the original genotype and one spore converted to the other allele.

Table 20-1   Correction of One or Both Heteroduplex DNA Molecules

DNA strands at start of meiosis		Strands at heteroduplex DNA stage		No correction	One heteroduplex DNA corrected to +	Both heteroduplex DNAs corrected to +
Two homologous chromosomes	$\{$ _m_ _m_ _m_ _m_ _+_ _+_ _+_ _+_ $\}$ — Four chromatids, two strands each	$\{$ _m_ _m_ $\}$ $\{$ _m_ _+_ $\}$ Crossover _m_ $\}$ strands _+_ $\{$ _+_ _+_ $\}$		$\}$ 3 m $\}$ 1 + $\}$ 1 m $\}$ 3 +	$\}$ 3 m $\}$ 5 +	$\}$ 2 m $\}$ 6 +
				Aberrant 4:4	5:3	6:2

3:5 ratio), only half the chromatid seems to have converted, so the process is called **half-chromatid conversion.** In half-chromatid conversions, different members of a spore pair have different genotypes. Recall that each spore pair is produced by mitosis from a single product of meiosis. Half-chromatid conversion implies that two strands of one double helix carry information for two different alleles at the conclusion of meiosis; that is, there is noncomplementarity within the helix. Following the next mitotic division to form a spore pair, the two strands segregate to separate nuclei.

It should be emphasized that alleles that are heterozygous at other loci in the same cross typically segregate 4:4, so the aberrant ratios are not the results of accidents of isolation. The process of conversion cannot be mutation, because the process is directional: the allele that is converted always changes to the other specific allele *involved in the cross,* not some other allele known for the locus but not a part of the cross. This specificity has been confirmed in molecular studies on the gene products of converted alleles.

Explanation. Aberrant ratios are explained as a consequence of the generation and subsequent mismatch repair of regions of heteroduplex DNA in the process of crossing-over (Figure 20-12). The various observed aberrant ratios can be produced by the events listed in Table 20-1, which deals only with correction (and hence conversion) from the heteroduplex *m* / + to +. The symbols *m* and + are used for simplicity instead of the nucleotides used in Figures 20-7 through 20-10. Similar ratios can be developed for the correction of heteroduplex DNA to *m*. It should be noted that the phenomenon of gene conversion and its association with crossing-over was a driving force for the formulation of the Holliday model.

2: Polarity. In genes for which accurate allele maps are available, we can compare the conversion frequencies of alleles at various positions within the gene. In almost every case, the sites closer to one end show higher frequencies than do the sites farther away from that end. In other words, there is a gradient, or **polarity,** of conversion frequencies along the gene (Figure 20-15).

Explanation. Polarity is explained by the fact that crossing-over produces mismatches (the "raw material" of conversions) only in heteroduplex DNA. Since heteroduplex DNA occurs between the break point and the branch point at which the Holliday structure is resolved, the farther a gene locus is from the breakage point, the more likely it is to be beyond the branch point and thus not be part of the heteroduplex. Figure 20-10 is helpful in seeing that heteroduplex DNA falls between the break and branch points but resolves to homoduplex outside this region.

3: Conversion and Crossing-Over. In heteroallelic crosses where the locus under study is closely flanked by other genetically marked loci, the conversion event is very often (about 50 percent of the time) accompanied by an exchange

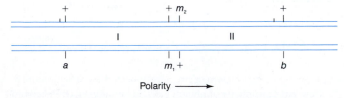

Figure 20-15 Diagram of chromatids involved in a cross. The arrow indicates the polarity of gene conversion in the *m* locus, pointing toward the end with lower conversion frequency.

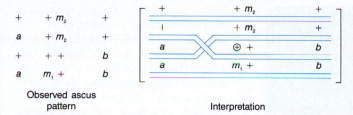

Observed ascus pattern			Interpretation
+	+ m_2	+	
a	+ m_2	+	
+	+ +	b	
a	m_1 +	b	

Figure 20-16 A specific ascus pattern can be explained by both crossover and a chromatid conversion. In this case, a conversion of $m_1 \rightarrow +$ is accomplished by a crossover in the region between a and m_1.

in one of the flanking regions. This exchange nearly always occurs on the side nearer the allele that has converted. Furthermore, it almost always involves the chromatid in which conversion has occurred.

For example, consider the chromatids diagrammed in Figure 20-15. Suppose the polarity is such that alleles toward the left end of the chromatid convert more often than those toward the right end. The cross diagrammed here is between $a^+ m_2 b^+$ and $a\, m_1\, b$, where m_1 and m_2 are different alleles of the m locus and a and b represent closely linked flanking markers. If we look at asci in which conversion has occurred at the m_1 site (the most frequent kind of conversion in this locus), we find that half these asci also will have a crossover in region I and half will have no crossover. In the smaller number of asci showing gene conversion at the m_2 site, half also will have a crossover in region II and half will have no crossover. Such events are detected in ascus genotypes like the one shown in Figure 20-16, which can be interpreted as a conversion of $m_1 \rightarrow +$, accompanied by a crossover in region I. If this is a nutritional locus (say, for arginine requirement), see if you can figure out how the genotypes $m_1\, +$, $+ m_2$, and $m_1 m_2$ can be distinguished from one another.

Explanation. The association of crossing-over with about half the cases of gene conversion is explained by the resolution of the exchange point in two equally likely ways. The ascus pattern used to illustrate the above points (see Figure 20-16) can be explained as shown in Figure 20-17. (Any con-

version of the m_2 allele must be explained by assuming that heteroduplex DNA sometimes "filters in" from some distant breakage point to the right.)

4: Co-conversion. In some asci, a single conversion event seems to include several sites at once. In a heteroallelic cross, this event is called a **co-conversion** (Figure 20-18). The frequency of co-conversion increases as the distance between alleles decreases.

Explanation. Co-conversion is explained by the location of both sites in the region of heteroduplex DNA and by the excision of both sites in the same excision-repair act. This double excision obviously converts both sites to the same parental type.

Message The phenomenon of gene conversion provides clues that lead to a heteroduplex DNA model to explain the mechanism of crossing-over. Mendelian (1:1) allele ratios are normally observed in crosses because it is only rarely that a heterozygous locus is the precise point of chromosome exchange. Remember that asci showing gene conversion at a heterozygous locus are relatively rare (on the order of 1 percent).

Visualization of Recombination Intermediates

Several of the individual steps that constitute the Holliday model have been demonstrated to occur in vivo or in vitro, such as nicking, strand displacement, branch migration, repair synthesis, and ligation. H. Potter and D. Dressler showed that DNA intermediates of the type predicted by the Holliday model can be found in recombining phages or plasmids. Figure 20-19 shows an electron micrograph of a recombinant molecule. It is formally equivalent to the central pair of DNA double helices shown in Figure 20-9, with two arms rotated to produce the central single-stranded "diamond," as in Figure 20-6c. The single strands correspond to the "elbows" in Figure 20-6c, which run between the sections of double-stranded DNA.

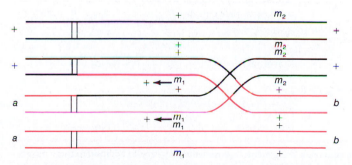

Figure 20-17 An explanation of the ascus in Figure 20-16 in terms of the heteroduplex DNA model. Arrows represent the direction of correction of the heteroduplex DNA.

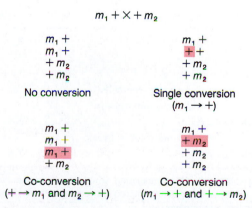

Figure 20-18 Sample ascus patterns obtained from a single conversion and co-conversions.

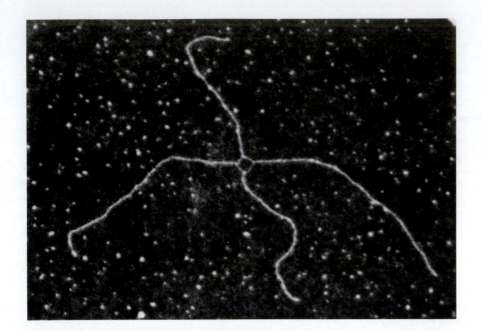

Figure 20-19 Recombination intermediate from recombination between plasmids. Shown are four double-helical arms and a single-stranded diamond uniting them. This is the same kind of intermediate as that postulated in the heteroduplex DNA model for recombination in eukaryotes and is formally equivalent to the structure shown in Figure 20-6c. (Photograph courtesy of H. Potter and D. Dressler.)

Enzymatic Mechanism of Recombination

By isolating mutants defective in some stage of recombination, much light has been shed on the enzymology of recombination. In *E. coli* the products of several genes involved in general recombination—the *recA, recB, recC,* and *recD* genes—have been well characterized, as has the single-stranded-DNA-binding (SSB) protein. Mutants deficient in any of these proteins have reduced levels of recombination. In fact three distinct recombination pathways have been identified. In addition to the major RecBCD pathway, two minor pathways, RecF and RecE, are activated in certain situations. All three pathways use the RecA protein. Table 20-2 summarizes information about these pathways.

Production of Single-Stranded DNA

An initial step in recombination is probably the nicking and unwinding of a DNA duplex by a protein complex consisting of the RecB, RecC, and RecD proteins. The complex has both helicase and nuclease activity. Figure 20-20 shows how this complex unwinds the DNA, driven by the hydrolysis of ATP as it moves and generates single strands from a duplex molecule. The nuclease activity recognizes an 8-bp sequence:

Table 20-2   Major Recombination Pathways in *E. coli*

	Pathway		
Property	RecBCD	RecF	RecE
Rec proteins required*	RecA, B, C, D	RecA, F, J, N, O, Q; Ruv	RecA, E, F, J, O, Q
Activating mutation	None	*sbcB, C*	*sbcA*
Responsible for conjugal recombination in wild-type cells	Yes	Marginally	No
Promotes plasmid recombination in wild-type cells	No	Yes	No
Requires double-stranded-DNA break	Yes	No	No
Activated by *chi* sites	Yes	No	No

* Assayed by conjugal recombination; when other recombination assays are employed, the genetic requirements may vary. In addition to these recombination factors, other gene products are required. The RecBCD pathway depends on SSB, gyrase, pol I, and ligase; the other pathways are likely to have similar requirements.

SOURCE: A. Kornberg and T. A. Baker, *DNA Replication,* 2d ed. Copyright 1992 by W. H. Freeman and Company. Modified from G. R. Smith, *Cell* 58, 1989, 807.

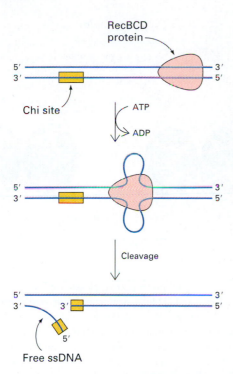

RecBCD
protein

Chi site

ATP

ADP

Cleavage

Free ssDNA

Figure 20-20 Model for the generation of single-stranded DNA by the RecBCD complex (red), an ATP-driven helicase and nuclease. Two single-stranded loops are formed as the complex moves forward. The enzyme cleaves one of the strands when it encounters a chi sequence (yellow) to form single-stranded DNA with a free end. Recombination can then take place. (After A. Taylor and G. R. Smith, *Cell* 22, 1980, 447; from L. Stryer, *Biochemistry*, 4th ed. Copyright ©1995 by Lubert Stryer.)

5′ G C T G G T G G 3′

called a **chi site,** which occurs approximately every 64 kb. As the complex unwinds the DNA, the free single-stranded DNA is generated, which can be used to initiate recombination (see below). The SSB protein, which also is involved in DNA replication (Chapter 11), can bind to and stabilize the single strands that are generated.

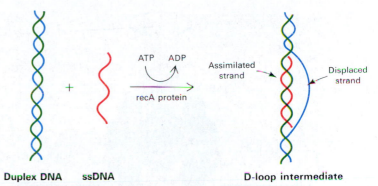

Duplex DNA   ssDNA   D-loop intermediate

Figure 20-21 The pairing of a single-stranded DNA molecule with the complementary strand of a duplex is catalyzed by the recA protein. The resulting structure is called a *D loop*. ATP hydrolysis releases recA protein from DNA. (From L. Stryer, *Biochemistry*, 4th ed. Copyright ©1995 by Lubert Stryer.)

RecA Protein–Mediated Single-Strand Exchange

The RecA protein, which also plays a role in the induction of the SOS repair system (see Chapters 18 and 19), can bind to single strands along their length, forming a nucleoprotein filament. RecA catalyzes single-strand invasion of a duplex and subsequent displacement of the corresponding strand from the duplex. This occurs in the presence of ATP, as shown in Figure 20-21. The displaced strand forms what is termed a **D loop.** Figure 20-22 depicts how this sequence can lead to Holliday junction.

Branch Migration

The movement of a Holliday junction (see, for instance, Figures 20-4e and 20-5), or branch migration, increases the length of heteroduplex DNA. The RuvA and RuvB proteins catalyze branch migration, driving the reaction by the hydrolysis of ATP. The RuvA protein binds to the crossover point, and then is flanked by two RuvB ATPase hexameric rings, as seen by electron microscopy. Figure 20-23 depicts a model for the action of these proteins in branch migration.

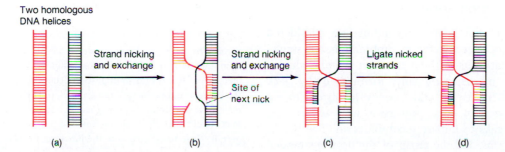

Two homologous
DNA helices

Strand nicking
and exchange

Strand nicking
and exchange

Site of
next nick

Ligate nicked
strands

(a)   (b)   (c)   (d)

Figure 20-22 A schematic view of some of the steps in recombination. (a) The pairing of two homologous duplexes. (b) A nick is made by the RecB,C nuclease. The helix is partially unwound, and the single-stranded region is extended and stabilized by the SSB protein. The RecA protein catalyzes the invasion by the single strand of the duplex. The SSB protein aids in keeping the single strand free. (c) After nicking by the RecB,C nuclease, the free single strand from the second duplex can anneal with the first duplex. (d) RNA ligase can seal this structure. (After B. Alberts et al., *Molecular Biology of the Cell.* Copyright © 1983 by Garland Publishing.)

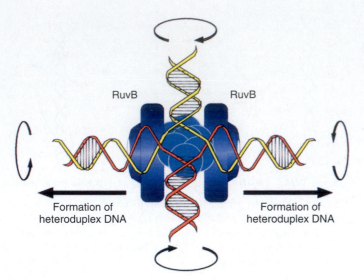

Figure 20-23 Model for RuvAB-mediated branch migration. The blue spheres indicate the RuvA protein, which binds to the crossover in the Holliday junction–RuvAB complex. Two hexameric rings of RuvB flank RuvA. The two RuvB ring motors lie in opposite orientation and affect branch migration by promoting the passage of DNA. (From Carol A. Parsons, Andrzej Stasiak, Richard J. Bennett, and Stephen C. West, "Structure of a Multisubunit Complex that Promotes DNA Branch Migration," *Nature* 374, 1995, 377.)

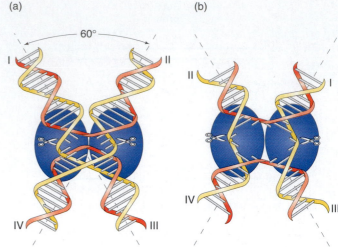

Figure 20-24 Model showing RuvC cleavage of two of the four strands of an antiparallel Holliday junction. The scissors depict sites where the nicking is symmetrical. (a) Here nicking resolves a stacked X-structure. The stars mark continuous strand. (b) Nicking resolves an unfolded junction. The formation of a twofold-symmetric unfolded structure shown here occurs by rotating arms I and II of the structure shown in (a) by 180°. (From Richard J. Bennett and Stephen C. West, "RuvC Protein Resolves Holliday Juctions via Cleavage of the Continuous (Noncrossover) Strands," *Proc. Natl. Acad. Sci. USA* 92, 1995, 5639.)

Resolution of Holliday Junctions

Several enzymatic pathways have been identified that are involved in cleaving across the point of strand exchange in a Holliday structure to yield two duplexes. RuvC is an endonuclease that resolves Holliday junctions by symmetric cleavage of the continuous (noncrossing) pair of DNA strands, as seen in Figure 20-24. In addition, the RecG and Rus proteins may provide alternative routes to cleavage. These reactions are summarized in Figure 20-25.

The Double-Strand Break-Repair Model for Recombination

In the two models for genetic recombination considered above, the initiation events for recombination are single-stranded nicks that result in the generation of heteroduplex DNA. Another model, originally formulated by Jack Szostak, Terry Orr-Weaver, and Rodney Rothstein to explain certain aspects of gene conversion in yeast, invokes double-strand breaks to initiate recombination. The breaks are enlarged to gaps, and the repair of the double-stranded gaps results in gene conversion. Figure 20-26 shows this model in detail. Note how the repair of the double-strand break and gap occurs by two rounds of single-strand repair, and that the DNA flanking the region of gap repair contains regions of heteroduplex DNA. The intermediate structure contains two Holliday junctions, which can be resolved in

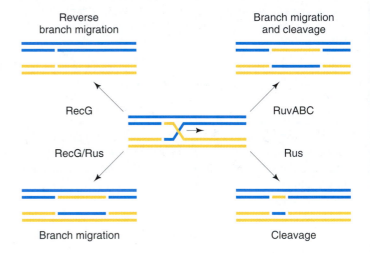

Figure 20-25 Possible pathways for processing Holliday junction in *E. coli*. The center of the figure shows a Holliday junction formed by homologous pairing and strand exchange by the RecA protein. In the presence of RecA, branch migration occurs in the same direction as the RecA strand exchange, and is promoted by RuvAB. RuvC cleavage resolves this structure (*upper right*). Also in the presence of RecA, RecG could drive the Holliday junction backward (*upper right*). However, if RecA dissociates from the junction, RecG could drive the reaction forward, with resolution coming from the Rus protein (*lower left*). The lower right shows the results of Rus cleavage alone, following strand exchange and branch migration by RecA. (From Gary Sharples, Sau Chan, Akeel Mahdi, Matthew Whitby, and Robert Lloyd, "Processing of Intermediates in Recombination and DNA Repair: Identification of a New Endonuclease That Specifically Cleaves Holliday Junctions," *EMBO Journal* 13 (24), 1994, 6140.)

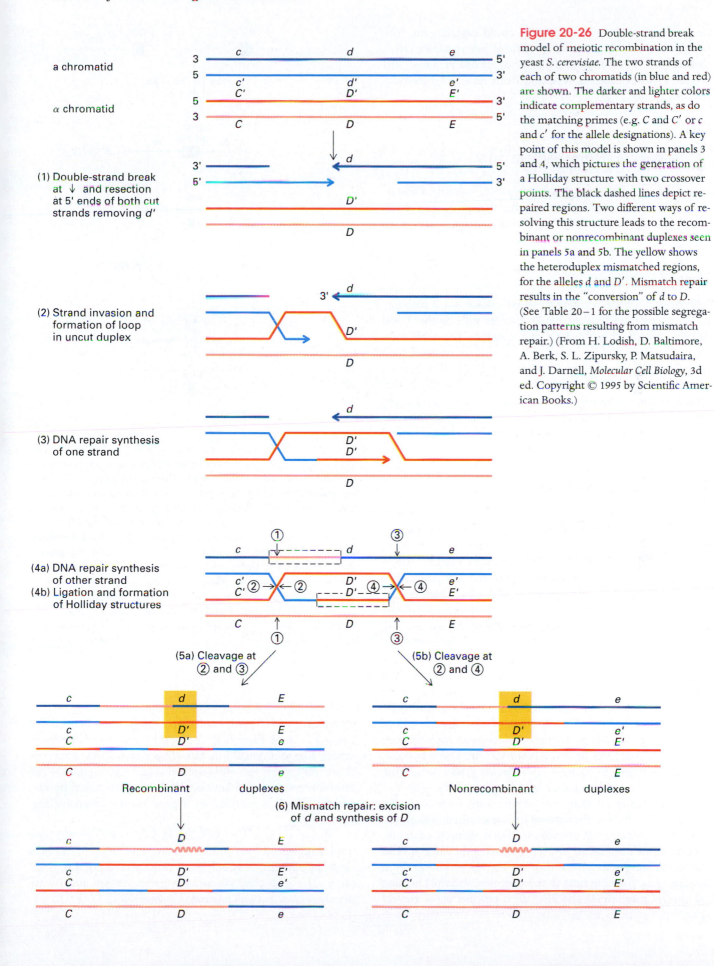

Figure 20-26 Double-strand break model of meiotic recombination in the yeast *S. cerevisiae*. The two strands of each of two chromatids (in blue and red) are shown. The darker and lighter colors indicate complementary strands, as do the matching primes (e.g. *C* and *C'* or *c* and *c'* for the allele designations). A key point of this model is shown in panels 3 and 4, which pictures the generation of a Holliday structure with two crossover points. The black dashed lines depict repaired regions. Two different ways of resolving this structure leads to the recombinant or nonrecombinant duplexes seen in panels 5a and 5b. The yellow shows the heteroduplex mismatched regions, for the alleles *d* and *D'*. Mismatch repair results in the "conversion" of *d* to *D*. (See Table 20–1 for the possible segregation patterns resulting from mismatch repair.) (From H. Lodish, D. Baltimore, A. Berk, S. L. Zipursky, P. Matsudaira, and J. Darnell, *Molecular Cell Biology*, 3d ed. Copyright © 1995 by Scientific American Books.)

several ways. Although the model could explain gene conversion without the mismatch repair of heteroduplex DNA, this in fact occurs frequently in yeast. (Consult Table 20-1 for segregation patterns resulting from repair of heteroduplexes.)

Site-Specific Recombination

In addition to generalized recombination, described above, there are also recombination systems that promote recombination only at specific sites. Some examples are given below.

Phage Integration

The bacteriophage λ integrates and excises via a pathway utilizing enzymes that recognize specific sites on the λ and on the *E. coli* chromosome and catalyze recombination events between them. These attachment, or *att,* sites contain short regions of homology (Figure 20-27). The homologous region of 15 base pairs is only part of each *att* site. The integration reaction is catalyzed by the λ-encoded **Int** protein, in cooperation with a host protein, while the excision reaction requires these two proteins plus the λ-encoded **Xis** enzyme.

Control of Gene Expression

Several genomes use site-specific recombination systems to control surface antigens or host-range phenotypes. One of the best-characterized systems involves phase variation in the bacterium *Salmonella.* The surface flagella can be one of two types in these strains, termed *H1* or *H2.* The conversion frequency of one type to another is 10^{-3}. This high rate of transition from one phase to another is caused by the unique arrangement of genes controlling the *H1* and *H2* loci (Figure 20-28). When the *H2* gene is actively expressed, a repressor protein, encoded by the *rH1* gene adjacent to *H2,* is also synthesized, which prevents transcription of the unlinked *H1* gene (Figure 20-28a). The promoter for the *H2* gene is situated on a roughly 1000-bp DNA segment adjacent to the operator for the *H2* gene. Forming each end of this segment is a 14-bp sequence identical with the sequence at the other end, but inverted relative to it. These two termini are referred to as **inverted repeats (IRs).** When the two IRs are brought together by a looping of the intervening DNA (Figure 20-28b), they are no longer inverted relative to each other (as the arrows demonstrate) and homologous recombination is possible between sections of them. As can be seen in Figure 20-28c, such recombination inverts the intervening segment, removing the *H2* promoter from its functional position next to the operator, and thus turning off the *H2* gene. Since this also stops production of the *H1*

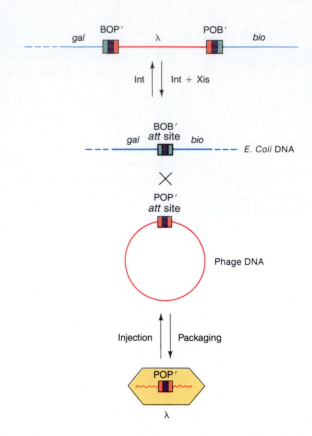

Figure 20-27 λ integration and excision. The λ and *E. coli* chromosomes contain attachment sites that are recognized by the λ integration and excision enzymes. These sites share a short region of homology, indicated by O, but also contain regions flanking the homology that are different. The flanking bacterial sequences are designated B and B′, and the flanking phage sequences are designated P and P′. Thus, the bacterial attachment site is designated BOB′, and the phage attachment site is designated POP′. The integration process proceeds from the bottom of the figure; the excision process, from the top. Looking at integration, the phage DNA is injected into the bacterium. The POP′ site splits and is inserted into the BOB′ site, under the mediation of the Int protein. Thus POP′ becomes flanked by BOB′ within the bacterial chromosome, with the phage DNA in between.

repressor, the *H1* gene, which is located elsewhere in the genome, can be expressed. A second inversion of the intervening segment repositions the *H2* promoter next to its operator and turns the *H2* gene on again. As Figure 20-28b shows, recombination between the IRs is mediated by Hin protein, which is encoded by a gene on the invertible segment itself.

Phage *mu* (see Chapter 21) and P1 (Chapter 10) have similar site-specific recombination systems that control host range by allowing a switch from one surface protein to another. The respective *gin* and *cin* genes for *mu* and P1 are closely related to the *hin* gene of *Salmonella.*

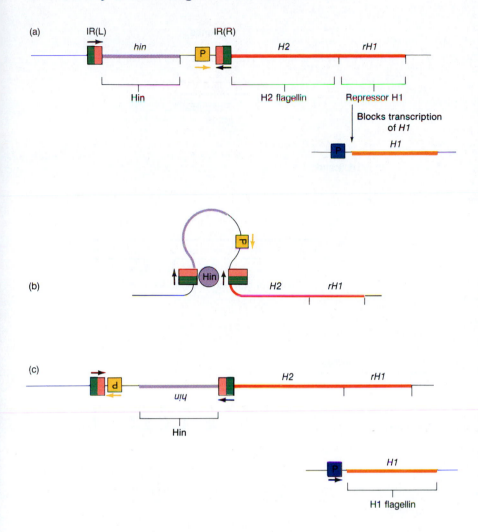

Figure 20-28 Site-specific recombination controls flagellar phase transition. Inversion of a region of DNA adjacent to the *H2* gene alternately couples and uncouples a promoter (P) for the *H2* gene. (a) When the promoter is positioned next to the *H2* operon, the H2 flagellin type is produced, along with a repressor of the *H1* gene, encoded by the *rH1* gene next to *H2*. The *H1* gene, which is not linked to *H2* is thus turned off when *H2* is expressed. (b) The invertible region includes at its right and left sides an inverted repeat [IR(R) and IR(L), respectively]. The arrows associated with the IRs show that they are oriented in opposite directions owing to their inversion relative to each other. However, when they are brought together, the arrows show that they are properly aligned, and Hin protein is able to mediate a recombination between them, which reverses the intervening segment. Hin is produced by a gene within this segment and is necessary for recombination to occur. (c) Inversion of the intervening segment moves the promoter away from the *H2* operon and thus turns off the *H2* and *rH1* genes. This derepresses *H1*, which has its own promoter, unrelated to the invertible promoter, and *H1* is then able to produce the H1 flagellin. (Note that the invertible promoter is able to promote expression of the gene on the left side of *H2* when in the "H2-off" orientation.) (After M. Silverman and M. Simon, *Mobile Genetic Elements*, J. A. Shapiro, ed. Copyright © 1983 by Academic Press.)

Recombination and Chromosomal Rearrangements

Rearrangement of chromosomes, such as deletions, duplications (see the following section), inversions, and translocations can result from recombinations, as discussed in Chapter 8. For instance, the bar-eye duplication in *Drosophila* results from an unequal crossover. Unequal crossovers can also be detected in human chromosomes. For example, some inherited thalassemias involve deletions that fuse two human globin genes (Figure 8-17), leading to anemic blood disease. Other diseases caused by recombination-driven rearrangements covered in Chapter 8 include cri du chat syndrome and Down syndrome.

Duplication and Evolution

It is clear that gene duplications generated by unequal crossing-over have played an important role in gene evolution. In Chapter 16 we discussed gene families, consisting of genes that are derived from a common ancestral gene that had duplicated, with the two or more resulting copies then undergoing sequence divergence. The globin family (Figure 16-20) is an example. Another example is illustrated in Figure 20-29, which shows a model for the evolution of mammalian albumin and α-fetoprotein. Each protein consists of three domains which are held together by disulfide cross-links (black bars) and which are very similar. Each domain is encoded by four exons, which have a similar arrangement for each of the three domains (Figure 20-29, *right side*). The arrangement of exons can be explained if there were an ancestral gene with four exons that underwent a triplication, to produce the gene for one of the two three-domain enzymes. This entire gene structure could then have been duplicated, to produce, after sequence divergence, the current genes for mammalian albumin and α-fetoprotein.

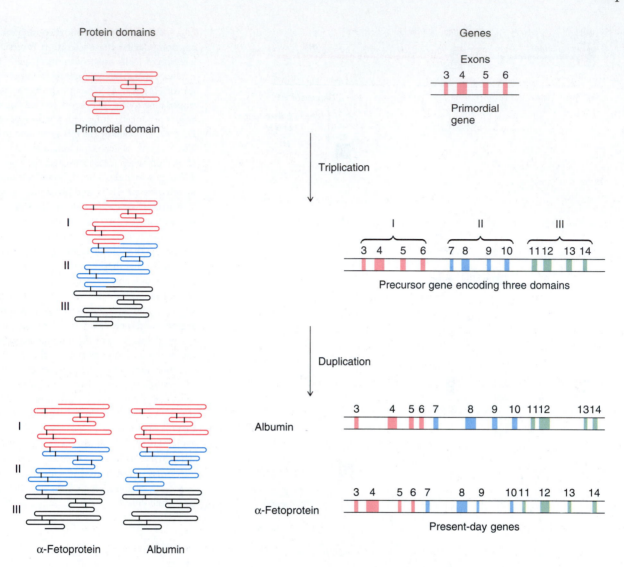

Figure 20-29 Model for evolution of mammalian albumin and α-fetoprotein. (From J. Darnell, H. Lodish, and D. Baltimore, *Molecular Cell Biology,* 2d ed. Copyright © 1990 by Scientific American Books.)

SUMMARY

We are now beginning to understand the molecular processes behind recombination, which produces new gene combinations by exchanging homologous chromosomes. Both genetics and physical evidence have led to the use of a heteroduplex DNA model to explain the mechanism of crossing-over. The process of recombination itself is under genetic control: there are genes that affect the efficiency of recombination. Several enzymes, including those encoded by the *recA*, *recB*, and *recC* genes, have been implicated in the recombination process in bacteria such as *E. coli*.

Concept Map

Draw a concept map interrelating as many of the following terms as possible. Note that the terms are listed in no particular order.

crossing-over / heteroduplex / chiasmata / gene conversion / correction / non-Mendelian ratio / recombination / 5:3 / base mismatch / heterozygote

CHAPTER INTEGRATION PROBLEM

In previous chapters we have learned about different mechanisms of generating mutations, both spontaneously and with mutagens. Describe gene conversion and how you can distinguish gene conversion from mutation.

Solution

Gene conversion is a meiotic process of directed change in which one allele directs the conversion of a partner allele to its own form. Instead of ending with equal numbers of both alleles during meiosis, gene conversion results in an excess of one allele and a deficiency of the other. Models of recombination employing heteroduplex DNA offer plausible mechanisms for gene conversion. In contrast, mutation is undirected change that can occur during both meiosis and mitosis.

(Solution from Diane K. Lavett.)

SOLVED PROBLEMS

1. In *Neurospora* an *ad3* double mutant consisted of two mutant sites within the *ad3* gene, site 1 on the left and site 2 on the right. This was crossed to wild type, using parental stocks heterozygous for two closely linked flanking loci *A* and *B*, as follows, where $+$ represents wild-type sequence at the mutant positions:

$$A \; 1 \; 2 \; B \times a + + b$$

Most asci were of the expected type showing regular Mendelian segregations, but there were also some unexpected types, of which several examples are represented here as I through III.

I	II	III
$A\,1\,2\,B$	$A\,1\,2\,B$	$A\,1\,2\,B$
$A\,1\,2\,B$	$A\,1\,2\,B$	$A\,1\,2\,B$
$A\,1\,2\,b$	$A\,1\,2\,B$	$A\,1\,2\,B$
$A++b$	$A\,1\,2\,B$	$A++B$
$a\,1\,2\,B$	$a++b$	$a+2\,b$
$a++B$	$a\,1\,2\,b$	$a++b$
$a++b$	$a++b$	$a++b$
$a++b$	$a++b$	$a++b$

Explain the likely origin of the rare types I through III according to molecular recombination models.

Solution

Ascus type I has two nonidentical sister spore pairs. Since the members of a spore pair are derived from a postmeiotic mitosis (see Chapter 6), this proves that the meiotic products in these cases must have contained both *1 2* and $+ +$ information; in other words, they must have contained heteroduplex DNA. Notice also that these spore pairs are recombinant for *A* and *B*. Therefore it is likely that a crossover occurred between the *A* and *B* loci, that the heteroduplex DNA that constituted the crossover spanned both the *ad3* mutant sites *1* and *2*, and that there was no correction of the heteroduplex at those sites.

Type II shows a 5:3 ratio of *1 2* doubles to $+ +$. Here again the heteroduplexes must have spanned both sites (in a noncrossover configuration), but this time there was correction of a $+ + / 1\,2$ heteroduplex to *1 2 / 1 2*, presumably by excision and repair of the $+ +$ information (co- or double conversion).

Type III reveals another noncrossover heteroduplex configuration, and this time correction occurred only at site 1; this is revealed as a 5:3 ratio for site 1, with $1 \rightarrow +$ conversion in ascospore 5.

2. In fungal crosses of the following general type,

$$
\begin{array}{ccc}
M & \boxed{+\ 2} & N \\
\hline
\times & & \\
\hline
m & \boxed{1\ +} & n
\end{array}
$$

where 1 and 2 are mutant sites of a nutritional gene and *M* and *N* are flanking loci, it is possible to select rare $+ +$ prototrophic recombinants by plating on minimal medium. These prototrophs are then examined for the alleles of the flanking loci. As might be expected, the combination

$$M + + n$$

is commonly encountered, but so, somewhat surprisingly, are

$$M + + N$$

and

$$m + + n$$

a. Explain the origin of these two genotypes in terms of molecular recombination models.

b. If $M + + N$ were more common than $m + + n,$ what would that mean?

Solution

a. It is likely that $M + + N$ arose from a non-crossover heteroduplex that spanned site 2, followed by a correction of $2 \rightarrow +.$

Similarly, $m + + n$ would be explained by a heteroduplex spanning site 1, and corrected $1 \rightarrow +.$

b. Since $M + + N$ arises from gene conversion at the right-hand site, it is likely that heteroduplex DNA is formed more commonly from the right than from the left, possibly because of a closer fixed break point. (NOTE: Prototrophs may also be formed by single-site correction in a heteroduplex spanning both sites, but then the inequality would require another explanation.)

PROBLEMS

1. Which of the following linear asci shows gene conversion at the *arg2* locus?

1	2	3	4	5	6
+	+	+	+	+	+
+	+	+	+	+	+
+	arg	+	+	arg	arg
+	arg	arg	arg	arg	arg
arg	arg	arg	+	+	arg
arg	arg	arg	arg	+	arg
arg	+	arg	arg	arg	arg
arg	+	arg	arg	arg	arg

2. At the light spore locus of *Ascobolus*, $1'$ mutant site is in the left portion of the gene and the $1''$ mutant site is more to the right. When crosses are made between $1'$- and $1''$-bearing strains,

Light spore gene   Light spore gene

asci with 6 light and 2 black spores can be selected visually. These are shown to be caused by gene conversion mostly of the following type:

$+ 1''$

$+ 1''$

$+ 1''$

$+ 1''$

$+ +$

$+ +$ black

$1' +$

$1' +$

but very rarely

$+ 1''$

$+ 1''$

$+ +$

$+ +$

$1' +$

$1' +$

$1' +$

$1' +$

What can account for this irregularity?

3. It has been proposed that the "fixed break point" of the heteroduplex recombination model might correspond to a promoter sequence. The following data relate to this idea. In the fungus *Podospora*, mutants were available in adjacent spore color cistrons 1 through 4, and a cross was made as follows:

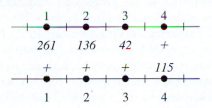

(261, 136, 42, and 115 are merely names for mutant sites.) Many asci were obtained showing gene conversion, but one that was relevant to the above suggestion was as follows:

Spore 1	261	136	42	+
2	+	+	+	+
3	+	+	+	115
4	+	+	+	115

Interpret this ascus in relation to the promoter idea.

4. Many mutagens increase the frequency of sister-chromatid exchange. Discuss possible explanations for this observation.

5. Mutations in locus *46* of the Ascomycete fungus *Ascobolus* produce light-colored ascospores (let's call them *a* mutants). In the following crosses between different *a* mutants, asci are observed for the appearance of dark wild-type spores. In each cross, all such asci were of the genotypes indicated:

$$
\begin{array}{ccc}
a_1 \times a_2 & a_1 \times a_3 & a_2 \times a_3 \\
\downarrow & \downarrow & \downarrow \\
a_1\ + & a_1\ + & a_2\ + \\
+\ + & a_1\ + & a_2\ + \\
+\ a_2 & +\ + & +\ + \\
+\ a_2 & +\ a_3 & +\ a_3
\end{array}
$$

Interpret these results in light of the models discussed in this chapter.

6. In the cross $A\ m_1\ m_2\ B \times a\ m_3\ b$, the order of the mutant sites m_1, m_2, and m_3 is unknown in relation to one another and to $A\,a$ and $B\,b$. One nonlinear conversion ascus is obtained:

$$
\begin{array}{c}
A\ m_1\ m_2\ B \\
A\ m_1\ b \\
a\ m_1\ m_2\ B \\
a\ m_3\ b
\end{array}
$$

Interpret this result in light of the heteroduplex DNA theory, and derive as much information as possible about the order of the sites.

7. In the cross $a_1 \times a_2$ (alleles of one locus) the following ascus is obtained:

$$
\begin{array}{c}
a_1\ + \\
a_1\ a_2 \\
a_1\ a_2 \\
+\ a_2
\end{array}
$$

Deduce what events may have produced this ascus (at the molecular level).

8. G. Leblon and J.-L. Rossignol have made the following observations in *Ascobolus*. Single-nucleotide-pair insertion or deletion mutations show gene conversions of the 6:2 or 2:6 type and only rarely of the 5:3, 3:5, or 3:1:1:3 type. Base-pair transition mutations show gene conversion of the 3:5, 5:3, or 3:1:1:3 type, and only rarely of the 6:2 or 2:6 type.

a. In terms of the hybrid DNA model, propose an explanation for these observations.

b. Leblon and Rossignol also have shown that there are far fewer 6:2 than 2:6 conversions for insertions and far more 6:2 than 2:6 conversions for deletions (where the ratios are $+:m$). Explain these results in terms of heteroduplex DNA. (You might also think about the excision of thymine photodimers.)

c. Finally, the researchers have shown that when a frameshift mutation is combined in a meiosis with a transition mutation at the same locus in a cis configuration, the asci showing *joint* conversion are all 6:2 or 2:6 for *both* sites (that is, the frameshift conversion pattern seems to have "imposed its will" on the transition site). Propose an explanation for this.

9. At the *grey* locus in the Ascomycete fungus *Sordaria*, the cross $+ \times g_1$ is made. In this cross, heteroduplex DNA sometimes extends across the site of heterozygosity, and two heteroduplex DNA molecules are formed (as discussed in this chapter). However, correction of heteroduplex DNA is not 100 percent efficient. In fact, 30 percent of all heteroduplex DNA is not corrected at all, whereas 50 percent is corrected to $+$ and 20 percent is corrected to g_1. What proportion of aberrant-ratio asci will be (a) 6:2? (b) 2:6? (c) 3:1:1:3? (d) 5:3? (e) 3:5?

10. Noreen Murray crossed α and β, two alleles of the *me-2* locus in *Neurospora*. Included in the cross were two markers, *trp* and *pan*, which each flank *me-2* at a distance of 5 m.u. The ascospores were plated onto a medium containing tryptophan and pantothenate but no methionine. The

Table 20-3

Cross	Genotype of *me-2*⁺ prototrophs			
	trp +	*+ pan*	*trp pan*	*+ +*
trp α + × + β pan	26	59	16	56
trp β + × + α pan	84	23	87	15

methionine prototrophs that grew were isolated and scored for the flanking markers, yielding the results shown in Table 20-3.

Interpret these results in light of the models presented in this chapter. Be sure to account for the asymmetries in the classes.

11. In *Neurospora,* the cross *A x × a y* is made, in which *x* and *y* are alleles of the *his-1* locus and *A* and *a* are mating-type alleles. The recombinant frequency between the *his-1* alleles is measured by the prototroph frequency when ascospores are plated on a medium lacking histidine; the recombinant frequency is measured as 10^{-5}. Progeny of parental genotype are backcrossed to the parents, with the following results. All *a y* progeny backcrossed to the *A x* parent show prototroph frequencies of 10^{-5}. When *A x* progeny are backcrossed to the *a y* parent, two prototroph frequencies are obtained: one half of the crosses show 10^{-5}, but the other half show the much higher frequency of 10^{-2}. Propose an explanation for these results, and describe a research program to test your hypothesis. (NOTE: Intragenic recombination is a *meiotic* function that occurs in a diploid cell. Thus, even though this is a haploid organism, dominance and recessiveness could be involved in this question.)

21

Mechanisms of Genetic Change III: Transposable Genetic Elements

Jumping gene in snapdragon (*Antirrhinum majus*). (Heinz Saedler, Max-Planck-Institut)

KEY CONCEPTS

► A series of genetic elements can occasionally move or transpose from one position on the chromosome to another position on the same chromosome or on a different chromosome.

► In bacteria, insertion sequences, transposons, and phage mu are examples of transposable genetic elements.

► Transposable elements can mediate chromosomal rearrangements.

► In higher cells, transposable elements have been extensively characterized in yeast, *Drosophila*, and maize and in mammalian systems.

► In eukaryotes, some transposable elements utilize an RNA intermediate during transposition, whereas in prokaryotes transposition occurs exclusively at the DNA level.

G enetic studies of maize, by Barbara McClintock and also Marcus Rhoades, have yielded results that greatly upset the classical genetic picture of genes residing only at fixed loci on the main chromosome. The research literature (see page 656) began to carry reports suggesting the existence of genetic elements of the main chromosomes that can somehow mobilize themselves and move from one location to another. These findings were viewed with skepticism for many years, but it is now clear that such mobile elements are widespread in nature.

A variety of colorful names (some of which help to describe their respective properties) have been applied to these genetic elements: controlling elements, cassettes, jumping genes, roving genes, mobile genes, mobile genetic elements, and transposons. We choose the term **transposable genetic elements,** which is formally most correct and embraces the entire family of types. The term *transposition* has long been used in genetics to describe transfer of chromosomal segments from one position to another in major structural rearrangements. In the present context, what is being transposed seems to be a gene, or a small number of linked genes, or a gene-sized fragment. Any genetic entity of this size can be called a *genetic element.*

Transposable genetic elements seem to be able to move to new positions within the same chromosome or even to move to a different chromosome. The normal genetic role of these elements is not known with certainty. They have been detected genetically through the abnormalities they produce in the activities and structures of the genes near the sites to which they move. A variety of physical techniques have been used to detect them as well. Transposable genetic elements have been found in phages, bacteria, fungi, higher plants, viruses, and insects.

Although transposable elements were first detected in eukaryotes, the molecular nature of transposable genetic elements was first understood in bacteria and phages. Therefore, we'll begin with the information derived from the original studies in these prokaryotes.

Insertion Sequences

Insertion sequences, or **insertion-sequence (IS) elements,** are now known to be segments of DNA that can move from one position on the chromosome to a different position on the same chromosome or on a different chromosome. When IS elements appear in the middle of genes, they interrupt the coding sequence and inactivate the expression of that gene. Owing to their size and in some cases the presence of transcription and translation termination signals, IS elements can also block the expression of other genes in the same operon if those genes are "downstream" from the promoter of the operon. This property was instrumental in their discovery, as we see in the following section.

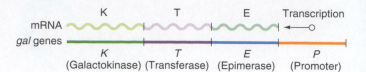

Figure 21-1 The *gal* genes of *E. coli.* Epimerase, transferase, and galactokinase are structural genes of the galactose gene set. The promoter controls their transcription into a single mRNA molecule.

IS elements were first found in *E. coli* in the *gal* operon—a set of three genes involved in the metabolism of the sugar galactose. These three adjacent genes—*E* (epimerase), *T* (transferase), and *K* (kinase)—are transcribed together from the same promoter (Figure 21-1).

Polar Mutations

The IS elements were found in mutants that had been selected by their deficiency in kinase activity. These mutations were found to map at various sites throughout the operon. In any particular mutant, the mutational site might map somewhere in the kinase gene itself or in the transferase or epimerase genes. Mutations in the transferase gene that have reduced kinase expression do not reduce epimerase expression. Each mutation obviously affects all structural gene functions transcriptionally downstream (but not upstream) in the operon. This directionality, or **polarity,** had been noted before for certain mutations, which are aptly termed **polar mutations.** For instance, certain chain-terminating nonsense mutations have polar effects. However, this particular kind of polar mutation was new. The mutants were observed to be capable of spontaneous reversion to wild type (showing that the mutation was not a deletion), but the reversion rate was not increased by any mutagen. Thus, they did not seem to be nonsense mutations, frameshifts, or any form of point mutation either. Then what were they? In the late 1960s, several experiments demonstrated that these IS elements were insertions of segments of DNA.

Physical Demonstration of DNA Insertion

Recall that the λ phage inserts next to the *gal* operon and that it is a simple matter to obtain λ*dgal* phage particles that have picked up the *gal* region (page 299). When the polar mutations in *gal* are incorporated into λ*dgal* phages and the buoyant density in a cesium chloride (CsCl) gradient of the phages is compared with that of the normal λ*dgal* phages, it is evident that the DNA carrying the polar mutation is longer than the wild-type DNA. This experiment clearly demonstrates that the mutations are caused by the insertion of a significant amount of DNA into the *gal* operon. Figure 21-2 depicts this experiment in more detail.

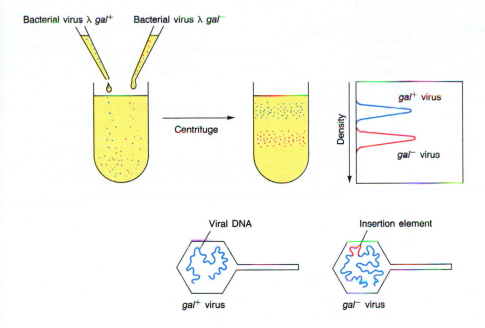

Bacterial virus λ *gal*⁺   Bacterial virus λ *gal*⁻

Centrifuge

Density

gal⁺ virus

gal⁻ virus

Viral DNA   Insertion element

gal⁺ virus   *gal*⁻ virus

Figure 21-2 Mutation by insertion is demonstrated with λ phage particles carrying the bacterial gene for galactose utilization (*gal*⁺) or the mutant gene *gal*⁻. The viruses are centrifuged in a cesium chloride (CsCl) solution. The *gal*⁻ particles are found to be denser. Because the virus particles all have the same volume and their outer shells all have the same mass, the increased density of *gal*⁻ particles shows that they must contain a larger DNA molecule: the *gal*⁻ mutation was caused by the insertion of DNA. (From S. M. Cohen and J. A. Shapiro, "Transposable Genetic Elements." Copyright © 1980 by Scientific American, Inc. All rights reserved.)

Direct Visualization of Inserted DNA

By hybridizing denatured λ*dgal* DNA containing the insertion mutation to denatured wild-type λ*dgal* DNA, the extra piece of DNA actually can be located under the electron microscope. In these experiments, some of the DNA molecules that form in the mixture are not the parent duplexes but heteroduplexes between one mutant and one wild-type strand. When point mutations are analyzed, the heteroduplexes are indistinguishable from the parental DNA molecules. However, in the case of the DNA involving the polar mutations, each heteroduplex shows a single-stranded buckle, or loop (Figure 21-3). This single-stranded buckle confirms the presence of an inserted sequence in the mutated DNA that has no complementary sequence in the wild-type DNA. The length of this single-stranded loop can be calibrated by including standardized marker DNA in the preparation. It proves to be approximately 800 nucleotides in length.

Identification of Discrete IS Elements

Are the segments of DNA that insert into genes merely random DNA fragments or are they distinct genetic entities? Hybridization experiments show that many different insertion mutations are caused by a small set of insertion sequences. In these experiments the λ*dgal* phages, which contain the *gal*⁻ gene, are isolated from the polar mutant bacteria, and their DNA is used to synthesize radioactive RNA in vitro. Certain fragments of this RNA are found to hybridize with the mutant DNA but not with wild-type DNA, reflecting the fact that the mutant contains an extra piece of DNA. These particular RNA fragments also hybridize to DNA from other polar mutants, showing that the

same bit of DNA is inserted in different places in the different polar mutants.

Based on their patterns of cross-hybridization, the insertion mutants are placed into categories. The first sequence, the 800-bp segment identified in *gal*, is termed *IS1*. A second sequence, termed *IS2*, is 1350 bp long. Table 21-1 lists some of the insertion sequences and their sizes. The "inverted repeats" listed in the table will be discussed shortly under "Transposons."

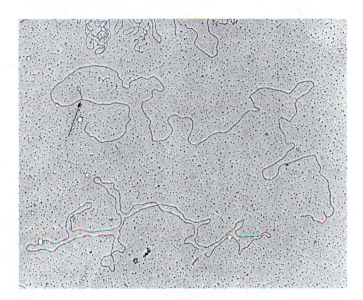

Figure 21-3 Electron micrograph of a λ*dgal*⁺/λ*dgal*ᵐ DNA heteroduplex. The single-stranded loop (arrow) is caused by the presence of an insertion sequence in λ*dgal*ᵐ. (From A. Ahmed and D. Scraba, "The Nature of the *gal3* Mutation of *Escherichia coli*," *Molecular and General Genetics* 136, 1975, 233.)

Table 21-1　**Prokaryotic Insertion Elements**

Insertion sequence	Normal occurrence in *E. coli*	Length (bp)	Inverted repeat (bp)
IS1	5–8 copies on chromosome	768	18/23
IS2	5 on chromosome; 1 on F	1327	32/41
IS3	5 on chromosome; 2 on F	1400	32/38
IS4	1 or 2 copies on chromosome;	1400	16/18
IS5	Unknown	1250	Short
γ-δ (TN1000)	1 or more on chromosome; 1 on F	5700	35
pSC101 segment	On plasmid pSC101	200	30/36

Source: M. P. Calos and J. H. Miller, *Cell* 20, 1980, 579–595.

We now know that the genome of the standard wild-type *E. coli* is rich in IS elements: it contains eight copies of IS1, five copies of IS2, and copies of other less well studied IS types as well. It should be emphasized that the sudden appearance of an insertion sequence at any given locus under study means that these elements are truly mobile, with a capability for transposition throughout the genome. Presumably, they produce a mutation or some other detectable alteration of normal cell function only when they happen to end up in an "abnormal" position, such as the middle of a structural gene. Insertion sequences also are commonly observed in the F factor. Figure 21-4 shows an example of an F-*lac* episome.

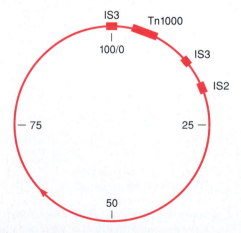

Figure 21-4 Genetic and physical map of the F factor. The positions of the resident insertion sequences, IS2, IS3, and Tn1000 (boxes), are shown relative to the map coordinates 0–100. The arrow indicates the origin and direction of F DNA transfer during conjugation.

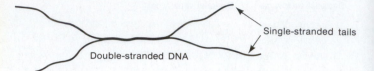

Figure 21-5 Diagrammatic representation of the appearance of a heteroduplex DNA molecule formed by annealing corresponding strands (that is, strands that have the same base sequence, excepting at the site of the IS) of λ*dgal*ᵐ DNA from two specific mutants caused by insertion sequences. This unexpected hybridization between corresponding strands (normally having the same rather than complementary base sequences) can be explained by the model shown in Figure 21-6.

Message　The bacterial genome contains segments of DNA, termed *IS elements,* that can move from one position on the chromosome to a different position on the same chromosome or on a different chromosome.

Orientation of IS Elements

Because of the base sequence, the two strands of λ*dgal*⁺ DNA happen to have different buoyant densities. After DNA denaturation, they can be recovered separately in the ultracentrifuge. In some cases, the *same* strands (that is, parallel strands) from two different IS1 mutants form an unexpected hybrid with each other. Under the electron microscope, these hybrids have a peculiar appearance: each is a double-stranded region with four single-stranded tails (Figure 21-5). This observation is explained by assuming that the IS1 elements are inserted in opposite directions in the two mutants (Figure 21-6).

Transposons

A frightening ability of pathogenic bacteria was discovered in Japanese hospitals in the 1950s. Bacterial dysentery is

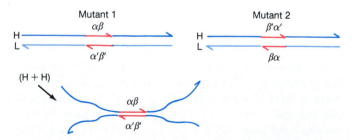

Figure 21-6 A model to explain the heteroduplex DNA structure shown in Figure 21-5. IS1 is represented by the red lines; α and β represent the ends of the IS1 molecule; H and L represent the strands of λ*dgal* DNA with high and low buoyant densities, respectively. The hybrid can be explained by assuming that the IS1 sequence is inserted in opposite directions in the two mutants.

caused by bacteria of the genus *Shigella*. This bacterium initially proved sensitive to a wide array of antibiotics that were used to control the disease. In the Japanese hospitals, however, *Shigella* isolated from patients with dysentery proved to be simultaneously resistant to many of these drugs, including penicillin, tetracycline, sulfanilamide, streptomycin, and chloramphenicol. This multiple drug-resistance phenotype was inherited as a single genetic package, and it could be transmitted in an infectious manner—not only to other sensitive *Shigella* strains, but also to other related species of bacteria. This talent is an extraordinarily useful one for the pathogenic bacterium, and its implications for medical science were terrifying. From the point of view of the geneticist, however, the situation is very interesting. The vector carrying these resistances from one cell to another proved to be a self-replicating element similar to the F factor. These **R factors** (for "resistance") are transferred rapidly on cell conjugation, much like the F particle in *E. coli* (Chapter 10).

In fact, these R factors proved to be just the first of many similar F-like factors to be discovered. These elements, which exist in the plasmid state in the cytoplasm, have been found to carry many different kinds of genes in bacteria. Table 21-2 shows just some of the characteristics that can be borne by plasmids. What is the mode of action of these plasmids? How do they acquire their new genetic abilities? How do they carry them from cell to cell? The discovery of *transposons* made it possible to answer some of these questions.

Table 21-2 Genetic Determinants Borne by Plasmids

Characteristic	Plasmid examples
Fertility	F, R1, Col
Bacteriocin production	Col E1
Heavy-metal resistance	R6
Enterotoxin production	Ent
Metabolism of camphor	Cam
Tumorigenicity in plants	T1 (in *Agrobacterium tumefaciens*)

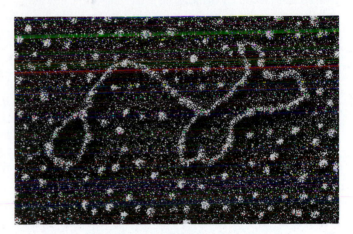

Figure 21-7 Peculiar structure formed when denatured plasmid DNA is reannealed. The double-stranded IR region separates a large circular loop from the small "lollipop" loop. (Photograph courtesy of S. N. Cohen.)

Physical Structure of Transposons

If the DNA of a plasmid conferring drug resistance (carrying the genes for kanamycin resistance, for example) is denatured to single-stranded forms and then allowed to renature slowly, some of the strands form an unusual shape under the electron microscope: a large, circular DNA ring is attached to a "lollipop"-shaped structure (Figure 21-7). The "stick" of the lollipop is double-stranded DNA, which has formed through the annealing of two **inverted repeat (IR) sequences** in the plasmid (Figure 21-8). Subsequent studies have shown that the IR sequences are a pair of IS elements in many cases. For instance, IS10 is present at the ends of the region carrying the genes for tetracycline resistance (Figure 21-9). In some cases, however, the IR sequences are much smaller.

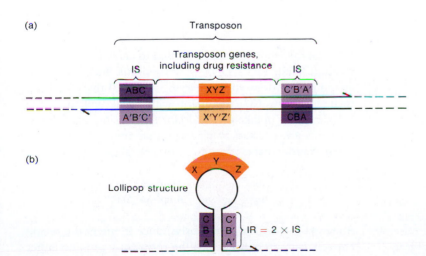

Figure 21-8 An explanation at the nucleotide level for the lollipop structure seen in Figure 21-7. The structure is called a *transposon*. (a) The transposon in its double-stranded form before denaturing. Note the presence of oppositely oriented copies of an insertion sequence (IS). (b) The lollipop structure formed by intrastrand annealing after denaturing. The two IS regions anneal to form the IR of the transposon; the transposon genes are carried in the lollipop loop.

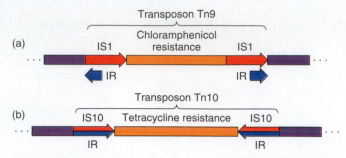

Figure 21-9 Two different transposons having different IR regions and carrying different drug-resistance genes. (a) Tn9 has a short IR region, because the two IS1 elements are in the same orientation and each element has a short inverted repeat. (b) Tn10 has a large IR region because the two IS10 components have opposite orientations, and the entire IS10 sequence constitutes the inverted repeat.

The genes for drug resistance or other genetic abilities carried by the plasmid are located between the IR sequences in the lollipop head. The IR sequences together with their contained genes have been collectively called a **transposon (Tn).** (Transposons are therefore longer than IS elements, since they contain extra protein-coding genes.) The remainder of the plasmid, bearing the genes coding for resistance-transfer functions (RTF), is called the **RTF region** (Figure 21-10). Table 21-3 lists some of the known transposons.

Movement of Transposons

A transposon can jump from one plasmid to another plasmid (Figure 21-11) or from a plasmid to a bacterial chromosome. Let's consider an actual experiment documenting this transposition.

A transposon (Tn3) containing an ampicillin-resistance gene (Amp^R) is carried in a large *E. coli* plasmid R64-1. This Tn3 is then transferred to a small plasmid RSF1010, which carries the genes for sulfonamide resistance (Su^R) and streptomycin resistance (Sm^R) in another transposon (Tn4):

$$\text{R64-1}\ (Amp^R \text{ in Tn3}) \longrightarrow \text{RSF1010}\ (Su^R Sm^R \text{ in Tn4})$$

Table 21-3

Transposon	Marker	Length (bp)	Inverted repeat
Tn1	Ampicillin	4,957	38
Tn2	Ampicillin		
Tn3	Ampicillin		
Tn4	Ampicillin streptomycin, sulfanilamide	20,500	Short
Tn5	Kanamycin	5,400	1500
Tn6	Kanamycin	4,200	Not detectable with electron microscopy
Tn7	Trimethoprim, streptomycin	14,000	Not detectable with electron microscopy
Tn9	Chloramphenicol	2,638	18/23
Tn10	Tetracycline	9,300	1400
Tn204	Chloramphenicol, fusidic acid	2,457	18/23
Tn402	Trimethoprim	7,500	Not detectable with electron microscopy
Tn501	Mercuric ions	7,800	38
Tn551	Erythromycin	5,200	35
Tn554	Erythromycin, spectinomycin	6,200	Not determined
Tn732	Gentamicin, tobramycin	11,000	Not determined
Tn903	Kanamycin	3,100	1050
Tn917	Erythromycin	5,100	Short
Tn951	*lac*	16,600	Short
Tn1681	Heat-stable enterotoxin	2,088	768 (IS1)
Tn1721	Tetracycline	10,900	Short

SOURCE: M. P. Calos and J. H. Miller, *Cell* 20, 1980, 579–595.

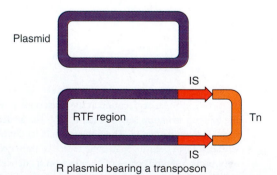

Figure 21-10 The insertion of a transposon (Tn) into a plasmid. RTF represents the resistance-transfer functional genes of the plasmid. Tn includes both the IS elements and the drug-resistance genes.

The following procedure is used to carry out this transposition. First, R64-1 DNA is isolated and added as transforming DNA to a strain carrying RSF1010 that had been treated with $CaCl_2$ (treatment with $CaCl_2$ enhances the probability of uptake of the donor DNA). The Amp^R transformants are selected by plating on an ampicillin medium. These transformants prove to be of several genotypes:

$$Amp^R\ Su^R\ Sm^R \qquad Amp^R\ Su^R\ Sm^S$$
$$Amp^R\ Su^S\ Sm^R \qquad Amp^R\ Su^S\ Sm^S$$

Apparently, the Amp^R transposon, once it enters the recipient cell, can insert itself into the recipient's genome either

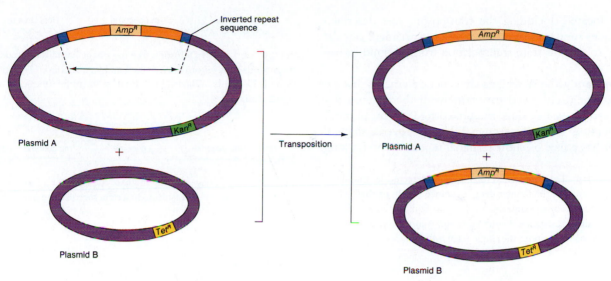

outside the $Su^R Sm^R$ transposon or within it. If it inserts within the recipient's original Tn4 transposon, it knocks out either one or both of the resistance functions originally possessed, depending on the precise point of insertion. New lollipop structures, corresponding to the acquisition of the new transposon, are observed in the transformed genotypes. Figure 21-12 shows a composite diagram of an R plasmid, indicating the various places at which transposons

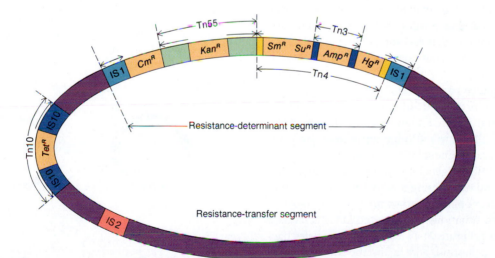

can be located, including the transposons just described. Techniques similar to those used in this experiment can also be used to demonstrate transposition from plasmid to bacterial DNA.

Apparently, the IS regions are genetic elements that can mobilize themselves and can carry with them genes that confer various traits. Thus, multiple drug-resistant plasmids are generated. As can be seen in Table 21-1, even a single IS element has a pair of short terminal repeats 20-50 bp long.

Message Transposons were originally detected as mobile genetic elements that confer drug resistance. Many of these elements consist of recognizable IS elements flanking a gene that encodes drug resistance. IS elements and transposons are now grouped together under the single term *transposable elements*.

Phage mu

Phage mu is a normal-appearing phage. We consider it here because, although it is a true virus, it has many features in common with IS elements. The DNA double helix of this phage is 36,000 nucleotides long—much larger than an IS element. However, it does appear to be able to insert itself anywhere in a bacterial or plasmid genome in either orientation. Once inserted, it causes mutation at the locus of insertion—again like an IS element. (The phage was named for this ability: *mu* stands for "mutator.") Normally, these mutations cannot be reverted, but reversion can be produced by certain kinds of genetic manipulation. When this reversion is produced, the phages that can be recovered show no deletion, proving that excision is exact and that the insertion of the phage therefore does not involve any loss of phage material either.

Each mature phage particle has on each end a piece of flanking DNA from its previous host (Figure 21-13). However, this DNA is not inserted anew into the next host. Its function is unclear. Phage mu also has an IR sequence, but neither of the repeated elements is at a terminus.

Mu can also act like a genetic snap fastener, mobilizing any kind of DNA and transposing it anywhere in a genome. For example, it can mobilize another phage (such as λ) or the F factor. In such situations, the inserted DNA is flanked by two mu genomes (Figure 21-14). It can also transfer bacterial markers onto a plasmid; here again, the transferred region is flanked by a pair of mu genomes (Figure 21-15). Finally, the phage mu can mediate various kinds of structural chromosome rearrangements (Figure 21-16).

Mechanism of Transposition

Several different mechanisms of transposition are employed by prokaryotic transposable elements. And, as we shall see later, eukaryotic elements exhibit still additional mechanisms of transposition.

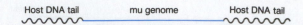

Figure 21-13 The DNA of a free mu phage has tails derived from its previous host. (From S. N. Cohen and J. A. Shapiro, "Transposable Genetic Elements." Copyright © 1980 by Scientific American, Inc. All rights reserved.)

Figure 21-14 Phage mu can mediate the insertion of phage λ into a bacterial chromosome, resulting in a structure like the one shown here.

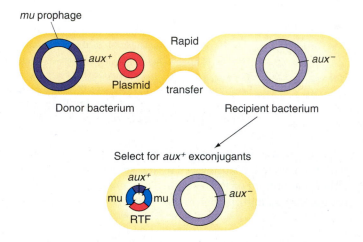

Figure 21-15 Phage mu can mediate the transposition of a bacterial gene into a plasmid. The selection procedure for detecting the transposition is indicated here. An auxotrophic mutant gene is indicated by *aux⁻*.

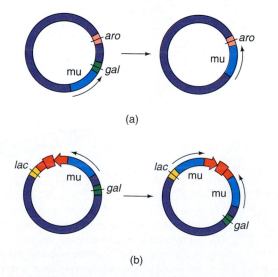

Figure 21-16 Phage mu can cause the deletion or inversion of adjacent bacterial segments. (a) The *gal* region is deleted by transposition of phage mu. (b) The F-factor region of an Hfr strain is inverted by transposition of phage mu.

In *E. coli,* we can identify **replicative** and **conservative** (nonreplicative) modes of transposition. In the replicative pathway, a new copy of the transposable element is generated during the transposition event. The results of the transposition are that one copy appears at the new site and one copy remains at the old site. In the conservative pathway, no replication occurs. Instead, the element is excised from the chromosome or plasmid and is integrated into the new site.

Replicative Transposition

When transposition occurs from one locus to a second locus for certain transposons, a copy of the transposable element is left behind at the first locus. Researchers used restriction-enzyme digestion patterns to show that when phage mu transposes, segments of mu phage remain at the original location in the *E. coli* chromosome and newly synthesized DNA appears at new locations in the chromosome. Therefore, in these cases the transposable elements do not strictly "jump" from one location to another by excising themselves from the chromosome and migrating through the cytoplasm; instead, they "replicate" into a new location, leaving one copy behind.

An analysis of transposon mutants revealed another interesting fact about the mechanism of transposition. Using the transposon Tn3 (Figure 21-17), researchers grouped the mutations that prevent transposition into two categories. A trans-recessive class maps in the gene that encodes the transposase enzyme, which is involved in transposition. A second class of cis-dominant mutations results in the buildup of an intermediate in the transposition process. Figure 21-18 diagrams the transposition pathway involved in Tn3 transposition from one plasmid to another. The intermediate is actually a double plasmid, with both donor and recipient plasmid being fused together. The combined circle resulting from the fusion of two circular elements is termed a **cointegrate.** Apparently, the mutations in this second class delete a region on the transposon at which a recombination event takes place that resolves cointegrates into two smaller circles. This region, called the **internal resolution site (IRS),** appears in Figure 21-17.

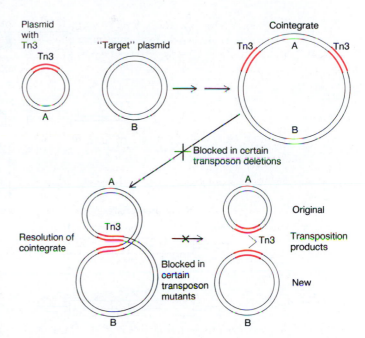

Figure 21-18 Transposition of Tn3 takes place via a cointegrate intermediate. Cointegrates in Tn3 transposition are observed for some internal deletions in the transposon. The correct explanation for this observation is that the cointegrates are intermediates in Tn3 transposition and that their resolution is blocked because the internal deletion has removed an internal resolution site (IRS), where recombination occurs. (From F. Heffron, in J. A. Shapiro, ed., *Mobile Genetic Elements,* pp. 223–260. Copyright © 1983 by Academic Press.)

The finding of a cointegrate structure as an intermediate in transposition helped establish a replicative mode of transposition for certain elements. In Figure 21-18, note how the transposable element is duplicated during the fusion event and how the recombination event that resolves the cointegrate into two smaller circles leaves one copy of the transposable element in each plasmid.

Conservative Transposition

Some transposons, such as Tn10, excise from the chromosome and integrate into the target DNA. In these cases, DNA replication of the element does not occur, and the element is lost from the site of the original chromosome. Researchers demonstrated this by constructing heteroduplexes of λTn10 derivatives containing the *lac* region of *E. coli.* The researchers used DNA from Tn10-*lacZ*$^+$ and Tn10-*lacZ*$^-$ derivatives. The heteroduplexes, therefore, contain one strand with the wild-type *lac* region and a second strand with the mutated (Z$^-$) *lac* region. Figure 21-19 diagrams this part of the experiment. The heteroduplex DNA is used to infect cells that have no *lac* genes, and transpositions of the Tet^R Tn10 are selected. Different types of colonies arise from the transposition of a heteroduplex Z$^-$/Z$^+$ carrying

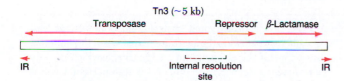

Figure 21-17 The structure of Tn3. Tn3 contains 4957 base pairs and codes for three polypeptides: the transposase is required for transposition, the repressor is a protein that regulates the transposase gene (see Chapter 18), and β-lactamase confers ampicillin resistance. Tn3 is flanked by inverted repeats (IR) of 38 base pairs and contains a site designated the *internal resolution site* that is necessary for the resolution of Tn3 cointegrates (discussed next).

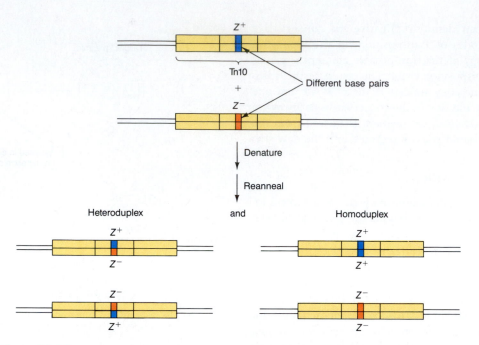

Figure 21-19 Generation of heteroduplex and homoduplex Tn10 elements. The denaturation and reannealing of a mixture of two parental λ phages carrying Tn10 elements that differ only at three single bases in the transposon yields a mixture of heteroduplex and homoduplex products. The base differences in the Z gene allow the ultimate determination of the heteroduplex or homoduplex nature of a cell that has received the Z gene via transposition. (Adapted from J. Bender and N. Kleckner, *Cell* 45, 1986, 801–815.)

transposon (Figure 21-20). If replication occurs (the replicative mode of transposition), all colonies are either completely Lac$^+$ or completely Lac$^-$, because the replication will convert the heteroduplex DNA into two homoduplex daughter molecules. The mechanism by which this conversion occurs will be examined in detail in the next section. However, if the transposition is conservative and does not involve replication, each colony arises from a *lacZ$^+$/lacZ$^-$* heteroduplex. Such colonies are partially Lac$^+$ and partially Lac$^-$. By using media that stain Lac$^+$ and Lac$^-$ cells different colors, researchers can observe the Lac$^+$ and Lac$^-$ sectors in colonies.

Therefore, the determination of whether replicative or conservative transposition occurs for Tn10 can be made by observing whether differently colored sectors occur within the same colony resulting from the transposition. Sectored colonies are observed in a majority of cases (Figure 21-21). Thus, Tn10—and perhaps other transposable elements in *E. coli*—transpose by excising themselves from the donor DNA and integrating directly into the recipient DNA.

Molecular Consequences of Transposition

The molecular consequences of transposition reveal an additional piece of evidence concerning the mechanism of transposition: on integration into a new target site, transposable elements generate a repeated sequence of the target DNA in both replicative and conservative transposition. Figure 21-22 depicts the integration of IS1 into a gene. In the example shown, the integration event results in the repeti-

tion of a 9-bp target sequence. Analysis of many integration events reveals that the repeated sequence does not result from reciprocal site-specific recombination (as is the case in λ phage integration; see page 630); rather, it is generated during the process of integration itself. The number of base pairs is a characteristic of each element. In bacteria, 9-bp and 5-bp repeats are most common.

The preceding observations have been incorporated into somewhat complicated models of transposition. Most models postulate that staggered cleavages are made at the target site and at the ends of the transposable element by a transposase enzyme that is encoded by the element. One end of the transposable element is then attached by a single strand to each protruding end of the staggered cut. Subsequent steps depend on which mode of transposition occurs (replicative or conservative). Figure 21-23 depicts this model of transposition.

Message In prokaryotes, transposition occurs by at least two different pathways. Some transposable elements can replicate a copy of the element into a target site, leaving one copy behind at the original site. In other cases, transposition involves the direct excision of the element and its reinsertion into a new site.

Rearrangements Mediated by Transposable Elements

Transposable elements generate a high incidence of deletions in their vicinity. These deletions emanate from one

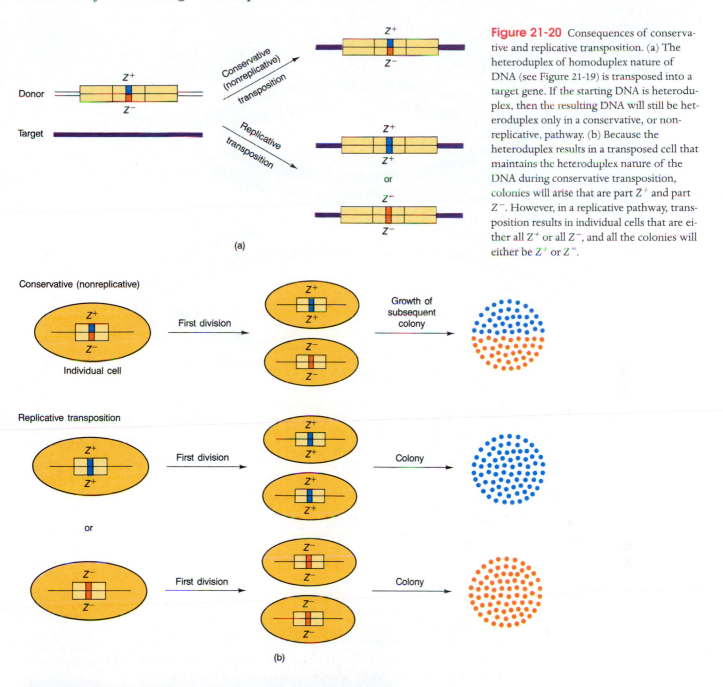

Figure 21-20 Consequences of conservative and replicative transposition. (a) The heteroduplex of homoduplex nature of DNA (see Figure 21-19) is transposed into a target gene. If the starting DNA is heteroduplex, then the resulting DNA will still be heteroduplex only in a conservative, or nonreplicative, pathway. (b) Because the heteroduplex results in a transposed cell that maintains the heteroduplex nature of the DNA during conservative transposition, colonies will arise that are part Z^+ and part Z^-. However, in a replicative pathway, transposition results in individual cells that are either all Z^+ or all Z^-, and all the colonies will either be Z^+ or Z^-.

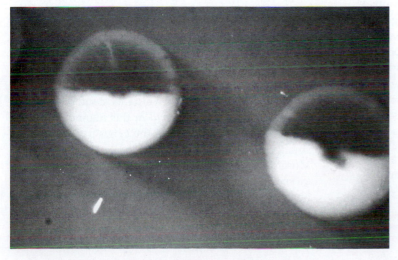

Figure 21-21 Colonies resulting from the experiment illustrated in Figures 21-19 and 21-20. A dye is used that stains Z^+ individuals blue. One half of this colony is Z^+ (dark area), and the other is Z^- (white). (Photo courtesy of N. Kleckner.)

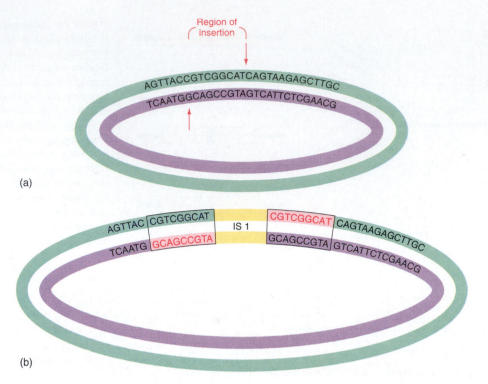

Figure 21-22 Duplication of a short sequence of nucleotides in the recipient DNA is associated with the insertion of a transposable element; the two copies bracket the inserted element. Here the duplication that attends the insertion of IS1 is illustrated in a way that indicates how the duplication may come about. IS1 insertion causes a nine-nucleotide duplication. If the two strands of the recipient DNA are cleaved (arrow) at staggered sites that are nine nucleotides apart, as shown in (a), followed by insertion of IS1 between the resulting single-stranded ends, then the subsequent filling in of single strands on each side of the newly inserted element, indicated by red letters in (b), with the right complementary nucleotides could account for the duplicated sequences (boxes). (From S. N. Cohen and J. A. Shapiro, "Transposable Genetic Elements." Copyright © 1980 by Scientific American, Inc. All rights reserved.)

end of the element into the surrounding DNA (Figure 21-24). Such events, as well as element-induced inversions, can be viewed as aberrant transposition events. Transposons also give rise to readily detectable deletions in which part of the element is deleted together with varying lengths of the surrounding DNA. This process of **imprecise excision** is now recognized as deletions or inversions emanating from the internal ends of the IR segments of the transposon. The process of **precise excision**—the loss of the transposable element and the restoration of the gene that was disrupted by the insertion—also occurs, although at very low rates compared with the frequencies of the events just described.

Message   Some DNA sequences in bacteria and phages act as mobile genetic elements. They are capable of joining different pieces of DNA and thus are capable of splicing DNA fragments into or out of the middle of a DNA molecule. Some naturally occurring mobile or transposable elements carry antibiotic-resistance genes.

Review of Transposable Elements in Prokaryotes

Let's examine what we have learned up to this point about prokaryotic transposable elements:

1. There are several different types of transposable elements, including insertion sequences (IS1, IS2, . . .), transposons (Tn1, Tn2, , . . .), and phage mu.

2. Two copies of a transposable element can act in concert to transpose the DNA segments in between them. Some of the antibiotic-resistance-conferring transposons are obviously formed in this manner, with two insertion sequences flanking the genes for antibiotic resistance.

3. Most of the transposable elements have recognizable inverted repeat (IR) structures, some of which can be observed under the electron microscope after denaturation and renaturation.

(a)

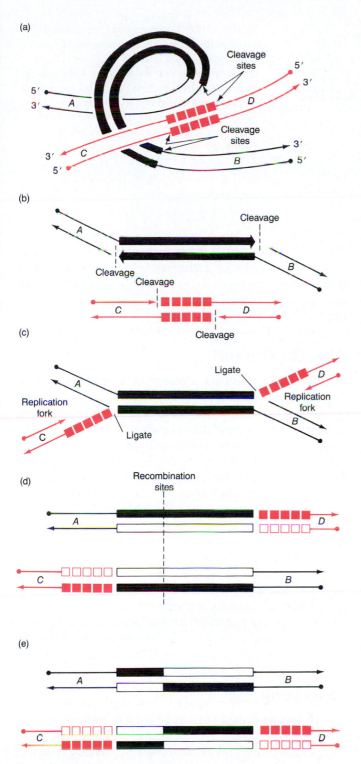

(b)

(c)

(d)

(e)

Figure 21-23 A possible molecular pathway is suggested to explain transposition and chromosomal rearrangements. The donor DNA, including the transposon (*thick bars*), is in black; the recipient DNA is in red. *A, B, C,* and *D* identify segments of the two DNA molecules. The pathway has five steps, beginning with single-stranded cleavage (a) at each end of the transposable element and at each end of the target nucleotide sequence (*red squares*) that will be duplicated. The cleavages (b) expose the chemical groups involved in the next step: the joining of DNA strands from donor and recipient molecules in such a way that the double-stranded transposable element has a DNA-replication fork at each end (c). At this point the figure shows alternative pathways, depending on whether transposition is replicative [(d) and (e)] or conservative [(f) and (g)]. In replicative transposition, DNA synthesis (d) replicates the transposon (*open bars*) and the target sequence (*open squares*), accounting for the observed duplication of the target sequence. This step forms two new complete double-stranded molecules; each copy of the transposable element unites a segment of the donor molecule with a segment of the recipient molecule. (The copies of the element thus serve as linkers for the recombination of two unrelated DNA molecules.) In the final step (e), reciprocal recombinations between the copies of the transposable element inserts a copy of the element at the new genetic site and regenerates the donor molecule. The mechanism of this recombination is not known; it does not require the proteins that are needed for homologous recombination; at least in Tn3, it is mediated by sequences within the element. Steps (f) and (g) show the alternative pathway from the intermediate step shown in (c), which leads to conservative (nonreplicative) insertions without cointegrates. This occurs if repair synthesis is initiated at the primer termini within the target DNA and if the displaced single strands that attach the transposon to the donor replicon are broken. Note that a short single-stranded segment of the transposon is lost at each end and is replaced by repair synthesis (indicated by the clear black boxes.) (Modified from S. N. Cohen and J. A. Shapiro, "Transposable Genetic Elements." Copyright © 1980 by Scientific American, Inc. All rights reserved.)

(f)

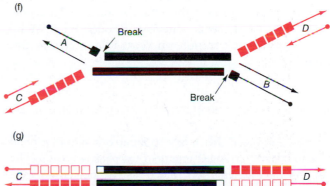

(g)

4. Transposable elements are found in bacterial chromosomes, as well as in plasmids.

5. After insertion into a new site on the DNA, transposable elements generate a short repeated sequence, commonly consisting of 9 or 5 bp.

6. The detailed mechanism of transposition is not known, but two different pathways for transposition have been identified. In some cases, transposition occurs by replicating a new copy of the element into the target site, leaving one copy behind at the original site. In other cases, transposition involves the excision of the element from the original site and its reintegration into a new site. These two modes of transposition are called *replicative* and *conservative*, respectively.

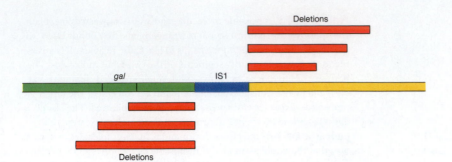

Figure 21-24 Deletion formation mediated by a transposable element. In this example, the transposable element IS1 is shown at a point in the *E. coli* chromosome near the *gal* genes. Deletions can be generated from each end of the IS1 element, extending into the neighboring DNA sequences. In cases where the deletions extend into the *gal* regions, they can be detected as a result of the Gal⁻ phenotype.

Transposable elements have been found in eukaryotes, and some have close similarities to those observed in bacteria, transposing via DNA intermediates. Interestingly, other mobile elements transpose via RNA intermediates, and in certain cases resemble mammalian retroviruses.

Ty Elements in Yeast

Figure 21-25 shows the structure of one of the **Ty elements** in yeast: the Ty1 sequence, which is present in approximately 35 copies in the yeast genome. The 330-bp-long termini (terminal sequences), called δ (*delta*) *sequences*, are present in about 100 copies of the genome. Yeast δ sequences, as well as Ty elements as a whole, show significant sequence divergence. The terminal δ sequences are present in direct-repeat orientation, in contrast to transposable elements in bacteria, which carry inverted repeat (IR) sequences. However, like prokaryotic transposons, Ty elements generate a repeated sequence of target DNA (in this case, 5 bp) during transposition. Also, Ty elements cause mutations by insertion into different genes in the yeast chromosome. It is now known that Ty elements transpose through an RNA intermediate (see the section on retroviruses later in this chapter).

Transposable Elements in *Drosophila*

It is now estimated that many spontaneous mutations and chromosomal rearrangements in *Drosophila* are caused by transposable elements. As much as 10 percent of the *Drosophila* genome may be composed of families of dispersed, repetitive DNA sequences that move as discrete elements. Three types of transposable elements have been characterized: the *copia*-like elements, the fold-back (FB) elements, and the P elements. Their structures are summarized in Figure 21-26.

Copia-like Elements

The **copia-like elements** comprise at least seven families, ranging in size from 5 to 8.5 kb. Members of each family appear at 10–100 positions in the *Drosophila* genome. Each member carries a long, direct terminal repeat and a short, imperfect inverted repeat (Figure 21-26) and is structurally similar to a yeast Ty element (Figure 21-25). Also, *copia*-like elements cause a duplication of a characteristic number of base pairs of *Drosophila* DNA on insertion. Certain classic *Drosophila* mutations result from the insertion of *copia*-like and other elements. For example, the white-apricot (w^a) mutation for eye color is caused by the insertion of an element from the *copia* family into the white locus. Some *copia*-like families have surprising properties. For instance, all the insertion mutations detected so far that result from the *gypsy* family of *copia*-like elements are suppressible by a specific allele at one particular outside locus. That is, the phenotypes resulting from the *gypsy* insertions are affected by unlinked genes. The mechanism of the effect is unknown.

FB Elements

The **FB elements** range in size from a few hundred to a few thousand base pairs. These elements have sequence homologies, but different elements have sequence differences also. Each carries long inverted repeats at its termini (see Figure 21-26). Sometimes the entire element consists of inverted repeats, but a central sequence separates the inverted repeats in other elements. In either case, FB elements can literally fold back on themselves owing to the inverted repeats (hence, their name). Several unstable mutations in *Drosophila* have been shown to be caused by the insertion of FB elements. Mutations can result either from the interruption of a gene-coding sequence by FB-element insertion or from the effects on gene expression due to FB-element in-

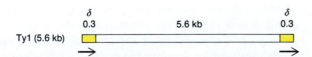

Figure 21-25 The structure of a yeast transposable element. The Ty1 sequence occurs approximately 35 times in the yeast genome. It contains two copies of delta (δ) sequence in direct orientation at each end. Delta occurs approximately 100 times in the yeast genome.

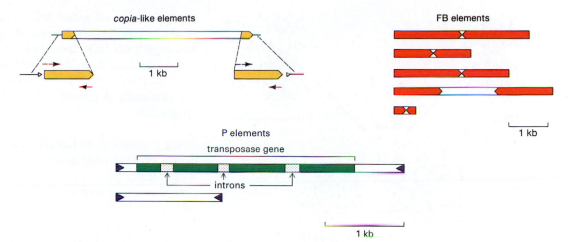

Figure 21-26 Summary of the structures of three classes of *Drosophila* transposable elements. The *copia*-like elements carry long direct terminal repeats. Each repeat makes up about 5 percent of the length of the element. These repeats are shown on an expanded scale below the element to illustrate the presence of short, imperfect inverted repeats (→) at the ends of each long direct repeat and the presence of a few base pairs of duplicate target sequence (▷) flanking the element after insertion. Such duplication of the target sequence is a virtually universal feature of transposable element insertion. The different genomic copies of the family elements are very similar in structure to one another.

The FB elements make up a family of heterogeneous but cross-homologous sequences, ranging in size from a few hundred base pairs to several kilobases. Each FB element carries long terminal inverted repeats. In some cases, the entire element consists of these inverted repeats. In other cases, a central sequence is located between the inverted repeats. The inverted repeat sequences themselves are internally repetitious, having a substructure made up primarily of 31-bp tandem repeats. The number of these 31-bp tandem repeats can differ not only between FB elements but also between the termini of a single FB element.

The P elements have a very different structure from that of the *copia*-like and FB elements. P elements carry perfect terminal inverted repeats of 31 bp. A fraction of the P elements (about one-third in the one strain examined) are very similar in sequence to one another and are 2.9 kb in length. The remainder of the P elements are more heterogeneous, but all appear to have structures that are consistent with their having been derived from the 2.9-kb element by one or more internal deletions. DNA sequence analysis of the 2.9-kb element reveals a gene coding for transposase, composed of four exons and three introns. (From G. Robin, in J. A. Shapiro, ed., *Mobile Genetic Elements*, pp. 329–361. Copyright © 1983 by Academic Press.)

sertion in or near a control region. The properties of some of the FB-insertion mutations suggest that FB elements can excise themselves from the genome and promote chromosomal rearrangements at high frequencies.

P Elements

Of all the transposable elements in *Drosophila*, the most intriguing and useful to the geneticist are the **P elements.** These elements were discovered as a result of studying **hybrid dysgenesis**—a phenomenon that occurs when females from laboratory strains of *Drosophila melanogaster* are mated to males derived from natural populations. In such crosses, the laboratory stocks are said to possess an **M cytotype** (cell type), and the natural stocks are said to possess a **P cytotype.** In a cross of M♀ × P♂, the progeny show a range of surprising phenotypes that are manifested in the germ line, including sterility, a high mutation rate, and a high frequency of chromosomal aberation and nondisjunction. These hybrid progeny are termed **dysgenic,** or biologically deficient (hence, the expression *hybrid dysgenesis*). Interestingly, the reciprocal cross P♀ × M♂ produces no dysgenic offspring. An important observation is that a large portion of the dysgenically induced mutations are unstable—that is, they revert to wild-type or to other mutant alleles at very high frequencies. This instability is generally restricted to the germ line of individuals possessing an M cytotype.

These findings led to the hypothesis that the mutations are caused by the insertion of foreign DNA into specific genes, thereby rendering them inactive. According to this hypothesis, reversion usually would result from the spontaneous excision of these inserted sequences. This hypothesis has been critically tested by isolating dysgenically derived unstable mutants at the eye-color locus *white*. A plasmid constructed to carry the white locus was used as a probe to recover the dysgenesis-mutated *white* genes. (This type of experiment is explained in Chapter 15). The majority of the mutations were found to be caused by the insertion of a genetic element into the middle of the *white*[+] gene. The element, called the *P element,* was found to be present in 30 to 50 copies per genome in P strains but to be completely absent in M strains. The P elements vary in size, ranging from 0.5 to 2.9 kb in length (this size difference reflects partially deleted elements derived from a single complete P element), but there is always a 31-bp perfect inverted repeat at their ends. The full-sized P element carries a gene containing four exons, coding for a transposase.

The current explanation of hybrid dysgenesis is based on the proposal that P elements encode both the trans-

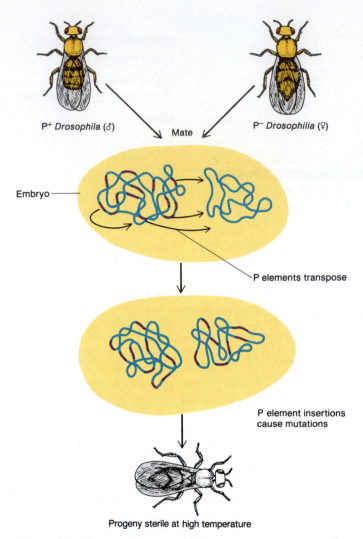

P+ *Drosophila* (♂)   Mate   P− *Drosophila* (♀)

Embryo

P elements transpose

P element insertions
cause mutations

Progeny sterile at high temperature

Figure 21-27 The phenomenon known as hybrid dysgenesis results
from the mobilization of DNA sequences called *P* elements in
Drosophila embryos. When a sperm from a P-carrying strain fertilizes
an egg from a non-P-carrying strain, the P elements transpose
throughout the genome, usually disrupting vital genes. (After J. D.
Watson, J. Tooze, and D. T. Kurtz, *Recombinant DNA: A Short Course.*
Copyright © 1983 by W. H. Freeman and Company.)

hand, as we noted earlier, P♀ × M♂ crosses do not result
in dysgenesis. This is because P cytotype females already
have high levels of P repressor.

Quite apart from their interest as a genetic phenomenon, the P elements have become major tools of the modern *Drosophila* geneticist. Two main analytical techniques
are possible.

Using P Elements to Locate Genes. P elements are used to
isolate any *Drosophila* gene of interest. First, the investigator
simply looks for mutants of that gene in progeny of dysgenic crosses. Then a vector is constructed in which a P element is inserted. This vector is used as a probe to identify
and isolate DNA segments containing P elements (see
Chapter 15 for more details of these types of experiments);
in a subset of these segments, P elements are found inserted
into the gene of interest. These genes can then be cloned
(Chapter 15) and studied.

Using P Elements to Insert Genes. The second major analytical technique stems from the discovery by Gerald Rubin
and Allan Spradling that P element DNA can be used as an
effective vehicle for transferring donor genes into the germ
line of a recipient fly. Rubin and Spradling devised the following experimental procedure (Figure 21-28). The recipient genotype is homozygous for the *rosy* (ry^-) mutation,
which confers a characteristic eye color and is of M cytotype. From this strain, embryos are collected at the completion of about nine nuclear divisions. At this stage, the embryo is one multinucleate cell, and the nuclei destined to
form the germ cells are clustered at one end. (P elements
mobilize only in germ line cells.) Two types of DNA are injected into embryos of this type. The first is a bacterial plasmid carrying a deleted P element into which the ry^+ gene
has been spliced. This deleted element is not able to transpose owing to the deletion, so a helper plasmid bearing a
complete element is also injected. Flies developing from
these embryos are phenotypically still *rosy* mutants, but
their offspring contain a large proportion of ry^+ individuals.
These ry^+ descendants show Mendelian inheritance of the
newly acquired ry^+ gene, suggesting that it is located on a
chromosome. This has been confirmed by in situ hybridization, which shows that the ry^+ gene, together with the
deleted P element, has been inserted into one of several distinct chromosome locations. None appears exactly at the
normal locus of the *rosy* gene. These new ry^+ genes are
found to be inherited in a stable fashion.

Mechanism of P Element Transposition. Figure 21-29
shows a model of P element transposition in which, following DNA synthesis, a P element transposes to a new position, leaving a double-stranded gap, which is increased by
nucleases. The gap is repaired with the aid of the sister-
chromatid P element that generates two Holliday junctions,
in a process analogous to the double-strand gap repair
model in yeast (Figure 20-26). Finally, the Holliday junctions

posase product and P-repressor products. According to this
model, which is depicted in Figure 21-27, the transposase is
responsible for mobilization of the P elements, whereas the
repressor prevents transposase production, thereby blocking
transposition of the element. (We consider repressor proteins in general in Chapter 18.) In the P cytotype, the high
copy number of P elements leads to the abundant production of repressor, so that the P elements are immobilized.
For some reason, most laboratory strains have no P elements; consequently, there is no repressor in the cytoplasm.
In hybrids from the cross M♀ × P♂, the P elements are in
a repressor-free environment and can transpose throughout
the genome, causing a variety of damage expressed as the
various manifestations of hybrid dysgenesis. On the other

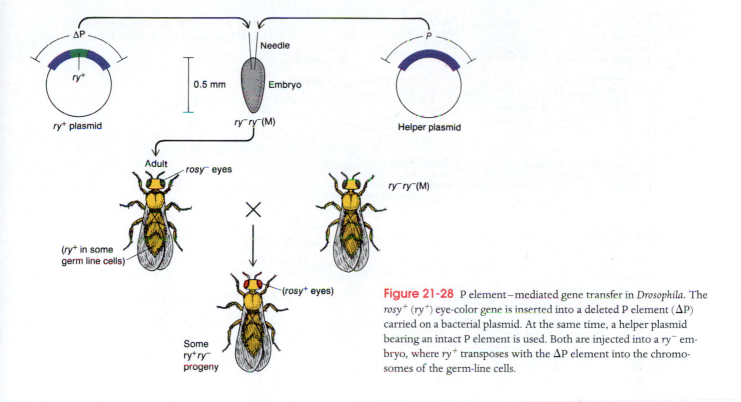

Figure 21-28 P element–mediated gene transfer in *Drosophila*. The *rosy*$^+$ (*ry*$^+$) eye-color gene is inserted into a deleted P element (ΔP) carried on a bacterial plasmid. At the same time, a helper plasmid bearing an intact P element is used. Both are injected into a *ry*$^-$ embryo, where *ry*$^+$ transposes with the ΔP element into the chromosomes of the germ-line cells.

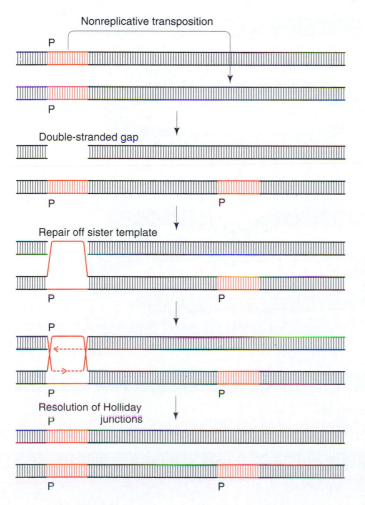

are resolved, and DNA synthesis restores a P element copy to both sister chromatids.

> **Message** P elements in *Drosophila* are a type of transposon that causes hybrid dysgenesis. They can be used through transposon mutagenesis to recover selectively any gene with a recognizable mutant phenotype. P elements in *Drosophila* can also be used as efficient vehicles for the transfer of specific genes to given recipient genotypes.

Retroviruses

Retroviruses are single-stranded RNA animal viruses that employ a double-stranded DNA intermediate for replication. The RNA is copied into DNA by the enzyme **reverse transcriptase.** The life cycle of a typical retrovirus is shown in Figure 21-30. Some retroviruses, such as mouse mammary tumor virus (MMTV) and Rous sarcoma virus (RSV), are responsible for the induction of cancerous tumors. When integrated into host chromosomes as double-

Figure 21-29 Model for P element transposition. A P element transposes, leaving a double-stranded gap that is repaired using the sister-chromatid copy and generating two Holliday junctions that are resolved into two intact molecules with a P element reinstated at the original position. (From William R. Engles, "The Origin of P Elements in *Drosophila melanogaster*," *BioEssays* 14(10), 1992, 681–686.)

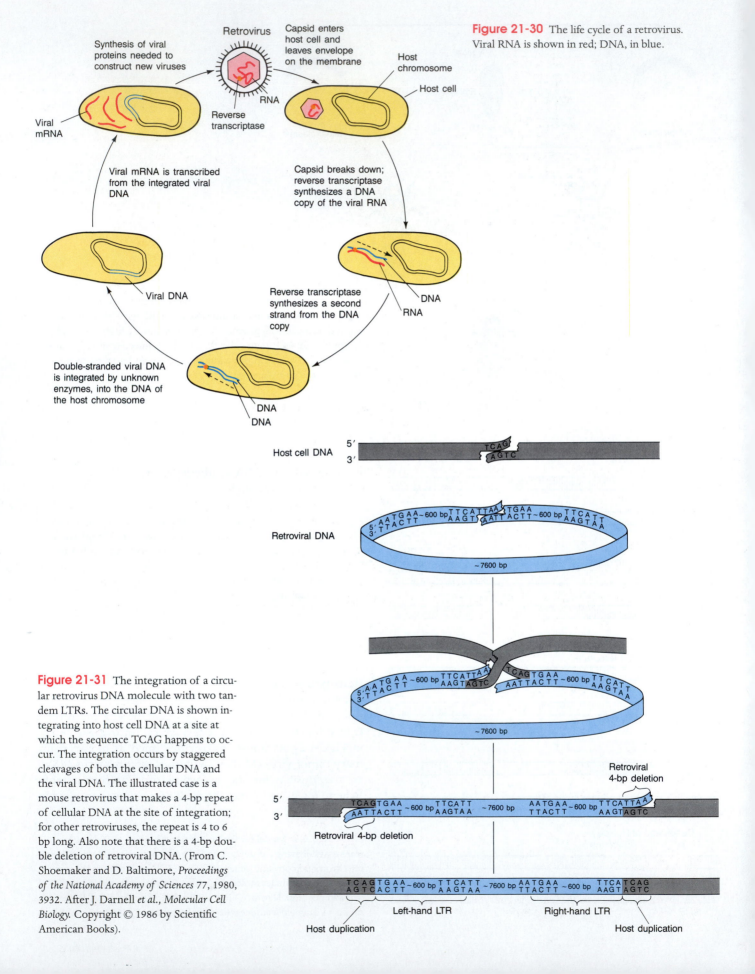

Figure 21-30 The life cycle of a retrovirus. Viral RNA is shown in red; DNA, in blue.

Figure 21-31 The integration of a circular retrovirus DNA molecule with two tandem LTRs. The circular DNA is shown integrating into host cell DNA at a site at which the sequence TCAG happens to occur. The integration occurs by staggered cleavages of both the cellular DNA and the viral DNA. The illustrated case is a mouse retrovirus that makes a 4-bp repeat of cellular DNA at the site of integration; for other retroviruses, the repeat is 4 to 6 bp long. Also note that there is a 4-bp double deletion of retroviral DNA. (From C. Shoemaker and D. Baltimore, *Proceedings of the National Academy of Sciences* 77, 1980, 3932. After J. Darnell *et al.*, *Molecular Cell Biology*. Copyright © 1986 by Scientific American Books).

stranded DNA, these retroviruses are termed **proviruses.** Proviruses, like the mu phage in bacteria, can be considered transposable elements, since they can, in effect, transpose from one location to another.

Retroviruses have structural similarities to some transposable elements from bacteria and other organisms. In particular, the ends of the proviruses have long terminal repeats (LTRs) reminiscent of the sequences of the Ty1 elements in yeast and the long terminal repeats of the *copia*-like elements in *Drosophila*. Also, integration results in the duplication of a short target sequence in the host chromosome. For example, in the case of the mouse retrovirus shown in Figure 21-31, a 4-bp sequence is duplicated on each side of the integrated provirus. (Note how the staggered cuts and the duplication of the short segment of the host DNA are similar to those shown in Figure 21-22 for the integration of bacterial transposable elements.) On the other hand, similarities between retroviruses and Ty1 elements in yeast and *copia*-like elements in *Drosophila* suggest that the Ty1 and *copia*-like elements might also be integrated forms of retrovirus-like elements (see Figure 21-32). This was nicely confirmed in 1985 by Jef Boeke and Gerald Fink and their co-workers in the experiments discussed in the next section.

Transposition via an RNA Intermediate

Figure 21-33 diagrams the experimental design used by Boeke and Fink and their colleagues to alter a yeast Ty element, cloned on a plasmid. First, a promoter was inserted near the end of the element that could be activated by the addition of galactose to the medium. The use of a galactose-sensitive promoter allows the manipulation of the expression of Ty RNA. In the presence of galactose, more transcription of Ty RNA occurs. Second, an intron (page 411) from another yeast gene was introduced into the Ty transposon coding region.

The addition of galactose greatly increases the frequency of transposition of the altered Ty element. This suggests the involvement of RNA, because galactose-stimulated transcription begins at the galactose-sensitive promoter and continues through the element (Figure 21-33). The key experimental result, however, is the fate of the transposed Ty DNA. When the researchers examined the Ty DNA resulting from transpositions, they found that the intron had been removed. Because introns are excised only during RNA processing (see Chapter 13), the transposed Ty DNA must have been copied from an RNA intermediate transcribed from the original Ty element and then

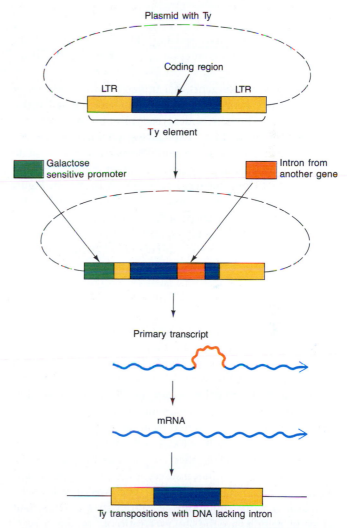

Figure 21-33 Demonstration of transposition through an RNA intermediate. A Ty element is altered by adding a promoter that can be activated by the addition of galactose. Activation of the promoter will increase transcription through the Ty element. Also, an intron from another gene is inserted into the Ty element. Because the final product of transposition contains no intron, the intron must have been spliced out from an RNA transcript (see Chapter 13). This must have occurred as shown here, where the primary transcript contains the intron but the final processed mRNA does not. This RNA is then copied by reverse transcriptase and integrated into the chromosomal DNA. (Modified from J. Darnell, H. Lodish, and D. Baltimore, *Molecular Cell Biology*, p. 450. Copyright © 1986 by Scientific American Books.)

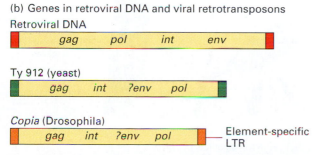

(b) Genes in retroviral DNA and viral retrotransposons

Retroviral DNA

gag   pol   int   env

Ty 912 (yeast)

gag   int   ?env   pol

Copia (Drosophila)

gag   int   ?env   pol   — Element-specific LTR

Figure 21-32 A comparison of the genes of integrated retrovirus DNA and the yeast Ty elements and *Drosophila copia* elements. The four functions encoded by the retroviral DNA have counterparts in the yeast and *Drosophila* elements shown here. The LTRs are represented by the colored ends of the elements. (From H. Lodish, D. Baltimore, A. Berk, S. L. Zipursky, P. Matsudaira, and J. Darnell, *Molecular Cell Biology*, 3d ed., p. 329. Copyright © 1995 by Scientific American Books.)

Viral retrotransposons

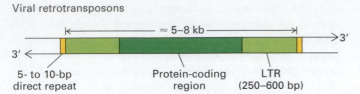

Figure 21-34 A schematic view of viral retrotransposons. These elements have two LTRs (long terminal repeats that flank a central region that encodes specific protein functions). (From H. Lodish, D. Baltimore, A. Berk, S. L. Zipursky, P. Matsudaira, and J. Darnell, *Molecular Cell Biology*, 3d ed., p. 329. Copyright © 1995 by Scientific American Books.)

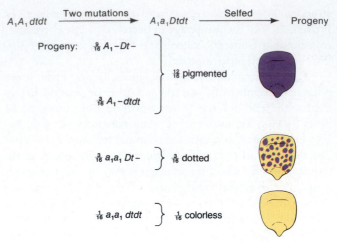

Figure 21-35 A formal genetic explanation of the appearance of the dotted phenotype in corn. The $A_1a_1\ Dtdt$ genotype is created by simultaneous mutations of $A_1 \rightarrow a_1$ and $dt \rightarrow Dt$. Upon selfing, this genotype yields the observed 12:3:1 ratio of kernel phenotypes.

processed by RNA-RNA splicing. The DNA copy of the spliced mRNA is then integrated into the yeast chromosome.

Transposable elements that utilize reverse transcriptase to transpose via an RNA intermediate are termed **retrotransposons.** They are widespread among eukaryotes and are generally divided into two classes. *Viral retrotransposons,* as exemplified by yeast Ty and *Drosophila copia* elements have properties similar to retroviruses (see Figure 21-32), including LTRs (long terminal repeats), as shown in Figure 21-34. *Nonviral retrotransposons,* which are the most frequently encountered transposable elements in mammals, include the LINES (long interspersed elements) and SINES (short interspersed elements) discussed in Chapter 16.

Controlling Elements in Maize

In 1938, Marcus Rhoades analyzed an ear of Mexican black corn. The ear came from a selfing of a pure-breeding pigmented genotype, but it showed a surprising modified Mendelian dihybrid segregation ratio of 12:3:1 among pigmented, dotted, and colorless kernels. Analysis showed that two events had occurred at unlinked loci. At one locus, a pigment gene A_1 had mutated to a_1, an allele for the colorless phenotype; at another locus, a dominant allele Dt (*Dotted*) had appeared. The effect of Dt was to produce pigmented dots in the otherwise colorless phenotype of a_1a_1 (Figure 21-35). Thus, the original line was very probably $A_1A_1\ dtdt$, and the mutations generated an $A_1a_1\ Dtdt$ plant, which on selfing gave the observed ratio of progeny.

But what was causing the dotted phenotype? A reverse mutation of $a_1 \rightarrow A_1$ in somatic cells would be an obvious possibility, but the large numbers of dots in the *Dotted* kernels would require extremely high reversion rates. Using special stocks, Rhoades was able to find anthers in the flowers of $a_1a_1\ Dt$ plants that showed patches of pigment (Figure 21-36). He reasoned that these anthers might contain pollen grains bearing the reverted pigment genotype, and he used the pollen from these anthers to fertilize a_1a_1 tested females. Sure enough, some of the progeny were com-

pletely pigmented, showing that each dot in the parental plants was in fact the phenotypic manifestation of a genetic reversion event. Thus, a_1 is one of the first known examples of an **unstable mutant allele**—an allele for which reverse mutation occurs at a very high rate. However, the allelic instability is dependent on the presence of the unlinked Dt gene. Once the reverse mutations occur, they are stable; the Dt gene can be crossed out of the line with no loss of the A_1 character. This would not be surprising if the a_1 phenotype arises from the insertion of a defective transposable element that by itself is unable to move. In the presence of a transfactor produced by the Dt locus, however, the element could move, yielding reversion to A_1. This would remain stable in the absence of Dt.

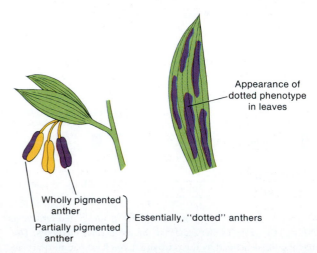

Figure 21-36 Rhoades used special stocks of $a_1a_1\ Dt$ corn plants carrying certain genes that allow pigmented sectors to be detected in tissue other than the kernels. (After M. M. Rhoades, *Genetics* 23, 1938, 382.)

McClintock's Experiments: The *Ds* Element

In the 1950s, Barbara McClintock demonstrated an analogous situation in another study of corn. She found a genetic factor *Ds* (*Dissociation*) that causes a high tendency toward chromosome breakage at the location at which it appears. These breaks can be located either cytologically (Figure 21-37a) or genetically by the uncovering of recessive genes (Figure 21-37b). As you will appreciate, this action of *Ds* is another kind of instability. Once again, this instability proves to be dependent on the presence of an unlinked gene *Ac* (*Activator*), in the same way that the instability of *a₁* is dependent on *Dt*.

McClintock found it impossible to map *Ac*. In some plants, it mapped to one position; in other plants of the same line, it mapped to different positions. As if this were not enough of a curiosity, the *Ds* locus itself (Figure 21-37) was constantly changing position on the chromosome arm, as indicated by the differing phenotypes of the variegated sections of the seeds (as different recessive gene combinations were uncovered in a system such as the one illustrated in Figure 21-37b).

The wanderings of the *Ds* element take on new meaning for us in the context of this chapter when we consider the results of the following cross:

$$\delta\ CC\ D\ Ds\ Ac\ Ac^+ \times cc\ Ds^+Ds^+\ Ac^+Ac^+\ \female$$

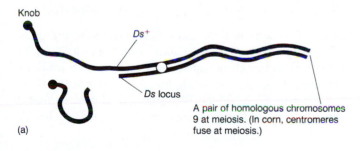

Knob

Ds⁺

Ds locus

A pair of homologous chromosomes 9 at meiosis. (In corn, centromeres fuse at meiosis.)

(a)

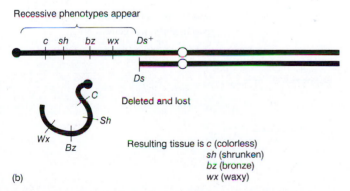

Recessive phenotypes appear

c sh bz wx *Ds⁺*

Ds

Deleted and lost

C

Sh

Wx

Bz

Resulting tissue is *c* (colorless)
sh (shrunken)
bz (bronze)
wx (waxy)

(b)

Figure 21-37 Detection of chromosomal breakage (instability) due to action of the *Ds* element in corn. (a) Cytological detection of the breakage. (*Ds⁺* indicates a lack of *Ds*.) (b) Genetic detection of the breakage.

Here *C* allows color expression and *Ds⁺* and *Ac⁺* indicate the lack of the element. Most of the kernels from this cross were of the expected types (Figure 21-38), but one exceptional kernel was very interesting. In the figure, the first seed shows the normal solid pigment pattern owing to the presence of the dominant *C* allele. The second seed shows the same basic background pigmentation, but with the expected white mottling caused by the loss of the *C* allele through chromosome breakage in some of the cell lines within the seed, with the resultant expression of the recessive *c*. Because of the clonal nature of cell growth in the seed, the size of a white patch is an indication of when in the seed's development the breakage occurred. A small white area suggests that the break came late in development, since it gave rise to only a small number of affected cells. A large patch suggests an early break, since many descendant cells are affected. The bottom seed in Figure 21-38 reflects expression of the *c* allele at the very beginning of development, since the background is white, not pigmented. However, the presence of pigmented blotches on a white background suggests a *reversible* process at work that allows the *C* allele to be reexpressed. Chromosomal breakage could not be the explanation in this case, since upon breakage the *C* alleles are left on acentric fragments that are lost during mitosis, and, therefore, the white phenotype is not reversible in such cases. The white coloration in the bottom seed appears to be the result of a second type of action in which the *Ds* gene transposes into the *C* allele, disrupting its function, but not inducing the chromosome to break. If the *Ds* allele then is excised and transposes elsewhere, the *C* allele regains its function, yielding the pigmented cells that form the patches. The *Ac* allele is still required, but in this case it has the effect of mediating a reversible instability at

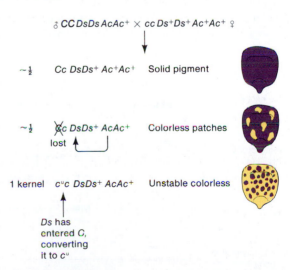

$$\delta\ CC\ DsDs\ AcAc^+ \times cc\ Ds^+Ds^+\ Ac^+Ac^+\ \female$$

~½   *Cc DsDs⁺ Ac⁺Ac⁺*   Solid pigment

~½   ~~*Cc*~~*c DsDs⁺ AcAc⁺*   Colorless patches
lost

1 kernel   *cᵘc DsDs⁺ AcAc⁺*   Unstable colorless

Ds has entered *C*, converting it to *cᵘ*

Figure 21-38 Results that indicate the transposition of *Ds* into the *C* gene in corn. (*C* allows color expression; *c* does not. *Ac* = activator.) The action of *Ds* is dependent on the presence of the unlinked gene *Ac*.

the *C* locus. If *Ac* is crossed out of the line, the *c*^{u} allele becomes a stable mutant.

The analogy of this system with the *a₁ Dt* system is obvious. Perhaps the earlier situation also is due to the insertion of a *Ds*-like element into the *A₁* gene. It is natural to ask whether *a₁* will respond to *Ac* or *c*^{u} will respond to *Dt*. The answer is no; some kind of specificity prevents this cross-activation of mutational instability.

The *wx* (*waxy*) Locus

The *Ds* element can wander not only into the middle of the *C* gene but also into other genes, rendering them unstable mutants dependent on *Ac*. One such locus, *wx* (*waxy*), has been the subject of an intense study on the effects of the *Ds* element. Oliver Nelson has paired many unstable *waxy* alleles in the absence of the *Ac* mutation. In such *wx*$^{m-1}$/*wx*$^{m-2}$ heterozygotes, he has looked for rare wild-type *Wx* recombinants by staining the pollen with KI-I₂ reagent, which stains *Wx* pollen black and *wx* pollen red. By counting the frequency of *Wx* pollen grains in each kind of heterozygote, Nelson has been able to do fine-structure recombination mapping of the *waxy* gene. He has showed that the different "mutable *waxy*" mutant alleles are in fact due to the insertion of the *Ds* element in different positions in the gene. Continuing the experiment, he allowed the *Wx*-bearing pollen to fertilize *wx* plants and produce rare *Wx* kernels, which could be detected by staining sliced-off slivers. The *Wx* kernels were then raised into plants, and the exchange of flanking markers expressed in the adults showed that the *Wx* pollen grains had arisen from chromosome exchange.

General Characteristics of Controlling Elements

Several systems like *a₁ Dt* and *Ds Ac* have now been found in corn. Each shows similar action, having a **target gene** that is inactivated, presumably by the insertion of some **receptor element** into it, and a distant **regulator gene** that maintains the mutational instability of the locus, presumably through its ability to "unhook" the receptor element from the target locus and return the locus to normal function. The receptor and the regulator are termed **controlling elements.**

In the examples discussed so far, the unstable allele is said to be **nonautonomous:** it can revert only in the presence of the regulator. Sometimes, however, a system such as the *Ac-Ds* system can produce an unstable allele that is **autonomous.** Such mutants are recognizable because they show Mendelian ratios (such as 3:1 for pigmented to dotted) that apparently are independent of any other element. In fact, such alleles appear to be caused by the insertion of *Ac* itself into the target gene. An allele of this type can subsequently be transformed into a nonautonomous allele. In such cases, the nonautonomy seems to result from the spontaneous generation of a *Ds* element from the inserted *Ac* element. In other words, *Ds* is in all likelihood an incomplete version of *Ac* itself.

Figure 21-39 summarizes the overall behavior of the controlling elements in maize as inferred from genetic data. Note that the mutation events that gave rise to Rhoades's original ratio are nicely explained by this model: a nonautonomous *Ds*-like element is generated from an *Ac*-like progenitor; transposition of the *Ds*-like element into the *A₁* gene produces an inactive but mutationally unstable allele *a₁*.

Molecular Analysis of Controlling Elements

Molecular studies in the last few years on *Ac, Ds,* and other controlling elements in maize have confirmed McClintock's genetic model in a very satisfying way. One such study focused on *Ac*- or *Ds*-containing unstable alleles at the *Waxy* locus. First, the wild-type *Wx* gene was cloned so that it could be used as a probe to retrieve the unstable alleles. To

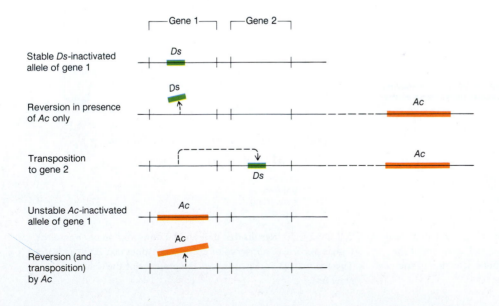

Figure 21-39 Summary of the main effects of controlling elements in corn. *Ac* and *Ds* are used as examples, acting on two hypothetical genes 1 and 2.

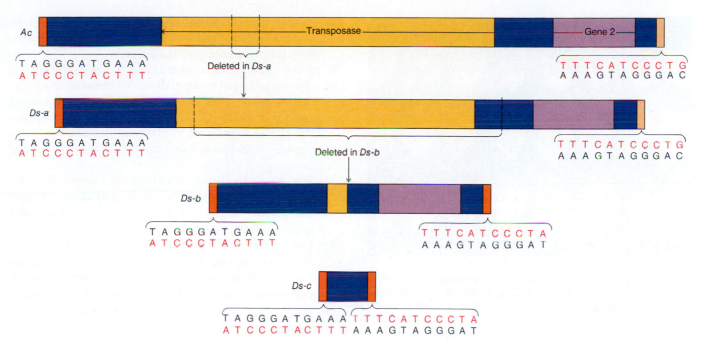

Figure 21-40 The structure of the *Ac* element of maize and several *Ds* elements. (From N. V. Federoff, "Transposable Genetic Elements in Maize." Copyright © 1984 by Scientific American, Inc. All rights reserved.)

do this, *Waxy* mRNA was isolated; it was identified as the endosperm mRNA which, on translation, produces a protein which is almost certainly the Waxy structural protein (as judged from the absence of this protein in *wx* mutants). A cDNA clone was made from this mRNA and used to fish out, from total genomic DNA, a *Wx* gene, a *wx* gene containing either *Ds* or *Ac* inserted, and *Wx* revertant alleles derived from unstable alleles. In all cases, the DNA was sequenced; some of the results are summarized in Figure 21-40. An *Ac* element is about 4500 bp long and has 11-bp imperfect inverted terminal repeats. It has two open reading frames, one of which seems like a good candidate for a

transposase. *Ds* elements prove to be deleted *Ac* elements, and the deletion can be small within the transposase, as in *Ds-a*, or large, as in *Ds-b* and *Ds-c* (Figure 21-40).

Flanking the inserted transposons are found 8-bp-long duplications of the *Wx* gene DNA. Commonly, the duplication is imperfect. In the revertants, the 8-bp duplication remains, although often with a slight modification that would seem to be necessary to retain the proper reading frame of the gene. These duplications are shown in Figure 21-41. *Ac* transposes via a DNA transposition mechanism, rather than via an RNA intermediate, and is similar in this respect and others to the *Drosophila* P elements.

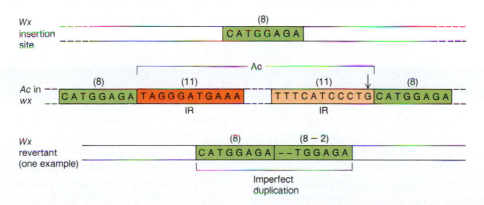

Figure 21-41 Target-insertion-site duplication produced by the transposon *Ac* at the *Wx* gene in maize.

Figure 21-42 Mosaicism through transposon mutagenesis in corn. These seeds represent genotypes in which a transposon has inserted into a gene that produces anthocyanin. Therefore, the cells of these seeds are predominantly lacking anthocyanin and are yellow; let us call the genotype $A^T A^T$. However, during development, the transposon can occasionally exit from the gene, forming a revertant cell of genotype $A A^T$. Cell division will result in a clone of revertant cells and hence a patch of pigmented cells. The three different rows represent corn lines in which the transposon exits early (large spots), late (small spots), or in between (intermediate spots). (A.J.F. Griffiths.)

Message Controlling elements in maize can inactivate a gene in which they reside, cause chromosome breaks, and transpose to new locations within the genome. Complete elements can perform these functions unaided; other forms with partial deletions can transpose only with the help of a complete element elsewhere in the genome. Figures 21-42 and 21-43 show examples of the effects of transposons in maize and similar effects in the snapdragon.

Review of Transposable Elements in Eukaryotes

Let's examine some of the essential points about eukaryotic transposable elements:

1. Transposable elements exist in all cells. Elements in yeast, *Drosophila,* and maize have been well studied, as have retroviruses in mammalian cells.

2. Some transposable elements can be used as tools for cloning and gene manipulation. For instance, the P elements of *Drosophila* can be employed to transfer genes into the germ line of a recipient fly. Another example—the T-DNA segment of Ti plasmids (described in Chapter 15)—can be used to introduce cloned genes into certain plants.

3. One similarity between eukaryotic transposable elements and their counterparts in prokaryotes is that transposition into a new site generates a short repeated sequence at the target site.

4. One difference between certain eukaryotic and prokaryotic transposable elements lies in the mechanism of transposition. Some eukaryotic transposable elements transpose via an RNA intermediate; prokaryotic elements do not use an RNA intermediate.

Figure 21-43 Mosaicism through transposon mutagenesis in snapdragon (*Antirrhinum*). As in the corn example (Figure 21-42), the plant has both copies of an anthocyanin gene inactivated by a transposon, and albino tissue results. Revertant sectors (red spots or stripes) occur when the transposon spontaneously exits from that gene. (Photo from Rosemary Carpenter and Enrico Coen.)

SUMMARY

Nature has devised many different ways of changing the genetic architecture of organisms. We are now beginning to understand the molecular processes behind some of these phenomena. Gene mutation, recombination between chromosomes (see Chapters 19 and 20), and transposition can all be reasonably explained at the DNA level. Far from merely producing genetic waste, these processes undoubtedly all have important roles in evolution. This idea is strengthened through the knowledge that the processes themselves are to a large extent under genetic control: there are genes that affect the efficiency of mutation, recombination, and transposition.

Although different mechanisms of transposition are sometimes used, the analogies between the transposable elements of phages, bacteria, and eukaryotes are striking. At present, it is not known if transposons are elements that normally play a role in the day-to-day transactions of the genome, as originally proposed by Barbara McClintock in the 1950s, or if they are pieces of "selfish DNA" that exist

for no purpose other than their own survival. Whatever the truth of this matter is, transposons certainly represent a completely unexpected element of chaos in the genome, which geneticists have already harnessed into their team of analytical procedures. At the evolutionary level, transposons may be important in the sudden leaps that characterize the fossil record.

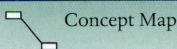

Concept Map

Draw a concept map interrelating as many of the following terms as possible. Note that the terms are listed in no particular order.

IS element / plasmids / transposons / inverted repeats / rearrangements / phenotype / tagging / unstable mutations / sectors

CHAPTER INTEGRATION PROBLEM

In Chapter 18 we studied the operon model. Note that for the *gal* operon the order of transcription of the genes in the operon is *E-T-K*. Suppose we have five different mutations in *galT*: *gal-1, gal-2, gal-3, gal-4,* and *gal-5*. The following table shows the expression of *galE* and *galK* in mutants carrying each of these mutations:

galT mutation	Expression of *galE*	Expression of *galK*
gal-1	+	−
gal-2	+	−
gal-3	+	−
gal-4	+	+
gal-5	+	+

In addition, the reversion patterns of these mutations with several mutagens we studied in Chapter 19 are shown in the following table. Here, a "+" indicates a high rate of reversion in the presence of each mutagen, a "−" depicts no reversion, and a "low" indicates a low rate of reversion.

	Reversion				
Mutation	Spon-taneous	2-Amino purine	ICR191	UV	EMS
gal-1	−	−	+	+	−
gal-2	−	−	+	−	−
gal-3	Low	Low	Low	Low	Low
gal-4	−	−	−	−	−
gal-5	Low	+	Low	+	+

Which mutation is most likely to result from the insertion of a transposable element such as IS1 and why? Can you assign the other mutations to other categories?

Solution

Transposable elements will cause polarity, preventing expression of downstream genes from the point of insertion, but not of upstream genes. Therefore, we would expect the insertion mutation to prevent expression of the *galK* gene. Three mutations are in this category, *gal-1, gal-2,* and *gal-3*. These could be frameshifts, nonsense mutations, or insertions, since each of these mutations can lead to polarity. If we examine the reversion data, however, we can distinguish among these possibilities. Transposable elements revert at low rates spontaneously, and this rate is not stimulated by base analogs, frameshift mutagens, alkylating agents, or UV. On the basis of these criteria, the *gal-3* mutation is most likely to result from an insertion, since it reverts at a low rate that is not stimulated by any of the mutagens. *gal-1* might be a frameshift, since it does not revert with 2-AP and EMS, but does with ICR191, a frameshift mutagen, and UV. (Refer to Chapter 19 for details of each mutagen.) Likewise, *gal-2* is probably a frameshift, since it reverts only with ICR191. The *gal-4* mutation is probably a deletion, since it is not stimulated to revert at all. The *gal-5* mutation appears to be a base substitution, since it reverts with 2-AP, but not above the spontaneous background rate with ICR191.

SOLVED PROBLEM

1. Transposable elements have been referred to as "jumping genes," since they appear to jump from one position to another, leaving the old locus and appearing at a new locus. In light of what we now know concerning the mechanism of transposition, how appropriate is the term "jumping genes" for bacterial transposable elements?

Solution

In bacteria, transposition occurs by two different modes. The conservative mode results in true jumping genes, since in this case the transposable element excises from its original position and inserts at a new position. A second mode is termed the *replicative mode*. In this pathway, transposable elements move to a new location by replicating into the target DNA, leaving behind a copy of the transposable element at the original site. When operating by the replicative mode, transposable elements are not really jumping genes, since a copy does remain at the original site.

PROBLEMS

1. Suppose that you want to determine whether a new mutation in the *gal* region of *E. coli* is the result of an insertion of DNA. Describe two physical experiments that would allow you to demonstrate the presence of an insertion.

2. What is a polar mutation?

3. Explain the difference between the replicative and conservative modes of transposition. Briefly describe an experiment demonstrating each of these modes in prokaryotes.

4. Describe the generation of multiple drug-resistance plasmids.

5. Briefly describe the experiment that demonstrates that the transposition of the Ty1 element in yeast occurs via an RNA intermediate.

6. Explain how the properties of P elements in *Drosophila* make possible gene-transfer experiments in this organism.

7. When Rhoades took pollen from wholly pigmented anthers on plants of genotype $a_1 a_1 Dt Dt$ and used this pollen to pollinate $a_1 a_1 dt dt$ tester females, he found wholly pigmented kernels and, in addition, some dotted kernels. Explain the origin of *both* phenotypes.

8. In yeast, the *his4* region has three cistrons—A, B, and C, in that order—each of which mediates an enzymatic step of histidine synthesis. A certain spontaneous mutation is mapped in the cistron A, but these mutants are defective for all three functions (A, B, and C). The mutation is not suppressible by nonsense or frameshift suppressors. Spontaneous reversion occurs quite frequently, but this rate is not enhanced by any mutagen. Discuss the possible nature of the mutation.

9. In *Drosophila*, M. Green found a *singed* allele (*sn*) with some unusual characteristics. Females homozygous for this X-linked allele have singed bristles, but they have numerous patches of sn^+ (wild-type) bristles on their heads, thoraxes, and abdomens. When these flies are bred to *sn* males, some females give only singed progeny, but others give both singed and wild-type progeny in variable proportions. Explain these results.

10. Crown gall tumors are found in many dicotyledonous plants infected by the bacterium *Agrobacterium tumefaciens*. The tumors are caused by the insertion of DNA from a large plasmid carried by the bacterium into the plant DNA. Suppose that a tobacco plant of type A (there are many types of tobacco plants) is infected, and it produces tumors. You remove tumor tissue and grow it on a synthetic medium. Some of these tumor cultures produce aerial shoots. You graft one of these shoots onto a normal tobacco plant of type B, and the graft grows to an apparently normal A-type shoot and flowers.

a. You remove cells from the graft and place them in synthetic medium, where they grow like tumor cells. Explain why the graft appears to be normal.

b. When seeds are produced by the graft, the resulting progeny are normal A-type plants. No trace of the inserted plasmid DNA remains. Propose a possible explanation for this "reversal."

11. Consider two maize plants:

a. Genotype $C c^m Ac Ac^+$, where c^m is an unstable allele caused by *Ds* insertion

b. Genotype $C c^m$, where c^m is an unstable allele caused by *Ac* insertion

What phenotypes would be produced and in what proportions when (1) each plant is crossed to a base-pair substitution mutant *cc* and when (2) plant (a) is crossed with plant (b)? Assume that *Ac* and *c* are unlinked, that the chromosome breakage frequency is negligible, and that mutant *cc* is Ac^+.

22

The Extranuclear Genome

► Chloroplasts and mitochondria each contain multiple copies of their own unique "chromosome" of genes.

► Generally, organelle DNA—and any variant phenotype coded therein—is inherited through the maternal parent in a cross.

► In mixtures of two genetically different mitochondrial DNAs or chloroplast DNAs, it is commonly observed that a sorting-out process results in descendant cells of one type or the other.

► In "dihybrid" organelle mixtures, recombination can be detected.

► Organelle genes code for organelle translation or transcription components, and components of energy-producing systems.

► Most organelle-encoded polypeptides unite with nucleus-encoded polypeptides to produce active proteins, and these function in the organelle.

A variegated mosaic of *Euonymus fortunei* caused by segregation during cell division of a mixture of two chloroplast DNA types, one normal and one that leads to albino tissue. The segregation results in variegated leaves or less commonly in whole branches that are green or albino. (Anthony Griffiths)

he notion that genes of eukaryotic organisms reside in the nucleus has become a firm part of the
general dogma of genetics. However, like most
fundamental truths, this one does have exceptions. The special inheritance patterns of some eukaryotic
genes reveal that these genes must be located *outside* the nucleus. Such **extranuclear genes** (*extra-* means "outside")
have now been found in most eukaryotes. The inheritance
patterns shown by extranuclear genes are less commonly

encountered than those of nuclear genes, and indeed there
are fewer extranuclear genes than nuclear genes. However,
it is now clear that extranuclear genes play normal, highly
specialized roles in the control of eukaryote phenotypes.

Extranuclear genes are found in mitochondria and
chloroplasts, membranous organelles found in the cytoplasm of eukaryotic cells (Figure 22-1). These organelles
serve the major functions of ATP synthesis and photosynthesis, respectively. The mitochondria and chloroplasts each

Figure 22-1 The heterogeneity of the cytoplasm.
The area called the *cytoplasm* (everything between
the nuclear membrane and
the cell membrane) em-
braces the cytosol (liquid
phase) plus a rich assortment of organelles, mem-
branes, and other struc-
tures. Extranuclear genes
are found in the mitochondria and in the chloroplasts
of green plants. (a) An animal cell. (b) A plant cell.
(Part a from S. Singer and
H. R. Hilgard, *The Biology
of People.* Copyright ©
1978 by W. H. Freeman
and Company; part b from
J. Janick et al., *Plant Science.*
Copyright © 1974 by W. H.
Freeman and Company.)

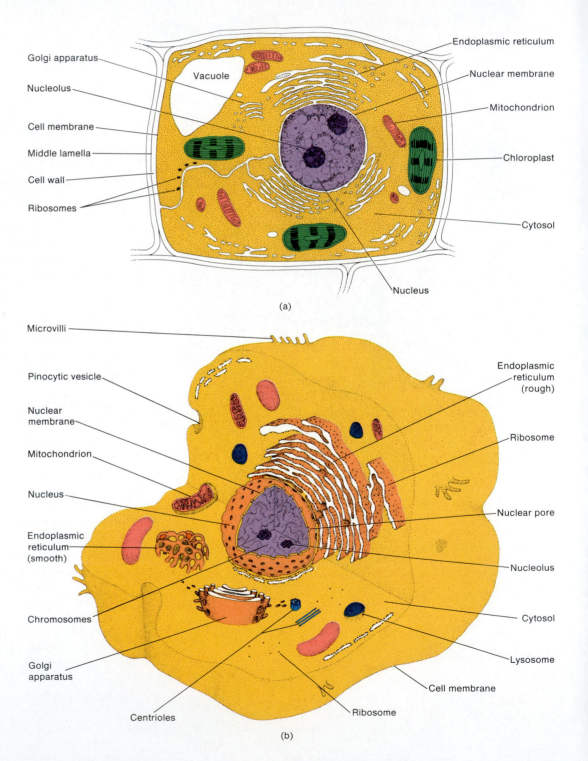

contain their own set of unique and autonomous genes, linked together in an organellar chromosome. Extranuclear genes often are called *extrachromosomal*—a term that is potentially confusing because the genes found in the mitochondria and chloroplasts in fact do comprise extranuclear chromosomes. Nevertheless, it must be emphasized that the organellar chromosomes show organizational and inheritance patterns that are quite different from those of nuclear chromosomes. Extranuclear genes have also been called *cytoplasmic genes,* based on their location in the cytoplasm.

A great deal is now known about the genetics and biochemistry of organellar genes. This chapter outlines some of the major steps in the evolution of our present understanding.

Message   The genes that have been called *cytoplasmic, extrachromosomal,* or *extranuclear* are in fact located on a unique kind of chromosome inside cytoplasmic organelles.

Extranuclear Inheritance in Higher Plants

In 1909, Carl Correns reported some surprising results from his studies on four-o'clock plants (*Mirabilis jalapa*). He noted that the blotchy leaves of these variegated plants show patches of green and white tissue, but some branches carry only green leaves and others carry only white leaves (Figure 22-2).

Flowers appeared on all types of branches. Correns intercrossed a variety of different combinations by transferring pollen from one flower to another. Table 22-1 shows the results of such crosses. Two features of these results are surprising. First there is a difference between reciprocal crosses: for example, white ♀ × green ♂ gives a different result from green ♀ × white ♂. Recall that Mendel never observed differences between reciprocal crosses. In fact, in conventional genetics, differences between reciprocal crosses are normally encountered only for sex-linked genes. However, the results of the four-o'clock crosses cannot be explained by sex linkage.

The second surprising feature is that the phenotype of the maternal parent is solely responsible for determining the phenotype of all progeny. The phenotype of the male parent appears to be irrelevant, and its contribution to the progeny appears to be zero. This pattern is known as **maternal inheritance.** Although the white progeny plants do not live long because they lack chlorophyll, the other progeny types do survive and can be used in further generations of crosses. In these subsequent generations, maternal inheritance always appears in the same patterns as the ones observed in the original crosses.

How can such curious results be explained? Whether a leaf is green or colorless depends on whether it has green or colorless chloroplasts. The inheritance patterns might be explained if these cytoplasmic organelles are somehow genetically autonomous and furthermore are never transmitted via the pollen parent. For an organelle to be genetically autonomous, it must have its own genetic determinants that are responsible for its phenotype. In this case, the chloroplasts would carry their own genetic determinants responsible for chloroplast color, which suggests that this organelle has its own genome. Its failure to be transmitted via the pollen parent is reasonable because the bulk of the cyto-

Table 22-1   Results of Crossing Flowers on Variegated Four-o'Clock Plants

Phenotype of branch bearing egg parent (♀)	Phenotype of branch bearing pollen parent (♂)	Phenotype of progeny
White	White	White
White	Green	White
White	Variegated	White
Green	White	Green
Green	Green	Green
Green	Variegated	Green
Variegated	Green	Variegated, green, or white
Variegated	Green	Variegated, green, or white
Variegated	Variegated	Variegated, green, or white

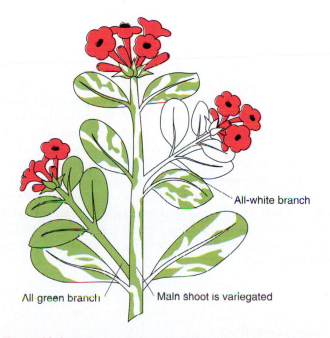

Figure 22-2 Leaf variegation in *Mirabilis jalapa,* the four-o'clock plant. Flowers may form on any branch (variegated, green, or white), and these flowers may be used in crosses.

All-white branch

All-green branch   Main shoot is variegated

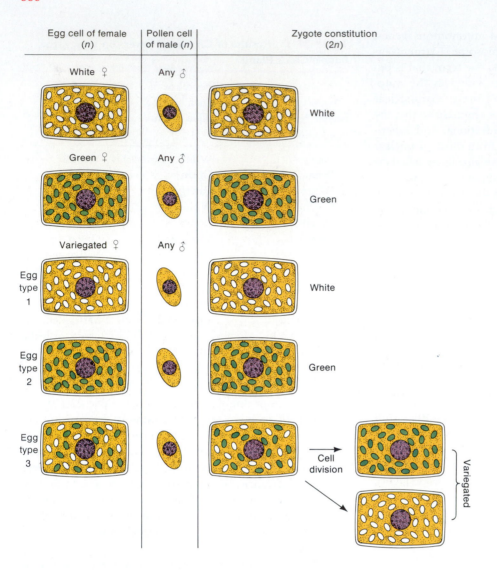

Egg cell of female (n)	Pollen cell of male (n)	Zygote constitution (2n)

Figure 22-3 A model explaining the results of the *Mirabilis jalapa* crosses in terms of autonomous chloroplast inheritance. The large, dark spheres are nuclei. The smaller bodies are chloroplasts, either green or white. Each egg cell is assumed to contain many chloroplasts, and each pollen cell is assumed to contain no chloroplasts. The first two crosses exhibit strict maternal inheritance. If the maternal branch is variegated, three types of zygotes can result, depending on whether the egg cell contains only white, only green, or both green and white chloroplasts. In the last case, the resulting zygote can produce both green and white tissue, so a variegated plant results.

plasm of the zygote is known to come from the maternal parent via the cytoplasm of the egg. Figure 22-3 diagrams a model that formally accounts for all the inheritance patterns in Table 22-1.

Variegated branches apparently produce three kinds of eggs: some contain only white chloroplasts, some contain only green chloroplasts, and some contain both kinds. The egg type containing both green and white chloroplasts produces a zygote that also contains both kinds of chloroplasts. In subsequent mitotic divisions, some cellular process causes like chloroplasts to segregate together in some cell lines, thus producing the variegated phenotype in that progeny individual.

This process of sorting might be described as "mitotic segregation." However, this is an extranuclear phenomenon, not to be confused with the mitotic segregation of nuclear genes, so a new term is needed. The term we shall use is **cytoplasmic segregation.** Throughout this chapter, cytoplasmic segregation crops up often; it appears to be a common behavior of extranuclear genomes. It must be stressed that the processes causing cytoplasmic segregation are tak-

ing place in the cytoplasm more or less simultaneously with the mitotic division of the nucleus. The parallel systems are shown schematically in Figure 22-4.

What causes cytoplasmic segregation? We might wonder if random chloroplast assortment can explain the green-white segregation in variegated plants. In a cell with, let's say, 40 chloroplasts, consisting of 20 green and 20 colorless chloroplasts, the production by random assortment of one daughter cell containing only green and one daughter cell containing only colorless chloroplasts would seem be a rare event at best. However, it has been calculated that the ratio of green to white in a cell lineage can drift randomly to 1 : 0 or 0 : 1 over several cell generations—so random assortment can explain this segregation pattern.

Extranuclear Inheritance in Fungi

Poky *Neurospora*

In 1952, Mary Mitchell isolated a mutant strain of *Neurospora* that she called *poky*. This mutant differs from the

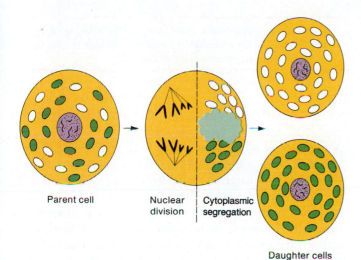

Figure 22-4 Two processes may be distinguished at cell division. Nuclear division (mitosis) is an observed physical process that parcels the genes on the nuclear chromosomes. In parallel to mitosis, a heterogeneous mixture of organelle genomes sometimes shows cytoplasmic segregation, resulting in cytoplasmically homogeneous daughter cells. The diagrams show cytoplasmic segregation generating two different organelle sets from a mixed parental cell. However, the cellular mechanisms of cytoplasmic segregation are not well understood.

But exactly where in the cytoplasm could the poky mutation be? Several aspects of the mutant phenotype seem to involve mitochondria. For instance, the cells' slow growth suggests a lack of energy, suggesting a deficiency of ATP which is normally produced by mitochondria. Also, mutants have abnormal amounts of cytochromes, which are also located in the mitochondria. These indications led geneticists to investigate mitochondrial autonomy.

David Luck labeled mitochondria with radioactive choline, a membrane component, and then followed their division autoradiographically in an unlabeled medium. He found that even after several doublings of mass, the radiation was distributed evenly among the mitochondria. Luck concluded that mitochondria propagate by division of previously existing mitochondria. If mitochondria were synthesized de novo (anew), some of the resulting population of mitochondria would be expected to be unlabeled and the original mitochondria should remain heavily labeled.

In 1965, a research group led by Edward Tatum extracted purified mitochondria from a mutant called *abnormal,* which is similar to *poky.* With an ultrafine needle and

wild-type fungus in a number of ways: it is slow-growing, it shows maternal inheritance of slow growth, and it has abnormal amounts of cytochromes. Cytochromes are mitochondrial electron-transport proteins necessary for the proper oxidation of foodstuffs to generate energy in the form of ATP. Like most organisms, wild-type *Neurospora* has three main types of cytochrome: *a*, *b*, and *c*. In poky, however, there is no cytochrome *a* or *b*, and there is an excess of cytochrome *c*.

How can maternal inheritance be demonstrated in a haploid organism? It is possible to cross some fungi in such a way that one parent contributes the bulk of the cytoplasm to the progeny; this cytoplasm-contributing parent is called the *maternal parent,* even though no true sex is involved. Mitchell demonstrated maternal inheritance for the poky phenotype in the following crosses:

poky ♀ × wild-type ♂ → all poky progeny

wild-type ♀ × poky ♂ → all wild-type progeny

In such crosses, any nuclear genes that differed between the parental strains were observed to segregate in the normal Mendelian manner and to produce 1 : 1 ratios in the progeny (Figure 22-5). All poky progeny behaved like the original poky strain, transmitting the poky phenotype down through many generations when crossed as females. Because poky does not behave like a phenotype arising from a nuclear mutation, the mutation responsible for poky was proposed to be extranuclear (cytoplasmic) in its location.

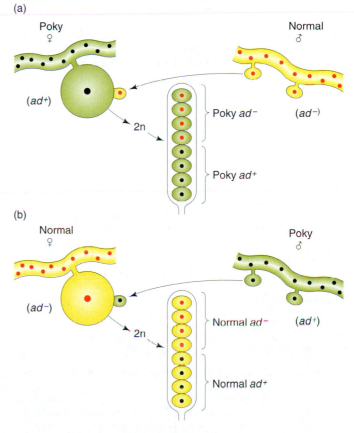

Figure 22-5 Explanation of the different results from reciprocal crosses of poky and normal *Neurospora.* The parent contributing most of the cytoplasm to the progeny cells is called female. Green shading represents cytoplasm with the poky determinants. The nuclear locus with the alleles *ad*⁺ and *ad*⁻ is used to illustrate the segregation of the nuclear genes in the expected 1 : 1 Mendelian ratio.

syringe, the experimenters injected the mitochondria into wild-type recipient cells, using appropriate controls. After several successive subcultures, the abnormal phenotype appeared! In transferring the mitochondria, the experimenters had transferred the hereditary determinants of the abnormal phenotype. Presumably, then, the extranuclear genes involved in this phenotype are located in the mitochondria. These inferences gained credibility with the discovery of DNA in the mitochondria of *Neurospora* and other species (but more of that story later).

The Heterokaryon Test

Maternal inheritance is one criterion for recognizing organelle-based inheritance. Another diagnostic test was originally used in filamentous fungi such as *Neurospora* and *Aspergillus,* but in principle, this test could be applied to any systems involving heterokaryons. A heterokaryon is made between the putative extranuclear mutant (say, a slow-growth mutant) and a strain carrying a known nuclear mutation. If the phenotype of the nuclear mutation can be recovered from the heterokaryon in combination with the phenotype of the slow-growth mutant being tested, it is likely that the growth mutant arose from an organelle-based mutation. This is because generally no diploidy occurs in a heterokaryon, so that there is no genetic exchange between nuclei. The slow-growth phenotype must have been transferred solely by cytoplasmic contact. The segregation of the pure extranuclear type from the mixed cytoplasm of the heterokaryon presumably involves the cytoplasmic segregation process (Figure 22-6).

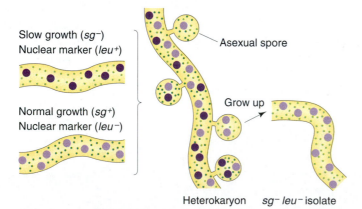

Figure 22-6 The heterokaryon test is used to detect extranuclear inheritance in filamentous (threadlike) fungi. A strain with a possible extranuclear mutation (here, *sg⁻*, causing slow growth) is combined with a strain having a nuclear mutation (*leu⁻*) to form a heterokaryon. Cultures having the phenotype caused by *leu⁻* can be derived from the heterokaryon. If some of these cultures also are *sg⁻* in phenotype, then *sg* is very likely to be an extranuclear gene, borne in an organelle. Since no nuclear recombination normally occurs in a heterokaryon, the *sg⁻* phenotype must have been acquired by cytoplasmic contact.

In figure: Slow growth (*sg⁻*) / Nuclear marker (*leu⁺*); Normal growth (*sg⁺*) / Nuclear marker (*leu⁻*); Asexual spore; Grow up; Heterokaryon; *sg⁻ leu⁻* isolate

Shell Coiling in Snails: A Red Herring

Does maternal inheritance always indicate extranuclear inheritance? Usually, but not always. It is possible for a maternal-inheritance pattern of reciprocal crosses to be generated by nuclear genes. In 1923, Alfred Sturtevant (whom you will remember from his studies on crossing-over in *Drosophila*) found a good example in the water snail *Limnaea*. Sturtevant analyzed the results of crosses between snails that differed in the direction of their shell coiling. On looking into the opening of the shells, an observer can see that some snails coil to the right (dextral coiling) and that others coil to the left (sinistral coiling). All the F₁ progeny of the cross dextral ♀ × sinistral ♂ were dextral, but all the F₁ progeny of the cross sinistral ♀ × dextral ♂ were sinistral. Thus far, the situation seems similar to the one observed with *poky* or with chloroplast inheritance. However, the F₂ progenies in both pedigrees were all dextral!

The F₃ generation in this study revealed that the inheritance of coiling direction involves nuclear genes rather than extranuclear genes. (The F₃ is produced by individually selfing the F₂ snails; this is possible because snails are hermaphroditic.) Sturtevant found that three-fourths of the F₃ snails were dextral and one-fourth were sinistral (Figure 22-7). This ratio reveals a Mendelian segregation in the F₂ generation. Apparently, dextral (s^+) is dominant to sinistral (s), but Sturtevant concluded that, strangely enough, the shell-coiling phenotype of any individual animal is determined by the genotype (not the phenotype) of its mother. We now know that this happens because the genotype of the mother's body determines the initial cleavage pattern of the developing embryo. Note that the segregation ratios shown in Figure 22-8 would never appear in the phenotypes of true organelle genes. This example shows that one generation of crosses is not enough to provide conclusive evidence that maternal inheritance is due to organelle-based inheritance. The term **maternal effect** can be used to describe the results in cases like the shell-coiling example to distinguish them from organelle-based inheritance.

Extranuclear Genes in *Chlamydomonas*

If an investigator could somehow design an ideal experimental organism with a simple life cycle, the result would be something like the unicellular freshwater alga *Chlamydomonas* (Figure 22-8). Like fungi, algae rarely have different sexes, but they do have mating types. In many algal and fungal species, two mating types are determined by alleles at one locus. A cross can occur only if the parents are of differ-

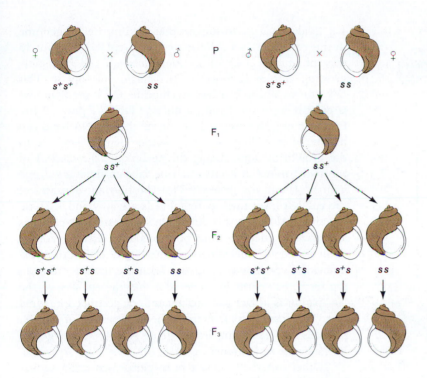

Figure 22-7 The inheritance of dextral (s^+) and sinistral (s) alleles of the nuclear gene for shell coiling in a species of water snail. The direction of coiling is determined by the nuclear genotype of the mother rather than the genotype of the individual itself. This explanation accounts for the initial difference between reciprocal crosses as well as the phenotypes of later generations. No organelle inheritance is involved (such a hypothesis would not explain the phenotypes of later generations).

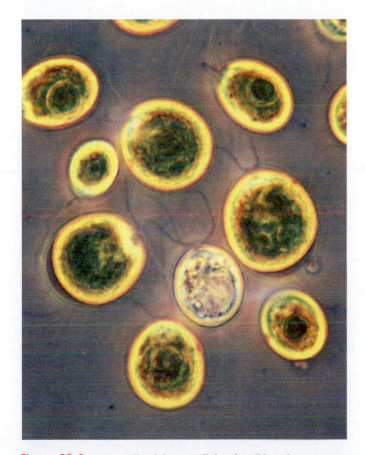

Figure 22-8 Living cells of the unicellular alga *Chlamydomonas reinhardtii*. Note the pair of flagella and the large single chloroplast. (M. I. Walker/Science Source/Photo Researchers.)

ent mating types. The mating types are physically identical but physiologically different. Such species are called *heterothallic* (literally, "different bodied"). In *Chlamydomonas*, the mating-type alleles are called mt^+ and mt^- (in *Neurospora* they are *A* and *a*; in yeast, *a* and *α*). Figure 22-9 diagrams the *Chlamydomonas* life cycle. This organism has played a central role in research on organelle genetics. As you might guess, tetrad analysis (see Chapter 6) is possible and in fact is routinely performed.

Uniparental Inheritance Demonstrated in *Chlamydomonas*

In 1954, Ruth Sager isolated a streptomycin-sensitive mutant of *Chlamydomonas* with a peculiar inheritance pattern. In the following crosses, *sm-r* and *sm-s* indicate streptomycin resistance and sensitivity, respectively, and *mt* is the mating type gene discussed earlier:

$$sm\text{-}r\ mt^+ \times sm\text{-}s\ mt^- \longrightarrow \text{all } sm\text{-}r \text{ progeny}$$

$$sm\text{-}s\ mt^+ \times sm\text{-}r\ mt^- \longrightarrow \text{all } sm\text{-}s \text{ progeny}$$

Here we see a difference in reciprocal crosses; all progeny cells show the streptomycin phenotype of the mt^+ parent. Like the maternal-inheritance phenomenon, this is a case of uniparental inheritance. In fact, Sager referred to the mt^+ mating type as the female, even though there is no observable physical distinction between the mating types, nor is there a difference in the contribution of cytoplasm as seen

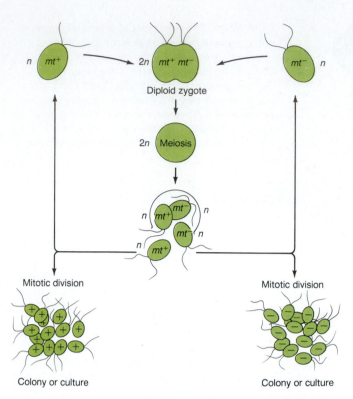

Figure 22-9 The life cycle of *Chlamydomonas*, a unicellular green alga. All diploid zygotes are heterozygous for the mating-type alleles *mt⁺* and *mt⁻* because only algae differing in these alleles can mate.

in *Neurospora*. In these crosses, the conventional nuclear marker genes (such as *mt* itself) all behave in a Mendelian manner and give 1:1 progeny ratios. For example, half the progeny of the cross *sm-r mt⁺* × *sm-s mt⁻* are *sm-r mt⁺* and half are *sm-r mt⁻*.

While performing these experiments, Sager observed that the drug streptomycin itself acts as a mutagen. If a pure strain of *sm-s* cells is plated onto streptomycin, quite a significant proportion of streptomycin-resistant colonies will appear. In fact, treatment with streptomycin will produce a whole crop of different kinds of mutant phenotypes, all of which show uniparental inheritance and hence presumably are phenotypes of extranuclear mutations. Furthermore, streptomycin treatment does not produce nuclear mutants. Most of the mutants produced by streptomycin treatment show either resistance to one of several different drugs or defective photosynthesis. The photosynthetic mutants are unable to make use of CO_2 from the air as a source of carbon, so they survive only if soluble carbon (in some form such as acetate) is added to their medium.

These experiments revealed to Sager the existence of a mysterious "uniparental genome" in *Chlamydomonas*—that is, a group of genes that all show uniparental transmission in crosses. Where is this genome located? What is the mechanism of the uniparental transmission? Since there seems to be no physical difference between mating types, why are these genes transmitted only by the *mt⁺* parent?

Evidence arose to suggest that the uniparental genome in this case is in fact chloroplast DNA (cpDNA). About 15 percent of the DNA of a *Chlamydomonas* cell in rapidly growing culture is found in the single chloroplast. This DNA forms a band in a cesium chloride (CsCl) gradient that is distinct from the band of nuclear DNA (Figure 22-10). Furthermore, the precise position of the cpDNA band can be altered by adding the heavy isotope of nitrogen, ¹⁵N, to the growth medium. Using this technique, the cpDNA of the two parents in a cross can be labeled differently: one light (¹⁴N) and one heavy (¹⁵N). The buoyant densities of the cpDNA from these parental cells are 1.69 and 1.70, respectively. Although this difference seems small, it provides appreciably different band positions in a CsCl gradient. With differently labeled parents, the zygote DNA can be examined to see how it compares with the parents. You can see from the results in Table 22-2 that the cpDNA of the *mt⁻* parent is in fact lost, inactivated, or destroyed in some way. This loss of cpDNA from the *mt⁻* parent, of course, parallels the loss of uniparental genes (such as the *sm* genes) borne by the *mt⁻* parent.

Other experiments have been performed using similar logic. For example, crosses can be made between *Chlamydomonas* strains having cpDNAs with distinctly different re-

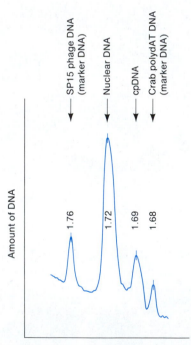

Figure 22-10 The chloroplast DNA (cpDNA) of Chlamydomonas can be detected in a CsCl density gradient as a band of DNA distinct from the nuclear DNA. In this particular experiment, two other DNA types were sedimented along with the *Chlamydomonas* DNA to act as known reference points in the gradient. The numbers represent buoyant densities (g/cm³) of the various DNA types. (After Ruth Sager, *Cytoplasmic Genes and Organelles*. Academic Press, 1972.)

Table 22-2 Buoyant Densities of cpDNA in Progeny from Various *Chlamydomonas* Crosses

Cross	Buoyant density of zygote cpDNA
^{14}N mt^+ × ^{14}N mt^-	1.69
^{15}N mt^+ × ^{15}N mt^-	1.70
^{15}N mt^+ × ^{14}N mt^-	1.70
^{14}N mt^+ × ^{15}N mt^-	1.69

SOURCE: Ruth Sager, *Cytoplasmic Genes and Organelles*, Academic Press, 1972.

striction-enzyme digest patterns on electrophoretic gels. Again, the specific cpDNA digest pattern passed on to the progeny is that of the mt^+ parent only.

> **Message** The cpDNA of *Chlamydomonas* is inherited uniparentally. Because the behavior of this DNA parallels the behavior of the uniparentally transmitted genes, it can be inferred that these genes are located in the cpDNA.

Mapping Chloroplast Genes in *Chlamydomonas*

Chlamydomonas has a large number of uniparentally inherited genes. Is there linkage among these genes? Are they all in one linkage group (on one chromosome), or are they arranged in several linkage groups? Or does each gene assort independently as if it were on its own separate piece of DNA? Of course, the way to approach this question in the true tradition of classical genetics is to perform a recombination analysis. But here we run into a problem. Both sets of parental DNA must be present for recombination to occur, and we have seen that the cpDNA of the mt^- parent is eliminated in the zygote cell. Luckily, there is a way out of this dilemma.

In crosses of mt^+ *sm-r* × mt^- *sm-s*, about 0.1 percent of the progeny zygotes are found to contain both *sm-r* and *sm-s*. The presence of both alleles can be inferred from the fact that the products of meiosis of such cells show both *sm-r* and *sm-s* phenotypes. Such zygotes are called **biparental zygotes,** for obvious reasons, and their genetic condition is described as a **cytohet** ("cytoplasmically heterozygous") or a **heteroplasmon.** It appears as though the inactivation of the mt^- parent's cpDNA fails in these rare zygotes. However, whether this is what happens or not, the cytohets provide the opportunity to study recombination because these cells contain both sets of parental cpDNA. The rarity of cytohet zygotes does pose a problem for research, but their frequency can be increased by treating the mt^+ parent with ultraviolet light before mating. After such treatment, 40 to 100 percent of the progeny zygotes are cytohets. (Note in passing that this observation implies that the mt^+ cell plays a normal role in actively eliminating the

mt^- cell's cpDNA. The ultraviolet irradiation of the mt^+ cell must inactivate such a function.)

> **Message** In *Chlamydomonas*, biparental zygotes (or cytohets) must be the starting point for all studies on the segregation and recombination of chloroplast genes.

Maps of the positions of uniparentally inherited genes on the cpDNA can be obtained by measuring recombination in heteroplasmons. Let's consider an example using uniparentally inherited genes for streptomycin resistance and kanamycin resistance. A "dihybrid" cross is made of the type mt^+ *sm-s kan-r* × mt^- *sm-r kan-s*. From this cross biparental zygotes are obtained. Cytoplasmic segregation of each gene will occur during the successive mitotic divisions of these zygotes. However, if genes are close together on the cpDNA there will be a tendency for the parental allele combinations to segregate together and few recombinants will arise. The recombination process has been quantified using standardized cell populations stemming from biparental zygotes.

It is now known that only genes that are clustered together on the cpDNA of Chlamydomonas will show linkage. One linkage group is shown in Figure 22-11. This linkage map can be derived on the basis of recombination data alone, but molecular analyses including sequencing have confirmed the positions as shown. Most of the uniparentally inherited drug resistance mutations in Chlamydomonas are in genes coding for chloroplast ribosomal RNA, and the positions of these genes are shown on the map.

Molecular studies have shown that the cpDNA is circular but the linked gene cluster shown in Figure 22-11, which is a small proportion of the cpDNA circle, gives a linear map. Furthermore, because of peculiarities of the cpDNA structure (see page 678) a circular linkage map of the entire molecule cannot be made.

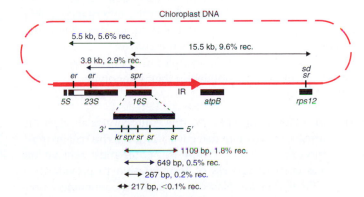

Figure 22-11 Linkage map of drug-resistance genes in Chlamydomonas. The symbols *5S, 23S,* and *16S* represent rRNA genes, *atp* stands for ATP synthase beta subunit, and *rps* a ribosomal protein. The thicker line with an arrowhead is one copy of a repeated sequence found in most cpDNAs. Recombination frequencies are shown together with DNA segment sizes measured physically. (Modified after E. H. Harris et al., 1989, *Genetics* 123: 281.)

Mitochondrial Genes in Yeast

Possibly the greatest success story in the clarification of extranuclear inheritance has been the development of the current view of the mitochondrial genome in bakers' yeast (*Saccharomyces cerevisiae*). This achievement combined genetic analysis in with modern molecular techniques. Let's start with the contributions of genetic analysis.

Three kinds of mutants have been of particular importance: the *petite*, *ant^R*, and *mit^−* mutants. In the 1940s, Boris Ephrussi and his colleagues first described some curious mutants in yeast. The wild-type cells of yeast form relatively large colonies on the surface of a solidified culture medium. Among these large, or "grande," colonies an occasional small, or "petite," colony is found. When isolated, the **petite mutants** prove to be of three types on the basis of their inheritance patterns. The first type is called **segregational petites** because on crossing to a grande strain, half the ascospores give rise to grande colonies and the other half give rise to petite colonies. This 1:1 Mendelian segregation indicates that the petite phenotype is due to a nuclear mutation in these cases. The second type is the **neutral petites,** which in crosses to a grande strain give ascospores that all grow into grande colonies—a clear case of uniparental inheritance. The third type of petite mutant is the **suppressive petites,** which give some ascospores that grow into grande colonies and some that grow into petite colonies. The ratio of grandes to petites is variable but strain-specific: some suppressive petites give exclusively petite offspring in such crosses. Thus the suppressive petites obviously show a non-Mendelian inheritance pattern, and some show uniparental inheritance.

In a yeast cross, the two parental cells fuse and apparently contribute equally to the cytoplasm of the resulting diploid cell (Figure 22-12). Furthermore, the inheritance of the neutral and suppressive petites is independent of mating type. In this sense, then, yeast is clearly quite different from *Chlamydomonas*. Nevertheless, since their inheritance is obviously extranuclear, the neutral and suppressive types have become known as the **cytoplasmic petites.**

Several properties of cytoplasmic petites point to the involvement of mitochondria in their phenotype:

1. In cytoplasmic petites, the mitochondrial electron-transport chain is defective. Because this chain is responsible for ATP synthesis, petites must rely on the less efficient process of fermentation to provide their ATP. If they are placed in a medium containing a nonfermentable energy source such as glycerol, they cannot grow.

2. These petites show no mitochondrial protein synthesis. Mitochondria normally possess their own unique protein-synthesizing apparatus, consisting of a unique set of tRNA molecules and unique ribosomes, all of which are quite different from those operating outside the mi-

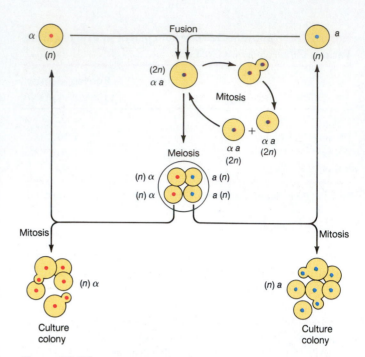

Figure 22-12 The life cycle of bakers' yeast (*Saccharomyces cerevisiae*). The nuclear alleles *a* and *α* determine mating type. Cell fusion between haploid *a* and *α* cells produces a diploid cell. Normally, the cell then goes through a diploid mitotic cycle (budding). However, the cell can be induced (by plating on a special medium) to undergo sporulation, producing haploid products. Meiosis occurs during sporulation. Note that budding involves the formation of a small growth on the side of the parent cell; this bud eventually enlarges and separates to become one of the daughter cells.

tochondrion in the cytosol, or nonorganellar, phase of the cytoplasm.

3. Mitochondria in all organisms have their own unique mitochondrial DNA (mtDNA), which, in addition to being smaller in amount, is quite different from nuclear DNA. The mtDNA of petites is radically different from that of wild types. Neutral petites totally lack mtDNA, whereas suppressive petites show altered base ratios compared with the grandes from which they spring.

The second major class of yeast mutants, the *ant^R* mutants, were initially recognized at the phenotypic level by their resistance to antibiotics supplied in the medium. For example, strains have been obtained that are resistant to chloramphenicol (cap^R), erythromycin (ery^R), spiromycin (spi^R), paramomycin (par^R), and oligomycin (oli^R). These mutations each show a non-Mendelian inheritance pattern similar to that in the suppressive petites; that is, in a cross such as *ery^R × ery^S*, a strain-specific non-Mendelian ratio is seen among the random ascospore progeny. However, tetrad analysis reveals several now-familiar phenomena. Let's trace the cross through its sequential stages (Figure 22-13). When the parental cells fuse, the fusion product, as well as being diploid, is a cytohet. The diploid cells can then

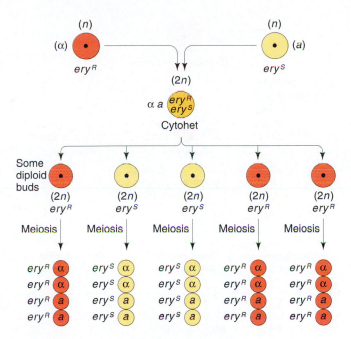

Figure 22-13 The special inheritance pattern shown by certain drug-resistant phenotypes in yeast. When diploid buds undergo meiosis, the products of each meiosis show uniparental inheritance. (Only a representative sample of the diploid buds is shown.) Note that the nuclear genes, represented here by the mating-type alleles *a* and *α*, segregate in a strictly Mendelian pattern. The alleles *ery*^{R} and *ery*^{S} determine erythromycin resistance and sensitivity, respectively.

be allowed to bud mitotically. During this mitotic division, because of cytoplasmic segregation, the daughter cells become either *ery*^{R} or *ery*^{S}. Therefore, when meiosis is induced by shifting the cells onto a special medium, all the haploid products of any single meiosis are identical with respect to erythromycin sensitivity or resistance. At the level of single meioses, then, we see uniparental inheritance at work. (Note that the nuclear mating-type alleles *a* and *α* always segregate in a 1 : 1 ratio.) Similar results have been shown by all the *ant*^{R} mutations, pointing once again to their location in an "extrachromosomal genome."

The third important class of mutants, the *mit*$^-$ mutants, was last to be discovered and required the development of special selective techniques. These mutants are similar to petite mutants in that they exhibit small colony size and abnormal electron-transport-chain functions, but they differ in that they have normal protein synthesis and are able to revert. In a way, *mit*$^-$ mutants are like point-mutation petites. The inheritance pattern of *mit*$^-$ mutants is comparable with that shown by *ant*^{R} types; that is, they show cytoplasmic segregation and also uniparental inheritance at meiosis.

Mapping the Mitochondrial Genome in Yeast

The demonstration of extranuclear genetic determinants immediately raises questions about their physical interrelationship: Are they all linked together on one mitochondrial

"chromosome," or are they located on separate structural units? Mapping the yeast mitochondrial genome has proceeded using many different approaches. A few representative analytical methods follow.

Recombination Mapping. We can begin to look for recombinants by crossing parents that differ in two extranuclear gene pairs (a kind of "dihybrid cross"). As shown in Figure 22-14, we can carry out the cross *ery*^{R} *spi*^{R} × *ery*^{S} *spi*^{S}, allow the resulting diploid cell to bud through several cell generations, and then induce the resulting cells to sporulate (go through meiosis). We can then identify the genotype of each bud cell by observing the phenotype common to all its ascospores.

Four genotypes can result from such a cross and all are in fact observed. An early cross yielded the following results:

ery^R *spi*^{R}	63 tetrads
ery^S *spi*^{S}	48 tetrads
ery^S *spi*^{R}	7 tetrads
ery^R *spi*^{S}	1 tetrad

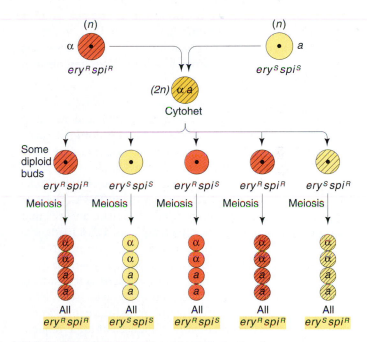

Totals of four genotypes

ery^R*spi*^{R}	63 tetrads		*ery*^{S} *spi*^{R}	7 tetrads
ery^S*spi*^{S}	48 tetrads		*ery*^{R} *spi*^{S}	1 tetrad

Figure 22-14 The study of inheritance in a cross between yeast cells differing with respect to two different drug-resistance alleles (*ery*^{R} = erythromycin resistance; *spi*^{R} = spiramycin resistance). Each diploid bud can be classified as parental or recombinant on the basis of these results. Note that the identity of the extranuclear genotype for all four products of meiosis confirms that cytoplasmic segregation and recombination must occur during the production of the diploid buds.

The genotypes $ery^S spi^R$ and $ery^R spi^S$, of course, represent cytoplasmic recombinants produced during the formation of diploid buds. (Note that these recombinants cannot have been produced by meiosis because the products of any given meiosis are identical with respect to the drug-resistance phenotypes.)

Message In yeast, there is cytoplasmic segregation and recombination during bud formation in diploid cytohets. The segregation and recombination may be detected directly in cultures of the diploid buds or may be detected by observing the products of meiosis that result when the buds are induced to sporulate.

We seem to be just a short step away from the development of a complete map of all the drug-resistance genes in yeast. Unfortunately, the technique of recombination mapping proved to be of only limited usefulness. For one thing, recombination involving mitochondria was shown to be a population phenomenon, similar to phage recombination. So many rounds of recombination are taking place that most genes appeared to be unlinked. Linkage is detectable only for genes that are very close together. Furthermore, the process of recombination has been shown to be strongly influenced by a specific genetic factor, ω (omega), that is present in some mitochondrial genomes and not in others. The most useful mapping developments came from less conventional analyses, examples of which are described in the following subsections.

Mapping by Petite Analysis. The *petite*, *ant^R*, and *mit$^-$* mutations are inherited on the mitochondrial genome of yeast. Some effective techniques for mapping that genome were developed through the combined study of these classes of mutants. Most of these approaches are based on the discovery that petites represent deletions of the mtDNA. This fact opens up a new and different kind of genetic analysis that has been combined with new techniques of DNA manipulation to produce a complete genetic map of yeast mtDNA.

The pivotal observation came in studies where drug-resistant grande strains (such as *ery^R*) were used as the starting material for the induction of petite mutants (see Figure 22-15). We have seen that petite mutants form spontaneously; however, they can also be induced at high frequencies by the use of various specific mutagens, notably ethidium bromide. It is of interest to determine the drug resistance of petite cultures obtained from grande *ery^R* strains. Are they still *ery^R*, or has the petite mutation caused them to become *ery^S*? The answer cannot be determined directly because drug resistance cannot be tested in petite cells.

However, genetic manipulation comes to the rescue. The induced petites are combined with grande *ery^S* cells (which, of course, lack resistance to erythromycin) to form diploids, and the diploids are allowed to form diploid buds. Depending on the type of petite used, varying amounts of petite and grande diploid buds are formed. It is the grande

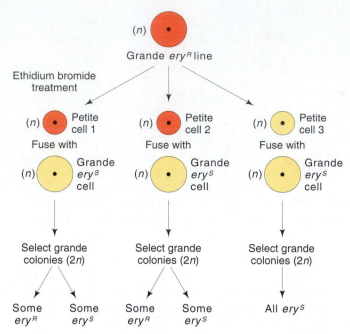

Figure 22-15 The retention or loss of drug resistance (*ery^R*) when petite mutants are induced from a drug-resistant grande culture. The petite cannot be tested directly for drug resistance, so any remaining drug resistance must be "rescued" through mating of petite cells with drug-sensitive grande cells. Diploid buds can then be tested for drug resistance. In the examples shown here, the drug resistance apparently is retained in petite cells 1 and 2, but the drug resistance apparently was lost, together with the determinant for the grande phenotype, during the mutation event giving rise to the petite phenotype in cell 3.

diploid buds that are of interest here because they *can* be tested for drug resistance. If the original petites retain the drug resistance, then some of these diploid grande cells may be expected to have acquired that resistance. In fact, for some petites, the derived grande diploid buds do prove to be of two phenotypes: *ery^S* (derived from the grande *ery^S* haploid) and *ery^R* (which must be derived from the petite haploid). This result indicates that only the genetic determinants for the grande phenotype are lost in the original mutation event that generates these particular petite mutants; the genetic determinants for drug resistance were not affected by the petite mutation.

Of particular interest, however, are those petites with derived grande diploids that are all *ery^S*. In this case, the original mutation event by which the petite phenotype arises apparently inactivates or destroys both the determinant for the grande phenotype and the determinant for drug resistance. The coincidental loss of several genetic determinants is characteristic of deletion mutations in general genetic analysis. In fact, it is now known that the petite phenotype is produced by large deletions of the mtDNA.

The discovery of marker loss in petites is interesting for two main reasons. First, because the petite mutation alters the mtDNA and causes the loss of a particular marker gene, the location of that marker can be assigned to the mtDNA

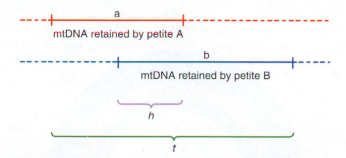

Figure 22-16 Measuring the extent of overlap between any two petite mtDNAs by hybridizing them to samples of grande mtDNA. A total of a units of petite A's mtDNA hybridizes to the grande mtDNA; b units of petite B's mtDNA hybridize. If there is some overlap (homology) between the mtDNAs retained by the two petites, the total amount (t units) of a mixture of the petite mtDNAs that hybridizes to the grande mtDNA must be less than the sum of the individual totals by an amount h, so that $h = a + b − t$. This value h is proportional to the amount of overlap. When many pairs of different petites are studied in this way, the relative sizes and positions of these different mtDNAs can be fitted together with information about the genes retained in common by each pair to produce a map of the mitochondrial genome.

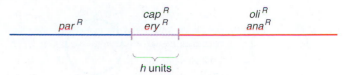

Figure 22-17 Genetic markers retained by specific petites can be correlated with regions of petite homology. Petite induction causes the deletion of a part of the mitochondrial DNA, and different petites retain different genes. In the example, one petite retains cap^R, ery^R, and par^R and another retains cap^R, ery^R, oli^R, and ana^R. The cap and ery loci must be within the overlap region of size h, as determined by DNA hybridization techniques. The loci par, oli, and ana must lie somewhere outside this region, and their relative positions can be assigned through study of other pairs of petites. In this manner, a map of genetic markers can be built up.

beyond any reasonable doubt. The procedure outlined in Figure 22-16 can be used as a routine test to detect a mitochondrial mutation.

> **Message** A good operational test showing that a yeast gene is located in mtDNA is to demonstrate that the gene can be deleted upon petite induction.

Second, the phenomenon of marker loss in petites has given rise to several mapping techniques. One technique begins with a grande cell line that is resistant to several drugs; such a line might have the genotype ery^R cap^R oli^R spi^R. Petites can be induced in this line and then tested to see which of the resistance markers are retained in the petite lines. (Of course, this test must be performed by "rescuing" the markers in a cross with a drug-sensitive grande cell, as we have seen.) It is then simple to compare the frequencies with which various pairs of resistance genes are either co-retained or lost and thus obtain a good idea of which genes are closely linked. This idea can be extended by another mapping technique, which combines genetic analysis with physical techniques. Two different petite strains are derived from a grande line that shows multiple drug resistance. For example, suppose that strain A has retained cap^R, ery^R, and par^R, whereas strain B has retained cap^R, ery^R, oli^R, and ana^R. DNA is extracted from each strain and denatured. This DNA then is hybridized to a standard sample of grande DNA, both in individual experiments using DNA from each strain and with a mixture of DNA from A and B. Suppose that the amounts hybridized are a units from A alone, b units from B alone, and t units from the mixture of A and B.

Anthony Linnane used these values to measure the degree of overlap (h) of different petite deletions. From Figure 22-16, we see that if the retained (nondeleted) DNA in the two petites has a region in common, then the amount of DNA hybridized from the mixture will be less than the sum of the amounts bound when the petite DNAs are hybridized individually. In fact, a study of the diagram should convince you that this difference equals the amount of overlap, so that $h = a + b − t$. Because cap^R and ery^R are the only two alleles retained in both of these petites, we know that these genes must be located in the overlap region. Therefore, the value h is proportional to the maximum possible distance between cap and ery. Thus, we can insert genetic markers into map segments of defined size, and we have begun a comprehensive mapping process (Figure 22-17). This kind of mapping was extended to many petites, and a library of petites with well-defined retained regions was established. These results were combined with the results of marker-retention analysis to produce a complete genetic map of the mtDNA. This map is a circle. Electron microscope observation of mtDNA molecules confirms that they are circular.

We have stated that the petite mtDNA is a small retained piece from the grande mtDNA. However, we must reconcile this view with the fact that a typical petite cell contains more or less the same amount of mtDNA as a typical grande cell, and many petite mtDNA circles are the same size as grande mtDNA circles. The answer to this puzzle is that the retained piece in the petite cell is present in as many tandem copies as it requires to make up an mtDNA of about the same length as the grande mtDNA (Figure 22-18). This model does not require any change in the interpretation of the petite-overlap experiments.

Once a defined set of petite deletions has been worked out, they are in subsequent mapping. For example, $mit^−$ mutants have been mapped by fusing $mit^−$ with several specific petites. If the $mit^−$ mutation is in the region of mtDNA retained by the petite cell, then recombination can substitute petite DNA in place of the $mit^−$ DNA and the recombinant cell will be able to grow on a nonfermentable energy source.

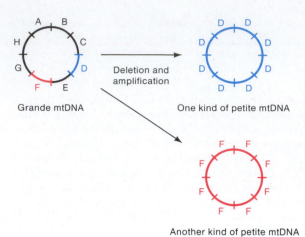

Figure 22-18 When a petite is produced from a grande cell, a large region of mtDNA may be deleted. The DNA region retained by the petite (D or F in these examples) is amplified through tandem duplication to provide a chromosome of approximately the normal size.

Message Marker loss or retention during petite induction and the overlap of petite deletions are the basis of novel genetic techniques used in mapping mitochondrial genomes.

Restriction Mapping. Both mtDNA and cpDNA can be mapped using restriction enzymes. The resulting map (Figure 22-19) shows the location of restriction-enzyme target sites. One obvious opportunity for using restriction analysis on yeast mtDNA is to map genes that code for the various unique mitochondrial RNAs, (tRNAs and rRNAs). A radioactively labeled RNA can be used as a probe in a Southern hybridization (see page 443) to a suitable restriction enzyme digest. The band that "lights up" in the autoradiogram represents the approximate location of the RNA gene under study. Further similar experiments using different restriction enzymes will narrow the region of hybridization down to a precise locus on the mtDNA.

What about the ant^R and mit^- genes? One approach is to correlate the retained restriction fragments in a number of different petites with, say, the antibiotic-resistance markers also retained in those strains. In this way, specific markers may be associated with specific regions of the mtDNA, as defined by the restriction-enzyme target sites. Another approach uses a petite that retains only a single drug-resistance gene—say, ery^R. The mtDNA is extracted from the petite and made radioactive for use as a probe. Its hybridization with the various mtDNA restriction fragments locates the portion of mtDNA that is retained by this particular petite—and hence the locus of the *ery* gene. For example, if the mtDNA from the ery^R petite hybridizes with fragment X from one restriction enzyme and with fragment Y from another enzyme, then the *ery* locus must be located within the overlap region of these two fragments.

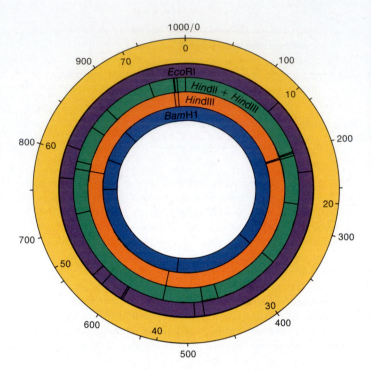

Figure 22-19 Map showing the target sites of various restriction enzymes on the yeast mtDNA. An arbitrary scale from 0 to 1000 is indicated on the outside of the outer circle; a scale indicating thousands of DNA base pairs is indicated on the inside of the outer circle. The inner four bands show the fragments produced by treatment with particular restriction enzymes: *Eco*RI, *Hind*II + *Hind*III, *Hind*III, and *Bam*HI. The position for the zero point of this map is arbitrary. (After J. P. M. Sanders et al., "The Organization of Genes in Yeast Mitochondrial DNA, III," *Molecular and General Genetics* 157, 1977.)

An Overview of the Mitochondrial Genome

The yeast mitochondrial genome has now been fully sequenced. The combined results of sequencing, genetic mapping, and biochemical analysis are shown in Figure 22-20. The human mtDNA map is also shown for comparison. The various genes in yeast discussed in this chapter are shown together with their protein products. We can see that cap^R, ery^R, and spi^R are alterations of the large mitochondrial rRNA genes, and that par^R is associated with changes in the small rRNA gene. Other ant^R mutations such as oli^R are associated with alterations in various subunits of the enzyme ATPase. On the other hand, mit^- mutations are lesions in several subunits of cytochrome oxidase (I–III) or in the cytochrome *b* gene. In addition, several other genes are indicated, such as those for the mitochondrial tRNAs and a gene for a ribosome-associated protein.

The map contains some surprises too. Most prominent are the introns in several genes. Subunit I of cytochrome oxidase contains nine introns. The discovery of introns in the

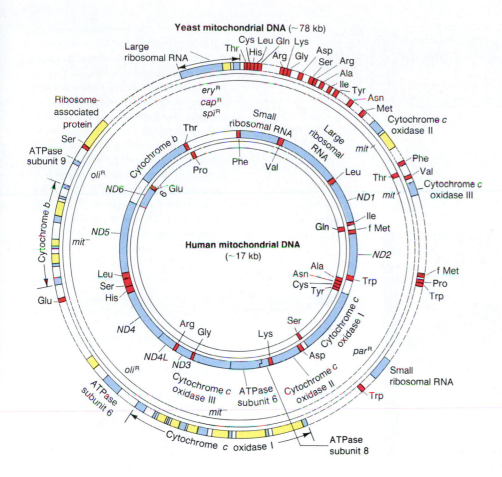

Figure 22-20 Maps of yeast and human mtDNAs. Each map is shown as two concentric circles corresponding to the two strands of the DNA helix. The human map has been produced exclusively by physical techniques. The yeast map has been produced by a combination of genetic and physical techniques, as discussed in the text. Note that the mutants used in the yeast genetic analysis are shown opposite their corresponding structural genes. Blue = exons and uninterrupted genes, red = tRNA genes, and yellow = URFs (unassigned reading frames). tRNA genes are shown by their amino acid abbreviations; ND genes code for subunits of NADH dehydrogenase. (Note that the human map is not drawn to the same scale as the yeast map.)

mitochondrial genes is particularly surprising because they are relatively rare in yeast nuclear genes. Another set of surprises was the unassigned reading frames (URFs). These are sequences that have correct initiation codons and are uninterrupted by stop codons. Some URFs within introns appear to specify proteins important in the splicing out of the introns themselves at the RNA level. Notice that the human mtDNA is by comparison much smaller and more compact. There seems to be much less spacer DNA between the genes.

The overall view of the mitochondrial genome shows it to have two main functions: it codes for some proteins that are actually in or associated with the electron-transport chain, and it codes for some proteins, all the tRNAs, and both rRNAs necessary for mitochondrial protein synthesis. But it is striking that the remaining necessary components for both these functions are encoded by nuclear genes with mRNA that is translated on cytosolic ribosomes followed by transport of the products to the mitochondrion (Figure 22-21). Why this peculiar division of labor exists between nuclear and mitochondrial DNA is not known. Another curiosity is that some specific subunits are encoded by mtDNA in one organism but by nuclear DNA in another. Evidently, an evolutionary transposition of information has occurred between these organelles. There is also evidence of transposition between mitochondria and chloroplasts. Further-

more, inactive pseudogenes are detectable in the nucleus, showing homology with mitochondrial genes.

Genes for 25 yeast and 22 human mitochondrial tRNAs are shown on the maps in Figure 22-20. These tRNAs carry out all the translation that occurs in mitochondria. These are far fewer than the minimum of 32 required to translate nucleus-derived mRNA. The economy is achieved by a "more wobbly" wobble pairing (see Figure 13-19) of tRNA anticodons. The tRNA specificities in human mtDNA are shown in Figure 22-22. Notice that the codon assignments are in some cases different from the nuclear code. It is also known that there is variation between the mitochondria of different species. Hence, the genetic code is obviously not universal, as had been supposed for many years.

Message Mitochondrial DNA has genes for mitochondrial translation components (mainly rRNAs and tRNAs) and for some subunits of the proteins associated with mitochondrial ATP production. Less understood regions include the introns, unassigned reading frames, and spacer DNA.

Now that yeast, human, and several other mtDNAs have been sequenced and intensively characterized, the study of mtDNA of other genetically less well studied organisms is made relatively simple. The technique is based

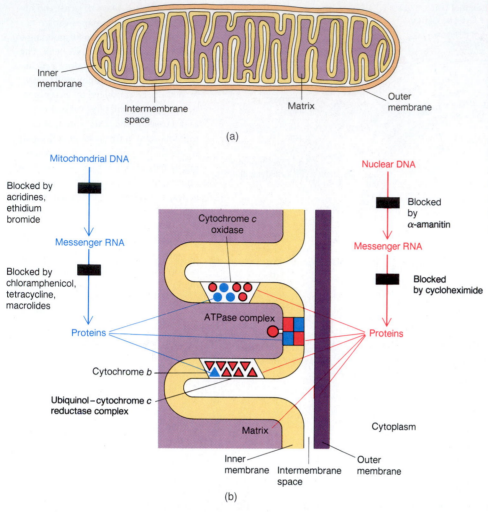

(a)

(b)

Figure 22-21 Cooperation of yeast mtDNA and nuclear DNA in coding for the protein components of the inner mitochondrial membrane. (a) Overview of mitochondrial structure. (b) Details of membrane constitution. Three functional proteins, cytochrome *c* oxidase, ATPase, and cytochrome *b*, show the cooperation. The mtDNA supplies six proteins, three subunits of cytochrome *c* oxidase, two subunits of ATPase, and one subunit of cytochrome *b*. The nuclear DNA supplies all the remaining subunits. Part of the dissection of this complex scheme was through the use of drugs that block specific reactions, as shown.

First letter	Second letter				Third letter
	U	C	A	G	
U	Phe	Ser	Tyr	Cys	U
	Phe	Ser	Tyr	Cys	C
	Leu	Ser	*Stop*	(Stop) Trp	A
	Leu	Ser	*Stop*	Trp	G
C	Leu	Pro	His	Arg	U
	Leu	Pro	His	Arg	C
	Leu	Pro	Gln	Arg	A
	Leu	Pro	Gln	Arg	G
A	Ile (Met)	Thr	Asn	Ser	U
	Ile	Thr	Asn	Ser	C
	(Ile) Met	Thr	Lys	(Arg) *Stop*	A
	Met	Thr	Lys	(Arg) *Stop*	G
G	Val	Ala	Asp	Gly	U
	Val	Ala	Asp	Gly	C
	Val	Ala	Glu	Gly	A
	Val	Ala	Glu	Gly	G

Figure 22-22 The genetic code of the human mitochondrion. The functions of the 22 tRNA types are shown by the 22 nonstop codon boxes.

on evolutionary conservation of DNA sequences. Cloned genes from yeast or human mtDNA can be used as probes to locate rapidly the corresponding gene on the restriction map of the mtDNA of any species.

An Overview of the Chloroplast Genome

Although many of the concepts behind chloroplast inheritance have been derived from the kind of genetic studies we have followed in *Mirabilis* and *Chlamydomonas*, the current perspective has come mainly from molecular studies. In fact, the cpDNA of several species now has been completely sequenced. As an example, Figure 22-23 shows the organization of the cpDNA from the liverwort *Marchantia polymorpha*.

Typically, cpDNA molecules range from 120 to 200 kb in different plant species. In *Marchantia,* the molecular size is 121 kb. Figure 22-24 shows the functions of most of the genes in the *Marchantia* cpDNA. The sequencing of 121,000 nucleotides does not automatically tell us what genes are

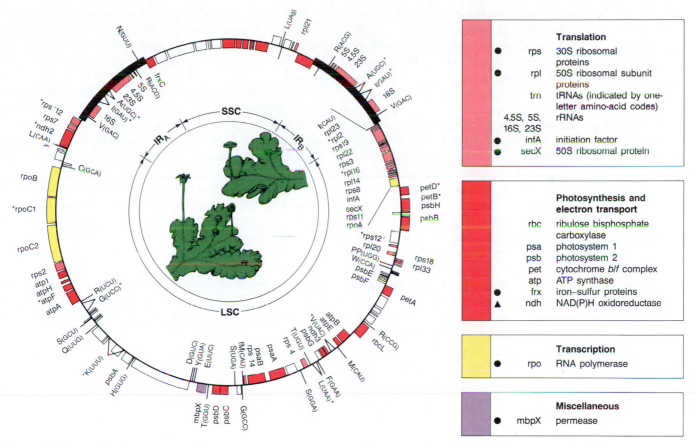

Figure 22-23 The chloroplast genome of the liverwort *Marchantia polymorpha*. IR_A and IR_B, LSC, and SSC on the inner circle indicate the inverted repeats, large single-copy and small single-copy regions, respectively. Genes shown inside the map are transcribed clockwise, and those outside are transcribed counterclockwise. Genes for rRNAs in the IR regions are represented by 16S, 23S, 4.5S, and 5S, respectively. Genes for tRNAs are indicated by the one-letter amino acid code with the unmodified anticodon. Protein genes identified are indicated by gene symbols, and the remaining open boxes represent unidentified ORFs, approximately to scale. Genes containing introns are marked with asterisks. The boxes to the right of the gene map summarize the functions of the genes identified to date; groups with related functions are shown in different shades. ORFs are shown in color. Genes identified on the basis of homology with genes from bacterial or mitochondrial genomes are marked ● and ▲, respectively. The central drawing depicts a male (*above*) and a female (*below*) *Marchantia* plant. The antheridia and archegonia are elevated on specialized stalks above the thallus, which contains the chloroplasts. *Machantia* can also reproduce asexually: disks of green tissue (gemmae) grow from the bottom of cup-shaped structures on the thallus surface. When mature, the gemmae separate from the thallus and grow to produce new gametophyte plants. (From K. Umesono and H. Ozeki, *Trends in Genetics 3*, 1987.)

Figure 22-24 Mutations in nuclear genes can result in white leaves. Here a green C*c* corn plant, heterozygous for a recessive allele causing albino leaves, was selfed and the progeny were $\frac{1}{4}$ *cc* and fully albino. Because they cannot photosynthesize, these albinos die as soon as the food reserve in the seed (laid down by the maternal plant) is exhausted. Such mutations can be in the synthesis of chlorophyll itself or in one of the other nuclear proteins that interact with the chloroplastencoded proteins to produce a functional photosynthetic light reaction. (Anthony Griffiths)

present. The presence of genes is inferred by the detection of open reading frames (ORFs)—long sequences that begin with a start codon but are uninterrupted by stop codons except at the termini. Determining what these specific genes code for is more difficult; although most ORFs have been assigned, some of the ORFs are still unassigned reading frames (URFs). Most of the protein-encoding genes have been assigned by comparing the predicted amino acid sequences with those of known chloroplast, mitochondrial, and bacterial proteins. Other genes, such as the rRNAs and tRNAs, have been assigned by hybridization experiments, using known genes from other organisms as probes.

There are about 136 genes in the *Marchantia* molecule, including four kinds of rRNA, 31 kinds of tRNA, and about 90 protein genes. Of the 90 protein genes, 20 code for photosynthesis and electron-transport functions. Genes coding for translation functions take up about half the chloroplast genome and include the proteins and RNA types necessary for translation in the organelle.

Also, notice the presence of a large inverted repeat in Figure 22-23. Such inverted repeats are found in the cpDNA of virtually all species of plants. However, there is some variation as to which genes are included in the inverted repeat region and therefore in the relative size of that region. One of the mysteries of the inverted repeat is that the duplicates have exactly the same sequence, yet to date no mechanism is known that ensures this complete identity.

Like mtDNA, cpDNA cooperates with nuclear DNA to provide subunits for functional proteins used inside the organelle. The nuclear components are translated outside in the cytosol and then transported into the chloroplast, where they are assembled, together with the components synthesized in the organelle. Mutations in the nuclear components of photosynthetic complexes are inherited as Mendelian alleles (Figure 22-24).

Mitochondrial Diseases in Humans

Several human diseases are known to be caused by mutations in the mitochondrial DNA. These mutations are all in the mitochondrial genes for protein subunits of the mitochondrial electron-transport chain. The mutations result in defects of mitochondrial ATP production. Humans that carry these mutations have mitochondrial cytopathies, diseases associated with various combinations of defects of the brain, heart, muscle, kidney, and liver.

The mutations themselves can be point mutations, deletions, or duplications of the mtDNA, as shown in Figure 22-25. As expected, such mitochondrial diseases show the inheritance features characteristic of mitochondrial mutations. First, they are passed on from one generation to the next only through maternal parents. Second, they show cytoplasmic segregation when present in heteroplasmic mixtures. Hence, a person with a mixture of normal and abnormal mitochondria can have cells with widely differing ratios

of these mitochondria because of cytoplasmic segregation. Similarly, a woman with mild expression of a mitochondrial disease can produce children with disease expression ranging from none to severe, again because of cytoplasmic segregation.

Incidentally, mutations in the nucleus-encoded subunits of the mitochondrial electron-transport chain (Figure 22-21) produce a similar spectrum of disease phenotypes. However, these mutations can be distinguished in pedigrees because they show the Mendelian inheritance typical of nuclear genes.

It has been proposed that even the aging process is influenced or perhaps caused by accumulated damage in mtDNA. Deleted mtDNA molecules accumulate in the brains of older people, and there is considerable variation between regions of the brain. In muscles, oxidative phosphorylation decreases with age, and mtDNA mutations have been observed to accumulate in localized bands in the muscle fiber. However, the proportions of damaged mtDNA molecules have been found not to exceed 12 percent, so the significance of these accumulations is poorly understood at present. Nevertheless, if these speculations are substantiated by further experiments, it could well emerge that normal aging is the most common mitochondrial "disease" of all.

Extragenomic Plasmids in Eukaryotes

We saw in Chapter 10 that plasmids are common in bacteria. However, plasmids are also regularly, although less commonly, encountered in eukaryotes. Because plasmids are generally not associated with nuclear chromosomes, they show various kinds of non-Mendelian inheritance.

Most eukaryotic plasmids are silent at the phenotypic level and can only be detected using molecular techniques. The best known of these is the "two-micron circle" (2μ) plasmid of yeast. This circular plasmid is located in the nucleoplasm. If a haploid strain containing the plasmid is mated with another haploid strain lacking the plasmid, the asexual or sexual descendants all tend to have the plasmid. Although mysterious biologically, the 2μ plasmid has assumed a prominent role in the molecular biology of yeast because it has been genetically engineered to act as a cloning vector for that organism (see Chapter 15). When nuclear DNA is spliced into the 2μ vector and used to transform a recipient cell type, the insert is carried to the nucleus, where it may become incorporated into the chromosomes via a variety of recombination mechanisms.

However, most eukaryotic plasmids are located in the mitochondria, and show inheritance mechanisms similar to those of mitochondria. Plasmids have been found in fungi and plants, but not in animals. An interesting and well-studied type of mitochondrial plasmid is associated with cytoplasmic male sterility in corn. Male sterility in plants is of great importance in agriculture. Plants with the male-sterile

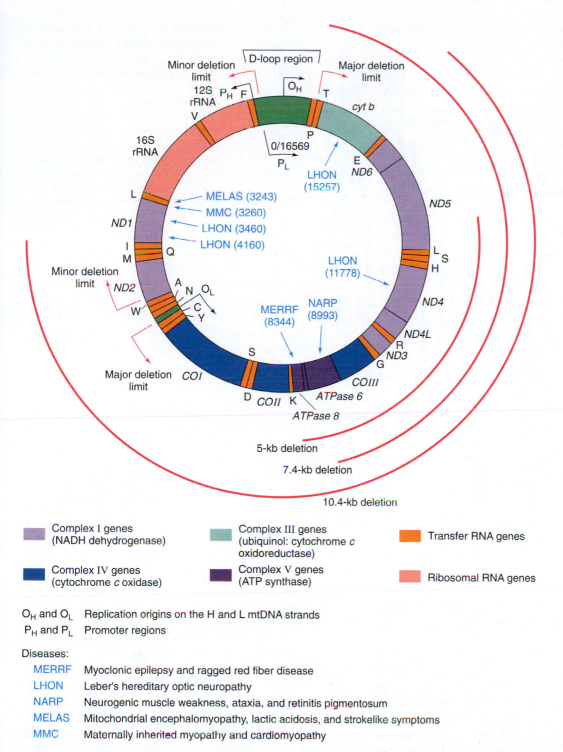

Figure 22-25 Locations of the major mitochondrial DNA mutations that result in human disease. Point mutations are shown by nucleotide number on the mtDNA map. The diseases CEOP (chronic external ophthalmoplegia plus), KSS (Kearns-Sayre syndrome), and Pearson's syndrome involve mtDNA deletions, and these occur within the major deletion limits indicated; others are found within the region called the *minor deletion limit*. Some examples of these deletions are shown as red arcs. Gene symbols are explained in Figure 22-20. (From D. C. Wallace, *Science* 256, 629, 1992.)

Complex I genes (NADH dehydrogenase)

Complex III genes (ubiquinol: cytochrome *c* oxidoreductase)

Transfer RNA genes

Complex IV genes (cytochrome *c* oxidase)

Complex V genes (ATP synthase)

Ribosomal RNA genes

O_H and O_L Replication origins on the H and L mtDNA strands

P_H and P_L Promoter regions

Diseases:

MERRF	Myoclonic epilepsy and ragged red fiber disease
LHON	Leber's hereditary optic neuropathy
NARP	Neurogenic muscle weakness, ataxia, and retinitis pigmentosum
MELAS	Mitochondrial encephalomyopathy, lactic acidosis, and strokelike symptoms
MMC	Maternally inherited myopathy and cardiomyopathy

phenotype produce no functional pollen. In the case of cytoplasmic male sterility, when male-sterile plants are crossed as female parents using normal fertile plants as males, all the progeny prove to be male-sterile and the inheritance pattern is clearly maternal. The best-studied type of cytoplasmic male sterility is in corn, where it is associated with two linear plasmids, S1 and S2, that are located within the mitochondria in addition to the mtDNA. The precise way in which the S1 and S2 plasmids are involved with male steril-

ity is not yet fully understood, although much is known about the molecular properties of these plasmids. One intriguing property is that S1 and S2 can recombine with the mtDNA.

Male sterility in corn and other crop plants is used to facilitate the production of hybrid seed issuing from a cross between genetically different lines; such seeds usually result in larger, more vigorous plants. The big problem is how to prevent self-pollination, which interferes with the produc-

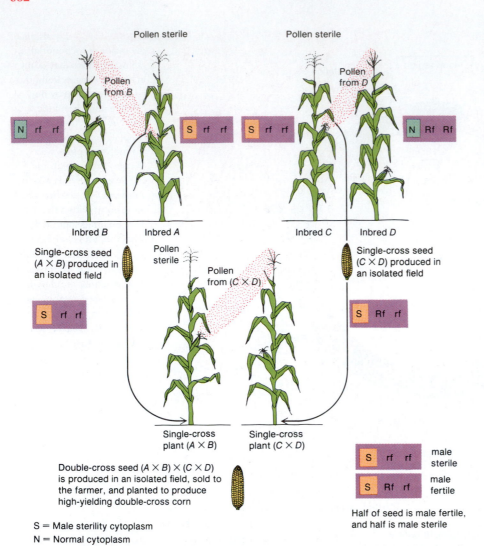

Figure 22-26 The use of cytoplasmic male sterility to facilitate the production of hybrid corn. In this scheme, the hybrid corn is generated from four pure parental lines: A, B, C, and D. Such hybrids are called *double-cross hybrids.* At each step, appropriate combinations of cytoplasmic genes and nuclear restorer genes ensure that the female parents will not self and that male parents will have fertile pollen. (From J. Janick et al., *Plant Science.* Copyright ©1974 by W. H. Freeman and Company.)

tion of hybrid seed. Male sterility is useful here because selfing is impossible, since there is no functional pollen. One breeding scheme is illustrated in Figure 22-26.

Our final example of eukaryotic plasmids is a type that determines senescence (aging) in the fungus *Neurospora.* Most *Neurospora* strains do not senesce and, given a large enough supply of medium, will keep growing forever. However, certain populations contain senescent individuals. One such population is from Hawaii, where the senescent strains are called *kalilo,* a Hawaiian word meaning "hovering between life and death." Kalilo strains will grow only a certain distance through the culture medium before they die. However, crossing these strains before they die reveals that the ability to die is inherited in a strict maternal fashion. It has been found that the senescence determinant is a linear 9-kb mitochondrial plasmid called *kalDNA.* This plasmid can exist autonomously—and apparently innocuously—in the cytoplasm, but the onset of death is preceded by the physical insertion of kalDNA into the mtDNA. At death, most mtDNA molecules contain the full-length kalDNA insert (Figure 22-27). Presumably, the strain dies because the

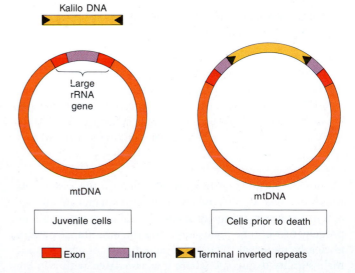

Figure 22-27 Senescence is *Neurospora* is associated with the insertion of a linear plasmid called *kalilo DNA* into the mtDNA. Here insertion is shown in the intron of a gene for rRNA, but other insertion sites are known.

kalDNA acts as an insertional mutagen and interferes with mitochondrial function. It can therefore be seen that kalDNA behaves like a molecular parasite.

Notice that the plasmids are associated with a definite phenotype in the examples from corn and *Neurospora*, whereas the yeast 2μ plasmid has no detectable phenotypic manifestation. The existence and evolutionary position of eukaryotic plasmids is still a mystery, but the study of the structure and behavior of such elements will undoubtedly provide important clues about the fundamental properties of the genetic material.

Message   Extragenomic eukaryotic plasmids are inherited in a non-Mendelian manner that often mimics the inheritance patterns of organelle DNA.

How Many Copies?

For the genetic and the biochemical approaches to the study of organelle genomes that we have examined, it does not matter much how many copies of organelle genome are present per cell. However, it is a question that occurs to most people interested in organelle inheritance.

The number of copies turns out to vary among species. More surprisingly, it also can vary within a single species. The leaf cells of the garden beet have about 40 chloroplasts per cell. The chloroplasts themselves contain specific areas that stain heavily with DNA stains; these areas are called **nucleoids,** and they are a feature commonly found in many organelles. Each beet chloroplast contains 4 to 18 cpDNA molecules. Thus, cells of a single beet leaf can contain as many as $40 \times 18 \times 8 = 5760$ copies of the chloroplast genome. Although *Chlamydomonas* has only one chloroplast per cell, the chloroplast contains 500 to 1500 cpDNA molecules, commonly observed to be packed in nucleoids.

What about mitochondria? A "typical" haploid yeast cell can contain 1 to 45 mitochondria, each having 10 to 30 nucleoids, with 4 or 5 molecules in each nucleoid. The mi-

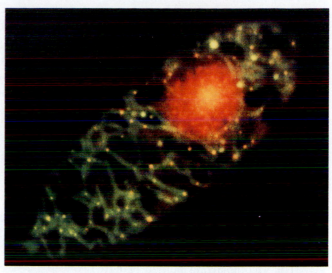

10 μm

Figure 22-28   Fluorescent staining of a cell of *Euglena gracilis*. With the dyes used, the nucleus appears red because of the fluorescence of large amounts of nuclear DNA. The mitochondria fluoresce green, and within mitochondria the concentrations of mtDNA (nucleoids) fluoresce yellow. (From Y. Huyashi and K. Veda, *Journal of Cell Science* 93, 565, 1989.)

tochondrial nucleoids of the unicellular organism *Euglena gracilis* are shown in Figure 22-28.

How does this genome duplication relate to cytoplasmic segregation and recombination? How many genomes are present at the beginning of this process? Do all copies of the genome actually become involved? Does recombination occur when organelles fuse, and must the nucleoids fuse also? How does the process physically segregate the many copies of the genome scattered through the cell? Few answers to such questions are available at present. The situation is rather like the one faced by the early geneticists, who knew about genes and linkage groups but knew nothing about their relation to the process of meiosis.

SUMMARY

The extranuclear genome supplements the nuclear genome of eukaryotic organisms, which was examined in earlier chapters. The existence of the extranuclear genome is recognized at the genetic level chiefly by uniparental transmission of the relevant mutant phenotypes. In fungi, heterokaryon tests can detect a cytoplasmically based phenotype by demonstrating transmission by cytoplasmic contact, in the absence of nuclear fusion. At the cellular level, a combination of genetic and biochemical techniques has demonstrated that the extranuclear genome is organelle DNA—either mitochondrial (mtDNA) or chloroplast (cpDNA). In heteroplasmons (mixed cytoplasms), the differ-

ent "alleles" of organellar genetic determinants regularly segregate to produce phenotypically different tissue sectors (for example, in variegated green and white plants) or phenotypically different cells.

The mtDNA of yeast is the best-understood organelle DNA. This DNA codes for unique mitochondrial translation components, and it also codes for some components of the respiratory enzymes found in the mitochondrial membranes. Mutations in the translational-component genes typically produce drug-resistant phenotypes; mutations in the respiratory-enzyme genes typically produce phenotypes involving respiratory insufficiency. Yeast petite mutants are caused by deletion of part of mtDNA. Yeast mtDNA can be mapped by recombination analysis in "dihybrid" heteroplasmons, marker co-retention in induced petites, and correlation of retained markers with retained restriction fragments in petites.

Chloroplast DNA is larger and more complex than mtDNA. Mutations in cpDNA typically produce photosynthetic defects or drug resistance.

Organelle genomes have been fully investigated and sequenced in only a few organisms, but these studies have produced general analytical techniques that have widespread application in the study of extranuclear genomes in other organisms. Now cloned genes from yeast and human mtDNA and Marchantia cpDNA can be used as probes to retrieve homologous genes from other less well studied organisms.

Some rare human diseases are caused by mutations or deletions in mtDNA. Aging itself may be a result of accumulated genetic damage in mtDNA.

Eukaryotic plasmids are extragenomic elements that are also inherited in a non-Mendelian manner. Most are mitochondrial in location.

CHAPTER INTEGRATION PROBLEM

1. The accompanying human pedigree concerns a rare visual abnormality in which the person affected loses central vision while retaining peripheral vision.

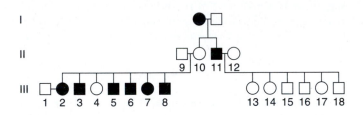

a. What inheritance pattern is shown? Can it be explained by nuclear inheritance? Mitochondrial inheritance? Molecular geneticists studied the mitochondrial DNA of the 18 members of generations II and III. A restriction fragment 212 base pairs long of the mtDNA from each person was digested with another restriction enzyme, *Sfa*N1, with the following results:

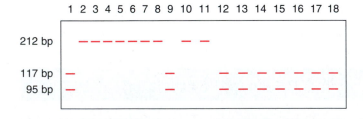

b. What inheritance pattern is shown by these restriction fragments?

c. How does the restriction pattern inheritance relate to the inheritance of the disease?

d. How can you explain individuals 4 and 10?

e. What is the likely nature of the mutation?

f. How would this analysis be useful in counseling this family?

Solution

a. Based on the pedigree alone, it is possible, but unlikely, that the disease is caused by a dominant nuclear allele. But we would have to invoke lack of penetrance in individual 10, who would have to carry the allele because it is passed on to her children. In addition, we have to explain the ratios in generation III. The matings 9 × 10 and 11 × 12 would have to be $Aa \times aa$, and the phenotypic ratio of affected to normal then expected among the children in each family is 1:1. So overall this is not an attractive model to explain the results.

The results can also be explained by maternal inheritance of the disease. Individuals 4 and 10, however, require special explanation. Once again, we can invoke incomplete penetrance. But alternatively, we could invoke cytoplasmic segregation; we have learned that cells can be mixtures of normal and abnormal cytoplasmic determinants (here, mitochondria), and cytoplasmic segregation can skew the ratio from cell to cell. The mother in the first generation would have to be a heteroplasmon and by chance pass along predominantly normal mitochondria to her daughter (10) who does not express the disease. Then, by a skewing in the other direction, 10 would pass along mainly abnormal mitochondria to six out of her seven children.

b. The restriction patterns clearly show maternal inheritance. This is expected because we are dealing with mtDNA.

c. There is obviously a close correlation between the presence of the large 212-bp fragment and the disease. If this same correlation were to be found in other similar pedigrees, one could formulate a model in which the mutation that causes the disease simultaneously causes the loss of a *Sfa*N1 restriction site.

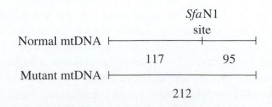

d. The possibility that 4 and 10 are heteroplasmons is now less attractive, because if there were mixtures, we would expect to see that the restriction-enzyme patterns of some persons in the family have all three bands—95, 117, and 212—but none were seen. Therefore the most likely explanation is the incomplete penetrance of a mitochondrial disease.

e. According to the model, the most likely kind of mutation would be a nucleotide pair substitution, because if the mutation is at the *Sfa*N1 site, no nucleotides are lost or gained, since 117 + 95 = 212.

f. If the model is upheld by other studies, appearance of the 212-bp fragment after *Sfa*N1 digestion would be a diagnostic marker for the mutation. All women with this marker could pass the disease on to their children, whereas men with the marker could not transmit the disease. Note that in answering this question we have combined concepts of Mendelian inheritance, cytoplasmic inheritance, mutation, and DNA restriction analysis. The question is based on patterns shown in a true pedigree for the disease Leber's hereditary optic neuropathy (LHON), which is believed to be mitochondrially based.

SOLVED PROBLEMS

1. In a strain of *Chlamydomonas* that carries the mt^+ allele, a temperature-sensitive mutation arises that renders cells unable to grow at higher temperatures. This mutant strain is crossed to a wild-type stock, and all the progeny of both mating types are temperature-sensitive. What can you conclude about the mutation?

Solution

We are told that the mutation arose in a mt^- stock. Therefore, the cross must have been

$$mt^+ \ ts \times mt^- \ ts^+$$

and the progeny must have been $mt^+ \ ts$ and $mt^- \ ts$. This is a clear-cut case of uniparental inheritance from the mt^+ parent to all the progeny. In *Chlamydomonas*, this type of inheritance pattern is diagnostic of genes in chloroplast DNA, so the mutation must have occurred in the chloroplast DNA.

2. Due to evolutionary conservation, organelle DNA shows homology across a wide range of organisms. Consequently, DNA probes derived from one organism often hybridize with the DNA of other species. Two probes derived from the cpDNA and mtDNA of a fir tree are hybridized to a Southern blot of the restriction digests of the cpDNA and mtDNA of two pine trees, R and S, that had been used as parents in a cross. The autoradiograms follow (numbers are in kb):

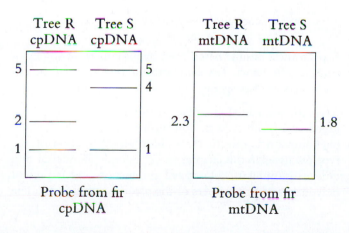

Probe from fir cpDNA

Probe from fir mtDNA

The cross R♀ × S♂ is made, and 20 progeny are isolated. They are all identical in regard to their hybridization to the two probes. The autoradiogram for each progeny is

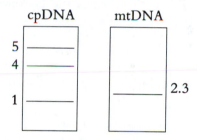

a. Explain the probe hybridization of parents and progeny.

b. Explain the progeny results. Compare and contrast them with the results in this chapter.

c. What do you predict from the cross S♀ × R♂?

NOTE: This question is based on results shown in several conifer species.

Solution

a. For both cpDNA and mtDNA, the total amount of DNA hybridized by the probe is different in plants R and S. Hence, we can represent the DNA something like this (other fragment arrangements are possible):

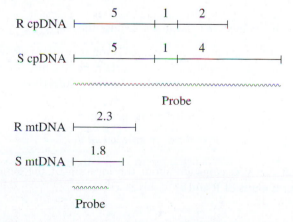

Thus, the probes reveal a (presumably neutral) restriction fragment length polymorphism of both the cpDNA and the mtDNA. These are useful organelle markers in the cross.

b. We can see that all the progeny have inherited their mtDNA from the maternal parent R, because all show the same R mtDNA fragment hybridized by the probe. This is what we might have predicted, based on the predominantly maternal inheritance encountered in this chapter. However, cpDNA is apparently inherited exclusively paternally, be-cause all progeny show the 5/4/1 pattern of the paternal plant S. This is surprising, but it is the only explanation of the data. In fact, all the gymnosperms studied so far show paternal inheritance of cpDNA. The cause is unknown, but the phenomenon contrasts with that in angiosperms.

c. From this cross, we can predict that all progeny will show the paternal 5/2/1 pattern for cpDNA and the maternal 1.8-kb band for mtDNA.

PROBLEMS

1. How do the nuclear and organelle genomes cooperate at the protein level?

2. Name and describe two tests for cytoplasmic inheritance.

3. What is the basis for the green-white color variegation in the leaves of *Mirabilis?* If the cross is made

variegated ♀ × green ♂

what progeny types can be predicted? What about the reciprocal cross?

4. In *Neurospora* the mutant *stp* exhibits erratic stop-start growth. The mutant site is known to be in the mitochondrial DNA. If a *stp* strain is used as the female parent in a cross to a normal strain acting as the male, what type of progeny can be expected? What about the progeny from the reciprocal cross?

5. If a yeast cell carrying an antibiotic-resistance mutation in its mtDNA is crossed to a normal cell and tetrads are produced, what ascus types can you expect with respect to resistance?

6. A new antibiotic mutation (ant^R) is discovered in a certain yeast. Cells of genotype ant^R are treated with ethidium bromide, and petite colonies are obtained. Some of these petites prove to have lost the ant^R determinant.

a. What can you conclude about the location of the ant^R gene?

b. Why didn't all the petites lose the ant^R gene?

7. Two corn plants are studied. One is resistant (R) and the other is susceptible (S) to a certain pathogenic fungus. The following crosses are made, with the results shown:

S♀ × R♂ → progeny all S

R♀ × S♂ → progeny all R

What can you conclude about the location of the genetic determinants of R and S?

8. In *Chlamydomonas,* a certain probe picks up a restriction fragment length polymorphism in cpDNA. There are two morphs, as follows:

Morph 1: two bands, sizes 2 and 3 kb

Morph 2: two bands, sizes 3 and 5 kb

If the following crosses are made:

mt^+ morph 1 × mt^- morph 2

mt^+ morph 2 × mt^- morph 1

what progeny types can be predicted from these crosses? Be sure to draw the DNA morphs with their restriction sites. Also draw a sketch of the autoradiogram.

9. In yeast, the following cross is dihybrid for two mitochondrial antibiotic genes:

MATa oli^R cap^R × *MATα* oli^S cap^S

(*MATa* and *MATα* are the mating-type alleles in yeast.) What types of tetrads can be predicted from this cross?

10. In the genus *Antirrhinum,* a yellowish leaf phenotype called prazinizans (pr) is inherited as follows:

normal ♀ × pr ♂ → 41,203 normal
+ 13 variegated

pr ♀ × normal ♂ → 42,235 pr + 8 variegated

Explain these results based on a hypothesis involving cytoplasmic inheritance. (Explain both the majority *and* the minority classes of progeny.)

11. You are studying a plant with tissue comprising both green and white sectors. You wish to decide whether this phenomenon is due to (1) a chloroplast mutation of the type discussed in this chapter, or (2) a dominant nuclear mutation that inhibits chlorophyll production and is present only in certain tissue layers of the plant as a mosaic. Outline the experimental approach you would use to resolve this problem.

12. A dwarf variant of tomato appears in a research line. The dwarf is crossed as female to normal plants, and all the F_1 progeny are dwarfs. These F_1 individuals are selfed, and the F_2 progeny are all normal. Each of the F_2 individuals is selfed, and the resulting F_3 generation is $\frac{3}{4}$ normal and $\frac{1}{4}$ dwarf. Can these results be explained by (1) cytoplasmic inheritance? (2) Cytoplasmic inheritance plus nuclear suppressor gene(s)? (3) Maternal effect on the zygotes? Explain your answers.

13. Assume that diploid plant A has a cytoplasm genetically different from that of plant B. To study nuclear-cytoplasmic relations, you wish to obtain a plant with the cytoplasm of plant A and the nuclear genome predominantly of plant B. How would you go about producing such a plant?

14. Two species of *Epilobium* (fireweed) are intercrossed reciprocally as follows:

    ♀ *E. luteum* × ♂ *E. hirsutum* ⟶ all very tall

    ♀ *E. hirsutum* × ♂ *E. luteum* ⟶ all very short

The progeny from the first cross are backcrossed as females to *E. hirsutum* for 24 successive generations. At the end of this crossing program, the progeny still are all tall, like the initial hybrids.

    a. Interpret the reciprocal crosses.

    b. Explain why the program of backcrosses was performed.

15. One form of male sterility in corn is maternally transmitted. Plants of a male-sterile line crossed with normal pollen give male-sterile plants. In addition, some lines of corn are known to carry a dominant nuclear restorer gene (*Rf*) that restores pollen fertility in male-sterile lines.

    a. Research shows that the introduction of restorer genes into male-sterile lines does not alter or affect the maintenance of the cytoplasmic factors for male sterility. What kind of research results would lead to such a conclusion?

    b. A male-sterile plant is crossed with pollen from a plant homozygous for gene *Rf*. What is the genotype of the F_1? The phenotype?

    c. The F_1 plants from part b are used as females in a testcross with pollen from a normal plant (*rf rf*). What would be the result of this testcross? Give genotypes and phenotypes, and designate the kind of cytoplasm.

    d. The restorer gene already described can be called *Rf-1*. Another dominant restorer, *Rf-2*, has been found. *Rf-1* and *Rf-2* are located on different chromosomes. Either or both of the restorer alleles will give pollen fertility. Using a

male-sterile plant as a tester, what would be the result of a cross where the male parent was:

    (i) Heterozygous at both restorer loci?

    (ii) Homozygous dominant at one restorer locus and homozygous recessive at the other?

    (iii) Heterozygous at one restorer locus and homozygous recessive at the other?

    (iv) Heterozygous at one restorer locus and homozygous dominant at the other?

16. Treatment with streptomycin induces the formation of streptomycin-resistant mutant cells in *Chlamydomonas*. In the course of subsequent mitotic divisions, some of the daughter cells produced from some of these mutant cells show the normal phenotype. Suggest a possible explanation of this phenomenon.

17. Cosegregation mapping is performed in *Chlamydomonas* on four chloroplast markers: *m1*, *m2*, *m3*, and *m4*. The markers are considered pairwise in heterozygous condition, and cosegregation frequencies are obtained as follows:

	m1	*m2*	*m3*	*m4*
m1	——	29.0	18.0	18.4
m2		——	10.9	26.2
m3			——	8.8
m4				——

(For example, *m1* and *m2* cosegregate in 29 percent of the cell divisions followed.) Draw a rough genetic map based on these results.

18. In *Aspergillus*, a "red" mycelium arises in a haploid strain. You make a heterokaryon with a nonred haploid that requires *para*-aminobenzoic acid (PABA). From this heterokaryon you obtain some PABA-requiring progeny cultures that are red, along with several other phenotypes. What does this information tell you about the gene determining the red phenotype?

19. On page 128 an experiment is described in which abnormal mitochondria are injected into normal *Neurospora*. The text mentions that "appropriate controls" are used. What controls would you use?

20. Adrian Srb crossed two closely related species, *Neurospora crassa* and *N. sitophila*. In the progeny of some of these crosses, a phenotype called aconidial (*ac*) appeared that lacks conidia (asexual spores). The observed inheritance was

    ♀ *N. sitophila* × ♂ *N. crassa* ⟶ $\frac{1}{2}$ ac, $\frac{1}{2}$ normal

    ♀ *N. crassa* × ♂ *N. sitophila* ⟶ all normal

    a. What is the explanation of this result? Explain all components of your model with symbols.

b. From which parent(s) did the genetic determinants for the ac phenotype originate?

c. Why were neither of the parental types ac?

21. Several crosses involving poky or nonpoky strains A, B, C, D, and E were made in *Neurospora*. Explain the results of the following crosses, and assign genetic symbols for each of the strains involved. (Note that poky strain D behaves just like poky strain A in all crosses.)

Cross	Progeny
a. nonpoky B♀ × poky A♂	all nonpoky
b. nonpoky C♀ × poky A♂	all nonpoky
c. poky A♀ × nonpoky B♂	all poky
d. poky A♀ × nonpoky C♂	$\frac{1}{2}$ poky, all identical (e.g., D); $\frac{1}{2}$ nonpoky, all identical (e.g., E)
e. nonpoky E♀ × nonpoky C♂	all nonpoky
f. nonpoky E♀ × nonpoky B♂	$\frac{1}{2}$ poky $\frac{1}{2}$ nonpoky

22. The mtDNA of seven cytoplasmic petites was characterized by restriction-enzyme analysis. The results showed that the mtDNA retained in each of these petites was as indicated by the arcs in the accompanying diagram. Ten *mit⁻* mutants were fused with each of the petites to make cytohets, and cells from these cytohet cultures were plated onto standard growth medium. Out of the 70 combinations, some showed only petite phenotype colonies on the plate (these are shown with a − in the grid below), but the remainder showed some grande colonies as well as petites (these are shown with a + in the grid). Use these results to locate the approximate positions in the mtDNA of the genes that mutated to give the original *mit⁻* cultures.

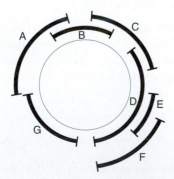

	1	2	3	4	5	6	7	8	9	10
A	+	−	−	−	−	−	−	−	+	+
B	−	+	+	−	−	−	−	−	+	−
C	−	−	+	+	+	−	−	−	−	−
D	−	−	−	−	+	−	+	+	−	−
E	−	−	−	−	−	−	+	−	−	−
F	−	−	−	−	−	−	+	−	−	−
G	−	−	−	−	−	−	−	−	−	−

 Unpacking the Problem

a. What is mtDNA?

b. Draw a mtDNA molecule, showing at least five specific genes found in mtDNA.

c. What is a cytoplasmic petite?

d. What is the nature of a cytoplasmic petite at the mtDNA level?

e. Why is it appropriate to represent petite DNA by an arc of a circle?

f. Briefly describe a restriction-enzyme analysis that might have been used to determine the extent of the DNA retained by a petite.

g. What is a *mit⁻* mutant? How does a *mit⁻* mutant compare with a petite mutant? Sketch a *mit⁻*, a petite, and a grande colony.

h. Are *mit⁻* mutants drug-resistant?

i. What is a cytohet? Give another example of a cytohet. Give another word for a cytohet, or make up a descriptive term for the concept yourself.

j. How would you make the cytohets in this experiment? Would auxotrophic markers be useful?

k. In what sense is the word *fused* used in this question?

l. What happens to yeast cells on growth medium?

m. Draw a typical plate representing a + result in the grid.

n. Draw a typical plate representing a − result in the grid.

o. What exactly do the + and − results mean? Is complementation or recombination involved?

p. Are all the *mit⁻* mutants different in their behavior?

q. Do all the petites show different behavior in combination with the mit^- mutants?

r. If some mit^- mutants show the same behavior, how is this possible if the petites are different according to the restriction analysis?

23. In yeast, an antibiotic-resistance haploid strain ant^R arises spontaneously. It is combined with a normal ant^S strain of opposite mating type to form a diploid culture that is then allowed to go through meiosis. Three tetrads are isolated:

Tetrad 1	Tetrad 2	Tetrad 3
$\alpha\ ant^R$	$\alpha\ ant^R$	$a\ ant^S$
$\alpha\ ant^R$	$a\ ant^R$	$a\ ant^S$
$a\ ant^R$	$a\ ant^R$	$\alpha\ ant^S$
$a\ ant^R$	$\alpha\ ant^R$	$\alpha\ ant^S$

a. Interpret these results.

b. Explain the origin of each ascus.

c. If an ant^R grande strain were used to generate petites, would you expect some of the petites to be ant^S? Explain your answer.

24. In yeast, two haploid strains are obtained that are both defective in their cytochromes; the mutants are designated $cyt1$ and $cyt2$. The following crosses are made:

$$cyt1^- \times cyt1^+$$
$$cyt2^- \times cyt2^+$$

One tetrad is isolated from each cross:

$cyt1^-$	$cyt2^-$
$cyt1^-$	$cyt2^-$
$cyt1^+$	$cyt2^-$
$cyt1^+$	$cyt2^-$

a. From these tetrad patterns, explain the differences in the two underlying mutations.

b. What other ascus types could be expected from each cross?

c. How might the two genes involved here interact at the functional level?

25. In a marker-retention analysis in yeast, a multiply resistant strain $apt^R\ bar^R\ cob^R$ is used to induce 500 petites. The table shows, for different pairs of markers, the number of petites in which only one marker of the pair is lost.

Genes	Petites in which first marker is lost	Petites in which second marker is lost
apt bar	87	120
apt cob	27	18
bar cob	48	69

(For example, 87 petites were $apt^S\ bar^R$, and 120 were $apt^R\ bar^S$.) Use these results to draw a rough map of these mitochondrial genes.

26. A grande yeast culture of genotype $cap^R\ ery^R\ oli^R\ par^R$ is used to obtain petites by treatment with ethidium bromide. The petites are tested for (1) their drug resistance and (2) the ability of their mtDNA to hybridize with various specific mtRNA types. In all, 12 petites are tested; Table 22-3

Table 22-3

Petite culture	Drug resistance (R) or sensitivity (S)				Ability (+ or −) of petite mtDNA to hybridize with mtRNAs						
	cap	ery	oli	par	rRNA$_{large}$	rRNA$_{small}$	tRNA$_1$	tRNA$_2$	tRNA$_3$	tRNA$_4$	tRNA$_5$
1	R	S	S	S	+	−	−	−	−	−	−
2	S	S	S	S	−	−	−	−	−	+	−
3	S	S	S	S	−	−	−	+	−	−	+
4	S	S	R	S	−	−	+	−	−	−	−
5	S	R	S	S	+	+	−	−	−	−	−
6	R	S	S	S	−	−	−	−	+	−	−
7	S	S	R	R	−	−	−	+	−	−	−
8	S	S	S	S	−	+	−	−	−	+	+
9	S	R	S	S	+	−	−	−	−	−	−
10	S	S	S	S	−	−	+	−	+	−	−
11	R	S	R	R	−	+	+	+	+	+	+
12	S	R	S	S	+	+	−	−	−	−	−

shows the results. Use this information to plot a map of the mtDNA, showing the sequence of the 11 genetic loci involved in these phenotypes. Be sure to state your assumptions and draw a complete map.

27. The mtDNA is compared from two haploid strains of baker's yeast. Strain 1 (mating type α) is from North America, and strain 2 (mating type a) is from Europe. A single restriction enzyme is used to fragment the DNAs, and the fragments are separated on an electrophoretic gel. The sample from strain 1 produces two bands, corresponding to one very large and one very small fragment. Strain 2 also produces two bands, but they are of more intermediate sizes. If a standard diploid budding analysis is performed, what results do you expect to observe in the resulting cells and in the tetrads derived from them? In other words, what kinds of restriction fragment patterns do you expect?

28. In yeast, some strains are found to have in their cytoplasm circular DNA molecules that are 2 micrometers (μm) in circumference. In some of these strains, this 2-μm DNA has a single *Eco*RI restriction site; in other strains, there are two such sites. A strain with one site is mated to a strain with two sites. All the resulting diploid buds are found to contain both kinds of 2-μm DNA.

a. Is the 2-μm DNA inherited in the same fashion as mtDNA?

b. If you used radioactive 2-μm DNA as a probe, predict the results of hybridizing it to a Southern blot of *Eco*RI-treated DNA of ascospores from these diploid cells.

29. Circular mitochondrial DNA is cut with two restriction enzymes A and B, with the following results:

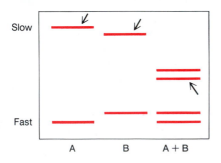

(The arrows indicate the bands that bound a radioactive mt rRNA-derived cDNA probe in a Southern blot.) Draw a rough map of the positions of the restriction site(s) of A and B, and also show approximately where the mt rRNA gene is located.

30. You are interested in the mitochondrial genome of a fungal species for which genetic analysis is very difficult but from which mtDNA can be extracted easily. How would you go about finding the positions of the major mitochondrially encoded genes in this species? (Assume some evolutionary conservation for such genes.)

31. Rare senescent cultures of a certain fungus have been found.

a. When these cultures are crossed as male with non-senescent partners, no senescent progeny are obtained. However, when they are crossed as female, some but not all progeny senesce; some senescent progeny die sooner than others and either sooner or later than the parental strain. Explain these findings.

b. When mtDNA from senescent cultures is cut with *Eco*RI, the large normal fragment 1 is missing. In its place are four new *Eco* fragments B, C, E, and G, totaling 10 kb more than fragment 1. A map of the normal fragment 1 follows, where *Eco* represents *Eco*RI sites, *Hi* represents *Hind*III sites, and two segments cloned as probes are also shown:

Probe pBH-10a binds to B but not to C, and probe pTVB-3 binds to C but not to B; neither probe binds to E or G. A *Hind*III digest shows the absence of fragment 20 and the appearance of two new large fragments, K1 and K2. K1 hybridizes to C, E, and G, and K2 hybridizes to B and E. Fragment C also hybridizes to *Hind*III fragment 15. What can you deduce from these data?

32. **a.** Early in the development of a plant, a mutation in cpDNA removes a specific *Bg*III restriction site (*B*) as follows:

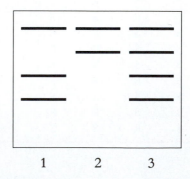

In this species, cpDNA is inherited maternally. Seeds from the plant are grown, and the resulting progeny plants are sampled for cpDNA. The cpDNAs are cut with *Bg*III, and Southern blots are hybridized with the probe P shown. The autoradiograms show three patterns of hybridization:

Explain the production of these three seed types.

b. In *Gryllus* (a cricket) and *Drosophila,* rare females have been found that show mixtures of two mtDNA types, differing by the presence or absence of one specific restriction site. In the progeny of these females, each and every individual also proves to be a mixture of the parental mtDNA types. Contrast these results with the results from part a and with other results in this chapter.

33. The mitochondrial genome of the turnip is a large, circular molecule 218 kb in size with a pair of 2-kb direct repeats located about 83 kb apart. However, when turnip mtDNA is examined carefully, three molecular types are seen: the 218-kb circle just described, a 135-kb circle bearing a single 2-kb repeat, and an 83-kb circle also bearing only one copy of the 2-kb repeat. Propose a model to explain the presence of the two smaller molecular types.

34. Reciprocal crosses and selfs were performed between the two moss species *Funaria mediterranea* and *Funaria hygrometrica.* The appearance of the sporophytes and the leaves of the gametophytes are shown in the accompanying diagram. The crosses are written with the female parent first.

a. Describe the results presented, summarizing the main findings.

b. Propose an explanation for the results.

c. Show how you would test your explanation; be sure to show how it could be distinguished from other explanations.

35. The accompanying pedigree shows a very unusual inheritance pattern that actually did occur. All progeny are shown, but the fathers in each mating have been omitted to draw attention to the remarkable pattern.

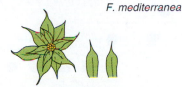

Gametophytes (leaves)   Sporophytes

F. mediterranea

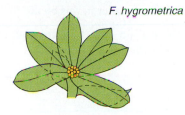

F. hygrometrica

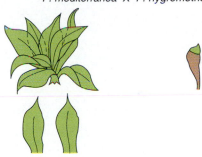

F. mediterranea X *F. hygrometrica*

F. hygrometrica X *F. mediterranea*

Diagrams after C. H. Waddington, *An Introduction to Modern Genetics,* Macmillan, 1939.

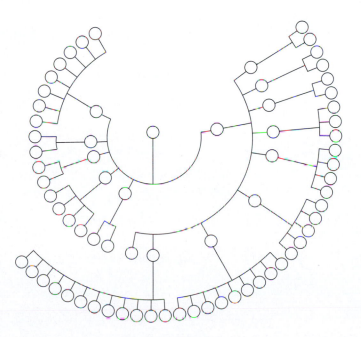

a. State concisely exactly what is unusual about this pedigree.

b. Can the pattern be explained by

(i) chance (if so, what is the probability)?

(ii) cytoplasmic factors?

(iii) Mendelian inheritance?

Explain.

36. Consider the following pedigree for a rare human muscle disease:

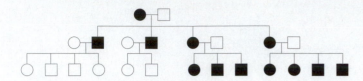

    a. What is the unusual feature that distinguishes this pedigree from those studied earlier in this book?

    b. Where in the cell do you think the mutant DNA resides that is responsible for this phenotype?

37. The following pedigree shows the recurrence of a rare neurological disease (large black symbols) and spontaneous fetal abortion (small black symbols) in one family. (Slashes mean the individual is deceased). Provide an explanation for this pedigree in terms of cytoplasmic segregation of defective mitochondria.

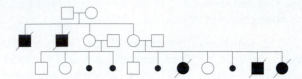

23

Gene Regulation during Development

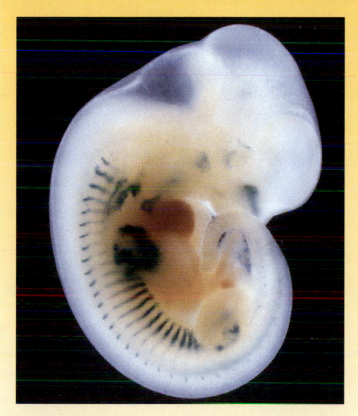

Tissue-specific expression of a transgenic reporter gene. An 11.5 transgenic mouse embryo contains recombinant DNA composed of a 258-base-pair mouse DNA sequence fused to the *E. coli lacZ* gene, which encodes the enzyme β-galactosidase. The 258-bp mouse DNA contains all of the cis-regulatory sequences necessary to direct expression of *lacZ* in muscle precursor cells, as revealed by the blue histochemical stain for β-galactosidase enzyme activity. The stained muscle precursors include the somites, limb buds, and branchial arches of the embryo. (Reproduced from D. J. Goldhamer, B. P. Brunk, A. Faerman, A. King, M. Shani, and C. P. Emerson, Jr. *Development* 121, 1995, 644.)

KEY CONCEPTS

▶ Regulation of protein synthesis can occur at many levels, including modulation of gene structure, regulation of transcription, processing of the nascent transcript into mRNA, or modification at post-transcriptional levels.

▶ Each level of control can contribute to tissue-specific protein production.

▶ Epigenetic inheritance implies that the expression of a gene can be regulated according to its ancestry.

▶ Modern transgenic techniques can be used to characterize the regulatory elements of eukaryotic genes.

▶ Misregulation of gene expression can lead to dominant mutant phenotypes.

▶ Developmental pathways are formed by the sequential implementation of various regulatory steps.

W hat distinguishes 150 pounds of *E. coli* from a person of similar weight? Indeed, many answers are possible. The one of particular importance here is that each one of the *E. coli* cells must perform all the functions necessary for life as a bacterium—every cell is on its own. By contrast, life for the cells of a higher organism is a cooperative venture—there is a division of labor. Cells and organ systems are specialized for many different roles. In the extreme, certain cells are in essence factories for the production of a single type of protein. For example, in humans, B islet cells of the pancreas make and secrete insulin, red blood cells synthesize hemoglobin for oxygen transport, and B lymphocytes produce antibodies. More often, however, a cell type is not characterized by the production of a single, highly abundant protein, but rather by an entire constellation of proteins expressed in that cell. Most proteins are expressed in many or all types of cell, and the amounts and variety of these proteins distinguish different classes of cell from one another.

As has been discussed in earlier chapters, higher organisms have much more DNA in their genome than do unicellular organisms. There appear to be two reasons for this, both relating to the multitissued nature of higher organisms. First, genes in higher organisms are bigger, principally because they need to contain many more cis-regulatory elements (the equivalent of bacterial operators) to control correctly the timing, location, and amount of expression of the proteins they encode. Second, there are many more genes in higher organisms. Whereas the bacterium *E. coli* has on the order of 5,000 genes, the fruit fly *Drosophila melanogaster* has about 20,000, and humans might have 100,000 or more. Because almost all genes encode proteins, this means that higher organisms have many more proteins than do unicellular systems. Some of the "extra" proteins in higher organisms contribute as structural components to the more complicated organization of their cells and tissues. However, we are also seeing that many of these extra genes serve regulatory functions, for example, by encoding proteins that act as transcription factors to regulate the complex expression patterns of other proteins. Here we will examine some of the many ways that the proteins encoded by such regulatory genes contribute to the differences between cell types in higher organisms.

Regulating the Synthesis of Proteins

How is synthesis of new proteins regulated, both qualitatively and quantitatively? In one gene or another, we find examples of regulation at each of the steps leading from the gene to the final protein (Figure 23-1). The structure and number of copies of a gene can be an important aspect of control. Transcription of a gene can be regulated. The processing of the nascent transcript into the mature mRNA can be altered in different tissues. The ability of a transcript to be translated can be modulated. Regulation at each of these

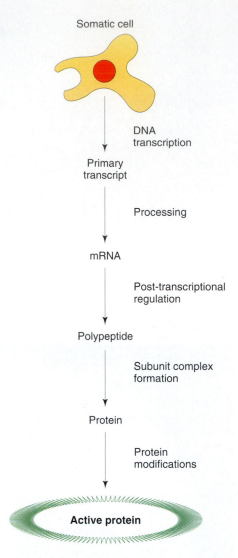

Figure 23-1 Regulation of protein activity. In a cell, there are many steps that occur between the gene and the active protein it encodes. During development, regulatory mechanisms at any of these levels can control protein expression.

levels has been shown to have an important impact on the elaboration of different cells or tissues during development. Just as understanding the triplet genetic code led to great insights into the process by which genetic information determines protein structure, unraveling the code for the regulation of gene expression will be crucial in understanding how cells adopt different developmental and physiological roles.

Tissue-Specific Regulation at the Level of DNA Structure

For many years it was believed that the structure of somatic cell and germ-line DNA was the same. This is indeed true for the bulk of DNA in a cell. However, some important exceptions have been observed in somatic cells of higher organisms. Because these somatic cells never contribute genetic material to offspring, regulatory mechanisms that alter the structure of the DNA in these dead-end cells have been able to evolve without altering germ-line gene organization.

Here, we will examine one instance of gene rearrangement that has important regulatory consequences. In this case, tissue-specific amplification of the number of copies of a gene leads to high levels of gene expression in that tissue. Later in this chapter, we will see how somatic DNA rearrangement plays a crucial role in a highly specialized and important developmental pathway—the production of antibodies by cells of the immune system.

Egg production (oogenesis) occurs in the ovary of the *Drosophila* adult female. Both a mature oocyte and a proteinaceous eggshell must be constructed during this process. Follicle cells, somatic cells that surround each oocyte, face the task of building the eggshell very rapidly and, hence, must synthesize large amounts of the eggshell proteins in a brief period of time. *Drosophila* has evolved an elegant solution to this problem: the copy number of the eggshell genes is increased through somatic DNA rearrangements that occur only in the follicle cells that make the eggshell. In *Drosophila*, there are only seven major eggshell genes in most cells. Specifically during follicle cell development, by a special replication mechanism (Figure 23-2), the eggshell genes in *Drosophila* are amplified 20-fold to 80-fold in the follicle cells just before they need to be transcribed. Thus, in *Drosophila*, extra DNA templates for the eggshell genes are present only when and where they are needed—in the follicle cells. We can imagine that this is much more efficient than having to carry around the multiple copies of the eggshell genes in every cell in the body.

Message   Somatic changes in gene structure or copy number can be used to regulate tissue-specific gene activity.

Regulation of Gene Transcription

By far, the most common mode of gene regulation appears to be at the level of transcription. This is an energy-efficient solution to gene regulation, since the organism doesn't have to expend energy to transcribe genes whose protein products will not be required in a given cell. Most genes have evolved mechanisms that can modulate when, where, and how much transcript is made in a given cell. These mechanisms depend on the presence of different DNA sequences that can be bound by proteins that can induce or repress transcription. Recall from Chapter 18 that promoters, enhancers, and silencers are DNA sequences that play a role in transcription. **Promoters** are the binding sites for RNA polymerase at the start of transcription. **Enhancers** are binding sites for proteins, transcription factors, that are thought to interact with the RNA polymerase in a way that causes it to initiate transcription. **Silencers** are sites that

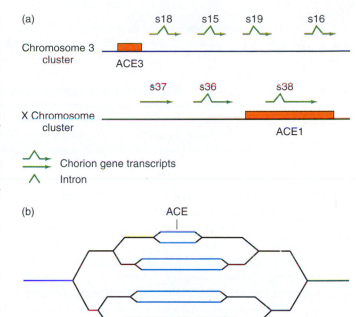

Figure 23-2   The structure of the two eggshell gene clusters in *Drosophila melanogaster*. (a) The cluster on chromosome 3 contains four eggshell genes, and the cluster on the X chromosome contains three. Each cluster is only about 10 kb long. Each ACE region contains an amplification control element. This region is thought to act as a special site for the initiation of tissue-specific DNA replication in follicle cells of the ovary. (b) A model of how amplification might occur through multiple DNA replication initiations at an ACE. Each line represents a double helix of DNA. (Adapted from T. Orr-Weaver, *BioEssays* 13, 1991, 97.)

bind transcription factors in a way that directly or indirectly blocks transcription initiation by RNA polymerase.

Tissue-Specific Regulation of Transcription

Many enhancer elements in higher eukaryotes activate transcription in a **tissue-specific** manner—that is, they induce expression of a gene in one or a few cell types. Tissue-specificity is conferred in one of two ways. As discussed previously in Chapter 18, enhancer elements are merely docking sites for transcription factors that interact with RNA polymerase at the promoter. An enhancer can act in a tissue-specific manner if the transcription factor that binds to it is present only in some types of cells, or if other proteins bind to other sites near the enhancer element, making the enhancer inaccessible to its transcription factor.

Properties of Tissue-Specific Enhancers

In some genes, regulation can be controlled by simple sets of enhancers. For example, in *Drosophila*, vitellogenins are large egg yolk proteins made in the female adult's ovary and fat body (an organ that is essentially the fly's liver) and trans-

Figure 23-3 A molecular map of a complex gene—the *dpp* gene of *Drosophila*. Units on the map are in kilobases. The basic transcription unit of the gene is shown below the map coordinate line. The abbreviations above the line mark the sites of a few of the many tissue-specific enhancers that regulate this transcript in different stages of development. Tissue-specific expression patterns conferred by these enhancers are shown in Figure 23-10, and the abbreviations are explained there.

ported into the developing oocyte. Two distinct enhancers located within a few hundred base pairs of the promoter regulate the vitellogenin gene, one driving expression in the ovaries and the other in the fat body.

The array of enhancers for a gene can be quite complex, reflecting similarly complex patterns of gene expression that must be controlled. The *dpp* (*decapentaplegic*) gene in *Drosophila*, for example, codes for a protein that mediates signals between cells (see Chapter 24), contains numerous enhancers, perhaps numbering in the tens or hundreds, dispersed along a 50-kb interval of DNA. Some of these are located 5′ to the promoter of *dpp*, others in introns, and still others 3′ to the polyadenylation site of the gene. Each of these enhancers regulates the expression of *dpp* in a different site in the developing animal (Figure 23-3).

In prokaryotes, cis-regulatory elements, such as operators, reside in the immediate neighborhood of the promoter. By contrast, as we have just seen, tissue-specific enhancers in eukaryotes can be quite far from the promoter, acting from a distance of 30 kb or more. How can a protein that is bound to a DNA element many kilobases away from the promoter interact with it? Under the right circumstances (for example, with certain DNA sequences or under the influence of certain DNA-binding proteins), DNA can bend so that sequences very far apart on the linear DNA molecule are actually quite close to each other in the three-dimensional nucleus.

The requirement for multiple enhancer elements to regulate tissue-specific expression helps to explain the large size of genes in higher eukaryotes. The tissue-specific regulation of a gene may be quite complex, requiring the action of numerous, distantly located enhancer elements.

Message Eukaryotic enhancers can act at great distance, regulating transcription by remote control.

Transcript Processing and Tissue-Specific Regulation

The transcripts of most higher eukaryotic genes are extensively processed during their maturation to mRNAs. Regulated mRNA splicing can be an important developmental control point. One striking example is the regulation of the

P element, a transposable element (Figure 23-4). Recall from Chapter 21 that the P element is a family of related DNA sequences found in many strains of *Drosophila melanogaster.*

The intact P element is a 3-kb-long piece of DNA that encodes its own transposase, an enzyme that catalyzes the transposition of the element from one site in the genome to another. In strains of flies carrying the P element, it transposes only in the germ line, not in somatic cells. This is because P transposase is found only in the germ line. P tran-

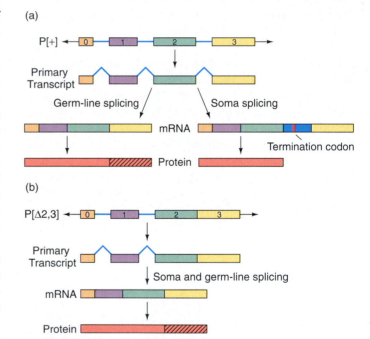

Figure 23-4 Germ-line-specific expression of *Drosophila* P elements: an example of tissue-specific regulation by RNA splicing. (a) The somatic and germ-line mRNA structures for a wild-type P element are shown. The germ-line mRNA is formed by splicing of four exons (numbered 0 to 3) and encodes P transposase. The somatic mRNA lacks the splice between exons 2 and 3; that is, the intron between these two exons is retained in the mRNA. Because the intron between exons 2 and 3 contains a termination codon, the resulting shorter somatic protein is defective, lacking the transposase activity. (b) A modified P element transgene, P{Δ2, 3}, that lacks the intron between exons 2 and 3 is depicted. In all tissues, P{Δ2, 3} produces an mRNA indistinguishable from that of the wild-type P element, resulting in P transposase activity in both germ line and soma.

scription, however, occurs in every cell. How can we reconcile ubiquitous P transcription with germ-line-specific transposase activity? The splicing of P transposase is different in germ line and soma. The germ-line P transposase mRNA contains four exons numbered 0, 1, 2, and 3 (Figure 23-4). Frank Laski, Don Rio, and Gerry Rubin hypothesized that in somatic cells the P transposase mRNA retains the intron between exons 2 and 3; in other words, during splicing of the nascent P transposase transcript in somatic cells, the 2,3 intron is not spliced out. A stop codon present in this intron creates a defective somatic transposase essentially encoded only by exons 0, 1, and 2. As a consequence, somatic P transposase is missing its normal carboxy terminus and is nonfunctional. To test this hypothesis, Laski, Rio, and Rubin constructed a P element in which the intron between exons 2 and 3 was removed in vitro by cutting and pasting of the P element DNA, and the resulting P{Δ2, 3} transgenic construct was introduced into flies by DNA transformation. The occurrence of clones of mutant tissue in flies carrying the P{Δ2, 3} transgene together with another P element containing a marker expressed in somatic tissue (the eye) did indeed prove able to induce somatic transposition of other P elements (Figure 23-5), confirming the idea that the

germ-line specificity of P transposase activity is at the level of regulated splicing.

Message The production of an active protein can be regulated by controlling the pattern of splicing of an initial transcript into an mRNA.

Post-transcriptional Regulation

A variety of mechanisms modulate the ability to translate mRNAs. Many of these operate through interactions of regulatory molecules with sequences in the 3′ ends of transcripts. An mRNA can be divided into three parts: a 5′ untranslated region (5′ UTR), the polypeptide coding region, sometimes called the **open reading frame (ORF),** and the 3′ untranslated region (3′ UTR). If certain sequences are present within the 3′ UTR of an mRNA, that mRNA will be very rapidly degraded. In other cases, sequences in the 3′ UTR do not cause mRNA degradation but nonetheless can lead to lower levels of translation. Such sequences have been identified in the 3′ UTRs of mRNAs encoded by genes involved in sex determination of the nematode *Caenorhabdi-*

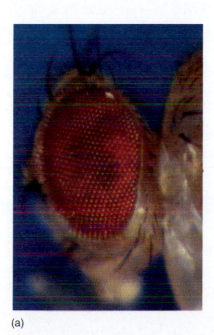 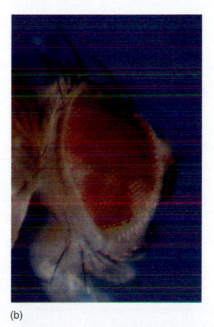

(a)   (b)

Figure 23-5 Somatic expression of P transposase in *Drosophila*. In both panels, the flies contain a source of P transposase and a second P element lacking transposase activity but capable of being transposed by a source of P transposase. This second P element is marked with the w^+ gene, which confers red eye color. The standard white genes in the fly contain w null mutations, so that any pigmentation depends on the presence of the P[w^+] transposon. (a) P transposase is derived from a wild-type P element (see Figure 23-4a). The resulting eye is a homogeneous red color. The absence of white eye tissue indicates that the P element is inactive in soma. Thus, the wild-type transposase mRNA was spliced to yield the inactive form of the enzyme. (b) Here, P transposase derives from the P{Δ2, 3} transgene. The resulting eyes are a mixture of red and white sectors. The white sectors represent tissue in which the P[w^+] element has "jumped out of" the chromosome, indicating that P transposase was active in those somatic cells. (Courtesy of S. Misra and G. Rubin, University of California, Berkeley.)

tis elegans. Mutations that alter these 3′ UTR sequences lead to higher-than-normal levels of synthesis of the proteins encoded by these genes. Presumably, these 3′ UTR sequences are the target sites for proteins that digest mRNA molecules or that block their translation.

A novel phenomenon, in which 3′ UTR sequences interact with a regulatory RNA molecule, has recently been uncovered in *C. elegans.* This phenomenon was discovered in studies of **heterochrony,** that is, in studies of mutations that alter the timing of development. In this nematode, the wild-type adult worm develops after four larval stages. Some heterochronic mutations cause the adult worm to develop prematurely, after only three larval stages. Others cause the adult worm to be delayed, with the adult arising only after an additional fifth larval stage occurs. The basis for the altered timing of development is that the normal chronology of cell divisions in the various cell lineages is altered such that certain divisions in a lineage either are skipped, causing premature adult development, or are reiterated, producing delayed adulthood. Two of the heterochronic genes are called *lin-4* and *lin-14* (*lin* is an abbreviation for *lineage-defective*). The product of the *lin-4* gene is known to repress translation of *lin-14* mRNA. It turns out that the repressor molecule encoded by the *lin-4* gene is not a protein but an RNA that has sequences complementary to sequences within the 3′ UTR of *lin-14*. It is thought that a double-stranded RNA structure is formed between the 3′ UTR of *lin-14* mRNA and the complementary *lin-4* RNA, and that this somehow leads to repression of translation. A 3′ UTR may also contain sequences that act as sites to anchor an mRNA to particular structures within a cell. The ability to localize an mRNA within a cell can lead to differences in the concentration of the protein product of that transcript in that cell and in its progeny. In Chapter 25, we will see how this localization mechanism contributes to the formation of one of the body axes of the *Drosophila* embryo.

Message Regulatory instructions are also contained within noncoding regions of mRNAs.

Post-translational Regulation

Once polypeptides are synthesized, there are still many opportunities for regulatory events to occur, and post-translational regulation is of major importance to the cell. Enzymatic modifications to proteins might change their biological activities. We will see examples of this in Chapter 24, where we discuss how phosphorylation of certain amino acids can change the activity of a protein. Interactions between different polypeptides to form multiprotein complexes might also produce changes in the activities of the constituent polypeptides. For example, in the discussion of the *Drosophila* sex determination pathway toward the end of this chapter, we will see that transcription factor activity is

controlled through competitive interactions between different possible subunits to form dimeric proteins.

Epigenetic Inheritance

We now have a general view of regulation of protein synthesis that can account for most observations that geneticists have made over the last century. However, there are still some phenomena that still beg for explanation. These phenomena have generally been termed **epigenetic inheritance** because they seem to reflect heritable alterations in which the DNA sequence itself is unchanged. Indeed, it is likely that these phenomena reflect another, poorly understood, level of gene control. We will consider some examples of epigenetic inheritance in which the activity state of a gene depends on its genealogical history.

Paramutation

One such phenomenon is paramutation, described in several plant species, most notably in corn (Figure 23-6). Paramutation has been observed only at a few genes in corn. In this phenomenon, certain special but seemingly normal alleles, called *paramutable alleles,* suffer irreversible changes after having been present in the same genome as another class of special alleles, called *paramutagenic alleles.* The *B-I* gene in corn encodes an enzyme in the pathway of anthocyanin pigments in various tissues in corn. Ordinary null *b* alleles lack these pigments, and these *b* alleles are completely recessive to *B-I.* There is a special paramutagenic allele, called B′, that confers the ability to make only a small amount of anthocyanin pigment. In crosses of *B-I* to B′ homozygotes, the resulting heterozygotes are weakly pigmented, thus appearing indistinguishable from the B′ homozygous plants. This result would seemingly suggest that *B-I* is recessive to B′. If this simple explanation were true, self-crosses of these heterozygous plants would generate homozygous *B-I* individuals. However, instead, only B′ alleles appear in the next (and subsequent) generations, indicating that the *B-I* allele has been paramutated. Somehow, by virtue of having been exposed to the paramutagenic B′ allele by being in the same genotype for but a single generation, the *B-I* allele has been permanently crippled in its activity.

Parental Imprinting

Another example of epigenetic inheritance, discovered about 10 years ago in mammals, is called **parental imprinting.** In parental imprinting, certain autosomal genes have seemingly unusual inheritance patterns. For example, the mouse *Igf2* gene is expressed in a mouse only if it was inherited from the mouse's father. It is said to be maternally imprinted, since a copy of the gene derived from the mother is inactive. Conversely, the mouse *H19* gene is expressed only if it was inherited from the mother; *H19* is pa-

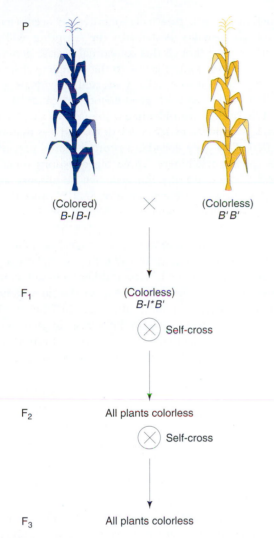

P

(Colored)
B-I B-I

×

(Colorless)
B' B'

F₁

(Colorless)
B-I*B'

⊗ Self-cross

F₂

All plants colorless

⊗ Self-cross

F₃

All plants colorless

Figure 23-6 A series of crosses depicting paramutation. The *B-I* mutation produces pigmented plants, whereas the *B'* mutation produces nearly unpigmented plants. Normally, when *B-I* is crossed to recessive colorless alleles of the *b* gene, the resulting plants are pigmented. However, when B-I and B' plants are intercrossed, the F_1 plants are essentially unpigmented, like the *B'* homozygotes. Thus, B-I is altered by being in the same genome as *B'*, indicated by the B-I* designation. If this were due simply to the dominance of *B'* to *B-I*, then a self-cross of the F_1 plants should generate *B-I* colored homozygotes as approximately 1/4 of the F_2 progeny. Instead, no F_2 are pigmented. Intercrosses of the F_2 and of further generations do not restore the pigmented phenotype. Thus, *B-I* is said to have been paramutated by virtue of being in the same nucleus with the B' allele.

ternally imprinted. The consequence of parental imprinting is that imprinted genes are expressed as if they are hemizygous, even though there are two copies of each of these autosomal genes in each cell. Furthermore, when these genes are examined at the molecular level, no changes in their DNA sequences are observed. Rather, the only changes that are seen are extra methyl (—CH₃) groups present on certain bases of the DNA of the imprinted genes. Occasional bases of the DNA of most higher organisms are methylated

(an exception being *Drosophila*). These methyl groups are enzymatically added and removed, through the action of special methylases and demethylases. The level of methylation generally correlates with the transcriptional state of a gene: active genes are less methylated than inactive genes. However, it is unknown if altered levels of DNA methylation cause epigenetic changes in gene activity, or if altered methylation levels arise as a consequence of such changes.

Imprinting and X-Inactivation. Some aspects of the phenomenon of X-inactivation are reminiscent of parental imprinting. Recall from Chapter 3 that X-inactivation is the way in which mammals cope with having differing numbers of X chromosomes in the two sexes. In organisms in which one sex is hemizygous for one of its chromosomes, a special mechanism called dosage compensation equalizes the gene activity of most X-linked genes in the two sexes. In female mammals, dosage compensation is accomplished by shutting off one of the X chromosomes, thereby maintaining only a single active X chromosome (Figure 23-7). The single X in an XY individual remains active. In an XX mammal, on the other hand, one of the two X chromosomes is inactivated and condensed into a highly compact Barr body (see Figure 3-25). In embryos of eutherian mammals, X-inactivation is a random event occurring early in development; that is, in XX individuals, the maternally derived X chromosome is inactivated in some cells, the paternally derived one in other cells (Figure 23-7). The mosaic hair pigmentation pattern in calico cats (see Figure 3-27) is due to random X-inactivation. However, in other instances, there is an imprinting event that determines which X chromosome is inactivated. In primitive (protherian) mammals, such as the kangaroo, it

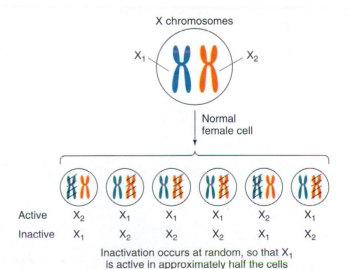

X chromosomes

X₁ — X₂

Normal
female cell

| Active | X₂ | X₁ | X₁ | X₁ | X₂ | X₁ |
| Inactive | X₁ | X₂ | X₂ | X₂ | X₁ | X₂ |

Inactivation occurs at random, so that X₁
is active in approximately half the cells

Figure 23-7 Inactivation of the X chromosome in mammals. X-inactivation occurs randomly in normal individuals. (Modified from J. D. Watson, M. Gilman, J. Witkowski, and M. Zoller, *Recombinant DNA*, 2d ed. Copyright © 1992 by Scientific American Books.)

is always the paternal X chromosome that is inactivated. Paternal X-inactivation also occurs in the extraembryonic tissues of all other (eutherian) mammals—the tissues that will form the membranes of the fluid-filled sacs that surround the fetus (see, for example, the discussion of amniocentesis in Chapter 15 and Figure 15-33). Thus, in these cases, the paternal X must somehow be identified in the zygote as different from the maternally derived X.

Parental Imprinting and Human Disease. Parental imprinting has been used to explain the seemingly bizarre genetic behavior of certain human diseases. Prader-Willi syndrome (PWS) is an inherited human disease in which affected individuals are characterized by short stature, obesity, small extremities, poor muscle tone, mental retardation, and other characteristic features. Typically, PWS patients have a paternally derived deletion for a specific part of human chromosome 15 (a region of three chromosome bands called *15q11-13*). Occasionally, PWS patients have a different abnormality of chromosome 15: unlike the usual situation where one copy of chromosome 15 comes from the mother and one from the father, both chromosome 15's in certain PWS patients are maternal in origin. Such an unusual circumstance must arise by a sperm lacking chromosome 15 (due to nondisjunction during spermatogenesis) fertilizing an ovum with two copies of chromosome 15 (due to independent nondisjunction during oogenesis).

Why do these two different abnormalities yield similar symptoms? It is thought that in all individuals there is a gene in chromosome 15q11-13 that is "maternally imprinted," meaning that the maternal copy (or copies, in the case of the second form of PWS) of the gene is inactivated, either during oogenesis or shortly after fertilization. Because of this, a normal phenotype requires that an individual inherit from the father a normal copy of the gene controlling the PWS trait (Figure 23-8). If a normal gene is paternally inherited, then the offspring is normal. If, on the other hand,

no functional gene is paternally inherited (as occurs in both types of abnormality described), the offspring will have PWS. Thus, even though this autosomal disease genetically behaves as a dominant, it is due to the complete absence of the PWS gene product in cells of affected individuals.

What might the PWS gene encode? A candidate gene called *SNRPN* has been identified. It maps to the 15q11-13 region, it is removed by PWS deletions, and it is maternally imprinted. This gene encodes a protein that is part of the mRNA-splicing machinery. Three big questions remain. Is the *SNRPN* gene actually the gene whose absence causes PWS? If so, how does the absence of *SNRPN* cause this particular genetic disease syndrome? Why is the *SNRPN* gene imprinted?

What do all these examples of epigenetic inheritance have in common? The main thread is that, somehow, a piece of a chromosome can be labeled as different based on its ancestry or on what other genes were in the same genome. For many of these examples, differences in DNA methylation have been associated with differences in gene activity. Nonetheless, the underlying mechanisms and rationales for why such systems evolved still seem rather mysterious.

Message   The ancestry of a gene can affect its ability to be expressed.

Dissecting Eukaryotic Regulatory Elements

In this chapter, we have seen that a gene's regulatory elements can come in many flavors, some acting to control aspects of DNA structure, to regulate transcription, to determine mRNA splicing patterns, and to determine properties of the mRNA such as its ability to be translated. An important part of modern genetics is the identification and char-

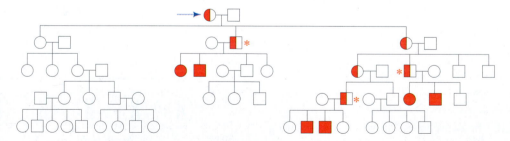

Figure 23-8 A pedigree depicting the inheritance of Prader-Willi syndrome (PWS). The individuals with PWS in this hypothetical family are denoted with the filled circles and squares. The fathers (noted with red asterisks) of the PWS offspring all derive from the first-generation mating shown in this pedigree. These fathers are heterozygous for the PWS mutant gene, and it is this mutant gene, lacking any normal gene activity, that they pass on to the PWS offspring. The unaffected sibs, on the other hand, inherit the normal allele of this PWS gene. In order to explain the inheritance of PWS in the pedigree, we must suppose that the PWS gene entered the pedigree through the female ancestor marked with the blue arrow.

acterization of such regulatory elements by means of transgenic constructs, in which recombinant DNA molecules are inserted into the genome of an organism (see Chapter 15). In these constructs, isolated pieces of a gene are incorporated to determine what tissue and temporal regulatory patterns it can control. By means of such slicing and dicing experiments, it is possible to home in on the locations of specific regulatory elements. We can also exploit these regulatory elements to develop new ways to identify genes of interest. In the following sections, we will discuss how such experiments are carried out, using studies of transcriptional enhancers as examples. Remember however that the same techniques and logic can be applied to all other classes of regulatory elements.

Using Reporter Genes to Find Enhancers

Enhancers of a cloned gene are typically identified by means of transgenic **reporter genes.** In reporter gene constructs, as described in Chapter 15, pieces of cis-regulatory DNA are fused (usually by restriction-enzyme-based cutting and pasting of recombinant DNA molecules) near a transcription unit that can express a reporter protein—that is, a protein whose presence can be monitored. This transcription unit contains a "weak" promoter—one that cannot initiate transcription without the assistance of an enhancer (Figure 23-9). The construct is then introduced by DNA transformation into the germ-line of a host organism and appropriate cells are histochemically assayed for the presence of the reporter protein (Figure 23-10). Our old friend, the *E. coli lacZ* gene, encoding β-galactosidase in the *lac* operon, is a very popular reporter gene. Reporter-protein

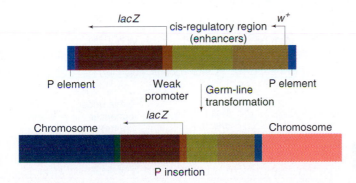

Figure 23-9 Use of a reporter-gene construct in *Drosophila* to identify enhancers. The top line represents a portion of a plasmid, bracketed by P element ends so that the material in between can be inserted into the genome by P element transformation (see Chapter 21). A region of DNA thought to contain one or more enhancers is inserted immediately adjacent to a "weak" promoter, that is, a promoter that by itself cannot initiate transcription. The promoter is joined to the *E. coli lacZ* structural gene (the reporter gene), which encodes β-galactosidase. If there are any enhancers in the construct, they will induce tissue-specific expression of *lacZ*. The embryos and imaginal disks in Figure 23-10 a, c, and d are examples of expression from such reporter-gene constructs.

expression in a tissue indicates the presence of an enhancer within the tested piece of cis-regulatory DNA. Once a piece of DNA has been found to act as an enhancer, the enhancer can be further localized by testing smaller and smaller pieces of the cis-regulatory DNA, using the same reporter-gene assay.

Ultimately, the DNA sequence of the enhancer can be identified by whittling down the piece of cis-regulatory DNA. With this sequence known, the next question of importance is the identity of the transcription factor proteins that bind to the enhancer. Methods now exist to identify enhancer-binding proteins and to clone the genes that encode these proteins. Once these genes are cloned, they can be characterized by genetic and molecular techniques. Using these approaches, it is possible to build detailed circuit diagrams of the genetic pathways that regulate gene expression in eukaryotes.

Message Reporter-gene techniques can be used to isolate individual regulatory elements of genes.

Gene Capture Using Enhancers

Knowing that regulatory elements exist allows us to identify genes in novel ways. For example, researchers studying the development of a specific organ might want to identify genes expressed in that organ but not in other tissues. For this purpose, we can exploit techniques that depend on the ability of certain cis-regulatory elements, such as enhancers, to act on adjacent genes. One of the most powerful such techniques is "enhancer trapping."

Enhancer traps are recombinant DNA constructs in which a weak promoter is connected to a gene whose expression will be assayed, such as the *E. coli lacZ* gene (Figure 23-11). However, unlike reporter-gene constructs, these enhancer trap constructs do *not* contain any cis-regulatory DNA. Rather, when these transgenic constructs are inserted into the genome, β-galactosidase will be expressed only if the construct "landed" near one or more enhancer elements. By histochemically staining a transgenic organism for β-galactosidase, the expression pattern conferred by nearby enhancers can be observed. Transgenes typically can insert anywhere in the genome, and so, by accumulating a great many independent transgenic lines, we can assay an entire genome for enhancer activities. Because the genome is replete with enhancers, many insertions will produce specific and reproducible tissue-specific patterns of expression (Figure 23-12). These patterns reflect the normal expression pattern of a nearby gene. Through recombinant DNA techniques, the neighboring gene that is normally regulated by these enhancers can be identified and analyzed. Enhancer trapping identifies genes based on their patterns of expression, without regard to the phenotypes elicited by any mutant alleles of the gene. For this reason, enhancer trapping has become a powerful complementary approach to standard mutagenesis techniques.

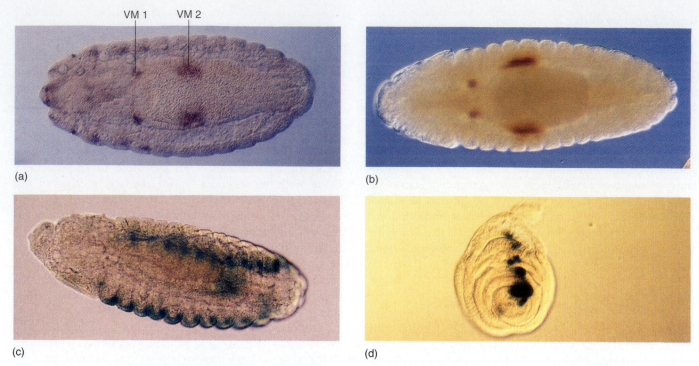

Figure 23-10 Some examples of the complex tissue-specific regulation of the *dpp* gene. In parts a, c, and d, the blue staining is due to a histochemical assay for *E. coli* β-galactosidase activity (the protein encoded by the *lacZ* reporter gene). The map positions of the *dpp* enhancers responsible for the staining patterns seen here are shown in Figure 23-3. (a) Reporter-gene assay for expression of the *Drosophila dpp* gene in two portions of the embryonic visceral mesoderm, the precursor of the gut musculature. The left-hand block of blue staining indicates the enhancer VM1, which drives anterior visceral mesoderm expression; the right-hand block of staining indicates the enhancer VM2 which drives posterior visceral mesoderm expression. (b) RNA in situ hybridization assay of *dpp* expression in the embryonic visceral mesoderm. Note that the blue reporter-gene expression pattern in part a is the same as the brown *dpp* RNA expression pattern shown here, confirming the reliability of the reporter-gene assay. (c) Reporter-gene expression driven by a different enhancer (LE) of *dpp* in the lateral ectoderm of an embryo. (d) Reporter-gene expression driven by ID, one of many enhancer elements driving imaginal disk expression of *dpp*. (An imaginal disk is a flat circle of cells in the larva that gives rise to one of the adult appendages.) A blue sector of *dpp* reporter-gene expression in a leg imaginal disk is shown. (Parts a and b courtesy of D. Hursh, part c courtesy of R. W. Padgett, and part d courtesy of R. Blackman and M. Sanicola.)

Figure 23-11 Use of an enhancer trap to capture tissue-specific genes in *Drosophila*. The essence of an enhancer trap is that it contains a weak promoter joined to a marker gene, in this case *lacZ*. In contrast to the reporter-gene construct (Figure 23-9), there are no enhancers within the transgene construct, and so the *lacZ* gene will be expressed only if the transgene has inserted near enhancers. Thus, insertions of the same transgene into different regions of the genome will be associated with different tissue patterns of *lacZ* expression. The triangle indicates the site at which the P transgene inserted in this example.

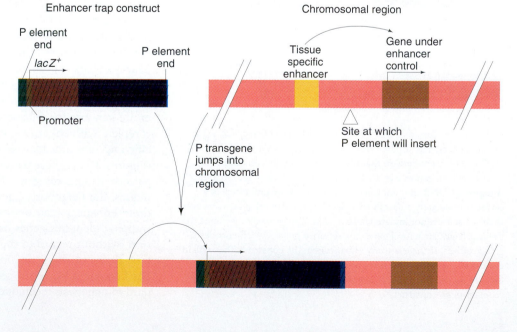

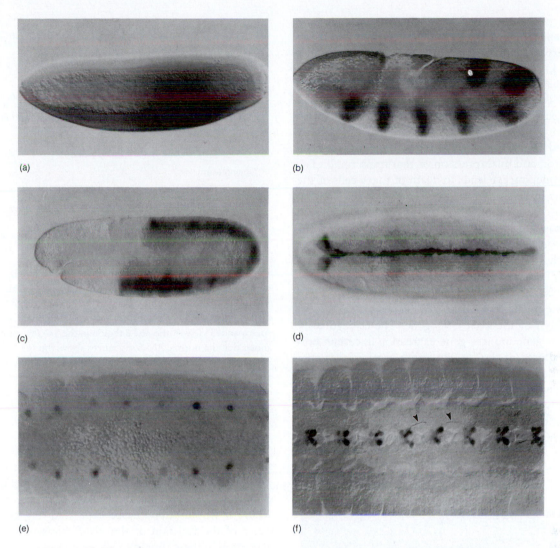

Figure 23-12 Examples of enhancer trap expression patterns during embryogenesis. (a–c) Patterns similar to expression patterns of genes involved in producing the segments of the larva. None of these genes had previously been identified through mutational screens. (d–f) Expression correlating with specific developmental steps in the central nervous system. (Y. Hiromi, Department of Molecular Biology, Princeton University.)

Message The properties of enhancers can be exploited to identify genes with interesting spatial expression patterns.

Regulatory Elements and Dominant Mutations

The properties of regulatory elements help us to understand certain classes of dominant mutations. We can think of dominant mutations in two general ways. In some cases, inactivation of one of the two copies of a gene reduces the gene product below some critical threshold for producing a normal phenotype; we can think of these as loss-of-function dominant mutations. In other cases, the dominant phenotype is due to some new property of the mutant gene, not to a reduction in its normal activity; these are **gain-of-function dominant mutations.**

A common origin of gain-of-function dominant mutations is through the fusion of parts of two genes to each other. Such fusions can occur at the breakpoints of chromosomal rearrangements such as inversions, translocations, duplications, or deletions (see Chapter 8). If these **gene fusions** join protein coding regions to each other, they can lead to the production of novel mutant proteins. Such fusion proteins can have special properties that cause dominant mutant phenotypes. In other cases, the cis-regulatory DNA of one gene is fused to the transcribed region of another gene. Because many cis-regulatory elements, such as enhancers, can act at long distance and can activate any promoter, this type of gene fusion places the transcription unit of the one gene under the control of the cis-regulatory DNA of the other gene. Often, this leads to the misexpression of the mRNA encoded by the transcription unit in question. Depending on the nature of the protein product of the misexpressed mRNA and the tissues in which it is misexpressed, such fusions can lead to gain-of-function dominant phenotypes. The classic *Bar* dominant mutation in *Drosophila* is an example of such a gene fusion. In the *Bar* mutation, cis-regulatory elements that promote expression in the developing eye are fused to a gene that is ordinarily not expressed in the eye. This latter gene encodes a transcription factor, and misexpression of that transcription factor in the developing eye leads to death of many cells of the eye and to the small eye Bar phenotype.

In a few cases, the basis for the misexpression in such gene fusions is quite well understood. One such case is the *Tab* (*Transabdominal*) mutation in *Drosophila*. *Tab* causes part of the thorax of the adult fly to develop instead as tissue normally characteristic of the sixth abdominal segment (A6) (Figure 23-13). *Tab* is associated with a chromosomal inversion. One breakpoint of the inversion is within the cis-regulatory DNA of the *sr* (*striped*) gene. The cis-regulatory DNA of the *sr* gene induces gene expression in certain portions of the thorax of the fly. The other breakpoint is near the transcription unit of the *Abd-B* (*Abdominal-B*) gene. The *Abd-B* gene encodes a transcription factor that normally is expressed only in posterior regions of the animal, and this *Abd-B* transcription factor is responsible for conferring an abdominal phenotype on any tissues that express it. (We'll have more to say about genes such as *Abd-B* in our discussion of homeotic genes in Chapter 25.) In the *Tab* inversion,

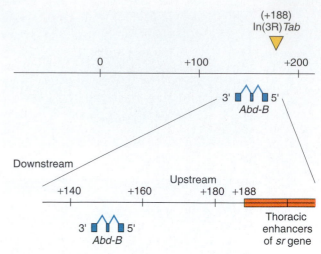

Figure 23-14 The *Tab* rearrangement. The top line is a molecular map of the chromosomal region of the *Abd-B* gene, with the coordinate units in kilobases. One inversion junction of *Tab*, just upstream of the Abd-B transcription unit, is indicated with the inverted triangle. An expanded view of the *Abd-B* region in the *Tab* inversion strain shows that the normal *Abd-B* sequences above the 188-kb position are replaced by sequences adjacent to the other breakpoint, which is in the *sr* gene. The *Tab* inversion brings the *Abd-B* transcription unit into contact with tissue-specific *sr* gene enhancers that drive expression in small blocks of the thorax.

the *sr* enhancer elements controlling thoracic expression are juxtaposed to the *Abd-B* transcription unit, causing the *Abd-B* gene to be activated in exactly those portions of the thorax where *sr* would ordinarily be expressed (Figure 23-14). Because of the function of the *Abd-B* transcription factor, its activation in these thoracic cells changes their fate to that of posterior abdomen. In this way, we can understand the molecular basis of a dominant mutation.

Gene fusions are an extremely important source of genetic variation. Through chromosomal rearrangements, novel proteins and novel patterns of gene expression can be generated. In addition to their developmental implications, as we will discuss in Chapter 24, such mutations can play a pivotal role in the formation and progression of certain cancers.

Message　The fusion of tissue-specific enhancers to genes not normally under their control can produce dominant mutant phenotypes.

Regulatory Cascades: Applying Regulatory Mechanisms to Developmental Decisions

In the previous sections, we have seen a variety of mechanisms that operate at different levels to control the production of specific proteins at the correct time, in the correct

Figure 23-13 The *Tab* mutation. The fly on the left is a wild-type male. The fly on the right is a *Tab*⁺ heterozygous mutant male. In the mutant fly, part of the thorax (the black tissue) is changed into tissue normally found in the dorsal portion of one of the posterior abdominal segments. The rest of the thorax is normal. (S. Celniker and E. B. Lewis, *Genes and Development* 1, 1987, 111.)

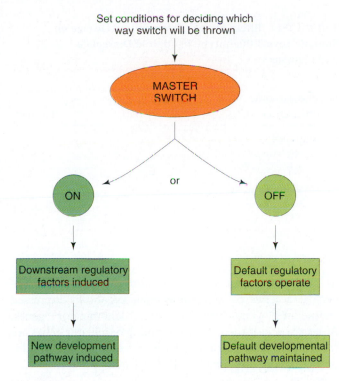

Set conditions for deciding which
way switch will be thrown

MASTER
SWITCH

or

ON   OFF

Downstream regulatory
factors induced

Default regulatory
factors operate

New development
pathway induced

Default developmental
pathway maintained

Figure 23-15 Decision making in developmental pathways. In developmental pathways, cells have to achieve different roles through a series of binary (on-off) decisions. Conditions within the cell allow a master switch to be regulated. Once the master switch is activated, it sets in motion a cascade of "downstream" regulatory events that eventually lead the cell down a new developmental pathway. In the absence of the activation of the master switch, a set of default signals remain in place and maintain the cell in the default pathway.

place, and in the correct amounts. Eukaryotes combine these regulatory devices in many different ways to control how cells develop and differentiate. A common scenario is that a series of regulatory events occur in sequential order and so form a **developmental pathway** (Figure 23-15). Part of such a pathway acts to set cellular conditions for making a crucial "on-off" decision for some key regulatory protein. Once these decisions are set, a switch is thrown that actually produces this active regulatory protein or that leaves it "off." Once this regulatory switch is thrown, then a cascade of other "downstream" regulatory proteins must be produced to execute the decision represented by that switch.

Developmental pathways can exploit all possible mechanisms that regulate the expression of an active protein. To exemplify the rich variety of mechanisms that have evolved to solve the problem of turning proteins on in the right place and the right time, we will look at three pathways. Two are the strikingly different solutions that *Drosophila* and mammals have taken to differentiate the two sexes. The third example comes from the vertebrate immune system, and explores the very elegant mechanism by which antibody production by B lymphocytes is regulated. In *Drosophila*, sex determination involves an interplay of regulation at the levels of transcription and mRNA splicing. In

mammals, sex determination largely relies on regulation of one transcription factor by another. In the antibody production pathway, antibody genes are regulated by special DNA sequence rearrangements in the context of transcriptional control of these genes to generate the diversity of antibodies required to sustain an immune response.

The Sex Determination Pathway in *Drosophila*

The presence of two sexes is quite a common theme in the biological world. In many organisms, sex determination is associated with the inheritance of a heteromorphic chromosome pair in one sex. XX-XY sex determination mechanisms appear to have arisen independently many times in evolution. Thus, molecular mechanisms underlying XX-XY sex determination can be quite different, as we will see by a comparison of *Drosophila* and mammals. In both flies and mammals, many areas of the body differentiate sexually dimorphic characteristics; that is, these areas differ in males and in females. For example, in *Drosophila*, the two sexes differ in structure of the sex organs themselves and in pigmentation of the abdomen. The male has special bristles on the foreleg that are lacking in females. Such secondary sexual characteristics will be the main focus of the discussions that follow.

Sexual differentiation in *Drosophila* requires several sequential regulatory steps (Figure 23-16). Let's look first at an overview of the pathway, and then examine it in detail.

The Basics of the Regulatory Pathway

The ratio of X chromosome to sets of autosomes (the X:A ratio) in the early embryo establishes whether a fly will become male or female. The pathway converting this ratio into phenotypic sex has been elegantly worked out by Thomas Cline, Bruce Baker, and their colleagues. This directive for sexual differentiation is carried out by a master regulatory switch and several downstream sex-specific genes, in a pathway that involves differential RNA splicing to produce different transcription factors for males and for females.

In the first step of the pathway, each cell in the early embryo calculates its X:A ratio, which is reflected in the concentration of certain X chromosome–encoded transcription factors. If the concentration is high enough (as in XX embryos, destined to become females), the gene serving as the master switch is turned on. If the concentration is too low, then the switch remains off. If the switch is on, RNA splicing regulates the production of a downstream transcription factor that represses male-specific genes. If the switch is off, the splicing pattern of the downstream transcription factor produces an alternative mRNA. The alternative mRNA produces a form of this transcription factor that can repress only female-specific genes.

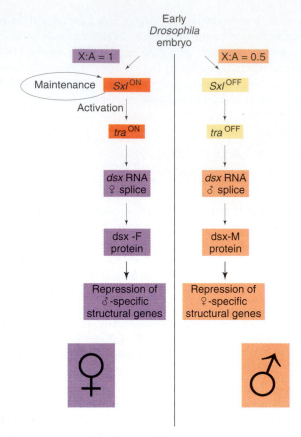

Figure 23-16 The pathway of sex determination and differentiation in *Drosophila*. The X:A ratio (described in the next section) is evaluated through the interaction of numerator and denominator transcription factors. This determines whether *Sxl* is to be permanently turned on or is to remain off. If *Sxl* is on, then the female sex differentiation pathway is turned on, ultimately causing splicing of a form of the *dsx* mRNA that produces a transcription factor that represses male-specific genes. If *Sxl* is off, then the sex differentiation pathway is not activated and the default *dsx* splicing pattern creates an mRNA encoding a female-repressing transcription factor.

The Role of the X:A Ratio

In *Drosophila,* the chromosomal basis of sex determination is due to the ratio of X chromosomes to *sets* of autosomes. (Recall that in *Drosophila,* $n = 4$—one sex chromosome and three different autosomes. Hence, one autosomal set, which we'll represent as A, comprises the three different autosomes, and in a diploid individual $A = 2$.) This X:A ratio effect can best be appreciated by examining sex chromosome aneuploids (see Table 23-1). A normal 2X *Drosophila* diploid (XX AA) has an X:A ratio of 1.0 and is phenotypically female. An XY diploid (XY AA) has an X:A ratio of 0.5 and is male; an XO diploid is also male (although sterile). Triploids with three X chromosomes (XXX AAA) are females, those with one X (XYY AAA) are male, and those with two X's (XXY AAA) are "in between" (intersexes).

The X:A ratio is established by an interaction of proteins made in the mother's ovary and in the early zygote. Genetic observations suggest that the molecular basis of the

Table 23-1 Effect of Sex Chromosome Dosage on Somatic Sexual Phenotype in Diploid *Drosophila* and Humans

Sex chromosome constitution	Somatic sexual phenotype	
	Drosophila	Humans
Euploidy		
XX	♀	♀
XY	♂	♂
Aneuploidy		
XXY	♀	♂
XO	♂	♀

X:A ratio is the relative concentrations of the protein products of a series of X-chromosomal, zygotically expressed "numerator" genes and autosomal "denominator" genes. At least some of the numerator and denominator genes are transcription factors of the basic helix-loop-helix (bHLH) protein family (see Chapter 18 for a description of bHLH proteins).

Two key features of bHLH proteins are important in thinking about how they might contribute to measuring the X:A ratio. First, it appears that all bHLH proteins function as dimers—proteins made up of two polypeptide chains. They can form homodimers of two identical polypeptide chains or heterodimers with other bHLH polypeptides. Sec-

Figure 23-17 (facing page) The initiation and maintenance of the *Sxl* switch. (a) A plausible mechanism for the molecular basis of the X:A ratio. During early embryogenesis, the numerator and denominator genes are expressed. Numerator polypeptide subunits (red circles) encoded by X chromosome genes and denominator polypeptide subunits (green circles) encoded by autosomal genes form dimers at random. Only numerator-numerator dimers form active transcription factor. When the X:A ratio is 1.0, high levels of these numerator-numerator dimers form. When the X:A ratio is 0.5, most numerator subunits dimerize with the more abundant denominator subunits. (b) The *Sxl* switch. High levels of the numerator-numerator transcription factor dimers activate transcription from the early promoter (P_E) of *Sxl.* (Activated promoters are represented by solid red or blue rectangles, inactivated promoters by clear rectangles.) Later in embryogenesis, P_E is turned off, and *Sxl* transcription occurs in all animals from the late promoter (P_L). Through binding to the primary P_L transcript and regulating its splicing pattern (preventing the male exon from being included in the final mRNA), preexisting active Sxl protein ensures the further production of active Sxl protein, which continues the splicing pattern leading to its own formation, thus creating an autoregulatory loop. On the other hand, when the X:A ratio is 0.5, then P_E is not activated. Thus, no Sxl protein is present in the early embryo, and the RNA splicing pattern of the P_L transcript generates an mRNA that includes the male exon. This exon contains a stop codon (UGA), terminating the Sxl polypeptide prematurely. The short Sxl protein made in males is completely nonfunctional. The AUG denotes the location of the translation initiation codon for the Sxl polypeptide.

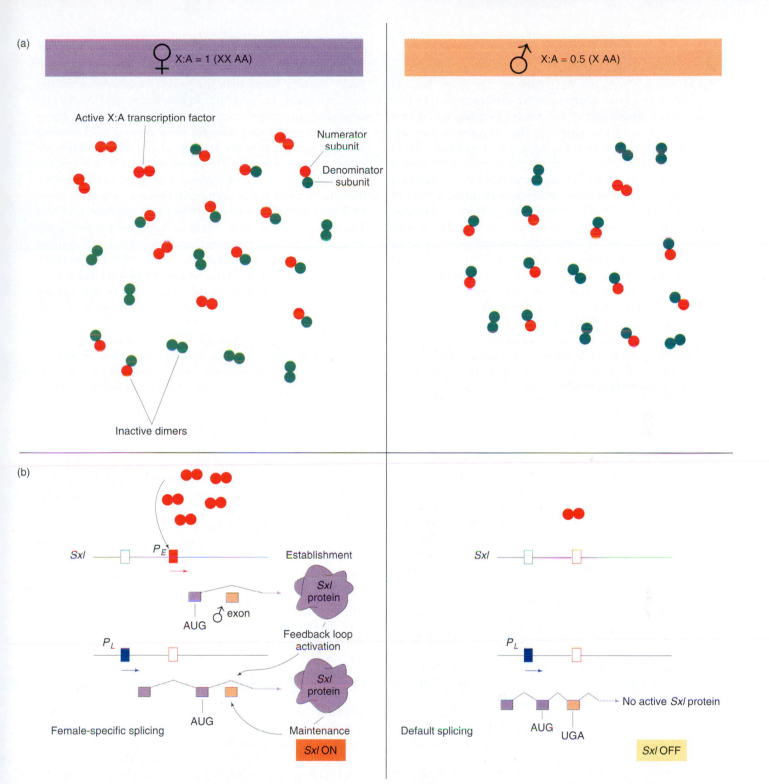

(a)

♀ X:A = 1 (XX AA)

Active X:A transcription factor

Numerator subunit

Denominator subunit

Inactive dimers

♂ X:A = 0.5 (X AA)

(b)

Sxl P_E — Establishment

Sxl protein

AUG ♂ exon

Feedback loop activation

Sxl protein

P_L

Female-specific splicing — AUG — Maintenance

Sxl ON

Sxl

P_L

No active *Sxl* protein

Default splicing — AUG UGA

Sxl OFF

ond, some polypeptides of the bHLH family lack parts of the protein necessary for DNA binding. Nevertheless, these proteins can dimerize with other bHLH proteins. These bHLH polypeptides lacking DNA-binding activity are thought to act as inhibitors of the bHLH proteins that do have binding activity, because in heterodimers between bHLH proteins of these two classes, the subunit without DNA-binding activity poisons the ability of the other subunit to bind effectively to its DNA targets.

The bHLH proteins very likely measure X:A ratio by competing with one another to form complexes that are ac-

tive transcription factors (Figure 23-17a). While we don't know for sure how this works, here is a plausible mechanism. First, certain numerator genes (on the X chromosomes) encode bHLH proteins with DNA-binding activity, whereas certain denominator genes (on the autosomes) encode bHLH proteins without DNA-binding activity. Second, each of the bHLH polypeptides is synthesized proportionately to the number of copies of each bHLH gene in the cell. In this way, embryos with an X:A ratio of 1.0 have twice as much numerator bHLH polypeptide as do embryos with an X:A ratio of 0.5. Third, all possible combinations of bHLH homodimers and heterodimers can form. Fourth, only numerator homodimers can act as transcription factors to turn on transcription of the master switch gene *Sxl* (see the next section of this chapter). The outcome of this scenario is that the higher the level of numerator bHLH polypeptide that is made (relative to denominator bHLH polypeptides), the more numerator bHLH homodimer that accumulates. Thus, in early embryos with an X:A ratio of 1.0, we can expect that considerably more numerator bHLH homodimers will accumulate than in embryos with an X:A ratio 0.5. Not all numerator and denominator genes encode bHLH proteins, but it is likely that this general mechanism, in which there is a competition between functional and poisoning subunits to form multimeric protein complexes, will be able to explain the contributions of other proteins to evaluating the X:A ratio.

Message   Protein-protein interactions, such as competition between several subunits for dimer formation, can serve as the triggers for controlling developmental switches.

The *Sxl* Gene: The Master Switch

The transcription factors such as the bHLH proteins produced in response to the X:A ratio appear to have only one role: to determine if a master regulatory switch gets flipped on. The *Sxl* gene is essentially a toggle switch that is permanently locked into an "on" position in females or an "off" position in males (see Figure 23-17b). To set the *Sxl* switch in the "on" position, the level of the active X:A transcription factors must be high (due to an X:A ratio of 1.0). With high (female) levels, the X:A transcription factors present in the early embryo bind to enhancers of the *Sxl* gene, activating its transcription from the *Sxl* early promoter. The transcript made from the early promoter then produces active *Sxl* protein. If the levels of these bHLH factors are too low, as is the case when the X:A ratio is 0.5, then there is insufficient transcription factor to activate *Sxl* transcription and, no *Sxl* protein is made.

Message   The essential question the cell faces in sex determination is very much like Hamlet's: To transcribe (*Sxl*) or not to transcribe?

For development to proceed along the female pathway, *Sxl* protein not only has to be made in the early embryo, but must continue to be made throughout the remainder of the lifetime of the fly. After early embryogenesis, the X:A transcription factors disappear and the *Sxl* early promoter is shut off. How then can *Sxl* transcription remain on in animals with an X:A ratio of 1.0? This is accomplished by requiring that a cell already have *Sxl* protein in order to produce more *Sxl* protein (Figure 23-17b). After early embryogenesis, constitutive *Sxl* transcription occurs from the late promoter. However, the primary transcript produced by transcription of the *Sxl* gene from the late pro-

moter is much larger than the primary transcript from the early promoter, and is subject to alternative mRNA splicing, depending on the presence or absence of preexisting active *Sxl* protein in the cell. The *Sxl* protein is an RNA-binding protein that can alter the splicing of the nascent *Sxl* transcript coming from this late promoter. The female-specific (X:A ratio of 1.0) splicing pattern of *Sxl* produces an mRNA that makes more active *Sxl* protein that is functionally identical with that produced from the early promoter mRNA. This *Sxl* protein, in turn, binds to more *Sxl* primary transcript from the late promoter, ensuring that the prevailing splicing pattern will produce the mRNA that makes more *Sxl* protein. Thus a feedback, or **autoregulatory loop,** controlled at the level of RNA splicing, maintains *Sxl* activity throughout development in individuals with an X:A ratio of 1.0.

Message   The autoregulatory loop explains how an early developmental decision can be "remembered" for the rest of development, even after the initial signals that established the decision have long disappeared.

In contrast, when the X:A ratio is 0.5, the *Sxl* switch is set in the "off" position. The early promoter is not activated early in embryogenesis and hence the early X:A = 0.5 embryo has no *Sxl* protein. As a consequence, in the absence of any active *Sxl* protein, the primary *Sxl* transcript of the late, constitutive *Sxl* promoter is processed in the "default" mRNA splicing pattern. This default *Sxl* mRNA is nonfunctional, in the sense that it encodes a stop codon shortly after the translation-initiation codon of its protein-coding region. The small protein produced from this male-specific spliced mRNA has no biological activity. Thus, in individuals with an X:A ratio of 0.5, the initial absence of active *Sxl* protein early in development predestines that *Sxl* will remain off throughout the remainder of development.

Controlling Sex-Specific Gene Expression

Once the decision is made to establish continuous *Sxl* protein production through the *Sxl* autoregulatory loop (or not to), this decision must then be communicated downstream to the genes that actually have to regulate the sexual phenotypes. This is accomplished through sequential sex-specific mRNA splicing events. The *Sxl* protein not only controls its own splicing, it also initiates the female differentiation pathway by altering the splicing of the *tra* (*transformer*) gene transcript (Figure 23-18a). As with *Sxl*, the female-specific *tra* mRNA encodes an active RNA-binding protein, while the male-specific mRNA encodes an inactive product. The active *tra* protein controls the female-specific splicing of the final gene transcript in the pathway, *dsx*. In the absence of active *tra* protein, by default the male-specific splice of *dsx* mRNA occurs. Both the female-specific and the male-specific spliced versions of the nascent *dsx* transcript produce proteins that have biological activity, which we can call

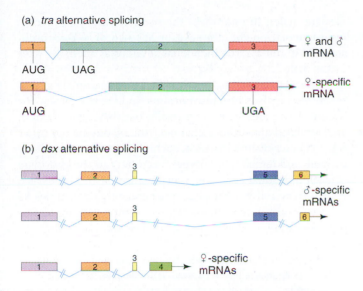

(a) *tra* alternative splicing

♀ and ♂ mRNA

♀-specific mRNA

(b) *dsx* alternative splicing

♂-specific mRNAs

♀-specific mRNAs

Figure 23-18 Alternative splicing of *tra* and *dsx* transcripts. (a) Two forms of *tra* mRNA are produced. One occurs in both sexes and, because of a stop codon (UAG) in exon 2, does not code for a functional protein. The other is female-specific and does code for the active *tra* polypeptide. (b) Different *dsx* mRNAs are produced in the two sexes. In males, exon 4 is not included in two related male-specific mRNAs. Both males mRNAs make related polypeptides dsx-M that act to repress transcription of female-specific genes. In females, exons 5 and 6 are not included in the *dsx* mRNA. The dsx-F polypeptide that is produced acts to repress transcription of male-specific genes. (Modified from M. McKeown, *Curr. Opinion Genet. Dev.* 2, 1992, 301.)

dsx-F and dsx-M proteins, respectively (Figure 23-18b). Without regulation by these dsx transcription factors, the downstream sex-specific genes would be constitutively transcribed. The dsx proteins intervene to repress the transcription of some of these sex-specific genes. Thus dsx-F protein represses transcription only of male-specific genes, thereby conferring a female phenotype. Conversely, dsx-M protein represses transcription only of female-specific genes, thereby conferring a male phenotype.

Mutations Affecting Sex Determination

The insights into *Drosophila* sex determination have emerged through molecular and genetic analysis of mutations altering the phenotypic sex of the fly. Now that we have discussed the pathway, we can briefly describe the phenotypic effects of mutations in the key genes in the pathway. In terms of the sex differentiated by mutant animals, the effects of null mutations in several of the genes in the pathway are to transform females into phenotypic males. Males homozygous for these mutations are completely normal. These include genes such as *sis-b* (sisterless-b), *Sxl* (Sex-lethal), and *tra* (transformer). These genes are dispensable in males because the male developmental pathway seems to be the default state of the developmental switch. In other words, the sex determination pathway in *Drosophila* is con-

structed so that the activities of several gene products are needed to shunt the animal from the default state into the female developmental pathway. The *sis-b* gene, a numerator gene encoding a bHLH protein, must be active to achieve an X:A ratio of 1.0. The mRNA splicing regulators, the RNA-binding proteins encoded by the *Sxl* and *tra* genes must be active for female development to ensue. They are ordinarily off in males anyway, so it is of no consequence to male development to have mutations knocking out these genes' functions. The exceptional gene is *dsx* (*doublesex*). The knockout of the *dsx* gene leads to the production of flies that simultaneously have male and female attributes. The reason for this is that each of the two alternative *dsx* proteins, dsx-F and dsx-M, represses the gene products that produce the phenotypic structures characteristic of the other sex. In the absence of repression, the gene products that build the structures characteristic of each of the two sexes operate simultaneously, and a fly that is simultaneously male and female develops.

Thus, we have seen in the *Drosophila* sex determination pathway an interplay at several different levels of control. Protein-protein interactions between transcription factors, reflecting the chromosomal X:A ratio, trigger a cascade of events, beginning with the major switch decision—whether or not to transcribe the *Sxl* gene early in development. This decision leads to the constitutive production of the *Sxl* RNA-binding protein that regulates the pattern of splicing of the *Sxl* late transcript. The continuous presence of active *Sxl* protein is then assured through the *Sxl* autoregulatory loop. The *Sxl* RNA-binding protein regulates *tra* mRNA splicing, which leads to the production of *tra* RNA-binding protein. The *tra* RNA-binding protein also controls mRNA splicing, in this case of *dsx* transcripts, leading to the production of the form of the *dsx* protein, dsx-F, that represses transcription of male-specific genes, thereby producing a female phenotype. In contrast, in the absence of active *Sxl* protein, no active *tra* protein is produced, and the default version of the *dsx* protein, dsx-M, represses female-specific transcription, producing a male phenotype.

Sex Determination in Mammals

Originally, because mammals and flies had a similar sex chromosome dimorphism, it was assumed that the X:A ratio mechanism would apply to mammals as well. However, once sex chromosome aneuploids were recognized, it was clear that the basis of mammalian sex determination and differentiation was completely different (Table 23-1). In humans, XXY individuals are phenotypically male, with a syndrome of moderate abnormalities (Klinefelter syndrome; see Figure 9-20). By comparison, XXY flies are phenotypically normal females. XO humans have a number of abnormalities (Turner syndrome; see Figure 9-17), including short stature, mental retardation, and mere traces of gonads, but they are clearly female in morphology. By comparison, XO

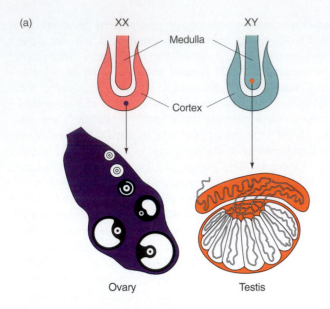

(a)

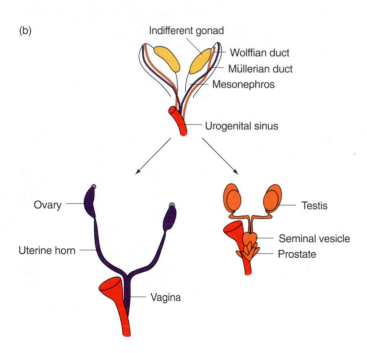

(b)

Figure 23-19 Development of the mammalian urogenital system.
(a) The embryonic genital ridge consists of a medulla surrounded by a
cortex. Female germ cells migrate into the cortex and become organized into an ovary. Male germ cells migrate into the medulla and become organized into a testis. (b) In the initial urogenital organization
at the indifferent gonad stage, precursors of both male (Wolffian) and
female (Müllerian) ducts are present. If a testis is present, it secretes
two hormones, testosterone and a polypeptide hormone called Mül-
lerian-inhibiting substance (MIS, or anti-Müllerian hormone). This
causes the Müllerian ducts to regress and the Wolffian ducts to develop into the male reproductive ducts. If an ovary is present, testosterone and MIS are absent and the opposite happens: the Wolffian
ducts regress and the Müllerian ducts develop into the female reproductive ducts. (From Ursula Mittwoch, *Genetics of Sex Differentiation.*
Copyright © 1973 by Academic Press.)

flies are males. In mammals, the sex determination mechanism is based on the presence or absence of a Y chromosome. Without a Y, the individual develops as a female; with
it, as a male.

Another difference between flies and mammals is that
sex in mammals is **nonautonomous.** Individuals mosaic for
XX and XY tissues arise occasionally, but they typically have
a generalized appearance characteristic of one or the other
sex. In *Drosophila,* by contrast, such mosaics show a mixture
of male and female phenotypes, according to the genotype
of each individual cell. The interpretation of this difference
is that every cell in *Drosophila* independently determines its
sex, whereas in mammals, sexual phenotypes are determined by some sort of global instruction mechanism.

Message Mammalian sex determination and differentiation are established by a pathway quite different from that in
Drosophila.

Mammalian Reproductive Development

The observation of nonautonomy in mammalian sex determination can be understood in view of the biology of the
reproductive system: secondary sexual phenotypes are driven by the presence or absence of the testes.

The gonad forms within the first two months of human gestation. Primordial germ cells migrate into the genital ridge, which sits atop the rudimentary kidney. The chromosomal sex of the germ cells determines whether they
will migrate superficially or deeply into the gonadal ridge
and whether they will organize into a testis or an ovary (Figure 23-19). If they form a testis, the Leydig cells of the testis
secrete testosterone, an androgenic (male-determining)
steroid hormone. This hormone binds to androgen receptors. These receptors are cytoplasmic proteins that can act
as transcription factors; their transcription factor activity
however depends on binding to their cognate hormone.
Thus, the androgen-receptor complex binds to androgen-re-
sponsive enhancer elements, leading to the activation of
male-specific gene expression. In chromosomally female individuals, no Leydig cells form in the gonad, no testosterone
is produced, androgen receptor is not activated, and individuals continue along a female pathway of development.
Hence, it is the presence or absence of a testis that determines the sexual phenotype, through the endocrine release
of testosterone. Indeed, in XY individuals lacking the androgen receptor, development proceeds along a completely female pathway even though these individuals have testes.
Mutations in the androgen receptor are the genetic basis for
testicular feminization (Tfm) syndrome (see Figure 3-22).

Role of the Y Chromosome
in Sex Determination

What initiates the sex determination pathway? Molecular
genetic analysis has focused on identifying the locus on the

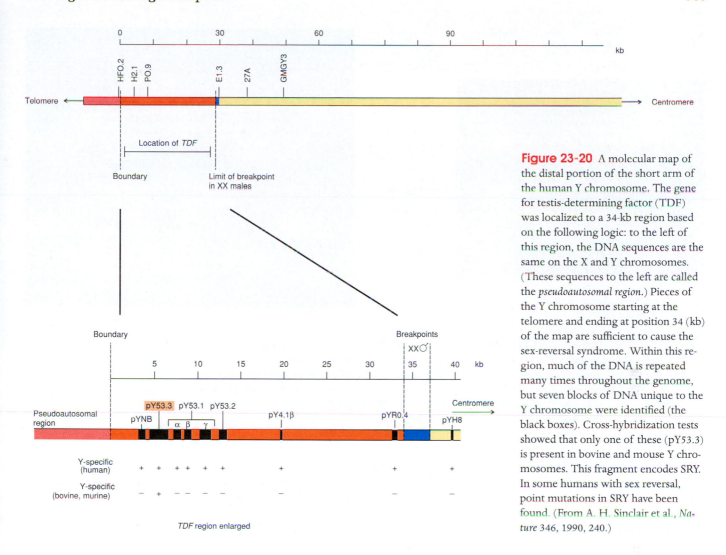

Figure 23-20 A molecular map of the distal portion of the short arm of the human Y chromosome. The gene for testis-determining factor (TDF) was localized to a 34-kb region based on the following logic: to the left of this region, the DNA sequences are the same on the X and Y chromosomes. (These sequences to the left are called the *pseudoautosomal region*.) Pieces of the Y chromosome starting at the telomere and ending at position 34 (kb) of the map are sufficient to cause the sex-reversal syndrome. Within this region, much of the DNA is repeated many times throughout the genome, but seven blocks of DNA unique to the Y chromosome were identified (the black boxes). Cross-hybridization tests showed that only one of these (pY53.3) is present in bovine and mouse Y chromosomes. This fragment encodes SRY. In some humans with sex reversal, point mutations in SRY have been found. (From A. H. Sinclair et al., *Nature* 346, 1990, 240.)

Y chromosome that drives testis formation. This hypothetical gene has been called the testis-determining factor on the Y chromosome (*TDF* in humans, *Tdy* in mice, and is now known to be the same gene as the *SRY/Sry* gene). *TDF* and *Tdy* have been identified through mapping and characterization by Robin Lovell-Badge and Peter Goodfellow of a genetic syndrome common to mice and humans that almost certainly affects this factor (Figure 23-20). This syndrome is called *sex reversal*. Sex-reversed XX individuals are phenotypic males and have been shown to carry a fragment of the Y chromosome in their genome. In general, these Y chromosome duplications arise by an illegitimate recombination between the X and Y chromosomes that fuses a piece of the Y chromosome to a tip of one of the X chromosomes. The portion of the Y chromosome that includes these duplications has been cloned in a **chromosome walk** (see Figure 14-17). Lovell-Badge and Goodfellow have identified from this region a transcript that is expressed in the appropriate location of the developing kidney capsule.

The gene encoding this transcript was named the *sex reversal on Y gene* (*SRY* in humans, *Sry* in mice), since it was identified on the basis of the sex reversal syndrome, but it is almost certainly the same gene as *TDF/Tdy*. Lovell-Badge and Goodfellow used a transgene to provide spectacular evidence in support of this identity (Figure 23-21). A cloned 14-kb genomic fragment of the mouse Y chromosome, including the *Sry* gene, was inserted into the mouse genome by germ-line transformation. An XX offspring containing this inserted *Sry* DNA (the transgene) was completely male in external and internal phenotype and, as predicted, possessed the somatic tissues of the testis—including the Leydig cells that make testosterone. (It should be noted, however, that this mouse was sterile. The sterility is probably a consequence of having two X chromosomes in a male germ cell, since XXY male mice are similarly sterile.) Thus, a single genetic unit was directly shown to alter the mammalian sexual phenotype profoundly, completely consistent with the role of *SRY/Sry* as the gene that determines testis development.

How does *SRY/Sry* contribute to sex determination? The SRY/Sry protein appears to be a transcription factor. The SRY/Sry protein has a polypeptide sequence, called an HMG (for high-mobility-group) domain, that is found in several other transcription factors. Some sex-reversed XY morphological females have point mutations in the HMG

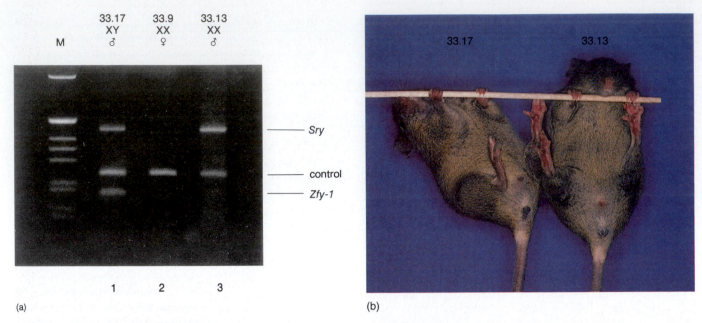

(a)

(b)

Figure 23-21 A transgenic mouse that proves *Sry* can cause the sex-reversal syndrome. (a) Amplification of genomic DNA by the polymerase chain reaction (PCR) shows that mouse 33.13 lacks a DNA marker for the presence of a Y chromosome (Zfy-1), but that it does contain the *Sry* transgene. M, marker bands. (b) The external genitalia of sex-reversed XX transgenic mouse 33.13 are indistinguishable from those of a normal XY male sib (33.17). (From P. Koopman, J. Gubbay, N. Vivian, P. Goodfellow, and R. Lovell-Badge, *Nature* 351, 1991, 117.)

domain of SRY/Sry; while recombinant SRY/Sry protein made from normal males is a strong DNA-binding protein, recombinant protein made from the *SRY/Sry* genes of these sex-reversed individuals shows reduced or no DNA-binding activity. Exactly how the SRY/Sry protein initiates testis formation is not understood. However, with the SRY/Sry protein sequence in hand, many avenues for answering this and other age-old questions about the biological basis of sexual phenotype can be pursued.

> **Message**   In mammals, a Y chromosome gene encodes a transcription factor that causes the gonad to become a testis. The testis produces androgens that are secreted throughout the body and activate the androgen receptor transcription factor.

The Pathway of Antibody Production: DNA Gene Rearrangement and Antibody Diversity

We have seen in the previous sections that regulation of protein activity at several different levels can contribute to the control of developmental pathways. Perhaps one of the most intricate pathways involves the vertebrate immune system, where many special needs of the system have been solved through the evolution of novel regulatory mechanisms. In this section, we will focus on one aspect of immune system development, namely, the elaborate changes

in DNA structure that are programmed as part of the process of building antibody proteins.

Antibodies and the Immune System

The immune system of a vertebrate organism provides the body's main line of defense against invasion by such disease-causing organisms as bacteria, viruses, and fungi; it probably also carries out surveillance for cancer cells and participates in their elimination. The immune system consists of many cell types that act in concert to produce a robust response to the presence of foreign substances (Figure 23-22).

Plasma cells and their precursors, B lymphocytes (B cells), are the cells of the immune system that make **antibodies,** or **immunoglobulins,** the proteins that recognize and bind foreign substances, called **antigens.** Each antibody is equipped to recognize a specific stereochemical shape that characterizes a particular antigen; this is known as *antibody specificity.* Antigens may be an entire molecule or a part of a molecule, and most large organic molecules can be antigenic. Hence, there are millions of potential antigens. The immune system must be able to recognize all these diverse substances. The immune system accomplishes this by making vast numbers of different antibody molecules, each with the ability to bind a different antigen.

An antibody molecule is made up of two identical subunits, each containing a light chain and a heavy chain, bound together by disulfide bridges (Figure 23-23). The major stem regions of all light-chain polypeptides are very sim-

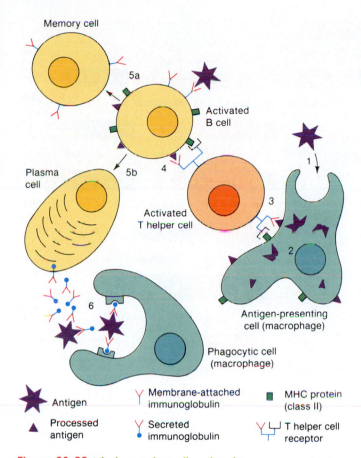

Antigen

Processed antigen

Y **Membrane-attached immunoglobulin**

Y **Secreted immunoglobulin**

■ **MHC protein (class II)**

Y **T helper cell receptor**

Figure 23-22 The humoral (B-cell-mediated) immune response to an infection. (1) The antigen (the infecting agent) is taken up by a cell such as a macrophage. (2) This cell processes the antigen, breaking it down into components, which are then displayed on the surface of the cell by the MHC class II protein. (3) The antigen is recognized by a T helper cell, which in turn (4) activates B cells that are also carrying pieces of the antigen. The activated B cells either (5a) differentiate into memory cells, which respond in future infections caused by the same agent, or (5b) become plasma cells that secrete antibody. (6) The antibodies bind to the antigen, thereby forming a complex that is destroyed by macrophages. (From S. Tonegawa, "The Molecules of the Immune System." Copyright © 1985 by Scientific American, Inc. All rights reserved.)

ilar in amino acid sequence; the same is true of the heavy chains. These sequences of similarity are called **constant regions.** At the amino-terminal ends of the antibody chains, there are variable regions, where the amino acid sequences differ from antibody to antibody, and within these variable regions, there are specific domains that are very different among different chains: the **hypervariable regions.** Antigens bind to antibodies within the variable regions of the antibody molecules, with the variable region of the heavy and the light chain both contributing to the antigen-binding capacity and to the differences in binding between different antibody molecules. Thus, thousands of different light and heavy chains can be mixed and matched to produce millions of different antibody molecules, each with unique antigen-binding properties.

The Genetic Source of Antibody Diversity

Each B lymphocyte makes only a single antibody molecule, and each different B lymphocyte makes a different antibody. This presents a genetic puzzle. An antibody's specificity reflects the primary amino acid sequences of its heavy- and light-chain subunits, which in turn are reflections of the DNA sequences that encode each unique antibody gene. However, if each antibody gene were present in every cell in the body, the vertebrate genome would have to be much larger than it is to accommodate all the genes encoding the thousands of different heavy- and light-chain proteins required by the immune system. In actuality, only a small number of genes actually code for all the millions of different antibody proteins: two for the light chains (the κ and λ light-chain genes) and one for the heavy chain. How do three genes generate such a vast number of different protein products?

The key is that separate DNA segments, each present in multiple versions but distinct in sequence from one another, code for different parts of the variable region of each chain

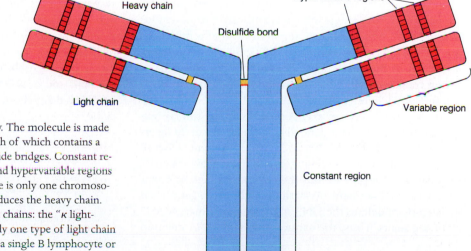

Figure 23-23 The structure of an antibody. The molecule is made up of two identical pairs of polypeptides, each of which contains a light chain and a heavy chain linked by disulfide bridges. Constant regions are located in the stem area; variable and hypervariable regions are located at the ends of the branches. There is only one chromosomal region, the "heavy-chain gene," that produces the heavy chain. Two chromosomal regions can produce light chains: the "κ light-chain gene" and the "λ light-chain gene." Only one type of light chain and one type of heavy chain are produced in a single B lymphocyte or plasma cell.

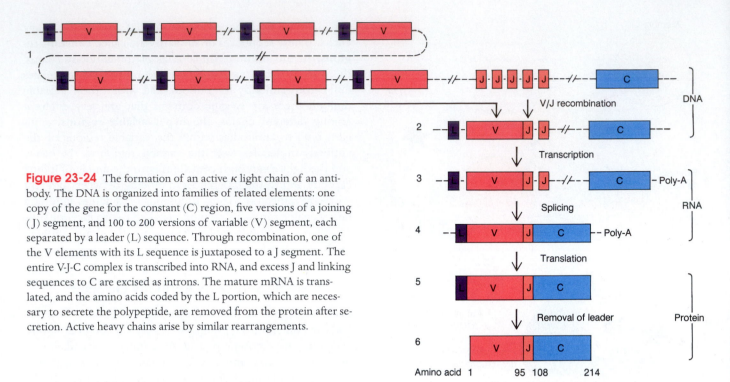

Figure 23-24 The formation of an active κ light chain of an antibody. The DNA is organized into families of related elements: one copy of the gene for the constant (C) region, five versions of a joining (J) segment, and 100 to 200 versions of variable (V) segment, each separated by a leader (L) sequence. Through recombination, one of the V elements with its L sequence is juxtaposed to a J segment. The entire V-J-C complex is transcribed into RNA, and excess J and linking sequences to C are excised as introns. The mature mRNA is translated, and the amino acids coded by the L portion, which are necessary to secrete the polypeptide, are removed from the protein after secretion. Active heavy chains arise by similar rearrangements.

(Figure 23-24). These separate DNA segments, called V and J for light-chain genes, can fuse in any combination. For example, any κ light-chain V segment can rearrange with any κ light-chain J segment. For the κ light-chain gene, in germline DNA, the constant (C) and variable (V) segments of the gene are not closely linked. Moreover, there are perhaps 100–200 different versions of the variable segment and about 5 short joining (J) segments. Thus, by combinatorial fusion, 500 to 1000 different κ light chains can be produced.

During B lymphocyte development a process of site-specific recombination deletes the DNA between a V segment and a J segment, bringing one V segment adjacent to a J segment. Such a deletion, removing many thousands of base pairs of DNA, brings the transcriptional promoter (located at the 5′ end of each V segment) into effective proximity of the light chain's tissue-specific enhancer (located in the vicinity of the C segment). This allows the enhancer to activate transcription of the rearranged V-J-C gene, producing a functional light-chain mRNA.

Heavy-chain gene rearrangement is quite similar, except that the heavy-chain gene also includes about 20 diversity (D) segments in addition to V and J segments (Figure 23-25). First there is a rearrangement of a heavy-chain V segment to a D segment, and then this combined segment rearranges to a J segment. As with the light-chain gene, each rearrangement deletes the DNA in between the merged V-D or D-J segments. Thus, the same process of DNA deletion

and RNA splicing acts to produce the final heavy-chain mRNA.

T lymphocytes have different antigen-binding proteins called *T-cell receptors*. These T-cell receptors also must be present in millions of different flavors. The diversity of T-cell receptors is produced by a very similar process of DNA rearrangement within the T-cell receptor genes.

Thus, by using a mechanism involving recombination between segments of individual DNA molecules in developing B lymphocytes, the immune system can generate a vast variety of B lymphocytes, each encoding one of millions of possible antibody molecules. A site-specific recombination system permits the production of thousands of different light chains and thousands of different heavy chains. A given light chain can be produced in a cell with any of the possible heavy chains. Because the antigen-binding site is formed by contributions from the variable regions of both the light and heavy chains, thousands of different light and heavy chains become millions of different antibody molecules.

Message Two combinatorial mechanisms—DNA rearrangement to produce many light chains and heavy chains and construction of the antigen-binding site from portions of both chains—allow a small set of genes to encode millions of different antibody molecules.

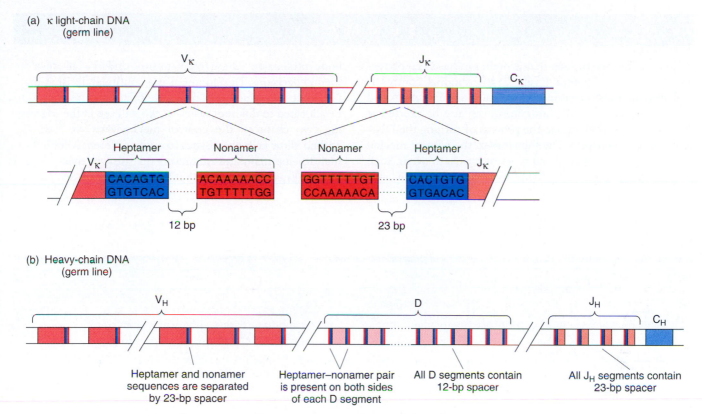

(a) κ light-chain DNA
(germ line)

V_κ J_κ C_κ

Heptamer   Nonamer   Nonamer   Heptamer

V_κ

| CACAGTG | ACAAAAACC | GGTTTTTGT | CACTGTG | J_κ |
| GTGTCAC | TGTTTTTGG | CCAAAAACA | GTGACAC | |

12 bp   23 bp

(b) Heavy-chain DNA
(germ line)

V_H D J_H C_H

Heptamer and nonamer sequences are separated by 23-bp spacer

Heptamer–nonamer pair is present on both sides of each D segment

All D segments contain 12-bp spacer

All J_H segments contain 23-bp spacer

Figure 23-25 The heptamer-nonamer sequences that align antibody gene rearrangements. (a) The top line shows a nonrearranged germ-line κ light-chain gene. Just 3′ to each variable (V) segment is the 28-bp structure shown below, with exact 7-bp and 9-bp sequences, separated by a 12-bp spacer that differs among the segments. Just 5′ to each joining (J) segment is the 39-bp structure shown below. In opposite orientation are the identical 9-bp and 7-bp sequences, but now separated by a 23-bp spacer. Recombination occurs only between segments with 12-bp and 23-bp spacers. Thus, a V can recombine only with a J: no V-V or J-J recombinants can form. A second light-chain gene, called the λ *light chain*, exhibits the same kind of V-J rearrangement. (b) A similar diagram for heavy-chain (H) rearrangement is shown. To build a functional heavy-chain gene, two recombination events must occur: V-D and D-J. These can be orchestrated properly because 23-bp spacer elements can recombine only with 12-bp spacers. (Adapted from J. D. Watson, M. Gilman, J. Witkowski, and M. Zoller, *Recombinant DNA*, 2d ed. Copyright © 1992 by Scientific American Books.)

How do the DNA recombination events occur at reliable positions to properly join the 3′ end of a V segment exactly to the 5′ end of a J segment? In the light-chain genes, at the 3′ end of each variable segment is a stretch of DNA that contains a 7-bp sequence and a 9-bp sequence, separated by a 12-bp spacer (see Figure 23-25). The 7- and 9-bp elements always have the same base sequences, but the 12-bp sequence appears to be random. 5′ to every joining region is almost the same structure, with the same 7- and 9-bp sequences, except that they are in opposite orientation and the spacer is 23 bp long. During antibody gene rearrangement, recombination always occurs between a segment with a 12-bp spacer and one with a 23-bp spacer. Thus, we

can envision that recombination involves forming a heteroduplex between the regions immediately adjacent to the variable and joining regions based on the complementarity of the 7-bp and 9-bp sequences. Two proteins, encoded by *RAG-1* and *RAG-2* (for recombination-activating genes 1 and 2), are thought to bind to these sequences and somehow mediate these site-specific recombination events.

Is this sort of programmed DNA rearrangement limited to the immune system? Perhaps not. Interestingly, the *RAG-1* gene is also expressed in the brain, suggesting that some neural functions may involve site-specific DNA rearrangements as well.

SUMMARY

In this chapter, we've seen that protein synthesis can be regulated at many different levels. Transcriptional control is probably the most common way of determining when and where a protein is made, but control can also be exerted at essentially any other step in the process of turning the DNA code into a polypeptide. In some cases, the control mechanisms permit very specialized types of events to occur, such as the production of an enormous array of different antibody molecules. These mechanisms underlie all developmental decisions. Development is a continual decision-making process in which cells of a developing organism need to be allocated to different tasks. As we will see in the succeeding two chapters, the control mechanisms we have described allow cells and tissues to adopt different roles in normal development, and mutations that disrupt these control mechanisms contribute to the formation of cancers.

Concept Map

Draw a concept map interrelating as many of the following terms as possible. Note that the terms are listed in no particular order.

transcription factors / enhancer / gain-of-function mutations / RNA splicing / polypeptides / 3′ UTR / DNA rearrangement / multimeric proteins / reporter genes / immunoglobulin heavy chains / autoregulatory loop

CHAPTER INTEGRATION PROBLEM

In developmental pathways, the crucial events seem to be the activation of master switches that set in motion a programmed cascade of regulatory responses. In the two examples of sex determination and differentiation discussed in this chapter—(a) sex determination in *Drosophila* and (b) sex determination in mammals—identify the master switches and how they operate.

Solution

a. In *Drosophila* sex determination, the master switch is the transcriptional regulation of the *Sxl* gene during early embryogenesis. The properties of the Sxl protein and the existence of a constitutive late promoter for *Sxl* ensure that once active Sxl protein is produced, an autoregulatory loop will continue to maintain Sxl protein activity in the cell. Sxl will also initiate a downstream set of regulated RNA splicing events, culminating in the production of *dsx-F* mRNA. In the absence of *Sxl,* the default RNA splicing machinery will produce *dsx-M* mRNAs. The key then is whether or not the *Sxl* switch is flipped on or stays off. This is controlled by the level of active numerator-encoded transcription factors.

The higher level of these transcription factors conferred by an X:A ratio of 1 is necessary to activate early *Sxl* transcription and set the female developmental pathway in motion. In the absence of these higher levels of numerator-encoded transcription factors, development continues along the default pathway leading to male development.

b. In mammalian sex determination, the master switch is the presence or absence of the *SRY/Sry* gene, which is ordinarily located on the Y chromosome. In the presence of the protein product of this gene, which acts as a DNA-binding protein, certain cells of the gonad (Leydig cells) synthesize androgens, male-inducing steroid hormones. These hormones are secreted into the bloodstream and act on target tissues to induce the transcription factor activity of the androgen receptors. In the absence of androgen receptor activation, development proceeds along the default pathway leading to female development. The factors that activate *SRY/Sry* expression in the testis are not understood. Since the master switch here is the actual presence or absence of the *SRY/Sry* gene itself, it is likely that the regulatory molecules that activate *SRY/Sry* are present in the indifferent gonad early in development.

SOLVED PROBLEMS

1. What would be the consequences if an individual were homozygous for any of the following mutations in the gene for the κ light-chain antibody gene?

    a. Deletion of the light-chain enhancer element

    b. Deletion of three of the five J segments

    c. Deletion of the entire constant region

Solution

    a. DNA rearrangements would still occur in this mutation, but, because the enhancer element is removed, transcription would not be activated in B lymphocytes and plasma cells. Therefore, no κ light-chain protein would be produced, and this individual would be unable to make any antibodies that included κ light chains. Any antibodies made by this individual would therefore include λ light chains. This would profoundly diminish the diversity of antibody molecules that this individual could produce.

    b. A mutation that deletes three of the five J segments would reduce the diversity of κ light chains that could be produced. Since there are 100–200 different V segments, in this mutant individual, only 200–400 different κ light chains could be produced. There would be an equivalent reduction then in the diversity of antibodies that could be produced.

    c. A mutation that deletes the constant region would produce a defective protein product of κ gene rearrangement. Rearrangements might still occur, but the resulting proteins would lack their carboxy-terminal sequences and would be unable to multimerize into the antibody tetramer. This would largely compromise the ability of that antibody to function in binding antigen.

2. In *Drosophila*, you have identified a new mutation, *mll* (*male-like*), that causes flies with an X:A ratio of 1 to develop as phenotypic males. You want to understand how the product of the *mll* gene acts in the developmental pathway for *Drosophila* sex determination and differentiation. You measure the presence of functional protein for *Sxl, tra,* and *dsx* in animals with an X:A ratio of 1. In each of the following cases, propose a role for mll protein in the sex determination and differentiation pathway.

    a. You observe that functional early Sxl, late Sxl, tra, and dsx-M proteins are produced.

    b. You observe that functional early Sxl protein and dsx-M are produced. No functional late Sxl or tra are produced.

    c. The only functional protein you observe is dsx-M.

Solution

It is actually through experiments like these that the sequential pathway of sex determination and differentiation in *Drosophila* was elucidated. More generally, it is through observations like these, where the effects of a mutant in one gene on the mRNA or protein expression of another gene are studied, that developmental pathways are pieced together.

    a. Given that Sxl and tra proteins are still operating normally in the context of an X:A ratio of 1, whereas the alternative splicing of *dsx* mRNA is leading to the male-specific form of the protein, it is likely that *mll* contributes to *dsx* RNA-splicing regulation.

    b. Here, the block seems to be between early and late *Sxl* protein production. This may indicate that mll plays a role in the autoregulation of late *Sxl* alternative splicing.

    c. In this example, the block is before *Sxl* expression in the early embryo. One possible explanation is that mll protein contributes to the numerator function in interpreting the X:A ratio.

PROBLEMS

1. Immunohistochemistry is a technique whereby specific antibodies can be used to stain tissues to identify the presence of a particular protein at a specific location in the body. You have acquired two antibodies, one that binds specifically to the amino acid sequence at the very carboxy terminus of the Sxl protein and another that binds specifically to the amino acid sequence at the very amino terminus of the protein.

    a. If, in separate experiments, you stained early embryos with each of these two antibodies, what would you expect to see?

b. What would you expect to see if you repeated the two experiments, but this time staining late embryos?

2. Several different levels of regulation contribute to the control of antibody production by the B lymphocytes.

    a. Explain how the specialized DNA rearrangements seen in antibody production and the multimeric nature of antibodies contribute to antibody diversity.

    b. It is important that the enhancer elements that contribute to antibody production are located near the constant segment of the gene. Why? What would be the consequence of having the enhancer element located instead near the V segment farthest from the C segment?

3. In *Drosophila*, individuals with two X chromosomes and three sets of autosomes (X:A ratio = 0.67) are intersexes. When examined more closely, it turns out that their intersexual phenotype is due to mosaicism. Some cells differentiate with an entirely male phenotype and other cells with an entirely female phenotype. Explain this observation in terms of how the *Sxl* on-off switch is established and maintained.

4. The P{Δ2, 3} transgene produces active P transposase in somatic as well as germ-line cells. Suppose that we want to develop a technique in which we can induce P element excisions in only one tissue of the fly, for example, the developing wing. Using the information discussed in this chapter, propose a way to accomplish this goal.

5. In the immune system of vertebrates, having a diverse set of antibodies is crucial if a robust immune response is to be generated. Further, cells producing an antibody that binds to an antigen present at a particular time must be able to proliferate rapidly (this is called *clonal expansion*) so that sufficient antibodies are available to fight an infection effectively.

    a. Once one light-chain gene rearranges successfully, all further light-chain rearrangements are prevented in a B lymphocyte. The same is true for the heavy-chain genes. This phenomenon is known as *allelic exclusion*. Why is this phenomenon crucial to the immune response?

    b. The mouse is estimated to have 300 variable segments and 4 joining segments for immunoglobulin light-chain genes, and 1000 variable, 4 joining, and 12 diversity segments for heavy-chain genes. If all light-chain and heavy-chain rearrangements occur randomly, how many possible antibody molecules could be generated by the immune system?

6. Dominant gain-of-function mutations, such as *Tab*, can be reverted to wild type at high frequency by treatment with a mutagen. Explain this observation.

7. The transposase protein of the P transposable element in *Drosophila* is encoded by 4 exons, numbered 0, 1, 2 and 3. The germ-line transcript is 2.1 kb long, contains these 4 exons, and encodes a protein with transposase activity. However, the somatic P element transcript, which is 2.5 kb long, contains all 4 exons and in addition does not remove the intron between exons 2 and 3.

    a. The longer somatic mRNA encodes a protein that is only 75 percent as large as the protein encoded by the shorter germ-line transcript. How can a longer transcript encode a shorter protein?

    b. The enzyme reverse transcriptase can be used to make a double-stranded DNA copy of an mRNA. This DNA copy of an mRNA is called a *cDNA* (see the description of cDNAs in Chapter 14). How could you make use of cDNAs and transgenes to produce a fly that has transposase activity in somatic cells?

    c. Suppose that you digest the P element with the restriction enzyme *Sal*I and you obtain eight fragments from the digest. You do a Southern blot on this digest and probe it with radioactive germ-line transposase cDNA. You find that all eight of the restriction fragments hybridize to the radioactive cDNA probe. What does this tell you about the positions of the *Sal*I restriction enzyme recognition sites?

8. You are studying a mouse gene that is expressed in the kidneys of male mice. You have previously cloned this gene. Now you wish to identify the segments of DNA that control the tissue-specific and sex-specific expression of this gene. Describe an experimental approach that would allow you to do so.

9. In *Drosophila*, homozygotes for mutations in the *tra* gene transform XX individuals into phenotypic males with regard to somatic secondary sexual characteristics. The gonad in *Drosophila* forms from somatic mesoderm tissue and germ-line cells. XX;*tra* homozygotes are sterile and have rudimentary gonads. You suspect that the reason for this sterility is that the somatic tissue is sexually transformed to male by *tra* but the germ-line cells remain female. Design an experiment that tests this prediction.

10. XYY humans are fertile males. XXX humans are fertile females. What do these observations reveal about the mechanisms of sex determination and dosage compensation?

11. Humans that are mosaics of XX and XY tissue occasionally occur. They generally exhibit a uniform sexual phenotype. Some individuals are phenotypically female, others male. Explain these observations in terms of the mechanism of sex determination in mammals.

12. There are dominant mutations of the *Sxl* gene, called *Sxl*^M alleles, that transform XY animals into females, but do not affect XX animals. Reversions of these *Sxl*^M alleles can be readily induced with mutagen treatment. These reversions, called *Sxl*^f alleles, yield normal individuals in XY animals, but in homozygotes, transform XX animals into phenotypic males. Provide a possible explanation for these observations, keeping in mind that *Sxl* ordinarily is dispensable for male development.

24

Genetics and Cellular Differentiation

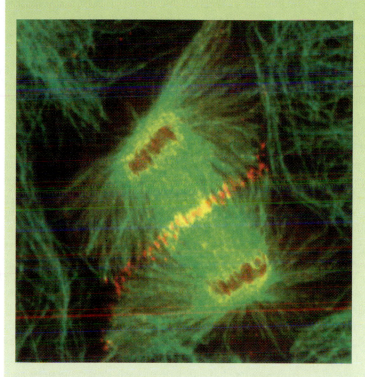

The spindle apparatus of a toad cell (*Xenopus laevis*) in metaphase of mitosis. In green are the spindle fibers, which are composed of cytoskeletal elements called microtubules, and in red are the locations of a motor protein called *kinesin* that moves along the microtubules, dragging cellular cargoes along with it. The kinesin is thought to haul the chromosomes along the spindle fibers to the poles during anaphase of mitosis. (Reproduced from I. Vernos, J. Raats, T. Hirano, J. Heasman, E. Karsenti, and C. Wylie. *Cell* 81, 1995, 120.

KEY CONCEPTS

▶ The cytoskeleton provides a structural meshwork and a series of conveyor belts for moving materials around the cell.

▶ Normal cell proliferation is modulated by cell cycle regulation.

▶ Signaling systems permit regulatory events to be coordinated between different cells and tissues.

▶ In cancer, cells proliferate out of control through the accumulation of a series of special mutations in the same somatic cell.

▶ Many of the classes of genes that are mutated to cause cancers are important components of the cell that directly or indirectly contribute to growth control and differentiation.

The many differentiated cell types in the body allow the different organs and tissues to carry out their specialized physiological roles. Single cells can be quite small or can extend long processes. They can be round, columnar, or highly stretched, or manifest highly branched and elongate processes. The human body has hundreds, perhaps thousands, of different cell types. Some cell types are fated never to divide again, others may divide rapidly, and still others will divide only under appropriate and highly regulated circumstances. Much research in cell biology seeks to understand the processes by which cells acquire and maintain their unique structures and functions, and how these processes can go awry. In the last several years, genetic approaches have provided additional tools for the researcher to understand and dissect the biology of normal and diseased cells. Of special interest to us is the study of cancer, which is clearly recognized to be a genetic disease of somatic cells. We cannot understand the biological basis of cancer without addressing the cell biology of differentiation and the genetics of normal somatic cells. In this chapter, we will explore some of the findings of genetics and cell biology as they impinge on one another, and we will examine how these findings affect our understanding of cancer. In addition, some of the cellular processes we will discuss here make important contributions to the mechanisms that establish complex biological patterns, as we will see in Chapter 25.

We will examine three important topics in the biology of normal cells: the cytoskeleton, the cell cycle, and intercellular communication. The cytoskeleton consists of the supporting rods within a cell that maintain its shape and along which "cargoes" are moved from one portion of the cell to another. The cell cycle consists of the mechanisms that govern progression through the cycle of cell division and interphase. Intercellular communication is the process by which chemical information and instructions are passed from one cell type to another.

While we will treat them as separate topics, we should be aware that they interrelate to one another. Consider mitosis. The spindle apparatus that moves chromosomes to the poles during anaphase is composed of those cytoskeletal elements called **microtubules,** and determination of the position of the cleavage furrow is dependent on another set of structural elements called **microfilaments.** The timing of the mitotic division and the ability of the cell to enter mitosis depend on cell cycle controls. Generally, cell numbers within a tissue are tightly governed, through intercellular communication pathways that either promote or repress progression of the cell cycle into mitosis.

Similarly, all these aspects of cell biology impact on the biology of cancer. The normal mechanisms that govern the cell cycle and that make it dependent on proper proliferative signals from other cells are abrogated in cancer cells. Many aspects of normal cell behavior, including the control of their shape and their ability to migrate, are abnormal in cancerous cells. In the latter half of this chapter, we will look at some of the ways that cancer-promoting mutations alter the normal regulation of the cell, focusing particularly on the mechanisms of cell cycle control and intercellular communication, where the connection to cancer is particularly well understood.

The Cytoskeleton and Cellular Architecture

As cells have been studied in more detail, hitherto unappreciated levels of organization and structure have been resolved. When scientists first saw cells through primitive microscopes, they thought of them as bags of "protoplasm," a more or less uniform liquid substance. As microscopes achieved better resolution, the distinctions of nucleus and cytoplasm became apparent. With further increases in resolution, the various organelles were identified. We now think of cells as highly compartmentalized entities, with many different organelles and other structural components playing specialized roles.

Among these structural components, we now appreciate that several classes of proteins form the networks of rods that act as the superstructure of a cell. This superstructure, the cytoskeleton, determines the cell's shape and mechanical properties, and provides highways for the trafficking of materials within the cell.

For our purposes, there are several reasons to be concerned with the cytoskeleton. The cytoskeleton provides the apparatus that directs the partitioning of chromosomal material to the poles during mitosis and meiosis. The location and orientation of the cleavage plane is also controlled by the cytoskeleton, offering the possibility that the cytoskeleton can determine specializations of particular somatic cell divisions. By determining the shape of a cell and its ability to migrate, differentiated roles for cells can be established and modulated. Coordinated changes in cell shape can lead to bulges and indentations in tissue layers that can ultimately produce outpocketings such as appendages or internalizations of cell layers during the morphogenetic movements characteristic of early embryogenesis and organ formation. Finally, while the molecular events are not well understood, it is clear that changes in cell shape and motility are an important part of the formation and growth of tumors.

The cytoskeleton consists of several networks of structural rods that run within each cell: **microfilaments, intermediate filaments,** and **microtubules** (Figure 24-1 and Table 24-1). Each has its own architecture, with a unique set of structural proteins, cross-linking proteins that connect individual rods to one another, and proteins that promote production or disassembly of the rods. While all their roles have not yet been discovered, it is clear that the various cytoskeletal systems make different contributions to the cell.

(a) Intermediate filaments (vimentin)

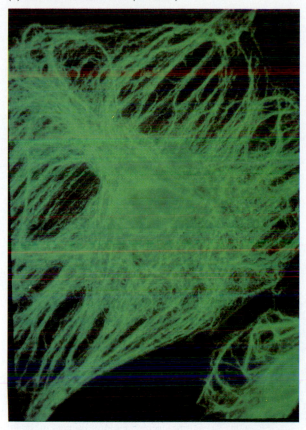

(b) Microtubules (tubulin)

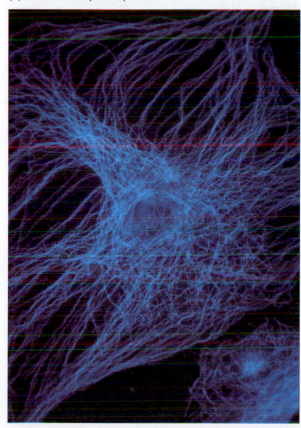

(c) Microfilaments (actin)

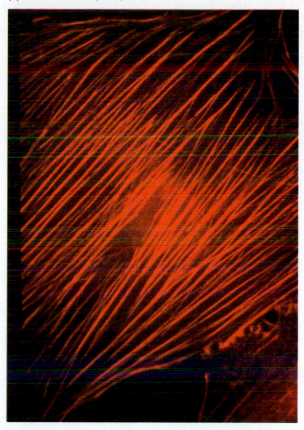

Figure 24-1 Different cytoskeletal systems in the same cell. The distribution of (a) the intermediate filament protein vimentin, (b) the microtubulin protein tubulin, and (c) the microfilament protein actin are shown. (Courtesy of V. Small. Reprinted from H. Lodish, D. Baltimore, A. Berk, S. L. Zipursky, P. Matsudaira, and J. Darnell, *Molecular Cell Biology*, 3d ed. Scientific American Books, 1995.)

Cell Shape

All three types of support rods probably contribute to the shape of a cell. Proteins within the plasma membrane of the cell can be attached to other proteins in the cytoplasm. In turn, these other proteins can be linked to still other proteins that can bind and cross-link cytoskeletal elements. In this manner, the plasma membrane can be tethered to the cytoskeleton such that the organization of the cytoskeleton will determine the outline of the cell.

Intracellular Transport

Microfilaments and microtubules also serve as highways for directed movement of molecules and organelles throughout the cell (Figure 24-2). Microfilaments and microtubules have polarity (conceptually like the 5′-to-3′ polarity of DNA and RNA strands, even though the molecular basis of polarity is

Table 24-1   **Some Properties of the Different Cytoskeletal Systems**

Cytoskeletal system	Diameter of fibers	Structural proteins	Motor proteins	Roles in cell
Microfilaments	7–9 nm	Actin	Myosin	Microvilli Cell polarity Ameboid movement Muscle contraction Cell furrow at mitosis
Intermediate filaments	10 nm	Vimentin, keratin, desmin, lamin	—	Structural support
Microtubules	24 nm	Tubulin	Kinesin, dynein	Motility Vesicle transport Protein transport Cilia and flagella Mitotic spindle apparatus

(a)

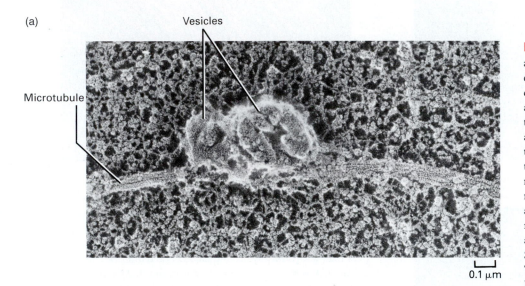

Figure 24-2 Movement of vesicles along microtubules. (a) A scanning electron micrograph of two small vesicles attached to a microtubule. (b) A diagram of how kinesin is thought to attach to cellular cargoes such as vesicles at its tail and to move the cargoes along the microtubule in the − to + direction, using the motor domain in the kinesin head. (c) A diagram of the kinesin protein showing the functions associated with various portions of the molecule. (Part a from B. J. Schnapp et al., *Cell* 40, 1985, 455. Courtesy of B. J. Schnapp, R. D. Valle, M. P. Sheetz, and T. S. Reese. All parts: Reprinted from H. Lodish, D. Baltimore, A. Berk, S. L. Zipursky, P. Matsudaira, and J. Darnell, *Molecular Cell Biology,* 3d ed. Scientific American Books, 1995.)

(b)

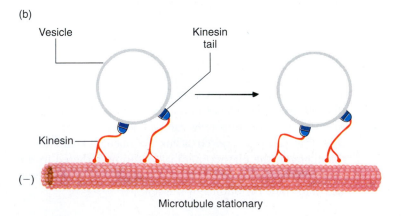

(c)

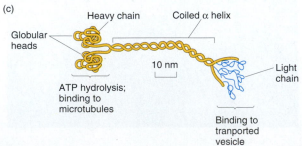

quite different) (Figure 24-3). Furthermore, the polarity of the cytoskeletal elements can be constant within a cell. Consider microtubules. There is a location near the center of a cell where all the − (minus) ends of the microtubules are found (Figure 24-4). This is called the **microtubule organizing center (MTOC)**. The + (plus) ends of the microtubules are located at the periphery of the cell. Very much as an automobile uses the combustion of gasoline to create energy that is then transduced into motion, special "motor" proteins hydrolyze ATP for energy that is turned into the force for movement along a microtubule. For example, a protein called *kinesin* is able to move in a − to + direction along microtubules carrying cargoes such as vesicles (see Figure 24-2b). The motor, the part of the kinesin protein that directly interacts with the microtubule rod is contained in the globular head of the protein (see Figure 24-2c). The tail of kinesin is thought to be where the cargo is attached. These cargoes might be individual molecules, organelles, or other subcellular particles to be towed from one part of the cell to another.

> **Message** The cytoskeleton serves both as the structural support system for the cell and as a highway system for the directed movement of subcellular particles and organelles.

What is the value of having multiple independent cytoskeletal systems? Having several systems of structural support for the cell may be important for maintaining the structural integrity of the cell. Certain compounds have the ability to induce disassembly of specific cytoskeletal elements (for example, cytochalasin, a natural plant product, can induce depolymerization of microfilaments). The cell may be able to survive such damage so long as other systems are in place to maintain at least partially the overall structure of the cell. Another part of the answer is probably division of labor. With multiple cytoskeletal systems, cargo can be moved from any particular location in the cell to any other location.

The Cytoskeleton and Cell Division

We can get an appreciation of how genetic and developmental mechanisms depend on the cytoskeleton by examining the process of cell division. During mitosis, chromosomes must be partitioned equally to the two daughter cells. The cellular machinery that carries out this task is one of the cytoskeletal systems: the microtubules. Another aspect of mitosis is cytokinesis, the process by which other subcellular components are distributed to daughter cells. Depending on the type of cell undergoing mitosis, cytokinesis might more or less equally distribute subcellular components, producing two equivalent daughter cells, or might asymmetrically distribute the components, thereby generating different types of daughter cells. The process of cytokinesis depends on the action of another cytoskeletal system: the microfilaments.

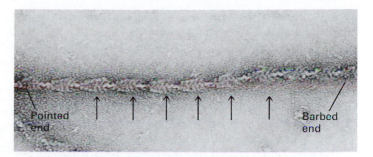

Figure 24-3 The polarity of subunits in an actin microfilament. An actin microfilament doesn't ordinarily have this appearance, but the microfilament has been coated with a protein that binds in a fashion that reveals the underlying polarity of the actin microfilament itself. (Courtesy of R. Craig. Reprinted from H. Lodish, D. Baltimore, A. Berk, S. L. Zipursky, P. Matsudaira, and J. Darnell, *Molecular Cell Biology*, 3d ed. Scientific American Books, 1995.)

The Spindle Apparatus

One aspect of the division process controlled by the cytoskeletal system is the spindle apparatus, which is responsible for proper chromosome separation at anaphase of mitosis and meiosis. The spindle fibers, some connecting the spindle poles to each other and others connecting centromeres to the spindle poles, are microtubule assemblies. Along these spindle fibers, microtubule-associated motor proteins are thought to pull chromosomal cargoes to the poles at anaphase. Indeed, in *Drosophila*, *ncd* and *nod*, two genes identified by mutant alleles defective in meiotic chromosome segregation, encode kinesinlike microtubule motor proteins. These are among the first concrete examples of how polarized motors might contribute to chromosome segregation.

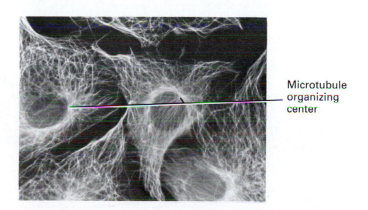

Figure 24-4 Fluorescence micrograph showing the distribution of tubulin in an interphase animal fibroblast. Notice that the microtubules radiate out from a microtubule organizing center (MTOC). The − ends of the microtubules are in the center, and the + ends are at the periphery of the cell. (Courtesy of M. Osborn. Reprinted from H. Lodish, D. Baltimore, A. Berk, S. L. Zipursky, P. Matsudaira, and J. Darnell, *Molecular Cell Biology*, 3d ed. Scientific American Books, 1995.)

Cytokinesis: Symmetric versus Asymmetric Cell Divisions

Depending on how it occurs, cytokinesis can generate equivalent or different daughter cell types. Which type of cell division—symmetric or asymmetric—occurs depends on the action of the cytoskeleton to determine the position of the metaphase plate which in turn determines the location of the cleavage furrow. To produce two equivalent daughter cells, typically, the metaphase plate is located at a central position in the cell. When mitosis occurs at this medial position, there is an equal distribution of cellular components to daughter cells. However, often during development, it is important to distribute the contents of a cell asymmetrically, so that the daughter cells acquire different amounts of key regulatory molecules and hence adopt different developmental roles. One way that a differential distribution of cellular contents is achieved is by locating the metaphase plate at an offset position, which in turn leads to a cleavage furrow at a similarly asymmetric position.

Such an asymmetrically located metaphase plate is typical of the divisions of undifferentiated **blast cells** or **stem cells.** For example, a stem cell divides asymmetrically, with one daughter product (typically the larger one) being another stem cell and the other, a specialized cell, undergoing differentiation. Through such blast or stem cell divisions, a cell can perpetuate its own cell type while spinning off numerous differentiating cells.

The first cell division of the zygote of the nematode *C. elegans* provides an example of an asymmetric division. One of the favorable properties of *C. elegans* as an experimental system is that the same pattern of cell divisions occurs from one animal to another, and that these can be readily followed under the microscope. A lineage tree can then be constructed that traces the descent of each of the thousand or so somatic cells of the worm. Part of this lineage tree is shown in Figure 24-5.

Figure 24-5 The development of the nematode *C. elegans.* (a) The *C. elegans* life cycle. The relative sizes of each stage are indicated, as well as the number of hours it takes to progress from one stage to the next. (b) The lineage of 13 cell lines present in the L1 stage of a *C. elegans* male. The symbols identifying the parental cells appear on the top row. (Note that the lineages of cells P3 to P9 are identical and therefore are represented as a single Pn family.) At each division, there is an anterior (a) and a posterior (p) cell. The history of each cell is indicated by its name. For example, P10.paap is a derivative of P10 that is the posterior cell in the first division, the anterior cell in the next two divisions, and the posterior cell in the last division. (From J. E. Sulston and H. R. Horwitz, *Developmental Biology* 56, 1977, 111.)

(a)

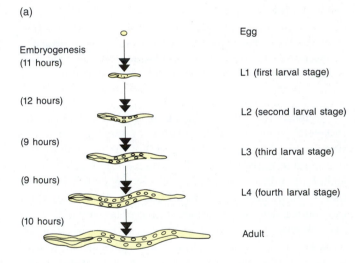

(b)

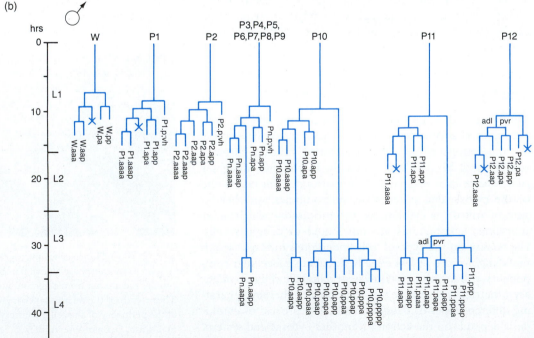

The one-cell zygote of *C. elegans* that is produced upon fertilization is called the P_0 cell. It divides asymmetrically across the long axis of the ellipsoidal P_0 cell to produce a larger, anterior AB cell and a smaller, posterior P_1 cell (Figure 24-6). This is a very important division in that it already sets up specialized roles for the descendants of these first two cells. The AB cell descendants will produce most of the skin cells of the worm (the hypoderm) and most of the neurons of the nervous system, while most of the muscles and all the digestive system and the germ-line cells will come from the P_1 cell (Figure 24-6a). The germ cell fate in this and future divisions seems to correlate with the distribution of certain fluorescent cytoplasmic particles called P granules. These granules are incorporated exclusively into the P_1 cell at the first division. When the P_1 cell divides, also asymmetrically, the P granules are incorporated into the progeny P_2 cell, and similarly at the next division into the P_3 cell, and so on. Only the P_x cell that has these P granules becomes the germ line of the worm—all other cells are somatic. Later on in development, once the germ line is established, this cell gives rise to numerous maturing germ cells to produce the oocytes and sperm.

How are the asymmetric divisions of the P cells controlled? It appears that microfilaments are involved. When applied at the right time to the P_0 cell, drugs such as cytochalasin that disrupt actin chain polymerization into microfilaments cause the first division to be equal. The consequences are also that the P granules are distributed equally to the daughter cells, and that all cells in the developing embryo adopt roles typical of the somatic cells produced by the P_1 lineage. Not surprisingly, these embryos are quite confused, and die as masses of cells that look nothing like a normal worm. The effects of cytochalasin can be mimicked by *par* (*partition defective*) mutations (Figure 24-6b), which act in the ovary to produce embryos in which the division of the P_0 cell is symmetric. The *par* genes likely encode products that contribute to the special cytoskeletal structures or functions that turn the division into an asymmetric one.

Message By attaching specific particles to the cytoskeleton, cell divisions can be unequal, giving rise to two different kinds of daughter cells.

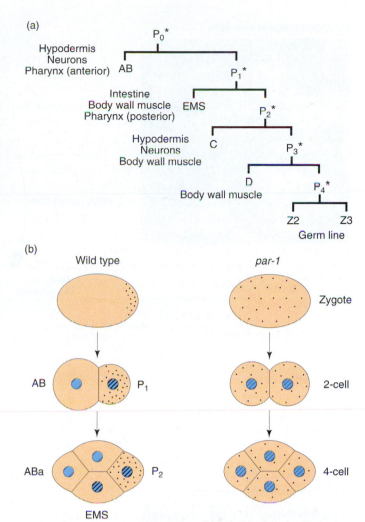

Figure 24-6 The early development of *C. elegans*. (a) The early lineage divisions of the zygote. The mature cell types that arise from the various daughter cells of the early divisions are indicated. Note that the entire germ line comes from the P4 cell. Each of the "P" cell divisions indicated with an asterisk is asymmetric, and each of the posterior daughter cells inherits all the P granules, which are thought to be germ-line determinants in the worm. The letters (for example, AB, EMS) are symbols for names of daughter cells. (b) The first two cell divisions of zygotes derived from wild-type and *par* mutant mothers. The stippling indicates the distribution of P granules at each of the divisions in wild-type and mutant embryos. Note that in the wild type, each of the four cells produced after the second mitotic division is distinct, whereas in the *par* mutant, the four cells are indistinguishable from one another. (After S. Guo and K. J. Kemphues, *Cell* 81, 1995, 612.)

The Cell Cycle

Mitotically dividing eukaryotic cells cycle through interphase and mitosis. However, the cell cycle is not a hardwired program that will proceed at a fixed pace. Rather, in complex eukaryotes, it proceeds at a pace characteristic of particular cell types and specific environmental conditions. In mature organisms, cell division rates in many tissues must be regulated so there are sufficient new cells to replace dying ones, but also so that excess cells are not produced.

As we saw in Chapter 11 (Figure 11-35), there are four main parts to the cell cycle: M phase—mitosis, the cell division process described in detail in Chapter 3—and three that are components of interphase: G_1, the *gap* period between the end of mitosis and the start DNA replication; S, the period during which DNA *synthesis* occurs; and G_2, the *gap* period following DNA replication and preceding the initiation of the mitotic prophase. In mammals, where the cell cycle is particularly well studied, differences in the rate of cell division are largely a consequence of the length of time

(a) *Saccharomyces cerevisiae*

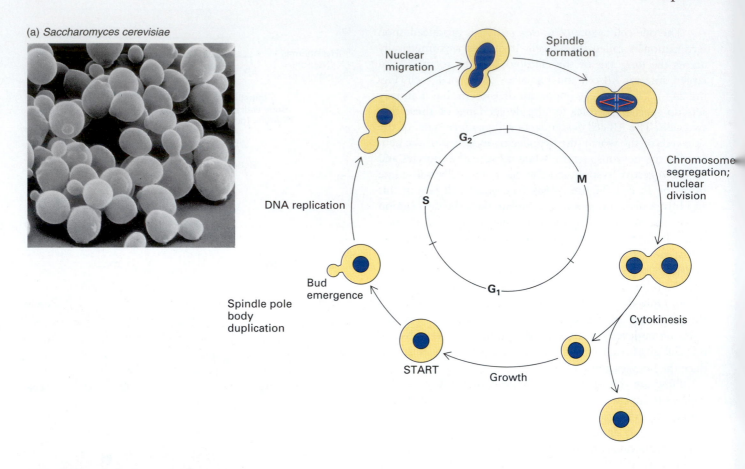

Nuclear migration

Spindle formation

DNA replication

G₂

M

S

Chromosome segregation; nuclear division

Spindle pole body duplication

Bud emergence

G₁

START

Growth

Cytokinesis

(b) *Schizosaccharomyces pombe*

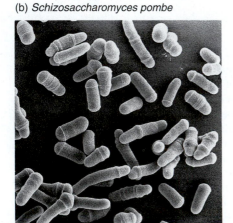

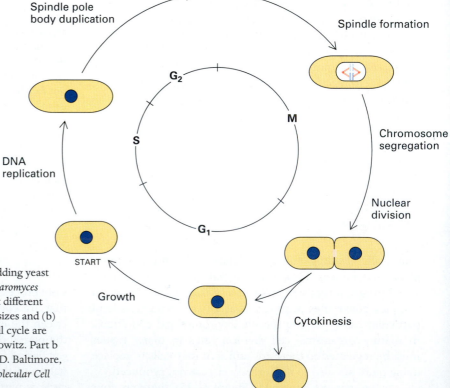

Chromosome condensation

Spindle pole body duplication

Spindle formation

DNA replication

G₂

M

S

Chromosome segregation

Nuclear division

G₁

START

Growth

Cytokinesis

Figure 24-7 The cell cycles of two yeasts. (a) The budding yeast *Saccharomyces cerevisiae.* (b) The fission yeast *Schizosaccharomyces pombe.* The scanning electron micrographs show cells at different points in the cell cycle, as indicated (a) by different bud sizes and (b) by the length of the cells. The principle events in the cell cycle are shown. (Part a courtesy of E. Schachtbach and I. Herskowitz. Part b courtesy of N. Hajibagheri. Reprinted from H. Lodish, D. Baltimore, A. Berk, S. L. Zipursky, P. Matsudaira, and J. Darnell, *Molecular Cell Biology,* 3d ed. Scientific American Books, 1995.)

a cell takes in G_1. This is accomplished by having an optional **G_0** resting phase into which G_1 phase cells can shunt and remain in for variable lengths of time, depending on the cell type and on environmental conditions.

Many aspects of the cell cycle normally seem to proceed on a fixed schedule. How is this schedule maintained? It would be suicidal if some events proceeded in the wrong order, for example, if mitosis began before chromosomes had been replicated. How are events kept in the correct order? The division rate of a cell must be appropriate for its circumstances. How does the cell receive environmental cues to determine whether it should be dividing or not?

Much of our knowledge of the cell cycle comes from a combination of genetic studies in two yeasts and from biochemical studies of cultured mammalian cells. Cell cycle genetics of the budding yeast *Saccharomyces cerevisiae* and the fission yeast *Schizosaccharomyces pombe* have revealed a large array of genetic functions that maintain the proper cell cycle (Figure 24-7). These functions are identified as a specific subset of temperature-sensitive (ts) mutations called *cdc* (*cell division cycle*) mutations. When grown at low temperature, yeasts with these *cdc* mutations would grow normally. When shifted to higher, restrictive temperatures, the yeasts would stop growing. What made these *cdc* mutations novel among the more general class of ts mutations was that a particular *cdc* mutant would stop growing at a specific time in the cell cycle, and all the yeast cells would look alike. Consider some examples of *S. cerevisiae*, a yeast that divides through budding, a process in which a mother cell develops a small outpocketing, a "bud." The bud grows and mitosis occurs such that one spindle pole is in the mother cell and the other in the bud. The bud continues to grow until it is as big as the mother cell, and the mother cell and bud then separate into two daughter cells. Any run-of-the-mill ts mutation in *S. cerevisiae*, when shifted to restrictive temperature, stops growth at varying times in the cycle of bud formation and growth. In contrast, after a shift to restrictive temperature, one *S. cerevisiae cdc* mutation would produce yeast cells only with tiny buds, whereas another would produce yeast cells exclusively with larger buds half the size of the mother cell. Such different *cdc* phenotypes reflect different defects in the machinery required to execute specific events in the progression of the cell cycle. We will summarize the salient features of how these genes and the proteins they encode contribute to the cell cycle.

Cyclins and Cyclin-Dependent Protein Kinases

The key triggers for progression from one step of the cell cycle to the next are two polypeptides that participate as subunits of a heterodimer: **cyclins** and **CDKs** (cyclin-dependent protein kinases). Cyclins are so named because they are transcribed and translated only at a particular time during the cell cycle. All cyclins are related in amino acid sequence, presumably reflecting similar three-dimensional structures. **Protein kinases** are enzymes that can phosphorylate cer-

tain amino acids in specific proteins. The protein substrates (the target proteins that are phosphorylated) are different for different protein kinases. The phosphorylation events can alter the activity of the substrate proteins. The CDK has the ability to phosphorylate certain serines or threonines on target proteins. In *S. cerevisiae*, there is a single CDK, encoded by the *cdc28* gene. (In mammals, as we will discuss below, there is a family of related CDKs.) The target proteins are determined by interactions with the associated cyclin. In other words, the cyclin tethers the target protein so that the CDK can phosphorylate it. Since different cyclins are present at different phases of the cell cycle (Figure 24-8), different phases of the cell cycle are characterized by the phosphorylation of different target proteins.

CDK Targets

How does the phosphorylation of some target proteins control the cell cycle? Phosphorylation initiates a chain of events that culminates in the activation of transcription factors that promote transcription of certain genes whose products are required for the next stage of the cell cycle. A

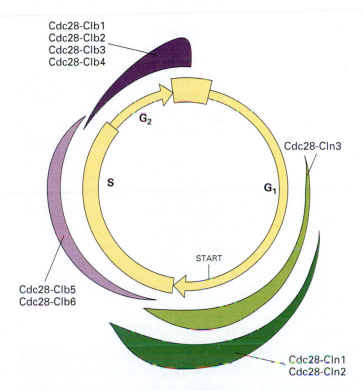

Figure 24-8 The variations in cyclin-CDK activities throughout the cell cycle in *S. cerevisiae*. The widths of the bands indicate the relative kinase activities of the various cyclin-CDK complexes. There are several different cyclins in this yeast (indicated with the Clb and Cln designations) but only one CDK (encoded by the *cdc28* gene). Restriction point is the stage at which the cell becomes irreversibly committed to mitosis. (Reprinted from H. Lodish, D. Baltimore, A. Berk, S. L. Zipursky, P. Matsudaira, and J. Darnell, *Molecular Cell Biology,* 3d ed. Scientific American Books, 1995.)

Figure 24-9 The contributions of the RB and E2F proteins in the regulation of the G_1 to S phase transition in a mammalian cell. (Reprinted from H. Lodish, D. Baltimore, A. Berk, S. L. Zipursky, P. Matsudaira, and J. Darnell, *Molecular Cell Biology*, 3d ed. Scientific American Books, 1995.)

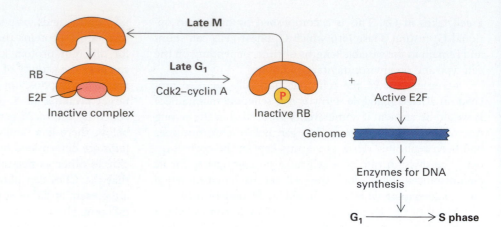

well-understood example is the RB-E2F pathway in mammalian cells; RB is the target of a CDK-cyclin heterodimer called Cdk2–cyclin A, and E2F is the transcription factor RB regulates (Figure 24-9). From late M phase through the middle of G_1, the RB and E2F proteins are combined in a protein complex that is inactive in promoting transcription. In late G_1, the Cdk2–cyclin A complex is produced and phosphorylates RB protein. This phosphorylation produces a conformational change in RB such that it can no longer bind to E2F protein. The free E2F protein is then able to promote transcription of certain genes that encode enzymes vital for DNA synthesis. This allows the next phase of the cell cycle—S phase—to proceed.

RB and E2F are in fact representatives of two families of related proteins. It is thought that different CDK-cyclin heterodimers selectively phosphorylate different proteins of the RB family, which in turn release the specific E2F family member to which they are bound. The different E2F transcription factors then have the ability to promote transcription of different genes that execute different aspects of the cell cycle. In this way, a cycle of activation of different CDK-cyclin heterodimers can produce a cell division cycle (Figure 24-10).

> **Message** Sequential activations of different CDK-cyclin heterodimers lead in turn to activation of different transcription factors and in turn to the progression of the cell cycle.

Negative Controls of CDK Activity

We can consider the cell cycle as an engine. We can think of the cyclins as the accelerator pedal to activate the CDK engine. To operate effectively, the system needs a way of braking as well, so that the cell cycle doesn't proceed under inappropriate conditions.

Through activation of CDK-cyclin-binding proteins that can inhibit the protein kinase activity of CDK-cyclin heterodimers, the cell cycle can be held in check until various monitoring mechanisms give a "greenlight," indicating that the cell is properly prepared to go on to the next phase of the cycle.

One example of how this system operates involves DNA damage (Figure 24-11). When DNA is damaged during G_1, for example, by X-irradiation, CDK activity of CDK-cyclin heterodimers is inhibited. The inhibition seems to be mediated by a protein called p53. Part of the p53 protein recognizes certain kinds of DNA mismatches. In the presence of such mismatches, p53 is able to activate another protein, p21. When its levels are high, p21 binds to the CDK-cyclin heterodimer and inhibits its protein kinase enzymatic activity. Through mechanisms we have just discussed, the inhibition of the CDK protein kinase activity blocks

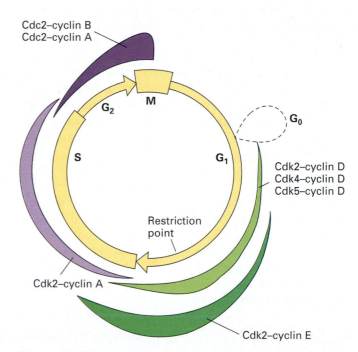

Figure 24-10 A current view of the variations in cyclin-CDK activities throughout the cell cycle of a mammalian cell. The widths of the bands indicate the relative kinase activities of the various cyclin-CDK complexes. Note that in contrast to yeast (see Figure 24-8), there are several different cyclins *and* several different CDKs. This increases the number of combinations of cyclin-CDK complexes that can form during the cell cycle. (Reprinted from H. Lodish, D. Baltimore, A. Berk, S. L. Zipursky, P. Matsudaira, and J. Darnell, *Molecular Cell Biology*, 3d ed. Scientific American Books, 1995.)

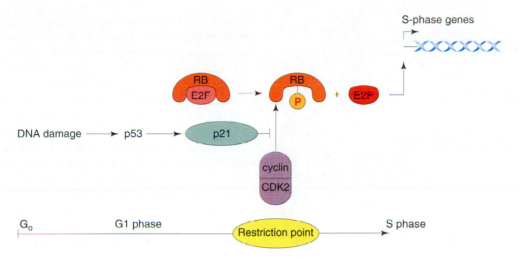

Figure 24-11 An example of inhibitory control of cell cycle progression. In mammals, the transition from G_1 to S phase requires the phosphorylation of Rb protein by the CDK2-cyclin complex. In the presence of damaged DNA, p53 protein is induced, which in turn induces p21 protein. The elevated levels of p21 protein inhibit the protein kinase activity of the CDK2-cyclin complex. Once the damaged DNA is repaired, p53 levels drop. In turn, then, p21 levels decrease, and the inhibition of the CDK2-cyclin protein kinase activity is relieved. This allows Rb to be phosphorylated and E2F to become an active transcription factor, permitting the cell to enter S phase. (Adapted from C. J. Sherr and J. M. Roberts, *Genes and Development* 9, 1995, 1150.)

the cell cycle from progressing until the DNA mismatches are repaired, leading to a drop in p53 levels and thus to the cessation of inhibition of the CDK-cyclin protein kinase activity.

In this manner, **checkpoints** that serve as monitors of the status of DNA replication, the spindle apparatus, and other key components of the cell cycle can operate as braking systems when necessary. The key is the existence of regulatory proteins that can modulate the protein kinase activity of the cyclin-CDK complex. In addition to monitoring intracellular conditions, the circuitry that controls the cell cycle is also sensitive to information originating as signals from other cells. Certain secreted proteins are known to inhibit cells from dividing. These proteins bind to receptors on target cells and thereby initiate a response in the target cell which leads to the activation of a protein which can inhibit the enzymatic activity of the CDK-cyclin heterodimer. How such signaling molecules and receptors operate will be described in the next section.

Message Fail-safe systems (checkpoints) ensure that the cell cycle does not progress until the cell is competent.

Intercellular Communication

The proper physiology and development of multicellular organisms depend on coordination of functions between different tissues and organs, and within cells of the same tissue. This is accomplished through a wide variety of differ-

ent systems for sending and receiving signals between cells. In the development of this complex pattern, we will see examples in which the transcriptional state of one cell can lead to the production of a signal that alters the transcriptional activity of neighboring cells. Such intercellular communication allows changes in the developmental state of the organism to be coordinated.

Components of Signaling Systems

All systems for intercellular communication have several components. A molecule called a **ligand** is released from signaling cells. Some ligands are proteins, while others are small molecules such as steroids or vitamin D. The ligand binds to **receptor** proteins on or within the target cells, forming a ligand-receptor complex. Some types of ligand-receptor complexes are able to alter transcription directly. Others need to operate through a series of intermediary steps, called a **signal transduction pathway,** which permit a signal to migrate from the cell surface to the nucleus, where it can affect transcription. Some ligands are nonpolar molecules and are able to pass directly through the cell membrane. For such ligands, the receptors are found in the cytoplasm or the nucleus. Other ligands, such as polypeptides and polar small molecules, cannot pass directly into the cell. For these ligands, their receptors are transmembrane proteins embedded in the cell membrane.

Some signals, called **hormones,** are transmitted long range throughout the body. They can act as master control switches for many different tissues, which can then respond

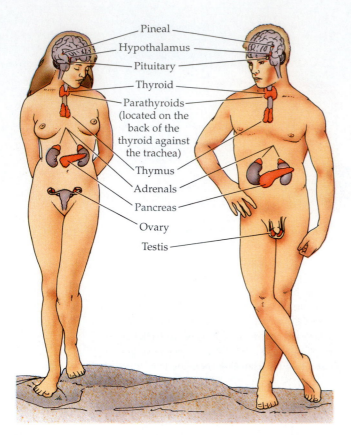

Pineal
Hypothalamus
Pituitary
Thyroid
Parathyroids
(located on the
back of the
thyroid against
the trachea)
Thymus
Adrenals
Pancreas
Ovary
Testis

Figure 24-12 The human endocrine organs. (Reprinted from W. K. Purves, G. H. Orians, and H. C. Heller, *Life: The Science of Biology,* 4th ed. Sinauer Associates, Inc., and W. H. Freeman and Company, 1995.)

be induced by interactions with small molecules, and that such changes were crucial to a transcriptional regulatory system that could rapidly respond to different environmental conditions. Similarly, the steps in ligand-receptor binding and in signaling within the cell depend on conformational changes. For example, conformational changes underlie the activation of signaling caused by the binding of ligands to receptors, of protein kinases that phosphorylate specific amino acids on specific proteins, and of certain proteins that have biological activity when they are bound to GTP. Not only do these conformational changes permit rapid response to an initial signal, but they also are readily reversible, enabling signals to be shut down rapidly and permitting recycling of the components of the signaling system so that they are ready to receive further signals.

The Steroid Hormone Receptor Superfamily

A diverse group of ligands, including various steroid hormones and other relatively small molecules, such as thyroid hormone and vitamin D, are able to pass through the plasma membrane of the cell directly because of their nonpolar structures. The receptors for these ligands (Figure 24-14a) are called the **steroid hormone receptor** superfamily, a group of structurally and evolutionarily related proteins named for the first ligands to be associated with such receptors. The ligands bind to these receptors, which are sequestered in the cytoplasm where they are bound to other proteins (Figure 24-14b). When ligand binds to its receptor,

in a coordinated fashion. Hormones are produced in organs of the **endocrine system,** a series of glands distributed throughout the body (Figure 24-12). Hormones are released into the circulatory system and can thus reach organs quite distant from their site of release (Figure 24-13a). Some endocrine signals act upon target tissues throughout the body. Examples of these are the androgens (masculinizing hormones) and estrogens (feminizing hormones) secreted by the testis and ovary, respectively. Other hormones act only upon one or a few target tissues, such as some of the hormones released by the pituitary gland that act upon the ovary.

Yet other secreted ligands act in a **paracrine** fashion (Figure 24-13b). That is, they do not enter the circulatory system, but act only locally, in some cases only on immediately adjacent cells. Paracrine signals are an extremely important part of the apparatus to control development, as will be discussed in Chapter 25.

Regulation of Signaling through Conformational Changes in Proteins

In discussions of the operon (Chapter 18), we saw that conformational changes in repressor or inducer proteins could

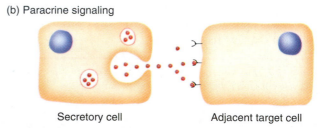

(a) Endocrine signaling

Blood vessel

Hormone secretion
into blood by endocrine gland

Distant target cells

(b) Paracrine signaling

Secretory cell   Adjacent target cell

Figure 24-13 Modes of intercellular signaling. (a) Endocrine signals enter the circulatory system and can be received by distant target cells. (b) Paracrine signals act locally and are received by nearby target cells. (Reprinted from H. Lodish, D. Baltimore, A. Berk, S. L. Zipursky, P. Matsudaira, and J. Darnell, *Molecular Cell Biology,* 3d ed. Scientific American Books, 1995.)

(a)

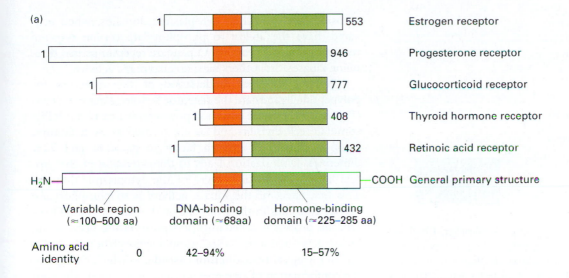

Variable region (≈100–500 aa)	DNA-binding domain (≈68aa)	Hormone-binding domain (≈225–285 aa)
Amino acid identity 0	42–94%	15–57%

(b)

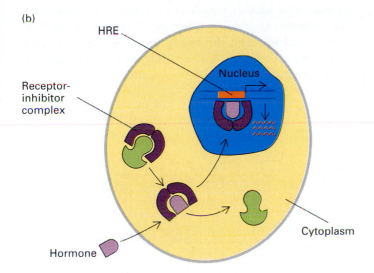

Figure 24-14 Action of steroid hormones and their receptors. (a) Some examples of steroid receptor superfamily members. In orange are the domains that bind to a specific HRE sequence (see text), and in green are the regions that bind the specific hormone. These DNA- and hormone-binding regions are the most conserved portions of the proteins, as indicated by the percentages of amino acid identity between different pairs of receptors. (b) The general model for activation of a steroid hormone receptor by its hormone. (c) Some examples of HRE sequences for different steroid hormone receptors. GRE, ERE, and VDRE are the glucocorticoid, estrogen, and vitamin D response elements, respectively. The arrows indicate identical sequences to which the various hormone receptor proteins bind. (Reprinted from H. Lodish, D. Baltimore, A. Berk, S. L. Zipursky, P. Matsudaira, and J. Darnell, *Molecular Cell Biology*, 3d ed. Scientific American Books, 1995.)

(c)

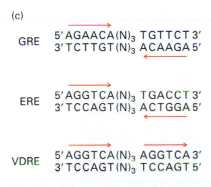

Mammalian sex determination occurs via the action of steroid hormones and receptors. For example, in human embryos, at about 6–8 weeks of gestation, sex is determined on the basis of whether or not testes have formed. If testes are present, they secrete testosterone—a masculinizing steroid hormone—that activates the testosterone receptor in target cells and induces transcription of male-specific genes. In the absence of testosterone, the embryo develops along a pathway that is otherwise in place, leading to female development. Thus, sex determination in mammals is a developmental on-off switch. In the absence of the activation of the testosterone receptor, the so-called default developmental program, which is in place, will lead to a female phenotype. The endocrine signal that activates the testosterone receptor is the "on" switch that shunts the organism into the male program of development. The biology of mammalian sex determination will be considered in more detail in Chapter 25.

changes in the structure of the receptor cause it to disassociate from the sequestering protein and move into the nucleus. In the nucleus, the hormone-receptor complex is then able, in concert with other transcription factors, to activate or repress transcription of target genes by binding to cis-regulatory sequences called *hormone response elements* (HREs) (Figure 24-14c).

As we saw in Chapter 23, this is an example of a common type of developmental decision, in which a cell is des-

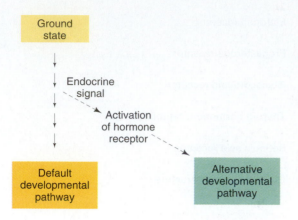

Figure 24-15 A general view of a developmental switch. Development proceeds along a default pathway unless, at a specific point in the pathway, a signal is received that redirects development along an alternative pathway. For example, in mammalian sex determination, a testosterone signal, by activating the androgen receptor, can redirect development from the female pathway into the alternative male pathway.

tined to go down a preset developmental pathway, but can be shunted into another pathway if the right signal is received (Figure 24-15). Further, for such endocrine signals, their effects can be global—a switch thrown by the action of one endocrine organ can coordinate changes in the developmental state (for example, male versus female) of all the tissues of the body.

Transmembrane Receptors and Signal Transduction Pathways

Many ligands are polypeptides or other polar molecules that cannot directly pass through the plasma membrane. Their targets are just outside the plasma membrane, where they interact with receptors anchored in the membrane. These ligand-receptor interactions initiate chemical signals in the cytoplasm just inside the plasma membrane of the cell. Such signals are passed through a series of intermediary molecules until they finally alter the structure of transcription factors in the nucleus, leading to the activation of transcription of some genes and repression of others.

Transmembrane receptors have one portion (the extracellular domain) outside of the cell, a middle portion that passes once or several times through the plasma membrane, and another portion located inside the plasma membrane (the cytoplasmic domain) (Figure 24-16).

The extracellular domain of the receptor is the site to which the ligand binds. Many polypeptide ligands are dimers and can simultaneously bind two receptor monomers. This simultaneous binding brings the cytoplasmic domains of the two receptor subunits into close proximity and activates the signaling activity of these cytoplasmic domains. Some receptors for polypeptide ligands are **receptor**

tyrosine kinases. Their cytoplasmic domains, when activated, have the ability to phosphorylate certain tyrosine residues on target proteins. Others are receptor serine/threonine kinases. Still others have no enzymatic activity.

Perhaps the best understood of the receptors for polypeptide ligands are the receptor tyrosine kinases (RTKs) (Figure 24-17). Some common polypeptide ligands for RTKs stimulate cell division and so are termed **growth factors.** Upon binding of a growth factor to its RTK, and RTK dimerizes. Dimerization leads to the activation of the protein kinase enzymatic activity of the cytoplasmic domain. This first effect of the kinase activity is to phosphorylate several tyrosines in the cytoplasmic domain of the RTK itself; this process is called **autophosphorylation** because the kinase acts upon itself. Once autophosphorylation occurs, it initiates a **signal transduction cascade,** in which changes in the conformation of one protein cause it to modify the conformation of another protein. One protein can modify the conformation of another in two possible ways: through direct contact between the two proteins, or by chemically modifying the other protein (typically through phosphorylation). Eventually, the signal transduction cascade leads to

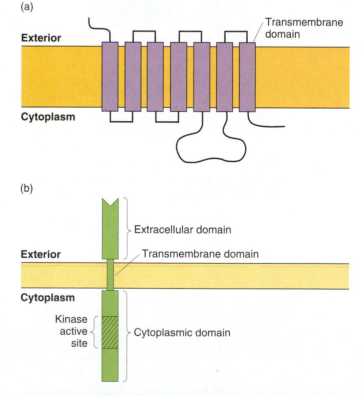

Figure 24-16 Examples of transmembrane receptors. (a) A receptor that passes through the cell membrane seven times. (b) A receptor tyrosine kinase (RTK). It has a single transmembrane domain. The extracellular domain binds to ligand. The active site of the tyrosine kinase is in the cytoplasmic domain. (Adapted from H. Lodish, D. Baltimore, A. Berk, S. L. Zipurksy, P. Matsudaira, and J. Darnell, *Molecular Cell Biology,* 3d ed. Scientific American Books, 1995.)

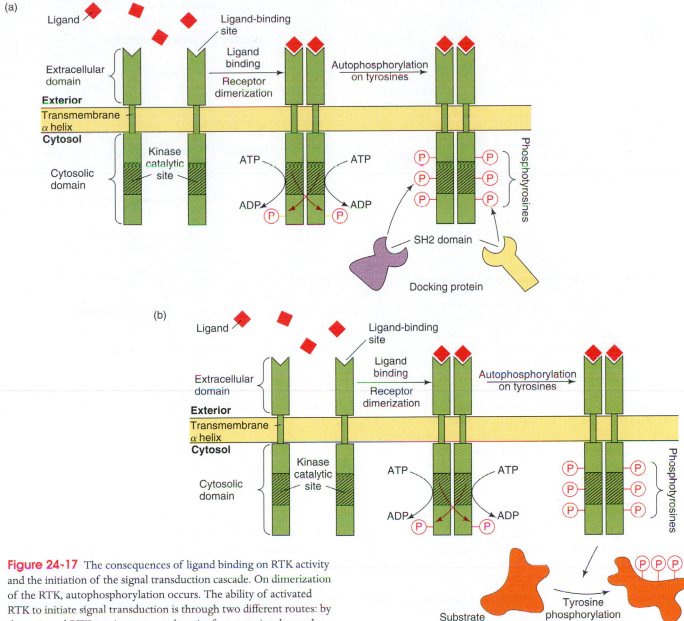

Figure 24-17 The consequences of ligand binding on RTK activity and the initiation of the signal transduction cascade. On dimerization of the RTK, autophosphorylation occurs. The ability of activated RTK to initiate signal transduction is through two different routes: by the activated RTK serving as an anchor site for some signal transduction proteins and by the activated RTK phosphorylating other proteins. (a) One consequence of autophosphorylation is to create binding sites for adaptor or "docking" proteins that bind to sites that include specific phosphorylated tyrosine residues. Through the docking of different adaptor proteins in close proximity to one another, protein-protein conformational changes lead to induction of the signal transduction activities of these proteins. (b) The other consequence of dimerization and autophosphorylation is that the RTK phosphorylates "substrate" proteins and, through this phosphorylation, induces their signal transduction activities. (Adapted from H. Lodish, D. Baltimore, A. Berk, S. L. Zipursky, P. Matsudaira, and J. Darnell, *Molecular Cell Biology*, 3d ed. Scientific American Books, 1995.)

the modifications of transcription factors and hence to changes in the activities of many genes in the target cell.

In the case of RTKs, there are two consequences of autophosphorylation of the receptor, both of which contribute to activation of the signal transduction cascades. Owing to conformational changes in the autophosphorylated RTK, the phosphorylated receptor binds to certain target proteins and phosphorylates them, causing conformational changes in these target proteins that allow them to propagate the signal. In addition, the phosphorylated RTK binds to various **adaptor proteins.** Interactions among these "docked" adaptor proteins then induce conformational changes in a signaling element, activating it and thereby propagating the signal.

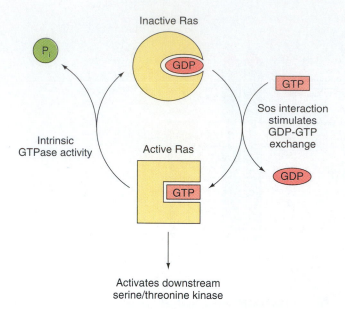

Inactive Ras

P_i

GDP

GTP

Sos interaction
stimulates
GDP-GTP
exchange

Intrinsic
GTPase activity

Active Ras

GDP

GTP

Activates downstream
serine/threonine kinase

Figure 24-18 An example of the G-protein activity cycle. Ras is a member of the G-protein family. When Ras binds GDP, it does not signal. Through direct interactions with Ras, another protein called *Sos* causes conformational changes in Ras so that it preferentially binds GTP. The Ras-GTP complex is able to interact in turn with a cytoplasmic serine/threonine kinase, activating its kinase activity and thus transmits the signal to the next step in the signal transduction pathway. When Ras-GTP is released from Sos, it hydrolyzes GTP to GDP and reassumes the inactive Ras-GDP state. (Adapted from J. D Watson, M. Gilman, J. Witkowski, and M. Zoller, *Recombinant DNA*, 2d ed. Copyright © 1992 by Scientific American Books.)

Growth factor

Receptor tyrosine kinase (RTK)

Adapter protein

Sos

GDP ← → GTP

Ras·GDP (inactive)   Ras·GTP (active)

P_i

Raf (serine/threonine kinase)

MEK (dual-specificity kinase)

MAP kinase (serine/threonine kinase)

Nuclear transcription factors

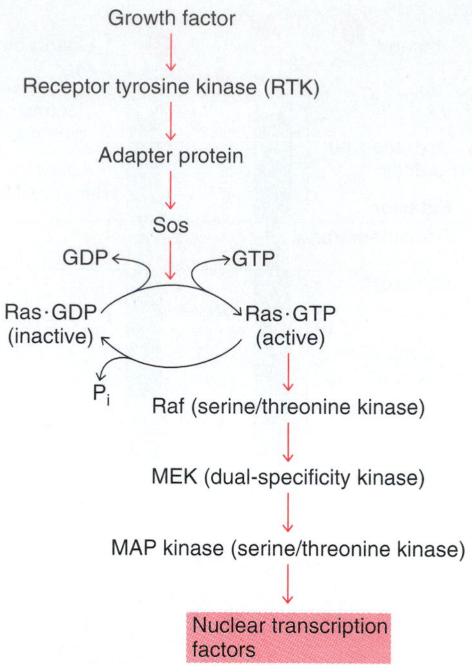

Figure 24-19 One pathway for RTK signaling. Raf, MEK, and MAP kinase are three cytoplasmic protein kinases that are sequentially activated in the signal transduction cascade. (Adapted from H. Lodish, D. Baltimore, A. Berk, S. L. Zipursky, P. Matsudaira, and J. Darnell, *Molecular Cell Biology*, 3d ed. Scientific American Books, 1995.)

Quite often, the next step in propagating the signal is to activate a **G-protein.** G-proteins cycle between being bound by GDP (the inactive state) or GTP (the activated state). The propagation of the signal from the RTK leads to the activation of a protein that binds to the inactive GDP-bound G-protein, changing its conformation so that it then binds to a molecule of GTP (Figure 24-18). The specific G-protein called *Ras* is of special importance in carcinogenesis, as discussed later.

The activated GTP-bound G-protein then binds to a cytoplasmic protein kinase, in turn changing its conformation and activating its protein kinase activity. This protein kinase then phosphorylates other proteins, including other protein kinases. The targets of some of these protein kinases are transcription factors. The phosphorylation of the transcription factors changes their conformations, leading to activation of transcription of some genes and repression of others (Figure 24-19).

One example of how paracrine signaling through an RTK contributes to developmental decisions is seen in the formation of the vulva of the nematode *C. elegans* (Figure 24-20), a process that has been studied in detail through the analysis of mutants that have either no vulva or too many. The vulva is the opening in the body wall of a worm that

permits eggs to pass out of the animal from the uterus. Within the body wall of the worm, called the *hypoderm*, several cells have the potential to build certain parts of the vulva. To make an intact vulva, one of the cells must become the primary vulva cell, and two others secondary vulva cells, while others become tertiary cells that contribute to the surrounding hypoderm (Figure 24-21).

Initially, all these cells can adopt any of these roles and so are called an **equivalence group.** The key to allocating the different roles to these cells is another single cell, called the *anchor,* which lies underneath the cells of the equivalence group. The anchor cell secretes a polypeptide ligand that binds to an RTK present on all the cells of the equivalence group. Only the cell that receives the highest level of this signal (the equivalence group cell nearest the anchor cell) activates the signal transduction pathway at a sufficient level to activate the transcription factors necessary for that cell to adopt the primary role. Having acquired its fate, the primary vulva cell sends out a different paracrine signal to its immediate neighbors in the equivalence group, inhibiting those cells from similarly interpreting the anchor cell signal to also adopt the primary role. This process of **lateral inhibition** leads these neighboring cells to adopt the secondary fate. The remaining cells of the equivalence group develop

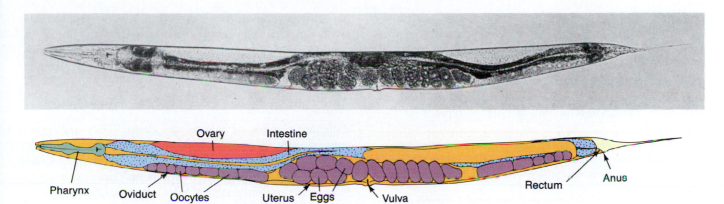

Figure 24-20 Adult *Caenorhabditis elegans.* Photomicrograph and drawing of an adult hermaphrodite, showing various organs and nuclei readily identified by their location. Note the position of the vulva midway along the anterior-posterior axis of the worm (From J. E. Sulston and H. R. Horvitz, *Developmental Biology* 56, 1977, 111.)

Figure 24-21 The production of the *C. elegans* vulva from the equivalence group by cell-cell interactions. (a) The primary, secondary, and tertiary cell types are distinguished by the cell division patterns that they undergo. (b) Early in development, no signal from the anchor cell has occurred, and all the equivalence group cells are in the default tertiary cell state. (c) Later in development, the anchor cell sends a signal that activates an RTK signal transduction cascade. The equivalence group cell nearest the anchor cell receives the strongest signal and becomes the primary vulva cell. It then sends out lateral inhibition signals to its neighbors, preventing them from also becoming primary vulva cells and shunting them into the secondary vulva cell pathway. (d) The portions of the vulva anatomy occupied by the descendants of the equivalence group cells. (Part a from R. Horvitz and P. Sternberg, *Nature* 351, 1991, 357. Parts b–d adapted from I. Greenwald, *Trends in Genetics 7,* 1991, 366.)

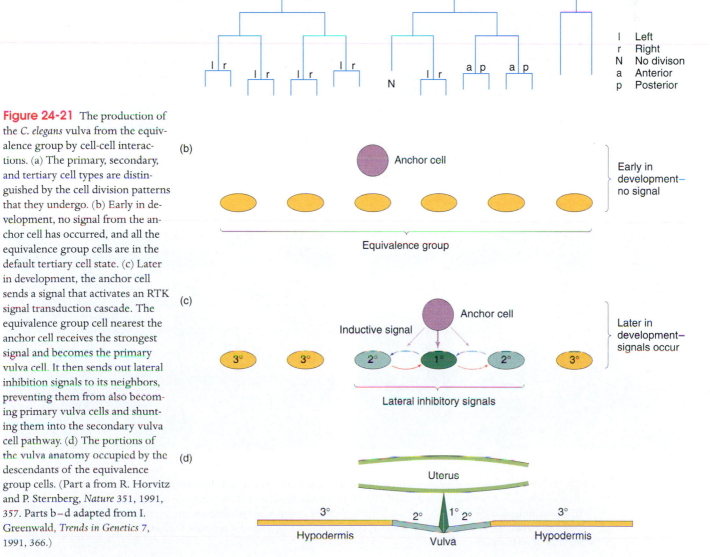

as tertiary cells and contribute to the hypoderm surrounding the vulva. For each of the three cell types that the equivalence group develops into, there is a specific constellation of transcription factors that are activated and that typify the state of the cell: primary, secondary or tertiary. Thus, through a series of paracrine intercellular signals, a group of equivalent cells can develop into distinct cell types.

Message A signal sent by certain cells or organs can change the status of other cells or tissues through alterations to the activities of transcription factors in the cells receiving the signal.

Cancer: A Genetic Disease of Somatic Cells

In the previous sections, we have discussed several areas of cell biology that impact on genetics and development. Two of these areas in particular—the cell cycle and intercellular communication—have had a tremendous impact on our understanding of abnormal cellular behavior in certain diseases, most notably cancer. Tumor cells are uncoupled from the regulatory mechanisms that normally keep proliferation in check. As we will see in the succeeding sections, this uncoupling occurs through an accumulation of mutations in somatic cells such that the normal brakes on the cell cycle—through checkpoint control and paracrine regulatory signals—are abrogated.

In genetics, we study the abnormal (mutations) in order to make inferences about how wild-type genes function and how normal processes operate. Similarly, by studying abnor-

mal cell proliferation (cancer), we can hope to gain insights into the mechanisms underlying normal growth control. In addition, we can hope to learn how to improve our diagnosis, treatment, and control of this major group of diseases.

How Cancer Cells Differ from Normal Cells

Malignant tumors, or cancers, are clonal. They are aggregates of cells, all derived from an initial aberrant founder cell that, although surrounded by normal tissue, is no longer integrated into that environment. Cancer cells often differ from their normal neighbors by a host of specific phenotypic changes, such as rapid division rate, invasion of new cellular territories, high metabolic rate, and altered shape. For example, when cells from normal epithelial cell sheets are placed in cell culture, they can grow only when anchored to the culture dish itself. In addition, normal epithelial cells in culture divide until they form a continuous monolayer. Then, they somehow recognize that they have formed an epithelial sheet and stop dividing. In contrast, malignant cells derived from epithelial tissue continue to proliferate, piling up on one another (Figure 24-22). Clearly, the factors regulating normal cell differentiation have been altered. What, then, is the underlying cause of cancer? Many different cell types can be converted to a malignant state. Is there a common theme to the ontogeny of these different types of cancer, or do they each arise in quite different ways? Indeed we can think about cancer in a general way: as occurring by the production of multiple mutations in a single cell that cause it to proliferate out of control. Some of those mutations may be transmitted from the parents through the germ line. Others arise *de novo* in the somatic cell lineage of a particular cell.

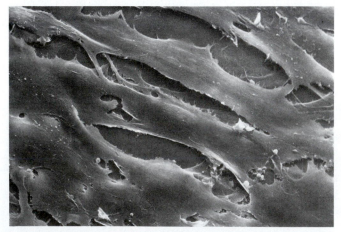

(a)

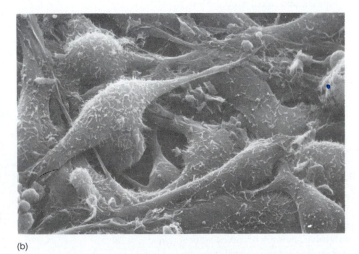

(b)

Figure 24-22 Scanning electron micrographs of (a) normal cells and (b) cells transformed with Rous sarcoma virus. (a) A normal cell line called 3T3. Note the organized monolayer structure of the cells. (b) A transformed derivative of 3T3. Note how the cells are rounder and piled up on one another. (From J. Darnell, H. Lodish, and D. Baltimore, *Molecular Cell Biology*, 2d ed. Copyright © 1990 by Scientific American Books.)

Evidence for the Genetic Origin of Cancers

As briefly discussed in Chapter 7, several lines of evidence have pointed to a genetic origin for the transformation of cells from the benign to the cancerous state. For many years, it has been known that most carcinogenic agents (chemicals and radiation) are also mutagenic. There are occasional instances in which certain cancers are inherited as highly penetrant single Mendelian factors; an example is familial retinoblastoma (see Figure 7-25). Perhaps reflecting the more general case are less penetrant susceptibility alleles that increase the probability of developing a particular type of cancer. In the last few years, several alleles have been recombinationally mapped and molecularly cloned and localized using RFLP mapping or related techniques (see Chapter 15). Oncogenes, dominant mutant genes that contribute to cancer in animals (see pages 482–486), have been isolated from tumor viruses, viruses that can transform normal cells in certain animals into tumor-forming cells. Such dominant oncogenes can also be isolated from tumor cells, using cell-culture assays that can distinguish between some types of benign and malignant cells. A tumor does not arise as a single genetic event, but rather involves a multiple-hit process, in which several mutations must arise within a single cell for it to become cancerous. In some of the best-studied cases, the progression of colon cancer and astrocytoma (a brain cancer) have each been associated with the sequential accumulation of several different mutations in the malignant cells (Figure 24-23). In the following sections, we will consider further the genetic origin of cancers and the nature of the proteins that are altered by cancer-producing mutations. We will see that many of these proteins are involved in intercellular communication and regulation of the cell cycle.

Message Many, perhaps all, tumors arise through a series of sequential mutational events that lead to a state of uncontrolled proliferation.

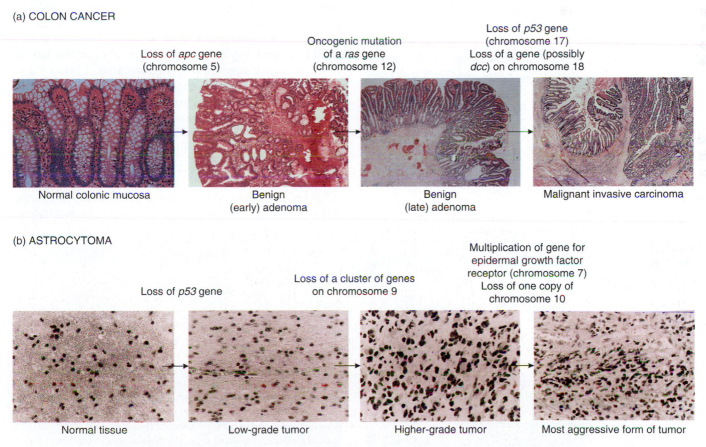

Figure 24-23 The multistep progression to malignancy in cancers of the colon and brain. Several histologically distinct stages can be distinguished in the progression of these tissues from the normal state to benign tumors to a malignant cancer. (a) A common sequence of mutational events in the progression to colon cancer. Note that the tissue becomes more disorganized as the tumor progresses toward malignancy. (b) A different characteristic series of mutations marks the progression toward a malignant astrocytoma, a form of brain cancer. (Micrographs by E. R. Fearon and K. R. Cho. From W. K. Cavenee and R. L. White, *Scientific American*, March 1995, 78–79.)

Mutations in Cancer Cells

Two general kinds of mutations are associated with tumors: dominant **oncogene** mutations and mutations in recessive **tumor suppressor genes.** Oncogenes are mutated in such a way that the protein they encode is "activated" in tumor cells. Typically, oncogenes are mutations in components of one of the intercellular communication pathways such that the cell behaves as if it is always receiving a signal to proliferate, even when it is not. Mutations in tumor suppressor genes inactivate proteins that normally contribute to the inhibition of cell proliferation. These genes are of two types. One type encodes proteins involved in inhibiting progression through the cell cycle. In recessive tumor-promoting mutations of these tumor suppressor genes, the inhibitory protein is inactivated. The other type of tumor suppressor gene encodes a protein involved in the repair of damaged DNA. Here, indirectly, mutations that inactivate these DNA repair genes induce tumor growth by elevating the rate at which oncogene and other recessive tumor-promoting mutations occur.

How have tumor-promoting mutations been identified? Several approaches have been used. With modern pedigree analysis techniques, familial tendencies toward certain kinds of cancer can be mapped relative to molecular markers such as RFLPs (see Chapter 15), and in several cases this has led to the successful identification of the mutated genes. Cytogenetic analysis of tumor cells themselves has also proved invaluable. Many types of tumors are typified by characteristic chromosomal translocations, or deletions of particular chromosomal regions. In some cases, these chromosomal rearrangements are so reliably a part of a particular cancer that they can be used for diagnostic purposes. For example, 95 percent of patients with chronic myelogenous leukemia (CML) have a characteristic translocation between chromosomes 9 and 22. This translocation, called the *Philadelphia*

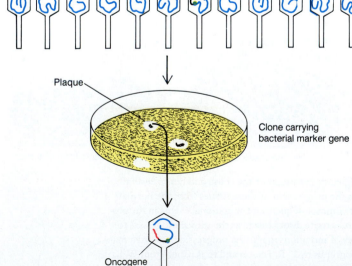

Human bladder carcinoma DNA

Cleave, link each fragment to bacterial marker gene

Marker gene

Transform mouse fibroblasts

Extract DNA, transform new mouse fibroblasts

Extract DNA

Surviving human DNA

Introduce into defective phage vector

Phage library

Plaque

Clone carrying bacterial marker gene

Oncogene

Figure 24-24 A method for identifying the DNA clone carrying a human oncogene. DNA from a human bladder cancer tumor is extracted and cleaved; each fragment is attached to a bacterial marker gene and then used to transform mouse fibroblast cells. A second round of transformation of mousefibroblasts enriches for those human DNA fragments that can induce rapid transformation of the fibroblasts (assayed through the formation of foci) as if those cells are now cancerous. DNA from cells that form foci is then cleaved and cloned in a defective phage that can grow only in the presence of the bacterial marker gene. When plated onto bacteria, the presence of the bacterial marker gene is indicated by phage plaques, which are then tested for the oncogene by the same mouse fibroblast transformation assay.

chromosome after the city where this translocation was first described, is a critical part of the CML diagnosis. We will discuss the Philadelphia chromosome in more detail later in this chapter. Other translocations characterize other sorts of tumors; diagnostic translocations are most often found associated with cancers of the white blood cells—leukemias and lymphomas. However, not all tumor-promoting mutations are specific to a given type of cancer. Rather, the same mutations seem to be tumor-promoting for a variety of cell types and thus are seen in many different cancers.

Some oncogenes have been identified through the use of tumor viruses. By mutational analysis of these viruses, usually a single gene within the small genome of such a virus can be associated with the ability to transform cells to the malignant state. It appears that these **viral transforming genes** are mutated versions of normal cellular genes. Another method, also used to identify oncogenes (Figure 24-24), relies on a logic parallel to that of Avery, McCarty, and McLeod (Chapter 11) in proving that DNA is the genetic material. Normal cells in culture are bathed in DNA isolated from tumor cells under conditions such that the cells will take up the DNA (this process is called *transfection*). If, through transfection, an oncogene from the tumor is incorporated into a chromosome of a cell, that cell will then be transformed and will form a raised colony (or focus) on soft agar plates, whereas normal cells are unable to form

foci. (Recall that normal cells cannot grow without being anchored and stop growing when they form a continuous monolayer of cells.) By recombinant DNA techniques, oncogene DNA from a focus is then joined to DNA from a bacterial plasmid or phage used as a cloning vector, which transfers the oncogene DNA into a host cell. By this means, it is possible to purify the oncogene by recombinant DNA techniques.

In the case of human bladder cancers, the oncogene turned out to be a mutated form of a gene called *ras*, whose wild-type protein product is a G-protein involved in RTK signal transduction (see Figure 24-18). The same *ras* oncogene was identified in bladder tumors from several patients. This finding, that many tumors of a particular type all possess similar sets of oncogenes, has been encountered for several different types of cancer. This finding is the basis of the motivational pathways outlined in Figure 24-23.

Message   Tumor-promoting mutations can be identified in a variety of ways. Once located, they can be cloned and studied to learn how they contribute to the malignant state.

Oncogenes

Roughly 100 different oncogenes have been identified (see some examples in Table 24-2; some of their chromosomal locations are shown in Figure 7-24). What are the normal

Table 24-2   Some Well-Characterized Oncogenes and the Proteins They Encode

Oncogenes*	How isolated†	Protein Location	Protein Function‡
Nuclear transcription factors			
jun	Retrovirus	Nucleus	Transcription factor (AP-1)
fos	Retrovirus	Nucleus	Transcription factor (AP-1)
myc	Retrovirus, tumor	Nucleus	DNA-binding protein
erbA	Retrovirus	Nucleus	Member of steroid-receptor family
Intracellular signal transducers			
src	Retrovirus	Cytoplasm	Protein-tyrosine kinase
abl	Retrovirus	Cytoplasm	Protein-tyrosine kinase
raf	Retrovirus	Cytoplasm	Protein-serine kinase
gsp	Tumor	Cytoplasm	G-protein α subunit
ras	Retrovirus, tumor	Cytoplasm	GTP/GDP-binding protein
Growth factors			
sis	Retrovirus	Secreted	Growth factor (PDGF)
Growth factor receptors			
erbB	Retrovirus	Membrane	Receptor for EGF
fms	Retrovirus	Membrane	Receptor for CSF-1
trk	Tumor	Membrane	Receptor for NGF

* Oncogenes can be divided into four categories based on the function of the proteins they encode. Virtually all known oncogenes encode constituents of growth-factor signal transduction pathways.

† Most oncogenes were first found as retroviral oncogenes, but some were first isolated or extensively characterized as oncogenes in human tumors.

‡ PDGF = Platelet-derived growth factor; EGF = epidermal growth factor; CSF-1 = colony-stimulating factor 1; NGF = nerve growth factor; GTP/GDP = guanosine tri- and diphosphate.

Source: After J. D. Watson, M. Gilman, J. Witkowski, and M. Zoller, Recombinant DNA, 2d ed. Copyright © 1992 by Scientific American Books.

(a)

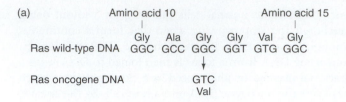

(b)

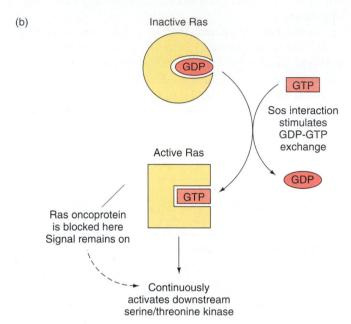

Figure 24-25 The Ras oncoprotein. (a) The *ras* oncogene differs from the wild type by a single base pair, producing a Ras oncoprotein that differs from the wild type in one amino acid, at position 12 in the *ras* open reading frame. (b) The effect of this missense mutation is to create a Ras oncoprotein that cannot hydrolyze GTP to GDP (compare with the normal Ras cycle depicted in Figure 24-18). Because of this defect, the Ras oncoprotein remains in the active Ras-GTP complex and continuously activates the downstream serine/threonine kinase (see Figure 24-19).

Message Oncogenes encode oncoproteins which are altered forms of proteins which participate in the intercellular communication pathways which normally control cell proliferation.

The change from normal protein to oncoprotein often involves structural modifications to the protein itself, such as those caused by simple point mutation. A single base-pair substitution that converts glycine to valine at amino acid number 12 of the Ras protein, for example, creates the oncoprotein found in human bladder cancer (Figure 24-25a). Recall that the normal Ras protein is a G-protein subunit involved in signal transduction and, as described earlier in this chapter, normally functions by cycling between the active GTP-bound state and the inactive GDP-bound state (Figure 24-18). The amino acid change caused by the *ras* oncogene missense mutation produces a oncoprotein that always binds GTP, even in the absence of the normal upstream signals that would permit such binding for a wild-type Ras protein (Figure 24-25b). In this way, the Ras oncoprotein continually propagates a signal that promotes cell proliferation.

Structural alterations can also occur by deletion of parts of a protein. The *v-erbB* oncogene encodes a mutated form of an RTK known as the *EGFR*, a receptor for the epidermal growth-factor (EGF) ligand (Figure 24-26). The mutant form of the EGFR lacks the extracellular, ligand-binding domain as well as some regulatory components of the cytoplasmic domain. The result of these deletions is that the truncated *v-erbB* EGFR oncoprotein is able to dimerize even in the absence of the EGF ligand. The constitutive EGFR oncoprotein dimer is always autophosphorylated through its tyrosine kinase activity, and thus continuously initiates a signal transduction cascade.

Perhaps the most remarkable type of structurally altered oncoprotein is one caused by a gene fusion. The classic example of this emerged from studies of the Philadelphia chromosome, which, as we have already discussed, is a translocation between chromosomes 9 and 22 that is a diagnostic feature of chronic myelogenous leukemia (CML). Recombinant DNA methods have shown that the breakpoints of the Philadelphia chromosome translocation in different CML patients are quite similar, and cause the fusion of two genes, *bcr1* and *abl* (Figure 24-27). The *abl* proto-oncogene encodes a cytoplasmic tyrosine-specific protein kinase. The *bcr1-abl* fusion oncoprotein has an activated protein kinase activity that is responsible for its oncogenic state.

Some oncogenes produce an oncoprotein that is identical in structure with the normal protein. In these cases, the oncogene mutation induces misexpression of the protein — that is, it is expressed in cell types from which it is ordinarily absent. Several such oncogenes that cause misexpression are also associated with chromosomal translocations diagnostic of various B lymphocyte tumors. (Recall from Chapter 23 that B lymphocytes and their descendants, plasma cells, are the cells that synthesize antibodies, or immunoglobulins.) In

functions of their cellular counterparts, sometimes called **proto-oncogenes**? Proto-oncogenes generally define a class of proteins that are selectively active only when the proper regulatory signals allow them to be activated, usually by some allosteric interaction. Indeed, many proto-oncogene products are elements of intercellular communication pathways, including growth-factor receptors, signal transduction proteins, and transcription factors. However, in oncogene mutations, the activity of the mutant **oncoprotein** has been uncoupled from the upstream regulatory pathway that ought to be controlling its activation, leading to continuous unregulated expression of the oncoprotein (Figure 24-25). Several categories of oncogenes have been identified that depict different ways in which the upstream regulatory functions have been uncoupled. We will look at examples of some of these.

Figure 24-26 An oncogenic mutation affecting signaling between cells. The EGFR, the normal receptor for epidermal growth factor (EGF) has a ligand-binding domain outside the cell, a transmembrane domain that allows the protein to span the plasma membrane, and an intracellular domain that has tyrosine-specific protein kinase activity. Normally, the kinase activity is activated only when EGF binds to the ligand-binding domain, and this activity is only transitory. The erythroblastosis tumor virus carries the *v-erbB* oncogene, which encodes a mutant form of the EGFR that lacks pieces at both ends. These deletions allow the mutant protein to constitutively dimerize, leading to continuous autophosphorylation, resulting in continuous transduction of a signal from the receptor. (Adapted from J. D. Watson, M. Gilman, J. Witkowski, and M. Zoller, *Recombinant DNA,* 2d ed. Copyright © 1992 by Scientific American Books.)

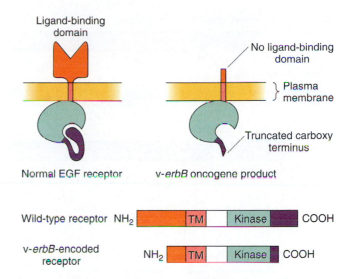

these B cell oncogene translocations, no protein fusion is produced, but, rather, the chromosomal rearrangement causes a gene near one breakpoint to be turned on in the wrong tissue. In Burkitt's lymphoma, for example, a translocation joins the *c-myc* proto-oncogene, located on human chromosome 8, to one of the three genes that encode the immunoglobulins *(Ig),* which are located on chromosomes 2, 14, and 22 (Figure 24-28). The *c-myc* protein is a transcription factor that is activated in many tissues in order to induce proliferation. The *Ig-c-myc* translocation puts the *c-myc* gene very near a tissue-specific immunoglobulin enhancer, which ordinarily functions to activate *Ig* gene transcription

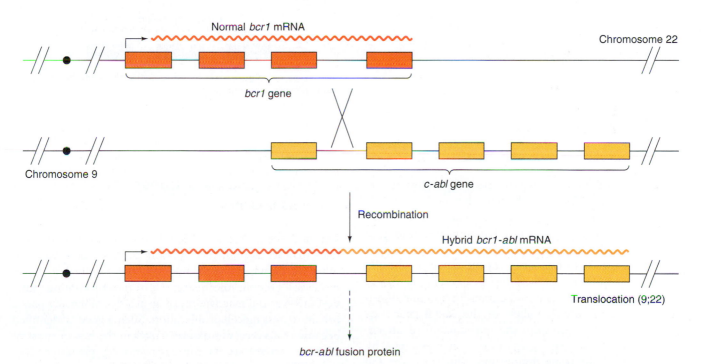

Figure 24-27 The chromosome rearrangement in CML. The Philadelphia chromosome, which is diagnostic of CML, is a translocation between chromosomes 9 and 22. The translocation breakpoints are in the middle of the *c-abl* gene, which encodes a cytoplasmic protein tyrosine kinase, and the *bcr1* gene, which has an unknown function. The translocation produces a hybrid *bcr1-abl* protein that lacks the normal controls repressing the protein tyrosine kinase activity. Only one of the two rearranged chromosomes of the reciprocal translocation is shown. (Adapted from J. D. Watson, M. Gilman, J. Witkowski, and M. Zoller, *Recombinant DNA,* 2d ed. Copyright © 1992 by Scientific American Books.)

(a)

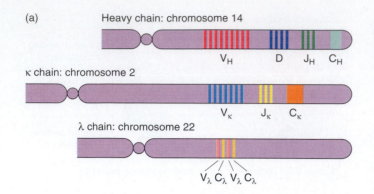

Figure 24-28 Fusion of an immunoglobulin gene to another gene, creating an oncogene. (a) The chromosomal locations of the human *heavy-chain Ig* gene and the two different versions of the *light-chain Ig* genes (κ and λ). (b) An example of a fusion between an Ig gene (in this example, the heavy-chain gene on chromosome 14) and the *c-myc* gene. In the fusion, the *c-myc* transcribed region is still intact (meaning that it will encode exactly the same protein as usual), but now it is turned on by the *heavy-chain Ig* enhancer (E). Thus, the c-myc onco-protein is expressed in B lymphocytes, leading to Burkitt's lymphoma. (Part a adapted from J. Darnell, H. Lodish, and D. Baltimore, *Molecular Cell Biology*. Copyright © 1986 by Scientific American Books. Part b adapted from J. D. Watson, M. Gilman, J. Witkowski, and M. Zoller, *Recombinant DNA*, 2d ed. Copyright © 1992 by Scientific American Books.)

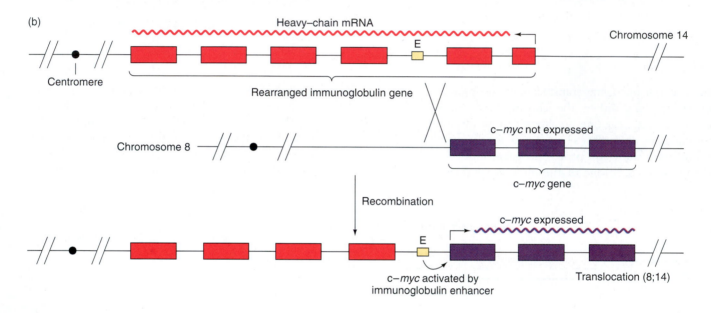

at high levels in B lymphocytes (see page 714). Normally, *c-myc* is not expressed in B lymphocytes, but in the Burkitt's lymphoma *Ig-c-myc* gene fusion, the *Ig* enhancer is co-opted to drive transcription of *c-myc* in B lymphocytes, helping to induce proliferation of these cells. There are strong parallels between this sort of dominant oncogene mutation and the dominant gain-of-function phenotypes caused by the fusion of the enhancer of one gene to the transcription unit of another in producing the *Tab* allele of the Abd-B gene (see pages 703–704). In each case, the introduction of an enhancer causes a dominant gain-of-function phenotype by misregulation of the transcription unit. Mutations like *Tab* arise in the germ line and are transmitted from one generation to the next, whereas most oncogene mutations arise in somatic cells and are not inherited by offspring.

Message Dominant oncogenes contribute to the oncogenic state by causing a protein to be expressed in an activated form or in the wrong cells.

Recessive Mutations in Tumor Suppressor Genes

The factors that are identified as oncogenes produce proteins that act as positive regulators of cell proliferation. In other words, when one of these proteins is activated, it promotes cell growth. Another type of tumor-promoting mutation involves the inactivation of genes that normally play a role in repressing cell proliferation. Such a gene is identified because of its recessive nature—only in the loss of most or all gene activity is its tumor-promoting phenotype observed. The genes altered in these recessive mutations are thought to act as **tumor suppressor genes** that normally function as negative regulators of cell proliferation. These tumor suppressor genes are turning out to be very numerous and to encode a diverse set of genetic functions. Some of the most important tumor suppressor genes identified thus far encode negative regulators of the cell cycle. Among these are the genes encoding the Rb and p53 proteins, discussed earlier in this chapter as important cell cycle regulators. In retinoblastoma, a cancer typically affecting young children (see pages 711–713), retinal cells lacking a functional *RB* gene proliferate out of control. These *RB* null cells either are homozygous for a single mutant *RB* allele or they are heterozygous for two different *RB* mutations. Most patients have one or a few tumors localized to one site in one eye, and the condition is sporadic—in other words, there is no previous history of retinoblastoma in the family and the affected individual does not transmit it to his or her offspring. This is because the *RB* mutation or mutations that inactivate both alleles of this autosomal gene arise in a somatic cell whose descendants populate the retina (see Figure 7-25). Presumably the mutations arise by chance at different times in development in the same cell lineage.

A few patients, however, have an inherited form of the disease, called *hereditary binocular retinoblastoma* (HBR). Such patients have many tumors, and the retinas of both eyes are involved. Paradoxically, even though *RB* is a recessive trait at the cellular level, the transmission of HBR is as an autosomal dominant (Figure 7-25). The paradox is resolved by realizing that in the presence of a germ-line mutation that knocks out one of the two copies of the *RB* gene, the mutation rate for *RB* makes it virtually certain that at least some of the retinal cells of HBR individuals will have acquired a mutation in the remaining *RB* gene, thereby producing cells with no functional RB protein.

Why does the absence of *RB* promote tumor growth? Recall from our discussion of the cell cycle that Rb protein functions to sequester the E2F transcription factor, preventing E2F from promoting transcription of genes whose products are needed for S phase functions such as DNA replication (see Figure 24-9 for a description of the RB-E2F interaction). Phosphorylation by active CDK-cyclin heterodimer causes the Rb protein to be inactivated. The inactive Rb is unable to bind E2F. In turn, this allows E2F to promote transcription of S phase genes. In homozygous null *RB* cells, Rb protein is permanently inactive. Thus, E2F is always able to promote S phase, and the arrest of normal cells in late G_1 does not occur in retinoblastoma cells.

Another very important recessive tumor-promoting mutation has identified the *p53* gene as a tumor suppressor gene. Mutations in *p53* are associated with many types of tumors. The active p53 protein serves as a monitor of DNA damage and prevents progression of the cell cycle until such damage is repaired. In the absence of a functional *p53* gene, cell division occurs even in the absence of DNA repair. This leads to an increase in mutations, chromosomal rearrangements, and aneuploidy, and increases the chances that other mutations promoting cell proliferation will occur. Other recessive tumor-promoting genes that have been identified recently also are implicated in the repair of DNA damage. Recent research suggests that genes which, when inactivated, produce the phenotype of elevated mutation rates are a very important contributor to the progression of tumors in humans. Such recessive tumor suppressor mutations that in-

terfere with DNA repair promote tumor growth indirectly, because their elevated mutation rates make it much more likely that a series of oncogene and recessive tumor-promoting mutations in genes encoding proteins of the intercellular communication and cell cycle control pathways will accumulate in a given cell.

The Complexities of Cancer

As we have discussed in this chapter, numerous mutations can occur that promote tumor growth. These mutations are thematically related and can be understood in terms of their cellular functions (Figure 24-29). In some instances such as colon cancer (Figure 24-23), we are even able to identify a series of independent mutations that contribute to the progression of a cell from a normal state through various stages of a benign tumor to a truly malignant state. The story doesn't stop there, however. Even among malignant tumors, their rates of proliferation and their abilities to invade other tissues—metastasize—are quite different. Undoubtedly, even after a malignant state is achieved, more mutations accumulate in the tumor cell that further promote its proliferation and invasiveness. Thus, there is a considerable way to go before we have a truly comprehensive view of how tumors arise and progress.

Message Some recessive mutations that contribute to tumor formation inactivate tumor suppressor genes whose normal function is to inhibit cell proliferation.

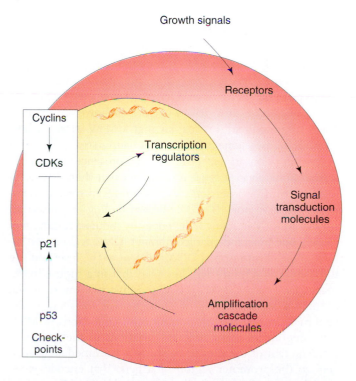

Figure 24-29 The types of molecules that control cell proliferation and that are altered by tumor-promoting mutations. (Adapted from A. Kamb, *Trends in Genetics* 11, 1995, 138.)

Higher eukaryotic cells have evolved mechanisms that control their structure and their ability to proliferate. These controls are all highly integrated and depend on the continual evaluation of the state of the cell, and continual communication of information among neighboring cells and between different tissues. Oncogenes encode oncoproteins, activated proteins of the intercellular communication pathways; these cause transcription factors whose role is to induce cell proliferation to be continuously activated. Recessive mutations in tumor suppressor genes promote tumor growth by knocking out the activity of proteins whose normal role is to inhibit progression through the cell cycle, or by generally elevating the mutation rate of the cell.

Concept Map

Draw a concept map interrelating as many of the following terms as possible. Note that the terms are listed in no particular order.

transcription factors / enhancer / oncogene / gain-of-function mutations / cyclin-dependent protein kinase / microtubules / checkpoints / signal transduction pathway / cell proliferation / RTK / tumor suppressor gene

CHAPTER INTEGRATION PROBLEM

MMTV, mouse mammary tumor virus, is an oncogenic retrovirus. It specifically causes tumors in the mammary glands of female mice and nowhere else. Unlike some other oncogenic retroviruses, it does not appear to produce its own oncogenic protein product (in contrast to the *v-erbB* oncogene carried by the erythroblastosis virus). Rather, MMTV encodes just the proteins necessary for its own reverse transcription into a DNA copy that integrates into the host genome, and for packaging into virion particles. It seems puzzling then that MMTV is an oncogenic virus.

Studies have provided two clues to how MMTV produces tumors. First, it turns out that MMTV carries a hormone response element (HRE) that causes strong increases in transcription in response to the presence of certain steroid hormones. Second, it turns out that there is usually just one MMTV insertion in the genome in mouse mammary gland tumors. The DNA surrounding the insertion sites of MMTV in many independently induced tumors was cloned out by standard recombinant DNA techniques, allowing the chromosomal DNA adjacent to the insertion site to be studied. This analysis revealed that in a mouse mammary gland tumor, MMTV is found integrated next to one of only a small number of sites (named *Int sites*) in the genome. (In contrast, in nontumorous infections, MMTV can integrate in many different locations in the genome.) *Int1*, the first of the sites to be studied, is immediately adjacent to the promoter region of a gene that encodes a secreted protein very similar to the *wg* protein of *Drosophila*, which is involved in cell-to-cell signaling during segmentation in the embryo.

a. Bearing in mind that female mammary gland development and lactation are dependent on certain female-specific steroid hormone signals, and based on the clues we have discussed, propose an explanation for how MMTV produces its oncogenic effects.

b. Propose an experimental approach for testing this hypothesis.

Solution

a. Because of the HRE that MMTV carries, it acts as a portable enhancer element. The steroid hormone receptor to which the HRE responds is probably expressed only in mammary glands, so that the HRE is a tissue-specific enhancer for female mammary glands. Thus, if the MMTV integrates near a gene that, when activated in female mammary glands cells, will deregulate cell proliferation, it has the potential to cause a tumor. In the case of *Int1*, for example, the protein is not normally expressed in mammary glands, but becomes expressed at high levels under the influence of the MMTV's HRE. In principle then, this is no different from dominant oncogenes arising from chromosomal rearrangements, as in Burkitt's lymphoma, in which an *Ig* enhancer is fused to the *c-myc* gene, activating *c-myc* in B lymphocytes. In the case of MMTV, however, a viral insertion rather than a chromosomal translocation causes the gene fusion.

b. This hypothesis postulates that the HRE is the only essential portion of MMTV with regard to oncogenesis, and that it acts via misregulation of the *Int1* gene. To test this, we could isolate a small DNA segment that included only the HRE of MMTV, and fuse this segment in vitro to a wild-type *Int1* gene. We could then directly inject this DNA into mouse blastocysts and integrate it randomly in the genome by germ-line transformation. Our prediction would be that each of these insertions should cause mammary gland tu-mors in females. We should also include two constructs as controls: first, germ-line transformation of a similar con-struct except that we place a nonsense mutation in the *Int1* coding sequences so that the protein product is inactive, and second, in another construct, we mutate the HRE so that it can no longer bind its steroid receptor. Neither of these control constructs should turn out to be oncogenic.

SOLVED PROBLEM

1. In the inherited form of retinoblastoma, an affected child is heterozygous for an *RB* mutation, which is either passed on from a parent or has newly arisen in the sperm or the oocyte nucleus that gave rise to the child. The heterozy-gous *RB RB⁺* cells are nonmalignant, however. The *RB⁺* al-lele of the heterozygote must be knocked out as well in the developing retinal tissue in order to create a tumorous cell. Such a knockout can occur through an independent muta-tion of the *RB⁺* allele or by mitotic crossing over such that the original *RB* mutation would now be homozygous.

a. If retinoblastoma is passed on to other siblings as well, could we determine if the original mutation was de-rived from the mother or the father? How?

b. Could we determine if the *RB* mutation was ma-ternally or paternally derived if it arose de novo in a germ cell of one parent?

Solution

a. If the trait is inherited, we can determine which parent it came from. The most straightforward approach is to identify DNA polymorphisms, such as restriction frag-ment length polymorphisms (RFLPs; Chapter 15), that map within or near the *RB* gene. *RB* has the curious property of being inherited as an autosomal dominant, even though it is recessive on a cellular basis. Given the dominant pattern of inheritance, by finding DNA differences in the parental

genomes that map near each of the four parental alleles, we should be able to determine which allele is passed on to all affected individuals. That allele would be the mutant one.

b. If the trait arose de novo in the sperm or the oocyte, we could still possibly determine which parent it came from, but with considerable difficulty. One way to do this would be to use recombinant DNA cloning techniques to isolate each of the two copies of the *RB* gene from nor-mal cells. Only one of these alleles should be mutant. Once the gene is cloned, by DNA sequencing of the two alleles we might be able to identify the mutation that inactivates *RB*. Since it arose de novo in the sperm or the oocyte, this mutation will not be present in the somatic cells of the par-ents. If sequencing also reveals some polymorphisms (for example in restriction-enzyme recognition sites) that distin-guish the alleles, we should then be able to go back to the parents' DNA and find out if the mother or the father car-ries the polymorphisms in the cloned mutant allele. (De-pending on the exact nature of the parental alleles and the mutation, this approach will not always work.)

If the mutation arises by mitotic crossing-over, we have additional tools available to us. In this case, the entire region around the *RB* gene will be homozygous for the mutant chromosome. By examining DNA polymorphisms known to map in this region, we should be able to determine if this chromosome derived from the mother or the father. This becomes a standard exercise in DNA fingerprinting similar to that described in part a.

PROBLEMS

1. Cancer is thought to be caused by the accumulation of two or more "hits"—that is, two or more mutations affect-ing cell proliferation and regulation—within the same cell. Many of these oncogenic mutations are dominant: one mu-tant copy of the pertinent gene is sufficient to change the proliferative properties of a cell. Which of the following general types of mutations have the potential to be domi-nant oncogenes? Justify each answer.

a. A mutation increasing the number of copies of a transcription factor

b. A nonsense mutation occurring shortly after the beginning of translation in a gene encoding a growth-factor receptor

c. A steroid-receptor mutant protein that blocks an inhibitory protein from masking the receptor's DNA-bind-ing domain

d. A mutation that disrupts the active site of a cyto-plasmic tyrosine-specific protein kinase

2. Some of the evidence that tumors derive from a single cell was based on the phenomenon of dosage compensation in humans. Explain how dosage compensation helped to reveal the clonal nature of tumors.

3. In *C. elegans* vulva development, one anchor cell in the gonad interacts with six equivalence group cells (cells with the potential to become parts of the vulva). The six equivalence group cells have three distinct phenotypic fates: primary, secondary, and tertiary. The equivalence group cell closest to the anchor cell develops the primary vulva phenotype. If the anchor cell is ablated, all six equivalence group lineages differentiate into the tertiary state.

a. Set up a model to explain these results.

b. The anchor cell and the six equivalence group cells can be isolated and grown in vitro; design an experiment to test your model.

4. There are two types of muscle cell in *C. elegans:* pharyngeal muscles and body wall muscles. They can be distinguished from each other, even as single cells. In the following figure, letters designate particular muscle precursor cells, black triangles (▲) are pharyngeal muscles, and white triangles (△) are body wall muscles.

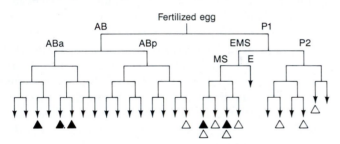

a. It is possible to move these cells around during development. When the positions of ABa and ABp are physically interchanged, the cells develop according to their new position. In other words, the cell that was originally in the ABa position now gives rise to one body wall muscle cell, while the cell that was originally in the ABp position now gives rise to three pharyngeal muscle cells. What does that tell us about the developmental processes controlling the fates of ABa and ABp?

b. If EMS is ablated (by heat inactivation with a laser beam aimed through a microscope lens), no AB descendants make muscles. What does that suggest?

c. If P2 is ablated, AB descendants all turn into muscle cells. What does that suggest?

(Figure from J. Priess and N. Thomson, *Cell* 48, 1987, 241.)

5. Some genes can be mutated to become oncogenes by increasing the copy number of the gene. This, for example, is true of the *c-myc* transcription factor. On the other hand, oncogenic mutations of *ras* are always point mutations that alter the protein structure of Ras. Rationalize these observations in terms of the roles of normal and oncogenic versions of *ras* and *c-myc.*

6. In humans, the male steroid hormone testosterone binds to the testosterone receptor. The hormone-receptor complex then activates transcription of male-specific genes. A mutation in the testosterone receptor prevents binding to testosterone. What would be the phenotype of an individual who is XY and is homozygous for such a mutation? Explain your conclusion in terms of how steroid receptors act.

7. Suppose that you had the ability to introduce normal copies of a gene into a tumor cell that had mutations in the gene that caused it to promote tumor growth.

a. If the mutations were in a recessive tumor suppressor gene, would you expect that these normal transgenes would block the tumor-promoting activity of the mutations? Why or why not?

b. If the mutations were of the dominant oncogene type, would you expect that the normal transgenes would block their tumor-promoting activity? Why or why not?

8. Insulin is a protein that is secreted by the pancreas (an endocrine organ) when blood sugar levels are high. Insulin acts on many distant tissues by binding and activating a receptor tyrosine kinase, leading to a reduction in blood sugar by appropriately storing the products of sugar metabolism.

Diabetes is a disease in which blood sugar levels remain high because some aspect of the insulin pathway is defective. One kind of diabetes, (let's call this type A) can be treated by giving the patient insulin. Another kind of diabetes (call it type B) is not ameliorated by insulin treatment.

a. Which type of diabetes is likely to be due to a defect in the pancreas, and which type is likely to be due to a defect in the target cells? Justify your answer.

b. Type B diabetes can be due to mutations in any of several different genes. Explain this observation?

9. Epinephrine is secreted by the adrenal gland that binds and activates the β-adrenergic receptor. This leads to the activation of adenylate cyclase, which in turn leads to the production of high levels of cyclic AMP (cAMP). The cAMP then binds to a regulatory subunit of a cAMP-dependent protein kinase (CPK). This binding causes the release of the regulatory subunit from the catalytic subunit of CPK, thereby allowing CPK to phosphorylate its protein substrate. What would be the effects on CPK activity in target cells with the following genotypes? Justify each answer.

a. Homozygosity for a mutation that leads to the constitutive production of epinephrine by the adrenal gland

b. Homozygosity for a mutation that destroys the active site of adenylate cyclase

c. Homozygosity for a mutation in the regulatory subunit of CPK that blocks cAMP binding

d. Double homozygosity for mutations in (a) *and* (b)

e. Double homozygosity for mutations in (b) *and* (c)

25

Developmental Genetics:
Cell Fate and Pattern Formation

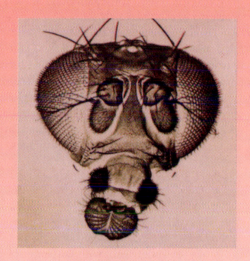

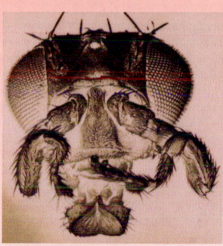

A homeotic mutation that alters the basic *Drosophila* body plan. Homeosis is the replacement of one body part by another. In place of the normal antennae (*above*), an *Antennapedia* mutation causes the antennal precursor cells to develop into a leg. (F. R. Turner/BPS)

KEY CONCEPTS

▶ A programmed set of instructions in the genome of a higher organism leads to the establishment of developmental fates of cells with respect to the major features of basic body plan.

▶ The zygote is totipotent, giving rise to every adult cell type; as development proceeds, successive decisions restrict each cell lineage to its particular fate.

▶ Gradients of maternally derived regulatory proteins establish polarity along the major body axes of the egg; these proteins control the local transcriptional activation of genes encoding master regulatory proteins in the zygote.

▶ Many proteins that act as master regulators of early development are transcription factors; others are components of pathways that mediate signaling between cells.

▶ Some fate decisions are made autonomously by individual cells; many fate decisions require communication and collaboration between cells.

▶ The same basic set of genes identified in *Drosophila* and the regulatory proteins they encode are conserved in mammals and appear to govern major developmental events in many—perhaps all—higher animals.

In all higher organisms, life begins with a single cell, the newly fertilized egg. It reaches maturity with thousands, millions, or even trillions of cells combined into a complex organism with many integrated organ systems. The goal of developmental biology is to unravel the fascinating and mysterious processes that achieve the transfiguration of egg into adult. Because we know more about complex pattern formation in these organisms, we will restrict most of our discussion to animal development.

The general body plan is common to all members of a species, and indeed is common to many very different species. All mammals have four legs, and all insects six. But mammals and insects must all, during their development, differentiate the anterior from the posterior end and the dorsal from the ventral side. Eyes and legs always appear in the appropriate places, and except for severe disturbances that interfere with development, appear to be quite independent of environmental variation. The study of the basic development of body plan can then be carried out by studying the internal genetic program of the organism without reference to the environment. We should not forget however that the study of the genetic determination of these basic developmental processes does not provide us with an explanation of the phenotypic differences between individuals of a species. This chapter will focus on the processes underlying pattern formation, the construction of complex form, and how these processes operate reliably to execute the developmental program for the basic body plan.

During the establishment of the body plan, cells adopt specific **cell fates,** that is, the capacity to differentiate into particular kinds of cells. The process of commitment to a particular fate, **determination,** is a gradual one. A cell does not go directly from being totally uncommitted (totipotent) to being earmarked for a single fate. Rather, as cells proliferate in the developing organism, decisions are made to specify more and more precisely the fate options of cells of a given lineage. Most such decisions are made by committee—that is, the fate of a cell becomes dependent on input from neighboring cells. In other words, in order to build a complex organism, cells must make developmental decisions in the context of the decisions being made by their neighbors.

In this chapter, we will consider several examples of the genetic dissection of cell fate decisions. Were this chapter written 15 years ago, we would have to have been satisfied with a phenomenological description of mutant phenotypes and with microsurgical manipulations of embryos. Not so today. The combination of genetics and recombinant DNA techniques has revolutionized our understanding of development. We are now in a position to identify the protein products contributing to these developmental events and to build a general framework of the molecular mechanisms involved. Perhaps the most exciting development is that recombinant DNA tools have allowed researchers to fish out related genes from very different organisms. From such studies, we are coming to realize that the same basic set of regulatory proteins—including certain classes of transcription factors and of molecules that pass information between cells—governs major developmental events in all higher animals.

Pattern Formation: The Establishment of the Basic Animal Body Plan

As with other biological phenomena, some systems are more favorable than others for studying pattern formation. Indeed, most of the insights in animal development have emanated from studies of two small invertebrates: the fruit fly *Drosophila melanogaster* and the roundworm *Caenorhabditis elegans*. What we learn from these systems provides the basis for examining similar developmental problems in other animals.

The most revealing information on early development has come from studies of *Drosophila*. There are a number of reasons why use of this fly has been so successful. The ease of culturing, short generation time, small genome (about 1.6×10^8 base pairs of DNA per haploid genome), low chromosome number (n = 4), polytene chromosomes, together with 85 years' worth of accumulated genetic lore, are the most important genetic advantages.

A Sneak Preview of Pattern Formation

The basic strategy for establishing the overall body plan is outlined in Figure 25-1. The body plan of most animals is bilaterally symmetrical, with an anterior-posterior (A/P) and a dorsal-ventral (D/V) axis. We can draw an analogy to geographic information on the surface of the earth, which is described by two coordinate systems, longitudes and latitudes. Similarly, developmental **positional information** is described along the two body axes, A/P and D/V, in the developing oocyte by differential concentrations of certain regulatory molecules (called **morphogens,** from *morphogenesis,* which means the "development of form"). We shall use the term *morphogen* to mean a molecule that controls cell fate in a concentration-dependent fashion. Different concentrations of the morphogen will commit cells to different fates. Somehow, during oogenesis, systems must be implemented to create asymmetric distributions of these morphogens within the single-celled oocyte. In the best-understood cases, including *Drosophila* (as discussed below), the asymmetric distributions of morphogens are accomplished by transporting and anchoring the mRNAs encoding morphogens themselves or encoding proteins that regulate morphogen synthesis to one end or the other of a polarized cytoskeletal element (see Chapter 24). The end result of this process is that the cells of the early embryo are exposed to different levels of morphogen along each axis and so shift into different developmental pathways.

Many concentrations of morphogen may be present in the early embryo, but embryonic cells appear able to distinguish only a few different concentration ranges of morphogen. The consequence of this is that the early embryo becomes divided into a few distinct domains; the expression of different regulatory molecules along each axis is characteristic for each domain. Once these domains are defined, cells are no longer totipotent; they are beginning the process of becoming restricted to specific fates.

Fate restriction occurs through a series of refinement steps, in which the allocation of cells to specific fates is coordinated through cell-cell interactions. Through inductive interactions between different cell populations and through communication of fate decisions as they are made, each domain coordinates the process of developmental specification so that sufficient numbers of cells are allocated to each fate. Some of the interactions are positive, so that one cell induces other cells to adopt a particular fate, whereas others are negative, so that a cell of a particular fate prevents its neighbors from adopting the identical fate. In higher animals, the A/P fate decisions lead to the production of repeating tissue units called **segments,** while D/V fate decisions distinguish the cell types of a domain (for example, neurons, epidermis, muscles, and gut).

The elegance of the approach of dividing up a population of cells into many organized cell types is flexibility. Because decisions are stepwise, and because many of the decisions subdivide populations by committee, through interplay of information among neighboring cells, it is not very important if exactly the right number of cells is assigned to each initial domain. The cell-cell interactions that take place during the refinement stage ensure that there are sufficient cells for each of the possible fates.

Another feature to keep in mind is that at every step of the process of pattern formation, a regulatory gene that is expressed in a particular domain activates other regulatory genes in different subsets of that domain. Thus, at each step, finer and finer subdivisions of the original **developmental field** are achieved.

Figure 25-1 The steps leading to the establishment of the basic body plan. The localization of precursors of the morphogen (see text) to poles of the embryo sets up conditions for the establishment of morphogen gradients. At different concentration levels, the morphogens activate different primary, zygotically active regulatory genes, dividing the embryo up into several broad domains. The domains of expression of these primary regulatory genes subdivide the embryo up into more refined domains, each with a different cell fate. The activation and maintenance of cell identity genes allow cells to retain their fates through the rest of development.

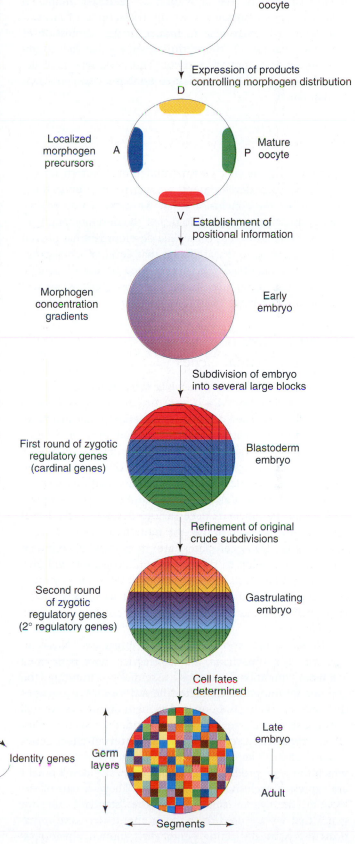

In many instances, once the final cell fates are established, cells must "remember" these fate decisions for the remainder of the lifetime of the organism. One way this is accomplished is by the activation of **feedback loops,** in which regulatory proteins—usually transcription factors—that define a fate are able to induce further expression of the genes that encode them, thus ensuring that the regulatory protein is continually present. Thus, two aspects of determination of cell fate are first to establish a fate and then to maintain it.

Investigating Early Development

In considering the mechanisms underlying pattern formation during early development, we will want to understand the basis for the conclusions that we draw. To do so, we need to discuss the methodologies of the developmental geneticist. The reason that *Drosophila* development has proved to be a gold mine to researchers is that developmental problems can simultaneously be approached by use of genetic and molecular techniques. In the next few sections, we will consider these various methodologies.

Identifying Mutations in Pattern-Formation Genes

The exoskeleton of the *Drosophila* larva is laid down as a mosaic in the embryo. Each structure of the exoskeleton is built by the epidermal cell or cells underlying that structure. With its intricate pattern of hairs, indentations, and other structures, the exoskeleton provides numerous landmarks to serve as indicators of the fates acquired by the many epidermal cells. In particular, there are many anatomical structures that are distinct along the A/P and D/V axes. Further, because all the nutrients necessary to develop to the larval stage are prepackaged in the egg, mutant embryos in which the A/P or D/V cell fates are drastically altered can nonetheless develop to the end of embryogenesis and produce a mutant larva. The exoskeleton of such a mutant larva mirrors the mutant fates assigned to subsets of the epidermal cells and can thus identify genes worthy of detailed analysis.

Researchers, most notably Christiane Nüsslein-Volhard, Eric Wieschaus, and colleagues, have performed extensive mutational screens, essentially saturating the genome for mutations that alter the A/P or D/V patterns of the larval exoskeleton. In our subsequent discussion, we will distinguish two classes of genes identified by such mutant alleles: **zygotically acting genes** and **maternal-effect genes** (Figure 25-2). The zygotically acting genes are those in which the gene products contributing to early development are expressed exclusively in the zygote. They are part of the DNA of the zygote itself and are the "standard" sorts of genes that we are used to thinking about. Recessive mutations in zygotically acting genes elicit mutant phenotypes

(a) Zygotically acting genes

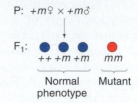

(b) Maternal-effect genes

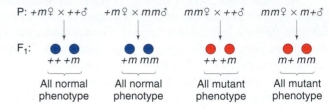

Figure 25-2 The genetic distinction between recessive (a) zygotically acting and (b) maternal-effect mutations. (a) The phenotypes of offspring are purely a reflection of their own genotype. (b) The phenotypes of offspring are purely a reflection of their mother's genotype.

only in homozygous mutant animals. The alternative category—the maternal-effect genes—affect early development through contributions of gene *products* from the ovary of the mother to the developing oocyte. As we will see, the initial determinants for the A/P and D/V axes act in the developing oocyte well in advance of fertilization. In maternal-effect mutations, the phenotype of the offspring depends on the genotype of the mother, not the offspring! A recessive maternal-effect mutation will produce mutant animals only when the mother is a mutant homozygote.

Message Phenotypic analysis of mutant larvae can identify genes controlling the basic body plan.

Molecular Analysis of Pattern-Formation Genes

Any *Drosophila* gene can be cloned if its chromosomal map location has been well established, using recombinant DNA techniques like those described in Chapters 14 and 15. The analysis of the cloned genes often provides valuable information on the function of the protein product—usually by identifying close relatives in amino acid sequence of the encoded polypeptide through comparisons with all the protein sequences stored in public databases. In addition, one can investigate the spatial and temporal pattern of expression of (1) an mRNA, using histochemically tagged single-stranded DNA sequences complementary to the mRNA to perform **RNA in situ hybridization,** or (2) a protein, using histochemically tagged antibodies that specifically bind to that protein to perform **immunohistochemistry.**

Extensive use is also made of in vitro mutagenesis techniques. P elements are used for germ-line transformation in *Drosophila* (see Chapter 21). A cloned pattern-formation gene is mutated in a test tube and put back in the fly. The mutated gene is then analyzed to see how the mutation alters the gene's function.

Message   We need to know only the chromosomal location of a gene in order to clone it. Recombinant DNA techniques allow us to physically purify genes, their transcripts, and their protein products and to study the spatial and temporal patterns of mRNA or protein localization in the embryo.

A Synopsis of Early *Drosophila* Development

Having considered the basic tools of the developmental geneticist, we are now ready to return to our discussion of early *Drosophila* development. First, we will consider the anatomy and organization of the developing oocyte and early embryo. In the context of this information, we will then discuss the major events that lead to the establishment of the A/P and D/V body axes.

Oogenesis

Each oocyte is part of an egg chamber, which is a cooperative venture of germ-line and somatic cells (Figure 25-3). The **oocyte** and 15 **nurse cells** are all the mitotic derivatives of a germ-line precursor cell. These 16 cells all share a common cytoplasm, with ring canals interconnecting the cells (Figure 25-4). These interconnections allow the nurse cells to provide most of the contents of the oocyte cytoplasm. The oocyte cytoplasm appears homogeneous except at the posterior pole, where small ribonucleoprotein (RNA plus protein) particles, the polar granules, are found. As we will

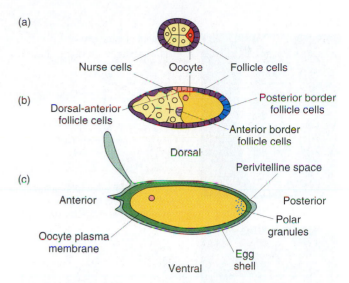

Figure 25-3 Three stages in the maturation of the *Drosophila* egg chamber. (a) A young egg chamber. Follicle cells surround a 16-cell cluster in which the nurse cells and oocyte are of equal size. (b) A midstage egg chamber. The nurse cells have begun to dump their contents through the ring canals (see text) into the oocyte cytoplasm. Follicle cells are beginning to lay down the eggshell. (c) A mature oocyte. The nurse cells and follicle cells have been discarded, and the plasma membrane of the oocyte is interior to the two layers of the eggshell: the vitelline coat and the chorion. Note the polar granules at the posterior end of the oocyte. (Modified from Christiane Nüsslein-Volhard, *Development,* Suppl. 1, 1991, 1.)

see, these polar granules participate in A/P axis formation as well as in determination of the germ line.

The somatic cells of each egg chamber are called **follicle cells.** Concurrent with the production of the oocyte cytoplasm (ooplasm) by the nurse cells, the 1000 follicle cells surrounding each oocyte–nurse cell complex contribute to oocyte maturation in two ways. First, it is the follicle cells

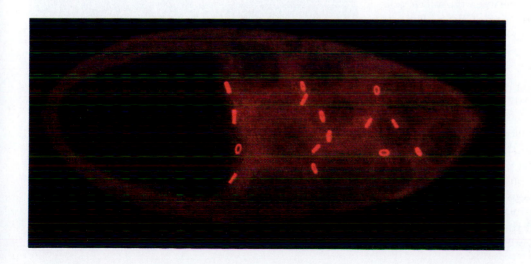

Figure 25-4 Photomicrograph of a midstage egg chamber, stained in a manner that highlights the outlines of the ring canals. The ring canals are the passageways for dumping cytoplasm from the nurse cells into the oocyte. (F. Xu and L. Cooley, Yale University School of Medicine)

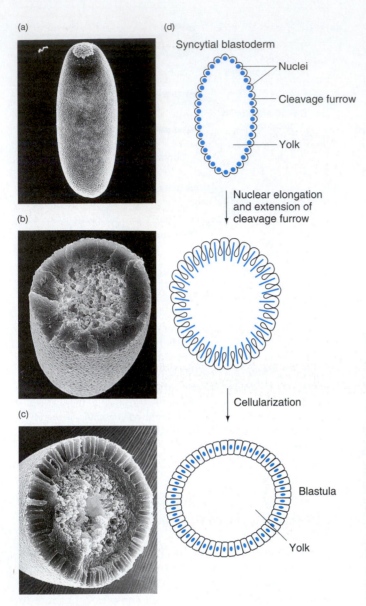

(a)

(b)

(c)

(d)

Syncytial blastoderm

Nuclei

Cleavage furrow

Yolk

Nuclear elongation
and extension of
cleavage furrow

Cellularization

Blastula

Yolk

Figure 25-5 Some aspects of nuclear multiplication and blastoderm formation in a *Drosophila melanogaster* embryo. The embryo begins as a syncytium; cellularization is not completed until there are about 6000 nuclei. (a) to (c) Scanning electron micrographs of embryos removed from the eggshell. (a) After the ninth nuclear division, the pole cells (the cap of cells on top of the embryo) lie outside the somatic syncytium. (b) A syncytial blastoderm–stage embryo, fractured to reveal the common cytoplasm toward the periphery and the central yolk. The bumps on the outside of the embryo are the beginning of cellularization, which moves from the outside inward. (c) A cellular blastoderm embryo, fractured to reveal the columnar cells that have formed by drawing cell membranes down between the nuclei to create some 6000 mononucleate cells. (d) Schematic diagram of the changes taking place in parts a to c, showing development from the syncytial to the cellularized form. (F. R. Turner and A. P. Mahowald, *Developmental Biology,* 50, 1976, 95.)

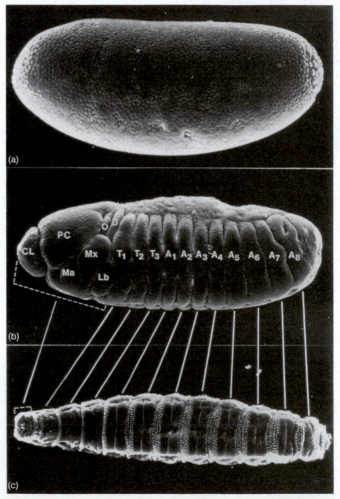

(a)

(b)

CL   PC   O   D
Mx   T₁ T₂ T₃ A₁ A₂ A₃ A₄ A₅ A₆ A₇ A₈
Ma   Lb

(c)

Figure 25-6 Scanning electron micrographs of (a) a 3-hour *Drosophila* embryo, (b) a 10-hour embryo, and (c) a newly hatched larva. Note that by 10 hours, the segmentation of the embryo is obvious. Lines are drawn to indicate that the segmental identity of cells along the A/P axis is already fated early in development. (T. C. Kaufman and F. R. Turner, Indiana University)

that are responsible for forming the eggshell that envelops the oocyte. Second, as we will discuss in subsequent sections, intercellular communication between the follicle cells and the oocyte helps establish the axes of the oocyte and the embryo. There are several important follicle cell subpopulations: the border, dorsal-anterior, and main-body follicle cells (Figure 25-3).

The Early Zygote

All of embryogenesis takes but one day. In this section, we will consider the first 3 hours—the rapid multiplication stage—of embryogenesis. The early *Drosophila* embryo is a single multinucleate syncytium: nuclei divide but the egg cell does not. Nine nuclear divisions occur in quick succession, with the nuclei remaining in the center of the egg. Then the nuclei migrate toward the periphery (cortex) of the egg (Figure 25-5). The first nuclei reach the posterior cortex and are cellularized (made into cells) by being captured in invaginating plasma membranes. The cytoplasm of these cells contains the polar granules, and these round "pole cells" ultimately form the entire germ line of the fly. The remaining nuclei undergo four more divisions and form a monolayer of nuclei in the egg cortex; this is the **syncytial blastoderm** stage. The plasma membrane of the egg then invaginates around each nucleus and, for the first time, each nucleus is contained within a different cell; this is the **cellular blastoderm** stage (Figure 25-6a).

Before the cellular blastoderm, soluble molecules can diffuse throughout the common egg cytoplasm. Once cellularization occurs, plasma membranes isolate the contents of each cell from one another, and interactions between cells must rely on the standard mechanisms of intercellular communication.

From Blastoderm to Larva

Shortly after the cellular blastoderm stage, gastrulation ensues, with its characteristic movements of tissue layers. Two furrows form: a ventral furrow forms along the ventral midline of the gastrulating embryo (Figure 25-7), and a slanted cephalic furrow forms about one-sixth of the way back

from the anterior end of the embryo. These furrows are the first morphological manifestations, other than pole-cell formation, of A/P and D/V asymmetries. Another manifestation of D/V fate differences is that by 4 hours, the embryo is divided into numerous cell cycle domains (Figure 25-8).

(a)

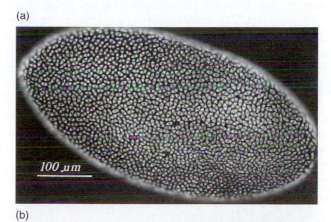

(b)

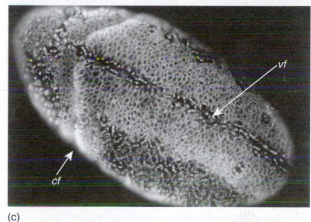

(c)

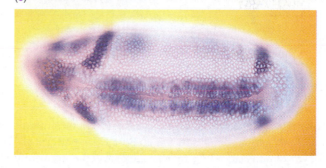

Figure 25-8 Cell cycle regulation in the early *Drosophila* embryo. (a) Before the cellular blastoderm stage is completed, all cell divisions occur in synchrony. (b) and (c) During gastrulation, the stage of the cell cycle varies in different geographic domains of the embryo. In part b, for example, the invaginated cells of the ventral furrow and mesectoderm cells, which form the lips of the ventral furrow (vf; see Figure 25-7), are in mitosis. In part c, the mRNA for *string*, a *Drosophila* gene that activates cyclin-CDK and hence induces cell cycle progression, is expressed in the cells, including the ventral furrow, that then undergo mitosis, as in part b. Parts a and b show staining of microtubules and prominently outline the mitotic spindles. Panel c shows mRNA distribution by in situ hybridization. (Parts a and b from V. E. Foe, *Development* 107, 1989, 1–22. Panel c from B. A. Edgar, D. A. Lehman, and P. H. O'Farrell, *Development* 120, 1994, 3133.)

(a)

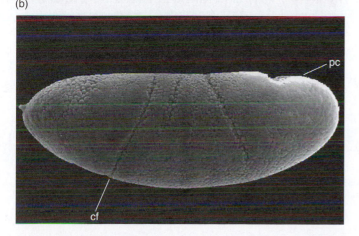

(b)

Figure 25-7 Invaginations of the early *Drosophila* embryo. (a) Ventral view. (b) Lateral view. vf = ventral furrow; cf = cephalic furrow; pc = pole cells. (Scanning electron micrographs courtesy of T. C. Kaufman and F. R. Turner, Indiana University.)

The clustered cells in the same domain pass through the cell cycle in synchrony. Cells in different domains are out of phase relative to one another. Many of these domains represent different cell populations along the D/V axis.

A few hours later, the first morphological manifestations of segmentation are apparent (Figure 25-6b). By 12 hours post-fertilization, organogenesis occurs; by 15 hours, the exoskeleton of the larva begins to form, with its specialized hairs and other external structures. About 24 hours after fertilization, a fully formed larva hatches out of the eggshell (Figure 25-6c).

The *Drosophila* larva is clearly highly differentiated along its head-to-tail (A/P) and top-to-bottom (D/V) axes. As with other animals, a major concern is how these two developmental axes are established. By 10 hours of development, the embryo is already externally divided into 14 segments from anterior to posterior—3 head, 3 thoracic, and 8 abdominal segments (Figure 25-6c). Furthermore, each of these segments has developed a unique set of anatomical structures, reflective of its special identity and role in the biology of the animal. In understanding the development of the A/P body axis, we want to keep in mind several questions. How is the segmental repeat pattern established? How does the correct number of segments always form? How is each segment assigned its unique identity?

Similarly, the D/V axis is highly differentiated. Along the D/V axis, commitments to different cell fates are also detectable quite early. The ventral furrow and the distribution of cell cycle domains are just two manifestations of different cell fates along the D/V axis as early as 4 hours post-fertilization. How does the embryo specialize D/V cell fates? In the following sections, we will see that the answers involve a collaborative effort of mother and offspring.

Message The basic body plan is established quite quickly after fertilization.

Establishing Positional Information

Through screens for maternal-effect mutations, mutations that show global defects in the A/P or D/V pattern of the oocyte identify about 30 maternal-effect genes contributing to body axis formation. Through the developmental and molecular analysis of these genes and their mutant alleles, we now have a good overview of how they contribute to the establishment of the chemical coordinates that define the two major body axes. Among these 30 genes, several classes can be distinguished based on their mutant phenotypes and on how they contribute to development (Figure 25-9 and Table 25-1). Most of the genes we are interested in alter the pattern of the oocyte and the resulting embryos, but do not disrupt the normal pattern of the somatic follicle cells or the eggshell they produce. The mutant phenotypes involve the loss of all anterior segments (for ex-

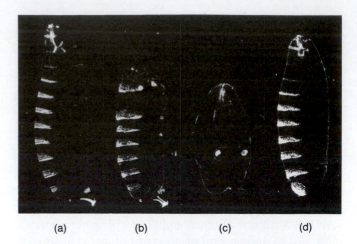

(a) (b) (c) (d)

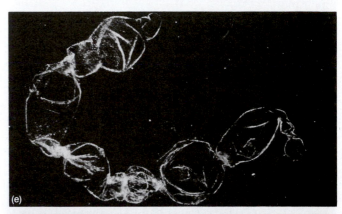

(e)

(f)

Figure 25-9 Photomicrographs of the exoskeletons of larvae derived from wild-type and maternal-effect-lethal mutant mothers. These photomicrographs are in darkfield, so that dense structures show up white, as in a photographic negative. Note the bright, segmentally repeated denticle bands present on the ventral side of the embryo. Maternal genotypes and larval phenotypes (and class of mutation) are as follows: (a) wild-type, normal phenotype; (b) *bcd* (*bicoid*), anterior head and thoracic structures missing (anterior); (c) *osk* (*oskar*), posterior (abdominal) segments missing; (d) *tsl* (*torso-like*), lacking the two unsegmented terminal regions; (e) *dl* (*dorsal*), lacking the ventral denticle bands; and (f) *cact* (*cactus*), lacking dorsal epidermis and having the ventral denticles wrapping around the embryo. (Parts a–d from C. H. Nüsslein-Volhard, G. Frohnhöfer, and R. Lehmann, *Science* 238, 1987, 1678. Parts e and f from Christiane Nüsslein-Volhard, *Development*, Suppl. 1, 1991, 1.)

Table 25-1   **Examples of *Drosophila* Maternal-Effect Genes Contributing to Positional Information**

Gene symbol	Gene name	Protein function	Contributions to positional information
A/P positional information genes:			
bcd	*bicoid*	Transcription factor—homeodomain protein	A/P morphogen
hb-m	*hunchback-maternal*	Transcription factor—Zn-finger protein	A/P morphogen
nos	*nanos*	RNA-binding protein (postulated)	Translational repressor of HB-M
pum	*pumilio*	RNA-binding protein	Contributes to HB-M repression by NOS
vas	*vasa*	Helicase (RNA or DNA unwinding protein)	Contributes to NOS localization
osk	*oskar*	Unknown	Protein *and* mRNA of *osk* gene must be localized to posterior pole for posterior localization of *nos* mRNA and NOS protein
exu	*exuperentia*	RNA-binding protein	Contributes to *bcd* mRNA localization
swa	*swallow*	RNA-binding protein	Contributes to *bcd* mRNA localization
D/V positional information genes:			
dl	*dorsal*	DL transcription factor	D/V morphogen
cact	*cactus*	Binds to DL protein in cytoplasm	Regulates regional activation of D/V morphogen
spz	*spätzle*	Signaling SPZ ligand	Regulates regional activation of D/V morphogen
Tl	*Toll*	Receptor for SPZ ligand	Regulates regional activation of D/V morphogen
snk	*snake*	Serine protease	Activates SPZ protein through protease cleavage enzymatic cascade
ea	*easter*		
gd	*gastrulation-defective*		
tube	*tube*	Unknown	Required for activation of DL protein
pll	*pelle*	Protein kinase	Required for activation of DL protein
A/P and D/V positional information genes:			
grk	*gurken*	Signaling GRK ligand	Signals from oocyte to follicle cells to establish posterior and anterior-dorsal fates
Egfr	*EGF receptor*	Receptor tyrosine kinase	Receives GRK protein signal in follicle cells

ample, *bicoid*), all posterior segments (for example, *nanos*), all ventral material (for example, *dorsal*), or all dorsal tissue (for example, *cactus*). These genes either encode the morphogens that define the two axes or encode gene products that contribute to the localization and asymmetric distribution of the morphogens. We will consider the maternal contributions to the two axes first.

The A/P Morphogens

There are two morphogens, BCD and HB-M, along the A/P axis that act in concert to provide A/P positional information. The BCD protein, encoded by the *bicoid* (*bcd*) gene is distributed in a steep gradient in the early embryo, while the HB-M protein, encoded by the *hunchback* (*hb*) gene, is

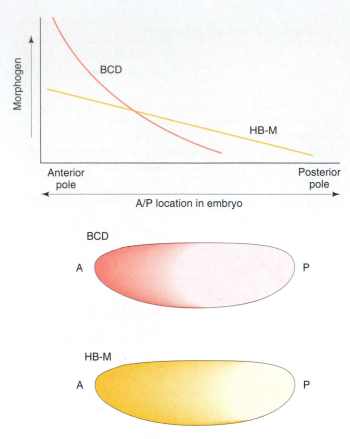

Figure 25-10 The gradients of BCD and HB-M proteins. The BCD gradient is steeper, and BCD protein is not detectable in the posterior half of the early *Drosophila* embryo. The HB-M gradient is shallower, and HB-M protein can be detected well into the posterior half of the embryo.

tured egg, anterior segments (head and thorax) are lost. Injection of anterior ooplasm from another egg into the anterior region of the anterior ooplasm-depleted egg restores normal anterior segment formation, and a normal larva is produced. Similarly, synthetic *bcd* mRNA can be made in a test tube and injected into the anterior region of an anterior ooplasm-depleted egg. Again, a normal larva is produced.

Third, *bcd* mRNA is localized to the anterior pole of the oocyte, as was predicted from ooplasm transplantation experiments. Transplanting cytoplasm from middle or posterior regions of an oocyte will not restore normal anterior formation. This predicted that the anterior determinant would be located at the anterior end of the egg. In situ localization of *bcd* mRNA reveals a tight localization at the anterior pole of the oocyte (Figure 25-13a).

Fourth, the distribution of BCD protein is asymmetric and graded along the A/P axis. Immunohistochemistry, using anti-BCD antibodies (antibodies that bind to BCD pro-

distributed in a shallow gradient (Figure 25-10). As we will discuss later, both these proteins are transcription factors, and they contribute jointly to the gene activation or repression of genes expressed in the early embryo.

The BCD Gradient

How do we know that these molecules are the A/P morphogens? Let's consider the example of BCD in detail. First, genetic changes in the *bcd* gene alter anterior fates. Embryos derived from *bcd* homozygous null mutant mothers lack anterior segments (Figure 25-9b). If, on the other hand, we overexpress *bcd* in the mother, by increasing the number of copies of the *bcd*+ gene from the normal 2 to 3, 4, or more, we "push" fates that ordinarily appear in anterior positions into increasingly posterior zones of the resulting embryos (Figure 25-11). These observations suggest that *bcd* exerts global control of anterior positional information.

Second, *bcd* mRNA can completely substitute for the anterior determinant activity of anterior ooplasm (Figure 25-12). If one removes the anterior cytoplasm from a punc-

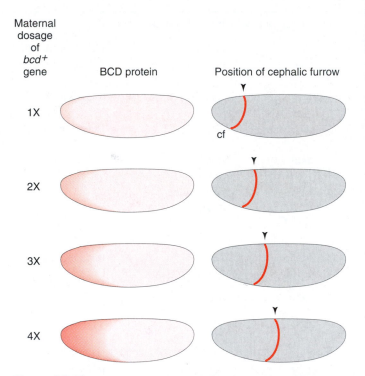

Figure 25-11 The concentration of BCD protein affects A/P cell fates. The amount of BCD protein can be changed by varying the number of copies of the *bcd*+ gene in the mother. Embryos derived from mothers carrying one to four copies of *bcd*+ have increasing amounts of BCD protein. (Two copies is the normal diploid gene dose.) Cells that invaginate to form the cephalic furrow (see Figure 25-7) are determined by a specific concentration of BCD protein. As embryos go from one maternal copy to four copies of *bcd*+, this specific concentration is present more and more posteriorly in the embryo. Thus, the position of the cephalic furrow (marked dorsally by the triangle) arises farther toward the posterior according to *bcd*+ gene dosage. (Adapted from W. Driever and C. Nüsslein-Volhard, *Cell* 54, 1988, 100–101.)

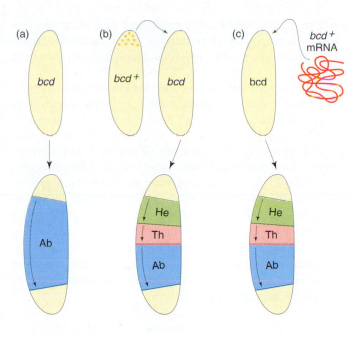

Figure 25-12 The *bcd* "anterior-less" mutant phenotype can be rescued by wild-type cytoplasm or purified *bcd*⁺ mRNA. (a) In embryos derived from *bcd* mothers, the anterior (head and thoracic) segments do not form. (b) If anterior cytoplasm from an early wild-type donor embryo is injected into the anterior of a recipient embryo derived from a *bcd* mutant mother, a fully normal embryo and larva are produced. Cytoplasm from any other part of the donor embryo does not rescue. (c) *bcd*⁺ mRNA injected into the anterior of an embryo derived from a *bcd* mutant mother also rescues the wild-type segmentation pattern. A *bicoid* larva is shown in Figure 25-9b. (Adapted from C. Nüsslein-Volhard, H. G. Fronhöfer, and R. Lehmann, *Science* 238, 1987, 1679.)

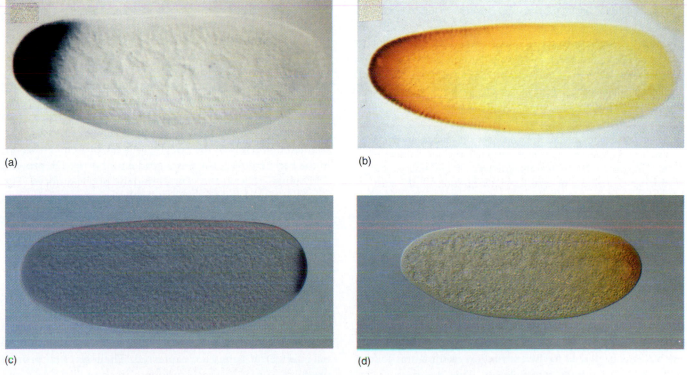

Figure 25-13 Photomicrographs showing the expression of localized A/P determinants in the embryo. (a) By in situ hybridization to RNA, the localization of *bcd* mRNA to the anterior (*left*) tip of the embryo can be seen. (b) By antibody staining, a gradient of BCD protein (brown stain) can be visualized, with its highest concentration at the anterior tip. (c) Similarly, *nos* mRNA localizes to the posterior (*right*) tip of the embryo, and (d) NOS protein is in a gradient with a high point at the posterior tip. (Parts a and b from Christiane Nüsslein-Volhard, *Development*, Suppl. 1, 1991, 1. Parts c and d provided by Elizabeth R. Gavis, Laura K. Dickinson, and Ruth Lehmann, Massachusetts Institute of Technology.)

tein), reveals that BCD protein is distributed in a gradient showing a high point at the anterior pole and dropping to background levels in the middle of the embryo (Figure 25-13b).

The HB-M Gradient

The HB-M protein is present in a shallower gradient, such that it extends more posteriorly than BCD (Figure 25-10). This is consistent with the important contributions of HB-M protein for fates in the middle and posterior regions of the embryo.

The HB-M protein gradient is produced by means of post-transcriptional regulation. The *hb-m* mRNA is uniformly distributed throughout the oocyte and the early embryo. However, translation of *hb-m* mRNA is blocked by a translational repressor protein [the NOS protein product of the *nanos* (*nos*) gene], which is itself distributed in an opposite gradient from BCD, showing a high point at the posterior pole and dropping down to background levels in the middle of the A/P axis. NOS protein has the ability specifically to block translation of *hb-m* mRNA. Indirectly then, the NOS translation repressor posterior-to-anterior gradient produces the shallow anterior-to-posterior gradient of HB-M protein. Consistent with this idea is the *nos* mutant phenotype. Homozygous *nos* null mothers produce embryos lacking posterior segments. How is the posterior-to-anterior gradient of NOS produced? The answer is quite similar to that for BCD. The *nos* mRNA is anchored at the posterior pole of the oocyte and early embryo (Figure 25-13c). Localized translation of *nos* mRNA produces the posterior-to-anterior NOS gradient (Figure 25-13d).

> **Message** The A/P morphogens demonstrate two ways to produce morphogen gradients: through the localization of morphogen mRNA at the anterior pole of the embryo and through a gradient of translational repression of the morphogen mRNA.

The Mechanism of mRNA Localization

Directly in one case and indirectly in the other, the anterior-to-posterior gradients of BCD and HB originate with localized mRNAs in the oocyte. How are the *bcd* and *nos* mRNAs localized respectively at the anterior and posterior poles of the oocyte? In both cases, the answer seems to be through interactions of the 3′ untranslated region (3′ UTR) of each mRNA (the region 3′ to the translation termination codon) with the microtubule cytoskeleton. This has been determined in part by "swapping" experiments. For example, by inserting in the fly genome a synthetic transgene that produces an mRNA with 5′ UTR and protein coding regions of the normal *nos* mRNA glued to the 3′ UTR of the normal *bcd* mRNA, this fused *nos-bcd* mRNA will be localized at the anterior pole of the oocyte. This causes a double gradient of NOS: one from anterior to posterior (due to the transgene mRNA) and one from posterior to anterior (due to the nor-

mal *nos* gene's mRNA). This produces a very weird embryo, with two mirror image posteriors and no anterior (Figure 25-14). (Can you explain why this double abdomen embryo is produced?)

How do the *bcd* and *nos* mRNAs get directed to opposite ends of the egg through their 3′ UTRs? Recall from Chapter 24 that polymerized microtubule chains have a − (minus) to + (plus) polarity. The microtubules of the late-stage oocyte and the zygote are all oriented with their − ends at the anterior pole and their + ends at the posterior. The *bcd* mRNA 3′ UTR sequences are bound by a protein that can also bind the − ends of the microtubules. Comparably, the 3′ UTR of *nos* mRNA can bind only to proteins that attach to the + ends of microtubules. In actuality, there are more intermediary steps in anchoring *nos* at the posterior end, with *nos* mRNA's being anchored to the molecules in the polar granules, which are in turn anchored to the + end of the microtubules.

> **Message** Localization of mRNAs is accomplished by anchoring the mRNAs to one end of polarized cytoskeleton chains.

The D/V Morphogen

The story of how the D/V morphogen gradient is laid down is quite different from the A/P examples and depends on a series of cell-cell interactions between subpopulations of follicle cells and the oocyte itself. The *dorsal* (*dl*) gene encodes the single D/V morphogen, the DL protein, a transcription factor that controls cell fate. The *dl* mRNA is distributed uniformly in the oocyte, as is DL protein. However, in the early embryo, there is a gradient of *active* DL protein, with a high point at the ventral midline of the embryo (Figure 25-15). In the absence of any DL activity, as in the case of embryos produced from homozygous *dl* null mothers, embryos completely lack ventral structures. For example, the ventral furrow that occurs at 4 hours of embryogenesis in wild-type embryos never occurs in embryos from *dl* mutant mothers.

How is the gradient of active DL protein produced? Inactive DL protein is located in the cytoplasm of the cell, by binding to the CACT protein encoded by the *cactus* (*cact*) gene. On the ventral side of the egg chamber, other maternal-effect D/V genes are expressed. Their secreted protein products, some of which originate in the follicle cells and others in the oocyte, activate the SPZ ligand (encoded by the *spaetzle* gene) on the ventral side only. SPZ binds to the TOLL transmembrane receptor (encoded by the *Toll* gene) that is present ubiquitously on the oocyte plasma membrane. The SPZ-TOLL complex triggers a signal transduction pathway that ends up phosphorylating the inactive DL protein. Phosphorylation of DL protein causes a conformational change that releases it from the CACT protein, which

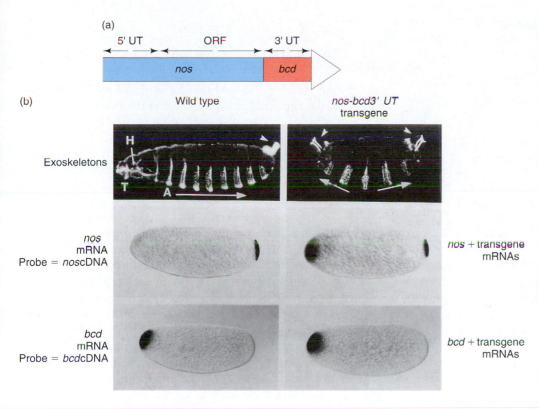

(a)

5′ UT   ORF   3′ UT

nos   bcd

(b)

Wild type   *nos-bcd3′ UT* transgene

Exoskeletons

nos mRNA
Probe = *nos*cDNA

nos + transgene mRNAs

bcd mRNA
Probe = *bcd*cDNA

bcd + transgene mRNAs

Figure 25-14 The effect of replacement of the 3′ UT of the *nanos* mRNA with the 3′ UT of the *bicoid* mRNA on mRNA localization and embryonic phenotypes. (a) The structure of the *nos-bcd3′UT* transgene. (b) The effects on embryonic development of the transgene. The embryos and larva in the left column are derived from wild-type mothers. Those in the right column are derived from transgenic mothers. All embryos are shown with anterior to the left and posterior to the right. The exoskeletons are shown for comparison. The transgene causes a perfect mirror-image double abdomen. In the embryo from a transgenic mother, mRNAs coding for NOS protein are now present at both poles of the embryo. NOS protein will inhibit translation of *hb-m* mRNA (and, actually, of *bcd* mRNA as well). (From E. R. Gavis and R. Lehmann, *Cell* 71, 1992, 303.)

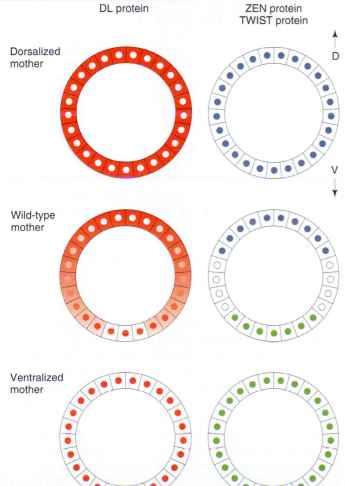

DL protein   ZEN protein
TWIST protein

Dorsalized mother

D

V

Wild-type mother

Ventralized mother

had sequestered DL in the cytoplasm (Figure 25-16). The free DL protein is now able to migrate into the nucleus, where it serves as a transcription factor for certain genes called cardinal genes whose products are required to determine the ventral fate. We will discuss cardinal genes in detail later in the chapter. Thus, cell-cell signaling between the oocyte and the follicle cells establishes the activity gradient of DL protein (Figure 25-15).

Message The D/V morphogen demonstrates a third way to produce a morphogen gradient: through the graded activation of a uniformly distributed morphogen.

Figure 25-15 The distribution of the morphogen DL and the ZEN and TWIST proteins of the D/V genes *zen* and *twist* in cross sections of blastoderm *Drosophila* embryos. The DL protein is normally present in a gradient of nuclear versus cytoplasmic localization around the circumference of the embryo, with the highest nuclear concentrations found ventrally. ZEN protein is normally found in nuclei located dorsally in the embryo, and TWIST is found in ventral nuclei. Both genes encode transcription factors and are regulated directly by the DL morphogen gradient. Note how the distributions of DL, ZEN, and TWIST are affected in dorsalized and ventralized embryos. (Adapted from S. Roth, D. Stein, and C. Nüsslein-Volhard, *Cell* 59, 1989, 1196.)

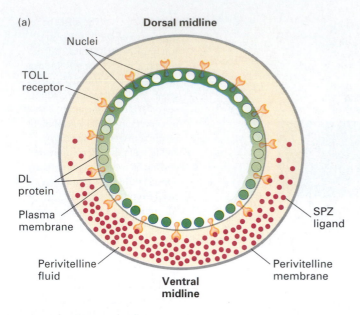

(a)

Dorsal midline

Nuclei

TOLL receptor

DL protein

Plasma membrane

Perivitelline fluid

SPZ ligand

Perivitelline membrane

Ventral midline

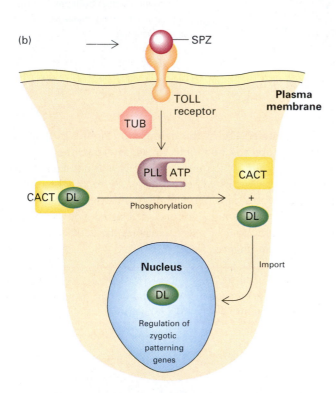

(b)

SPZ

TOLL receptor

Plasma membrane

TUB

PLL | ATP

CACT

CACT DL

CACT

+

DL

Phosphorylation

Nucleus

DL

Regulation of zygotic patterning genes

Import

Figure 25-16 The signaling pathway that leads to the gradient of nuclear versus cytoplasmic localization of DL protein. (a) A cross section of a *Drosophila* embryo showing the blastoderm cells inside the plasma membrane and the space (perivitelline space) between the inside boundary of the eggshell (perivitelline membrane) and the plasma membrane, where the SPZ ligand is produced on the ventral side of the embryo. (b) The SPZ ligand binds to the TOLL receptor, activating a signal transduction cascade through two proteins called TUB and PLL, leading to the phosphorylation of DL and its release from CACT. DL then is able to migrate into the nucleus, where it serves as a transcription factor for D/V cardinal genes. (Adapted from H. Lodish, D. Baltimore, A. Berk, S. L. Zipursky, P. Matsudaira, and J. Darnell, *Molecular Cell Biology,* 3d ed. Scientific American Books, 1995.)

The Establishment of Polarity in the Oocyte

What we have discussed thus far leaves us with the chicken and the egg problem. Why is it that the microtubule polarity runs along the A/P axis? Why is the SPZ ligand produced only on the ventral side of the oocyte? We have only a partial answer, but it seems that an earlier set of intercellular signals from the oocyte to subpopulations of follicle cells first distinguishes posterior from anterior and then dorsal from ventral.

Early in oogenesis, the microtubule structure of the developing oocyte is like that of a typical somatic cell, with the − ends of the microtubules in the center of the oocyte (at the microtubule organizing center, or MTOC; see Chapter 24) and the + ends in the periphery of the cytoplasm. In this early oocyte, there are already different sets of follicle cells at the ends of the oocyte, the border follicle cells, than are in the middle.

The key is that the oocyte nucleus transcribes an mRNA that encodes a protein required for posterior and dorsal fate of both the oocyte and the follicle cells (Figure 25-17). The oocyte nucleus has in its vicinity localized *grk* mRNA, encoded by the *gurken* (*grk*) gene. The oocyte nucleus moves nearer one end of the oocyte, perhaps by chance, and thus more GRK protein is thereby produced and secreted in the vicinity of one of the two sets of border cells. The GRK protein is a ligand for the *Drosophila* EGFR, a receptor tyrosine kinase (see Chapter 24). The EGFR protein, which is located in the cell membrane of all the follicle cells, is activated in the nearby border follicle cells. In a way that we don't understand, this causes a reorientation of the microtubules in the in the oocyte so that now all the + ends are located near the border cells which now have an activated EGFR kinase activity, and this leads them to organize the posterior end of the oocyte. This microtubule rearrangement moves the oocyte nucleus more anteriorly, and, again perhaps by chance, it happens to move closer to one side of the main-body follicle cells. These follicle cells are then exposed to a high level of secreted EGF, and this converts these follicle cells into dorsal-anterior follicle cells. One consequence (and again, we don't understand all the steps involved) is that the dorsal-anterior cells prevent SPZ ligand from being activated in their vicinity, limiting SPZ activity to the ventral side of the egg follicle.

We have devoted considerable space to the elaboration of positional information in *Drosophila*. This is because the gradients of the positional information molecules are really the key to understanding how development works. We have seen how different aspects of the normal biology of a cell—the cytoskeleton and the cell-cell signaling systems—are exploited by the oocyte to accomplish the creation of morphogen gradients. Once the gradients are established, the remainder of development proceeds by local activation of different gene products by the different local concentrations

Figure 25-17 The current model for how the same signaling system sets up both A/P and D/V polarities in the developing *Drosophila* oocyte. GRK is a ligand for the EGFR, a receptor tyrosine kinase. GRK is produced by mRNA located near the oocyte nucleus. First the oocyte nucleus is located near the posterior pole of the oocyte and, through the GRK signal to EGFR, causes the posteriormost follicle cells to adopt a different fate from the anteriormost ones. This sets up a reorganization of the microtubules and the differential localization of *bcd* and *nos* mRNAs along the A/P axis. Then, the oocyte nucleus migrates and, again through the GRK signal to the EGFR, locally activates the nearby cells to adopt the dorsal-anterior fate, and thereby creates the D/V axis. (Adapted from A. González-Reyes, H. Elliott, and D. St. Johnston, *Nature 375*, 1995, 657.)

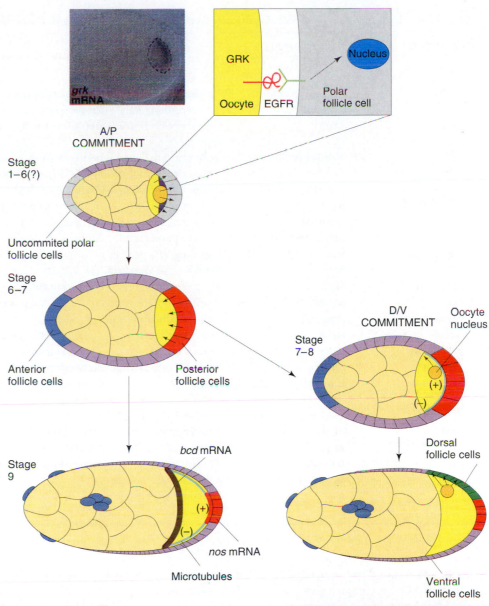

of the BCD, HB-M, and DL protein activities. We will consider these processes in the next section.

Message Gradients of protein products establish polarity—different geographic positions—along the A/P and D/V axes of the *Drosophila* embryo.

The Initial Response to Positional Information

How do BCD, HB-M, and DL proteins act as morphogens to provide positional information? We will see that they initiate the localized expression of regulatory proteins (mostly transcription factors), which in turn control the localized expression of other subsidiary regulatory proteins. Through several steps of regulation, the localized patterns of expression of the final regulatory proteins become more and more

fine-grained, until essentially every cell type is defined by the expression of a unique constellation of regulatory proteins.

BCD, HB-M, and DL are all transcription factors and induce, in a concentration-dependent fashion, the expression of a small set of zygotically expressed regulatory genes, collectively known as the **cardinal genes** because they are the

Table 25-2 Examples of *Drosophila* Zygotically Acting Genes Contributing to Pattern Formation

Gene symbol	Gene name	Protein function	Role(s) in early development
A/P axis genes:			
hb-z	*hunchback-zygotic*	Transcription factor—Zn-finger protein	Gap gene
Kr	*Krüppel*	Transcription factor—Zn-finger protein	Gap gene
kni	*knirps*	Transcription factor—steroid receptor type protein	Gap gene
eve	*even-skipped*	Transcription factor—homeodomain protein	Pair-rule gene
ftz	*fushi tarazu*	Transcription factor—homeodomain protein	Pair-rule gene
opa	*odd-paired*	Transcription factor—Zn-finger protein	Pair-rule gene
prd	*paired*	Transcription factor—PHOX protein	Pair-rule gene
en	*engrailed*	Transcription factor—homeodomain protein	Segment-polarity gene
ci	*cubitus-interruptus*	Transcription factor—Zn-finger protein	Segment-polarity gene
wg	*wingless*	Signaling WG protein	Segment-polarity gene
hh	*hedgehog*	Signaling HH protein	Segment-polarity gene
fu	*fused*	Cytoplasmic serine-threonine kinase	Segment-polarity gene
ptc	*patched*	Transmembrane protein	Segment-polarity gene
arm	*armadillo*	Cell-cell junction protein	Segment-polarity gene
lab	*labial*	Transcription factor—homeodomain protein	Segment-identity gene
Dfd	*Deformed*	Transcription factor—homeodomain protein	Segment-identity gene
Antp	*Antennapedia*	Transcription factor—homeodomain protein	Segment-identity gene
Ubx	*Ultrabithorax*	Transcription factor—homeodomain protein	Segment-identity gene
D/V axis genes:			
twi	*twist*	Transcription factor—bHLH protein	D/V cardinal gene
sna	*snail*	Transcription factor—Zn-finger protein	D/V cardinal gene
zen	*zerknüllt*	Transcription factor—homeodomain protein	D/V cardinal gene
dpp	*decapentaplegic*	Signaling DPP protein	D/V cardinal gene
spi	*spitz*	Signaling SPI protein	D/V 2° regulatory gene
S	*Star*	Transmembrane protein	D/V 2° regulatory gene

first genes to respond to the maternally supplied positional information. These cardinal genes (Table 25-2) were first identified through recessive mutant alleles in which homozygous mutant animals lack a large continuous piece of the A/P or D/V embryonic and larval anatomy (Figures 25-18 and 25-19). The promoters of the cardinal genes each have enhancer elements that are bound by one or another of the morphogen proteins, thereby activating or repressing transcription of the cardinal gene in question. Consider some of the so-called **gap genes,** cardinal genes of the A/P axis. Expression of one cardinal gene, *Krüppel (Kr),* is repressed by high levels of the BCD transcription factor but is promoted by low levels of BCD and HB-M. In contrast, the *knirps* (*kni*) gene is repressed by the presence of any BCD protein but does require low levels of the HB-M transcription factor for its expression. These enhancer and promoter properties thereby ensure that the *kni* gene is expressed more posteriorly than is *Kr* (Figure 25-20a). By having promoters that are differentially sensitive to the concentrations of one or more morphogens, the cardinal genes can be expressed in a distribution that begins to restrict the fates of their constituent cells.

Refining the Initial Patterning Decisions

The cardinal genes lay out a patchwork quilt of gene expression that anticipates the basic outlines of the body plan. However, the patterns of expression are too coarse to allocate all the cell fates that need to be assigned in order to achieve the final pattern of the larva. All the cardinal genes encode regulatory proteins (mostly transcription factors, the one exception being a ligand for a transmembrane receptor serine/threonine kinase). Through interactions of the protein products of adjacently expressed cardinal genes with each other, their domains of expression are refined and narrowed. For example, in the dorsal region of the embryo, the cardinal genes *decapentaplegic* (*dpp*) and *zerknüllt* (*zen*) are expressed in exactly the same domain at cellular blastoderm. Through interactions that are not completely understood, but that are dependent on different levels of DPP protein being expressed in different dorsal cells, *dpp* and *zen* become expressed in mutually exclusive sets of cells as gastrulation begins. In this manner, *zen* fates the amnioserosa cells, the

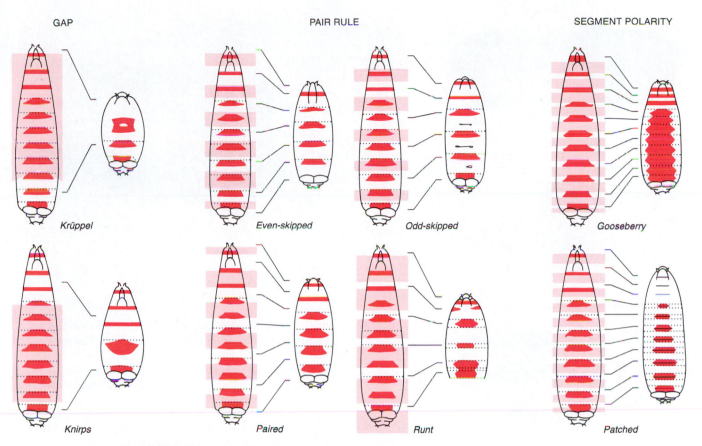

Figure 25-18 A diagram depicting the mutant larval phenotypes due to mutations in the three classes of zygotically acting genes controlling segment number in *Drosophila,* with representative mutants in each class. The denticle belts, shown as red trapezoids, are segmentally repeating swatches of dense projections on the ventral surface of the larval exoskeleton. The boundary of each segment is the dotted line. On the left, the pink regions indicate, on a diagram of a wild-type larva, the A/P domains of the larva that are missing in each mutant. The diagrams to the right are the resulting mutant phenotypes.

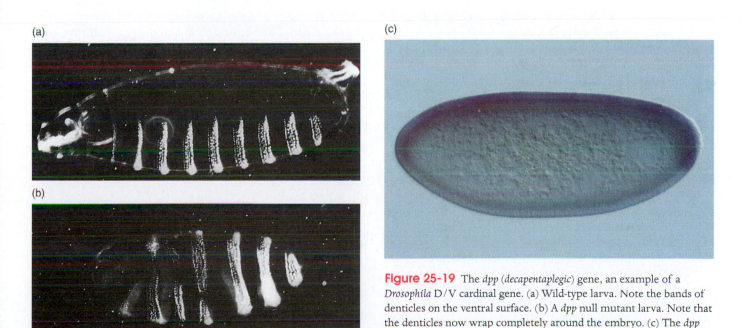

Figure 25-19 The *dpp* (*decapentaplegic*) gene, an example of a *Drosophila* D/V cardinal gene. (a) Wild-type larva. Note the bands of denticles on the ventral surface. (b) A *dpp* null mutant larva. Note that the denticles now wrap completely around the embryo. (c) The *dpp* mRNA blastoderm expression pattern in the dorsal half of the embryo.

(b)

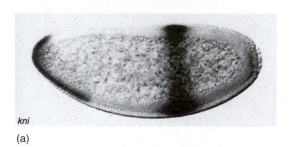

(a)

Figure 25-20 Photomicrographs showing the early embryonic expression patterns of gap and pair-rule genes. (a) Early blastoderm expression patterns of proteins from three gap genes: *hb-z, Kr,* and *kni.* (b) Late blastoderm expression patterns of proteins from two pair-rule genes, *ftz* (stained gray) and *eve* (stained brown). Note the localized gap gene expression patterns compared with the reiterated pair-rule gene expression pattern. (Part a from M. Hulskamp and D. Tautz, *BioEssays* 13, 1991, 261. Part b from Peter Lawrence, *The Making of a Fly.* Copyright © 1992 by Blackwell Scientific Publications.)

most dorsal cells of the embryo, whereas *dpp* fates the dorsolateral cells of the dorsal epidermis.

After narrowing their own expression domains, the proteins expressed by the cardinal genes cause localized expression of **secondary regulatory genes.** As with the morphogens and the cardinal genes, the first hints as to the nature of the genes involved in refinement of pattern come from mutant analysis. Along the D/V axis, these genes, which we will term **secondary D/V patterning genes,** are identified by mutant phenotypes in which a subset of cell types produced from one of the cardinal gene domains is missing. In the A/P axis, the next events are split along two paths. One set of secondary A/P patterning genes is involved in determining the correct number of segments. The other set is involved in determining the identity of each segment (that is, the kind of segment that will develop).

These segment-number genes are recognized by mutations in which there is a deletion of segmental material in a repeating fashion (Figure 25-18). Homozygous mutants for some segment-number genes have half as many segments as normal and are called **pair-rule** mutants. Homozygous mutants for others are missing subsets of cells from each

segment and are called **segment-polarity** mutants. The pair-rule genes are expressed in seven-stripe patterns, with each "on" and "off" stripe domain being approximately the width of one segment (Figure 25-20b). These pair-rule genes participate in defining the locations and widths of the segments. The segment-polarity gene products are expressed in 14 stripes, with each stripe corresponding to a subset of cells along the A/P axis of each segment. The segment-polarity genes contribute to the definition of the discrete positions of the various cells within each segment.

A Cascade of Regulatory Events

We see in the establishment of the A/P segments and the D/V cell types that the previous action of one set of regulatory molecules that are expressed in one type of domain leads to the progressive restriction of the embryo into finer and finer subdivisions (Figure 25-21 summarizes this progression for the A/P axis). Most of the elements of the initial patterning decisions are either transcription factors themselves or are cell-cell signals that themselves lead to dif-

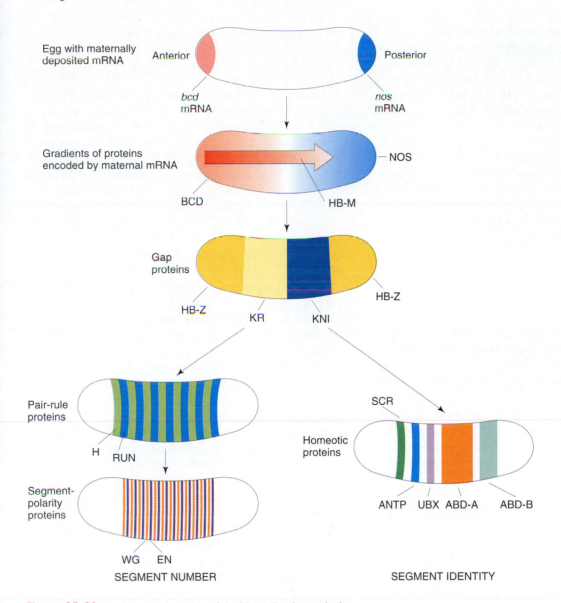

Figure 25-21 A schematic depiction of the hierarchical cascade that activates the elements forming the A/P segmentation pattern in *Drosophila*. The maternally derived *bcd* and *nos* mRNAs are located at the anterior and posterior poles, respectively. Early in embryogenesis, these mRNAs are translated to produce a steep anterior → posterior gradient of BCD transcription factor. The posterior → anterior gradient of NOS inhibits translation of *hb-m* mRNA, thereby creating a shallow anterior → posterior gradient of HB-M transcription factor (shown as an arrow). The gap genes, which are the A/P cardinal genes, are activated in different portions of the embryo in response to the anterior → posterior gradients of the two factors BCD and HB-M. (The posterior band of HB-Z expression gives rise to certain internal organs, not to segments.) The correct number of segments is determined by activation of the pair-rule genes in a "zebra-stripe" pattern in response to the gap gene–encoded transcription factors. The segment-polarity genes are then activated in response to the activities of the several pair-rule proteins, leading to further refinement of the organization within each segment. The correct identities of the segments are determined by expression of the homeotic genes due to direct regulation by the transcription factors encoded by the gap genes. (Modified from J. D. Watson, M. Gilman, J. Witkowski, and M. Zoller, *Recombinant DNA*, 2d ed. Copyright © 1992 by Scientific American Books.)

ferential activation of transcription factors. This theme seems to hold throughout the processes that we have described. All the way through to the segment-polarity genes, all the encoded functions are implicated in transcriptional regulation or in cell-cell signaling. In this regard, we can view intercellular signaling as a mechanism that allows different cells to coordinate their transcriptional activities.

Peculiarly, during early *Drosophila* pattern formation, we see that a repeating pattern of gene expression—the activation of the seven stripes of expression of each pair-rule gene—is somehow determined by the asymmetric and localized expression of the several gap genes. As is often true, we have little idea of the forces that led to the evolution of such a system, but here's how it works. We can define two subclasses of pair-rule genes: primary and secondary. The primary pair-rule genes—the gene even-skipped is one of these—are directly activated by the gap gene-encoded transcription factors. The secondary pair-rule genes are activated by transcription factors encoded by the primary pair-rule genes or by other secondary pair-rule genes. How do the primary pair-rule genes get activated in a stripe pattern by the gap genes? The key is that the cis-regulatory enhancer elements for the primary pair-rule genes are quite complex. For primary genes such as even-skipped, there are separate enhancer elements that regulate the activation of each even-skipped stripe. The stripe 1 enhancer is activated by high levels of the HB-Z gap gene-encoded transcription factor, the stripe 2 enhancer by low levels of HB-Z but high levels of the KR transcription factor, and so forth. Thus, through enhancer-specific activation of the primary pair-rule genes, the gap protein domains of expression direct the production of the pair-rule striping.

Determination of Segmental Identity

The pattern of A/P cardinal gene expression has two eventual consequences. One, as we just discussed, is to establish the correct number of segments in the larva. The other, which we will now address, is to confer the correct A/P identity to each of the segments. This topic is especially important because, as we shall see, it underscores that many of the mechanisms of pattern formation are ancient inventions that have been conserved in animals as divergent as mammals and flies through several hundred million years of evolution.

Even a brief look at the exoskeleton of an adult fly demonstrates the differences in the structures of the appendages, hairs, and pigmentation of the various segments of the fly's head, thorax, and abdomen. While more subtle, differences in the anatomy of the larval exoskeleton are also clear. Even in internal tissues, such as the central nervous system and the gut, there are segmental subdivisions. The ability to analyze the phenotypes of mutations in genes that control segmental identity has been an invaluable research tool.

The **homeotic genes** are the primary determinants of segmental identity. Mutations in these homeotic genes change the identity of one or more segments, but do not alter their number. In animals exhibiting homeotic mutant phenotypes, the segment number is exactly the same as in wild-type, but one or more of the segments have transformed their pattern of structures into extra copies of a segment that is more anterior or more posterior in the wild-type body plan. The homeotic genes all encode transcription factors. We will have more to say about the function and evolution of the homeotic genes and their proteins toward the end of this chapter.

Different homeotic genes are expressed at different positions along the A/P axis of the embryo. The expression patterns of the several gap genes determines the spatial patterns of homeotic gene expression (Figure 25-21).

Homeotic Mutations and the Establishment of Segment Identity

At the turn of the century, William Bateson collected examples of monstrosities in which one body part was turned into another; he called this phenomenon **homeosis.** Early in the history of *Drosophila* genetics, researchers uncovered mutations that regularly produced certain homeotic transformations. To cite two extreme examples, the bithorax class of mutations can cause the entire third thoracic segment (T3) to be transformed into a second thoracic segment (T2), giving rise to flies with four wings instead of the normal two (Figure 25-22), and dominant Antennapedia (Antp) class of mutations transforms antennas into legs (see the photograph on the first page of this chapter).

Most of the pattern-formation genes that we described in previous sections are dispersed throughout the genome. In contrast, the principal genes involved in establishing segment identity, the homeotic genes, map in two clusters (or complexes) on *Drosophila* chromosome 3. The bithorax gene complex (BX-C), studied in depth by E. B. Lewis and colleagues, contains the homeotic genes controlling the segmental identity of the third thoracic segment and all the abdominal segments throughout development. The Antennapedia gene complex (ANT-C), similarly characterized in Thomas Kaufman's laboratory, contains a cluster of homeotic genes controlling the identity of the head and thoracic segments. The linear order of genes within the ANT-C and BX-C is reflective of the A/P locations of their expression domains. As we go from one end of the ANT-C to the other end of the BX-C, genes are expressed in ever more posterior domains (Figure 25-23).

Once they were cloned, it was recognized that all homeotic genes share a very similar 180-bp DNA sequence, the **homeobox.** The homeobox encodes a 60 amino-acid polypeptide segment—the **homeodomain.** The term *homeodomain* is used, even though we now know that it is a more general motif, also present in many proteins that are not encoded by homeotic genes. The homeodomain is the region of the protein encoded by a homeotic gene protein that binds to DNA in a sequence-specific manner and thereby activates or represses the transcription of specific target genes (see Chapter 18 for a discussion of how homeodomains bind to DNA).

There is a stepwise progression in the phenotypes elicited by ANT-C and BX-C mutations (Figure 25-24). For example, a larva homozygous for a null mutation of *Abd-B* shows a transformation of what would ordinarily by the fifth to eighth abdominal segments (A5–A8) into reiterated copies of the fourth segment (A4). A double homozygous null for both *abd-A* and *Abd-B* shows a stronger transformation, in which segments A2–A8 are each transformed into segments indistinguishable from the normal A1. Finally, a triple homozygous null for the entire BX-C, which lacks the

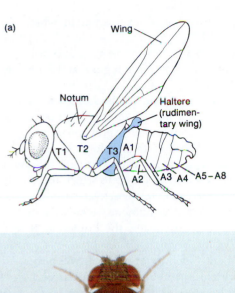

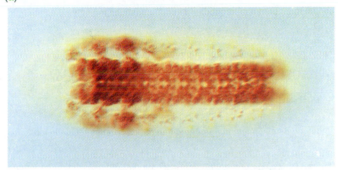

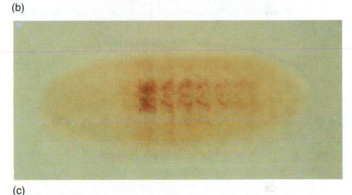

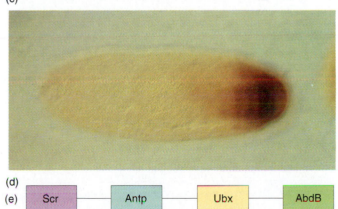

(e)

Scr	Antp	Ubx	AbdB

Figure 25-22 The homeotic transformation of the third thoracic segment (T3) of *Drosophila* into an extra second thoracic segment (T2). (a) Diagram showing the normal thoracic and abdominal segments; note the rudimentary wing structure normally derived from T3. Most of the thorax of the fly, including the wings and the dorsal part of the thorax, comes from T2. (b) A wild-type fly with one copy of T2 and one of T3. (c) A bithorax triple mutant homozygote completely transforms T3 into a second copy of T2. Note the second dorsal thorax and second pair of wings (T2 structures) and the absence of the halteres (T3 structures). (From E. B. Lewis, *Nature* 276, 1978, 565. Photographs courtesy of E. B. Lewis. Reprinted by permission of *Nature.* Copyright © 1978 by Macmillan Journals Ltd.)

Figure 25-23 Photomicrographs of embryos that exhibit homeotic gene protein expression patterns in *Drosophila*. (a) SCR; (b) ANTP; (c) UBX; (d) ABD-B. Note that the anterior boundary of homeotic gene expression is ordered from SCR (most anterior) to ANTP, UBX, and ABD-B (most posterior). (e) This order is matched by the linear arrangement of the corresponding genes along chromosome 3. (Parts a and b from T. Kaufman, Indiana University. Parts c and d from S. Celniker and E. B. Lewis, California Institute of Technology.)

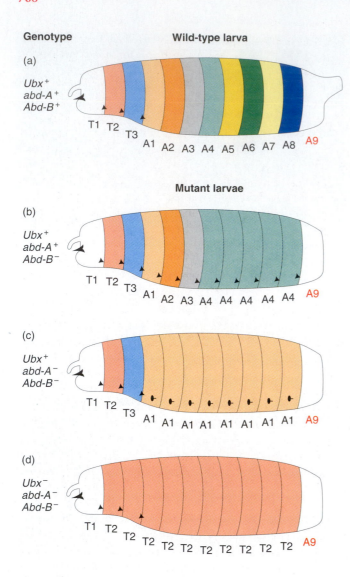

Genotype

Wild-type larva

(a)

Ubx^+
$abd-A^+$
$Abd-B^+$

T1 T2 T3 A1 A2 A3 A4 A5 A6 A7 A8 A9

Mutant larvae

(b)

Ubx^+
$abd-A^+$
$Abd-B^-$

T1 T2 T3 A1 A2 A3 A4 A4 A4 A4 A4 A9

(c)

Ubx^+
$abd-A^-$
$Abd-B^-$

T1 T2 T3 A1 A1 A1 A1 A1 A1 A1 A1 A9

(d)

Ubx^-
$abd-A^-$
$Abd-B^-$

T1 T2 T2 T2 T2 T2 T2 T2 T2 T2 T2 A9

Figure 25-24 The effects of mutations in the BX-C on the segmental identity of *Drosophila* larvae. Note that as more and more of the homeotic genes are inactivated, increasing numbers of posterior segments are transformed to increasingly anterior segmental fates. (Adapted from H. Lodish, D. Baltimore, A. Berk, S. L. Zipursky, P. Matsudaira, and S. Darnell, *Molecular Cell Biology,* 3d ed. Scientific American Books, 1995.)

three homeotic genes *Ubx, abd-A,* and *Abd-B,* transforms T3–A8 into copies of T2.

How do the ANT-C and BX-C genes normally get turned on in the right places? As we discussed earlier in this chapter, they appear to be regulated by the A/P cardinal genes; moreover, the homeotic genes are able to regulate one another. These multiple layers of regulation produce a

stable pattern of gene expression in which, for example, only the *Abd-B* gene is expressed in the most posterior abdominal segments, *Ubx* in the most anterior abdominal segments, *Antp* in the thoracic region, and *Scr* in the head (Figure 25-23).

Typically, only one of the homeotic genes is continuously expressed in a given cell throughout the lifetime of an animal. This continuous expression is known to be extremely important in their role in development. In some tissues, the continuous expression of a homeotic gene is mediated by a feedback loop in which the encoded homeodomain transcription factor activates its own promoter. If at any time during development we create a mutation in a somatic cell that inactivates the homeotic gene that should normally function in that cell, a homeotic transformation of the segmental identity of that cell will occur. A large set of accessory chromatin-binding proteins (perhaps 100) somehow contributes to the maintenance of the active state of the one expressed homeotic gene and the repression of all the other homeotic genes in a given cell. These accessory proteins are encoded by groups of genes named collectively the *Polycomb-group* and *trithorax-group* genes after the weak homeotic transformations that leaky, adult viable mutants in these genes can produce. It is possible that the clustering of the homeotic genes somehow contributes to ensuring that proper maintenance of expression of one homeotic gene and repression of all others occurs. How this is accomplished, however, is not clear. Nonetheless, the maintenance functions for the homeotic genes and for other genes at the ends of the cascades of interactions in D/V patterning and in the establishment of segment number are crucial in understanding development. Not only do "fates" have to be established early on in development, but cells must maintain these fates through the rest of development. The maintenance functions that have been outlined here serve this role and are an important part of the molecular basis of cell fate determination.

What is it that the proteins of the homeotic genes do? Each of them is a transcription factor with the ability to bind to different tissue-specific enhancers, in some cases to promote transcription of a specific gene and in others to repress gene expression. The genes regulated by the homeotic genes genes execute the functions required to build the structures of the various segments. Only a few of the direct targets of homeotic gene regulation have been identified. Most of these are themselves regulatory proteins involved in transcriptional control or intercellular signaling. Thus, it is likely that there are further regulatory events that occur after homeotic gene function to subdivide the fates of all the cells of a given segment.

The homeotic genes are expressed in many different tissues throughout the lifetime of the fly. In these different tissues, the various homeodomain proteins are thought to activate and repress different batteries of genes. This is one of many examples of an important developmental principle—

the same regulatory gene is reutilized throughout development in many different roles. Exactly what that regulatory protein will accomplish depends on its developmental context. That is, the action of many regulatory proteins including the homeodomain proteins is dependent on coordinated or cooperative interactions with other transcription factors to drive a cell down a particular developmental pathway. Thus, the particular set of genes regulated by the homeotic gene-encoded protein will depend on the array of other transcription factors expressed in the cell.

Message Once cell identity is established, it must be remembered. The properties of the homeotic genes and their proteins provide a model for the mechanisms underlying this maintenance of the determined state.

Applying the Fly Lesson to Other Organisms: Mouse Molecular Genetics

How universal are the developmental principles uncovered in *Drosophila?* Until recently, the type of genetic analysis possible in *Drosophila* has not been feasible in most other organisms, at least not without a huge investment to develop comparable genetic tools. However, in the last few years, recombinant DNA technology has provided the tools for addressing the generality of the *Drosophila* findings. Some of the most spectacular advances have come from studying early mouse development.

Finding Cognates of *Drosophila* Developmental Genes

With the discovery that there were numerous homeobox genes within the *Drosophila* genome, similarities among the DNA sequences of these genes could be exploited in treasure hunts for other members of the homeotic gene family. These hunts depend on the DNA base-pair complementarity. For this purpose, DNA hybridizations were carried out under "moderate stringency conditions," in which some mismatch of bases between the hybridizing strands could occur without disrupting the proper hydrogen bonding of nearby base pairs. Some of these treasure hunts were carried out in the *Drosophila* genome itself, looking for *more* family members. Others searched for homeobox genes in other animals, by means of "zoo blots" (Southern blots of restriction-enzyme-digested DNA from different animals), using radioactive *Drosophila* homeobox DNA as the probe. This approach has led to the discovery of homologous homeobox sequences in many different animals, including humans and mice. (Indeed, it is a very powerful approach to go "fishing" for relatives of most any gene in your favorite organism.) Some of these mammalian homeobox genes are very similar in sequence to the *Drosophila* genes. Perhaps the most striking case is the similarity between the clusters of mammalian homeobox genes called the *Hox complexes* and the insect ANT-C and BX-C homeotic gene clusters, now collectively called the HOM-C (homeotic gene complex) (Figure 25-25).

The ANT-C and BX-C clusters, which are far apart on chromosome 3 of *Drosophila melanogaster,* are tightly linked in more primitive insects such as the flour beetle, *Tribolium castaneum.* The tight linkage of the HOM-C cluster is considered the general case in insects. Moreover, the genes of the HOM-C cluster are arranged on the chromosome in an order that is colinear with their spatial pattern of expression: the genes at the left-hand end of the complex are transcribed near the anterior end of the embryo; going rightward along the chromosome, the genes are transcribed progressively more posteriorly (see Figure 25-25a).

The molecular anatomy of the mouse (and the human) Hox clusters has similar organizational features to the insect HOM-C cluster (Figure 25-25a). The map of the Hox genes also is colinear with the spatial organization of expression. The major difference is that, while there is only one HOM-C cluster in the insect genome, there are four Hox clusters, each located on a different chromosome, in mammals. These four Hox clusters are "paralogous," meaning that the structure (order of genes) in each cluster is very similar, as if the entire cluster had been quadruplicated during vertebrate evolution. The genes near the left end of each Hox cluster are quite similar not only to each other but also to one of the insect HOM-C genes at the left end of the cluster. Similar relationships hold throughout the clusters. Finally, and most notably, the Hox genes are expressed in a segmental fashion in the developing somites and central nervous system of the mouse (and presumably the human) embryo. Each Hox gene is expressed in a continuous block beginning at a specific anterior limit and running posteriorly to the end of the developing vertebral column (Figure 25-26). The anterior limit differs for different Hox genes. Within each Hox cluster, the leftmost genes have the most anterior limits. These limits proceed more and more posteriorly as we move rightward in each Hox cluster. Thus, the Hox gene clusters appear to be arranged and expressed in an order that is strikingly similar to that of the insect HOM-C genes (Figure 25-25b).

How can such disparate organisms—fly, mouse, human—have such similar gene sequences? The simplest interpretation is that the Hox and HOM-C genes are the vertebrate and insect descendants of a homeobox gene cluster present in a common ancestor some 600 million years ago. The evolutionary conservation of the HOM-C and Hox genes is not a singular occurrence. Many examples of strong evolutionary and functional conservation (such as the mouse *aniridia* and *Drosophila eyeless* genes depicted on this book's cover) of genes and entire pathways have been uncovered. As just one example, the pathway for activation

(a)

Directions of transcription of HOM-C genes

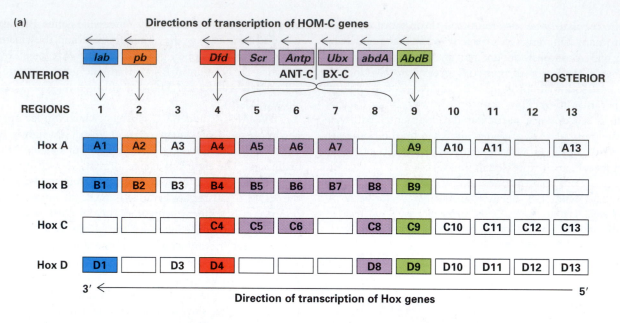

Figure 25-25 Comparisons of the structures and functions of the insect and mammalian homeotic genes. (a) The comparative anatomy of the HOM-C and Hox gene clusters. The genes of the HOM-C are shown on top. Each of the four paralogous (see text) Hox clusters maps on a different chromosome. Genes shown in the same color are most closely related to one another in structure and function. (b) The expression domains and regions of the *Drosophila* and mouse embryos that require the various HOM-C and Hox genes. The color scheme parallels that in part a. Note that the order of domains in the two embryos is the same. (Adapted from H. Lodish, D. Baltimore, A. Berk, S. L. Zipursky, P. Matsudaira, and S. Darnell, *Molecular Cell Biology*, 3d ed. Scientific American Books, 1995.)

of the DL morphogen in *Drosophila* has clear evolutionary and functional parallels to the pathway for activation of NF-κB, the transcription factor that binds to the immunoglobulin gene enhancers described in Chapter 23 (Figure 25-27). Other examples of conservation among genes important to building the body plan are listed in Table 25-3.

Message From HOM-C and Hox and many other similarities uncovered in the last few years, it is clear that many genes with important developmental roles are highly conserved during animal evolution.

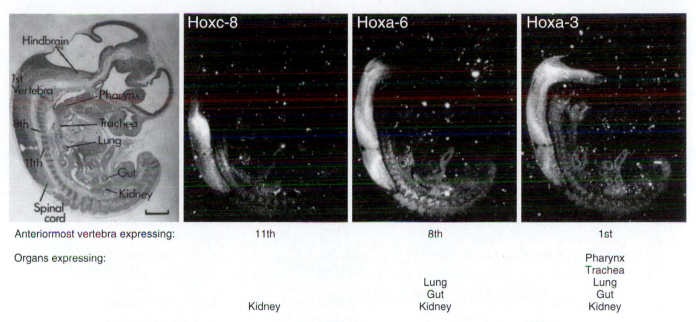

Anteriormost vertebra expressing:	11th	8th	1st
Organs expressing:	Kidney	Lung Gut Kidney	Pharynx Trachea Lung Gut Kidney

Figure 25-26 Photomicrographs showing the RNA expression patterns of three mouse Hox genes in the vertebral column of a sectioned 12.5-day-old mouse embryo. Note that the anterior limit of each of the expression patterns is different. (From S. J. Gaunt and P. B. Singh, *Trends in Genetics, 6,* 1990, 208.)

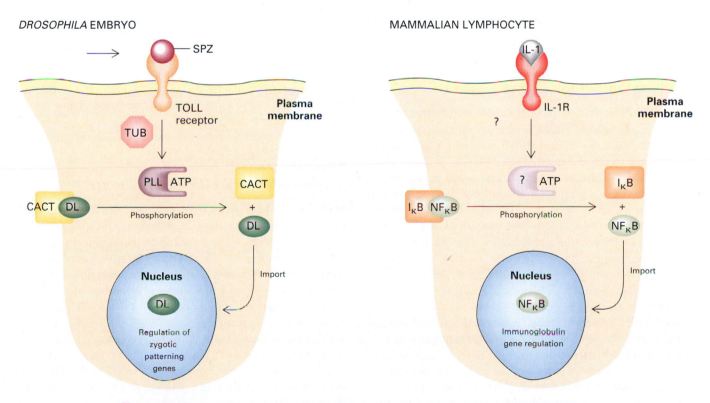

Figure 25-27 The signaling pathway for activation of the *Drosophila* DL morphogen parallels a mammalian signaling pathway for activation of NF-$_\kappa$B, the transcription factor that activates the immunoglobulin gene enhancers (see Chapter 23 for a discussion of immunoglobulin gene regulation). There are structural protein similarities between SPZ and IL-1, TOLL and IL-1R, CACT and I$_\kappa$B, and DL and NF-$_\kappa$B. (Adapted from H. Lodish, D. Baltimore, A. Berk, S. L. Zipursky, P. Matsudaira, and J. Darnell, *Molecular Cell Biology,* 3d ed. Scientific American Books, 1995.)

Table 25-3 Examples of Evolutionarily Conserved Genes Contributing to *Drosophila* Pattern Formation

Drosophila gene symbol	*Drosophila* gene name	Mammalian counterpart(s)	Protein function	Role(s) in early development
dl	*dorsal*	NF-$_\kappa$B	DL transcription factor	D/V morphogen
cact	*cactus*	I$_\kappa$B	Binds to DL protein in cytoplasm	Regulates D/V morphogen
spz	*spätzle*	IL-1	Signaling SPZ ligand	Regulates D/V morphogen
Tl	*Toll*	IL-1 receptor	Receptor for SPZ ligand	Regulates D/V morphogen
snk	*snake*	Many	Serine protease	Activates *spz* protein
dpp	*decapentaplegic*	BMP2, BMP4	Signaling DPP ligand	Cardinal gene for dorsal ectoderm
sax	*saxophone*		Receptor serine/	Contribute to dorsal ectodermal
tkv	*thick veins*	BRK	threonine kinases	pattern formation
pnt	*punt*		for DPP ligand	
sog	*short gastrulation*	chordin	Signaling SOG ligand	Contributes to ectodermal patterning
spi	*spitz*	TGF-α	Signaling SPI ligand	Secondary regulatory gene for ventral ectoderm
grk	*gurken*	EGF	Signaling GRK ligand	D/V and A/P polarity signals
Egfr	*EGF receptor*	EGFR	Receptor tyrosine kinase for SPI and GRK ligands	D/V and A/P polarity, and ventral ectodermal pattern formation
en	*engrailed*	En1, En2	Homeodomain protein	Segment-polarity gene
hh	*hedgehog*	Sonic hedgehog	Signaling HH ligand	Segment-polarity gene
wg	*wingless*	Wnt1, Wnt2, etc.	Signaling WG ligand	Segment-polarity gene
arm	*armadillo*	plakoglobin	Cell-cell junctions	Segment-polarity gene

NOTE: Many other conservations, especially among families of transcription factors such as the homeotic gene products are not listed here.

Knocking Out Mouse Cognate Genes

The above analysis sidesteps the question of whether or not the Hox genes and the HOM-C genes perform a similar developmental role. Some very clever technology is allowing this question to be addressed. Techniques for mixing different cell types in cultured mouse embryos and reimplanting these embryos in surrogate mothers have been in use for about 30 years. More recently, techniques have been developed to grow mouse germ-line precursors (**embryonic stem cells,** or ES cells) under conditions in which they can successfully be injected into the inner cell mass of host embryos and successfully repopulate the germ line of the resulting chimeric mouse (Figure 25-28; also see pages 477–480). Furthermore, individual cultured ES cells can be injected with altered, defective versions of standard mouse genes, which can replace their normal counterparts by homologous recombination that exploits DNA repair enzymes present in the ES cells. Hence, researchers can now produce a **targeted knockout** of a specific gene. ES cells containing a knockout are grown into a colony and injected into host embryos, which are in turn implanted into surrogate mothers carrying genetic markers (such as coat color) that differentiate them from both the host and the ES cells. The resulting adult mice are back-crossed to appropriate testers, and F$_1$ individuals that carry the genetic markers of the ES cells are therefore heterozygous for the knockout mutation. Such F$_1$ mice are then interbred to observe the phenotype of knockout homozygotes in the F$_2$.

Applying Knockout Technology to the Hox Gene Clusters

Many of the Hox genes have now been knocked out, and the striking result is that the phenotypes of the homozygous knockout mice are thematically parallel to the phenotypes of homozygous null HOM-C flies. For example, the *Hoxc-8* knockout causes a striking anatomical transformation: the production of ribs on the first lumbar vertebra, L1, which ordinarily is the first nonribbed vertebra behind those vertebrae-bearing ribs (Figure 25-29). Thus, the L1 vertebra is homeotically transformed to the segmental identity of a more anterior vertebra. To use geneticists' jargon, *Hox-c8*$^-$ has caused a fate shift toward anterior. Clearly, this Hox gene seems to control segmental fate in a manner quite similar to the HOM-C genes, since, for example, the absence of the *Drosophila Ubx* gene also causes a fate shift toward anterior.

Knockout genetics is also contributing to our understanding of how the four separate paralogous Hox clusters contribute to A/P segmental identity in mammals (Figure

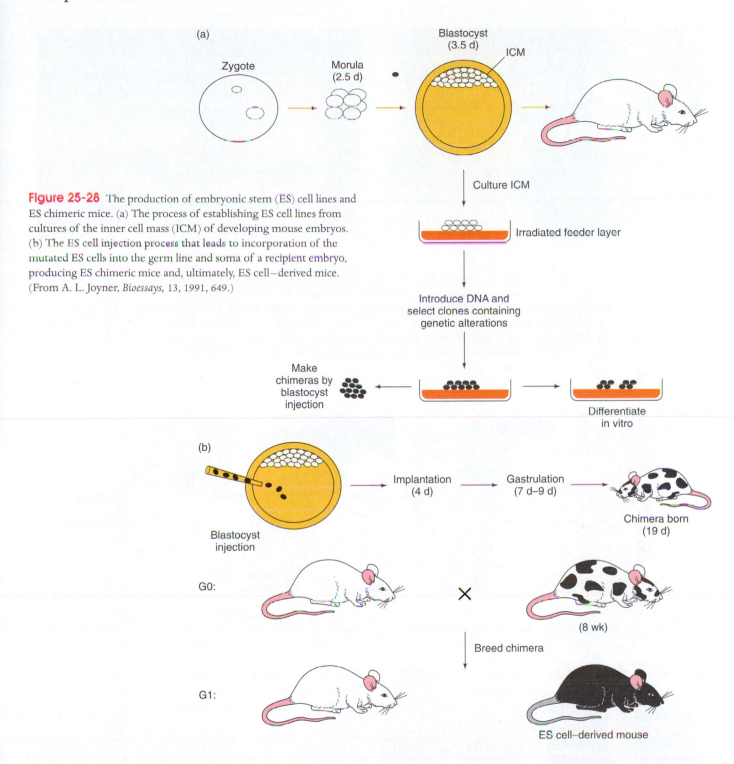

Figure 25-28 The production of embryonic stem (ES) cell lines and ES chimeric mice. (a) The process of establishing ES cell lines from cultures of the inner cell mass (ICM) of developing mouse embryos. (b) The ES cell injection process that leads to incorporation of the mutated ES cells into the germ line and soma of a recipient embryo, producing ES chimeric mice and, ultimately, ES cell–derived mice. (From A. L. Joyner, *Bioessays,* 13, 1991, 649.)

25-30). The group 4 Hox genes—*Hoxa-4, Hoxb-4, Hoxc-4* and *Hoxd-4*—are thought to be paralogous, that is, the equivalent members of each of the four Hox clusters. Knockouts of *Hoxb-4* and *Hoxd-4* have now been made and affect development of the seven cervical vertebrae of the spinal column. Either of the homozygous knockouts has very weak homeotic effects, causing a slight transformation of the second cervical vertebra (C2) toward the first (C1). The *Hoxa-4* knockout has even weaker phenotypic defects

and is essentially wild-type. The double knockout homozygote for *Hoxb-4* and *Hoxd-4* has a much more severe phenotype, fully transforming C2 and partially transforming C3 into C1. Further, C6 now looks very much like C5, and C7 like C6. A triple knockout homozygote of *Hoxa-4, Hoxb-4,* and *Hoxd-4* is even more dramatically affected, with C2 and C3 fully, and C4 and C5 partially, transformed into phenotypically C1 vertebrae, and both C6 and C7 transformed into C5 types of vertebrae. Thus, the aggregate effect of re-

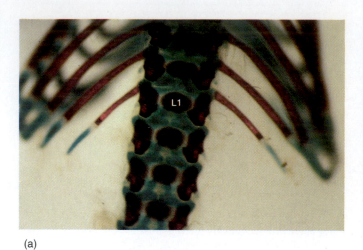

(a)

(b)

Figure 25-29 The phenotype of a homeotic mutant mouse. Mice homozygous for a targeted knock-out of the *Hoxc-8* gene were generated by using cultured ES cells, as shown in Figure 25-28. (a) A close-up of the thoracic and lumbar vertebrae of a homozygous *Hoxc-8⁻* mouse. Note the ribs coming from L1, the first lumbar vertebra. L1 in wild-type mice has no ribs. (b) An unexpected second phenotype of the *Hoxc-8⁻* knockout. Note that the homozygous mutant mouse on the right has clenched fingers, while the wild-type mouse on the left has normal fingers. (From H. Le Mouellic, Y. Lallemand, and P. Brulet, *Cell* 69, 1992, 251.)

(a)

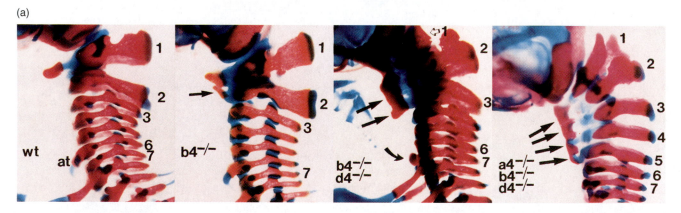

(b)

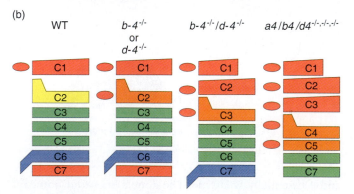

Figure 25-30 The combined effects of targeted knockout mutations of the *Hoxa-4, Hoxb-4,* and *Hoxd-4* genes on development of the seven cervical vertebrae (numbered 1–7 from anterior to posterior) in the mouse. (a) Stained vertebrae of wild-type, and single, double, and triple mutant homozygotes, as indicated. The arrows indicate some of the structures that have undergone homeotic transformations. (b) A diagram highlighting the transformations within the cervical vertebrae C1–C7. (Adapted from G. S. B. Horan, R. Ramírez-Solis, M. S. Featherstone, D. J. Wolgemuth, A. Bradley, and R. R. Behringer, *Genes & Development* 9, 1995, 1670 and 1672. Photos in part (b) from G. S. B. Horan and R. R. Behringer, *Genes and Development* 9, 1995, 1667–1677.)

moving the paralogous genes from each of the Hox clusters is much more pronounced than the removal of just one of these genes. This suggests that there is a certain level of overlap or redundancy of function between the paralogous genes of the four clusters, and that they may participate jointly in fating the various segments of the vertebrate body plan.

A final type of experiment that speaks to the level of conservation of the various homeotic genes is through heterologous gene replacement. In these experiments, the protein coding sequences of a gene from one of the vertebrate Hox clusters is introduced into a *Drosophila* mutant genotype lacking the function of the supposed homolog of that vertebrate gene. In several cases, the vertebrate protein is able to replace the missing function of the *Drosophila* gene. Thus, not only are the sequences of these homeodomain proteins quite conserved, and their expression patterns quite analogous, but they are still able to function in the proper context to regulate the segmental identity of very distantly related creatures. *Plus ça change, plus c'est le même chose!*

All the work to date suggests that the Hox clusters not only have strong evolutionary parallels to the HOM-C genes, but may have considerable functional parallels as well. A strong message is emerging that developmental strategies in animals are quite ancient and highly conserved, and that a mammal and a fly are put together using the same regulatory devices.

SUMMARY

The details of how animal development proceeds will undoubtedly differ from species to species. However, the examples that we have looked at here do portray general themes.

Early in life, each cell is totipotent: it has the potential to differentiate into a number of cell types. But in later stages of development, the type of cell it can become is increasingly restricted. In other words, the cell's determination—its commitment to a specific fate—is successively restricted to every-narrowing possibilities.

The egg must quickly establish polarity—the orientations of its body axes. Flies create polarity through maternally contributed localized determinants, exploiting properties of the cytoskeleton and cell-cell signaling systems to establish positional information.

Positional information is then used to activate regional specific expression of the cardinal genes, and the properties of the cardinal genes then produce proper fating along the A/P and D/V axes.

The homeotic genes and the homeodomain transcription factors they encode are highly conserved between insects and mammals—in sequence, chromosomal organization, expression pattern, and function.

Many of the regulatory genes of the fly have cognates in mammals; in many cases, these cognates appear to serve analogous roles during mammalian development.

Concept Map

Draw a concept map interrelating as many of the following terms as possible. Note that the terms are listed in no particular order.

targeted gene knockouts / homeotic genes / totipotency / EGFR / A/P axis / microtubules / gap genes / segment-polarity genes / morphogen gradients / paralogous genes

CHAPTER INTEGRATION PROBLEM

Comparative studies of the homeotic gene clusters in *Drosophila* and in the mouse have given us the view that these gene clusters are ancient entities, predating the evolutionary split of the lineages that gave rise to insects and mammals. You want to see if other genes involved in pattern formation are also highly conserved in structure and function. In *Drosophila*, *wg* (*wingless*) is one of the segment-polarity class of segmentation genes.

a. You wish to identify the gene that is equivalent to *wg* (that is, a *wg* homolog) in mice. Design experiments using the recombinant DNA techniques described in previous chapters to clone DNA corresponding to a mouse *wg* homolog (which we will call *mwg* for *mouse wingless*).

b. Assume that you are successful in cloning a complementary DNA (cDNA) corresponding to *mwg*. Outline a

protocol for characterizing this gene and for determining if it has developmental effects that parallel those of *wg*.

Solution

a. One message that comes through from the homeotic gene clusters is that parts of the protein-coding regions of the genes, the homeodomains, are highly conserved in sequence compared with other parts of the cluster. Therefore, if there is a mouse homolog of *wg* (that is, a gene presumably derived from a common ancestor), the DNA sequence of the homolog is much more likely to be conserved in some or all its protein-coding region than in nontranscribed regions or introns. For that reason, you would do best to use a *wg* cDNA to try to identify clones corresponding to *mwg*, because the cDNA contains the protein-coding region of the gene, minus all nontranscribed regions and introns. Using one classic approach, you could radioactively label and denature the *wg* cDNA and use it as a probe. You would then hybridize your probe to nitrocellulose filters carrying bound DNA from a genomic library of mouse DNA fragments cloned in bacteriophage vectors. The mouse gene, if it exists, will certainly have many differences from the fly *wg* gene in the DNA sequence for its protein. For example, even if *mwg* and *wg* produced the same amino acid, because the genetic code is degenerate, the amino acid might be encoded by a different codon, usually different in the third position of the codon. To compensate for such differences, this kind of cross-species hybridization is performed at "lower stringency"; in other words, conditions of hybridization are adjusted so that hybridization will occur even if there is some base mismatching.

Next, if you expose the hybridized filter to X-ray film, any phage plaques that contain some of the *mwg* gene will show up as black spots. You would then go back to the bacteriophage plate corresponding to that filter, pick the phage from the location on the plate that gave rise to the black spot, and retest the phage by hybridization to make sure that you picked the right one. You would then purify a restriction fragment of the putative *mwg* gene that hybridized on Southern blots with a *wg* probe and use that *mwg* fragment as a probe to identify mouse cDNAs containing some sequences corresponding to that *mwg* genomic fragment. Having obtained the *mwg* cDNA, you would then want to sequence it to determine if it had an open reading frame (that is, a protein-coding region) similar in amino acid sequence to *wg*. If it did, you would have a potential *mwg* homolog in hand.

b. Once you knew that you had an *mwg* gene, you would want to characterize it in several ways. For one thing,

you would want to know when and where it was expressed. One way to determine this is by Northern blots, in which you use radioactively labeled, denatured *mwg* cDNA to hybridize to filters containing electrophoretically separated RNAs from different developmental stages or tissues of the mouse. Another way is by RNA in situ hybridization, using an *mwg* probe to hybridize to transcripts in histological sections of mice. A third way is by making antibodies against recombinant *mwg* protein and using those antibodies as probes for immunohistochemistry.

Once you knew where and when the gene was expressed, you would have some ideas about its possible functional significance that you would like to test. The best way to test these ideas nowadays is by making a mouse-targeted knockout mutation. For this, you would first construct an altered *mwg* gene that contained some nonsense DNA within the gene. By using the technique of transformation into cultured ES cells, you could select for homologous gene replacement, effectively knocking out one of the copies of *mwg* in the cell. Then you would transplant the ES cell clone containing the *mwg* knockout into recipient mouse embryos, reimplant the embryos in a foster mother host, recover the resulting mosaic adults, and test their germ lines for transmission of the knockout homolog of *mwg* by probing the DNA of the F_1 for presence of the nonsense DNA sequence used in the knockout. After two more generations of inbreeding of such F_1 individuals, you would have generated mice homozygous for the *mwg* knockout. You would then determine the phenotypes of these individuals and try to relate the phenotypes to the expression patterns of the gene and to your expectation from *wg* in flies.

This scenario has been used many times now and is turning out to be a highly profitable way to approach problems of mammalian development.

SOLVED PROBLEMS

1. In the embryogenesis of mammals, the inner cell mass, or ICM (the prospective fetus), quickly separates from the cells that will serve as enclosing membranes and respiratory, nutritive, and excretory channels between the mother and the fetus.

a. Design experiments using mosaics in mice to determine when the two fates are decided.

b. How would you trace the formation of different fetal membranes?

Solution

a. We must have markers that enable us to distinguish different cell lineages. This can be done with mice by using strains that differ in chromosomal or biochemical

markers. (Other ways would be to use differences in sex chromosomes of XX and XY cells or to induce chromosome loss or aberrations by irradiating embryos.)

Once you have decided on the marker difference to be used, one way to answer the question is to inject a single cell from one strain into embryos of the other at various developmental stages. Another approach is to fuse embryos of defined cell numbers from the two strains. In either case, you would inspect the embryos when the ICM and membranes are distinct and recognizable. When cell insertion or fusion results in membranes and ICM that are exclusively made up of one cell type and never a mosaic of the two, the two developmental fates have been set.

b. Carry out the same injection or fusion experiment on early embryos. Now look for the pattern of mosaicism. Correlate the occurrence of cells of similar genotype in different membranes. It should be possible to determine the cell lineage of cells in each set of membranes.

2. *Drosophila* maternal-effect lethal mutagenesis screens require that mutant homozygous daughters derived from heterozygous parents survive to adulthood, since it is these daughters that will be assayed for the production of nonviable offspring. However, it may be that genes which provide important maternal products for A/P and D/V axis formation also perform other important functions in some cell types later in development, such that null mutants in these genes will be homozygous lethal. How might one get around the homozygous lethality to test the contributions of a gene to maternal A/P and D/V axis formation?

HINTS: (1) Think about the genotypes of mice that are recipients of transplanted ES cells. (2) Recall that the germ line and soma of the fly separate very early in development.

Solution

The key to this problem is to generate a germ-line mosaic individual. If female germ-line precursor cells of the homozygous mutant genotype are viable and can develop oocytes, and if cells of this genotype can be placed in a female that is somatically wild-type, then we can test this mosaic for its ability to produce offspring with A/P or D/V patterning defects.

Such a mosaic could be created by transplantation of the germ-line precursors, the pole cells, by techniques analogous to ES cell transplantation. For example, we could remove pole cells from donor blastoderm-stage embryos derived from a strain heterozygous for the recessive lethal mutation. We would transplant these cells into differentially marked recipient embryos. (We would have to be careful to transplant pole cells from only one donor embryo into each recipient.) One-quarter of the donor embryos will be homozygous for the recessive lethal mutation, and half of those will be female. For the recipient embryos, half the recipient embryos will be female. It turns out that in order for pole cell transplants to succeed, the sex of the pole cells and the recipient must be the same. Thus, we expect that one-sixteenth of our transplants will be potential successes. The others either will be sexually incompatible (half the injected embryos) and give rise to no donor-derived offspring, or will be derived from pole cells carrying at least one wild-type allele of the gene in question. Among these, we would look for cases where we obtain some progeny embryos with A/P or D/V patterning defects.

This type of germ-line mosaic experiment has been extremely valuable in identifying maternally acting genes that could not have been recovered in a standard mutant screen. Additional genetic tricks are used to kill off the germ-line cells of the recipient and to allow the recessive mutant chromosome to be marked in the embryo so that the genotype of the donor germ cells can be unambiguously ascertained.

PROBLEMS

1. In his book *In His Image: The Cloning of a Man,* David Rorvik claimed that a baby had been cloned from the cells of an elderly man. Few (if any) geneticists believe this story, but many scientists do feel that human cloning is a definite possibility in the future, since naturally occurring human clones (identical twins) do exist. How would the ability to produce human clones in the laboratory be useful in studying the relative importance of heredity and environment in determining the final phenotype of a human individual?

2. For many of the mammalian Hox genes, it has been possible to determine that some of them are more similar to one of the insect HOM-C genes than to the others. Describe an experimental approach using the tools of molecular biology that would allow such a determination.

3. **a.** When you remove the anterior 20 percent of the cytoplasm of a newly formed *Drosophila* embryo, you can cause a bicaudal phenotype, in which there is a mirror image duplication of the abdominal segments. In other words, from the anterior tip of the embryo to the posterior tip, the order of segments is A8-A7-A6-A5-A4-A4-A5-A6-A7-A8. Explain this phenotype in terms of the action of the anterior and posterior determinants and how they affect gap gene expression.

b. Females homozygous for the mutation *oskar* (*osk*) produce embryos with an absence of the abdominal segments and in which the head and thoracic segments are broader. In terms of the action of the anterior and posterior determinants and gap gene action, explain how *oskar* produces this mutant phenotype. In your answer, explain why there is a loss of segments rather than a mirror-image duplication of anterior segments.

4. The three homeodomain proteins *Abd-B*, *abd-A*, and *Ubx* are encoded by genes within the BX-C of *Drosophila*. In wild-type embryos, the *Abd-B* gene is expressed in the posterior abdominal segments, *abd-A* in the middle abdominal segments, and *Ubx* in the anterior abdominal and posterior thoracic segments. When the *Abd-B* gene is deleted, *abd-A* is expressed in both the middle and posterior abdominal segments. When *abd-A* is deleted, *Ubx* is expressed in the posterior thorax and in the anterior and middle abdominal segments. When *Ubx* is deleted, the patterns of *abd-A* and *Abd-B* expression are unchanged from wild-type. When both *abd-A* and *Abd-B* are deleted, *Ubx* is expressed in all segments from the posterior thorax to the posterior end of the embryo. Explain these observations, taking into consideration the fact that the gap genes control the initial expression patterns of the homeotic genes.

5. When an embryo is homozygous mutant for the gap gene *Kr,* the fourth and fifth stripes of the pair-rule gene *ftz* (counting from the anterior end) do not form normally. When the gap gene *kni* is mutant, the fifth and sixth *ftz* stripes do not form normally. Explain these results in terms of how segment number is established in the embryo.

6. In discussing the formation of the A/P and D/V axes in *Drosophila,* we noted that for mutations like *bcd,* homozygous mothers uniformly produce mutant offspring with segmentation defects. This is true regardless of whether the offspring themselves are *bcd⁺ bcd* or *bcd bcd*. Some other maternal-effect lethal mutations are different, in that the mutant phenotype can be "rescued" by introducing a wild-type allele of the gene from the father. In other words, for such rescuable maternal-effect lethals, *mut⁺ mut* animals are normal while *mut mut* animals have the mutant defect. Rationalize the difference between rescuable and nonrescuable maternal-effect lethal mutations.

7. The anterior determinant in the *Drosophila* egg is *bcd*. A mother heterozygous for a *bcd* deletion has only one copy of the *bcd* gene. Using P elements to insert copies of the cloned *bcd* gene into the genome by transformation, it is possible to produce mothers with extra copies of the gene.

Shortly after blastoderm formation is over, the *Drosophila* embryo develops an indentation that is more or less perpendicular to the longitudinal body axis, called the *cephalic* furrow. In the progeny of *bcd* monosomics, this furrow is very close to the anterior tip, lying at position $\frac{1}{6}$ of the distance from the anterior to the posterior tip. In the progeny of standard wild-type diploids (disomic for *bcd⁺*), the cephalic furrow arises more posteriorly, at a position $\frac{1}{5}$ of the distance from the anterior to the posterior tip of the embryo. In the progeny of *bcd* trisomics, it is even more posterior. As additional gene doses are added, it moves more and more posteriorly, until in the progeny of hexasomics, the cephalic furrow is midway along the A/P axis of the embryo.

a. Explain the gene dosage effect of *bcd* on cephalic furrow formation in terms of *bcd*'s contribution to A/P pattern formation.

b. Diagram the relative expression patterns of mRNAs from the gap genes *Kr* and *kni* in blastoderm embryos derived from *bcd* monosomic, trisomic, and hexasomic mothers.

8. In the *Drosophila* embryo, the 3′ untranslated regions (abbreviated 3′ UT) of the mRNAs [the regions between the translation termination codons and the poly(A) tails] are responsible for localizing *bcd* and *nos* to the anterior and posterior poles, respectively. Experiments have been done in which the 3′ UTs of *bcd* and *nos* have been swapped. Supposing that we make P element transformation constructs with both swaps (*nos* mRNA with *bcd* 3′ UT and *bcd* mRNA with *nos* 3′ UT) and transform them into the *Drosophila* genome. We then make a female which is homozygous mutant for *bcd* and *nos,* and which also carries both swap constructs. What phenotype would you expect for her embryos in terms of A/P axis development?

26

Population Genetics

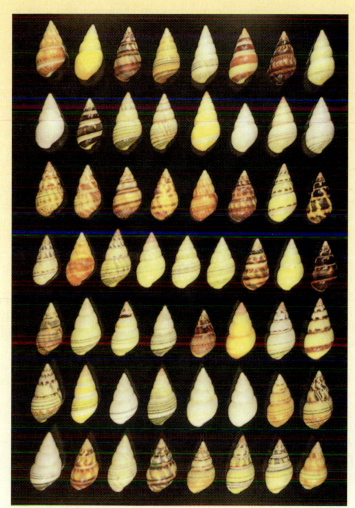

Shell color polymorphism in *Liguus fascitus*. (From David Hillis, *Journal of Heredity*, July–August 1991.)

KEY CONCEPTS

▶ The goal of population genetics is to understand the genetic composition of a population and the forces that determine and change that composition.

▶ In any species, a great deal of genetic variation within and between populations arises from the existence of various alleles at different gene loci.

▶ A fundamental measurement in population genetics is the frequency at which the alleles can occur at any gene locus of interest.

▶ The frequency of a given allele in a population can be changed by recurrent mutation, selection, or migration, or by random sampling effects.

▶ In an idealized population, in which no forces of change are acting, a randomly interbreeding population would show constant genotypic frequencies for a given locus.

Mendel's investigations of heredity—indeed all the interest in heredity in the nineteenth century—arose from two related problems: how to breed improved crops and how to understand the nature and origin of species. What is common to these problems (and differentiates them from the problems of transmission and gene action) is that they are concerned with *populations* rather than with *individuals*. Studies of gene replication, protein synthesis, development, and chromosome movement focus on processes that go on within the cells of individual organisms. But the transformation of a species, either in the natural course of evolution or by the deliberate intervention of human beings, is a change in the properties of a collectivity—of an entire population or a set of populations.

Message   Population genetics relates the heritable changes in populations of organisms to the underlying individual processes of inheritance and development.

Darwin's Revolution

The modern theory of evolution is so completely identified with the name of Charles Darwin (1809–1882) that many people think that the concept of organic evolution was first proposed by Darwin, but that is certainly not the case. Most scholars had abandoned the notion of fixed species, unchanged since their origin in a grand creation of life, long before publication of Darwin's *The Origin of Species* in 1859. By that time, most biologists agreed that new species arise through some process of evolution from older species; the problem was to explain *how* this evolution could occur.

Darwin's theory of the mechanism of evolution begins with the variation that exists among organisms within a species. Individuals of one generation are qualitatively different from one another. Evolution of the species as a whole results from the differential rates of survival and reproduction of the various types, so that the relative frequencies of the types change over time. Evolution, in this view, is a sorting process.

For Darwin, evolution of the group resulted from the differential survival and reproduction of individual variants *already existing* in the group—variants arising in a way unrelated to the environment, but whose survival and reproduction does depend on the environment.

Message   Darwin proposed a new explanation to account for the accepted phenomenon of evolution. He argued that the population of a given species at a given time includes individuals of varying characteristics. The population of the next generation will contain a higher frequency of those types that most successfully survive and reproduce under the existing environmental conditions. Thus, the frequencies of various types within the species will change over time.

There is an obvious similarity between the process of evolution as Darwin described it and the process by which the plant or animal breeder improves a domestic stock. The plant breeder selects the highest-yielding plants from the current population and (as far as possible) uses them as the parents of the next generation. If the characteristics causing the higher yield are heritable, then the next generation should produce a higher yield. It was no accident that Darwin chose the term **natural selection** to describe his model of evolution through differential rates of reproduction of different variants in the population. As a model for this evolutionary process, he had in mind the selection that breeders exercise on successive generations of domestic plants and animals.

We can summarize Darwin's theory of evolution through natural selection in three principles:

1. **Principle of variation.** Among individuals within any population, there is variation in morphology, physiology, and behavior.
2. **Principle of heredity.** Offspring resemble their parents more than they resemble unrelated individuals.
3. **Principle of selection.** Some forms are more successful at surviving and reproducing than other forms in a given environment.

Clearly, a selective process can produce change in the population composition only if there are some variations to select among. If all individuals are identical, no amount of differential reproduction of individuals can affect the composition of the population. Furthermore, the variation must be in some part heritable if differential reproduction is to alter the population's genetic composition. If large animals within a population have more offspring than small ones but their offspring are no larger on average than those of small animals, then no change in population composition can occur from one generation to another. Finally, if all variant types leave, on average, the same number of offspring, then we can expect the population to remain unchanged.

Message   Darwin's principles of variation, heredity, and selection must hold true if there is to be evolution by a variational mechanism.

Variation and Its Modulation

Population genetics is the translation of Darwin's three principles into precise genetic terms. As such, it deals with the description of genetic variation in populations and with the experimental and theoretical determination of how that variation changes in time and space.

Message   Population genetics is the study of inherited variation and its modulation in time and space.

Observations of Variation

Population genetics necessarily deals with genotypic variation, but by definition, only phenotypic variation can be observed. The relation between phenotype and genotype varies in simplicity from character to character. At one extreme, the phenotype may be the observed DNA sequence of a stretch of the genome. In this case, the distinction between genotype and phenotype disappears, and we can say that we are, in fact, directly observing the genotype. At the other extreme lie the bulk of characters of interest to plant and animal breeders and to most evolutionists—the variations in yield, growth rate, body shape, metabolic ratio, and behavior that constitute the obvious differences between varieties and species. These characters have a very complex relation to genotype, and we must use the methods introduced in Chapter 27 to say anything at all about the genotypes. But as we will see in Chapter 27, it is not possible to make very precise statements about the genotypic variation underlying quantitative characters. For that reason, most of the study of experimental population genetics has concentrated on characters with simple relations to the genotype, much like the characters studied by Mendel. A favorite object of study for human population geneticists, for example, has been the various human blood groups. The qualitatively distinct phenotypes of a given blood group—say, the MN group—are coded for by alternative alleles at a single locus, and the phenotypes are insensitive to environmental variations.

The study of variation, then, consists of two stages. The first is a description of the phenotypic variation. The second is a translation of these phenotypes into genetic terms and the redescription of the variation genetically. If there is a perfect one-to-one correspondence between genotype and phenotype, then these two steps merge into one, as in the case of the MN blood group. If the relation is more complex—for example, as the result of dominance, so that heterozygotes resemble homozygotes, it may be necessary to carry out experimental crosses or to observe pedigrees to translate phenotypes into genotypes. This is the case for the human ABO blood group (see page 93).

The simplest description of Mendelian variation is the frequency distribution of genotypes in a population. Table 26-1 shows the frequency distribution of the three genotypes at the MN blood group locus in several human populations. Note that there is variation between individuals in each population, as there are different genotypes present, and there is variation in the frequencies of these genotypes from population to population. More typically, instead of the frequencies of the diploid genotypes, the frequencies of the alternative alleles are used. If f_{AA}, f_{Aa}, and f_{aa} are the proportions of the three genotypes at a locus with two alleles, then the frequency $p(A)$ of the A allele and the frequency $q(a)$ of the a allele are obtained by counting alleles. Since each homozygote AA consists only of A alleles and only half the alleles of each heterozygote Aa are type A, the total frequency (p) of A alleles in the population is

Table 26-1 Frequencies of Genotypes for Alleles at MN Blood Group Locus in Various Human Populations

Population	Genotype			Allele frequencies	
	MM	*MN*	*NN*	*p(M)*	*q(N)*
Eskimo	0.835	0.156	0.009	0.913	0.087
Australian aborigine	0.024	0.304	0.672	0.176	0.824
Egyptian	0.278	0.489	0.233	0.523	0.477
German	0.297	0.507	0.196	0.550	0.450
Chinese	0.332	0.486	0.182	0.575	0.425
Nigerian	0.301	0.495	0.204	0.548	0.452

SOURCE: W. C. Boyd, *Genetics and the Races of Man*. D. C. Heath, 1950.

$$p = f_{AA} + \tfrac{1}{2} f_{Aa} = \text{frequency of } A$$

Similarly, the frequency q of a alleles is given by

$$q = f_{aa} + \tfrac{1}{2} f_{Aa} = \text{frequency of } a$$
$$p + q = f_{AA} + f_{aa} + f_{Aa} = 1.00$$

So,
$$q = 1 - p$$

If there are multiple alleles, the frequency for each allele is simply the frequency of its homozygote plus half the sum of the frequencies for all the heterozygotes in which it appears. Table 26-1 shows the values of p and q for each of the MN blood group populations.

As an extension of p, which represents the **gene frequency** or **allele frequency,** or **gamete frequency,** we can describe variation at more than one locus simultaneously, in terms of the multilocus gametic frequencies, or **haplotype** frequencies.

Locus S (the secretor factor, determining whether the M and N proteins are also contained in the saliva) is closely linked to the MN locus in humans. Table 26-2 shows the frequencies, commonly symbolized by g's, of the four haplotypes ($M S$, $M s$, $N S$, and $N s$) in various populations. The haplotype frequency is obtained by summing all the contributions of the different heterozygotes and homozygotes to the total pool of gametes. For example, the frequency of the $M S$ haplotype, denoted by $g(M S)$ is given by

$$
\begin{aligned}
g(M S) = \; &\text{frequency of } M S/M S \\
&+ \tfrac{1}{2} \text{ frequency of } M S/N S + \tfrac{1}{2} \text{ frequency of } M S/M s \\
&+ \tfrac{1}{2} \text{ frequency of } M S/N s
\end{aligned}
$$

Note that the last term in the sum involves the frequency of the double heterozygote $M S/ N s$. There is no contribution from the other double heterozygote $M s / N S$, because it produces no $M S$ gametes. If there were recombination, the $M s / N S$ heterozygote would produce $M S$ gametes at a rate

Table 26-2 Frequencies of Gametic Types for MNS System in Various Human Populations

| Population | Gametic type | | | | Heterozygosity (*H*) | |
	MS	*Ms*	*NS*	*Ns*	From gametes	From alleles
Ainu	0.024	0.381	0.247	0.348	0.672	0.438
Ugandan	0.134	0.357	0.071	0.438	0.658	0.412
Pakistan	0.177	0.405	0.127	0.291	0.704	0.455
English	0.247	0.283	0.080	0.290	0.700	0.469
Navaho	0.185	0.702	0.062	0.051	0.467	0.286

SOURCE: A. E. Mourant, *The Distribution of the Human Blood Groups.* Blackwell Scientific Pub., 1954.

 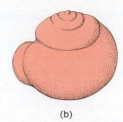

(a)   (b)

Figure 26-1 Shell patterns of the snail *Cepaea nemoralis:* (a) banded yellow; (b) unbanded pink.

proportional to the recombination fraction. Thus, to give a gametic frequency description of a population at more than one locus simultaneously, we need to be able to distinguish coupling from repulsion double heterozygotes and to know the recombination fraction between the genes. For the human MNS system, there is essentially no recombination, and the different types of heterozygotes can be distinguished by pedigree analysis.

A *measure* of genetic variation (as opposed to its *description* by gene frequencies) is the amount of **heterozygosity** at a locus in a population, which is given by the total frequency of heterozygotes at a locus. If one allele is in very high frequency and all others are near zero, then there will be very little heterozygosity because, by necessity, most individuals will be homozygous for the common allele. We expect heterozygosity to be greatest when there are many alleles at a locus, all at equal frequency. In Table 26-1, the heterozygosity is simply equal to the frequency of the MN genotype in each population. When more than one locus is considered, there are two possible ways of calculating heterozygosity. First, we can average the frequency of heterozygotes at each locus separately. Alternatively, we can consider each haplotype as a unit as in Table 26-2 and calculate the proportion of all individuals who carry two different haplotypic forms. This form of heterozygosity is also referred to as *haplotype diversity.* The results of both calculations are given in Table 26-2. Note that the haplotype diversity is always greater than the average heterozygosity of the separate loci because an individual is a haplotypic heterozygote if *either* of its loci is heterozygous. (See the discussion of Hardy-Weinberg equilibrium on page 790 for the calculation of heterozygosity.)

Simple Mendelian variation can be observed within and between populations of any species at various levels of phenotype, from external morphology down to the amino acid sequence of enzymes and other proteins. Indeed, with the new methods of DNA sequencing, variations in DNA sequence (such as third-position variants that are not differen-

tially coded in amino acid sequences and even variations in nontranslated intervening sequences) have been observed. Every species of organism ever examined has revealed considerable genetic variation, or **polymorphism,** reflected at one or more levels of phenotype, within populations, between populations, or both. A gene or a phenotypic trait is said to be *polymorphic* if there is more than one form of the gene or trait in a population. Genetic variation that might be the basis for evolutionary change is ubiquitous. The tasks for population geneticists are to describe that ubiquitous variation quantitatively and to build a theory of evolutionary change that can use these observations in prediction.

It is impossible in this text to provide an adequate picture of the immense richness of genetic variation that exists in species. We can consider only a few examples of the different kinds of Mendelian variation to gain a sense of the genetic diversity within species. Each of these examples can be multiplied many times over in other species and with other traits.

Morphological Variation. The shell of the land snail *Cepaea nemoralis* may be pink or yellow, depending on two alleles at a single locus, with pink dominant to yellow. Also, the shell may be banded or unbanded (Figure 26-1) as a result of segregation at a second linked locus, with unbanded dominant to banded. Table 26-3 shows the variation of these two loci in several European colonies of the snail. The populations also show polymorphism for the number of bands and the

Table 26-3 Frequencies of Snails (*Cepaea nemoralis*) with Different Shell Colors and Banding Patterns in Three French Populations

| Population | Yellow | | Pink | |
	Banded	Unbanded	Banded	Unbanded
Guyancourt	0.440	0.040	0.337	0.183
Lonchez	0.196	0.145	0.564	0.095
Peyresourde	0.175	0.662	0.100	0.062

SOURCE: Maxime Lamotte, *Bulletin Biologique de France et Belgique,* suppl. 35, 1951.

Table 26-4 Frequencies of Plants with Supernumerary Chromosomes and of Translocation Heterozygotes in a Population of *Clarkia elegans* from California

No supernumeraries or translocations	Supernumeraries	Translocations	Both translocations and supernumeraries
0.560	0.265	0.133	0.042

SOURCE: H. Lewis, *Evolution* 5, 1951, 142–157.

height of the shells, but these characters have complex genetic bases.

Examples of naturally occurring morphological variation within plant species are *Plectritis* (see Figure 1-8), *Collinsia* (blue-eyed Mary, page 48), and clover (see Figure 4-5).

Chromosomal Polymorphism. Although the karyotype is often regarded as a distinctive characteristic of a species, in fact, numerous species are polymorphic for chromosome number and morphology. Extra chromosomes (supernumeraries), reciprocal translocations, and inversions segregate in many populations of plants, insects, and even mammals.

Table 26-4 gives the frequencies of supernumerary chromosomes and translocation heterozygotes in a population of the plant *Clarkia elegans* from California. The "typical" species karyotype would be hard to identify.

Immunological Polymorphism. A number of loci in vertebrates code for antigenic specificities such as the ABO blood types. Over 40 different specificities on human red cells are known, and several hundred are known in cattle. Another major polymorphism in humans is the HLA system of cellular antigens, which are implicated in tissue graft compatibility (Chapter 17). Table 26-5 gives the allelic frequencies for the ABO blood group locus in some very different human populations. The polymorphism for the HLA system is vastly greater. There appear to be two main loci, each with five distinguishable alleles. Thus, there are $5^2 = 25$ different possible gametic types, making 25 different homozygous forms and $(25)(24)/2 = 300$ different heterozygotes. All

genotypes are not phenotypically distinguishable, however, so only 121 phenotypic classes can be seen. L. L. Cavalli-Sforza and W. F. Bodmer report that, in a sample of only 100 Europeans, 53 of the 121 possible phenotypes were actually observed!

Protein Polymorphism. In recent years, studies of genetic polymorphism have been carried down to the level of the polypeptides coded by the structural genes themselves. If there is a nonredundant codon change in a structural gene (say, GGU to GAU), this will result in an amino acid substitution in the polypeptide produced at translation (in this case, glycine to aspartic acid). If a specific protein could be purified and sequenced from separate individuals, then it would be possible to detect genetic variation in a population at this level. In practice, this is tedious for large organisms and impossible for small ones unless a large mass of protein can be produced from a homozygous line.

There is, however, a practical substitute for sequencing that makes use of the change in the physical properties of a protein when an amino acid is substituted. Five amino acids (glutamic acid, aspartic acid, arginine, lysine, and histidine) have ionizable side chains that give a protein a characteristic net charge, depending on the pH of the surrounding medium. Amino acid substitutions may directly replace one of these charged amino acids, or a noncharged substitution near one of them in the polypeptide chain may affect the degree of ionization of the charged amino acid, or a substitution at the joining between two α helices may cause a slight shift in the three-dimensional packing of the folded polypeptide. In all these cases, the net charge on the polypeptide will be altered because the net charge on a protein is not simply the sum of all the individual charges on its amino acids but is dependent on their exposure to the liquid medium surrounding them.

To detect the change in net charge, protein can be subjected to the method of gel electrophoresis. Figure 26-2 shows the outcome of such an electrophoretic separation of variants of an esterase enzyme in *D. pseudoobscura*, where each track is the protein from a different individual. Figure 26-3 shows a similar gel for different variant human hemoglobins. In this case, most individuals are heterozygous for the variant and normal hemoglobin A. Table 26-6 shows the frequencies of different alleles for three enzyme-coding loci in *D. pseudoobscura* in several populations: a nearly monomorphic locus (malic dehydrogenase), a moderately poly-

Table 26-5 Frequencies of the Alleles I^A, I^B, and i at the ABO Blood Group Locus in Various Human Populations

Population	I^A	I^B	i
Eskimo	0.333	0.026	0.641
Sioux	0.035	0.010	0.955
Belgian	0.257	0.058	0.684
Japanese	0.279	0.172	0.549
Pygmy	0.227	0.219	0.554

SOURCE: W. C. Boyd, *Genetics and the Races of Man*. D. C. Health, 1950.

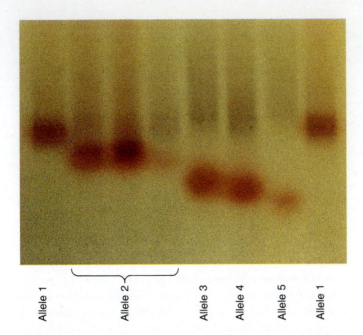

Figure 26-2 Electrophoretic gel showing homozygotes for five different alleles at the *esterase-5* locus in *Drosophila pseudoobscura*. Repeated samples of the same allele are identical, but there are repeatable differences between alleles.

all structural genes in the genome of a species is polymorphic and what the average heterozygosity is in a population. Very large numbers of species have been sampled by this method, including bacteria, fungi, higher plants, vertebrates, and invertebrates. The results are remarkably consistent over species. About one-third of structural-gene loci are polymorphic, and the average heterozygosity in a population over all loci sampled is about 10 percent. This means that scanning the genome in virtually any species would show that about 1 in every 10 loci is in heterozygous condition and that about one-third of all loci have two or more alleles segregating in any population. This represents an immense potential of variation for evolution. The disadvantage of the electrophoretic technique is that it detects variation only in structural genes. If most of the evolution of shape, physiology, and behavior rests on changes in regulatory genetic elements, then the observed variation in structural genes would be beside the point.

DNA Sequence Polymorphism

DNA analysis makes it possible to examine variation among individuals and between species in their DNA sequences. There are two levels at which such studies can be done. Studying variation in the sites recognized by restriction enzymes provides a coarse view of base-pair variation. At a finer level, methods of DNA sequencing allow variation to be observed base pair by base pair.

Restriction-Site Variation. A restriction enzyme that recognizes six-base sequences (a "six-cutter") will recognize an appropriate sequence approximately once every $4^6 = 4096$ base pairs along a DNA molecule [determined from the probability that a specific base (of which there are four) will

morphic locus (α-amylase), and a highly polymorphic locus (xanthine dehydrogenase).

The technique of gel electrophoresis (or sequencing) differs fundamentally from other methods of genetic analysis in allowing the study of loci that are not segregating, because the presence of a polypeptide is prima facie evidence of a structural gene, that is, a DNA sequence coding for a protein. Thus, it has been possible to ask what proportion of

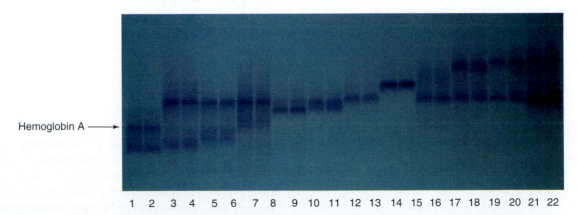

Figure 26-3 Electrophoretic gel showing heterozygotes of normal hemoglobin A and a number of variant hemoglobin alleles. One of the dark-staining bands is marked as hemoglobin A. The second dark-staining band in each track (seen most clearly in tracks 3 and 4) represents the second protein derived from the second allele of the heterozygote. Hemoglobin A is missing from tracks 9 and 10 because these are homozygotes for the variant allele.

Table 26-6   Frequencies of Various Alleles at Three Enzyme-Coding Loci in Four Populations of *Drosophila pseudoobscura*

Locus (enzyme-encoded)	Allele	Population			
		Berkeley	Mesa Verde	Austin	Bogotá
Malic dehydrogenase	A	0.969	0.948	0.957	1.00
	B	0.031	0.052	0.043	0.00
α Amylase	A	0.030	0.000	0.000	0.00
	B	0.290	0.211	0.125	1.00
	C	0.680	0.789	0.875	0.00
Xanthine dehydrogenase	A	0.053	0.016	0.018	0.00
	B	0.074	0.073	0.036	0.00
	C	0.263	0.300	0.232	0.00
	D	0.600	0.581	0.661	1.00
	E	0.010	0.030	0.053	0.00

SOURCE: R. C. Lewontin, *The Genetic Basis of Evolutionary Change.* Columbia University Press, 1974.

be found at each of the six positions]. If there is polymorphism in the population for one of the six bases at the recognition site, then there will be a restriction fragment length polymorphism (RFLP) in the population, because in one variant the enzyme will recognize and cut the DNA, while in the other variant it will not (see pages 456–457). A panel of, say, eight enzymes will then sample every $4096/8 \cong 500$ base pairs for such polymorphisms. Of course, when one is found, we do not know which of the six base pairs at the recognition site is polymorphic.

If we use enzymes that recognize four-base sequences ("four-cutters"), there is a recognition site every $4^4 = 256$ base pairs, so a panel of eight different enzymes can sample about once every 32 base pairs along the enzyme. In addition to single base-pair changes that destroy restriction-enzyme recognition sites, there are insertions and deletions of stretches of DNA that also cause restriction fragment lengths to vary.

Extensive samples have been made for several regions of the genome in a number of species of *Drosophila* using both four-cutting and six-cutting enzymes. The result of one such study of the xanthine dehydrogenase gene in *Drosophila pseudoobscura* is shown in Figure 26-4. The figure shows, symbolically, the restriction pattern of 53 chromosomes (haplotypes) sampled from nature, polymorphic for 78 restriction sites along a sequence 4.5 kb in length. Among the 53 haplotypes there are 48 different ones. (Try to find the identical pairs.) Clearly there is an immense amount of nucleotide variation at the xanthine dehydrogenase locus in nature.

Twenty studies of different regions of the X chromosome and the two large autosomes of *Drosophila melanogaster* by restriction enzymes have found between 0.1 and 1.0 percent heterozygosity per nucleotide site, with an average of 0.4 percent. A study of the very small fourth chromosome, however, found no polymorphism at all.

Tandem Repeats. Another form of DNA sequence variation that can be revealed by restriction fragment surveys arises from the occurrence of multiply repeated DNA sequences. In the human genome, there are a variety of different short DNA sequences dispersed through the genome, each one of which is multiply repeated in a tandem row. The number of repeats may vary from a dozen to more than 100 in different individual genomes. Such sequences are known as **variable number tandem repeats (VNTR).** If the restriction enzymes cut sequences that flank either side of such a tandem array, a fragment will be produced whose size is proportional to the number of repeated elements. The different-sized fragments will migrate at different rates in an electrophoretic gel. Unfortunately, the individual elements are too short to allow distinguishing between, say, 64 and 68 repeats, but size classes (*bins*) can be established, and a population can be assayed for the frequencies of the different classes. Table 26-7 shows the data for two different VNTRs sampled in two American Indian groups from Brazil. In one case, D14S1, the Karitiana are nearly homozygous while the Surui are very variable, while in the other, D14S13, both populations are variable but with different frequency patterns.

Complete Sequence Variation. Studies of variation at the level of single base pairs by DNA sequencing can provide information of two kinds. First, by translating the sequences of coding regions obtained from different individuals in a population or from different species, the exact amino acid sequence differences can be determined. Electrophoretic studies show only that there is variation in amino acid sequences, but cannot identify how many or which amino acids differ between individuals. So, when DNA sequences were obtained for the various electrophoretic variants of esterase-5 in *Drosophila pseudoobscura* (see Figure 26-2), it was found that electrophoretic classes differ from each other by

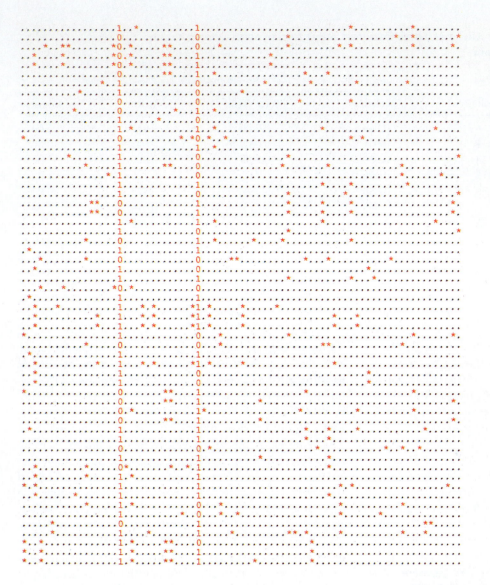

Figure 26-4 The result of a four-cutter survey of 53 chromosomes, probed for the xanthine dehydrogenase gene in *Drosophila pseudoobscura*. Each line is a chromosome (haplotype) sampled from a natural population. Each position along the line is a polymorphic restriction rate along the 4.5-kb sequence studied. Where an asterisk appears, the haplotype differs from the majority, either cutting where most haplotypes are not cut, or not cutting where most haplotypes are cut. At two sites there is no clear majority type, so a 0 or 1 is used to show whether the site is absent or present.

an average of 8 amino acids, and the 20 different kinds of amino acids were involved in polymorphisms at about the frequency that they were represented in the protein. Such studies also show that different regions of the same protein have different amounts of polymorphism. For the esterase-5 protein, consisting of 545 amino acids, 7 percent of amino acid positions are polymorphic, but the last amino acids at the carboxyl terminus of the protein are totally invariant between individuals, probably because of functional constraints on these amino acids.

Second, DNA base-pair variation can also be studied for those base pairs that do not determine or change the protein sequence. This includes DNA in introns, in 5'-flanking sequences that may be regulatory, in nontranscribed DNA 3' to the gene, and in those nucleotide positions within codons (usually third positions) whose variation does not result in amino acid substitutions. Within coding sequences, these so-called silent or synonymous base-pair polymorphisms are much more common than changes that result in amino acid polymorphism, presumably because many

amino acid changes interfere with normal function of the protein and are eliminated by natural selection. An examination of the codon translation table (page 397) shows that approximately 25 percent of all random base-pair changes would be synonymous, giving an alternative codon for the same amino acid, while 75 percent of random changes will change the amino acid coded. For example, a change from AAT to AAC still codes for asparagine, but a change to ATT, ACT, AAA, AAG, AGT, TAT, CAT or GAT, all single-base-pair changes from AAT, changes the amino acid coded. So, if mutations of base pairs are at random and if the substitution of an amino acid made no difference to function, we would expect a 3:1 ratio of amino acid replacement to silent polymorphisms. The actual ratios found in *Drosophila* vary from 2:1 to 1:10. Clearly, there is a great excess of synonymous polymorphism, showing that most amino acid changes are subject to natural selection. It should not be supposed, however, that silent sites in coding sequences are entirely free from constraints. Different alternative triplet codings for the same amino acid may differ in speed and ac-

Table 26-7 Size Class Frequencies for Two Different VNTR Sequences, D14S1 and D14S13, in the Karitiana and Surui of Brazil

	D14S1		D14S13	
Size class	Karitiana	Surui	Karitiana	Surui
3–4	105	4	0	0
4–5	0	3	3	14
5–6	0	11	1	4
6–7	0	2	1	2
7–8	0	1	1	2
8–9	3	3	8	16
9–10	0	11	28	9
10–11	0	2	22	0
11–12	0	4	18	8
12–13	0	0	13	18
13–14	0	0	13	3
>14	0	0	0	2
	108	78	108	78

SOURCE: Data of J. Kidd and K. Kidd, *American Journal of Physical Anthropology* 81, 1992, 249.

curacy of transcription, and the mRNA corresponding to different alternative triplets may have different accuracy and speed of translation because of limitations on the pool of tRNAs available. Evidence for this latter effect is that alternative synonymous triplets for an amino acid are not used equally, and the inequality of use is much more pronounced for genes that are transcribed at a very high rate.

There are also constraints on $5'$ and $3'$ noncoding sequences and on intron sequences. Both $5'$ and $3'$ noncoding DNA contain signals for transcription, and introns may contain enhancers of transcription (see Chapter 17).

Message Within species there is great genetic variation. This is manifest at the morphological level of chromosome form and number and at the level of DNA segments that may have no observable developmental effects.

Variation within and between Populations

The various examples just given show that there are genetic differences between individuals within a population and also that the allelic frequencies differ between populations. The relative amounts of variation within and between populations vary from species to species, depending on history and environment. In humans, some gene frequencies (for example, those for skin color or hair form) obviously are well differentiated between populations and major geographical groups (so-called geographical races). If, however, we look at single structural genes identified immunologically or by electrophoresis rather than these outward phenotypic characters, the situation is rather different. Table 26-8 gives the allelic frequencies in a random sample of enzyme loci (chosen only for experimental convenience) in a European and an African population. Except for phosphoglucomutase-3, where the allelic frequencies are reversed between the populations, blacks and whites are very similar in their allelic distributions. In contrast to this random sample of loci, Table 26-9 shows the three loci for which Caucasians, Negroids, and Mongoloids are known to be most different from one another (Duffy and Rhesus blood groups and the P antigen) compared with the three polymorphic loci for which the races are most similar (Auberger blood group and Xg and secretor factors). Even for the most divergent loci, no race is homozygous for one allele that is absent in the other two races.

In general, different human populations show rather similar frequencies for polymorphic genes. Figure 26-5 is a **triallelic diagram** for the three main allelic classes I^A, I^B,

Table 26-8 Allelic Frequencies at Seven Polymorphic Loci in Europeans and Black Africans

	Europeans			Africans		
Locus	Allele 1	Allele 2	Allele 3	Allele 1	Allele 2	Allele 3
Red-cell acid phosphatase	0.36	0.60	0.04	0.17	0.83	0.00
Phosphoglucomutase-1	0.77	0.23	0.00	0.79	0.21	0.00
Phosphoglucomutase-3	0.74	0.26	0.00	0.37	0.63	0.00
Adenylate kinase	0.95	0.05	0.00	1.00	0.00	0.00
Peptidase A	0.76	0.00	0.24	0.90	0.10	0.00
Peptidase D	0.99	0.01	0.00	0.95	0.03	0.02
Adenosine deaminase	0.94	0.06	0.00	0.97	0.03	0.00

SOURCE: R. C. Lewontin, *The Genetic Basis of Evolutionary Change.* Columbia University Press, 1974. Adapted from H. Harris, *The Principles of Human Biochemical Genetics.* North Holland, Amsterdam and London, 1970.

Table 26-9 Examples of Extreme Differentiation and Close Similarity in Blood Group Allelic Frequencies in Three Racial Groups

Gene	Allele	Population		
		Caucasoid	Negroid	Mongoloid
Duffy	Fy	0.0300	0.9393	0.0985
	Fy^a	0.4208	0.0000	0.9015
	Fy^b	0.5492	0.0607	0.0000
Rhesus	R_0	0.0186	0.7395	0.0409
	R_1	0.4036	0.0256	0.7591
	R_2	0.1670	0.0427	0.1951
	r	0.3820	0.1184	0.0049
	r'	0.0049	0.0707	0.0000
	Others	0.0239	0.0021	0.0000
P	P_1	0.5161	0.8911	0.1677
	P_2	0.4839	0.1089	0.8323
Auberger	Au^a	0.6213	0.6419	No data
	Au	0.3787	0.3581	No data
Xg	Xg^a	0.67	0.55	0.54
	Xg	0.33	0.45	0.46
Secretor	Se	0.5233	0.5727	No data
	se	0.4767	0.4273	No data

Source: A summary is provided in L. L. Cavalli-Sforza and W. F. Bodmer, *The Genetics of Human Populations.* W. H. Freeman and Company, 1971, pp. 724–731. See L. L. Cavalli-Sforza, P. Menozzi, and A. Piazza, *The History and Geography of Human Genes* (Princeton University Press, 1994), for detailed data.

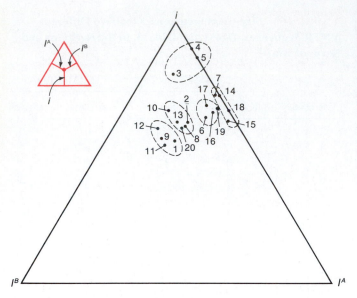

Figure 26-5 Triallelic diagram of the ABO blood group allelic frequencies for human populations. Each point represents a population; the perpendicular distances from the point to the sides represent the allelic frequencies, as indicated in the small triangle. Populations 1 to 3 are African, 4 to 7 are American Indian, 8 to 13 are Asian, 14 to 15 are Australian aborigine, and 16 to 20 are European. Dashed lines enclose arbitrary classes with similar gene frequencies; these groupings do not correspond to "racial" classes. (After A. Jacquard, *Structures génétiques des populations.* Copyright © 1970 by Masson et Cie.)

and i of the ABO blood group. Each point represents the allelic composition of a population, where the three allelic frequencies can be read by taking the lengths of the perpendiculars from each side to the point. The diagram shows that all human populations are bunched together in the region of high i, intermediate I^A, and low I^B frequencies. Moreover, neighboring points (enclosed by dashed lines) do not correspond to geographical races, so that such races cannot be distinguished from one another by characteristic allelic frequencies for this gene. The study of polymorphic blood groups and enzyme loci in a variety of human populations has shown that about 85 percent of total human genetic diversity is found within local populations, about 8 percent is found among local populations within major geographical races, and the remaining 7 percent is found among the major geographical races. Clearly, the genes influencing skin color, hair form, and facial form that are well differentiated among races are not a random sample of structural gene loci.

> **Message** In general, the genetic variation between individuals within human races is much greater than the average variation between races.

Quantitative Variation

Not all variation in traits can be described in terms of allelic frequencies, because many characteristics, like height, vary continuously over a range rather than falling in a few qualitatively distinct classes. There is no allele for being 5′8″ or 5′4″ tall. Such characters, if they are varying as a consequence of genetic variation, will be influenced by several or many genes and by environmental variation as well. Special techniques are needed for the study of such *quantitative traits,* and we devote Chapter 27 to those techniques. For the moment we confine ourselves to the question of whether genetic differences between individuals influence the trait at all. In experimental organisms a simple way to ask this question is to choose two groups of parents that differ markedly in the trait and to raise offspring from both groups in the same environment. If the offspring of the two groups are different, then the trait is said to be *heritable* (see Chapter 27 for a more detailed discussion of the concept and estimation of heritability). A simple measure of the degree of heritability of the variation is the ratio of the difference between the offspring groups to the difference between the parental groups. So, if two groups of *Drosophila* parents differed by, say, 0.1 mg in weight while the offspring groups, raised in identical environments, differed by 0.03 mg, the heritability of weight difference would be estimated as 30 percent. When this technique has been applied to morphological variation in *Drosophila,* virtually every variable trait is found to have some heritability, so evolution of the trait could occur. It is important to note that this method cannot be applied to organisms where there is no rigorous control possible over developmental environment. In humans, for example, children of different parental groups differ from one another not only because their genetic makeup is different, but also because the environments of different families, social classes, and nations are different. Japanese are, on the average, shorter than Europeans, but the difference between children of Japanese ancestry and children of European ancestry, both born in North America, is less and becomes even less in the second generation, presumably because of diet. It is not clear whether all the differences in height would disappear or even be reversed if the family environments were identical.

It should not be supposed that all variable traits are heritable, however. Certain metabolic traits (such as resistance to high salt concentrations in *Drosophila*) show individual variation but no heritability. In general, behavioral traits have lower heritabilities than morphological traits, especially in organisms with more complex nervous systems that exhibit immense individual flexibility in central nervous states. Before any judgment can be made about the evolution of a particular quantitative trait, it is essential to determine if there is genetic variance for it in the population whose evolution is to be predicted. Thus, suggestions that such traits in the human species as performance on IQ tests, temperament, and social organization are in the process of evolving or have evolved at particular epochs in human history depend critically on evidence about whether there is genetic variation for these traits.

One of the most important findings in evolutionary genetics has been the discovery of substantial genetic variation underlying characters that show no morphological variation! These are called **canalized characters,** because the final outcome of their development is held within narrow bounds despite disturbing forces. Development is such that all the different genotypes for canalized characters have the same constant phenotype over the range of environments that is usual for the species. The genetic differences are revealed if the organisms are put in a stressful environment or if a severe mutation stresses the developmental system. For example, all wild-type *Drosophila* have exactly four scutellar bristles (Figure 26-6). If the recessive mutant *scute* is present, the number of bristles is reduced, but, in addition, there is variation from fly to fly. This variation is heritable, and lines with zero or one bristle and lines with three or four bristles can be obtained by selection in the presence of the *scute* mutation. When the mutation is removed, these lines now have two and six bristles, respectively. Similar experiments have been performed using extremely stressful environments in place of mutants. A consequence of such hidden genetic variation is that a character that is phenotypically uniform in a species may nevertheless undergo rapid evolution if a stressful environment uncovers the genetic variation.

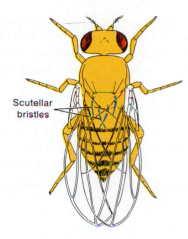

Figure 26-6 The scutellar bristles of the adult *Drosophila,* shown in blue. This is an example of a canalized character; all wild-type *Drosophila* will have four scutellar bristles in a very wide range of environments.

Message Even characters with no apparent phenotypic variant may evolve when developmental conditions are changed drastically if genetic variation for the character is hidden by developmental canalization.

The Effect of Sexual Reproduction on Variation

The evolutionary theorists of the nineteenth century encountered a fundamental difficulty in dealing with Darwin's theory of evolution through natural selection. The possibility of continued evolution by natural selection is limited by the amount of genetic variation. But biologists of the nineteenth century, including Darwin, believed in one form or another of **blending inheritance,** a model postulating that the characteristics of each offspring are some intermediate mixture of the parental characters. Such a model of inheritance has fatal implications for a theory of evolution that depends on variation.

Suppose that some trait (say, height) has a distribution in the population and that individuals mate more or less at random. If intermediate individuals mated with each other, they would produce only intermediate offspring according to a blending model. The mating of a tall with a short individual also would produce only intermediate offspring. Only the mating of tall with tall individuals and short with short individuals would preserve extreme types. The net result of all matings would be an increase in intermediate types and a decrease in extreme types. The variance of the distribution would shrink, simply as a result of sexual reproduction. In fact, it can be shown that the variance is *cut in half* in each generation, so that the population would be essentially uniformly intermediate in height before very many generations had passed. There then would be no variation on which natural selection could operate. This was a very serious problem for the early Darwinists; it made it necessary for Darwin to assume that new variation is generated at a very rapid rate by the inheritance of characters acquired by individuals during their lifetimes.

The rediscovery of Mendelism changed this picture completely. Because of the discrete nature of the Mendelian genes and the segregation of alleles at meiosis, a cross of intermediate with intermediate individuals does *not* result in all intermediate offspring. On the contrary, extreme types (homozygotes) segregate out of the cross. To see the consequence of Mendelian inheritance for genetic variation, consider a population in which males and females mate with each other at random with respect to some gene locus *A*; that is, individuals do not choose their mates preferentially with respect to the partial genotype at the locus. Such random mating is equivalent to mixing all the sperm and all the eggs in the population together and then matching randomly drawn sperm with randomly drawn eggs.

If the frequency of allele *A* is p in both the sperm and the eggs and the frequency of allele *a* is $q = 1 - p$, then the consequences of random unions of sperm and eggs are shown in Figure 26-7. The probability that both the sperm and the egg will carry *A* is $p \times p = p^2$, so this will be the frequency of *AA* homozygotes in the next generation. In like manner, the chance of heterozygotes *Aa* will be $(p \times q) + (q \times p) = 2pq$, and the chance of homozygotes *aa* will be $q \times q = q^2$. The three genotypes, after a generation of random mating, will be in the frequencies $p^2 : 2pq : q^2$. As the figure shows, the allelic frequency of *A* has not changed and is still p. Therefore, in the second generation, the frequencies of the three genotypes will again be $p^2 : 2pq : q^2$, and so on, forever.

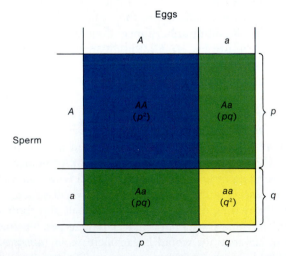

Figure 26-7 The Hardy-Weinberg equilibrium frequencies that result from random mating. The frequencies of *A* and *a* among both eggs and sperm are *p* and *q* (= 1 − *p*), respectively. The total frequencies of the zygote genotypes are p^2 for *AA*, $2pq$ for *Aa*, and q^2 for *aa*. The frequency of the allele *A* in the zygotes is the frequency of *AA* plus half the frequency of *Aa*, or $p^2 + pq = p(p + q) = p$.

The equilibrium distribution

$$
\begin{array}{ccc}
AA & Aa & aa \\
p^2 & 2pq & q^2
\end{array}
$$

is called the **Hardy-Weinberg equilibrium** after those who independently discovered it. (A third independent discovery was made by the Russian geneticist Sergei Chetverikov.)

The Hardy-Weinberg equilibrium means that sexual reproduction does not cause a constant reduction in genetic variation in each generation; on the contrary, the amount of variation remains constant generation after generation, in the absence of other disturbing forces. The equilibrium is the direct consequence of the segregation of alleles at meiosis in heterozygotes.

Numerically, the equilibrium shows that irrespective of the particular mixture of genotypes in the parental generation, the genotypic distribution after one round of mating is completely specified by the allelic frequency p. For example, consider three hypothetical populations:

	$f(AA)$	$f(Aa)$	$f(aa)$
I	0.3	0.0	0.7
II	0.2	0.2	0.6
III	0.1	0.4	0.5

The allele frequency p of *A* in the three populations is

$$
\begin{array}{lll}
\text{I} & p = f(AA) + \tfrac{1}{2}f(Aa) = 0.3 + \tfrac{1}{2}(0) & = 0.3 \\
\text{II} & p = & 0.2 + \tfrac{1}{2}(0.2) = 0.3 \\
\text{III} & p = & 0.1 + \tfrac{1}{2}(0.4) = 0.3
\end{array}
$$

So, despite their very different genotypic compositions, they have the same allele frequency. After one generation of random mating, however, each of the three populations will have the same genotypic frequencies:

AA	*Aa*	*aa*
$(0.3)^2 = 0.09$	$2(0.3)(0.7) = 0.42$	$(0.7)^2 = 0.49$

and they will remain so indefinitely.

One consequence of the Hardy-Weinberg proportions is that rare alleles are virtually never in homozygous condition. An allele with a frequency of 0.001 occurs in homozygotes at a frequency of only one in a million; most copies of such rare alleles are found in heterozygotes. In general, since two copies of an allele are in homozygotes but only one copy of that allele is in each heterozygote, the relative

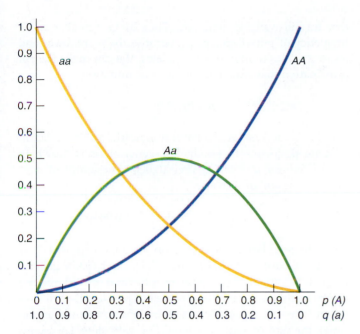

Figure 26-8 Curves showing the proportions of homozygotes AA (blue line), homozygotes aa (yellow line), and heterozygotes Aa (green line) in populations of different allelic frequencies if the populations are in Hardy-Weinberg equilibrium.

Table 26-10 Comparison between Observed Frequencies of Genotypes for the MN Blood Group Locus and the Frequencies Expected from Random Mating

Population	Observed			Expected		
	MM	MN	NN	MM	MN	NN
Eskimo	0.835	0.156	0.009	0.834	0.159	0.008
Egyptian	0.278	0.489	0.233	0.274	0.499	0.228
Chinese	0.332	0.486	0.182	0.331	0.488	0.181
Australian aborigine	0.024	0.304	0.672	0.031	0.290	0.679

NOTE: The expected frequencies are computed according to the Hardy-Weinberg equilibrium, using the values of p and q computed from the observed frequencies.

frequency of the allele in heterozygotes (as opposed to homozygotes) is, from the Hardy-Weinberg equilibrium frequencies,

$$\frac{2pq}{2q^2} = \frac{p}{q}$$

which for $q = 0.001$ is a ratio of 999:1. The general relation between homozygote and heterozygote frequencies as a function of allele frequencies is shown in Figure 26-8.

In our derivation of the equilibrium, we assumed that the allelic frequency p is the same in sperm and eggs. The Hardy-Weinberg equilibrium theorem does not apply to sex-linked genes if males and females start with unequal gene frequencies (see Problem 7 at the end of the chapter).

The Hardy-Weinberg equilibrium was derived on the assumption of "random mating," but we must carefully distinguish two meanings of that process. First, we may mean that individuals do not choose their mates on the basis of some heritable character. Human beings are random-mating with respect to blood groups in this first sense, because they generally do not know the blood type of their prospective mates, and even if they did, it is unlikely that blood type would be used as a criterion for choice. In the first sense, random mating will occur with respect to genes that have no effect on appearance, behavior, smell, or other characteristics that directly influence mate choice.

There is a second sense of random mating that is relevant when there is any subdivision of a species into subgroups. If there is genetic differentiation between subgroups so that the frequencies of alleles differ from group to group and if individuals tend to mate within their own subgroup (**endogamy**), then with respect to the species as a whole, mating is not at random and frequencies of genotypes will depart more or less from Hardy-Weinberg frequencies. In this sense, human beings are not random-mating, because ethnic and racial groups differ from one another in gene frequencies and people show high rates of endogamy, not only within major races but also within local ethnic groups. Spaniards and Russians differ in their ABO blood group frequencies, Spaniards marry Spaniards and Russians marry Russians, so there is unintentional nonrandom mating with respect to ABO blood groups. Table 26-10 shows random mating in the first sense and nonrandom mating in the second sense for the MN blood group. Within Eskimos, Egyptian, Chinese, and Australian subpopulations, females do not choose their mates by MN type, and, thus, Hardy-Weinberg equilibrium exists *within* the subpopulations. But Egyptians do not mate with Eskimos or Australian aborigines, so the nonrandom associations in the human species *as a whole* result in large differences in genotype frequencies and departure from Hardy-Weinberg equilibrium.

The Sources of Variation

The variational theory of evolution has a peculiar self-defeating property. If evolution occurs by the differential reproduction of different variants, we expect the variant with the highest rate of reproduction eventually to take over the population and all other genotypes to disappear. But then there is no longer any variation for further evolution. The possibility of continued evolution therefore is critically dependent on renewed variation.

For a given population, there are three sources of variation: mutation, recombination, and immigration of genes. However, recombination by itself does not produce variation unless alleles are segregating already at different loci;

Table 26-11 Some Point-Mutation Rates in Different Organisms

Organism	Gene	Mutation rate per generation
Bacteriophage	Host range	2.5×10^{-9}
Escherichia coli	Phage resistance	2×10^{-8}
Zea mays (corn)	R (color factor)	2.9×10^{-4}
	Y (yellow seeds)	2×10^{-6}
Drosophila melanogaster	Average lethal	2.6×10^{-5}

SOURCE: T. Dobzhansky, *Genetics and the Origin of Species*, 3d ed., rev. Columbia University Press, 1951.

otherwise there is nothing to recombine. Similarly, immigration cannot provide variation if the entire species is homozygous for the same allele. Ultimately, the source of all variation must be mutation.

Variation from Mutations

Mutations are the *source* of variation, but the *process* of mutation does not itself drive evolution. The rate of change in gene frequency from the mutation process is very low because spontaneous mutation rates are low (Table 26-11). Let μ be the **mutation rate** from allele *A* to some other allele *a* (the probability that a gene copy *A* will become *a* during the DNA replication preceding meiosis). If p_t is the frequency of the *A* allele in generation *t*, if $q_t = 1 - p_t$ is the frequency of the *a* allele, and if there are no other causes of gene frequency change (no natural selection, for example), then the change in allelic frequency in one generation is

$$\Delta p = p_t - p_{t-1} = -\mu p_{t-1}$$

where p_{t-1} is the frequency in the preceding generation. This tells us that the frequency of *A* decreases (and the frequency of *a* increases) by an amount that is proportional to the mutation rate μ and to the proportion *p* of all the genes

that are still available to mutate. Thus Δp gets smaller as the frequency of *p* itself decreases, because there are fewer and fewer *A* alleles to mutate into *a* alleles. We can make the approximation that after *n* generations of mutation

$$p_n = p_0 e^{-n\mu}$$

where *e* is the base of the natural logarithms. This relation of gene frequency to number of generations is shown in Figure 26-9 for $\mu = 10^{-5}$. After 10,000 generations of continued mutation of *A* to *a*

$$p = p_0 e^{-(10^4)(10^{-5})} = p_0 e^{-0.1} = 0.904 p_0$$

So, if the population starts with only *A* alleles ($p_0 = 1.0$), it would still have only 10 percent *a* alleles after 10,000 generations at a rather high mutation rate and would require 60,000 additional generations to reduce *p* to 0.5. Even if mutation rates were doubled (say, by environmental mutagens), the rate of evolution would be very slow. For example, radiation levels of sufficient intensity to double the mutation rate over the reproductive lifetime of an individual human are at the limit of occupational safety regulations, and a dose of radiation sufficient to increase mutation rates by an order of magnitude would be lethal, so rapid genetic change in the species would not be one of the effects of increased radiation. Although we have many things to fear from environmental radiation pollution, turning into a species of monsters is not one of them.

If we look at the mutation process from the standpoint of the increase of a particular new allele rather than the decrease of the old form, the process is even slower. Most mutation rates that have been determined are the sum of all mutations of *A* to any mutant form with a detectable effect. Any *specific* base substitution is likely to be at least two orders of magnitude lower in frequency than the sum of all changes. So, precise reverse mutations ("back mutations") to the original allele *A* are unlikely, although many mutations may produce alleles that are *phenotypically* similar to the original.

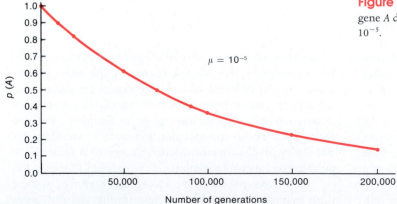

Figure 26-9 The change over generations in the frequency of a gene *A* due to mutation from *A* to *a* at a constant mutation rate (μ) of 10^{-5}.

$\mu = 10^{-5}$

It is not possible to measure locus-specific mutation rates for continuously varying characters, but the rate of accumulation of genetic variance can be determined. Beginning with a completely homozygous line of *Drosophila* derived from a natural population, $\frac{1}{1000}$ to $\frac{1}{500}$ of the genetic variance in bristle number in the original population is restored each generation by spontaneous mutation.

Variation from Recombination

The creation of genetic variance by recombination can be a much faster process than its creation by mutation. When just two chromosomes with "normal" survival, taken from a natural population of *Drosophila,* are allowed to recombine for a single generation, they produce an array of chromosomes with 25 to 75 percent as much genetic variation in survival as was present in the entire natural population from which the parent chromosomes were sampled. This is simply a consequence of the very large number of different recombinant chromosomes that can be produced even if we take into account only single crossovers. If a pair of homologous chromosomes is heterozygous at n loci, then a crossover can take place in any one of the $n-1$ intervals between them, and since each recombination produces two recombinant products, there are $2(n-1)$ new unique gametic types from a single generation of crossing-over, even considering only single crossovers. If the heterozygous loci are well spread out on the chromosomes, these new gametic types will be frequent and a considerable variance will be generated. Asexual organisms or organisms like bacteria that very seldom undergo sexual recombination do not have this source of variation, so that new mutations are the only way in which a change in gene combinations can be achieved. As a result, asexual organisms may evolve more slowly under natural selection than sexual organisms.

Variation from Migration

A further source of variation is migration into a population from other populations with different gene frequencies. If p_t is the frequency of an allele in the recipient population in generation t and P is the allelic frequency in a donor population (or the average over several donor populations), and if m is the proportion of the recipient population that is made up of new migrants from the donor population, then the gene frequency in the recipient population in the next generation, p_{t+1}, is the result of mixing $1-m$ genes from the recipient with m genes from the donor population. Thus

$$p_{t+1} = (1-m)p_t + mP = p_t + m(P-p_t)$$

and

$$\Delta p = p_{t+1} - p_t = m(P-p_t)$$

The change in gene frequency is proportional to the difference in frequency between the recipient population and the average of the donor populations. Unlike the mutation rate, the migration rate (m) can be large, so the change in frequency may be substantial.

We must understand *migration* as meaning any form of the introduction of genes from one population into another. So, for example, genes from Europeans have "migrated" into the population of African origin in North America steadily since the Africans were introduced as slaves. We can determine the amount of this migration by looking at the frequency of an allele that is found only in Europeans and not in Africans and comparing its frequency among blacks in North America.

We can use the formula for the change in gene frequency from migration if we modify it slightly to account for the fact that several generations of admixture have taken place. If the rate of admixture has not been too great, then (to a close order of approximation) the sum of the single-generation migration rates over several generations (let's call this M) will be related to the total change in the recipient population after these several generations by the same expression as the one used for changes due to migration. If, as before, P is the allelic frequency in the donor population and p_0 is the original frequency among the recipients, then

$$\Delta p_{\text{total}} = M(P-p_0)$$

so

$$M = \frac{\Delta p_{\text{total}}}{P-p_0}$$

For example, the Duffy blood group allele Fy^a is absent in Africa but has a frequency of 0.42 in whites from the state of Georgia. Among blacks from Georgia, the Fy^a frequency is 0.046. Therefore, the total migration of genes from whites into the black population since the introduction of slaves in the eighteenth century is

$$M = \frac{\Delta p_{\text{total}}}{P-p} = \frac{0.046-0}{0.42-0} = 0.1095$$

When the same analysis is carried out on American blacks from Oakland (California) and Detroit, M is 0.22 and 0.26, respectively, showing either greater admixture rates in these cities than in Georgia or differential movement into these cities by American blacks who have more European ancestry. In any case, the genetic variation at the Fy locus has been increased by this admixture.

The Origin of New Functions

Point mutations or chromosomal rearrangements are themselves a limited source of variation for evolution because they can only alter a function or change one kind of function into another. To add quite new functions requires expansion in the total repertoire of genes through duplication and polyploidy, followed by a divergence between the duplicated genes, presumably by the usual process of mutation. Expansion of the genome by polyploidy has clearly been a frequent process, at least in plants. Figure 26-10 shows the frequency distribution of haploid chromosome numbers among dicotyledonous plant species. Even numbers are much more common than odd numbers—a consequence of frequent polyploidy (see Chapter 9).

Once an expansion of the total DNA of the genome has occurred, it may require only a few base substitutions in a gene to provide it with a new function. For example, B. Hall has experimentally changed a gene to a new function in *Escherichia coli*. In addition to the *lacZ* genes specifying the usual lactose-fermenting β-galactosidase activity in *E. coli*, another structural gene locus *ebg* specifies another β-galactosidase that does not ferment lactose, although it is induced by lactose. The natural function of this second enzyme is unknown. Hall was able to alter this gene into one specifying an enzyme that ferments another substrate, lactobionate. To do so, it was necessary to alter the regulatory element to a constitutive state and to produce three successive structural-gene mutations.

> **Message** Evolution would come to a stop by running out of variation if new genetic variation were not added to populations by mutation, recombination, and migration. Ultimately, all new variation is derived from gene and chromosome mutations.

Inbreeding and Assortative Mating

Random mating with respect to a locus is common, but it is not universal. Two kinds of deviation from random mating must be distinguished. First, individuals may mate with each other nonrandomly because of their degree of common ancestry, that is, their degree of genetic relationship. If mating between relatives occurs more commonly than would occur by pure chance, then the population is **inbreeding.** If mating between relatives is less common than would occur by chance, then the population is said to be undergoing **enforced outbreeding,** or **negative inbreeding.**

Second, individuals may tend to choose each other as mates, not because of their degree of genetic relationship but because of their degree of resemblance to each other at some locus. Bias toward mating of like with like is called **positive assortative mating.** Mating with unlike partners is called **negative assortative mating.** Assortative mating is never complete.

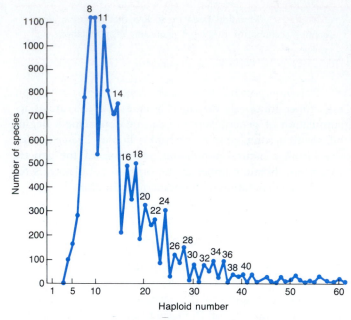

Figure 26-10 Frequency distribution of haploid chromosome numbers in dicotyledonous plants. (From Verne Grant, *The Origin of Adaptation.* Copyright © 1963 by Columbia University Press.)

Inbreeding and assortative mating are not the same. Close relatives resemble each other more than unrelated individuals on the average, but not necessarily for any particular trait in particular individuals. So inbreeding can result in the mating of quite dissimilar individuals. On the other hand, individuals who resemble each other for some trait may do so because they are relatives, but unrelated individuals also may have specific resemblances. Brothers and sisters do not all have the same eye color, and blue-eyed people are not all related to one another.

Inbreeding levels in natural populations are a consequence of geographical distribution, the mechanism of reproduction, and behavioral characteristics. If close relatives occupy adjacent areas, then simple proximity may result in inbreeding. The seeds of many plants, for example, fall very close to the parental source and the pollen is not widely spread, so a high frequency of sib mating occurs. Some plants (such as corn) can be self-pollinated as well as cross-pollinated, so that wind pollination results in some very close inbreeding. Yet other plants, like the peanut, are obligatorily selfed. Many small mammals (such as house mice) live and mate in restricted family groups that persist generation after generation. Humans, on the other hand, generally have complex mating taboos and proscriptions that reduce inbreeding.

Assortative mating for some traits is common. In humans, there is a positive assortative mating bias for skin color and height, for example. An important difference between assortative mating and inbreeding is that the former is specific to a trait whereas the latter applies to the entire

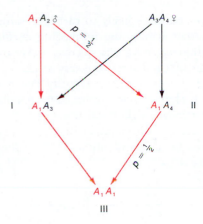

Figure 26-11 Calculation of homozygosity by descent for an offspring (III) of a brother-sister (I-II) mating. The probability that II will receive A_1 from its father is $\frac{1}{2}$; if it does, the probability that II will pass A_1 on to the generation producing III is $\frac{1}{2}$. Thus, the probability that III will receive an A_1 from II is $\frac{1}{2} \times \frac{1}{2} = \frac{1}{4}$.

genome. Individuals may mate assortatively with respect to height but at random with respect to blood group. Cousins, on the other hand, resemble each other genetically on the average to the same degree at all loci.

For both positive assortative mating and inbreeding, the consequence to population structure is the same: there is an increase in homozygosity above the level predicted by the Hardy-Weinberg equilibrium. If two individuals are related, they have at least one common ancestor. Thus, there is some chance that an allele carried by one of them and an allele carried by the other are both descended from the identical DNA molecule. The result is that there is an extra chance of **homozygosity by descent,** to be added to the chance of homozygosity ($p^2 + q^2$) that arises from the random mating of unrelated individuals. The probability of homozygosity by descent is called the **inbreeding coefficient (F).** Figure 26-11 illustrates the calculation of the probability of homozygosity by descent. Individuals I and II are full sibs because they share both parents. We label each allele in the parents uniquely to keep track of them. Individuals I and II mate to produce individual III. If individual I is $A_1 A_3$ and the gamete that it contributes to III contains the allele A_1, then we would like to calculate the probability that the gamete produced by II is also A_1. The chance is $\frac{1}{2}$ that II will receive A_1 from its father, and if it does, the chance is $\frac{1}{2}$ that II will pass A_1 on to the gamete in question. Thus, the probability that III will receive an A_1 from II is $\frac{1}{2} \times \frac{1}{2} = \frac{1}{4}$, and this is the chance that III—the product of a full-sib mating—will be homozygous by descent.

Such close inbreeding can have deleterious consequences. Let's consider a rare deleterious allele a that, when homozygous, causes a metabolic disorder. If the frequency of the allele in the population is p, then the probability that a random couple will produce a homozygous offspring is

only p^2 (from the Hardy-Weinberg equilibrium). Thus, if p is, say, $\frac{1}{1000}$, the frequency of homozygotes will be 1 in 1,000,000. Now suppose that the couple are brother and sister. If one of their common parents is a heterozygote for the disease, they may both receive it and may both pass it on to the offspring they produce. The probability of a homozygous aa offspring is

probability one or the other grandparent is Aa

$\times$ probability a is passed to male sib

$\times$ probability a is passed to female sib

$\times$ probability of a homozygous aa
offspring from $Aa \times Aa$

$= (2pq + 2pq) \times \frac{1}{2} \times \frac{1}{2} \times \frac{1}{4}$

$= \dfrac{pq}{4}$

We assume that the chance that both grandparents are Aa is negligible. If p is very small, then q is nearly 1.0 and the chance of an affected offspring is close to $p/4$. For $p = \frac{1}{1000}$, there is 1 chance in 4000 of an affected child, compared to the one-in-a-million chance from a random mating. In general, for full sibs, the ratio of risks will be

$$\frac{p/4}{p^2} = \frac{1}{4p}$$

so the rarer the gene, the worse the *relative* risk of a defective offspring from inbreeding. For more distant relatives the chance of homozygosity by descent is, of course, less but still substantial. For first cousins, for example, the relative risk is $1/16p$ compared with random mating.

The population consequences of inbreeding depend on its intensity and form. Next we consider some examples. In experimental genetics (especially in plant and animal breeding), generation after generation of systematic selfing, full-sib, parent-offspring, or some other form of mating between relatives may be used to increase homozygosity. Such systematic inbreeding between close relatives eventually leads to complete homozygosity of the population, but at different rates. Referring to Table 27-2 in the next chapter, we can see the amount of heterozygosity left within lines after various numbers of generations of inbreeding. Which allele is fixed within a line is a matter of chance. If, in the original population from which the inbred lines are taken, allele A has frequency p and allele a has frequency $q = 1 - p$, then a proportion p of the homozygous lines established by inbreeding will be homozygous AA and a proportion q of the lines will be aa. What inbreeding does is take the genetic variation present *within* the original population and convert it into variation *between* homozygous inbred lines sampled from the population (Figure 26-12).

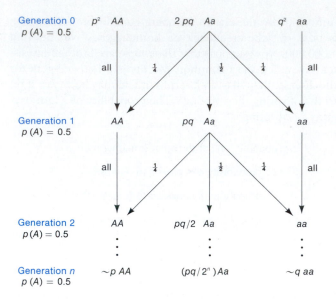

Generation 0
$p(A) = 0.5$ — p^2 AA $2pq$ Aa q^2 aa

all $\frac{1}{4}$ $\frac{1}{2}$ $\frac{1}{4}$ all

Generation 1
$p(A) = 0.5$ — AA pq Aa   aa

all $\frac{1}{4}$ $\frac{1}{2}$ $\frac{1}{4}$ all

Generation 2
$p(A) = 0.5$ — AA $pq/2$ Aa   aa

Generation n
$p(A) = 0.5$ — $\sim p$ AA $(pq/2^n)$ Aa $\sim q$ aa

Figure 26-12 Repeated generations of self-fertilization (or inbreeding) will eventually split a heterozygous population into a series of completely homozygous lines. The frequency of *AA* lines among the homozygous lines will be equal to the frequency of allele *A* in the original heterozygous population.

In a natural population, there will be some fraction of mating between relatives (or even selfing if that is possible) because of spatial proximity. However, there is no continuity of inbreeding within any specific family. If some proportion of wind-pollinated plants are selfed in a particular generation, these are not necessarily the progeny of selfed plants in the previous generation; they are distributed at random over selfed and outcrossed progeny. A consequence of such random inbreeding is that there is an equilibrium frequency of homozygotes and heterozygotes similar to the Hardy-Weinberg equilibrium, but with more homozygotes. Thus, genetic variation is still preserved, in contrast to the result of systematic experimental inbreeding.

Suppose that a population is founded by some small number of individuals who mate at random to produce the next generation. Also assume that no further immigration into the population ever occurs again. (For example, the rabbits now in Australia probably have descended from a single introduction of a few animals in the nineteenth century.) In later generations, then, everyone is related to everyone else, because their family trees have common ancestors here and there in their pedigrees. Such a population is then inbred, in the sense that there is some probability of a gene's being homozygous by descent. Since the population is, of necessity, finite in size, some of the originally introduced family lines will become extinct in every generation, just as family names disappear in a closed human population because, by chance, no male offspring are left. As original family lines disappear, the population comes to be made up of descendants of fewer and fewer of the original founder individuals, and all the members of the population

become more and more likely to carry the same alleles by descent. In other words, the inbreeding coefficient F increases, and the heterozygosity decreases over time until finally F reaches 1.00 and heterozygosity reaches 0.

The rate of loss of heterozygosity per generation in such a closed, finite, randomly breeding population is inversely proportional to the total number $(2N)$ of haploid genomes, where N is the number of diploid individuals in the population. In each generation, $1/2N$ of the remaining heterozygosity is lost, so that

$$H_t = H_0 \left(1 - \frac{1}{2N}\right)^t \cong H_0 e^{-t/2N}$$

where H_t and H_0 are the proportions of heterozygotes in the tth and original generations, respectively. As the number t of generations becomes very large, H_t approaches zero.

Which of the original alleles becomes fixed at each locus is a chance matter. If allele A_i has frequency p_i in the original population, then the probability is p_i that eventually the population will become homozygous $A_i A_i$. Suppose that a number of isolated island populations are founded from some large, heterozygous mainland population. Eventually, if these island populations remain completely isolated from one another, each will become homozygous for one of the alleles at each locus. Some will be homozygous $A_1 A_1$, some $A_2 A_2$, and so on. Thus, the result of this form of inbreeding is to cause genetic differentiation between populations.

Message Once again, we see that inbreeding is a process that converts genetic variation within a population into difference between populations by making each separate population homozygous for a randomly chosen allele.

Within each population, there is a change in allelic frequency from the original p_i to either 1 or 0, depending on which allele is fixed, but the average allelic frequency over all such populations remains p_i. Figure 26-13 shows the distribution of allelic frequencies among islands in successive generations, where $p(A_1) = 0.5$. In generation 0, all populations are identical. As time goes on, the gene frequencies among the populations diverge and some become fixed. After about $2N$ generations, every allelic frequency except the fixed classes ($p = 0$ and $p = 1$) is equally likely, and about half the populations are totally homozygous. By the time $4N$ generations have gone by, 80 percent of the populations are fixed, being homozygous AA and half being homozygous aa.

The process of differentiation by inbreeding in island populations is slow, but not on an evolutionary or geological time scale. If an island can support, say, 10,000 individuals of a rodent species, then after 20,000 generations (about 7000 years, assuming 3 generations per year), the population will be homozygous for about half of all the loci that were initially at the maximum of heterozygosity. Moreover, the island will be differentiated from other similar islands in

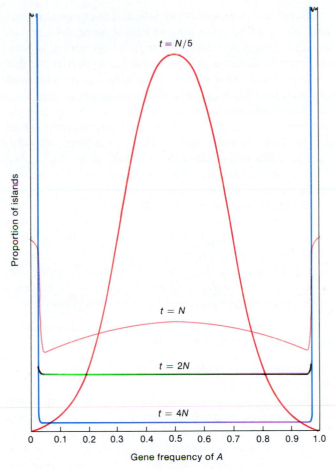

Figure 26-13 Distribution of gene frequencies among island populations after various numbers of generations of isolation, where the number of generations that have passed (t) is given in multiples of the population size (N).

this or a higher rate. If m is the migration rate into a given population and μ is the rate of mutation to new alleles, then roughly (to an order of magnitude) a population will retain most of its heterozygosity and will not differentiate much from other populations by local inbreeding if

$$m \geq \frac{1}{N} \quad \text{or} \quad \mu \geq \frac{1}{N}$$

or if

$$Nm \geq 1 \quad \text{or} \quad N\mu \geq 1$$

For populations of intermediate and even fairly large size, it is unlikely that $N\mu \geq 1$. For example, if the population size is 100,000, then the mutation rate must exceed 10^{-5}, which is somewhat on the high side for known mutation rates, although it is not an unknown rate. On the other hand, a migration rate of 10^{-5} per generation is not unreasonably large. In fact

$$m = \frac{\text{number of migrants}}{\text{total population size}} = \frac{\text{number of migrants}}{N}$$

Thus, the requirement that $Nm \geq 1$ is equivalent to the requirement that

$$Nm = N \times \frac{\text{number of migrants}}{N} \geq 1$$

or that

$$\text{number of migrant individuals} \geq 1$$

irrespective of population size! For many populations, more than a single migrant individual per generation is quite likely. Human populations (even isolated tribal populations) have a higher migration rate than this minimal value, and, as a result, no locus is known in humans for which one allele is fixed in some populations and an alternative allele is fixed in others (see Table 26-9).

two ways. For the loci that are fixed, many of the other islands will still be segregating, and others will be fixed at a different allele. For the loci that are still segregating in all the islands, there will be a large variance in gene frequency from island to island, as shown in Figure 26-13.

The Balance between Inbreeding and New Variation

Any population of any species is finite in size, so all populations should eventually become homozygous and differentiated from one another as a result of inbreeding. Evolution would then cease. In nature, however, new variation is always being introduced into populations by mutation and by some migration between localities. Thus, the actual variation available for natural selection is a balance between the introduction of new variation and its loss through local inbreeding. The rate of loss of heterozygosity in a closed population is $1/2N$, so any effective differentiation between populations will be negated if new variation is introduced at

Selection

Fitness and the Struggle for Existence

Darwin recognized that evolution consists of two processes, both of which must be explained. One is the origin of the *diversity* of organisms, and the second is the origin of the *adaptation* of these same organisms. Evolution is not simply the origin and extinction of different organic forms; rather, it is also a process that creates some kind of match between the phenotypes of species and the environments in which they live. Darwin's explanation of that match was that there is a constant *struggle for existence.* Organisms with phenotypes that are better-suited to the environment have a greater probability of surviving the struggle and will leave more offspring. Presumably, the better an organism can see, the better chance it has of locating food, defending itself, finding mates, and so on, and the greater will be its chance of survival and reproduction. Darwin called the process of differential survival and reproduction of different types **natural selection** by analogy with the **artificial selection** carried out by animal and plant breeders when they deliberately select some individuals of a preferred type.

The relative probability of survival and rate of reproduction of a phenotype or genotype is now called its **Darwinian fitness.** Although geneticists sometimes speak loosely of the fitness of an individual, the concept of fitness really applies to the average survival and reproduction of individuals in a phenotypic or genotypic class. Because of chance events in the life histories of individuals, even two organisms with identical genotypes and identical environments will differ in their survival and reproduction rates. It is the fitness of a genotype on average over all its possessors that matters.

Fitness is a consequence of the relationship between the phenotype of the organism and the environment in which the organism lives, so the same genotype will have different fitnesses in different environments. In part, this is because the exposure to different environments during development will result in different phenotypes for the same genotypes. But even if the phenotype is the same, the success of the organism depends on the environment. Having webbed feet is fine for paddling in water but a positive disadvantage for walking on land, as a few moments spent observing a duck walk will reveal. No genotype is unconditionally superior in fitness to all others in all environments.

Furthermore, the environment is not a fixed situation that is experienced passively by the organism. The environment of an organism is defined by the activities of the organism itself. Dry grass is part of the environment of a junco, so juncos that are most efficient at gathering it may waste less energy in nest building and thus have a higher reproductive fitness. But dry grass is part of a junco's environment *because juncos gather it to make nests.* The rocks among which the grass grows are not part of the junco's environment, although the rocks are physically present there. But the rocks are part of the environment of thrushes; these birds use the rocks to break open snails. Moreover, the environment that is defined by the life activities of an organism evolves as a result of those activities. The structure of the soil that is in part determinative of the kinds of plants that will grow is altered by the growth of those very plants. Environment is both the cause and the result of the evolution of organisms. As primitive plants evolved photosynthesis, they changed the earth's atmosphere from one that had had essentially no free oxygen and a high concentration of carbon dioxide to the atmosphere that we know today, which contains 21 percent oxygen and only 0.03 percent carbon dioxide. Plants that evolve today must do so in an environment created by the evolution of their own ancestors.

Darwinian or reproductive fitness is not to be confused with "physical fitness" in the everyday sense of the term, although they may be related. No matter how strong, healthy, and mentally alert the possessor of a genotype may be, that genotype has a fitness of zero if for some reason its possessors leave no offspring. Thus such statements as the "unfit are outreproducing the fit so the species may become extinct" are meaningless. The fitness of a genotype is a consequence of all the phenotypic effects of the genes involved. Thus, an allele that doubles the fecundity of its carriers but at the same time reduces the average lifetime of its possessors by 10 percent will be more fit than its alternatives, despite its life-shortening property. The most common example is parental care. An adult bird that expends a great deal of its energy gathering food for its young will have a lower probability of survival than one that keeps all the food for itself. But a totally selfish bird will leave no offspring, because its young cannot fend for themselves. As a consequence, parental care is favored by natural selection.

Two Forms of the Struggle for Existence

Darwin saw the "struggle for existence" as having two quite different forms, with different consequences for fitness. In one form, the organism "struggles" with the environment directly. Darwin's example was the plant that is struggling for water at the edge of a desert. The fitness of a genotype in such a case does not depend on whether it is frequent or rare in the population, because fitness is not mediated through the interactions of individuals but is a direct consequence of the individual's physical relationship to the external environment. Fitness is then **frequency-independent.**

The other form of struggle is between organisms competing for a resource in short supply or otherwise interacting so that their relative abundances do determine fitness. An example is *Mullerian mimicry* in butterflies. Some species of brightly colored butterflies (such as monarchs and viceroys) are distasteful to birds, which learn, after a few trials, to avoid attacking butterflies with that pattern. If two species differ in their pattern, there will be selection to make them more similar (Figure 26-14), since both will be protected and they share the burden of the birds' initial learning period. The less frequent pattern will be at a disadvantage with respect to the more frequent one, since birds will less often learn to avoid them. Within a species, rarer patterns will be selected against for the same reason. The rarer the pattern, the greater is the selective disadvantage, because birds will be unlikely to have had a prior experience of a low-frequency pattern and therefore will not avoid it. This selection to blend in with the crowd is an example of **frequency-dependent fitness.**

For reasons of mathematical convenience, most models of natural selection are based on frequency-independent fitness. In actual fact, however, a very large number of selective processes (perhaps most) are frequency-dependent. The kinetics of the evolutionary process depend on the exact form of frequency dependence, and, for that reason alone, it

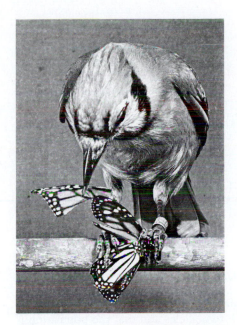

Figure 26-14 (a) A blue jay eating a monarch butterfly, which (b) induces vomiting in the jay. Because of this experience, the jay later will refuse to eat a viceroy butterfly that is similar in appearance to the monarch, although jays that have never tried monarchs will eat the viceroys. (Photographs courtesy of Lincoln Brower.)

is difficult to make any generalizations. The result of *positive* frequency dependence (such as competing predators, where fitness increases with increasing frequency) is quite different from the case of *negative* frequency dependence (such as the butterfly mimics, where fitness of a genotype declines with increasing frequency). For the sake of simplicity and as an illustration of the main qualitative features of selection, we deal only with models of frequency-independent selection in this chapter, but convenience should not be confused with reality.

Measuring Fitness Differences

For the most part, the differential fitness of different genotypes can be most easily measured when the genotypes differ at many loci. In very few cases (except for laboratory mutants, horticultural varieties, and major metabolic disorders) does the effect of an allelic substitution at a single locus make enough difference to the phenotype to be reflected in measurable fitness differences. Figure 26-15 shows the probability of survival from egg to adult—that is, the **viability**—of a number of second-chromosome homozygotes of *Drosophila pseudoobscura* at three different temperatures. As is generally the case, the fitness (in this case, a component of the total fitness, viability) is different in different environments. A few homozygotes are lethal or nearly so at all three temperatures, whereas a few have consistently high viability. Most genotypes, however, are not consistent in viability between temperatures, and no genotype is unconditionally the most fit at all temperatures. The fitness of

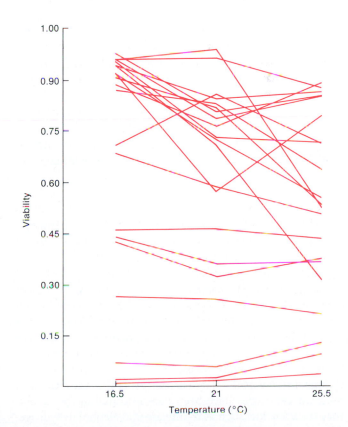

Figure 26-15 Viabilities of various chromosomal homozygotes of *Drosophila pseudoobscura* at three different temperatures.

Table 26-12 Comparison of Fitnesses for Inversion Homozygotes and Heterozygotes in Laboratory Populations of *Drosophila pseudoobscura* when Measured in Different Competitive Combinations

Experiment	Homozygotes			Heterozygotes		
	ST / ST	*SR / AR*	*CH / CH*	*ST / AR*	*ST / CH*	*AR / CH*
ST and *AR* alone	0.8	0.5	—	1.0	—	—
ST and *CH* alone	0.8	—	0.4	—	1.0	—
AR and *CH* alone	—	0.86	0.48	—	—	1.0
ST, *AR*, and, *CH* together	0.83	0.15	0.36	1.0	0.77	0.62

these chromosomal homozygotes was not measured in competition with each other; all are measured against a common standard, so we do not know whether they are frequency-dependent. An example of frequency-dependent fitness is shown in the estimates for inversion homozygotes and heterozygotes of *Drosophila pseudoobscura* in Table 26-12.

Examples of clear-cut fitness differences associated with single-gene substitutions are the many "inborn errors of metabolism," where a recessive allele interferes with a metabolic pathway and causes lethality of the homozygotes. An example in humans is phenylketonuria, where tissue degeneration is the result of the accumulation of a toxic intermediate in the pathway of tyrosine metabolism. A case that illustrates the relation of fitness to environment is sickle-cell anemia. An allelic substitution at the structural-gene locus for the β chain of hemoglobin results in substitution of valine for the normal glutamic acid at chain position 6. The abnormal hemoglobin crystallizes at low oxygen pressure, and the red cells deform and hemolyze. Homozygotes $Hb^S Hb^S$ have a severe anemia, and survivorship is low. Heterozygotes have a mild anemia and under ordinary circumstances exhibit the same or only slightly lower fitness than normal homozygotes $Hb^A Hb^A$. However, in regions of Africa with a high incidence of falciparum malaria, heterozygotes ($Hb^A Hb^S$) have a *higher* fitness than normal homozygotes because the presence of some sickling hemoglobin apparently protects them from the malaria. Where malaria is absent, as in North America, the fitness advantage of heterozygosity is lost.

It has not been possible to measure fitness differences for most single-locus polymorphisms. The evidence for differential net fitness for different ABO or MN blood types is shaky at best. The extensive enzyme polymorphism present in all sexually reproducing species has for the most part not been connected with measurable fitness differences, although in *Drosophila* clear-cut differences in the fitness of different genotypes have been demonstrated in the laboratory for a few loci such as α-amylase and alcohol dehydrogenase.

How Selection Works

Suppose that a population is mating at random with respect to a given locus with two alleles and that the population is so large that (for the moment) we can ignore inbreeding. Just after the eggs have been fertilized, the zygotes will be in Hardy-Weinberg equilibrium:

Genotype	*AA*	*Aa*	*aa*
Frequency	p^2	$2pq$	q^2

and $p^2 + 2pq + q^2 = (p + q)^2 = 1.0$, where p is the frequency of A.

Further suppose that the three genotypes have probabilities of survival to adulthood (viabilities) of $W_{AA} : W_{Aa} : W_{aa}$. For simplicity, let us also assume that all selective differences are differences in survivorship between the fertilized egg and the adult stage. Differences in fertility give rise to much more complex mathematical formulations. Then among the progeny once they have reached adulthood, the frequencies will be

Genotype	*AA*	*Aa*	*aa*
Frequency	$p^2 W_{AA}$	$2pq W_{Aa}$	$q^2 W_{aa}$

These adjusted frequencies do not add up to unity since the W's are all fractions smaller than 1. However, we can readjust them so that they do, without changing their relation to each other, by dividing each frequency by the sum of the frequencies after selection ($\overline{W}$):

$$\overline{W} = p^2 W_{AA} + 2pq W_{Aa} + q^2 W_{aa}$$

So defined, $\overline{W}$ is called the **mean fitness** of the population because it is, indeed, the mean of the fitnesses of all individuals in the population. After this adjustment, we have

Genotype	*AA*	*Aa*	*aa*
Frequency	$p^2 \dfrac{W_{AA}}{\overline{W}}$	$2pq \dfrac{W_{Aa}}{\overline{W}}$	$q^2 \dfrac{W_{aa}}{\overline{W}}$

We can now determine the frequency p' of the allele A in the next generation by summing up genes:

$$p' = AA + (\tfrac{1}{2})Aa = p^2\frac{W_{AA}}{\overline{W}} + \frac{pqW_{Aa}}{\overline{W}} = p\frac{pW_{AA} + qW_{Aa}}{\overline{W}}$$

Finally, we note that the expression $pW_{AA} + qW_{Aa}$ is the mean fitness of A alleles because, from the Hardy-Weinberg frequencies, a proportion p of all A alleles occur in homozygotes with another A and then they have a fitness of W_{AA}, while a proportion q of all the A alleles occur in heterozygotes with a and have a fitness of W_{Aa}. Using $\overline{W}_A$ to denote $pW_{AA} + qW_{Aa}$ yields the final new gene frequency:

$$p' = p\frac{\overline{W}_A}{\overline{W}}$$

In other words, after one generation of selection, the new value of the frequency of A is equal to the old value (p) multiplied by the ratio of the average fitness of A alleles to the fitness of the whole population. If the fitness of A alleles is greater than the average fitness of all alleles, then $\overline{W}_A/\overline{W}$ is greater than unity and p' is larger than p. Thus, the allele A increases in the population. Conversely, if $\overline{W}_A/\overline{W}$ is less than unity, A decreases. But the mean fitness of the population ($\overline{W}$) is the average fitness of the A alleles and of the a alleles. So if $\overline{W}_A$ is greater than the mean fitness of the population, it must be greater than $\overline{W}_a$, the mean fitness of a alleles.

Message   The allele with the higher average fitness increases in the population.

It should be noted that the fitnesses W_{AA}, W_{Aa}, and W_{aa} may be expressed as absolute probabilities of survival and absolute reproduction rates, or they may all be rescaled relative to one of the fitnesses, which is given the standard value of 1.0. This rescaling has absolutely no effect on the formula for p' because it cancels out in the numerator and denominator.

Message   The course of selection depends only on relative fitnesses.

An increase in the allele with the higher fitness means that the average fitness of the population as a whole increases, so that selection can also be described as a process that *increases mean fitness*. This rule is strictly true only for frequency-independent genotypic fitnesses, but it is close enough to a general rule to be used as a fruitful generalization. This maximization of fitness does not necessarily lead to any optimal property for the species as a whole because fitnesses are only defined relative to each other within a population. It is relative (not absolute) fitness that is increased by selection. The population does not necessarily become larger or grow faster, nor is it less likely to become extinct.

The Rate of Change of Gene Frequency

An alternative way to look at the process of selection is to solve for the *change* in allelic frequency in one generation:

$$\Delta p = p' - p = \frac{p\overline{W}_A}{\overline{W}} - p = \frac{p(\overline{W}_A - \overline{W})}{\overline{W}}$$

But $\overline{W}$, the mean fitness of the population, is the average of the allelic fitnesses $\overline{W}_A$ and $\overline{W}_a$, so that

$$\overline{W} = p\overline{W}_A + q\overline{W}_a$$

Substituting this expression for $\overline{W}$ in the formula for Δp and remembering that $q = 1 - p$, we obtain (after some algebraic manipulation)

$$\Delta p = \frac{pq(\overline{W}_A - \overline{W}_a)}{\overline{W}}$$

which is the general expression for a change in allelic frequency as a result of selection. This general expression is particularly illuminating. It says that Δp will be positive (A will increase) if the mean fitness of A alleles is greater than the mean fitness of a alleles, as we saw before. But it also shows that the speed of the change depends not only on the difference in fitness between the alleles but also on the factor pq, which is proportional to the frequency of heterozygotes ($2pq$). For a given difference in fitness of alleles, gene frequency will change most rapidly when the alleles A and a are in intermediate frequency, so that pq is large. If p is near 0 or 1 (that is, if A or a is nearly fixed), then pq is nearly 0 and selection will proceed very slowly. Figure 26-16 shows the **S**-shaped curve that represents the course of selection of a new favorable allele A that has recently entered a population of homozygotes aa. At first, the change in frequency is very small because p is still close to 0. Then it accelerates as A becomes more frequent, but it slows down again as A takes over and a becomes very rare. This is precisely what is expected from a selection process. When most of the population is of one type, there is nothing to select. For evolution by natural selection to occur, there must be genetic variance; the more variance, the faster the process.

An important consequence of the dependence of selection on genetic variance can be seen in a case of artificial selection. In the early part of this century, it became fashionable to advocate a program of **negative eugenics.** It was proposed that individuals with certain undesirable genetic traits (say, metabolic or nervous disorders) should be prevented from having any offspring. By this means, it was thought, the frequency of the trait in the population would be lowered, and the trait could eventually be eradicated. Suppose that such a program is completely efficient, so that every homozygote aa is prevented from reproducing. The fitnesses of the genotypes then are

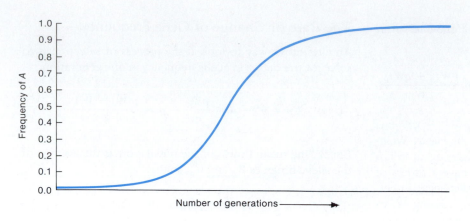

Genotype	AA	Aa	aa
Fitness	1.0	1.0	0

If p is the frequency of A and q is the frequency of a, then

$$\overline{W}_A = pW_{AA} + qW_{Aa} = p(1) + q(1) = 1.0$$

$$\overline{W}_a = pW_{Aa} + qW_{aa} = p(1) + q(0) = p$$

$$\overline{W} = p^2W_{AA} + 2pqW_{Aa} + q^2W_{AA} = p\overline{W}_A + q\overline{W}_a$$
$$= p(1) + q(p) = p(1 + q)$$

so that

$$q' = q\frac{\overline{W}_a}{\overline{W}} = q\frac{p}{p(1 + q)} = \frac{q}{1 + q}$$

If we iterate this formula over generations, we obtain

$$q'' = \frac{q'}{1 + q'} = \frac{q/(1 + q)}{1 + q/(1 + q)} = \frac{q}{1 + 2q}$$

$$q''' = \frac{q}{1 + 3q}$$

$$q^n = \frac{q}{1 + nq} = \frac{1}{n + 1/q}$$

From this sequence, we can see what the fate of a negative eugenics program would be. A deleterious gene will already be rare in a population. Suppose that $q = \frac{1}{100}$, for example. Then after one generation, the frequency would be $\frac{1}{101}$; after two generations, $\frac{1}{102}$; and so on. It would take 100 generations to reduce the frequency to $\frac{1}{200}$. But a human generation is 25 years, so it would require 2500 years (the time since the founding of the Roman republic) with perfectly efficient selection against the recessive just to reduce the frequency from $\frac{1}{100}$ to $\frac{1}{200}$. The negative eugenics plan clearly is impractical.

Of course, if the heterozygote for the deleterious genes (as, for example, in sickle-cell anemia) could be detected, then all copies of the gene could be removed from the pop-

ulation in a single generation if the heterozygotes were all prevented from having offspring. The only trouble with this suggestion is that every human being is heterozygous for many different deleterious recessive genes, so no one would be allowed to breed.

The objection to such eugenic measures, of course, cannot rest on their impracticality, since it could always be asserted that some genetic effects are worse than others, and it is these that should be singled out for efficient eugenic elimination. The real problem is one of power and rights. Eugenic problems are the problems of the conflict of rights between individuals and the state, and between different individuals. Indeed, they involve the very definition of an individual. Who should have the power to decide what constitutes a "life not worth living"? And since a newborn or a fetus clearly cannot have such power, how should the conflict between rights be resolved? More pressing than the problem of how they *should* be resolved is the question of how they *will* be resolved in actual social practice.

When alternative alleles are not rare, selection can cause quite rapid changes in allelic frequency. Figure 26-17 shows the course of elimination of a malic dehydrogenase allele in a laboratory population of *Drosophila melanogaster.* The fitnesses in this case are

$$W_{AA} = 1.0 \qquad W_{Aa} = 0.75 \qquad W_{aa} = 0.40$$

Of course, the frequency of a is not reduced to 0, and further reduction in frequency will require longer and longer times, as shown in the negative eugenics case.

Message Unless alternative alleles are present in intermediate frequencies, selection (especially against recessives) is quite slow. Selection is dependent on genetic variation.

Balanced Polymorphism

Let's reexamine the general formula for allelic frequency change (page 801).

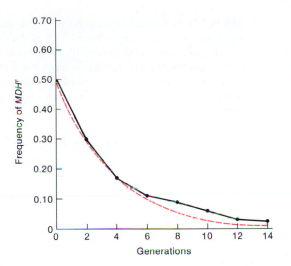

Figure 26-17 The loss of an allele of the malic dehydrogenase locus MDH^F due to selection in a laboratory population of *Drosophila melanogaster*. The red dashed line shows the theoretical curve of change computed for fitnesses $W_{AA} = 1.0$, $W_{Aa} = 0.75$, $W_{aa} = 0.4$. (From R. C. Lewontin, *The Genetic Basis of Evolutionary Change.* Copyright © 1974 by Columbia University Press. Data courtesy of E. Berger.)

$$\Delta p = pq \frac{\overline{W}_A - \overline{W}_a}{\overline{W}}$$

Under what conditions will the process stop? When is $\Delta p = 0$? Two answers are when $p = 0$ or when $q = 0$ (that is, when either allele A or allele a, respectively, has been eliminated from the population). One of these events will eventually occur if $\overline{W}_A - \overline{W}_a$ is consistently positive or negative, so that Δp is always positive or negative irrespective of the value of p. The condition for such unidirectional selection is that the heterozygote fitness be somewhere between the fitnesses of the two homozygotes: If $W_{AA} \geq W_{Aa} \geq W_{aa}$, then A alleles are more fit than a alleles in both the heterozygous and the homozygous condition. Then the mean allelic fitness of A, $\overline{W}_A$, is larger than the mean allelic fitness of a, $\overline{W}_a$, no matter what the frequencies of the genotypes may be. In this case $\overline{W}_A - \overline{W}_a$ is positive, and A always increases until it reaches $p = 1$. If, on the other hand, $W_{AA} \leq W_{Aa} \leq W_{aa}$, then $\overline{W}_A - \overline{W}_a$ is negative, and a always increases until it reaches $q = 1$.

But there is another possibility for $\Delta p = 0$, even when p and q are not 0:

$$\overline{W}_A = \overline{W}_a$$

which can occur if the heterozygote is not intermediate between the homozygotes but has a fitness that is more extreme than either homozygote. Then, for part of the range of values of p, $\overline{W}_A > \overline{W}_a$, whereas $\overline{W}_A < \overline{W}_a$ for the rest of the range. Just between those ranges is a value of p (denoted by $\hat{p}$) for which the mean fitnesses of the two alleles are equal. A little algebraic manipulation of

$$\overline{W}_A - \overline{W}_a = 0 = (\hat{p}W_{AA} + \hat{q}W_{Aa}) - (\hat{p}W_{Aa} + \hat{q}W_{aa})$$

gives us the solution for $\hat{p}$:

$$\hat{p} = \frac{W_{aa} - W_{Aa}}{(W_{aa} - W_{Aa}) + (W_{AA} - W_{Aa})}$$

The equilibrium value is a simple ratio of the differences in fitness between the homozygotes and the heterozygote. As an example, suppose the fitnesses are

$\dfrac{W_{AA}}{0.9}$	$\dfrac{W_{Aa}}{1.0}$	$\dfrac{W_{aa}}{0.8}$

The equilibrium will be $\hat{p} = \frac{2}{3}$.

There are, in fact, two qualitatively different possibilities for $\hat{p}$. One possibility is that $\hat{p}$ is an *unstable* equilibrium. There will be no change in frequency if the population has exactly this value of p, but the frequency will move *away* from the equilibrium (toward $p = 0$ or $p = 1$) if the slightest perturbation of frequency occurs. This unstable case will exist when the heterozygote is *lower* in fitness than either homozygote; such a condition is an example of **underdominance.** The alternative possibility is a *stable* equilibrium, or **balanced polymorphism,** in which slight perturbations from the value of $\hat{p}$ will result in a return to $\hat{p}$. The conditions for this balance is that the heterozygote be *greater* in fitness than either homozygote—a condition termed **overdominance.**

In nature, the chance that a gene frequency will remain balanced on the knife edge of an unstable equilibrium is negligible, so we should not expect to find naturally occurring polymorphisms in which heterozygotes are less fit than homozygotes. On the contrary, the observation of a long-lasting polymorphism in nature might be taken as evidence of a superior heterozygote.

Unfortunately, life confounds theory. The *Rh* locus (rhesus blood group) in humans has a widespread polymorphism with Rh^+ and Rh^- alleles. In Europeans, the frequency of the Rh^- allele is about 0.4, whereas in Africans it is about 0.2. Thus, this must be a very old human polymorphism, antedating the origin of modern geographical races. But this polymorphism causes a maternal fetal incompatibility when an RH^- mother (homozygous Rh^-Rh^-) produces an RH^+ fetus (heterozygous Rh^-Rh^+). This incompatibility results in hemolytic anemia (from a destruction of red blood cells) and the death of the fetus in a moderate proportion of cases if the mother has been previously sensitized by an earlier pregnancy with an incompatible fetus. Thus, there is selection against heterozygotes, although it is frequency-dependent, since it occurs only when the mother is a homozygous recessive. This polymorphism is unstable and

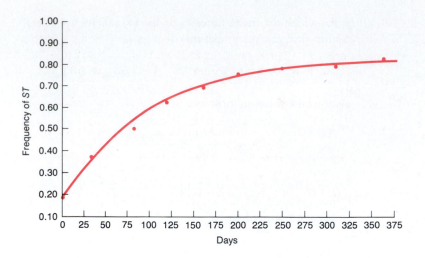

should have disappeared from the species, yet it exists in most human populations. There have been many hypotheses proposed to explain its apparent stability, but the mystery remains.

In contrast, no fitness difference at all can be demonstrated for many polymorphisms of blood groups (and for the ubiquitous polymorphism of enzymes revealed by electrophoresis). It has been suggested that such polymorphisms are not under selection at all but that

$$W_{AA} = W_{Aa} = W_{aa}$$

This situation of **selective neutrality** would, of course, also satisfy the requirement that $\overline{W}_A = \overline{W}_a$, but instead of a stable equilibrium, it gives rise to a **passive** (or **neutral**) **equilibrium** such that any allele frequency p is as good as any other. This leaves unanswered the problem of how the populations became highly polymorphic in the first place. The best case of overdominance for fitness at a single locus remains that of sickle-cell anemia, where the two homozygotes are at a disadvantage relative to the heterozygote for quite different reasons.

The best-studied cases of balanced polymorphism in nature and in the laboratory are the inversion polymorphisms in several species of *Drosophila*. Figure 26-18 shows the course of frequency change for the inversion *ST* in competition with the alternative chromosomal type *CH* in a laboratory population of *Drosophila pseudoobscura*. The inversions *ST* and *CH* are part of a chromosomal polymorphism in natural populations of this species. The fitnesses estimated for the three genotypes in the laboratory are

$$W_{ST/ST} = 0.89 \qquad W_{ST/CH} = 1.0 \qquad W_{CH/CH} = 0.41$$

Applying the formula for the equilibrium value $\hat{p}$, we obtain $\hat{p} = 0.85$, which agrees quite well with the observations in Figure 26-18.

Multiple Adaptive Peaks

We must avoid taking an overly simplified view of the consequences of selection. At the level of the gene—or even at the level of the partial phenotype—the outcome of selection for a trait in a given environment is not unique. Selection to alter a trait (say, to increase size) may be successful in a number of ways. In 1952, F. Robertson and E. Reeve successfully selected to change wing size in *Drosophila* in two different populations. However, in one case the *number* of cells in the wing changed, whereas in the other case the *size* of the wing cells changed. Two different genotypes had been selected, both causing a change in wing size. The initial state of the population at the outset of selection determined which of these selections occurred.

The way in which the same selection can lead to different outcomes can most easily be illustrated by a simple hypothetical case. Suppose that the variation of two loci (there will usually be many more) influences a character and that (in a particular environment) intermediate phenotypes have the highest fitness. (For example, newborn babies have a higher chance of surviving birth if they are neither too big nor too small.) If the alleles act in a simple way in influencing the phenotype, then the three genetic constitutions *A a B b*, *A A b b*, and *a a B B* will produce a high fitness because they will all be intermediate in phenotype. On the other hand, very low fitness will characterize the double homozygotes *A A B B* and *a a b b*. What will the result of selection be? We can predict the result by using the mean fitness $\overline{W}$ of a population. As previously discussed, selection acts in most simple cases to increase $\overline{W}$. Therefore, if we calculate $\overline{W}$ for every possible combination of gene frequencies at the two loci, we can determine which combinations yield high values of $\overline{W}$. Then we should be able to predict the course of selection by following a curve of increasing $\overline{W}$.

The surface of mean fitness for all possible combinations of allelic frequency is called an **adaptive surface,** or an

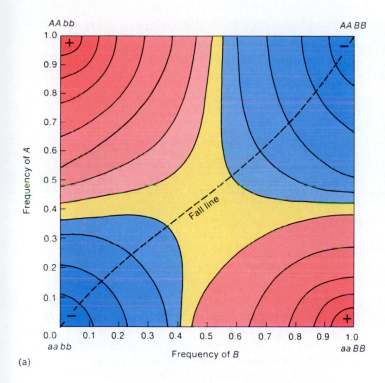

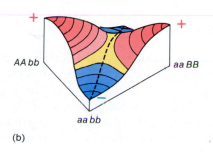

(b)

Figure 26-19 An adaptive landscape with two adaptive peaks (red), two adaptive valleys (blue), and a topographic saddle in the center of the landscape. The topographic lines are lines of equal mean fitness. If the genetic composition of a population always changes in such a way as to move the population "uphill" in the landscape, then the final composition will depend on where the population began with respect to the fall (dashed) line. (a) Topographic map of the adaptive landscape. (b) A perspective sketch of the surface shown in the map.

adaptive landscape (Figure 26-19). The figure is like a topographic map. The frequency of allele A at one locus is plotted on one axis, and the frequency of allele B at the other locus is plotted on the other axis. The height above the plane (represented by topographic lines) is the value of $\overline{W}$ that the population would have for a particular combination of frequencies of A and B. According to the rule of increasing fitness, selection should carry the population from a low-fitness "valley" to a high-fitness "peak." However, Figure 26-19 shows that there are two adaptive peaks, corresponding to a fixed population of $AAbb$ and a fixed population of $aaBB$,

with an adaptive valley between them. Which peak the population will ascend—and therefore what its final genetic composition will be—depends on whether the initial genetic composition of the population is on one side or the other of the dashed "fall line" shown in the figure.

Message   Under identical conditions of natural selection, two populations may arrive at two different genetic compositions as a direct result of natural selection.

The existence of multiple adaptive peaks for a selective process means that some differences between species are the result of history and not of environmental differences. For example, African rhinoceroses have two horns, and Indian rhinoceroses have one (Figure 26-20). We need not invent a special story to explain why it is better to have two horns on the African plains and one in India. It is much more plausible that the trait of having horns was selected, but that two long, slender horns and one short, stout horn are simply alternative adaptive features, and that historical accident differentiated the species. Explanations of adaptations by natural selection do not require that every difference between species be differentially adaptive.

It is important to note that nothing in the theory of selection requires that the different adaptive peaks be of the same height. The kinetics of selection is such that $\overline{W}$ increases, not that it necessarily reaches the highest possible peak in the field of gene frequencies. Suppose, for example, that a population is near the peak $AAbb$ in Figure 26-19 and that this peak is lower than the $aaBB$ peak. Selection alone cannot carry the population to $aaBB$, because that would require a temporary decrease in $\overline{W}$ as the population descended the $AAbb$ slope, crossed the saddle, and ascended the other slope. Thus, the force of selection is myopic. It drives the population to a *local* maximum of $\overline{W}$ in the field of gene frequencies—not to a *global* one.

Artificial Selection

In contrast to the difficulties of finding simple, well-behaved cases in nature that exemplify the simple formulas of natural selection, there is a vast record of the effectiveness of artificial selection in changing populations phenotypically. These changes have been produced by laboratory selection experiments and by selection of animals and plants in agriculture (as examples, for increased milk production in cows and for rust resistance in wheat). No analysis of these experiments in terms of allelic frequencies is possible because individual loci have not been identified and followed. Nevertheless, it is clear that genetic changes have occurred in the populations and that some analysis of selected populations has been carried out according to the methods described in Chapter 27. Figure 26-21 shows, as an example, the large changes in average bristle number achieved in a selection ex-

(a)

(b)

Figure 26-20 Differences in horn morphology in two geographically separated species of rhinoceros: (a) the African rhinoceros; (b) the Indian rhinoceros. (Part a from Anthony Bannister/NHPA; part b from K. Ghani/NHPA.)

periment with *Drosophila melanogaster.* Figure 26-22 shows the changes in the number of eggs laid per chicken as a consequence of 30 years of selection.

The usual method of selection is **truncation selection.** The individuals in a given generation are pooled (irrespective of their families), a sample is measured, and only those individuals above (or below) a given phenotypic value (the truncation point) are chosen as parents for the next genera-

tion. This phenotypic value may be a fixed value over successive generations; then selection is by **constant truncation.** More commonly, a fixed percent of the population representing the highest (or lowest) value of the selected character is chosen; then selection is by **proportional truncation.** With constant truncation, the intensity of selection decreases with time, as more and more of the population exceeds the fixed truncation point. With proportional truncation, the intensity of selection is constant, but the truncation point moves upward as the population distribution moves. Figure 26-23 illustrates these two types of truncation.

Figure 26-21 Changes in average bristle number obtained in two laboratory populations of *Drosophila melanogaster* through artificial selection for high bristle number in one population and for low bristle number in the other. The dashed segment in the curve for the upwardly selected line indicates a period of five generations during which no selections were performed. (From K. Mather and B. J. Harrison, "The Manifold Effects of Selection," *Heredity* 3, 1949, 1–52.)

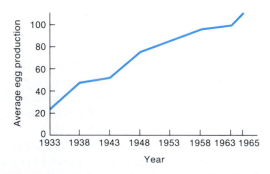

Figure 26-22 Changes in average egg production in a chicken population selected for its increase in egg-laying rate over a period of 30 years. (From I. M. Lerner and W. J. Libby, *Heredity, Evolution, and Society,* 2d ed. Copyright © 1976 by W. H. Freeman and Company. Data courtesy of D. C. Lowry.)

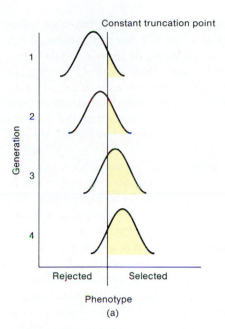

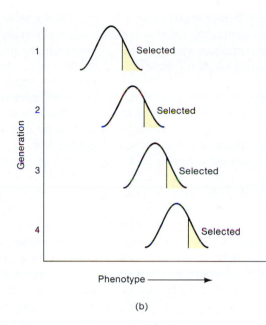

Figure 26-23 Two schemes of truncation selection for a continuously varying trait: (a) constant truncation; (b) proportional truncation.

A common experience in artificial selection programs is that as the population becomes more and more extreme, its viability and fertility decrease. As a result, eventually no further progress under selection is possible, despite the presence of genetic variance for the character, because the selected individuals do not reproduce. The loss of fitness may be a direct phenotypic effect of the genes for the selected character, in which case nothing much can be done to improve the population further. Often, however, the loss of fitness is tied to linked sterility genes that are carried along with the selected loci. In such cases, a number of generations without selection allow recombinants to be formed, and selection can then be continued, as in the upwardly selected line in Figure 26-21.

We must be very careful in our interpretation of long-term agricultural selection programs. In the real world of agriculture, changes in cultivation methods, machinery, fertilizers, insecticides, herbicides, and so on are occurring along with the production of genetically improved varieties. Increases in average yields are consequences of all of these changes. For example, the average yield of corn in the United States increased from 40 bushels to 80 bushels per acre between 1940 and 1970. But experiments comparing old and new varieties of corn in common environments show that only about half this increase is a direct result of new corn varieties (the other half being a result of improved farming techniques). Furthermore, the new varieties are superior to the old ones only at the high densities of modern planting for which they were selected.

Random Events

If a population is finite in size (as all populations are) and if a given pair of parents has only a small number of offspring, then even in the absence of all selective forces, the frequency of a gene will not be exactly reproduced in the next generation because of sampling error. If in a population of 1000 individuals, the frequency of *a* is 0.5 in one generation, then it may by chance be 0.493 or 0.505 in the next generation because of the chance production of a few more or a few less progeny of each genotype. In the second generation, there is another sampling error based on the new gene frequency, so the frequency of *a* may go from 0.505 to 0.511 or back to 0.498. This process of random fluctuation continues generation after generation, with no force pushing the frequency back to its initial state, because the population has no "genetic memory" of its state many generations ago. Each generation is an independent event. The final result of this random change in allelic frequency is that the population eventually drifts to $p = 1$ or $p = 0$. After this point, no further change is possible; the population has become homozygous. A different population, isolated from the first, also undergoes this **random genetic drift,** but it may become homozygous for allele *A*, whereas the first population has become homozygous for allele *a*. As time goes on, isolated populations diverge from each other, each losing heterozygosity. The variation originally present *within* populations now appears as variation *among* populations.

One form of genetic drift occurs when a small group breaks off from a larger population to found a new colony. This "acute drift," called the **founder effect,** results from a single generation of sampling, followed by several generations during which the population remains small. The founder effect is probably responsible for the virtually complete lack of blood group B in Native Americans, whose ancestors arrived in very small numbers across the Bering Strait during the end of the last ice age, about 20,000 years ago.

The process of genetic drift should sound familiar. It is, in fact, another way of looking at the inbreeding effect in small populations discussed earlier. Whether regarded as inbreeding or as random sampling of genes, the effect is the same. Populations do not exactly reproduce their genetic constitutions; there is a random component of gene frequency change.

One result of random sampling is that most new mutations, even if they are not selected against, never succeed in entering the population. Suppose that a single individual is heterozygous for a new mutation. There is some chance that the individual in question will have no offspring at all. Even if it has one offspring, there is a chance of $\frac{1}{2}$ that the new mutation will not be transmitted. If the individual has two offspring, the probability that neither offspring will carry the new mutation is $\frac{1}{4}$, and so on. Suppose that the new mutation is successfully transmitted to an offspring. Then the lottery is repeated in the next generation, and again the allele may be lost. In fact, if a population is of size N, the chance that a new mutation is eventually lost by chance is $(2N - 1)/2N$. (For a derivation of this result, which is beyond the scope of this book, see Chapters 2 and 3 of Hartl and Clark, *Principles of Population Genetics.*) But if the new mutation is not lost, then the only thing that can happen to it in a finite population is that eventually it will sweep through the population and become fixed! This event has the probability of $1/2N$. In the absence of selection, then, the history of a population looks like Figure 26-24. For some period of time, it is homozygous; then a new mutation appears. In most cases, the new mutant allele will be lost immediately or very soon after it appears. Occasionally, however, a new mutant allele drifts through the population, and the population becomes homozygous for the new allele. The process then begins again.

Even a new mutation that is slightly favorable selectively will usually be lost in the first few generations after it appears in the population, a victim of genetic drift. If a new mutation has a selective advantage of S in the heterozygote in which it appears, then the chance is only $2S$ that the mutation will ever succeed in taking over the population. So a mutation that is 1 percent better in fitness than the standard allele in the population will be lost 98 percent of the time by genetic drift.

The fact that occasionally an unselected mutation will, by chance, be incorporated in a population has given rise to a theory of *neutral evolution,* according to which unselected mutations are being incorporated in populations at a steady rate, which we can calculate. If the mutation rate per locus is μ, and the size of the population is N, so that there are $2N$ copies of each gene, then the absolute *number* of mutations that will appear in a population per generation at a given locus is $2N\mu$. But the probability that any given mutation is eventually incorporated is $1/2N$, so that the absolute number of new mutations that will be incorporated per generation per locus is $(2N\mu)(1/2N) = \mu$. If there are k loci mutating, then in each generation there will be $k\mu$ newly incorporated mutations in the genome. This is a very powerful result, for it predicts a regular, clocklike rate of evolution that is independent of external circumstances and that depends only on the mutation rate, which we assume to be constant over long periods of time. The total genetic divergence between species should, on this theory, be proportional to the length of time since their separation in evolution. It has been proposed that much of the evolution of amino acid sequences of proteins has been without selection, and that evolution of synonymous bases and other DNA that neither codes for proteins nor regulates protein synthesis should behave like a *molecular clock* with a constant rate over all evolutionary lineages. Of course, different proteins will have different clock rates, depending on what portion of their amino acids are free to be substituted without selection.

Message New mutations can become established in a population even though they are not favored by natural selection simply by a process of random genetic drift. Even new favorable mutations are often lost.

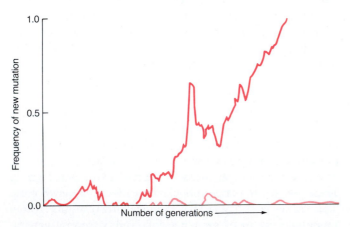

Figure 26-24 The appearance, loss, and eventual incorporation of new mutations during the life of a population. If random genetic drift does not cause the loss of a new mutation, then it must eventually cause the entire population to become homozygous for the mutation (in the absence of selection). In the figure, 10 mutations have arisen, of which 9 increased slightly in frequency and then died out. Only the fourth mutation eventually spread into the population. (After J. Crow and M. Kimura, *An Introduction to the Population Genetics Theory.* Copyright © 1970 by Harper & Row.)

A Synthesis of Forces

Variation and Divergence of Populations

The genetic variation within and between populations is a result of the interplay of the various evolutionary forces (Figure 26-25). Generally, as Table 26-13 shows, forces that increase or maintain variation within populations prevent the differentiation of populations from each other, whereas the divergence of populations is a result of forces that make each population homozygous. Thus, random drift (or in-

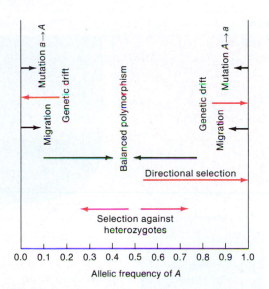

Figure 26-25 The effects of gene frequency of various forces of evolution. The black arrows show a tendency toward increased variation within the population; the red arrows, decreased variation.

Table 26-13 How the Forces of Evolution Increase (+) or Decrease (−) Variation within and between Populations

Force	Variation within populations	Variation between populations
Inbreeding or genetic drift	−	+
Mutation	+	−
Migration	+	−
Selection		
Directional	−	+/−
Balancing	+	−
Incompatible	−	+

breeding) produces homozygosity while causing different populations to diverge. This divergence and homozygosity are counteracted by the constant flux of mutation and the migration between localities, which introduce variation into the populations again and tend to make them more like each other.

The effects of selection are more variable. **Directional selection** pushes a population toward homozygosity, rejecting most new mutations as they are introduced but occasionally (if the mutation is advantageous) spreading a new allele through the population to create a new homozygous state. Whether or not such directional selection promotes differentiation of populations depends on the environment and on chance events. Two populations living in very similar environments may be kept genetically similar by directional selection, but if there are environmental differences, selection may direct the populations toward different compositions.

A particular case of interest, especially in human populations, is the interaction between mutation and directional selection in a very large population. New deleterious mutations are constantly arising spontaneously or as the result of the action of mutagens. These mutations may be completely recessive or partly dominant. Selection removes them from the population, but there will be an equilibrium between their appearance and removal. If we let q be the frequency of the deleterious allele a and $p = 1 − q$ be the frequency of the normal allele, then the change in allelic frequency due to the mutation rate μ is

$$\Delta q_{\text{mut}} = \mu p$$

A simple way to express the fitness for a recessive deleterious gene is $W_{AA} = W_{Aa} = 1.0$ and $W_{aa} = 1 − s$, where s is

the loss of fitness in homozygotes. We now can substitute these fitnesses in our general expression for allelic frequency change (see page 801) and obtain

$$\Delta q_{\text{sel}} = - \frac{pq(sq)}{1 - sq^2}$$

Equilibrium means that the increase in the allelic frequency due to mutation must exactly balance the decrease in the allelic frequency due to selection so that

$$\Delta \hat{q}_{\text{mut}} + \Delta \hat{q}_{\text{sel}} = 0$$

Remembering that $\hat{q}$ at equilibrium will be quite small, so that $1 - s\hat{q}^2 \simeq 1$, and substituting the terms for $\Delta \hat{q}_{\text{mut}}$ and $\Delta \hat{q}_{\text{sel}}$ in the formula above, we have

$$\mu \hat{p} - \frac{s\hat{p}\hat{q}^2}{1 - s\hat{q}^2} \simeq \mu \hat{p} - s\hat{p}\hat{q}^2 = 0$$

or

$$q^2 = \frac{\mu}{s} \qquad \text{and} \qquad \hat{q} = \sqrt{\frac{\mu}{s}}$$

So, for example, a recessive lethal ($s = 1$) mutating at the rate of $\mu = 10^{-6}$ will have an equilibrium frequency of 10^{-3}. Indeed, if we knew that a gene was a recessive lethal and had no heterozygous effects, we could estimate its mutation rate as the square of the allelic frequency. But the basis for such calculations must be firm. Sickle-cell anemia was once thought to be a recessive lethal with no heterozygous effects, which led to an estimated mutation rate in Africa of 0.1 for this locus.

If we let the fitnesses be $W_{AA} = 1.0$, $W_{Aa} = 1 - hs$, and $W_{aa} = 1 - s$ for a partly dominant gene, where h is the degree of dominance of the deleterious allele, then a similar calculation gives us

$$\hat{q} = \frac{\mu}{hs}$$

Thus, if $\mu = 10^{-6}$ and the lethal is not totally recessive but has a 5 percent deleterious effect in heterozygotes ($s = 1.0$, $h = 0.05$), then

$$\hat{q} = \frac{10^{-6}}{5 \times 10^{-2}} = 2 \times 10^{-5}$$

which is smaller by two orders of magnitude than the equilibrium frequency for the purely recessive case. In general, then, we can expect deleterious, completely recessive genes to have much higher frequencies than partly dominant genes.

Selection favoring heterozygotes (balancing selection) will, for the most part, maintain more or less similar polymorphisms in different populations. However, again, if the environments are different enough between them, then the populations will show some divergence. The opposite of balancing selection is selection against heterozygotes, which produces unstable equilibria. Such selection will cause homozygosity and divergence between populations.

The Exploration of Adaptive Peaks

Random and selective forces should not be thought of as simple antagonists. Random drift may counteract the force of selection, but it can enhance it as well. The outcome of the evolutionary process is a result of the simultaneous operation of these two forces. Figure 26-26 illustrates these possibilities. Note that there are multiple adaptive peaks in this landscape. Because of random drift, a population under selection does not ascend an adaptive peak smoothly. Instead, it takes an erratic course in the field of gene frequencies, like an oxygen-starved mountain climber. Pathway I shows a population history where adaptation has failed. The random fluctuations of gene frequency were sufficiently great that the population by chance became fixed at an unfit genotype. In any population, some proportion of loci are fixed at a selectively unfavorable allele because the intensity of selection is insufficient to overcome the random drift to fixation. The existence of multiple adaptive peaks and the random fixation of less fit alleles are integral features of the evolutionary process. Natural selection cannot be relied on to produce the best of all possible worlds.

Pathway II in Figure 26-26, on the other hand, shows how random drift may improve adaptation. The population was originally in the sphere of influence of the lower adaptive peak; however, by random fluctuation in gene frequency, its composition passed over the adaptive saddle, and the population was captured by the higher, steeper adaptive peak. This passage from a lower to a higher adaptive stable state could never have occurred by selection in an infinite population, because by selection alone, $\overline{W}$ could never decrease temporarily in order to cross from one slope to another.

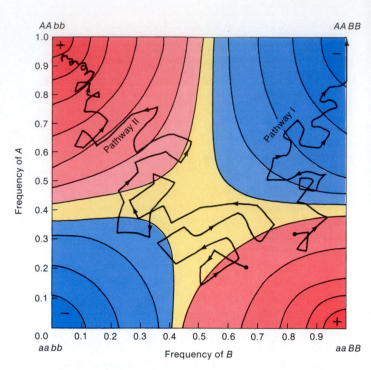

Figure 26-26 Selection and random drift can interact to produce different changes of gene frequency in an adaptive landscape. Without random drift, both populations would have moved toward *aa BB* as a result of selection alone.

Message The interaction of selection and random drift makes possible the attainment of higher fitness states than are obtainable when natural selection is operating alone.

Another consequence of the interaction of random and selective forces is that the effectiveness of the selective force in driving population composition depends on population size. The magnitude of the random effect is proportional to the reciprocal of population size, $1/N$, while the magnitude of a deterministic force depends on the migration rate m, or mutation rate μ, or selection coefficient s. We have already seen (page 797) that migration and mutation are effective if

$$m \geq \frac{1}{N} \quad \text{or} \quad \mu > \frac{1}{N}$$

that is, if

$$Nm \geq 1 \quad \text{or} \quad N\mu > 1$$

The same is true of selection; selection is effective only if $Ns \geq 1$. When Ns is small, because selection is weak or population size is small, then mutations are *effectively neutral*, even though there is some selection on them. Small populations will be less affected by selection than large populations even under otherwise identical conditions. For example, human populations were very small for nearly all the history of our species, having grown large only during the last few hundred generations. Thus, we may expect to find that many mutations that are now under selection were effectively neutral for a long time and may have reached high frequency by chance.

The Origin of Species

By a *species* (at least in sexually reproducing organisms), we mean a group of individuals that are biologically capable of interbreeding yet isolated genetically from other groups. **Speciation**—the origin of a new species—is the origin of a group of individuals capable of making a living in a new way and at the same time acquiring some barrier to genetic exchange with the species from which it arose. The genetic differentiation of a population by inbreeding, genetic drift, and differential selection is always threatened by the reintroduction of genes from other groups by migration. The reduction of gene migration to a very low value is therefore a prerequisite for speciation.

Generally, this reduction is the result of the geographical isolation of the population as a consequence of chance historical events: A few long-distance migrants may reach a new island; a part of the mainland may be cut off by a rise in sea level; an insect vector that formerly passed pollen from one population to another may become locally extinct; the grassy plain that connected the feeding grounds of two grazers may, by a slight change in rainfall pattern, become a desert. Once the population is isolated physically, the processes of genetic differentiation will go on unimpeded until the genetic constitution of the isolated population is so different from its parental group that there is real difficulty in interbreeding.

If migration is reestablished before this critical period, speciation will not occur and the divergent populations will once again converge. This has already happened in the human species, in which the genetic differentiation of geographical populations never has proceeded beyond some superficial physical traits and a mixed differentiation of frequencies at polymorphic loci. On the other hand, if populations are very divergent before they come back in contact with each other, hybrid offspring will have genotypes with such low fitness that they do not survive or are sterile. At this stage of differentiation, there is a definite selective advantage for the newly forming species to avoid mating with each other and so avoid the wastage of gametes. New (secondary) barriers to interbreeding may then be selected, thereby completing the speciation process.

Beyond this generalized sketch of speciation, remarkably little can be said with certainty. Because species do not interbreed, it is difficult to analyze their differences genetically. A great deal must be made of the few cases in which some hybrid offspring can be produced in the laboratory or the garden. The methods of electrophoresis, immunology, and protein sequencing have made it possible to describe the differences in the proteins of species, but there are very few cases of species that have just recently separated. Thus, we do not know how much of the genome—or what part of it—is involved in the first divergence, nor do we know whether that divergence is often a consequence of selection in opposite directions or of random drift. A detailed genetic analysis of the process of speciation remains one of the most important tasks for population genetics.

SUMMARY

Charles Darwin's *The Origin of Species* revolutionized biology. He constructed a theory of evolution based on the principles that variation existed within populations, that variation was heritable, and that the phenotype of the individuals in the population changed through generations because of natural selection. These basic tenets of evolution—put forward in 1859, before any knowledge of Mendelian genetics—have required only minor modification since that time. The study of changes within a population, or population genetics, relates the heritable changes in populations or organisms to the underlying individual processes of inheritance and development. Population genetics is the study of inherited variation and its modification in time and space.

Identifiable inherited variation within a population can be studied by observing morphological differences between individuals, examining the differences in specific amino acid sequences of proteins, or even examining, most recently, the differences in nucleotide sequences within the DNA. These kinds of observations have led to the conclusion that there is considerable polymorphism at many loci within a population. A measure of this variation is the amount of heterozygosity in a population. Population studies have shown that, in general, the genetic differences between individuals within human races are much greater than the average differences between races.

The ultimate source of all variation is mutation. However, within a population, the quantitative frequency of specific genotypes can be changed by recombination, immigration of genes, continued mutational events, and chance.

One property of Mendelian segregation is that random mating results in an equilibrium distribution of genotypes after one generation. However, inbreeding is one process that converts genetic variation within a population into differences between populations by making each separate population homozygous for a randomly chosen allele. On the other hand, for most populations, a balance is reached for any given environment between inbreeding, mutation from one allele to another, and immigration.

"Directed" changes of allelic frequencies within a population occur through the natural selection of a favored genotype. In many cases, such changes lead to homozygosity at a particular locus. On the other hand, the heterozygote may be more suited to a given environment than either of the homozygotes, leading to a balanced polymorphism.

Environmental selection of specific genotypes is rarely this simple, however. More often than not, phenotypes are determined by several interacting genes, and alleles at these different loci will be selected for at different rates. Furthermore, closely linked loci, unrelated to the phenotype in question, may have specific alleles carried along during the selection process. In general, genetic variation is the result of the interaction of evolutionary forces. For instance, a recessive, deleterious mutant may never be totally eliminated from a population, because mutation will continue to resupply it to the population. Immigration can also reintroduce the undesirable allele into the population. And, indeed, a deleterious allele may, under environmental conditions of which we are unaware (including the remaining genetic makeup of the individual), be selected for.

Unless alternative alleles are in intermediate frequencies, selection (especially against recessives) is very slow, requiring many generations. In many populations, especially those of small size, new mutations can become established even though they are not favored by natural selection, simply by a process of random genetic drift. Such slow changes in different allelic frequencies throughout the genome can lead to the eventual formation of new species.

Concept Map

Draw a concept map interrelating as many of the following terms as possible. Note that the terms are listed in no particular order.

allelic frequency / heterozygosity / polymorphism / mutation / selection / Hardy-Weinberg equilibrium / immigration / inbreeding / genetic drift / Mendelian ratios

CHAPTER INTEGRATION PROBLEM

The polymorphisms for shell color (yellow or pink) and for the presence or absence of shell banding in the snail *Cepaea nemoralis* (see page 782) are each the result of a pair of segregating alleles at a separate locus. Design an experimental program that would reveal the forces that determine the frequency and geographical distribution of these polymorphisms.

Solution

a. A description is made of the frequencies of the different morphs in a large number of populations covering the geographic and ecological range of the species. Each snail must be scored for *both* polymorphisms. At the same time a description of the habitat of each population is recorded. In addition, an estimate is made of the number of snails in a population.

b. By marking a sample of snails with a spot of paint on the shell, replacing them in the population, and then resampling at a later date, migration distances can be measured.

c. By raising broods from individual snails, the genotype of male parents can be inferred, so nonrandom mating patterns can be observed. The segregation frequencies *within* each family will reveal differences between genotypes in probability of survivorship of early developmental stages.

d. Further evidence of selection is sought from (a) geographical clines in the frequencies of the alleles; (b) correlation between allele frequency and ecological variables, including population density; (c) correlation between the frequencies of the two different polymorphisms (are populations with, say, high frequencies of pink shells also characterized by, say, high frequencies of banded shells?); (d) nonrandom associations *within* populations of the alleles at the two loci, indicating that certain combinations may have a higher fitness.

e. Evidence of the importance of random genetic drift is sought by comparing the variation in allele frequencies among small populations with the variation among large populations. If small populations vary more from each other than large ones, then random drift is implicated.

SOLVED PROBLEMS

1. About 70 percent of all white North Americans can taste the chemical phenylthiocarbamide, and the remainder cannot. The ability to taste is determined by the dominant allele T, and the inability to taste is determined by the recessive allele t. If the population is assumed to be in Hardy-Weinberg equilibrium, what are the genotypic and allelic frequencies in this population?

Solution

Since 70 percent are tasters (TT), 30 percent must be non-tasters (tt). This homozygous recessive frequency is equal to q^2, so to obtain q, we simply take the square root of 0.30:

$$q = \sqrt{0.30} = 0.55$$

Since $p + q = 1$, we can write

$$p = 1 - q = 1 - 0.55 = 0.45$$

Now we can calculate

$$p^2 = 0.45^2 = 0.20(TT)$$

$$2pq = 0 \times 0.45 \times 0.55 = 0.50(Tt)$$

$$q^2 = 0.3(tt)$$

2. In a large natural population of *Mimulus guttatus* one leaf was sampled from each of a large number of plants. The leaves were crushed and run on an electrophoretic gel. The gel was then stained for a specific enzyme X. Six different banding patterns were observed as shown in the following frequencies:

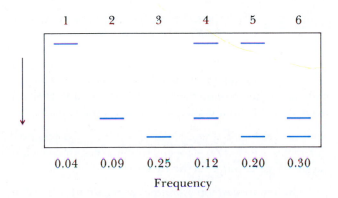

Frequency

a. Assuming that these patterns are produced by a single locus, propose a genetic explanation for the six types.

b. How can you test your idea?

c. What are the allelic frequencies in this population?

d. Is the population in Hardy-Weinberg equilibrium?

Solution

a. Inspection of the gel reveals that there are only three band positions: we will call them slow, intermediate, and fast. Furthermore, any individual can show either one band or two. The simplest explanation of this is that there are three alleles of one locus (let's call them A^S, A^I, and A^F) and that the individuals with two bands are heterozygotes. Hence, $1 = SS$, $2 = II$, $3 = FF$, $4 = SI$, $5 = SF$, and $6 = IF$.

b. The hypothesis can be tested by making controlled crosses. For example, from a self of type 5, we can predict $\frac{1}{4}SS, \frac{1}{2}SF$, and $\frac{1}{4}FF$.

c. The frequencies can be calculated by a simple extension of the two-allele formulas. Hence

$$f(S) = 0.04 + \tfrac{1}{2}(0.12) + \tfrac{1}{2}(0.20) = 0.20$$

$$f(I) = 0.09 + \tfrac{1}{2}(0.12) + \tfrac{1}{2}(0.30) = 0.30$$

$$f(F) = 0.25 + \tfrac{1}{2}(0.20) + \tfrac{1}{2}(0.30) = 0.50$$

d. The Hardy-Weinberg genotypic frequencies are

$$(p + q + r)^2 = p^2 + q^2 + r^2 + 2pq + 2pr + 2qr$$

$$= 0.04 + 0.09 + 0.25 + 0.12 + 0.20 + 0.30$$

which are precisely the observed frequencies, so it appears that the population is in equilibrium.

3. In a large experimental *Drosophila* population, the fitness of a recessive phenotype is calculated to be 0.90, and the mutation rate to the recessive allele is 5×10^{-5}. If the population is allowed to come to equilibrium, what allelic frequencies can be predicted?

Solution

Here mutation and selection are working in opposite directions, so an equilibrium is predicted. Such an equilibrium is described by the formula

$$\hat{q} = \sqrt{\frac{\mu}{s}}$$

In the present question, $\mu = 5 \times 10^{-5}$ and $s = 1 - W = 1 - 0.9 = 0.1$. Hence

$$\hat{q} = \sqrt{\frac{5 \times 10^{-5}}{10^{-1}}} = 2.2 \times 10^{-2} = 0.022$$

$$\hat{p} = 1 - 0.022 = 0.978$$

4. In a population, q is currently 0.2. If the fitness of $aa = 0$, what will q be after 95 generations?

$$q^n = \frac{1}{n + 1/q}$$

Thus

Solution

The formula needed here is

$$q^n = \frac{1}{95 + 1/0.2} = \frac{1}{100} = 0.01$$

PROBLEMS

1. What are the forces that can change the frequency of an allele in a population?

2. In a population of mice, there are two alleles of the A locus ($A1$ and $A2$). Tests showed that in this population there are 384 mice of genotype $A1A1$, 210 of $A1A2$, and 260 of $A2A2$. What are the frequencies of the two alleles in the population?

3. In a randomly mating laboratory population of *Drosophila,* 4 percent of the flies have black bodies (black is the autosomal recessive b), and 96 percent have brown bodies (the normal color B). If this population is assumed to be in Hardy-Weinberg equilibrium, what are the allelic frequencies of B and b and the genotypic frequencies of BB and Bb?

4. In a population, the $D \rightarrow d$ mutation rate is 4×10^{-6}. If $p = 0.8$ today, what will p be after 50,000 generations?

5. You are studying protein polymorphism in a natural population of a certain species of a sexually reproducing haploid organism. You isolate many strains from various parts of the test area and run extracts from each strain on electrophoretic gels. You stain the gels with a reagent specific for enzyme X and find that in the population there is a total of, say, five electrophoretic variants of enzyme X. You speculate that these variants represent various alleles of the structural gene for enzyme X.

a. How could you demonstrate that this is so, both genetically and biochemically? (You can make crosses, make diploids, run gels, test enzyme activities, test amino acid sequences, and so on). Outline the steps and conclusions precisely.

b. Name at least one other possible way of generating the different electrophoretic variants, and say how you would distinguish this possibility from the one described here.

6. A study made in 1958 in the mining town of Ashibetsu in the Hokkaido province of Japan revealed the frequencies of MN blood type genotypes (for individuals and for married couples) shown in the following table:

Genotype	Number of individuals or couples
Individuals	
$L^M L^M$	406
$L^M L^N$	744
$L^N L^N$	332
Total	1482
Couples	
$L^M L^M \times L^M L^M$	58
$L^M L^M \times L^M L^N$	202
$L^M L^N \times L^M L^N$	190
$L^M L^M \times L^N L^N$	88
$L^M L^N \times L^N L^N$	162
$L^N L^N \times L^N L^N$	41
Total	741

a. Show whether or not the population is in Hardy-Weinberg equilibrium with respect to the MN blood types.

b. Show whether mating is random with respect to MN blood types.

(Problem 6 is from J. Kuspira and G. W. Walker, *Genetics: Questions and Problems.* Copyright © 1973 by McGraw-Hill.)

7. For a sex-linked character is a species in which the male is the heterogametic sex, suppose that the allelic frequencies at a locus are different for males and females.

a. Letting the frequencies of A be 0.8 in males and 0.2 in females, show what happens to the allelic frequencies in successive generations in the two sexes.

b. Try to develop a general expression for the difference in the allelic frequencies in males (p) and females (P) in the nth generation, given that the initial values were p_0 and P_0, respectively.

8. Consider the populations that have the genotypes shown in the following table.

Population	AA	Aa	aa
1	1.0	0.0	0.0
2	0.0	1.0	0.0
3	0.0	0.0	1.0
4	0.50	0.25	0.25
5	0.25	0.25	0.50
6	0.25	0.50	0.25
7	0.33	0.33	0.33
8	0.04	0.32	0.64
9	0.64	0.32	0.04
10	0.986049	0.013902	0.000049

a. Which of the populations are in Hardy-Weinberg equilibrium?

b. What are p and q in each population?

c. In population 10, it is discovered that the $A \rightarrow a$ mutation rate is 5×10^{-6} and that reverse mutation is negligible. What must be the fitness of the aa phenotype?

d. In population 6, the a allele is deleterious; furthermore, the A allele is incompletely dominant, so that AA is perfectly fit, Aa has a fitness of 0.8, and aa has a fitness of 0.6. If there is no mutation, what will p and q be in the next generation?

9. Colorblindness results from a sex-linked recessive allele. One in every ten males is colorblind.

a. What proportion of all women are colorblind?

b. By what factor is colorblindness more common in men (or, how many colorblind men are there for each colorblind woman)?

c. In what proportion of marriages would colorblindness affect half the children of each sex?

d. In what proportion of marriages would all children be normal?

e. In a population that is not in equilibrium, the frequency of the allele for colorblindness is 0.2 in women and 0.6 in men. After one generation of random mating, what proportion of the female progeny will be colorblind? What proportion of the male progeny?

f. What will the allelic frequencies be in the male and in the female progeny in (e)?

(Problem 9 courtesy of Clayton Person.)

10. In a wild population of beetles of species X, you notice that there is a 3:1 ratio of shiny to dull wing covers. Does this prove that *shiny* is dominant? (Assume that the two states are caused by the alleles of one gene.) If not, what does it prove? How would you elucidate the situation?

11. It seems clear that most new mutations are deleterious. Why?

12. Most mutations are recessive to the wild type. Of those rare mutations that are dominant in *Drosophila*, for example, the majority turn out either to be chromosomal aberrations or to be inseparable from chromosomal aberrations. Can you explain why the wild type is usually dominant?

13. Ten percent of the males of a large and randomly mating population are colorblind. A representative group of 1000 from this population migrates to a South Pacific island, where there are already 1000 inhabitants and where 30 percent of the males are colorblind. Assuming that Hardy-Weinberg equilibrium applies throughout (in the two original populations before emigration and in the mixed population immediately following immigration), what fraction of males and females can be expected to be colorblind in the generation immediately following the arrival of the immigrants?

14. Using pedigree diagrams, find the probability of homozygosity by descent of the offspring of (a) parent-offspring matings; (b) first-cousin matings; (c) aunt-nephew or uncle-niece matings.

15. In a survey of Indian tribes in Arizona and New Mexico, it was found that albinos were completely absent or very rare in most groups (there is one albino per 20,000 North American Caucasians). However, in three Indian populations, albino frequencies are exceptionally high: one per 277 Indians in Arizona; one per 140 Jemez Indians in New Mexico; and one per 247 Zuni Indians in New Mexico. All three of these populations are culturally but not linguistically related. What possible factors might explain the high incidence of albinos in these three tribes?

16. In an animal population, 20 percent of the individuals are AA, 60 percent are Aa, and 20 percent are aa. What are the allelic frequencies? In this population, mating is always with *like phenotype* but is random within phenotype. What genotypic and allelic frequencies will prevail in the next generation? Such *assortative mating* is common in animal populations. Another type of assortative mating is that which occurs only between *unlike* phenotypes: answer the above question with this restriction imposed. What will the end result be after many generations of mating of both types?

17. In *Drosophila*, a stock isolated from nature has an average of 36 abdominal bristles. By selectively breeding only those flies with more bristles, the mean is raised to 56 in 20 generations. What is the source of this genetic flexibility? The 56-bristle stock is very infertile, so selection is relaxed for several generations and the bristle number drops to about 45. Why doesn't it drop back to 36? When selection is reapplied, 56 bristles are soon attained, but this time the stock is *not* sterile. How can this situation arise?

18. The fitnesses of three genotypes are $W_{AA} = 0.9$, $W_{Aa} = 1.0$, and $W_{aa} = 0.7$.

a. If the population starts at the allelic frequency $p = 0.5$, what is the value of p in the next generation?

b. What is the predicted equilibrium allelic frequency?

19. AA and Aa individuals are equally fertile. If 0.1 percent of the population is aa, what selection pressure exists against aa if the $A \rightarrow a$ mutation rate is 10^{-5}?

20. Gene B is a deleterious autosomal dominant. The frequency of affected individuals is 4.0×10^{-6}. The reproductive capacity of these individuals is about 30 percent that of normal individuals. Estimate μ, the rate at which b mutates to its deleterious allele B.

21. Of 31 children born of father-daughter matings, six died in infancy, 12 were very abnormal and died in childhood, and 13 were normal. From this information, calculate roughly how many recessive lethal genes we have, on average, in our human genomes. For example, if the answer is 1, then a daughter would stand a 50 percent chance of carrying the gene, and the probability of the union's producing a lethal combination would be $\frac{1}{2} \times \frac{1}{4} = \frac{1}{8}$. (So obviously 1 is not the answer.) Consider also the possibility of undetected fatalities in utero in such matings. How would they affect your result?

22. If we define the **total selection cost** to a population of deleterious recessive genes as the loss of fitness per individual affected (s) multiplied by the frequency of affected individuals (q^2), then

$$\text{Genetic cost} = sq^2$$

a. Suppose that a population is at equilibrium between mutation and selection for a deleterious recessive gene, where $s = 0.5$ and $\mu = 10^{-5}$. What is the equilibrium frequency of the gene? What is the genetic cost?

b. Suppose that we start irradiating individual members of the population, so that the mutation rate doubles. What is the new equilibrium frequency of the gene? What is the new genetic cost?

c. If we do not change the mutation rate but we lower the selection intensity to $s = 0.3$ instead, what happens to the equilibrium frequency and the genetic cost?

27

Quantitative Genetics

Quantitative variation in flower color, flower diameter, and number of flower parts in the composite flowers of *Gaillardia pulchella* (J. Heywood, *Journal of Heredity*, May/June 1986)

KEY CONCEPTS

▶ In natural populations, variation in most characters takes the form of a continuous phenotypic range rather than discrete phenotypic classes. In other words, the variation is quantitative, not qualitative.

▶ Mendelian genetic analysis is extremely difficult to apply to such continuous phenotypic distributions, so statistical techniques are employed instead.

▶ A major task of quantitative genetics is to determine the ways in which genes interact with the environment to contribute to the formation of a given quantitative trait distribution.

▶ The genetic variation underlying a continuous character distribution can be the result of segregation at a single genetic locus or at numerous interacting loci that produce cumulative effects on the phenotype.

▶ The estimated ratio of genetic to environmental variation is *not* a measure of the relative contribution of genes and environment to phenotype.

▶ Estimates of genetic and environmental variance are specific to the single population and the particular set of environments in which the estimates are made.

U ltimately, the goal of genetics is the analysis of the genotype of organisms. But the genotype can be identified—and therefore studied—only through its phenotypic effect. We recognize two genotypes as different from each other because the phenotypes of their carriers are different. Basic genetic experiments, then, depend on the existence of a simple relationship between genotype and phenotype. That is why studies of DNA sequences are so important, because we can read off the genotype directly from this most basic of all phenotypes. In general, we hope to find a uniquely distinguishable phenotype for each genotype and only a simple genotype for each phenotype. At worst, when one allele is completely dominant, it may be necessary to perform a simple genetic cross to distinguish the heterozygote from the homozygote. Where possible, geneticists avoid studying genes that have only partial penetrance and incomplete expressivity (page 108) because of the difficulty of making genetic inferences from such traits. Imagine how difficult (if not impossible) it would have been for Benzer to study the fine structure of the gene in phage, if the only effect of the rII mutants had been to be able to grow on *E. coli* K almost as well as wild-type phage could (see page 361). For the most part, then, the study of genetics presented in the previous chapters has been the study of allelic substitutions that cause *qualitative* differences in phenotype.

However, the actual variation between organisms is usually quantitative, not qualitative. Wheat plants in a cultivated field or wild asters at the side of the road are not neatly sorted into categories of "tall" and "short," any more than humans are neatly sorted into categories of "black" and "white." Height, weight, shape, color, metabolic activity, reproductive rate, and behavior are characteristics that vary more or less continuously over a range (Figure 27-1). Even when the character is intrinsically countable (such as eye facet or bristle number in *Drosophila*), the number of distinguishable classes may be so large that the variation is nearly continuous. If we consider extreme individuals—say, a corn plant 8 feet tall and another one 3 feet tall—a cross between them will not reproduce a Mendelian result. Such a corn cross will produce plants about six feet tall, with some clear variation among siblings. The F_2 from selfing the F_1 will not fall into two or three discrete height classes in ratios of 3:1 or 1:2:1. Instead, the F_2 will be continuously distributed in height from one parental extreme to the other. This behavior of crosses is not an exception, but is the rule, for most characters in most species. Mendel obtained his simple results because he worked with horticultural varieties of the garden pea that differed from one another by single allelic differences that had drastic phenotypic effects. Had Mendel conducted his experiments on the natural variation of the weeds in his garden, instead of abnormal pea varieties, he would never have discovered Mendel's laws. In general, size, shape, color, physiological activity, and behavior do not assort in a simple way in crosses.

The fact that most phenotypic characters vary continuously does not mean that their variation is the result of some genetic mechanisms different from the Mendelian genes we have been dealing with. The continuity of phenotype is a result of two phenomena. First, each genotype does not have a single phenotypic expression but a norm of reaction (see Chapter 1) that covers a wide phenotypic range. As a result, the phenotypic differences between genotypic classes become blurred, and we are not able to assign a particular phenotype unambiguously to a particular genotype. Second, many segregating loci may have alleles that

Figure 27-1 Quantitative inheritance of bract color in Indian paintbrush (*Castilleja hispida*). The left photograph shows the extremes of the color range, and the right one shows examples from throughout the phenotypic range.

make a difference to the phenotype being observed. Suppose, for example, that five equally important loci affect the number of flowers that will develop in an annual plant and that each locus has two alleles (call them + and −). For simplicity, also suppose that there is no dominance and that a + allele adds one flower whereas a − allele adds nothing. Thus, there are $3^5 = 243$ different possible genotypes [three possible genotypes (++, +−, and −−) at each of five loci], ranging from

$$
\begin{array}{ccc}
+\;+\;+\;+\;+ & +\;+\;+\;+\;+ & -\;-\;-\;-\;- \\
+\;+\;+\;+\;+ & -\;-\;-\;-\;- & -\;-\;-\;-\;- \\
\text{through} & \text{to} &
\end{array}
$$

but there are only 11 phenotypic classes (10, 9, 8, . . . , 0) because many of the genotypes will have the same numbers of + and − alleles. For example, although there is only one genotype with 10 + alleles and therefore an average phenotypic value of 10, there are 51 different genotypes with 5 + alleles and 5 − alleles, for example,

$$
\begin{array}{ccc}
+\;+\;+\;+\;- & & +\;+\;-\;+\;- \\
+\;-\;-\;-\;- & \text{and} & +\;+\;-\;-\;- \\
\end{array}
$$

Thus, many different genotypes may have the same average phenotype. At the same time, because of environmental variation, two individuals of the same genotype may not have the same phenotype. This lack of a one-to-one correspondence between genotype and phenotype obscures the underlying Mendelian mechanism. If we cannot study the behavior of the Mendelian factors controlling such traits directly, then what can we learn about their genetics?

Using current experimental techniques, geneticists can answer the following questions about the genetics of a continuously varying character in a population (say, height in a human population). These questions constitute the study of **quantitative genetics**—the study of the genetics of continuously varying characters:

1. Is the observed variation in the character influenced *at all* by genetic variation? Are there alleles segregating in the population that produce some differential effect on the character, or is all the variation simply the result of environmental variation and developmental noise (pages 14–17)?
2. If there is genetic variation, what are the norms of reaction of the various genotypes?
3. How important is genetic variation as a source of total phenotypic variation? Are the norms of reaction and the environments such that nearly all the variation is a consequence of environmental difference and developmental instabilities, or does genetic variation predominate?

4. Do many loci (or only a few) vary with respect to the character? How are they distributed over the genome?
5. How do the different loci interact with one another to influence the character? Is there dominance, and is there any epistasis (interaction between genes at different loci)?
6. Is there any nonnuclear inheritance (for example, any maternal effect)?

In the end, the purpose of asking these questions is to be able to predict what kinds of offspring will be produced by crosses of different phenotypes.

The precision with which these questions can be framed and answered varies greatly. In experimental organisms on the one hand, it is relatively simple to determine whether there is any genetic influence at all, but extremely laborious experiments are required to localize the genes (even approximately). In humans, on the other hand, it is extremely difficult to answer even the question of the presence of genetic influence for most traits because it is almost impossible to separate environmental from genetic effects in an organism that cannot be manipulated experimentally. As a consequence, we know a relatively large amount about the genetics of bristle number in *Drosophila* but virtually nothing about the genetics of complex human traits; a few (such as skin color) clearly are influenced by genes, whereas others (such as the specific language spoken) clearly are not. It is the purpose of this chapter to develop the basic statistical and genetic concepts needed to answer these questions and to provide some examples of the applications of these concepts to particular characters in particular species.

Some Basic Statistical Notions

In order to consider the answers to these questions about the most common kinds of genetic variation, we must first examine a number of statistical tools that are essential in the study of quantitative genetics.

Distributions

The outcome of a cross for a Mendelian character can be described in terms of the proportions of the offspring that fall into several distinct phenotypic classes or often simply in terms of the presence or absence of a class. For example, a cross between a red-flowered plant and a white-flowered plant might be expected to yield all red-flowered plants or, if it were a backcross, $\frac{1}{2}$ red-flowered plants and $\frac{1}{2}$ white-flowered plants. However, we require a different mode of description for quantitative characters. The basic concept is that of the **statistical distribution.** If the heights of a large number of male undergraduates are measured to the nearest 5 centimeters (cm), they will vary (say, between 145 and 195 cm), but many more individuals will fall into the middle categories (say, between 170 and 180 cm) than at the extremes.

Representing each measurement class as a bar, with its height proportional to the number of individuals in each class, we can graph the result as shown in Figure 27-2a. Such a graph of numbers of individuals observed against measurement class is a **frequency histogram.** Now suppose that five times as many individuals are measured, each to the nearest centimeter. The classes in Figure 27-2a are now subdivided to produce a histogram like the one shown in Figure 27-2b. If we continue this process, refining the measurement but proportionately increasing the number of individuals measured, then the histogram eventually takes on the continuous appearance of Figure 27-2c, which is the **distribution function** of heights in the population.

Of course, this continuous curve is an idealization, because no measurement can be taken with infinite accuracy or on an unlimited number of individuals. Moreover, the measured variate itself may be intrinsically discontinuous because it is the count of some number of discrete objects

such as eye facets or bristles. It is sometimes convenient, however, to develop concepts using this slightly idealized picture as a shorthand for the more cumbersome observed frequency histogram (Figure 27-2a).

The Mode. Most distributions of phenotypes look roughly like those in Figure 27-2: a single-most-frequent class, the **mode,** is located near the middle of the distribution, with frequencies decreasing on either side. There are exceptions to this pattern, however. Figure 27-3a shows the very asymmetric distribution of seed weights in the plant *Crinum longifolium.* Figure 27-3b shows a **bimodal** (two-mode) distribution of larval survival probabilities for different second-chromosome homozygotes in *Drosophila willistoni.*

A bimodal distribution may indicate that the population being studied could better be considered as a mixture of two populations, each with its own mode. In Figure 27-3b, the left-hand mode probably represents a subpopulation of severe single-locus mutations that are extremely deleterious when homozygous but whose effects are not felt in the heterozygous state in which they usually exist in natural populations. The right-hand mode is part of the distribution of "normal" viability modifiers of small effect.

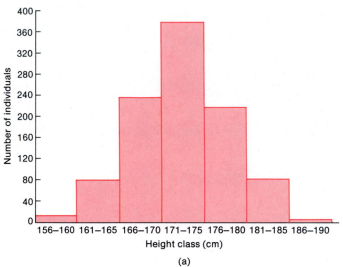

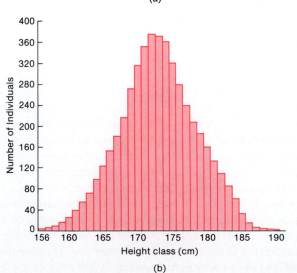

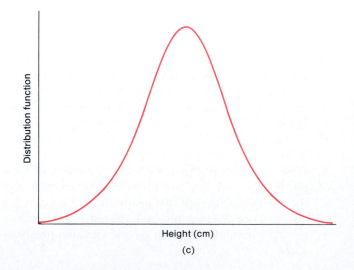

Figure 27-2 Frequency distributions for height of males. (a) A histogram with 5-cm class intervals. (b) A histogram with 1-cm class intervals. (c) The limiting continuous distribution.

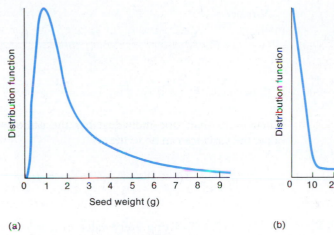

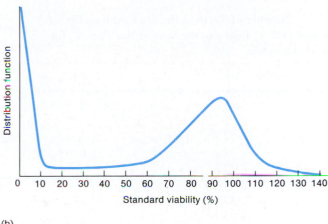

(a)

(b)

Figure 27-3 Asymmetric distribution functions. (a) Asymmetric distribution of seed weight in *Crinum longifolium*. (b) Bimodal distribution of survival of *Drosophila willistoni* expressed as a percentage of standard survival. (Adapted from S. Wright, *Evolution and the Genetics of Populations*, Vol. 1. Copyright © 1968 by University of Chicago Press, 1968.)

The Mean. Complete information about the distribution of a phenotype in a population can be given only by specifying the frequency of each measured class, but a great deal of information can be summarized in two statistics. First, we need some measure of the location of the distribution along the axis of measurement (for example, do the individual measurements tend to cluster around 100 cm or 200 cm)? One possibility is to give the measurement of the most common class, the mode. In Figure 27-2b, the mode is 172 cm (for females, the mode would be about 6 cm less). A more common measure of location is the arithmetic average, or the **mean.** The mean of the measurement ($\bar{x}$) is simply the sum of all the measurements (x_i) divided by the number of measurements in the sample (N):

$$\text{Mean} = \bar{x} = \frac{x_1 + x_2 + x_3 + \cdots + x_N}{N} = \frac{1}{N}\Sigma x_i$$

where Σ represents summation and x_i is the ith measurement.

In a typical large sample, the same measured value will appear more than once, because several individuals will have the same value within the accuracy of the measuring instrument. For example, many individuals will be 170 cm tall. In such a case, $\bar{x}$ can be rewritten as the sum of all measurement values, each weighted by how frequently it occurs in the population. Suppose that there are k distinct measurement classes into which all the measurements fall. From a total of N individuals measured, suppose that n_1 fall in the class with value x_1, that n_2 fall in the class with value x_2, and so on, so that $\Sigma n_i = N$. If we let f_i be the **relative frequency** of the ith measurement class, so that

$$f_i = \frac{n_i}{N}$$

then we can rewrite the mean as

$$\bar{x} = f_1 x_1 + f_2 x_2 + \cdots + f_k x_k = \Sigma f_i x_i$$

where x_i equals the value of the ith measurement class.

Let us apply these calculation methods to the data of Table 27-1, the numbers of toothlike bristles in the sex combs on the right (x) and left (y) front legs and the sum of both legs ($T = x + y$) of 20 *Drosophila*. Looking for the moment only at the sum of the two legs T, the mean number of teeth $\bar{T}$ is

$$\bar{T} = \frac{11 + 12 + 12 + 12 + 13 + \cdots + 15 + 16 + 16}{20}$$

$$= \frac{274}{20}$$

$$= 13.7$$

Alternatively, using the relative frequencies of the different measurement values,

$$\bar{T} = 0.05(11) + 0.15(12) + 0.20(13) + 0.35(14)$$
$$+ 0.15(15) + 0.10(16)$$
$$= 13.7$$

The Variance. A second characteristic of a distribution is the width of its spread around the central class. Two distributions with the same mean might differ very much in how closely the measurements are concentrated around the mean. Figure 27-4 shows two distributions, A and B, with the same mean, but different amounts of variation. The most common measure of variation around the center is the **variance,** which is defined as the average squared deviation of the observations from the mean, or

Table 27-1   Number of Teeth in the Sex Comb on the Right (x) and Left (y) Legs and the Sum of the Two (T) for 20 Males of *Drosophila*

x	y	T	n_i	$f_i = n_i/N$
6	5	11	1	$\frac{1}{20} = 0.05$
6	6	12		
5	7	12	3	$\frac{3}{20} = 0.15$
6	6	12		
7	6	13		
5	8	13		
6	7	13	4	$\frac{4}{20} = 0.20$
7	6	13		
8	6	14		
6	8	14		
7	7	14		
7	7	14	7	$\frac{7}{20} = 0.35$
7	7	14		
6	8	14		
8	6	14		
8	7	15		
7	8	15	3	$\frac{3}{20} = 0.15$
6	9	15		
8	8	16	2	$\frac{2}{20} = 0.10$
7	9	16		

$N = 20$
$\bar{x} = 6.25$ $s_x^2 = 0.8275$ $s_x = 0.9096$
$\bar{y} = 7.05$ $s_y^2 = 1.1475$ $s_y = 1.0722$
$\bar{T} = 13.70$ $s_T^2 = 1.71$ $s_T = 1.308$
$\text{cov } xy = -0.1325$
$r_{xy} = -0.1360$

$\text{Variance} = s^2$

$$= \frac{(x_1 - \bar{x})^2 + (x_2 - \bar{x})^2 + \cdots + (x_N - \bar{x})^2}{N}$$

$$= \frac{1}{N} \Sigma (x_i - \bar{x})^2$$

When more than one individual has the same measured value, the variance can be written

$$s^2 = f_1(x_1 - \bar{x})^2 + f_2(x_2 - \bar{x})^2 + \cdots + f_k(x_k - \bar{x})^2$$
$$= \Sigma f_i (x_i - \bar{x})^2$$

To avoid subtracting every value of x separately from the mean, we can use an alternative computing formula that is algebraically identical with the preceding equation:

$$s^2 = \frac{1}{N} \Sigma x_i^2 - \bar{x}^2$$

Because the variance is in squared units (square centimeters, for example), it is common to take the square root of the variance, which then has the same units as the measurement itself. This square-root measure of variation is called the **standard deviation** of the distribution:

$$\text{Standard deviation} = s = \sqrt{\text{variance}} = \sqrt{s^2}$$

The data in Table 24-1 can be used to exemplify these calculations:

$$s_T^2 = \frac{(11 - 13.7)^2 + (12 - 13.7)^2 + (12 - 13.7)^2}{20}$$
$$+ \frac{\cdots + (15 - 13.7)^2 + (16 - 13.7)^2}{20}$$
$$= \frac{34.20}{20} = 1.71$$

or, alternatively,

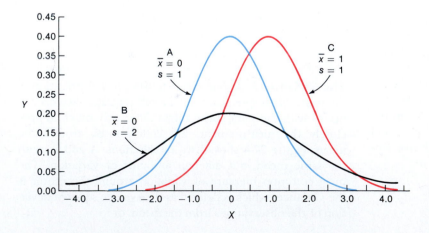

Figure 27-4 Three distribution functions, two of which have the same mean and two of which have the same standard deviation.

$$s_T^2 = 0.05(11 - 13.7)^2 + 0.15(12 - 13.7)^2$$
$$+ 0.20(13 - 13.7)^2 + 0.35(14 - 13.7)^2$$
$$+ 0.15(15 - 13.7)^2 + 0.10(16 - 13.7)^2$$
$$= 1.71$$

Finally, we can use the computing formula that avoids taking individual deviations:

$$s_T^2 = \frac{1}{N} \sum T_i^2 - \overline{T}^2 = \frac{3788}{20} - 187.69 = 1.71$$

and

$$s = \sqrt{1.71} = 1.308$$

Figure 27-4 shows two distributions having the same mean but different standard deviations (curves A and B) and two distributions having the same standard deviation but different means (curves A and C).

The mean and the variance of a distribution do not describe it completely, of course. They do not distinguish a symmetric distribution from an asymmetric one, for example. We can even construct symmetric distributions that have the same mean and variance but still have somewhat different shapes. Nevertheless, for the purposes of dealing with most quantitative genetic problems, the mean and variance suffice to characterize a distribution.

Covariance and Correlation

Another statistical notion that is of use in the study of quantitative genetics is the association, or **correlation,** between variables. As a result of complex paths of causation, many variables in nature vary together but in an imperfect or approximate way. Figure 27-5a provides an example, showing the lengths of two particular teeth in several individual specimens of a fossil mammal, *Phenacodus primaevis*. The

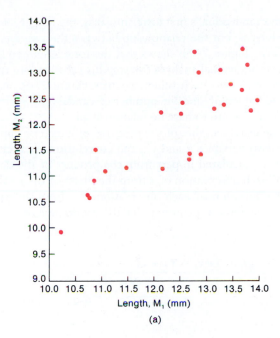

(a)

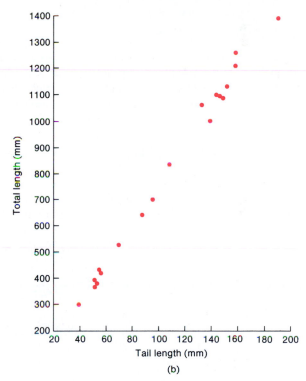

(b)

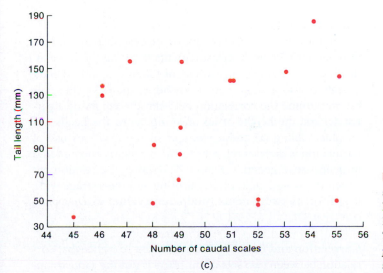

(c)

Figure 27-5 Scatter diagrams of relationships between pairs of variables. (a) Relationship between the lengths of the first and second lower molars (M_1 and M_2) in the extinct mammal *Phenacodus primaevis*. Each point gives the M_1 and M_2 measurements for one individual. (b) Tail length and body length of 18 individuals of the snake *Lampropeltis polyzona*. (c) Number of caudal scales and tail length related in the same 18 snakes in part b.

longer an individual's first lower molar is, the longer its second molar is, but the relationship between the two teeth is imprecise. Figure 27-5b shows that the total length and tail length in individual snakes (*Lampropeltis polyzona*) are quite closely related to each other, whereas Figure 27-5c shows that the tail length and the number of caudal (tail) scales in these snakes seem to have no relation at all.

The usual measure of the precision of a relationship between two variables x and y is the **correlation coefficient,** (r_{xy}). It is calculated in part from the product of the deviation of each observation of x from the mean of the x values and the deviation of each observation of y from the mean of the y values—a quantity called the **covariance** of x and y (cov xy):

$$\text{cov } xy = \frac{(x_1 - \bar{x})(y_1 - \bar{y}) + (x_2 - \bar{x})(y_2 - \bar{y}) + \cdots}{N}$$

$$+ \frac{(x_N - \bar{x})(y_N - \bar{y})}{N}$$

$$= \frac{1}{N} \Sigma (x_i - \bar{x})(y_i - \bar{y})$$

This formula for the covariance is rather awkward computationally because it requires subtracting every value of x and y from the respective means $\bar{x}$ and $\bar{y}$. A formula that is exactly algebraically equivalent but that makes computation easier is

$$\text{cov } xy = \frac{1}{N} \Sigma x_i y_i - \bar{xy}$$

Using these formulas, we can calculate the covariance between the right (x) and the left (y) leg counts in Table 27-1.

cov $xy =$

$$\frac{(6 - 6.65)(5 - 7.05) + (6 - 6.65)(6 - 7.05) + \cdots}{20}$$

$$+ \frac{(8 - 6.65)(8 - 7.05) + (7 - 6.65)(9 - 7.05)}{20}$$

cov $xy = -0.1325$

or, alternatively, from the computing formula

$$\text{cov } xy = \frac{1}{N} \Sigma xy - \bar{xy}$$

$$= \frac{(6)(5) + (6)(6) + \cdots + (8)(8) + (7)(9)}{20}$$

$$- (6.65)(7.05)$$

$$= -0.1325$$

$$\text{Correlation} = r_{xy} = \frac{\text{cov } xy}{s_x s_y}$$

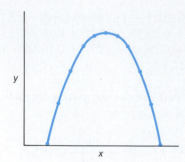

Figure 27-6 A parabola. Each value of y is perfectly predictable from the value of x, but there is no linear correlation.

In the formula for correlation, the products of the deviations are divided by the product of the standard deviations of x and y (s_x and s_y). This normalization by the standard deviations has the effect of making r_{xy} a dimensionless number that is independent of the units in which x and y are measured. So defined, r_{xy} will vary from -1, which signifies a perfectly linear negative relation between x and y, to $+1$, which indicates a perfectly linear positive relation between x and y. If $r_{xy} = 0$, there is no linear relation between the variables. It is important to notice, however, that sometimes when there is no *linear* relation between two variables but there is a regular *nonlinear* relationship between them, one variable may be perfectly predicted from the other. Consider, for example, the parabola shown in Figure 27-6. The values of y are perfectly predictable from the values of x; yet $r_{xy} = 0$, because on average over the whole range of x values, larger x values are not associated with either larger or smaller y values. The data in parts (a), (b), and (c) in Figure 27-5 have r_{xy} values of 0.82, 0.99, and 0.10, respectively. In the example of the sex comb teeth of Table 27-1, the correlation between left and right legs is

$$r_{xy} = \frac{\text{cov } xy}{\sqrt{s_x^2 s_y^2}} = \frac{-0.1325}{\sqrt{(0.8275)(1.1475)}} = -0.1360$$

a very small value.

In these examples, correlations are described as the relation between a pair of measurements taken on the same individual, but a single measurement taken on pairs of individuals is also a subject for correlation analysis. Thus, we can determine the correlation between the height of a parent (x) and the height of an offspring (y) or the heights of an older sibling (x) and a younger sibling (y). This use of correlation is directly relevant to the problems encountered in quantitative genetics. Figure 27-7 shows the relation between the wing length of an offspring and the average wing length of its two parents (**midparent value**) in *Drosophila* ($r = +0.83$).

Correlation and Equality. It is important to notice that correlation between two sets of numbers is not the same as numerical identity. For example, two sets of values can be per-

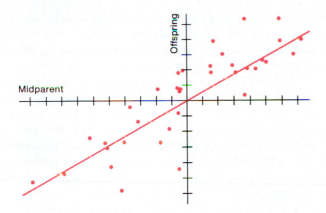

Figure 27-7 Relation between the wing lengths of individual *Drosophila* and the mean wing lengths of their two parents. (From D. Falconer, *Quantitative Genetics.* Copyright © 1981 by Longman Group Limited.)

fectly *correlated,* even though the values in one set are very much larger than the values in the other set. Consider the following pairs of values:

x	y
1	22
2	24
3	26

The variables x and y in the pairs are perfectly correlated ($r = +1.0$), although each value of y is about 20 units greater than the corresponding value of x. Two variables are perfectly correlated if, for a unit increase in one, there is a constant increase in the other (or a constant decrease if r is negative). The importance of the difference between correlation and identity arises when we consider the effect of environment on heritable characters. Parents and offspring can be perfectly correlated in some trait such as height, yet because of an environmental difference between generations, every child can be taller than the parents. This phenomenon appears in adoption studies, in which children may be correlated with their biological parents but, on the average, may be quite different from the parents as a result of a change in social situation.

Covariance and the Variance of a Sum. In Table 27-1 the variances of the left and right legs are 0.8275 and 1.1475, which adds up to 1.975, but the variance of the sum of the two legs T is only 1.71. That is, the variance of the whole is less than the sum of the variances of the parts. This discrepancy is a consequence of the negative correlation between left and right sides. Larger left sides are associated with smaller right sides and vice versa, so the sum of the two sides varies less than each side separately. If, on the other hand, there were a positive correlation between sides, then larger left and right sides would go together and the varia-

tion of the sum of the two sides would be larger than the sum of the two separate variances. In general, if $x + y = T$ then

$$s_T^2 = s_x^2 + s_y^2 + 2 \text{ cov } xy$$

For the data of Table 27-1

$$s_T^2 = 1.71 = 0.8275 + 1.1475 - 2(0.1325)$$

Regression

The measurement of correlation provides us with only an estimate of the *precision* of relationship between two variables. A related problem is predicting the value of one variable given the value of the other. If x increases by two units, by how much will y increase? If the two variables are linearly related, then that relationship can be expressed as

$$y = bx + a$$

where b is the slope of the line relating y to x and a is the y intercept of that line.

Figure 27-8 shows a scatter diagram of points for two variables, y and x, together with a straight line expressing the general linear trend of y with increasing x. This line, called the **regression line of y on x,** has been positioned so that the deviations of the points from the line are as small as possible. Specifically, if Δy is the distance of any point from the line in the y direction, then the line has been chosen so that

$$\Sigma (\Delta y)^2 = \text{a minimum}$$

Any other straight line passed through the points on the scatter diagram will have a larger total squared deviation of the points from it.

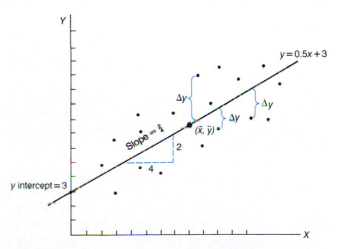

Figure 27-8 A scatter diagram showing the relation between two variables, x and y, with the regression line of y on x. This line, with a slope of $\frac{2}{4}$ minimizes the squares of the deviations (Δy).

Obviously, we cannot find this **least-squares line** by trial and error. It turns out, however, that if the slope b of the line is calculated by

$$b = \frac{\text{cov } xy}{s_x^2}$$

and if a is then calculated from

$$a = \bar{y} - b\bar{x}$$

so that the line passes through the point $\bar{x}, \bar{y}$, then these values of b and a will yield the least-squares prediction equation.

Note that the prediction equation cannot predict y exactly for a given x, because there is scatter around the line. The equation predicts the *average* y for a given x, if large samples are taken.

Samples and Populations

The preceding sections have described the distributions and some statistics of particular assemblages of individuals that have been collected in some experiments or sets of observations. For some purposes, however, we are not really interested in the particular 100 undergraduates or 27 snakes that have been measured. Instead, we are interested in a wider world of phenomena, of which those particular individuals are representative. Thus, we might want to know the average height *in general* of undergraduates or the average seed weight *in general* of plants of the species *Crinum longifolium*. That is, we are interested in the characteristics of a **universe,** of which our small collection of observations is only a **sample.** The characteristics of any particular sample are, of course, not identical with those of the universe but vary from sample to sample. Two samples of undergraduates drawn from the universe of all undergraduates will not have exactly the same mean ($\bar{x}$) or variance (s^2), nor will these values for a sample typically be exactly equal to the mean and variance of the universe. We distinguish between the statistics of a sample and the values in the universe by using Roman letters, such as $\bar{x}$, s^2, and r, for sample values and Greek letters μ (mean), σ^2 (variance), and ρ (correlation) for the values in the universe.

The sample mean $\bar{x}$ is an approximation of the true mean μ of the universe. It is a statistical *estimate* of that true mean. So, too, s^2, s, and r are estimates of σ^2, σ, and ρ in the universe from which respective samples have been taken.

If we are interested in a particular collection of individuals not for its own sake but as a way of obtaining information about a universe, then we want the sample statistics such as $\bar{x}$, s^2, and r to be good estimates of the true values of μ, σ^2, and ρ, and so on. There are many criteria of what a "good" estimate is, but the one that seems clearly desirable is that if we take a very large number of samples, then

the average value of the estimate over these samples should be the true value in the universe; that is, the estimate should be **unbiased.** If a very large number of samples are taken and $\bar{x}$ is calculated in each one, then the average of these $\bar{x}$ values will indeed be μ, so $\bar{x}$ is an unbiased estimate of μ. Unfortunately, s^2, as we have defined it, is not an unbiased estimate of σ^2. The calculation of s^2 is based on the squared deviations of each observation from the *sample* mean $\bar{x}$, and not from the true mean μ. But the sample mean is calculated from the sample observations, so the deviations of the observations from the sample mean are not free to vary as much as they do from the true mean. The result is that s^2 tends to be a little too small, so that the average of many s^2 values is less than σ^2. The amount of bias is precisely related to the size N of each sample, and it can be shown that $[N/(N-1)]s^2$ is an unbiased estimate of σ^2. Thus, whenever we are interested in the variance of a set of measurements—not as a characteristic of the particular collection but as an estimate of a universe that the sample represents—then the appropriate quantity to use, rather than s^2 itself, is $[N/(N-1)]s^2$. Note that this new quantity is equivalent to dividing the sum of squared deviations by $N-1$ instead of N in the first place, so that

$$\left(\frac{N}{N-1}\right)s^2 = \left(\frac{N}{N-1}\right)\frac{1}{N}\sum(x_i - \bar{x})^2$$

$$= \frac{1}{N-1}\sum(x_i - \bar{x})^2$$

All these considerations about bias also apply to the sample covariance. In the formula for the correlation coefficient (pages 823–824), however, the factor $N/(N-1)$ would appear in both the numerator and the denominator and therefore cancel out, so we can ignore it for the purposes of computation.

Genotypes and Phenotypic Distribution

Using the concepts of distribution, mean, and variance, we can understand the difference between quantitative and Mendelian genetic traits. Suppose that a population of plants contains three genotypes, each of which has some differential effect on growth rate. Further, assume that there is some environmental variation from plant to plant because of inhomogeneity in the soil in which the population is growing and that there is some developmental noise (see page 16). For each genotype, there will be a separate distribution of phenotypes with a mean and a standard deviation that depend on the genotype and the set of environments. Suppose that these distributions look like the three height distributions in Figure 27-9a. Finally, assume that the population consists of a mixture of the three genotypes but in the unequal proportions 1:2:3 ($aa:Aa:AA$). Then the

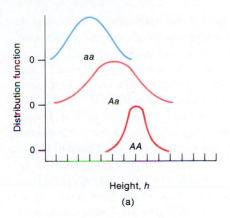

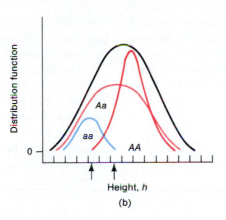

Figure 27-9 (a) Phenotypic distributions of three genotypes. (b) A population phenotypic distribution results from mixing individuals of the three genotypes in a proportion 1:2:3 ($aa:Aa:AA$).

phenotypic distribution of individuals in the population as a whole will look like the black line in Figure 27-9b, which is the result of summing the three underlying separate genotypic distributions, weighted by their frequencies in the population. This weighting by frequency is indicated in Figure 26-9b by the different heights of the component distributions that add up to the total distribution. The mean of this total distribution is the average of the three genotypic means, again weighted by the frequencies of the genotypes in the population. The variance of the total distribution is produced partly by the environmental variation within each genotype and partly by the slightly different means of the three genotypes.

Two features of the total distribution are noteworthy. First, there is only a single mode. Despite the existence of three separate genotypic distributions underlying it, the population distribution as a whole does not reveal the separate modes. Second, any individual whose height lies between the two arrows could have come from any one of the three genotypes, because they overlap so much. The result is that we cannot carry out any simple Mendelian analysis to determine the genotype of an individual organism. For example, suppose that the three genotypes are the two homozygotes and the heterozygote for a pair of alleles at a locus. Let aa be the short homozygote and AA be the tall one, with the heterozygote being of intermediate height. Because there is so much overlap of the phenotypic distributions, we cannot know to which genotype a given individual belongs. Conversely, if we cross a homozygote aa and a heterozygote Aa, the offspring will not fall into two discrete classes in a 1:1 ratio but will cover almost the entire range of phenotypes smoothly. Thus, we cannot know that the cross is in fact $aa \times Aa$ and not $aa \times AA$ or $Aa \times Aa$.

Suppose we grew the hypothetical plants in Figure 27-9 in an environment that exaggerated the differences between genotypes, for example, by doubling the growth rate of all genotypes. At the same time, we were very careful to provide all plants with exactly the same environment. Then, the phenotypic variance of each separate genotype would be reduced because all the plants are grown under identical conditions; at the same time the differences between geno-

types would be exaggerated by the more rapid growth. The result (Figure 27-10) would be a separation of the population as a whole into three nonoverlapping phenotypic distributions, each characteristic of one genotype. We could now

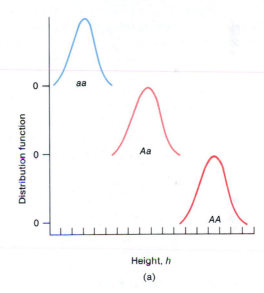

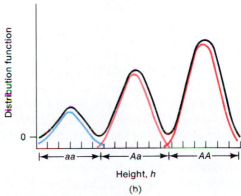

Figure 27-10 When the same genotypes as those in Figure 27-9 are grown in carefully controlled stress environments, the result is a smaller phenotypic variation in each genotype and a greater difference between genotypes. The heights of the individual distributions in part b are proportional to the frequencies of the genotypes in the mixture.

carry out a perfectly conventional Mendelian analysis of plant height. A "quantitative" character has been converted into a "qualitative" one. This conversion has been accomplished by finding a way to make the differences between the means of the genotypes large compared with the variation within genotypes.

Message A **quantitative** character is one for which the average phenotypic differences between genotypes are small compared with the variation between individuals within genotypes.

It is sometimes assumed that continuous variation in a character is necessarily caused by a large number of segregating genes, so that continuous variation is taken as evidence for control of the character by many genes. But, as we have just shown, this is not necessarily true. If the difference between genotypic means is small compared with the environmental variance, then even a simple one-gene–two-allele case can result in continuous phenotypic variation.

Of course, if the range of a character is limited and if there are many segregating loci influencing it, then we expect the character to show continuous variation, because each allelic substitution must account for only a small difference in the trait. This **multiple-factor hypothesis** (that large numbers of genes, each with a small effect, are segregating to produce quantitative variation) has long been the basic model of quantitative genetics, although there is no convincing evidence that such groups of genes really exist. A special name, **polygenes,** has been coined for these hypothetical factors of small-but-equal effect as opposed to the genes of simple Mendelian analysis.

It is important to remember, however, that the *number* of segregating loci that influence a trait is not what separates quantitative and qualitative characters. Even in the absence of large environmental variation, it takes only a few genetically varying loci to produce variation that is indistinguishable from the effect of many loci of small effect. As an example, we can consider one of the earliest experiments in quantitative genetics, that of Wilhelm Johannsen on pure lines. By **inbreeding** (mating close relatives), Johannsen produced 19 homozygous lines of bean plants from an originally genetically heterogeneous population. Each line had a characteristic average seed weight ranging from 0.64 gram per seed for the heaviest line down to 0.35 gram per seed for the lightest line. It is by no means clear that all these lines were genetically different (for example, five of the lines had seed weights of 0.450, 0.453, 0.454, 0.454, and 0.455 g), but let's take the most extreme position—that the lines *were* all different. Obviously, these observations would be incompatible with a simple one-locus–two-allele model of gene action. In that case, if the original population were segregating for the two alleles *A* and *a*, all inbred lines derived from that population would have to fall into one of two classes: *AA* or *aa*. If, in contrast, there were, say, 100 loci, each of small effect, segregating in the original popula-

tion, then a vast number of different inbred lines could be produced, each with a different combination of homozygotes at different loci.

However, we do not need such a large number of loci to obtain the result observed by Johannsen. If there were only five loci, each with three alleles, then $3^5 = 243$ different kinds of homozygotes could be produced from the inbreeding process. If we make 19 inbred lines at random, there is a good chance (about 50 percent) that each of the 19 lines will belong to a different one of the 243 classes. So Johannsen's experimental results can easily be explained by a relatively small number of genes. Thus, there is no real dividing line between polygenic traits and other traits. It is safe to say that no phenotypic trait above the level of the amino acid sequence in a polypeptide is influenced by only one gene. Moreover, traits influenced by many genes are not equally influenced by them all. Some genes will have major effects on the trait; some, minor effects.

Message The critical difference between Mendelian and quantitative traits is not the number of segregating loci but the size of phenotypic differences between genotypes compared with the individual variation within genotypic classes.

Norm of Reaction and Phenotypic Distribution

The phenotypic distribution of a trait, as we have seen, is a function of the average differences between genotypes and of the variation between genotypically identical individuals. But both of these are in turn functions of the sequence of environments in which the organisms develop and live. For a given genotype, each environment will result in a given phenotype (for the moment, ignoring developmental noise). Then a *distribution of environments* will be reflected biologically as a *distribution of phenotypes*. The way in which the environmental distribution is transformed into the phenotypic distribution is determined by the **norm of reaction,** as shown in Figure 27-11. The horizontal axis is environment (say, temperature) and the vertical axis is phenotype (say, plant height). The norm of reaction curve for the genotype shows how each particular temperature results in a particular plant height. So, the dashed lines from the 18°C point on the temperature axis is reflected off the norm of reaction curve to a corresponding plant height on the vertical phenotype axis, and so on for each temperature. If a large number of individuals develop at, say, 20°C, then a large number of individuals will have the phenotype that corresponds to 20°C as shown by the dashed line, and if only small numbers develop at 18°C, few plants will have the corresponding plant height. Then the frequency distribution of developmental environments will be reflected as a frequency distribution of phenotypes as determined by the shape of the norm of reaction curve. It is as if an observer, standing at

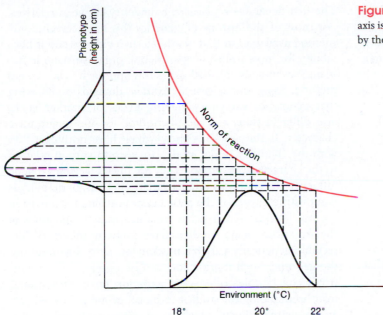

Figure 27-11 The distribution of environments on the horizontal axis is converted to the distribution of phenotypes on the vertical axis by the norm of reaction of a genotype.

the vertical phenotype axis, were seeing the environmental distribution, not directly, but reflected in the curved mirror of the norm of reaction. The shape of the curvature will determine how the environmental distribution is distorted on the phenotype axis. So, the norm of reaction in Figure 27-11 falls very rapidly at lower temperatures (the phenotype changes rapidly with small changes in temperature) but flattens out at higher temperatures, so that the plant height is much less sensitive to temperature differences at the higher temperatures. The result is that the symmetric environmental distribution is converted into an asymetric

phenotype distribution with a long tail at the larger plant heights, corresponding to the lower temperatures.

By means of the same analysis, Figure 27-12 shows how a population consisting of two genotypes with different norms of reaction has a phenotypic distribution that depends on the distribution of environments. If the environments are distributed as shown by the black distribution curve, then the resulting population of plants will have a unimodal distribution, because the difference between genotypes is very small in this range of environments compared with the sensitivity of the norms of reaction to small

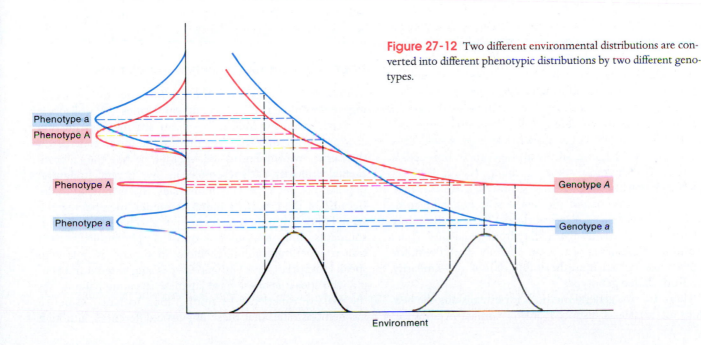

Figure 27-12 Two different environmental distributions are converted into different phenotypic distributions by two different genotypes.

changes in temperature. If the distribution of environments is shifted to the right, however, as shown by the gray distribution curve, a bimodal distribution of phenotypes results, because the norms of reaction are nearly flat in this environmental range but very different from each other.

> **Message** A distribution of environments is reflected biologically as a distribution of phenotypes. The transformation of environmental distribution into phenotypic distribution is determined by the norm of reaction.

The Heritability of a Trait

The most basic question to be asked about a quantitative trait is whether or not the observed variation in the character is influenced by genes at all. It is important to note that this is not the same as asking whether or not genes play any role in the character's development. Gene-mediated developmental processes lie at the base of every character, but *variation* from individual to individual is not necessarily the result of *genetic variation*. Thus, the possibility of speaking any language at all depends critically on the structures of the central nervous system as well as of the vocal cords, tongue, mouth, and ears, which depend in turn on the nature of the human genome. There is no environment in which cows will speak. But although the particular language that is spoken by humans varies from nation to nation, that variation is totally nongenetic.

> **Message** The question of whether or not a trait is heritable is a question about the role that differences in genes play in the phenotypic differences between individuals or groups.

Familiality and Heritability

In principle, it is easy to determine whether any genetic variation influences the phenotypic variation among organisms for a particular trait. If genes are involved, then (on average) biological relatives should resemble one another more than unrelated individuals do. This resemblance would be reflected as a positive correlation between parents and offspring or between siblings (offspring of the same parents). Parents who are larger than the average would have offspring who are larger than the average; the more seeds a plant produces, the more seeds its siblings would produce. Such correlations between relatives, however, are evidence for genetic variation only if the relatives do not share common environments *more than nonrelatives do*. It is absolutely fundamental to distinguish *familiality* from *heritability*. Traits are **familial** if members of the same family share them, for whatever reason. Traits are **heritable** only if the similarity arises from shared genotypes.

There are two general methods for establishing the heritability of a trait as distinct from its familial occurrence.

The first depends on *phenotypic similarity* between relatives. For most of the history of genetics this has been the only method available, so that nearly all the evidence about heritability for most traits in experimental organisms and in humans has been established using this approach. The second method, using *marker-gene segregation*, depends on showing that genotypes carrying different alleles of marker genes also differ in their average phenotype for the quantitative character. If the marker genes (which have nothing to do with the character under study) are seen to vary in relation to the character, presumably they are linked to genes that *do* influence the character and its variation. Thus, heritability is demonstrated, even if the actual genes causing the variation are not known. This method requires that the genome of the organism being studied have large numbers of detectable genetically variable marker loci spread throughout the genome. Such marker loci can be observed from electrophoretic studies of protein variation or, in vertebrates, from immunological studies of blood group genes. For example, within flocks, chickens of different blood groups show some difference in egg weight.

Since the introduction of molecular methods for the study of DNA sequence variation, very large numbers of variable nucleotide positions have been discovered in a great variety of organisms. This molecular variation includes both single nucleotide replacements and insertions and deletions of longer nucleotide sequences. These variations are usually detected by the gain or loss of sites of cleavage of restriction enzymes or by length variation of DNA sequences between two fixed restriction sites, both of which are a form of restriction fragment length polymorphisms (RFLPs). In tomatoes, for example, strains carrying different RFLP variants differ in fruit characteristics.

However, because so much of what is known or claimed about heritability still depends on phenotypic similarity between relatives, especially in human genetics, we will begin the examination of the problem of heritability by analyzing phenotypic similarity.

Phenotypic Similarity between Relatives

In experimental organisms, there is no problem in separating environmental from genetic similarities. The offspring of a cow producing milk at a high rate and the offspring of a cow producing milk at a low rate can be raised together in the same environment to see whether, despite the environmental similarity, each resembles its own parent. In natural populations, and especially in humans, this is difficult to do. Because of the nature of human societies, members of the same family not only share genes but also have similar environments. Thus, the observation of simple familiality of a trait is genetically uninterpretable. In general, people who speak Hungarian have Hungarian-speaking parents and people who speak Japanese have Japanese-speaking parents. Yet the massive experience of immigration to North America has demonstrated that these linguistic differences, although familial, are nongenetic. The highest correlations between parents and offspring for any social traits in the United States are those for political party and religious sect, but they are not heritable. The distinction between familiality and heredity is not always so obvious. The Public Health Commission, which originally studied the vitamin-deficiency disease pellagra in the southern United States in 1910 came to the conclusion that it was genetic because it ran in families.

To determine whether a trait is heritable in human populations, we must use adoption studies to avoid the usual environmental similarity between biological relatives. The ideal experimental subjects are identical twins reared apart, because they are genetically identical but environmentally different. Such adoption studies must be so contrived that there is no correlation between the social environment of the adopting family and that of the biological family. These requirements are exceedingly difficult to meet, so that in practice we know very little about whether human quantitative traits that are familial are also heritable. Skin color is clearly heritable, as is adult height—but even for these traits we must be very careful. We know that skin color is affected by genes from studies of cross racial adoptions and observations that the offspring of black African slaves were black even when they were born and reared in Canada. But are the differences in height between Japanese and Europeans affected by genes? The children of Japanese immigrants who are born and reared in North America are taller than their parents but shorter than the North American average, so we might conclude that there is some influence of genetic difference. However, second-generation Japanese Americans are even taller than their American-born parents. It appears that some environmental-cultural influence, or perhaps a maternal effect, is still felt in the first generation of births in North America. We cannot yet say whether genetic differences in height distinguish North Americans of, say, Japanese and Swedish ancestry.

Personality traits, temperament, and cognitive performance (including IQ scores), and a whole variety of behaviors like alcoholism and mental disorders like schizophrenia, have been the subject of heritability studies in human populations. Many show familiality. There is indeed a positive correlation between the IQ scores of parents and the scores of their children (the correlation is about 0.5 in white American families), but the correlation does not distinguish familiality from heritability. To make that distinction requires that the environmental correlation between parents and children be broken, so adoption studies are common. Because it is difficult to randomize the environments, even in cases of adoption, evidence of heritability for human personality and behavior traits remains equivocal despite the very large number of studies that exist. Prejudices about the causes of human differences are widespread and deep, and as a result, the canons of evidence adhered to in studies of the heritability of IQ, for example, have been much more lax than in studies of milk yield in cows.

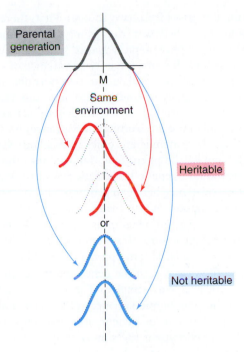

Figure 27-13 Standard method for testing heritability in experimental organisms. Crosses are performed within two populations of individuals selected from the extremes of the phenotypic distribution in the parental generation. If the phenotypic distributions of the two groups of offspring are significantly different from each other, then the trait is heritable. If the offspring distributions both resemble the distribution for the parental generation, then the trait is not heritable.

Figure 27-13 summarizes the usual method for testing heritability in experimental organisms. Individuals from both extremes of the distribution are mated with their own kind, and the offspring are raised in a common controlled environment. If there is an average difference between the two offspring groups, the trait is heritable. Most morphological traits in *Drosophila*, for example, turn out to be heritable—but not all of them. If flies with right wings that are slightly longer than their left wings are mated together, their offspring have no greater tendency to be "right-winged" than do the offspring of "left-winged" flies. As we shall see later (page 844) this method can also be used to obtain quantitative information about heritability.

Message In experimental organisms, environmental similarity often can be readily distinguished from genetic similarity (or heritability). In humans, however, it is very difficult to determine whether a particular trait is heritable.

Quantifying Heritability

If a trait is shown to have some heritability in a population, then it is possible to quantify the degree of heritability. In Figure 27-9, we see that the variation between phenotypes in a population arises from two sources. First, there are average differences between the genotypes; second, each genotype exhibits phenotypic variance because of environmental variation. The total phenotypic variance of the population (s_p^2) can then be broken into two portions: the variance between genotypic means (s_g^2) and the remaining variance (s_e^2). The former is called the **genetic variance,** and the latter is called the **environmental variance;** however, as we shall see, these names are quite misleading. Moreover, the breakdown of the phenotypic variance into the sum of environmental and genetic variance leaves out the possibility of some covariance between genotype and environment. For example, suppose it were true (we do not know) that there are genes that influence musical ability. Parents with such genes might themselves be musicians, who would create a more musical environment for their children, who would then have both the genes and the environment promoting musical performance. The result would be an increase in the phenotypic variances of musical ability and an erroneous estimate of genetic and environmental variances. If the phenotype is the sum of a genetic and an environmental effect, $P = G + E$, then as we showed on page 825,

$$s_p^2 = s_g^2 + s_e^2 + 2 \operatorname{cov} ge$$

If genotypes are not distributed randomly across environments, there will be some covariance between genotype and environmental values and the covariance will be hidden in the genetic and environmental variances.

The degree of heritability can be defined as the portion of the total variance that is due to genetic variance:

$$H^2 = \frac{s_g^2}{s_p^2} = \frac{s_g^2}{s_g^2 + s_e^2}$$

H^2, so defined, is called the **broad heritability** of the character.

It must be stressed that this measure of "genetic influence" tells us what portion of the population's *variation* in phenotype can be assigned to *variation* in genotype. It does not tell us what portions of an *individual's* phenotype can be ascribed to its heredity and to its environment. This latter distinction is not a reasonable one. An individual's phenotype is a consequence of the interaction between its genes and its sequence of environments. It clearly would be silly to say that you owe 60 inches of your height to genes and 10 inches to environment. All measures of the "importance" of genes are framed in terms of the proportion of variance ascribable to their variation. This approach is a special application of the more general technique of the **analysis of variance** for apportioning relative weight to contributing causes. The method was, in fact, invented originally to deal with experiments in which different environmental and genetic factors were influencing the growth of plants. (For a sophisticated but accessible treatment

of the analysis of variance written for biologists, see R. Sokal and J. Rohlf, *Biometry,* 2d ed. W. H. Freeman and Company, 1980.)

Methods of Estimating H^2

Genetic variance and heritability can be estimated in several ways. Most directly, we can obtain an estimate of s_e^2 by making a number of homozygous lines from the population, crossing them in pairs to reconstitute individual heterozygotes, and measuring the phenotypic variance *within* each heterozygous genotype. Because there is no genetic variance within a genotypic class, these variances will (when averaged) provide an estimate of s_e^2. This value then can be subtracted from the value of s_p^2 in the original population to give s_g^2. Using this method, any covariance between genotype and environment in the original population will be hidden in the estimate of genetic variance and inflate it.

Other estimates of genetic variance can be obtained by considering the genetic similarities between relatives. Using simple Mendelian principles, we can see that half the genes of full siblings will (on average) be identical. For identification purposes, we can label the alleles at a locus carried by the parents differently, so that they are, say, $A_1 A_2$ and $A_3 A_4$. Now the older sibling has a probability of $\frac{1}{2}$ of getting A_1 from its father, as does the younger sibling, so the two siblings have a chance of $\frac{1}{2} \times \frac{1}{2} = \frac{1}{4}$ of both carrying A_1. On the other hand, they might both have received an A_2 from their father, so again, they have a probability of $\frac{1}{4}$ of carrying a gene in common that they inherited from their father. Thus, the chance is $\frac{1}{4} + \frac{1}{4} = \frac{1}{2}$ that both siblings will carry an A_1 or that both siblings will carry an A_2. The other half of the time, one sibling will inherit an A_1 and the other will inherit an A_2. So, as far as paternally inherited genes are concerned, full siblings have a 50 percent chance of carrying the same allele. But the same reasoning applies to their maternally inherited gene. Averaging over their paternally and maternally inherited genes, half the genes of full siblings are identical between them. Their **genetic correlation,** which is equal to the chance that they carry the same allele, is $\frac{1}{2}$.

If we apply this reasoning to half-siblings, say, with a common father but with different mothers, we get a different result. Again, the two siblings have a 50 percent chance of inheriting an identical gene from their father, but this time they have no way of inheriting the same gene from their mothers because they have two different mothers. Averaging the maternally inherited and paternally inherited genes thus gives a probability of $(\frac{1}{2} + 0)/2 = \frac{1}{4}$ that these half-siblings will carry the same gene.

We might be tempted to use the theoretical correlation between, say, siblings to estimate H^2. If the observed phenotypic correlation were, for example, 0.4, and we expect on purely genetic grounds a correlation of 0.5, then by the formula $H^2 = s_g^2/s_p^2$, an estimate of heritability would be 0.4/0.5 = 0.8. But such an estimate fails to take into account the fact that siblings also may be environmentally correlated. Unless we are careful to raise the siblings in independent environments, the estimate of H^2 would be too large, and could even exceed 1 if the observed phenotypic correlation were greater than 0.5. To get around this problem, we use the *differences* between phenotypic correlations of different relatives. For example, the difference in genetic correlation between full and half-siblings is $\frac{1}{2} - \frac{1}{4} = \frac{1}{4}$. Let's contrast this with their **phenotypic correlations.** If the environmental similarity is the same for half- and full siblings—a very important condition for estimating heritability—then environmental similarities will cancel out if we take the difference in correlation between the two kinds of siblings. This difference in phenotypic correlation will then be proportional to how much of the variance is genetic. Thus

$$\begin{bmatrix} \text{Genetic correlation} \\ \text{of full siblings} \end{bmatrix} - \begin{bmatrix} \text{genetic correlation} \\ \text{of half-siblings} \end{bmatrix} = \tfrac{1}{4}$$

but

$$\begin{bmatrix} \text{Phenotypic} \\ \text{correlation} \\ \text{of full siblings} \end{bmatrix} - \begin{bmatrix} \text{phenotypic} \\ \text{correlation} \\ \text{of half-siblings} \end{bmatrix} = H^2 \times \tfrac{1}{4}$$

so an estimate of H^2 is

$$H^2 = 4 \left(\begin{array}{cc} \text{correlation} & \text{correlation} \\ \text{of full siblings} & - \text{of half-siblings} \end{array} \right)$$

where the correlation here is the *phenotypic* correlation.

We can use similar arguments about genetic similarities between parents and offspring and between twins to obtain two other estimates of H^2:

$$H^2 = 4 \left(\begin{array}{c} \text{correlation} \\ \text{of full siblings} \end{array} \right) - 2 \left(\begin{array}{c} \text{parent-offspring} \\ \text{correlation} \end{array} \right)$$

or

$$H^2 = 2 \left(\begin{array}{cc} \text{correlation of} & \text{correlation of} \\ \text{monozygotic twins} & - \text{dizygotic twins} \end{array} \right)$$

These formulas are derived from considering the genetic similarities between relatives. They are only approximate and depend on assumptions about the ways in which genes act. The first two formulas, for example, assume that genes at different loci add together in their effect on the character. The last formula also assumes that the alleles at each locus show no dominance (see the discussion of components of variance on pages 842–843).

All these estimates, as well as others based on correlations between relatives, depend *critically* on the assumption that environmental correlations between individuals are the same for all degrees of relationship. If closer relatives have more similar environments, as they do in humans, the estimates of heritability are biased. It is reasonable to assume that most environmental correlations between relatives are positive, in which case the heritabilities would be overestimated. Negative environmental correlations can also exist. For example, if the members of a litter must compete for food that is in short supply, negative correlations in growth rates among siblings could occur.

The difference in correlation between monozygotic and dizygotic twins is commonly used in human genetics to estimate H^2 for cognitive or personality traits. Here the problem of degree of environmental similarity is very severe. Identical (monozygotic) twins are generally treated more similarly to each other than are fraternal (dizygotic) twins. People often give their identical twins names that are similar, dress them alike, treat them identically, and in general, accent their similarities. As a result, heritability is overestimated.

The Meaning of H^2

Attention to the problems of estimating broad heritability distracts from the deeper questions about the meaning of the ratio when it can be estimated. Despite its widespread use as a measure of how "important" genes are in influencing a trait, H^2 actually has a special and limited meaning.

There are two conclusions that can be drawn from a properly designed heritability study. First, if there is a nonzero heritability, then in the population measured and in the environments in which the organisms have developed, genetic differences have influenced the variation between individuals, so genetic differences do matter to the trait. This is not a trivial finding and is a first step in a more detailed investigation of the role of genes. It is important to notice that the reverse is not true. If no heritability for the trait is found, that is not a demonstration that genes are irrelevant, but only that in the particular population studied there is no genetic variation at the relevant loci, or that the environments in which the population developed were such that different genotypes had the same phenotype. In other populations or other environments, the trait might be heritable.

Message In general, the heritability of a trait is different in each population and in each set of environments; it cannot be extrapolated from one population and set of environments to another.

Moreover, we must distinguish between *genes* being relevant to a trait, and *genetic differences* being relevant to *differences* in the trait. The experiment of immigration to North America has proved that the ability to pronounce the sounds of North American English, as opposed to French, Swedish, or Russian, is not a consequence of genetic differences between our immigrant ancestors. But without the appropriate genes, we could not speak any language at all.

Second, the value of the H^2 provides a limited prediction of the effect of environmental modification under particular circumstances. If all the relevant environmental variation is eliminated *and the new constant environment is the same as the mean environment in the original population,* then H^2 estimates how much phenotypic variation will still be present. So, if the heritability of performance on an IQ test were, say, 0.4, then if all children had the same developmental and social environment as the "average child," about 60 percent of the variation in IQ test performance would disappear and 40 percent would remain.

The requirement that the new constant environment be at the mean of the old environmental distribution is absolutely essential to this prediction. If the environment is shifted toward one end or the other of the environmental

distribution, or a new environment is introduced, nothing at all can be predicted. In the example of IQ performance, the heritability gives us no information at all about how variable performance would be if children's developmental and social environments were generally enriched. To understand why this is so, we must return to the concept of the norm of reaction.

The separation of variance into genetic and environmental components s_g^2 and s_e^2 does not really separate the genetic and environmental causes of variation. Consider Figure 27-14, which shows two genotypes in a population with their two norms of reaction. In Figure 27-14a, the population is assumed to consist of the two genotypes in equal frequency, and the distribution of environments (shown on the horizontal axis) is centered toward the right. The phenotypic distribution (shown on the vertical axis) is composed of two underlying distributions that are very different in their means and exhibit little environmental variation. H^2 has a high value. In Figure 27-14b, the environmental distribution has been shifted to the left. As a result, the average difference between genotypes is very much less. The important point here is that the so-called *genetic variance* has been changed by shifting the *environment.* In Figure 27-14c, we suppose that the environments are the same as in Figure 27-14b, but now genotype II has become extremely common in the population and genotype I is rare. Then the phenotypic distribution is almost entirely a reflection of norm-of-reaction II. The population has a smaller environmental variance (s_e^2) than does the population of Figure 27-14b because it has become enriched for the more developmentally stable genotype. In this case, the *environmental variance* has been changed by a change in *genotypes.* In general, genetic variance depends on the environments to which the population is exposed, and environmental variance depends on the frequencies of the genotypes.

Message Because genotype and environment interact to produce phenotype, no partition of variation can actually separate causes of variation.

As a consequence of the argument just given, knowledge of the heritability of a trait does not permit us to predict how the distribution of that trait will change if either genotypic frequencies or environmental factors change markedly.

Message A high heritability does not mean that a trait is unaffected by its environment.

Let's compare Figure 27-14a and b, for example. All that high heritability means is that for the particular population developing in the particular distribution of environments in which the heritability was measured, average differences between genotypes are large compared with environmental variation within genotypes. If the environment is changed, large differences in phenotype may occur.

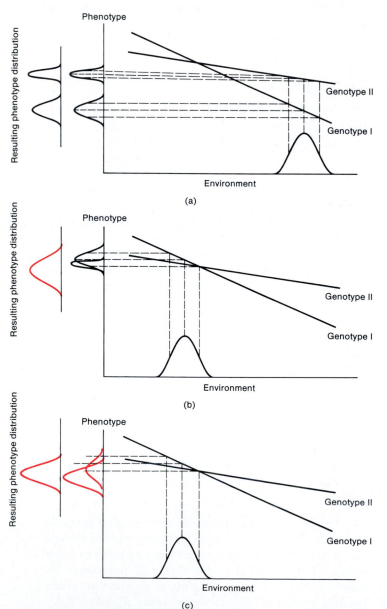

Figure 27-14 Genetic variance can be changed by changing the environment, and environmental variance can be changed by changing the relative frequencies of genotypes.

Perhaps the most famous example of an erroneous use of heritability arguments to make claims about the changeability of a trait is the case of human IQ performance and social success. In 1969 an educational psychologist, A. R. Jensen, published a long paper in the *Harvard Educational Review*, asking the question (in its title) "How much can we boost IQ and scholastic achievement?" Jensen's conclusion was "not much." As an explanation and evidence of this unchangeability, he offered a claim of high heritability for IQ performance. A great deal of criticism has been made of the evidence offered by Jensen for the high heritability of IQ scores. But irrespective of the correct value of H^2 for IQ performance, the real error of Jensen's argument lies in his equation of high heritability with unchangeability. In fact, the heritability of IQ is *irrevalent* to the question raised in the title of his article.

To see why this is so, let us consider the results of adoption studies in which children are separated from their biological parents in infancy and reared by adoptive parents. Although results may vary quantitatively from study to study, there are three characteristics in common. First, adopting parents generally have higher IQ scores than the biological parents. Second, the adopted children have higher IQ scores than their biological parents. Third, the adopted children show a higher correlation of IQ scores with their biological parents than with their adoptive families. The following table is a hypothetical data set that shows all these characteristics, in idealized form, to illustrate the concepts. The scores given for parents are meant to be the average of mother and father.

Children	Biological parents	Adoptive parents
110	90	118
112	92	114
114	94	110
116	96	120
118	98	112
120	100	116
Mean 115	95	115

First, we can see that the children have a high correlation with their biological parents but a low correlation with their adoptive parents. In fact, in our hypothetical example, the correlation of children with biological parents is $r = 1.00$, but with adoptive parents it is $r = 0$. (Remember from page 824 that correlation between two sets of numbers does not mean that the two sets are identical but that for each unit increase in one set, there is a constant proportion increase in the other set.) This perfect correlation with biological parents and zero correlation with adoptive parents means that $H^2 = 1$, given the arguments developed on page 833. All the variation in IQ score between the children is explained by the variation between the biological parents.

Second, however, we notice that each of the IQ scores of the children is 20 points higher than the IQ scores of their respective biological parents and that the mean IQ of the children is equal to the mean IQ of the adoptive parents. Thus, adoption has raised the average IQ of the children 20 points higher than the average IQ of their biological parents, so that as a *group* the children resemble their adoptive parents. So we have perfect heritability, yet high environmental plasticity.

An investigator who is seriously interested in knowing how genes might constrain or influence the course of development of any trait in any organism must study directly the norms of reaction of the various genotypes in the population over the range of projected environments. No less detailed information will do. Summary measures such as H^2 are not first steps toward a more complete analysis and therefore are not valuable in themselves.

Message   Heritability is not the opposite of phenotypic plasticity. A character may have perfect heritability in a population and still be subject to great changes resulting from environmental variation.

Determining Norms of Reaction

Remarkably little is known about the norms of reaction for any quantitative traits in any species. This is partly because it is difficult in most sexually reproducing species to replicate a genotype so that it can be tested in different environments. It is for this reason, for example, that we do not have a norm of reaction for any genotype for any human quantitative trait.

In Domesticated Plants and Animals

A few norm of reaction studies have been carried out with plants that can be clonally propagated. The results of one of these experiments are discussed on page 15. It is possible to replicate genotypes in sexually reproducing organisms by the technique of mating close relatives, or inbreeding. By selfing (where possible) or by mating brother and sister repeatedly generation after generation, a **segregating line** (one that contains both homozygotes and heterozygotes at a locus) can be made homozygous.

The purpose of creating homozygous lines is to produce groups of organisms within which all individuals are genetically identical. These can then be allowed to develop in different environments to produce a norm of reaction. Alternatively, two different homozygous lines can be crossed and the F_1 offspring, all genetically identical with one another, can be characterized in different environments.

Ideally for a norm of reaction study all the individuals should be absolutely identical genetically, but the process of inbreeding only increases the homozygosity of the group slowly, generation after generation, depending on the closeness of the relatives that are mated. In corn, for example, a single individual is chosen and self-pollinated. Then in the next generation, a single one of its offspring is chosen and self-pollinated. In the third generation, a single one of *its* offspring is chosen and self-pollinated, and so on. Suppose that the original individual in the first generation is already a homozygote at some locus. Then all of its offspring from self-pollination will also be homozygous and identical at the locus. Future generations of self-pollination will simply preserve the homozygosity. If, on the other hand, the original individual is a heterozygote, then the selfing $Aa \times Aa$ will produce $\frac{1}{4}$ AA homozygotes and $\frac{1}{4}$ aa homozygotes. If a single offspring is chosen in this subsequent generation to propagate the line, then there is a 50 percent chance that it is now a homozygote. If, by bad luck, the chosen individual should still be a heterozygote, there is another 50 percent chance that the selected individual in the third generation is homozygous, and so on. Of the ensemble of all heterozygous loci, then, after one generation of selfing, only $\frac{1}{2}$ will still be heterozygous; after two generations, $\frac{1}{4}$; after three, $\frac{1}{8}$. In the nth generation

$$\mathrm{Het}_n = \frac{1}{2^n} \mathrm{Het}_0$$

where Het_n is the proportion of heterozygous loci in the nth generation and Het_0 is the proportion in the 0th generation. When selfing is not possible, brother-sister mating will accomplish the same end, although more slowly. Table 27-2 is a comparison of the amount of heterozygosity left after n generations of selfing and brother-sister mating.

In Natural Populations

To carry out a norm of reaction study of a natural population, a large number of lines are sampled from the population and inbred for a sufficient number of generations to

Table 27-2 **Heterozygosity Remaining after Various Generations of Inbreeding for Two Systems of Mating**

Generation	Remaining heterozygosity	
	Selfing	Brother-sister mating
0	1.000	1.000
1	0.500	0.750
2	0.250	0.625
3	0.125	0.500
4	0.0625	0.406
5	0.03125	0.338
10	0.000977	0.114
20	1.05×10^{-6}	0.014
n	$\mathrm{Het}_n = \frac{1}{2}\mathrm{Het}_{n-1}$	$\mathrm{Het}_n = \frac{1}{2}\mathrm{Het}_{n-1} + \frac{1}{4}\mathrm{Het}_{n-2}$

guarantee that each line is virtually homozygous at all its loci. Each line is then homozygous at each locus for a randomly selected allele present in the original population. The inbred lines themselves cannot be used to characterize norms of reaction in the natural population, because such totally homozygous genotypes do not exist in the original population. Each inbred line can be crossed to every other inbred line to produce heterozygotes that reconstitute the original population, and an arbitrary number of individuals from each cross can be produced. If inbred line 1 has the genetic constitution $AA\,BB\,cc\,dd\,EE$ . . . and inbred line 2 is $aa\,BB\,CC\,dd\,ee$ . . . , then a cross between them will produce a large number of offspring, all of whom are identically $Aa\,BB\,Cc\,dd\,Ee$ . . . and can be raised in different environments.

Inbreeding by mating of close relatives for many generations results in total homozygosity for the entire genome. In species like *Drosophila* in which the necessary dominant markers and crossover suppressors are available, it is possible to produce lines that are homozygous for only a single chromosome, rather than for the whole set, as shown for an autosome in Figure 27-15. A single male from the population to be sampled is crossed to a female carrying a chromosome with a crossover suppressor C (usually a complex inversion), a recessive lethal l, and a dominant visible marker M_1 heterozygous with a second dominant visible M_2. In the F_1, a *single* male carrying the ClM_1 chromosome is chosen. This male, which is also carrying a wild-type chromosome from the population, is again crossed to the marker stock. In the F_2, all flies showing the M_1 trait but not M_2 are necessarily all heterozygotes for copies of the original wild-type chromosome because ClM_1/ClM_1 is lethal, and no crossovers have taken place. In the F_3, all wild-type flies are identically homozygous for the wild-type chromosome and are now available to make a stock for norm of reaction studies for crosses. (See Chapter 7 for another use of this technique.)

Figure 27-16 shows the norms of reaction of abdominal bristle number as a function of temperature for second-chromosome homozygotes of *Drosophila pseudoobscura*. Like the growth rate norms of *Achillea* (Figure 1-18), the norms cross each other, with different genotypes having different temperature maxima. The heterozygotes (the natural genotypes) are more similar to each other than are the chromosomal homozygotes, which seldom, if ever, occur in a natural population.

Results of Norm of Reaction Studies

Very few norm of reaction studies have been carried out for quantitative characters for the normally heterozygous genotypes found in natural populations. Only easily inbred species such as corn or *Drosophila* and clonal species such as strawberries have been studied to any degree. The outcomes of such studies resemble Figure 27-16. No genotype is consistently above or below other genotypes, instead,

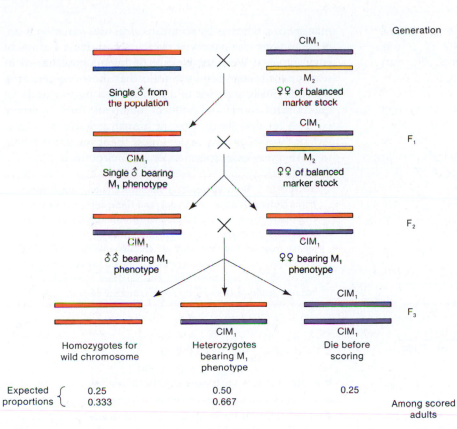

Figure 27-15 Method for making autosomes homozygous using dominant marker genes M_1 and M_2, a crossover suppressor C, and a lethal gene l.

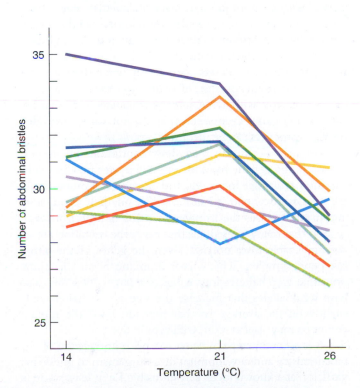

Figure 27-16 The number of abdominal bristles in different homozygous genotypes of *Drosophila pseudoobscura* at three different temperatures. (Data courtesy of A. P. Gupta.)

there are small differences between genotypes, and the direction of these differences is not consistent over a wide range of environments.

These factors have two important consequences. First, the selection of "superior" genotypes in domesticated animals and cultivated plants will result in very specifically adapted varieties that may not show their superior properties in other environments. To some extent, this problem is overcome by deliberately testing genotypes in a range of environments (for example, over several years and in several locations). It would be even better, however, if plant breeders could test their selections in a variety of controlled environments in which different environmental factors could be separately manipulated. The consequences of actual plant-breeding practices can be seen in Figure 27-17, where the yields of two varieties of corn are shown as a function of different farm environments. Variety 1 is an older variety of hybrid corn; variety 2 is a later "improved" hybrid. These performances are compared at a low planting density, which prevailed when variety 1 was developed, and at a high planting density characteristic of farming practice when hybrid 2 was selected. At the high density, the new variety is clearly superior to the old variety in all environments (Figure 27-17a). At the low density, however, the situation is quite different. First, note that the new variety is less sensitive to environment than the older hybrid, as evidenced by its flatter norm of reaction. Second, the new "improved" variety

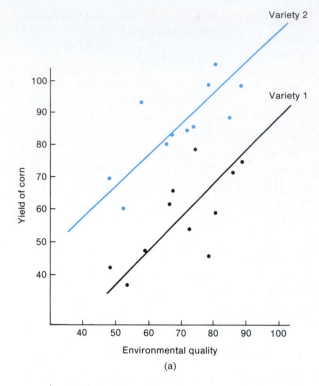

(a)

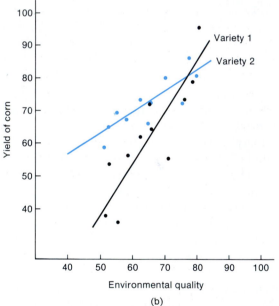

(b)

Figure 27-17 Yields of grain of two varieties of corn in different environments: (a) at a high planting density; (b) at a low planting density. (Data courtesy of W. A. Russell, *Proceedings of the 29th Annual Corn and Sorghum Research Conference,* 1974.)

is actually poorer under the best farm conditions. Third, the yield improvement of the new variety is not apparent under the low densities characteristic of earlier agricultural practice.

The second consequence of the nature of reaction norms is that even if it should turn out that there is genetic variation for various mental and emotional traits in the hu-

man species, which is by no means clear, this variation is unlikely to favor one genotype over another across a range of environments. We must beware of hypothetical norms of reaction for human cognitive traits that show one genotype unconditionally superior to another. Even putting aside all questions of moral and political judgment, there is simply no basis for describing different human genotypes as "better" or "worse" on any scale, unless the investigator is able to make a very exact specification of environment.

Message Norm of reaction studies show only small differences between natural genotypes, and these differences are not consistent over a wide range of environments. Thus, "superior" genotypes in domesticated animals and cultivated plants may be superior only in certain environments. If it should turn out that humans exhibit genetic variation for various mental and emotional traits, this variation is unlikely to favor one genotype over another across a range of environments.

Locating the Genes

It is not possible with purely genetic techniques to identify all the genes that influence the development of a given trait. This is true even for simple qualitative traits—for example, the genes involved in determining the total antigenic configuration of the membrane of the human red blood cell. About 40 loci determining human blood groups are known at present; each has been discovered by finding at least one person with an immunological specificity that differs from the specificities of other people. Many other loci that determine red-cell membrane structure may remain undiscovered, because all the individuals studied are genetically identical. *Genetic* analysis detects genes only when there is some allelic variation. In contrast, of course, *molecular* analysis, by dealing directly with DNA and its translated information, can identify genes even when they do not vary—provided the gene products can be identified.

Even though a trait may show continuous phenotypic variation, the genetic basis for the differences may be allelic variation at a single locus. Most of the classical mutations in *Drosophila* are phenotypically variable in their expression, and in many cases the mutant class differs little from wild type so that many individuals that carry the mutation are indistinguishable from normal. Even the genes of the bithorax gene complex, which have dramatic homeotic mutations that turn halteres into wings (see pages 766–769), also have weak alleles that increase the size of the haltere only slightly on the average so that individuals of the mutant genotype may appear to be wild-type.

It is sometimes possible to use prior knowledge of the biochemistry and development of an organism to guess that variation at a known locus is responsible for at least some of the variation in phenotype. This locus then is a *candidate gene* for investigation of continuous phenotypic variation. An example is the variation in activity of the enzyme acid phosphatase in human red blood cells. Because we are dealing with variation in enzyme activity, a good hypothesis would be that there is allelic variation at the locus that codes for this enzyme. When H. Harris and D. Hopkinson sampled an English population, they found that there were, indeed, three allelic forms, A, B, and C, with different activities. Table 27-3 shows the mean activity, the variance in activity, and the population frequency of the six genotypes. Figure 27-18 shows the distribution of activity in the entire population and how it is composed of the distributions of the different genotypes. The table shows that, of the variance of activity in the total distribution (607.8), about half is explained by the average variance within genotypes (310.7), so half (607.8 − 310.7 = 297.1) is accounted for by the variance between the means of the six genotypes. While much of the variation in activity is explained by the mean differences between the genotypes, there remains variation within each genotype that may be the result of environmental influences or of the segregation of other, as yet unidentified, genes. This partial explanation of variation by alleles at a single identified locus is typical of what is found by the candidate gene method, and the proportion of variance associated with the single locus is usually less than what was found for acid phosphatase. For example, the three common alleles for the gene *apoE* that codes for the protein apolipoprotein E account for only about 16 percent of the variance in blood levels of low-density lipoproteins that carry cholesterol and are implicated in excess cholesterol levels.

Marker Gene Segregation

The genes segregating for a quantitative trait, so-called *quantitative trait loci*, or QTLs, cannot be individually identified in most cases. It is possible, however, to localize those regions of the genome in which the relevant loci lie and to estimate how much of the total variation is accounted for by QTL variation in each region. This analysis is done in experimental organisms by crossing two lines that differ markedly in the quantitative trait and also differ in alleles at well-known loci, *marker genes*, where the different genotypes can be distinguished by criteria such as some visible phenotypic effect that is not confused with the quantitative trait, say, eye color in *Drosophila*, or by the electrophoretic mobility of the proteins they code for, or by the DNA sequence of the genes themselves. The F_1 between the two lines is then crossed with itself to make a segregating F_2, or it may be backcrossed to one of the parental lines. If there are QTLs closely linked to a marker gene, then the different marker genotypes in the segregating generation will also carry the QTL alleles that were linked to them in the original parental lines. Thus different marker genotypes in the F_2 or backcross will have different average phenotypes for the quantitative character.

The method is most easily illustrated by an application of the principle to localizing QTLs to particular chromo-

somes. We assume that the species being investigated has a known array of gene mutations that mark each chromosome in the genome, as, for example, exist in *Drosophila melanogaster*. We could then extend the method to finding the gene regions within chromosomes by the use of the large amount of DNA sequence variation that exists in all organisms (see the section on linkage analysis, page 841).

To find the chromosomes on which the QTLs lie in a genetically well marked species such as *Drosophila melanogaster*, we begin with two populations that are very divergent for the character. These may be two natural populations, but more frequently they are two subpopulations

Table 27-3 Red Blood Cell Activity of Different Genotypes of Red-Cell Acid Phosphatase in the English Population

Genotype	Mean activity	Variance of activity	Frequency in population
AA	122.4	282.4	.13
AB	153.9	229.3	.43
BB	188.3	380.3	.36
AC	183.8	392.0	.03
BC	212.3	533.6	.05
CC	~240	—	.002
Grand average	166.0	310.7	
Total distribution	166.0	607.8	

NOTE: Averages are weighted by frequency in population.

SOURCE: H. Harris, *The Principles of Human Biochemical Genetics*, 3d ed. North-Holland, 1980.

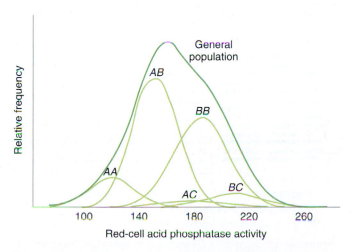

Figure 27-18 Red-cell acid phosphatase activity of different genotypes and the distribution of activity in an English population made up of a mixture of these genotypes. (H. Harris, *The Principles of Human Biochemical Genetics*, 3d ed. Copyright © 1970 by North Holland.)

Figure 27-19 Scheme of mating to produce individuals with different combinations of chromosomes from a high-selection line (+) and a low-selection line (−). A_1, B_1, . . . are dominant marker, crossover suppressors.

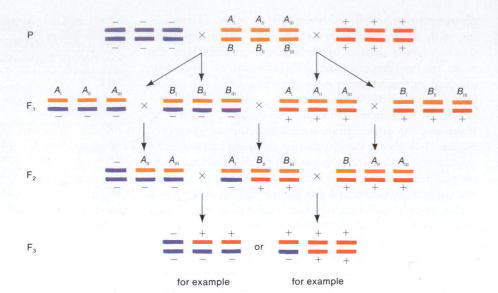

that have been created by artificial selection. For example, in one subpopulation, the largest individuals are chosen as parents in each generation; in the other subpopulation, the smallest individuals are chosen as parents. As generations pass, the upwardly selected line will become enriched for alleles that lead to larger size, whereas the alternative alleles will accumulate in the downwardly selected line. If selection is carried on for a long enough time, the two populations will become virtually homozygous for "high" and "low" alleles. Alternatively, many inbred lines could be made from the original population, and the most divergent lines then could be chosen as the "high" and "low" populations. Whichever method is used, the two divergent populations will probably differ only for those loci that were somewhat heterozygous in the original population.

Once the divergent lines have been established, each chromosome in one line can be separately substituted into the other line by using dominant marker stocks with crossover suppressors. An idealization of the method is shown in Figure 27-19, where A_I, A_{II}, A_{III}, B_I, . . . are dominant marker systems for chromosomes I, II, and III. Lines homozygous and heterozygous for various combinations of "+" and "−" chromosomes are measured for the trait; in this way, the contribution of each chromosome to the genetic differences between the lines is determined. An example of the application of this technique by J. Crow to the study of DDT resistance in *Drosophila melanogaster* is shown in Figure 27-20. Crow did not manufacture lines that were homozygous for "+" and "−" chromosomes, so we see only the effect of chromosomes when lines are heterozygous. The figure shows that every chromosome seems to have some genes that differ between the resistant and the susceptible line.

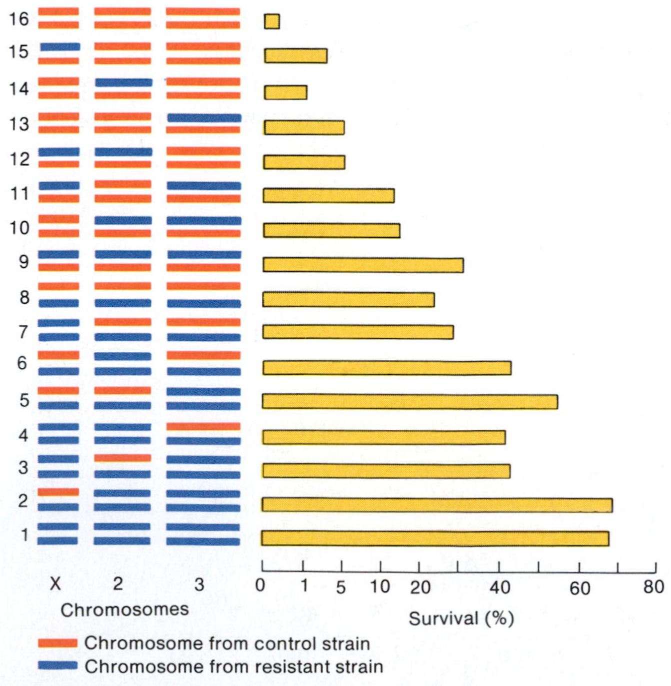

Gene Action

The problem of detecting gene loci influencing quantitative characters is closely tied to the magnitude of the effects of different allelic substitutions, to the degree of dominance of alleles at each locus, and to the amount of epistasis among loci. The methods for locating genes described in the previous section, for example, are biased toward the detection of

Figure 27-20 Survival of *Drosophila melanogaster* genotypes with different combinations of chromosomes from a selected line (resistant strain) and a nonselected line (control strain). Resistant chromosomes are from strains that have been selected for resistance to DDT. (From J. Crow, *Annual Review of Entomology* 2, 1957, 228.)

loci where allelic differences are of large effect. There may be many loci which, in the aggregate, contribute a large amount to the genetic variation of a trait, but which individually cannot be located and characterized because of the small individual effect of each of their allelic substitutions.

Even without being able to localize the genes, we can obtain some information about their actions and interactions. We can judge average dominance at the chromosomal level by comparing homozygotes and heterozygotes in chromosomal substitution experiments. In Figure 27-20, a comparison of lines 1 and 2 shows that the X chromosomes do not differ, whereas lines 15 and 16 show that they do. An explanation of this discrepancy would be the complete dominance of resistance alleles over susceptibility alleles.

There is also evidence of specific epistatic interactions. Line 5 in the figure has both the substitutions of lines 2 and 3 and should be at least susceptible as line 3; yet it is intermediate between 2 and 3. It would seem that heterozygosity for a susceptible X chromosome actually decreases the susceptibility when in combination with a susceptible second-chromosome heterozygote. Such evidence of dominance and epistasis of whole chromosomes throws only indirect light on individual gene action because with dominance or strong epistasis, a single gene of major effect would hide the effects of several other genes of small effect.

Evidence for nonnuclear effects can also be deduced from the chromosomal substitution experiments. For example, in Figure 27-20, lines 8 and 9, which differ in their resistance, have the same chromosomal constitution but are the result of reciprocal crosses and so have different cytoplasm.

Linkage Analysis

The localization of QTLs to small regions within chromosomes requires that there be closely spaced marker loci along the chromosome. Moreover, it must be possible to have parental lines that differ from each other in the alleles carried at these loci. For most of the history of genetics these requirements could not be met, even in a genetically well known species like *Drosophila*, because most marker loci were known from severe morphological mutants that had deleterious effects on the viability and fecundity of their carriers. As a result it was not possible to create a line that carried large numbers of mutant alleles that would distinguish it from an alternative line carrying the wild-type alleles. With the advent of molecular techniques that can detect genetic polymorphism at the DNA level (see pages 784–787), a very high density of variant loci has been discovered along the chromosomes of all species. Especially useful are restriction-site polymorphisms and tandem repeats in DNA (see pages 484–486, 505, and pages 784–785). Such polymorphisms are so common that any two lines selected for a difference in quantitative traits are also sure to differ from each other at known molecular marker loci spaced a few cross-over units from each other along each chromosome.

The experimental protocol is similar to that illustrated for the chromosomal experiment. Two lines are divergently selected for the quantitative trait. It is not necessary, however, to introduce dominant markers in this case, because the lines will differ in detectable DNA markers. A cross is made between the two lines, and the F_1 is crossed with itself to produce a segregating F_2 or is crossed back to one of the parental lines to produce a segregating backcross. A large number of offspring from the segregating generation are then measured for the quantitative trait and characterized for their genotype at the marker loci. A marker locus that is unlinked or very loosely linked to any QTLs will have the same average value of the quantitative trait for all its genotypes, while one that is closely linked to some QTLs will differ in its mean phenotype from one of its genotypes to another. How much difference there is in the mean phenotype between the marker locus genotypes depends both on the strength of the effect of the QTLs and on the tightness of linkage between the QTLs and the marker locus. Suppose, for example that there are two selected lines that differ by a total of 100 units in some quantitative character and that the high line is homozygous $+ +$ at a QTL, while the low line is homozygous $- -$ and that each $+$ allele at this QTL accounts for 5 units of the total difference between the lines. Further suppose that the high line is MM and the low line is mm at a marker locus 10 crossover units away from the QTL. Then, as shown in Figure 27-21, there are 2 units of difference between the average gamete carrying an M al-

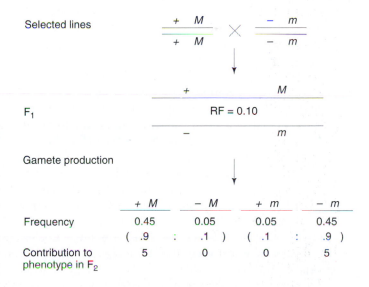

Average phenotypic effect of *M* class = 5(.9) + 0(.1) = 4.5
Average phenotypic effect of *m* class = 5(.1) + 0(.9) = 0.5
Difference between *M*-carrying gametes and *m*-carrying gametes =
4.5 − 0.5 = 4
Difference between *MM* homozygotes and *mm* homozygotes = 8

Figure 27-21 Result of a cross between two selected lines that differ at a QTL and in a molecular marker 10 crossover units away from the QTL. The QTL $+$ allele adds 5 units to the phenotype.

lele and an average gamete carrying an *m* allele in the segregating F_2, or 8 units of the difference between the two original homozygous lines. Thus we have accounted for 8 percent of the average difference between the lines although the QTL genes actually account for 10 percent of the difference. The discrepancy comes from the recombination between the marker gene and the QTL.

This technique has been used to locate segments of chromosome associated with such traits as fruit weight in tomatoes and bristle number in *Drosophila*. Typically any marker region will account for a few percent of the effect, but taken altogether it is possible to account for as much as half of the total difference between lines. Unfortunately, in human genetics, although marker gene segregation can be used to attempt to localize single-gene disorders, the small size of human pedigree groups makes the marker-segregation technique inapplicable for quantitative trait loci.

More on Analyzing Variance

Knowledge of the broad heritability (H^2) of a trait in a population is not very useful in itself, but a finer subdivision of phenotypic variance can provide important information for plant and animal breeders. The genetic variation and the environmental variation can themselves each be further subdivided to provide information about gene action and the possibility of shaping the genetic composition of a population. The concepts involved in subdividing the genetic and environmental variance were first introduced by the founder of modern statistical genetics, R. A. Fisher, in a classic paper in 1918.

Additive and Dominance Variance

Our previous consideration of gene action suggests that the phenotypes of homozygotes and heterozygotes ought to have a simple relation. If one of the alleles codes for a less active gene product or one with no activity at all and if one unit of gene product is sufficient to allow full physiological activity of the organism, then we would expect complete dominance of one allele over the other, as Mendel observed for flower color in peas. If, on the other hand, physiological activity is proportional to the amount of active gene product, we would expect the heterozygote phenotype to be exactly intermediate between the homozygotes (show no dominance).

For many quantitative traits, however, neither of these simple cases is the rule. In general, heterozygotes are not exactly intermediate between the two homozygotes but are closer to one or the other (show partial dominance), even though there is an equal mixture of the primary products of the two alleles in the heterozygote. Indeed, in some cases, the heterozygote phenotype may lie outside the phenotypic range of the homozygotes altogether—a feature termed **overdominance.** For example, newborn babies who are intermediate in size have a higher chance of survival than

very large or very small newborns. Thus, if survival were the phenotype of interest, heterozygotes for genes influencing growth rate would show overdominance for fitness although not for growth rate.

Suppose that two alleles, *a* and *A*, segregate at a locus influencing height. In the environments encountered by the population, the mean phenotypes (heights) and frequencies of the three genotypes might be:

	a a	*A a*	*A A*
Phenotype	10	18	20
Frequency	0.36	0.48	0.16

There is genetic variance in the population; the phenotypic means of the three genotypic classes are different. Some of the variance arises because there is an average effect on phenotype of substituting an allele *A* for an allele *a*; that is, the average height of all individuals with *A* alleles is greater than that of all individuals with *a* alleles. By defining the average effect of an allele as the average phenotype of all individuals that carry it, we necessarily make the average effect of the allele depend on the frequencies of the genotypes.

The average effect is calculated by simply counting the *a* and *A* alleles and multiplying them by the heights of the individuals in which they appear. Thus, 0.36 of all the individuals are homozygous *a a*, each *a a* individual has two *a* alleles, and the average height of *a a* individuals is 10 cm. Heterozygotes make up 0.48 of the population, each has only one *a* allele, and the average phenotypic measurement of *A a* individuals is 18 cm. The total "number" of *a* alleles is $2(0.36) + 1(0.48)$. Thus, the average effect of all the *a* alleles is

$$\bar{a} = \text{average effect of } a = \frac{2(0.36)(10) + 1(0.48)(18)}{2(0.36) + 1(0.48)}$$
$$= 13.20 \text{ cm}$$

and, by a similar argument

$$\bar{A} = \text{average effect of } A = \frac{2(0.16)(20) + 1(0.48)(18)}{2(0.16) + 1(0.48)}$$
$$= 18.80 \text{ cm}$$

This average difference in effect between *A* and *a* alleles of 5.60 cm accounts for some of the variance in phenotype—but not for all of it. The heterozygote is not exactly intermediate between the homozygotes; there is some dominance.

We would like to separate the so-called **additive effect** caused by substituting *a* alleles for *A* alleles, from the variation caused by dominance. The reason is that the effect of selective breeding depends on the additive variation and not on the variation caused by dominance. Thus, for purposes of plant and animal breeding or for making predictions about evolution by natural selection, we must determine the additive variation. An extreme example will illustrate the principle. Suppose that there is overdominance and that the phenotypic means and frequencies of three genotypes are:

	AA	Aa	aa
Phenotype	10	12	10
Frequency	0.25	0.50	0.25

It is apparent (and a calculation like the preceding one will confirm) that there is no average difference between the a and A alleles, because each has an effect of 11 units. So there is no *additive* variation although there is obviously variation in phenotype between the genotypes. The largest individuals are heterozygotes. If a breeder attempts to increase height in this population by selective breeding, mating these heterozygotes together will simply reconstitute the original population. Selection will be totally ineffective. This illustrates the general law that the effect of selection depends on the *additive* genetic variation and not on genetic variation in general.

We partition the total genetic variance in a population into **additive genetic variation** s_a^2, the variance that arises because there is an average difference between the carriers of a alleles and the carriers of A alleles, and a component called the **dominance variance** s_d^2, which results from the fact that heterozygotes are not exactly intermediate between the monozygotes. Thus

$$s_g^2 = s_a^2 + s_d^2$$

The components of variance in the first example, where $aa = 10$, $Aa = 18$, and $AA = 20$, can be calculated using the definitions of mean and variance developed earlier in this chapter. Remembering that a mean is the sum of the values of a variable, each weighted by the frequency with which that value occurs (see page 821), we can calculate the mean phenotype to be

$$\bar{x} = \Sigma f_i x_i = (0.36)(10) + (0.48)(18) + (0.16)(20)$$
$$= 15.44 \text{ cm}$$

The total genetic variance that arises from the variation among the mean phenotypes of the three genotypes is

$$s_g^2 = \Sigma f_i (x_i - \bar{x})^2 = (0.36)(10 - 15.44)^2$$
$$+ (0.48)(18 - 15.44)^2$$
$$+ (0.16)(20 - 15.44)^2$$
$$= 17.13 \text{ cm}^2$$

The frequency of allele a is (by counting alleles)

$$f_a = \frac{2(aa) + 1(Aa)}{2}$$
$$= \frac{2(0.36) + 1(0.48)}{2} = 0.60$$

and the frequency of the A allele is

$$f_A = \frac{2(AA) + 1(Aa)}{2}$$
$$= \frac{2(0.16) + 1(0.48)}{2} = 0.40$$

The variance of allelic means is then

$$s^2 = f_a(\bar{a} - \bar{x})^2 + f_A(\bar{A} - \bar{x})^2$$
$$= (.60)(13.20 - 15.44)^2 + (.40)(18.80 - 15.44)^2$$
$$= 7.525 \text{ cm}^2$$

But we want the variance among diploid individuals that results from the allelic effects, and every diploid individual carries two alleles, so

$$s_a^2 = (2)(7.525) = 15.05 \text{ cm}^2$$

and

$$s_d^2 = s_g^2 - s_a^2 = 17.13 - 15.05 = 2.08 \text{ cm}^2$$

The total phenotypic variance can now be written as

$$s_p^2 = s_g^2 + s_e^2 = s_a^2 + s_d^2 + s_e^2$$

We define a new kind of heritability, the **heritability in the narrow sense (h^2)**, as

$$h^2 = \frac{s_a^2}{s_p^2} = \frac{s_a^2}{s_a^2 + s_d^2 + s_e^2}$$

It is this heritability, not to be confused with H^2, that is useful in determining whether a program of selective breeding will succeed in changing the population. The greater the h^2 is, the greater the difference is between selected parents and the population as a whole that will be preserved in the offspring of the selected parents.

Message The effect of selection depends on the amount of *additive* genetic variance and not on the genetic variance in general. Therefore, the narrow heritability h^2, not the broad heritability H^2, is relevant for a prediction of response to selection.

What has been described as the "dominance" variance is really more complicated. It is all the genetic variation that cannot be explained by the average effect of substituting A for a. If there is more than one locus affecting the character, then any epistatic interactions between loci will appear as variance not associated with the average effect of substituting alleles at the A locus. In principle, we can separate this **interaction variance** (s_i^2) from the dominance variance s_d^2. In practice, however, this cannot be done with any semblance of accuracy, so all the nonadditive variance appears as "dominance" variance.

Estimating Genetic Variance Components

Genetic components of variance can be estimated from covariance between relatives, but the derivation of these estimates is beyond the scope of an elementary text.

There is, however, another way to estimate h^2 that provides an insight into its real meaning. If we plot the phenotypes of the offspring against the average phenotypes of their two parents (the midparent value), we may observe a relationship like the one illustrated in Figure 27-22. The regression line will pass through the mean of all the parents and the mean of all the offspring, which will be equal to each other because no change has occurred in the population between generations. Moreover, taller parents have taller children and shorter parents have shorter children, so that the slope of the line is positive. But the slope is not unity; very short parents have children who are somewhat taller and very tall parents have children who are somewhat shorter than they themselves are. This slope of less than unity for the regression line arises because heritability is less than perfect. If the phenotype were additively inherited with complete fidelity, then the height of the offspring would be identical with the midparent value and the slope of the line would be 1. On the other hand, if the offspring had no heritable similarity to their parents, all parents would have offspring of the same average height and the slope of the line would be 0. This suggests that the slope of the regression line of the offspring value on the midparent value is an estimate of additive heritability. In fact, the relationship is precise.

The fact that the slope equals the additive heritability now allows us to use h^2 to predict the effects of artificial selection. Suppose that we select parents for the next generation who are on the average 2 units above the general mean of the population from which they were chosen. If $h^2 = 0.5$, then the offspring who form the next, selected generation will lie $0.5(2.0) = 1.0$ unit above the mean of the present population, since the regression coefficient predicts how much increase in y will result from a unit increase in x. We can define the **selection differential** as the difference between the selected parents and the unselected mean, and the **selection response** as the difference between their offspring and the previous generation. Then

$$\text{Selection response} = h^2 \times \text{selection differential}$$

or

$$h^2 = \frac{\text{selection response}}{\text{selection differential}}$$

The second expression provides us with yet another way to estimate h^2: by selecting for one generation and comparing the response with the selection differential. Usually this is carried out for several generations, and the average response is used.

Remember that any estimate of h^2, just as for H^2, depends on the assumption of no greater environmental correlation between closer relatives. Moreover, h^2 in one population in one set of environments will not be the same as h^2 in a different population at a different time. Figure 27-23

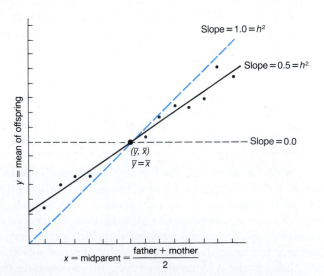

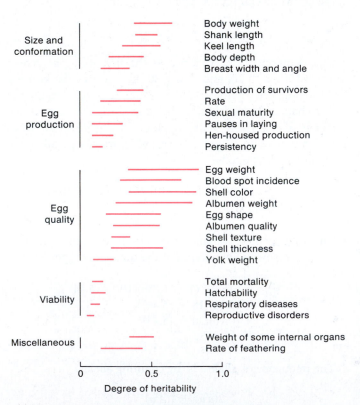

Figure 27-22 The regression (black line) of offspring measurements (y) on midparents (x) for a trait with narrow heritability (h^2) of 0.5. The blue line shows the regression slope if the trait were perfectly heritable.

Figure 27-23 Ranges of heritabilities (h^2) reported for a variety of characters in chickens. (From I. M. Lerner and W. J. Libby, *Heredity, Evolution and Society.* Copyright © 1976 by W. H. Freeman and Company.)

Table 27-4 Partition of Total Phenotypic Variance for Two Characters in Population of *Drosophila melanogaster*

Source of variation			Percentage of variance	
			Number of abdominal bristles	Ovary size
Additive genetic	s_a^2	s_g^2	52	30
Dominance + epistatic variance	s_d^2		9	40
Environmental variance	s_e^2	s_e^2	1	3
Developmental noise	s_n^2		38	27
Total		s_p^2	100	100

SOURCE: D. Falconer, *Quantitative Genetics*. Longman Group Limited. Copyright © 1981.

shows the range of heritabilities reported in various studies for a number of traits in chickens. The very small ranges are generally close to zero. For most traits for which a substantial heritability has been reported in some population, there are big differences from study to study.

Partitioning Environmental Variance

Environmental variance, like genetic variance, can also be further subdivided. In particular, developmental noise (see page 16) is usually confounded with environmental variance, but when a character can be measured on the left and right sides of an organism or over repeated body segments, it is possible to separate noise from environment. Table 27-4 shows the complete partitioning of variation for two characters in a population of *Drosophila melanogaster* raised under standard laboratory conditions. For each character, there is a substantial h^2 ($h^2 = s_a^2/s_p^2$), so we might expect selective breeding to increase or decrease bristle number and ovary size. Furthermore, nearly all the nongenetic variation is due to developmental noise. These values, however, are a consequence of the relatively rigorously controlled environment of the laboratory. Presumably s_e^2 in nature would be considerably larger, with a consequent diminution in the relative sizes of h^2 and of s_a^2. As always, such studies of variation are applicable only to a particular population in a given distribution of environments.

The Use of h^2 in Breeding

Even though h^2 is a number that applies only to a particular population and a given set of environments, it is still of great practical importance to breeders (Figure 27-24). A poultry geneticist interested in increasing, say, growth rate, is not concerned with the genetic variance over all possible flocks and all environmental distributions. Given a particular flock (or a choice between a few particular flocks) under the environmental conditions approximating present husbandry practice, the question becomes: Can a selection scheme be devised to increase growth rate and, if so, how fast? If one flock has a lot of genetic variance and another

only a little, the breeder will choose the former to carry out selection. If the heritability in the chosen flock is very high, then the mean of the population will respond quickly to the selection imposed, because most of the superiority of the selected parents will appear in the offspring. The higher h^2

Figure 27-24 Quantitative genetic theory has been extensively applied to poultry breeding. (Larry Lefever/Grant Heilman Photography, Inc.)

is, the higher the parent-offspring correlation is. If, on the other hand, h^2 is low, then only a small fraction of the increased growth rate of the selected parents will be reflected in the next generation.

If h^2 is very low, some alternative scheme of selection or husbandry may be needed. In this case, H^2 together with h^2 can be of use to the breeder. Suppose that h^2 and H^2 are both low. This means that there is a lot of environmental variance compared with genetic variance. Some scheme of reducing s_e^2 must be used. One method is to change the husbandry conditions so that environmental variance is lowered. Another is to use **family selection.** Rather than choosing the best individuals, the breeder allows pairs to produce several progeny, and the *mating* is selected on the basis of the average performance of the progeny. By averaging over progeny, uncontrolled environmental and developmental noise variation is canceled out and a better estimate of the genotypic difference between pairs can be made so that the best pairs can be chosen as parents of the next generation.

If, on the other hand, h^2 is low but H^2 is high, then there is not much environmental variance. The low h^2 is the result of a small amount of additive genetic variance compared with dominance and interaction variance. Such a situation calls for special breeding schemes that make use of nonadditive variance. One such scheme is the **hybrid-in-bred method,** which is used almost universally for corn. A large number of inbred lines are created by selfing. These are then crossed in many different combinations (all possible combinations, if this is economically feasible), and the cross that gives the best hybrid is chosen. Then new inbred lines are developed from this best hybrid, and again crosses are made to find the best second-cycle hybrid. This scheme selects for dominance effects, because it takes the best heterozygotes; it has been the basis of major genetic advances in hybrid maize yield in North America since 1930. Yield in corn does not appear to have large amounts of nonadditive genetic variance, so it is debatable whether this technique *ultimately* produces higher-yielding varieties than those that would have resulted from years of simple selection techniques based on additive variance.

The hybrid method has been introduced into the breeding of all kinds of plants and animals. Tomatoes and chickens, as examples, are now almost exclusively hybrids. Attempts also have been made to breed hybrid wheat, but thus far the wheat hybrids obtained do not yield consistently better than the nonhybrid varieties now used.

Message   The subdivision of genetic variation and environmental variation provides important information about gene action that can be used in plant and animal breeding.

SUMMARY

Many—perhaps most—of the phenotypic traits we observe in organisms vary continuously. In many cases, the variation of the trait is determined by more than a single segregating locus. Each of these loci may contribute equally to a particular phenotype, but it is more likely that they contribute unequally. The measurement of these phenotypes and the determination of the contributions of specific alleles to the distribution must be made on a statistical basis in these cases. Some of these variations of phenotype (such as height in some plants) may show a normal distribution around a mean value; others (such as seed weight in some plants) will illustrate a skewed distribution around a mean value.

In other characters, the variation in one phenotype may be correlated with the variation in another. A correlation coefficient may be calculated for these two variables.

A quantitative character is one for which the average phenotypic differences between genotypes are small compared with the variation between the individuals within the genotypes. This situation may be true even for characters that are influenced by alleles at one locus. The distribution of environments is reflected biologically as a distribution of phenotypes. The transformation of environmental distribution into phenotypic distribution is determined by the norm of reaction. Norms of reaction can be characterized in organisms in which large numbers of genetically identical individuals can be produced.

By the use of genetically marked chromosomes, it is possible to determine the relative contributions of different chromosomes to variation in a quantitative trait, to observe dominance and epistasis from whole chromosomes, and, in some cases, to map genes that are segregating for a trait.

Traits are familial if members of the same family share them, for whatever reason. Traits are heritable, however, only if the similarity arises from shared genotypes. In experimental organisms, environmental similarities may be readily distinguished from genetic similarities, or heritability. In humans, however, it is very difficult to determine whether a particular trait is heritable. Norm of reaction studies show only small differences between genotypes, and these differences are not consistent over a wide range of environments. Thus, "superior" genotypes in domesticated animals and cultivated plants may be superior only in certain environments. If it should turn out that humans exhibit genetic variation for various mental and emotional traits, this variation is unlikely to favor one genotype over another across a range of environments.

The attempt to quantify the influence of genes on a particular trait has led to the determination of heritability in the broad sense (H^2). In general, the heritability of a trait is different in each population and each set of environments and cannot be extrapolated from one population and set of environments to another. Because H^2 characterizes present populations in present environments only, it is fundamentally flawed as a predictive device. Heritability in the narrow sense, h^2, measures the proportion of phenotypic variation that results from substituting one allele for another. This quantity, if large, predicts that selection for a trait will succeed rapidly. If h^2 is small, special forms of selection are required.

Concept Map

Draw a concept map interrelating as many of the following terms as possible. Note that the terms are listed in no particular order.

quantitative variation / polygenes / Mendel's law / norm of reaction / environment / heritability / variance / selection response / additive variance / QTL

CHAPTER INTEGRATION PROBLEM

In some species of songbirds, populations living in different geographical regions sing different "local dialects" of the species song. Some people believe that this is the result of genetic differences between populations, while others believe these differences arose from purely individual idiosyncracies in the founders of these populations and have been passed on from generation to generation by learning. Outline an experimental program that would determine the importance of genetic and nongenetic factors and their interaction in this dialect variation. If there is evidence of genetic difference, what experiments could be done to provide a detailed description of the genetic system, including the number of segregating genes, their linkage relations, and their additive and nonadditive phenotypic effects?

input from their own ancestors and in various combinations of auditory environments of other populations. This is done by raising birds from the egg

(1) in isolation

(2) surrounded by hatchlings consisting only of birds derived from the same population

(3) surrounded by hatchlings consisting of birds derived from other populations

(4) in the presence of singing adults from other populations

(5) in the presence of singing adults from their own population (as a control on the rearing conditions)

Solution

This example has been chosen because it illustrates the very considerable experimental difficulties that arise when we try to examine claims that observed differences in quantitative characters in some species have a genetic basis. To be able to say anything at all about the roles of gene and developmental environment requires, at minimum, that the organisms can be raised from fertilized eggs in a controlled laboratory environment. To be able to make more detailed statements about the genotypes underlying variation in the character requires, further, that the results of crosses between parents of known phenotype and known ancestry be observable, and that the offspring of some of those crosses be, in turn, crossed to other individuals of known phenotype and ancestry. Very few animal species can satisfy this requirement, although it is much easier to carry out controlled crosses in plants. We will assume that the songbird species in question can indeed be raised and crossed in captivity, but that is a big assumption.

a. To determine whether there is any genetic difference underlying the observed phenotypic difference in dialect between the populations, we need to raise birds of each population, from the egg, in the absence of auditory

If there are no genotypic differences, and all dialect differences are learned, then birds from group 5 will sing their population dialect and those from group 4 will sing the foreign dialect. Groups 1, 2, and 3 may not sing at all; they may sing a generalized song not corresponding to any of the dialects; or they may all sing the same song dialect—this dialect would then represent the "intrinsic" developmental program unmodified by learning.

If dialect differences are totally determined by genetic differences, birds from groups 4 and 5 will sing the same dialect, that of their parents. Birds from groups 1, 2, and 3, if they sing at all, will each sing the song dialect of their parent population, irrespective of the other birds in their group. There are then the possibilities of less clear-cut results, indicating that both genetic and learned differences influence the trait. For example, birds in group 4 might sing a song with both population elements. Note that if the birds in the control group 5 do not sing their normal dialect, the rest of the results are uninterpretable, because the conditions of artificial rearing are interfering with the normal developmental program.

b. If the results of the first experiments show some heritability in the broad sense, then a further analysis is possible. This requires a genetically segregating population, made from a cross between two dialect populations, say I and II. A cross between males from population I and females from population II and the reciprocal cross will give an estimate of the average degree of dominance of genes influencing the trait and whether there is any sex linkage. (Remember that in birds the female is the heterogametic sex.) The offspring of this cross and all subsequent crosses *must* be raised in conditions that do not confuse the learned and the genetic component of the differences, as revealed in the experiments in part a. If learned effects cannot be separated out, this further genetic analysis is impossible.

c. To localize genes influencing dialect differences would require a large number of segregating genetic markers. These could be morphological mutants or molecular variants such as restriction-site polymorphisms. Families segregating for the quantitative trait differences would be examined to see if there were cosegregation of any of the marker loci with the quantitative trait. These would then be candidates for loci linked to the quantitative trait loci. Further crosses between individuals with and without mutant markers, and measurement of the quantitative trait values in F_2 individuals, would establish whether there was actual linkage between the marker and quantitative trait loci. In practice it is very unlikely that such experiments could be carried out on a songbird species because of the immense time and effort required to establish lines carrying the large number of different marker genes and molecular polymorphisms.

SOLVED PROBLEMS

1. Two inbred lines of beans are intercrossed. In the F_1, the variance in bean weight is measured at 1.5. The F_1 is selfed; in the F_2, the variance in bean weight is 6.1. Estimate the broad heritability of bean weight in this experiment.

Solution

The key here is to recognize that all the variance in the F_1 population must be environmental because all individuals must be of identical genotype. Furthermore, the F_2 variance must be a combination of environmental and genetic components, because all the genes that are heterozygous in the F_1 will segregate in the F_2 to give an array of different genotypes that relate to bean weight. Hence, we can estimate

$$s_e^2 = 1.5$$

$$s_e^2 + s_g^2 = 6.1$$

Therefore

$$s_g^2 = 6.1 - 1.5 = 4.6$$

and broad heritability is

$$H^2 = \frac{4.6}{6.1} = 0.75 \ (75\%)$$

2. In an experimental population of *Tribolium* (flour beetles), the body length shows a continuous distribution with a mean of 6 mm. A group of males and females with body lengths of 9 mm are removed and interbred. The body lengths of their offspring average 7.2 mm. From these data, calculate the heritability in the narrow sense for body length in this population.

Solution

The selection differential is $9 - 6 = 3$ mm, and the selection response is $7.2 - 6 = 1.2$ mm. Therefore, the heritability in the narrow sense is

$$h^2 = \frac{1.2}{3} = 0.4 \ (40\%)$$

PROBLEMS

1. Distinguish between continuous and discontinuous variation in a population, and give some examples of each.

2. In a large herd of cattle, three different characters showing continuous distribution are measured, and the variances in the following table are calculated:

Variance	Characters		
	Shank length	Neck length	Fat content
Phenotypic	310.2	730.4	106.0
Environmental	248.1	292.2	53.0
Additive genetic	46.5	73.0	42.4
Dominance genetic	15.6	365.2	10.6

a. Calculate the broad- *and* narrow-sense heritabilities for each character.

b. In the population of animals studied, which character would respond best to selection? Why?

c. A project is undertaken to decrease mean fat content in the herd. The mean fat content is currently 10.5 percent. Animals of 6.5 percent fat content are interbred as parents of the next generation. What mean fat content can be expected in the descendants of these animals?

3. Suppose that two triple heterozygotes $Aa\,Bb\,Cc$ are crossed. Assume that the three loci are in different chromosomes.

a. What proportions of the offspring are homozygous at one, two, and three loci, respectively?

b. What proportions of the offspring carry 0, 1, 2, 3, 4, 5, and 6 alleles (represented by capital letters), respectively?

4. In Problem 3, suppose that the average phenotypic effect of the three genotypes at the A locus is $AA = 4$, $Aa = 3$, $aa = 1$ and that similar effects exist for the B and C loci. Moreover, suppose that the effects of loci add to each other. Calculate and graph the distribution of phenotypes in the population (assuming no environmental variance).

5. In Problem 4, suppose that there is a threshold in the phenotypic character so that when the phenotypic value is above 9, the individual *Drosophila* has three bristles; when it is between 5 and 9, the individual has two bristles; and when the value is 4 or less, the individual has one bristle. Discuss the outcome of crosses within and between bristle classes. Given the result, could you infer the underlying genetic situation?

6. Suppose that the general form of a distribution of a trait for a given genotype is

$$f = 1 - \frac{(x - \bar{x})^2}{s_e^2}$$

over the range of x where f is positive.

a. On the same scale, plot the distributions for three genotypes with the following means and environmental variances:

Genotype	$\bar{x}$	s_e^2	Approximate range of phenotype
1	0.20	0.3	$x = 0.03$ to $x = 0.37$
2	0.22	0.1	$x = 0.12$ to $x = 0.24$
3	0.24	0.2	$x = 0.10$ to $x = 0.38$

b. Plot the phenotypic distribution that would result if the three genotypes were equally frequent in a population. Can you see distinct modes? If so, what are they?

7. The following table shows a distribution of bristle number in *Drosophila*:

Bristle number	Number of individuals
1	1
2	4
3	7
4	31
5	56
6	17
7	4

Calculate the mean, variance, and standard deviation of this distribution.

8. The following sets of hypothetical data represent paired observations on two variables (x, y). Plot each set of data pairs as a scatter diagram. Look at the plot of the points, and make an intuitive guess about the correlation between x and y. Then calculate the correlation coefficient for each set of data pairs, and compare this value with your estimate.

a. (1, 1); (2, 2); (3, 3); (4, 4); (5, 5); (6, 6).

b. (1, 2); (2, 1); (3, 4); (4, 3); (5, 6); (6, 5).

c. (1, 3); (2, 1); (3, 2); (4, 6); (5, 4); (6, 5).

d. (1, 5); (2, 3); (3, 1); (4, 6); (5, 4); (6, 2).

9. A book on the problem of heritability of IQ makes the following three statements. Discuss the validity of each statement and its implications about the authors' understanding of h^2 and H^2.

a. "The interesting question then is . . . 'How heritable?' The answer [0.01] has a very different theoretical and practical application from the answer [0.99]." [The authors are talking about H^2.]

b. "As a rule of thumb, when education is at issue, H^2 is usually the more relevant coefficient, and when eugenics and dysgenics (reproduction of selected individuals) are being discussed, h^2 is ordinarily what is called for."

c. "But whether the different ability patterns derive from differences in genes . . . is not relevant to assessing discrimination in hiring. Where it could be relevant is in deciding what, in the long run, might be done to change the situation."

(From J. C. Loehlin, G. Lindzey, and J. N. Spuhler, *Race Differences in Intelligence.* Copyright © 1975 by W. H. Freeman and Company.)

10. Using the concepts of norms of reaction, environmental distribution, genotypic distribution, and phenotypic distribution, try to restate the following statement in more exact terms: "80 percent of the difference in IQ performance between the two groups is genetic." What would it mean to talk about the heritability of a difference between two groups?

11. Describe an experimental protocol involving studies of relatives that could estimate the broad heritability of alcoholism. Remember that you must make an adequate observational definition of the trait itself.

12. A line selected for high bristle number in *Drosophila* has a mean of 25 sternopleural bristles, whereas a low-selected line has a mean of only 2. Marker stocks involving the two large autosomes II and III are used to create stocks with various mixtures of chromosomes from the high (h) and low (l) lines. The mean number of bristles for each chromosomal combination is as follows:

$$\frac{h\ h}{h\ h}\ 25.1 \qquad \frac{h\ h}{l\ h}\ 22.2 \qquad \frac{l\ h}{l\ h}\ 19.0$$

$$\frac{h\ h}{h\ l}\ 23.0 \qquad \frac{h\ h}{l\ l}\ 19.9 \qquad \frac{l\ h}{l\ l}\ 14.7$$

$$\frac{h\ l}{h\ l}\ 11.8 \qquad \frac{h\ l}{l\ l}\ 9.1 \qquad \frac{l\ l}{l\ l}\ 2.3$$

What conclusions can you reach about the distribution of genetic factors and their actions from these data?

13. Suppose that number of eye facets is measured in a population of *Drosophila* under various temperature conditions. Further suppose that it is possible to estimate total genetic variance s_g^2 as well as the phenotypic distribution. Finally, suppose that there are only two genotypes in the population. Draw pairs of norms of reaction that would lead to the following results:

a. An increase in mean temperature decreases the phenotypic variance.

b. An increase in mean temperature increases H^2.

c. An increase in mean temperature increases s_g^2 but decreases H^2.

d. An increase in temperature *variance* changes a unimodal into a bimodal phenotypic distribution (one norm of reaction is sufficient here).

14. Francis Galton compared the heights of male undergraduates with the heights of their fathers, with the results shown in the following graph:

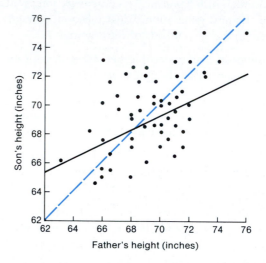

The average height of all fathers is the same as the average height of all sons, but the individual height classes are not equal across generations. The very tallest fathers had somewhat shorter sons, whereas the very short fathers had somewhat taller sons. As a result, the best line that can be drawn through the points on the scatter diagram has a slope of about 0.67 (*solid line*) rather than 1.00 (*dashed line*). Galton used the term *regression* to describe this tendency for the phenotype of the sons to be closer than the phenotype of their fathers to the population mean.

a. Propose an explanation for this regression.

b. How are regression and heritability related here?

(Graph after W. F. Bodmer and L. L. Cavalli-Sforza, *Genetics, Evolution, and Man.* Copyright © 1976 by W. H. Freeman and Company.)

Further Readings

Students interested in pursuing genetics further should start reading original research articles in scientific journals. Some important journals are *Cell, Current Genetics, Evolution, Gene, Genetic Research, Genetics, Heredity, Human Genetics, Journal of Medical Genetics, Journal of Molecular Biology, Molecular and General Genetics, Mutation Research, Nature, Plasmid, Proceedings of the National Academy of Sciences of the United States of America,* and *Science.* Useful review articles may be found in *Annual Review of Genetics, Advances in Genetics,* and *Trends in Genetics.*

Some particularly useful references, mostly general reviews, are listed below under the chapters to which they relate.

Chapter 1

Clausen, J., D. D. Keck, and W. W. Hiesey. 1940. *Experimental Studies on the Nature of Species,* Vol. 1: *The Effect of Varied Environments on Western North American Plants.* Carnegie Institute of Washington, Publ. No. 520, 1–452. This publication and the following one by the same authors are the classic studies of norms of reaction of plants from natural populations.

Clausen, J., D. D. Keck, and W. W. Hiesey. 1958. *Experimental Studies on the Nature of Species,* Vol. 3: *Environmental Responses of Climatic Races of* Achillea. Carnegie Institute of Washington, Publ. No. 581, 1–129.

Milunsky, A., and G. J. Annas, eds. 1975. *Genetics and the Law.* New York: Plenum Press. Interesting accounts of the ramifications of genetics in the lives of individuals.

Moore, J. A. 1985. *Science As a Way of Knowing—Genetics. American Zoologist* 25: 1–165. One short book from an excellent series that focuses on the modus operandi of science.

Schmalhausen, I. I. 1949. *Factors of Evolution: The Theory of Stabilizing Selection.* Philadelphia: Blakiston. The most general discussion of the relation of genotype and environment in the formation of phenotypic variation.

Chapter 2

Carlson, E. A. 1966. *The Gene: A Critical History.* Philadelphia: W. B. Saunders. A readable history of genetics.

Grant, V. 1975. *Genetics of Flowering Plants.* New York: Columbia University Press. One of the few texts on this subject.

Harpstead, D. 1971. "High-Lysine Corn." *Scientific American* (August). An account of the breeding of lines with increased amounts of normally limiting amino acids.

Hutt, F. B. 1964. *Animal Genetics.* New York: Ronald Press. A standard text on the subject with many interesting examples.

Jennings, P. R. 1976. "The Amplification of Agricultural Production." *Scientific American* (September). A discussion of genetics and the green revolution.

McKusick, V. A., et al. 1990. *Mendelian Inheritance in Man,* 9th ed. Baltimore and London: Johns Hopkins Press. The "Bible" of medical genetics; a 2000-page compendium of all the known human inherited disorders, both autosomal and X-linked.

Mange, A. P., and E. J. Mange. 1990. *Genetics: Human Aspects,* 2d ed. Sunderland, Mass.: Sinauer. A useful and up-to-date text for a course in human genetics.

Olby, R. C. 1966. *Origins of Mendelism.* London: Constable. An enjoyable account of Mendel's work and the intellectual climate of his time.

Rousseau, F., et al. 1994. "Mutations in the Gene Encoding Fibroblast Growth Factor Receptor in Achondroplasia." *Nature* 371 253–254. A short research paper describing the molecular identification of the abnormal allele that causes achondroplasia.

Stern, C., and E. R. Sherwood. 1966. *The Origin of Genetics: A Mendel Source Book.* New York: W. H. Freeman and Co. A short collection of important early papers, including Mendel's papers and correspondence.

Sturtevant, A. H. 1965. *A History of Genetics.* New York: Harper & Row. Another useful historical text.

Thompson, J. S., and M. W. Thompson. 1980. *Genetics in Medicine.* Philadelphia: Saunders. One of the classic textbooks in medical genetics. Short and easy to read.

Todd, N. B. 1977. "Cats and Commerce." *Scientific American* (November). Includes some genetics of domestic cat coat colors and the use of this information to study cat migration throughout history.

Vogel, F. and A. G. Motulsky. 1986. *Human Genetics: Problems and Approaches,* 2d ed. New York: Springer-Verlag. Another classic textbook useful as a reference work.

Chapter 3

Crow, J. F. 1983. *Genetics Notes,* 8th ed. New York: Macmillan. A short and concise review of genetics from Mendel to populations.

McIntosh, J. R., and K. L. McDonald. 1988. "The Mitotic Spindle." *Scientific American* (October).

McLaren, A. 1988. "Sex Determination in Mammals." *Trends in Genetics* 4: 153–157.

McLeish, J., and B. Snoad. 1958. *Looking at Chromosomes,* New York: Macmillan. A short classic book consisting of many superb photos of mitosis and meiosis.

Rick, C. M. 1978. "The Tomato." *Scientific American* (August). Includes an account of tomato genes and chromosomes and their role in breeding.

Stern, C. 1973. *Principles of Human Genetics.* 3d ed. New York: W. H. Freeman and Co. A standard text including many examples of the inheritance of human traits.

von Wettstein, D., et al. 1984. "The Synaptonemal Complex in Genetic Segregation." *Annual Review of Genetics* 18: 331–414. An up-to-date review of the structure and function of the complex.

Chapter 4

Bodmer, W. F., and L. L. Cavalli-Sforza. 1976. *Genetics, Evolution, and Man.* New York: W. H. Freeman and Co. A very readable, well-illustrated book, including a clear account of HLA genetics.

Griffiths, A. J. F., and F. R. Ganders. 1984. *Wildflower Genetics.* Vancouver: Flight Press. A field guide to plant variation in natural populations and its genetic basis, including examples relevant to this chapter.

Griffiths, A. J. F., and J. McPherson. 1989. *One Hundred + Principles of Genetics.* New York: W. H. Freeman and Co. A short book that distills genetics down to its basic key concepts; useful for rapid review of the whole subject.

Hutt, W. B. 1979. *Genetics for Dog Breeders.* New York: W. H. Freeman and Co. A short book that will make genetics immediately relevant to all dog owners and breeders.

Searle, A. G. 1968. *Comparative Genetics of Coat Color in Mammals.* New York: Academic Press. A classic treatment of the subject with many examples relevant to this and other chapters.

Silvers, W. K. 1979. *The Coat Colors of Mice.* New York: Springer-Verlag. A standard handbook on the subject, including many examples of gene interaction.

Wright, M., and S. Walters, eds. 1981. *The Book of the Cat.* New York: Summit Books. A fascinating book that has an excellent chapter on gene interaction in determining the coat colors of domestic cats.

Chapter 5

O'Brien, S. J., ed. 1984. *Genetic Maps.* Cold Spring Harbor Press. A compendium of the detailed maps of 80 well-analyzed organisms.

Peters, J. A., ed. 1959. *Classic Papers in Genetics.* Englewood Cliffs, N. J.: Prentice-Hall. A collection of important papers in the history of genetics.

White, R., et al. 1985. "Construction of Linkage Maps with DNA Markers for Human Chromosomes." *Nature* 313: 101–104. An extension of the techniques of this chapter to DNA markers.

Chapter 6

Finchman, J. R. S., P. R. Day, and A. Radford. 1979. *Fungal Genetics,* 3d ed. London: Blackwell. A large, standard technical work. Good for tetrad analysis.

Kemp, R. 1970. *Cell Division and Heredity.* London: Edward Arnold. A short, clear introduction to genetics. Good for map functions and tetrad analysis.

Murray, A. W., and J. W. Szostak. 1983. "Construction of Artificial Chromosomes in Yeast." *Nature* 305: 189–193. The first creation of new chromosomes by splicing together known telomere, centromere, replicator, and then DNA fragments by recombinant DNA technology. Includes tetrad analysis of markers on the new chromosomes.

Puck, T. T., and F-T. Kao. 1982. "Somatic Cell Genetics and Its Application to Medicine." *Annual Review of Genetics* 16: 225–272. A technical but readable review.

Ruddle, F. H., and R. S. Kucherlapati. 1974. "Hybrid Cells and Human Genes." *Scientific American* (July). A popular account of the use of cell hybridization in mapping human genes.

Stahl, F. W. 1969. *The Mechanics of Inheritance,* 2d ed. Englewood Cliffs, N. J.: Prentice-Hall. A short introduction to genetics, including some advanced material presented with a novel approach.

Chapter 7

Induced Mutations—A Tool in Plant Research. 1981. Vienna: International Atomic Energy Agency. A collection of papers by eminent workers in agricultural genetics illustrating the practical uses of mutations in plant breeding.

Lawrence, C. W. 1971. *Cellular Radiobiology.* London: Edward Arnold. A short standard text.

Lindsley, D. L., and E. H. Grell. 1972. *Genetic Variations of* Drosophila melanogaster. Washington, D.C.: Carnegie Institute of Washington. A reference book for fruitfly researchers, containing all the thousands of known *Drosophila* mutations. Fascinating browsing for anyone interested in genetics generally and in the major contributions made by *Drosophila* research.

Neuffer, M. G., L. Jones, and M. S. Zuber. 1968. *The Mutants of Maize.* Madison: Crop Science Society of America. A color catalog of the many, and often bizarre, mutants used by corn geneticists.

Schull, W. J., et al. 1981. "Genetic Effect of the Atomic Bombs: A Reappraisal." *Science* 213: 1220–1227. A summary of all the indicators of potential genetic effects of the Hiroshima and Nagasaki explosions, concluding that "In no instance is there a statistically significant effect of parental exposure; but for all indicators the observed effect is in the direction suggested by the hypothesis that genetic damage resulted from the exposure."

deSerres, F. J., and A. Hollaender, eds. 1982. *Chemical Mutagens: Principles and Methods for Their Detection,* Vol. 7. New York: Plenum Press. One of a set of useful volumes on this important class of mutagens that is particularly relevant to human mutation.

Chapters 8 and 9

Dellarco, V. L., P. E. Voytek, and A. Hollaender. 1985. *Aneuploidy: Etiology and Mechanisms.* New York: Plenum. A collection of research summaries on aneuploidy in humans and experimental organisms.

Epstein, C. J., et al. 1983. "Recent Developments in Prenatal Diagnosis of Genetic Diseases and Birth Defects." *Annual Review of Genetics* 17: 49–83. Includes amniocentesis.

Feldman, M. G., and E. R. Sears. 1981. "The Wild Gene Resources of Wheat." *Scientific American* (January). A general discussion of the genomes of wheat and its relatives, and how new genes can be introduced.

Friedmann, T. 1971. "Prenatal Diagnosis of Genetic Disease." *Scientific*

American (November). An early article on amniocentesis and its uses.

Fuchs, F. 1980. "Genetic Amniocentesis." *Scientific American* (August).

German, J., ed. 1974. *Chromosomes and Cancer.* New York: Wiley. A large technical work, but readable, describing the relation of chromosome changes and cancer.

deGrouchy, J., and C. Turleau. 1984. *Clinical Atlas of Human Chromosomes.* New York: Wiley. A systematic examination of all the human chromosomes and the aberrations associated with them.

Hassold, T. J., and P. A. Jacobs, 1984. "Trisomy in Man." *Annual Review of Genetics* 18: 69–98. A comprehensive summary of trisomy, including a discussion of the maternal age effect.

Hulse, J. H., and D. Spurgeon. 1984. "Triticale." *Scientific American* (August). An account of the development and possible benefits of this wheat-rye amphidiploid.

Lawrence, W. J. C. 1968. *Plant Breeding.* London: Edward Arnold (*Studies in Biology,* No. 12). A short introduction to the subject.

Mangelsdof, P. C. 1986. "The Origins of Corn." *Scientific American* (August).

Maniatis, T. E., et al. 1980. "The Molecular Genetics of Human Hemoglobins." *Annual Review of Genetics* 14: 145–178. A useful summary, which could be profitably read at this point in the course or after reading the material on molecular genetics.

Patterson, D. 1987. "The Causes of Down Syndrome." *Scientific American* (August).

Shepherd, J. F. 1982. "The Regeneration of Potato Plants from Protoplasts." *Scientific American* (May). A review by one of the leaders in this field.

Swanson, C. P., T. Mertz, and W. J. Young, 1967. *Cytogenetics.* Englewood Cliffs, N. J.: Prentice-Hall.

Chapter 10

Adelberg, E. A. 1966. *Papers on Bacterial Genetics.* Boston: Little, Brown.

Brock, T. D. 1990. *The Emergence of Bacterial Genetics.* Cold Spring Harbor, N.Y.: Cold Spring Harbor Laboratory Press. The definitive treatise on the beginnings and development of the field of bacterial genetics.

Hayes, W. 1968. *The Genetics of Bacteria and Their Viruses,* 2d ed. New York: Wiley. The standard and classic text, written by a pioneer in the subject.

Lewin, B. 1977. *Gene Expression,* Vol. 1: *Bacterial Genomes.* New York: Wiley. An excellent set of volumes, all of which are relevant to various sections of this text.

Lewin, B. 1977. *Gene Expression,* Vol. 3: *Plasmids and Phages.* New York: Wiley.

Miller, J. H. 1992. *A Short Course in Bacterial Genetics.* Cold Spring Harbor, N.Y.: Cold Spring Harbor Laboratory Press. A laboratory manual with extensive historical introductions to each area of bacterial genetics.

Stent, G. S., and R. Calendar. 1978. *Molecular Genetics,* 2d ed. New York: W. H. Freeman and Co. A lucidly written account of the development of our present understanding of the subject, based mainly on experiments in bacteria and phage.

Chapter 11

Dickerson, R. E. 1983. "The DNA Helix and How It is Read." *Scientific American* (December). An article with some beautiful color models of DNA structures.

Kornberg, A., and T. Baker. 1992. *DNA Replication,* 2d ed. New York: W. H. Freeman and Co.

Wang, J. C. 1982. "DNA Topoisomerases." *Scientific American* (July). Diagrams different topological forms of DNA.

Watson, J. D. 1981. *The Double Helix.* New York: Atheneum. An enjoyable personal account of Watson and Crick's discovery, including the human dramas involved.

Chapter 12

Benzer, S. 1962. "The Fine Structure of the Gene." *Scientific American* (January). A popular version of the author's pioneer experiments.

Dressler, D., and H. Potter. 1991. *Discovering Enzymes.* New York: Scientific American Library. A delightful book that describes the discovery of enzymes and shows many of their features in well-planned easy-to-read figures.

Felsenfeld, G. 1985. "DNA." *Scientific American* (October).

Prusiner, S. B. 1995. "The Prion Diseases." *Scientific American* (January). A review of an article describing the fascinating phenomenon of infectious protein particles.

Radman, M., and R. Wagner. 1988. "The High Fidelity of DNA Duplication." *Scientific American* (August).

Watson, J. D., et al. 1987. *The Molecular Biology of the Gene,* 4th ed. Menlo Park, Calif.: Benjamin/Cummings. A superb development of the subject, written in a highly readable style and well illustrated.

Yanofsky, C. 1967. "Gene Structure and Protein Structure." *Scientific American* (May). This article gives the details of colinearity at the molecular level.

Chapter 13

Crick, F. H. C. 1962. "The Genetic Code." *Scientific American* (October). This article and the following one are popular accounts of code-cracking experiments.

Crick, F. H. C. 1966. "The Genetic Code: III." *Scientific American* (October).

Darnell, J. E., Jr. 1985. "RNA." *Scientific American* (October).

Doolittle, R. F. 1985. "Proteins." *Scientific American* (October).

Lake, J. A. 1981. "The Ribosome." *Scientific American* (August). Three-dimensional model of the ribosome.

Lane, C. 1976. "Rabbit Haemoglobin from Frog Eggs." *Scientific American* (August). This article describes experiments illustrating the universality of the genetic system.

Lawn, R. M., and G. A. Vehar. 1986. "The Molecular Genetics of Hemophilia." *Scientific American* (March).

Miller, O. L. 1973. "The Visualization of Genes in Action." *Scientific American* (March). A discussion of electron microscopy of transcription and translation.

Moore, P. B. 1976. "Neutron-Scattering Studies of the Ribosome." *Scientific American* (October). This article gives the details of ribosome substructure.

Nirenberg, M. W. 1963. "The Genetic Code: II." *Scientific American* (March). Another account of early code-cracking experiments.

Radman, M., and R. Wagner. 1988. "The High Fidelity of DNA Duplication." *Scientific American* (August).

Rich, A., and S. H. Kim. 1978. "The Three-Dimensional Structure of Transfer RNA." *Scientific American* (January). A presentation of the experimental evidence behind the structure described in this chapter.

Weinberg, R. A. 1985. "The Molecules of Life." *Scientific American* (October).

Chapter 14

Britten, R. J., and D. Kohne. 1968. "Repeated Sequences in DNA." *Science* 161: 529–540. One of the important summaries of the theoretical basis for distinguishing DNAs by renaturation.

Broda, P. 1979. *Plasmids*. New York: W. H. Freeman and Co. One of the few technical books on the subject.

Brown, D. D. 1973. "The Isolation of Genes." *Scientific American* (August). Illustrates the power of focusing molecular techniques on one specific locus with special properties.

Cohen, S. 1975. "The Manipulation of Genes." *Scientific American* (July). A summary of recombinant DNA techniques by one of the main innovators.

Fiddes, J. C. 1977. "The Nucleotide Sequence of a Viral DNA." *Scientific American* (December). This is a review of a landmark, the DNA sequence of an entire virus genome with an unexpected discovery.

Gilbert, W., and L. Villa-Komaroff. 1980. "Useful Proteins from Recombinant Bacteria." *Scientific American* (April). A description of the method of DNA sequencing used most extensively. Also discusses the potential application of recombinant DNA techniques.

Itakura, K., et al. 1977. "Expression in *E. coli* of a Chemically Synthesized Gene for the Hormone Somatostatin." *Science* 198: 1056–1063. A technical paper well worth reading for its historical significance. It represents the start of bioengineering—using DNA manipulation to modify cells to produce a medically useful human protein.

Khorana, H. G., et al. 1972. "Studies on Polynucleotides. CIII. Total Synthesis of the Structural Gene for an Alanine Transfer Ribonucleic Acid from Yeast." *Journal of Molecular Biology* 72: 209–217. A classic technical paper.

Nathans, D., and H. O. Smith. 1975. "Restriction Endonucleases in the Analysis and Restructuring of DNA Molecules." *Annual Review of Biochemistry* 44: 273–293. A technical review of restriction enzymes by two pioneers in the field.

Sinsheimer, R. L. 1977. "Recombinant DNA." *Annual Review of Biochemistry* 46: 415–438. A provocative article by a leading molecular biologist who has expressed concern about potential hazards of DNA manipulation.

Varmus, H. 1987. "Reverse Transcription." *Scientific American* (September).

Watson, J. D., M. Gilman, J. Witkowski, and M. Zoller. 1992. *Recombinant DNA*, 2d ed. New York: W. H. Freeman and Co. A well-written account of recombinant DNA and its applications, with many wonderful illustrations.

Chapter 15

Capecchi, M. R. 1994. "Targeted Gene Replacement." *Scientific American* (March). A popular article describing details of the production of knockout mice by gene replacement.

Mertens, T. R. 1975. *Human Genetics: Readings on the Implications of Genetic Engineering*. New York: Wiley. A collection of popular articles.

Moses, P. B., and N.-H. Chua. 1988. "Light Switches for Plant Genes." *Scientific American* (April).

Murray, A. W., and J. W. Szostak. 1987. "Artificial Chromosomes." *Scientific American* (November).

Watson, J. D., M. Gilman, J. Witkowski, and M. Zoller. 1992. *Recombinant DNA*, 2d ed. New York: W. H. Freeman and Co. A well-written account of recombinant DNA and its applications, with many wonderful illustrations.

White, R., and J. M. Lalouel. 1988. "Chromosome Mapping with DNA Markers." *Scientific American* (February).

Chapter 16

Britten, R. J., and D. E. Kohne, 1970. "Repeated Segments of DNA." *Scientific American* (April).

Brown, S. W. 1966. "Heterochromatin." *Science* 151: 417–425. A nice review of the classic cytological observations.

Chambon, P. 1981. "Split Genes." *Scientific American* (May). The discovery of intervening sequences is described.

Darnell, J., H. Lodish, and D. Baltimore. 1990. *Molecular Cell Biology,* 2d ed. New York: Scientific American Books. A useful book that does a good job of translating many of the concepts of genetics into cellular processes, including a particularly strong section on chromosome structure and types of repetitive DNA.

Davidson, E., and R. Britten. 1973. "Organization, Transcription and Regulation in the Animal Genome." *Quarterly Review of Biology* 48: 565–613. The analysis of renaturation kinetics of DNA fragments provides insights into chromosome structure.

Dupraw, E. J. 1970. *DNA and Chromosomes*. New York: Holt, Rinehart & Winston. A useful book on chromosome substructure, containing excellent photographs by the author.

Gasser, S. M., and U. K. Laemmli. 1987. "A Glimpse at Chromosomal Order." *Trends in Genetics* 3: 16–22. A review that focuses on the various levels of chromosome organization.

Greider, C. W. 1990. "Telomeres, Telomerase, and Senescence." *BioEssays* 12: 363–369. A good review written for nonexperts.

Grigliatti, T. A. 1991. "Position Effect Variegation—An Assay for Nonhistone Chromosomal Proteins and Chromatin Assembly and Modification Factors." *Methods in Cell Biology* 35: 587–627. A readable account of research progress on this topic.

Hayashi, S., et al. 1980. "Hybridization of tRNAs of *Drosophila melanogaster*." *Chromosoma* 76: 65–84. A technical report showing how specific genes can be located cytologically by hybridization of labeled RNA to chromosomes in situ.

Hilliker, A. J., and C. B. Sharp. "New Perspectives on the Genetics and Molecular Biology of Constitutive Heterochromatin." In *Chromosome Structure and Function,* J. P. Gustaffson and R. Appels, eds. New York: Plenum. A summary of research progress focusing on *Drosophila* heterochromatin. The book generally is useful on the subject matter of this chapter.

Jeffreys, A. J., V. Wison, and S. L. Thein. 1985. "Hypervariable 'Minisatellite' Regions in Human DNA." *Nature* 314: 67–73. One of the first papers on the subject of DNA fingerprinting.

Jeffreys, A. J., V. Wison, S. L. Thein, D. J. Weatherall, and B. A. J. Ponder. 1986. "DNA Fingerprints and Segregation Analysis of Multiple Markers in Human Pedigrees." *American Journal of Human Genetics* 39: 25–37. A technical article on the inheritance patterns of DNA fingerprints.

Manuelidis, L. 1990. "A View of Interphase Chromosomes." *Science* 250: 1533–1540. An easily readable review that relates metaphase chromosome structure to structures in interphase.

Chapter 17

Coulson, A., et al. 1991. "YACs and the *C. Elegans* Genome." *BioEssays* 13: 413–417. A review describing the use of YACs to build up contigs for this nematode genome.

Fleischmann, M. D., et al. 1995. "Whole-Genome Random Sequencing and Assembly of *Haemophilus influenzae* Rd." *Science* 269: 496–512. The article reports the first full genomic sequence for a free-living organism.

Foote, S., D. Vollrath, A. Hilton, and D. C. Page. 1992. "The Human Y Chromosome: Overlapping DNA Clones Spanning the Euchroatic Region." *Science* 258: 60–66. The use of sequence-tagged sites in producing the physical map of the Y chromosome.

Green, E. D., and M. V. Olson. 1990. "Chromosomal Region of the Cystic Fibrosis Gene in Yeast Artificial Chromosomes: A Model for Human Genome Mapping." *Science* 250: 94–98. An easy-to-understand article on the principles of genome mapping using YACs.

Murray, J. C., et al. 1994. "A Comprehensive Human Linkage Map with Centimorgan Density." *Science* 265: 2049–2054. Each year *Science* publishes a genome map issue describing progress in mapping the human genome. This issue describes the use of molecular markers and other methods in obtaining a fine scale map of the human genome.

Walter, M., et al. 1994. "A Method for Constructing Radiation Hybrid Maps of Whole Genomes." *Nature Genetics* 7: 22–28. An easy-to-understand research paper describing the technique of radiation hybrid mapping.

Chapter 18

Hendrix, R. W., J. W. Roberts, F. W. Stahl, and R. A. Weisberg. 1983. *Lambda II.* Cold Spring Harbor, N. Y.: Cold Spring Harbor Laboratory Press. A series of incisive reviews about phage lambda.

Herskowitz, I. 1973. "Control of Gene Expression in Bacteriophage Lambda." *Annual Review of Genetics* 7: 289–324. A technical review of the complex and well-analyzed regulation of lambda genes.

Jacob, F., and J. Monod. 1961. "Genetic Regulatory Mechanisms in the Synthesis of Proteins." *Journal of Molecular Biology* 3: 318–356. A classic paper setting forth the elements of an operon and the experimental evidence.

Maniatis, T., S. Goodbourn, and J. A. Fischer. 1987. "Regulation of Inducible and Tissue-Specific Gene Expression." *Science* 236: 1237–1244. A current review of regulatory elements in eukaryotes.

Maniatis, T., and M. Ptashne. 1976. "A DNA Operator-Repressor System." *Scientific American* (January). This article discusses the molecular structures of the components of the *lac* operon.

Miller, J. H., and W. S. Reznikoff, eds. 1978. *The Operon.* Cold Spring Harbor, N. Y.: Cold Spring Harbor Laboratory Press. A valuable set of reviews on gene regulation in bacteria.

Moses, P. B., and N.-H. Chua. 1988. "Light Switches for Plant Genes." *Scientific American* (April).

Ptashne, M., and W. Gilbert. 1970. "Genetic Repressors." *Scientific American* (June). The exciting story of how repressors were identified and purified, thereby confirming the predictions of Jacob and Monod.

Steitz, J. A. 1988. "Snurps." *Scientific American* (June).

Trends in Biochemical Sciences. November 1991. "Transcription." A whole issue of this journal devoted to reviews about eukaryotic transcription regulation.

Tijan, R. 1995. "Molecular Machines that Control Genes." *Scientific American* (February). An up-to-date review of eukaryotic gene transcription regulation.

Weber, I. T., D. B. McKay, and T. A. Steitz. 1982. "Two Helix DNA Binding Motifs of CAP Found in *lac* Repressor and *gal* Repressor." *Nucleic Acids Research* 10: 5085–5102. Discussion of models for repressor-operator recognition.

Chapter 19

Auerbach, C. 1976. *Mutation Research.* London: Chapman & Hall. Standard text by a pioneer researcher.

Cairns, J. 1978. *Cancer: Science and Society.* New York: W. H. Freeman and Co. A fascinating discussion of the origins of human cancers, and of the role played by the environment.

Croce, C. M., and H. Koprowski. 1978. "The Genetics of Human Cancer." *Scientific American* (February). An excellent popular account.

Devoret, R. 1979. "Bacterial Tests for Potential Carcinogens." *Scientific American* (August). This article describes the use of mutation tests to screen for carcinogens and includes some details of DNA reactions.

Drake, J. W. 1970. *The Molecular Basis of Mutation.* San Francisco: Holden-Day. One of the few standard texts on the subject.

Eisenstadt, E. 1987. "Analysis of Mutagenesis." In Escherichia coli *and* Salmonella typhimurium. Cellular and Molecular Biology, F. C. Neidhardt, ed. Washington, D.C.: American Society for Microbiology. A comprehensive review of studies of mutagenesis in bacteria.

Friedberg, E. C., G. C. Walker, and W. Siede. 1995. *DNA Repair and Mutagenesis.* Washington, D.C.: American Society for Microbiology. A complete review text covering mutagenesis and repair in both bacteria and higher cells. An excellent source for human genetic diseases and cancer in relation to repair genes.

Miller, J. H. 1983. "Mutational Specificity in Bacteria." *Annual Review of Genetics* 17: 215–238. A discussion of the specificity of mutagens in bacteria.

Walker, G. C. 1984. "Mutagenesis and Inducible Responses to Deoxyribonucleic Acid Damage in *Escherichia coli.*" *Microbiological Reviews* 48: 60–93. A review of mutagenesis and repair, with an excellent discussion of the SOS system.

Walker, G. C. 1987. "The SOS Response of *Escherichia coli.*" In Escherichia coli *and* Salmonella typhimurium: Cellular and Molecular Biology. F. C. Neidhardt, ed. Washington, D.C.: American Society for Microbiology. A detailed account of the model for SOS regulation.

Chapter 20

Alberts, B., D. Bray, J. Lewis, M. Raff, K. Roberts, and J. D. Watson. 1989. *Molecular Biology of the Cell,* 2d ed. New York: Garland. See pages 239 to 248 for an excellent description of recombination at the molecular level, with nice figures.

Stahl, F. W. 1979. *Genetic Recombination: Thinking About It in Phage and Fungi.* New York: W. H. Freeman and Co. A rather technical short book on recombination models.

Stahl, F. W. 1994. "The Holliday Junction on its Thirtieth Anniversary." *Genetics* 138: 241–246. An easy-to-read review of the Holliday model and subsequent variations that are used today to explain recombination.

Whitehouse, H. L. K. 1973. *Towards an Understanding of the Mechanism of Heredity.* 3d ed. London: Edward Arnold. An excellent general introduction to genetics stressing the historical approach and the pivotal experiments. It includes a good section on recombination models.

Chapter 21

Baltimore, D. 1985. "Retroviruses and Retrotransposons: The Role of Reverse Transcription in Shaping the Eukaryotic Genome." *Cell* 40: 481–482. A short review of work on RNA intermediates in transposition.

Berg, D. E., and M. M. Howe, eds. 1989. *Mobile DNA*. Washington, D.C.: American Society for Microbiology. A collection of incisive reviews on transposable elements from bacteria to humans.

Bukhari, A. I., J. A. Shapiro, and S. L. Adhya, eds. 1977. *DNA Insertion Elements, Plasmids and Episomes*. Cold Spring Harbor, N. Y.: Cold Spring Harbor Laboratory Press. An excellent large collection of short summary papers involving lower and higher life forms.

Cohen, S. N., and J. A. Shapiro. 1980. "Transposable Genetic Elements." *Scientific American* (February). A popular account stressing bacteria and phages.

Fedoroff, N. V. 1984. "Transposable Genetic Elements in Maize." *Scientific American* (June). An account of the early experiments in maize, with more recent results on the molecular basis of transposition.

Shapiro, J. A., ed. 1983. *Mobile Genetic Elements*. New York: Academic Press. A comprehensive set of reviews covering the latest developments on transposition in later prokaryotes and eukaryotes.

Chapter 22

Borst, P., and L. A. Grivell. 1978. "The Mitochondrial Genome of Yeast." *Cell* 15: 705–723. A nontechnical review of the molecular biology of mtDNA.

Dujon, B. 1981. "Mitochondrial Genetics and Functions." In *The Molecular Biology of the Yeast* Saccharomyces. J. N. Strathern, E. W. Jones, and J. R. Broach, eds. Cold Spring Harbor, N. Y.: Cold Spring Harbor Laboratory Press. The most comprehensive review available. This book is useful for many other aspects of yeast genetics and molecular biology.

Gillham, N. W. 1978. *Organelle Heredity*. New York: Raven Press. A rather technical work on the subject, especially strong on genetic analysis.

Griffiths, A. J. F. 1992. "Fungal Senescence." *Annual Review of Genetics* 26: 349–370. A review article that shows how most cases of fungal senescence are caused by mitochondrial mutations or mitochondrial plasmids.

Grivell, L. A. 1983. "Mitochondrial DNA." *Scientific American* (March). An excellent and up-to-date review including a discussion of the unique mitochondrial translation system and of splicing of mitochondrial introns.

Linnane, A. W., and P. Nagley. 1978. "Mitochondrial Genetics in Perspective: The Derivation of a Genetic and Physical Map of the Yeast Mitochondrial Genome." *Plasmid* 1: 324–345. An excellent review of the genetics and molecular biology of yeast mtDNA.

Sager, R. 1965. "Genes Outside Chromosomes." *Scientific American* (January). A popular description of early *Chlamydomonas* experiments.

Sager, R. 1972. *Cytoplasmic Genes and Organelles*. New York: Academic Press.

Wallace, D. C. 1992. "Mitochondrial Genetics: A Paradigm for Aging and Degenerative Diseases?" *Science* 256: 628–632. A review article on the various types of mtDNA mutations in humans, and their disease outcomes.

Chapter 23

Affara, N. A. 1991. "Sex and the Single Y." *BioEssays* 13: 475–478. Mammalian sex determination.

Ballabio, A., and H. F. Willard. 1992. "Mammalian X-Chromosome Inactivation and the XIST Gene." *Current Opinion in Genetics & Development* 2: 439–448.

Barlow, D. P. 1994. "Imprinting: A Gamete's Point of View." *Trends in Genetics* 10: 194–199.

Curtis, D., R. Lehmann, and P. D. Zamore. 1995. "Translational Regulation in Development." *Cell* 81: 171–178.

Eden, S., and H. Cedar. 1994. "Role of DNA Methylation in the Regulation of Transcription." *Current Opinions in Genetics and Development* 4: 255–259.

Efstratiadis, A. 1994. "Parental Imprinting of Autosomal Mammalian Genes." *Current Opinions in Genetics and Development* 4: 265–280.

Ehlich, A., and R. Kppers. 1995. "Analysis of Immunoglobulin Rearrangements in Single B Cells." *Current Opinions in Immunology* 4: 281–284.

Engelhard, V. H. 1994. "How Cells Present Antigen." *Scientific American* (August).

Jorgensen, R. A. 1995. "Co-suppression, Flower Color Patterns, and Metastable Gene Expression States." *Science* 268: 686–691.

Kingston, R. E., and M. R. Green. 1994. "Modeling Eukaryotic Transcriptional Activation." *Current Biology* 4: 325–332.

Lovell-Badge, R. 1992. "Testis Determination: Soft Talk and Kinky Sex." *Current Opinion in Genetics & Development* 2: 596–601.

Marshall-Graves, J. A. 1995. "The Origin and Function of the Mammalian Y Chromosome and Y-Borne Genes—An Evolving Understanding." *BioEssays* 17: 311–320.

McKeown, M. 1992. "Sex Differentiation: The Role of Alternative Splicing." *Current Opinion in Genetics & Development* 2: 299–304.

Moore, T., and D. Haig. 1991. "Genomic Imprinting in Mammalian Development: A Parental Tug-of-War." *Trends in Genetics* 7: 45–49.

Newman, A. J. 1994. "Pre-mRNA Splicing." *Current Opinions in Genetics and Development* 4: 298–304.

Parkhurst, S. M., and P. M. Meneely. 1994. "Sex Determination and Dosage Compensation: Lessons from Flies and Worms." *Science* 264: 924–932.

Pfeifer, K., and S. M. Tilghman. 1994. "Allele-Specific Gene Expression in Mammals: The Curious Case of the Imprinted RNAs." *Genes & Development* 8: 1867–1874.

Tijan, R. 1995. "Molecular Machines that Control Genes." *Scientific American* (February).

Weissman, I. L., and M. D. Cooper. 1993. "How the Immune System Develops." *Scientific American* (September). This issue is entirely devoted to the biology of the immune system and contains several other interesting papers.

Chapter 24

Baserga, R. 1994. "Oncogenes and the Strategy of Growth Factors." *Cell* 79: 927–930.

Bowerman, B. 1995. "Determinants of Blastoderm Identity in the Early *C. Elegans* Embryo." *BioEssays* 17: 405–414.

Cavenee, W. K., and R. L. White. 1995. "The Genetic Basis of Cancer." *Scientific American* (March).

Cohen, G. B., R. Ren, and D. Baltimore. 1995. "Modular Binding Domains in Signal Transduction Proteins." *Cell* 80: 237–248.

Cox, L. S., and D. P. Lane. 1995. "Tumour Suppressors, Kinases and Clamps: How p53 Regulates the Cell Cycle in Response to DNA Damage." *BioEssays* 17: 501–508.

Draetta, G. F. 1994. "Mammalian G1 Cyclins." *Current Opinions in Cell Biology* 6: 842–846.

Edgar, B. A. 1994. "Cell-Cycle Control in a Developmental Context." *Current Biology* 4: 522–524.

Elledge, S. J., and J. W. Harper. 1994. "Cdk Inhibitors: On the Threshold of Checkpoints and Development." *Current Opinions in Cell Biology* 6: 847–852.

Fishel, R., and R. D. Kolodner. 1995. "Identification of Mismatch Repair Genes and Their Role in the Development of Cancer." *Current Opinions in Genetics and Development* 5: 382–396.

Haffner, R., and M. Oren. 1995. "Biochemical Properties and Biological Effects of p53." *Current Opinions in Genetics and Development* 5: 84–91.

Hartwell, L. H., and M. B. Kastan. 1994. "Cell Cycle Control and Cancer." *Science* 266: 1821–1828.

Jiricny, J. 1994. "Colon Cancer and DNA Repair: Have Mismatches Met Their Match?" *Trends in Genetics* 10: 164–168.

Kamb, A. 1995. "Cell-Cycle Regulators and Cancer." *Trends in Genetics* 11: 136–140.

Karin, M., and T. Hunter. 1995. "Transcriptional Control by Protein Phosphorylation: Signal Transmission from the Cell Surface to the Nucleus." *Current Biology* 5: 747–757.

Kenyon, C. 1995. "A Perfect Vulva Every Time: Gradients and Signaling Cascades in *C. Elegans*." *Cell* 82: 171–174.

Maruta, H., and A. W. Burgess. 1994. "Regulation of the Ras Signaling Network." *BioEssays* 16: 489–496.

Murray, A. W. 1995. "The Genetics of Cell Cycle Checkpoints." *Current Opinions in Genetics and Development* 5: 5–11.

Neer, E. J. 1995. "Heterotrimeric G Proteins: Organizers of Transmembrane Signals." *Cell* 80: 249–257.

Nigg, E. A. 1995. "Cyclin-Dependent Protein Kinases: Key Regulators of the Eukaryotic Cell Cycle." *BioEssays* 17: 471–480.

Picksley, S. M., and D. P. Lane. 1994. "p53 and Rb: Their Cellular Roles." *Current Opinions in Cell Biology* 6: 853–858.

Rhyu, M. S., and J. A. Knoblich. 1995. "Spindle Orientation and Asymmetric Cell Fate." *Cell* 82: 523–525.

Sohn, R. H., and P. J. Goldschmidt-Clermont. 1994. "Profilin: At the Crossroads of Signal Transduction and the Actin Cytoskeleton." *BioEssays* 16: 465–472.

Sullivan, W., and W. E. Theurkauf. 1995. "The Cytoskeleton and Morphogenesis of the Early *Drosophila* Embryo." *Current Opinions in Cell Biology* 7: 18–22.

Thomas, B. J., and S. L. Zipursky. 1994. "Early Pattern Formation in the Developing *Drosophila* Eye." *Trends in Cell Biology* 4: 389–394.

Weinberg, R. A. 1995. "The Retinoblastoma Protein and Cell Cycle Control." *Cell* 81: 323–330.

Zipursky, S. L., and G. M. Rubin. 1994. "Determination of Neuronal Cell Fate: Lessons from the R7 Neuron of *Drosophila*." *Annual Review of Neuroscience* 17: 373–397.

Chapter 25

Capecchi, M. R. 1994. "Targeted Gene Replacement." *Scientific American* (March). A description of approaches to the alteration of gene structure exploiting the cellular machinery for DNA repair and homologous recombination by one of the innovators of the technology.

DiNardo, S., J. Heemskerk, S. Dougan, and P. H. O'Farrell. 1994. "The Making of a Maggot: Patterning the *Drosophila* Embryonic Epidermis." *Current Opinion in Genetics and Development* 4: 529–534.

Duboule, D., and G. Morata. 1994. "Colinearity and Functional Hierarchy among the Genes of the Homeotic Complexes." *Trends in Genetics* 10: 358–364.

Gonzalez Crespo, S., and M. Levine. 1994. "Related Target Enhancers for Dorsal and NF-kappa B-Signaling Pathways." *Science* 264: 255–258.

Hlskamp, M., and D. Tautz. 1991. "Gap Genes and Gradients—The Logic Behind the Gaps." *BioEssays* 13: 261–268.

Ip, Y. T., and M. Levine. 1994. "Molecular Genetics of *Drosophila* Immunity." *Current Opinion in Genetics and Development* 4: 672–677.

Kenyon, C. 1994. "If Birds Can Fly, Why Can't We? Homeotic Genes and Evolution." *Cell* 78: 175–180.

Kornberg, T. B., and T. Tabata. 1993. "Segmentation of the *Drosophila* Embryo." *Current Opinions in Genetics and Development* 3: 585–594.

Klingler, M. 1994. "Segmentation in Insects: How Singular is *Drosophila?*" *BioEssays* 16: 391–392.

Krumlauf, R. 1994. "*Hox* Genes in Vertebrate Development." *Cell* 78: 191–201.

Lawrence, P. A., and G. Morata. 1994. "Homeobox Genes: Their Function in *Drosophila* Segmentation and Pattern Formation." *Cell* 78: 181–189.

McGinnis, W., and M. Kuziora. 1994. "The Molecular Architecture of Body Design." *Scientific American* (February).

Meyerowitz, E. M. 1994. "Genetics of Flower Development." *Scientific American* (November). Similar families of transcription factors are found to be responsible for homeotic mutations during plant pattern formation as well.

Orlando, V., and R. Paro. 1995. "Chromatin Multiprotein Complexes Involved in the Maintenance of Transcription Patterns." *Current Opinions in Genetics and Development* 5: 174–179.

Perrimon, N., and C. Desplan. 1994. "Signal Transduction in the Early *Drosophila* Embryo: When Genetics Meets Biochemistry." *Trends in Biochemical Sciences* 19: 509–513.

Quiring, R., U. Walldorf, U. Kloter, and W. Gehring. 1994. "Homology of the Eyeless Gene of *Drosophila* to the Small Eye Gene in Mice and Aniridia in Humans. *Science* 265: 785–789.

Schpbach, T., and S. Roth. 1994. "Dorsoventral Patterning in *Drosophila* Oogenesis." *Current Opinion in Genetics and Development* 4: 502–507.

St. Johnston, R. D. 1994. "RNA Localization: Getting to the Top." *Current Opinion in Genetics and Development* 4: 54–56.

St. Johnston, R. D. 1995. "The Intracellular Localization of Messenger RNAs." *Cell* 81: 161–170.

Tautz, D., and R. J. Sommer. 1995. "Evolution of Segmentation Genes in Insects." *Trends in Genetics* 11: 23–27.

Wolpert, L. 1994. "Positional Information and Pattern Formation in Development." *Developmental Genetics* 15: 485–490.

Chapter 26

Beadle, G. W. 1980. "The Ancestry of Corn." *Scientific American* (January). A popular account of the various clues that led to the modern version.

Bodmer, W. F., and L. L. Cavalli-Sforza. 1976. *Genetics, Evolution, and Man.* New York: W. H. Freeman and Co.

Cavalli-Sforza, L. L., P. Menozzi, and A. Piazza. 1994. *The History and Geography of Human Genes.* Princeton, N.J.: Princeton University Press. A complete discussion of human genetic variation, its history and the forces molding it, with the most recent and complete tables of human gene frequencies.

Clarke, B. 1975. "The Causes of Biological Diversity." *Scientific American* (August). A popular account emphasizing genetic polymorphism.

Crow, J. F. 1979. "Genes That Violate Mendel's Rules." *Scientific American* (February). An interesting article on a topic called segregation distortion (not treated in this text) and its effects in populations.

Dobzhansky, T. 1951. *Genetics and the Origin of Species*. New York: Columbia University Press. The classic synthesis of population genetics and the processes of evolution. The most influential book on evolution since Darwin's *Origin of Species*.

Ford, E. B. 1971. *Ecological Genetics,* 3d ed. London: Chapman & Hamm. A nonmathematical treatment of the role of genetic variation in nature, stressing morphological variation.

Futuyma, D. J. 1986. *Evolutionary Biology,* 2d ed. Sunderland, Mass.: Sinauer. The best modern discussion of population genetics, ecology, and evolution from both a theoretical and an experimental point of view.

Hartl, D. L., and A. G. Clark. 1989. *Principles of Population Genetics,* 2d ed. Sunderland, Mass.: Sinauer.

Lerner, I. M., and W. J. Libby. 1976. *Hereditary, Evolution and Society,* 2d ed. New York: W. H. Freeman and Co. A text meant for non-science students.

Lewontin, R. C. 1974. *The Genetic Basis of Evolutionary Change.* New York: Columbia University Press. A discussion of the prevalence and role of genetic variation in natural populations. Both morphological and protein variations are considered.

Lewontin, R. C. 1982. *Human Diversity.* New York: Scientific American Books. Applies concepts of population genetics to human diversity and evolution.

Li, W. H., and D. Graur. 1991. *Fundamentals of Molecular Evolution.* Sunderland, Mass.: Sinauer. An excellent review of what is known about the forces and rates of evolution of proteins and DNA sequences.

Scientific American. September 1978. "Evolution." This volume contains several articles relevant to population genetics.

Wilson, A. C. 1985. "The Molecular Basis of Evolution." *Scientific American* (October).

Chapter 27

Bodmer, W. F., and L. L. Cavalli-Sforza. 1970. "Intelligence and Race." *Scientific American* (October). A popular treatment, including a discussion of heritability.

Falconer, D. S. 1981. *Introduction to Quantitative Genetics,* 2d ed. New York: Ronald Press. A widely read text with a strong mathematical emphasis.

Feldman, M. W., and R. C. Lewontin. 1975. "The Heritability Hangup." *Science* 190: 1163–1168. A discussion of the meaning of heritability and its limitations, especially in relation to human intelligence.

Lewontin, R. C. 1974. "The Analysis of Variance and the Analysis of Causes." *American Journal of Human Genetics.* 26: 400–411. A discussion of the meaning of the analysis of variance in genetics as a method for determining the roles of heredity and environment in determining phenotype.

Lewontin, R. C. 1982. *Human Diversity.* New York: Scientific American Books. Includes a discussion of quantitative variation in human populations.

Glossary

A Adenine, or adenosine.

abortive transduction The failure of a transducing DNA segment to be incorporated into the recipient chromosome.

acentric chromosome A chromosome having no centromere.

achondroplasia A type of dwarfism in humans, inherited as an autosomal dominant phenotype.

acrocentric chromosome A chromosome having the centromere located slightly nearer one end than the other.

active site The part of a protein that must be maintained in a specific shape if the protein is to be functional—for example, in an enzyme, the part to which the substrate binds.

adaptation In the evolutionary sense, some heritable feature of an individual's phenotype that improves its chances of survival and reproduction in the existing environment.

adaptive landscape The surface plotted in a three-dimensional graph, with all possible combinations of allele frequencies for different loci plotted in the plane, and mean fitness for each combination plotted in the third dimension.

adaptive peak A high point (perhaps one of several) on an adaptive landscape; selection tends to drive the genotype composition of the population toward a combination corresponding to an adaptive peak.

adaptive surface *See* **adaptive landscape.**

adaptor protein A protein that binds to certain specific phosphorylated amino acid sequences on second protein, often a transmembrane receptor, and also associates with still other proteins, thereby allowing a complex of proteins to "dock" to the receptor. This docking brings proteins of this complex in proximity to one another and by doing so, permits propagation of an intracellular signal in a signal transduction pathway.

additive genetic variance Genetic variance associated with the average effects of substituting one allele for another.

adenine A purine base that pairs with thymine in the DNA double helix.

adenosine The nucleoside containing adenine as its base.

adenosine triphosphate *See* **ATP.**

adjacent segregation In a reciprocal translocation, the passage of a translocated and a normal chromosome to each of the poles.

ADP Adenosine diphosphate.

Ala Alanine (an amino acid).

albino A pigmentless "white" phenotype, determined by a mutation in a gene coding for a pigment-synthesizing enzyme.

alkylating agent A chemical agent that can add alkyl groups (for example, ethyl or methyl groups) to another molecule; many mutagens act through alkylation.

allele One of the different forms of a gene that can exist at a single locus.

allele frequency A measure of the commonness of an allele in a population; the proportion of all alleles of that gene in the population that are of this specific type.

allopolyploid *See* **amphidiploid.**

allosteric transition A change from one conformation of a protein to another.

alternate segregation In a reciprocal translocation, the passage of both normal chromosomes to one pole and both translocated chromosomes to the other pole.

alternation of generations The alternation of gametophyte and sporophyte stages in the life cycle of a plant.

alternative splicing The process by which different mRNAs are produced from the same primary transcript, through variations in the splicing pattern of the transcript. Multiple mRNA "isoforms" can be produced in a single cell, or the different isoforms can display different tissue-specific patterns of expression. If the alternative exons fall within the open reading frames of the mRNA isoforms, different proteins will be produced by the alternative mRNAs.

amber codon The codon UAG, a nonsense codon.

amber suppressor A mutant allele coding for a tRNA whose anticodon is altered in such a way that the tRNA inserts an amino acid at an amber codon in translation.

Ames test A widely used test to detect possible chemical carcinogens; based on mutagenicity in the bacterium *Salmonella*.

amino acid A peptide; the basic building block of proteins (or polypeptides).

amniocentesis A technique for testing the genotype of an embryo or fetus in utero with minimal risk to the mother or the child.

AMP Adenosine monophosphate.

amphidiploid   An allopolyploid; a polyploid formed from the union of two separate chromosome sets and their subsequent doubling.

amplification   The production of many DNA copies from one master region of DNA.

anaphase   An intermediate stage of nuclear division during which chromosomes are pulled to the poles of the cell.

aneuploid cell   A cell having a chromosome number that differs from the normal chromosome number for the species by a small number of chromosomes.

animal breeding   The practical application of genetic analysis for development of lines of domestic animals suited to human purposes.

annealing   Spontaneous alignment of two single DNA strands to form a double helix.

antibody   A protein (immunoglobulin) molecule, produced by the immune system, that recognizes a particular substance (antigen) and binds to it.

anticodon   A nucleotide triplet in a tRNA molecule that aligns with a particular codon in mRNA under the influence of the ribosome, so that the amino acid carried by the tRNA is inserted in a growing protein chain.

antigen   A molecule that is recognized by antibody (immunoglobulin) molecules. Generally, multiple antibody molecules can recognize a given antigen.

antiparallel   A term used to describe the opposite orientations of the two strands of a DNA double helix; the 5′ end of one strand aligns with the 3′ end of the other strand.

AP sites   Apurinic or apyrimidinic sites resulting from the loss of a purine or pyrimidine residue from the DNA.

Arg   Arginine (an amino acid).

ascospore   A sexual spore from certain fungus species in which spores are found in a sac called an ascus.

ascus   In fungi, a sac that encloses a tetrad or an octad of ascospores.

asexual spores   *See* **spore.**

Asn   Asparagine (an amino acid).

Asp   Aspartate (an amino acid).

ATP (adenosine triphosphate)   The "energy molecule" of cells, synthesized mainly in mitochondria and chloroplasts; energy from the breakdown of ATP drives many important cell reactions.

attached X   A pair of *Drosphila* X chromosomes joined at one end and inherited as a single unit.

attenuator   A region adjacent to the structural genes of the *trp* operon; this region acts in the presence of tryptophan to reduce the rate of transcription from the structural genes.

autonomous controlling element   A controlling element that seems to have both regulator and receptor functions combined in the single unit, which enters a gene and causes an unstable mutation.

autonomous phenotype   A genetic trait in multicellular organisms in which only genotypically mutant cells exhibit the mutant phenotype. Conversely, a *nonautonomous* trait is one in which genotypically mutant cells cause other cells (regardless of their genotype) to exhibit a mutant phenotype.

autonomous replication sequence (ARS)   A segment of a DNA molecule needed for the initiation of its replication; generally a site recognized and bound by the proteins of the replication system.

autophosphorylation   The process by which a protein kinase phosphorylates specific amino acid residues on itself.

autopolyploid   A polyploid formed from the doubling of a single genome.

autoradiogram   A pattern of dark spots in a developed photographic film or emulsion, in the technique of autoradiography.

autoradiography   A process in which radioactive materials are incorporated into cell structures, which are then placed next to a film or photographic emulsion, thus forming a pattern on the film corresponding to the location of the radioactive compounds within the cell.

autoregulatory loop   The process by which the expression of a gene is controlled by it's own gene product.

autosome   Any chromosome that is not a sex chromosome.

auxotroph   A strain of microorganisms that will proliferate only when the medium is supplemented with some specific substance not required by wild-type organisms.

BAC   Bacterial artificial chromosome; an F plasmid engineered to act as a cloning vector that can carry large inserts.

Back mutation   *See* **reversion.**

bacteriophage (phage)   A virus that infects bacteria.

balanced polymorphism   Stable genetic polymorphism maintained by natural selection.

Balbiani ring   A large chromosome puff.

Barr body   A densely staining mass that represents an inactivated X chromosome.

base analog   A chemical whose molecular structure mimics that of a DNA base; because of the mimicry, the analog may act as a mutagen.

bead theory   The disproved hypothesis that genes are arranged on the chromosome like beads on a necklace, indivisible into smaller units of mutation and recombination.

bimodal distribution   A statistical distribution having two modes.

binary fission   The process in which a parent cell splits into two daughter cells of approximately equal size.

biparental zygote   A *Chlamydomonas* zygote that contains cpDNA from both parents; such cells generally are rare.

blast cell   A cell that divides, generally asymmetrically, to give rise to two different progeny cells. One is a blast cell just like the parental cell and the other is a cell that enters a differentiation pathway. In this manner, a continuously propagating cell population can maintain itself and spin off differentiating cells.

blastoderm   In an insect embryo, the layer of cells that completely surrounds an internal mass of yolk.

blastula   An early developmental stage of lower vertebrate embryos, in which the embryo consists of a single layer of cells surrounding the central yolk.

blending inheritance   A discredited model of inheritance suggesting that the characteristics of an individual result from the smooth blending of fluidlike influences from its parents.

brachydactyly   A human phenotype of unusually short digits, generally inherited as an autosomal dominant.

branch migration   The process by which a single "invading" DNA strand extends its partial pairing with its complementary strand as it displaces the resident strand.

bridging cross A cross made to transfer alleles between two sexually isolated species by first transferring the alleles to an intermediate species that is sexually compatible with both.

broad heritability (H^2) The proportion of total phenotypic variance at the population level that is contributed by genetic variance.

bud A daughter cell formed by mitosis in yeast; one daughter cell retains the cell wall of the parent, and the other (the bud) forms a new cell wall.

buoyant density A measure of the tendency of a substance to float in some other substance; large molecules are distinguished by their differing buoyant densities in some standard fluid. Measured by density-gradient ultracentrifugation.

Burkitt's lymphoma A cancer of the lymphatic system manifested by tumors in the jaw, associated with a chromosomal translocation bringing a specific oncogene next to regulatory elements of one of the immunoglobulin genes.

C Cytosine, or cytidine.

callus An undifferentiated clone of plant cells.

cAMP (cyclic adenosine monophosphate) A molecule that plays a key role in the regulation of various processes within the cell.

cancer The class of disease characterized by rapid and uncontrolled proliferation of cells within a tissue of a multi-tissued eukaryote. Cancers are generally thought to be genetic diseases of somatic cells, arising through sequential mutations that create oncogenes and inactivate tumor suppressor genes.

candidate gene A sequenced gene of previously unknown function that, because of its chromosomal position or some other property, becomes a candidate for a particular function such as disease determination.

CAP (catabolite activator protein) A protein whose presence is necessary for the activation of the *lac* operon.

carbon source A nutrient (such as sugar) that provides carbon "skeletons" needed in the organism's synthesis of organic molecules.

carcinogen A substance that causes cancer.

cardinal gene Those pattern formation genes in Drosophila that are the zygotically acting genes directly responding to the gradients of anterior-posterior and dorsal-ventral positional information created by the maternally expressed pattern formation genes.

carrier An individual who possesses a mutant allele but does not express it in the phenotype because of a dominant allelic partner; thus, an individual of genotype *Aa* is a carrier of *a* if there is complete dominance of *A* over *a*.

cassette model A model to explain mating-type interconversion in yeast. Information for both *a* and *α* mating types is assumed to be present as silent "cassettes"; a copy of either type of cassette may be transposed to the mating-type locus, where it is "played" (transcribed).

catabolite activator protein *See* **CAP.**

catabolite repression The inactivation of an operon caused by the presence of large amounts of the metabolic end product of the operon.

cation A positively charged ion (such as K^+).

CDK *See* **cyclin-dependent protein kinase.**

cDNA *See* **complementary DNA.**

cDNA library A library composed of cDNAs, not necessarily representing all mRNAs.

cell autoninous A genetic trait in multicellular organisms in which only genotypically mutant cells exhibit the mutant phenotype. Conversely, a *nonautonomous* trait is one in which genotypically mutant cells cause other cells (regardless of their genotype) to exhibit a mutant phenotype.

cell cycle The set of events that occur during the divisions of mitotic cells. The cell cycle oscillates between mitosis (M phase) and interphase. Interphase can be subdivided in order into G_1, S phase, and G_2. DNA synthesis occurs during S phase. The length of the cell cycle is regulated through a special option in G_1, in which G_1 cells can enter a resting phase called G_0.

cell division The process by which two cells are formed from one.

cell fate The ultimate differentiated state to which a cell has become committed.

cell lineage A pedigree of cells related through asexual division.

cellular blastoderm The stage of blastoderm in insects after the nuclei have each been packaged in an individual cellular membrane.

centimorgan (cM) *See* **map unit.**

central dogma The hypothesis that information flows only from DNA to RNA to protein; although some exceptions are now known, the rule is generally valid.

centromere A kinetochore; the constricted region of a nuclear chromosome, to which the spindle fibers attach during division.

character Some attribute of individuals within a species for which various heritable differences can be defined.

character difference Alternative forms of the same attribute within a species.

chase *See* **pulse-chase experiment.**

checkpoints Stages of the cell cycle at which the completion of certain events of the cell cycle, such as chromosome replication, must have been successfully completed in order for the cell cycle to progress to the next stage.

chiasma (plural, **chiasmata**) A cross-shaped structure commonly observed between nonsister chromatids during meiosis; the site of crossing-over.

chimera *See* **mosaic.**

chimeric DNA *See* **recombinant DNA.**

chi-square (χ^2) **test** A statistical test used to determine the probability of obtaining the observed results by chance, under a specific hypothesis.

chloroplast A chlorophyll-containing organelle in plants that is the site of photosynthesis.

chromatid One of the two side-by-side replicas produced by chromosome division.

chromatid conversion A type of gene conversion that is inferred from the existence of identical sister-spore pairs in a fungal octad that shows a non-Mendelian allele ratio.

chromatid inference A situation in which the occurrence of a crossover between any two nonsister chromatids can be shown to affect the probability of those chromatids being involved in other crossovers in the same meiosis.

chromatin The substance of chromosomes; now known to include DNA, chromosomal proteins, and chromosomal RNA.

chromocenter The point at which the polytene chromosomes appear to be attached together.

chromomere A small beadlike structure visible on a chromosome during prophase of meiosis and mitosis.

chromosome A linear end-to-end arrangement of genes and other DNA, sometimes with associated protein and RNA.

chromosome aberration Any type of change in the chromosome structure or number.

chromosome loss Failure of a chromosome to become incorporated into a daughter nucleus at cell division.

chromosome map *See* **linkage map.**

chromosome mutation Any type of change in the chromosome structure or number.

chromosome puff A swelling at a site along the length of a polytene chromosome; the site of active transcription.

chromosome rearrangement A chromosome mutation involving new juxtapositions of chromosome parts.

chromosome set The group of different chromosomes that carries the basic set of genetic information of a particular species.

chromosome theory of inheritance The unifying theory stating that inheritance patterns may be generally explained by assuming that genes are located in specific sites on chromosomes.

chromosome walking A method for the dissection of large portions of DNA, in which a cloned portion of DNA, usually eukaryotic, is used to screen recombinant DNA clones from the same genome bank for other clones containing neighboring sequences.

cis conformation In a heterozygote involving two mutant sites within a gene or within a gene cluster, the arrangement $a_1 a_2 / ++$.

cis dominance The ability of a gene to affect genes next to it on the same chromosome.

cis-trans test A test to determine whether two mutant sites of a gene are in the same functional unit or gene.

cistron Originally defined as a functional genetic unit within which two mutations cannot complement. Now equated with the term gene, as the region of DNA that encodes a single polypeptide (or functional RNA molecule such as tRNA or rRNA).

clone (1) A group of genetically identical cells or individuals derived by asexual division from a common ancestor. (2) (*colloquial*) An individual form by some asexual process so that it is genetically identical to its "parent." (3) *See* **DNA clone.**

cM (centimorgan) *See* **map unit.**

code dictionary A listing of the 64 possible codons and their translational meanings (the corresponding amino acids).

codominance The situation in which a heterozygote shows the phenotypic effects of both alleles equally.

codon A section of DNA (three nucleotide pairs in length) that codes for a single amino acid.

coefficient of coincidence The ratio of the observed number of double recombinants to the expected number.

cohesive ends Ends of DNA which are cut in a staggered pattern, and then can hydrogen-bond with complementary base sequences from other similarly formed ends.

cointegrate The product of the fusion of two circular elements to form a single, larger circle.

colinearity The correspondence between the location of a mutant site within a gene and the location of an amino acid substitution within the polypeptide translated from that gene.

colony A visible clone of cells.

compartmentalization The existence of boundaries within the organism beyond which a specific clone of cells will never extend during development.

competent Able to take up exogenous DNA and therby be transformed.

complementary DNA (cDNA) Synthetic DNA transcribed from a specific RNA through the action of the enzyme reverse transcriptase.

complementary RNA (cRNA) Synthetic RNA produced by transcription from a specific DNA single-stranded template.

complementation The production of a wild-type phenotype when two different mutations are combined in a diploid or a heterokaryon.

complementation test *See* **cis-trans test.**

conditional mutation A mutation that has the wild-type phenotype under certain (permissive) environmental conditions and a mutant phenotype under other (restrictive) conditions.

conjugation The union of two bacterial cells, during which chromosomal material is transferred from the donor to the recipient cell.

conjugation tube *See* **pilus.**

conservative replication A disproved model of DNA synthesis suggesting that one-half of the daughter DNA molecules should have both strands composed of newly polymerized nucleotides.

constant region A region of an antibody molecule that is nearly identical with the corresponding regions of antibodies of different specificities.

constitutive Always expressed in an unregulated fashion (when referring to gene control).

constitutive heterochromatin Specific regions of heterochromatin always present and in both homologs of a chromosome.

contig A set of ordered overlapping clones that comprise a chromosomal region or a genome.

controlling element A mobile genetic element capable of producing an unstable mutant target gene; two types exist, the regulator and the receptor elements.

copy-choice model A model of the mechanism for crossing-over, suggesting that crossing-over occurs during chromosome division and can occur only between two supposedly "new" nonsister chromatids; the experimental evidence does not support this model.

correction The production (possibly by excision and repair) of a properly paired nucleotide pair from a sequence of hybrid DNA that contains an illegitimate pair.

correlation coefficient A statistical measure of the extent to which variations in one variable are related to variations in another.

cosegregation In *Chlamydomonas*, parallel behavior of different chloroplast markers in a cross, due to their close linkage on cpDNA.

cosmid A cloning vector that can replicate autonomously like a plasmid and also be packaged into phage.

cotransduction The simultaneous transduction of two bacterial marker genes.

contransformation The simultaneous transformation of two bacterial marker genes.

coupling conformation Linked heterozygous gene pairs in the arrangement *A B / a b.*

covariance A statistical measure used in computing the correlation coefficient between two variables; the covariance is the mean of $(x - \bar{x})(y - \bar{y})$ overall pairs of values for the variables x and y, where $\bar{x}$ is the mean of the x values and $\bar{y}$ is the mean of the y values.

cpDNA Chloroplast DNA.

cri du chat syndrome A lethal human condition in infants caused by deletion of part of one homolog of chromosome 5.

crisscross inheritance Transmission of a gene from male parent to female child to male grandchild—for example, X-linked inheritance.

cRNA *See* **complementary RNA.**

cross The deliberate mating of two parental types of organisms in genetic analysis.

crossing-over The exchange of corresponding chromosome parts between homologs by breakage and reunion.

crossover suppressor An inversion (usually complex) that makes pairing and crossing-over impossible.

cruciform configuration A region of DNA with palindromic sequences in both strands, so that each strand pairs with itself to form a helix extending sideways from the main helix.

culture Tissue or cells multiplying by asexual division, grown for experimentation.

CVS Chorionic villus sampling, a placental sampling procedure for obtaining fetal tissue for chromosome and DNA analysis to assist in prenatal diagnosis of genetic disorders.

cyclic adenosine monophosphate *See* **cAMP.**

cyclin A family of labile proteins that are synthesized and degraded at specific times within each cell cycle and which regulate cell cycle progression through their interactions with specific cyclin-dependent protein kinases.

cyclin-dependent protein kinase A family of protein kinases that, upon activation by cyclins and an elaborate set of positive and negative regulatory proteins, phosphorylate certain transcription factors whose activity is necessary for a particular stage of the cell cycle.

Cys Cysteine (an amino acid).

cystic fibrosis A potentially lethal human disease of secretory glands; the most prominent symptom is excess secretion of lung mucus; inherited as an autosomal recessive.

cytidine The nucleoside containing cytosine as its base.

cytochromes A class of proteins, found in mitochondrial membranes, whose main function is oxidative phosphorylation of ADP to form ATP.

cytogenetics The cytological approach to genetics, mainly involving microscopic studies of chromosomes.

cytohet A cell containing two genetically distinct types of a specific organelle.

cytoplasm The material between the nuclear and cell membranes; includes fluid (cytosol), organelles, and various membranes.

cytoplasmic inheritance Inheritance via genes found in cytoplasmic organelles.

cytosine A pyrimidine base that pairs with guanine.

cytoskeleton The protein cable systems and associated proteins that together form the architecture of a eukaryotic cell.

cytosol The fluid portion of the cytoplasm, outside the organelles.

Darwinian fitness The relative probability of survival and reproduction for a genotype.

default state The developmental state of a cell (or group of cells) in the absence of activation of a developmental regulatory switch.

degenerate code A genetic code in which some amino acids may be encoded by more than one codon each.

deletion Removal of a chromosomal segment from a chromosome set.

denaturation The separation of the two strands of a DNA double helix, or the severe disruption of the structure of any complex molecule without breaking the major bonds of its chains.

denaturation map A map of a stretch of DNA showing the locations of local denaturation loops, which correspond to regions of high AT content.

deoxyribonuclease *See* **DNase.**

deoxyribonucleic acid *See* **DNA.**

determinant A spatially localized molecule that causes cells to adopt a particular fate or set of related fates.

determination The process of commitment of cells to particular fates.

development The process whereby a single cell becomes a differentiated organism.

developmental field A set of cells that together interact to form a developing structure (e.g., an embryo, a tissue, an organ, a limb, etc.).

developmental pathway The chain of molecular events that take a set of equivalent cells and produce the assignment of different fates among those cells.

dicentric chromosome A chromosome with two centromeres.

differentiation The changes in cell shape and physiology associated with the production of the final cell types of a particular organ or tissue.

dihybrid cross A cross between two individuals identically heterozygous at two loci—for example, $Aa\,Bb \times Aa\,Bb$.

dimorphism A "polymorphism" involving only two forms.

dioecious plant A plant species in which male and female organs appear on separate individuals.

diploid A cell having two chromosome sets, or an individual having two chromosome sets in each of its cells.

directed mutagenesis Altering some specific part of a cloned gene and reintroducing the modified gene back into the organism.

directional selection Selection that changes the frequency of an allele in a constant direction, either toward or away from fixation for that allele.

dispersive replication Disproved model of DNA synthesis suggesting more-or-less random interspersion of parental and new segments in daughter DNA molecules.

distribution *See* **statistical distribution.**

distribution function A graph of some precise quantitative measure of a character against its frequency of occurrence.

DNA (deoxyribonucleic acid) A double chain of linked nucleotides (having deoxyribose as their sugars); the fundamental substance of which genes are composed.

DNA clone   A section of DNA that has been inserted into a vector molecule, such as a plasmid or a phage chromosome, and then replicated to form many copies.

DNA fingerprint   The largely individual-specific autoradiographic banding pattern produced when DNA is digested with a restriction enzyme that cuts outside a family of VNTRs, and a Southern blot of the electrophoretic gel is probed with a VNTR-specific probe.

DNA polymerase   An enzyme that can synthesize new DNA strands using a DNA template; several such enzymes exist.

DNase (deoxyribonuclease)   An enzyme that degrades DNA to nucleotides.

dominance variance   Genetic variance at a single locus attributable to dominance of one allele over another.

dominant allele   An allele that expresses its phenotypic effect even when heterozygous with a recessive allele; thus if *A* is a dominant over *a*, then *AA* and *Aa* have the same phenotype.

dominant phenotype   The phenotype of a genotype containing the dominant allele; the parental phenotype that is expressed in a heterozygote.

donor DNA   Any DNA to be used in cloning.

dosage compensation   The process in organisms using a chromosomal sex determination mechanism (such as XX versus XY) that allows standard structural genes on the sex chromosome to be expressed at the same levels in females and males, regardless of the number of sex chromosomes. In mammals, dosage compensation operates by maintaining only a single active X chromosome in each cell; in *Drosophila,* it operates by hyperactivating the male X chromosome.

dose   *See* **gene dose.**

double crossover   Two crossovers occuring in a chromosomal region under study.

double helix   The structure of DNA first proposed by Watson and Crick, with two interlocking helices joined by hydrogen bonds between paired bases.

double infection   Infection of a bacterium with two genetically different phages.

Down syndrome   An abnormal human phenotype, including mental retardation, due to a trisomy of chromosome 21; more common in babies born to older mothers.

drift   *See* **random genetic drift.**

Duchenne muscular dystrophy   A lethal muscle disease in humans caused by mutation in a huge gene coding for the muscle protein dystrophin; inherited as an X-linked recessive phenotype.

duplicate genes   Two identical allele pairs in one diploid individual.

duplication   More than one copy of a particular chromosomal segment in a chromosome set.

dyad   A pair of sister chromatids joined at the centromere, as in the first division of meiosis.

ecdysone   A molting hormone in insects.

ectopic expression   The occurrence of gene expression in a tissue in which it is normally not expressed. Such ectopic expression can be caused by the juxtaposition of novel enhancer elements to a gene.

ectopic integration   In a transgenic organism, the insertion of an introduced gene at a site other than its usual locus.

electrophoresis   A technique for separating the components of a mixture of molecules (proteins, DNAs, or RNAs) in an electric field within a gel.

embryonic (or tissue) polarity   The production of axes of asymmetry in a developing embryo or tissue primordium.

embryonic stem cells   Cultured cell lines that are established from very early embryos and that are essentially totipotent. That is, these cells can be implanted into a host embryo and populate many or all tissues of the developing animal, most importantly including the germ line. Manipulations of these embryonic stem cells (ES cells) are used extensively in mouse genetics to produce targeted gene knockouts.

endocrine system   The organs in the body that secrete hormones into the circulatory system.

endogenote   *See* **merozygote.**

endonuclease   An enzyme that cleaves the phosphodiester bond within a nucleotide chain.

endopolyploidy   An increase in the number of chromosome sets caused by replication without cell division.

endosperm   Triploid tissue in a seed, formed from fusion of two haploid female and one haploid male nucleus.

enforced outbreeding   Deliberate avoidance of mating between relatives.

enhancer   A cis-regulatory sequence that can elevate levels of transcription from an adjacent promoter. Many *tissue-specific enhancers* can determine spatial patterns of gene expression in higher eukaryotes. Enhancers can act on promoters over many tens of kilobases of DNA and can be 5′ or 3′ to the promoter they regulate.

enhancer trap   A transgenic construction inserted in a chromosome which is used to identify tissue-specific enhancers in the genome. In such a construct, a promoter sensitive to enhancer regulation is fused to a reporter gene, such that expression patterns of the reporter gene identify the spatial regulation conferred by nearby enhancers.

enucleate cell   A cell having no nucleus.

environment   The combination of all the conditions external to the genome that potentially affect its expression and its structure.

environmental variance   The variance due to environmental variation.

enzyme   A protein that functions as a catalyst.

epigenetic inheritance   Processes by which heritable modifications in gene function occur but that are not due to changes in the base sequence of the DNA of the organism. Examples of epigenetic inheritance are paramutation, X-chromosome inactivation, and parental imprinting.

episome   A genetic element in bacteria that can replicate free in the cytoplasm or can be inserted into the main bacterial chromosome and replicate with the chromosome.

epistasis   A situation in which the differential phenotypic expression of genotypes at one locus depends upon the genotype at another locus.

epitope   The portion of an antigen molecule that is recognized by a specific immunoglobulin.

equational division   A nuclear division that maintains the same ploidy level of the cell.

equivalence group   A set of immature cells that all have the same developmental potential. In many cases, cells of an equivalence group end up adopting different fates from one another.

ES cells   *See* **embryonic stem cells.**

ethidium   A molecule that can intercalate into DNA double helices when the helix is under torsional stress.

euchromatin   A chromosomal region that stains normally; thought to contain the normally functioning genes.

eugenics   Controlled human breeding based on notions of desirable and undesirable genotypes.

eukaryote   An organism having eukaryotic cells.

eukaryotic cell   A cell containing a nucleus.

euploid   A cell having any number of complete chromosome sets, or an individual composed of such cells.

excision repair   The repair of a DNA lesion by removal of the faulty DNA segment and its replacement with a wild-type segment.

exconjugant   A female bacterial cell that has just been in conjugation with a male and that contains a fragment of male DNA.

exogenote   *See* **merozygote.**

exon   Any nonintron section of the coding sequence of a gene; together, the exons constitute the mRNA and are translated into protein.

exonuclease   An enzyme that cleaves nucleotides one at a time from an end of a polynucleotide chain.

expression library   A library in which the vector carries transcriptional signals to allow any cloned insert to produce mRNA and ultimately a protein product.

expressivity   The degree to which a particular genotype is expressed in the phenotype.

F^- **cell**   In *E. coli*, a cell having no fertility factor; a female cell.

F^+ **cell**   In *E. coli*, a cell having a free fertility factor; a male cell.

F factor   *See* **fertility factor.**

F' **factor**   A fertility factor into which a portion of the bacterial chromosome has been incorporated.

F_1 **generation**   The first filial generation, produced by crossing two parental lines.

F_2 **generation**   The second filial generation, produced by selfing or intercrossing the F_1.

facultative heterochromatin   Heterochromatin located in positions that are composed of euchromatin in other individuals of the same species, or even in the other homolog of a chromosome pair.

FACS   Fluorescence-activated chromosome sorting. Using specific fluorescence signals of stained chromosomes in droplets to activate deflector plates that sort them into individual tubes of uniform types.

familial trait   A trait shared by members of a family.

family selection   A breeding technique of selecting a pair on the basis of the average performance of their progeny.

fate map   A map of an embryo showing areas that are destined to develop into specific adult tissues and organs.

feedback loop   *See* **autoregulatory loop.**

fertility factor (F factor)   A bacterial episome whose presence confers donor ability (maleness).

filial generations   Successive generations of progeny in a controlled series of crosses, starting with two specific parents (the P generation) and selfing or intercrossing the progeny of each new (F_1, F_2, . . .) generation.

filter enrichment   A technique for recovering auxotrophic mutants in filamentous fungi.

fingerprint   The characteristic spot pattern produced by electrophoresis of the polypeptide fragments obtained through denaturation of a particular protein with a proteolytic enzyme.

first-division segregation pattern   A linear pattern of spore phenotypes within the ascus for a particular allele pair, produced when the alleles go into separate nuclei at the first meiotic division, showing that no crossover has occurred between that allele pair and the centromere.

FISH   Fluorescent in situ hybridization. In situ hybridization using a probe coupled to a fluorescent molecule.

fitness   *See* **Darwinian fitness.**

fixed allele   An allele for which all members of the population under study are homozygous, so that no other alleles for this locus exist in the population.

fixed breakage point   According to the heteroduplex DNA recombination model, the point from which unwinding of the DNA double helices begins, as a prelude to formation of heteroduplex DNA.

flow sorting   *See* **FACS.**

fluctuation test   A test used in microbes to establish the random nature of mutation, or to measure mutation rates.

fMET   *See* **formylmethionine.**

focus map   A fate map of areas of the *Drosophila* blastoderm destined to become specific adult structures, based on the frequencies of specific kinds of mosaics.

foreign DNA   DNA from another organism.

formylmethionine (fMet)   A specialized amino acid that is the very first one incorporated into the polypeptide chain in the synthesis of proteins.

forward mutation   A mutation that converts a wild-type allele to a mutant allele.

frame-shift mutation   The insertion or deletion of a nucleotide pair or pairs, causing a disruption of the translational reading frame.

frequency-dependent fitness   Fitness differences whose intensity changes with changes in the relative frequency of genotypes in the population.

frequency-dependent selection   Selection that involves frequency-dependent fitness.

frequency histogram   A "step curve" in which the frequencies of various arbitrarily bounded classes are graphed.

frequency-independent selection   Selection in which the fitnesses of genotypes are independent of their relative frequency in the population.

frequency-independent fitness   Fitness that is not dependent upon interactions with other individuals of the same species.

fruiting body   In fungi, the organ in which meiosis occurs and sexual spores are produced.

functional complementation   The use of a cloned fragment of wild type DNA to transform a mutant to wild type; used in identifying a clone containing one specific gene.

G   Guanine, or guanosine.

G$_0$ phase   The resting phase that can occur during G$_1$ of interphase of the cell cycle.

G$_1$ phase   The portion of interphase of the cell cycle that precedes S phase.

G$_2$ phase   The portion of interphase of the cell cycle that follows S phase.

gain-of-function mutation   A mutation that results in a new functional ability for a protein, detectable at the phenotypic level.

gamete   A specialized haploid cell that fuses with a gamete from the opposite sex or mating type to form a diploid zygote; in mammals, an egg or a sperm.

gametophyte   The haploid gamete-producing stage in the life cycle of plants; prominent and independent in some species but reduced or parasitic in others.

gap gene   In Drosophila, the class of cardinal genes that are activated in the zygote in response to the anterior-posterior gradients of positional information. Through regulation of the pair-rule and homeotic genes, the patterns of expression of the various gap gene products lead to the specification of the correct number and types of body segments. Gap mutations cause the loss of several adjacent body segments.

gastrulation   The first process of movements and infoldings of the cell sheet in early animal embryos, usually immediately following blastula (or blastoderm).

gene   The fundamental physical and functional unit of heredity, which carries information from one generation to the next; a segment of DNA, composed of a transcribed region and a regulatory sequence that makes possible transcription.

gene amplification   The process by which the number of copies of a chromosomal segment is increased in a somatic cell.

gene conversion   A meiotic process of directed change in which one allele directs the conversion of a partner allele to its own form.

gene disruption   Inactivation of a gene by the integration of a specially engineered introduced DNA fragment.

gene dose   The number of copies of a particular gene present in the genome.

gene family   A set of genes in one genome all descended from the same ancestral gene.

gene frequency   *See* **allele frequency.**

gene fusion   The accidental joining of DNA of two genes, such as can occur in a translocation. Gene fusions can give rise to hybrid proteins or to the misregulation of the transcription unit of one gene by the cis-regulatory elements (enhancers) of another.

gene interaction   The collaboration of several different genes in the production of one phenotypic character (or related group of characters).

gene locus   The specific place on a chromosome where a gene is located.

gene map   A linear designation of mutant sites within a gene, based upon the various frequencies of interallelic (intragenic) recombination.

gene mutation   A point mutation that results from changes within the structure of a gene.

gene pair   The two copies of a particular type of gene present in a diploid cell (one in each chromosome set).

gene replacement   The insertion of a genetically engineered transgene in place of a resident gene; often occurs by a double crossover.

gene rearrangement   The process of programmed changes in the DNA structure of somatic cells, leading to changes in gene number or in the structural and functional properties of the rearranged gene.

gene therapy   The correction of a genetic deficiency in a cell by the addition of new DNA and its insertion into the genome.

generalized transduction   The ability of certain phages to transduce any gene in the bacterial chromosome.

genetic code   The set of correspondences between nucleotide pair triplets in DNA and amino acids in protein.

genetic dissection   The use of recombination and mutation to piece together the various components of a given biological function.

genetic markers   Alleles used as experimental probes to keep track of an individual, a tissue, a cell, a nucleus, a chromosome, or a gene.

genetic variance   Phenotypic variance resulting from the presence of different genotypes in the population.

genetics   (1) The study of genes. (2) The study of inheritance.

genome   The entire complement of genetic material in a chromosome set.

genomic library   A library encompassing an entire genome.

genomics   The cloning and molecular characterization of entire genomes.

genotype   The specific allelic composition of a cell—either of the entire cell or, more commonly, for a certain gene or a set of genes.

germinal mutations   Mutations occurring in the cells that are destined to develop into gametes.

germ line   The cell lineage in a multi-tissued eukaryote from which the gametes derive.

Gln   Glutamine (an amino acid).

Glu   Glutamate (an amino acid).

Gly   Glycine (an amino acid).

G-protein   A member of a family of proteins that contribute to signal transduction through protein-protein interactions that occur when the G-protein binds GTP but not when the G-protein binds GDP.

gradient   A gradual change in some quantitative property over a specific distance.

ground state   The developmental state of a cell (or group of cells) in the absence of activation of a developmental regulatory switch.

growth factors   Paracrine signaling molecules, usually secreted polypeptides, that induce cell division in cells receiving these signals.

guanine   A purine base that pairs with cytosine.

guanosine   The nucleoside having guanine as its base.

gynandromorph   An individual that is a mosaic of male and female structures. The underlying cause is frequently sex chromosome mosaicism, such that some cells are chromosomal females while others are chromosomal males.

half-chromatid conversion A type of gene conversion that is inferred from the existence of nonidentical sister spores in a fungal octad showing a non-Mendelian allele ratio.

haploid A cell having one chromosome set, or an organism composed of such cells.

haploidization Production of a haploid from a diploid by progressive chromosome loss.

Hardy-Weinberg equilibrium The stable frequency distribution of genotypes AA, Aa, and aa, in the proportions p^2, $2pq$, and q^2, respectively (where p and q are the frequencies of the alleles A and a), that is a consequence of random mating in the absence of mutation, migration, natural selection, or random drift.

harlequin chromosomes Sister chromatids that stain differently, so that one appears dark and the other light.

helix-loop-helix (HLH) protein A protein, a portion of which forms α-helices separated by a loop that acts as a sequence-specific DNA binding domain. HLH proteins are thought to act as transcription factors.

hemizygous gene A gene present in only one copy in a diploid organism—for example, X-linked genes in a male mammal.

hemoglobin (Hb) The oxygen-transporting blood protein in most animals.

hemophilia A human disease in which the blood fails to clot, caused by a mutation in a gene coding for a clotting protein; inherited as an X-linked recessive phenotype.

Hereditary Nonpolyposis Colorectal Cancer (HNPCC) One of the most common predispositions to cancer.

heredity The biological similarity of offspring and parents.

heritability in the broad sense *See* **broad heritability.**

heritability in the narrow sense (h^2) The proportion of phenotypic variance that can be attributed to additive genetic variance.

hermaphrodite (1) A plant species in which male and female organs occur in the same flower of a single individual (*compare* **monoecious plant**). (2) An animal with both male and female sex organs.

heterochromatin Densely staining condensed chromosomal regions, believed to be for the most part genetically inert.

heteroduplex A DNA double helix formed by annealing single strands from different sources; if there is a structural difference between the strands, the heteroduplex may show such abnormalities as loops or buckles.

heteroduplex DNA model A model that explains both crossing-over and gene conversion by assuming the production of a short stretch of heteroduplex DNA (formed from both parental DNAs) in the vicinity of a chiasma.

heterogametic sex The sex that has heteromorphic sex chromosomes (for example, XY) and hence produces two different kinds of gametes with respect to the sex chromosomes.

heterogeneous nuclear RNA (HnRNA) A diverse assortment of RNA types found in the nucleus, including mRNA precursors and other types of RNA.

heterokaryon A culture of cells composed of two different nuclear types in a common cytoplasm.

heterokaryon test A test for cytoplasmic mutations, based on new associations of phenotypes in cells derived from specially marked heterokaryons.

heteromorphic chromosomes A chromosome pair with some homology but differing in size, shape, or staining properties.

heteroplasmon A cell containing a mixture of genetically different cytoplasms, generally different mitochondria or different chloroplasts.

heterothallic fungus A fungus species in which two different mating types must unite to complete the sexual cycle.

heterozygosity A measure of the genetic variation in a population; with respect to one locus, stated as the frequency of heterozygotes for that locus.

heterozygote An individual having a heterozygous gene pair.

heterozygous gene pair A gene pair having different alleles in the two chromosome sets of the diploid individual—for example, Aa or A^1A^2.

hexaploid A cell having six chromosome sets, or an organism composed of such cells.

high-frequency recombination (Hfr) cell In *E. coli*, a cell having its fertility factor integrated into the bacterial chromosome; a donor (male) cell.

Himalayan A mammalian temperature-dependent coat phenotype, generally albino with pigment only at the cooler tips of the ears, feet, and tail.

His Histidine (an amino acid).

histocompatibility antigens Antigens that determine the acceptance or rejection of a tissue graft.

histocompatibility genes The genes that code for the histocompatibility antigens.

histone A type of basic protein that forms the unit around which DNA is coiled in the nucleosomes of eukaryotic chromosomes.

HLH proteins A family of proteins in which a portion of the polypeptide forms two α-helices separated by a loop (the HLH domain); this domain acts as a sequence-specific DNA-binding domain. HLH proteins are thought to act as transcription factors.

homeobox (homeotic box) A family of quite similar 180-base-pair DNA sequences that encode a polypeptide sequence called a homeodomain, a sequence-specific DNA-binding sequence. While the homeobox was first discovered in all homeotic genes, it is now known to encode a much more widespread DNA-binding motif.

homeodomain A highly conserved family of sequences 60 amino acids in length found within a large number of transcription factors that can form a helix-turn-helix structure and bind DNA in a sequence-specific manner.

homeologous chromosomes Partially homologous chromosomes, usually indicating some original ancestral homology.

homeosis The replacement of one body part by another. Homeosis can be caused by environmental factors leading to developmental anomalies, or by mutation.

homeotic gene Genes that control the fate of segments along the anterior-posterior axis of higher animals.

homeotic mutations Mutations that can change the fate of an imaginal disk.

homogametic sex The sex with homologous sex chromosomes (for example, XX).

homolog A member of a pair of homologous chromosomes.

homologous chromosomes Chromosomes that pair with each other at meiosis or chromosomes in different species that have retained most of the same genes during their evolution from a common ancestor.

homothallic fungus A fungus species in which a single sexual spore can complete the entire sexual cycle (*compare* **heterothallic fungus**).

homozygote An individual having a homozygous gene pair.

homozygous gene pair A gene pair having identical alleles in both copies—for example, AA or A^1A^1.

hormone A molecule secreted by an endocrine organ into the circulatory system and that acts as a long-range signaling molecule through activating receptors on or within target cells.

hormone response element (HRE) For hormones that act by binding to receptors that can act as transcription factor, an HRE is a cis-regulatory DNA sequence that is a binding site for a hormone-receptor complex.

host range The spectrum of strains of a given bacterial species that a given strain of phage can infect.

hot spot A part of a gene that shows a very high tendency to become a mutant site, either spontaneously or under the action of a particular mutagen.

Huntington's disease A lethal human disease of nerve degeneration, with late-age onset. Inherited as an autosomal dominant phenotype; new mutations rare.

hybrid (1) A heterozygote. (2) A progeny individual from any cross involving parents of differing genotypes.

hybrid dysgenesis A syndrome of effects including sterility, mutation, chromosome breakage, and male recombination in the hybrid progeny of crosses between certain laboratory and natural isolates of *Drosophila*.

hybridization in situ Finding the location of a gene by adding specific radioactive probes for the gene and detecting the location of the radioactivity on the chromosome after hybridization.

hybridize (1) To form a hybrid by performing a cross. (2) To anneal nucleic acid strands from different sources.

hydrogen bond A weak bond involving the sharing of an electron with a hydrogen atom; hydrogen bonds are important in the specificity of base pairing in nucleic acids and in the determination of protein shape.

hydroxyapatite A form of calcium phosphate that binds double-stranded DNA.

hyperploid Aneuploid containing a small number of extra chromosomes.

hypervariable region The part of a variable region that actually determines the specificity of an antibody.

hypha (plural, **hyphae**) A threadlike structure (composed of cells attached end to end) that forms the main tissue in many fungus species.

hypoploid Aneuploid with a small number of chromosomes missing.

Ig *See* **immunoglobulin.**

Ile Isoleucine (an amino acid).

imago An adult insect.

immune system The animal cells and tissues involved in recognizing and attacking foreign substances within the body.

immunoglobulin *See* **antibody.**

immunohistochemistry The use of antibodies or antisera as histological tools for identifying patterns of protein distribution within a tissue or an organism. An antibody (or mixture of antibodies) that binds to a specific protein is tagged with an enzyme that can convert a substrate to a visible dye. The tagged antibody is incubated with the tissue, unbound antibody is washed off, and the enzymatic substrate is then added, revealing the pattern of protein (actually, antigen) localization.

in situ "In place"; *see* **hybridization in situ.**

in vitro In an experimental situation outside the organism (literally, "in glass").

in vitro mutagenesis The production of either random or specific mutations in a piece of cloned DNA. Typically, the DNA will then be repackaged and introduced into a cell or an organism to assess the results of the mutagenesis.

in vivo In a living cell or organism.

inbreeding Mating between relatives.

inbreeding coefficient The probability of homozygosity that results because the zygote obtains copies of the *same* ancestral gene.

incomplete dominance The situation in which a heterozygote shows a phenotype quantitatively (but not exactly) intermediate between the corresponding homozygote phenotypes. (Exact intermediacy is no dominance.)

independent assortment *See* **Mendel's second law.**

inducer An environmental agent that triggers transcription from an operon.

induction The relief of repression for a gene or set of genes under negative control.

inductive interaction The interaction between two groups of cells in which a signal passed from one group of cells causes the other group of cells to change their developmental state (or fate).

infectious transfer The rapid transmission of free episomes (plus any chromosomal genes they may carry) from donor to recipient cells in a bacterial population.

inosine A rare base that is important at the wobble position of some tRNA anticodons.

insertion sequence (IS) A mobile piece of bacterial DNA (several hundred nucleotide pairs in length) that is capable of inactivating a gene into which it inserts.

insertional translocation The insertion of a segment from one chromosome into another nonhomologous one.

intercalating agent A chemical that can insert itself between the stacked bases at the center of the DNA double helix, possibly causing a frame-shift mutation.

interchromosomal recombination Recombination resulting from independent assortment.

interference A measure of the independence of crossovers from each other, calculated by subtracting the coefficient of coincidence from 1.

intermediate filaments A heterogeneous class of cytoskeletal elements, characterized by an intermediate cable diameter, larger than that of microfilaments but smaller than microtubules.

interphase The cell cycle stage between nuclear divisions, when chromosomes are extended and functionally active.

interrupted mating A technique used to map bacterial genes by determining the sequence in which donor genes enter recipient cells.

interstitial region The chromosomal region between the centromere and the site of a rearrangement.

intervening sequence An intron; a segment of largely unknown function within a gene. This segment is initially transcribed, but the transcript is not found in the functional mRNA.

intrachromosomal recombination Recombination resulting from crossing-over between two gene pairs.

intron *See* **intervening sequence.**

inversion A chromosomal mutation involving the removal of a chromosome segment, its rotation through 180 degrees, and its reinsertion in the same location.

inverted repeat (IR) sequence A sequence found in identical (but inverted) form, for example, at the opposite ends of a transposon.

IR *See* **inverted repeat sequence.**

IS *See* **insertion sequence.**

isoaccepting tRNAs The various types of tRNA molecules that carry a specific amino acid.

isotope One of several forms of an atom having the same atomic number but differing atomic masses.

karyotype The entire chromosome complement of an individual or cell, as seen during mitotic metaphase.

kinetochore *See* **centromere.**

Klinefelter syndrome An abnormal human male phenotype involving an extra X chromosome (XXY).

knockout Inactivation of one specific gene. Same as gene disruption.

lagging strand In DNA replication, the strand that is synthesized apparently in the 3′ to 5′ direction, by ligating short fragments synthesized individually in the 5′ to 3′ direction.

λ (lambda) phage One kind ("species") of temperate bacteriophage.

λdgal A λ phage carrying a *gal* bacterial gene and defective (*d*) for some phage function.

lampbrush chromosomes Large chromosomes found in amphibian eggs, with lateral DNA loops that produce a brushlike appearance under the microscope.

lateral inhibition The signal produced by one cell that prevents adjacent cells from acquiring its fate.

lawn A continuous layer of bacteria on the surface of an agar medium.

leader sequence The sequence at the 5′ end of an mRNA that is not translated into protein.

leading strand In DNA replication, the strand that is made in the 5′ to 3′ direction by continuous polymerization at the 3′ growing tip.

leaky mutant A mutant (typically, an auxotroph) that results from a partial rather than a complete inactivation of the wild-type function.

leaky mutation A mutation that confers a mutant phenotype but still retains a low but detectable level of wild type function.

lesion A damaged area in a gene (a mutant site), a chromosome, or a protein.

lethal gene A gene whose expression results in the death of the individual expressing it.

Leu Leucine (an amino acid).

library A collection of DNA clones obtained from one DNA donor.

ligand-receptor interaction The interactions between a molecule (usually of an extracellular origin) and a protein on or within a target cell. One type of ligand-receptor interaction can be between steroid hormones and their cytoplasmic or nuclear receptors. Another can be between secreted polypeptide ligands and transmembrane receptors.

ligase An enzyme that can rejoin a broken phosphodiester bond in a nucleic acid.

line A group of identical pure-breeding diploid or polyploid organisms, distinguished from other individuals of the same species by some unique phenotype and genotype.

LINE Long interspersed element; a type of large repetitive DNA segment found throughout the genome.

linear tetrad A tetrad that results from the occurrence of the meiotic and postmeiotic nuclear divisions in such a way that sister products remain adjacent to one another (with no passing of nuclei).

linkage The association of genes on the same chromosome.

linkage group A group of genes known to be linked; a chromosome.

linkage map A chromosome map; an abstract map of chromosomal loci, based on recombinant frequencies.

locus (plural, **loci**) *See* **gene locus.**

Lys Lysine (an amino acid).

lysis The rupture and death of a bacterial cell upon the release of phage progeny.

lysogen *See* **lysogenic bacterium.**

lysogenic bacterium A bacterial cell capable of spontaneous lysis due, for example, to the uncoupling of a prophage from the bacterial chromosome.

M phase The mitotic phase of the cell cycle.

macromolecule A large polymer such as DNA, a protein, or a polysaccharide.

Manx Tailless phenotype in cats, caused by an autosomal dominant mutation that is lethal when homozygous.

map unit (m.u.) The "distance" between two linked gene pairs where 1 percent of the products of meiosis are recombinant; a unit of distance in a linkage map.

mapping function A formula expressing the relationship between distance in a linkage map and recombinant frequency.

Marfan's syndrome A human disorder of the connective tissue expressed as a range of symptoms, including very long limbs and digits, and heart defects; inherited as an autosomal dominant phenotype.

marker *See* **genetic markers.**

marker retention A technique used in yeast to test the degree of linkage between two mitochondrial mutations.

maternal effect The environmental influence of the mother's tissues on the phenotype of the offspring.

maternal effect lethal A mutation that is viable in zygotes, but in which mutant mothers produce inviable offspring.

maternal inheritance A type of uniparental inheritance in which all progeny have the genotype and phenotype of the parent acting as the female.

maternally expressed gene A gene that contributes to the phenotype of an offspring based on its expression in the mother.

mating types The equivalent in lower organisms of the sexes in higher organisms; the mating types typically differ only physiologically and not in physical form.

matroclinous inheritance Inheritance in which all offspring have the nucleus-determined phenotype of the mother.

mean The arithmetic average.

medium Any material on (or in) which experimental cultures are grown.

meiocyte Cell in which meiosis occurs.

meiosis Two successive nuclear divisions (with corresponding cell divisions) that produce gametes (in animals) or sexual spores (in plants and fungi) have one-half of the genetic material of the original cell.

meiospore Cell that is one of the products of meiosis in plants.

melting Denaturation of DNA.

Mendelian ratio A ratio of progeny phenotypes reflecting the operation of Mendel's laws.

Mendel's first law The two members of a gene pair segregate from each other during meiosis; each gamete has an equal probability of obtaining either member of the gene pair.

Mendel's second law The law of independent assortment; unlinked or distantly linked segregating gene pairs assort independently at meiosis.

merozygote A partially diploid *E. coli* cell formed from a complete chromosome (the endogenote) plus a fragment (the exogenote).

messenger RNA *See* **mRNA.**

Met Methionine (an amino acid).

metabolism The chemical reactions that occur in a living cell.

metacentric chromosome A chromosome having its centromere in the middle.

metameres Segmental repeat units in higher animals.

metaphase An intermediate stage of nuclear division when chromosomes align along the equatorial plane of the cell.

methylation Modification of a molecule by the addition of a methyl group.

microfilaments The smallest diameter cable system of the cytoskeleton. Microfilament cables are composed of actin polymers.

microsatellite DNA A type of repetitive DNA based on very short repeats such as dinucleotides.

microtubule organizing center The portion of the microtubule cytoskeleton in which all of the minus ends of the microtubules are clustered. Ordinarily, this is near the center of the cell.

microtubules The largest diameter cable system of the cytoskeleton. Microtubules are composed of polymerized tubulin subunits forming a hollow tube.

midparent value The mean of the values of a quantitative phenotype for two specific parents.

minimal medium A medium containing only inorganic salts, a carbon source, and water.

minisatellite DNA A type of repetitive DNA sequence based on short repeat sequences with a unique common core; used for DNA fingerprinting.

misexpression The occurence of gene expression in a tissue in which it is normally not expressed. Such ectopic expression can be caused by the juxtaposition of novel enhancer elements to a gene.

missense mutation A mutation that alters a codon so that it encodes a different amino acid.

mitochondrial cytopathies Human disorders caused by point mutations or deletions in mitochondrial DNA; inherited maternally.

mitochondrion A eukaryotic organelle that is the site of ATP synthesis and of the citric acid cycle.

mitosis A type of nuclear division (occurring at cell division) that produces two daughter nuclei identical to the parent nucleus.

mitotic crossover A crossover resulting from the pairing of homologs in a mitotic diploid.

mobile genetic element *See* **transposable genetic element.**

mode The single class in a statistical distribution having the greatest frequency.

modifier gene A gene that affects the phenotypic expression of another gene.

molecular genetics The study of the molecular processes underlying gene structure and function.

monocistronic mRNA An mRNA that codes for one protein.

monoecious plant A plant species in which male and female organs are found on the same plant but in different flowers (for example, corn).

monohybrid cross A cross between two individuals identically heterozygous at one gene pair—for example, $Aa \times Aa$.

monoploid A cell having only one chromosome set (usually as an aberration), or an organism composed of such cells.

monosomic A cell or individual that is basically diploid but that has only one copy of one particular chromosome type and thus has chromosome number $2n - 1$.

morphogen A molecule that can induce the acquisition of different cell fates according to the level of morphogen to which a cell is exposed.

mosaic A chimera; a tissue containing two or more genetically distinct cell types, or an individual composed of such tissues.

motor protein Proteins that are able to move unidirectionally along a specific type of cytoskeletal cable. Kinesins and dyneins are microtubule-based and myosins are microfilament-based motor proteins. By attaching to other subcellular components, motor proteins are capable of directed movement of these components within the cell.

mRNA (messenger RNA) An RNA molecule transcribed from the DNA of a gene, and from which a protein is translated by the action of ribosomes.

mtDNA Mitochondrial DNA.

m.u. *See* **map unit.**

mu **phage** A kind ("species") of phage with properties similar to those of insertion sequences, being able to insert, transpose, inactivate, and cause rearrangements.

multimeric structure A structure composed of several identical or different subunits held together by weak bonds.

multiple allelism The existence of several known alleles of a gene.

multiple-factor hypothesis A hypothesis to explain quantitative variation by assuming the interaction of a large number of genes (polygenes), each with a small additive effect on the character.

multiple-hit hypothesis The proposal that a single cell must receive a series of mutational events in order to become malignant or cancerous.

multiplicity of infection The average number of phage particles that infect a single bacterial cell in a specific experiment.

mutagen An agent that is capable of increasing the mutation rate.

mutant An organism or cell carrying a mutation.

mutant allele An allele differing from the allele found in the standard, or wild type.

mutant hunt The process of collecting different mutants showing abnormalities in a certain structure or in a certain function, as a preparation for mutational dissection of that function.

mutant site The damaged or altered area within a mutated gene.

mutation (1) The process that produces a gene or a chromosome set differing from the wild-type. (2) The gene or chromosome set that results from such a process.

mutation breeding Use of mutagens to develop variants that can increase agricultural yield.

mutation event The actual occurrence of a mutation in time and space.

mutation frequency The frequency of mutants in a population.

mutation rate The number of mutation events per gene per unit of time (for example, per cell generation).

mutational dissection The study of the components of a biological function through a study of mutations affecting that function.

muton The smallest part of a gene that can be involved in a mutation event; now known to be a nucleotide pair.

myeloma A cancer of the bone marrow.

nanometer 10^{-9} meters.

narrow heritability *See* **heritability in the narrow sense.**

negative assortative mating Preferential mating between phenotypically unlike partners.

negative control Regulation mediated by factors that block or turn off transcription.

neurofibromatosis A human disease with tumors of nerve cells and café au lait spots, both in the skin. The allele generally arises from germinal mutation, but it is inherited as an autosomal dominant.

Neurospora A pink mold, commonly found growing on old food.

neutral mutation (1) A mutation that has no effect on the Darwinian fitness of its carriers. (2) A mutation that has no phenotypic effect.

neutral petite A petite that produces all wild-type progeny when crossed with wild type.

neutrality *See* **selective neutrality.**

nicking Nuclease action to sever the sugar-phosphate backbone in one DNA strand at one specific site.

nitrocellulose filter A type of filter used to hold DNA for hybridization.

nitrogen bases Types of molecules that form important parts of nucleic acids, composed of nitrogen-containing ring structures; hydrogen bonds between bases link the two strands of a DNA double helix.

nonautonomous *See* **autonomous phenotype.**

nondisjunction The failure of homologs (at meiosis) or sister chromatids (at mitosis) to separate properly to opposite poles.

nonlinear tetrad A tetrad in which the meiotic products are in no particular order.

non-Mendelian ratio An unusual ratio of progeny phenotypes that does not reflect the simple operation of Mendel's laws; for example, mutant : wild ratios of $3:5$, $5:3$, $6:2$, or $2:6$ in tetrads indicate that gene conversion has occurred.

nonparental ditype (NPD) A tetrad type containing two different genotypes, both of which are recombinant.

nonsense codon A codon for which no normal tRNA molecule exists; the presence of a nonsense codon causes termination of translation (ending of the polypeptide chain). The three nonsense codons are called amber, ocher, and opal.

nonsense mutation A mutation that alters a gene so as to produce a nonsense codon.

nonsense suppressor A mutation that produces an altered tRNA that will insert an amino acid during translation in response to a nonsense codon.

norm of reaction The pattern of phenotypes produced by a given genotype under different environmental conditions.

Northern blot Transfer of electrophoretically separated RNA molecules from a gel onto an absorbent sheet, which is then immersed in a labeled probe that will bind to the RNA of interest.

NPD *See* **nonparental ditype.**

nu body *See* **nucleosome.**

nuclease An enzyme that can degrade DNA by breaking its phosphodiester bonds.

nucleoid A DNA mass within a chloroplast or mitochondrion.

nucleolar organizer A region (or regions) of the chromosome set physically associated with the nucleolus and containing rRNA genes.

nucleolus An organelle found in the nucleus, containing rRNA and amplified multiple copies of the genes coding for rRNA.

nucleoside A nitrogen base bound to a sugar molecule.

nucleosome A nu body; the basic unit of eukaryotic chromosome structure; a ball of eight histone molecules wrapped about by two coils of DNA.

nucleotide A molecule composed of a nitrogen base, a sugar, and a phosphate group; the basic building block of nucleic acids.

nucleotide pair A pair of nucleotides (one in each strand of DNA) that are joined by hydrogen bonds.

nucleotide-pair substitution The replacement of a specific nucleotide pair by a different pair; often mutagenic.

null allele An allele whose effect is either an absence of normal gene product at the molecular level or an absence of normal function at the phenotypic level.

null mutation A mutation that results in complete absence of function for the gene.

nullisomic A cell or individual with one chromosomal type missing, with a chromosome number such as $n - 1$ or $2n - 2$.

nurse cells   The sister cells of the oocyte in insects. The nurse cells produce the bulk of the cytoplasmic contents of the mature oocyte.

ocher codon   The codon UAA, a nonsense codon.

octad   An ascus containing eight ascospores, produced in species in which the tetrad normally undergoes a postmeiotic mitotic division.

oligonucleotide   A short segment of synthetic DNA.

oncogene   A gene that contributes to the production of a cancer. Oncogenes are generally mutated forms of normal cellular genes.

opal codon   The codon UGA, a nonsense codon.

open reading frame   *See* **ORF.**

operator   A DNA region at one end of an operon that acts as the binding site for repressor protein.

operon   A set of adjacent structural genes whose mRNA is synthesized in one piece, plus the adjacent regulatory signals that affect transcription of the structural genes.

ORF (open reading frame)   A section of a sequenced piece of DNA that begins with a start codon and ends with a stop codon; it is presumed to be the coding sequence of a gene.

organogenesis   The production of organ systems during animal embryogenesis.

organelle   A subcellular structure having a specialized function—for example, the mitocondrion, the chloroplast, or the spindle apparatus.

origin of replication   The point of specific sequence at which DNA replication is initiated.

overdominance   A phenotypic relation in which the phenotypic expression of the heterozygote is greater than that of either homozygote.

P element   A *Drosophila* transposable element that has been used as a tool for insertional mutagenesis and for germ-line transformation.

pair-rule gene   In *Drosophila,* a member of a class of zygotically-expressed genes that act at an intermediary stage in the process of establishment of the correct numbers of body segments. Pair-rule mutations have half of the normal number of segments, due to the loss of every other segment.

paracentric inversion   An inversion not involving the centromere.

paracrine signaling   The process by which a secreted molecule binds to receptors on or within nearby cells, thereby inducing a signal transduction pathway within the receiving cell.

paramutation   An epigenetic phenomenon in plants, in which the genetic activity of a normal allele is heritably reduced by virtue of that allele having been heterozygous with a special "paramutagenic" allele.

parental ditype (PD)   A tetrad type containing two different genotypes, both of which are parental.

parental imprinting   An epigenetic phenomenon in which the activity of a gene is dependent upon whether it was inherited from the father or the mother. Some genes are maternally imprinted, other paternally.

paralogous genes   Two genes in the same species that have evolved by gene duplication.

parthenogenesis   The production of offspring by a female with no genetic contribution from a male.

partial diploid   *See* **merozygote.**

particulate inheritance   The model proposing that genetic information is transmitted from one generation to the next in discrete units ("particles"), so that the character of the offspring is not a smooth blend of essences from the parents (*compare* **blending inheritance**).

pathogen   An organism that causes disease in another organism.

patroclinous inheritance   Inheritance in which all offspring have the nucleus-based phenotype of the father.

pattern formation   The developmental processes by which the complex shape and structure of higher organisms occurs.

PD   *See* **parental ditype.**

pedigree   A "family tree," drawn with standard genetic symbols, showing inheritance patterns for specific phenotypic characters.

penetrance   The proportion of individuals with a specific genotype who manifest that genotype at the phenotype level.

peptide   *See* **amino acid.**

peptide bond   A bond joining two amino acids.

pericentric inversion   An inversion that involves the centromere.

permissive conditions   Those environmental conditions under which a conditional mutant shows the wild-type phenotype.

petite   A yeast mutation producing small colonies and altered mitochondrial functions. In cytoplasmic petites (neutral and suppressive petites), the mutation is a deletion in mitochondrial DNA; in segregational petites, the mutation occurs in nuclear DNA.

phage   *See* **bacteriophage.**

Phe   Phenylalanine (an amino acid).

phenocopy   An environmentally induced phenotype that resembles the phenotype produced by a mutation.

phenotype   (1) The form taken by some character (or group of characters) in a specific individual. (2) The detectable outward manifestations of a specific genotype.

phenotypic sex determination   Sex determination by nongenetic means.

phenylketonuria (PKU)   A human metabolic disease caused by a mutation in a gene coding for a phenylalanine-processing enzyme, which leads to mental retardation if not treated; inherited as an autosomal recessive phenotype.

Philadelphia chromosome   A translocation between the long arms of chromosomes 9 and 22, often found in the white blood cells of patients with chronic myeloid leukemia.

phosphodiester bond   A bond between a sugar group and a phosphate group; such bonds form the sugar-phosphate backbone of DNA.

physical mapping   Mapping the positions of cloned genomic fragments.

piebald   A mammalian phenotype in which patches of skin are unpigmented because of lack of melanocytes; generally inherited as an autosomal dominant.

pilus (plural, **pili**)   A conjugation tube; a hollow hairlike appendage of a donar *E. coli* cell that acts as a bridge for transmission of donor DNA to the recipient cell during conjugation.

plant breeding   The application of genetic analysis to development of plant lines better suited for human purposes.

plaque A clear area on a bacterial lawn, left by lysis of the bacteria through progressive infections by a phage and its descendants.

plasmid Autonomously replicating extrachromosomal DNA molecule.

plate (1) A flat dish used to culture microbes. (2) To spread cells over the surface of solid medium in a plate.

pleiotropic mutation A mutation that has effects on several different characters.

point mutation A mutation that can be mapped to one specific locus.

Poisson distribution A mathematical expression giving the probability of observing various numbers of a particular event in a sample when the mean probability of an event on any one trial is very small.

poky A slow-growing mitochondrial mutant in *Neurospora*.

polar gene conversion A gradient of conversion frequency along the length of a gene.

polar granules Cytoplasmic granules localized at the posterior end of a Drosophila oocyte and early embryo. These granules are associated with the germ-line and posterior determinants.

polar mutation A mutation that affects the transcription or translation of the part of the gene or operon on only one side of the mutant site—for example, nonsense mutations, frame-shift mutations, and IS-induced mutations.

poly(A)tail A string of adenine nucleotides added to mRNA after transcription.

polyacrylamide A material used to make electrophoretic gels for separation of mixtures of macromolecules.

polycistronic mRNA An mRNA that codes for more than one protein.

polydactyly More than five fingers and/or toes. Inherited as an autosomal dominant phenotype.

polygenes *See* **multiple-factor hypothesis.**

polymerase chain reaction (PCR) A method for amplifying specific DNA segments which exploits certain features of DNA replication.

polymorphism The occurrence in a population (or among populations) of several phenotypic forms associated with alleles of one gene or homologs of one chromosome.

polypeptide A chain of linked amino acids; a protein.

polyploid A cell having three or more chromosome sets, or an organism composed of such cells.

polysaccharide A biological polymer composed of sugar subunits—for example, starch or cellulose.

polytene chromosome A giant chromosome produced by an endomitotic process in which the multiple DNA sets remain bound in a haploid number of chromosomes.

position effect Used to describe a situation where the phenotypic influence of a gene is altered by changes in the position of the gene within the genome.

position-effect variegation Variegation caused by the inactivation of a gene in some cells through its abnormal juxtaposition with heterochromatin.

positional information The process by which chemical cues that establish cell fate along a geographic axis are established in a developing embryo or tissue primordium.

positive assortative mating A situation in which like phenotypes mate more commonly than expected by chance.

positive control Regulation mediated by a protein that is required for the activation of a transcription unit.

primary structure of a protein The sequence of amino acids in the polypeptide chain.

prion Proteinaceous infectious particle that causes degenerative disorders of the central nervous system, such as "scrapie" in sheep and Creutzfeldt-Jacob disease in humans.

Pro Proline (an amino acid).

probe Define nucleic acid segment that can be used to identify specific DNA molecules bearing the complementary sequence, usually through autoradiography.

product of meiosis One of the (usually four) cells formed by the two meiotic divisions.

product rule The probability of two independent events occurring simultaneously is the product of the individual probabilities.

proflavin A mutagen that tends to produce frame-shift mutations.

prokaryote An organism composed of a prokaryotic cell, such as bacteria and blue-green algae.

prokaryotic cell A cell having no nuclear membrane and hence no separate nucleus.

promoter A regulator region a short distance from the 5' end of a gene that acts as the binding site for RNA polymerase.

prophage A phage "chromosome" inserted as part of the linear structure of the DNA chromosome of a bacterium.

prophase The early stage of nuclear division during which chromosomes condense and become visible.

propositus In a human pedigree, the individual who first came to the attention of the geneticist.

protein kinase An enzyme that phosphorylates specific amino acid residues on specific target proteins. One major class of protein kinases phosphorylates tyrosines, and the other phosphorylates serines and threonines on target proteins.

proto-oncogene The normal cellular counterpart of genes that can be mutated to become dominant oncogenes.

protoplast A plant cell whose wall has been removed.

prototroph A strain of organisms that will proliferate on minimal medium (*compare* **auxotroph**).

provirus A virus "chromosome" integrated into the DNA of the host cell.

pseudodominance The sudden appearance of a recessive phenotype in a pedigree, due to deletion of a masking dominant gene.

pseudogene An inactive gene derived from an ancestral active gene.

puff *See* **chromosome puff.**

pulse-chase experiment An experiment in which cells are grown in radioactive medium for a brief period (the pulse) and then transferred to nonradioactive medium for a longer period (the chase).

pulsed-field gel electrophoresis An electrophoretic technique in which the gel is subjected to electrical fields alternating between different angles, allowing very large DNA fragments to "snake" through the gel, and hence permitting efficient separation of mixtures of such large fragments.

Punnett square A grid used as a graphic representation of the progeny zygotes resulting from different gamete fusions in a specific cross.

pure-breeding line or strain A group of identical individuals that always produce offspring of the same phenotype when intercrossed.

purine A type of nitrogen base; the purine bases in DNA are adenine and guanine.

pyrimidine A type of nitrogen base; the pyrimidine bases in DNA are cytosine and thymine.

quantitative variation The existence of a range of phenotypes for a specific character, differing by degree rather than by distinct qualitative differences.

quaternary structure of a protein The multimeric constitution of the protein.

R plasmid A plasmid containing one or several transposons that bear resistance genes.

radiation hybrid A type of human/mouse hybrid cell in which human chromosomes have been fragmented by radiation to determine which markers are inherited together and therefore linked.

random genetic drift Changes in allele frequency that result because the genes appearing in offspring do not represent a perfectly representative sampling of the parental genes.

random mating Mating between individuals where the choice of a partner is not influenced by the genotypes (with respect to specific genes under study).

RAPD Randomly amplified polymorphic DNA. A set of several genomic fragments amplified by single PCR primer; somewhat variable from individual to individual; $+/-$ heterozygotes for individual fragments can act as markers in genome mapping.

reading frame The codon sequence that is determined by reading nucleotides in groups of three from some specific start codon.

realized heritability The ratio of the single-generation progress of selection to the selection differential of the parents.

reannealing Spontaneous realignment of two single DNA strands to re-form a DNA double helix that had been denatured.

receptor *See* **ligand-receptor interactions.**

receptor element A controlling element that can insert into a gene (making it a mutant) and can also excise (thus making the mutation unstable); both of these functions are nonautonomous, being under the influence of the regulator element.

receptor tyrosine kinase A transmembrane receptor whose cytoplasmic domain includes a tyrosine kinase enzymatic activity. In normal situations, the kinase is only activated upon binding of the appropriate ligand to the receptor.

recessive allele An allele whose phenotypic effect is not expressed in a heterozygote.

recessive phenotype The phenotype of a homozygote for the recessive allele; the parental phenotype that is not expressed in a heterozygote.

reciprocal crosses A pair of crosses of the type genotype A ♀ × genotype B ♂ and genotype B ♀ × genotype A ♂.

reciprocal translocation A translocation in which part of one chromosome is exchanged with a part of a separate nonhomologous chromosome.

recombinant An individual or cell with a genotype produced by recombination.

recombinant DNA A novel DNA sequence formed by the combination of two nonhomologous DNA molecules.

recombinant frequency (RF) The proportion (or percentage) of recombinant cells or individuals.

recombination (1) In general, any process in a diploid or partially diploid cell that generates new gene or chromosomal combinations not found in that cell or in its progenitors. (2) At meiosis, the process that generates a haploid product of meiosis whose genotype is different from either of the two haploid genotypes that constituted the meiotic diploid.

recombinational repair The repair of a DNA lesion through a process, similar to recombination, that uses recombination enzymes.

recon A region of a gene within which there can be no crossing-over; now known to be a nucleotide pair.

reduction division A nuclear division that produces daughter nuclei each having one-half as many centromeres as the parental nucleus.

redundant DNA *See* **repetitive DNA.**

regression A term coined by Galton for the tendency of the quantitative traits of offspring to be closer to the population mean than are their parents' traits. It arises from dominance, gene interaction, and nongenetic influences on traits.

regression coefficient The slope of the straight line that most closely relates two correlated variables.

regulator element *See* **receptor element.**

regulatory genes Genes that are involved in turning on or off the transcription of structural genes.

repetitive DNA Redundant DNA; DNA sequences that are present in many copies per chromosome set.

replication DNA synthesis.

replication fork The point at which the two strands of DNA are separated to allow replication of each strand.

replicon A chromosomal region under the influence of one adjacent replication-initiation locus.

reporter gene A gene whose phenotypic expression is easy to monitor and which is used to study tissue-specific promoter and enhancer activities in transgenes.

repressor protein A molecule that binds to the operator and prevents transcription of an operon.

repulsion conformation Two linked heterozygous gene pairs in the arrangement *A b / a B.*

resolving power The ability of an experimental technique to distinguish between two genetic conditions (typically discussed when one condition is rare and of particular interest).

restriction enzyme An endonuclease that will recognize specific target nucleotide sequences in DNA and break the DNA chain at those points; a variety of these enzymes are known, and they are extensively used in genetic engineering.

restriction map A map of a chromosomal region showing the positions of target sites of one or more restriction enzymes.

restrictive conditions Environmental conditions under which a conditional mutant shows the mutant phenotype.

retinoblastoma A childhood cancer of the human retina.

retrotransposon Transposable elements that utilize reverse transcriptase to transpose via an RNA intermediate.

retrovirus An RNA virus that replicates by first being converted into double-stranded DNA.

reverse genetics The experimental procedure that begins with a cloned segment of DNA, or a protein sequence, and uses this (through directed mutagenesis) to introduce programmed mutations back into the genome in order to investigate function.

reverse transcriptase An enzyme that catalyzes the synthesis of a DNA strand from an RNA template.

reversion The production of a wild-type gene from a mutant gene.

RF *See* **recombinant frequency.**

RFLP Restriction fragment length polymorphism. At some chromosomal locations a probe sometimes detects different sizes or different numbers of restriction fragments (often as a result of presence and absence of restriction sites) and this situation is an RFLP. If an individual is heterozygous for such a chromosomal difference that region can be used as a marker in chromosome mapping.

RFLP mapping A technique in which DNA restriction fragment length polymorphisms are used as reference loci for mapping in relation to known genes or other RFLP loci.

rho A protein factor required to recognize certain transcription termination signals in *E. coli.*

ribonucleic acid *See* **RNA.**

ribosomal RNA *See* **rRNA.**

ribosome A complex organelle that catalyzes translation of messenger RNA into an amino acid sequence. Composed of proteins plus rRNA.

ribozymes RNAs with enzymatic activities, for instance, the self-splicing RNA molecules in *Tetrahymena.*

RNA (ribonucleic acid) A single-stranded nucleic acid similar to DNA but having ribose sugar rather than deoxyribose sugar and uracil rather than thymine as one of the bases.

RNA in situ hybridization A technique that is used to identify the spatial pattern of expression of a particular transcript (usually an mRNA). In this technique, the DNA probe is labeled, either radioactively or by chemically attaching an enzyme that can convert a substrate to a visible dye. A tissue or organism is soaked in a solution of single-stranded labeled DNA under conditions that allow the DNA to hybridize to complementary RNA sequences in the cells; unhybridized DNA is then removed. Radioactive probe is detected by autoradiography. Enzyme-labeled probe is detected by soaking the tissue in the substrate; the dye develops in sites where the transcript of interest was expressed.

RNA polymerase An enzyme that catalyzes the synthesis of an RNA strand from a DNA template. In eukaryotes, there are several classes of RNA polymerase. Structural genes for proteins are transcribed by RNA polymerase II.

rRNA (ribosomal RNA) A class of RNA molecules, coded in the nucleolar organizer, that have an integral (but poorly understood) role in ribosome structure and function.

S (Svedberg unit) A unit of sedimentation velocity, commonly used to describe molecular units of various sizes (because sedimentation velocity is related to size).

SARs Scaffold attachment regions; the positions along DNA where it is anchored to the central scaffold of the chromosome.

satellite A terminal section of a chromosome, separated from the main body of the chromosome by a narrow constriction.

satellite chromosomes Chromosomes that seem to be additions to the normal genome.

satellite DNA Any type of highly repetitive DNA; previously defined as DNA forming a satellite band after cesium chloride density gradient centrifugation.

saturation mutagenesis Induction and recovery of large numbers of mutations in one area of the genome, or in one function, in the hope of identifying all the genes in that area, or affecting that function.

scaffold The central framework of a chromosome to which the DNA solenoid is attached as loops; composed largely of topoisomerase.

SCE *See* **sister-chromatid exchange.**

secondary sexual characteristics The sex-associated phenotypes of somatic tissues in sexually dimorphic animals.

secondary structure of a protein A spiral or zigzag arrangement of the polypeptide chain.

second-division segregation pattern A pattern of ascospore genotypes for a gene pair showing that the two alleles separate into different nuclei only at the second meiotic division, as a result of a crossover between that gene pair and its centromere; can only be detected in a linear ascus.

second-site mutation The second mutation of a double mutation within a gene; in many cases, the second-site mutation suppresses the first mutation, so that the double mutant has the wild-type phenotype.

sector An area of tissue whose phenotype is detectably different from the surrounding tissue phenotype.

sedimentation The sinking of a molecule under the opposing forces of gravitation and buoyancy.

segment polarity gene In *Drosophila,* a member of the class of genes that contribute to the final aspects of establishing the correct number of segments. Segment polarity mutations cause a loss of a comparable portion of each of the body segments.

segmentation The process by which the correct number of segments are established in a developing segmented animal.

segregation (1) Cytologically, the separation of homologous structures. (2) Genetically, the production of two separate phenotypes, corresponding to two alleles of a gene, either in different individuals (meiotic segregation) or in different tissues (mitotic segregation).

segregational petite A petite that in a cross with wild type produces $\frac{1}{2}$ petite and $\frac{1}{2}$ wild-type progeny; caused by a nuclear mutation.

selection coefficient (s) The proportional excess or deficiency of fitness of one genotype in relation to another genotype.

selection differential The difference between the mean of a population and the mean of the individuals selected to be parents of the next generation.

selection progress The difference between the mean of a population and the mean of the offspring in the next generation born to selected parents.

selective neutrality   A situation in which different alleles of a certain gene confer equal fitness.

selective system   An experimental technique that enhances the recovery of specific (usually rare) genotypes.

self   To fertilize eggs with sperms from the same individual.

self-assembly   The ability of certain multimeric biological structures to assemble from their component parts through random movements of the molecules and formation of weak chemical bonds between surfaces with complementary shapes.

semiconservative replication   The established model of DNA replication in which each double-stranded molecule is composed of one parental strand and one newly polymerized strand.

semisterility (half sterility)   The phenotype of individuals heterozygotic for certain types of chromosome aberration; expressed as a reduced number of viable gametes and hence reduced fertility.

Sequence-tagged site   A relatively small sequenced region of a cloned genomic fragment that can be used by a computer to align the cloned fragment into a contig.

Ser   Serine (an amino acid).

sex chromosome   A chromosome whose presence or absence is correlated with the sex of the bearer; a chromosome that plays a role in sex determination.

sex determination   The genetic or environmental process by which the sex of an individual is established.

sex linkage   The location of a gene on a sex chromosome.

sex reversal   A syndrome known in humans and mice in which chromosomally XX individuals develop as males. In some cases, sex reversal is now known to be due to the translocation of the testis-determining region of the Y chromosome to the tip of the X chromosome in such individuals.

sexduction   Sexual transmission of donor *E. coli* chromosomal genes on the fertility factor.

sexual spore   *See* **spore.**

shotgun technique   Cloning a large number of different DNA fragments as a prelude to selecting one particular clone type for intensive study.

shuttle vector   A vector (e.g. a plasmid) constructed in such a way that it can replicate in at least two different host species, allowing a DNA segment to be tested or manipulated in several cellular settings.

sickle-cell anemia   Potentially lethal human disease caused by a mutation in a gene coding for the oxygen-transporting molecule hemoglobin. The altered molecule causes red blood cells to be sickle shaped. Inherited as an autosomal recessive.

signal sequence   The N-terminal sequence of a secreted protein, which is required for transport through the cell membrane.

signal transduction cascade   A series of sequential events, such as protein phosphorylations, that pass a signal received by a transmembrane receptor through a series of intermediate molecules until final regulatory molecules, such as transcription factors, are modified in response to the signal.

silent mutation   Mutation in which the function of the protein product of the gene is unaltered.

SINE   Short interspersed element. A type of small repetitive DNA sequence found throughout a eukaryotic genome.

sister-chromatid exchange (SCE)   An event similar to crossing-over that can occur between sister chromatids at mitosis or at meiosis; detected in harlequin chromosomes.

site-specific recombination   Recombination occurring between two specific sequences that need not be homologous; mediated by a specific recombination system.

S-9 mix   A liver-derived supernatant used in the Ames test to activate or inactivate mutagens.

solenoid structure   The supercoiled arrangement of DNA in eukaryotic nuclear chromosomes produced by coiling the continuous string of nucleosomes.

somatic cell   A cell that is not destined to become a gamete; a "body cell," whose genes will not be passed on to future generations.

somatic mutation   A mutation occurring in a somatic cell.

somatic-cell genetics   Asexual genetics, involving study of somatic mutation, assortment, and crossing-over, and of cell fusion.

somatostatin   A human growth hormone.

SOS repair   The error-prone process whereby gross structural DNA damage is circumvented by allowing replication to proceed past the damage through imprecise polymerization.

Southern blot   Transfer of electrophoretically separated fragments of DNA from the gel to an absorbent sheet such as paper. This sheet is then immersed in a solution containing a labeled probe that will bind to a fragment of interest.

spacer DNA   DNA found between genes; its function is unknown.

specialized (restricted) transduction   The situation in which a particular phage will transduce only specific regions of the bacterial chromosome.

specific-locus test   A system for detecting recessive mutations in diploids. Normal individuals treated with mutagen are mated to testers that are homozygous for the recessive alleles at a number of specific loci; the progeny are then screened for recessive phenotypes.

S phase   The portion of interphase of the cell cycle in which DNA synthesis occurs.

spindle   The set of microtubular fibers that appear to move eukaryotic chromosomes during division.

splicing   The reaction that removes introns and joins together exons in RNA.

spontaneous mutation   A mutation occurring in the absence of mutagens, usually due to errors in the normal functioning of cellular enzymes.

spore   (1) In plants and fungi, sexual spores are the haploid cells produced by meiosis. (2) In fungi, asexual spores are somatic cells that are cast off to act either as gametes or as the initial cells for new haploid individuals.

sporophyte   The diploid sexual-spore-producing generation in the life cycle of plants—that is, the stage in which meiosis occurs.

SSLP   Short sequence length polymorphisms; the presence of different numbers of short repetitive elements (mini- and microsatellite DNA) at one particular locus in different homologous chromosomes; heterozygotes represent useful markers for genome mapping.

stacking   The packing of the flattish nitrogen bases at the center of the DNA double helix.

staggered cuts   The cleavage of two opposite strands of duplex DNA at points near one another.

standard deviation   The square root of the variance.

statistic   A computed quantity characteristic of a population, such as the mean.

statistical distribution   The array of frequencies of different quantitative or qualitative classes in a population.

stem cells   *See* **blast cells.**

steroid hormone   A class of hormones synthesized by glands of the endocrine system that, by virtue of their non-polar nature, are able to pass directly through the plasma membrane of cells. Steroid hormones act by binding to and activating transcription factors called steroid hormone receptors.

steroid hormone receptor   A family of related proteins that act as transcription factors when bound to their cognate hormones. Not all members of this family actually bind to steroids; the name derives from the first family member that was discovered, which was indeed a steroid hormone receptor.

strain   A pure-breeding lineage, usually of haploid organisms, bacteria, or viruses.

structural gene   A gene encoding the amino acid sequence of a protein.

subvital gene   A gene that causes the death of some proportion (but not all) of the individuals that express it.

sum rule   The probability that one or the other of two mutually exclusive events will occur is the sum of their individual probabilities.

supercoil   A closed double-stranded DNA molecule that is twisted on itself.

superinfection   Phage infection of a cell that already harbors a prophage.

supersuppressor   A mutation that can suppress a variety of other mutations; typically a nonsense suppressor.

suppressive petite   A petite that in a cross with wild type produces progeny of ehich variable non-Mendelian proportions are petite.

suppressor   A secondary mutation that can cancel the effect of a primary mutation, resulting in wild-type phenotype.

suppressor mutation   A mutation that counteracts the effects of another mutation. A suppressor maps at a different site from the mutation it counteracts, either within the same gene or at a more distant locus. Different suppressors act in different ways.

synapsis   Close pairing of homologs at meiosis.

synaptonemal complex   A complex structure that unites homologs during the prophase of meiosis.

syncytial blastoderm   In insects, the stage of blastoderm preceding the formation of cell membranes around the individual nuclei of the early embryo.

syncytium   A single cell with many nuclei.

T   (1) Thymine, or thymidine. (2) *See* **tetratype.**

tagging   The use of a piece of foreign DNA or a transposon to tag a gene so that a clone of that gene can be identified readily in a library.

tandem duplication   Adjacent identical chromosome segments.

targeted gene knockout   The introduction of a null mutation in a gene by a designed alteration in a cloned DNA sequence that is then introduced into the genome through homologous recombination and replacement of the normal allele.

tautomeric shift   The spontaneous isomerization of a nitrogen base to an alternative hydrogen-bonding condition, possibly resulting in a mutation.

T-DNA   A portion of the Ti plasmid that is inserted into the genome of the host plant cell.

telocentric chromosome   A chromosome having the centromere at one end.

telomere   The tip (or end) of a chromosome.

telophase   The late stage of nuclear division when daughter nuclei re-form.

temperate phage   A phage that can become a prophage.

temperature-sensitive mutation   A conditional mutation that produces the mutant phenotype in one temperature range and the wild-type phenotype in another temperature range.

template   A molecular "mold" that shapes the structure or sequence of another molecule; for example, the nucleotide sequence of DNA acts as a template to control the nucleotide sequence of RNA during transcription.

teratogen   An agent that interferes with normal development.

terminal redundancy   In phage, a linear DNA molecule with single-stranded ends that are longer than is necessary to close the DNA circle.

tertiary structure of a protein   The folding or coiling of the secondary structure to form a globular molecule.

testcross   A cross of an individual of unknown genotype or a heterozygote (or a multiple heterozygote) to a tester individual.

tester   An individual homozygous for one or more recessive alleles; used in a testcross.

testicular feminization syndrome   A human condition, caused by a mutation in a gene coding for androgen receptors, in which XY males develop into phenotypic females.

tetrad   (1) Four homologous chromatids in a bundle in the first meiotic prophase and metaphase. (2) The four haploid product cells from a single meiosis.

tetrad analysis   The use of tetrads (definition 2) to study the behavior of chromosomes and genes during meiosis.

tetraparental mouse   A mouse that develops from an embryo created by the experimental fusion of two separate blastulas.

tetraploid   A cell having four chromosome sets; an organism composed of such cells.

tetratype (T)   A tetrad type containing four different genotypes, two parental and two recombinant.

Thr   Threonine (an amino acid).

three-point testcross   A testcross involving one parent with three heterozygous gene pairs.

tymidine   The nucleoside having thymine as its base.

thymine   A pyrimidine base that pairs with adenine.

thymine dimer   A pair of chemically bonded adjacent thymine bases in DNA; the cellular processes that repair this lesion often make errors that create mutations.

Ti plasmid   A circular plasmid of *Agrobacterium tumifaciens* that enables the bacterium to infect plant cells and produce a tumor (crown gall tumor).

tissue-specific gene expression   The expression of a gene in a higher eukaryote in a specific and reproducible subset of tissues and cells during development.

totipotency The ability of a cell to proceed through all the stages of development and thus produce a normal adult.

trans conformation In a heterozygote involving two mutant sites within a gene or gene cluster, the arrangement $a_1 + / + a_2$.

transcription The synthesis of RNA using a DNA template.

transcription factor A protein that binds to a cis-regulatory element (e.g., an enhancer) and thereby, directly or indirectly, affects the initiation of transcription.

transduction The movement of genes from a bacterial donor to a bacterial recipient using a phage as the vector.

transfection The process by which exogenous DNA in solution is introduced into cultured cells.

transgene A gene that has been modified by externally applied recombinant DNA techniques and reintroduced into the genome by germ-line transformation.

transmembrane receptor A protein that spans the plasma membrane of a cell, with the extracellular portion of the protein having the ability to bind to a ligand and the intracellular portion having an activity (such as protein kinase) that can be induced upon ligand binding.

transfer RNA *See* **tRNA.**

transformation (1) The directed modification of a genome by the external application of DNA from a cell of different genotype. (2) Conversion of normal higher eukaryotic cells in tissue culture to a cancerlike state of uncontrolled division.

transgenic organism One whose genome has been modified by externally applied new DNA.

transient diploid The stage of the life cycle of predominantly haploid fungi (and algae) during which meiosis occurs.

transition A type of nucleotide-pair substitution involving the replacement of a purine with another purine, or of a pyrimidine with another pyrimidine — for example, GC → AT.

translation The ribosome-mediated production of a polypeptide whose amino acid sequence is derived from the codon sequence of an mRNA molecule.

translocation The relocation of a chromosomal segment in a different position in the genome.

transmission genetics The study of the mechanisms involved in the passage of a gene from one generation to the next.

transposable genetic element A general term for any genetic unit that can insert into a chromosome, exit, and relocate; includes insertion sequences, transposons, some phages, and controlling elements.

transposition *See* **translocation.**

transposon A mobile piece of DNA that is flanked by terminal repeat sequences and typically bears genes coding for transposition functions.

transversion A type of nucleotide-pair substitution involving the replacement of a purine with a pyrimidine, or vice versa — for example, GC → TA.

triplet The three nucleotide pairs that compose a codon.

triplet expansion The expansion of a 3-bp repeat from a relatively low number of copies to a high number of copies, that is responsible for a number of genetic diseases, such as Fragile X Syndrome.

triploid A cell having three chromosome sets, or an organism composed of such cells.

trisomic Basically a diploid with an extra chromosome of one type, producing a chromosome number of the form $2n + 1$.

tritium A radioactive isotope of hydrogen.

tRNA (transfer RNA) A class of small RNA molecules that bear specific amino acids to the ribosome during translation; the amino acid is inserted into the growing polypeptide chain when the anticodon of the tRNA pairs with a codon on the mRNA being translated.

Trp Tryptophan (an amino acid).

true-breeding line or strain *See* **pure-breeding line or strain.**

truncation selection A breeding technique in which individuals in whom quantitative expression of a phenotype is above or below a certain value (the truncation point) are selected as parents for the next generation.

tumor suppressor gene A gene encoding a protein that suppresses tumor formation. The wild-type alleles of tumor suppressor genes are thought to function as negative regulators of cell proliferation.

tumor virus A virus that is capable of inducing a cancer.

Turner syndrome An abnormal human female phenotype produced by the presence of only one X chromosome (XO).

twin spot A pair of mutant sectors within wild-type tissue, produced by a mitotic crossover in an individual of appropriate heterozygous genotype.

2-μm (2-micrometer) plasmid A naturally occurring extragenomic circular DNA molecule found in some yeast cells, with a circumference of 2 μm. Engineered to form the basis for several types of gene vectors in yeast.

Tyr Tyrosine (an amino acid).

U Uracil, or uridine.

underdominance A phenotypic relation in which the phenotypic expression of the heterozygote is less than that of either homozygote.

unequal crossover A crossover between homologs that are not perfectly aligned.

uniparental inheritance The transmission of certain phenotypes from one parental type to all the progeny; such inheritance is generally produced by organelle genes.

unstable mutation A mutation that has a high frequency of reversion; a mutation caused by the insertion of a controlling element, whose subsequent exit produces a reversion.

uracil A pyrimidine base that appears in RNA in place of thymine found in DNA.

URF Unassigned reading frame. An open reading frame (ORF) whose function has not yet been determined.

uridine The nucleoside having uracil as its base.

Val Valine (an amino acid).

variable A property that may have different values in various cases.

variable region A region in an immunoglobulin molecule that shows many sequence differences between antibodies of different specificities. The variable regions of the light and heavy chains of an immunoglobulin bind antigen.

variance A measure of the variation around the central class of a distribution; the average squared deviation of the observations from their mean value.

variant An individual organism that is recognizably different from an arbitrary standard type in that species.

variate A specific numerical value of a variable.

variation The differences among parents and their offspring or among individuals in a population.

variegation The occurrence within a tissue of sectors with differing phenotypes.

vector In cloning, the plasmid or phage chromosome used to carry the cloned DNA segment.

viability The probability that a fertilized egg will survive and develop into an adult organism.

viral transforming gene A gene within a viral genome that can induce abnormal proliferation of cells in culture, and similarly, can induce tumors in infected whole animals.

virulent phage A phage that cannot become a prophage; infection by such a phage always leads to lysis of the host cell.

VNTR (variable number tandem repeat) A chromosomal locus at which a particular repetitive sequence is present in different numbers in different individuals or in the two different homologs in one diploid individual.

wild type The genotype or phenotype that is found in nature or in the standard laboratory stock for a given organism.

wobble The ability of certain bases at the third position of an anticodon in tRNA to form hydrogen bonds in various ways, causing alignment with several possible codons.

X : A ratio The ratio between the X chromosome and the number of sets of autosomes.

X chromosome inactivation The process by which the genes of an X chromosome in a mammal can be completely repressed as part of the dosage compensation mechanism (*see* **dosage compensation** and **Barr body**).

X hyperactivation In *Drosophila*, the process by which the structural genes of the male X chromosome are transcribed at the same rate as the two X chromosomes of the female combined.

X linkage The inheritance pattern of genes found on the X chromosome but not on the Y.

X-and-Y linkage The inheritance pattern of genes found on both the X and Y chromosomes (rare).

X-ray crystallography A technique for deducing molecular structure by aiming a beam of X rays at a crystal of the test compound and measuring the scatter of rays.

Y linkage The inheritance pattern of genes found on the Y chromosome but not on the X (rare).

zygote The cell formed by the fusion of an egg and a sperm; the unique diploid cell that will divide mitotically to create a differentiated diploid organism.

zygotic induction The sudden release of a lysogenic phage from an Hfr chromosome when the prophage enters the F⁻ cell, and the subsequent lysis of the recipient cell.

zygoticallyacting gene A gene whose product is expressed only in the zygote and not included in the maternal contribution to the oocyte.

Answers to Selected Problems

Chapter 1

3. DNA determines all the specific attributes of a species (shape, size, form, behavioral characteristics, biochemical processes, etc.) and sets the limits for possible variation that is environmentally induced.

4. Two properties of DNA that are vital to its being the hereditary molecule are its ability to replicate and its stability that nevertheless contains within it the ability to change (mutate). Alien life forms might utilize RNA, just as some viruses do, as a hereditary molecule. However, of the types of molecules that can exist on earth, only the nucleic acids possess the characteristics necessary for a hereditary molecule.

6. Phenotypic variation within a species can be due to genotype, environmental effects, and pure chance (random noise).

9. The goal of genetic dissection is to understand the functioning of all the genes that affect a specific trait, to know the DNA sequence of each gene, and to understand the regulation during development of each of the genes.

Chapter 2

2. Do a testcross.

4. a. 9

   b. $1:2:1:2:4:2:1:2:1$

   c. Genotypes $= 3^n$

      Phenotypes $= 2^n$

   d. Selfing and testcross

5. a. $\left(\frac{1}{6}\right)^3$

   b. $\left(\frac{1}{6}\right)^3$

   c. $\left(\frac{1}{6}\right)^3$

   d. $\left(\frac{5}{6}\right)^3$

   e. $3\left(\frac{1}{6}\right)^3$

   f. $2\left(\frac{1}{6}\right)^3$

   g. $\left(\frac{1}{6}\right)^2$

   h. $\frac{5}{9}$

7. a. $\frac{1}{8}, \frac{3}{8}, \frac{3}{8}$

   b. $\frac{9}{16}$

10. $\frac{5}{8}$

16. b. 0.0069

21. b. 0.005

   c. $\frac{3}{4}$

25. 1 dwarf : 1 normal : 1 neurofibromatosis : 1 dwarf, neurofibromatosis

28. a. 1. $\frac{9}{128}$ **b.** 1. $\frac{1}{32}$

      2. $\frac{9}{128}$ 2. $\frac{1}{32}$

      3. $\frac{9}{64}$ 3. $\frac{1}{16}$

      4. $\frac{55}{64}$ 4. $\frac{15}{16}$

32. a. Three years of selfing produces $0.725\ AA : 0.05\ Aa : 0.225\ aa$.

   b. 0.05 remain heterozygous.

   c. $\left(\frac{1}{2}\right)^n X$, where $n =$ number of generations and $X =$ initial percentage of heterozygosity.

Chapter 3

2. $Aa\,Bb\,Cc$

6. a. 92 chromatids

   b. 92 chromatids

   c. 92 chromatids

   d. 92 chromatids

   e. 46 chromatids

7. e. Chromosome pairing

8. $\frac{1}{8}$

9. $\left(\frac{1}{2}\right)^{n-1}$

11. Female is G, male is gg.

13. e. 0

15. a. $\frac{1}{8}$

 b. $\frac{1}{4}$

 c. 0

17. a. $X^C X^c, x^c x^c$

 b. $\frac{1}{4}$

 c. 1 normal : 1 colorblind

 d. 1 normal : 1 colorblind

19. e. $\frac{1}{4}$

23. $\frac{3}{8}$ red, long $\frac{1}{8}$ brown, short

  $\frac{3}{8}$ red, short $\frac{1}{8}$ brown, long

28. Autosomal dominant with expression limited to males

Chapter 4

1. The mating is $AO \times AB$. The children are

Genotype	Phenotype
1 AA	A
1 AB	AB
1 AO	A
1 BO	B

4. e. 0 percent

5. a. Order of dominance: $b > s > c > a$

 b. 1 bb black   1 bc black

  1 bs black   1 sc sepia

7. Husband fathered child 1; lover fathered child 2; either could have fathered child 3.

11. The first cross was $Aa \times aY$. The expected progeny are
  1 Aa short-bristled females
  1 aa long-bristled females
  1 aY long-bristled males
  1 AY nonviable males

13. P $AABB$ (disc) $\times aabb$ (long)

 F_1 $AaBb$ disc

 F_2 $9A-B-$ disc

  3 $aaB-$ sphere

  3 $A-bb$ sphere

  1 $aabb$ long

15. a. Frizzle is Aa. Normal and woolly are homozygotes.

 b. Cross normal $\times$ woolly.

18. a. In line 1, the wild-type allele makes pisatin, and the recessive deviation does not produce pisatin. In line 2, the wild-type allele allows the expression of the pisatin that is normally made by the first gene, while the dominant deviation blocks expression of that wild-type product.

 b.

	Cross 1:	P	$aabb \times AAbb$
		F_1	$Aabb$
		F_2	$3A-bb$
			1 $aabb$
	Cross 2:	P	$AABB \times AAbb$
		F_1	$AABb$
		F_2	$3AAB-$
			1 $AAbb$
	Cross 3:	P	$aabb \times AABB$
		F_1	$AaBb$
		F_2	$9A-B-$ no pisatin
			$3A-bb$ pisatin
			$3aaB-$ no pisatin
			1 $aabb$ no pisatin

 c. Line 1 does not make pisatin, whereas line 2 blocks the expression of pisatin.

19. It is possible to produce black offspring from two pure-breeding recessive albino parents if albinism results from alleles of two different genes. If the cross is designated

$$AAbb \times aaBB$$

and epistasis is assumed, then all the offspring would be

$$AaBb$$

and they would have a black phenotype because of complementation.

23. b. $\frac{9}{32}$

26. a. P AA (agouti) $\times aa$ (nonagouti)

  Gametes A and a

 F_1 Aa (agouti)

  Gametes A and a

 F_2 1 AA (agouti) : 2 Aa (agouti) : 1 aa (nonagouti)

 b. P BB (wild type) $\times bb$ (cinnamon)

  Gametes B and b

F$_1$ Bb (wild type)

Gametes B and b

F$_2$ $1\,BB$ (wild type) : $2\,Bb$ (wild type) : $1\,bb$ (cinnamon)

c. P $AA\,bb$ (cinnamon or brown agouti) $\times$ $aa\,BB$

(black nonagouti)

Gametes $A\,b$ and $a\,B$

F$_1$ $Aa\,Bb$ (wild type or black agouti)

d. $9\,A-B-$ black agouti

$3\,aa\,B-$ black nonagouti

$3\,A-bb$ cinnamon

$1\,aa\,bb$ chocolate

e. P $AA\,bb$ (cinnamon) $\times$ $aa\,BB$ (black nonagouti)

Gametes $A\,b$ and $a\,B$

F$_1$ $Aa\,Bb$ (wild type)

Gametes AB, Ab, aB, and ab

F$_2$ $9\,A-B-$ wild type

$1\,AA\,BB$

$2\,Aa\,BB$

$2\,AA\,Bb$

$4\,Aa\,Bb$

$3\,aa\,B-$ black nonagouti

$1\,aa\,BB$

$2\,aa\,Bb$

$3\,A-bb$ cinnamon

$1\,AA\,bb$

$2\,Aa\,bb$

$1\,aa\,bb$ chocolate

f. P $Aa\,Bb \times AA\,bb$ $Aa\,Bb \times aa\,BB$

(wild type) (cinnamon)   (wild type) (black nonagouti)

F$_1$ $1\,AA\,Bb$ wild type $1\,Aa\,BB$ wild type

$1\,Aa\,Bb$ wild type $1\,Aa\,Bb$ wild type

$1\,AA\,bb$ cinnamon $1\,aa\,BB$ black nonagouti

$1\,Aa\,bb$ cinnamon $1\,aa\,Bb$ black nonagouti

g. P $Aa\,Bb \times aa\,bb$

(wild type) (chocolate)

F$_1$ $1\,Aa\,Bb$ wild type

$1\,Aa\,bb$ cinnamon

$1\,aa\,Bb$ black nonagouti

$1\,aa\,bb$ chocolate

h. Cross 1:   The F$_1$ parent is AA, and the original albino must have been $cc\,AA\,bb$.

Cross 2:   The albino parent must be $cc\,AA\,BB$.

Cross 3:   The F$_1$ must be $Cc\,Aa\,BB$, and the albino parent must be $cc\,aa\,BB$.

Cross 4:   The albino parent must be $cc\,aa\,bb$.

29. The cross is

P $AA\,pp \times aa\,P/Y$

F$_1$ $Aa\,Pp$ purple-eyed females

$Aa\,p/Y$ red-eyed males

F$_2$

Females		Males	
$\frac{3}{8}\,A-Pp$	purple	$\frac{3}{8}\,A-P/Y$	purple
$\frac{3}{8}\,A-pp$	red	$\frac{3}{8}\,A-p/Y$	red
$\frac{1}{8}\,aa\,Pp$	white	$\frac{1}{8}\,aa\,P/Y$	white
$\frac{1}{8}\,aa\,pp$	white	$\frac{1}{8}\,aa\,p/Y$	white

31. The seed is $Aa\,CC\,Rr$.

33. a. The cross is

P $td\,su$ (wild type) $\times$ $td^+\,su^+$ (wild type)

F$_1$ $1\,td\,su$ wild type

$1\,td\,su^+$ requires tryptophan

$1\,td^+\,su^+$ wild type

$1\,td^+\,su$ wild type

b. 1 tryptophan-dependent : 3 tryptophan-independent

37. Pedigrees like this are quite common. They indicate lack of penetrance due to epistasis or environmental effects. Individual A must have the dominant autosomal gene.

39. a. This is a dihybrid cross with only one phenotype in the F$_2$ being colored. The ratio of white to red indicates that the double recessive is not the colored phenotype. Instead, the general formula for color is represented by $X-yy$.

Let line 1 be $AA\,BB$ and line 2 be $aa\,bb$. The F$_1$ is $Aa\,Bb$. Assume that A blocks color in line 1 and bb blocks color in line 2. The F$_1$ will be white because of the presence of A. The F$_2$ are

$9\,A-B-$ white because of A

$3\,A-bb$ white because of A

$3\,aa\,B-$ red

$1\,aa\,bb$ white because of bb

b. Cross 1: $AA\,BB \times Aa\,Bb \rightarrow$ all $A-B-$ white

Cross 2: $aa\,bb \times Aa\,Bb \rightarrow \frac{1}{4}\,Aa\,Bb$ white

$\frac{1}{4}\,Aa\,bb$ white

$\frac{1}{4}\,aa\,bb$ white

$\frac{1}{4}\,aa\,Bb$ red

41. a.

Cross	Parents	Progeny
1	$A^S A^S \times A^D A^D$	$A^S A^D$

CONCLUSION: A^S is dominant to A^D

2	$A^S A^D \times A^S A^D$	$3\, A^S- : 1\, A^D A^D$

CONCLUSION: supports conclusion from cross 1

3	$A^D A^D \times A^{Sd} A^D$	$1\, A^{Sd} A^D : 1\, A^D A^D$

CONCLUSION: A^{Sd} is dominant to A^D

4	$A^S A^S \times A^{Sd} A^D$	$1\, A^{Sd} A^S : 1\, A^S A^D$

CONCLUSION: A^{Sd} is dominant to A^S

5	$A^D A^D \times A^{Sd} A^S$	$1\, A^{Sd} A^D : 1\, A^D A^S$

CONCLUSION: supports conclusion of heterozygous superdouble

6	$A^D A^D \times A^S A^D$	$1\, A^D A^D : 1\, A^D A^S$

CONCLUSION: supports conclusion of heterozygous superdouble

b. While this explanation does rationalize all the crosses, it does not take into account either the female sterility or the origin of the superdouble plant from a double-flowered variety.

A number of genetic mechanisms could be proposed to explain the origin of superdouble from the double-flowered variety. Most of the mechanisms will be discussed in later chapters and so will not be mentioned here. However, it can safely be assumed at this point that, whatever the mechanism, it was aberrant enough to block the proper formation of the complex structure of the female flower. Because of female sterility, no homozygote for superdouble can be observed.

43. a. A trihybrid cross would give a 63:1 ratio. Therefore, there are three R loci segregating in this cross.

b. P $\quad R_1 R_1\ R_2 R_2 R_3 R_3 \times r_1 r_1 r_2 r_2 r_3 r_3$

F_1 $\quad R_1 r_1\ R_2 r_2 R_3 r_3$

F_2		
27 $R_1- R_2- R_3-$	red	
9 $R_1- R_2- r_3 r_3$	red	
9 $R_1- r_2 r_2 R_3-$	red	
9 $r_1 r_1 R_2- R_3-$	red	
3 $R_1- r_2 r_2 r_3 r_3$	red	
3 $r_1 r_1 R_2- r_3 r_3$	red	
3 $r_1 r_1 r_2 r_2 R_3-$	red	
1 $r_1 r_1 r_2 r_2 r_3 r_3$	white	

c. 1. In order to obtain a 1:1 ratio, only one of the genes can be heterozygous. A representative cross would be $R_1 r_1 r_2 r_2 r_3 r_3 \times r_1 r_1 r_2 r_2 r_3 r_3$.

2. In order to obtain a 3 red:1 white ratio, two alleles must be segregating, and they cannot be within the same gene. A representative cross would be $R_1 r_1 R_2 r_2 r_3 r_3 \times r_1 r_1 r_2 r_2 r_3 r_3$.

3. In order to obtain a 7 red:1 white ratio, three alleles must be segregating, and they cannot be within the same gene. The cross would be $R_1 r_1 R_2 r_2 R_3 r_3 \times r_1 r_1 r_2 r_2 r_3 r_3$.

d. The formula is $1 - \left(\frac{1}{2}\right)^N$, where $N =$ the number of loci that are segregating in the representative crosses above. In F_1 crosses, the formula is $1 - \left(\frac{1}{2}\right)^{2N}$, where N is the number of loci.

45. a. The gene causing blue sclera and brittle bones is pleiotropic with variable expressivity.

b. The allele appears to be an autosomal dominant.

c. Both incomplete penetrance and variable expressivity are demonstrated in the pedigree. Individuals II-4, II-14, III-2, and III-14 have descendants with the disorder although they do not themselves express the disorder. Therefore, 4 out of 20 people have the gene but it is not penetrant in them. That is an 80 percent penetrance. Of the 16 individuals who have the allele expressed in their phenotype, 9 do not have brittle bones. Usually, expressivity is put in terms of none, variable, and highly variable, rather than expressed as percentages.

47. a. The gene is on the X chromosome. This is a multiple allelic series with oval > sickle > round.

Let $W^O =$ oval, $W^S =$ sickle and $W^r =$ round. The three crosses are

Cross 1: $\quad W^s W^s \times W^r Y \rightarrow W^s W^r : W^s Y$

Cross 2: $\quad W^r W^r \times W^s Y \rightarrow W^s W^r : W^r Y$

Cross 3: $\quad W^s W^s \times W^o Y \rightarrow W^o W^s : W^s Y$

b. $W^o W^s \times W^r Y \rightarrow 1\ W^o W^r : 1\ W^s W^r : 1\ W^o Y : 1\ W^s Y$, or 1 female oval:1 female sickle:1 male oval:1 male sickle

49. a. Let platinum be a, aleutian be b, and wild type be $A- B-$.

Cross 1:	P	$AA\,BB \times aa\,BB$
	F_1	$Aa\,BB$
	F_2	$3\,A-BB : 1\,aa\,BB$
Cross 2:	P	$AA\,BB \times AA\,bb$
	F_1	$AA\,Bb$
	F_2	$3\,AA\,B- : 1\,AA\,bb$
Cross 3:	P	$aa\,BB \times AA\,bb$
	F_1	$Aa\,Bb$
	F_2	$9\,A-B-$ wild type
		$3\,A-bb$ aleutian
		$3\,aa\,B-$ platinum
		$1\,aa\,bb$ sapphire

b. Sapphire × Platinum   Sapphire × Aleutian

P	$aa\,bb \times aa\,BB$		$aa\,bb \times AA\,bb$	
F$_1$	$Aa\,Bb$	platinum	$Aa\,bb$	aleutian
F$_2$	$3\,aa\,B{-}$	platinum	$3\,A{-}\,bb$	aleutian
	$1\,aa\,bb$	sapphire	$1\,aa\,bb$	sapphire

51. a. Let the gene be E and assume the following designations:

$$E^1 = \text{black}$$
$$E^2 = \text{brown}$$
$$E^3 = \text{eyeless}$$

If you next assume, based on the various crosses, that black $>$ brown $>$ eyeless, genotypes in the pedigree become

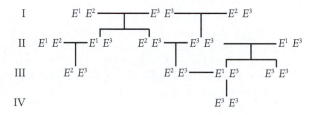

b. The genotype of individual II-3 is $E^2 E^3$.

Chapter 5

For the following, *CO* is used to designate single recombinants, and *DCO* is used to designate double recombinants.

1. 45 percent of the progeny will be $Aa\,Bb$.

2.

P	$A\,d/A\,d \times a\,D/a\,D$
F$_1$	$A\,d/a\,D$
F$_2$	$1\,A\,d/A\,d$
	$2\,A\,d/a\,D$
	$1\,a\,D/a\,D$

4. The two genes are 33.3 map units (m.u.) apart.

5. Because only parental types were recovered, the two genes must be quite close to each other, making recombination quite rare.

8. a. 4 percent

b. 4 percent

c. 46 percent

d. 8 percent

11. a. All four genes are linked.

b. and c. The map is

```
      B          A                    C
      +----------+--------------------+
        10 m.u.        30 m.u.
```

The parental chromosomes actually were $B\,(A,\,d)\,c/b\,(a,\,D)\,C$, where the parentheses indicate that the order of the genes within is unknown.

d. Interference = 0.5

12. a. Males must be heterozygous for both genes, and the two must be closely linked: $M\,F\,/\,m\,f$.

b. $m\,f\,/\,m\,f$

c. Sex is determined by the male contribution. The two parental gametes are $M\,F$, determining maleness ($M\,F\,/\,m\,f$), and $m\,f$, determining femaleness ($m\,f\,/\,m\,f$). Occasional recombination would yield $M\,f$, determining a hermaphrodite ($M\,f\,/\,m\,f$), and $m\,F$, determining total sterility ($m\,F\,/\,m\,f$).

d. Recombination in the male yielding $M\,f$

e. Hermaphrodites are rare because the genes are tightly linked.

15. a. $a{-}b$:   100% (91 + 9) / 1001 = 10 m.u.
       $b{-}c$:   100% (171 + 9) / 1001 = 18 m.u.

b. Coefficient of coincidence = 0.5

17. $v{-}b$: 18.0 m.u.
$b{-}lg$: 28.0 m.u.
Coefficient of coincidence = 0.79

20. 1. $b\,a\,c$
2. $b\,a\,c$
3. $b\,a\,c$
4. $a\,c\,b$
5. $a\,c\,b$

22. a. The hypothesis is that the genes are not linked. Therefore, a 1:1:1:1 ratio is expected.

b. $\chi^2 = 0.76$

c. With 3 degrees of freedom, the p value is between 0.50 and 0.90.

d. Between 50 percent and 90 percent of the time values this extreme from the prediction would be obtained by chance alone.

e. Accept the initial hypothesis.

f. Because the χ^2 value was insignificant, the two genes are assorting independently. The genotypes of all individuals are

P	$dp^+dp^+ee \times dpdp\,e^+e^+$
F$_1$	$dp^+dp\,e^+e$
Tester	$dpdp\,ee$
Progeny	long ebony $dp^+dp\,ee$
	long gray $dp^+dp\,e^+e$
	short gray $dpdp\,e^+e$
	short ebony $dpdp\,ee$

29. Let A = resistance to rust 24, a = susceptibility to rust 24, B = resistance to rust 22, b = susceptibility to rust 22.

   a. P $AAbb$ (770B) $\times$ $aaBB$ (Bombay)

   F$_1$ $AaBb \times AaBb$

   F$_2$ 184 $A-B-$

   63 $A-bb$

   58 $aaB-$

   15 $aabb$
   ———
   320

   b. Expect:

   180 $A-B-$

   60 $A-bb$

   60 $aaB-$

   20 $aabb$

   $\chi^2 = 1.555$, nonsignificant; accept the hypothesis.

31. a. Blue sclerotics appears to be an autosomal dominant disorder. Hemophilia appears to be an X-linked recessive disorder.

   b. If the individuals in the pedigree are numbered as generations I through IV and the individuals in each generation are numbered clockwise, starting from the top right-hand portion of the pedigree, their genotypes are

   I: $bbHh, BbHY$

   II: $BbHY, BbHY, bbHY, BbH-, bbHY, BbHh, BbH-, bbH-$

   III: $bbH-, BbH-, bbhY, bbHY, BbHY, BbH-, BbHY, BbhY, BbH-, bbHY, BbH-, bbHY, BbH-, BbHY, BbhY, bbHY, bbHY, bbH-, bbHY, bbHY, BbH-, BbHY, BbhY$

   IV: $bbH-, BbH-, BbH-, bbHh, bbHh, bbHY, bbHH, bbHY, bbHh, bbH-, bbH-, bbHY, bbHY, bbH-, bbHY, bbHY, BbHY, bbHY, bbHH, bbHY, bbHY, bbHH, bbH-, bbH-, bbH-, bbHY, bbHY, bbHY, bbHh, BbH-, BbHY, bbHY, BbHY, bbH-$

   c. There is no evidence of linkage between these two disorders.

   d. The two genes exhibit independent assortment.

   e. No individual could be considered intrachromosomally recombinant. However, a number show interchromosomal recombination: all individuals in generation III that have both disorders.

34. a.

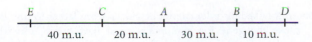

   b. No interference

36. a. 0.0336

   b. 0.182

Chapter 6

1. a. +, +, al-2, al-2, +, +, al-2, al-2
       al-2, al-2, +, +, al-2, al-2, +, +

   b. 4 percent

3. a. $f(0) = e^{-2}2^0/0! = e^{-2} = 0.135$

   b. $f(1) = e^{-2}2^1/1! = e^{-2}(2) = 0.27$

   c. $f(2) = e^{-2}2^2/2! = e^{-2}(2) = 0.27$

6. a. 25.5 m.u.

   b. $\chi^2 = 2.52$

   With 3 degrees of freedom, the probability is greater than 10 percent that the genes are not linked. Therefore, the hypothesis of no linkage can be accepted.

7. a. The parents were

   $\underline{ad^-\ nic^+\ leu^+\ arg^-} \times \underline{ad^+\ nic^-\ leu^-\ arg^+}$

   b. Culture 16 resulted from a crossover between ad and nic. The reciprocal did not show up in the small sample.

8. a. 0.28

   b. 0.35

   c. 0.22

   d. 0.72

9. a. 0.704

   b. 0

   c. 0.176

   d. 0.012

   e. 0.096

   f. 0

   g. 0.012

11. a.

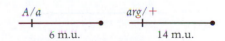

   b. Class 6 can be obtained if a single crossover occurred between chromatids 2 and 3 between each gene and its centromere.

13. Cross 1:

Cross 2:

Cross 3:

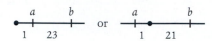

The first diagram is the better interpretation of the data.

Cross 4:

Cross 5: The genes are considered unlinked to their centromeres in tetrad analysis.

Cross 6:

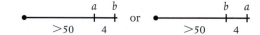

Cross 7:

Cross 8: Same as cross 5.

Cross 9:

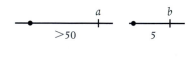

Cross 10:

or

or

Cross 11:

15. Cross 1: recombinant frequency = 26.5%

uncorrected map distance = 26.5 m.u.

corrected map distance = 34.5 m.u.

Cross 2: recombinant frequency = 19%

uncorrected map distance = 19 m.u.

corrected map distance = 29 m.u.

Cross 3: recombinant frequency = 30%

uncorrected map distance = 30 m.u.

corrected map distance = 40 m.u.

17. a. A 4:0 ascus is a PD ascus. The 3:1 ascus must represent a T
ascus. The 2:2 ascus is an NPD ascus.

b. PD = NPD = 40, and NPD/T = 40/20 = 2.00

22. a. The final map is

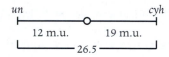

23. a. 0%

b. 84%

c. 16%

d–g. 0%

26. The resulting daughter cells would be GG (yellow) and gg
(green).

27.

Chapter 7

1. The petal will now be Ww, or blue, either in whole or in part,
depending on the timing of the reversion.

3. Plate the cells on medium lacking proline. Nearly all colonies
will come from revertants. The remainder will be second-site
suppressors.

6. $\dfrac{1}{10^{-6}}$ cell divisions

8. Stain pollen grains, which are haploid, from a homozygous
Wx parent with iodine. Look for red pollen grains, indicating
mutations to wx, under a microscope.

10. An X-linked disorder cannot be passed from father to son. Because the gene for hemophilia must have come from the
mother, the nuclear power plant cannot be held responsible. It
is possible that the achondroplastic gene mutation was caused
by exposure to radiation.

13. a. Reddish all over

b. Reddish all over

c. Many small, red spots

d. A few large, red spots

e. Like c, but with fewer reddish patches

f. Like d, but with fewer reddish patches

g. Some large spots and many small spots

16. Mutant 1: An auxotroph capable of growing only on complete medium

Mutant 2: A dominant temperature-sensitive mutation

Mutant 3: A conditional auxotrophic mutation

Mutant 4: A dominant conditional temperature-sensitive mutation

Mutant 5: A conditional auxotrophic temperature-sensitive mutation

18. a. The experiment was designed to detect the number of genes that are involved with locomotion.

b.

Gene	Mutants
A	1, 5
B	2, 6, 8, 10
C	3, 4
D	7, 11, 12

c. Mutant 1: $AA\ B^+B^+\ C^+C^+\ D^+D^+$

Mutant 2: $A^+A^+\ BB\ C^+C^+\ D^+D^+$

Mutant 5: $AA\ B^+B^+\ C^+C^+\ D^+D^+$

Hybrid $\frac{1}{2}$: $AA\ B^+B^+\ C^+C^+\ D^+D^+ \times$
$A^+A^+\ BB\ C^+C^+\ D^+D^+ \rightarrow$
$AA^+\ B^+BC^+C^+\ D^+D^+$

Hybrid $\frac{1}{5}$: $AA\ B^+B^+\ C^+C^+\ D^+D^+ \times$
$AA\ B^+B^+\ C^+C^+\ D^+D^+ \rightarrow$
$AA\ B^+B^+\ C^+C^+\ D^+D^+$

In hybrid $\frac{1}{2}$, there is complementation. In hybrid $\frac{1}{5}$, the mutations are in the same gene, and complementation cannot occur.

21. The cross is $arg^r \times arg^+$, where arg^r is the revertant. However, it might also be $arg^-\ su \times arg^+su^+$, where su^+ has no effect on the arg gene.

a. If the revertant is a precise reversal of the original change that produced the mutant allele, 100 percent of the progeny would be arginine independent.

b. If a suppressor mutation on a different chromosome is involved, then the cross is $arg^-\ su \times arg^+su^+$. Independent assortment would lead to the following:

1 $arg^-\ su$	arginine-independent
1 $arg^+\ su^+$	arginine-independent
1 $arg^-\ su^+$	arginine-dependent
1 arg^+su	arginine-independent

c. If a suppressor mutation 10 map units from the arg locus occurred, then the cross is $arg^-\ su \times arg^+su^+$, but it is now necessary to write the diploid intermediate as $arg^-\ su\,/\,arg^+su^+$.

The two parental types would occur 90 percent of the time, and the two recombinant types would occur 10 percent of the time. The progeny would be

45% $arg^-\ su$	arginine-independent
45% $arg^+\ su^+$	arginine-independent
5% $arg^-\ su^+$	arginine-dependent
5% $arg^+\ su$	arginine-independent

Chapter 8

2. a. 27%

b. 36%

5. A deletion occurred in the left arm of the large chromosome in nucleus 2, and leu^+, mating type a, and un^+ were lost.

8. The order is $b\ a\ c\ e\ d\ f$.

Allele	Band
b	1
a	2
c	3
e	4
d	5
f	6

10. a. Eighteen maps units (m.u.) were either deleted or inverted. A large inversion would result in semisterility, whereas a large deletion would most likely be lethal. Thus, an inversion is more likely.

b. Normal

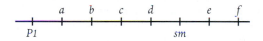

Inversion

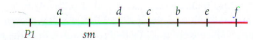

Alternatively, sm could be located external to the e locus on both the normal and inversion chromosomes.

c. The semisterility is the result of crossing-over in the inverted region. All products of crossing-over would have both duplications and deletions.

13. The most likely explanation is that one or both break points were located within essential genes.

20. a. The inversion must involve *b* through *x*.

b. To obtain recombinant progeny when an inversion is involved, either a double crossover occurred within the inverted region or single crossovers occurred between *f* and the point of inversion, which occurred someplace between *f* and *x*.

22. a. and b.

Class 1: parental

Class 2: parental

Class 3: DCO *y-cv* and *B-car*

Class 4: reciprocal of class 3

Class 5: DCO *cv-v* and *v-f*

Class 6: reciprocal of class 5

Class 7: DCO *cv-v* and *f-car*

Class 8: reciprocal of class 7

Class 9: DCO *v-cv* and *v-f*

Class 10: reciprocal of class 9

Class 11: This class is identical to the male parent's X chromosome and could not have come from the female parent. Thus, the male sperm must have donated it to the offspring. In *Drosophila*, sex is determined by the ratio of X chromosomes to the number of sets of autosomes. The ratio in males is 1X:2A, where A stands for the autosomes contributed by one parent (the ratio in females is 2X:2A). Thus, this class of males must have arisen from the union of an X-bearing sperm with an egg that was the product of nondisjunction for X and contained only autosomes.

c. Class 11 should have only one sex chromosome, which could be checked cytologically.

32. a. The gray male 1 was crossed with a yellow female, yielding yellow females and gray males, which is reversed sex linkage. If the e^+ allele were translocated to the Y chromosome, the gray male would be $X^e Y^{-e+}$, or gray. When crossed with yellow females, the results would be

$X^e Y^{-e+}$ gray males $X^e X^e$ yellow females

b. The gray male 2 was crossed with a yellow female, yielding gray and yellow males and females in equal proportions. If the e^+ allele were translocated to an autosome, the progeny would be as below, where "A" indicates an autosome involved in the translocation and the "/" separates male and female contributions:

P $A^{e+}A\ XY \times AA\ X^e X^e$
F$_1$ $A^{e+}X / AX^e$ gray female
      $A^{e+}Y / AX^e$ gray male
      AX / AX^e yellow female
      AY / AX^e yellow male

38. The percent degeneration seen in the progeny of the exceptional rat is 51 percent larger than that seen in the progeny of the normal rat. One explanation could be a chromosomal inversion. This could be verified by cytological observation of meiotic cells in the rat. An alternative possibility would be a translocation that could also be checked cytologically.

39. a. Heterozygous reciprocal translocations lead to duplications and deletions. Therefore, in asci in which crossing-over occurs within the translocated region, each crossing-over event would lead to two white ascospores that abort and two viable dark ascospores. In asci in which no crossing-over occurs within the translocated region, the ascospores would be normal color. Alternate segregation is assumed.

No CO   CO

●   ●
●   ○
●   ○
●   ●

b. Heterozygous pericentric inversions lead to duplications and deletions. Therefore, in asci in which crossing-over occurs within the pericentric inversion, each crossing-over event would lead to two white ascospores that abort and two viable dark ascospores. In asci in which no crossing-over occurs within the pericentric inversion, the ascospores would be normal color. Alternate segregation is assumed.

No CO   CO

●   ●
●   ○
●   ○
●   ●

c. Heterozygous paracentric inversions result in an acentric fragment that has lost some genetic material (deletion) and a dicentric chromosome that has gained some genetic material (duplication) if crossing-over occurs within the paracentric inversion. Therefore, in asci in which crossing-over occurs within the paracentric inversion, each crossing-over event would lead to two white ascospores that abort and two viable dark ascospores. In asci in which no crossing-over occurs within the paracentric inversion, the ascospores would be normal color. Alternate segregation is assumed.

No CO   CO

●   ●
●   ○
●   ○
●   ●

40. Species B is probably the "parent" species. A paracentric inversion in this species would give rise to species D. Species E could then occur by a translocation of *z x y* to *k l m*. Next, species A could result from a translocation of *a b c* to *d e f*. Finally, species C could result from a pericentric inversion of *b c d e*.

Chapter 9

1. Klinefelter syndrome   XXY male
   Down syndrome   trisomy 21
   Turner syndrome   XO female

3. **a.** 3, 3, 3, 3, 3/3, 3/3, 0, 0

   b. 7, 7, 7, 7, 8, 8, 6, 6

5. b

7. If there is no pairing between chromosomes from the same parent, then all pairs are *Aa*. The gametes are 1*AA* : 2*Aa* : 1*aa*. With selfing, the progeny are $\frac{1}{16}$ *AAAA*, $\frac{4}{16}$ *AAAa*, $\frac{6}{16}$ *AAaa*, $\frac{4}{16}$ *Aaaa*, and $\frac{1}{16}$ *aaaa*. If there is no pairing between chromosomes from different parents, then all pairs are *AA* and *aa* and all gametes are *Aa*. Thus 100% of the progeny are *AAaa*.

11. b, f, and h, and sometimes c

12. Colorblindness appears to be an X-linked recessive disorder. That means that the individual with Turner syndrome had to have obtained her sole X from her mother. She did not obtain a sex chromosome from her father, which indicates that nondisjunction occurred in him. The nondisjunction could have occurred at either M_I or M_{II}.

17. **a.** Loss of one X in the developing fetus after the two-celled stage.

    b. Nondisjunction leading to Klinefelter syndrome (XXY) followed by a nondisjunctive event in one cell for the Y chromosome after the two-celled stage, leading to XX and XXYY.

    c. Nondisjunction for X at the one-celled stage.

    d. Either fused XX and XY zygotes or fertilization of an egg and polar body by one sperm bearing an X and another bearing a Y, followed by fusion.

    e. Nondisjunction of X at the two-celled stage or later.

19. Type a: The extra chromosome must be from the mother. Because the chromosomes are identical, nondisjunction had to have occurred at M_{II}.
    Type b: The extra chromosome must be from the mother. Because the chromosomes are not identical, nondisjunction had to have occurred at M_I.
    Type c: The mother correctly contributed one chromosome, but the father did not contribute any chromosome 4. Therefore, nondisjunction occurred in the male during either meiotic division.
    Type d: One cell line lacks a maternal contribution while the other has a double maternal contribution. Because the two lines are complementary, the best explanation is that nondisjunction occurred in the developing embryo during mitosis.

Type e: Each cell line is normal, indicating that nondisjunction did not occur. The best explanation is that the second polar body was fertilized and was subsequently fused with the developing embryo.

20. **a.** The common progeny are be^+/be and b^+e/be.

    b. The rare female could have come from crossing-over, which would have resulted in a gamete that was b^+e^+. The rare female also could have come from nondisjunction that resulted in a gamete that was be^+/b^+e. Such a gamete might give rise to viable progeny.

    c. The female was a product of nondisjunction.

23. Radiation could have caused point mutations or induced recombination, but nondisjunction is a more likely explanation.

27. Aneuploidy is the result of nondisjunction. Therefore, any system that will detect nondisjunction will work. One of the easiest follows. Cross white-eyed females with red-eyed males and look for the reverse of X linkage in the progeny. Compare populations unexposed to any environmental pollutants with those exposed to different suspect agents.

Chapter 10

1. An Hfr strain has the fertility factor, F, integrated into the chromosome. An F$^+$ strain has the fertility factor free in the cytoplasm. An F$^-$ strain lacks the fertility factor.

3. **a.** Hfr cells involved in conjugation transfer host genes in a linear fashion. The genes transferred depend on both the Hfr strain and the length of time during which the transfer occurred. Therefore, a population containing several different Hfr strains will appear to have an almost random transfer of host genes. This is similar to generalized transduction, in which the viral protein coat forms around a specific amount of DNA rather than specific genes. In generalized transduction, any gene can be transferred.

    b. F$'$ factors arise from improper excision of an Hfr from the bacterial chromosome. They can have only specific bacterial genes on them because the integration site is fixed for each strain. Specialized transduction resembles this in that the viral particle integrates into a specific region of the bacterial chromosome and then, on improper excision, can take with it only specific bacterial genes. In both cases, the transferred gene exists as a second copy.

6.   —
   M Z X W C

       W C N A L

         A L B R U

           B R U M Z
             —

   The regions with the bars above or below are identical in sequence. The order on the circular map is

       MZXWCNALBRUMZ

7. An F⁻ strain will respond differently to an F⁺ (L) or an Hfr (M) strain, while Hfr × Hfr, Hfr × F⁺, F⁺ × F⁺, and F⁻ × F⁻ will give 0. Thus strains 2, 3, and 7 are F⁻. Strains 1 and 8 are F⁺, and strains 4, 5, and 6 are Hfr.

10. a. The gene order is *arg bio leu*.

    b. *arg-bio:* RF = 12.76 m.u.

      bio-leu: RF = 2.12 m.u.

12. The sequence is *pro*-4-5-7-1-6-3-2-8-*ade*.

13. The most straightforward way would be to put an Hfr at both ends of the same sequence and measure the time of transfer between two specific genes. For example,

15. The best explanation is that the integrated *pro⁺* was incorporated onto an F′ factor that was transferred into recipients early in the mating process. These cells now carry the F factor and are able to transmit F⁺ in the second cross as part of the F′ factor, which still carries *pro⁺*.

16. The high rate of integration and the preference for the same site originally occupied by the sex factor suggest that the F′ contains some homology with the original site. The source of homology could be a fragment of the sex factor or it could be the chromosomal copy of the bacterial gene (most likely).

21. The expected number of double recombinants is 2. Interference = −1.5. By definition, the interference is negative.

24. The instability of *gal⁺* transductants comes from the fact that they are partial diploids and have a tendency to lose a *gal* gene. If the *gal⁺* gene is lost, a *gal⁻* clone will develop. The stable transductants are not partial diploids and, therefore, do not have a tendency to segregate *gal*. They are not partial diploids because there was an exchange of *gal* alleles rather than an insertion of the second *gal* allele, as with partial diploids. This occurred when lambda looped out between the two *gal* alleles and one *gal⁺* allele was reconstructed.

25. a. Specialized transduction is at work here. It is characterized by the transduction of one to a few markers.

    b. The prophage is located in the *cys-leu* region, which is the only region that gave rise to colonies when tested against the six nutrient markers.

28. Recognize that if a compound is not added and growth occurs, the *E. coli* has received the genes for it by transduction. Thus, the BCE culture must have received *a⁺* and *d⁺*. The BCD culture received *a⁺* and *e⁺*. The ABD culture received *c⁺* and *e⁺*. The order is thus *d a e c*. Notice that *b* is never cotransduced and is therefore distant from this group of genes.

30. a. The colonies are all *cys⁺* and either + or − for the other two genes.

    b. (1) *cys⁺ leu⁺ thr⁺* and *cys⁺ leu⁺thr⁻*
        (2) *cys⁺ leu⁺thr⁺* and *cys⁺leu⁻ thr⁺*
        (3) *cys⁺ leu⁺ thr⁺*

    c. Because none grew on minimal medium, no colony was *leu⁺ thr⁺*. Therefore, medium (1) had *cys⁺ leu⁺ thr⁻*, and medium (2) had *cys⁺ leu⁻ thr⁺*. The remaining cultures were *cys⁺ leu⁻ thr⁻*, and this genotype occurred in 39 percent of the colonies.

    d.

Chapter 11

1. The DNA double helix has two types of bonds, covalent and hydrogen. Covalent bonds occur within each linear strand and strongly bond nucleotides, sugars, and phosphate groups (both within each component and between components). Hydrogen bonds occur between the two strands and involve a nucleotide from one strand with a nucleotide from the second in complementary pairing. These hydrogen bonds are individually weak but collectively quite strong.

3. A primer is a short segment of RNA that is synthesized by RNA polymerase using DNA as a template during DNA replication. Once the primer is synthesized, DNA polymerase then adds DNA to the 3′ end of the RNA. Primers are required because the major DNA polymerase involved with DNA replication is unable to initiate DNA synthesis and, rather, requires a 3′ end. The RNA is subsequently removed and replaced with DNA so that no gaps exist in the final product.

5. Because the DNA polymerase is capable of adding new nucleotides only at the 3′ end of a DNA strand, and because the two strands are antiparallel, at least two molecules of DNA polymerase must be involved in the replication of any specific region of DNA. When a region becomes single-stranded, the two strands have an opposite orientation. Imagine a single-stranded region that runs from left to right. At the left end, the 3′ end of one strand points to the right, and synthesis can initiate and continue toward the right end of that region. The other strand has a 5′ end pointing toward the right, and synthesis cannot initiate and continue toward the right end of the single-stranded region at the 5′ end. Instead, synthesis must initiate somewhere to the right of the left end of the single-stranded region and move toward the left end of the region. As the first strand continues synthesis (continuous synthesis), the single-stranded region extends toward the right. This now leaves the second strand unreplicated in this new region of single-strandedness, and there must be a second initiation of DNA synthesis moving from the current right end of the single-stranded region toward the first initiation point on that strand. This results in discontinuous synthesis along that strand.

7. The frequency of both A and T is $\frac{1}{2}$ (52%) = 26%.

9. The results suggest that the DNA is replicated in short segments that are subsequently joined by enzymatic action (DNA ligase). Because DNA replication is bidirectional, because there are multiple points along the DNA where replication is initiated, and because DNA polymerases work only in a $5' \rightarrow 3'$ direction, one strand of the DNA is always in the wrong orientation for the enzyme. This requires synthesis in fragments.

12. Chargaff's rules are that A = T and G = C. Because this is not observed, the most likely interpretation is that the DNA is single-stranded. The phage would first have to synthesize a complementary strand before it could begin to make multiple copies of itself.

13. Remember that there are two hydrogen bonds between A and T, while there are three hydrogen bonds between G and C. Denaturation involves the breaking of these bonds, which requires energy. The more bonds that need to be broken, the more energy that must be supplied. Thus the temperature at which a given DNA molecule denatures is a function of its base composition. The higher the temperature of denaturation, the higher the percentage of G–C pairs.

18. The data suggest that each chromosome is composed of one long, continuous molecule of DNA and that chromosomes are fragmented during translocation.

Chapter 12

1. The primary structure of a protein is the sequence of amino acids along its length. It is held together by covalent bonds. The secondary structure of a protein is caused by hydrogen bonding between CO and NH groups on different amino acids. Frequently, α helices and β pleated sheets result. The tertiary structure of a protein is caused by electrostatic, hydrogen, and Van der Waals bonds between the R groups of amino acids. If the final, functional protein is composed of more than one polypeptide, then the protein has a quaternary structure.

3. An auxotroph is a strain that requires at least one nutrient for growth beyond that normally required for the organism.

5. Yanofsky analyzed mutations in the *trpA* gene and ordered them using transduction. He also determined the amino acid sequence of the altered gene products. By this, he demonstrated an exact correlation between the sequence of mutation sites and the sequence of altered amino acids.

7.

	B	K
rII	large plaques	no plaques
rII⁺	small plaques	plaques

9. Lactose is composed of one molecule of galactose and one molecule of glucose. A secondary cure would result if all galactose and lactose were removed from the diet. The disorder would be expected not to be dominant, because one good copy of the gene should allow for at least some, if not all, breakdown of galactose. In fact, the disorder is recessive.

11. a. The main use is in detecting carrier parents and in diagnosing a disorder in the fetus.

b. Because the values for normal individuals and carriers overlap for galactosemia, there is ambiguity if a person has 25 to 30 units as a result. That person could be either a carrier or normal.

c. These wild-type genes are phenotypically dominant but are incompletely dominant at the molecular level. A minimal level of enzyme activity apparently is enough to ensure normal function and phenotype.

12. One less likely possibility is a germ-line mutation. More likely is that each parent was blocked at a different point in a metabolic pathway. If one were *AA bb* and the other were *aa BB*, then the child would be *Aa Bb* and would have sufficient levels of both resulting enzymes to produce pigment.

14. a. Complementation refers to genes within a cell, which is not what is happening here. Most likely, what is known as cross-feeding is occurring, whereby a product made by one strain diffuses to another strain and allows growth of the second strain. This is equivalent to supplementing the medium. Because cross-feeding seems to be taking place, the suggestion is that the strains are blocked at different points in the metabolic pathway.

b. For cross-feeding to occur, the growing strain must have a block that occurs earlier in the metabolic pathway than does the block in the strain from which it is obtaining the product for growth.

c. *E-D-B*

d. Without some tryptophan, no growth at all would occur, and the cells would not have lived long enough to produce a product that could diffuse.

18. a. 60 percent of the meioses will not have a crossover.

b. There are 20 m.u. between the two genes.

c. 10 percent

d. $\longrightarrow$ B $\longrightarrow$ A $\longrightarrow$ valine
   val-1   val-2

20. a. White

b. Blue

c. Purple

d. P   *bb DD* × *BB dd*

     F_1 *Bb Dd* × *Bb Dd*

     F_2 9 *B— D—*   purple

         3 *bb D—*   white

         3 *B— dd*   blue

         1 *bb dd*   white

         or 9 : 3 : 4

21. The cis and trans burst size should be the same if the mutants are in different cistrons, and if they are in the same cistron, the trans burst size should be zero. Therefore, assuming *rV* is in A, *rW* also is in A, and *rU*, *rX*, *rY*, and *rZ* are in B.

28.

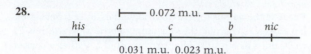

29. a. The allele s^n will show dominance over s^f because there will be only 40 units of square factor in the heterozygote.

   b. Here, the functional allele is recessive.

   c. The allele s^f may become dominant over time in two ways: (1) it could mutate slightly, so that it produces more than 50 units, or (2) other modifying genes may mutate to increase the production of s^f.

30. 129.15

33. a. There are three cistrons:

           Cistron 1: mutants 1, 3, and 4

           Cistron 2: mutants 2 and 5

           Cistron 3: mutant 6

   b. The order is *A/a* 6 (1, 3, 4) (2, 5) *B/b*.

Chapter 13

1. Because RNA can hybridize to both strands, the RNA must be transcribed from both strands. This does not mean, however, that both strands are used as a template *within each gene*. The expectation is that only one strand is used within a gene but that different genes are transcribed in different directions along the DNA. The most direct test would be to purify a specific RNA coding for a specific protein, and then hybridize it to the lambda genome. Only one strand should hybridize to the purified RNA.

3. A single nucleotide change should result in three adjacent amino acid changes in a protein. One and two adjacent amino acid changes would be expected to be much rarer than the three changes. This is directly the opposite of what is observed in proteins.

7. The codon for amber is UAG. Listed below are the amino acids that would need to have been inserted to continue the wild-type chain and their codons:

glutamine	CAA, CAG*
lysine	AAA, AAG*
glutamic acid	GAA, GAG*
tyrosine	UAU*, UAC*
tryptophan	UGG*
serine	AGU, AGC, UCU, UCC, UCA, UCG*

In each case, the codon that has an asterisk by it would require a single base change to become UAG.

8. a. $\frac{1}{8}$

   b. $\frac{1}{4}$

   c. $\frac{1}{8}$

   d. $\frac{1}{8}$

10. a. $(GAU)_n$ codes for Asp (GAU), Met (AUG), and stop (UGA). $(GUA)_n$ codes for Val (GUA), Ser (AGU), and stop (UAG). One reading frame contains a stop codon.

   b. Each of the three reading frames contains a stop codon.

11. Mutant 1: A simple substitution of Arg for Ser exists, suggesting a nucleotide change. Two codons for Arg are AGA and AGG, and one codon for Ser is AGU. The final U for Ser could have been replaced by either an A or a G.

Mutant 2: The Trp codon (UGG) changed to a stop codon (UGA or UAG).

Mutant 3: Two frameshift mutations occurred:

5′-GCN CCN (−U)GGA GUG AAA AA(+U or C) UGU/C CAU/C-3′

Mutant 4: An inversion occurred after Trp and before Cys. The DNA original sequence was

   3′-CGN GGN ACC TCA CTT TTT ACA/G GTA/G-5′

Therefore, the complementary RNA sequence was

   5′-GCN CCN UGG AGU GAA AAA UGU/C CAU/C-3′

The DNA inverted sequence became

   3′-CGN GGN ACC∧AAA AAG TGA ACA/G G∧TA/G-5′

Therefore, the complementary RNA sequence was

   5′-GCN CCN UGG U∧UU UUC ACU UGU/C C∧AU/C-3′

12. a.

   old: ⁻AAA/G AGU CCA UCA CUU AAU GCN GCN AAA/G

   new: AAA/G GUC CAU CAC UUA AUG GCN GCN AAA/G
                            +

   b. Plus (+) and minus (−) are indicated in the appropriate strands.

14. 3′ CGT ACC ACT GCA 5′

   5′ GCA TGG TGA CGT 3′

5′ GCA UGG UGA CGU 3′

3′ CGU ACC ACU GCA 5′

NH$_3$-Ala-Trp stop Arg

16. f, d, j, e, c, i, b, h, a, g

17. Cells in long-established culture lines usually are not fully diploid. For reasons that are currently unknown, adaptation to culture frequently results in both karyotypic and gene dosage changes. This can result in hemizygosity for some genes, which allows for the expression of previously hidden recessive alleles.

Chapter 14

2. GTTAAC occurs, on average, every 4^6 bases, which is 4.096 kb. GGCC occurs, on average, every 4^4 bases, which is 0.256 kb.

4.

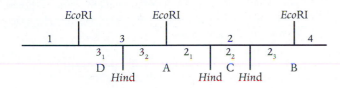

6. Reading from the top the sequence is

*Hin*d-*Hae*-*Hae*-*Hin*d-*Hae*-*Hae*-*Hae*-*Hin*d-*Hin*d-*Hae*-*Hin*d-*Hae*-*Eco*RI

7. Reading from the bottom up,

left column: GGTACAACTATATATCAATTATAAAC

right column: GGATCTATTCTTATGATTATATAG

9. *Alu*I $\left(\frac{1}{4}\right)^4$ = every 256 nucleotide pairs

*Eco*RI $\left(\frac{1}{4}\right)^6$ = every 4096 nucleotide pairs

*Acy*I $\left(\frac{1}{4}\right)^4\left(\frac{1}{2}\right)^2$ = every 1024 nucleotide pairs

10. a.

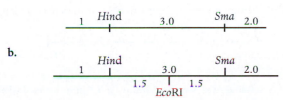

b.

18. a. and b.

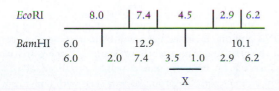

Chapter 15

2. Pulsed field gel electrophoresis involves the movement through a gel of chromosome-sized pieces of DNA. Unless the overall size of a chromosome is changed, it should always migrate to the same relative position.

 a. 7 bands identical to wild type

 b. 7 bands identical to wild type

 c. The largest and smallest bands would be expected to disappear. Two new intermediate bands would appear, unless they happened to co-migrate with one of the other five wild-type bands.

 d. One band would be larger than expected and one would be smaller than expected, when compared to the wild type.

 e. 7 bands identical to wild type

 f. 7 bands

 g. The largest wild-type band would be missing and an even larger new band would be present.

6. One approach would be to "knock out" wild-type function and then observe the phenotype. To do this, follow the one-step protocol described for Figure 15–13. Insert a selectable gene into the cloned gene of interest. Linearize the plasmid and use it for transformation of yeast cells. Select for the selectable gene. Another possibility is site-directed mutagenesis (pp. 461–462).

7. After electrophoresis, Southern blot the gel and probe with radioactive copies of the cloned gene.

9. a. Let B = bent tail, b = normal tail. The two genes are linked with 40 m.u. between them.

 b. The original cross is $B\,r_2/b\,r_1 \times b\,r_1/b\,r_1$. The wild-type progeny for tail conformation are

60%	$b\,r_1/b\,r_1$
40%	$b\,r_2/b\,r_1$

10. a. 1: *A1, B2*, recombinant
    2: *A2, B1*, recombinant
    3: *A1, B1*, parental
    4: *A2, B2*, parental

 b.
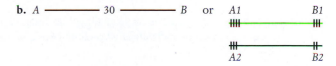

12. a. Recall that yeast plasmid vectors can exist free in the cytoplasm, producing multiple copies that are distributed randomly to daughter cells, or they can be integrated into yeast chromosomes, in which case they replicate with the host chromosome and are distributed as any other yeast gene during mitosis and meiosis.

YP1 produces 100 percent *leu*⁺ progeny, suggesting that the *leu*⁺ gene is not integrated into a chromosome and that enough copies exist of the plasmid so that all progeny obtained at least one copy.

YP2 produces 50 percent *leu*⁺ progeny, which is indicative of orderly distribution and suggests that the *leu*⁺ gene is integrated into a yeast chromosome.

b. The "insert," which is not cut with the enzyme, is the *leu*⁺ gene inserted into the plasmid.

Following digestion, electrophoresis, blotting, and probing, YP1 will produce one band that is somewhat heavier than the plasmid without an insert (that is, it migrates less distance than the plasmid during electrophoresis). YP2 will produce two bands that contain a portion of the integrated plasmid, plus genomic DNA. The location of these bands relative to the plasmid cannot be predicted.

Chapter 16

1. a. F2 is among the group of men who could be a father to the child in question.

b. If the bands in the child are numbered from the top to the bottom, there are a total of 22 bands. The following bands are from the mother: 1, 4, 5, 10, 12, 13, 14, 15, 16, 17, 18, 19, 20, and 21. The remaining bands are from the presumed father.

c. No

d. A mutation occurred that altered, eliminated, or made a restriction site.

e. The cellular DNA is cut with a restriction enzyme and electrophoresed. This results in a continuous "smear" of DNA along the length of the gel. Discrete bands are seen because the probe is detecting regions that are complementary to it. The multiple bands on the gel indicate a repetitive sequence that is of variable distance from a restriction site from repeat to repeat. The probe most likely is detecting a subsection of the piece of DNA that is seen in the band.

3. a. One locus on each of the homologous chromosomes would be indicated.

b. Both ends of all chromatids would be indicated.

c. The chromosomal satellite region on all chromosomes containing NORs would be indicated.

d. Many small regions on many chromosomes would be indicated.

e. Centromeres, telomeres, and NORs would be indicated.

5. a. The pattern of suspect 1 is compatible with including him among the group of individuals who could have committed the rape.

b. Suspects 2 and 3 could not have committed the rape.

8. a. Isolate DNA from the cells, cut it with a restriction enzyme, and electrophorese the fragments. After Southern blotting, probe with radioactive transposon sequences.

b. You could disrupt the transposon and look for a loss of the abnormal phenotype.

c. The transposon could have inserted in spacer DNA or in nonessential genes.

10. a. and b. One explanation is that the gene is surrounded by some type or types of repeating sequences and that the probe contained at least a portion of that repeat. Alternatively, the *sp* locus is not a single, unique locus. Although the three rare cutters each produced different multiple cuts, the repeating sequence, or the *sp* gene family, was located in at least four different regions that were separated by the cuts in each case, producing four bands.

c. It is unlikely that each enzyme cut the gene in three places to produce four bands because the enzymes all recognized rare target sequences.

Chapter 17

1. a. The sequence is 2 1 3 (or 3 1 2).

b. YAC A spans at least a portion of both 1 and 3. YAC B is within region 1. YAC C spans at least a portion of regions 1 and 2. YAC D is contained within region 2. YAC E is contained within region 3. A diagram of these results is shown below. In the diagram, there is no way to know the exact location of the ends of each YAC.

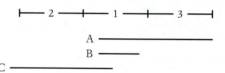

3. The cross is

$$cys\text{-}1\ RFLP\text{-}1^O\ RFLP\text{-}2^O \times cys\text{-}1^+\ RFLP\text{-}1^M\ RFLP\text{-}2^M$$

A parental type will have the genotype of either strain and be in a majority while a recombinant type will have a mixed genotype and be in a minority of the ascospores. Clearly, the first two ascospore types are parental, with the remaining being recombinant.

a. The *cys-1* locus is in this region of chromosome 5. If it were not in this region, linkage to either of the RFLP loci would not be observed.

b. The entire region is approximately 17 map units in length.

```
 ├──5 m.u.──┼────────12 m.u.──┤
RFLP-1       cys-1              RFLP-2
```

c. A suitable next step in cloning *cys-1* might be to do PFGE to obtain a pure sample of chromosome 5, followed by cloning of chromosome fragments. The clones could be screened by the *RFLP* probes.

5. Remember that a gene is one small region of a long strand of DNA and that a cloned gene will contain the entire sequence of the gene under normal circumstances. If there are two cuts within a gene, three fragments will be produced, all of which will interact with the probe, as was seen with enzyme 1. Cuts external to a gene will produce one fragment that will interact with the probe, which was seen with enzyme 2. One cut within a gene will produce two fragments that will interact with the probe, as was seen with enzyme 3.

8. a. RAPDS are formed by being bracketed by two inverted copies of the primer sequence. Below, the primer is indicated by X's, and the amplified region appears in brackets. For convenience, the two amplified regions are placed on the same lengthy piece of DNA for strain 1.

3'-XXX-5'   3'XXX 5'

5' ⎯ [⎯⎯⎯⎯] ⎯⎯⎯ [⎯⎯⎯⎯] ⎯ 3'

3' ⎯ [⎯⎯⎯⎯] ⎯⎯⎯ [⎯⎯⎯⎯] ⎯ 5'

5'-XXX-3'   5'-XXX-3'

Strain 2 lacks one or two regions complementary to the primer.

b. Progeny 1 and 6 are identical to the strain 1 parent. Progeny 4 and 7 are identical to the strain 2 parent. Progeny 2 and 5 received the chromosome holding the upper band from the strain 1 parent and the chromosome holding the lower band from the strain 2 parent (resulting in no second band). Progeny 3 received the opposite: the chromosome holding the lower band from the strain 1 parent and the chromosome holding the upper band from the strain 2 parent (resulting in no second band). Therefore, bands 1 and 2 are unlinked.

c. Recall that a nonparental ditype has two types only, both of which are recombinant. Therefore, the tetrad would be composed of four progeny like progeny 2 and four progeny like progeny 3.

9. a. Gene *x* is located in region 5.

b. The location of gene *y* is region 8.

c. Both probes are able to hybridize with cosmid E because cosmid E is long enough to contain part of genes *x* and *y*.

13. a. DNA from each individual was obtained. It was restricted, electrophoresed, blotted, and then probed with the five probes. After each probing, an autoradiograph was produced.

b. The RFLP locus closest to the disease allele is 4°.

c. Now that the RFLP locus closest to the gene has been identified, the disease gene can be cloned using probe 4 as an indicator of the region of interest. Shotgun cloning could be done first, and clones then screened with the probe. Transcripts from the clones positive for 4° could be compared and used to "fish out" the DNA region containing the gene of interest. Ultimately, DNA sequencing could be done, and the mutations causing the disease could be identified.

Chapter 18

2. O^c mutants have impaired binding of the repressor product of the *I* gene to the operator, and therefore, the *lac* operon asso-

ciated with the O^c operator cannot be turned off. Because an operator controls only the genes on the same DNA strand, it is cis (on the same strand) and dominant (cannot be turned off).

7. Nonpolar Z^- mutants cannot convert lactose to allolactose, and, thus, the operon is never induced.

9. An operon is turned off or inactivated by the mediator in negative control, and the mediator must be removed for transcription to occur. An operon is turned on by the mediator in positive control, and the mediator must be added or activated for transcription to occur.

10. The *lacY* gene produces a permease that transports lactose into the cell. A *lacY⁻* gene could not transport lactose into the cell, so β-galactosidase will not be induced.

13. The bacterial operon consists of a promoter region that extends approximately 35 bases upstream of the site where transcription is initiated. Within this region is the promoter. Inducers and repressors, both of which are trans-acting proteins that bind to the promoter region, regulate transcription of associated cistrons in cis only.

The eukaryotic cistron has the same basic organization. However, the promoter region is somewhat larger. Also, enhancers up to several thousand nucleotides upstream or downstream can influence the rate of transcription. A major difference is that eukaryotes have not been demonstrated to have polycistronic messages.

Chapter 19

5. Depurination results in the loss of adenine or guanine from the nucleotide. Since the resulting apurinic site cannot specify a complementary base, replication is blocked. Under certain conditions, replication proceeds with a random insertion of a base opposite the apurinic site. In three-fourths of these insertions, a mutation will result.

Deamination of cytosine yields uracil. If left unrepaired, uracil will be paired with adenine during replication, ultimately resulting in a transition mutation.

8-OxodG can pair with adenine, resulting in a transversion.

6. 5-Bromouracil is an analog of thymine. It undergoes tautomeric shifts at a higher frequency than thymine and, therefore, is more likely to pair with G than is thymine during replication. At the next replication this will lead to a GC pair rather than the original AT pair.

Ethyl methanesulfonate is an alkylating agent that produces O-6-ethylguanine. This alkylated guanine will mispair with thymine, which leads from a GC pair to an AT pair at the next replication.

8. Mismatch repair occurs if a mismatched nucleotide is inserted during replication. The new, incorrect base is removed, and the proper base is inserted. The enzymes involved can distinguish between new and old strands because, in *E. coli*, the old strand is methylated.

Recombination repair occurs if lesions such as AP sites and UV photodimers block replication (there is a gap in the complementary strand). Recombination fills this gap with the corresponding segment from the sister DNA molecule, which is normal in both strands. This produces one DNA molecule with a gap across from a correct strand, which can then be filled by complementation, and one with a photodimer across from a correct strand.

9. Leaky mutants are mutants with an altered protein product that retains a low level of function. Enzyme activity may, for instance, be reduced rather than abolished by a mutation.

11. **a.** Because 5′-UAA-3′ does not contain G or C, a transition to a GC pair in the DNA cannot result in 5′-UAA-3′. 5′-UGA-3′ and 5′-UAG-3′ have the DNA antisense-strand sequence of 3′-ACT-5′ and 3′-ATC-5′, respectively. A transition to either of these stop codons occurs from the nonmutant 3′-ATT-5′, respectively. However, a DNA sequence of 3′-ATT-5′ results in an RNA sequence of UAA, itself a stop codon.

    b. Yes. An example would be 5′-UGG-3′, which codes for Trp, to 5′-UAG-3′.

    c. No. In the three stop codons the only base that can be acted on is G (in UAG, for instance). Replacing the G with an A would result in 5′-UAA-3′, a stop codon.

13. To understand these data, recall that half of the progeny should come from the wild-type parent.

    a. A lack of revertants suggests either a deletion or an inversion within the gene.

    b. *Prototroph A:* Because 100 percent of the progeny are prototrophic, a reversion at the original mutant site may have occurred.
    Prototroph B: Half of the progeny are parental prototrophs, and the remaining prototrophs, 28 percent, are the result of the new mutation. Notice that 28 percent is approximately equal to the 22 percent auxotrophs. The suggestion is that an unlinked suppressor mutation occurred, yielding independent assortment with the *nic* mutant.
    Prototroph C: There are 496 "revertant" prototrophs (the other 500 are parental prototrophs) and 4 auxotrophs. This suggests that a suppressor mutation occurred in a site very close [100% (4 × 2)/1000 = 0.8 m.u.] to the original mutation.

Chapter 20

1. 3, 4, 6

5. First notice that gene conversion is occurring. In the first cross, a_1 converts (1:3). In the second cross, a_3 converts. In the third cross, a_3 converts. Polarity is obviously involved. The results can be explained by the following map, where hybrid DNA enters only from the left.

$$a_3 \rule{1.5em}{0.4pt} a_1 \rule{1.5em}{0.4pt} a_2$$

7. The ratios for a_1 and a_2 are both 3:1. There is no evidence of polarity, which indicates that gene conversion as part of recombination is occurring. The best explanation is that two separate excision-repair events occurred and, in both cases, the repair retained the mutant rather than the wild type.

8. **a.** and **b.** A heteroduplex that contains an unequal number of bases in the two strands has a larger distortion than does a simple mismatch. Therefore, the former would be more likely to be repaired. For such a case, both heteroduplex molecules are repaired (leading to 6:2 and 2:6) more often than one (leading to 5:3 or 3:5) or none (leading to 3:1:1:3). The preference in direction (i.e., adding a base rather than subtracting) is analogous to thymine dimer repair. In thymine dimer repair, the unpaired, bulged nucleotides are treated as correct and the strand with the thymine dimer is excised.
      A mismatch more often than not escapes repair, leading to a 3:1:1:3 ascus.
      Transition mutations would not cause as large a distortion of the helix, and each strand of the heteroduplex should have an equal chance of repair. This would lead to 4:4 (two repairs each in the opposite direction), 5:3 (1 repair), 3:1:1:3 (no repairs or two repairs in opposite directions), and, less frequently, 6:2 (two repairs in the same direction).

    c. Because excision repair excises the strand opposite the larger buckle (i.e., opposite the frameshift mutation), the cis transition mutation will also be retained. The nearby genes are converted because of the length of the excision repair.

9. **a.** 6:2 = 31.25%

    b. 2:6 = 5%

    c. 3:1:1:3 = 11.25%

    d. 5:3 = 37.5%

    e. 3:5 = 15%

Chapter 21

2. Polar mutations affect the transcription or translation of the part of the gene or operon on only one side of the mutant site, usually described as *downstream*. Examples are nonsense mutations, frameshift mutations, and IS-induced mutations.

4. R plasmids are the main carriers of drug resistance. They acquire these genes by transposition of drug-resistance genes located between IR (inverted repeat) sequences. Once in a plasmid, the transposon carrying drug resistance can be transferred upon conjugation if it stays in the R plasmid, or it can insert into the host chromosome.

6. P elements are transposons (genes flanked by inverted repeats, allowing for great mobility). Because they are transposons, they can insert into chromosomes. By inserting specific DNA between the inverted repeats of the P elements and injecting the altered transposons into cells, a high frequency of gene transfer will occur.

8. The best explanation is that the mutation is due to an insertion of a transposable element.

10. **a.** The expression of the tumor is blocked in plant B. This suggests either that plant B can suppress the functioning of

5. Pro
mei

8. The

Chapte

1. Sel
net

2. *A1*

4. 0.6

6. a.

b.

8.

10.

the plasmid that causes the tumor, or that plant A provides something to the tissue with the tumor-causing plasmid that plant B does not provide.

b. Tissue carrying the plasmid, when grafted to plant B, appears normal, but the graft produces tumor cells in synthetic medium. This indicates that the plasmid sequences are present and capable of functioning in the right environment. However, the production of normal type A plants from seeds from the graft suggests a permanent loss of the plasmid during meiosis.

Chapter 22

3. Maternal inheritance of chloroplasts results in the green-white color variegation observed in *Mirabilis*.

Cross 1: variegated female × green male ⟶ variegated progeny

Cross 2: green female × variegated male ⟶ green progeny

5. Both yeast parents contribute mitochondria to the cytoplasm of the resulting diploid cell. Subsequent meiosis shows uniparental inheritance for mitochondria. Therefore, 4:0 and 0:4 asci will be seen.

7. The genetic determinants of R and S are cytoplasmic and are showing maternal inheritance.

9. Both yeast parents contribute mitochondria to the cytoplasm of the resulting diploid cell. Because of cytoplasmic segregation, the asci will be of four types:

$$oli^R \, cap^R$$

$$oli^S \, cap^S$$

$$oli^R \, cap^S$$

$$oli^S \, cap^R$$

11. If the mutation is in the chloroplast, reciprocal crosses will give different results, while if it is in the nucleus and dominant, reciprocal crosses will give the same results.

13. After the initial hybridization, a series of backcrosses using pollen from B will result in the desired combination of cytoplasm A and nucleus B. With each cross the female contributes all of the cytoplasm and half the nuclear contents, while the male contributes half the nuclear contents.

16. The suggestion is that the mutants were cytoheterozygotes for streptomycin resistance.

17. Realize that the closer two genes are, the higher the rate of cosegregation. A rough map of the results is as follows:

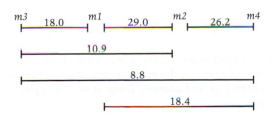

18. The red phenotype in the heterokaryon indicates that the red phenotype is caused by a cytoplasmic organelle allele.

23. a. and **b.** Each meiosis shows uniparental inheritance, suggesting cytoplasmic inheritance.

c. Because *ant*[R] is probably mitochondrial and because petites have been shown to result from deletions in the mitochondrial genome, *ant*[R] may be lost in some petites.

25. Consider the following hypothetical situation of three genes:

If a deletion of one of them occurred, *b* by itself would be least likely to be the only gene deleted if the genes were closely linked. The closer *a* is to *b*, the less likely it is that *a* will be deleted if *b* is retained.

Applying the above logic to the data, the *apt-cob* pair had the lowest rate of loss (45 total), and the *apt-bar* pair had the highest rate of loss (207 total). This puts *cob* between *apt* and *bar*. The relative rate of loss is

apt ——————————————— *cob* ——————————————— *bar*
      1   2.6

27. Some tetrads will show strain-1 type, some will show strain-2 type, and some will be recombinant.

28. a. No; during diploid budding all the progeny receive one type of mtDNA. Most likely it is a plasmid or episome.

b. 3 bands, one at 2μ, the others totaling 2μ.

30. First, prepare a restriction map of the mtDNA using various restriction enzymes. Using the assumption of evolutionary conservation, on a Southern blot hybridize equivalent fragments from yeast or another organism in which the genes have already been identified.

31. a. The cytoplasm from senescent cultures is a mixture of normal and abnormal mitochondria. The mitochondrial types are distributed in different ratios to different spores. The abnormal mitochondria appear to have a replicative advantage over the normal since senescence seems, ultimately, to "win out" over normal nonsenescence. The rapidity of the onset of senescence seems to be related to the ratio of normal-to-abnormal mitochondria.

b. The mutation is a 10-kb insertion into fragment 20. It carries bands E and G and splits the original fragment into two fragments, B and C.

Chapter 23

1. a. Female embryos will react to both antibodies, but male embryos will not react to either.

b. As in early development, the female will react to both probes. The male will react to the probe for the amino end only.

2. a. If only one gene coded for antibodies, then the number of different alleles would limit antibody production to two

7. Mean = 4.7

    Variance = 0.2619

    Standard deviation = 0.5117

9. a. H^2 has meaning only with respect to the population that was studied in the environment in which it was studied. Otherwise, it has no meaning.

    b. Neither H^2 nor h^2 is a reliable measure that can be used to generalize from a particular sample to a "universe" of the human population. They certainly should not be used in social decision making (as implied by the terms *eugenics* and *dysgenics*).

    c. Again, H^2 and h^2 are not reliable measures, and they should not be used in any decision making with regard to social problems.

11. First, define *alcoholism* in behavioral terms. Next, realize that all observations must be limited to the behavior you used in the definition and that all conclusions from your observations are applicable only to that behavior. In order to do your data gathering, you must work with a population in which familiarity is distinguished from heritability. In practical terms, this means using individuals who are genetically close but who are found in all environments possible.

14. a. If you assume that individuals at the extreme of any spectrum are homozygous, then their offspring are more likely to be heterozygous than the original individuals. That is, they will be less extreme.

    b. For Galton's data, regression is an estimate of heritability (h^2), assuming that there were few environmental differences between father and son.

Index